Physical Chemistry

P. W. ATKINS

Second Edition

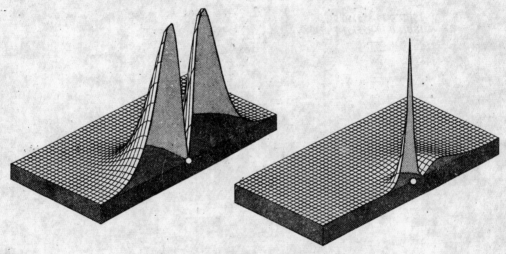

OXFORD UNIVERSITY PRESS

1982

Oxford University Press, Walton Street, Oxford OX2 6DP

London Glasgow New York Toronto
Delhi Bombay Calcutta Madras Karachi
Kuala Lumpur Singapore Hong Kong Tokyo
Nairobi Dar es Salaam Cape Town
Melbourne Auckland

and associate companies in
Beirut Berlin Ibadan Mexico City

© P. W. Atkins 1978, 1982

First edition 1978
Second edition 1982

ISBN 0-19-855150-9
ISBN 0-19-855151-7 Pbk

Printed in Great Britain
at the University Press, Oxford
by Eric Buckley
Printer to the University

Preface to the Second Edition

In preparing the second edition of this book I have taken into account the advice I have received from readers throughout the world. The principal changes include a revised introduction to thermodynamics, a strengthening of the sections on quantum theory and molecular structure, and an entirely new chapter on macromolecules. I have also brought up to date the sections on X-ray diffraction, magnetic resonance, and surface chemistry. Wherever possible I have simplified the notation and the presentation, particularly in the sections on electrochemistry. The chapters on statistical thermodynamics have been restructured into a simpler form. Every section of the original edition has been reconsidered in order to improve the presentation, introduce modern material, or to simplify the notation without loss of rigour. The illustrations have been revised and the visual presentation of the material has been improved in a variety of ways, including the use of computer graphics in a novel format. All the Tables, Problems, Examples, and Further Reading entries have been reconsidered in detail. Throughout the revision, however, I have aimed to retain the level of presentation of the first edition, and to preserve its style and approach.

I owe a considerable debt to all those who wrote to me with comments either on the first edition or on the draft of the second. Extensive sections of the latter were commented on by W. J. Albery (London), A. D. Buckingham (Cambridge), A. J. B. Cruickshank (Bristol), A. A. Denio (Wisconsin), R. A. Dwek (Oxford), A. H. Francis and T. M. Dunn (Michigan), D. A. King (Liverpool), G. Lowe (Oxford), M. L. McGlashan (London), J. Murto (Helsinki), M. J. Pilling (Oxford), C. K. Prout (Oxford), H. Reiss (UCLA), H. S. Rossotti (Oxford), J. S. Rowlinson (Oxford), W. A. Wakeham (London), D. H. Whiffen (Newcastle), J. S. Winn (Berkeley), and M. Wolfsberg (Irvine) and I am grateful to them all. I should also like to thank the following, who made particularly helpful comments on the first edition, and whose remarks have been built into this: M. D. Archer (Cambridge), L. Brewer (Berkeley), D. H. Everett (Bristol), D. Hussein (Cambridge), R. M. Lynden-Bell (Cambridge), I. M. Mills (Reading), and A. D. Pethybridge (Reading), as well as those others whom I have acknowledged privately. The correspondence with my translators, K. P. Butin (Moscow), G. Chambaud (Paris), H. Chihara (Osaka), M. Guardo (Bologna), and A. Höpfner (Heidelberg), has been a particularly fruitful source of advice.

Finally, I should like to thank Judith Adam, who helped to prepare this edition, Caron Crisp, who typed it accurately and always on time, Daniel James, who provided the computer graphics, and the officers of both Oxford University Press and W. H. Freeman and Co. who, as always, have done so much by way of support and encouragement.

Oxford 1981   P.W.A.

Preface to the First Edition

Authors should not preach to teachers. Textbooks should be flexible and adaptable, yet have a strong story-line. I have tried to conform to these demands by dividing the text into three parts, *Equilibrium*, *Structure*, and *Change*. Each part begins in an elementary way, drawing on the others only weakly. Of course they rapidly get tangled up with each other—as they should because the subject is a unity—but teachers will be able to match the text to their own needs without unduly burdening the student. The student, I hope, will be enticed to read his way into chemistry's web of interdependencies, and will find that he can master them without getting confused.

Physical chemistry possesses its mathematics for a purpose: there has to be enough mathematical spine in the subject to enable our ideas on the behaviour of molecules and systems to stand up to experimental verification. Ideas that cannot be tested do not belong to science. Nevertheless, in an introductory treatment the ideas must not be overborne by the mathematics. In this text I show how physical ideas can be developed mathematically, and I take care to interpret the mathematical statements I make. Only where the mathematics and the chemistry lose sight of each other is physical chemistry a difficult subject, so I try never to let that happen.

These views have led me to a further organization of the text. In several places I have treated a subject in two parts, as 'Concepts' and as 'Machinery'. The former establishes the ideas, while the latter extends and develops them more mathematically. This is the arrangement I have adopted for the First and Second Laws of thermodynamics and, later on in Part 2, for statistical thermodynamics. The 'Concepts' chapters emphasize the underlying physics and let the reader understand the conceptual basis of the subject; the 'Machinery' chapters let him discover the ramifications of these ideas and show him how to apply them to chemical problems.

Throughout the text I have used a series of worked *Examples*. These serve two purposes. The first is to show how calculations are actually done: an example with all its detailed working can save pages of explanation and give the reader a much clearer impression of what is involved, and a sense of reality. Their second purpose is to introduce a remark to extend the text, or to stimulate the reader's imagination and interest. Apart from the *Examples* there are the *Problems* at the end of each chapter. I have included a lot of simple ones as well as a number that require more time and effort, and occasionally access to a small computer. Many are based on recent literature. I do not expect readers to do all of them, but I have provided a large number so that teachers can be selective. The answers to most Problems will be found at the end of the book. Virtually all necessary data are given in the Tables, and the Table Index on p. 1083 should permit quick location of any item. SI units are used throughout, but I have sprinkled a selection of others through the text in order to keep older literature accessible. *Boxes* serve to collect results of arguments or to summarize handy information. *Appendices* contain detailed developments or background that would have encumbered the main text with too much detail or too many equations.

A book such as this could not have appeared without the sustained help of a number of people. Chapters from an early draft were read and criticized by Professor G. Allen (University of London), Dr M. H. Freemantle (University of Jordan), Professor P. J. Gans (New York University), Dr R. J. Hunter (University of Sydney), Professor L. G. Pedersen (University of North Carolina), Professor D. W. Pratt (University of Pittsburg), Dr D. J. Waddington (University of York), Dr S. M. Walker (University of Liverpool), and Professor R. W. Zuehlke (University of Bridgeport). From their remarks grew the second draft, which in turn owes a considerable debt to others. W. H. Freeman and Company of San Francisco played a key role in having the book extensively reviewed and in making sure that it would fit the requirements of courses in North America. They obtained comments and advice from Professor H. C. Andersen (Stanford University), Professor J. Simons (University of Utah), and Professor R. C. Stern (Lawrence Livermore Laboratory). I owe a particular debt of gratitude to Professor Stern who made penetrating remarks on almost every word, or so it seemed; to Miss A. J. MacDermott (University of Oxford) who read the whole and criticized acutely; and to Mr S. P. Keating (University of Oxford), who worked hard and thoughtfully on many of the Problems. After such global assistance there appears to be little left to which the author can attach his name, except to the apologies for the errors that may remain.

I would also like to pay tribute to the officers of the Oxford University Press, who have suffered without public complaint the intrusions of a pernickity and local author. The sustained good humour of our relations made the whole exercise most agreeable. In particular, though, I must thank Dr M. G. Rodgers, whose wise advice, in the form of enthusiasm tempered by selective discouragement, is in no small part responsible for the eventual appearance of this book.

Finally, I thank my wife and my daughter for suffering, again without complaint, the brutishness to which authors sink in order to create something they hope is worthwhile.

Oxford 1977   P.W.A.

Contents

Units and notation

SI units are used throughout, but the atmosphere (atm) is retained as a very convenient unit of pressure. The adoption of SI leads to the appearance of unfamiliar units in only two cases. First, pm (picometer, 10^{-12} m or 0.01 ångström) is used as the unit of length for molecules: with this choice molecular dimensions and bond lengths are of the order of 100 pm and decimal points conveniently disappear. Secondly, dm^3 (1000 cm^3, 1 litre) is used as the unit of volume, unless cm^3 or m^3 are more convenient. The replacement of the litre by dm^3 may take a little getting used to, but it simplifies the numerical working of equations. Simply remember that 1 dm^3 is exactly the same as 1 l.

Some interconversions, and some commonly encountered notational differences are listed below. Others are listed inside the front cover.

Units

$1 l \equiv 1 \, dm^3 \equiv 1000 \, cm^3 \equiv 10^{-3} \, m^3$

1 molar, $1M \equiv 1 \, mol \, l^{-1} \equiv 1 \, mol \, dm^{-3}$

$100 \, pm \equiv 1.00 \, \text{ångström} \equiv 10^{-8} \, cm \equiv 10^{-10} \, m$

$1 \, atm = 760 \, Torr = 760 \, mmHg = 1.013 \, 25 \times 10^5 \, N \, m^{-2}$.

Notation

U, internal energy. Some texts use E.

G, Gibbs function. Some texts use F.

A, Helmholtz function. Some texts use F.

L, Avogadro's constant. Some texts use N_A, $\mathcal{N}_A$, or N_0.

$p^{\ominus}$ denotes a pressure of 1 atm (101.325 kPa).

$m^{\ominus}$ denotes a molality of 1 mol kg^{-1}.

USEFUL RELATIONS

At 298.15 K

$$RT = 2.4789 \text{ kJ mol}^{-1}$$

$$RT/F = 0.025693 \text{ V}$$

$$2.3026 \, RT/F = 0.05915 \text{ V}$$

$$kT/hc = 207.223 \text{ cm}^{-1}$$

$$V_m^\ominus = RT/p^\ominus = 2.4465 \times 10^{-2} \text{ m}^3 \text{ mol}^{-1} = 24.465 \text{ dm}^3 \text{ mol}^{-1}$$

T/K	100.00	298.15	500.00	1000	1500	2000
$(kT/hc)/\text{cm}^{-1}$	69.50	207.223	347.51	695.03	1042.54	1390.06

$$p^\ominus = 101.325 \text{ kPa} = 1.01325 \times 10^5 \text{ N m}^{-2}$$
$$p^\ominus \triangleq 1 \text{ atm} \triangleq 760 \text{ mmHg} \qquad \qquad 1 \text{ mmHg} \triangleq 133.322 \text{ N m}^{-2}$$

$$1 \text{ eV} \triangleq 1.602189 \times 10^{-19} \text{ J} \qquad \qquad 1000 \text{ cm}^{-1} \triangleq 1.986 \times 10^{-20} \text{ J}$$
$$96.485 \text{ kJ mol}^{-1} \qquad \qquad \qquad 11.96 \text{ kJ mol}^{-1}$$
$$8065.5 \text{ cm}^{-1} \qquad \qquad \qquad 0.1240 \text{ eV}$$

$$hc = 1.98648 \times 10^{-23} \text{ J cm} \qquad \qquad hc/k = 1.43879 \times 10^{-2} \text{ m K}$$
$$g/\text{m s}^{-2} = 9.8064 - 0.0259 \cos\{2(\text{latitude})\} \approx 9.811 \text{ at } 50°$$
$$1 \text{ cal} = 1 \text{ cal}_{\text{th}} \triangleq 4.184 \text{ J} \qquad \qquad 1 \text{ debye} = 1 \text{ D} \triangleq 3.33564 \times 10^{-30} \text{ C m}$$

$$N = J \text{ m}^{-1} \triangleq 10^5 \text{ dyn} \qquad W = J \text{ s}^{-1} \qquad T = J \text{ C}^{-1} \text{ s m}^{-2}$$
$$J \triangleq 10^7 \text{ erg} \qquad \qquad \quad A = C \text{ s}^{-1} \qquad J = A \text{ V s}$$

MATHEMATICAL INFORMATION

$$\ln \equiv \log_e \qquad \lg \equiv \log_{10} \qquad \ln x = (\ln 10) \lg x = 2.302585 \lg x$$
$$\pi = 3.14159265359 \qquad e = 2.71828182846$$

$=$ equals	$\triangleq$ corresponds to
$\approx$ approximately equals	$\overset{\text{def}}{=}$ equal by definition to
$\equiv$ identical to	$\sim$ asymptotically equal to

GENERAL DATA

Speed of light	c	$2.997\,925 \times 10^8$ m s^{-1}
Charge of proton (Charge on the electron is $-e$)	e	$1.602\,19 \times 10^{-19}$ C
Faraday constant	$F = eL$	$9.648\,46 \times 10^{-19}$ C mol^{-1}
Boltzmann constant	k	$1.380\,66 \times 10^{-23}$ J K^{-1}
Gas constant	$R = kL$	$8.314\,41$ J K^{-1} mol^{-1}
		$1.987\,17$ cal K^{-1} mol^{-1}
		$8.205\,75 \times 10^{-2}$ dm^3 atm K^{-1} mol^{-1}
Planck constant	h	$6.626\,18 \times 10^{-34}$ J s
	$\hbar = h/2\pi$	$1.054\,59 \times 10^{-34}$ J s
Avogadro constant	L	$6.022\,05 \times 10^{23}$ mol^{-1}
Atomic mass unit	$u = 10^{-3}$ kg/(L mol)	$1.660\,56 \times 10^{-27}$ kg
Mass of electron	m_e	$9.109\,53 \times 10^{-31}$ kg
proton	m_p	$1.672\,65 \times 10^{-27}$ kg
neutron	m_n	$1.674\,95 \times 10^{-27}$ kg
nuclide	$m = M_r u$	$1.660\,56 \times 10^{-27} \times M_r$ kg
Vacuum permittivity	ε_0	$8.854\,188 \times 10^{-12}$ J^{-2} C^{-2} m^{-1}
	$4\pi\varepsilon_0$	$1.112\,650 \times 10^{-10}$ J^{-1} C^2 m^{-1}
Vacuum permeability (Note that $\varepsilon_0\mu_0 = 1/c^2$)	μ_0	$4\pi \times 10^{-7}$ J s^2 C^1 m^{-1}
Bohr magneton	$\mu_B = e\hbar/2m_e$	$9.274\,08 \times 10^{-24}$ J T^{-1}
Nuclear magneton	$\mu_N = e\hbar/2m_p$	$5.050\,82 \times 10^{-27}$ J T^{-1}
Bohr radius	$a_0 = 4\pi\varepsilon_0\hbar^2/m_e e^2$	$5.291\,77 \times 10^{-11}$ m
Rydberg constant	$R_\infty = m_e e^4/8h^2\varepsilon_0^2$	$2.179\,908 \times 10^{-23}$ J
	R_∞/hc	$1.097\,373 \times 10^5$ cm^{-1}
Gravitational constant	G	6.6720×10^{-11} N m^2 kg^{-2}

PREFIXES

p	n	μ	m	c	d	k	M	G
pico	nano	micro	milli	centi	deci	kilo	mega	giga
10^{-12}	10^{-9}	10^{-6}	10^{-3}	10^{-2}	10^{-1}	10^3	10^6	10^9

Introduction

The nature of matter: orientation and background

Learning objectives

This chapter presents in general outline a number of important results and concepts. The explanation of the basis of the assertions made in this chapter, and the way of using the results, will all be examined later in the book. After careful study of this chapter you should be able to

(1) Indicate the relative sizes of atoms and molecules (p. 5).

(2) Describe the effects of the *quantization of energy* and state the size of the quantum associated with translational, rotational, vibrational, and electronic motion (p. 8).

(3) Describe the formation of *spectral lines* and state their *frequencies* and *wavenumbers* by application of the *Bohr frequency condition* (eqn (1.1)).

(4) Define *mole* and define and interrelate *molar mass, molecular mass,* and *relative molar mass* (Box 0.1).

(5) State the formula for the *Boltzmann distribution* (eqn (1.2)) and calculate the relative populations of translational, rotational, vibrational, and electronic energy levels at different temperatures (p. 12).

(6) State the *equipartition theorem* (p. 15).

(7) Classify bonds as *ionic, polar,* and *covalent* (p. 16) and describe the forces holding together the molecules of liquids and solids (p. 16).

(8) State the *perfect gas law* (eqn (2.2)) and describe the basis of the *kinetic theory* of gases (p. 22).

(9) Deduce expressions for the *mean speed* of gas molecules (eqn (2.3)), and define *collision frequency* and *mean free path* (p. 22).

(10) Write an expression for the *temperature dependence of reaction rates* (eqn (2.4)) and explain the term *activation energy*.

(11) State the units used to express *force, pressure,* and *energy* and make interconversions between SI units and others (pp. 23 and 24).

Introduction

We know that atoms and molecules exist because we can see them, Figs.
0.1 and 0.2. Their existence had been inferred long before the invention
of the techniques used to obtain these pictures, and estimates of their size
and shape were made in the nineteenth century. Now we can measure the
masses of molecules, atoms, and subatomic particles with great precision
and we can determine the shapes of molecules as complicated as enzymes
and proteins, Fig. 0.3. Whenever we want an explanation of some
observation we shall build one in terms of atoms and molecules. If this is
to be done successfully we must know something about their structure
and properties, and how individual molecules contribute to the samples
we normally encounter.

Fig. 0.1. An image of a platinum tip about 150 nm radius (E. W. Müller). The technique
employed for this photograph is called *field ionization microscopy*, p. 1009.

Fig. 0.2. The density of electrons in a molecule of anthracene. The picture was obtained by X-ray diffraction. This technique is described on p. 744. (V. L. Sinclair, J. M. Robertson, and A. McL. Mathieson, *Acta Crystallogr.* **3**, 254 (1950).)

This chapter gives an outline of the concepts that are encountered throughout chemistry. All the ideas introduced here are developed with more explanation in later chapters, but this chapter gives a general background to the rest of the book.

0.1 The microscopic world

At the centre of an atom lies the nucleus. Almost the whole of the mass of the atom is concentrated there even though it accounts for only a minute proportion of the atom's total volume. Around the nucleus cluster the electrons. They contribute very little to the total mass of the atom but occupy an appreciable volume, and are responsible for the atom's bulk.

How big are atoms? When we attempt to picture things happening on a molecular scale it is important to know *relative* sizes. For example, a proton is about 2000 times as massive as an electron, an atom of carbon is about 12 times as massive as an atom of hydrogen, and in a molecule a chlorine atom has a diameter about $1\frac{1}{2}$ times that of an oxygen atom, Fig. 0.4. Wherever possible we shall deal with relative sizes. In this way we shall develop an insight into the behaviour of atoms and molecules without being troubled by our inability to visualize objects as small as 10^{-10} m.

The *absolute* magnitudes of atoms and molecules do, however, play one very important role. The laws of mechanics made familiar to us by everyday objects are not applicable to species the size of atoms. One of the biggest surprises of modern physics was the discovery that conventional mechanics is only an approximation: classical mechanics is inapplicable to atomic objects and has to be replaced by a new theory of matter, the *quantum theory*. If we want to understand atomic and molecular processes, we must do so on the basis of the rules of this theory.

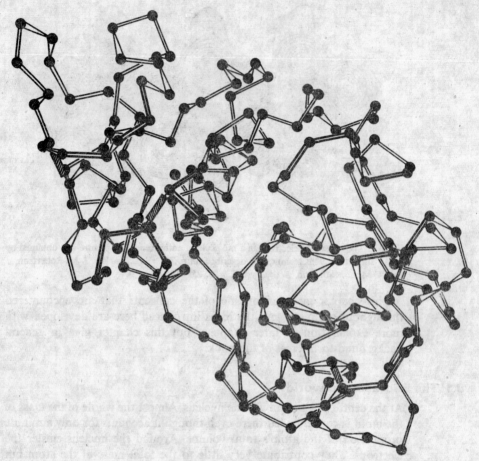

Fig. 0.3. Perspective drawing of the main chain conformation of the enzyme papain (which is related to digestive enzymes). The filled circles are the α-carbon atoms of the amino acid residues. (J. Drenth, J. N. Jansonius, R. Koekoek, H. N. Swen, B. G. Wolphers. *Nature*, **218**, 929 (1968), with permission.)

The absolute magnitude of atoms compels us to think in terms of quantum theory, but with that established we can attempt to visualize the microscopic world on the basis of the relative sizes of its components.

The quantum rules. According to the quantum theory, the energy of an object cannot be changed by an arbitrary amount. The nature of the object determines what energies it may possess, and it is impossible to make it acquire intermediate values.

These remarks can be illustrated by a pendulum. According to classical physics a pendulum can be made to swing with any energy. A small impulse sets it swinging with its natural frequency ν (nu) which is determined by its length, but its amplitude is small. A bigger impulse sets it swinging with the same frequency, but its amplitude is bigger because more energy has been transferred. According to classical physics an impulse

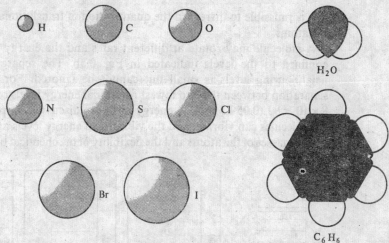

Fig. 0.4. The relative sizes of some atoms and molecules.

of any magnitude can set the pendulum in motion, and by governing the impulse any amplitude of swing can be established.

Things are quite different when we turn to quantum theory. The pendulum, this theory states, can possess one of a precisely defined, discrete set of energies. The energy of the pendulum must be one of the values $\frac{1}{2}h\nu$, $\frac{3}{2}h\nu$, $\frac{5}{2}h\nu$, and so on, where ν is the natural frequency and h, *Planck's constant*, is a fundamental constant of nature. Put another way, in a sense only some amplitudes of swing of the pendulum are permitted, and it is impossible to set it swinging with an intermediate amplitude. Furthermore, since the minimum energy is $\frac{1}{2}h\nu$, according to quantum theory a pendulum can never be absolutely still.

Why did it take until the twentieth century to discover this behaviour? The answer lies in the magnitude of Planck's constant. Experiment gives its value as 6.626×10^{-34} J s. This implies that a pendulum of natural frequency 1 Hz can accept energy in steps of about 6.6×10^{-34} J, which is so small that for all practical purposes it appears that energy can be transferred continuously. Only when very high-frequency motions are involved does the quantum $h\nu$ become large enough to give rise to significant effects. A mass on a spring oscillates with simple harmonic motion, and its natural frequency increases as the mass decreases. When the mass is that of an atom the frequency is extremely high, and so we must be prepared for significant quantum effects in vibrating molecules.

The energy associated with every type of motion is quantized. When we come to think about the rotation of molecules and the energy of electrons in atoms and molecules, we have to take into account the role of the quantum theory in governing their behaviour. Figure 0.5 shows some examples of the arrangement of energy levels in typical systems. An object free to move in a region of space possesses translational energy, and can be accelerated into any of its translational energy levels. The separation between neighbouring levels is so small, Fig. 0.5a, that for most purposes it is permissible to disregard the quantization of translational energy even for atoms.

A molecule may rotate at different rates and the energy of rotation is confined to the levels indicated in Fig. 0.5b. The separation between neighbouring levels is small but cannot be ignored. For instance, the separation between the two lowest rotational energy levels of CO is about 8×10^{-23} J (0.05 kJ mol^{-1}, energy units are discussed on p. 24).

Molecules can vibrate, and the vibrational energy levels are determined by the masses of the atoms and the flexibility of the chemical bonds between

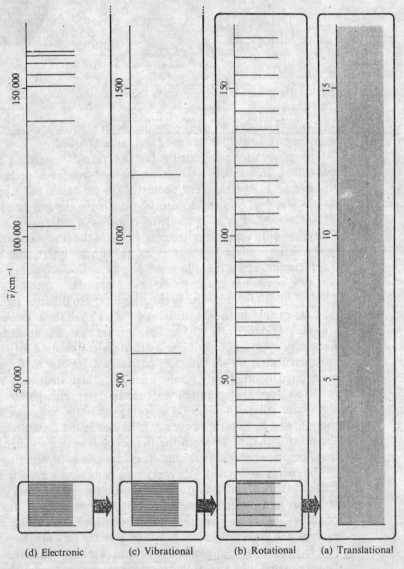

(d) Electronic   (c) Vibrational   (b) Rotational   (a) Translational

Fig. 0.5. An indication of the relative separations of the energy levels of atoms and molecules. The scales are marked in cm^{-1}, see below.

them. This is indicated in Fig. 0.5c. The separation of neighbouring levels in CO is as much as 4×10^{-20} J, or 25 kJ mol^{-1}, and the effects of quantization are very important.

The energy of an electron in an atom like hydrogen is restricted to the levels illustrated in Fig. 0.5d. Because the electron is so light we can expect its energy levels to be much more widely spaced than the vibrational energy levels of a molecule. This is the case, as the figure shows. The separation between the lowest two energy levels of atomic hydrogen is as much as 1.5×10^{-18} J, or 1000 kJ mol^{-1}.

The results just quoted are theoretical conclusions, but there is plenty of experimental evidence for the quantization of energy. Compelling evidence comes from direct visual observation. Figure 0.6 shows the spectrum of light emitted by an atom excited to a high energy state (for example, in a flame or a spark). You can see that the light is emitted at a series of definite frequencies, just as we might expect for a system that drops between different, discrete energy levels of the sort illustrated in Fig. 0.5d, and emits the excess energy as radiation. The frequency v emitted or absorbed when a transition occurs can always be calculated from the *Bohr frequency condition*:

(0.1.1)
$$hv = E_{\text{upper}} - E_{\text{lower}}.$$

For example, when a small molecule drops from one rotational state to another (and then rotates more slowly) the energy loss is about 0.1 kJ mol^{-1}, corresponding to an emission of radiation of frequency 2×10^{11} Hz or wavelength λ (lambda) $\approx$ 1.5 mm ($\lambda = c/v$ in a vacuum, or $\lambda = v/v$ in general, where v is the speed of propagation of the radiation). This is the wavelength of microwave radiation, and so rotational transitions occur in the microwave region of the electromagnetic spectrum. Instead of frequency or wavelength, the *wavenumber* $\tilde{v}$ (nu tilde), of the radiation is often quoted (as in Fig. 0.5): this is defined as $\tilde{v} = v/c$. In a vacuum (where the radiation has a speed c) it follows that $\tilde{v} = 1/\lambda$, and so the wavenumber may then be envisaged as the number of waves per unit length. In the present example, $\tilde{v} \approx 7$ cm^{-1} (7 wavelengths per centimetre).

Vibrational transitions give rise to, and can be caused by, infrared radiation. Electronic transitions, when electrons are shifted from one region of an atom or molecule to another, occur in the visible and ultraviolet part of the spectrum.

If the tightly bound inner electrons of heavy atoms can be involved in a transition, we can expect very short-wavelength radiation. This is the source of X-rays. Even shorter-wavelength radiations constitute the gamma-rays (γ-rays). For their source we have to look beyond the atomic

Hg

Fig. 0.6. The spectrum of light emitted by excited mercury atoms.

electrons and into the nucleus itself: transitions of the components of the nucleus between their extremely widely spaced quantum levels generate this high-energy, penetrating radiation.

The contributions made to the spectrum by transitions of different types are indicated in Fig. 0.7.

Assemblies of molecules. Most chemical systems (for example, a flask of solution, or a human body) are built from very large numbers of atoms or molecules. A very common measure of the amount of substance is the *mole* (abbreviated mol). A mole of substance contains 6×10^{23} molecules (see Box 0.1 for a more precise statement). Finding the number of moles present in a sample is easy once we know its mass and the molar mass of the components. For instance, if we have a 16 g sample of oxygen, since the molar mass is 32 g mol^{-1}, the amount of oxygen molecules present is 0.5 mol. This, in turn, signifies that the sample contains 3×10^{23} molecules of oxygen.

Constant thermal agitation ensures that the molecules of a sample are distributed among the energy levels available to them. Some molecules occupy the lowest energy levels, others are in the excited levels. This is particularly easy to see in the case of vibrational motion, and for this purpose we refer again to Fig. 0.5c.

At very low temperatures almost every molecule in a sample occupies its lowest vibrational energy level, which corresponds to the virtual absence of vibrational motion. Warming the sample stimulates some of the molecules to vibrate. Transitions up the ladder of energy levels correspond to acceleration into more violent motion (larger amplitude of vibration). The warmer the sample, the higher up the ladder the molecules are driven. Not all reach the higher levels: there is a predominance of molecules in the lower, less energetic levels, but the tail of the distribution gets longer and penetrates further into the high energy region as the temperature is raised, Fig. 0.8.

The formula for calculating the population of the available levels is known as the *Boltzmann distribution*. This asserts that the ratio of the numbers of atoms or molecules in the states with energies E_i and E_j is

(0.1.2) $$N_i/N_j = \exp\left[-(E_i - E_j)/kT\right].$$

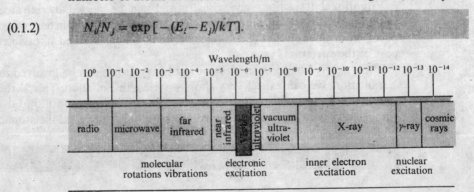

Fig. 0.7. The wavelength and type of electromagnetic radiation.

Box 0.1   Moles and molecular masses

1 mole (1 mol) of substance contains as many elementary entities (e.g. atoms, molecules, or other specified species) as there are atoms in 12 g of ^{12}C, which is 6.02205×10^{23}. Thus, if a sample contains N elementary entities, *the amount of substance is* $n = N/L$, where L is *Avagadro's constant*:

$$L = 6.02205 \times 10^{23} \text{ mol}^{-1}.$$

Likewise, if the amount of substance is n (e.g. 2 mol of O_2) the number of specified elementary entities (O_2 molecules) is nL (e.g. $2 \text{ mol} \times 6.02 \times 10^{23} \text{ mol}^{-1} = 12.04 \times 10^{23}$ oxygen molecules).

If the *mass of the sample* is $\mathcal{M}$ and the *relative molar mass* (R.M.M., the 'molecular weight') is M_r, the amount of substance is

$$n = \mathcal{M}/(M_r \text{ g mol}^{-1})$$

and the number of molecules is nL, or $L\mathcal{M}/(M_r \text{ g mol}^{-1})$. The *molar mass*, M_m, is the mass of unit amount of substance (e.g. 1 mol):

$$M_m = M_r \text{ g mol}^{-1}$$

The *molecular mass* is m, and so

$$m = M_m/L = (M_r \text{ g mol}^{-1})/L = (M_r \text{ kg mol}^{-1})/1000 L$$
$$= M_r \text{ u}$$

where u is the *atomic mass unit*:

$$1 \text{ u} = 1 \text{ kg mol}^{-1}/1000 L = 1.66056 \times 10^{-27} \text{ kg}.$$

*   *   *

L has been determined by a variety of methods. These include the determination of the gas constant and comparison with Boltzmann's constant (p. 12), the determination of the Faraday constant and comparison with the electronic charge (p. 332), and the determination by X-ray diffraction of the number of ions in a crystal of known density (p. 732). There are also several old-fashioned methods involving Brownian motion and the distribution of particles in gravitational fields.

*   *   *

The following examples indicate how much substance constitutes 1 mol at room temperature and pressure.

(i) 1 mol of atoms of a perfect gas, or of any gas behaving perfectly, occupies 24.5 l (24.5 dm³). 24 l is about 1 ft³.

(ii) 1 mol of H_2O (18 g) occupies 18 cm³, and 1 mol of C_2H_5OH (46 g) occupies 58 cm³.

(iii) 2 mol of Fe (112 g) is about 1 cubic inch.

(iv) 1 mol of NaCl (58 g) is a conical pyramid 2 cm high and 4 cm in diameter. A pinch of salt is about 10^{-3} mol.

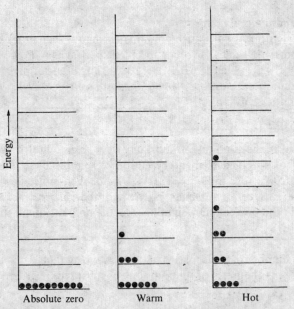

Fig. 0.8. The dependence of the populations of energy levels on the temperature.

In this expression, the fundamental constant k is *Boltzmann's constant*, with the value 1.381×10^{-23} J K^{-1}. T is the *thermodynamic temperature*, which is related to the temperature t on the centigrade (Celsius) scale by $T/K = 273.15 + t/°C$, (T is sometimes called the *absolute temperature*; K denotes *kelvin*). The multiple kL, where L is Avogadro's constant, is denoted R and called the *gas constant* ($R = 8.314$ J K^{-1}mol^{-1}) for it also appears, as we shall shortly see, in the description of the properties of gases. When the Boltzmann equation is expressed in terms of R in place of k, the energies in the exponential are molar energies (e.g. kJ mol^{-1}).

Consider first the distribution of atoms among their electronic states. In a sample of atomic hydrogen at room temperature (conventionally 25 °C, corresponding to 298.15 K), what proportion of the atoms are in the first excited electronic state? At 25 °C, $RT \approx 2.48$ kJ mol^{-1}.

The first electronically excited state of hydrogen lies about 1000 kJ mol^{-1} above the ground state. It follows that

$$\frac{N(\text{first excited electronic state})}{N(\text{electronic ground state})} = \exp\{-1000\,\text{kJ mol}^{-1}/2.48\,\text{kJ mol}^{-1}\}$$
$$= e^{-403} \approx 10^{-175}.$$

Consequently, essentially the whole sample is in the electronic ground state at room temperature. The population of the excited state rises to about 1 per cent only when the temperature is raised to 10 000 K. You should not, however, draw the conclusion that it is never necessary to consider the chemistry of electronically excited atoms. Very high temperatures are reached in rocket exhausts and explosions, and the absorption of light and the fracture of bonds can be used to generate significant concentrations of excited atoms. *The Boltzmann distribution applies only to systems in thermal equilibrium.*

The vibrational levels of molecules are closer together than the electronic levels, but normal molecules are almost entirely in their ground vibrational states at normal temperatures. Take CO, for example. The energy spacing is 25 kJ mol^{-1} and so at room temperature the populations in the first excited and ground states are in the ratio

$$\frac{N(\text{first excited vibrational state})}{N(\text{vibrational ground state})} = \exp\{-25\,\text{kJ mol}^{-1}/2.48\,\text{kJ mol}^{-1}\}$$

$$\approx e^{-10} \approx 5 \times 10^{-5}.$$

This means that only 0.005 per cent of any sample, or 3×10^{19} molecules in each mole, would be in the first excited state.

When we turn to rotational states the spacing is much smaller, and at room temperature a significant proportion of molecules are rotating at high speed. Direct application of the Boltzmann distribution, evaluated for CO, gives

$$\frac{N(\text{first excited rotational state})}{N(\text{rotational ground state})} = \exp\{-0.05\,\text{kJ mol}^{-1}/2.48\,\text{kJ mol}^{-1}\}$$

$$\approx e^{-0.02} \approx 0.98.$$

This calculation is not quite right because the first rotationally excited level is really three states of the same energy. This complication comes about because quantum theory shows that in this state the molecule can rotate in planes orientated in three different directions in space. Therefore, since all three orientations are equally likely, the ratio of the populations is 3×0.98, or 2.9. There are *more* rotating molecules than non-rotating molecules. The next rotational level is really five equal-energy states because quantum theory shows that five planes of rotation are permitted. As it lies at an energy of 0.14 kJ mol^{-1} above the ground state its relative population is

$$\frac{N(\text{second excited rotational level})}{N(\text{rotational ground state})} = 5 \exp(-0.14/2.48) \approx 4.7.$$

This result, and the population distribution at room temperature, are illustrated in Fig. 0.9.

When we come to the distribution of molecules among their translational states the Boltzmann distribution takes a special form. In the case of a gas of non-interacting particles, each energy level corresponds to a different kinetic energy, and therefore to a different speed. This means that the Boltzmann formula can be used to predict the numbers of molecules having different speeds at some temperature T. The distribution of speeds is called the *Maxwell–Boltzmann distribution*, and it has the features shown in Figs. 0.10 and 0.11. Figure 0.10 shows the distribution of molecular speeds at two widely different temperatures: notice how the tail towards high speeds is much larger at the higher temperature. Figure 0.11 shows how the molecular speeds depend on the mass: the figure shows the

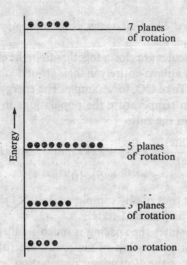

7 planes
of rotation

5 planes
of rotation

Energy ⟶

͓ planes
of rotation

no rotation

Fig. 0.9. The populations of rotational energy
levels of a linear molecule.

distribution of speeds of some constituents of air at 25 °C. The lighter
molecules move, on the average, much faster than the heavy ones, but
even the most probable speed of carbon dioxide molecules is about
330 m s^{-1}, or over 750 m.p.h.

An important deduction from the Maxwell–Boltzmann distribution is
the mean kinetic energy of a gas of particles each of mass m. The kinetic
energy of one particle is $\frac{1}{2}mv^2$. Different molecules travel at different
speeds, and to evaluate the average kinetic energy of 1 mol of particles we
must know what proportion of the sample has each speed v. This can be
calculated from the Maxwell–Boltzmann distribution, and the final result
is

(0.1.3) $$\left. \begin{array}{c} \text{mean translational} \\ \text{kinetic energy} \end{array} \right\} = \tfrac{3}{2}RT.$$

This result is an aspect of the *equipartition theorem* of classical physics.
This states, roughly, that *the average energy of each different mode of*

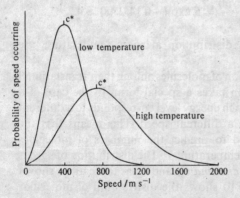

Fig. 0.10. The Maxwell–Boltzmann distribution
of molecular speeds at two temperatures.
c* denotes the most probable speed.

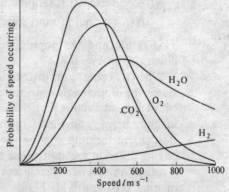

Fig. 0.11. How the distribution of molecular
speeds depends on the molecular mass.
Speeds at room temperature are shown.

motion is $\frac{1}{2}RT$. In the case of a gas there are three translational modes (corresponding to motion along three perpendicular axes), and so the total average kinetic energy is expected to be $\frac{3}{2}RT$, as stated above. The theorem also indicates that a rotating molecule may have an energy of $\frac{1}{2}RT$ associated with each of its axes of rotation, and so the mean energy of rotation of methane, with three modes, is expected to be $\frac{3}{2}RT$. The theorem also suggests that the mean vibrational energy of a bond ought to be RT ($\frac{1}{2}RT$ for the kinetic energy and $\frac{1}{2}RT$ for the potential energy), but this is hardly ever found in practice because the equipartition theorem, which is based on classical physics, fails when quantum effects are important.

Although the equipartition theorem sometimes fails we can often draw on the result that average thermal energies are of the order of RT, and $RT \approx 2.5$ kJ mol^{-1} will be our yardstick for assessing the magnitude of thermal effects at room temperature.

The only tricky point about the application of the Boltzmann distribution is the determination of the number of states corresponding to a particular energy level. A set of rules can be constructed and they are dealt with later (Part 2). The essential features of the Boltzmann distribution formula should not be obscured by this minor problem. The formula indicates that the distribution of atoms and molecules among their states is an exponential function of energy and temperature, and that more states are populated if they are close together in comparison with RT (like rotational and translational states) than if they are far apart (like vibrational and electronic states); also more states are occupied at high temperatures than at low. The exponential form occurs very widely in chemistry. For instance, many reactions proceed at a rate that depends on temperature through a term of the form $\exp(-E/RT)$, and the appearance of this factor can always be traced back to the Boltzmann distribution.

Figure 0.12 summarizes the general implications of the Boltzmann distribution for some typical energy level arrays. This figure, and Fig. 0.10, carry all the information necessary at this stage.

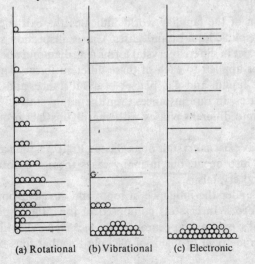

(a) Rotational (b) Vibrational (c) Electronic

Fig. 0.12. The Boltzmann distribution for different types of energy level.

The association of atoms and molecules. Atoms stick together in molecules on account of electrostatic interactions.

The simplest type of bond between atoms is the *ionic bond*. This is formed when one or more electrons are transferred from one atom to another, and the resulting ions adhere by direct Coulombic attraction. A typical example is the structure of NaCl formed from the union of Na^+ cations and Cl^- anions. As in this example, ionic bonding often leads to extensive aggregates of ions and the formation of an ionic crystal.

The other extreme type of bond is the *covalent bond*. Its basis is the sharing of a pair of electrons, and the simplest example is the bond between hydrogen atoms in molecular hydrogen. When the electrons are shared equally the molecule is *non-polar* (like hydrogen, or chlorine). When the electron pair is located nearer one atom than another the molecule is *polar* (H_2O, with negative charge predominantly on the oxygen and positive on the hydrogens, is an example).

Covalently bonded molecules are often discrete units, like N_2 or H_2O, but nevertheless may be very large (as in the case of the one illustrated in Fig. 0.3). In some cases, covalent bonds extend virtually indefinitely, and a crystal of diamond may be regarded as a colossal single, covalent molecule.

Covalent molecules congregate to form liquids and solids. This indicates that there must be forces of attraction between them even though their normal valencies are satisfied. These *van der Waals forces* have an electromagnetic origin and arise in a variety of ways. In some cases molecules stick together because they are polar. Sometimes this is assisted by the formation of hydrogen bonds (as in water). In the case of non-polar molecules, like N_2 or benzene, the rapid fluctuations of the electron clouds can give rise to instantaneous electric dipoles strong enough to result in stable solids at low enough temperatures.

0.2 The states of matter

Casual inspection of the familiar world indicates the existence of three states of matter: solids, liquids, and gases. Closer inspection shows that some solids can exist in different crystal forms (e.g. diamond and graphite). The term *phase* is applied to each of these different forms. Then we can talk of the solid, liquid, and gas phases of a substance, and also of its various solid phases. In rare instances even the liquid state of a material may be divided into different phases with sharply distinct properties.

The solid state. Pure solids may exist as crystals or as glasses, Fig. 0.13. Fine amorphous ('formless') dusts also occur, but in many cases these are nothing more than finely ground crystals.

Crystals form under the influence of ionic and covalent bonding and van der Waals forces. In crystals composed of ions, the structure (and hence the crystal's external appearance, or *morphology*) is determined largely by the geometrical problem of packing together ions of different

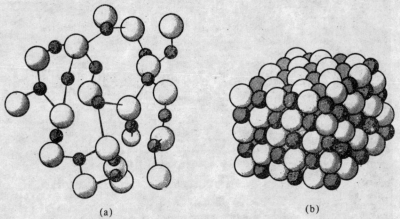

(a)   (b)

Fig. 0.13. The arrangements of atoms in (a) glassy and (b) crystalline solids.

sizes into an almost infinite, symmetrical array. The rigidity of such solids can be traced to the difficulty of disrupting large numbers of bonds simultaneously. In all real crystals there are breaks, called *defects*, in the uniformity of the array, and crystals fracture when defects are made to spread under stress.

Aggregates of covalent molecules (which are called *molecular crystals* when the molecules form periodic, symmetrical arrays, but might also be amorphous conglomerates, like butter) can often be disrupted by the application of a little heat. Warmth may cause enough molecular motion to overcome the weak intermolecular bonds, and such aggregates have low melting and boiling points.

In some cases van der Waals attractions are so weak that the crystals possess virtually no rigidity. An example of this is found among *liquid crystals*, which are swarms of aligned, often rod-like, molecules able to flow like liquids.

Metals are a special type of crystalline array because they are built from an assembly of cations (e.g. Cu^{2+}) immersed in a sea of electrons. Their electric and thermal conductivities arise from the mobility of the electron sea. Their malleability and ductility (the ability of the cations to move under stress, and then to remain in their new positions) can be traced to the ease with which the sea of electrons adjusts to permit different bonding arrangements.

The liquid state. When a solid is heated the molecules, atoms, or ions that constitute it vibrate with a greater amplitude, and at some temperature, the *melting point*, are able to move from site to site in the lattice. Figure 0.14 shows a computer simulation of the melting process in the case of a two-dimensional crystal. Above the melting point the molecules move so much that the crystal lattice ceases to be significant, and the whole sample becomes a mobile and almost structureless fluid.

The ordered structure of the solid may not be wholly lost. In the case

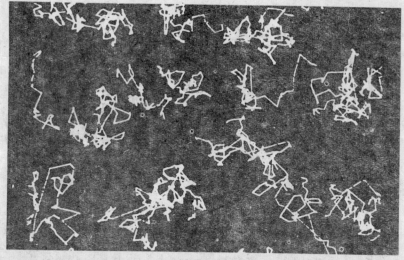

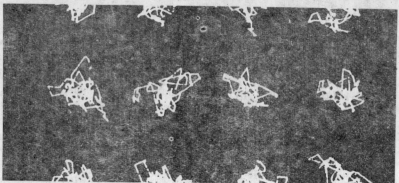

Fig. 0.14. Calculated paths of molecules in the solid and just above its melting point (B. J. Alder). Note how the molecules move from site to site in the latter case.

of water, for example, the liquid can be pictured as a collection of ice-like regions separated by structureless zones. These structures are continually forming and dispersing, and at one moment a water molecule may be in an ice-like environment, and at another in a structureless zone. This is illustrated in Fig. 0.15.

A major characteristic of liquids is their ability to flow. Highly viscous liquids, such as glass and molten polymers, flow only very slowly because their large molecules get entangled. Mobile liquids like benzene have low viscosities. Water has a higher viscosity than benzene because its molecules bond together more strongly and this hinders the flow.

We can expect viscosities to decrease with increasing temperature because the molecules then move more energetically and can escape from their neighbours more easily. Since moving a molecule from site to site involves breaking weak van der Waals bonds we can guess that the number of molecules with enough energy to move follows a Boltzmann distribution. This suggests that the ability of the liquid to flow ought to behave as

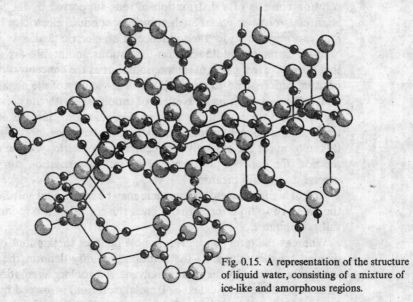

Fig. 0.15. A representation of the structure of liquid water, consisting of a mixture of ice-like and amorphous regions.

fluidity $\propto \exp\left(-\Delta E/RT\right)$

where ΔE is the energy to be overcome. The viscosity is the inverse of the fluidity, and so we can expect that

(0.2.1) viscosity $\propto \exp\left(\Delta E/RT\right)$.

Experimental observation of the temperature-dependence of the viscosity follows this exponential form, and shows that the value of ΔE is of the same order as the intermolecular binding energy (a few kJ mol^{-1}; e.g. 11 kJ mol^{-1} for benzene and 3 kJ mol^{-1} for methane).

Ionic crystals often dissolve in solvents that can form an electrostatic association with the ions. For instance, water is composed of polar molecules which are able to associate with the ions and to break up the crystals by *solvation*, Fig. 0.16. When an ionic crystal has dissolved, the

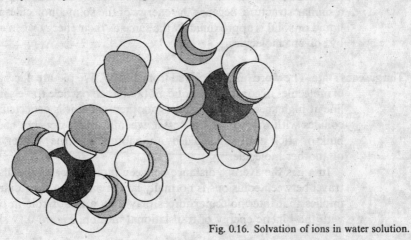

Fig. 0.16. Solvation of ions in water solution.

solution consists of a distribution of ions supported by the solvent; this is an *electrolyte solution*. Such solutions conduct electricity because the ions can migrate under the influence of an electric field.

In an extremely dilute solution the cations and anions are so far apart that they have insignificant interactions, but as the concentration increases positive cations tend to congregate in the vicinity of the negative anions, and vice versa. This has the effect of modifying both the conductivity of the ions and their ability to take part in reactions. Instead of talking in terms of the concentration of ions it then becomes more significant to talk in terms of their effective concentration, or *activity*. Later on we shall see how the concept of activity of an ion in a solution can be given a precise and useful meaning.

At still higher concentrations there may be insufficient solvent to solvate the ions and to hold them apart. Then the crystal lattice re-forms and the salt precipitates.

Whereas the term *concentration* now denotes the amount of substance divided by the volume of the solution, *molality* denotes the amount of substance divided by the mass of solvent. A 1 mol kg^{-1} solution (formerly and often colloquially called a '1 molal' solution) is formed by dissolving 1 mol of substance in 1 kg of *solvent*. A 1 mol dm^{-3} solution (formerly and often colloquially called a '1 molar' solution) is formed by dissolving 1 mol of substance in enough solvent to produce 1 dm^3 of *solution*. You should note that concentration depends on the temperature but molality does not.

It is useful to be able to picture the structure of solutions of various strengths. In the case of a 1 mol dm^{-3} solution of sodium chloride in water the average distance between oppositely charged ions is then about 1 nm, enough to accommodate about 3 water molecules. A 'dilute solution' often means a concentration of about 0.01 mol dm^{-3} (1/100 molar) or less, and in this the ions are separated by about 10 water molecules.

Non-ionic molecules often dissolve in non-polar or weakly polar solvents to form *solutions of non-electrolytes*. A typical example is toluene dissolved in benzene. The dissolution proceeds well if the solvent and solute have a similar structure, because the energy of the solute molecules in the pure liquid or solid is approximately the same as their energy when surrounded by solvent molecules.

The gaseous state. The word 'gas' is derived from 'chaos'. We picture a gas as a swarm of molecules in constant, chaotic motion. Each particle travels in a straight line at high speed until it reaches another, when it is deflected: or until it collides with the wall of the vessel, when it might ricochet back into the bulk or stick until dislodged by the vibration of the wall or the impact of another molecule.

In a gas the average distance between molecules and the distance they travel between collisions is normally large relative to their diameter. This implies that intermolecular forces play only a minor role in comparison with the kinetic energy of translational motion. Figure 0.17 shows three

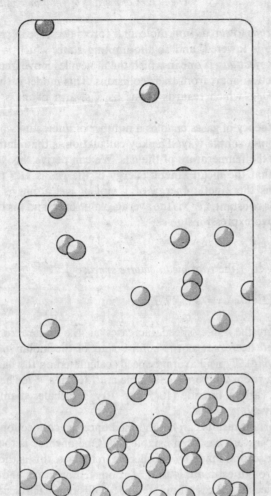

Fig. 0.17. Samples of argon gas at 1 atm (*top*), 10 atm (*middle*), 30 atm (*bottom*). The drawings represent the number of molecules in a slab 5 nm thick and are scaled to the atomic size of argon.

scale drawings of a sample of argon at room temperature and three different pressures. The drawing suggests that intermolecular forces are likely to be negligible at the lowest pressure (1 atm), becoming significant in the region of pressures somewhat above normal (10 atm), and very important at higher pressures (30 atm).

It is found by experiment that, when the pressure is sufficiently low, all gases satisfy the *perfect gas law*:

(0.2.2)
$$pV = nRT.$$

p is the pressure, V the volume, n the amount of substance of gas, and T the absolute temperature. Deviations from this relation occur at high pressures, but it is a law to which real gases conform increasingly closely as their pressure is lowered. Intermolecular forces decrease in significance as the pressure is lowered, and so the limiting state, which is called an *ideal gas* or a *perfect gas*, is one in which the molecules move freely, totally without interaction apart from their collisions. This model is the basis of the *kinetic theory*, which regards a gas as a swarm of mass points in constant motion.

The kinetic theory of gases enables a number of interesting conclusions to be deduced in a simple way. The key calculation is the relation of the mean speed to the temperature of the gas. We can derive this very easily from the fact that the mean kinetic energy of the gas is $\frac{3}{2}RT$ (p. 15). In molecular terms the kinetic energy of 1 mol of gas atoms is $\frac{1}{2}m\langle v^2\rangle L$ (L is Avogadro's constant, $\langle v^2\rangle$ is the average value of v^2, and v is the speed). Equating the two expressions gives

$$\langle v^2\rangle = 3RT/mL.$$

We shall call $\sqrt{\langle v^2\rangle}$ the *root mean square speed*:

(0.2.3)
$$\sqrt{\langle v^2\rangle} = \sqrt{(3RT/mL)} = \sqrt{(3RT/M_m)}.$$

This shows that the r.m.s. speed increases as the square root of the temperature and decreases as the square root of the molar mass (from Box 0.1, $M_m = mL$). Typical average speeds calculated on the basis of this expression for a room temperature sample are 1360 m s^{-1} for helium and 411 m s^{-1} for carbon dioxide (10^{13} and 10^{12} molecular diameters per second respectively).

Several important results can be obtained once the connection between speed and temperature has been established. One of these is the number of collisions per second made by a molecule in the gas: this is the *collision frequency z*. A rough estimate is that at room temperature and pressure an oxygen molecule makes about 6×10^9 collisions each second. The distance a molecule flies between collisions is called its *mean free path λ* (lambda). Since a molecule collides with a frequency z and travels with a speed $\sqrt{\langle v^2\rangle}$, we have immediately that $\lambda \approx (\sqrt{\langle v^2\rangle})/z$. For oxygen the value of λ comes out as 70 nm (200 collision diameters) at room temperature and pressure.

Transformations of matter. Two kinds of question arise in connection with the physical and chemical transformations of matter. One is 'Can it occur?', and the other is 'How fast does it occur?'.

The second question is explored in *chemical kinetics* and, at a more fundamental level, in *molecular reaction dynamics*. It is possible at this stage, though, to construct a rudimentary model of a reaction as involving the collision of two species with sufficient energy to break and form bonds. An expression for the rate of such a process is obtained as follows.

The frequency of collisions is z. The probability that when the collision occurs it does so with sufficient energy to react is determined by the Boltzmann distribution. Therefore if E_a is the necessary energy (the *activation energy* of the reaction) the rate is given by

rate of reaction $\approx$ (rate of collision) $\times$

(probability that collision carries enough energy);

so that

(0.2.4)

$$\text{rate} \approx z(T)\exp(-E_a/RT).$$

The collision frequency varies with temperature, but its variation is normally dominated by the exponential temperature dependence coming from the Boltzmann factor. The last expression, the *Arrhenius Law*, is often obeyed experimentally, and is a fair first approximation to the actual experimental results.

, The answer to the first question, whether a change *can* occur, is the domain of *thermodynamics*. Thermodynamics deals with the possible transformations of energy, and embodies its conclusions in three laws.

The *First Law* states that energy can be neither created nor destroyed. Therefore any change must conserve the total amount of energy.

The *Second Law* takes the argument a stage further. It introduces another property of the system, the *entropy*, and uses it to set up a criterion for assessing whether or not a transformation of the system has a natural tendency to occur (like the flow of heat from hot to cold, or the rusting of iron).

The entropy of a system is determined by the way its molecules are distributed over the available energy levels, and it can be calculated once these are known: This important connection between the ideas of quantum theory and the laws of thermodynamics is the domain of *statistical thermodynamics*.

0.3 Force, pressure, and energy

The concepts of force, pressure, and energy occur throughout chemistry, and it is important to be familiar with their units.

If you can imagine the force necessary to move a mass of 1 kg so that it accelerates at the rate of 1 m s^{-2} (if the mass is at rest initially, that means it should have a velocity of 1 m s^{-1} after 1 s, and 2 m s^{-1} after 2 s, and so on), then you will have some appreciation of the basic unit of force, which is the *newton* ($1 \text{ N} = 1 \text{ kg m s}^{-2}$). For instance, this book, which has a mass of about 1.5 kg, would, if released, fall downward with an acceleration of 9.8 m s^{-2} (the acceleration due to gravity, g) and so it is subject to a force of 15 N. Holding the book enables you to experience that force. A small apple on a tree experiences a force of about 1 newton.

Imagining a force of 1 N applied to an area of 1 m^2 gives an idea of the basic unit of pressure. When we come to do quantitative calculations we shall quote pressures in N m^{-2}. This unit is also called the *pascal* ($1 \text{ Pa} = 1 \text{ N m}^{-2}$). An important unit of pressure is the *atmosphere* (atm). There is international agreement that 1 atm is exactly $101\,325 \text{ N m}^{-2}$

(101.325 kPa), but remembering that 1 atm $\approx 10^5$ N m^{-2} is often sufficiently accurate for estimating the magnitude of a physical quantity.

One way of appreciating the magnitude of a pressure of 1 atm is to
imagine the pressure exerted by a mercury column in a barometer: a
column of mercury 760 mm high exerts a pressure of 1 atm. The use of
mercury in barometers is responsible for two other measures of pressure.
The first is *millimetres of mercury*, written mmHg; in these units 1 atm $\triangleq$
760 mmHg (the sign $\triangleq$ means 'corresponds to'). The other is the *Torr*
(named after Torricelli, who invented the barometer). The Torr is virtually
identical to the mmHg, and so you will sometimes see 1 atm $\triangleq$ 760 Torr.
Pressure is also quoted in *bar*: 1 bar $\triangleq 10^5$ N m^{-2}, and so a rough approximation is that 1 bar $\approx$ 1 atm. This unit and its derivative the kilobar
(kbar) are used extensively in high-pressure chemistry and geochemistry.

The unit for measuring energy is the *joule* (J). The formal definition of
1 J is that it is the energy needed to push against a force of 1 N for 1 m
(1 J = 1 N m). There are several ways of envisaging the size of this unit.
For instance, raising this book 1 m in the air requires the expenditure of
about 15 J of energy. Each pulse of a human heart consumes about 1 J
of energy. Alternatively, when 1 kJ of energy is dissipated as heat in
50 cm^3 of water, its temperature rises by about 5 °C. This means it requires
about 15 kJ of energy to make a small cup of coffee.

Other units of energy are often encountered. One of the most common
is the thermochemical calorie (cal or cal$_{th}$). This is defined as 4.184 J, and
so it is easy to make the conversion. Much of the literature of chemistry
quotes energies in kcal mol^{-1} (kilocalories per mole) even though the
measurements were originally made in joules, and the data can easily be
translated into kJ mol^{-1} by multiplication by 4.184.

Another unit is the *electron volt* (eV). This is the energy acquired by an
electron when it is accelerated through a potential difference of 1 V. For
chemical applications it is better to think in terms of 1 mol of electrons,
then the conversion reads 1 eV $\triangleq$ 100 kJ mol^{-1}, or 96.485 kJ mol^{-1} to be
precise.

In the first section of this chapter we recognized that spectral lines arise
from transitions between energy levels. Since the energy of a transition
can be expressed in terms of its frequency through the Bohr frequency
condition (p. 9), it is appropriate in some cases to express energies as an
equivalent frequency, or as the *wavenumber* $\tilde{v} = v/c$. This conversion reads
1000 cm^{-1} $\triangleq$ 11.96 kJ mol^{-1}.

These units, and their interconversions, are collected in the Tables on
the first endpapers of the book.

Part 1 · Equilibrium

1 The properties of gases

Learning objectives

After careful study of this chapter you should be able to

(1) Write the *equation of state* for a perfect gas (eqn (1.1.1)) and use it to predict changes in pressure, volume, and temperature (p. 30).

(2) Explain the meaning of the term *limiting law* (p. 29).

(3) Describe the application of a constant volume gas thermometer to the setting up of a *temperature scale* (p. 32).

(4) Define *partial pressure* of a gas in a mixture and relate it to the *mole fraction* of the component (p. 34).

(5) State how *real gases* differ from the perfect gas (p. 36) and interpret the *pressure–volume* behaviour of gases (p. 38).

(6) Define the terms *isotherm* (p. 30), *compression factor* (p. 36), *vapour pressure* (p. 39), and *critical constants* (p. 39).

(7) Write the form of a *virial expansion* (eqns (1.3.1) and (1.3.2)) and define the *Boyle temperature* (p. 38).

(8) Write down and explain the basis of the *van der Waals equation of state* (eqn (1.4.1)), and relate its parameters to the critical constants of the gas (p. 43).

(9) Define the *reduced variables* of a gas (p. 44) and state and use the principle of *corresponding states*.

Introduction

The first purpose of this chapter is to see how the state of a gas can be described, and how its properties depend on its condition. The ideas developed here lie at the heart of the calculations encountered when we try to predict and describe the rates and courses of chemical reactions.

The second purpose is to begin to show how to construct models of molecular behaviour. This is important because understanding the behaviour of molecules makes it possible to suggest explanations for newly discovered phenomena and to judge the plausibility of others' explanations without getting trapped in lengthy calculations. Gases are so simple that they provide an excellent introduction to this technique.

1.1 Equations of state: the perfect gas

The basic quantities for the study of gases are pressure and temperature. Pressures may be measured with a *manometer*, which in its simplest form is a U-tube filled with some liquid of low volatility (mercury is commonly used). The pressure of the system is given by the difference in height of the two columns of mercury (plus the external pressure if one tube is open to the atmosphere). That is the basis of the unit mmHg for the measurement of pressures. More sophisticated techniques are used at lower pressures. The *McLeod gauge*, for instance, works on the basis of withdrawing a known volume of the sample gas, compressing it into a known smaller volume, and measuring the pressure of the compressed sample. The compression magnifies the pressure and makes it readily measurable; the pressure within the vessel itself is then calculated on the basis of the gas laws (which we develop in this chapter). Methods that avoid the complication of having to account for the intrusion of vapour from the manometer fluid are also readily available. These include monitoring the deflection of a diaphragm, either mechanically or electrically, or the change in some other pressure sensitive electrical property.

The existence of the 'temperature' of a sample, and its measurement, depends upon the validity of a generalization which is often referred to as the *Zeroth Law of thermodynamics*. The Zeroth Law states that if a system A is in thermal equilibrium with a system B (in the sense that no change of state occurs when the two are in thermal contact), and if B is in equilibrium with C, then C is also in equilibrium with A whatever the detailed nature of the systems. This indicates the existence of a property of systems that is independent of their nature and which signifies the existence of a condition of thermal equilibrium. This property we refer to as the system's *temperature*. The Zeroth Law also guarantees that we can construct a device from any material and be confident that a particular property (such as the length of a column of mercury or the resistance of a length of wire) will give the same reading when it is in contact with any of the systems A, B, C, . . . that are in mutual thermal equilibrium (that 'have the same temperature').

The relation of the numerical value of the temperature to the property selected for monitoring it is arbitrary, but the common sense of practical workers in laboratories has resulted in the adoption of a scale which can be readily used and also sharpened into precise significance. In the early days of thermometry temperatures were related to the length of a column of liquid, and the difference of length shown when the thermometer was first in thermal equilibrium with melting ice and then in contact with water boiling at atmospheric pressure was divided into 100 steps, the lowest mark being labelled 0. That essentially constituted the *Celsius scale*. Different liquids, though, expand at different rates, and so thermometers constructed from different materials showed different numerical values of the temperature. Thus, while the systems A, B, C, . . . may all have been ascribed the temperature 28.7 °C when a mercury-in-glass thermometer was used, they may have been ascribed 28.8 °C when an alcohol-in-glass thermometer was used. The variation in quoted temperatures is even greater when electrical measurements are included. One class of materials, however, gives rise to a temperature scale that is almost independent of the nature of the monitoring substance: these are the gases, the monitored property being the pressure at constant volume. Furthermore, this approximate uniformity tends towards exact uniformity as the densities of the measuring gases are reduced, and becomes exact in the limit of zero density. This lets us establish a scale of temperature known as the *perfect gas temperature* scale, and we shall return to its precise specification below.

The suspicion will arise that the measurement of temperature is based on a cyclic argument: temperature measurement depends on the properties of the idealization we refer to as a perfect gas; perfect gases show a simple temperature dependence. That is indeed the case: the argument is cyclic. Nevertheless, in due course we shall see that a temperature scale can be established wholly independently of the properties of any substance, and of perfect gases in particular; happily this *thermodynamic temperature scale* coincides with the perfect gas scale, and its introduction breaks the cycle of argument. For the present, therefore, we shall regard the perfect gas scale as a common sense refinement of practical temperature scales, and return to the rigorous discussion in Chapter 6.

The experiments of Boyle, Gay-Lussac, and their successors show that the pressure (p), volume (V), temperature (T), and amount of substance (n) of gases are related by the expression

$(1.1.1)°$
$$pV = nRT$$

and that this relation is obeyed increasingly closely as the density is decreased or the temperature is increased. R, the *gas constant*, is a fundamental constant, independent of the nature of the gas. A gas that obeys eqn (1.1.1) exactly is called *perfect* or *ideal*. All gases conform ever more closely to this equation as the density is reduced, and conform exactly in the limit of zero density: for this reason eqn (1.1.1) is termed a *limiting law* for the description of real gases. When an equation relates to a perfect gas (by which we shall also mean a real gas behaving perfectly) a superscript ° will be attached to the equation number.

The perfect gas equation is one example of an *equation of state*. It is impossible to force a given amount of a perfect gas into a state of pressure, volume, and temperature that does not satisfy this relation.

The perfect gas equation of state is an idealization of experimental observations. We shall look briefly at its implications and interpret them in terms of molecular behaviour.

Response to pressure: Boyle's Law. In 1661 Boyle, acting on the suggestion of his assistant Townley, verified that at constant temperature *the volume of a given amount of gas is inversely proportional to the pressure*:

(1.1.2)°   *Boyle's Law*: $V \propto 1/p$, or $pV = $ constant (at constant n and T).

The p, V relation is illustrated in Fig. 1.1. Each curve corresponds to a single temperature, and is called an *isotherm* ('iso' means 'the same'). The early experiments were crude, and we now know that gases obey this law strictly only in the limit of $p \rightarrow 0$ or of $T \rightarrow \infty$.

Boyle's Law is used to predict the pressure of a gas when its volume is changed (or vice versa). If the initial values of pressure and volume are p_i and V_i, the final values p_f and V_f must satisfy

(1.1.3)° $p_f V_f = p_i V_i$ (constant n and T).

The explanation of the law is based on the view that the pressure exerted by a gas arises from the impact of its molecules on the walls of the vessel. If the volume is halved the density of molecules is doubled. In a given interval of time twice as many molecules strike the walls, and so the pressure is doubled. Therefore, halving the volume doubles the pressure, in accord with Boyle's Law. At very low densities the effects of inter-molecular forces are negligible, and so it is also easy to see why the law is *universal* in the sense that it applies to all gases (so long as they are behaving perfectly) without reference to their chemical composition.

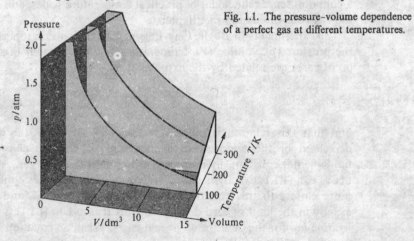

Fig. 1.1. The pressure-volume dependence of a perfect gas at different temperatures.

Response to temperature: The law of Gay-Lussac and Charles. The quantitative study of the thermal expansion of gases was made first by Charles (1787), the inventor of the hydrogen balloon (once known as the *Charlière*). He measured the effect of temperature on the volume of a fixed amount of gas, but did not publish his results. Then Gay-Lussac (1802) studied the effect in greater detail.

Gay-Lussac's observations led him to conclude that, at constant pressure, *the volume of a given amount of gas increases in proportion to the temperature*, Fig. 1.2. An alternative form is that at constant volume *the pressure of a given amount of gas is proportional to the temperature*. The mathematical expression of these observations is

$$Gay\text{-}Lussac\text{'}s\ Law:\ V \propto T \text{ (at constant } n \text{ and } p)$$

(1.1.4)° $$p \propto T \text{ (at constant } n \text{ and } V).$$

Gay-Lussac's Law enables us to predict the volume of a perfect gas when a fixed amount is heated under constant pressure. Equation (1.1.4) gives

(1.1.5)° $$V_f = (T_f/T_i)V_i \text{ (constant } n \text{ and } p).$$

The alternative version allows us to predict the pressure when a fixed amount is heated in a constant volume:

(1.1.6)° $$p_f = (T_f/T_i)p_i \text{ (constant } n \text{ and } V).$$

Example (Objective 1). In an industrial process nitrogen has to be heated to 500 K in a vessel of constant volume. If it enters the vessel at a pressure of 100 atm and a temperature of 300 K, what pressure does it exert at the working temperature?

• *Method.* In the absence of more detailed information we assume perfect behaviour and use eqn (1.1.6).

• *Answer.* $p_f = (500 \text{ K}/300 \text{ K}) \times (100 \text{ atm}) = 167 \text{ atm}.$

• *Comment.* Experiment shows that the pressure exerted is 183 atm under these circumstances, and so the assumption of perfect behaviour is not very good.

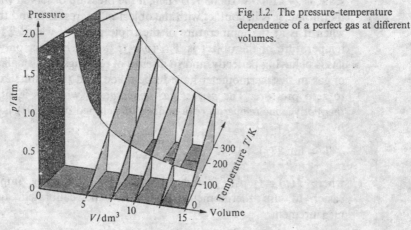

Fig. 1.2. The pressure–temperature dependence of a perfect gas at different volumes.

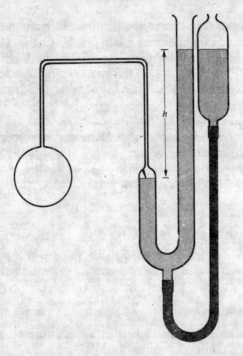

Fig. 1.3. A constant volume gas thermometer. The pressure of the enclosed gas is measured in terms of the height h.

The temperature dependence of the pressure of a gas allows us to establish a temperature scale (and to measure temperature) without having to rely on the vagaries of the behaviour of liquids in glass capillaries. Since a real gas behaves perfectly in the limit of zero pressure, the temperature can be measured with a *constant volume gas thermometer* (Fig. 1.3) by comparing the pressures when the thermometer is in thermal contact first with the object of interest and then with a standard system. The latter is taken as the *triple point of water* (the unique condition of pressure and temperature when ice, liquid water, and water vapour can coexist in equilibrium), and its temperature is defined as 273.16 K exactly (the freezing temperature under a pressure of 1 atm, the zero on the Celsius scale, is 273.1500 ± 0.0003 K). Then, if the pressure observed when the gas thermometer is in contact with the object of interest is p, and the pressure when it is at the temperature of the triple point of water, T_3^*, is p_3, the temperature of the object is $T \approx (p/p_3)T_3^*$. This is exact only when the gas is behaving perfectly, and so a series of readings with smaller amounts of gas in the thermometer have to be taken, and the results extrapolated to zero pressure. The *perfect gas temperature*, which is identical to the *thermodynamic temperature* (a concept developed on p. 201), is then given by

$$T = \lim_{p \to 0} T(p)$$

where $T(p) = (p/p_3)T_3^*$, and $T_3^* = 273.16$ K. Ordinary, imperfect, but easier to use thermometers may then be calibrated against this measurement.

The explanation of Gay-Lussac's Law recognizes that raising the temperature of a gas increases the average molecular speed. If the confining volume is kept constant, the molecules collide with the walls more frequently and with a greater impact and therefore exert on them a stronger net force. If the volume and number of molecules are held constant, this appears as an increase in the pressure exerted by the gas.

Why does the pressure increase *linearly* with the temperature? The key lies in the dependence of the average molecular speed on the square root of the temperature (p. 22). The pressure arises from collisions of the molecules with the walls and the frequency of collisions increases in proportion to the molecular speed. But the *momentum* of a molecule, which also depends linearly on its speed, determines the force it exerts on the wall during a collision. Therefore both the frequency of collisions and their effectiveness increase with the speed. Consequently we find that the pressure increases as the square of the average speed, and therefore in proportion to the temperature. This qualitative argument is made quantitative in the opening chapter of Part 3.

Collecting the fragments: the gas constant. The two sets of experimental observations described above, $V \propto 1/p$ and $V \propto T$ can be combined into $pV \propto T$, or $pV_m \propto T$, where V_m is the *molar volume* (the volume occupied by unit amount of gas, $V_m = V/n$, n being the amount of substance (usually in moles, Box 0.1)). All we need now is the constant of proportionality, the *gas constant*, R, in the full equation, eqn (1.1.1). This may be obtained from the value of pV/nT for a real gas in the limit of zero pressure (so that perfect behaviour is ensured), and so measurements of the pressure and volume of a known amount of gas are made at a series of pressures and their product is extrapolated to zero. The result is that $R = 0.082\,057\,5\ \mathrm{dm^3\ atm\ K^{-1}\ mol^{-1}}$ or $8.314\,41\ \mathrm{J\ K^{-1}\ mol^{-1}}$. The implication of this result is that the molar volume of a perfect gas at $25\,^\circ\mathrm{C}$ ($298.15\ \mathrm{K}$) and 1 atm ($101.325\ \mathrm{kPa}$, $1.013\,25 \times 10^5\ \mathrm{N\ m^{-2}}$) is

$$V_m = RT/p$$
$$= \frac{(8.314\,41\ \mathrm{J\ K^{-1}\ mol^{-1}}) \times (298.15\ \mathrm{K})}{(1.013\,25 \times 10^5\ \mathrm{N\ m^{-2}})}$$
$$= 2.446\,52 \times 10^{-2}\ \mathrm{m^3\ mol^{-1}},\ \text{or}\ 24.465\,2\ \mathrm{dm^3\ mol^{-1}}.$$

This is about 1 cubic foot.

1.2 Mixtures of gases: partial pressures

So far we have dealt only with pure gases. This is too restrictive for chemical considerations, and so we shall now look at mixtures.

Suppose an amount of substance n_A of some gas A is injected into a container of volume V; then according to the perfect gas equation of state the pressure exerted is $p_A = n_A(RT/V)$. Had we injected an amount n_B of another gas B into a container of the same volume and at the same temperature, it would have exerted a pressure $p_B = n_B(RT/V)$. But suppose that gas B is injected into the container already containing gas A: what is the total pressure?

In the nineteenth century Dalton* made observations which answer this question. *Dalton's law of partial pressures* states that *the pressure exerted by a mixture of gases behaving perfectly is the sum of the pressures exerted by the individual gases occupying the same volume alone.* This means that in the present example the total pressure is simply

$(1.2.1)°$ $$p = p_A + p_B = (n_A + n_B)(RT/V).$$

Dalton's Law is not confined to a mixture of only two gases: if the mixture consists of the gases A, B, C, ... present with the amounts n_A, n_B, n_C, ... respectively, then the total pressure is simply

$(1.2.2)°$ $$p = p_A + p_B + p_C + \ldots$$

where the *partial pressure* p_J of the component J is given by

$(1.2.3)°$ $$p_J = n_J(RT/V), \quad J = A, B, C \ldots$$

Example (Objective 2). 1 mol of nitrogen and 3 mol of hydrogen are injected into a container of volume 10 dm³ at 298 K. What are the partial pressures and the total pressure?

- *Method.* We assume perfect behaviour and calculate p_J on the basis of eqn (1.2.3) and the total pressure from eqn (1.2.2).

- *Answer.* $p(N_2) = (1 \text{ mol}) \times (0.0821 \text{ dm}^3 \text{ atm K}^{-1} \text{ mol}^{-1}) \times (298 \text{ K})/(10 \text{ dm}^3)$
  $$= 2.45 \text{ atm.}$$
  $$p(H_2) = 3 \times (2.45 \text{ atm}) = 7.35 \text{ atm.}$$
  $$p = 2.45 \text{ atm} + 7.35 \text{ atm} = 9.80 \text{ atm.}$$

- *Comment.* Below we shall see that the definition of the partial pressure of a *real* gas is $p_J = x_J p$; in the present case $p(N_2) = 0.25p$ and $p(H_2) = 0.75p$ irrespective of whether the gases are real or perfect, but for the real gases the combined pressure is different from the value $p = 9.80$ atm calculated by assuming perfect behaviour.

Another way of expressing the law is to introduce the *mole fraction* of the component J. Suppose we have a mixture of gases A, B, ... with the amounts n_A, n_B, ..., and a total amount $n = n_A + n_B + \ldots$ of the mixture. Then the fraction n_J/n is the relative amount of component J present: this is called the *mole fraction* of J, and denoted x_J. It follows that the partial pressure of component J in an amount n of perfect gas occupying a volume V at a temperature T is

$(1.2.4)°$ $$p_J = n_J(RT/V) = x_J nRT/V = x_J p.$$

---

* Dalton is the Dalton of the atomic hypothesis, and also the Dalton of 'daltonism', or colour-blindness (from which he suffered and which he described). He was described as 'an indifferent experimenter, and singularly wanting in the language and power of illustration'. He paid another price: 'Into society he rarely went, and amusement he had none, with the exception of a game of bowls on Thursday afternoons'. Was that the source of the atomic hypothesis?

That is, the *partial pressure is proportional to the mole fraction*. Since the sum of the mole fractions is unity, the total pressure is

$$p = p_A + p_B + \ldots = (x_A + x_B + \ldots)nRT/V = nRT/V.$$

This is just the perfect gas law for an amount n of a gas of unspecified and irrelevant composition. Figure 1.4 illustrates how the partial pressures contribute to the total pressure of a two-component system as the mole fraction of one component varies from 0 to 1. The relation $p_J = x_J p$ also serves as the *definition* of the partial pressure of a *real* gas even when it is not under conditions of perfect behaviour. Then, however, we cannot assume that the partial pressure of J is the pressure it would exert if it were alone in the container.

Example (Objective 2). The composition of dry air at sea level, in mass per cent, is approximately as follows: N_2: 75.52; O_2: 23.15; Ar: 1.28; CO_2: 0.046. What is the partial pressure of each component when the total pressure is 1 atm?

● *Method.* Use eqn (1.2.4) in the form $p_J = x_J p$: this means that we must find the mole fraction of each component. Percentage composition by mass can be converted into mole fractions from a knowledge of the molar masses $M_m(J)$ of the components. If the component J contributes w_J per cent of the mass of the sample, 100 g of sample contain w_J g of that component, and therefore the amount of J in the sample is $n_J = w_J g/M_m(J)$. The total amount present is $n = w(N_2)g/M_m(N_2) + w(O_2)g/M_m(O_2) + \ldots$, and so the mole fraction x_J of the component J is $w_J/M(J)$ divided by this sum.

● *Answer.*
$$w(N_2)g/M_m(N_2) = (75.52\ g)/(28.0\ g\ mol^{-1}) = 2.70\ mol$$
$$w(O_2)g/M_m(O_2) = (23.15\ g)/(32.0\ g\ mol^{-1}) = 0.72\ mol$$
$$w(Ar)g/M_m(Ar) = (1.28\ g)/(39.9\ g\ mol^{-1}) = 0.032\ mol$$
$$w(CO_2)g/M_m(CO_2) = (0.046\ g)/(44.0\ g\ mol^{-1}) = 0.001\ mol$$

total amount in 100 g = 3.45 mol
mole fractions: $x(N_2) = 0.782$, $x(O_2) = 0.208$, $x(Ar) = 0.009$, $x(CO_2) = 0.0003$
partial pressures: $p(N_2) = 0.782$ atm, $p(O_2) = 0.208$ atm, etc.

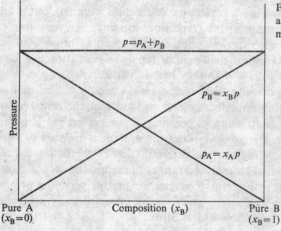

Fig. 1.4. The total pressure and partial pressures of a mixture of perfect gases.

$p = p_A + p_B$

$p_B = x_B p$

$p_A = x_A p$

Pressure

Pure A
$(x_B = 0)$

Composition (x_B)

Pure B
$(x_B = 1)$

• *Comment.* Perfect gas behaviour has not been assumed: the partial pressures are defined as $p_J = x_J p$ for any gas.

1.3 Incorporating imperfections

If a gas does not behave perfectly, we should expect its equation of state to differ from the simple $pV = nRT$ characteristic of a perfect gas. We can guess that we cannot cool a gas to zero volume, or squash it out of existence, and we know that in some regions of pressure and temperature gases become liquids and solids.

The molecular basis of the deviations from ideality is the interaction between individual molecules. Molecules are small, but they are not infinitesimal; therefore we can expect them to show resistance to compression when they are squeezed together as the gas is compressed. We might be tempted to draw the conclusion that it is more difficult to compress a real gas because the molecules resist being pressed together. But we must be careful: we must also remember that molecules attract each other. Attractions are responsible for the loose but significant cohesion of gaseous molecules into a mobile liquid. Attractions between molecules have the effect of favouring compression.

Which process triumphs? Is a real gas easier to compress than a perfect gas on account of the molecular attractions, or is it more difficult because of the molecular repulsions?

The key to the problem is the knowledge that the repulsive forces come into operation only when the molecules are almost in contact: *repulsive forces are short-range forces,* even on a scale measured in molecular diameters. Consequently they can be expected to dominate when the molecules are squashed together (which means at high density and so at high pressure). Attractive forces have a relatively long range and are effective over several molecular diameters. Therefore we can expect them to dominate when the molecules spend most of their time fairly close to one another, but not squashed together (which means at moderate pressures). This suggests that at moderate pressures a real gas is compressed more easily than a perfect gas (attraction winning), but at high pressures it is less compressible (repulsion winning). At very low pressures the molecules are rarely in contact, and spend so much of their time so far from one another that neither repulsive nor attractive forces play any significant role; and so they behave perfectly.

Is there evidence that real gases do behave in this manner? One way of examining imperfect behaviour is to plot the *compression factor* $Z = pV_m/RT$ as a function of pressure. For a perfect gas Z is unity under all conditions, and so deviation from unity is a measure of imperfection.

Some results from the determination of the compression factors of a number of real gases over a range of pressures but at constant temperature are shown in Fig. 1.5. You can see that at very low pressures all these gases have Z close to unity; as we expected, they are behaving almost perfectly. At high pressures all the gases have $Z > 1$ (the product pV_m is

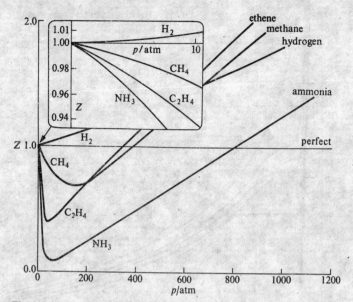

Fig. 1.5. The compression factor $Z = pV_m/RT$ for a variety of gases at $0\,°C$.

greater than RT) which means they are harder to compress than a perfect gas. This fits the argument that repulsive forces are dominant in this range. At lower pressures some of the gases have $Z < 1$, indicating that the attractive long-range forces are dominating and are favouring compression.

Note that for some gases (such as methane) marked deviations from perfect behaviour occur only at high pressures. This is pleasing because it shows how well the perfect gas equations apply to a number of real gases. But you can also see how some common gases (for example, ammonia) show large deviations even at low pressures: this confirms that we must be prepared to discuss gas imperfections if we want to make useful deductions.

The next step in the analysis is to turn from the compression factor and look at the whole p, V, T-behaviour of a real gas. One way of doing this is to select a series of temperatures and illustrate how p depends on V for each one. In other words, we draw a number of experimental isotherms. Figure 1.6 shows this done on the basis of experimental data for carbon dioxide.

At low pressures the real and perfect isotherms (drawn in Fig. 1.1) do not differ very much. The isothermal compression of the real gas—which corresponds to moving along one of the lines in Fig. 1.6 from right to left (the pressure is raised and the volume contracts)—is in reasonable accord with Boyle's Law. Nevertheless there are differences, and we might guess that $pV_m = RT$ is only the first term of a more complicated expression of the form

(1.3.1) $$pV_m = RT\{1 + B'(T)p + C'(T)p^2 + \ldots\}.$$

In many applications a more convenient expansion is

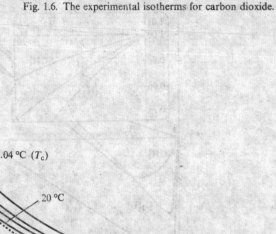

Fig. 1.6. The experimental isotherms for carbon dioxide.

$$pV_m = RT\{1 + B(T)/V_m + C(T)/V_m^2 + \ldots\}.$$

(1.3.2)

This is known as the *virial equation of state* (virial comes from the Latin word for force). The coefficient $B(T)$ is called the *second virial coefficient*; $C(T)$, which is usually less important, is the third, and so on. We need to be aware of this way of writing an equation of state, but apart from some illustrations in the Problems, it will not appear in any detail until Part 2. Note, however, that the coefficients are temperature-dependent, and there may be a temperature where $B'(T)$ is zero: in that case $pV_m \approx RT$ over an extended pressure range because $C'(T)p^2$ and the higher terms are negligibly small. This temperature, at which perfect behaviour is shown over an extended range of pressures, is called the *Boyle temperature*, and is denoted T_B. For helium $T_B = 22.64$ K, for air $T_B = 346.81$ K. (Further values are given in Table 1.2 on p. 44.)

Considerable deviations from perfect gas isotherms occur at low temperatures, and we have to investigate the significance of the sharp deflections of the curves in Fig. 1.6.

Consider what happens when the volume of a sample of gas in the state A in Fig. 1.6 is decreased at constant temperature (e.g. by pushing in a piston). The pressure rises in approximate accord with Boyle's Law in the region close to A, for there the pressure is low. Serious deviations from Boyle's Law begin to appear when the volume has been reduced to the point represented by B: the pressure required to get to this volume might be wildly different from the Boyle's Law prediction.

At point C, which lies at about 60 atm for carbon dioxide, all similarity to perfect behaviour is lost, for suddenly we find that the piston slides in easily without resulting in any rise in pressure: this is represented by the horizontal portion of the line, CDE. Inspecting the contents of the vessel explains this behaviour. Just to the left of C a liquid appears, and we can see two phases separated by a sharply defined surface (which, in a narrow tube, is the *meniscus*). As the volume is diminished, the gas condenses and the fraction of liquid increases. Because the gas can respond by condensing there is no resistance to further insertion of the piston. The pressure corresponding to the line CDE is called the *vapour pressure* of the liquid at the temperature of the experiment: at this pressure and temperature the gas and liquid can coexist in equilibrium.

When point E is reached all the gas has disappeared, the sample is entirely liquid, and the piston is resting on its surface. Any further reduction of volume of the sample requires the exertion of considerable pressure (liquids are very difficult to compress, which is why they can be used in hydraulic systems), and this is reflected by the sharply rising line to the left of E. Even a small reduction of volume from E to F requires a great increase in pressure.

The isotherm at the temperature T_c plays a very special role in the theory of states of matter. An isotherm corresponding to a temperature a fraction of a kelvin below T_c behaves in the manner already described: at a particular volume and pressure a liquid condenses out of the gas, and is distinguishable from it by the presence of a visible surface. The range of volumes over which the liquid and gas coexist is very short at this temperature, and only a small reduction of volume is required to produce a wholly liquid sample. If the compression takes place at T_c, the condensation point and the wholly liquid point occur at the same volume, and there is no appearance of a surface separating the two phases. At this *critical temperature* the gas phase and the liquid phase are continuous and we cannot point to one part of the sample and call it gas, and the other part liquid. At and above this temperature a gas changes into a liquid without the appearance of an interface. The pressure and molar volume at the critical point are called the *critical pressure* p_c and *critical molar volume* $V_{m,c}$. Taken together, p_c, $V_{m,c}$, and T_c are referred to as the *critical constants* of the gas.

.4 Imperfect gases: an equation of state

Conclusions can be drawn from the virial equation of state only by inserting specific values of the coefficients and taking note of their temperature dependence. That is too specialized for our purposes, and would tell us about only one gas at a time. We want a broader, if less precise, view of all gases. We shall use the equation of state introduced by van der Waals as an example of an exercise in thinking scientifically about a mathematically complicated but physically simple problem. Van der Waals himself proposed it on the basis of experimental evidence available to him in conjunction with rigorous thermodynamic arguments.

Our aim is to find a simple expression which can serve as an approximate equation of state of a real gas. This equation of state must take into account the repulsive and attractive interactions between the molecules. We shall take the repulsive interactions into account by supposing that they cause the molecules to behave like small impenetrable spheres each of volume v_{mol}. We shall take the attractive forces into account by supposing that they reduce the pressure exerted by the gas.

The effect of molecular volume. The non-zero volume of the molecules implies that instead of having an empty space of volume V to move in, the space is partly filled by molecules and reduced to $V - nb$, where nb is approximately the volume occupied by the molecules themselves. This suggests that the volume in the perfect gas equation of state should be replaced by the free volume $V - nb$:

$$p(V - nb) = nRT,$$

or

$$p = nRT/(V - nb).$$

The effect of attractive forces. Attractive forces tend to hold the gas together, and so reduce the pressure it exerts. We have seen that the pressure depends on *both* the frequency of molecular collisions with the walls *and* the impulse delivered by each collision. Both contributions are diminished by the attractive forces, and the strength with which these operate is roughly proportional to the density of molecules. Therefore the average pressure exerted by the molecules is decreased in proportion to the square of the density, or

$$\text{reduction of pressure} = an^2/V^2$$

where a is a constant for each gas. It follows that

(1.4.1a) $$p = nRT/(V - nb) - an^2/V^2$$

This is often reorganized into a form resembling $pV = nRT$:

(1.4.1b) $$(p + an^2/V^2)(V - nb) = nRT.$$

Noting that $V_m = V/n$, another form is

(1.4.2) $$p = RT/(V_m - b) - a/V_m^2.$$

The extra term a/V_m^2 is often called the *internal pressure of the gas*.

Example (Objective 3). In an industrial process nitrogen has to be heated to 500 K at constant volume. If it enters the system at 300 K and 100 atm, what pressure does it exert at its final working temperature? Treat it as a van der Waals gas.

● *Method.* Use eqn (1.4.2) since V_m is a constant (because V and n are both constant during the heating). Then

$$p_f - p_i = \{RT_f/(V_m - b) - a/V_m^2\} - \{RT_i/(V_m - b) - a/V_m^2\}$$
$$= R\{(T_f - T_i)/(V_m - b)\}.$$

Find V_m by solving eqn (1.4.2) written in the form

$$V_m^3 - (b + RT/p)V_m^2 + (a/p)V_m - (ab/p) = 0$$

for the initial conditions and with a, b taken from Table 1.1. This can be done either analytically, or quite quickly simply by estimating a value of V_m (on the basis $V_m \approx RT/p$) and changing it until the last equation is satisfied.

● *Answer.*

$$RT/p_i = \frac{(0.0821 \text{ dm}^3 \text{ atm K}^{-1} \text{ mol}^{-1}) \times (300 \text{ K})}{(100 \text{ atm})} = 0.246 \text{ dm}^3 \text{ mol}^{-1}.$$

$$b + RT/p = 3.913 \times 10^{-2} \text{ dm}^3 \text{ mol}^{-1} + 0.246 \text{ dm}^3 \text{ mol}^{-1} = 0.285 \text{ dm}^3 \text{ mol}^{-1}.$$

$$ab/p = (1.390 \text{ dm}^6 \text{ atm mol}^{-2}) \times (3.913 \times 10^{-2}. \text{dm}^3 \text{ mol}^{-1})/(100 \text{ atm})$$
$$= 5.439 \times 10^{-4} \text{ (dm}^3 \text{ mol}^{-1})^3.$$

Solving $(V_m/\text{dm}^3 \text{ mol}^{-1})^3 - 0.285 (V_m/\text{dm}^3 \text{ mol}^{-1})^2 + 1.39 \times 10^{-2} (V_m/\text{dm}^3 \text{ mol}^{-1})$
$$-5.439 \times 10^{-4} = 0$$

starting with $V_m \approx 0.246 \text{ dm}^3 \text{ mol}^{-1}$ gives $V_m \approx 0.2358 \text{ dm}^3 \text{ mol}^{-1}$. Then

$$p_f - p_i = \frac{(0.0821 \text{ dm}^3 \text{ atm K}^{-1} \text{ mol}^{-1}) \times (500 \text{ K} - 300 \text{ K})}{(0.2358 \text{ dm}^3 \text{ mol}^{-1} - 0.03913 \text{ dm}^3 \text{ mol}^{-1})} = 83.5 \text{ atm}.$$

Therefore $p_f = 183.5$ atm.

● *Comment.* The assumption of perfect gas behaviour would have given $p_f \approx 167$ atm. The experimental value is 183 atm.

Table 1.1. Van der Waals constants for gases

	$a/\text{dm}^6 \text{ atm mol}^{-2}$	$100b/\text{dm}^3 \text{ mol}^{-1}$
He	0.034 12	2.370
Ne	0.210 7	1.709
Ar	1.345	3.219
Kr	2.318	3.978
Xe	4.194	5.105
H_2	0.244 4	2.661
O_2	1.360	3.183
N_2	1.390	3.913
Cl_2	6.493	5.622
CO_2	3.592	4.267
H_2O	5.464	3.049
NH_3	4.170	3.707
CH_4	2.253	4.278
C_2H_4	4.471	5.714
C_2H_6	5.489	6.380
C_6H_6	18.00	11.54

Source: *Handbook of chemistry and physics*, Chemical Rubber Co.

Equation (1.4.1) is the *van der Waals equation of state*. We built it on the basis of quite vague arguments about molecular volume and intermolecular forces. There are other ways of deriving the equation, but the present method has the advantage that it shows how to force the form of an equation out of a general idea about what is going on at a molecular level, and that this forcing can be done without knowing a lot of details or mathematics. The derivation also has the advantage of keeping the significance of the coefficients a and b imprecise: they are much better regarded as adjustable parameters ra.her than as precisely defined molecular properties.

We must see whether the van der Waals equation bears any resemblance to the true equation of state (whatever that may be). This can be done by seeing whether it predicts isotherms anything like the experimental ones shown in Fig. 1.6. A selection of curves drawn on the basis of this equation are shown in Fig. 1.7, and apart from the peculiar oscillations (which disappear at high temperatures) they do resemble the experimental isotherms. The oscillations, the *van der Waals loops*, are obviously unrealistic—they suggest that under some conditions an increase in pressure increases the volume—and so we cut them out and replace this part of the curve by a horizontal line drawn so that the loop defines equal areas above and below the line (this is the *Maxwell construction*).

The principal features of this equation of state are as follows.

(1) At high temperatures RT may be so large that the first term of eqn (1.4.2) greatly exceeds the second. The equation then differs insignificantly from the perfect gas equation, apart from the reduction in the available volume from V_m to $V_m - b$. Therefore at high temperatures we should expect perfect gas isotherms, but displaced to the region of higher volume (so that the left legs of the curves rise to infinity at $V_m = b$ rather than at $V_m = 0$). This is the behaviour shown at the top left of Fig. 1.7.

(2) The loops—which we replaced by straight lines and identified with the region where gas and liquid can coexist—occur where both terms in

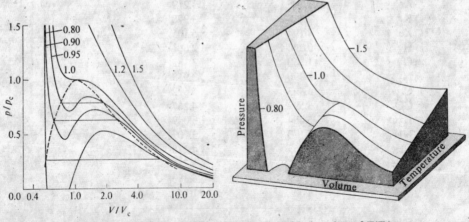

Fig. 1.7. The van der Waals isotherms. (Isotherms are labelled with the values of T/T_c.)

eqn (1.4.2) have about equal magnitude. The first term arises from the kinetic energy of the molecules and their repulsive interactions, and the second is due to the attractive interactions. We expect liquids and gases to coexist when the attractive forces are about equal to the forces favouring dispersal.

(3) We can find the position of the critical point. Below the critical temperature the curve oscillates and passes through a minimum and then a maximum: these get closer together as T approaches T_c, and at T_c they coincide. Therefore, at the critical temperature the curve has a flat inflexion rather than a maximum or a minimum. From the properties of curves we know that an inflexion of this type occurs where both the slope and the curvature are zero, and so we look for the values of V_m, p, and T for which the first and second derivatives of p with respect to V_m disappear.

$$\left.\begin{array}{ll} \text{slope:} & \dfrac{dp}{dV_m} = \dfrac{-RT}{(V_m - b)^2} + \dfrac{2a}{V_m^3} = 0 \\[3mm] \text{curvature:} & \dfrac{d^2p}{dV_m^2} = \dfrac{2RT}{(V_m - b)^3} - \dfrac{6a}{V_m^4} = 0 \end{array}\right\} \text{ at } T = T_c, \ p = p_c, \ V_m = V_{m,c}.$$

Solving these equations leads to

(1.4.3)
$$\begin{aligned} V_{m,c} &= 3b \\ p_c &= a/3V_{m,c}^2 = a/27b^2 \\ T_c &= 8p_c V_{m,c}/3R = 8a/27Rb. \end{aligned}$$

Taken together these relations suggest that the values of p_c, $V_{m,c}$, and T_c of a gas should be related by

$$Z_c = p_c V_{m,c}/RT_c = \tfrac{3}{8} = 0.375.$$

These conclusions can be tested by seeing whether the critical compression factor (Z_c) is equal to $\tfrac{3}{8}$. Table 1.2 lists the experimental critical constants of a variety of gases. You can see that although Z_c is less than 0.375 the discrepancy is gratifyingly small.

Another general conclusion that may be drawn from the van der Waals equation of state relates to its low pressure limit, for it can be expressed in the form of the virial expansion, eqn (1.3.1), and $B'(T)$ can be identified as

$$B'(T) = \{b - (a/RT)\}/RT.$$

It follows that the Boyle temperature (the temperature at which $B'(T)$ vanishes, p. 38) of a van der Waals gas is the temperature at which $b = a/RT$. Therefore

(1.4.4)
$$T_B = a/bR = 8T_c/27.$$

An important point should now be appreciated. For a perfect gas

$$(d/dp)(pV_m/RT) = 0$$

but for a real gas

Table 1.2. Critical constants of gases

	p_c/atm	$V_{m,c}$/cm^3 mol^{-1}	T_c/K	Z_c	T_B/K
He	2.26	58	5.2	0.307	22.64
Ne	26.9	41.7	44.4	0.308	122.11
Ar	48.0	75.2	150.7	0.292	411.52
Xe	58.0	119	289.7	0.290	768.03
H_2	12.77	65.5	32.99	0.309	110.04
O_2	50.1	78.0	154.8	0.308	405.88
N_2	33.5	90.1	126.2	0.291	327.22
F_2	55	—	144	—	—
Cl_2	76.1	124	417	0.276	—
Br_2	102	135	584	0.286	—
CO_2	72.8	94.0	304.2	0.274	714.81
H_2O	218.3	59.1	647.3	0.243	—
NH_3	111.5	72.5	405.4	0.243	—
CH_4	45.6	98.7	190.6	0.288	509.66
C_2H_4	50.5	127.4	282.4	0.278	—
C_2H_6	48.2	148	305.4	0.284	—
C_6H_6	48.4	254	562.2	0.266	—

Source: G. W. C. Kaye and T. H. Laby, *Tables of physical and chemical constants*, Longmans; *American Institute of Physics handbook*, McGraw Hill.

$$\lim_{p \to 0} (d/dp)(pV_m/RT) = RTB'(T),$$

and so, although the *equation of state* may coincide with the perfect gas form as $p \to 0$, the gas's *properties* (which in general may depend on derivatives) might not. We shall see examples later.

Comparing gases. When the properties of objects are compared, an important technique is to choose a characteristic fundamental feature and set up a scale of measurement on that basis. This technique was applied in a simple way when we expressed the distance between molecules in terms of a scale based on molecular diameters. We have seen that the critical pressure, critical molar volume, and critical temperature are characteristic properties of gases, and so it may be that a scale of pressure, volume, and temperature can be set up using them as yardsticks. In order to investigate this point of view we define the *reduced variables* of a gas as the actual variables divided by the corresponding critical constant:

reduced pressure: $p_r = p/p_c$

reduced volume: $V_r = V_m/V_{m,c}$

reduced temperature: $T_r = T/T_c$.

Van der Waals, who first tried this, hoped that the same reduced volume of different real gases at the same reduced temperature would exert the same reduced pressure. The hope was largely fulfilled. Figure 1.8 shows the compression factor $Z = pV_m/RT$ for a number of gases expressed as

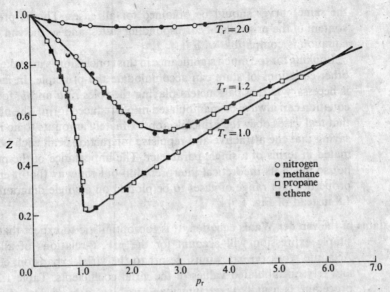

Fig. 1.8. The compression factor Z plotted as a function of reduced pressure.

a function of their reduced pressures at various reduced temperatures. It is strikingly clear that the values of Z are approximately the same for a wide variety of gases under a wide range of conditions. Compare this with Fig. 1.5 where similar data are plotted without using reduced variables.

The observation that real gases in the same state of reduced volume and reduced temperature exert approximately the same reduced pressure is called the *principle of corresponding states*. The principle is approximately valid for gases composed of spherical non-polar molecules, but it breaks down when the molecules are non-spherical or polar.

We now see whether the van der Waals equation is able to shed any light on the principle. In order to do this we write the p, V_m, and T of eqn (1.4.2) in terms of the reduced variables, and then express the latter in terms of the relations in eqn (1.4.3). First we get

$$p = p_r p_c = \frac{RT_r T_c}{(V_r V_{m,c} - b)} - \frac{a}{V_r^2 V_{m,c}^2}$$

and then

$$\frac{a p_r}{27b^2} = \frac{8aT_r}{27b(3bV_r - b)} - \frac{a}{9b^2 V_r^2}$$

which reorganizes into

(1.4.5)
$$p_r = \frac{8T_r}{3V_r - 1} - \frac{3}{V_r^2}.$$

This equation has the same form as the original equation, but the constants a and b, which differ from gas to gas, have disappeared. It follows that if the isotherms are plotted in terms of the reduced variables, the same curves should be obtained for all gases. This is precisely the content of the principle of corresponding states, and so the van der Waals equation is compatible with it.

Looking for too much significance in this conclusion is mistaken, because other equations of state can accommodate the principle. In fact all that is necessary is two parameters playing the roles of a and b, for then the equation can always be manipulated into a reduced form. The observation that real gases obey the principle approximately amounts to no more than saying that the attractive and repulsive interactions can each be approximated in terms of a single parameter. The importance of the principle is not so much its theoretical interpretation but the way that it enables the properties of a range of gases to be plotted on a single diagram (e.g. Fig. 1.8 instead of Fig. 1.5).

The status of the van der Waals equation. It is too optimistic to expect that a single, simple expression will account for the p, V, T-relations of all systems. Accurate work on gases must resort to the virial expansion, eqn (1.3.2), and rely on tabulated values of the virial coefficients, Table 1.3. Such a procedure is usually very clumsy and involves a good deal of numerical analysis. The advantage of the van der Waals expression is that it is analytical and enables us to draw some general conclusions about the behaviour of real systems. But we have to be circumspect: we have to remember that under all circumstances it is an approximation, and under many quite common conditions it is a poor approximation. When it fails (at high pressures and densities) we have to use one of the other equations of state that are available (some are listed in Table 1.4), invent a new one, or resort to the virial expansion.

Table 1.3. Second virial coefficients of some gases, $B/\text{cm}^3 \text{ mol}^{-1}$*

	100 K	273 K	373 K	600 K
He	11.4	12.0	11.3	10.4
Ne	−4.8	10.4	12.3	13.8
Ar	−187.0	−21.7	−4.2	11.9
Kr		−62.9	−28.7	2.0
Xe		−153.7	−81.7	−19.6
H_2	−2.5	13.7	15.6	
N_2	−160.0	−10.5	6.2	21.7
O_2	−197.5	−22.0	−3.7	12.9
CO_2		−149.7	−72.2	−12.4
CH_4		−53.6	−21.2	8.1
Air	−167.3	−13.5	3.4	19.0

Data: *American Institute of Physics handbook*, McGraw-Hill.

* The values of B quoted here relate to the expansion

$$pV_m/RT = 1 + B/V_m + C/V_m^2 + \ldots$$

The relations between B, C of this expansion and the B', C' of the expansion

$$pV_m/RT = 1 + B'p + C'p^2 + \ldots \text{ are } B' = B/RT, \ C' = (C - B^2)/R^2T^2.$$

Table 1.4. Equations of state

Many people have proposed equations of state for real gases. The criteria of success are accurate representation of observed p, V, T-relations over moderate ranges of conditions, a simplicity of form (so that they can be differentiated and integrated reasonably easily), and the use of only a few adjustable parameters. The following small selection conforms to the first criterion with increasing success on passing down the list, but the number of parameters increases from zero (in (i)) to five (in (v)) and to an indefinite number (in (vi)).

(i) *Perfect gas equation*

$$p = RT/V_m$$

(ii) *Van der Waals equation*

$$p = RT/(V_m - b) - a/V_m^2$$

Critical constants:

$$p_c = a/27b^2, \qquad V_{m,c} = 3b, \qquad T_c = 8a/27Rb$$
$$Z_c = \tfrac{3}{8} = 0.375.$$

Reduced form:

$$p_r = 8T_r/(3V_r - 1) - 3/V_r^2.$$

(iii) *Berthelot equation*

$$p = RT/(V_m - b) - a/TV_m^2$$

Critical constants:

$$p_c = \tfrac{1}{12}(2aR/3b^3)^{1/2}, \qquad V_{m,c} = 3b, \qquad T_c = \tfrac{2}{3}(2a/3bR)^{1/2}$$
$$Z_c = \tfrac{3}{8} = 0.375.$$

Reduced form:

$$p_r = 8T_r/(3V_r - 1) - 3/T_r V_r^2.$$

(iv) *Dieterici equation*

$$p = \{RT/(V_m - b)\} \exp(-a/RTV_m)$$

Critical constants:

$$p_c = a/4e^2b^2, \qquad V_{m,c} = 2b, \qquad T_c = a/4bR$$
$$Z_c = 2/e^2 = 0.2706 \ldots$$

Reduced form:

$$p_r = \{e^2 T_r/(2V_r - 1)\} \exp(-2/T_r V_r)$$

(e is the exponential e, not a parameter).

(v) *Beattie–Bridgeman equation*

$$p = (1 - \gamma)RT(V_m + \beta)/V_m^2 - \alpha/V_m^2$$
$$\alpha = a_0(1 + a/V_m)$$
$$\beta = b_0(1 - b/V_m)$$
$$\gamma = c_0/V_m T^3.$$

(vi) *Virial equation (Kammerlingh Onnes)*

$$p = (RT/V_m)\{1 + B(T)/V_m + C(T)/V_m^2 + \ldots\}.$$

Further Reading

Properties of matter. B. H. Flowers and E. Mendoza; Wiley, London, 1970.

Gases, liquids, and solids. D. Tabor; Cambridge University Press, 1979.

Thermodynamics for chemical engineers. K. E. Bett, J. S. Rowlinson, and G. Saville; Athlone Press, London, 1975.

Chemical thermodynamics. M. L. McGlashan; Academic Press, London, 1979.

Heat and thermodynamics. M. W. Zemansky; McGraw-Hill, Tokyo, 1968.

Determination of pressure and volume. G. W. Thomson and D. R. Douslin; In *Techniques of chemistry.* (A. Weissberger and B. W. Rossiter, eds.) V, 23, Wiley-Interscience, New York, 1971.

Temperature. J. F. Swindells; NBS Special Publication 300, 1968.

Comparisons of equations of state. J. B. Ott, J. R. Coates, and H. T. Hall; *J. chem. Educ.* **48**, 515 (1971).

The molecular theory of gases and liquids. J. O. Hirschfelder, C. F. Curtiss, and R. B. Bird; Wiley, New York, 1954.

The virial coefficients of pure gases and mixtures. J. H. Dymond and E. B. Smith; Oxford University Press, 1980.

International critical tables, Vol. 3 (p, V, T data). McGraw-Hill, New York, 1928.

Problems

1.1. A sample of air occupies 1 dm^3 at room temperature and pressure. What pressure is needed to compress it so that it occupies only 100 cm^3 at that temperature?

1.2. At sea level, where the pressure was 755 mmHg, the gas in a balloon occupied 2 m^3. What volume will the balloon expand to when it has risen to an altitude where the pressure is (a) 100 mmHg, (b) 10 mmHg? Assume that the material of the balloon is infinitely extensible.

1.3. A diving bell has an air space of 3 m^3 when on the deck of a boat. What is the volume of the air space when it has been lowered to a depth of 50 m? Take the mean density of sea water as 1.025 g cm^{-3} and assume that the temperature is the same at 50 m as at the surface.

1.4. What pressure difference must be generated across the length of a vertical drinking straw of length 15 cm in order to drink a water-like liquid? Estimate the expansion of the lungs that is needed in order to create the appropriate partial vacuum at the upper end of the straw.

1.5. To what temperature must a 1 dm^3 sample of a perfect gas be cooled from room temperature in order to reduce its volume to 100 cm^3?

1.6. A car tyre (i.e. an automobile tire) was inflated to a pressure of 24 lb in^{-2} on a winter's day when the temperature was $-5\,°C$. What pressure will be found, assuming no leaks to have occurred, on a subsequent summer's day when the temperature is 35 $°C$? What complications should be taken into account in practice?

1.7. A meteorological balloon had a radius of 1 m when released from sea level, and expanded to a radius of 3 m when it had risen to its maximum altitude, where the temperature was $-20\,°C$. What is the pressure inside the balloon at that altitude?

1.8. The perfect gas law is a *limiting law* in the sense described on p. 29. First deduce the relation between *pressure* and *density* (ρ) of a perfect gas, and then confirm on the basis of the following data for dimethyl ether, $CH_3.O.CH_3$, at 25 $°C$

that perfect behaviour is approached at low pressures. Find the relative molar mass (R.M.M.) of the gas.

p/mmHg	91.74	188.98	277.3	452.8	639.3	760.0
$10^3 \rho/\text{g cm}^{-3}$	0.2276	0.4695	0.6898	1.1291	1.5983	1.9029

1.9. Investigate some of the technicalities of ballooning on the basis of the perfect gas law. Suppose your balloon has a radius of 3 m and that it is a sphere when inflated. How much hydrogen is needed to inflate it to a pressure of 1 atm at an ambient temperature of 25 °C at sea level? What mass can the balloon lift at sea level, where the density of air is 1.22 kg m^{-3}? What would be the payload if helium were used instead of hydrogen? With you and a companion on board, the balloon ascends to 30 000 ft where the pressure is 0.28 atm, the temperature -43 °C, and the density of air 0.43 kg m^{-3}. Can you in fact attain that height with both hydrogen and helium? Do you need more hydrogen to rise further? (Leaks are encountered in Problem 25.22.)

1.10. In an experiment to determine the R.M.M. of ammonia, 250 cm^3 of the gas was confined to a glass vessel. The pressure was 152 mmHg at 25 °C and after correcting for buoyancy effects, the mass of the gas was 33.5 mg. What is (a) the molecular mass, (b) the R.M.M. of the gas?

1.11. The R.M.M. of a newly synthesized fluorocarbon gas was measured with a *gas microbalance*. This consists of a glass bulb forming one end of a beam, the whole surrounded by a closed container. The beam is pivoted, and the balance point is attained by raising the pressure of gas in the container, and so increasing the buoyancy of the enclosed bulb. In one experiment the balance point was reached when the fluorocarbon pressure was 327.10 mmHg, and for the same setting of the pivot a balance was achieved when CHF_3 was introduced at a pressure of 423.22 mmHg. A repeat of the experiment with a different setting of the pivot required a pressure of 293.22 mmHg of the fluorocarbon and 427.22 mmHg of the CHF_3. What is the R.M.M. of the fluorocarbon? Suggest a molecular formula. (Take $M_r = 70.014$ for trifluoromethane.)

1.12. A constant-volume ideal gas thermometer indicates a pressure of 50.2 mmHg at the triple point of water (273.16 K). What change of pressure indicates a change of 1 K at this temperature? What pressure indicates a temperature of 100 °C (373.15 K)? What change of pressure indicates a change of 1 K at the latter temperature?

1.13. The synthesis of ammonia is an important process technologically, and it has features which make it useful for emphasizing and illustrating points made in the text. A simple problem is the following. A vessel of volume 22.4 dm^3 contains 2 mol of hydrogen and 1 mol of nitrogen at 273.15 K. What is the mole fraction of each component, their partial pressures, and the total pressure?

1.14. The question arose as to what the partial and total pressures would be if the whole of the hydrogen in the last Problem were converted to ammonia by reaction with the appropriate amount of nitrogen. What are these pressures?

1.15. Could 131 g of xenon in a vessel of capacity 1 dm^3 exert a pressure of 20 atm at 25 °C if it behaved as a perfect gas? If not, what pressure would it exert?

1.16. Now assume that the xenon in the last Problem behaves as a van der Waals gas (with the constants given in Table 1.1). What pressure will 131 g exert under the same conditions?

1.17. Calculate the pressure exerted by 1 mol of ethene behaving as (a) a perfect gas, (b) a van der Waals gas, when it is confined under the following conditions: (i) at 273.15 K in 22.414 dm³, (ii) at 1000 K in 100 cm³. Use the data in Table 1.1 even though that data refers to temperatures around 25 °C.

1.18. Use the data in Table 1.2 to suggest the pressure and temperature that 1 mol of (a) ammonia, (b) xenon, (c) helium will have in states corresponding to 1 mol of hydrogen at 25 °C at 1 atm.

1.19. Estimate the critical constants (p_c, $V_{m,c}$, T_c) of a gas with van der Waals parameters $a = 0.751$ atm dm⁶ mol⁻², $b = 0.0226$ dm³ mol⁻¹.

1.20. The critical constants of methane are $p_c = 45.6$ atm, $V_{m,c} = 98.7$ cm³ mol⁻¹, and $T_c = 190.6$ K. Calculate the van der Waals parameters and estimate the size (volume and radius) of the gas molecules.

1.21. Estimate the radii of the rare-gas atoms on the basis of their critical volumes and the Dieterici equation of state, Table 1.4.

1.22. Estimate the coefficients a and b in the Dieterici equation of state for the critical constants of xenon. Calculate the pressure exerted by 1 mol of the gas when it is confined to 1 dm³ at 25 °C. (Compare Problem 1.16.)

1.23. Express the van der Waals equation of state as a virial expansion in powers of $1/V_m$ and obtain expressions for $B(T)$ and $C(T)$ in terms of the parameters a and b.

1.24. Repeat the last Problem for a Dieterici gas.

1.25. The virial equation of state is often a convenient method of expressing both theoretical and experimental results on gases. For instance, measurements on the deviation of argon from ideality give the following virial expansion at 273 K:

$$pV_m/RT = 1 - (21.7 \text{ cm}^3 \text{ mol}^{-1}/V_m) + (1200 \text{ cm}^6 \text{ mol}^{-2}/V_m^2) + \ldots$$

Use the parameters in this expansion to predict the critical constants of argon on the basis of the results in the last two Problems and the relations set out in Table 1.4.

1.26. A scientist with a simple view of life proposes the following equation of state for a gas:

$$p = RT/V_m - B/V_m^2 + C/V_m^3.$$

Demonstrate that critical behaviour is accommodated by this equation. Express p_c, $V_{m,c}$, and T_c in terms of B and C, and find an expression for the critical compression factor Z_c.

1.27. The virial expansion can be expressed either as a series in powers of $1/V_m$ or in powers of p, see eqns (1.3.1) and (1.3.2) on p. 37. Often it is convenient to express the coefficients B', C' in terms of B, C. This can be done by inverting the $1/V$ expansion and expressing the result in powers of p. The inversion need be done only to p^2 in order to find B' and C'. Express B' and C' in terms of B and C.

1.28. Find an expression for the Boyle temperature T_B in terms of the van der Waals parameters of a gas. Find the temperature at which 1 mol of xenon in a 5 dm³ vessel has a compression factor of unity.

1.29. Express the Boyle temperature in terms of reduced variables of (a) a van der Waals gas, (b) a Dieterici gas.

1.30. The second virial coefficient B can be obtained from measurements of the density of a gas at a series of pressures. Show that the graph of p/ρ should be a straight line with slope proportional to B. Use the data in Problem 1.8 to find B for dimethyl ether. (The data relate to 25 °C.)

1.31. The *barometric formula* relates the pressure of a gas at some height h to its pressure p_0 at sea level (or any other base line). It can be derived in various ways, one involving the Boltzmann distribution (p. 11). Here we approach it from another viewpoint, and start from the change in pressure dp for an infinitesimal change in height dh, where the density is ρ: $dp = -\rho g\,dh$. Prove this relation, and then show that it integrates to $p(h) = p_0 \exp(-M_m gh/RT)$ for a perfect gas, where M_m is the molar mass of the gas molecules. The consequences of this expression are explored in the next few Problems.

1.32. One of the consequences of the barometric formula is the extra complication that no gas in any terrestrial laboratory has a uniform pressure. But how serious is the effect of the gravitational field? Find the pressure difference between top and bottom of (a) a laboratory vessel of height 15 cm and (b) the World Trade Center, 1350 ft.

1.33. That (Problem 1.32) is not the only problem. If the gas is a mixture, its local composition depends on the altitude: the heavier molecules tend to sink to the bottom of a column because Mgh depends more strongly on height when M is big. As a first step in unravelling this effect, show that for a mixture of ideal gases the partial pressure of a component J follows $p_J = p_{J,0} \exp(-M_{J,m}gh/RT)$. Then assess the magnitude of the effect on the distortion of the composition of the atmosphere. Is life faster at the top of the World Trade Center? Does helium abound at the top of Everest? Deduce the composition of the atmosphere at heights of (a) 1350 ft, (b) 29 000 ft, (c) 100 km in order to answer this sort of question. The sea-level composition is given on p. 35. Disregard the extra complication of temperature variation over this range of altitudes, and take it as 20 °C throughout.

1.34. The barometric formula also found application in the early determination of Avogadro's constant. Although the method has been superseded, there is some interest in seeing how simple macroscopic measurements can give values of atomic constants. As a first step, show that the numbers of particles of mass m at two heights separated by h are in the ratio $\exp(-m'gh/kT)$, where m' is their effective mass in the solvent of density ρ. In an actual experiment, spheres of rubber latex of radius 2.12×10^{-4} mm and density 1.2049 g cm^{-3} were distributed in water at 20 °C (when its density is 0.9982 g cm^{-3}). The average numbers of spherical particles at various heights were as follows:

h/mm	0	0.05	0.07	0.09	0.10	0.15	0.20
N	1000	400	280	190	160	60	25

Estimate Boltzmann's constant, and hence Avogadro's constant L from the known value of the gas constant R.

2 The First Law: the concepts

After careful study of this chapter you should be able to

(1) Define *thermodynamic system, surroundings, closed system,* and *isolated system* (p. 53).

(2) Define *energy, heat,* and *work* (p. 53).

(3) State the *First Law of thermodynamics* (pp. 54, 55).

(4) Define *internal energy* (p. 54) and express how it changes under the influence of heat and work (eqn (2.1.1)).

(5) Calculate the work done when a gas expands against an external pressure (eqn (2.2.5)).

(6) Define thermodynamic *reversibility* (p. 62).

(7) Calculate the work done and the change of internal energy during the *isothermal reversible expansion* of a perfect gas (eqn (2.2.11)).

(8) State the connection between *reversibility, equilibrium, quasi-static process,* and *maximum work* (p. 62).

(9) Define the *heat capacity* of a system (p. 66).

(10) Relate the changes of internal energy in a system to the heat transferred at constant volume (eqn (2.3.4)).

(11) Define and explain the significance of *enthalpy* (eqn (2.3.9)).

(12) Relate changes of enthalpy in a system to the heat transferred at constant pressure (eqn (2.3.7)).

(13) Establish a connection between the constant volume and constant pressure heat capacities of a perfect gas (eqn (2.3.10)).

(14) Distinguish between the effects of work and heat at a molecular level (p. 70).

Introduction

Energy is stored by molecules, and its release can be used to provide heat when a fuel burns, mechanical work when a fuel burns in an engine, and electrical work when a chemical reaction pumps electrons through a circuit. In chemistry we encounter reactions that can be harnessed to provide heat and work, reactions whose liberated energy might be squandered but whose products are required, and reactions that constitute the processes of life. Thermodynamics, the study of the transformations of energy, enables us to discuss all these matters rationally.

2.1 Heat, work, and the conservation of energy

In the introduction on p. 5 we began with the atomic nucleus because there our aim was to see how the familiar world is based on the microscopic world of the atom. In thermodynamics we take the opposite viewpoint, and begin with the universe.

The universe can be thought of as consisting of two parts. One is the *system*: this is the part in which we have a special interest, and which may be a reaction vessel, an engine, an electric cell, and so on. The *surroundings* are the rest of the universe, and are where we make observations. The two parts may be in contact. When matter can be transferred between the system and its surroundings we call it an *open system*; when such transfer is not possible the system is *closed*. An *isolated system* is a closed system with neither mechanical nor thermal contact with its surroundings. Normally a fairly small part of a laboratory in the vicinity of the system of interest is sufficiently isolated from the rest of the actual universe to be regarded as the full extent of the surroundings.

The basic concepts of thermodynamics are work, heat, and energy. We all think we know what we mean by the terms, but thermodynamics is a very precise subject, and demands precise definitions. There is no point in erecting an elaborate, rigorous superstructure on vague foundations, and so we have to spend a little time on sharpening the everyday meaning of these terms.

Work is done if the process can be used to bring about a change in the height of a weight somewhere in the surroundings. Work has been done *by* the system if a weight has been raised somewhere in the surroundings; work has been done *on* the system if a weight has been lowered.

Energy is the capacity to do work. When we do work on an otherwise isolated system its capacity to do work is increased, and so its energy has been increased. When the system does work its energy is reduced because it is then capable of doing less work.

Experiments have shown that the energy of a system, its capacity to do work, may be changed by means other than work itself. When there is a temperature difference between the system and its surroundings the energy of the system will change if they are in thermal contact. *When the energy changes as a result of a temperature difference we say that there has been a flow of heat.* Containers that permit energy transfer as heat are termed *diathermic.* Containers that do not are termed *adiabatic*: a Dewar flask is a good approximation. We can detect whether heat has been transferred by making an observation in the surroundings. For instance, in an ice calorimeter (an adiabatic container surrounding the diathermic system, and containing a mixture of ice and water) the amount of ice will change as a result of the flow of heat.

Experiments have established the crucial point that, whereas *we* know what mode of transfer of energy was employed in a particular case (because we can see whether a weight was raised or lowered in the surroundings, or whether ice melted), the *system* is blind to the mode employed. Heat and work are equivalent ways of changing its energy: energy is energy, however it is acquired or lost. The system is like a bank: it accepts deposits through its walls in either currency, but stores its reserves as energy. The experimental evidence for this equivalence is abundant. In an adiabatic system, for instance, the same rise in temperature is brought about by the same amount of any kind of work we choose to do on the system (stirring it with rotating paddles, electrical work, and so on). The temperature rise is also the same if the same amount of heat is passed into the same system, but one now enclosed in a diathermic vessel.

The *First Law of thermodynamics* is essentially a summary of the preceding remarks. It can be expressed in a number of ways. An involved but rich statement is:

> *When a system changes from one state to another along an adiabatic path, the amount of work done is the same irrespective of the means employed.*

The richness of this statement can be unfolded as follows. Suppose an amount w_{ad} of work is done on the system in some adiabatic process taking it from an initial state i to a final state f. The work may be of any kind—compression work, rotating paddles, electrical, or mixtures of different processes, and so on, and therefore we ought to label w_{ad} with the path (the specification of the process). But the statement implies that the value of w_{ad} is the same for all such paths, and depends only on the initial and final states of the system. This immediately suggests that *there is a property of the state of the system* such that w_{ad} may be expressed as the difference of its values for the initial and final states:

$$w_{ad} = U_f - U_i.$$

The property U is called the *internal energy*.

The statement should also be understood as implying the following. Suppose the system changes between the same initial and final states as before, but that the change occurs along a *non-adiabatic path*. The same change of internal energy has to occur, because the internal energy depends on the state and not the path, but the work involved might not be the same as before. The difference between the work done and the change in internal energy produced is *defined* as the heat absorbed in the process:

$$q \overset{\text{def}}{=} \Delta U - w,$$

where ΔU is the difference $U_f - U_i$ (Δ, Greek capital delta, is commonly used to denote the non-infinitesimal, or measurable difference between two quantities). You can see that this gives a *mechanical* and therefore truly fundamental definition of heat, because ΔU for the change $i \rightarrow f$ can be measured by measuring w_{ad} for a process between the specified states; the w in the last equation can be measured for the non-adiabatic change between the same two states, and the heat is then simply the difference of the two amounts of work of the two processes:

$$q = w_{ad} - w.$$

Finally, the statement of the First Law implies that the internal energy of an *isolated system* cannot change. This is because in an isolated system there can be no transfers of heat or work, and so $q = 0$, $w = 0$; which implies that $\Delta U = 0$ (as $\Delta U = q + w$). (An isolated system can, of course, undergo a change of state; a pendulum will come to rest in an isolated container; but the energy of the enclosed, isolated system remains constant.) This leads to a very succinct expression of the First law:

First Law: *The energy of an isolated system is constant.*

But to comprehend the full content of this statement, so that it is regarded as a *thermodynamic* statement and not merely a statement about mechanics, all that has gone before has to be borne in mind. The evidence for the law in this form is the impossibility of constructing perpetual motion machines, for if energy could arise spontaneously, an engine could be constructed which would run without fuel. Experience has shown this to be unrealizable, and the First Law is based on the gloomy acceptance of its impossibility. That is not to say that claims have not been made to the contrary: the amount of fraud that has been perpetrated is summarized in articles referred to in the bibliography.

The next step in the development involves switching attention to infinitesimal transfers of heat and work: we shall soon see how this opens up powerful methods of calculation.

Let the work done on a system be the infinitesimal amount dw and the heat added be the infinitesimal amount dq. Then in place of $\Delta U = q + w$, which is appropriate when the transfers are measurable, in the case of infinitesimal changes the internal energy of the system changes by the amount dU, where

(2.1.1) $$dU = dq + dw.$$

We have had to choose a convention about the significance of signs. Look at the discussion in Box 2.1 before going any further.

Box 2.1   **The sign convention**

We use the convention that dq denotes the heat supplied to the system and dw denotes the work *done on* the system.

When dq *is positive* it signifies that heat has been transferred to the system and that it has contributed to an increase in the internal energy.

When dq *is negative* (so that a negative amount has been added to the system), it signifies that heat has flowed out of the system and has contributed to a decrease in the internal energy.

If the system gains an amount of heat dq, the surroundings lose that amount (or gain $-dq$).

When dw *is positive* it signifies that work has been done on the system and has contributed to an increase in the internal energy.

When dw *is negative* (so that a negative amount of work has been done on the system), it signifies that work has been done by the system on the outside world, and has contributed to a decrease in the internal energy.

If the system does an amount of work on the outside world, it is incorporated into the scheme by saying that the negative of that amount of work was done *on* the system.

*   *   *

The convention applies to measurable changes as well as to infinitesimal ones. Thus if $q = 10\,kJ$ it means that $10\,kJ$ of heat has been injected into the system; but if $q = -10\,kJ$ it means that $10\,kJ$ of heat has left the system. If $w = 10\,kJ$ then that amount of work has been done on the system; if $w = -10\,kJ$ it means that the system has done $10\,kJ$ of work on the outside world.

Always think clearly about signs: you can keep the convention clear by thinking about whether the internal energy of the system increases or decreases during the change.

Equation (2.1.1) is the mathematical expression of the First Law. It expresses the observation that the internal energy of a closed system changes by the amounts of work and heat that pass through its walls.

2.2 Work

In order to do anything useful with eqn (2.1.1) we must be able to write dq and dw in terms of things actually done to the system. We begin by discussing *mechanical work* and aim at getting expressions for dw. Mechanical work includes the work required to compress gases, and the work that gases do when they expand and drive back the atmosphere at the walls of the containing vessel. Many chemical reactions involve the

Fig. 2.1. The basic nature of mechanical work.

generation of or the reaction of gases, and the extent of reaction depends on the work the system is allowed to do.

We base the discussion of work on its definition given in elementary physics and develop this approach to include the various types of work of chemical significance.

When an object is displaced along a path through a distance dz against a force $F(z)$ the amount of work that has to be done on it, Fig. 2.1, is

$$ (2.2.1) \qquad dw = -F(z)\,dz. $$

The force has a sign: if it pushes towards $+z$ (opposing motion downwards) it is positive; if it pushes towards $-z$ (opposing motion upwards, as in Fig. 2.1), it is negative.

If the force is constant (independent of position) the work done on the object as it is pushed through dz is simply $dw = -F\,dz$. The work spent in moving the object from z_i to z_f is the sum of the work required to move it through all the segments dz of the path. Since the segments are infinitesimal the sum is really an integral:

$$ (2.2.2) \qquad w = -\int_{z_i}^{z_f} F\,dz = -(z_f - z_i)F. $$

In this case the work is proportional to the distance moved along the chosen path.

If the constant force opposes motion upwards, F is negative, $F = -|F|$, and

$$ (2.2.3) \qquad w = +(z_f - z_i)|F|. $$

If the final position z_f lies above z_i ($z_f > z_i$) w is positive: work has to be done *on* the object to move it *against* the force pulling it downwards. This shows how taking note of signs leads directly to the right result.

Example Calculate the work required to raise a mass of 1.5 kg (this book, approximately) through a height of 10 cm.

- *Method.* At the surface of the earth the force on a mass m is of magnitude mg and directed towards the centre of the earth (g is the acceleration due to gravity, 9.81 m s^{-2}). If z increases vertically we can write $F = -mg$. The force is constant (for small displacements) and so eqn (2.2.2) can be used. Therefore $w = -(z_f - z_i)(-mg)$, or $w = mgh$, with $h = z_f - z_i$, the vertical displacement.

- *Answer.* $w = (1.5 \text{ kg}) \times (9.81 \text{ m s}^{-2}) \times (0.10 \text{ m})$
  $$= 1.47 \text{ kg m}^2 \text{ s}^{-2} = 1.47 \text{ J} \approx 1.5 \text{ J}.$$

- *Comment.* This result is positive, indicating that work has to be done on the book. If it fell from a height of 10 cm the book could do 1.5 J of work or generate 1.5 J of heat on impact with the table top.

If the force is not constant but varies from point to point along the path, the work done on the object in each segment of the path depends on the force acting there, and this work must be calculated from eqn (2.2.1). The total work involved in getting the object from z_i to z_f is still the sum of all the individual efforts, and so in this case

(2.2.4)
$$w = -\int_{z_i}^{z_f} F(z)\,dz.$$

We cannot integrate until we know the dependence of the force on position. As soon as we know $F(z)$ we can evaluate the integral and get a useful formula for the work.

Example Calculate the work required to raise a rocket of mass 10^5 kg from the surface of the earth to an altitude where the gravitational attraction is negligible.

- *Method.* In this calculation we have to take into account the decrease of gravitational force with distance. If z is the distance from the centre of the earth, Newton's law of gravity reads $F = -GmM/z^2$, where M is the earth's mass (5.98×10^{24} kg), m the object's, and G the gravitational constant ($G = 6.672 \times 10^{-11} \text{ kg}^{-1} \text{ m}^3 \text{ s}^{-2}$). We can use eqn (2.2.3) to calculate the work by setting $z_i = R$, the radius of the earth (6.38×10^6 m), and $z_f = \infty$, a point infinitely far away. The work involved in raising the mass to that distance is therefore
  $$w = -\int_R^\infty (-GmM/z^2)\,dz = GmM \int_R^\infty (1/z^2)\,dz = GmM/R.$$

- *Answer.* $w = (6.672 \times 10^{-11} \text{ kg}^{-1} \text{ m}^3 \text{ s}^{-2}) \times (10^5 \text{ kg})$
  $$\times (5.98 \times 10^{24} \text{ kg})/(6.38 \times 10^6 \text{ m})$$
  $$= 6.25 \times 10^{12} \text{ kg m}^2 \text{ s}^{-2} = 6.25 \times 10^{12} \text{ J}.$$

- *Comment.* The calculation neglects air resistance. In practice the mass of the rocket decreases with altitude, much of the mass being lost as the fuel burns during the initial stage of ascent. Can you recast the solution to take the effect of this into account?

The next step is to use eqn (2.2.4) to discuss the expansion and the compression of a gas. The simplest way of dealing with the problem is to consider the rectangular piston illustrated in Fig. 2.2. We choose the piston to be massless, frictionless, rigid, and a perfect fit. Choosing these properties simply means that we are focusing our minds on the properties of the gas rather than on any technological or economic deficiencies. This is the way to concentrate on the essentials of the problem.

Inside the container is a gas. (This need not necessarily be a perfect gas. It does not even have to be a gas: any substance, liquid or solid, would do, but it is easier to think about gases.) Its pressure is p_{in}. The external pressure is p_{ex}. The force exerted by the gas on the inner face of the piston is $+p_{in}A$ (pushing up) where A is the piston's area. The force on the outside face of the piston is $-p_{ex}A$ (pushing down).

Now arrange matters so that p_{ex} is less than p_{in}. Then the force driving the piston outwards is the greater, and the tendency is to expansion. If the piston moves through a distance dz into the outside world it does so against a constant force $-p_{ex}A$. The work done on the piston is just $-(\text{force}) \times (\text{distance})$ which in the present case is $+p_{ex}A\,dz$. But $A\,dz$ is the volume swept out during the infinitesimal expansion (we write it dV), and therefore the work done *by* the system *on* the atmosphere, as represented by the piston, is $+p_{ex}\,dV$. But we have agreed that dw should be used to denote the work done *on* the system (the confined gas). The

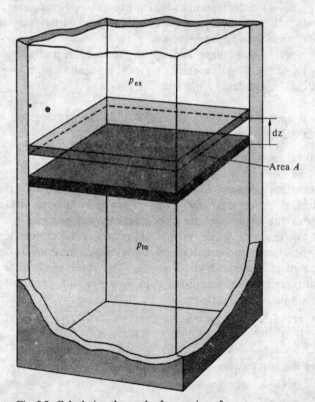

Fig. 2.2. Calculating the work of expansion of a gas.

work done *on* the system is the same in magnitude, but opposite in sign to the work done *by* the system on the outside world (see Box 2.1). We conclude that *the work done on the system when the gas volume changes by* dV *against an external pressure* p_{ex} *is*

(2.2.5)
$$dw = -p_{ex}\, dV.$$

Always bear in mind the fact that work is an effect detectable in the surroundings as a change in the height of a weight (p. 53). In the case of expansion work, the effect of the external pressure can be reproduced by a weight resting on the piston. This weight is raised by the expansion process, and so work is done *by* the system on the surroundings. This emphasis on the interpretation of work may seem pedantic, but its significance becomes clear when we turn to the work of compression. Is the work of compression proportional to the internal pressure? After all, it is the pressure inside that is opposing the pressing in of the piston. Or is it proportional to the external pressure? The answer is that it is the *external* pressure that determines the quantity of work that is done. This is because during compression a weight in the outside world is lowered. It follows that it is the external pressure (to which the external weight is equivalent) that occurs in the compression work, and the appropriate expression is d$w = -p_{ex}\, dV$ for compression as well as expansion. Notice how the signs are still rational: for compression dV is negative, and so dw is positive; work is done *on* the system in compression.

Other types of work have analogous expressions, and some are collected in Table 2.1. For the present we shall continue with p,V-work, and see what can be extracted from eqn (2.2.5).

Free expansion into a vacuum. If $p_{ex} = 0$ it follows that the gas does no work on expansion—the piston has nothing to push against:

(2.2.6)   d$w = 0$ for each step: therefore $w = 0$.

Expansion against constant pressure. In this case the gas expands until it meets a mechanical stop, or until the internal pressure falls to the external. Throughout the expansion the opposing external pressure p_{ex} remains constant (for example, when outside the piston is the atmosphere of the outside world) and the work done on the system as it passes through each successive displacement dV is $-p_{ex}\, dV$. The total work done on the system during its expansion from V_i to V_f is the sum (integral) of these successive, equal contributions:

$$w = -\int_{V_i}^{V_f} p_{ex}\, dV = -p_{ex} \int_{V_i}^{V_f} dV = -p_{ex}(V_f - V_i).$$

Therefore the work done on the system is

(2.2.7)
$$w = -p_{ex}\Delta V,$$

Table 2.1. Varieties of work

In general the work done on a system can be expressed in the form $dw = -F(z)\,dz$, where dz is a 'generalized displacement' and $F(z)$ is a 'generalized force'. The table below shows some examples of this expression.

Type of work	dw	Comment	Units (for dw in J)
Change of volume	$-p_{ex}\,dV$	p_{ex} is the external pressure and dV the change of volume	p_{ex}: Nm^{-2} V: m^3
Change of surface area	$\gamma\,d\sigma$	γ is the surface tension and $d\sigma$ the change in area	γ: Nm^{-1} σ: m^2
Change of length	$f\,dl$	f is the tension, and dl the change in length	f: N l: m
Electrical work	$\phi\,dQ$	ϕ is the electric potential dQ is the charge increment.	ϕ: V Q: C

Only the first type of work concerns us in this chapter, but the techniques developed are applicable to the others too. The expression for electrical work will be explained in more detail in Chapter 11.

and is proportional to the constant external pressure and the change of volume $\Delta V = V_f - V_i$. This result can be illustrated graphically, Fig. 2.3: w is equal, apart from sign, to the area beneath the line $p = p_{ex}$ between $V = V_i$ and V_f.

Note that the work done depends on the *external* pressure. The pressure of the gas inside the piston might change as the volume changes but the work done is independent of that change. The only way the internal pressure enters the calculation is in the requirement that it exceed the external pressure in order that the system expand rather than contract.

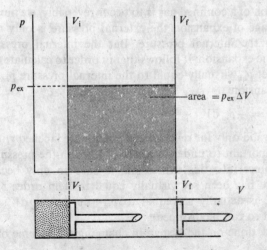

Fig. 2.3. The work of expansion of a gas against a constant external pressure. The magnitude of the work is given by the shaded area. This type of graph is called an *indicator diagram*.

Example (Objective 5). A simplified model of an internal combustion engine is as follows. At the start of the power stroke the ignited gases exert a pressure of 20 atm, and drive the piston back against a constant force which is equivalent to an external pressure of 5 atm. In so doing the piston sweeps out a volume of 250 cm³. How much work is done during the power stroke? What is the power output of an engine containing six cylinders working at 2000 r.p.m. (one power stroke from each cylinder every second revolution)?

- *Method.* The external pressure is constant and so eqn (2.2.7) can be used directly. Power is energy per unit time $(J s^{-1}$, or watts, W).

- *Answer.* $w = -(5 \times 1.013 \times 10^5 \, N \, m^{-2}) \times (2.50 \times 10^{-4} \, m^3) = -127 \, J.$

  Six such cylinders working at 2000 r.p.m. deliver the following amount of power:

  $$\text{power} = (127 \, J) \times 6 \times \tfrac{1}{2} \times (2000/60 \, s^{-1}) = 12.7 \, kW.$$

- *Comment.* The calculation is extremely crude because it neglects all the features that impair the efficiency of engines, nevertheless it does give some idea of the energy and power output of a car engine.

Reversible expansion. In thermodynamics *a reversible change is one that can be reversed by an infinitesimal modification of a variable.* The key word is 'infinitesimal', and its appearance in the definition sharpens the common-sense interpretation of the word 'reversible'. What does it signify?

Suppose the gas is confined by a piston, and the external pressure is only infinitesimally less than the internal. The gas expands. If, however, the external pressure is increased infinitesimally, it will rise above the internal pressure and the gas will be compressed. The process is therefore reversible in the full thermodynamic sense, for an infinitesimal change has brought about a change of direction: instead of expansion there is compression. In contrast, suppose the external pressure differs by a measurable amount from the internal pressure. Changing p_{ex} by an infinitesimal amount will not decrease it below the internal pressure, and so expansion under these circumstances is an *irreversible change* in the thermodynamic sense.

If the expansion of a confined gas is to occur reversibly, we must ensure that at every stage of expansion the external pressure is only infinitesimally less than the internal pressure. But the internal pressure may change during the expansion. It follows that in order to calculate the work done we must set p_{ex} virtually equal to the internal pressure p_{in} at every stage of the expansion:

(2.2.8)ᵣ $$dw = -p_{ex} \, dV = -p_{in} \, dV.$$

(The equations valid only for reversible processes are labelled with a subscript ᵣ.) It is important to understand that, although the pressure of the confined gas comes into this expression, it does so only because the external pressure has been set virtually equal to it in order to ensure reversibility (p_{ex} is actually set equal to $p_{in} \pm dp$, but $dp \, dV$ is doubly infinitesimal and so can be dropped).

The total work done on the system in changing the volume of the gas from V_i to V_f is obtained by summing all the infinitesimal contributions of the form given in the last equation. This means that

(2.2.9)$_r$
$$w = -\int_{V_i}^{V_f} p_{in}\, dV.$$

p_{in} is not necessarily constant: it might drop as the piston comes out. In other words, p_{in} might be a function of the volume V. We can evaluate this integral only if we know how the pressure of the confined gas depends on its volume. This gives a major link with the material covered in Chapter 1: if we know the equation of state of the confined gas, we can express p_{in} as a function of V, and, with luck, do the integral.

The procedure can be illustrated by considering an isothermal expansion and using a perfect gas as the working substance. The expansion can be made isothermal by keeping the system in thermal contact with its surroundings. In practice that means immersing it in a constant-temperature bath. The equation of state of the gas is

(2.2.10)$^\circ$
$$pV = nRT.$$

In an isothermal expansion T is held constant, and so it is independent of V. It follows that the work done on expanding from V_i to V_f at a temperature T can be found by substituting eqn (2.2.10) in (2.2.9):

$$w = -\int_{V_i}^{V_f} \left(\frac{nRT}{V}\right) dV = -nRT \int_{V_i}^{V_f} \left(\frac{1}{V}\right) dV = -nRT \ln(V_f/V_i).$$

(2.2.11)$_r^\circ$
$$w = -nRT \ln(V_f/V_i).$$

This result is limited to reversible processes ($_r$) and perfect ($^\circ$) gases; the logarithms are natural logarithms.

The expression conforms to common sense. When the final volume of the gas is greater than the initial volume (as in an expansion) V_f/V_i is greater than unity; this means that its logarithm is positive, and so w is negative. A negative value of the work done on the system indicates that the system has done work on the outside world.

The equation also shows that the system does more work for a given change of volume if the temperature is raised. This is plausible, and can be traced to the greater pressure of the confined gas at higher temperatures (and therefore to the need for a higher opposing force to ensure reversibility).

We can interpret eqn (2.2.11) in terms of a graph (an *indicator diagram*). The equation shows that the work done when the system expands reversibly and isothermally is, apart from the sign, the area under the isotherm $p = nRT/V$, Fig. 2.4. Superimposed on the figure is the rectangular area obtained in the case of irreversible expansion against an external pressure fixed at the same final value (i.e. Fig. 2.3).

It is quite clear from these graphs that we get more work out of the system in the reversible than in the irreversible case. This is because

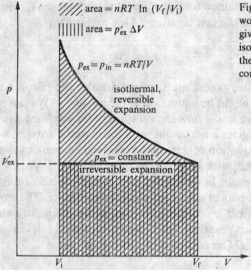

area $= nRT \ln (V_f/V_i)$

area $= p'_{ex} \Delta V$

$p_{ex} = p_{in} = nRT/V$

isothermal, reversible expansion

$p_{ex} = $ constant
irreversible expansion

V_i V_f V

Fig. 2.4. The reversible isothermal work of expansion of a perfect gas is given by the area beneath the isotherm. Note that it is greater than the irreversible work done against constant pressure.

matching the external pressure to the internal pressure ensures that none of the pushing power of the system is wasted. More work than this cannot be obtained because increasing the opposing force even infinitesimally further leads to contraction.

We may infer from this discussion that the *maximum work* available from a system operating between specified initial and final states, and passing along a specified path, is obtained when it is operating *reversibly*. The maximum work can be extracted when at every stage along the path the system is in *equilibrium* with its surroundings. In a reversible process the system passes through a sequence of states only infinitesimally removed from equilibrium states (where, for example, expansion and compression tendencies just balance). Unfortunately, in practice this normally means that the path has to be traversed infinitely slowly, and so reversible processes are also *quasi-static*. Nevertheless, this is not quite as devastating a restriction as at first sight it might appear. The principal point is that the use of reversible paths as *formal* ways of taking a system from one state to another is of extreme importance for calculating the difference between thermodynamic functions even though they may not be realizable in practice. A secondary point is that even some practical processes may be regarded as occurring effectively reversibly, for what really matters is not the absolute rate, but the rate relative to the rapidity of response of the molecules constituting the system. For example, the propagation of sound through a gas is a sequence of very rapid compressions and contractions; but since the molecules follow the changes the propagation can be treated as a sequence of reversible steps. On the other hand, if we do arrange for a process to occur infinitely slowly, then we can be sure that the molecules do have time to respond, and that the process is reversible (if the conditions are right, such as matching internal and external pressures). In other words, granted the correct balance of forces, infinite slowness is a sufficient condition for total reversibility, but not a necessary one for effective reversibility.

The connections between reversibility, equilibrium, and maximum work have been introduced for the particular case of a perfect gas doing expansion work, but the connection applies to all substances and to all kinds of work, as we shall confirm when we turn to the consequences of the Second Law.

Example Although we have been dealing with chemically remote concepts like pistons and cylinders, these ideas do have chemical applications. For instance, calculate the work done when 50 g of iron dissolves in hydrochloric acid in (a) a closed vessel and (b) an open beaker at 25 °C.

• *Method.* The reaction is

$$Fe(s) + 2HCl(aq) \rightarrow FeCl_2(aq) + H_2(g).$$

In the course of reaction 1 mole of gas is generated for each mole of iron consumed. The gas drives back the surrounding atmosphere and thereby does the work $-p\Delta V$. Treat the hydrogen as perfect and ignore the initial volume of the system. For (a) $\Delta V = 0$ because the vessel is closed and cannot expand. For (b) $\Delta V \approx V(g)$, and the system expands until its pressure is p (the pressure of the atmosphere). Then use $pV(g) = n_{H_2} RT$. Note also that if an amount n_{Fe} of iron is consumed, an amount n_{H_2} of H_2 is produced: $n_{H_2} = n_{Fe}$ is

(a) $w = -p\Delta V = 0$

(b) $w = -p\Delta V \approx -n_{H_2} RT = -n_{Fe} RT.$

• *Answer.* (a) $w = 0$.

$$\text{(b)} \quad w = -(50 \, g/55.85 \, g \, mol^{-1}) \times (8.314 \, J \, K^{-1} \, mol^{-1}) \times (298.15 \, K)$$
$$= 2.22 \, kJ.$$

• *Comment.* This is not the maximum amount of work done because we have considered expansion against constant external pressure.

2.3 Heat

When energy is transferred to a system as heat there is a change of state which may appear as a rise in its temperature. For an infinitesimal transfer of heat the increase in temperature is proportional to the amount of heat supplied, and so

$$dT \propto dq, \quad \text{or} \quad dT = \text{coefficient} \times dq,$$

the magnitude of the coefficient depending on the extent, composition, and state of the system. It proves to be more convenient to invert this relation and to write it as

(2.3.1) $dq = C \, dT.$

The coefficient C is called the *heat capacity*. The *molar heat capacity* is $C_m = C/n$.

Heat Capacity. The heat capacity enables the energy supplied to a system to be measured in terms of the resulting temperature rise, a property which is easily monitored. For instance, the heat capacity of the local surroundings of a system (such as a water bath) is an essential piece of information in thermochemistry because the heat absorbed or evolved by the system can be monitored by observing the change of its temperature.

When C is large the transfer of a given amount of energy to a system as heat leads to only a small rise in temperature (the system has a large capacity for heat); but when C is small the transfer of the same amount of energy as heat can lead to a large temperature rise. Water has a large heat capacity: a lot of energy is needed to make it hot (i.e. to raise its temperature), and central heating systems take advantage of this property because a lot of energy can be transported by a slow flow of hot water. Similarly, ponds freeze only slowly.

The heat capacity depends on the conditions under which heat transfer is done. Suppose the system is constrained to have a constant volume and is not able to do any kind of work (e.g. neither mechanical nor electrical work): the energy required to bring about a rise in temperature dT is some amount $C_V dT$, where the subscript V denotes the specified constraint. If instead the system is subjected to constant pressure, and is allowed to expand (or to contract) as the energy is transferred, the amount required to bring about the same rise in temperature is $C_p dT$, where the subscript p denotes the constraint. In the second example, but not in the first, the system has changed its volume, and therefore has done expansion work. It follows that C_p is not the same as C_V because it has to take account of the extra work done by the system. In this case heat is turned into work of expansion and is not used exclusively to raise the system's temperature: energy has been returned to the surroundings as work.

The heat capacities at constant volume and constant pressure are both special cases of the general definition in eqn (2.3.1). This means that both may be defined by adding a label denoting the constraint:

(2.3.2) $dq = C_V dT$, at constant volume, no work

$dq = C_p dT$, at constant pressure, no work other than pV-work.

C_V has been defined in terms of the heat supplied to a body under specified conditions (constant volume, no other forms of work involved). It can, however, be related to the increase in internal energy, because when the system is doing no work $dw = 0$ and so $dU = dq$; therefore $C_V = dU/dT$, volume constant. When one or more variables are held constant during the change of another, the derivatives are called *partial derivatives* with respect to the changing variable. The d is replaced by ∂, and the constant quantities are added as a suffix. It follows that

(2.3.3) $C_V = (\partial U/\partial T)_V$

for systems performing no work. This manipulation opens up roads to useful calculations in thermodynamics because it shifts attention from the type of energy transfer (heat) to the change in a property of the state of the system (its internal energy).

A more formal deduction of the last relation proceeds as follows. The First Law can be written

$$dU = dq + dw = dq + dw_e - p_{ex}\,dV$$

where $-p_{ex}\,dV$ is expansion work and dw_e is any other work (e for extra, or electrical if we want to be explicit). If no volume change is allowed the last term is zero, and if no non-pV work is allowed the middle term is also zero. Hence

(2.3.4) $$dU = (dq)_V \qquad \text{for } dV = 0, \qquad dw_e = 0.$$

Substitution of this result into eqn (2.3.1) leads immediately to the expression for C_V in terms of the internal energy of the system.

Example A function of two variables which has already been met is the van der Waals equation of state, p. 40, which reads

$$p(V, T) = nRT/(V-nb) - an^2/V^2.$$

Find the partial derivatives of p with respect to V and T.

- *Method.* In order to find $(\partial p/\partial T)_V$ differentiate $p(V, T)$ with respect to T, regarding V as a constant (just like $R, a,$ and b). To find $(\partial p/\partial V)_T$ differentiate with respect to V, holding T constant.

- *Answer.* $(\partial p/\partial T)_V = nR/(V-nb).$
  $$(\partial p/\partial V)_T = -nRT/(V-nb)^2 + 2an^2/V^3.$$

- *Comment.* The two partial derivatives are themselves functions of V and T, and so we could go on to find the four second derivatives. Later we shall see that it is particularly important to note that $\partial^2 p/\partial V \partial T = \partial^2 p/\partial T \partial V$, as may be checked here.

Enthalpy. But what of C_p? Is there a thermodynamic property of the system that can be identified with $(dq)_p$, the heat absorbed at constant pressure? It turns out that it is easy to construct one simply by adding the product pV to the internal energy U. This gives a new quantity denoted H:

$$H = U + pV.$$

We now confirm that dH can be identified with the amount of heat added to a system subjected to a constant pressure p (for example, a system open to the atmosphere) and doing no work except expansion work.

The first step is to consider what happens to H when the pressure and volume change. U changes to $U+dU$, p changes to $p+dp$, V changes to $V+dV$. H changes to

$$H + dH = (U+dU) + (p+dp)(V+dV)$$
$$= U + pV + dU + Vdp + pdV + dpdV.$$

As we are considering only infinitesimal changes, the last term, being doubly infinitesimal, can be ignored. Since $U + pV$ on the right is simply H, it follows that

(2.3.5) $dH = dU + pdV + Vdp.$

We can get this result more quickly. The differential of $H = U + pV$ is $dH = dU + d(pV)$; the second term is the differential of a product, and so it can be written $(dp)V + p(dV)$ or $Vdp + pdV$. (We shall use this method from now on, and avoid the clumsy method, on which it is based.)

Now consider a system that is in equilibrium with its surroundings at a pressure p. If this is so, we can substitute $dU = dq + dw$ in the form

$$dU = dq + dw_e - pdV$$

(because $p_{ex} = p$ under the specified conditions), cancel the pdV terms, and arrive at

(2.3.6) $dH = dq + dw_e + Vdp.$

Now impose the conditions that (a) there is no non-pV work, (b) there is no pressure change during the heating. Both dw_e and Vdp are then zero, and so

(2.3.7) $dH = (dq)_p,$ for $dp = 0,$ $dw_e = 0.$

This confirms that the increase in the property H of the system is equal to the amount of energy added as heat at constant pressure so long as no other types of work are involved.

The result in eqn (2.3.7) can be used in the definition of C_p to give

(2.3.8) $C_p = (\partial H / \partial T)_p$

for systems performing no non-pV work.

It may be objected that all we seem to have done is to replace q by something called H. This is not so, for what we have actually done is to replace something that is measured in the surroundings and which depends on the mode of energy transfer (heat, dq) by a change in a property of the state of the system (H depends only on the state of the system, because U depends only on the state, and so do the pressure and volume). It will be seen shortly that H is a very useful quantity, and inventing it has not merely been an exercise in generating equations. In fact H plays a central role in chemistry because we are so often concerned with processes at constant pressure (reactions occurring in open vessels, including the body, are typical examples). Because H is so useful it carries its own name, the *enthalpy* of the system:

(2.3.9)   Enthalpy: $H = U + pV.$

Note that p is to be taken as the pressure of the *system*, and the form pV is a part of the definition of the enthalpy for any substance and does not imply a restriction to perfect gases.

Example (Objectives 10 and 12). Water is brought to the boil under a pressure of 1 atm. When an electric current of 0.5 A from a 12 V source is passed for 5 minutes through a resistance in thermal contact with it, it is found that 0.798 g of water is distilled and then condensed. Calculate the molar internal energy change and molar enthalpy of vaporization of water, ΔU_{vap}, ΔH_{vap} at the boiling point (373.15 K).

• *Method.* The change $H_2O(l) \rightarrow H_2O(g)$ is accompanied by a change in internal energy $\Delta U_{vap} = q_{vap} + w_{vap}$. The heat absorbed during the vaporization is equal to the amount of electrical work (amps × volts × time) done on the resistance, or $q_{vap} = IVt$. The work done on vaporization is the expansion work done when water liquid is turned into water vapour, or $-p\Delta V$, ΔV being the volume change. If we assume perfect behaviour for the gas its volume is nRT/p, and since the gas volume is so much greater than that of the same amount of liquid we can write $\Delta V = V(g) - V(l) \approx V(g)$. Hence $p\Delta V \approx nRT$. Therefore

$$\Delta U = IVt - nRT.$$

Furthermore, since $\Delta H = \Delta U + \Delta(pV) = \Delta U + p\Delta V$ because $\Delta p = 0$,

$$\Delta H = IVt.$$

The amount n can be obtained from the mass evaporated and the molar mass of water ($18.02 \, \text{g mol}^{-1}$).

• *Answer.* $n = (0.798 \, \text{g})/(18.02 \, \text{g mol}^{-1}) = 0.0443 \, \text{mol}$.

$$\begin{aligned}
\Delta U &= (0.5 \, \text{A}) \times (12 \, \text{V}) \times (300 \, \text{s}) \\
&\quad - (0.0443 \, \text{mol}) \times (8.314 \, \text{J K}^{-1} \, \text{mol}^{-1}) \times (373.15 \, \text{K}) \\
&= 1800 \, \text{J} - 137.4 \, \text{J} = 1663 \, \text{J}.
\end{aligned}$$

Hence

$$\Delta U_{vap,m}(373 \, \text{K}) = (1.663 \, \text{kJ})/(0.0443 \, \text{mol}) = 37.5 \, \text{kJ mol}^{-1}.$$

$$\Delta H_{vap,m}(373 \, \text{K}) = (1.800 \, \text{kJ})/(0.0443 \, \text{mol}) = 40.6 \, \text{kJ mol}^{-1}.$$

• *Comment.* Notice that the enthalpy change is greater than the internal energy change because it takes into account the amount of work that has to be done in order for the change to proceed and to drive back the atmosphere. Both ΔU and ΔH are positive because energy has to be supplied to the system in order to bring about evaporation.

Taking stock. We set out to express transfers of heat in terms of changes of temperature. In the course of the discussion we found that the amount of heat transferred at constant volume can be identified with the change of internal energy of the system, and the amount transferred at constant pressure can be identified with the change of a new property, the enthalpy, there being no work other than pV-work involved in the process.

How does enthalpy differ from internal energy? When heat passes into a system at constant pressure the increased internal energy does not entirely remain inside the system; some is passed back to the surroundings as work. It follows that the changes in enthalpy and internal energy differ by the amount of work involved in changing the dimensions of the system. In many problems of chemistry we are interested in the heat evolved under conditions of constant pressure, and so it is important to know the enthalpy of a system rather than its internal energy. This is the subject matter of Chapter 4.

The heat capacities at constant volume and pressure also differ on account of the work involved in changing the size of the system. Since work has to be done in the case of C_p it follows that the temperature rise will be less for a given heat transfer, and therefore that C_p is greater than C_V. The difference is greater for gases than for either liquids or solids, because the latter change their volumes only slightly when heated, and therefore do little work. We shall see an important qualification of this conclusion in the next chapter, but it remains true that C_p is greater than C_V.

The difference between C_p and C_V can be calculated readily for a perfect gas because the connection between enthalpy and internal energy depends only on the temperature:

$$H = U + pV = U + nRT.$$

When the temperature of the system is raised from T to $T + dT$, the corresponding enthalpy change dH is related to the energy change dU by

$$dH = dU + nRdT.$$

We have seen that $dH = C_p\,dT$ and $dU = C_V\,dT$, and so it follows that

$(2.3.10)°$ $C_p - C_V = nR.$

Molar heat capacities of gases are of the order of R itself (see Table 2.2), and so the difference between C_p and C_V is very significant.

Table 2.2. Heat capacities at 25 °C and 1 atm.

	$C_{V,m}/\mathrm{J\,K^{-1}\,mol^{-1}}$	$C_{p,m}/\mathrm{J\,K^{-1}\,mol^{-1}}$	$\gamma = C_{p,m}/C_{V,m}$
He, Ne, Ar, Kr, Xe	12.48	20.79	1.666
H_2	20.53	28.84	1.405
O_2	21.06	29.37	1.395
N_2	20.81	29.12	1.399
CO_2	28.81	37.12	1.288
NH_3	27.32	35.63	1.304
CH_4	27.33	35.64	1.304

Source: G. W. C. Kaye and T. H. Laby, *Tables of physical and chemical constants*, Longmans.

2.4 What is work, and what is heat?

In this section we introduce some simple and qualitative interpretations of thermodynamics, drawing on a knowledge of molecular behaviour no deeper than we needed for our discussion of gases. The description can be sharpened and made quantitative: that is the job of *statistical thermodynamics*, a subject treated in Part 2.

We identify thermodynamic internal energy with molecular energy, energy stored in molecular bonds and in molecular translation, vibration, and rotation.

Heat is a way of increasing internal energy because it stimulates molecular motion. When an object is heated the molecular motion occurs in random directions. Heating a gas causes the molecules to move

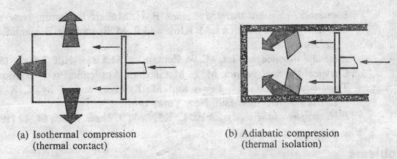

(a) Isothermal compression
(thermal contact)

(b) Adiabatic compression
(thermal isolation)

Fig. 2.5. The heat flows involved in isothermal and adiabatic compressions.
Perpetual Motion Machines. S. W. Angrist; *Scientific American*, **218**(1), 114 (1968).

faster in all directions at random. When energy is transferred to a solid by heating the atoms oscillate around their sites, but do so in random directions (or at least the displacements to the right occur as frequently as displacements to the left so that no net displacement can be identified). *Heat stimulates random motion*. We call this random motion *thermal motion*.

Work involves *organized* motion. When a spring is compressed the atoms move closer together in a definite direction, and they move apart in a definite direction when it unwinds. When a piston compresses a gas the initial effect is to accelerate the molecules in the direction of the piston's movement (what happens next we consider in a moment). *Work stimulates organized motion*.

When a gas is compressed *isothermally* (so that the initial temperature is maintained by a heat flow) the accelerated molecules strike the conducting walls of the vessel, and its atoms are excited into vibration. This jostling is handed on to the outside world. By then the motion is random, and so the work of compression has been degraded into thermal motion of the surroundings, Fig. 2.5a.

When a gas is compressed *adiabatically* (so that no heat enters or leaves the system) the accelerated molecules strike insulating walls, which are unable to transmit energy out of the system. Collisions with the other molecules of the sample ensure that the accelerated motion of one molecule is rapidly randomized in direction and shared between all the other molecules. Thus the rise in internal energy appears as a rise in the temperature of the compressed gas, and is stored as thermal motion of the system, Fig. 2.5b.

We can begin to see what is meant by saying that work *degrades* into heat: we mean that the directional characteristics of the molecular motion are being randomized. The Second Law (Chapter 5) is a commentary on this chaos.

Further reading

Basic chemical thermodynamics (2nd ed.). E. B. Smith; Oxford University Press, 1977.

Elementary chemical thermodynamics. B. H. Mahan; Benjamin, New York, 1963.
Chemical thermodynamics. I. M. Klotz and R. M. Rosenberg; Benjamin, New York, 1972.
Heat and thermodynamics. M. W. Zemansky; McGraw-Hill, Tokyo, 1968.
Chemical thermodynamics. M. L. McGlashan; Academic Press, London, 1979.
Thermodynamics. G. N. Lewis and M. Randall; revised by K. S. Pitzer and L. Brewer; McGraw-Hill, New York, 1961.
Bibliography of thermodynamics. L. K. Nash; *J. chem. Educ.*, **64**, 42 (1965).

Problems

2.1.   Calculate the work necessary to raise a mass of 1 kg through a height of 1 m on the surface of (a) the earth, where $g = 9.8\,\mathrm{m\,s^{-2}}$, (b) the moon, where $g = 1.6\,\mathrm{m\,s^{-2}}$.

2.2.   How much work must a man do to climb a flight of stairs? Take his mass as 150 lb and suppose he climbs through 10 ft.

2.3.   Calculate the work necessary to compress a spring by 1 cm if its force constant is $2 \times 10^5\,\mathrm{N\,m^{-1}}$. What work must be done to stretch the spring by the same amount?

2.4.   In a spacecraft designed to land on the surface of Mars there was a sampling device that incorporated a spring of force constant $2 \times 10^6\,\mathrm{N\,m^{-1}}$ attached to a cam which alternately compresses and expands it through a displacement of 1 cm from equilibrium. Power supplies had to be planned very carefully, and the question arose as to how much power would be needed to drive the sampler. How much work would have to be done on the spring to compress and extend it 1000 times? What is the power (in watts) of an electric motor that can accomplish these 1000 oscillations in 1000 s? The spring is in contact with a thermally insulated mass of metal of heat capacity $4.2\,\mathrm{kJ\,K^{-1}}$: if its initial temperature is 20 °C what will be the metal's temperature after the 1000 oscillations?

2.5.   In a machine of a particular design, the force acting on a mass of 2 g varies as $-F\sin(\pi x/a)$. Calculate the work required in order to move the mass (a) from $x = 0$ to $x = a$, (b) from $x = 0$ to $x = 2a$.

2.6.   A chemical reaction occurs in a vessel of cross-section 100 cm² fitted with a loosely fitted seal. During the reaction the seal is pushed out through 10 cm against the external pressure of 1 atm. How much work does the reaction do on the outside world?

2.7.   Using the same system as in the last Problem, calculate the work done when the pressure of the atmosphere is replaced by a mass of 5 kg acting downwards on the vertical piston. What is the work done when the same mass stands on a piston of cross-section 200 cm² and moves through the same distance.

2.8.   In another reaction occurring within the same apparatus employed in the last Problem, a contraction of 10 cm occurs. Discuss the work involved in the change.

2.9.   A 5 g lump of solid carbon dioxide is allowed to evaporate in a 100 cm³ vessel maintained at room temperature. Calculate the work done when the system is then allowed to expand at 1 atm pressure (a) isothermally against a pressure of 1 atm, (b) isothermally and reversibly.

2.10.   1 mol of $CaCO_3$ was heated to 700 °C, when it decomposed. The operation was carried out in a container closed by a piston which was initially resting on the sample and was restrained throughout by the atmosphere. How much work was done during complete decomposition?

2.11. The same experiment as in the last Problem was repeated, the only difference being that the carbonate was heated in an open vessel. How much work was done during decomposition?

2.12. A strip of magnesium of mass 15 g is dropped into a beaker of dilute hydrochloric acid. What work is done on the surrounding atmosphere (1 atm pressure, 25 °C) by the subsequent reaction?

2.13. We have to be able to deal with the work involved in the expansion of real gases. In the case of reversible expansion or compression this involves knowing how the pressure varies with the volume of the enclosed system. Information of this kind is contained in the equation of state, and the next few problems explore the consequences of employing some of the approximate equations that are available. In the first place, calculate the work done in an isothermal, reversible expansion of a gas that satisfies the virial equation of state, $pV_m = RT(1 + B/V_m + ...)$.

2.14. The virial equation that represents the behaviour of argon at 273 K has $B = -21.7 \text{ cm}^3 \text{ mol}^{-1}$ and $C = 1200 \text{ cm}^6 \text{ mol}^{-2}$. Calculate (a) the work of reversible, isothermal expansion at this temperature, (b) the work of expansion against a constant pressure of 1 atm, (c) the work of expansion on the assumption that argon behaves perfectly. Take $V_{m,i} = 500 \text{ cm}^3$ and $V_{m,f} = 1000 \text{ cm}^3$.

2.15. The van der Waals gas is a useful model of a real gas, and we know the significance of the parameters a and b. It is instructive to see how these parameters affect the work of isothermal, reversible expansion. Calculate this work, and account physically for the way that a and b appear in the final expression.

2.16. Plot on the same graph the indicator diagrams for the isothermal expansion of (a) a perfect gas, (b) a van der Waals gas for which $b = 0$ and $a = 4.2 \text{ dm}^6$ atm mol^{-1}, (c) the same but with $a = 0$ and $b = 5.105 \times 10^{-2} \text{ dm}^3 \text{ mol}^{-1}$. This will show how the a and b parameters modify the area under the isotherm, and therefore the work. The values of a and b selected exaggerate imperfections for normal conditions, but the exaggeration is instructive because of the large distortions of the indicator diagram that result. Take $V_i = 1 \text{ dm}^3$, $n = 1 \text{ mol}$, and $T = 298 \text{ K}$.

2.17. Show that the work of isothermal, reversible expansion of a van der Waals gas can be expressed in reduced variables, and that by defining the *reduced work* as $w_r = 3bw/a$ an expression can be obtained that is independent of the nature of the gas.

2.18. The technique of partial differentiation, p. 66, is quite straightforward once one realizes that only the variable stated explicitly, e.g. the x in $\partial f/\partial x$, is varied, all others being regarded as constants. In order to get further practice, take (a) the ideal gas equation, (b) the Dieterici equation, Table 1.4, and evaluate $(\partial p/\partial T)_V$ and $(\partial p/\partial V)_T$. Go on to confirm that $\partial^2 p/\partial V \partial T = \partial^2 p/\partial T \partial V$.

2.19. Partial derivatives can (and, in thermodynamics, do) have real physical meaning. For instance take the expressions for the two partial derivatives of p for an ideal gas obtained in the last Problem and answer the following. In an estimation of the pressure exerted by a gas, the temperature was uncertain by 1 per cent and the volume was uncertain by 2 per cent. What is the uncertainty in the prediction of the pressure arising from (a) the temperature uncertainty, (b) the volume uncertainty, (c) both.

2.20. The *equipartition principle*, p. 15, lets us calculate the internal energy of an ideal gas. In the case of a monatomic gas it gives $U = (3/2)nRT$. This result (and others like it for more complicated molecules) gives a very quick way of predicting the heat capacity of these materials. Deduce the value of the heat capacity of (a) a monatomic perfect gas, (b) a rotating, translating, nonlinear, polyatomic molecule.

2.21. When 1 calorie of heat is transferred to 1 g of water at 14.5 °C under constant atmospheric pressure its temperature rises to 15.5 °C. What is the molar heat capacity of water at this temperature? Express your answer in $cal K^{-1} mol^{-1}$ and in $J K^{-1} mol^{-1}$.

2.22. The constant-pressure heat capacity of helium is $20.79 J K^{-1} mol^{-1}$. How much heat is required to raise the temperature of a sample of 1 mol by 10 K at 25 °C (a) when it is in a constant volume vessel at a pressure of 10 atm, (b) when it is in a vessel fitted with a piston subjected to a constant external pressure of 10 atm? How much work is done in each case?

2.23. The heat capacity of air at room temperature and 1 atm pressure is approximately $21 J K^{-1} mol^{-1}$. How much heat is required to heat an otherwise empty room through 10 K at room temperature? Let the room measure 5 m × 5 m × 3 m. Neglecting losses, how long will a 1 kW heater take to heat the room by that amount?

2.24. Assume that the heat capacity of water has the value calculated in Problem 2.21 over its whole liquid range. How much heat must be supplied to raise 1 kg of water from room temperature to its boiling point? For how long must a 1 kW heater be operated in order to supply this energy?

2.25. A kettle containing 1 kg of boiling water is heated until evaporation is complete. Calculate (a) w, (b) q, (c) ΔU, (d) ΔH for the process. ($\Delta H_{vap,m}$ (373 K) = $40.6 kJ mol^{-1}$.)

2.26. The same amount of water as in the last Problem is evaporated in a large container where the pressure is only 0.1 atm. Calculate the same four quantities for the process. ($\Delta H_{vap,m}$ at the temperature of boiling water (46 °C) at that pressure is $44 kJ mol^{-1}$.)

2.27. In each of the last two Problems the heat was supplied electrically. How much heat has to be supplied in each case, and what power of heater (in watts) is required to bring about total evaporation in 10 minutes in each case? From what height must a 10 kg mass drop to supply the required energy?

2.28. The heat capacity of water in the range 25–100 °C is $C_{p,m} = 75.48 J K^{-1} mol^{-1}$. How much heat must be supplied to 1 kg of water initially at 25 °C in order to bring it to its boiling point at atmospheric pressure?

2.29. A piston exerting a pressure of 1 atm rests on the surface of water at 100 °C. The pressure is reduced infinitesimally, and as a result 10 g of water evaporate. This process absorbs 22.2 kJ of heat. What are the values of q, w, ΔU, ΔH, and ΔH_m for the vaporization?

2.30. One of the factors in the selection of refrigerant fluids is their enthalpy of vaporization. A new fluorocarbon of R.M.M. 102 was synthesized and a small amount was placed in an electrically heated vessel. Under a pressure of 650 mmHg the liquid boiled at 78 °C. When it was boiling it was found that when a current of 0.232 A from a 12 V supply was passed for 650 s the quantity of distillate was 1.871 g. What are the values of the molar enthalpy and internal energy of vaporization of the fluorocarbon?

3 The First Law: the machinery

Learning objectives

After careful study of this chapter you should be able to

(1) Define *extensive* and *intensive* properties (p. 76), *state function* and *path function* (p. 76), and *complete (exact) differentials* (p. 78).

(2) Express changes of internal energy in terms of changes in volume and temperature (eqn (3.1.3)).

(3) Use the properties of *partial derivatives* to deduce expressions for the dependence of the internal energy on the temperature at constant pressure (eqn (3.2.11)) and the dependence of the enthalpy on the temperature at constant volume (eqn (3.2.13)).

(4) Define and use *isobaric expansivity* (p. 84) and *isothermal compressibility* (p. 86).

(5) Explain the thermodynamic significance of the *Joule–Thomson experiment*, define the *Joule–Thomson coefficient*, and describe its significance for the *liquefaction of gases* (p. 87).

(6) Derive the relation between heat capacities at constant volume and constant pressure (eqn (3.2.19)) and apply it to solids, gases, and liquids.

(7) Calculate the work done on a perfect gas during an *adiabatic change* (eqn (3.2.30)).

(8) Calculate the final volume, pressure, and temperature of a perfect gas after a reversible, adiabatic change of volume (eqns (3.2.29), (3.2.31), and (3.2.33)).

Introduction

In the last chapter we established what we meant by work and heat, and indicated how they could be dealt with in simple cases. We saw that the work of expansion can be calculated by working with the expression $-p_{ex} dV$, and that the heat transferred to a single phase system can be related to the change in temperature through the expression $dq = C dT$. We also saw a mathematical expression of the First Law in the form $dU = dq + dw$.

This chapter develops these elementary relations. The first section is important because it is essential to know what is meant by a 'state function'. The next section is for those who want to go a little further in seeing how to manipulate thermodynamic expressions. That section can be returned to as later work indicates its relevance.

3.1 State functions and differentials

A common classification of properties is based on whether or not they depend on the amount of substance present in the system. The internal energy of a system depends on the amount of material it contains: doubling the amount of material doubles the internal energy. This is an example of an *extensive property*. Other examples are the mass, the heat capacity, and the volume. Other properties are independent of the amount of material present. These are called *intensive properties*, and include temperature, density, pressure, viscosity, and the molar properties (molar volume, molar heat capacity, etc.).

Some quantities are properties of the state of the system, but others are related to what is happening to the system when changes are in progress. The internal energy is an example of the first kind of quantity because a system in a particular state possesses a particular amount of internal energy. The work done on a system, w, is an example of the second kind of quantity: we do not speak of a system as possessing a particular amount of 'work', or of its 'work' having a particular value. Properties of the first kind, properties of the state of the system, are called *state functions*. Other examples of state functions are volume, pressure, temperature, density, refractive index, and so on. Properties that depend on the path we call *path functions*.

The classification of properties of a system into state functions and path functions is very important in thermodynamics. We can begin to appreciate why this is so by considering the implications of the classification for the First Law. The central point in this connection is that *a state function, being a property of the present state of the system, is independent of the way the state was prepared.*

The distinction can be illustrated by referring to the changes accompanying the expansion and compression of a gas (or any material for that matter). Look at Fig. 3.1. The initial state of the system is p_i, V_i, T_i, and in this state the internal energy is U_i. Work is done on the system so that it is compressed adiabatically to a volume V_f, a pressure p_f, and a tempera-

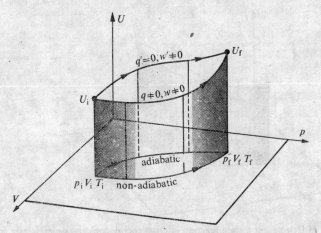

Fig. 3.1. q and w depend on the path, but U depends only on the state of the system.

ture T_f. In the new state the internal energy is U_f, and the work done on the system as it traversed the path is some amount w'. Notice our use of language: U is a property of the state; w' is a property of the path.

Now imagine another process in which the initial conditions are also p_i, V_i, T_i and the final conditions are p_f, V_f, T_f, but in which the compression is not adiabatic. The internal energy of the initial state is U_i, the same as before (because it is a state function) and the internal energy of the final state is U_f, also the same as before (for the same reason). But as an amount of heat q is allowed to escape from the system, the work done on the system, w, is different from w'. Here we see the difference between a state function and a path function: U_f is unchanged, so is U_i, hence U is a state function; w differs from w' and q differs from q', hence these are not state functions.

How do we know that U is a state function? The First Law implies that it is a state function, because if it were not we could get work out of nothing: we could have perpetual motion. To see this, suppose that U is not a state function. It will then have a value that depends on the path the system is taken through. Take it along path 1 (Fig. 3.2): U changes from U_i to U_f. Now take it from this state back to the initial state via path 2: U changes from U_f to U_i', and if U is not a state function U_i' may be different from U_i. Therefore, simply by changing the state of the system from p_i, V_i, T_i to p_f, V_f, T_f and back to p_i, V_i, T_i we have changed the internal energy from U_i to U_i'. If we choose the paths so that U_i' is larger than U_i, the cycle of operations generates an amount of energy $\delta U = U_i' - U_i$, which we could turn into useful work.

A regrettable fact of life is that experience has shown that perpetual motion machines cannot be constructed. From this experience we are forced to conclude that energy cannot be created; which implies that U is a state function.

An infinitesimal change in U is represented by dU. If a system is transported along a path (e.g. by compressing it isothermally) U will

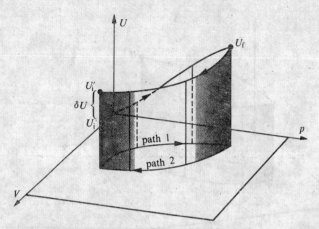

Fig. 3.2. If U were not a state function internal energy could be generated by going round the cycle of path 1 followed by path 2.

change from U_i to U_f, and the change in U can be calculated by adding together all the individual dUs:

(3.1.1) $$\Delta U = \int_{U_i}^{U_f} \mathrm{d}U = U_f - U_i.$$

The value of ΔU so calculated depends on the initial and final states but is independent of the path: all paths connecting them give the same value of ΔU. We summarize this path independence of the integral of dU by saying that dU is an *exact differential*.

An infinitesimal transfer of heat is denoted dq. If a system is transported along a path the total heat transferred may be obtained by adding all the individual dqs:

(3.1.2) $$q = \int_{\text{path}} \mathrm{d}q.$$

But notice the difference between this and the preceding equation. First, we do not write Δq, because q is not a state function and the heat added cannot be expressed in the form $q_f - q_i$. Secondly we have to specify the path of integration because q depends on the path selected. We summarize this by saying that dq is an *inexact* differential ('inexact', because to integrate it we must also specify the path). Often dq is written đq in order to emphasize that it is inexact. This looks a little mysterious, and we shall not do it, but it is essential to bear in mind the distinction between the derivatives of state functions and path functions. The latter have meaning *only* when the path is specified.

Example   Consider a perfect gas inside a cylinder fitted with a piston. Let the initial state of the system be T, V_i and the final state be T, V_f, so that the net change corresponds to isothermal expansion. The change of state can be brought about in many ways, of which the two simplest are the following.
(i) *Path 1*. Free, irreversible expansion against zero external pressure, accompanied by whatever influx of heat is required to maintain constant temperature.

(ii) *Path 2.* Reversible, isothermal expansion accompanied by the appropriate influx of heat.
Find w, q, and ΔU for each process.

- *Method.* Use the fact (Table 3.3, p. 99) that U does not change in an isothermal expansion of a perfect gas: $\Delta U = 0$ since $(\partial U/\partial V)_T = 0$. Since this is so, and since $\Delta U = q + w$, we have $q = -w$. The work for Path 1 can be calculated from eqn (2.2.7), p. 60, and that for Path 2 from eqn (2.2.11), p. 63.

- *Answer.* Path 1: $\Delta U = 0$, $w = 0$ (because $p_{\mathrm{ex}} = 0$), $q = 0$ (because $q = -w$).
  Path 2: $\Delta U = 0$, $w = -nRT \ln (V_f/V_i)$, $q = nRT \ln (V_f/V_i)$.

- *Comment.* The work and heat terms depend on the paths, but their sum, ΔU, is independent. This shows that q and w are path functions.

An infinitesimal amount of work done on the system is denoted dw. We know that the work done depends on the path selected; therefore we know immediately that dw is an inexact differential. It is often written đw.

We are now almost at the point where we can develop the idea of an exact differential into some powerful machinery for deducing quantitative conclusions about heat, work, and energy. In order to do this, consider the state function U as being a function of volume and temperature, so that it can be written $U(V, T)$. (U can be regarded as a function of V, T, and p; but since there is an equation of state it is possible to express p as a function of V and T, and so p is not an independent variable. We could choose p, T or p, V as the independent variables, but V, T fit our purpose. Throughout this section we are dealing only with a closed system, and one in which the composition remains the same. Later we shall have to allow U to depend on the composition too.) Now let V change to $V + \mathrm{d}V$: by how much will U change? If we know the slope of U with respect to V, with T held constant, we can find the change. Denoting this slope as $(\partial U/\partial V)_T$ we can write

$$U(V + \mathrm{d}V, T) = U(V, T) + (\partial U/\partial V)_T \, \mathrm{d}V.$$

The coefficient $(\partial U/\partial V)_T$, in the language introduced on p. 66, is the partial derivative of U with respect to V, see Fig. 3.3a.

If T were to change from T to $T + \mathrm{d}T$ at constant V, the internal energy would change to

$$'U(V, T + \mathrm{d}T) = U(V, T) + (\partial U/\partial T)_V \, \mathrm{d}T$$

and we encounter another partial derivative. This is illustrated in Fig. 3.3b.

Suppose now that both V and T change by an infinitesimal amount, Fig. 3.3c. Neglecting second-order infinitesimals, the new U is

$$U(V + \mathrm{d}V, T + \mathrm{d}T) = U(V, T) + (\partial U/\partial V)_T \, \mathrm{d}V + (\partial U/\partial T)_V \, \mathrm{d}T.$$

The internal energy at $(V + \mathrm{d}V, T + \mathrm{d}T)$ differs from that at (V, T) by an infinitesimal amount, which we write dU. Then from the last equation we can write the important expression

(3.1.3) $$\mathrm{d}U = (\partial U/\partial V)_T \, \mathrm{d}V + (\partial U/\partial T)_V \, \mathrm{d}T.$$

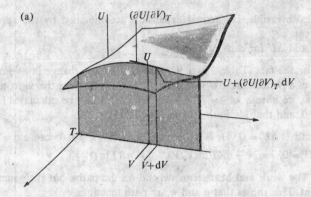

(a)

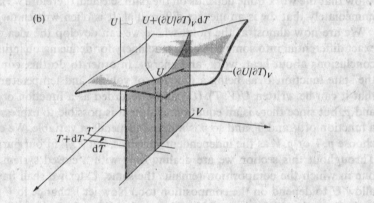

(b)

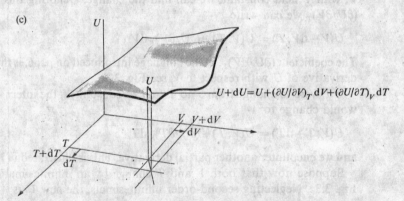

(c)

Fig. 3.3. The partial derivatives of U with respect to (a) V and (b) T, and (c) the total change (the total derivative) of U when V and T both change.

The interpretation of this equation is quite straightforward: it simply says that (in a closed system of constant composition) any infinitesimal change in U is proportional to the infinitesimal changes in volume and temperature, the coefficients of proportionality being the partial derivatives. Very often these partial derivatives have an easily discernible physical significance, and thermodynamics gets shapeless and difficult only when this meaning is not kept in sight. In the present case $(\partial U/\partial T)_V$ has already been encountered on p. 66; we saw that it is the constant-volume heat capacity C_V. The other coefficient, $(\partial U/\partial V)_T$, we have not met before. It is the rate of change of the internal energy as the volume of the system is changed isothermally. This quantity is sufficiently important to deserve closer attention, and we consider it below. Other partial derivatives will be met in the following pages, and all can be interpreted in physical terms.

Example (Objective 2). Measurements of $(\partial U_m/\partial V)_T$ for ammonia give the value 840 J m^{-3} mol^{-1} at 300 K (how this value can be estimated from the van der Waals parameters is explained on p. 164). The value of $(\partial U_m/\partial T)_V$, which is $C_{V,m}$, is 27.32 J K^{-1} mol^{-1}. What is the change in molar internal energy of ammonia when it is heated through 2 K and compressed through 100 cm^3?

- *Method.* An infinitesimal change in volume and temperature leads to the infinitesimal change in U given by eqn (3.1.3). The changes in the present problem are small and may be regarded as virtually infinitesimal. Then the equation becomes

$$\Delta U_m \approx (\partial U_m/\partial V)_T \, \Delta V + (\partial U_m/\partial T)_V \, \Delta T.$$

- *Answer.* $\Delta U_m = (840 \text{ J m}^{-3} \text{ mol}^{-1}) \times (-1.00 \times 10^{-4} \text{m}^3)$
$$+ (27.32 \text{ J K}^{-1} \text{ mol}^{-1}) \times (2 \text{ K})$$
$$= -0.084 \text{ J mol}^{-1} + 54.64 \text{ J mol}^{-1} = 54.56 \text{ mol}^{-1}.$$

- *Comment.* Note how the change of internal energy is dominated by the temperature change. If ammonia behaved perfectly, the coefficient $(\partial U/\partial V)_T$ would be zero and the volume change would have no influence on U. In a more precise calculation it would be necessary to integrate dU properly in order to calculate an accurate value of ΔU.

Partial derivatives have many useful properties and in the development of thermodynamics we shall draw on them frequently. The kind of property we need to know about is that the partial derivatives $(\partial V/\partial T)_p$ and $(\partial T/\partial V)_p$ can be related by an expression as simple as

(3.1.4) $(\partial V/\partial T)_p = 1/(\partial T/\partial V)_p.$

There are a number of relations like this, and they are collected in Box 3.1. Adroit use of them can turn some unfamiliar quantity into something we can recognize or interpret. Note that the relations are built entirely on mathematical principles, and so are independent of the nature of the system.

Box 3.1   **Partial-derivative relations**

If f is a function of x and y, when x and y are changed by dx and dy, f itself changes by df:

$$df = (\partial f/\partial x)_y\, dx + (\partial f/\partial y)_x\, dy.$$

Partial derivatives may be taken in either order:

$$\partial^2 f/\partial x\, \partial y = \partial^2 f/\partial y\, \partial x.$$

In more detail this reads

$$\left[\frac{\partial}{\partial x}\left(\frac{\partial f}{\partial y}\right)_x\right]_y = \left[\frac{\partial}{\partial y}\left(\frac{\partial f}{\partial x}\right)_y\right]_x.$$

1. Suppose x and y depend on a variable z (for example, x might be pressure, y might be volume; then z could be temperature). How does f vary when x is changed under conditions of constant z? *Relation No. 1* generates the answer:

If $df = (\partial f/\partial x)_y\, dx + (\partial f/\partial y)_x\, dy,$ x and y depend on z,

then $(\partial f/\partial x)_z = (\partial f/\partial x)_y + (\partial f/\partial y)_x(\partial y/\partial x)_z.$

2. Suppose x, y, and z are related (for example, as p, V, and T are related by an equation of state; or when z can be written as a function of x and y). Then relation No. 2 can be used to turn partial differentials upside-down:
Relation No. 2, The Inverter:

$$(\partial x/\partial y)_z = 1/(\partial y/\partial x)_z.$$

3. Suppose x, y, and z are related. Can we relate $(\partial x/\partial y)_z$ to either $(\partial x/\partial z)_y$ or $(\partial z/\partial y)_x$?
Relation No. 3, The Permuter, generates the answer:

$$(\partial x/\partial y)_z = -(\partial x/\partial z)_y(\partial z/\partial y)_x.$$

By connecting relation Nos. 2 and 3 we get *Euler's chain relation*, which is a useful way of remembering the manipulations performed by relation 3:

$$(\partial x/\partial y)_z(\partial y/\partial z)_x(\partial z/\partial x)_y = -1.$$

4. Suppose we encounter some differential df and we want to know whether it is an exact differential. How do we find out?
Relation No. 4 generates the answer:

For $df = a(x, y)\, dx + b(x, y)\, dy,$

df is exact if $(\partial a/\partial y)_x = (\partial b/\partial x)_y.$

If df is exact, then we also know that it may be integrated without needing to specify the path.

3.2 Using the machinery: manipulating the First Law

As a first step in developing thermodynamics we collect together some of the information met so far. Everything we do will be based on the First Law written in the form

(3.2.1) $dU = dq + dw.$

The First Law states essentially that U is a state function (or alternatively that dU is an exact differential); it follows that, for a closed system of constant composition, any change in U can be expressed in terms of changes in V and T according to

(3.2.2) $dU = (\partial U/\partial T)_V \, dT + (\partial U/\partial V)_T \, dV.$

Analogous expressions in terms of p, T or p, V would also be acceptable, but they are often less convenient.

The work done on a system can be written

(3.2.3) $dw = -p_{ex} \, dV + dw_e$

where the first term is work accompanying a change of volume and the second is any other kind of work (e.g. electrical work).

The enthalpy H was introduced as a property which would be useful for the discussion of constant-pressure changes:

(3.2.4) $H = U + pV.$

U is a state function, and so are p and V; therefore H is a state function and consequently dH is an exact differential. It turns out to be convenient to regard H as a function of T and p (instead of T and V which we selected for U), and so for a closed system of constant composition,

(3.2.5) $dH = (\partial H/\partial T)_p \, dT + (\partial H/\partial p)_T \, dp.$

The heat capacities at constant volume and constant pressure were introduced on pp. 66 and 68 and expressed as

(3.2.6) $C_V = (\partial U/\partial T)_V$

(3.2.7) $C_p = (\partial H/\partial T)_p.$

Insertion of these into the expressions for dU and dH gives

(3.2.8) $dU = C_V \, dT + (\partial U/\partial V)_T \, dV$

(3.2.9) $dH = C_p \, dT + (\partial H/\partial p)_T \, dp.$

How the internal energy varies with temperature. Suppose we want to find how the internal energy depends on the temperature when the *pressure*, rather than the volume, of the system is maintained constant. Can we use the relations in Box 3.1 to extract from eqn (3.2.8) an expression for $(\partial U/\partial T)_p$?

Relation No. 1 does this: it takes an expression like eqn (3.2.8), divides through by dT (to give dU/dT on the left as an intermediate) and then imposes the constant pressure condition (to restrict dU/dT to constant-pressure changes, and to shape it into $(\partial U/\partial T)_p$:

$$(\partial U/\partial T)_p = C_V + (\partial U/\partial V)_T(\partial V/\partial T)_p.$$

At this point always inspect the output of the relation to determine whether it contains any recognizable physical quantity. The final differential coefficient in this equation is the rate of change of volume with increase of temperature (at constant pressure). This is a readily accessible physical property which is normally tabulated as the *isobaric thermal expansivity* or *isobaric coefficient of thermal expansion* defined as

(3.2.10) $\alpha = (1/V)(\partial V/\partial T)_p$

(i.e., the rate of change of volume with temperature, per unit volume). On introducing this definition we find

(3.2.11) $(\partial U/\partial T)_p = C_V + \alpha V(\partial U/\partial V)_T.$

Example (Objective 4). At 300 K the isobaric coefficient of thermal expansion of neon is 3.3×10^{-3} K^{-1} and of copper 5.01×10^{-5} K^{-1}. What volume changes will occur when 50 cm^3 samples of both materials are heated through 5 K?

• *Method.* For a change of temperature the volume changes by

$$dV = (\partial V/\partial T)_p \, dT = \alpha V dT.$$

Assume that the temperature range is so small as to be virtually infinitesimal, and write

$$\Delta V \approx \alpha V \, \Delta T.$$

• *Answer.* For neon $\Delta V \approx (3.3 \times 10^{-3}$ K$^{-1}) \times (50$ cm$^3) \times (5$ K$) = 0.83$ cm^3.
For copper $\Delta V \approx (5.01 \times 10^{-5}$ K$^{-1}) \times (50$ cm$^3) \times (5$ K$) = 12.5$ mm^3.

• *Comment.* Note that the relations do not assume anything about the perfect behaviour, or otherwise, of the materials.

Equation (3.2.11) is an entirely general result for a closed system of constant composition, and it enables us to find how the internal energy of any material depends on the temperature under constant pressure conditions. We need to know C_V, which can be measured in one experiment, the coefficient of expansion α, which can be measured in another, Table 3.1, and the ubiquitous quantity $(\partial U/\partial V)_T$, which, if we knew what it was, could probably be measured in a third.

What, then, is this $(\partial U/\partial V)_T$? At this point we can give no more than a crude interpretation of it, but it is sufficient for our present understanding. When a material is compressed isothermally the interactions between the molecules are enhanced. These interactions contribute to the internal energy of the system, and so the coefficient $(\partial U/\partial V)_T$ *is a measure of how the interactions change when the volume of the sample is changed*. This is supported by noting that in the case of a van der Waals gas, $(\partial U/\partial V)_T = a/V^2$ (a result we obtain in Chapter 6). We might suspect that the coefficient is smaller for gases than for solids because we know that molecular interactions are not very important in gases. We might further suspect that $(\partial U/\partial V)_T$ is zero for a perfect gas. J. P. Joule thought he could measure $(\partial U/\partial V)_T$ by attempting to determine the change in temperature of a gas when it was allowed to expand into a vacuum. He used two vessels immersed in a water bath. One vessel was filled with air at 22 atm, the other was evacuated. Using the same sensitive thermometer that he had used in his experiments on the 'mechanical equivalent of heat' he attempted to measure the change in temperature of the water in the bath when a stopcock was opened and the air expanded into the vacuum. He observed no change.

Table 3.1. Isobaric volume expansivities α and isothermal compressibilities κ†

	Substance	$10^4\alpha/\text{K}^{-1}$	$10^6\kappa/\text{atm}^{-1}$
Liquids	Water	2.1	49.4
	CCl$_4$	12.4	90.5
	Benzene	12.4	92.1
	Ethanol	11.2	76.8
	Mercury	1.82	3.87
Solids	Pb	0.861	2.21
	Cu	0.501	0.735
	C (diamond)	0.030	0.187
	Fe	0.354	0.597

† Compressibility is often expressed in kilobars^{-1}; 1 bar $= 10^5\,\text{N m}^{-2}$, and so 1 kbar $= 986.9$ atm. The values in the table are appropriate to about 20 °C.
Source: *American Institute of Physics handbook*, McGraw-Hill (α); G. W. C. Kaye and T. H. Laby, *Tables of physical and chemical constants*, Longmans (κ).

The thermodynamic implication of the experiment is as follows. No work was done in the expansion into the vacuum; therefore $w = 0$. No heat entered or left the system (the gas) because the bath temperature did not change; therefore $q = 0$. Consequently $\Delta U = 0$. It follows that U does not change when a gas expands isothermally. Therefore, at constant temperature U is independent of volume. Or, expressed mathematically, $(\partial U/\partial V)_T = 0$.

Joule's experiment was crude; in particular the heat capacity of the water bath was so large that the temperature change that gases do in fact cause was too small to measure. His experiment was on a par with Boyle's: he extracted an essential limiting feature of a gas (a feature of a perfect gas) without detecting the small deviations characteristic of real gases.

How the enthalpy varies with temperature. We have seen how to find the temperature dependence of U under conditions of constant pressure; now we shall see how to find the temperature dependence of H under conditions of constant volume. At the end of this piece of work, which gives practice in manipulating thermodynamic expressions, we shall possess expressions for $(\partial U/\partial T)_V$, $(\partial U/\partial T)_p$, $(\partial H/\partial T)_V$, and $(\partial H/\partial T)_p$ and we shall be able to predict how U and H change when the temperature of any closed, constant composition system is changed.

We start in the same way as before, and act on eqn (3.2.9) with relation No. 1. In the present case we impose the condition of constant volume (previously we imposed constant pressure): the output is

$$(\partial H/\partial T)_V = C_p + (\partial H/\partial p)_T(\partial p/\partial T)_V.$$

Now inspect the equation. The final differential coefficient looks like something we ought to recognize, but it would be better if it could be turned into $(\partial V/\partial T)_p$, the thermal expansion coefficient encountered on p. 84. Relation No. 3 does just this kind of transformation: it shuffles p, V, and T around inside partial differentials, which is what we want. Acting with it on $(\partial p/\partial T)_V$ gives $-1/(\partial T/\partial V)_p(\partial V/\partial p)_T$. Unfortunately $(\partial T/\partial V)_p$ appears instead of $(\partial V/\partial T)_p$, but relation No. 2 inverts partial differentials and so application of it leads to

$$(\partial p/\partial T)_V = -(\partial V/\partial T)_p/(\partial V/\partial p)_T.$$

Is this an improvement? The numerator we recognize as αV; but is the denominator significant? $(\partial V/\partial p)_T$ is the change of volume under the influence of pressure at constant temperature. The *isothermal compressibility* is given the symbol κ (kappa) and defined as

(3.2.12) $$\kappa = -(1/V)(\partial V/\partial p)_T.$$

The negative sign is incorporated into the definition so that κ is positive. (When the pressure is increased the volume of any substance decreases. Positive values of dp therefore correspond to negative values of dV. Consequently $(\partial V/\partial p)_T$ is always negative.)

Example (Objective 4). The isothermal compressibility of neon at room temperature and 1 atm pressure is 1.00 atm^{-1} and that of copper is 0.735×10^{-6} atm^{-1}. What are the changes of volume when 50 cm^3 samples of both materials are subjected to a change of pressure from 1 atm to 0.5 atm?

● *Method.* For an infinitesimal isothermal change of pressure dp the volume changes by
$$dV = (\partial V/\partial p)_T\, dp = -\kappa V dp.$$

If the changes in pressure are small enough to be regarded as virtually infinitesimal this relation becomes
$$\Delta V \approx -\kappa V \Delta p.$$

● *Answer.* For neon $\Delta V \approx -(1.00 \text{ atm}^{-1}) \times (50 \text{ cm}^3) \times (-0.5 \text{ atm}) = 25 \text{ cm}^3.$
        For copper $\Delta V \approx -(0.735 \times 10^{-6} \text{ atm}^{-1}) \times (50 \text{ cm}^3) \times (-0.5 \text{ atm})$
$$= 0.018 \text{ mm}^3.$$

● *Comment.* The assumption that Δp is infinitesimal is very poor for neon because of its large compressibility, but good for copper because hardly any change in its volume occurs. If we assume perfect behaviour for neon the change in volume is $+50$ cm^3. If a temperature change of 5 K (as in the last *Example*) occurs as well as the pressure change, the overall change is the sum $(\partial V/\partial T)_p\, dT + (\partial V/\partial p)_T\, dp$. For copper this can be replaced by

$$\Delta V \approx \alpha V \Delta T - \kappa V \Delta p = 12.5 \text{ mm}^3 + 0.018 \text{ mm}^3 \approx 12.5 \text{ mm}^3.$$

1,3-Cyclohexandione

(ketone)

Collecting all the fragments leads to

(3.2.13) $\qquad (\partial H/\partial T)_V = C_p + (\alpha/\kappa)(\partial H/\partial p)_T.$

The coefficient representing the dependence of the enthalpy on the pressure, $(\partial H/\partial p)_T$, is clearly the analogue of $(\partial U/\partial V)_T$ in eqn (3.2.11)—V and p play analogous roles in U and H respectively—and we might enquire whether it can be measured in a similar fashion. The next few paragraphs explore this point.

Can the coefficient $(\partial H/\partial p)_T$ be changed into something recognizable? Shuffling H, p, and T leads to something containing $(\partial T/\partial p)$, the change in temperature on a change in pressure; and that looks as though it is a measurable quantity. Relation No. 3 effects the transformation

$\qquad (\partial H/\partial p)_T = -1/(\partial p/\partial T)_H (\partial T/\partial H)_p$

and a double use of the inverter gives

$\qquad (\partial H/\partial p)_T = -(\partial T/\partial p)_H (\partial H/\partial T)_p$

(3.2.14) $\qquad\qquad = -(\partial T/\partial p)_H C_p.$

We seem to be on the right track because the other term generated is simply the heat capacity C_p, and so we have not generated something more complicated than we started with (an occupational hazard in thermodynamics: when it happens be prepared to start again).

All would be well if the rise in temperature for a change in pressure could be measured under conditions of constant enthalpy: but how is that constraint imposed? The cunning required is the basis of the *Joule–Thomson* experiment, which measures the change in temperature of a gas as it squirts through a valve into a region of lower pressure, the whole system being thermally insulated so that the expansion is adiabatic. The fault with the Joule experiment was that the heat capacity of the water bath was too great. Joule and Thomson (later Kelvin) had the good idea to use the gas as its own heat bath and to set up a steady-state flow. They decided to let gas expand from one constant pressure through a valve to another constant pressure. On either side of the valve they measured the temperature. The whole apparatus was lagged so that no heat entered or left the system. They observed a drop in temperature when the gas passed the valve, and the drop was proportional to the pressure difference maintained: $\Delta T \propto \Delta p$.

The thermodynamic implication of the experiment is as follows. No heat enters or leaves the system; therefore $q = 0$. What work is done on the gas as it passes the valve? Consider the passage of a particular amount through the valve. On one side the pressure and temperature are p_i and T_i, and the amount occupies a volume V_i. On the other side the same amount will be at a pressure p_f, a temperature T_f, and will occupy a volume V_f. The gas on the first side is compressed isothermally by the up-stream gas acting as a piston, Fig. 3.4. The relevant pressure is p_i, the volume changes from V_i to 0; and so the work done on the gas is $-p_i(0 - V_i)$, or $p_i V_i$. The

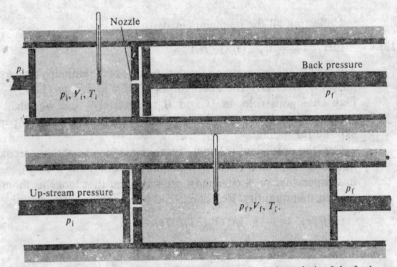

Fig. 3.4. Diagrammatic representation of the thermodynamic analysis of the Joule–Thomson experiment. The pistons represent the up-stream and down-stream gases.

gas is allowed to expand isothermally (but possibly at a different constant temperature) against the constant pressure p_f on the right of the valve provided by the down-stream gas acting as a piston to be driven out. The volume changes from 0 to V_f and so the work done on the gas in this stage is $-p_f(V_f - 0)$, or $-p_f V_f$. The total work done on the gas is therefore the sum of these two amounts, or $p_i V_i - p_f V_f$. It follows that the change in internal energy across the valve is $U_f - U_i = w = p_i V_i - p_f V_f$. Reorganizing this gives $U_f + p_f V_f = U_i + p_i V_i$, or $H_f = H_i$. Therefore the expansion, occurs without change of enthalpy (it is *isenthalpic*). The property observed is the temperature drop per unit drop in pressure: this is $\Delta T / \Delta p$. Adding the constraint of constant enthalpy, and taking the limit of small Δp, implies that the thermodynamic quantity measured is $(\partial T / \partial p)_H$; this is called the *isenthalpic Joule–Thomson coefficient*:

(3.2.15)
$$\mu_{JT} = (\partial T / \partial p)_H.$$

The modern method of determining μ_{JT} is indirect and involves measuring the *isothermal Joule–Thomson coefficient*, the quantity $(\partial H / \partial p)_T$. According to eqn (3.2.14) the two coefficients are related by

(3.2.16)
$$(\partial H / \partial p)_T = -\mu_{JT} C_p.$$

The procedure is to pump the gas continuously at a steady pressure through a heat exchanger (which brings it to the required temperature), and then through a throttle valve inside a thermally insulated container. The sharp pressure drop is measured, and the cooling effect is exactly offset by an electric heater placed immediately after the throttle. The energy provided by the heater is also monitored. Since the heat q can be identified with the value of ΔH for the gas, and the pressure drop Δp is known, the value of $(\partial H/\partial p)_T$ can be obtained from the limiting value of $\Delta H/\Delta p$ as $\Delta p \to 0$.

Real gases have non-zero coefficients (even in the limit of zero pressure). The sign of the coefficient may be positive or negative. A positive sign implies that dT is negative when dp is negative, in which case the gas cools on expansion. A negative sign implies that dT is positive when dp is negative, and so the gas is heated by expansion. The sign and magnitude of μ_{JT} depend on the gas and the conditions. Gases showing a heating effect $(\mu_{JT} < 0)$ show a cooling effect $(\mu_{JT} > 0)$ when their temperature has been lowered beneath their *inversion temperature*. A list of inversion temperatures and Joule–Thomson coefficients can be found in Table 3.2 and are summarized by the curves in Fig. 3.5. In the case of a van der Waals gas, $T_I = 2a/b$, and so we expect $T_I \approx 2T_B$ where T_B is the Boyle temperature (p. 38). This is in broad agreement with the data in the Table.

The technological importance of the Joule–Thomson experiment lies in its application to the cooling and liquefaction of gases. The *Linde refrigerator* works on the principle that below its inversion temperature a gas is cooled on expansion. For a sufficiently large pressure drop the cooling may drop the temperature below the condensation temperature corresponding to the exit pressure of the gas, when the liquid will form. The principle is illustrated in Fig. 3.6. Note the importance of working beneath the inversion temperature, and therefore of the importance of cooling some gases by other means before passing them into the refrigerator: using helium at room temperature would turn the refrigerator into an expensive oven.

For a perfect gas the Joule–Thomson coefficient is zero: its temperature is unchanged in a Joule–Thomson expansion. This points clearly to the involvement of intermolecular forces in determining the magnitude of the effect. Later we shall look at this in more detail. You should note that the Joule–Thomson coefficient of a real gas does not necessarily become zero

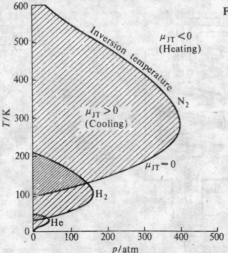

Fig. 3.5. Inversion curves for three gases.

Table 3.2. Inversion, freezing, and boiling temperatures of some gases at 1 atm

	$T_i/K\dagger$	T_m/K	T_b/K	$\mu_{JT}/K\ atm^{-1}$
He	40	—	4.6	−0.060
Ne	231	24.5	27.2	
Ar	723	83.8	87.3	
Kr	1090	115.8	119.9	
H_2	202	14.0	20.4	
N_2	621	63.2	77.4	0.25
O_2	764	54.4	90.2	0.31
CO_2	1500		194.6	1.11 (300 K)
CH_4	968	90.7	109.2	
Air	603			0.189 (50 °C)

† Maximum value.
Source: *American Institute of Physics handbook*, McGraw-Hill and M. W. Zemansky, *Heat and thermodynamics*, McGraw-Hill.

as the pressure is reduced: μ_{JT} is an example of a property which does not approach the perfect gas value at zero pressure (p. 44): it depends on a derivative and not simply on p, V, T themselves.

We return to the development of eqn (3.2.13). On introducing eqn (3.2.16) we find

(3.2.17) $(\partial H/\partial T)_V = (1 - \alpha\mu_{JT}/\kappa)C_p.$

The last equation is the final one for $(\partial H/\partial T)_V$: everything in it can be measured in suitable experiments, and so we can calculate how H varies with T. Remember that this expression applies to any substance, not only to a gas.

Take note of the kind of result we have obtained in eqn (3.2.16): a property at constant temperature has been expressed in terms of properties related to changes of temperature. This illustrates both the power and the

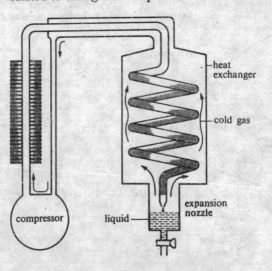

Fig. 3.6 The Linde refrigerator for liquefying gases.

heat exchanger

cold gas

expansion nozzle

liquid

compressor

poverty of thermodynamics. The power is that we are able to relate one quantity, which might be difficult to measure directly, to other properties, which might be easily measured. The poverty is that the results are merely mathematical identities: they rely on mathematical manipulations which, being independent of the physics, give little information about the microscopic nature of matter.

The relation between C_V and C_p. The development so far has shown that properties defined and measured under one set of conditions may be related to properties defined under a different set. The heat capacities at constant volume and constant pressure are examples of such properties, and we are now in a position to derive an expression relating them. The approach adopted illustrates the way of tackling similar problems in thermodynamics.

C_p differs from C_V on account of the work needed to change the volume of the system when the pressure, and not the volume, is held constant. This work arises in two ways. One is the work done on driving back the atmosphere. The other is the work needed to stretch bonds in the material, including the weak intermolecular (van der Waals) interactions. In the case of a perfect gas, the second effect makes no contribution. In the present case we shall derive an entirely general expression, and show that it reduces to the perfect gas result in the absence of intermolecular forces.

When seeking a way into a problem in thermodynamics a very useful rule is to *go back to first principles*. In the present problem we need to do this twice, first by expressing C_V and C_p in terms of their definitions

$$C_p - C_V = (\partial H/\partial T)_p - (\partial U/\partial T)_V$$

and then by inserting the definition $H = U + pV$

$$C_p - C_V = (\partial U/\partial T)_p + (\partial pV/\partial T)_p - (\partial U/\partial T)_V.$$

The difference of the first and third quantities has already been calculated in eqn (3.2.11):

$$(\partial U/\partial T)_p - (\partial U/\partial T)_V = \alpha V(\partial U/\partial V)_T.$$

αV gives the change of volume when the temperature is raised, and $(\partial U/\partial V)_T$ converts this change of volume to a change of internal energy.

The $(\partial pV/\partial T)_p$ term is simplified by noting that, since p is constant,

$$(\partial pV/\partial T)_p = p(\partial V/\partial T)_p = \alpha pV.$$

The middle term of this expression identifies it as the contribution to the work of pushing back the atmosphere: $(\partial V/\partial T)_p$ is the change of volume caused by a rise of temperature, and multiplication by p converts this into work.

Collecting the two contributions gives

(3.2.18) $$C_p - C_V = \alpha \{ p + (\partial U/\partial V)_T \} V.$$

This is an entirely general relation applicable to all materials. It contains the ubiquitous quantity $(\partial U/\partial V)_T$. At this point we can go further by anticipating a result which will be proved in Chapter 6 (p. 164). This states that

$$(\partial U/\partial V)_T = T(\partial p/\partial T)_V - p.$$

If this is inserted into the present equation we obtain a very neat expression:

$$C_p - C_V = \alpha TV(\partial p/\partial T)_V.$$

We can do even better than this. In the previous section we also encountered the coefficient $(\partial p/\partial T)_V$: the use of relations Nos. 3 and 2 turned it into

$$(\partial p/\partial T)_V = \alpha/\kappa,$$

see eqn (3.2.12). Therefore,

(3.2.19) $$C_p - C_V = (\alpha^2/\kappa)TV.$$

This is a thermodynamic expression, which means that it is universally true.

What does it tell us? Since the thermal expansivities, α, of liquids and solids are small it is tempting to say for them that C_p must be virtually equal to C_V. But this may be erroneous, because we would be forgetting that the compressibility κ might also be small, and so α^2/κ might be appreciable. Put another way, we might look at eqn (3.2.18), and get as far as α, which may be small, but not look as far as $p + (\partial U/\partial V)_T$, which might be big. In some cases, in fact, C_p and C_V differ by up to about 30 per cent.

Example (Objective 6). The molar heat capacity of water at constant volume is 74.8 J K^{-1} mol^{-1} at 25°C. What is its molar heat capacity at constant pressure?

- *Method.* The exact thermodynamic relation between the two heat capacities is given by eqn (3.2.19); for molar quantities it reads

$$C_{p,m} = C_{V,m} + (\alpha^2/\kappa)TV_m.$$

In order to use it we need to know the isobaric coefficient of expansion, α (Table 3.1 gives 2.1×10^{-4} K^{-1}), the isothermal compressibility, κ (Table 3.1 gives 4.94×10^{-5} atm^{-1}), and the molar volume (18.07 cm^3 mol^{-1}, the density is 0.99704 at 25°C and $M_m = 18.02$ g mol^{-1}). The value of κ must be expressed in (N m^{-2})$^{-1}$ to obtain a set of coherent units.

- *Answer.* $\kappa = (4.94 \times 10^{-5} \text{ atm}^{-1})/(1.0133 \times 10^5 \text{ N m}^{-2} \text{ atm}^{-1})$
  $$= 4.88 \times 10^{-10} \text{ N}^{-1} \text{ m}^2.$$

Then

$$C_{p,m} = 74.8 \text{ J K}^{-1} \text{ mol}^{-1}$$
$$+ \left\{ \frac{(2.1 \times 10^{-4} \text{ K}^{-1})^2 \times (298.15 \text{ K}) \times (1.807 \times 10^{-5} \text{ m}^3 \text{ mol}^{-3})}{(4.88 \times 10^{-10} \text{ N}^{-1} \text{ m}^2)} \right\}$$

$$= 74.8 \text{ J K}^{-1} \text{ mol}^{-1} + 0.487 \text{ N m K}^{-1} \text{ mol}^{-1} = 75.3 \text{ J K}^{-1} \text{ mol}^{-1}$$

● *Comment*. Remember that 1 N m = 1 J. This example shows that there is a small but significant difference (in this case 0.5 per cent) between the constant volume and constant pressure heat capacities of water.

Gases expand enormously when heated, and for them we expect α to be large. So long as the compressibility κ is not too large we expect $C_p - C_V$ to be large, and on p. 70 we did in fact find that $C_p - C_V = nR$ in the case of a perfect gas. The same result can be deduced from eqn (3.2.19) once we have expressions for α and κ.

It is a simple matter to insert $pV = nRT$ into the definitions of α and κ, and to deduce that

(3.2.20)° $\qquad \alpha = (1/V)(\partial V/\partial T)_p = 1/T$

(3.2.21)° $\qquad \kappa = -(1/V)(\partial V/\partial p)_T = 1/p.$

It follows immediately that

(3.2.22)° $\qquad C_p - C_V = (1/T^2)pVT = nR$

as we deduced before. Of course, if α and κ can be calculated from other equations of state, $C_p - C_V$ can be calculated for real gases: this extension is examined in the Problems.

Work of adiabatic expansion. On p. 63 we discovered that the work done on a system when it expands reversibly is

$$w = -\int_{V_i}^{V_f} p(V)\,\mathrm{d}V$$

(see eqn 2.2.9). We dealt with that in the case of an isothermal change using a perfect gas because we were able to relate p to V via $p = nRT/V$, and T was a constant. In the case of an adiabatic change the same substitution is valid, but now T is no longer a constant because it changes during the expansion, and hence depends on V. How, then, can we calculate w for an adiabatic change?

We use another useful rule in thermodynamics: *instead of calculating what you are asked to calculate, set it equal to a state function, and calculate that by the most convenient path.*

In the present problem, because the change is adiabatic, the heat flow into the system is zero during every stage of expansion: $\mathrm{d}q = 0$. Consequently

$$\mathrm{d}U = \mathrm{d}w.$$

Therefore, to calculate the work done during the expansion, why not calculate the change in internal energy instead? The last equation implies

$$w = \int_{\text{initial state}}^{\text{final state}} \mathrm{d}U.$$

Already we have a number of expressions for dU which might be useful in dealing with this expression. We try using the basic expression, eqn (3.2.8).

The easiest case to consider is a perfect gas as the working substance. Then $(\partial U/\partial V)_T$ is zero and the work is simply

$$w = \int_{T_i}^{T_f} C_V \, dT.$$

For many gases C_V is almost independent of temperature. (Accept this for now—it will be discussed in greater detail in Part 2.) This implies that the integration may be performed trivially:

(3.2.23)° $w = C_V \int_{T_i}^{T_f} dT = C_V(T_f - T_i) = C_V \Delta T.$

This is a very simple result: it shows that the work done on the system when it expands adiabatically is proportional to the temperature difference.

But what is the temperature difference? That is one of the items left out of the calculation: a few other things have been left out as well. First we have not specified whether the change in the system is an expansion or contraction: we have calculated the change in internal energy, and thence the work involved, between some initial state and an unspecified final state. Even more important, we have not even specified whether the work is done under conditions of reversibility. And yet eqn (3.2.23) is true for all adiabatic expansions or contractions involving a perfect gas, reversible or irreversible!

Before sorting out these points we can draw a general conclusion. If w has a negative value (which means that we have made the system do work), the equation shows that T_f is less than T_i irrespective of the manner of extracting the work (reversibly or irreversibly). Such a conclusion is not very surprising: in an adiabatic process no heat passes into the system and if work is done the temperature must fall as internal energy is withdrawn as work.

Equation (3.2.23) applies to both reversible and irreversible processes: any difference in the amount of work between these conditions must arise from different values of T_f and T_i. So we must examine the different values of ΔT that arise when the expansion is carried out irreversibly and reversibly, but in each case in a thermally isolated system.

If the expansion occurs against zero outside pressure it does no work. (This is a conclusion we drew on p. 60 before we had distinguished between adiabatic and isothermal processes; the opposing pressure is zero, irrespective of the rest of the arrangement.) Therefore $w = 0$. Consequently $\Delta T = 0$ for eqn (3.2.23) to be satisfied. We have the peculiar case of an expansion being both isothermal and adiabatic.

If the expansion occurs against a fixed external pressure the work done *on* the system is $w = -p_{ex} \Delta V$. This too is a general conclusion from our original discussion of work, and does not depend on the expansion being adiabatic or isothermal. This result can be used to calculate the drop in temperature that occurs during an adiabatic expansion. By setting $-p_{ex}\Delta V$ equal to $C_V\Delta T$ (which is also the expression for the work) we find ΔT for this irreversible expansion:

(3.2.24)° $$\Delta T = -p_{ex}\Delta V/C_V.$$

Notice that if ΔV is positive (expansion) the temperature of the system falls: the signs are looking after themselves, but we should learn their language. Also notice that if the external pressure is zero the temperature drop is also zero, in accord with the first example. (It is always sensible to check an equation by seeing whether it reduces to a simpler, known result.)

Example (Objective 8). A 2 mol sample of argon in a cylinder of 5 cm² cross-section at a pressure of 5 atm is allowed to expand adiabatically against an external pressure of 1 atm. During the expansion it pushes the piston through (a) 10 cm, (b) 10 m. If the initial temperature is 300.0 K, what is the final temperature of the gas in each case?

• *Method.* The expansion is adiabatic and irreversible against constant external pressure; therefore we use eqn (3.2.24). We need to know C_V for argon, and from Table 2.2 we have $C_{V,m} = 12.48$ J K⁻¹ mol⁻¹. In order to convert this to C_V, we require knowledge of the amount of gas in the system (2 mol). Since the temperature change is proportional to the volume change, part (b) can be obtained from part (a) simply by multiplication by 100.

• *Answer.* (a) $$\Delta T = \frac{-(1.0133 \times 10^5 \text{ N m}^{-2}) \times (5 \text{ cm}^2 \times 10 \text{ cm})}{(2 \text{ mol}) \times (12.48 \text{ J K}^{-1} \text{ mol}^{-1})}$$

$$= -0.203 \text{ K}, \quad \text{implying } T_f = 299.8 \text{ K}.$$
$$\text{(b)} \quad T = -20.3 \text{ K}, \quad \text{implying } T_f = 279.7 \text{ K}.$$

• *Comment.* The calculation assumes perfect behaviour. In each case the temperature falls on expansion, for energy is extracted as work. The change in internal energy of the gas is equal to the amount of work done (since $q = 0$, $\Delta U = w$). In (a) $\Delta U = -5.1$ J, in (b) $\Delta U = -510$ J.

Now consider reversible expansion. In this case at every step the external pressure is matched to the internal, the only difference from the isothermal, reversible expansion being that the gas cannot draw any heat from the surroundings. We know that the work done is equal to $C_V\Delta T$, and so our job is to find ΔT. We proceed first to calculate T_f by setting the work calculated from the expression $C_V\,dT$ equal to the work calculated from the expression $-p_{ex}\,dV$. In this case p_{ex} is matched to the pressure of the confined gas, and depends on both volume and the temperature of each step of the expansion.

First, because $dq = 0$, we can write

(3.2.25)ᵣ $$dU = dw = -p\,dV.$$

Then we can recall that for a perfect gas

(3.2.26)° $dU = C_v$.

Combining the two

(3.2.27)$_r^\circ$ $C_V\,dT = -p\,dV$.

The pressure of the working gas depends on its volume and temperature in accord with the perfect gas equation of state. On inserting $pV = nRT$ the last equality becomes

$$C_V\frac{dT}{T} = -nR\frac{dV}{V}.$$

Since C_V may be regarded as independent of temperature both sides may be integrated easily between the final and initial values:

$$C_V\int_{T_i}^{T_f}\frac{dT}{T} = -nR\int_{V_i}^{V_f}\frac{dV}{V}$$

or

$$C_V\ln(T_f/T_i) = -nR\ln(V_f/V_i).$$

If we introduce the quantity $c = C_V/nR = C_{V,m}/R$ and do a little rearranging, we get

$$\ln(T_f/T_i)^c = \ln(V_i/V_f),$$

which means that

(3.2.28)$_r^\circ$ $V_f T_f^c = V_i T_i^c$.

This is our goal. It means that we can predict the temperature of a gas that has expanded (or contracted) adiabatically and reversibly from a volume V_i and temperature T_i to a volume V_f:

(3.2.29)$_r^\circ$ $T_f = (V_i/V_f)^{1/c}T_i$.

This is the end of the main calculation, for the work done on the perfect gas as it expands adiabatically and reversibly from V_i to V_f is obtained simply by substituting the last equation into $w = C_V\,\Delta T$:

(3.2.30)$_r^\circ$ $w = C_V(T_f - T_i) = C_V T_i\{(V_i/V_f)^{1/c} - 1\}$.

This is normally expressed in terms of the ratio of heat capacities

(3.2.31) $\gamma = C_p/C_V$.

Since $C_{p,m} - C_{V,m} = R$ it follows that $1/c = \gamma - 1$, and so

(3.2.32)$_r^\circ$ $w = C_V T_i\{(V_i/V_f)^{\gamma-1} - 1\}$.

Example (Objective 7). A sample of argon ($C_{V,m} = 12.48$ J K^{-1} mol^{-1}) at 1 atm pressure expands reversibly and adiabatically from 0.5 dm^3 to 1.0 dm^3. Initially its temperature is 25 °C. What is its final temperature, how much work is done during the expansion, and what is the change in internal energy?

• *Method.* The final temperature can be calculated from eqn (3.2.29) and the work from eqn (3.2.30). In order to use these equations we need to know $c = C_{V,m}/R$ and the amount of gas present (from $n = pV/RT$, using the initial values of p, V, and T). The change is adiabatic, therefore $q = 0$, and so $\Delta U = w$.

• *Answer.* $\quad c = (12.48 \text{ J K}^{-1} \text{ mol}^{-1})/(8.314 \text{ J K}^{-1} \text{ mol}^{-1}) = 1.501$

$\qquad 1/c = 0.6662.$

$\qquad\qquad n = (1 \text{ atm}) \times (0.5 \text{ dm}^3)/(0.0821 \text{ dm}^3 \text{ atm K}^{-1} \text{ mol}^{-1}) \times (298.15 \text{ K})$

$\qquad\qquad\quad = 0.0204 \text{ mol}.$

From eqn (3.2.29):

$\quad T_f = (0.5 \text{ dm}^3/1.0 \text{ dm}^3)^{0.6662} \times (298.15 \text{ K}) = 187.88 \text{ K}.$

From eqn (3.2.30):

$\qquad w = (0.0204 \text{ mol}) \times (12.48 \text{ J K}^{-1} \text{ mol}^{-1}) \times (187.88 \text{ K} - 298.15 \text{ K}) = -28.1 \text{ J}$

$\qquad \Delta U = -28.1 \text{ J}.$

• *Comment.* Had the work been done reversibly but *isothermally* an amount $-nRT \ln 2$, or -35.1 J would have been done: more is done on the outside world because the influx of heat helps to raise the pressure inside the system. In this case $\Delta U = 0$, in contrast to the decrease of internal energy in the adiabatic case.

At this point it is easy to find the relation between p and V when a perfect gas undergoes reversible adiabatic change. From the perfect gas equation of state,

$$p_i V_i / p_f V_f = T_i / T_f,$$

while from eqn (3.2.29)

$$T_i / T_f = (V_f / V_i)^{1/c} = (V_f / V_i)^{\gamma - 1}.$$

By combining these two expressions it follows that

$(3.2.33)_r^\circ \qquad p_f V_f^\gamma = p_i V_i^\gamma.$

This relation is often expressed in the form

$(3.2.34)_r^\circ \qquad pV^\gamma = \text{constant},$

where the constant is simply the initial value $p_i V_i^\gamma$. On making use of these results in eqn (3.2.32), it follows that

$(3.2.35)_r^\circ \qquad w = C_V T_i \{(p_f/p_i)^{(\gamma-1)/\gamma} - 1\}.$

For all gases γ exceeds unity. (For a perfect monatomic gas $\gamma = \frac{5}{3}$; this value is obtained in Part 2, but see Problem 2.28.) This means that p falls off faster with volume ($p \propto 1/V^\gamma$: the curve is called an *adiabat*) than in the case of isothermal expansion ($p \propto 1/V$), Fig. 3.7. The physical reason for this difference is easy to find. In an isothermal expansion heat continuously flows into the system, and so the pressure does not fall as much as in a thermally isolated, adiabatic expansion.

Example (Objective 8). Using the same data as in the last Example, find the final pressure of the gas and the change in enthalpy during the expansion.

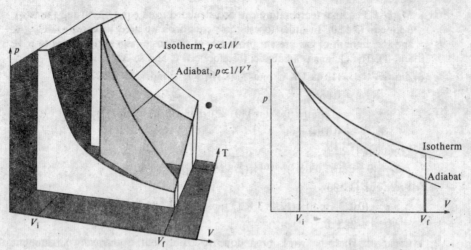

Fig. 3.7. The indicator diagram for isothermal and adiabatic reversible expansions of a perfect gas.

● *Method.* The final pressure is given by eqn (3.2.33). In order to use it we need $\gamma = C_p/C_V$; C_p can be obtained from $C_p - C_V = nR$ if perfect behaviour is assumed. There are two ways of obtaining ΔH. The first recognizes that since $H = U + pV$ and $pV = nRT$, the change in H is related to the change in U by

$$\Delta H = U_f - U_i + nRT_f - nRT_i = \Delta U + nR(T_f - T_i),$$

and ΔU was calculated in the last *Example*. The second way notes that in an adiabatic change $dq = 0$, and so from eqn (2.3.6) $dH = V\,dp$. We know from eqn (3.2.34) how V depends on the pressure, and so we can set about integrating $V\,dp$ from p_i to p_f.

● *Answer.* $\gamma = C_p/C_V = C_{p,m}/C_{V,m} = (C_{V,m} + R)/C_{V,m}$

$$= (12.48 \text{ J K}^{-1} \text{ mol}^{-1} + 8.314 \text{ J K}^{-1} \text{ mol}^{-1})/(12.48 \text{ J K}^{-1} \text{ mol}^{-1})$$

$$= 1.667.$$

From eqn (3.2.33):

$$p_f = (0.5 \text{ dm}^3/1.0 \text{ dm}^3)^{1.667} \times (1 \text{ atm}) = 0.315 \text{ atm}.$$

The enthalpy change is given by

$$\Delta H = \Delta U + nR\,\Delta T = nC_{V,m}\,\Delta T + nR\,\Delta T = nC_{p,m}\,\Delta T.$$

Since

$$\Delta T = -110.3 \text{ K (from the last } Example),$$

$$\Delta H = (0.0204 \text{ mol}) \times (20.79 \text{ J K}^{-1} \text{ mol}^{-1}) \times (-110.3 \text{ K}) = 46.8 \text{ J}.$$

● *Comment.* Had the expansion been isothermal the final pressure would have been 0.5 atm and the enthalpy change zero. Calculating ΔH by integrating $V\,dp$ is left as an exercise (Problem 3.26).

3.3 Comments on isothermal and adiabatic, and reversible and irreversible processes

The algebraic results for the different types of change are collected in Table 3.3. It will be helpful to remember the following points.

Table 3.3. Work done on a perfect gas

For any kind of expansion of any kind of material the work done on the system as it changes from V_i to V_f is given by

$$w = -\int_{V_i}^{V_f} p_{ex} \, dV.$$

p_{ex}, the external pressure, depends upon how the change is organized. Entries marked ° relate only to a perfect gas, while the rest apply to anything.

Type of work	w	q	ΔU	ΔT
Expansion against vacuum				
(1) Isothermal	0	$0°$	$0°$	0
(2) Adiabatic	0	0	0	$0°$
Expansion against constant pressure p_{ex}				
(1) Isothermal	$-p_{ex}\Delta V$	$+p_{ex}\Delta V°$	$0°$	0
(2) Adiabatic	$-p_{ex}\Delta V$	0	$-p_{ex}\Delta V$	$-p_{ex}\Delta V/C_V°$
Reversible expansion or compression				
(1) Isothermal	$-nRT\ln(V_f/V_i)°$	$+nRT\ln(V_i/V_f)°$	$0°$	0
(2) Adiabatic	$C_V\Delta T°$	0	$C_V\Delta T°$	$T_i\{V_i/V_f)^{1/c}-1\}°$ $c = C_{V,m}/R$

$\ln x$ is the natural logarithm of x; to express it in terms of logs to the base 10, use $\ln x = 2.303 \lg x$. Note that we use the notation $\ln x$ to mean $\log_e x$, and $\lg x$ to mean $\log_{10} x$.

(i) *Work done on a system* is always given by $-p_{ex}\,dV$, where p_{ex} is the pressure acting on the system.

(ii) In a *reversible change* the external pressure is always matched to the internal pressure, and so $p_{ex} = p$, and p depends on the volume and the temperature of the gas in the system.

(iii) In an *irreversible expansion* the external pressure is smaller than the internal pressure. The work done by the expansion is less than in the reversible case.

(iv) In an *isothermal expansion* the temperature of the working gas is maintained by leakage of heat into the system from the thermostat representing the outside world. For a *perfect gas* the internal energy is independent of the volume at constant temperature (we have seen that $(\partial U/\partial V)_T = 0$), which implies that as much heat seeps into the system as work is drawn out: $dq = -dw$.

(v) In an *adiabatic expansion* no heat enters the system, but work is done. Therefore $dq = 0$, and the internal energy changes by the amount of work done: $dU = dw$.

(vi) In the *reversible expansion* of a perfect gas through a given volume from the same initial state, the isothermal process yields more work than the adiabatic because the internal energy is continuously replenished by the energy that enters and maintains the temperature at its initial value.

Further reading

An introduction to chemical thermodynamics. E. F. Caldin; Clarendon Press, Oxford, 1961.

Thermodynamics. G. N. Lewis and M. Randall, revised by K. S. Pitzer and L. Brewer; McGraw-Hill, New York, 1961.

Heat and thermodynamics. M. W. Zemansky; McGraw-Hill, New York, 1957.

Thermodynamics. J. G. Kirkwood and I. Oppenheim; McGraw-Hill, New York, 1961.

Thermodynamics. E. A. Guggenheim; North-Holland, Amsterdam, 1977.

Thermodynamics for chemical engineers. K. E. Bett, J. S. Rowlinson, and G. Saville; Athlone Press, London, 1975.

Chemical thermodynamics. M. L. McGlashan, Academic Press, London, 1979.

Problems

3.1. Which of the following properties are *extensive*? Density, pressure, mass, temperature, enthalpy, refractive index, magnetic susceptibility, heat capacity.

3.2. The equipartition principle (p. 15) gives $(\frac{3}{2})nRT$ for the internal energy of an ideal monatomic gas. What are the values of $(\partial U/\partial V)_T$ and $(\partial H/\partial V)_T$ for an ideal gas? These two quantities occur frequently in thermodynamic discussions, and this is one of several ways of arriving at their value for an ideal gas.

3.3. The isothermal compressibility of lead is 2.3×10^{-6} atm^{-1}. Express this in N^{-1} m^2. A cube of lead of side 10 cm was to be incorporated into the keel of an underwater exploration TV camera, and the designers needed to know the stresses in the equipment. What change of volume will such a cube undergo on the floor of the sea at depths of (a) 100 ft, (b) 5000 fathoms? Take the mean density of sea water as $\rho \approx 1.03$ g cm^{-3}.

3.4. The isobaric coefficient of thermal expansion of lead is 8.61×10^{-5} K^{-1}. The temperature of the water where the camera will operate is $-5\,°C$. What effect will this temperature reduction have on the volume of the block? What effect will both pressure and temperature have?

3.5. In this chapter we encountered various properties of partial and complete differentials, and the next few Problems give some practice in dealing with them. In the first place, we consider a case in which the volume and temperature of a perfect gas depend on the *time t*. Show that the rate of change of the pressure can be related to the rate of change of the volume and the temperature by

$$d \ln p/dt = d \ln T/dt - d \ln V/dt.$$

Do this by finding an expression for dp, with p regarded as a function of V and T.

3.6. As an illustration of the result in the last Problem, consider a system in which the temperature is cooling exponentially towards absolute zero with a time-constant τ_T: exponential cooling is termed *Newtonian cooling*. At the same time a piston is driven into the vessel containing the gas, and the design of the cam controlling the piston is based on a logarithmic spiral, so that the volume is also decreased exponentially towards zero with some time-constant τ_V. How rapidly does the pressure change? What is the time-dependence of the pressure when $\tau_T = \tau_V$?

3.7. The pressure of a given amount of a van der Waals gas may also be regarded as a function of V and T, but its dependence is more complicated. Find an expression for dp for infinitesimal changes in V and T. If we wanted to repeat the time-dependence calculation of the last Problem, what modification to the time-dependence of V would be appropriate?

3.8. Some of the general relations between partial derivatives can be illustrated explicitly by taking the van der Waals equation of state. As a first step, express T as a function of p and V. This then permits us to calculate $(\partial T/\partial p)_V$: confirm the relation $(\partial y/\partial x)_z = 1/(\partial x/\partial y)_z$, and then go on to confirm Euler's chain relation (Box 3.1).

3.9. Find expressions for the isothermal compressibility κ, p. 86, and the isobaric expansivity α, p. 84, of a van der Waals gas. Show from Euler's chain relation that κ and α should be related by $\kappa R = \alpha(V_m - b)$ and confirm explicitly that this is so. In Chapter 1 we saw that the introduction of reduced variables (p. 44) lets us express relations in a way that is independent of the identity of the gas: show that the reduced compressibility and expansivity of a van der Waals gas are related by $8\kappa_r = \alpha_r(3V_r - 1)$.

3.10. Later on (p. 178) we shall see that the Joule–Thomson coefficient μ_{JT} can be determined from the expression $\mu_{JT}C_p = T(\partial V/\partial T)_p - V$. Devise an expression for the value of μ_{JT} in terms of the parameters of a van der Waals gas, and estimate the value of μ_{JT} for xenon at room temperature and pressure.

3.11. The Joule–Thomson coefficient changes sign at the *inversion temperature* T_i. This is an important quantity, because it indicates the conditions at which a real gas will switch from heating to cooling on adiabatic expansion. Use the information in the last Problem to derive an expression for T_i of a van der Waals gas, and express your answer in both ordinary and reduced variables. Estimate the temperatures at which (a) hydrogen, (b) carbon dioxide switch to cooling from heating when their pressures are 50 atm.

3.12. When you design a refrigerator you have to know what temperature drop is brought about by adiabatic expansion of the refrigerant fluid. The value for one type of freon is about 1.2 K atm^{-1}: what pressure drop will you have to ensure in order to bring about a 5 K drop in temperature of the circulating fluid?

3.13. Another fluorocarbon was investigated with a view to using it in a commercial refrigerating system. The temperature drop for a series of initial pressures p_i expanding into 1 atm pressure were measured at 0 °C, with the following results:

p_i/atm	32	24	18	11	8	5
$-\Delta T/K$	22	18	15	10	7.4	4.6

What is the value of μ_{JT} for the gas at this temperature? Is it more suitable than the first freon for the job? What other factors have to be taken into account?

3.14. One *thermodynamic equation of state* is given on p. 92. Find its partner, based on $(\partial H/\partial p)_T$, from it and the general relations between partial differentials.

3.15. Show that the value of μ_{JT} for a gas that is described by the virial equation of state $pV_m/RT = 1 + B/V_m$ is related to B when $4pB/RT \ll 1$ by

$$\mu_{JT} = (T^2/C_{p,m})\left\{\frac{d}{dT}\left(\frac{B}{T}\right)\right\}.$$

Estimate the value of μ_{JT} for argon at room temperature, using the information in Table 1.3.

3.16.   Estimate the changes in volume that occur when 1 cm^3 blocks of (a) mercury, (b) diamond are heated through 5 K at room temperature. Use the data in Table 3.1.

3.17.   In eqn (3.2.19) we possess an entirely general expression for the difference in heat capacities $C_p - C_V$ for any substance. What is the value of $C_{p,m} - C_{V,m}$ for a van der Waals gas? Show, using the result of Problem 3.9, that for such a gas, $C_{p,m} - C_{V,m} = \lambda R$, with $1/\lambda = 1 - (3V_r^2 - 1)^2/4V_r^3 T_r$, and estimate the value of the difference for xenon at room temperature and 10 atm.

3.18.   The heat capacities of (a) copper, (b) benzene at constant pressure and volume are related by eqn (3.2.19). Calculate this difference at room temperature and pressure using the data in Table 3.1 and estimate the difference in heat that is required to raise the temperature of a 500 g sample of each through 50 K at constant pressure and at constant volume.

3.19.   Combine the results of the last Problem with eqn (3.2.18) to find a value of $(\partial U/\partial V)_T$ for copper and benzene.

3.20.   By how much does (a) the internal energy, (b) the enthalpy, of water change when 1 mol is heated through 10 K at room temperature and pressure? Data for the problem will be found in the *Example* on p. 92. Account for the difference between the two quantities.

3.21.   Whereas the material of Chapter 2 let us deal with isothermal changes, this chapter has shown how to deal with another important class, the adiabatic changes. The next few Problems look at some examples of this kind of process. A straightforward application of the methods described in the text will orientate our ideas. Consider a system consisting of a 2 mol room-temperature sample of carbon dioxide (to be regarded as a perfect gas at this stage) confined to a cylinder of 10 cm^2 cross-section and at a pressure of 10 atm. It is allowed to expand adiabatically against a pressure of 1 atm. During the expansion the piston is pushed through 20 cm. Find the values of w, q, ΔU, ΔH, and ΔT for the expansion.

3.22.   65 g of xenon is held in a container at 2 atm and room temperature. It is then allowed to expand adiabatically (a) reversibly to 1 atm, (b) against 1 atm pressure. If the initial temperature is 298 K, what will the final temperature be in each case?

3.23.   A 1 mol sample of a fluorocarbon was allowed to expand reversibly and adiabatically to twice its volume. In the expansion the temperature dropped from 298.15 K to 248.44 K. On the basis of assuming that the gas behaves perfectly, estimate the value of $C_{V,m}$. Does the value you obtain conflict with the assumption?

3.24.   We have just seen that the heat capacity of a gas C_V can be measured by observing the temperature drop on adiabatic expansion. If the pressure drop is also monitored we can use the same experiment to find γ, and hence C_p of the gas. In the experiment just described the initial pressure was 1522.2 mmHg, and after expansion the pressure was 613.85 mmHg. What are the values of γ and $C_{p,m}$ for the fluorocarbon? Continue to assume perfect gas behaviour.

3.25.   Find the change in the internal energy and the enthalpy of the fluorocarbon during the expansion described in Problem 3.23. Continue to assume that the gas behaves perfectly.

3.26.   In the *Example* on p. 98 the value of ΔH for an adiabatic change was found by equating pV with nRT, but it was pointed out that $dH = Vdp$ could be integrated directly so long as we knew V as a function of p. Use this method to find ΔH for the reversible, adiabatic expansion of an ideal gas.

3.27. We have seen (Problem 2.20) that the equipartition theorem lets us write the value of $C_{V,m}$ for a perfect monatomic gas as $(3/2)R$. Predict a value for γ of such a gas. What is the value of γ for a translating, rotating, non-linear, polyatomic molecule?

3.28. In Problem 3.17 we found an expression relating C_p and C_V of a van der Waals gas. Find an expression for γ of a monatomic gas and of γ for a non-ideal polyatomic gas, both being regarded as van der Waals gases.

3.29. Estimate γ for (a) xenon, (b) water vapour, both at $100\,°C$ and 1 atm pressure, on the assumption that they are van der Waals gases.

3.30. The speed of sound in a gas is related to the ratio of heat capacities γ by $c_s = \sqrt{(RT\gamma/M_m)}$ where M_m is the molar mass. Show that this can be written $c_s = \sqrt{(\gamma p/\rho)}$ where p is the pressure and ρ the density. Calculate the speed of sound (a) in helium, (b) in air, both at $25\,°C$.

3.31. Measuring the speed of sound in a gas is a method of determining the heat capacity. The speed of sound in ethene at $0\,°C$ was measured as 317 m s^{-1}. Find the value of γ and, assuming ideal behaviour, the value of $C_{V,m}$ at this temperature.

3.32. An old man begins to blow an ocarina, and generates the note A (440 Hz). He goes on blowing, and as he blows the air he exhales turns more and more to pure carbon dioxide. His last gasp is pure carbon dioxide: what is his dying frequency? You need to know that the frequency of the ocarina is given by Kc_s, where K depends upon its dimensions, and c_s is the speed of sound.

4 The First Law in action: thermochemistry

Learning objectives

After careful study of this chapter you should be able to

(1)   Define the terms *endothermic* and *exothermic* reaction, and *reaction enthalpy* (p. 106).

(2)   Define *standard state* of a substance, *standard enthalpy of reaction* (p. 106), and *standard enthalpy of formation* (p. 107).

(3)   State, derive, and use *Hess's Law* of constant heat summation (p. 108).

(4)   Derive and use *Kirchhoff's Law* for the temperature dependence of the reaction enthalpy (eqn (4.1.2)).

(5)   Relate the reaction enthalpy to the change in internal energy (p. 111).

(6)   Describe the measurement of ΔU and ΔH by *calorimetry* (p. 112).

(7)   Define and use the *enthalpy of combustion* (p. 113), *bond enthalpy* (p. 113), *bond dissociation energy* (p. 114), and *enthalpy of atomization* (p. 116).

(8)   Define and use the *enthalpy of sublimation* (p. 116), *enthalpy of phase transition* (p. 116), and *enthalpy of hydrogenation* (p. 118).

(9)   Define, deduce, and use the *enthalpy of solution* (p. 118), *enthalpy of solvation* (p. 118) of ions, and *enthalpy of formation* of ions (p. 118).

(10)   Construct *Born–Haber cycles* to determine enthalpies from other data (p. 120).

(11)   Define *ionization energy, electron affinity,* and *lattice enthalpy* (p. 119).

Introduction

Thermochemistry is the study of the energy changes that occur during a chemical reaction. The designer of a new chemical plant needs to know how much energy each reaction generates or absorbs, and must make sure that the plant can accommodate them. If the plant is to work with maximum economy the designer has to know how much heat to supply to one region and how much can be taken from another. Elsewhere we meet questions about reactions in biological systems. Can a sequence of reactions release enough energy to drive a biological process? Can a proposed reaction scheme supply enough power to keep a cell alive? Such questions are not solely a matter of the heat available from the reactions, as we shall see later, but thermochemistry does give a first indication of the energy resources available in a particular fuel.

4.1 Heat in chemical reactions

A reaction vessel can be regarded as a thermodynamic system in the sense used in Chapter 2. The amount of heat transferred as heat during a reaction is q. In Chapter 3 we saw that q is a path function. It is much more convenient to discuss the energy changes accompanying reactions in terms of state functions because they are independent of the way the reactions are carried out. In particular, they are independent of whether the reaction was carried out reversibly or not. The next sections explore this shift of emphasis.

The reaction enthalpy. If the transfer of heat occurs at constant volume, and if no other forms of work are permitted, $(\mathrm{d}q)_V$ is equal to the change of internal energy (p. 66):

$$(\Delta U)_V = q_V.$$

For a specified change of state ΔU is independent of how the change is brought about; therefore the subscript V can be dropped from U:

$$\Delta U = q_V.$$

The significance of this equation is that if we measure the heat transferred at constant volume we can identify it with the change in a thermodynamic state function.

Another important result of Chapter 2 is the analogous result for changes occurring at constant pressure. The heat transferred to a system at constant pressure can be identified with the change of enthalpy H. Enthalpy is a state function, and therefore, by the same argument as before, when there is no work other than pV-work,

$$\Delta H = q_p.$$

It follows that if the reaction occurs at constant pressure, the observed heat change can be identified with the change of enthalpy.

We have established an identification of a *thermochemical observable* (q_V or q_p) with the change in a *thermodynamic state function* (ΔU or ΔH). This is the central link between thermochemistry and thermodynamics.

Reactions for which $\Delta H > 0$ are called *endothermic*; those for which $\Delta H < 0$ are *exothermic*. When the reaction takes place in an adiabatic container an endothermic reaction results in a lowering of temperature and an exothermic reaction results in a rise of temperature. If the reaction mixture is maintained at a constant temperature by thermal contact with the surroundings then an endothermic reaction results in a transfer of heat into the system. These points are summarized as follows:

	Endothermic $\Delta H > 0$	Exothermic $\Delta H < 0$
Isothermal conditions, p constant	Heat enters $q_p > 0$	Heat exits $q_p < 0$
Adiabatic conditions	T falls	T rises

One aspect of this discussion must be emphasized. Under constant pressure conditions, and when no work other than pV-work is involved, the whole of the ΔH of a reaction may be extracted as *heat* ($q_p = \Delta H$); but this does not imply that a simple modification of the apparatus will allow the same amount of *work* to be obtained. In some cases the amount of work may even exceed the value of ΔH for the reaction. This is a matter for the Second Law, and is dealt with in the next Chapter.

Standard enthalpy changes. Tables of reaction enthalpies refer to the case where the reactants and the products are in their *standard states*. The standard state of a substance is the stable form at a pressure of 1 atm (101.325 kPa) and the temperature specified (which is often, but not necessarily 25 °C, 298.15 K). The precise definition of the standard state of a real gas involves a correction for non-ideality: this is discussed in Chapter 6 and need not confuse matters at this stage. The *standard enthalpy of reaction* is denoted $\Delta H^\ominus(T)$ and is the difference $H(\text{products}) - H(\text{reactants})$, $H(\text{products})$ and $H(\text{reactants})$ being the sums of the enthalpies of the product and reactant species respectively, all substances being in their standard states at the temperature T. As an example, consider the reaction

$$C(s) + O_2(g) \rightarrow CO_2(g); \qquad \Delta H_m^\ominus(298.15\ \text{K}) = -393.51\ \text{kJ mol}^{-1}.$$

The subscript m indicates that the enthalpy change refers to unit amount of the reaction as written (e.g., to the formation of 1 mol of CO_2 from 1 mol of C and 1 mol of O_2). The standard states at 25 °C are graphite (the most stable form of carbon at 1 atm and 25 °C), and pure oxygen and carbon dioxide, each at 1 atm. Normally there are several reactants or products: the standard enthalpy change then refers to the overall process (unmixed reactants) → (unmixed products), but except in the case of ionic reactions in solution the enthalpy changes accompanying mixing and un-mixing are insignificant in comparison with the contribution from the reaction itself.

The reaction quoted above is an example of a *formation reaction*, in which a compound is formed from its elements. *Standard molar enthalpies of formation* of compounds are the enthalpy changes that occur when unit amount of the compound in its standard state is formed from its elements in their standard states. By definition the standard enthalpies of formation of the elements are zero. Standard molar enthalpies of formation, which are often denoted $\Delta H_f^\ominus(T)$, of a variety of compounds, are given in Tables 4.1 and 4.2.

Table 4.1. Standard molar enthalpies of formation at 298.15 K

	M_r	$\Delta H_f^\ominus/\text{kJ mol}^{-1}$		M_r	$\Delta H_f^\ominus/\text{kJ mol}^{-1}$
$H_2O(g)$	18.015	−241.8	$O_3(g)$	47.998	+142.7
$H_2O(l)$	18.015	−285.8	$NO(g)$	30.006	+90.2
$H_2O_2(l)$	34.015	−187.8	$NO_2(g)$	46.006	+33.2
$NH_3(g)$	17.031	−46.1	$N_2O_4(g)$	92.012	+9.2
$N_2H_4(l)$	32.045	+50.6	$SO_2(g)$	64.063	−296.8
$N_3H(l)$	43.028	+264.0	$H_2S(g)$	34.080	−20.6
$N_3H(g)$	43.028	+294.1	$SF_6(g)$	146.054	−1209
$HNO_3(l)$	63.013	−174.1	$HF(g)$	20.006	−271.1
$NH_2OH(s)$	33.030	−114.2	$HCl(g)$	36.461	−92.3
$NH_4Cl(s)$	53.492	−314.4	$HCl(aq)$	36.461	−167.2
$HgCl_2(s)$	271.50	−224.3	$HBr(g)$	80.917	+36.4
$H_2SO_4(l)$	98.078	−814.0	$HI(g)$	127.912	+26.5
$H_2SO_4(aq)$	98.078	−909.3	$CO_2(g)$	44.010	−393.5
$NaCl(s)$	58.443	−411.0	$CO(g)$	28.011	−110.5
$NaOH(s)$	39.997	−426.7	$Al_2O_3(\alpha, s)$	101.945	−1675.7
$KCl(s)$	74.555	−435.9	$SiO_2(s)$	60.085	−910.9
$KBr(s)$	119.011	−392.2	$FeS(s)$	87.91	−100.0
$KI(s)$	166.006	−327.6	$FeS_2(s)$	119.975	−178.2
			$AgCl(s)$	143.323	−127.1

Source: G. W. C. Kaye and T. H. Laby, *Tables of physical and chemical constants*, Longmans; M_r from *American Institute of Physics handbook*, McGraw-Hill.

Hess's Law and reaction enthalpies. The standard enthalpy change for any reaction can be calculated from tables of standard enthalpies of formation. This is because enthalpy is a state function, and its value is independent of the path used to go from a specified initial state (reactants) to a specified final state (products). Therefore the reaction enthalpy for a reaction such as

$$2HN_3(l) + 2NO(g) \rightarrow H_2O_2(l) + 4N_2(g)$$

can be expressed as the appropriate sum and differences of the formation enthalpies of all the components, because we can *formally* regard the reaction as proceeding by 'unforming' the reactants into their elements and then forming those elements into the products. It follows that

$$\Delta H_m^{\ominus}(298.15\,K) = [(-187.8\,kJ\,mol^{-1}) + 4(0)] - [2(264.0\,kJ\,mol^{-1})$$
$$+ 2(90.25\,kJ\,mol^{-1})]$$

$$= -896.3\,kJ\,mol^{-1}.$$

The m subscript refers to the generation of unit amount of H_2O_2.

Table 4.2. Standard molar enthalpies of formation and combustion at 298.15 K

	M_r	$\Delta H_f^{\ominus}/kJ\,mol^{-1}$	$-\Delta H_c^{\ominus}/kJ\,mol^{-1}$
$CH_4(g)$	16.043	−74.81	890.4
$C_2H_2(g)$	26.038	+226.8	1300
$C_2H_4(g)$	28.054	+52.30	1411
$C_2H_6(g)$	30.070	−84.64	1560
C_3H_6 cyclopropane(g)	42.081	53.35	2091
C_3H_6 propene(g)	42.081	20.5	2058
C_4H_{10} n-butane(g)	58.124	−126.11	2877
C_5H_{12} n-pentane(g)	72.151	−146.4	3536
C_6H_{12} cyclohexane(l)	84.163	−156.2	3920
C_6H_{14} n-hexane(l)	86.178	−198.7	4163
C_6H_6 benzene(l)	78.115	+48.99	3268
C_8H_{18} n-octane(l)	114.233	−249.8	5471
$C_{10}H_8$ naphthalene(s)	128.175	+78.53	5157
$CH_3OH(l)$	32.042	−239.0	726.1
$CH_3CHO(g)$	44.054	−166.0	1193
$CH_3CH_2OH(l)$	46.070	−277.0	1368
$CH_3COOH(l)$	60.053	−484.2	874.5
$CH_3COOC_2H_5(l)$	88.107	−486.6	2231
$C_6H_5OH(s)$	94.114	−165.0	3054
$C_6H_5NH_2(l)$	93.129	−31.1	3393
$NH_2CO.NH$, urea(s)	60.056	−333.0	632.2
$CH_2(NH_2)CO_2H$, glycine(s)	75.068	−537.2	964.4
$C_6H_{12}O_6$, α-D-glucose(s)	180.159	−1274	2802
$C_6H_{12}O_6$, β-D-glucose(s)	180.159	−1268	2808
$C_{12}H_{22}O_{11}$, sucrose(s)	342.303	−2222	5645
$CH_3CH(OH)COOH$, lactic acid(s)	90.079	−694.0	1344

Source: *American Institute of Physics handbook*, McGraw-Hill.

Hess's Law of constant heat summation is a generalization of the calculation just described. It states that *the standard enthalpy change in any reaction can be expressed as the sum of the standard enthalpy changes, at the same temperature, of a series of reactions into which the overall reaction may formally be divided.* The thermodynamic basis of the law is that ΔH is independent of the path taken between specified states, in this case the path between reactants and products. The data in Tables 4.1 and 4.2 are therefore capable of providing thermochemical information on a wide variety of reactions. Note that the tabulated data refer to 1 atm, the standard pressure. The reactions in rockets and factories, however, do not normally take place at 1 atm and 25 °C; we must therefore examine the pressure and temperature dependence of reaction enthalpies.

Example (Objective 3). The standard enthalpy of the hydrogenation of propene in the reaction $CH_2:CH.CH_3(g) + H_2(g) \rightarrow CH_3.CH_2.CH_3(g)$ is $-124\,kJ\,mol^{-1}$. The standard enthalpy of the oxidation of propane in the reaction $CH_3.CH_2.CH_3(g) + 5O_2(g) \rightarrow 3CO_2(g) + 4H_2O(l)$ is $-2220\,kJ\,mol^{-1}$. Find the standard enthalpy of the combustion reaction of propene.

- *Method.* The enthalpy of the reaction may be found by adding and subtracting the appropriate reactions and using Hess's Law. You will find that you also need to know the standard enthalpy of the reaction $H_2(g) + \frac{1}{2}O_2(g) \rightarrow H_2O(l)$: it is $-285.8\,kJ\,mol^{-1}$.

- *Answer.* We require the enthalpy of the reaction

$$C_3H_6(g) + \tfrac{9}{2}O_2(g) \rightarrow 3CO_2(g) + 3H_2O(l), \qquad \Delta H_m^{\ominus}.$$

The reactions given are

- (a) $C_3H_6(g) + H_2(g) \rightarrow C_3H_8(g)$, $\qquad \Delta H_m^{\ominus}(a) = -124\,kJ\,mol^{-1}$,
- (b) $C_3H_8(g) + 5O_2(g) \rightarrow 3CO_2(g) + 4H_2O(l)$, $\qquad \Delta H_m^{\ominus}(b) = -2220\,kJ\,mol^{-1}$.

Note that (a) + (b) is the reaction

$$C_3H_6(g) + H_2(g) + 5O_2(g) \rightarrow 3CO_2(g) + 4H_2O(l),$$

and so if we also include

- (c) $H_2(g) + \tfrac{1}{2}O_2(g) \rightarrow H_2O(l)$ $\qquad \Delta H_m^{\ominus}(c) = -285.8\,kJ\,mol^{-1}$

we can obtain the required reaction as (a) + (b) − (c). It follows that

$$\Delta H_m^{\ominus} = \Delta H_m^{\ominus}(a) + \Delta H_m^{\ominus}(b) - \Delta H_m^{\ominus}(c) = (-124 - 2220 + 285.8)\,kJ\,mol^{-1}$$
$$= -2058\,kJ\,mol^{-1}.$$

- *Comment.* This is the value in Table 4.2. The only imagination required in this kind of application is in deciding on the right set of reaction steps; then keep the signs right.

The temperature-dependence of reaction enthalpies. For many important reactions the standard enthalpy changes have been measured and are listed for a variety of temperatures. In the absence of this information, the enthalpy change at some other temperature may be estimated in the following way.

The heat capacity at constant pressure of some pure species, C_p, is the rate of change of enthalpy with respect to temperature at constant pressure. Therefore, when the temperature is changed from T_1 to T_2, the enthalpy of a single species changes from $H(T_1)$ to

$$H(T_2) \approx H(T_1) + (T_2 - T_1)(\partial H/\partial T)_p = H(T_1) + (T_2 - T_1)C_p.$$

(This assumes that C_p is independent of temperature in the range T_1 to T_2: we do a more precise calculation below.) This equation applies to the enthalpy of every species involved in a reaction, and so for the difference $\Delta H = H(\text{products}) - H(\text{reactants})$ we find

$$\Delta H(T_2) \approx \Delta H(T_1) + (T_2 - T_1)[C_p(\text{products}) - C_p(\text{reactants})].$$

If the difference of the products' and the reactants' heat capacities is written ΔC_p, and $\delta T = T_2 - T_1$, this expression simplifies to

(4.1.1) $$\Delta H(T_2) \approx \Delta H(T_1) + \delta T \Delta C_p.$$

Therefore in order to calculate the temperature-dependence of the reaction enthalpy, all we need to know are the molar heat capacities of the species involved in the reaction.

Example (Objective 4). The constant-pressure heat capacities of gaseous hydrogen, oxygen, and water are $28.84\,\text{J K}^{-1}\text{mol}^{-1}$, $29.37\,\text{J K}^{-1}\text{mol}^{-1}$, and $33.58\,\text{J K}^{-1}\text{mol}^{-1}$. The enthalpy of formation of gaseous water at $25\,°\text{C}$ is $-241.82\,\text{kJ mol}^{-1}$; what is its value at $100\,°\text{C}$?

- *Method.* In the absence of other information, take the heat capacities to be independent of temperature and use eqn (4.1.1).

- *Answer.* $\Delta C_{p,m} = C_{p,m}(H_2O, g) - C_{p,m}(H_2, g) - \frac{1}{2} C_{p,m}(O_2, g)$

  $\qquad\qquad = (33.58 - 28.84 - 14.68)\,\text{J K}^{-1}\text{mol}^{-1}$

  $\qquad\qquad = -9.94\,\text{J K}^{-1}\text{mol}^{-1}.$

  $\delta T \Delta C_{p,m} = (75.0\,\text{K}) \times (-9.94\,\text{J K}^{-1}\text{mol}^{-1})$

  $\qquad\qquad = -0.746\,\text{kJ mol}^{-1}.$

  Then from eqn (4.1.1),

  $\Delta H_m^{\ominus}(373\,\text{K}) = -241.82\,\text{kJ mol}^{-1} - 0.746\,\text{kJ mol}^{-1} = -242.57\,\text{kJ mol}^{-1}.$

- *Comment.* The way to take temperature dependence of the heat capacities into account is described below.

The assumption that all the values of C_p are independent of temperature may be poor when δT is large. In the general case we use the following technique. The enthalpy changes by dH when the temperature changes by dT, and at constant pressure $dH = C_p(T)dT$. Integrating this expression between T_1 and T_2 gives

$$H(T_2) - H(T_1) = \int_{T_1}^{T_2} C_p(T)\,dT.$$

This expression applies to every species involved in the reaction, and so the reaction enthalpies at T_2 and T_1 are related by

(4.1.2) $$\textit{Kirchhoff's Law: } \Delta H(T_2) = \Delta H(T_1) + \int_{T_1}^{T_2} \Delta C_p(T)\,dT,$$

where $\Delta C_p(T)$ is the difference of the heat capacities of the individual (unmixed) products and reactants at the temperature T.

Kirchhoff's Law can be applied when the temperature dependences of all the heat capacities are known in the temperature range of interest. Occasionally numerical tables or theoretical expressions are available, and the integration can be carried out. Often the molar heat capacity of some species is expressed in the form

(4.1.3) $$C_{p,m}(T) = a + bT + cT^{-2}$$

and the temperature-independent coefficients a, b, and c listed: a selection of values is given in Table 4.3. This expression can be inserted into the integral in eqn (4.1.2) in the appropriate molar combination and the integration performed (see Problem 4.18).

Table 4.3. Temperature-dependence of heat capacities, $C_{p,m} = a + bT + cT^{-2}$

	$a/\mathrm{J\,K^{-1}\,mol^{-1}}$	$b/10^{-3}\mathrm{J\,K^{-2}\,mol^{-1}}$	$c/10^5\mathrm{J\,K\,mol^{-1}}$
Gases (298–2000 K)			
He, Ne, Ar, Kr, Xe	20.78	0	0
H_2	27.28	3.26	0.50
O_2	29.96	4.18	−1.67
N_2	28.58	3.77	−0.50
F_2	34.56	2.51	−3.51
Cl_2	37.03	0.67	−2.85
Br_2	37.32	0.50	−1.26
CO_2	44.22	8.79	−8.62
H_2O	30.54	10.29	0
NH_3	29.75	25.10	−1.55
CH_4	23.64	47.86	−1.92
Liquids ($T_{\mathrm{melt}} \to T_{\mathrm{boil}}$)			
H_2O	75.48	0	0
$C_{10}H_8$ (naphthalene)	79.5	0.4075	0
Solids			
C (graphite)	16.86	4.77	−8.54
Cu	22.64	6.28	0
Al	20.67	12.38	0
Pb	22.13	11.72	0.96
I_2	40.12	49.79	0
NaCl	45.94	16.32	0
$C_{10}H_8$ (naphthalene)	−115.9	3920	0

Source: G. N. Lewis and M. Randall, *Thermodynamics* (revised by K. S. Pitzer and L. Brewer), McGraw-Hill.

The relation between ΔH and ΔU. The enthalpy of a substance differs from its internal energy by an amount pV. It follows that the reaction enthalpy and reaction energy are related by

$$\Delta H = \Delta U + [pV](\text{products}) - [pV](\text{reactants}).$$

ΔH and ΔU differ because at constant pressure, but not at constant volume, the system does work on the outside world in the course of the reaction. In reactions involving only solids or liquids the volumes of the products and reactants are approximately the same, and except under some geophysical conditions where pressures are very large

for reactions involving only solids and liquids: $\Delta H \approx \Delta U$.

In the case of reactions involving gases ΔH and ΔU may be quite different because the volume may change markedly. In elementary applications it is adequate to assume perfect behaviour for all the gases involved in the reaction, and then the product pV may be replaced by nRT for each gas present. If we write $\Delta n_{gas} = n_{gas}(\text{products}) - n_{gas}(\text{reactants})$, where n_{gas} is the total amount of gas present as products or as reactants, then

$(4.1.4)°$

$$\Delta H = \Delta U + \Delta n_{gas} RT.$$

This expression provides a simple way of interconnecting reaction enthalpies and reaction internal energies.

Example (Objective 5). The standard enthalpy of combustion of propene was calculated in the *Example* on p. 99. What is the value of $\Delta U_m^{\ominus}(298.15\,\text{K})$?

- *Method.* The reaction is $C_3H_6(g) + \frac{9}{2}O_2(g) \rightarrow 3CO_2(g) + 3H_2O(l)$, and so $\Delta n_{gas}/\text{mol} = 3 - 1 - \frac{9}{2} = -\frac{5}{2}$. Use eqn (4.1.4) and $RT = 2.48\,\text{kJ mol}^{-1}$ at 298.15 K.

- *Answer.* From eqn (4.1.4),
$$\Delta U_m^{\ominus}(298\,\text{K}) = -2058\,\text{kJ mol}^{-1} - (-\tfrac{5}{2}) \times (2.48\,\text{kJ mol}^{-1})$$
$$= -2052\,\text{kJ mol}^{-1}.$$

- *Comment.* In a very careful measurement it would be important to note that $\Delta H_m^{\ominus}$ in eqn (4.1.4) is really the value for a constant volume reaction, and an additional small correction (which is zero in the case of perfect gases) should be applied.

Calorimetric measurements of ΔU and ΔH. The calorimetric measurements of ΔU and ΔH are based on the identification $\Delta U = q_V$ and $\Delta H = q_p$, which are valid when the system is doing no work other than pV-work. The actual observation involved is the change in temperature brought about by the heat produced or absorbed: if the reaction is run in a thermally insulated system of heat capacity C and the temperature changes by δT, the value of q is obtained from $C\delta T$; this q is then identified with ΔU or ΔH, depending on the conditions of the reaction.

The most important device for measuring q_V is the *adiabatic bomb calorimeter*, Fig. 4.1a. The reaction is initiated inside a constant volume container (which, if the reaction is a combustion, contains oxygen at about 30 atm), and the temperature is monitored. The temperature of the surrounding water bath is monitored simultaneously and adjusted to the same value: this eliminates the temperature gradient between the calorimeter itself and its surroundings (the bath), and ensures that there is no loss of heat (hence the name *adiabatic* calorimeter). The temperature change produced by the completed reaction is then converted to q_V using the known heat capacity of the calorimeter. The apparatus is often calibrated against a standard (e.g. the combustion of benzoic acid), or electrically. The ΔU so obtained is converted to ΔH, if that is required, using eqn (4.1.4).

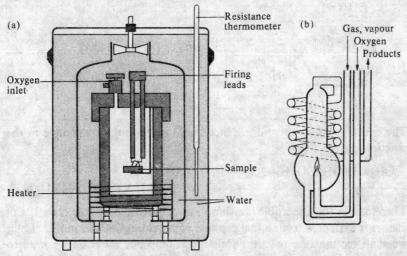

Fig. 4.1. (a) Constant volume bomb calorimeter, (b) constant pressure flame calorimeter insert, which replaces the constant volume bomb.

The *adiabatic flame calorimeter*, Fig. 4.1b, works in a similar way but the reaction proceeds at constant pressure, and so ΔH is obtained directly.

Reaction enthalpies and energies may also be determined by non-calorimetric methods: these are described in Chapters 9 and 12.

4.2. Varieties of enthalpy

In chemistry we are often interested in the enthalpy changes in a variety of different types of reaction. The $\Delta H^{\ominus}$ then carry special names, and a few of these are reviewed in this section.

The enthalpies of chemical and physical change. The *enthalpy of combustion* is the change of enthalpy accompanying total oxidation of a material. Some values are listed in Table 4.2. A particularly important example is the enthalpy of combustion of glucose:

$$(glucose) + 6O_2 \rightarrow 6CO_2 + 6H_2O;$$
$$\Delta H^{\ominus}_{m}(298.15\,K) = -2808\,kJ\,mol^{-1}.$$

This large amount of energy is the basis of much cellular activity and its value reveals something about the efficiency of the cells of the more highly developed organisms. In *aerobic respiration* glucose is consumed in accord with the last equation. In *anaerobic fermentation* (the method of obtaining energy used by some microorganisms) a basic reaction is glycolysis, in which glucose is broken down into lactic acid:

CH$_2$OH
O
HO   OH   OH
OH

$\xrightarrow{\text{enzymes}}$ 2CH$_3$.CH(OH).COOH.

The enthalpy of combustion of lactic acid is the enthalpy change in the reaction

$$CH_3.CH(OH).CO_2H + 3O_2 \rightarrow 3CO_2 + 3H_2O;$$
$$\Delta H_m^{\ominus}(298.15\,K) = -1344\,kJ\,mol^{-1}.$$

Therefore, in the anaerobic consumption of glucose only $120\,kJ\,mol^{-1}$ of energy is extracted from a fuel capable of providing $2808\,kJ\,mol^{-1}$. Later we shall see that the reaction enthalpy is only one criterion we need to consider when we examine the efficient use of resources in the respiration process, but this example suggests that aerobic respiration is a more sophisticated process than the more primitive anaerobic fermentation.

Enthalpies of formation of compounds were introduced on p. 107. Since the values refer to the overall enthalpy changes that occur when the elements in their standard states are broken up into atoms and re-assembled into molecules, they do not give direct information about the strengths of bonds in the compound. That information can be deduced from the enthalpy changes in reactions such as

$$AB(g) \rightarrow A(g) + B(g).$$

The enthalpy change in this process is called the *bond dissociation enthalpy*. A and B may be *groups* of atoms, and g denotes that the species are in the gas phase. Quite often the molar bond enthalpy is denoted $DH_m^{\ominus}(A-B)$ where D is a signal that we are talking about the dissociation of a molecule.

As an example of this type of measurement, the bond dissociation enthalpy of molecular nitrogen is $DH_m^{\ominus}(N\equiv N) = 945\,kJ\,mol^{-1}$, and that of the HO—H bond in water is $DH_m^{\ominus}(HO-H) = 492\,kJ\,mol^{-1}$. Removing the second hydrogen atom from water requires a different amount of energy because $DH_m^{\ominus}(O-H) = 428\,kJ\,mol^{-1}$. Some experimental values are given in Table 4.4.

The bond dissociation enthalpy has to be distinguished from two related quantities. First there is the *bond dissociation energy*: this is strictly the change of internal energy that accompanies bond cleavage at $T = 0$, but it normally differs only slightly from the bond dissociation enthalpy at 298 K. Then there is the *mean bond dissociation enthalpy*, or simply the *bond enthalpy*, $EH_m^{\ominus}(A-B)$. This is the *average* value of bond dissociation enthalpies of the A—B bond in a series of different compounds, Table 4.5. For instance, the O—H bond enthalpy is the mean of $DH_m^{\ominus}(O-H)$ and $DH_m^{\ominus}(HO-H)$ for water (and other compounds, e.g. $DH_m^{\ominus}(CH_3O-H)$ in methanol), and is $463\,kJ\,mol^{-1}$. The bond enthalpy is useful because it lets us make estimates of enthalpy changes in reactions where data might not be available (or to hand): see below.

Table 4.4. Bond dissociation enthalpies, $DH^{\ominus}_m(A-B)/\text{kJ mol}^{-1}$ at 298.15 K

Diatomic molecules				Polyatomic molecules			
H—H	436	O=O	497	H—CH$_3$	435	CH$_3$—OH	377
O—H	428	C=O	1074	H—NH$_2$	431	CH$_3$—Cl	452
F—H	565	N≡N	945	H—OH	492	CH$_3$—Br	293
Cl—H	431	F—F	155	H—C$_6$H$_5$	469	CH$_3$—I	234
Br—H	366	Cl—Cl	242	H$_3$C—CH$_3$	368	O=CO	531
I—H	299	Br—Br	193	H$_2$C=CH$_2$	699	HO—OH	213
		I—I	151	HC≡CH	962	O$_2$N—NO$_2$	57

Source: *Handbook of chemistry and physics*, Chemical Rubber Co.; G. W. C. Kaye and T. H. Laby, *Tables of physical and chemical constants*, Longmans.

Table 4.5. Single bond enthalpies, $EH^{\ominus}_m(A-B)/\text{kJ mol}^{-1}$

	H	C	N	O	F	Cl	Br	I	S
H	436								
C	413	348							
N	391	292	161						
O	463	351	(157)	139					
F	563	441	270	185	153				
Cl	432	328	200	203	254	243			
Br	366	276				219	193		
I	299	240				210	178	151	
S	399	259				250	212		213

Source: L. Pauling, *The nature of the chemical bond*, Cornell University Press, 3 edn., 1960.

Table 4.6. Standard enthalpies of atomization and sublimation, $\Delta H^{\ominus}_m(298.15\,\text{K})/\text{kJ mol}^{-1}$

H	C	N	O	F
217.97[a]	716.682[s]	472.70[a]	249.17[a]	78.99[a]
Na	Si	P	S	Cl
108.4[s]	455.6[s]	314.6[s]	278.81[s]	121.68[a]
K	Ag	Cu	Pb	Br
89.8[s]	286.2[s]	339.3[s]	195.8[s]	111.88[a]
Hg	Fe	Zn		I
60.84[a]	404.5[s]	130.5[s]		106.84[s+a]
				62.43[s]

[a] atomization $\frac{1}{2}X_2(g) \to X(g)$; [s] sublimation $X(s) \to X(g)$
Source: *American Institute of Physics handbook*, McGraw-Hill. G. N. Lewis and M. Randall, *Thermodynamics* (revised by K. S. Pitzer and L. Brewer), McGraw-Hill.

The enthalpy change on shattering a molecule entirely into its component atoms is called the *enthalpy of atomization*. It follows from Table 4.4 that the enthalpy of atomization of H_2O is $(492\,kJ\,mol^{-1})+(428\,kJ\,mol^{-1}) = 920\,kJ\,mol^{-1}$. Note that this is positive because energy must be injected into water if it is to be atomized. A special case of an enthalpy of atomization is given by

$$\text{metal (crystal)} \rightarrow \text{metal (monatomic gas)}$$

where the enthalpy change is the *enthalpy of sublimation*. The enthalpy of sublimation of carbon (which is also the enthalpy of atomization of graphite) proved to be a notoriously difficult quantity to obtain, but it is so important for the discussion of the energetics of organic molecules that a great deal of effort went into its determination. A reasonable value appears to be $716.68\,kJ\,mol^{-1}$.

Example (Objective 7). Estimate the standard enthalpy of formation of methanol.

- *Method.* The reaction is

$$C(\text{graphite}) + 2H_2(g) + \tfrac{1}{2}O_2(g) \rightarrow CH_3OH(l), \qquad \Delta H_f^\ominus$$

Analyse this as follows:

(i) $C(\text{graphite}) \rightarrow C(g)$ $\Delta H_m^\ominus(i)$
(ii) $2H_2(g) \rightarrow 4H(g)$ $\Delta H_m^\ominus(ii)$
(iii) $\tfrac{1}{2}O_2(g) \rightarrow O(g)$ $\Delta H_m^\ominus(iii)$
(iv) $C(g) + O(g) + 4H(g) \rightarrow CH_3OH(g)$ $\Delta H_m^\ominus(iv)$
(v) $CH_3OH(g) \rightarrow CH_3OH(l)$ $\Delta H_m^\ominus(v)$

$\Delta H_m^\ominus(i)$ is the enthalpy of atomization of carbon (Table 4.6), $\Delta H_m^\ominus(ii)$ is twice the bond dissociation enthalpy of $H_2(g)$ (Table 4.4), $\Delta H_m^\ominus(iii)$ half that of $O_2(g)$ (Table 4.4), $\Delta H_m^\ominus(iv)$ is (approximately) the negative of the bond enthalpies:

$$\Delta H_m^\ominus(iv) = -\{3EH_m^\ominus(C\!-\!H) + EH_m^\ominus(C\!-\!O) + EH_m^\ominus(O\!-\!H)\}$$

(Table 4.5), and $\Delta H_m^\ominus(v)$ is the negative of the standard enthalpy of vaporization of methanol (Table 4.7).

- *Answer.* $\Delta H_m^\ominus(i) = 717\,kJ\,mol^{-1}$

$\Delta H_m^\ominus(ii) = 2 \times 436\,kJ\,mol^{-1} = 872\,kJ\,mol^{-1}$

$\Delta H_m^\ominus(iii) = \tfrac{1}{2}(497\,kJ\,mol^{-1}) = 248\,kJ\,mol^{-1}$

$\Delta H_m^\ominus(iv) \approx -\{3 \times 413 + 351 + 463\}\,kJ\,mol^{-1} = -2053\,kJ\,mol^{-1}$

$\Delta H_m^\ominus(v) = -37.99\,kJ\,mol^{-1}$

$\Delta H_f^\ominus \approx \{717 + 872 + 248 - 2053 - 38\}\,kJ\,mol^{-1} = -254\,kJ\,mol^{-1}$

- *Comment.* The measured value is $-239.0\,kJ\,mol^{-1}$ (Table 4.2). The source of the discrepancy is in the use of the average bond enthalpies in step (iv).

The enthalpy of sublimation is also a special case of an *enthalpy of phase transition*. Phase transitions include evaporation (ΔH_{vap}, enthalpy of vaporization, also called the 'latent heat' of vaporization), melting (or fusion, ΔH_{melt}), and changes of crystal form. For example, the evaporation of water corresponds to the process

Table 4.7. Enthalpies of fusion and evaporation $\Delta H_m/\text{kJ mol}^{-1}$ at the transition temperature

	T_f/K	Fusion[a]	T_b/K	Evaporation[b]
He	3.5	0.021	4.22	0.084
Ar	83.81	1.188	87.29	6.506
H_2	13.96	0.117	20.38	0.9163
N_2	63.15	0.719	77.35	5.586
O_2	54.36	0.444	90.18	6.820
Cl_2	172.12	6.406	239.05	20.410
Br_2	265.90	10.573	332.35	29.45
I_2	386.75	15.52	458.39	41.80
Hg	234.29	2.292	629.73	59.296
Ag	1234	11.30	2436	250.63
Na	370.95	2.601	1156	98.01
CO_2	217.0	8.33	194.64	25.23[†]
H_2O	273.15	6.008	373.15	40.656 (44.016 at 298.15 K)
NH_3	195.40	5.652	239.73	23.351
H_2S	187.61	2.377	212.80	18.673
CH_4	90.68	0.941	111.66	8.18
C_2H_6	89.85	2.86	184.55	14.7
C_6H_6	278.65	10.59	353.25	30.8
CH_3OH	175.25	3.159	337.22	35.27 (37.99 at 298.15 K)

[†] Sublimation; [a] various pressures; [b] at 1 atm.
Source: *American Institute of Physics handbook*, McGraw-Hill.

$$H_2O(l) \rightarrow H_2O(g) \qquad \Delta H_m^{\ominus}(298\ \text{K}) = 44.01\ \text{kJ mol}^{-1}$$

(at 100 °C the enthalpy change is $\Delta H_m^{\ominus}(373\ \text{K}) = 40.67\ \text{kJ mol}^{-1}$). This indicates that evaporation is an endothermic process, which accords with common sense. Table 4.7 gives a selection of values. In a phase change there may be a substantial change of volume, and so ΔU and ΔH may differ significantly. For example, at 100 °C and 1 atm, 1 mol of water liquid occupies about $18\ \text{cm}^3$ but 1 mol of water vapour occupies about $30\,000\ \text{cm}^3$. Treating this in accord with eqn (4.1.4) with $\Delta n_{gas} = -1\ \text{mol}$ gives

$$\Delta U_m^{\ominus} = (40.67\ \text{kJ mol}^{-1}) - (3.10\ \text{kJ mol}^{-1}) = 37.57\ \text{kJ mol}^{-1}$$

($RT = 3.10\ \text{kJ mol}^{-1}$ at 100 °C), and so ΔH and ΔU differ by 8 per cent.

Example (Objective 8). A man sits still in a warm room and eats $\frac{1}{2}$ lb of cheese (an energy intake of 4000 kJ). Supposing that none of the energy is stored in the body, what mass of water would he need to perspire in order to maintain his original temperature?

● *Method.* Perspiration cools the body because water requires energy in order to evaporate. We take the molar enthalpy of vaporization of water $\Delta H_{vap,m}(298\ \text{K})$ as sufficiently precise for the present problem. Evaporation proceeds at constant pressure, and so ΔH can be equated to q, the amount of heat to be dissipated. We therefore have to find n in the expression $n\Delta H_{vap,m} = q$.

- *Answer.* $n = (4000\,kJ)/(44.0\,kJ\,mol^{-1}) \approx 91\,mol$.

  91 mol of water has a mass of $(18.02\,g\,mol^{-1}) \times (91\,mol) = 1.6\,kg$.

- *Comment.* When calculating thermochemical aspects of digestion it is important to remember that the 'calories' of a diet are the kilocalories of scientific usage: therefore multiply by 4184 to obtain the energy content in joules.

The *enthalpy of hydrogenation* is the change of enthalpy when an unsaturated organic compound becomes fully saturated. Specially important cases are the hydrogenation of ethene and benzene:

$$CH_2{:}CH_2 + H_2 \rightarrow CH_3.CH_3 \qquad \Delta H_m^{\ominus}(298\,K) = -132\,kJ\,mol^{-1}$$

$+ 3H_2 \rightarrow$ $\qquad \Delta H_m^{\ominus}(298\,K) = -246\,kJ\,mol^{-1}.$

The interest in these two values lies in the observation that the second is not three times the first, as might have been expected on the assumption that benzene contains three double bonds. The enthalpy of hydrogenation of benzene is less than $3 \times (-132\,kJ\,mol^{-1})$ by $150\,kJ\,mol^{-1}$, which therefore represents a *thermochemical stabilization* of benzene (so that the molecule is closer in energy than expected to the fully hydrogenated, stable form). The reason for this stabilization will be dealt with in Part 2.

The enthalpies of ions in solution. The *enthalpy of solution* of a substance is the change in enthalpy when it dissolves in a stated amount of solvent. The *enthalpy of solution to infinite dilution* is the enthalpy change when the substance dissolves in an infinite amount of solvent. In the case of water as solvent this is denoted by the label 'aq'. In the case of HCl we have

$$HCl(g) \rightarrow HCl(aq) \qquad \Delta H_m^{\ominus}(298.15\,K) = -75.14\,kJ\,mol^{-1}.$$

Once enthalpies of solution have been established we are able to obtain values of *enthalpies of formation of species in solution* by combining them with standard enthalpies of formation. These are of importance because so many reactions take place in solution. The enthalpy of formation of HCl(aq) for instance, is the sum of the enthalpies for the following two processes:

$$\tfrac{1}{2}H_2(g) + \tfrac{1}{2}Cl_2(g) \rightarrow HCl(g) \qquad \Delta H_f^{\ominus}(298.15\,K) = -92.31\,kJ\,mol^{-1}$$
$$HCl(g) \rightarrow HCl(aq) \qquad \Delta H_m^{\ominus}(298.15\,K) = -75.14\,kJ\,mol^{-1}$$

$$\tfrac{1}{2}H_2 + \tfrac{1}{2}Cl_2(g) \rightarrow HCl(aq) \qquad \Delta H_m^{\ominus}(298.15\,K) = -167.45\,kJ\,mol^{-1}$$

Likewise the enthalpy of formation of NaCl(s) can be combined with its enthalpy of solution to give the enthalpy of formation of NaCl(aq).

The enthalpy of formation of an ionic compound in solution can be regarded as being the sum of the enthalpies of formation of the constituent solvated ions, such as $H^+(aq)$ and $Cl^-(aq)$. It follows that a simple way of tabulating the enthalpies of formation is to ascribe the value zero to the enthalpy of formation of one ion and to refer all the other values to that.

The convention has been established that the standard enthalpy of formation of $H^+(aq)$ is to be set equal to zero:

$$\tfrac{1}{2}H_2(g) \rightarrow H^+(aq) + e^-(aq); \quad \Delta H^{\ominus}_m(298.15\,K) = 0 \text{ by convention.}$$

On this basis it follows that the enthalpy of formation of $Cl^-(aq)$ is $-167.45\,kJ\,mol^{-1}$. Combining this value with the enthalpy of formation of NaCl(aq) then leads to a value for $Na^+(aq)$, and so on. This procedure leads to the data in Table 4.8.

Table 4.8. Standard enthalpies of formation of ions in solution at infinite dilution, $\Delta H^{\ominus}_f(298.15\,K)/kJ\,mol^{-1}$

Cations		Anions	
H^+	0†	OH^-	-230.0
Li^+	-278.5	F^-	-332.6
Na^+	-239.7	Cl^-	-167.2
K^+	-251.2	Br^-	-121.5
NH_4^+	-132.5	I^-	-55.2
Ag^+	$+105.6$	CO_3^{2-}	-677.1
Mg^{2+}	-462.0	NO_3^-	-207.4
Cu^{2+}	$+64.8$	SO_4^{2-}	-909.3
Zn^{2+}	-153.9	ClO_4^-	-129.3
Ca^{2+}	-543.0	PO_4^{3-}	-1277.4
Al^{3+}	-531.4	HPO_4^{2-}	-1292.1
		$CH_3CO_2^-$	-486.0

† Zero by definition.
Source: G. W. C. Kaye and T. H. Laby, *Tables of physical and chemical constants*, Longmans.

Enthalpies of formation of ions in solution may be regarded as the sum of various contributions. For example, the formation of NaCl(aq) can be considered as the following sequence of steps.

(i) $Na(s) \rightarrow Na(g)$. This process is the sublimation of the sodium metal, and involves an enthalpy change $\Delta H^{\ominus}_{sub,m}$.

(ii) $Na(g) \rightarrow Na^+(g) + e^-$. This is the ionization of the sodium atom, and involves an *ionization energy*, I, which is obtained from spectroscopic observations, Chapter 14: values for various atoms are given in Table 4.9. (We should really deal with the ionization enthalpy, $I + RT$, since $\Delta n_{gas} = +1\,mol$; but not only is the difference small because $RT \approx 2.5\,kJ\,mol^{-1}$ but the RT is in fact cancelled by a similar term, as we see in a moment.)

(iii) $\tfrac{1}{2}Cl_2(g) \rightarrow Cl(g)$. This is the bond dissociation enthalpy of the chlorine molecule, $DH^{\ominus}_m(Cl-Cl)$: it too can be obtained spectroscopically.

(iv) $Cl(g) + e^- \rightarrow Cl^-(g)$. The energy of electron attachment is measured by the *electron affinity*, E_A and is defined so that a positive electron affinity corresponds to the ion X^- having a lower energy than the separated species $X + e^-$. The energy of the reaction $X + e^- \rightarrow X^-$ is then $-E_A$. Values for various atoms are listed in Table 4.10. Some can be obtained spectroscopically, some are calculated; sometimes the present procedure is used to determine the electron affinity if it is the only unknown. (In this step too we should deal with the enthalpy, which in this case is $-E_A - RT$ because $\Delta n_{gas} = -1$ mol: this is the RT that cancels the RT mentioned in step (ii).)

(v) $Na^+(g) + Cl^-(g) \rightarrow NaCl(s)$ is one possible subsequent step: it corresponds to the formation of the NaCl crystal, and the enthalpy change is the *lattice enthalpy*, $\Delta H^{\ominus}_{lattice,m}$. An alternative step is $Na^+(g) + Cl^-(g) \rightarrow Na^+(aq) + Cl^-(aq)$, or NaCl(aq). The enthalpy change is the sum of the solvation enthalpies of the two species of ions, $\Delta H^{\ominus}_{solv,m}$.

Table 4.9. First and second ionization energies of some elements $I/\text{kJ mol}^{-1}$

H							He
1312							2372.4
							5251
Li	Be	B	C	N	O	F	Ne
520.3	899.5	800.6	1086.4	1401.9	1314.0	1681.1	2080.7
7298	1757	2427	2353	2856	3388	3375	3952
Na	Mg	Al	Si	P	S	Cl	Ar
495.8	737.7	577.6	786.6	1011.9	999.6	1251.1	1520.6
4562	1451	1817	1577	1903	2258	2296	2665
K	Ca					Br	Kr
418.9	589.8					1139.9	1350.8
3069	1145					2084	2370

Source: *American Institute of Physics handbook*, McGraw-Hill; $1\,\text{eV} \triangleq 96.485\,\text{kJ mol}^{-1}$.

Table 4.10. Electron affinities of atoms $E_A/\text{kJ mol}^{-1}$

H						He
74 ± 2						18
Li	B	C	N	O	F	
[56]	[29]	121 ± 3	[−26]	141.4 ± 0.5	332.7 ± 0.5	
Na	Al	Si	P	S	Cl	Ar
[75]	[47]	[134]	[75]	200 ± 6	348.6 ± 0.3	−29
K					Br	
79					328	

Source: Principally C. A. McDowell, *Physical chemistry* Vol. III (Eyring, Henderson, and Jost, editors). Academic Press, 1969: $1\,\text{eV} \triangleq 96.485\,\text{kJ mol}^{-1}$.

The sequence of steps is depicted in Fig. 4.2a. When the step $Na + \frac{1}{2}Cl_2 \rightarrow NaCl(s)$ is included we see that a cycle is completed, and the complete diagram is known as a *Born–Haber cycle*. The importance of the Born–Haber cycle arises from the fact that, since H is a state function, *the sum of the enthalpy changes around a cycle must be zero*. Thus starting at $Na(s) + \frac{1}{2}Cl_2(g)$ and going round the cycle depicted brings the system back to

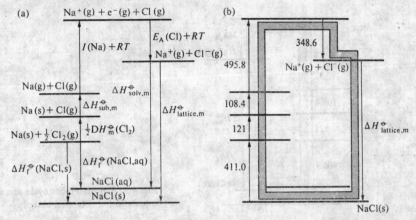

Fig. 4.2. A Born–Haber cycle summarizing the calculation of lattice enthalpy.

the same state; therefore the sum of the enthalpy changes around the cycle must be zero. It follows that if all but one of the enthalpy changes on a cycle are known, the unknown may be found.

Example (Objective 10). Find the lattice enthalpy of NaCl(s) at 298.15 K on the basis of the Born–Haber cycle illustrated in Fig. 4.2a.

- *Method.* Find the enthalpy change associated with the step

$$Na^+(g) + Cl^-(g) \rightarrow NaCl(c), \quad \Delta H_m^{\ominus}(lattice).$$

From the illustration

$$\tfrac{1}{2}DH_m^{\ominus}(Cl—Cl) + \Delta H_m^{\ominus}(Na\ sublimation) + \{I(Na) + RT\}$$
$$- \{E_A(Cl) + RT\} + \Delta H_m^{\ominus}(lattice) - \Delta H_f^{\ominus}(NaCl) = 0$$

Alternatively: *distance up on the left of diagram = distance down on right.* Then insert the following data:

$$\tfrac{1}{2}DH_m^{\ominus}(Cl—Cl) = 242\ kJ\ mol^{-1}\ (Table\ 4.4)$$

$$\Delta H_m^{\ominus}(Na\ sublimation) = 108.4\ kJ\ mol^{-1}\ (Table\ 4.6)$$

$$I(Na) = 495.8\ kJ\ mol^{-1}\ (Table\ 4.9)$$

$$E_A(Cl) = 348.6\ kJ\ mol^{-1}\ (Table\ 4.10)$$

$$\Delta H_f^{\ominus}(NaCl) = -411.0\ kJ\ mol^{-1}\ (Table\ 4.1)$$

- *Answer.* $\Delta H_m^{\ominus}(lattice) = \Delta H_f^{\ominus}(NaCl) + E_A(Cl) - I(Na) - \Delta H_m^{\ominus}(Na\ sublimation)$
$$- \tfrac{1}{2}DH_m^{\ominus}(Cl—Cl)$$
$$= \{-411.0 + 348.6 - 495.8 - 108.4 - 121\}\ kJ\ mol^{-1}$$
$$= -788\ kJ\ mol^{-1}.$$

- *Comment.* The lattice enthalpy would be quoted as $788\ kJ\ mol^{-1}$, which is the enthalpy change required to bring about the disruption NaCl(c) $\rightarrow$ Na$^+$(g) + Cl$^-$(g) at 298.15 K. The internal energy change for the same process is smaller by an amount $2RT$; that is, $783\ kJ\ mol^{-1}$.

Another cycle is obtained when the process NaCl(s) $\rightarrow$ NaCl(aq) is incorporated, with its enthalpy change, the enthalpy of solution of NaCl, Fig.

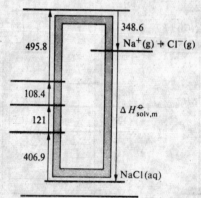

Fig. 4.3. A Born–Haber cycle summarizing the
calculation of solvation enthalpy.

4.3. This cycle gives the *enthalpy of solvation* of the gaseous ions Na^+ and
Cl^- since all the other steps in it can be measured separately. The sum of
enthalpies around the cycle is zero, and so the distance up on the left
($1132.1 \, kJ \, mol^{-1}$) is equal to the distance up on the right ($348.6 \, kJ \, mol^{-1} -$
$\Delta H^{\ominus}_{solv,m}$):

$$\Delta H^{\ominus}_{solv,m} = -(1132.1 - 348.6) \, kJ \, mol^{-1} = -783.5 \, kJ \, mol^{-1}.$$

Hydration enthalpies are listed in Table 4.11 for the alkali halides.

Table 4.11. Standard enthalpies of hydration at infinite dilution,
$-\Delta H^{\ominus}_m (298.15 \, K)/kJ \, mol^{-1}$

	Li^+	Na^+	K^+	Rb^+	Cs^+
F^-	1026	911	828	806	782
Cl^-	884	784	685	664	640
Br^-	856	742	658	637	613
I^-	815	701	617	596	572

Source: Principally J. O'M. Bockris and A. K. N. Reddy, *Modern electrochemistry*, Vol. I,
Plenum.

The solvation enthalpy may be regarded as the sum of the solvation
enthalpies of the $Na^+(g)$ and $Cl^-(g)$ ions separately. This time, however,
we cannot adopt an arbitrary convention about the solvation enthalpy
of some ion, such as the proton, because it would conflict with the choice
already made about its enthalpy of formation: we have already chosen
zero for the standard enthalpy of the step $\frac{1}{2}H_2(g) \rightarrow H^+(aq)$ and we cannot
simultaneously select zero for the step $H^+(g) \rightarrow H^+(aq)$. We are therefore
forced to estimate the actual value of the enthalpy change in the step, and
there is some agreement that its value is about $-1090 \, kJ \, mol^{-1}$. Once
this value has been established the value for $Cl^-(g) \rightarrow Cl^-(aq)$ can be
obtained from data on HCl, and then combined with the value of $\Delta H^{\ominus}_{solv}$
for NaCl as obtained above, to arrive at the value of about $-400 \, kJ \, mol^{-1}$

for $Na^+(g) \rightarrow Na^+(aq)$. The hydration enthalpies listed in Table 4.12 have been established in this way.

Table 4.12. Standard ion hydration energies $-\Delta H_m^{\ominus}(298.15\,K)/kJ\,mol^{-1}$

Li^+	520	F^-	506
Na^+	405	Cl^-	364
K^+	321	Br^-	337
Rb^+	300	I^-	296
Cs^+	277		

Source: Based on J. O'M. Bockris and A. K. N. Reddy, *Modern electrochemistry*, Vol. I, Plenum, with adjustment to $1090\,kJ\,mol^{-1}$ for $H^+(g) \rightarrow H^+(aq)$.

The data in the Table conform with common sense (and with calculation, as we shall see in due course, Chapter 12). Thus small ions of high charge have the largest enthalpy decrease on hydration because they attract the solvent so strongly. The hydration enthalpy of the electron itself, the enthalpy change in the process $e^-(g) \rightarrow e^-(aq)$, is only about $-170\,kJ\,mol^{-1}$; which suggests that when it is trapped in water (e.g. when water is irradiated with high energy radiation) it is spread over a volume about the size of one or two water molecules.

Further reading

Thermodynamics. G. N. Lewis and M. Randall, revised by K. S. Pitzer and L. Brewer; McGraw-Hill, New York, 1961.

Calorimetry. J. M. Sturtevant, in *Techniques of chemistry* (A. Weissberger and B. W. Rossiter, Eds.), V, 347, Wiley-Interscience, New York, 1971.

Energy changes in biochemical reactions. I. Klotz; Academic Press, New York, 1967.

Bioenergetics. A. L. Lehninger; Benjamin, New York, 1965.

The strengths of chemical bonds. T. L. Cottrell; Butterworths, London, 1958.

Tables of physical and chemical constants. G. W. C. Kaye and T. H. Laby; Longmans, London, 1973.

American Institute of Physics handbook. D. E. Gray (ed.); McGraw-Hill, New York, 1972.

Bond energies, ionization potentials, and electron affinities. V. I. Vedeneyev, L. V. Gurvich, V. N. Kondrat'yev, V. A. Mendaredev, and Y. L. Frankevich; Arnold, London, 1966.

Physico-chemical constants of pure organic compounds. J. Timmermans; Elsevier, Amsterdam, 1956.

Selected values of chemical thermodynamic properties. National Bureau of Standards Technical Note 270, 1965–71 (six parts).

Selected values of the thermodynamic properties of metals and alloys. R. Hultgren, R. L. Orr, P. D. Anderson, and K. K. Kelley; Wiley, New York, 1963.

The chemical thermodynamics of organic compounds. D. R. Stull, E. F. Westrum Jr., and G. C. Sinke; Wiley, New York, 1969.

Thermodynamic properties of pure substances. V. P. Glushko (ed.), Nauka, Moscow, 1978 et seq.

Problems

4.1. Which of the following reactions are *exothermic* and which *endothermic*?

(a) $CH_4 + 2O_2 \rightarrow CO_2 + 2H_2O$ $\quad \Delta H_m^{\ominus}(298\,K) = -890\,kJ\,mol^{-1}$

(b) $C_2 + H_2 \rightarrow C_2H_2$ $\qquad\qquad \Delta H_m^{\ominus}(298\,K) = 227\,kJ\,mol^{-1}$

(c) $NaCl(s) \rightarrow NaCl(aq)$ $\qquad \Delta H_m^{\ominus}(298\,K) = 3.9\,kJ\,mol^{-1}$.

4.2. Use the tables of standard molar enthalpies, Tables 4.1 and 4.2, to find the standard molar enthalpies of the following reactions at 25 °C.

(a) $2NO_2(g) \rightarrow N_2O_4(g)$

(b) $NH_3(g) + HCl(g) \rightarrow NH_4Cl(s)$

(c) cyclopropane (g) → propene (g).

(d) $HCl(aq) + NaOH(aq) \rightarrow NaCl(aq) + H_2O(l)$

4.3. A bomb calorimeter rose in temperature by 1.617 K when a current of 3.200 A was passed for 27 s from a 12 V source. What is its heat capacity?

4.4. The same calorimeter was used as in the last Problem. When 0.3212 g of glucose was oxidized at 25 °C under conditions of constant volume the temperature rose 7.793 K. What is (a) the standard molar enthalpy of combustion of glucose, (b) ΔU for the combustion, and (c) the standard molar enthalpy of formation of glucose at 25 °C?

4.5. The standard enthalpy of formation of the sandwich compound bis(benzene)chromium, where benzene is the bread and a chromium atom the meat, was measured in a microcalorimeter (J. A. Connor, H. A. Skinner, and Y. Virmani, *J. chem. Soc. Faraday Trans.* 1, 1218 (1973)). The change in internal energy for the reaction $Cr(C_6H_6)_2(c) \rightarrow Cr(c) + 2C_6H_6(g)$ was found to be $8.0\,kJ\,mol^{-1}$ at 583 K. Find $\Delta H^{\ominus}$ for the reaction, and estimate the standard molar enthalpy of formation of the sandwich compound at this temperature. Take the molar heat capacity of benzene as $140\,J\,K^{-1}\,mol^{-1}$ in its liquid range and at $28\,J\,K^{-1}\,mol^{-1}$ in its vapour phase.

4.6. A 0.727 g sample of the sugar D-ribose was weighed into a calorimeter and then ignited in the presence of excess oxygen. The temperature of the calorimeter rose by 0.910 K from 25 °C. In a separate experiment in the same apparatus, 0.825 g of benzoic acid, which has the accurately known value of $-3251\,kJ\,mol^{-1}$ for the internal energy of combustion, was ignited, and the rise in temperature was 1.940 K. Calculate the molar internal energy of combustion, the molar enthalpy of combustion, the molar enthalpy of formation of D-ribose.

4.7. The standard enthalpy of combustion of naphthalene is $-5157\,kJ\,mol^{-1}$ at 25 °C. What is its standard enthalpy of formation at that temperature?

4.8. The standard enthalpy of decomposition of the yellow solid adduct $NH_3.SO_2 \cdot$ is $40\,kJ\,mol^{-1}$ at 25 °C. What is its standard enthalpy of formation at that temperature?

4.9. By how much do the standard molar internal energy and standard molar enthalpy of combustion of diphenyl differ at (a) room temperature, (b) 99 °C, (c) 101 °C?

4.10. Geophysical conditions are sometimes so extreme that quantities neglected in normal laboratory experiments assume overriding importance. For example, consider the formation of diamond under geophysically typical pressures. The densities of the two allotropes are $2.25\,g\,cm^{-3}$(graphite) and $3.52\,g\,cm^{-3}$(diamond) at the temperature of the environment. By how much does ΔU_m for the transition differ from ΔH_m in a region where the pressure is 500 kbar (1 bar ≈ 1 atm)?

4.11. Several aspects of a hydrocarbon have to be considered when a choice is being made about what fuel to employ. Among them are the amount of heat evolved for every gram of fuel consumed: the advantage of a high molar enthalpy of combustion may be eliminated if a large mass of fuel has to be transported. Use Table 4.2 to find the following information on butane, pentane, and n-octane: (a) the amount of heat evolved at 25 °C when 1 mol of each hydrocarbon is burnt at constant pressure, (b) the heat evolved per gram, (c) the cost per kilojoule of heat (for the last, use your initiative to find the current cost of the hydrocarbons).

4.12. What amount of heat is available per mole and per gram of the three hydrocarbons referred to in the last Problem when combustion takes place in a constant-volume vessel at the same temperature?

4.13. The enthalpy of combustion of glucose is $-2808 \, \text{kJ mol}^{-1}$ at 25 °C. How many grams of glucose do you need to consume (a) to climb a flight of stairs rising through 3 m, (b) to climb a mountain of altitude 3000 m? Assume that 25 per cent of the enthalpy can be converted to useful work.

4.14. Calculate the standard enthalpy of formation of N_2O_5 at 25 °C on the basis of the following data:

$$2NO(g) + O_2(g) \rightarrow 2NO_2(g) \qquad \Delta H_m^{\ominus} = -114.1 \, \text{kJ mol}^{-1}$$
$$4NO_2(g) + O_2(g) \rightarrow 2N_2O_5(g) \qquad \Delta H_m^{\ominus} = -110.2 \, \text{kJ mol}^{-1}$$
$$N_2(g) + O_2(g) \rightarrow 2NO(g) \qquad \Delta H_m^{\ominus} = +180.5 \, \text{kJ mol}^{-1}.$$

4.15. The enthalpy of combustion of graphite at 25 °C is $-393.51 \, \text{kJ mol}^{-1}$ and that of diamond is $-395.41 \, \text{kJ mol}^{-1}$. What is the enthalpy of the graphite $\rightarrow$ diamond phase transition at this temperature?

4.16. Quite often the heat capacity of a substance can be expressed with adequate accuracy by the expression $C(T) = a + bT + c/T^2$, and coefficients for some materials are reported in Table 4.3. Find an expression for the enthalpy of the reaction $a'A + b'B \rightarrow c'C + d'D$ at a temperature T_2 in terms of the enthalpy at T_1 and the coefficients a, b, and c for the heat capacities of the species involved.

4.17. Use the result of the last Problem to predict the standard reaction enthalpy of $2NO_2(g) \rightarrow N_2O_4(g)$ at 100 °C given its value at 25 °C, Problem 4.2. (Estimate a, take $b \approx c \approx 0$.)

4.18. Make use of the data in Tables 4.1, 4.3 and 4.5 to predict the standard molar enthalpy of formation of water at (a) -0.1 °C, (b) 100.1 °C.

4.19. When the coefficients a, b, c, of the heat capacity formula are not available, or when only a rough answer is required, the assumption is made that the heat capacities of all the species are constant over the temperature range of interest. What is the error introduced by this assumption in the case of the enthalpy of formation of water at 99 °C? Calculate $\Delta H^{\ominus}(372 \, \text{K})$ using the room-temperature heat capacities, and also using the a, b, c coefficients.

4.20. Devise an expression for the reaction internal energy at a temperature T_2 in terms of its known value at T_1.

4.21. Samples of D-arabinose and α-D-glucose were burnt to completion in separate experiments in the same constant pressure microcalorimeter at 25 °C. 88 mg of D-arabinose ($M_r = 150.1$) led to a temperature rise of 0.761 K; 102 mg of the glucose ($M_r = 180.2$) led to a rise of 0.881 K. The standard enthalpy of formation of α-D-glucose is $-1274 \, \text{kJ mol}^{-1}$; what are the standard enthalpies of (a) formation, (b) combustion of D-arabinose at 25 °C.

4.22. The standard enthalpy of combustion of sucrose at 25 °C is $-5645\,\mathrm{kJ\,mol^{-1}}$. What is the advantage of complete aerobic oxidation as compared with incomplete anaerobic hydrolysis to lactic acid at this temperature (and under standard conditions). Use the data in Table 4.2.

4.23. A typical sugar cube has a mass of 1.5 g. What is its standard enthalpy of combustion at 25 °C? To what height can you climb assuming 25 per cent of the enthalpy is available for such work?

4.24. Damp clothes can be fatal on a mountain. Suppose the clothing you were wearing had absorbed 1 kg of water, and a cold wind dried it. What heat loss does the body have to make good? How much glucose would have to be consumed to replace that loss? Suppose your body did not make good the heat loss, what would your temperature be at the end of the evaporation? (Assume your heat capacity is the same as that of water.)

4.25. The standard enthalpy of combustion of propane gas at 25 °C is $-2220\,\mathrm{kJ\,mol^{-1}}$ and the standard molar enthalpy of vaporization of liquid propane at this temperature is $15\,\mathrm{kJ\,mol^{-1}}$. What is the standard enthalpy of combustion of the liquid at this temperature? What is the value of $\Delta U^{\ominus}$ for the combustion?

4.26. The constant pressure heat capacities of liquid propane and water are $39.0\,\mathrm{J\,K^{-1}\,mol^{-1}}$ and $75.5\,\mathrm{J\,K^{-1}\,mol^{-1}}$ respectively, and gaseous O_2 and CO_2 have the values $29.3\,\mathrm{J\,K^{-1}\,mol^{-1}}$ and $37.1\,\mathrm{J\,K^{-1}\,mol^{-1}}$ respectively. Combine this information with that in the last Problem to find $\Delta H_m^{\ominus}(308\,\mathrm{K})$ and $\Delta U_m^{\ominus}(308\,\mathrm{K})$ for the combustion of liquid propane.

4.27. The standard enthalpies of hydrogenation of ethene and benzene are $-132\,\mathrm{kJ\,mol^{-1}}$ and $-246\,\mathrm{kJ\,mol^{-1}}$ respectively at 25 °C. Calculate the thermochemical stabilization energy (the 'resonance energy') of benzene at this temperature.

4.28. In an experiment to measure the enthalpy of solution of potassium fluoride in glacial acetic acid (J. Emsley, *J. chem. Soc.* 2702 (1971)), a known weight of the anhydrous salt was added to a known weight of acid in a Dewar flask fitted with a heating coil, stirrer, and thermometer. The heat capacity of the system was determined by supplying a known amount of electricity and monitoring the temperature rise. The experiment was repeated for the salt KF.AcOH, where AcOH stands for acetic acid (ethanoic acid). The following is a reconstruction of an experiment:

KF.	Heat capacity $4.168\,\mathrm{kJ\,K^{-1}\,mol^{-1}}$				
	molality/(mol KF/kg AcOH):	0.194	0.590	0.821	1.208
	$\Delta T/\mathrm{K}$	1.592	4.501	5.909	8.115
KF.AcOH.	Heat capacity $4.203\,\mathrm{kJ\,K^{-1}\,mol^{-1}}$				
	molality/(mol KF/kg AcOH):	0.280	0.504	0.910	1.190
	$\Delta T/\mathrm{K}$	-0.227	-0.432	-0.866	-1.189

Calculate the enthalpies of solvation of the two species at these molalities and at infinite dilution, and find the best straight line of the form $\Delta H/\mathrm{kJ\,mol^{-1}} = a + bm$, where m is the molality. Account for the difference between the enthalpies of the two salts.

4.29. The bond dissociation enthalpy of $H_2(g)$ is $436\,\mathrm{kJ\,mol^{-1}}$ and that of $N_2(g)$ is $945\,\mathrm{kJ\,mol^{-1}}$. What is the atomization enthalpy of ammonia?

4.30. The table below gives the enthalpies of solvation of several alkali metal halides. Complete the table from the information it contains.

	KCl	KI	RbCl	RbI
Lattice energy/kJ mol^{-1}	702.5	637.6	672.0	608.0
$\Delta H^{\ominus}_{soln}$/kJ mol^{-1}	17.2	21.3	17.1	?

4.31. Set up the Born–Haber cycle for determining the enthalpy of solvation of Mg^{2+} ions by water given the following data. Enthalpy of sublimation of Mg(s), 167.2 kJ mol^{-1}; first ionization potential of Mg, 7.646 eV; second ionization potential, 15.035 eV; dissociation enthalpy of $Cl_2(g)$, 241.6 kJ mol^{-1}; electron affinity of Cl(g), 3.78 eV; enthalpy of formation of $MgCl_2(s)$, -639.5 kJ mol^{-1}; enthalpy of solution of $MgCl_2(s)$, -150.5 kJ mol^{-1}; enthalpy of hydration of $Cl^-(g)$, -383.7 kJ mol^{-1}.

4.32. The hydrogen bond between F^- and acetic acid has been found to be exceptionally strong (J. Emsley, *J. chem. Soc.* 2702 (1971)) and an indication of this was encountered in Problem 4.28. In this Problem we set up the Born–Haber cycle which enables the strength of the bond to be ascertained. Use the following data. Lattice energy of KF, 797 kJ mol^{-1}; lattice energy of KF.AcOH, *ca.* 734 kJ mol^{-1}; enthalpy of vaporization of acetic acid, 20.8 kJ mol^{-1}; enthalpy of solution of KF, 35.2 kJ mol^{-1}; enthalpy of solution of KF.AcOH, -3.1 kJ mol^{-1}. Find the hydrogen bond energy between F^- and acetic acid in the gas phase.

5   The Second Law: the concepts

Learning objectives

After careful study of this chapter you should be able to

(1)   State the criterion for the direction of *spontaneous change* (p. 130).

(2)   Explain the basis of the *statistical definition* of entropy (p. 131).

(3)   Derive an expression for the *change of the entropy* on isothermal expansion of a perfect gas (eqn (5.1.2)) by statistical arguments.

(4)   Distinguish between *reversible and irreversible processes* in terms of the generation of entropy (p. 135).

(5)   State the *Second Law of thermodynamics* (p. 136).

(6)   Define the entropy change in terms of processes in a reference system (eqn (5.1.3)).

(7)   Use the reference system to define the entropy change in a system of interest (p. 137).

(8)   Express an entropy change in terms of a reversible transfer of heat (eqn (5.1.4)).

(9)   Calculate the change of entropy when a perfect gas expands isothermally (p. 138).

(10)   Derive the *Clausius inequality* (eqn (5.1.7)) and use it to show that spontaneous processes are accompanied by an increase of entropy (p. 139).

(11)   State the *Second Law of thermodynamics* in terms of the entropy (p. 140).

(12)   Calculate the change of entropy when a system is heated (eqns (5.2.1) and (5.2.2)).

(13)   Calculate the changes of entropy in the surroundings of a system (eqn (5.2.4)).

(14)   Calculate the change of entropy during a phase transition (eqn (5.2.5)).

(15)   Calculate the change of entropy during *irreversible processes* (p. 144).

(16)   Define the *Helmholtz function* and the *Gibbs function* of a system (p. 146) and use them as criteria for the direction of spontaneous change.

(17)   Relate the Helmholtz function to the *maximum amount of work* available from a changing system (p. 146).

(18)   Relate the Gibbs function to the *maximum amount of non-pV work* available from a changing system (p. 149).

(19)   Evaluate the entropy of a system from the thermochemical data (p. 150).

(20) State the *Third Law of thermodynamics* (p. 152), and explain the

experimental evidence for it.

(21) Explain what is meant by the *standard entropy of reaction* (p. 154) and *standard Gibbs function of formation* (p. 155).

Introduction

Some things happen spontaneously, some things don't. A gas expands to fill the available volume; it does not spontaneously contract into something smaller. A hot body cools to the temperature of its surroundings; it does not spontaneously get hotter at their expense. A chemical reaction runs in one direction rather than another: burning diamonds gives hot carbon dioxide; hot carbon dioxide does not spontaneously collapse into diamonds. Some aspect of the world determines the direction of spontaneous changes, the changes things tend to undergo when they are not interfered with by our technology. We can in fact compress a gas into a smaller volume, refrigerators do refrigerate, and we can make diamonds. But none of these things happens spontaneously, they happen only when an outside agent does work.

What determines the direction of spontaneous change? It is not the total energy. The First Law states that energy is conserved in any process, and we cannot disregard it now and say that everything tends towards a state of lowest energy. The energy of the universe (in a thermodynamic sense) is constant: every change occurs without changing its total energy.

Is it, then, the energy of the system itself that tends towards a minimum? Two arguments show that this cannot be the criterion. First, a perfect gas expands spontaneously into a vacuum, yet its internal energy does not change (see pp. 81 and 85). Second, if the energy of a system does happen to decrease during a spontaneous change, the energy of the rest of the universe must increase by the same amount (in order to satisfy the First Law). The increase in energy of the surroundings is just as spontaneous a process as the decrease in energy of the system. Why should we favour one part of the universe over another?

When a change occurs the total energy remains constant, but it is parcelled out in different ways. Can the direction of the spontaneous change be related to some aspect of the *distribution* of energy? We shall see that this is so. Spontaneous changes are always accompanied by a reduction in the 'quality' of energy, in the sense that it is degraded into a more dispersed, chaotic form. Spontaneous, natural changes are simply manifestations of the natural tendency of the universe towards greater chaos.

The role of the distribution of energy can be illustrated by thinking about a ball bouncing on a floor. It is common experience that after each bounce the ball does not rise as high. This is because there are frictional losses in both the rubber of the ball and the material of the floor. Both the ball and the floor get a little hotter on each bounce, and the ordered motion of the ball is reduced. The direction of natural, spontaneous change is towards a state in which the ball is at rest with all its energy degraded into the chaotic thermal motion of the virtually infinite floor, Fig. 5.1.

A ball resting quietly on a warm floor has never been observed to start bouncing. For bouncing to start, something rather special would have to happen. In the first place, some of the thermal motion of the floor would have to accumulate in a single, small object, the ball. This requires a spontaneous localization of energy from the myriad of vibrations of the atoms that constitute the floor into the much smaller number that constitute the ball. Furthermore, whereas thermal motion is chaotic motion, in order for the ball to shoot off upwards the motions of its atoms must all be in the same direction. This localization of uniform motion is such an unlikely event that we can dismiss it as wholly improbable.

Now it is possible to understand, in the present example at least, why the direction of change is as we have described. The energy of the ball is dissipated into the enormous number of modes of random vibration of the floor, which acts as a thermal reservoir; this is a natural, spontaneous process. The reverse process is unnatural because the chaotic distribution of energy is extremely unlikely to coordinate itself into local uniform motion.

This is the signpost for the direction of spontaneous change: *we look for the direction of change that leads to chaotic dispersal of the total energy.*

Although this criterion has been introduced on the basis of a single example, it can easily be developed to account for the observed directions of spontaneous change in the three examples of the opening paragraph. A gas does not spontaneously contract, because in order to do so the motion of the molecules would have to take them all simultaneously into the same region of the container, and the probability of this occurring is so small as to be virtually zero. A cool object does not spontaneously

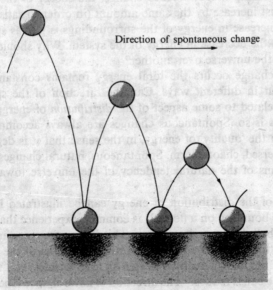

Direction of spontaneous change

Fig. 5.1. The direction of spontaneous change for a ball bouncing on a floor.

become warmer than its surroundings because it is extremely improbable that the jostling of randomly vibrating atoms will lead to the accumulation in it of thermal motion. Burnt diamonds turn into carbon dioxide because the energy locked in the highly localized, small diamonds is carried away in the highly unlocalized gas molecules and dissipated into the rest of the universe as hot molecules collide with cooler ones.

Our experience of everyday phenomena has led to the idea that chaotic dispersal determines direction. In order to develop the idea we must make it quantitative.

5.1 Measuring dispersal: the entropy

The First Law led to the introduction of the internal energy, a property of state. Since the internal energy of an isolated system is constant, we can regard U as being a function that lets us assess whether a proposed change is feasible: only those changes occur for which the total energy of the isolated system (the 'universe') remains constant. The Second Law also leads to a property, the *entropy*, which lets us assess whether a state is accessible from another by a *spontaneous* change: the entropy of the universe is always greater after the occurrence of a spontaneous change. The First Law uses the internal energy to identify the *permissible* changes: the Second Law uses the entropy to identify the *natural* changes among these permissible changes.

One way of introducing the entropy develops the view that the extent of the dispersal of energy can be calculated: this leads to the *statistical* definition of entropy. Another develops the view that the dispersal can be related to the amount of heat involved in a process: this leads to the *thermodynamic* definition.

The statistical view of the entropy. We concentrate on an isolated, perfect monatomic gas because it has the important simplifying feature that the only energy to consider is the chaotic, kinetic energy of translation of the atoms. Wherever an atom happens to be, there kinetic energy will be found. By thinking about the dispersal of the atoms, we automatically think about the dispersal of the energy.

Figure 5.2 shows two connected vessels, one having a volume V_i and the other chosen so that the combined volume is V_f. They are isolated from the rest of the universe. The direction of spontaneous change when the partition is opened is from an arrangement where the gas is initially entirely within V_i to one where it fills V_f. The entropy will be defined so that its value for the final state is greater than that for the less dispersed initial state.

Consider the case when the system is able to explore all its states. In the case of the isolated perfect gas, exploring its states means allowing the atoms to go everywhere. In some cases they will be spread uniformly throughout the volume V_f; in others they will be accumulated in various regions; in some they will be spread uniformly through the volume V_i.

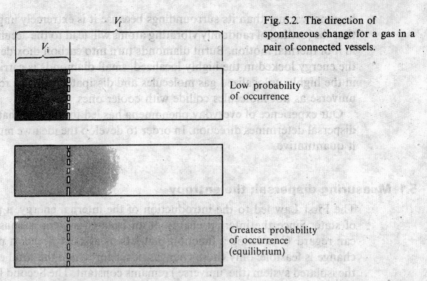

Fig. 5.2. The direction of spontaneous change for a gas in a pair of connected vessels.

Low probability of occurrence

Greatest probability of occurrence (equilibrium)

Intuitively it is obvious that the state in which the atoms are spread uniformly throughout V_f is much more likely to occur than any other, and in particular much more probable than the state in which all the atoms are within the smaller region V_i. Therefore we can say that *the direction of spontaneous change is away from a state with a low intrinsic probability of occurring and towards one of greater intrinsic probability.* Alternatively, we can say that *the equilibrium state is the most probable state*, it being understood that the system is able to explore all its available states.

The probabilities can be expressed quantitatively. Concentrate for the moment on one atom. The probability that it is actually in V_i as a result of its explorations is proportional to V_i. Denoting this probability by $w(V_i)$ we can write

probability of one atom being in V_i, $w(V_i) = \tau V_i$

where τ (tau) is some constant. The probability that two perfect gas atoms are in the same volume is the product of probabilities that each is there separately: this is $w(V_i)^2$. (This point is illustrated in Fig. 5.3.) If the system contains N atoms, the probability $W(V_i)$ that every one of them is in the same volume V_i is the product of the independent, individual probabilities:

probability that N atoms are in V_i, $W(V_i) = w(V_i)^N = \tau^N V_i^N$.

The same calculation repeated for the volume V_f gives

probability that N atoms are in V_f, $W(V_f) = \tau^N V_f^N$.

Since V_f is larger than V_i, it follows that the probability $W(V_f)$ is larger than $W(V_i)$.

It is worth pausing to consider the stage we have reached. We have inferred that systems tend to change spontaneously in the direction corresponding to the greater chaotic dispersal of energy. We have also

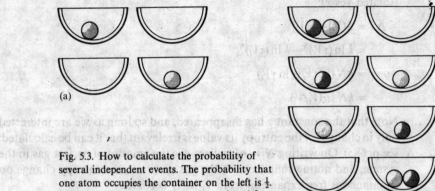

(a)

(b)

Fig. 5.3. How to calculate the probability of several independent events. The probability that one atom occupies the container on the left is $\frac{1}{2}$. The probability that two occupy it is $\frac{1}{4} = (\frac{1}{2})^2$.

established, in the case of an isolated sample of a perfect gas, that the direction of spontaneous change is from a state of low intrinsic probability to one of higher intrinsic probability. We are seeking a property, the entropy S, which indicates the direction of spontaneous change. Since we have obtained a measure of the probability, $W(V)$, it is tempting to set $S = W$, for then we could go on to say that systems have a natural tendency to change in the direction of greater entropy, and evaluate S by equating it to W.

The simple idea of identifying entropy and intrinsic probability would lead, however, to a serious difficulty: the magnitude of the entropy would then depend on the *power* of the amount of material present, and would not be *linearly* proportional to the amount, as is the case for the other extensive properties. For example, because W is proportional to V^N, doubling the amount of material in the given volume ($N \rightarrow 2N$) squares the value of W. This would lead to severe complications if we were trying to talk about both the internal energy and the entropy of a changing system.

Fortunately there is an easy way out of the difficulty. Instead of equating S to W we set it proportional to its logarithm:

$(5.1.1)°$ $\qquad S(V) \propto \ln W(V), \text{ or } S(V) = k \ln W(V)$

where k is some constant (which for historical reasons has been chosen to be Boltzmann's constant). This turns the entropy into an extensive property of the usual kind. For example, if the amount of matter changes from N to $2N$, W changes to W^2, and the entropy changes to

$$S(V) = k \ln W(V)^2 = 2k \ln W(V),$$

and the entropy doubles when the amount of material in the system is doubled. Furthermore, taking logarithms does not upset the relative order of quantities: if $W < W'$, $\ln W < \ln W'$. This means that as W increases so too does S, and so the present definition preserves the idea of increasing entropy as the indicator of the direction of spontaneous change.

We can now calculate the amount by which the entropy of the universe changes when a perfect gas undergoes free expansion from V_i to V_f in an isolated vessel:

$$\Delta S = S(V_f) - S(V_i)$$

$$= k \ln (\tau V_f)^N - k \ln (\tau V_i)^N$$

$$= kN (\ln \tau V_f - \ln \tau V_i)$$

$$= kN \ln (V_f/V_i).$$

Note that the constant τ has disappeared, and so long as we are interested only in changes of the entropy its value is irrelevant (but it can be calculated: see p. 685). On writing $N = nL$, where n is the amount of perfect gas in the system, and noting that $Lk = R$, the gas constant, the entropy change on spontaneous free expansion from V_i to V_f becomes

$(5.1.2)°$ $$\Delta S = nR \ln (V_f/V_i),$$

a positive quantity if $V_f > V_i$.

Example (Objective 3). A sample of hydrogen is confined to a cylinder fitted with a piston of 50 cm^2 cross-section. It occupies 500 cm^3 and at 25 °C it exerts a pressure of 2 atm. What is the change of entropy of the gas when the piston is withdrawn isothermally through 10 cm?

- *Method.* We assume perfect behaviour and use eqn (5.1.2). This requires that we know the amount of gas, and so we use $n = pV/RT$.

- *Answer.* $n = (2 \text{ atm}) \times (0.500 \text{ dm}^3)/(0.0821 \text{ dm}^3 \text{ atm K}^{-1} \text{ mol}^{-1}) \times (298.15 \text{ K})$

    $= 0.049$ mol.

    From eqn (5.1.2):

    $\Delta S = (0.0409 \text{ mol}) \times (8.314 \text{ J K}^{-1} \text{ mol}^{-1}) \ln (1000 \text{ cm}^3/500 \text{ cm}^3)$

    $= (0.340 \text{ J K}^{-1}) \ln 2 = 0.236 \text{ J K}^{-1}.$

- *Comment.* Note the units for entropy. Sometimes entropies are quoted in 'entropy units', e.u., which means cal/deg/mol. These are not part of SI: 1 e.u. $\triangleq 4.184 \text{ J K}^{-1}$ mol^{-1}. It is helpful to remember that the dimensions of S_m, molar entropy, are the same as those both of R and of molar heat capacity.

The restrictions of the calculation that led to eqn (5.1.2) should be appreciated. In the first place it applies only to an isolated, perfect gas, because the chaotic dispersal of energy has been identified with the spatial dispersal of the atoms. This is valid only when the energy can be ascribed to individual particles. The calculation is much more difficult when the constituent particles of a system interact with each other, because then their energies depend on their separations. Complications also arise when the particles can rotate and vibrate. In the second place, the calculation applies only to changes occurring at constant temperature (the temperature of a perfect gas is unchanged by free expansion, p. 85): the same amount of energy is dispersed over a greater volume by the expansion. If the temperature were to change, a different amount of energy would be carried around by the atoms at the end of the expansion, and the calculation has not taken that possibility into account. If energy were transferred to the surroundings, the calculation would also have to take that dispersal into account.

The calculation of entropies for systems more complicated than a perfect gas depends on a detailed knowledge of the structure, properties, and interactions of the constituent molecules. For this reason these calculations are postponed until later (Part 2) and we concentrate on another way of dealing with the entropy. Nevertheless we shall relate what follows to what has been established in this section: *entropy measures the chaotic dispersal of energy* and *the natural tendency of spontaneous change is towards states of higher entropy*.

The thermodynamic view of the entropy. The First Law can be discussed without knowing anything about the properties or even the existence of individual atoms and molecules. If the entropy is a true thermodynamic quantity, we ought to be able to discuss it in the same way. In this section we set up a purely thermodynamic definition of entropy, and show that it coincides with the statistical entropy.

The most basic type of spontaneous, irreversible process is the generation of heat in a reservoir as a result of dropping a weight. In such a process the 'quality' of the energy is degraded, in the sense that it becomes less available for doing work. All spontaneous processes are accompanied by this kind of degradation. Furthermore, if we could define the entropy change for this 'reference system', of a falling weight and a thermal reservoir, then we could use it to find the entropy change in any system of interest. This is because we can drive a system back to its initial state by coupling it to the reference system; if the system is restored to its initial state its overall entropy change is zero (entropy will be defined so as to be a state function), and if the restoration is performed in such a way that no more entropy is generated, then all the entropy generated in the original change will have been transferred to the reference system, where we shall know how to measure it.

There are a few preliminary remarks. First, we must draw a distinction between *irreversible* and *reversible* processes. The first are the spontaneous processes that we have been discussing: they cause a degradation in the quality of energy and hence an increase in the entropy of the universe. *Irreversible processes generate entropy*. In contrast, reversible processes are finely balanced changes, where the system is in equilibrium with its environment at every stage. Every infinitesimal step along a reversible path is reversible; reversible processes occur without degrading the quality of energy, without dispersing energy chaotically, and therefore without increasing the entropy of the universe. *Reversible processes do not generate entropy* (but they may transfer it from one part of the universe to another).

Next, we need to invoke an observation based on experience. No spontaneous change has ever been reversed by some agency *without* there being, somewhere, a degradation of energy. In other words, while it is certainly possible to restore a temperature difference between two blocks of metal, it can be done only at the expense of degrading energy elsewhere in the universe, for example by running a refrigerator on electricity generated by a falling weight. In particular, the reverse of the reference process, a weight rising at the expense of heat from a reservoir, with no other change elsewhere, has never been observed, Fig. 5.4. This last remark is promoted to a general principle:

Second Law: *No process is possible in which the sole result is the absorption of heat from a reservoir and its conversion into work.*

Note that the law does permit natural processes in which the sole result is the conversion of work into heat, but it is unsymmetrical: it cuts off one side of all possible processes allowed by the First Law.

Now we move on to making all these matters quantitative. We do this by making explicit what is meant by the entropy change in the reference system.

Suppose the falling weight is coupled to the reservoir (for instance, by driving a generator connected to a heater, as in Fig. 5.4), and that when it falls it results in the transfer of an amount of heat $q^{\dagger}$ to the reservoir ($\dagger$ denotes 'reference system'). The degradation of the quality of the energy of the reference system is related to this amount of heat, but we cannot simply set $\Delta S^{\dagger} = q^{\dagger}$. The reason is that it is necessary to take into account the *temperature* of the reservoir: the energy is degraded more completely if it is transferred directly to a cold reservoir than if it is transferred to a hot one. For instance, in the latter case we could go on to extract work from an engine operated by a flow of $q^{\dagger}$ from the hot reservoir to the cold, but direct transfer to the cold eliminates this possibility. We want there to be less entropy produced when the heat is transferred to a hot reservoir than when it is transferred to one that is cold. The simplest way of taking the temperature into account is to define the entropy change of the reference system as

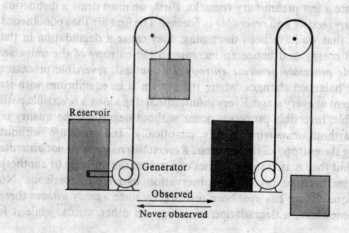

Fig. 5.4. The fundamental spontaneous process.

Reservoir

Generator

Observed

Never observed

(5.1.3)

$$\Delta S^\dagger \overset{\text{def}}{=} q^\dagger / T^\dagger$$

where $T^\dagger$ is the reservoir's temperature (which, in common with the temperature of any reservoir, remains constant however much heat is transferred). In the case of an infinitesimal change, the corresponding expression is:

(5.1.3a)

$$dS^\dagger = dq^\dagger / T^\dagger$$

The definition of entropy often seems both mysterious and arbitrary. If it does, it is helpful to recall that it has been introduced as a measure of the extent to which a high quality form of energy (a weight at some elevation) has been degraded into a low quality form (the thermal motion of a reservoir), taking account of the temperature of the reservoir and hence its ability to supply energy down some other temperature gradient. Big changes of entropy occur when there is a lot of degradation, when a lot of thermal motion is generated at low temperature. Note, too, how the definition of entropy depends on observations in the *surroundings* (as modelled by the standard system): this mirrors the standpoint of the definition of heat and work for the First Law, which were distinguished by observations in the surroundings.

The next step is to see how the definition of ΔS in the reference system can be used to find the ΔS of any system undergoing any change. There are two basic points. First, we can use the standard system as both a thermal reservoir and a source of work to restore the system of interest to its initial state. Second, if this restoration is carried out reversibly, there is no net entropy production, only transfer between the two coupled systems.

These two points are used as follows. If the change in the system of interest produces an amount of entropy dS^{sys}, when it is restored the change is $-dS^{sys}$. In the reversible restoration step an amount of heat $dq_{rev}^\dagger$ enters the standard system, and so its entropy change is $dq_{rev}^\dagger / T^\dagger$. The total entropy change for the restoration step is zero because it is carried out reversibly:

$$(-dS^{sys}) + (dq_{rev}^\dagger / T^\dagger) = 0.$$

This expression can be rearranged to $dS^{sys} = dq_{rev}^\dagger / T^\dagger$, but there are two useful simplifications arising from the reversibility of the restoration step. First, the temperature of the reference system has to be the same as that of the system of interest, $T^\dagger = T^{sys}$. Therefore $dS^{sys} = dq_{rev}^\dagger / T^{sys}$. Second, $dq_{rev}^\dagger$ is the heat transferred *to* the reference system during the restoration; that heat comes reversibly from the system of interest, and so it is equal to the heat transferred *from* the system of interest during the restoration, $-dq_{rev}^{sys}$ (restoration). But, on account of the reversibility, this is equal to $+dq_{rev}^{sys}$ for the same change in the original forward direction. Collecting these remarks leads to

(5.1.4) $dS^{sys} = dq_{rev}^{sys}/T^{sys}.$

That is, we can determine the entropy change when a system changes from state i to state f by finding the amount of heat it is necessary to transfer in order to take it along a *reversible* path between the same two states. For a measurable change the entropy change is the sum (integral) of the infinitesimal changes:

(5.1.5) $\Delta S^{sys} = S_f - S_i = \int_i^f dq_{rev}^{sys}/T^{sys}.$

All this may seem remote from the statistical definition of entropy, and so it is time to check that it leads to the same value of the entropy change for the free expansion of a perfect gas, eqn (5.1.2).

When an isolated perfect gas expands freely its state changes from (V_i, T) to (V_f, T). In order to apply eqn (5.1.5) we have to find some *reversible* path between the same two states, and calculate dq_{rev}^{sys} for each segment of the path. Since the expansion is isothermal the temperature T^{sys} remains constant (at T); therefore,

$$\Delta S^{sys} = \int_i^f dq_{rev}^{sys}/T^{sys} = (1/T)\int_i^f dq_{rev}^{sys} = q_{rev}^{sys}/T.$$

The work of Chapter 2 led to the result that the amount of heat transferred to a perfect gas expanding reversibly and isothermally from V_i to V_f is

$$q_{rev}^{sys} = nRT\ln(V_f/V_i).$$

(Table 3.3 on p. 99). Therefore the entropy change accompanying the change of state is

$$\Delta S^{sys} = nR\ln(V_f/V_i),$$

in accord with the statistical result, eqn (5.1.2).

The crucial point to note is that *whatever* the type of change, the entropy change of the system can be calculated from eqn (5.1.5) by finding some *reversible* path between the same two specified initial and final states.

Natural events. Consider a system in thermal and mechanical contact with a reference system at the same temperature, but not necessarily in mechanical balance. Any change will be accompanied by a change of entropy dS^{sys} in the system and $dS^\dagger$ in the standard system; but the overall change of entropy will be greater than zero because there is dissipation:

(5.1.6) $dS^{sys} + dS^\dagger \geqslant 0$, or $dS^{sys} \geqslant -dS^\dagger$.

(The equality applies if we happen to arrange matters so that the two systems are in equilibrium.) Since $dS^\dagger = dq^\dagger/T^\dagger$ (by definition), $dq^\dagger = -dq^{sys}$ (because the heat entering the standard system comes from the other system), and $T^\dagger = T^{sys}$ (by arrangement); it follows that for *any* change

(5.1.7)

$$dS^{sys} \geqslant dq^{sys}/T^{sys}$$

which is the *Clausius inequality*.

The Clausius inequality has the following important consequence for changes taking place in isolated systems. Since then $dq^{sys} = 0$, for any change $dS^{sys} \geqslant 0$. A measurable change is the sum of infinitesimal changes, and so it follows that in an isolated system $\Delta S^{sys} \geqslant 0$. Furthermore, a universe (in the thermodynamic sense) is itself an isolated system, and so for any change inside a universe

(5.1.8)

$$\Delta S \geqslant 0,$$

where S is the total entropy of all parts of the universe. If there are changes going on inside an isolated system (including the universe) they must be spontaneous, natural processes (because our technology cannot penetrate into an isolated system). This inequality therefore shows that these spontaneous processes must lead to an increase in the entropy of the universe. Only when the universe is at equilibrium, so that every change is thermodynamically reversible, does the equality $\Delta S = 0$ apply.

A further consequence of the Clausius inequality is that *the integral of* dq/T *for a system taken round any closed cycle cannot exceed zero*. This remarkable result is proved as follows. Consider a cycle in which a system changes from a state i to a state f along an irreversible path and is then brought back to state i along a reversible path, Fig. 5.5. At each stage of the irreversible path an amount of heat dq is transferred to the system at a temperature T (which may be different at different stages). At each stage of the reversible path an amount dq_{rev} is transferred. The integral of dq/T round the cycle is therefore

$$\int_{cycle} dq/T = \int_i^f dq/T + \int_f^i dq_{rev}/T.$$

The limits on the second integral may be reversed (the path is reversible), and so

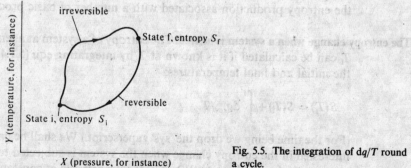

Fig. 5.5. The integration of dq/T round a cycle.

$$\int_{\text{cycle}} dq/T = \int_i^f dq/T - \int_i^f dq_{rev}/T = \int_i^f dq/T - \int_i^f dS$$

where dS is the entropy change dq_{rev}/T. According to the Clausius inequality, which now comes into play, $dq/T \leqslant dS$, implying

$$\int_i^f dq/T \leqslant \int_i^f dS.$$

It follows that the integral round the cycle is

(5.1.9) $$\int_{\text{cycle}} dq/T \leqslant 0,$$

as was to be proved. We shall make use of this result shortly.

It will be recalled that the First Law can be presented in a variety of forms. One (p. 54) was quite involved, and consisted of a report on direct experience. Another (p. 55, 'the energy of an isolated system is conserved') was much more succinct, but it had the same implications so long as the meaning of energy was comprehended. The Second Law can be expressed similarly. The form on p. 136 is of the 'report on experience' type. We are now in a position to give a succinct form, but one with the same implications so long as we comprehend the meaning of entropy:

> Second Law: *The entropy of an isolated system increases during any natural process.*

5.2 Entropy changes in the universe

By 'universe' we mean the system together with its surroundings. This is an important emphasis, because the entropy of the system itself may *decrease* in the case of a spontaneous change so long as the entropy change in the surroundings compensates for it. For example, living things are highly ordered, relatively low entropy, structures, but they grow and are sustained because their metabolism generates excess entropy in their surroundings. Ultimately we shall wish to discuss entropy changes in complex systems, but as a first step we shall examine how to determine the entropy production associated with a number of basic processes.

The entropy change when a system is heated. The entropy of a system at a temperature T_f can be calculated if it is known at T_i by integrating eqn (5.1.4) between the initial and final temperatures:

$$S(T_f) = S(T_i) + \int_i^f dq_{rev}/T.$$

(For the time being we drop the 'sys' superscript.) We shall be particularly interested in the entropy change when the system is subjected to constant pressure during the heating. When heat is transferred at constant pressure

dq_{rev} is written $(dq_{rev})_p$. But $(dq)_p = dH$ (p. 68) so long as the system is doing no non-pV work; therefore

$$S(T_f) = S(T_i) + \int_i^f dH/T \qquad \text{(const. } p; \text{ no non-}pV \text{ work).}$$

The 'rev' subscript has been dropped because H is a state function and has the same value independent of path. The integral over dH/T can be evaluated provided we know how the enthalpy increase is related to the temperature increase; but from Chapter 3 (p. 83) we know that $dH = C_p dT$ when the pressure is constant. It follows that

(5.2.1)
$$S(T_f) = S(T_i) + \int_{T_i}^{T_f} (C_p/T) dT \qquad \text{(const. } p).$$

This expression enables the entropy at any temperature to be obtained from measurements of the heat capacity in the temperature range of interest. In the case of many gases, C_p is fairly independent of temperature over moderate ranges, and so then

$$S(T_f) \approx S(T_i) + C_p \int_{T_i}^{T_f} (1/T) dT$$

(5.2.2)
$$\approx S(T_i) + C_p \ln(T_f/T_i) \qquad \text{(const. } p).$$

Example (Objectives 9 and 12). Calculate the entropy change when argon at 25 °C and 1 atm pressure in a container of volume 500 cm³ is allowed to expand to 1000 cm³ and simultaneously is heated to a temperature of 100 °C.

• *Method.* The initial state is (500 cm³, 298.15 K); the final state is (1000 cm³, 373.15 K). The sample can be taken from the initial to the required final state in two steps. The first is an isothermal expansion from V_i to V_f; this is followed by heating from T_i to T_f at constant volume. The entropy change in the first step is

$$\Delta S = nR \ln(V_f/V_i).$$

The entropy change in the second step can be obtained as follows:

$$\Delta S = \int_i^f (dq_{rev})_V/T = \int_i^f dU/T = \int_{T_i}^{T_f} C_V dT/T$$

$$= C_V \ln(T_f/T_i).$$

(The same steps that led to eqn (5.2.2) have been used, with the change confined to constant volume instead of constant pressure.) The total entropy change for the overall change is the sum

$$\Delta S = nR \ln(V_f/V_i) + nC_{V,m} \ln(T_f/T_i).$$

We find n from $n = p_i V_i/RT_i$. $C_{V,m} = 12.48$ J K⁻¹ mol⁻¹ (Table 2.2).

• *Answer.* $n = (1 \text{ atm}) \times (0.500 \text{ dm}^3)/(0.0821 \text{ atm dm}^3 \text{ K}^{-1} \text{ mol}^{-1}) \times (298.15 \text{ K})$

$$= 0.020 \text{ mol.}$$

$$\Delta S = (0.020 \text{ mol}) \times (8.314 \text{ J K}^{-1} \text{ mol}^{-1}) \ln 2 + (0.020 \text{ mol}) \times$$

$$\times (12.48 \text{ J K}^{-1} \text{ mol}^{-1}) \times \ln(373.15 \text{ K}/298.15 \text{ K})$$

$$= 0.115 \text{ J K}^{-1} + 0.056 \text{ J K}^{-1} = 0.171 \text{ J K}^{-1}.$$

● *Comment.* Notice that we have not needed to know how the expansion and heating is actually carried out: S is a state function, and therefore independent of the processes used to change the system from one state to another. For the same reason we have been able to break the overall change $(V_i, T_i) \rightarrow (V_f, T_f)$ into two steps, both of which can be dealt with very simply. Since S is a state function we are at liberty to choose the simplest path or set of paths from the initial to the final state.

Entropy changes in the surroundings. The surroundings of a system are just a huge reservoir, and so themselves constitute a reference system. It follows that we can use the definition eqn (5.1.3a) to write down the entropy change that occurs when an amount of heat dq^{surr} enters them:

(5.2.3) $dS^{surr} = dq^{surr}/T^{surr}$.

The same change of entropy occurs however the transfer of heat takes place, reversibly or irreversibly (because the definition, eqn (5.1.3a), is for a general process) so long as the reservoir remains at internal equilibrium. Furthermore, the temperature of the surroundings does not change however much heat is injected. Therefore the last expression can be integrated with T^{surr} a constant,

(5.2.4) $\Delta S^{surr} = q^{surr}/T^{surr}$,

where q^{surr} is the total heat injected into the surroundings by the system.
   This argument makes it very simple to calculate the changes in entropy of the surroundings for any process. For instance, in *any* adiabatic change, since $q^{surr} = 0$, the entropy change of the surroundings is zero. When a chemical reaction occurs at constant pressure with an enthalpy change ΔH, the entropy change of the surroundings is $-\Delta H/T^{surr}$.

The entropy change arising from a phase transition. Some change of order occurs when a substance freezes or boils, and so the statistical interpretation of the entropy suggests that such phase transitions will be accompanied by a change of entropy. Since a liquid is more chaotic than a solid we can anticipate that a substance's entropy increases when it melts and when it boils. The thermodynamic definition of entropy lets us calculate the entropy change very simply.
   Under the conditions of constant pressure the 'latent heat' of a phase transition is its enthalpy of transition, ΔH_t (p. 116). At the temperature of the transition, T_t, the system is in equilibrium (for instance, at the boiling point liquid and vapour are in equilibrium and the system's temperature is the same as its surroundings), and so heat is transferred reversibly from the surroundings. We can therefore use eqn (5.1.3) to deduce immediately that the *entropy of transition* is

(5.2.5) $\Delta S^{sys} = \Delta H_t/T_t$.

Since the transition is an equilibrium process the overall entropy generation is zero; consequently the entropy change in the surroundings (which are at the same temperature T_t) is $-\Delta H_t/T_t$.

Some experimental transition entropies are listed in Table 5.1. Both melting (fusion) and boiling (evaporation) are endothermic processes ($\Delta H_t > 0$), and so each is accompanied by an increase in the system's entropy, in accord with the statistical interpretation.

Table 5.1. Entropies and temperatures (in brackets) of phase transitions at 1 atm, $\Delta S_m^{\ominus}/J \ K^{-1} \ mol^{-1}$

	Melting (T_f/K)	Boiling (T_b/K)
He	6.0 (3.5)	19.9 (4.22)
Ar	14.17 (83.81)	74.53 (87.29)
H_2	8.38 (13.96)	44.96 (20.38)
N_2	11.39 (63.15)	75.22 (77.35)
O_2	8.17 (54.39)	75.63 (90.18)
Cl_2	37.22 (172.12)	85.38 (239.05)
Br_2	39.76 (265.90)	88.61 (332.35)
H_2O	22.00 (273.15)	108.95 (373.15)
H_2S	12.67 (187.61)	87.75 (212.0)
NH_3	28.93 (195.40)	97.41 (239.73)
CH_3OH	18.03 (175.25)	104.59 (337.22)
CH_3COOH	40.4 (289.76)	61.9 (391.45)
C_6H_6	38.00 (278.65)	87.19 (353.25)

Source: *American Institute of Physics handbook*, McGraw-Hill, using $\Delta S_m = \Delta H_m/T$.

Example (Objective 14). The following table gives the molar enthalpies of vaporization and the boiling points of several simple liquids. Calculate the entropy of vaporization for each one.

	$\Delta H_{vap,m}/kJ \ mol^{-1}$	$T_b/^{\circ}C$
methane	8.18	−161.5
carbon tetrachloride	30.00	76.7
cyclohexane	30.1	80.7
benzene	30.8	80.1
hydrogen sulphide	18.7	−60.4
water	40.7	100.0

- *Method.* The entropy of vaporization is given by eqn (5.2.5): but remember that boiling points must be expressed in kelvin ($t + 273.15$ K).

- *Answer.* $\Delta S_{vap,m}$(methane) $= (8.18 \times 10^3 \ J \ mol^{-1})/(111.7 \ K) = 73.2 \ J \ K^{-1} \ mol^{-1}$.

Likewise for the others:

carbon tetrachloride (85.8 J K^{-1} mol^{-1}), cyclohexane (85.1 J K^{-1} mol^{-1}), benzene (87.2 J K^{-1} mol^{-1}), hydrogen sulphide (87.9 J K^{-1} mol^{-1}), water (109.1 J K^{-1} mol^{-1}).

- *Comment.* A wide range of liquids give approximately the same molar entropy of vaporization (about 85 J K^{-1} mol^{-1}): this is *Trouton's rule*. This is because a comparable amount of disorder is generated when 1 mol of any liquid evaporates. Some liquids, however, deviate sharply from the rule. This is often because the liquids have structure, and so a greater amount of disorder is introduced when they evaporate. An example is water, where the relatively large entropy change reflects the presence of hydrogen bonds between the molecules. These bonds tend to organize the molecules in the liquid so that they are less random than, for example, molecules in liquid hydrogen sulphide.

How to determine the entropy change in an irreversible process. We have already emphasized that as entropy is a state function, any change in the entropy of a system is independent of the path between the specified initial and final states. In contrast, the overall entropy change (of the system and its surroundings) will depend on the path, for if the change is reversible the overall entropy change is zero, but if it is irreversible the overall change is non-zero (and positive). Our concern here is to see how to calculate the overall entropy change for various processes.

Consider the case of the expansion of a perfect gas from (V_i, T) to (V_f, T). The entropy change of the system is $nR \ln(V_f/V_i)$ irrespective of whether the expansion is reversible or irreversible. If it is reversible, the entropy change of the surroundings (which are thermally and mechanically balanced against the system) must compensate to give a net entropy change of zero. Therefore in this case the entropy of the surroundings changes by $-nR \ln(V_f/V_i)$. On the other hand, suppose the expansion occurs freely, and therefore irreversibly. In this case, since no work is done, and the internal energy remains constant, there is no transfer of heat between the system and its surroundings. Consequently the entropy of the surroundings is unchanged, the system's entropy change is still $nR \ln(V_f/V_i)$, and so the overall entropy change is $nR \ln(V_f/V_i)$: a positive amount of entropy has been generated in the universe. This amount of entropy, when multiplied by T, is also equal to the quantity of work that failed to materialize.

Another example of entropy generation in an irreversible process is provided by the flow of heat from a hot body to a cold. Suppose the two temperatures are T_h and T_c: what is the overall change in entropy when an infinitesimal amount of heat $|dq|$ passes from T_h to T_c? We can guess that entropy is generated, because the flow is spontaneous, and we can use the reference system to find its quantitative value.

Instead of allowing the heat to flow directly from one block to another, the same result can be achieved in three steps. First $|dq|$ is transferred reversibly to a reference system at T_h. There is no overall entropy production in this reversible step. Then $|dq|$ is allowed to leak from the reference system to another at a temperature T_c. From the definition of entropy, the change in entropy of the hot reference system is $-|dq|/T_h$ and the change in entropy of the cold reference system is $|dq|/T_c$. The total entropy produced in this step is $|dq|\{(1/T_c)-(1/T_h)\}$. Finally, the heat is transferred reversibly to the cold block (at a temperature T_c). There is no generation of entropy in this step, only transfer. The overall change in entropy of the universe is therefore

$$dS = |dq|\left(\frac{1}{T_c} - \frac{1}{T_h}\right).$$

This is a positive quantity because $T_c < T_h$: heat flows spontaneously down a temperature gradient. Since entropy is a state function, the same change of entropy is obtained if the blocks are put in direct contact. Note that if the blocks are finite, the spontaneous transfer of heat reduces the temperature gradient as it proceeds, and so the total entropy change has to allow for changing T_c and T_h; this is a more involved calculation than the one just done, but involves nothing different in principle (see Problem 5.27).

Where is the new entropy? The two reference systems are back in their original states (dq went into the hot one, then left it for the cold one, and left that for the cold system). Therefore, in this case, the whole of the newly generated entropy is in the cold block.

5.3 Concentrating on the system

Entropy is the basic concept for discussing the direction of natural changes. Nevertheless, in order to use it we have to investigate both the system and its surroundings. We have seen that it is always very simple to calculate the entropy change in the surroundings, and now we shall show that it is possible to devise a method of taking the surroundings into account automatically. This focuses attention on the system, and simplifies thermodynamic discussions.

Consider a system in thermal equilibrium with its surroundings (so that $T^{sys} = T^{surr} = T$); then the Clausius inequality, eqn (5.1.7), reads

(5.3.1) $\qquad dS^{sys} - dq^{sys}/T \geq 0.$

The importance of this inequality is that it expresses the criterion for natural, spontaneous change solely in terms of the properties of the system. *From now on the superscript denoting the system will be dropped, and everything we do will relate to the system* (unless explicitly stated otherwise). This convention greatly simplifies the notation.

The last equation can be developed in two ways. First take the case where the heat is lost from a constant volume system. Then $(dq)_V$ can be identified with dU if there is no non-pV work involved. Putting this into eqn (5.3.1) gives

(5.3.2) $\qquad dS - dU/T \geq 0, \quad \text{or} \quad TdS \geq dU \qquad \text{(const. } V, \text{ no non-}pV \text{ work).}$

At constant energy ($dU = 0$) and entropy ($dS = 0$) respectively, these become

(5.3.2a) $\qquad (dS)_{U,V} \geq 0 \text{ and } (dU)_{S,V} \leq 0$

as the changes that accompany natural processes. Remember that dS is the change in entropy of the system, dU the change of its internal energy, T its temperature, and V its volume. When the heat is lost from the system under conditions of constant pressure and when no non-pV work is involved, the only modification is to identify $(dq)_p$ with dH, the change of enthalpy of the system. Then

(5.3.3) $dS - dH/T \geqslant 0,$ or $T\,dS \geqslant dH$ (const. p, no non-pV work)

are the appropriate criteria at constant pressure. At constant entropy and pressure this criterion becomes

(5.3.3a) $(dH)_{S,p} \leqslant 0.$

These expressions can be simplified even more by introducing two new thermodynamic functions. These are the *Helmholtz function* and the *Gibbs function* (or *free energy*). They are defined as follows:

(5.3.4)   Helmholtz function: $A = U - TS.$

(5.3.5)   Gibbs function: $G = H - TS.$

All the symbols in these definitions refer to the *system*.

In order to see how G and A simplify the expressions in eqns (5.3.2) and (5.3.3), consider what happens to them when the state of a system changes at constant temperature:

$$dA = dU - T\,dS \qquad \text{(const. } T\text{)}$$

$$dG = dH - T\,dS \qquad \text{(const. } T\text{)}.$$

Now introduce $T\,dS \geqslant dU$ (const. V) into the first, and $T\,dS \geqslant dH$ (at const. p) into the second; then the criteria for a change being spontaneous are

(5.3.6) $dA \leqslant 0$ (const. V, T; no non-pV work), or $(dA)_{T,V} \leqslant 0$

(5.3.7) $dG \leqslant 0$ (const. p, T; no non-pV work), or $(dG)_{T,p} \leqslant 0.$

These inequalities are the most important conclusions from thermodynamics in chemistry. They will be developed in subsequent sections.

Some remarks on the Helmholtz function. When the changes in a system are constrained to occur under conditions of constant temperature and volume eqn (5.3.6) determines whether they can occur spontaneously. A change corresponding to a decrease in A can occur spontaneously. Systems tend to move naturally towards states of lower A, and the criterion of equilibrium is $(dA)_{T,V} = 0$.

Sometimes the expression $dA = dU - T\,dS$ is interpreted as follows. A negative value of dA is favoured by a negative value of dU and a positive value of $T\,dS$. This suggests that the tendency of A to move towards lower values can be ascribed to a tendency of the system to move towards lower values of U and to higher values of S. In other words, a system has a natural tendency towards states of lower internal energy and to states of higher entropy.

This argument is a false interpretation (even though it is a good rule of thumb for remembering the expression for dA). This is because *the basis of the reduction of A is solely a tendency towards states of greater universal entropy.* Systems move spontaneously in one direction or the other solely because that corresponds to an increase of the entropy of the universe, not because one component is moving towards a state of lower energy. The form of dA gives the impression that systems favour lower energies, but the impression is misleading. ΔS is the system's entropy change, and $-\Delta U/T$ is that of the surroundings: the total tends to a maximum.

It turns out that A carries a greater significance than being simply a signpost. If we know the value of ΔA for a change we can also state the maximum amount of work the system can do. This relation is the reason why A is sometimes called the *maximum work function*, or the *work function* (*Arbeit* is the German for work, hence A).

We shall now demonstrate the connection between ΔA and w_{max}, and at the same time prove that *a system does maximum work when it is working reversibly*. (The latter point was illustrated in Chapter 2 by the perfect gas expansion: here we prove its universal validity.) The proof depends on combining the Clausius inequality, d$S \geqslant$ dq/T (quantities now refer to the system), or T d$S \geqslant$ dq, with the First Law, d$U =$ d$q +$ dw. This gives d$U \leqslant T$ d$S +$ dw (because T dS is bigger than dq in general), which rearranges to d$w \geqslant$ d$U - T$ dS. The work done *by* the system is $-$dw, and so (remembering that a large negative quantity is algebraically *less* than a more positive quantity)

(5.3.8)
$$-dw \leqslant -dU + T\,dS.$$

It follows from the fact that T, S, and U are all state functions that the amount of work a system can do for a specified change of state cannot exceed some maximum value, $-$d$w_{\text{max}} = T$ d$S -$ dU, and that this work can be obtained only when the path is traversed reversibly, as was to be proved.

Furthermore, when the path is traversed reversibly, T d$S =$ dq_{rev}, and d$w =$ dw_{rev}; and so, for an isothermal change,

$$dA = dU - T\,dS = (dq_{\text{rev}} + dw_{\text{rev}}) - T\,dS = dw_{\text{rev}} = dw_{\text{max}}.$$

That is,

(5.3.9)
$$-dA = -dw_{\text{max}} \quad \text{(const. } T\text{)}.$$

It follows that if we know ΔA for a process, then we also know the maximum amount of work the system can do when it changes between the same two states ($-w_{\text{max}} = -\Delta A$).

In the case of an isothermal change between specified initial and final states

5.3.10) $-w_{max} = -\Delta A = -\Delta U + T\Delta S.$

This shows that in some cases, depending on the sign of $T\Delta S$, not all the change in the internal energy (ΔU) can be extracted as work, but that in others more may be. If the change occurs with a large production of entropy in the system the maximum work available exceeds $-\Delta U$ (because $T\Delta S$ is then positive). The explanation for this apparent paradox is that the system is not isolated, and so heat may flow in and fuel the work if ΔS (of the system) is positive. If ΔS is negative, on the other hand, some heat must flow out of the system in order to lead to an overall increase in the entropy of the universe: not all the change in internal energy is available for doing work, and as a result $-w_{max}$ is less than $-\Delta U$.

Example (Objective 17). When glucose is oxidized to carbon dioxide and water according to the reaction

$$glucose + 6O_2 \rightarrow 6CO_2 + 6H_2O$$

calorimetric measurements give $\Delta U_m = -2810$ kJ mol^{-1} and $\Delta S_m = 182.4$ J K^{-1} mol^{-1}. How much of this energy change can be extracted as heat, and how much can be extracted as work?

● *Method.* The whole of the energy change can be extracted as heat if the sample is burnt in a closed container. The amount available to do work is given by $\Delta A = \Delta U - T\Delta S$.

● *Answer.* The amount of heat available is $q_V = \Delta U_m = -2810$ kJ mol^{-1}. The amount of work available is determined by

$$\Delta A = (-2810 \text{ kJ mol}^{-1}) - (298.15 \text{ K}) \times (182.4 \text{ J K}^{-1} \text{ mol}^{-1})$$
$$= -2864 \text{ kJ mol}^{-1}.$$

Therefore the reversible, slow oxidation of glucose can be used to do 2864 kJ of work.

● *Comment.* The work available is greater than the change in internal energy because the favourable entropy change of the oxidation (partly due to the generation of a large number of small molecules from one big one) can suck in heat from the surroundings and make it available for doing work.

Some remarks on the Gibbs function. The Gibbs function is more common in chemistry than the Helmholtz function. This is because we are usually more interested in the conditions of equilibrium and the direction of chemical change when systems are at constant pressure rather than constant volume.

When we want to know whether a reaction will tend to go in a particular direction (the pressure and temperature being constant) we have to determine ΔG for the reaction:

initial state (reactants) $\rightarrow$ final state (products)

$$\Delta G = G_{products} - G_{reactants}.$$

If this ΔG is negative, then the reaction has a natural tendency to move spontaneously from reactants to products. If ΔG is positive, then the reaction as written will not proceed spontaneously, but the reverse reaction will be the spontaneous one.

The interpretation of reactions as tending to sink down the slope of the Gibbs function until they attain equilibrium at the minimum is the same as for the Helmholtz function. In this case the apparent driving force is a tendency to move towards lower enthalpy and a higher entropy. But the real interpretation is towards maximum entropy of the universe, which can be achieved by maximizing the sum of the system's entropy and the entropy of the surroundings by pumping enthalpy into them.

An illustration of the importance of the Gibbs function in determining the direction of spontaneous chemical change is provided by the existence of endothermic reactions (p. 106). In such reactions $\Delta H > 0$, and so the system rises spontaneously to states of higher enthalpy. The spontaneity implies that ΔG is negative, and for this to be so even though ΔH is positive requires $T\Delta S$ to be positive and larger than ΔH. They are therefore driven by a large increase in entropy of the system, and this entropy change overcomes the negative entropy change in the surroundings that arises when enthalpy is drawn in ($\Delta S^{surr} = -\Delta H/T$).

The analogue of the maximum work interpretation of ΔA can be found for ΔG. In a reversible change dH can be replaced by $dq_{rev} + dw_{rev} + d(pV)$. Therefore at constant temperature,

$$dG = dq_{rev} + dw_{rev} + V\,dp + p\,dV - T\,dS$$
$$= dw_{rev} + V\,dp + p\,dV.$$

Now dw_{rev} is the maximum work of the system, dw_{max}, and it consists of expansion work ($-p\,dV$) and possibly some other kind of work (for example, the work of pumping electrons through a circuit). Replacing dw_{rev} by $-p\,dV + dw_{e,max}$ (where p is the pressure exerted by the system, because the change is occurring reversibly) and imposing the condition of constant pressure (d$p = 0$) leaves the last equation as

(5.3.11) $$dG = dw_{e,max}. \quad \text{(const. } p, T\text{).}$$

This means that under conditions of constant pressure and temperature the change of the Gibbs function between specified states gives the maximum work available from that process *other* than expansion work ($-w_{e,max} = -\Delta G$).

Example (Objective 18). Animals operate under conditions of constant pressure (diving species are an exception), and most of the processes that maintain life are electrical (in a broad sense). How much energy is available for sustaining this type of nervous and muscular activity from 1 mol of glucose?

- *Method.* The amount of non-pV work available is determined from ΔG for the combustion reaction. From experiment $\Delta H = -2808$ kJ mol^{-1}, $\Delta S = 182.4$ J K^{-1} mol^{-1}; we use $\Delta G = \Delta H - T\Delta S$. Use $T = 310.0$ K (blood heat).

● *Answer.* $\Delta G = (-2808 \text{ kJ mol}^{-1}) - (310.0 \text{ K}) \times (182.4 \text{ K}^{-1} \text{ mol}^{-1})$
$$= -2865 \text{ kJ mol}^{-1}.$$

Therefore the digestion of 1 mol (180.2 g) of glucose can do 2865 kJ of non-pV work.

● *Comment.* A 70 kg man would need to do 2.1 kJ of work to climb vertically through 3 m; therefore he would need at least 0.13 g of glucose to complete the task (and in practice significantly more unless he moved infinitely slowly).

5.4 Evaluating the entropy and the Gibbs function

The entropy of a system at a temperature T can be related to its entropy at absolute zero by determining the heat capacity C_p as a function of temperature and calculating the integral in eqn (5.2.2). At some temperatures between 0 and T the material may change its phase and absorb heat in the process. Therefore the entropy of each phase-transition must be added in order to find $S(T)$. For example, if a material melts at T_f and boils at T_b, its entropy at some temperature higher than T_b is given by

(5.4.1)
$$S(T) = S(0) + \int_0^{T_f} (C_p^{\text{solid}}/T)\,dT + \Delta H_{\text{melt}}/T_f + \int_{T_f}^{T_b} (C_p^{\text{liq}}/T)\,dT$$

$$+ \Delta H_{\text{evap}}/T_b + \int_{T_b}^{T} (C_p^{\text{gas}}/T)\,dT.$$

All the quantities involved (except $S(0)$) can be measured calorimetrically, and the integrations can be carried out graphically. This is illustrated in Fig. 5.6: the area under the curve of C_p/T against T is the integral required. Since $dT/T = d \ln T$, an alternative procedure is to plot C_p against $\ln T$ and to determine the area under it up to the temperature of interest. Examples of this procedure are given in the Problems.

One problem of this approach is the difficulty of measuring heat capacity in the vicinity of absolute zero. The normal procedure is to measure it to as low a temperature as possible, and then to extrapolate. There are good theoretical grounds for assuming that the heat capacity varies as T^3 when T is small, and this is the basis of the Debye extrapolation (examined in Part 2). In this method C_p is measured to a low temperature and a curve of the form aT^3 is fitted to it. This determines the value of a, and the expression $C_p = aT^3$ is assumed valid down to absolute zero. The entropy is then evaluated by integration. It may be noted that in the vicinity of absolute zero:

$$S(T) - S(0) = \int_0^T (C_p/T)\,dT = \int_0^T aT^2\,dT$$

$$= \tfrac{1}{3}aT^3 = \tfrac{1}{3}C_p(T)$$

and since heat capacities are very small at low temperatures, only small errors arise from this extrapolation. In the case of metals there is also a contribution to the heat capacity from the electrons: this varies linearly with temperature when T is small.

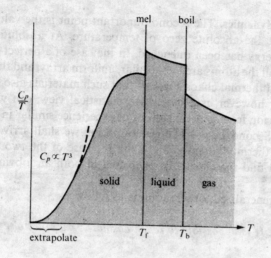

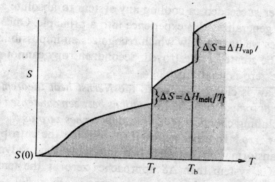

Fig. 5.6. The calculation of entropy from heat capacity data.

Example (Objective 19). The Third Law entropy of nitrogen gas at 298.15 K and 1 atm has been calculated from the following data.

	$\Delta S_m/\text{J K}^{-1}\text{ mol}^{-1}$
T^3 extrapolation, 0–10 K	1.92
graphical integration (eqn (5.2.1))	25.25
phase transition at 35.61 K	6.43
graphical integration (eqn (5.2.1))	23.38
phase transition: fusion at 63.14 K	11.42
graphical integration (eqn (5.2.1))	11.41
phase transition: vaporization at 77.32 K	72.13
perfect gas behaviour (eqn (5.2.2)) from 77.32–298.15 K	39.20
correction for non-ideality	0.92
Total entropy change:	192.06

By setting the entropy at absolute zero to zero we get the Third Law standard entropy (see below) of nitrogen gas as $S_m^{\ominus}$ (298.15 K) = 192.06 J K^{-1} mol^{-1}.

The Third Law of thermodynamics. The second important point is the value of $S(0)$, the entropy at the absolute zero of temperature. At absolute zero all quenchable energy has been quenched. In the case of a perfect crystal at absolute zero all the atoms are in a regular, uniform array, and the absence of disorder and thermal chaos suggests that such materials also have zero entropy. That, however, is a molecular, statistical view; can we draw a similar conclusion from generalizations of experience similar in nature to the First and Second Laws? The generalization we shall arrive at is the *Third Law* of thermodynamics, and as in the case of the two preceding Laws we shall find that it can be expressed in both an 'obvious' and a succinct way.

The basic, generalized, observation is as follows:

> Third Law: *It is impossible to reach absolute zero in a finite number of steps.*

No one has ever succeeded in cooling any system to absolute zero, and the Third Law generalizes this experience into a principle. Once more in thermodynamics we have a Law which recognizes an impossibility (First: energy cannot be created or destroyed; Second: entropy cannot decrease; Third: absolute zero is unattainable).

The Third Law is closely related to the *Nernst heat theorem* that *the entropy change accompanying transformations between condensed phases in equilibrium*, including chemical reactions, *approaches zero as the temperature approaches zero.** In other words, if we consider the entropy change ΔS for the transformation between two internally thermodynamically stable states of a system then ΔS approaches zero as the temperature approaches zero. It follows that if the value zero is *arbitrarily* ascribed to the entropies of the elements (in the perfect crystalline form stable at $T = 0$), then all perfect crystalline compounds also have zero entropies at absolute zero.

Since entropies at higher temperatures are given by expressions like eqn (5.4.1), and heat capacities are never negative, these conclusions can be expressed in an alternative form:

---

* Whereas the Third Law (in the form of its expression about the unattainability of absolute zero) is implied by the Nernst heat theorem, the opposite is not true except in restricted circumstances. This is a point of some subtlety, and is discussed by A. Münster (*Statistical thermodynamics*, Vol. 12, Springer, 1974, p. 79). Hence the unattainability statement and the heat theorem are not truly logically equivalent. Indeed, a precise and general formulatio₋ of the Third Law—one that implies the Nernst theorem—is still awaited, and it appears that when it is formulated it will necessarily involve statistical concepts. Hence the Third Law appears to be intrinsically different from the other laws of thermodynamics. We shall ignore these subtle points, and regard the Third Law in the form quoted above and in the form to be quoted in the next paragraph as 'equivalent' for our simple purposes. The question, though, is real and deep, and shows that thermodynamics is still alive. The implication of the Nernst theorem that absolute zero is unattainable is taken up again on p. 202.

> Third Law: *If the entropy of every element in the state stable at $T = 0$ is taken as zero, every substance has a positive entropy which at $T = 0$ may become zero, and does become zero for all perfect crystalline substances, including compounds.*

Note that a non-crystalline perfect state, such as the superfluid state of helium, is captured by the opening remark. Note, too, that the Third Law does not imply that entropies are zero at absolute zero; it merely implies that all perfect materials have the same entropy: choosing this common value as zero is then merely a convenience. Note finally that although, as expressed, the law appears to entail referring to molecules (since a perfect crystal is a regular array of individual particles), this could be avoided by replacing 'perfect crystal' by circumlocutions referring to 'internally thermodynamically stable states'.

Evidence for the Third Law comes from a variety of sources. One is the measurement of heat capacities down to very low temperatures. For instance, the entropy of the transition *monoclinic sulphur → rhombic sulphur* can be measured by determining its enthalpy -401.7 kJ mol^{-1} at the transition temperature (368.5 K):

$$\Delta S_m = S_m' (\text{rh}; 368.5 \text{ K}) - S_m (\text{monocl}; 368.5 \text{ K})$$

$$= (-401.7 \text{ kJ mol}^{-1})/368.5 \text{ K} = -1.09 \text{ J K}^{-1} \text{ mol}^{-1}.$$

But the two individual entropies can also be determined by measuring the heat capacities from $T = 0$ to $T = 368.5$ K (the rate of transformation is so slow at low temperatures that the relative thermodynamic instability of the monoclinic form is unimportant—they are both *internally* thermodynamically stable, i.e. perfect crystals). It is then found that

$$S_m (\text{rh}; 368.5 \text{ K}) = S_m (\text{rh}; 0) + (36.86 \pm 0.20) \text{ J K}^{-1} \text{ mol}^{-1}$$

$$S_m (\text{monocl}; 368.5 \text{ K}) = S_m (\text{monocl}; 0) + (37.82 \pm 0.40) \text{ J K}^{-1} \text{ mol}^{-1},$$

implying that at the transition temperature

$$\Delta S_m = S_m (\text{rh}; 0) - S_m (\text{monocl}; 0) - (0.96 \pm 0.60) \text{ J K}^{-1} \text{ mol}^{-1}.$$

On comparing this expression with the value obtained above we conclude that, within experimental error,

$$S_m (\text{rh}; 0) - S_m (\text{monocl}; 0) \approx 0;$$

therefore the entropies of the two materials are identical at $T = 0$. We then *ascribe* the value zero to both.

The choice $S(0) = 0$ will henceforth be made, and entropies reported on that basis will be referred to as *Third Law entropies*. When the substance is in its standard state at a temperature T the Third Law entropy is denoted $S^{\ominus}(T)$; a list of values at 25 °C is given in Table 5.2.

Third Law entropies are sometimes referred to as *absolute entropies*. This is a misnomer on two grounds. First, the value at absolute zero is merely a convention. Second, even on the statistical view there may be non-zero entropy at absolute zero on account of the existence of a random distribution of isotopes through the sample. This randomness is preserved in chemical reactions, but it contributes to the 'absolute' value of the entropy of any sample.

Table 5.2. Third Law entropies at 298.15 K, $S_m^\ominus/J\ K^{-1}\ mol^{-1}$

Solids		Liquids		Gases	
Ag	42.6	Hg	76.0	He	126.0
C(gr)	5.7	Br_2	152.2	H_2	130.6
C(d)	2.4	H_2O	69.9	N_2	192.1
Cu	33.1	HNO_3	155.6	O_2	205.0
Zn	41.6	C_2H_5OH	160.7	Cl_2	223.0
I_2	116.1	CH_3OH	126.8	CO_2	213.6
S(rh)	31.8	C_6H_6	173.3	HCl	186.8
AgCl	96.2	CH_3COOH	159.8	H_2S	205.7
AgBr	107.1	C_6H_{12}	204.3	NH_3	192.3
$CuSO_45H_2O$	300.4			CH_4	186.2
$HgCl_2$	146.0			C_2H_6	229.5
Sucrose	360.2			CH_3CHO	250.2

Source: Principally G. W. C. Kaye and T. H. Laby, *Tables of physical and chemical constants*, Longmans.

A brief inspection of Table 5.2 gives some familiarity with the magnitudes of entropies, and indicates how $S^\ominus$ varies from one substance to another. Notice, for example, that the standard entropies of gases (except hydrogen) are all much the same, and generally bigger than the entropies of liquids and solids composed of molecules of comparable complexity. This fits in well with the view that gases are more chaotic than either liquids or solids. Note the low value of the entropy of water in comparison with benzene. This accords with the view that water retains some ice-like structure because of the hydrogen bonds. Note the exceptionally low value of the entropy of diamond, and the high value for a much more complex solid, such as hydrated copper sulphate.

If the Third Law entropies of compounds are known, it is a simple matter to calculate the *standard entropy of reaction*. This is defined, like the enthalpy of reaction, as the difference of entropies between the pure, separated products and the pure, separated reactants, all in their standard states at the temperature of interest. Thus, for the reaction

$$H_2(g) + \tfrac{1}{2}O_2(g) \rightarrow H_2O(l)$$

at 25 °C the standard entropy of reaction is

$$\Delta S_m^\ominus = S_m^\ominus(H_2O, l) - S_m^\ominus(H_2, g) - \tfrac{1}{2}S_m^\ominus(O_2, g)$$

$$= (70.00 - 130.6 - 102.6)\ J\ K^{-1}\ mol^{-1}$$

$$= -163.2\ J\ K^{-1}\ mol^{-1}.$$

Standard molar Gibbs functions. Once the entropies of reactions have been determined, they can be combined with reaction enthalpies in order to obtain reaction Gibbs functions using $\Delta G = \Delta H - T\,\Delta S$. The value of ΔG for the formation of unit amount of a compound from its elements, with all the components in their standard states (p. 106), is then called the *standard molar Gibbs function of formation* of that compound, and written $\Delta G_f^\ominus$. It follows that the standard molar Gibbs functions of formation of elements are zero. A selection of values is given in Table 5.3. With these values available it is a simple matter to obtain the standard Gibbs function for any reaction by taking the appropriate combinations. For instance, the standard molar Gibbs function for a reaction such as

$$CO(g) + \tfrac{1}{2}O_2(g) \rightarrow CO_2(g)$$

is obtained from

$$\Delta G_m^\ominus = \Delta G_f^\ominus(CO_2, g) - \Delta G_f^\ominus(CO, g) - \tfrac{1}{2}\Delta G_f^\ominus(O_2, g)$$

$$= \{-394.38 - (-137.27) - 0\} \text{ kJ mol}^{-1}$$

$$= -257.11 \text{ kJ mol}^{-1}.$$

Table 5.3. Standard Gibbs function of formation at 298.15 K, $\Delta G_f^\ominus/\text{kJ mol}^{-1}$

Solids					
NaCl	-384.0	CaO	-604.2	SiO_2	-856.7
NH_4Cl	-203.0	$CaCO_3$	-1128.8	FeS	-100.4
KCl	-408.3	Al_2O_3	-1582.4	FeS_2	-166.9
KOH	-374.5	C (diamond)	2.9	AgCl	-109.8
Liquids					
H_2O	-237.2	CH_3CH_2OH	-174.1	C_6H_6	124.3
H_2O_2	-105.6	HNO_3	-80.8	$H_2SO_4(aq)$	-744.6
CH_3OH	-166.4	CS_2	$+65.3$	HCl(aq)	-131.3
Gases					
NH_3	-16.5	N_2O_4	$+97.8$	HCN	$+124.7$
NO	$+86.6$	CH_4	-50.8	HCl	-95.3
NO_2	$+51.3$	C_2H_2	$+209.2$	HBr	-53.4
O_3	$+163.2$	C_2H_4	$+68.1$	H_2S	-33.6
CO	-137.2	C_2H_6	-32.9	N_3H	328.0
CO_2	-394.4	C_4H_{10}	-17.2		

Source: G. W. C. Kaye and T. H. Laby, *Tables of physical and chemical constants*, Longmans.

Note that this is the change of molar Gibbs function for the overall process of starting with the pure, separated gases, then carrying out the reaction proper, and so it includes a contribution from the Gibbs function of mixing of the reactants (and in more complicated reactions, from the Gibbs function of unmixing of the products): these contributions are analysed in Chapter 7, but they are generally small in the case of gas-phase reactions.

Calorimetry (for ΔH, and for ΔS via heat capacities) is only one of the ways of determining the value of the Gibbs function of a reaction. $\Delta G^\ominus$

can also be obtained by determining the equilibrium constant for a reaction (Chapter 9), from electrochemical measurements (Chapter 12), and by calculations on the basis of data obtained from spectroscopic observations (Chapter 21).

Further reading

Thermodynamics. G. N. Lewis and M. Randall, revised by K. S. Pitzer and L. Brewer; McGraw-Hill, New York, 1961.
The second law. H. A. Bent; Oxford University Press, New York, 1965.
Entropy. J. D. Fast; McGraw-Hill, New York, 1963.
The third law of thermodynamics. J. Wilks; Clarendon Press, Oxford, 1961.
Heat and thermodynamics. M. W. Zemansky; McGraw-Hill, Tokyo, 1968.
Bibliography of thermodynamics. L. K. Nash; *J. chem. Educ.* **42**, 71 (1965).

Problems

5.1. Find the probability that all the molecules of a gas will be found in one half of a container when the sample consists of (a) 4 molecules, (b) 10 molecules, (c) 6×10^{23} molecules.

5.2. We have seen that the change of entropy of a system can be calculated from $dS = dq_{rev}/T$. It is important to appreciate some of the magnitudes involved, and the next few problems give some experience. As a first example, find the change of entropy when 25 kJ of heat is transferred reversibly and isothermally to a system at (a) $0\,°C$, (b) $100\,°C$.

5.3. A block of copper of mass 500 g initially at $20\,°C$ is in thermal contact with an electric heater of resistance $1000\,\Omega$ and negligible mass. An electric current of 1 A is passed for 15 s. What is the change of entropy of the copper? Assume $C_{p,m} = 24.4\ \mathrm{J\ K^{-1}\ mol^{-1}}$ throughout.

5.4. The same block of copper as in the last Problem is immersed in a stream of water which maintains its temperature at $20\,°C$. What is the change of entropy of (a) the metal, (b) the water when the same amount of electricity is passed through the resistance?

5.5. Find the molar entropy of neon at 500 K and 1 atm given that its entropy at 298 K is $146.22\ \mathrm{J\ K^{-1}\ mol^{-1}}$. Take constant volume conditions.

5.6. The standard molar entropy of ammonia at 298 K is $192.5\ \mathrm{J\ K^{-1}\ mol^{-1}}$. We have seen that its heat capacity follows $C_{p,m} = a + bT + c/T^2$, with the coefficients given in Table 4.3. What is the standard molar entropy of ammonia at (a) $100\ °C$, (b) $500\ °C$?

5.7. What is the change of entropy when 50 g of hot water at $80\ °C$ is poured into 100 g of cold water at $10\,°C$ in an insulated vessel? Take $C_{p,m} = 75.5\ \mathrm{J\ K^{-1}\ mol^{-1}}$.

5.8. Calculate the change of entropy when 200 g of water at $0\,°C$ is added to 200 g of water at $90\,°C$ in an insulated vessel. (Data in Table 4.7.)

5.9. In calculating entropy changes we have to be careful to take into account phase transitions. Calculate the change of entropy when 200 g of ice at $0\,°C$ is added to 200 g of water at $90\,°C$ in an insulated vessel.

5.10. Calculate the changes of entropy in the system, the surroundings, and the universe when a 14 g sample of nitrogen gas at room temperature doubles its volume in (a) an isothermal reversible expansion, (b) an isothermal irreversible expansion, (c) an adiabatic reversible expansion.

5.11. Calculate the differences in molar entropy (a) between water liquid at $-5\,°C$ and ice at $-5\,°C$, (b) between water liquid and water vapour at $95\,°C$ and 1 atm pressure. Distinguish between entropy differences of the system and entropy differences of the universe, and discuss the spontaneity of transitions between phases at these temperatures. (Difference of heat capacities on melting: 37.3 J K^{-1} mol^{-1}; difference on vaporization: -41.9 J K^{-1} mol^{-1}.)

5.12. We saw that the Gibbs function concentrates attention on the system and automatically carries around the entropy changes in the surroundings. Calculate the difference in Gibbs function between water and ice at $-5\,°C$ and between water and water vapour at $95\,°C$, and see how ΔG incorporates the information of the last Problem.

5.13. Calculate the entropy change when 1 mol of (a) water, (b) benzene are evaporated at their boiling points under 1 atm pressure. What are the entropy changes in (i) the system, (ii) the surroundings, (iii) the universe?

5.14. Suppose now that the evaporated water and benzene of the last Problem are compressed to half their volume and *simultaneously* heated to twice the absolute temperature at which they originally boiled. What change of entropy occurs? ($C_{V,m} = 25.3$ J K^{-1} mol^{-1} for water vapour, and $C_{V,m} = 130$ J K^{-1} mol^{-1} for benzene vapour.)

5.15. One of the advantages of introducing the Helmholtz and Gibbs functions is that the maximum amount of work available from processes may be stated very simply. As an example, calculate (a) the maximum amount of work, (b) the maximum amount of non-pV work that can be obtained from the freezing of supercooled water at $-5\,°C$ under atmospheric pressure. What are the corresponding values under 100 atm pressure? Use $\rho(\text{water}) = 0.99$ g cm^{-3} and $\rho(\text{ice}) = 0.917$ g cm^{-3} at $-5\,°C$ to calculate ΔV.

5.16. In Part 3 we shall see that *fuel cells* are designed to extract electrical work from the direct reaction of readily available fuels. Even at this stage we can make a thermodynamic assessment of the maximum amount of electrical work available from various reactions. Methane and oxygen are readily available, and the standard molar Gibbs function for the reaction $CH_4(g) + 2O_2(g) \rightarrow CO_2(g) + 2H_2O(l)$ is -802.8 kJ mol^{-1} at $25\,°C$. What is the maximum amount of work and the maximum amount of electrical work available from this reaction under these standard conditions?

5.17. In order to use entropies and Gibbs functions we need to be able to measure their values. Succeeding chapters will bring forward a number of methods for their measurement, but some were encountered in this chapter and can be illustrated here. The following table gives the heat capacity of lead over a range of temperatures. What is the standard molar Third Law entropy of lead at (a) $0\,°C$, (b) $25\,°C$?

T/K	10	15	20	25	30	50
$C_{p,m}/\text{J K}^{-1}\text{ mol}^{-1}$	2.8	7.0	10.8	14.1	16.5	21.4
T/K	70	100	150	200	250	298
$C_{p,m}/\text{J K}^{-1}\text{ mol}^{-1}$	23.3	24.5	25.3	25.8	26.2	26.6

5.18. In a thermochemical study of nitrogen the following heat capacity data were found

$$\int_0^{T_t} (C_{p,m}/T)\,dT = 27.2 \text{ J K}^{-1} \text{ mol}^{-1};$$

$$\int_{T_t}^{T_f} (C_{p,m}/T)\,dT = 23.4 \text{ J K}^{-1} \text{ mol}^{-1};$$

$$\int_{T_f}^{T_b} (C_{p,m}/T)\,dT = 11.4 \text{ J K}^{-1} \text{ mol}^{-1}; \qquad T_t = 35.61 \text{ K},$$

the enthalpy of transition being 0.229 kJ mol^{-1}; $T_f = 63.14$ K, $\Delta H_{melt,m}^{\ominus} = 0.721$ kJ mol^{-1}; $T_b = 77.32$ K, $\Delta H_{vap,m} = 5.58$ kJ mol^{-1}. What is the Third Law entropy of nitrogen gas at its boiling point?

5.19. We encountered a special sequence of changes called a *cycle*. One important example of a cycle, which was used a great deal in classical thermodynamics and still plays a central role in the discussion of the efficiency of heat engines and refrigerators, is called the *Carnot cycle*. We now have enough material at our command to set it up and use it to illustrate some of the remarks made in the text. The initial state is that of a perfect gas with pressure, volume, and temperature p_i, V_i, T_h. The first step in the cycle is reversible, isothermal expansion to p', V', T_h. The next is reversible, adiabatic expansion to p'', V'', T_c. The third is isothermal, reversible compression to p''', V''', T_c, and the fourth, final step is adiabatic, reversible compression to p_i, V_i, T_h, the initial state. Calculate (a) the work done on each step of the cycle, (b) the heat absorbed at each stage, (c) the entropy change at each step, (d) the change in internal energy at each stage. Confirm that ΔS and ΔU are both zero for the entire cycle, but that q and w differ from zero.

5.20. In order to demonstrate that the general relation $\oint dq/T \leqslant 0$ (p. 140) holds in a special case, repeat the calculation of q/T used in the last Problem, but with the isothermal reversible expansion step replaced by an irreversible isothermal expansion against a pressure p'.

5.21. Develop the last Problem a little further by calculating the entropy difference between the states p', V', T_h and p_i, V_i, T_h, and show that the Clausius inequality eqn (5.1.7), is satisfied.

5.22. The *coefficient of performance* of a heat engine is defined as the ratio of the work produced to the heat absorbed: $c = -w/q_{absorbed}$. Show that the efficiency of a Carnot engine working perfectly is determined only by T_h and T_c, through $c_0 = (T_h - T_c)/T_h$. In fact this result applies to any working substance, not just a perfect gas, and is the maximum thermodynamic efficiency of any heat engine. This point is developed in Chapter 7.

5.23. Find the thermodynamic coefficient of performance of a primitive steam engine operating on steam at $100\,^{\circ}$C and discharging at $60\,^{\circ}$C. A modern steam turbine operates on steam at $300\,^{\circ}$C and discharges at $80\,^{\circ}$C. What is its coefficient of performance?

5.24. Using the data in the last Problem, calculate the minimum amount of oil that must be burnt in order to raise a mass of 1000 kg through 50 m. Take the work content of the fuel as 4.3×10^4 kJ kg^{-1} and neglect all losses. Heat engines work at less than their thermodynamic coefficient of performance, but at least thermodynamics points to the ceiling of performance for a particular design, and indicates how and how far the performance can be improved.

5.25.   The internal combustion engine is not noted for its efficiency, but some of its deficiencies are due to thermodynamic rather than technological constraints. Take the fuel to be an octane having an enthalpy of combustion of -5512 kJ mol^{-1}, and use 1 gallon ≈ 3.03 kg. Then examine the coefficient of performance from the point of view of an engine working with a cylinder temperature of 2000 °C and an exit temperature of 800 °C. What is the maximum height, neglecting all forms of friction, to which a 2500 lb car can be driven on one gallon of fuel?

5.26.   The Carnot cycle can be represented on a pV-indicator diagram of the kind encountered in Chapter 2. Draw an indicator diagram showing the sequence of isotherms and adiabats described in Problem 5.19. Show that the same cycle can be represented in a much simpler way if the pV diagram is replaced by one with coordinates representing temperature and entropy.

5.27.   Two equal blocks of the same metal, one at a temperature T_h and the other at T_c, are placed in contact and come to thermal equilibrium. Assuming the heat capacity to be constant over the temperature range at 24.4 J K^{-1} mol^{-1}, calculate the change of entropy. Calculate the value for the case of two 500 g blocks of copper, with $T_h = 500$ K and $T_c = 250$ K.

5.28.   The enthalpy of the graphite $\rightarrow$ diamond phase transition, which under a pressure of 100 kbar occurs at 2000 K, is $+1.90$ kJ mol^{-1}. What is the entropy of the phase transition at the transition temperature? Does that fit your idea of the different crystal structures of graphite and diamond?

5.29.   Solid hydrogen chloride undergoes a phase transition at 98.36 K and the enthalpy change involved is 1.19 kJ mol^{-1}. Calculate the molar entropy of transition. The sample is maintained in contact with a block of copper, the whole being thermally isolated. What is the change of entropy of the copper at the phase transition, and the change of entropy in the universe?

5.30.   Use the Third Law entropies, Tables 5.2 and 12.2, to deduce the standard molar entropy changes for the following reactions at 25 °C.
   (a) $Hg(l) + Cl_2(g) \rightarrow HgCl_2(s)$
   (b) $Zn(s) + CuSO_4(aq) \rightarrow Cu(s) + ZnSO_4(aq)$
   (c) sucrose $+ 12O_2(g) \rightarrow 12CO_2 + 11H_2O(l)$.

5.31.   Combine the results of the last Problem with the reaction enthalpies given in Table 4.1 and 4.2, and find the standard molar Gibbs functions for the reactions at 25 °C.

5.32.   We shall encounter several methods for determining the Gibbs functions of reactions in later chapters, but even at this stage it is easy to use the information contained in Tables 4.1, 4.2, and 5.2 to draw conclusions about the spontaneity of chemical reactions. As a first example of this type of calculation, determine $\Delta S_m^{\ominus}$, $\Delta H_m^{\ominus}$, and $\Delta G_m^{\ominus}$ for the reactions at 298 K, and state whether they are spontaneous in the forward or reverse directions.
   (a) $H_2(g) + \frac{1}{2}O_2(g) \rightarrow H_2O(l)$
   (b) $3H_2(g) + C_6H_6(l) \rightarrow C_6H_{12}(l)$
   (c) $CH_3CHO(g) + \frac{1}{2}O_2(g) \rightarrow CH_3COOH(l)$.

5.33.   The standard molar Gibbs function for the reaction

$$K_4Fe(CN)_6.3H_2O \rightarrow 4K^+(aq) + Fe(CN)_6^{4-}(aq) + 3H_2O(l)$$

is 26.120 kJ mol^{-1} (I. R. Malcolm, L. A. K. Staveley, and R. D. Worswick, *J. chem. Soc. Faraday Trans.* **1**, 1532 (1973)). The enthalpy of solution of the trihydrate is

55.000 kJ mol^{-1}. Find the standard molar entropy of solution at 298.15 K and use the values in Table 12.2 on p. 362 to deduce a value of the standard molar entropy of the ferrocyanide ion in aqueous solution. ($S^{\ominus}$(ferrocy.) = 599.7 J K^{-1} mol^{-1}.)

5.34. The same authors referred to in the last Problem also measured the heat capacity (C_p) of the anhydrous potassium ferrocyanide (potassium hexacyano-ferrate (II)) over a wide temperature range, and obtained the following results:

T/K	10	20	30	40	50	60	70
$C_{p,m}$/J K^{-1} mol^{-1}	2.09	14.43	36.44	62.55	87.03	111.0	131.4

T/K	80	90	100	110	120	130	140
$C_{p,m}$/J K^{-1} mol^{-1}	149.4	165.3	179.6	192.8	205.0	216.5	227.3

T/K	150	160	170	180	190	200
$C_{p,m}$/J K^{-1} mol^{-1}	237.6	247.3	256.5	265.1	273.0	280.3

What is the entropy of the salt at these temperatures?

5.35. Integrate the data in the last Problem to find the enthalpy of the salt relative to its enthalpy at absolute zero; that is, find $H_m(T) - H_m(0)$. In Chapter 9 we shall meet the *Giauque function* $[G_m(T) - H_m(0)]/T$. This function varies more slowly than $G(T)$ itself, and so it is useful for recording the Gibbs function of substances at various temperatures. Calculate both $G_m(T) - G_m(0)$, which is the same as $G_m(T) - H_m(0)$, and the Giauque function for anhydrous potassium ferrocyanide using the data in the last Problem, and plot both as a function of temperature.

5.36. 1,3,5-trichloro-2,4,6-trifluorobenzene is an intermediate in the conversion of hexachlorobenzene to hexafluorobenzene. Its thermodynamic properties have been examined by measuring its heat capacity over a wide temperature range (R. L. Andon and J. F. Martin, *J. chem. Soc. Faraday Trans.* 1, 871 (1973)). A selection of the results are as follows:

T/K	14.14	16.33	20.03	31.15	44.08	64.81
$C_{p,m}$/J K^{-1} mol^{-1}	9.492	12.70	18.18	32.54	46.86	66.36

T/K	100.90	140.86	183.59	225.10	262.99	298.06
$C_{p,m}$/J K^{-1} mol^{-1}	95.05	121.3	144.4	163.7	180.2	196.4

Find the Third Law entropy, the molar enthalpy, and the Giauque function at 298 K.

6   The Second Law: the machinery

Learning objectives

After careful study of this chapter you should be able to

(1)   State how the internal energy changes when the entropy and the volume change (eqn (6.1.2)).

(2)   Derive the *Maxwell relations* between thermodynamic variables and use them to derive the *thermodynamic equation of state* (p. 164).

(3)   Deduce how the Gibbs function depends on the pressure and the temperature (eqns (6.2.1) to (6.2.3)).

(4)   Derive and use the *Gibbs–Helmholtz equation* (eqn (6.2.4)).

(5)   State how the Gibbs functions of solids and liquids vary with pressure (eqn (6.2.8)).

(6)   Derive an expression for the pressure dependence of the Gibbs function and the chemical potential of a perfect gas (eqns (6.2.9), (6.2.10)).

(7)   Define the *fugacity* of a gas (p. 169) and relate it to the pressure of the gas (eqn (6.2.12)).

(8)   Define the *standard state* of a real gas (p. 173).

(9)   State how the Gibbs function changes when the composition of a system changes (p. 175).

(10)   Define the *chemical potential* of a species in terms of the composition dependence of the thermodynamic functions (p. 175).

Introduction

The main aim of this chapter is to obtain the dependence of the Gibbs function on temperature, pressure, and the composition of the system. The most important conclusions are eqn (6.2.4) for the temperature dependence of G, eqn (6.2.9) for the pressure dependence, and eqn (6.3.3) for the composition dependence. This chapter also introduces the important concept of *chemical potential* upon which the whole of the subsequent applications of thermodynamics are based.

6.1 Combining the First and Second Laws

The First Law may be written

(6.1.1) $dU = dq + dw.$

If a change of a closed system is brought about reversibly, then in the absence of any kind of work other than pV-work, dw_{rev} may be replaced by $-p\,dV$ and dq_{rev} may be replaced by $T\,dS$. Therefore

$$dU_{rev} = T\,dS - p\,dV.$$

By introducing the entropy into the First Law expression, we have combined the two laws.

The last equation applies when the changes are reversible. But dU is an exact differential and is independent of the path. Therefore, $dU_{rev} = dU_{irrev}$. It follows that the equation is also true if the 'irrev' subscript is applied to dU. Therefore it applies irrespective of whether the change is brought about reversibly or irreversibly, and so *for any change* of a closed system, and where there is no change of composition,

(6.1.2) $dU = T\,dS - p\,dV.$

This we call the *fundamental equation*.

The fact that this important equation applies to both reversible and irreversible changes may be puzzling at first sight. The reason is that only in the case of a reversible change may $T\,dS$ be identified with dq and $-p\,dV$ with dw. When the change is irreversible $T\,dS$ is bigger than dq (the Clausius inequality, p. 139) and $p\,dV$ bigger than dw. The sum of dq and dw, however, is equal to the sum of $T\,dS$ and $-p\,dV$ if the composition is constant: this must be the case because U is a state function.

There are two ways of using the fundamental equation and both make use of the fact that dU is an exact differential. Since dU is exact it can be developed by the techniques described in Chapter 3 and collected in Box 3.1.

One way of looking at the fundamental equation. The appearance of eqn (6.1.2) indicates that U changes in a simple way when S and V are changed ($dU \propto dS$, and $dU \propto dV$). This suggests that the most sensible way of regarding U is as a function of S and V. Therefore we write it $U(S, V)$. We are at liberty to regard it instead as a function of S and p, or of T and V, or of some other pair of variables, because all these properties are interrelated, but the simplicity of the master equation suggests that $U(S, V)$ is the most sensible choice.

The *mathematical* consequence of U being a function of S and V is that dU can be related to changes in S and V by the relation

(6.1.3) $$dU = (\partial U/\partial S)_V \, dS + (\partial U/\partial V)_S \, dV.$$

This can be compared to the thermodynamic expression (6.1.2). It follows that for systems of constant composition,

(6.1.4) $$(\partial U/\partial S)_V = T$$

(6.1.5) $$-(\partial U/\partial V)_S = p.$$

Both these expressions, and especially the first, are remarkable. The first enables a temperature to be expressed solely in terms of extensive thermodynamic quantities. If the volume and composition are constant, the relation states, that the ratio of the change in energy (a First Law concept) to the corresponding change in entropy (a Second Law concept) is equal to the temperature of the system, whatever the latter's nature.

The coefficient $(\partial U/\partial V)_T$ played a central role in the manipulation of the First Law, and on p. 92 we used the relation $(\partial U/\partial V)_T = T(\partial p/\partial T)_V - p$, which was to be proved in this chapter. The coefficient required can be obtained from eqn (6.1.3) by using relation No. 1 (of Box 3.1). Dividing by dV and imposing constant temperature gives

(6.1.6) $$(\partial U/\partial V)_T = (\partial U/\partial S)_V(\partial S/\partial V)_T + (\partial U/\partial V)_S$$
$$= T(\partial S/\partial V)_T - p.$$

This is beginning to look like the expression we want, but we cannot yet do anything useful with $(\partial S/\partial V)_T$. The next section will permit further progress to be made.

Another way of looking at the fundamental equation. Since the fundamental equation, eqn (6.1.2), is for an exact differential, the coefficients of dS and dV must pass the test of relation No. 4 in Box 3.1; with $a = T$ and $b = -p$ it follows that

(6.1.7) $$(\partial T/\partial V)_S = -(\partial p/\partial S)_V.$$

In this simple way a relationship has been generated between two quantities which, at first sight, would not seem to be related.

The equation just derived is an example of a *Maxwell relation*. Apart from being unexpected, it does not look particularly interesting. Nevertheless, it does suggest that there may be other similar relations and that the coefficient $(\partial S/\partial V)_T$ encountered above might be related to something recognizable. It turns out that the fact that H, G, and A are state functions can be used to derive three more Maxwell relations. The argument to obtain each one runs in the same way: since H, G, and A are state functions the coefficients in the expressions for dH, dG, and dA satisfy relation No. 4 in Box 3.1.

The Maxwell relations are collected in Box 6.1. In the next section we see how to derive another of them, but as no new principles are involved we shall not derive them all.

Box 6.1

The Maxwell relations

$$(\partial T/\partial V)_S = -(\partial p/\partial S)_V \qquad (\partial p/\partial T)_V = (\partial S/\partial V)_T$$
$$(\partial T/\partial p)_S = (\partial V/\partial S)_p \qquad (\partial V/\partial T)_p = -(\partial S/\partial p)_T$$

One of the Maxwell relations does the job of turning $(\partial S/\partial V)_T$ into something else:

$$(\partial S/\partial V)_T = (\partial p/\partial T)_V.$$

Using this in eqn (6.1.6) completes the proof of the relation

(6.1.8)
$$(\partial U/\partial V)_T = T(\partial p/\partial T)_V - p$$

which was used in Chapter 3. This equation is called a *thermodynamic equation of state* because it relates V, T, and p.

Example (Objective 1). Show thermodynamically that $(\partial U/\partial V)_T$ is zero for a perfect gas, and compute its value for a van der Waals gas.

- *Method.* Showing a result thermodynamically means basing it entirely on general thermodynamic results and equations of state, and without drawing on arguments about molecular behaviour, for example the absence of intermolecular forces. Eqn (6.1.8) is a thermodynamic result, and so we consider it in connection with the equations of state of a perfect gas and a van der Waals gas (p. 47).

- *Answer.* For a perfect gas $p = nRT/V$ and so $(\partial p/\partial T)_V = nR/V$. From eqn (6.1.8):
$$(\partial U/\partial V)_T = nRT/V - p = 0.$$
The equation of state of a *van der Waals gas* is
$$p = nRT/(V-nb) - n^2 a/V^2$$
Therefore
$$(\partial p/\partial T)_V = nR/(V-nb)$$
and so
$$(\partial U/\partial V)_T = nRT/(V-nb) - [nRT/(V-nb) - n^2 a/V^2]$$
$$= n^2 a/V^2 = a/V_m^2.$$

- *Comment.* The last result shows that the internal energy of a van der Waals gas increases when it expands isothermally, and that the increase is related to the parameter a, which is related to the attractive interactions between the molecules: a larger volume implies less strong average attraction between the species. Working backwards from an assumed form for the energy-dependence on the volume and the number of molecules present, through eqn (6.1.8), is a very good way of constructing an equation of state.

6.2 Properties of the Gibbs function

Having seen the application of some simple arguments to the fundamental equation for the internal energy we can apply the same techniques to the Gibbs function. This is important because our discussion of chemical equilibria will be based on the properties of G.

First the fundamental equation is combined with the definition of G, and then the new equation is examined in the same way as the original fundamental equation for U.

The definition of the Gibbs function was given on p. 146, as

$$G = H - TS.$$

When the system changes, G might change because H, T, and S may change. For infinitesimal changes in each property,

$$dG = dH - T\,dS - S\,dT.$$

Since $H = U + pV$, we have

$$dH = dU + p\,dV + V\,dp$$

(an expression first seen on p. 68). For a closed system changing without change of composition dU can be replaced by the fundamental equation. Combining all the parts gives

$$dG = (T\,dS - p\,dV) + p\,dV + V\,dp - T\,dS - S\,dT$$

or

(6.2.1)
$$dG = V\,dp - S\,dT.$$

This is the new fundamental equation.

This new equation suggests that the most sensible way of regarding G is as a function of p and T, and writing it $G(p, T)$. This confirms that it is a very important quantity in chemistry because the pressure and temperature are variables that are usually under our control. *G, it appears, carries around the combined consequences of the First and Second Laws in a way that makes it particularly suitable for chemical applications.*

Once G is regarded in this way some conclusions can be drawn very quickly. The same argument that led to eqns (6.1.4) and (6.1.5) now gives

(6.2.2)
$$(\partial G/\partial T)_p = -S$$

(6.2.3)
$$(\partial G/\partial p)_T = V$$

which show how the Gibbs function varies with temperature and pressure.

The second conclusion from the new fundamental equation follows when dG is examined with the relation that tests for an exact differential. dG is exact, and so the coefficients $a = V$ and $b = -S$ pass the test of Relation 4:

$$(\partial V/\partial T)_p = -(\partial S/\partial p)_T.$$

This is another of the Maxwell relations in Box 6.1.

How the Gibbs function depends on the temperature. Equation (6.2.2) implies that, as S is always a positive quantity, G must decrease when the temperature is raised at constant pressure (and constant composition). Furthermore it decreases more when the entropy of the state is large. Therefore the Gibbs functions of gases are likely to be more sensitive to temperature than those of liquids and solids.

The temperature dependence given by eqn (6.2.2) can be expressed in another way that needs a knowledge only of the enthalpy of the system. In order to obtain this result replace S by

$$S = (H-G)/T$$

(which follows from the definition of G) then

$$(\partial G/\partial T)_p = (G-H)/T, \qquad \text{or} \qquad (\partial G/\partial T)_p - G/T = -H/T.$$

This simple replacement can be taken further. First note that $(\partial G/\partial T)_p - G/T$ can be written in terms of $\partial(G/T)/\partial T$. This follows from

$$\left(\frac{\partial(G/T)}{\partial T}\right)_p = \frac{1}{T}\left(\frac{\partial G}{\partial T}\right)_p + G\left(\frac{\partial}{\partial T}\left(\frac{1}{T}\right)\right)_p = \frac{1}{T}\left\{\left(\frac{\partial G}{\partial T}\right)_p - \frac{G}{T}\right\}.$$

This leads immediately to the

(6.2.4)   *Gibbs–Helmholtz equation* $\left(\dfrac{\partial(G/T)}{\partial T}\right)_p = -H/T^2.$

($G-H$ is a helpful way of remembering what this equation relates.) It shows that if the enthalpy of the system is known, then the temperature-dependence of G is also known.

The Gibbs–Helmholtz equation takes its most useful form when it is applied to a chemical reaction

initial state (reactants) → *final state* (products)

$$\Delta G = G_f - G_i.$$

The equation applies to both G_f and G_i, and on subtraction

$$\left\{\frac{\partial}{\partial T}\left(\frac{G_f}{T} - \frac{G_i}{T}\right)\right\}_p = -\left(\frac{H_f}{T^2} - \frac{H_i}{T^2}\right)$$

or

(6.2.5) $$\left\{\frac{\partial(\Delta G/T)}{\partial T}\right\}_p = -\Delta H/T^2.$$

ΔH is the enthalpy of reaction. This equation is important because it shows that from the enthalpy of reaction we can predict whether a rise in temperature will favour one direction of a reaction or the other. Chapter 9 develops this point. $\Delta G/T$ may seem to be a clumsy quantity, but in Chapter 9 we shall see that it is exactly what is needed for discussing equilibrium constants.

How the Gibbs function depends on the pressure. Equation (6.2.3) indicates that knowing the volume of the system is sufficient to predict how G depends on the pressure. Because V must be positive, G invariably increases when the pressure of the system is increased with the temperature (and composition) held constant.

Equation (6.2.3) can be used to find G at any pressure given its value at some other pressure. Integration of the equation leads to

(6.2.6)
$$G(p_2) = G(p_1) + \int_{p_1}^{p_2} V(p)\,dp.$$

There are two cases where the pressure dependence of the volume is very simple and allows this expression to be evaluated.

In the case of a liquid or a solid the volume depends only very weakly on the pressure, and enormous pressures are needed to bring about significant changes in volume. This suggests that the pressure dependence of V may be neglected and that V may be taken outside the integral. Then for molar amounts

(6.2.7)
$$G_m(p_2) = G_m(p_1) + (p_2 - p_1)V_m.$$

Except at very high pressures $(p_2 - p_1)V_m$ is very small and virtually no error is introduced if it is neglected. This implies that

(6.2.8)
   for solids, liquids: $G_m(p_2) \approx G_m(p_1).$

In other words, G is virtually independent of pressure.

Is this always true? If we are interested in geophysical problems we have to consider reactions going on deep inside the earth. Pressures in the earth's bowels are huge, and their effect on the Gibbs function cannot be ignored. In normal laboratory experiments and in biochemical applications, however, the pressures are atmospheric and G is insensitive to the small variations of pressure. For water $V_m \approx 18\,\text{cm}^3\,\text{mol}^{-1}$; therefore, on going from 1 atm to 10 atm pressure, the Gibbs function changes by $(10-1) \times (1.013 \times 10^5\,\text{N m}^{-2}) \times (18 \times 10^{-6}\,\text{m}^3\,\text{mol}^{-1})$, which is only 0.02 kJ mol^{-1}.

Example (Objective 5). The pressure at the centre of the earth is probably greater than 3×10^6 atm, and the temperature there is about 4000 °C. What is the change in the Gibbs function of reaction on going from crust to core for a reaction in which $\Delta V_m = 1.0\,\text{cm}^3\,\text{mol}^{-1}$ and $\Delta S_m = 2.1\,\text{J K}^{-1}\,\text{mol}^{-1}$?

● *Method.* Since $(\partial\Delta G_m/\partial p)_T = \Delta V_m$ and $(\partial\Delta G_m/\partial T)_p = -\Delta S_m$ we can make a rough estimate from

$$\Delta G_m(\text{core}) - \Delta G_m(\text{crust}) \approx \Delta V_m(p_{\text{core}} - p_{\text{crust}}) - \Delta S_m(T_{\text{core}} - T_{\text{crust}}).$$

● *Answer.* $\Delta V_m(p_{\text{core}} - p_{\text{crust}}) \approx (1.0 \times 10^{-6}\,\text{m}^3\,\text{mol}^{-1}) \times (3.00 \times 10^6 \times 1.0133$
$$\times 10^5\,\text{N}\,\text{m}^{-2})$$
$$\approx 3.04 \times 10^5\,\text{J}\,\text{mol}^{-1} \text{ or } 304\,\text{kJ}\,\text{mol}^{-1}.$$
$$\Delta S_m(T_{\text{core}} - T_{\text{crust}}) \approx (2.1\,\text{J}\,\text{K}^{-1}\,\text{mol}^{-1}) \times (4273\,\text{K} - 298\,\text{K})$$
$$\approx 8300\,\text{J}\,\text{mol}^{-1} \approx 8.3\,\text{kJ}\,\text{mol}^{-1}.$$

Then

$$\Delta G_m(\text{core}) - \Delta G_m(\text{crust}) \approx 304\,\text{kJ}\,\text{mol}^{-1} - 8.3\,\text{kJ}\,\text{mol}^{-1} \approx 296\,\text{kJ}\,\text{mol}^{-1}.$$

● *Comment.* The pressure effect dominates, and even for such small volume changes as used in this example, the change in Gibbs function is dramatic. This is the thermodynamic basis of why materials change their structure at great depths in the earth's interior.

The other simple case is a perfect gas. The volume of a gas is generally large, and so the correction term may also be large even though the pressure range is small. In order to find an explicit expression for the pressure dependence the equation of state $V = nRT/p$ is put into the integral:

$$G(p_2) = G(p_1) + nRT \int_{p_1}^{p_2} (1/p)\,\mathrm{d}p.$$

Then

(6.2.9)°
$$G(p_2) = G(p_1) + nRT \ln(p_2/p_1).$$

This shows that when the pressure is increased ten-fold the molar Gibbs function increases by about $6\,\text{kJ}\,\text{mol}^{-1}$ at room temperature.

The chemical potential of a perfect gas. The standard state of a perfect gas is established at a pressure of 1 atm. This standard pressure is denoted $p^\ominus$; henceforth we shall normally write $p^\ominus$ to denote 1 atm (or 101.325 kPa). The Gibbs function is then $G^\ominus$, and so at any other pressure p the Gibbs function is

$$G(p) = G^\ominus + nRT \ln(p/p^\ominus).$$

When the Gibbs function refers to 1 mol of material the last equation becomes

$$G_m(p) = G_m^\ominus + RT \ln(p/p^\ominus).$$

A special symbol is given to the molar Gibbs function of a pure substance. Henceforth we write $\mu = G_m$, and in this notation

(6.2.10)°
$$\mu(p) = \mu^\ominus + RT \ln(p/p^\ominus).$$

The molar Gibbs function μ is usually called the *chemical potential.*

An interpretation of this name is as follows. In mechanical systems the direction of spontaneous change can be predicted simply by examining the potential in the vicinity of the particle. Particles that are stationary initially have a natural tendency to slip down potentials. When energy can be dissipated as heat, the situation is not as simple. Nevertheless we know that the direction of spontaneous change when T and p are held constant is towards a minimum of the Gibbs function for the system. The Gibbs function therefore plays a role analogous to the potential, and for this reason the molar Gibbs function (the Gibbs function for unit amount of molecules in the assembly) is called the *chemical potential*. Later we shall have to generalize its definition in order to take into account changes of composition; but the general significance will remain the same: chemical potential indicates potential for chemical change.

In the remaining chapters of Part 1 the chemical potential plays a central role. We have to determine how it varies with pressure, temperature, and composition. Then we shall apply it to the discussion of equilibria in a variety of chemically important systems.

Real gases: the fugacity. Equation (6.2.10) provides a moderately good description of the pressure-dependence of real gases so long as the conditions are not too severe. Nevertheless, deviations from perfect behaviour are significant under quite common conditions, and so we ought to be equipped to cope with them. A straightforward way of calculating $G(p)$ would be to determine the pressure dependence of the volume of a sample of real gas and to calculate $G(p)$ from $\int V(p)\,dp$ by numerical integration. We know, however, that gas imperfections are normally not very large, and so it might be sensible to preserve the form of eqn (6.2.10) and to write

(6.2.11)
$$\mu(p) = \mu^{\ominus} + RT \ln (f/p^{\ominus}).$$

The quantity f plays the role of the pressure, but it has a value which ensures that the chemical potential is given by the last equation whatever the pressure. f is a kind of equivalent pressure. It is called the *fugacity* (from the Latin for fleetness).

In order to find μ for some real gas at a pressure p a calibration is required that relates the value of the fugacity to the pressure. This is established as follows.

Equation (6.2.6) is true for all substances. Expressing it in terms of molar quantities and then using eqn (6.2.11) (which is also true, by definition) gives

$$\int_{p_1}^{p_2} V_m\,dp = \mu(p_2, T) - \mu(p_1, T) = RT \ln (f_2/f_1)$$

(f_2 is the fugacity when the pressure is p_2, f_1 the fugacity when it is p_1). If the gas were perfect we could write

$$\int_{p_1}^{p_2} V_m^\circ \, dp = \mu^\circ(p_2, T) - \mu^\circ(p_1, T) = RT \ln(p_2/p_1).$$

where the $^\circ$ denotes quantities relating to a perfect gas. The difference of the two equations is

$$\int_{p_1}^{p_2} \{V_m(p) - V_m^\circ(p)\} \, dp = RT \ln(f_2/f_1) - RT \ln(p_2/p_1)$$

or

$$\ln\{(f_2/p_2)/(f_1/p_1)\} = \frac{1}{RT} \int_{p_1}^{p_2} \{V_m(p) - V_m^\circ(p)\} \, dp.$$

This untidy expression can be improved. In the first place one of the pressures, p_1, is allowed to approach some very low value, and finally go to zero. Under these conditions the real gas behaves perfectly and so in this limit the fugacity f_1 can be identified with the pressure p_1. Therefore f_1/p_1 approaches 1 as p_1 approaches zero. If we drop the subscripts from p_2 and f_2, the last equation becomes in this limit

$$\ln(f/p) = \frac{1}{RT} \int_0^p \{V_m(p) - V_m^\circ(p)\} \, dp.$$

Now consider the integral. The perfect gas volume–pressure dependence is $V_m^\circ = RT/p$. The real volume–pressure dependence can be written in terms of the compression factor Z (p. 36) because $Z = pV_m/RT$ and so $V_m = RTZ/p$. (For a real gas Z depends on p and T.) Collecting all these remarks leads to

$$\ln(f/p) = \int_0^p \left\{ \frac{Z(p,T) - 1}{p} \right\} dp$$

or

(6.2.12) $$f = p \exp \int_0^p \left\{ \frac{Z(p,T) - 1}{p} \right\} dp.$$

The last equation is the recipe for determining the fugacity of a gas at any pressure. Experimental data are needed on the compression factor from very low pressures up to the pressure of interest. Sometimes these are available in numerical tables, in which case the integral may be evaluated graphically. Sometimes an analytical expression for $Z(p,T)$ is available (for example, from one of the equations of state encountered in Chapter 1) and it may be possible to evaluate the integral analytically. For instance, if we have the virial equation coefficients in the form of eqn (1.3.1), we can obtain the fugacity simply by inserting the appropriate pressure into

$$f = p \exp(B'p + \tfrac{1}{2}C'p^2 + \ldots),$$

which is obtained by explicit evaluation of the integral in eqn (6.2.12).

Example (Objective 7). Consider two simplified forms of the van der Waals equation of state (p. 47), one in which $b = 0$, the other with $a = 0$, and find an explicit expression for the fugacity of the gas in each case when its pressure is 10 atm. Use the data for ammonia (Table 1.1) to assess the contributions to the fugacity of the two aspects of imperfection.

- *Method.* When $b = 0$ the equation of state is

$$p = RT/V_m - a/V_m^2,$$

and when $a = 0$

$$p = RT/(V_m - b).$$

In order to use eqn (6.2.12) we need to know $Z = pV_m/RT$ as a function of p. This involves inverting these two equations to find V_m. With Z found, the integration in eqn (6.2.12) can be carried out without difficulty.

- *Answer.* When $b = 0$, $V_m = (RT/2p)\{1 + [1 - (4ap/R^2T^2)]^{1/2}\}$ (a result obtained by solving the quadratic equation for V_m). If deviation from perfect behaviour is small, the term $4ap/R^2T^2$ is small and the square root can be approximated by $(1 - 4ap/R^2T^2)^{1/2} \approx 1 - 2ap/R^2T^2$. The compression factor is then $Z \approx (p/RT) \times (RT/2p)(1 + 1 - 2ap/R^2T^2) = 1 - ap/R^2T^2$. The fugacity is then given by

$$f \approx p\exp\int_0^p \left\{\frac{(Z-1)}{p}\right\}dp = p\exp\int_0^p (-a/R^2T^2)dp = p\exp(-ap/R^2T^2).$$

When $a = 0$, $V_m = (RT/p) + b$ and so $Z = 1 + pb/RT$. Therefore

$$f = p\exp\int_0^p (b/RT)dp = p\exp(pb/RT).$$

Since for ammonia $a = 4.170 \, \text{dm}^6 \, \text{atm} \, \text{mol}^{-2}$ and $b = 3.707 \times 10^{-2} \, \text{dm}^3 \, \text{mol}^{-1}$ at room temperature (298.15 K) and 10 atm, the fugacity corrections depend on

neglect of b:

$$f = (10\,\text{atm})\exp\left\{\frac{-(4.170\,\text{dm}^6\,\text{atm}\,\text{mol}^{-2}) \times (10\,\text{atm})}{(0.082\,\text{dm}^3\,\text{atm}\,\text{K}^{-1}\,\text{mol}^{-1})^2 \times (298.15\,\text{K})^2}\right\}$$

$$= (10\,\text{atm})\exp(-0.0696)$$

$$= 9.32\,\text{atm};$$

neglect of a:

$$f = (10\,\text{atm})\exp\left\{\frac{(3.707 \times 10^{-2}\,\text{dm}^3\,\text{mol}^{-1}) \times (10\,\text{atm})}{(0.082\,\text{dm}^3\,\text{atm}\,\text{K}^{-1}\,\text{mol}^{-1}) \times (298.15\,\text{K})}\right\}$$

$$= (10\,\text{atm})\exp(0.01514)$$

$$= 10.15\,\text{atm}.$$

- *Comment.* We see that the effect of a is to reduce the fugacity and the effect of b is to increase it. This conforms to the relation of these parameters to the attractive and repulsive interactions respectively.

From Fig. 1.5 on p. 37 it is clear that for some gases the compression factor is initially less than unity, but it always becomes greater than unity at higher pressures. If Z is less than unity throughout the range of the integral from 0 to p, the exponent in eqn (6.2.12) is negative and the fugacity is less than the pressure (the molecules tend to stick together). At higher pressures the region where Z exceeds unity dominates the region

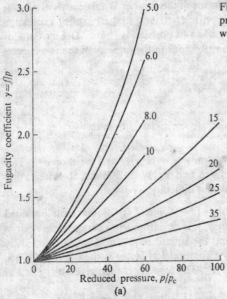

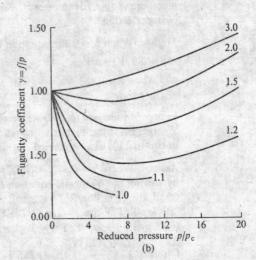

Fig. 6.1. How the fugacity of real gases depends on the pressure at different temperatures. (Lines are labelled with the reduced temperature T/T_c.)

where it is less than unity; the integral is then positive and the fugacity is greater than the pressure (the repulsive part of the' intermolecular potential is dominating and tending to cause the molecules to fly apart). Figure 6.1 indicates how the fugacity depends on the pressure for gases in terms of their reduced variables. Table 6.1 gives explicit values for nitrogen.

Table 6.1. The fugacity of nitrogen at 273 K, f/atm

p/atm	f/atm	p/atm	f/atm
1	0.99955	200	194.4
10	9.9560	300	301.7
50	49.06	400	424.8
100	97.03	600	743.4
150	145.1	1000	1839

Source: G. N. Lewis and M. Randall, *Thermodynamics* (revised by K. S. Pitzer and L. Brewer), McGraw-Hill.

The importance of the fugacity. We shall derive thermodynamically exact expressions for equilibrium constants in terms of the chemical potentials of the species involved in the reaction. The fugacity is simply another way of expressing the chemical potential because $\mu = \mu^{\ominus} + RT \ln f$ is no more than a definition of f. Therefore we shall be able to express equilibrium constants in terms of the fugacities of the components. Only if these fugacities can be related to pressures of the reacting species will the equations be of practical use. Thermodynamic expressions in terms of the fugacity are exact; the relation of the fugacity to the pressure is sometimes approximate, but gives results of practical utility.

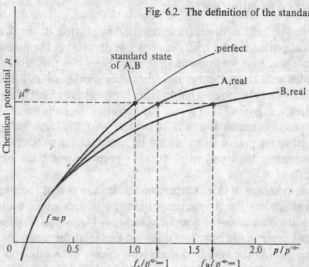

Fig. 6.2. The definition of the standard state of a real gas.

Standard states of real gases. The standard state of a perfect gas is established at 1 atm or $p = p^{\ominus}$. In the case of a real gas the chemical potential has its standard value when the fugacity is 1 atm, eqn (6.2.11). At 25 °C, $f = 1$ atm occurs at about $p = 1$ atm, and in all but the most accurate calculations it is normally adequate to identify $f = 1$ atm with $p = 1$ atm. Note, however, that we have not said that a real gas is *in* its standard state when its fugacity is 1 atm. This is because the definition of the standard state itself is quite subtle. This can be understood as follows. First write

$$(6.2.13) \qquad f = \gamma p.$$

The *fugacity coefficient* γ depends on the conditions, but approaches unity at very low pressures where the fugacity becomes equal to the pressure. Introducing this expression into the equation for the chemical potential gives

$$\mu(p) = \mu^{\ominus} + RT \ln(p/p^{\ominus}) + RT \ln \gamma.$$

The first two terms on the right have the form of the chemical potential of a perfect gas, and so $RT \ln \gamma$ measures the deviations from ideality. If, however, the *whole* of the deviations from ideality are to be ascribed to γ, the $\mu^{\ominus}$ must be characteristic of a perfect gas. We see that it is not sufficient to say that in the expression $\mu(p) = \mu^{\ominus} + RT \ln(f/p^{\ominus})$ the standard state is established at $f = 1$ atm. We should say that *the standard state is some hypothetical state in which the gas is behaving perfectly*.

One way of imagining this hypothetical (and unattainable) state is as one in which the real gas is at a pressure of 1 atm but in which all the molecular interactions have been extinguished. The concept is mysterious at first sight, but it is well defined, and Fig. 6.2 will help to make it clear. If we had a perfect gas its chemical potential could be related to its standard value by

$$\mu(p) = \mu^{\ominus} + RT \ln(p/p^{\ominus}).$$

This well-defined mathematical function is plotted in the diagram. The chemical potential of a real gas does not depend on the pressure in such a simple way, but we know that at sufficiently low pressures the fugacity coincides with the pressure, and so the chemical potential then coincides with that of a perfect gas. This equality is exact in the limit of zero pressure. The chemical potential of a real gas therefore depends on the pressure in the way indicated in Fig. 6.2. It is clear that the chemical potential at some arbitrary pressure can always be referred back to the standard value by descending the 'real' curve to zero pressure and then ascending the 'perfect' curve.

The advantage of this cumbersome definition is that the standard state of a real gas has the simple properties of a perfect gas: if we had defined the standard state as the one for which $f = 1$ atm, the standard states of different gases would have had different and relatively complex properties; for one gas (A) the standard state would have been attained at 1.2 atm, for another (B) it would have been attained at 1.7 atm, and so on. The choice of a hypothetical standard state literally standardizes the interactions between the molecules.

6.3 Open systems and changes of composition

The formulation presented so far is obviously incomplete because it has been tacitly assumed that composition remains constant. Furthermore, the expression $dG = Vdp - SdT$ suggests that G is a function of only p and T, and that when these are held constant $dG = 0$; this conflicts with the result derived on p. 149 that at constant p and T we have $dG = dw_{e,max}$. The two points are connected, for we shall now see that taking composition variation into account allows us to identify a source of non-pV work.

The Gibbs function is a function of composition as well as pressure and temperature, and should therefore be written $G(p, T, n_1, n_2, \ldots)$ and not simply $G(p, T)$. A general change can therefore be written

(6.3.1) $$dG = (\partial G/\partial p)_{T,n_1,n_2,\ldots} \, dp + (\partial G/\partial T)_{p,n_1,n_2,\ldots} \, dT$$
$$+ (\partial G/\partial n_1)_{p,T,n_2,\ldots} \, dn_1 + (\partial G/\partial n_2)_{p,T,n_1,\ldots} \, dn_2.$$

This fearsome expression simply says that G may change because any of its variables may change. How G depends on its variables depends on its composition and the intensive variables (p, T) that determine the state of the system, and so all the coefficients in the expression may depend on the composition, the pressure, and the temperature.

The first job is to simplify the appearance of the last expression. As a first step, consider how G responds to infinitesimal changes in the pressure and temperature, its composition being held constant:

$$dG = (\partial G/\partial p)_{T,n_1,n_2,\ldots} \, dp + (\partial G/\partial T)_{p,n_1,n_2,\ldots} \, dT.$$

We already know that $dG = Vdp - SdT$ under the same conditions, and since dG is an exact differential we may identify the coefficients:

$$(\partial G/\partial p)_{T,n_1,n_2,\ldots} = V$$

$$(\partial G/\partial T)_{p,n_1,n_2,\ldots} = -S.$$

These are the same as eqns (6.2.2) and (6.2.3), the only difference being that they are dressed more elaborately because the constant composition is specified explicitly. In passing, note that V and S both depend on the amounts of the components.

Now suppose that only substance 1 is present. G can be written as $n_1 G_{1m}$, which is $n_1\mu_1$, where μ_1 is the chemical potential of pure component 1. It follows that at constant temperature and pressure, $dG = \mu_1 \, dn_1$, and consequently

$$(\partial G/\partial n_1)_{p,T} = \mu_1.$$

Expressed in this way it is seen that the chemical potential carries the information about how G changes when species 1 is added.

..The last remark can be generalized as follows. Henceforth we define the chemical potential of a species in terms of the effect it has on the Gibbs function of a system of arbitrary composition. That is, we *define* the chemical potential of species 1 as the coefficient

(6.3.2)
$$\mu_1 \overset{\text{def}}{=} (\partial G/\partial n_1)_{p,T,n2\ldots}$$

This reduces to the earlier definition in the case of a single component because then $G = nG_m$. Notice that the chemical potential of a substance depends not only on the pressure and temperature but also on the composition. Thus adding 10^{-3} mol of methanol (which is almost an infinitesimal amount) to $1 \, \text{dm}^3$ of a 20 per cent methanol/water mixture leads to a change in Gibbs function of the total system which is different from the change when the same amount is added to $1 \, \text{dm}^3$ of an 80 per cent mixture. It is also amusing to note that a species may have a chemical potential even when it is absent! This is because the Gibbs function of a system will change when species 1 is added even though it contains no species 1 originally, and so $(\partial G/\partial n_1)_{p,T,n_2,\ldots}$ is non-zero (negatively infinite in fact) even at $n_1 = 0$.

We can now introduce the definition of the chemical potential into the general expression for dG and obtain the much simpler form

(6.3.3)
$$dG = Vdp - SdT + \mu_1 \, dn_1 + \mu_2 \, dn_2 + \ldots$$

All the coefficients depend, in general, on the conditions and the composition, but that will be taken for granted so that the appearance of the equation remains simple.

Under conditions of constant pressure and temperature the last expression reduces to

$$dG = \mu_1 \, dn_1 + \mu_2 \, dn_2 + \ldots$$

Now we see the source of the non-pV work in $dG = dw_{e,max}$ at constant pressure and temperature: it arises from the variation of the composition, such as may occur in a chemical reaction inside an electrochemical or a biological cell.

This idea can be taken further, because we know how G is related to U: $G = H - TS$, implying that $G = U + pV - TS$. A general change in U can therefore be written

$$dU = dG - pdV - Vdp + SdT + TdS$$

$$= (Vdp - SdT + \mu_1 dn_1 + \mu_2 dn_2 + \ldots) - pdV - Vdp + SdT + TdS$$

$$(6.3.4) \qquad = \mu_1 dn_1 + \mu_2 dn_2 + \ldots - pdV + TdS,$$

which is the generalization of eqn (6.1.2) to systems in which the composition may change. When the change is constrained to occur reversibly TdS is dq_{rev}, and so the remaining part of the expression is dw_{rev}. When the system is at constant volume the only work is non-pV work, and so

$$dU = dw_{e,rev} = \mu_1 dn_1 + \mu_2 dn_2 + \ldots,$$

which identifies how changing composition is related to non-pV work. Furthermore, eqn (6.3.4) lets us write

$$(\partial U / \partial n_1)_{V,S,n_2,\ldots} = \mu_1;$$

and so not only does the chemical potential show how G depends on the amount of a particular species present, it also shows how the internal energy varies (but under a different set of conditions). In the same way it is a simple matter to deduce that

$$(\partial H / \partial n_1)_{p,S,n_2,\ldots} = \mu_1 \quad \text{and} \quad (\partial A / \partial n_1)_{V,T,n_2,\ldots} = \mu_1.$$

Thus we see that μ_J are indeed *potentials* in as much as they reveal how all the extensive thermodynamic properties U, H, G, and A depend on the composition.

There remains one more item of business. We stressed on p. 146 that a central result for the whole of chemical thermodynamics was the result $dG \leqslant 0$ for a system at constant pressure and temperature, and one not doing any non-pV work. How does the possibility that composition may change affect this conclusion? The point to appreciate is that even when a system is changing its composition it need not be doing any work: for instance, the reaction characteristic of an electrochemical cell may also take place outside the cell. Therefore $dG \leqslant 0$ indicates the direction of spontaneous change for *any* change. The difference, though, is that the changing composition *may* be harnessed to do non-pV work, and *if* it is harnessed the maximum amount of work that can be obtained for a specified infinitesimal change of composition at constant pressure and temperature is dG, or $\mu_1 dn_1 + \mu_2 dn_2$.

Equation (6.3.3) is the *fundamental equation of chemical thermodynamics*:

its implications and consequences are explored and developed in the next six chapters.

Further reading

Basic chemical thermodynamics (2nd ed.). E. B. Smith; Oxford University Press, 1977.

Elementary chemical thermodynamics. B. H. Mahan; Benjamin, New York, 1963.

An introduction to chemical thermodynamics. E. F. Caldin; Clarendon Press, Oxford, 1961.

Chemical thermodynamics. I. Klotz and R. M. Rosenberg; Benjamin, New York, 1972.

Thermodynamics. G. N. Lewis and M. Randall, revised by K. S. Pitzer and L. Brewer; McGraw-Hill, New York, 1961.

Thermodynamics. J. G. Kirkwood and I. Oppenheim; McGraw-Hill, New York, 1961.

Chemical thermodynamics. M. L. McGlashan, Academic Press, London, 1979.

Thermodynamics. E. A. Guggenheim; North-Holland, Amsterdam, 1967.

Problems

6.1. Show that $(\partial U/\partial S)_V = T$ and $(\partial U/\partial V)_S = -p$ in the case of a perfect gas.

6.2. Two of the four Maxwell relations were derived in the text, but two were not. Complete their derivation by showing that $(\partial S/\partial V)_T = (\partial p/\partial T)_V$ and $(\partial T/\partial p)_S = (\partial V/\partial S)_p$.

6.3. One of the ways of employing these relations can be illustrated by using the first of the two just derived to show that in the case of a perfect gas the entropy depends on the volume as $S = \text{const} + R \ln V$, as was seen by explicit calculation in Chapter 5.

6.4. One of the *thermodynamic equations of state*, that for $(\partial U/\partial V)_T$, was derived on p. 164, eqn (6.1.8). In Chapter 3, Problem 3.14, the corresponding equation involving $(\partial H/\partial p)_T$ was derived in the case of a perfect gas. Now deduce a general expression for $(\partial H/\partial p)_T$ applicable to all substances.

6.5. Find an expression for the dependence of the enthalpy on the pressure $(\partial H/\partial p)_T$, for (a) a perfect gas, (b) a van der Waals gas. Estimate its value for 1 mol of argon at 10 atm and 25 °C. By how much does the enthalpy of the argon change when the pressure is changed to 11 atm without change of temperature.

6.6. Deviations from perfect behaviour are often listed in terms of the virial equation of state $pV_m/RT = 1 + B/V_m + \dots$. Show that the important quantity $(\partial U/\partial V)_T$ can be obtained from tables of the coefficient B as a function of temperature, and if $\Delta B = B(T_2) - B(T_1)$ and $\Delta T = T_2 - T_1$, that $(\partial U/\partial V)_T \approx (RT^2/V_m^2)(\Delta B/\Delta T)$, where T is the mean of T_1 and T_2. In the case of argon, the second virial coefficient is $-28.0 \, \text{cm}^3 \, \text{mol}^{-1}$ at 250 K and $-15.6 \, \text{cm}^3 \, \text{mol}^{-1}$ at 300 K. Estimate $(\partial U/\partial V)_T$ at 275 K and (a) 1 atm, (b) 10 atm.

6.7. Prove that the heat capacities of a perfect gas are independent of both volume and pressure. May they depend on the temperature?

6.8. Deduce an expression for the volume dependence of $C_{V,m}$ of a gas that is described by $pV_m/RT = 1 + B/V_m$, and calculate the change of heat capacity when a 1 mol sample of argon at 10 atm is reduced isothermally to 1 atm. You will need the following values of the virial coefficients: $-15.49\,cm^3\,mol^{-1}$ at 298 K, $-11.06\,cm^3\,mol^{-1}$ at 323 K, and $-7.14\,cm^3\,mol^{-1}$ at 348 K.

6.9. Confirm the relation used in Problem 3.10 that the Joule–Thomson coefficient for a gas is given by $\mu_{JT}C_p = T(\partial V/\partial T)_p - V$, and show that this may be expressed in terms of the isobaric expansivity α as $\mu_{JT}C_p = V(\alpha T - 1)$.

6.10. The *Joule coefficient* for the free expansion of a gas is defined by $\mu_J = (\partial T/\partial V)_U$. Show that it may be expressed in terms of C_V, α, and κ the isothermal compressibility.

6.11. The parameter a in the van der Waals equation of state, Table 1.4 on p. 47, represents the role of attractive interactions between the molecules of a gas. We should therefore expect it to govern the magnitude of $(\partial U/\partial V)_T$ and that this coefficient should disappear when the molar volume is very large. Relate the coefficient to the parameter a and confirm that it shows this behaviour. Evaluate $(\partial U/\partial V)_T$ for argon at (a) 1 atm, (b) 10 atm. Express your answer in $kJ\,mol^{-1}\,m^{-3}$.

6.12. Find an expression for $(\partial U/\partial V)_T$ for a Dieterici gas, and express the results in reduced variables by defining the reduced internal energy by $U = (a/e^2b)U_r$.

6.13. We have already met the isothermal compressibility κ. If compression takes place adiabatically the relevant quantity is the *adiabatic compressibility* κ_S. Why is the subscript S? Show that $p\gamma\kappa_S = 1$ for a perfect gas, where $\gamma = C_p/C_V$.

6.14. A very useful expression for $T\,dS$ can be deduced on the basis of the Maxwell relations and the properties of partial derivatives, and so this Problem has the double job of giving some experience in these manipulations and at the same time finding a valuable result. The work is spread over this and the next few Problems. First, show that $T\,dS = C_V\,dT + T(\partial p/\partial T)_V\,dV$ by starting at the statement 'S may be regarded as a function of T and V'. Hence calculate how much heat has to be transferred to 1 mol of a van der Waals gas when it is expanded reversibly and isothermally from V_i to V_f.

6.15. An alternative expression for $T\,dS$ can be derived on the basis that S is a function of T and p. Show that $T\,dS$ may be written in terms of $C_p\,dT$ and dp. Hence find the heat involved in the reversible, isothermal compression of a liquid or solid of constant isobaric expansivity α when the pressure is increased from p_i to p_f. Evaluate the heat involved in the compression of $100\,cm^3$ of mercury at $0\,°C$ when the pressure on the sample is increased from zero to 1000 atm. ($\alpha = 1.82 \times 10^{-4}\,K^{-1}$.)

6.16. The properties of materials change quite markedly when they are subjected to high pressures, and there is now a great deal of interest in their behaviour. In one experiment 100 g of benzene at 298 K was subjected to a pressure that was increased reversibly and isothermally from zero to 4000 atm. Calculate (a) the heat transferred to the sample, (b) the work done, (c) the change in its internal energy. Take $\alpha = 1.24 \times 10^{-3}\,K^{-1}$, $\rho = 0.879\,g\,cm^{-3}$, $\kappa = 9.6 \times 10^{-5}\,atm^{-1}$.

6.17. Assuming that water is incompressible, estimate the change in Gibbs function of $100\,cm^3$ of water at $25\,°C$ when the pressure is changed from 1 atm to 100 atm.

6.18. In fact the volume of a sample does change when it is subjected to pressure, and so we ought to see how to take the effect into account, and then judge whether it is significant. The volume varies in a way that can be determined by specifying the isothermal compressibility $\kappa = -(1/V)(\partial V/\partial p)_T$, and we may assume that this is virtually constant over a pressure range of interest. (That is still an approximation, but it is a better approximation than assuming that the volume itself is constant.) Deduce an expression for the Gibbs function at p_f in terms of its value at p_i, the original volume of the sample, and the compressibility.

6.19. Now we are in a position to estimate the change of G with pressure, and to judge the significance of assuming that the volume remains unchanged. The isothermal compressibility of copper is $0.8 \times 10^{-6} \text{ atm}^{-1}$. What is the change in its molar Gibbs function when it is subjected to an extra pressure of (a) 100 atm, (b) 10 000 atm? What error is introduced by assuming that the block is incompressible? (The density of copper is 8.93 g cm^{-3}.)

6.20. Repeat the calculation in the last Problem, but for water. The isothermal compressibility is $4.94 \times 10^{-5} \text{ atm}^{-1}$ at 25 °C and the density is 0.997 g cm^{-3}.

6.21. Much simpler expressions generally result when approximations are made, but it is always important to know whether they are plausible. In order to find the temperature dependence of the Gibbs function of a reaction it is sometimes supposed that ΔH is independent of temperature. In fact we know that ΔH is temperature dependent, and Kirchhoff's law is that $\Delta H(T) = \Delta H(T_i) + (T-T_i)\Delta C_p$, p. 110. Find an approximate expression for $\Delta G(T_i)$ on the basis that ΔH is independent of temperature. Then take the temperature dependence into account by integrating eqn (6.2.5) with the Kirchhoff law expression for $\Delta H(T)$. Assume that ΔC_p is independent of temperature.

6.22. The full expression just found is much more inconvenient than the simplification; but how important are the additional terms? The standard enthalpy of the reaction $H_2 + \frac{1}{2}O_2 \rightarrow H_2O$ is $-285.8 \text{ kJ mol}^{-1}$ at 298 K, and the heat capacities are H_2: $28.8 \text{ J K}^{-1}\text{ mol}^{-1}$; O_2: $29.4 \text{ J K}^{-1}\text{ mol}^{-1}$, and $H_2O(l)$: $75.3 \text{ J K}^{-1} \text{ mol}^{-1}$. The Gibbs function for the reaction is $-237.2 \text{ kJ mol}^{-1}$ at 298 K; what is its value at 330 K? What is the error of ignoring the heat capacities of the components (i.e., of assuming that ΔH is independent of temperature)?

6.23. The standard enthalpy of the reaction $N_2 + 3H_2 \rightarrow 2NH_3$ is $-92.2 \text{ kJ mol}^{-1}$ of ammonia and the standard Gibbs function is $-31.0 \text{ kJ mol}^{-1}$, both at 298 K. Estimate the Gibbs function at (a) 500 K, (b) 1000 K. Is the reaction spontaneous at room temperature? Is the formation of ammonia favoured or disfavoured by a rise in temperature?

6.24. At 25 °C the standard enthalpy of combustion of sucrose is $-5645 \text{ kJ mol}^{-1}$ and the Gibbs function $-5797 \text{ kJ mol}^{-1}$. What is the extra amount of non-pV work that can be obtained per mole by raising the temperature from 25 °C to 35 °C?

6.25. Calculate the change in Gibbs function of 1 mol of hydrogen, regarded as a perfect gas, when it is compressed isothermally from 1 atm to 100 atm at 298 K.

6.26. The volume of a newly synthesized plastic was found to depend exponentially on the pressure, it being V_0 when there was no excess pressure, and decreased towards zero as $V(p) = V_0 \exp(-\bar{p}/p^*)$, where $\bar{p}$ is the excess pressure and p^* some constant. Deduce an expression for the Gibbs function of the plastic as a function of excess pressure. What is the natural direction of change of the compressed material?

6.27. The *fugacity* f of a gas at a pressure p is given by eqn (6.2.12). In order to evaluate it we have to know how the compression factor varies with pressure all the way up to the pressure of interest. This information can be given in several ways, and the next few Problems give some examples. One way of arriving at the fugacity is to measure the compression factor Z over a range of pressures, and then to do the integration numerically or graphically. Find the fugacity of oxygen at 100 atm and 200 K on the basis of the following data:

p/atm	1	4	7	10	40	70	100
Z_m	0.99701	0.98796	0.97880	0.96956	0.8734	0.7764	0.6871

6.28. At such low temperatures and high pressures as in the last Problem, it is not surprising that the fugacity differs markedly from the pressure. When the conditions are not so extreme some of the equations of state may be used and the integration performed explicitly. For instance, suppose that the gas obeys the virial equation of state, $pV_m/RT = 1 + B/V_m + C/V_m^2$. Find an expression for the fugacity and $\ln \gamma$ in terms of a power expansion in the pressure (Problem 1.27).

6.29. On the basis of the expression just obtained, estimate the fugacity of argon at 1 atm pressure at (a) 100 K, (b) 273 K. Use the following data. At 273 K: $B = -21.13 \, \text{cm}^3 \, \text{mol}^{-1}$, $C = 1054 \, \text{cm}^6 \, \text{mol}^{-2}$; at 373 K: $B = -3.89 \, \text{cm}^3 \, \text{mol}^{-1}$, $C = 918 \, \text{cm}^6 \, \text{mol}^{-2}$.

6.30. In order to see how the fugacity differs from the pressure over a typical range, plot the fugacity of argon at 273 K for the range $0{-}10^3$ atm pressure, and on the same graph plot the fugacity of a perfect gas. At what pressure does argon have unit fugacity at 273 K?

6.31. The second virial coefficient of ammonia has the value $-261 \, \text{cm}^3 \, \text{mol}^{-1}$ at 298 K. What is its fugacity when its pressure is (a) 1 atm, (b) 100 atm? Use the condition $(C - B^2/2R^2T^2)p^2 \ll Bp/RT$.

6.32. A third way of arriving at the fugacity is to use one of the approximate equations of state introduced in Chapter 1. What is the fugacity of a gas that obeys the equation of state $pV_m/RT = 1 + BT/V_m$?

7   Changes of state: physical transformations of pure materials

Learning objectives

After careful study of this chapter you should be able to

(1) State how the chemical potential depends on temperature and pressure (eqns (7.1.1) and (7.1.2)).

(2) Describe how changes of pressure and temperature affect the boiling and freezing characteristics of solids, liquids, and gases (p. 184).

(3) Define *vapour pressure* (p. 185) and derive and use the *Clapeyron equation* (eqn (7.2.1)) for the temperature dependence of the vapour pressure.

(4) Represent the states of a system in terms of a *phase diagram* (p. 188) and derive expressions for the *phase boundaries* (pp. 187–91).

(5) Derive and use the *Clausius–Clapeyron equation* (eqn (7.2.3)) relating vapour pressure to temperature.

(6) State the *phase rule* for a system with one component (eqn (7.3.1)).

(7) Draw and interpret the phase diagrams for water, carbon dioxide, carbon, and helium.

(8) Define *first-order* and *second-order* phase transitions, and indicate what is meant by a *λ-transition*.

(9) Define *coefficient of performance* (p. 199), deduce an expression for it in terms of the temperature (eqn (7.6.1)), and calculate the minimum work necessary to bring about a specified amount of cooling (p. 200).

(10) Describe the method of attaining very low temperatures using *adiabatic demagnetization*.

(11) Define *surface tension* (p. 203) and write an expression for the dependence of the Gibbs function on surface area (7.7.2)).

(12) Derive the *Laplace equation* (eqn (7.7.3)) for the pressure difference across a curved surface, and use it to derive the *Kelvin equation* (eqn (7.7.5)) for the vapour pressure of droplets and bubbles.

(13) Indicate the meaning and importance of *nucleation* (p. 207).

(14) Derive and use an expression for the height to which a liquid can rise by capillary action (eqn (7.7.7)).

Introduction

Boiling, freezing, and the onset of ferromagnetism are all aspects of changes of phase without change of chemical composition. This chapter describes these processes in terms of the language of thermodynamics. The guiding principle is the tendency of systems to slide down the Gibbs hill to lower chemical potential because that corresponds to increasing overall entropy at constant temperature and pressure. If at a particular temperature and pressure a vapour has a lower chemical potential than the liquid, then the system has a natural tendency to evaporate and attain the lower chemical potential. If a solid lies lower in chemical potential than a liquid, then the solid is the stable phase under those conditions, and the liquid will have a natural tendency to freeze.

In this chapter we deal with the thermodynamics of phase transitions, and not with their rates. This restriction is of considerable practical importance. For instance, at normal temperatures graphite has a lower chemical potential than diamond, yet diamonds do not crumble into graphite. There is certainly a thermodynamic tendency for diamonds to decay, but in order to do so the atoms have to rearrange their crystal lattice, and this is an insignificantly slow process except at very high temperatures. The rate of attainment of equilibrium is a *kinetic* problem, and is outside the range of thermodynamics. In gases, liquids, and liquid solutions the molecular mobilities allow equilibria to be obtained rapidly, but in solids thermodynamic instability may be frozen in.

7.1 The stability of phases

The chemical potential of a sample of matter must be uniform when it is at equilibrium. In order to see why this must be true, consider a system in which the chemical potential varies from place to place. Suppose that at some point the potential is $\mu(1)$ and at another it is $\mu(2)$. When an amount of substance dn is transferred from the first point to the second at constant pressure and temperature the Gibbs function changes by $-\mu(1)\,dn$ on the first operation, and by $+\mu(2)\,dn$ on the second. The overall change is $dG = \{\mu(2) - \mu(1)\}\,dn$. If the chemical potential at 1 is higher than at 2, the transfer is accompanied by a decrease in the Gibbs function, and so it has a natural tendency to occur. Only if $\mu(1) = \mu(2)$ does no change occur in the Gibbs function, and so only then is the distribution of matter in equilibrium.

This *principle of uniform chemical potential* applies however many phases are in equilibrium. For instance, if a liquid and its vapour are in equilibrium, then the chemical potential is the same throughout the vapour, throughout the liquid, and it is the same in the vapour as in the liquid.

The chemical potentials of the solid, liquid, and gas (or vapour) phases will be written $\mu(s)$, $\mu(l)$, and $\mu(g)$. If at the temperature and pressure of interest $\mu(l)$ is smaller than the others, then the liquid is the stable phase. On lowering the temperature $\mu(l)$ might rise above $\mu(s)$ (see Fig. 7.1), in which case the solid becomes more stable, and the natural process is

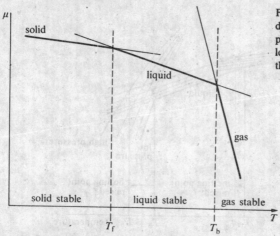

Fig. 7.1 The temperature dependence of the chemical potential. The phase with the lowest μ is the stable phase at that temperature.

freezing. Raising the temperature might lower the gas-phase potential beneath the liquid-phase potential, and the system will roll down that slope and turn spontaneously into a vapour.

Do the chemical potentials behave in this way? In the last chapter we saw that at constant pressure the chemical potential varies with temperature according to

(7.1.1)
$$(\partial\mu/\partial T)_p = -S_m$$

(see eqn (6.2.2) on p. 165). We immediately conclude that as the temperature rises the chemical potential of a pure substance falls (because S_m is always positive), and the gradient is steeper for gases than for liquids $(S_m(g) > S_m(l))$, and steeper for a liquid than the corresponding solid $(S_m(l) > S_m(s))$.

These deductions enable the rough form of the temperature dependence of the chemical potentials to be sketched for three possible phases of a pure system, see Fig. 7.1. This diagram shows the regions of temperature where each phase is the most stable. At high temperatures the vapour is the most stable, but at a temperature T_b the vapour's potential rises above the liquid's, and the liquid is then more stable. On cooling the liquid a point is reached, T_f, where the solid becomes most stable, and the system freezes. The temperature at which the liquid and solid are able to coexist in equilibrium, which is when their chemical potentials are equal, is the *melting point*. The temperature (at a given pressure) at which the chemical potentials of gas and liquid are equal, which corresponds to gas and liquid existing together in equilibrium, is the *boiling point* at that pressure. Conversely, the pressure corresponding to liquid–vapour equilibrium at a specified temperature is the *vapour pressure* at that temperature.

Figure 7.1 can be adapted to show the pressure dependence of the boiling and freezing points. The chemical potential depends on the pressure according to

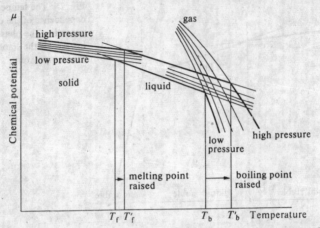

Fig. 7.2. The pressure dependence of the chemical potential, and its effect on melting and boiling points.

(7.1.2) $\qquad (\partial\mu/\partial p)_T = V_\mathrm{m}$

(see p. 165). An increase in pressure raises the chemical potential of a pure substance (because V is certainly positive) and increases it much more for gases than either liquids or solids (because the molar volume of a gas is typically about 1000 times larger than the molar volume of either a liquid or solid). In most, but not all, cases the volume of a solid increases when it melts, and so at the melting point the molar volume of a liquid is greater than the molar volume of the solid. The equation then predicts that an increase in pressure increases the chemical potential of a liquid slightly more than that of the solid. These changes are shown in Fig. 7.2. The effect is to raise the boiling point quite appreciably, and to raise the melting point, but only slightly.

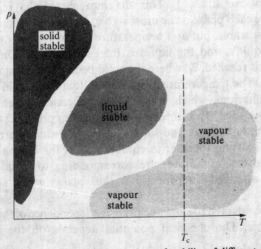

Fig. 7.3. The general p, T regions of stability of different phases.

7.2 How to depict phase equilibria: phase diagrams

Figure 7.3 indicates in a general way the regions of pressure and temperature where each phase is the most stable.

The boundaries between the regions lie at the values of p and T where the two phases coexist. Since the phases are in equilibrium on the phase-diagram boundaries, their chemical potentials are then equal. Therefore, for the phases α and β to be in equilibrium, the pressure and temperature must be such as to satisfy $\mu(\alpha; p, T) = \mu(\beta; p, T)$. By solving this equation for p as a function of T we shall find the equation of the boundary curve. In the case of the liquid/vapour and solid/vapour boundaries the p versus T curve is the dependence of the *vapour pressure* on the temperature.

Example (Objective 3). What the vapour pressure means in practice can be explained by considering the system shown in Fig. 7.4a. The vessel contains water at room temperature and the pressure exerted by the piston can be controlled (e.g., by adding weights to it, or by pulling it out without allowing any air to enter). Suppose the pressure exerted is 1 atm and that we monitor the volume of the

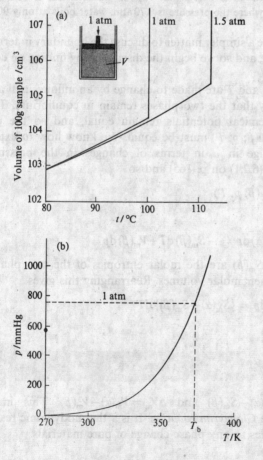

Fig. 7.4. Vapour pressure.

system (the water) as the temperature is raised. The volume expands smoothly with temperature, Fig. 7.4a, until the temperature reaches 100 °C. At that point there is a massive increase in volume, and although heat might be injected into the system the temperature does not rise until all the liquid has evaporated. When both liquid and vapour are present the temperature stays at 100 °C, and the proportions of the two phases can be modified arbitrarily either by withdrawing heat (which causes condensation, the temperature remaining at 100 °C) or by heating (which causes evaporation at 100 °C). At this temperature (100 °C) and pressure (1 atm) the liquid and vapour are in equilibrium: we say that the vapour pressure of water at 100 °C is 1 atm. When all the liquid has evaporated the addition of heat raises the temperature of the vapour, and it expands.

If weights are added to the piston sufficient to exert a pressure of 1.5 atm, the same heating experiment gives a sharp increase in volume, and an equilibrium between liquid and vapour, when the temperature is 112 °C. Therefore we say that the vapour pressure at 112 °C is 1.5 atm.

The same sequence can be repeated for any external pressure, and the *vapour pressure curve* drawn in Fig. 7.4b. Note that this curve can be used to predict the boiling point of water under any pressure conditions. The boiling point is the temperature at which liquid and vapour are in equilibrium. Therefore to find the boiling point when the pressure is 1.5 atm we simply have to read off the temperature at which water exerts that vapour pressure: this is 112 °C. At an altitude of 10 000 ft where the pressure is 0.70 atm, water boils at only 90 °C.

It turns out to be a simpler matter to discuss the boundary in terms of its slope at any point, and so we begin the discussion by finding an equation for dp/dT.

Suppose both p and T are made to change by an infinitesimal amount, but in such a way that the two phases remain in equilibrium. The new values of the chemical potentials remain equal, and so the changes $d\mu(\alpha; p, T)$ and $d\mu(\beta; p, T)$ must be equal. We know how to express an infinitesimal change in μ in terms of changes in the pressure and temperature (eqn (6.2.1) on p. 165) and so

$$d\mu(\alpha; p, T) = d\mu(\beta; p, T)$$

implies

$$-S_m(\alpha)\,dT + V_m(\alpha)\,dp = -S_m(\beta)\,dT + V_m(\beta)\,dp$$

where $S_m(\alpha)$ and $S_m(\beta)$ are the molar entropies of the two phases and $V_m(\alpha)$ and $V_m(\beta)$ their molar volumes. Rearranging this gives

$$\{V_m(\alpha) - V_m(\beta)\}\,dp = \{S_m(\alpha) - S_m(\beta)\}\,dT$$

or

(7.2.1) $$dp/dT = \Delta S_m/\Delta V_m$$

where $\Delta S_m = S_m(\alpha) - S_m(\beta)$ and $\Delta V_m = V_m(\alpha) - V_m(\beta)$. This important result is called the *Clapeyron equation*. It is a thermodynamic result, it is exact, and it applies to any phase change of pure materials.

The solid–liquid boundary. Melting is accompanied by a molar enthalpy of fusion $\Delta H_{melt,m}$ and occurs at some temperature T_f. The molar entropy of melting is therefore $\Delta H_{melt,m}/T_f$ and so the Clapeyron equation becomes

$$dp/dT = \Delta H_{melt,m}/T_f\Delta V_{melt,m},$$

where $\Delta V_{melt,m}$ is the molar volume change on melting. The enthalpy of melting is positive, and the volume change is positive except in a number of odd but important systems, but it is always small. This indicates that the slope dp/dT is large and usually positive (but sometimes negative). The last equation can be integrated immediately if ΔH_{melt} and ΔV_{melt} are assumed to be independent of both pressure and temperature (because solids and liquids respond strongly to neither). The equation of the solid–liquid equilibrium curve, the boundary in the phase diagram, is then

(7.2.2)
$$p = p^* + (\Delta H_{melt,m}/\Delta V_{melt,m})\ln(T/T^*)$$

where p^* and T^* are the pressure and temperature on some point of the line. This equation was originally obtained by yet another Thomson— this time James the brother of Lord Kelvin. When T is close to T^*, the logarithm can be approximated using $\ln(T/T^*) = \ln\{1+(T-T^*)/T^*\} \approx (T-T^*)/T^*$, and then

(7.2.2a)
$$p \approx p^* + (\Delta H_{melt,m}/T^*\Delta V_{melt,m})(T-T^*),$$

which is linear in T. This equation allows us to sketch in the boundary shown in Fig. 7.5a.

Example (Objective 4). Construct the ice/liquid equilibrium curve for water for temperatures, between $-1\,°C$ and $+1\,°C$. What is the freezing point of water under a pressure of 1500 atm $(1.5 \times 10^8\,N\,m^{-2})$?

- *Method.* Both parts of the question can be answered on the basis of eqn (7.2.2). For the fixed points (p^*, T^*) we take the triple point $(6.03 \times 10^{-3}\,atm, 273.16\,K)$, and for the enthalpy of fusion we take $\Delta H_{melt,m}(273\,K) = 6.01\,kJ\,mol^{-1}$, Table 4.7, and assume it to be constant over the range of interest. For $\Delta V_{melt,m}$ we take $-1.7\,cm^3\,mol^{-1}$ from density measurements. For the second part we take (p^*, T^*) as the point $(1\,atm, 273.15\,K)$, the conventional freezing point, and use the equation to find T when $p = 1500\,atm$.

- *Answer.* Evaluation of the equation over the stated range gives the following values of the equilibrium pressure:

$t/°C$:	-1.0	-0.5	0.0	0.5	1.0
p/atm:	129	65.1	1.28	-62.6	-126

This is plotted in Fig. 7.6 as line EF. For the second part we have

$$T = T^* \exp\{(p-p^*)\Delta V_{melt,m}/\Delta H_{melt,m}\}$$

$$= (273.15\,K)\exp\left\{\frac{(1500-1) \times (1.0133 \times 10^5\,N\,m^{-2}) \times (-1.7\,cm^3\,mol^{-1})}{(6.01\,kJ\,mol^{-1})}\right\}$$

$$= (273.15\,K)\exp(-0.043)$$

$$= 261.7\,K, \text{ or } -11.5\,°C.$$

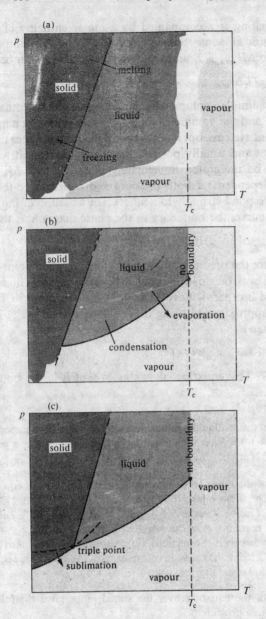

Fig. 7.5. General form of the phase boundaries.

● *Comment.* Note the decrease in freezing point with pressure: the water molecules occupy less space when packed together as a liquid, and so ice responds to pressure by melting. We have asssumed ΔV_m to be constant.

The liquid–gas boundary. The enthalpy of vaporization ΔH_{vap} determines the appropriate value of the entropy change. If $\Delta V_{vap,m}$ is the molar volume change on vaporization, the Clapeyron equation becomes

$$dp/dT = \Delta H_{vap,m}/T\Delta V_{vap,m}.$$

$\Delta H_{vap,m}$ is positive; $\Delta V_{vap,m}$ is also large and always positive. Therefore dp/dT is much smaller than in the previous case, but is always positive. A rough estimate of its value can be obtained by replacing $\Delta H_{vap,m}/T$ at the boiling point by Trouton's constant (p. 143), $85\,J\,K^{-1}\,mol^{-1}$ and using for $\Delta V_{vap,m}$ the molar volume of a perfect gas (c. $30\,dm^3\,mol^{-1}$ at 1 atm). This gives a slope of $4\times10^{-2}\,atm\,K^{-1}$, or a dT/dp slope of $25\,K\,atm^{-1}$, which means that a change of pressure of 0.1 atm changes the boiling point by about 2.5 K.

The slope depends on $\Delta V_{vap,m}$, and the volume of the vapour is sensitive to pressure. Increasing pressure decreases $V(g)$ and therefore ΔV_{vap} also. It follows that dp/dT increases as p increases, and this is why the line drawn in Fig. 7.5b curls upwards.

We can be more definite about the shape of the vapour-pressure curve by recognizing that $\Delta V_{vap,m}$ can be replaced by $V_m(g)$ itself because $V_m(g)$ is so much bigger than $V_m(l)$. If we also assume that the vapour behaves perfectly $V_m(g)$ may be replaced by RT/p. These approximations turn the Clapeyron equation into the

(7.2.3)° | *Clausius–Clapeyron equation*: $d(\ln p)/dT = \Delta H_{vap,m}/RT^2$.

(We have used the relation $dx/x = d(\ln x)$.) If we also assume that $\Delta H_{vap,m}$ does not depend on the temperature this equation integrates to

$$\ln p = \text{const} - \Delta H_{vap,m}/RT$$

or

(7.2.4)°
$$p = p^* \exp\left\{\frac{-\Delta H_{vap,m}}{R}\left(\frac{1}{T} - \frac{1}{T^*}\right)\right\},$$

where p^* is the vapour pressure at some temperature T^*. This formula gives the p versus T curve representing the states of the system where liquid and vapour coexist and is plotted in Fig. 7.5b. The line does not extend beyond the critical temperature T_c because above this temperature a liquid phase does not exist (recall Chapter 1). Instead of regarding the liquid/vapour boundary as a plot of vapour pressure against temperature, we can also regard it as a plot of the boiling point against applied pressure. The *normal boiling point* T_b is the temperature when the vapour pressure is 1 atm.

Example (Objective 5). Construct the vapour pressure curve for water between $-5\,°C$ and $100\,°C$.

- *Method.* Use eqn (7.2.4) taking as the fixed point (p^*,T^*) the triple point $(60.3\times10^{-2}\,atm, 273.16\,K)$ for temperatures in that vicinity, and the boiling point $(1\,atm, 373.15\,K)$ for temperatures nearby. Use $\Delta I_{vap,m}(273\,K) = 45.05\,kJ\,mol^{-1}$ and $\Delta H_{vap,m}(373\,K) = 40.66\,kJ\,mol^{-1}$ for the respective ranges.

- *Answer.* Evaluation of eqn (7.2.4) for several temperatures gives the following values:

$t/^\circ C$	-5	0	5	10	20	30	70	80	90	100
p/atm	0.004	0.006	0.008	0.012	0.023	0.042	0.318	0.476	0.697	1.00

The curve is plotted in Fig. 7.6 (AB and CD) and compared there with the experimental one. (The negative curvatures comes from the use of a log scale.)

• *Comment.* The deviation of the calculated curve from the experimental one is a result of neglecting the temperature dependence of the enthalpy of vaporization. Note that we could also use eqn (7.2.3) in conjunction with the experimental curve to obtain the enthalpy of vaporization at any temperature.

The gas–solid boundary. The only difference between this case and the last is the replacement of the enthalpy of vaporization by the enthalpy of sublimation. The approximations that led to the Clausius–Clapeyron equation give the following expressions for the sublimation vapour pressure:

$$d(\ln p)/dT = \Delta H_{sub,m}/RT^2$$

or

$(7.2.5)^\circ$
$$p = p^* \exp\left\{ \frac{-\Delta H_{sub,m}}{R}\left(\frac{1}{T} - \frac{1}{T^*}\right) \right\}.$$

Since the molar enthalpy of sublimation $\Delta H_{sub,m}$ is greater than the molar enthalpy of vaporization this equation predicts a steeper slope. The line drawn in Fig. 7.5c reflects this observation.

Example (Objective 5). Construct the ice/vapour equilibrium phase line over the range $-10\,^\circ C$ to $+5\,^\circ C$, using the information that $\Delta H_{vap,m}(273\,K) = 45.05\ kJ\ mol^{-1}$ and $\Delta H_{melt,m}(273\,K) = 6.01\ kJ\ mol^{-1}$.

• *Method.* In order to use eqn (7.2.5) we need to know the molar enthalpy of sublimation, $\Delta H_{sub,m}$. The change solid-to-gas can be expressed as the sequence solid-to-liquid-to-gas, and so, because enthalpy is a state function, we are able to write

$$\Delta H_{sub,m} = H_m(g) - H_m(s) = H_m(g) - H_m(l) + H_m(l) - H_m(s)$$
$$= \Delta H_{vap,m} + \Delta H_{melt,m}.$$

The phase line can then be drawn on the basis of eqn (7.2.5).

• *Answer.* The equilibrium vapour pressure for a range of temperatures is as follows:

$t/^\circ C$	-10	-5	0	5
p/atm	0.0026	0.0039	0.0060	0.0090

These are plotted as GH in Fig. 7.6 and compared with the experimental curve.

• *Comment.* The small discrepancy is a result of assuming that the enthalpy of sublimation is a constant.

The solid–liquid–gas equilibrium. There is generally some point where the solid, liquid, and vapour phases all coexist in equilibrium. It is given by the values of p and T for which all three chemical potentials are equal. Geometrically it occurs at the intersection of the three boundary curves, Fig. 7.5: this is called the *triple point*. It is very important to note that the position of the

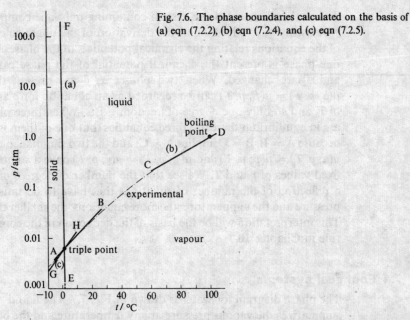

Fig. 7.6. The phase boundaries calculated on the basis of (a) eqn (7.2.2), (b) eqn (7.2.4), and (c) eqn (7.2.5).

triple point of a pure material is completely outside our control. It occurs at a single definite pressure and temperature. For instance, in the case of water it occurs at a temperature of 273.16 K (the ice point at 1 atm is 0.0100 K lower) and a pressure of 4.58 mmHg. The three phases coexist at equilibrium at no other combination of pressure and temperature. This property makes the triple point a good invariant point for defining a temperature scale.

7.3 A first look at the phase rule

When only *one phase* is present the values of p and T can both be varied independently over a range of values without causing a phase change. The system is said to be *bivariant* because the two variables p and T can be varied independently. Other ways of expressing this are to say that there are two *degrees of freedom*, or the *variance* $F = 2$.

When two phases are present in equilibrium only one of the variables may be varied independently. That is the meaning of the boundary lines in Fig. 7.5. If a temperature is selected, then there is no freedom in the choice of pressure if two phases are required to coexist. The system is univariant when two phases are present; there is only one degree of freedom; $F = 1$.

When three phases are present there is freedom to choose neither the pressure nor the temperature. The system is invariant; there are no degrees of freedom; $F = 0$.

If the number of phases is denoted P, these observations are summarized by

(7.3.1) $$F = 3 - P.$$

This is the *phase rule* for a system containing only one component.

A useful way of looking at the derivation of the phase rule is in terms of the equations relating the chemical potentials of the phases. When only one phase is present the chemical potential of the phase can vary as p and T are changed. When two phases α, β are present the equality $\mu(\alpha; p, T) = \mu(\beta; p, T)$ can be regarded as an equation for p as a function of T, and so only one degree of freedom exists. When three phases α, β, γ are in equilibrium there are three equalities (but one of them is redundant because A = B, B = C implies A = C) and the two simultaneous equations $\mu(\alpha; p, T) = \mu(\beta; p, T)$ and $\mu(\beta; p, T) = \mu(\gamma; p, T)$ have a solution only for fixed values of p and T. We see that the number of degrees of freedom is a reflection of the number of equations that have to be satisfied by the pressure and the temperature: the more equations, the smaller the variance. This interpretation will be the basis of the deduction of the complete phase rule in Chapter 10.

7.4 Four real systems

The phase diagram for water is drawn in Fig. 7.7. The liquid–vapour line summarizes the vapour pressure at any temperature and the boiling point at any pressure. The solid–liquid boundary shows how the freezing point depends on the temperature and it indicates that enormous pressures are needed to make any significant change. Notice that the line slopes backwards (the gradient is negative), meaning that the freezing point drops as the pressure is raised. The thermodynamic reason for this is that ice contracts on melting, and so ΔV is negative (see the discussion in the *Example* on p. 187). The molecular reason for the decrease in volume on melting is the very open structure of the ice crystal—the water molecules are held apart (as well as together) by the hydrogen bonds, but the structure partially collapses on melting. At higher pressures other phases come into evidence as the applied pressure modifies the water–water bonds. Some of these ice forms (they are called ice-II, III, V, VI, and VII: ice-IV was an illusion) have high melting points. For example, ice-VII melts at $100\,°C$, but it exists only when the pressure exceeds $25\,000$ atm.

A short time ago it was thought that another form of liquid water existed. A number of laboratories reported the existence of this new phase, and if their claims had been true it would have introduced another phase boundary in the liquid region of Fig. 7.7. Unfortunately it has been shown that the experimenters were misled by impurities, and polywater has been dismissed as another illusion.

Figure 7.8 is the phase diagram for carbon dioxide. The features to notice include the orthodox slope of the solid–liquid boundary, which indicates that the freezing point increases as the pressure is raised. Notice also that at 1 atm the liquid–vapour equilibrium cannot exist whatever the temperature. This means that solid carbon dioxide sublimes when left open to the atmosphere. In order to form liquid carbon dioxide it is necessary to exert a pressure of at least 5.11 atm. Cylinders of carbon

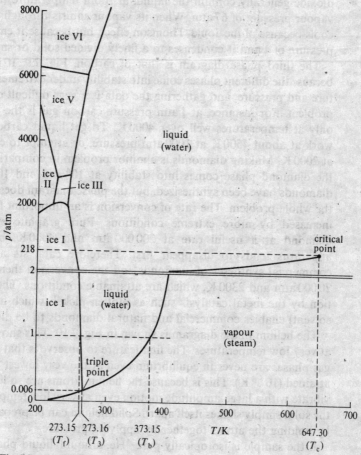

Fig. 7.7. The phase diagram for water.

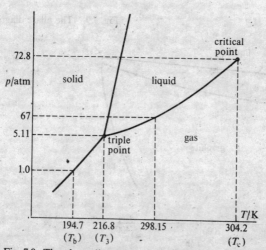

Fig. 7.8. The phase diagram for carbon dioxide.

dioxide generally contain the liquid; at room temperature this implies a vapour pressure of 67 atm. When its vapour squirts through the nozzle it cools because of the Joule–Thomson effect, but because it emerges into a pressure of 1 atm it condenses to a finely divided solid or snow.

The third phase diagram is that of carbon, Fig. 7.9. It is ill-defined because the different phases come into stability under extremes of temperature and pressure, and gathering the data is a very difficult experimental problem. For instance, at 1 atm pressure carbon gas is the stable phase only at temperatures well over 4000 K. To get liquid carbon one must work at about 4500 K at 1000 atm pressure, or attempt to exert 10^6 atm at 2000 K. Making diamonds is a minor problem in comparison, because the diamond phase comes into stability at 10^4 atm and 1000 K. Small diamonds have been synthesized, but the phase diagram does not indicate the whole problem. The rate of conversion is an important factor, and is increased by more extreme conditions. Pure graphite switches into diamond at a useful rate at 200 000 atm and 4000 K, but then the apparatus tends to disappear first. Therefore catalysts are added in commercial syntheses of diamonds and the conversion then proceeds at 70 000 atm and 2300 K, which are attainable conditions. The contamination by the metal catalysts such as molten nickel (which also acts as a solvent) enables commercial and natural diamonds to be distinguished.

The helium phase diagram is shown in Fig. 7.10. This shows anomalies at very low temperatures. The first feature to observe is that the solid and gas phases are never in equilibrium even at the very lowest temperatures attained (10^{-6} K). This is because the helium atoms are so light that they vibrate with a large-amplitude motion even at very low temperatures, and the solid simply shakes itself apart. Solid helium can be prepared, but only by holding the atoms together by applying pressure.

If the sample is isotopically pure ^{4}He, a liquid–liquid phase transition occurs on the λ-line (the *lambda-line*; we shall see the reason for this name in the next section). The liquid phase marked He-I behaves like a

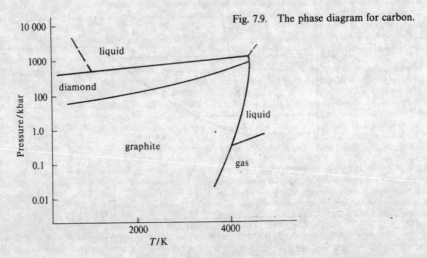

Fig. 7.9. The phase diagram for carbon.

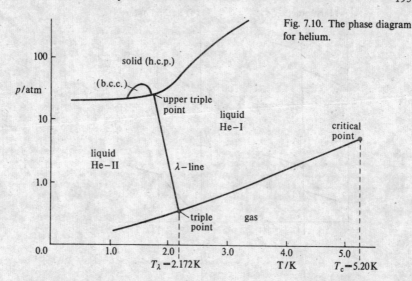

Fig. 7.10. The phase diagram for helium.

normal liquid. The other liquid phase, He-II, is a superfluid, so called because it flows without viscosity.

The isotope ^{3}He has a phase diagram quite similar to the one shown for ^{4}He, but until very recently it was thought that it lacked the λ-line and had only one liquid phase. Recent observations have indicated that a superfluid phase does exist, and evidence is still being accumulated.

7.5 More remarks about phase transitions

Familiar phase transitions, like melting and boiling, are accompanied by changes of enthalpy and volume. The slope of the chemical potential with respect to temperature or pressure is the molar entropy or the molar volume respectively (p. 165), and so at the transition point

$$(\partial\mu(\beta)/\partial p)_T - (\partial\mu(\alpha)/\partial p)_T = (\partial\Delta\mu/\partial p)_T = \Delta V_m \neq 0$$

$$(\partial\mu(\beta)/\partial T)_p - (\partial\mu(\alpha)/\partial T)_p = (\partial\Delta\mu/\partial T)_p = -\Delta S_m = -\Delta H_m/T \neq 0.$$

Since both ΔV_m and ΔH_m are non-zero it follows that $(\partial\mu/\partial p)_T$ and $(\partial\mu/\partial T)_p$ are different on either side of the transition; in other words, the first derivatives of the chemical potential are discontinuous at the transition. This is the basis of the term *first-order phase transition*, Fig. 7.11a.

The discontinuity in the slope of μ has a further implication. The heat capacity C_p is the slope of H with respect to temperature. If H changes discontinuously at the transition temperature, its slope there must be infinite (H has to step up a finite amount in an infinitesimal distance, Fig. 7.11a). The physical reason for this is that the addition of energy to a system at its transition temperature is used in driving the transition rather than in raising the temperature (e.g., water boiling in a kettle). It follows that a first-order transition is characterized by an infinite heat capacity at the transition point.

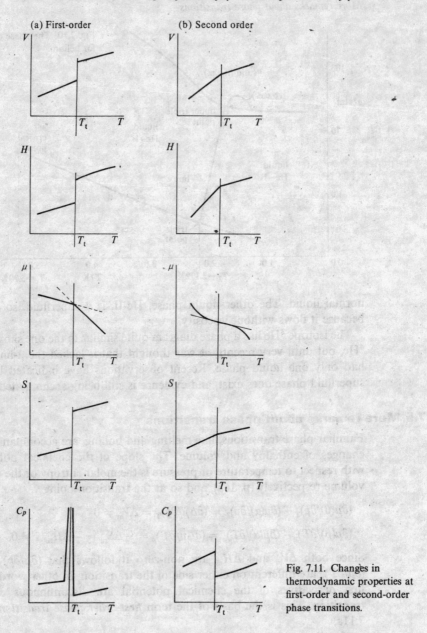

Fig. 7.11. Changes in thermodynamic properties at first-order and second-order phase transitions.

A first-order transition has a discontinuous first-derivative of the chemical potential. This suggests that there might exist a *second-order phase transition* in which the first derivative is continuous but the second derivative is discontinuous. If the gradient of μ is continuous the entropy and the volume of the system do not change when the transition occurs. If there is no entropy of phase transition, then nor is there any enthalpy of transition. Thus a second-order transition is not accompanied by a transition enthalpy, Fig. 7.11b. Nevertheless the heat capacity may still change discontinuously from one side of the transition to the other. This is shown in Fig. 7.11b.

It used to be thought that many examples of second-order transitions existed, but the closer measurements were made to the transition point, the greater the heat capacity became. Only in the conducting-superconducting transition does the heat capacity appear to change through a finite step. The transitions that are not first order yet approach an infinite heat capacity at the transition point include order–disorder transitions in alloys, the onset of ferromagnetism, and the fluid–superfluid transition in helium. Although the heat capacity goes to infinity ('has a singularity') they differ from first-order processes because the heat capacity does not change abruptly, but begins to get large well before the transition point is reached. Compare Figs. 7.11 and 7.12. The shape of the heat capacity curve, Fig. 7.12, resembles the Greek letter λ (lambda), and so this type of transition is called a *λ-transition*. Thermodynamically, λ-transitions are characterized by a continuous gradient of μ, which means that the entropy and volume remain constant in the transition, and that there is no enthalpy change.

λ-transitions appear to be easier to understand than first-order transitions. The rapid but continuous rise of the heat-capacity curve indicates that the system is reorganizing itself well before the actual transition temperature. The process can be pictured as a co-operative process where regions of the new phase begin to form in different parts of the old phase, and the presence of an ordered domain in one region favours the formation of more ordered domains. Fig. 7.13 shows three stages of the process.

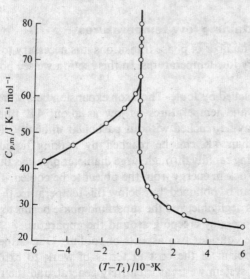

Fig. 7.12. The λ curve for helium.

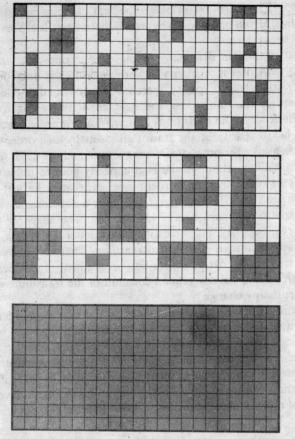

Fig. 7.13. The microscopic structure of an order–disorder transition.

7.6 A practical matter: attaining low temperatures

In order to study a number of phase transitions it is necessary to cool the system, often to very low temperatures. In this section we see how this is achieved.

Gases may be liquefied by Joule–Thomson expansion below their inversion temperatures, and temperatures as low as about 4 K (the boiling point of helium) may be reached without particular difficulty. Temperatures even lower than 4 K can be reached by making liquid helium evaporate by pumping rapidly through large diameter pipes: as the helium evaporates it must suck in energy from the object to be cooled. Temperatures of about 1 K can be obtained, but below this temperature the vapour pressure of helium is negligible, and the superfluid phase begins to interfere with the cooling process by creeping around the apparatus.

In this section we consider two aspects of cooling. One is the general thermodynamic problem of the cost in terms of work to extract heat from a system. The system we treat is entirely general, and the refrigeration may be the final step of cooling to 10^{-6} K from 10^{-5} K, or the problem of air-conditioning a normal room. The other aspect is the technique of getting to exceptionally low temperatures, and virtually to absolute zero.

The energetics of refrigeration. Transferring heat from a cool body to a warmer sink is not a natural, spontaneous process, and so to bring it about requires work. The amount of work to extract a given amount of heat q_c can be expressed in terms of the *coefficient of performance* $c = q_c/w$. In this section we calculate the best possible coefficient of performance c_0, which corresponds to the least possible work required to withdraw a given amount of heat.

If an amount of heat q_c is extracted from a body, and an amount of work w is done in the process, then the amount of heat to be dissipated in the hot sink (e.g. the room) must be $q_c + w$. This is illustrated in Fig. 7.14. Therefore $w = q_h - q_c$, where q_h is the amount of heat delivered to the hot sink, and so

$$1/c = w/q_c = (q_h - q_c)/q_c = (q_h/q_c) - 1.$$

The refrigerator is most efficient when it is working reversibly because w is then a minimum. The best coefficient of performance is therefore given by

$$1/c_0 = (q_h/q_c)_{rev} - 1$$

where the 'rev' subscript indicates that q_h and q_c are being transferred reversibly.

The second law is now used to relate $(q_h/q_c)_{rev}$ to the temperatures T_h and T_c of the hot and cold parts of the apparatus. The entropy of the cold object changes by an amount $(-q_c/T_c)_{rev}$ when the heat is withdrawn, and the entropy of the hot sink increases by $(q_h/T_h)_{rev}$. But as the whole process is proceeding reversibly there is no net entropy production. Therefore

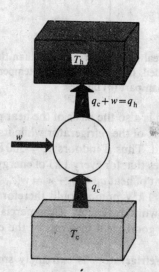

Fig. 7.14. The flow of work and heat in a refrigerator.

$$\Delta S = (q_h/T_h)_{rev} - (q_c/T_c)_{rev} = 0$$

or

$$(q_h/q_c)_{rev} = (T_h/T_c).$$

It follows that the coefficient of performance of a perfect refrigerator working between the temperatures T_c and T_h is

(7.6.1)$_r$ $c_0 = T_c/(T_h - T_c).$

The ideal coefficient of performance is determined by the conditions (T_h and T_c) and makes no reference to the type of refrigerator other than the requirement that it is operating reversibly. Available refrigerators have efficiencies less than ideal, and so their coefficients of performance are smaller than the c_0 just calculated.

Example (Objective 9). Assuming perfect thermodynamic efficiency, calculate the amount of work needed (a) to freeze 100 g of water at 0 °C, the temperature of the surroundings being 25 °C; (b) to withdraw the same amount of heat from a body at 10^{-5} K, the surroundings being at 1 K.

- *Method.* The coefficient of performance c_0 can be calculated from eqn (7.6.1) simply on the basis of the temperatures of the hot and cold sinks. The amount of work to be done in order to remove an amount q of heat is $w_0 = q/c_0$. The amount of heat to extract in order to freeze 100 g of water at 0 °C is the enthalpy of fusion, $n\Delta H_{melt,m}$; $\Delta H_{melt,m} = 6.01$ kJ mol^{-1}.

- *Answer.* In (a), $T_h = 298$ K, $T_c = 273$ K, therefore
$$c_0 = (273 \text{ K})/[(298 \text{ K}) - (273 \text{ K})] = 10.9.$$
In (b), $T_h = 1$ K, $T_c = 10^{-5}$ K: therefore $c_0 \approx 10^{-5}$.

The amount of heat to remove is $q = (100 \text{ g}/18.02 \text{ g mol}^{-1}) \times (6.01 \text{ kJ mol}^{-1}) = 33.4$ kJ. Therefore the minimum work is

(a) $w_0 = (33.4 \text{ kJ})/10.9 = 3.06$ kJ,

(b) $w_0 = (33.4 \text{ kJ})/10^{-5} = 33.4 \times 10^5$ kJ.

- *Comment.* The work w_0 is the minimum: real refrigerators are less than thermodynamically perfect and more work is needed. In practice very low temperatures are attained with much smaller samples than part (b) implies.

The thermodynamics of refrigeration is also the basis of the heat pump, where warmth is obtained from the back of the refrigerator while its front is being used to cool the outside world. Thus if indoors $T_h = 300$ K and outside $T_c = 290$ K, $c_0 = 29$; this signifies that for every 1 kJ of energy used to drive the pump ($w = 1$ kJ), an amount of heat $q_h = q_c + w = w(c_0 + 1) = 30$ kJ will be supplied to the house. A 1 kW pump would therefore be a supplier of 30 kW of heat if it could work reversibly. Commercial heat pumps have $c \approx 5$, which still gives a good yield in terms of the energy consumed to drive the equipment.

The coefficient of performance of refrigerators is extremely small at very low temperatures, and at absolute zero disappears completely. For instance, if the hot sink is at 1 K and the object to be cooled is at 10^{-5} K, the ideal coefficient of performance is only 10^{-5}, and so to remove only 1 J of heat involves doing 100 kJ of work. Whatever the temperature of the sink (so long as it is not itself at absolute zero), the coefficient of performance disappears as the temperature of the cold body approaches 0 K. Therefore an infinite amount of work is needed to reach absolute zero: absolute zero is unattainable.

A final consequence of this discussion is that we have arrived at a fundamental basis of a temperature scale, and one that is independent of the nature of the thermometric material. This is because the coefficient of performance c_0 is *measureable* in terms of the observables q_h and q_c and *expressible* in terms of a property (T) that is independent of the nature of the working substance. The *thermodynamic temperature scale* is the scale based on eqn (7.6.1) with the size of the divisions chosen so as to coincide with the perfect gas scale described earlier (p. 32).

Adiabatic demagnetization: on the track of absolute zero. Although absolute zero cannot be reached, the world record stands at 5×10^{-8} K. The principal method of attaining such very low temperatures is adiabatic demagnetization.

In Part 2 we shall see that magnetic properties arise because electrons behave as tiny magnets. In normal circumstances these magnets are orientated at random, but in a magnetic field a high proportion of them align along the field (Fig. 7.15). In thermodynamic terms the application of a magnetic field lowers the entropy of the system because of the ordering it induces.

The refrigeration process works as follows. A sample of some magnetic

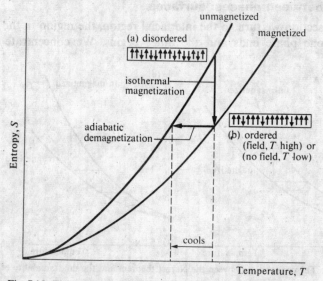

Fig. 7.15. The adiabatic demagnetization technique for attaining very low temperatures.

transition-metal complex (e.g., gadolinium sulphate) is cooled to about 1 K in the way already described. Then it is magnetized by the application of a strong field. This magnetization is done while the sample is in contact with the cold bath, and so it proceeds isothermally. The electronic magnetic moments are aligned and so the entropy falls, Fig. 7.15b. Next the thermal contact between sample and bath is broken, and then the magnetic field is reduced to zero. Since there is no heat flow in this last, reversible, adiabatic step the entropy remains constant. At the end of this step the magnetic field is absent but the magnets still have their earlier ordered arrangement. This arrangement corresponds to a lower temperature, Fig. 7.15. Therefore the adiabatic demagnetization step has cooled the system.

Even lower temperatures can be reached if instead of the electronic magnetic moments the magnetic moments of the nuclei of the ions are used. The process of adiabatic nuclear demagnetization works on the same principle as the electronic method, and has been the method of establishing the world record.

From the Third Law (p. 152) we know that the entropy curves coincide as T approaches absolute zero: hence the adiabatic demagnetization procedure cannot be used (in a finite number of steps) to cool an object to absolute zero. In order to appreciate this point, suppose that the Nernst heat theorem (p. 152) were false. Then the entropy might behave as in Fig. 7.16a. The last step indicated could cool the system to $T = 0$. As far as we know experimentally (and on statistical grounds we can be confident theoretically) the curves actually coincide at $T = 0$, as in Fig. 7.16b, and so no finite sequence of steps can cool the system to $T = 0$. By this method (as for every method), absolute zero is unattainable.

7.7 The region between phases: surfaces

In this section we turn to the interfacial region, the region in the sample where one phase ends and the other begins. We concentrate on the

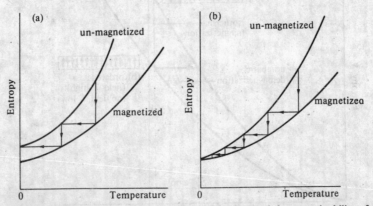

Fig. 7.16. The connection between the Nernst theorem and the unattainability of absolute zero.

vapour–liquid surface which is interesting because it is so mobile.

Liquids assume a shape that minimizes their surface area because that enables the maximum number of molecules to be 'bulk' molecules rather than 'surface molecules'. For this reason droplets are spherical, because a sphere is the solid object with the smallest surface/volume ratio. There may be other forces in operation which destroy this ideal shape, and in particular gravitational forces may flatten spheres into puddles or oceans. Nevertheless the shape is always affected by the tendency to acquire the minimum surface area.

Surface effects can be expressed in the language of chemical potentials. Changing the amount of surface enclosing a system involves doing work, and so it can be incorporated as another contribution to dw. The work involved in forming a surface of area dσ is proportional to the area of the surface formed, and so

(7.7.1) $$dw = \gamma\, d\sigma$$

where γ is a coefficient known as the *surface tension* of the material (a list is given in Table 7.1).

Table 7.1. Surface tensions of liquids at 20 °C

	$\gamma / N\,m^{-1}$		$\gamma / N\,m^{-1}$
Water	7.275×10^{-2}	CH_3OH	2.26×10^{-2}
	7.20×10^{-2} at 25 °C	CH_3CH_2OH	2.28×10^{-2}
	5.80×10^{-2} at 100 °C	$C_2H_5OC_2H_5$	1.696×10^{-2}
Benzene	2.886×10^{-2}	Hg	47.2×10^{-2}
n-Hexane	1.84×10^{-2}		
CCl_4	2.70×10^{-2}		

Source: G. W. C. Kaye and T. H. Laby, *Tables of physical and chemical constants*, Longmans.

The work of forming a surface is additional to pV-work, and so we should regard it as a contribution to the Gibbs function of the system (remember that dG is to be identified with the maximum amount of net work, see p. 149). Therefore in the presence of a single component system with a variable surface area

(7.7.2) $$dG = -S\,dT + V\,dp + \gamma\, d\sigma.$$

This is the link with thermodynamics, and the reasoning about systems tending to smaller Gibbs functions (or, if divided by amount of substance, chemical potentials) can be applied to this equation. For instance, at constant temperature and pressure a smaller value of the Gibbs function can be attained if the area is allowed to diminish. The condition for a natural change, d$G < 0$, implies d$\sigma < 0$, which means that surfaces have a natural tendency to contract. This is a more formal way of expressing what we have already described.

Bubbles and drops. By bubbles we mean either ordinary bubbles, in which air and vapour are trapped by a thin film, or cavities full of vapour in a liquid. Drops are spheres of liquid in equilibrium with their vapour. The principal difference between ordinary bubbles and cavities is that the former have two surfaces whereas cavities have only one. The treatment of both is much the same, but the factor of 2 in the case of ordinary bubbles (to take into account the doubled surface area) must be remembered.

Bubbles are at equilibrium because the tendency to decrease their surface area is balanced by the rise in internal pressure. The easiest way of finding out how the internal pressure depends on the radius is to find the condition of balance between the forces tending to expand the bubble and those tending to contract it. If the pressure inside the bubble is p_{in} and the radius is r, then the total outwards force is $4\pi r^2 p_{in}$. The force inwards is due to the sum of the outside pressure p_{out} and the surface tension. We know that the energy of the surface of area σ is $\gamma\sigma$, or $4\pi r^2\gamma$ if the bubble is a spherical cavity, and to find the force we calculate the work required to stretch this surface through dr; it is just $d(\gamma\sigma)$, or $8\pi\gamma r\,dr$. But work is (force) × (distance), and so the force opposing stretching through the distance dr at this radius is $8\pi\gamma r$. Balancing the forces gives

$$4\pi r^2 p_{in} = 4\pi r^2 p_{out} + 8\pi\gamma r,$$

or

(7.7.3)
$$p_{in} - p_{out} = 2\gamma/r.$$

This simple relation is the *Laplace equation*.

We see that the pressure inside a curved surface is always greater than the pressure outside, but the difference drops to zero as the radius of curvature becomes infinite (when the surface is flat). Small bubbles have very small radii of curvature and so the pressure difference across them is quite large. For instance, a bubble of radius 0.1 mm in champagne implies the existence of a pressure differential of

$$p_{in} - p_{out} = 2(7.3 \times 10^{-2}\,\mathrm{N\,m^{-1}})/(10^{-4}\,\mathrm{m}) = 1460\,\mathrm{N\,m^{-2}},$$

or enough to sustain a 15 cm column of water.

The thermodynamic consequences of bubbles and drops can be established by exploring the consequences of the increased pressure inside the curved surface. As a first step we show that whenever the pressure on a liquid is increased, then its vapour pressure is also increased. Thermodynamically, increasing the pressure corresponds to an increase in the enthalpy of the liquid (through the pV term in H), and therefore to an increase in its chemical potential. In practice an increase in pressure can be brought about by curving the surface of the liquid (as in a bubble) or by introducing an inert gas into a sealed container. The latter method suffers from the complication that the pressurizing gas might dissolve in the liquid.

At equilibrium the chemical potentials of a vapour and its liquid are equal. Therefore, any change in one is equal to a change in the other: $d\mu(g) = d\mu(l)$. Now let the pressure on the liquid be increased. It follows from $d\mu = V_m dp$ that if the pressure on the liquid changes by $dp(l)$ then the change in the pressure of its vapour, $dp(g)$ will be such that

$$V_m(g) \, dp(g) = V_m(l) \, dp(l) \qquad \text{(const. } T\text{)}.$$

As usual the vapour is assumed to be perfect, in which case $V_m(g) = RT/p(g)$. Then

$$dp(g)/p(g) = V_m(l) \, dp(l)/RT.$$

This expression can be integrated once the limits of integration have been established. When the pressure exerted on the liquid is the normal vapour pressure p^*, $p(g) = p^*$ and $p(l) = p^*$. When there is an additional pressure on the liquid it experiences the pressure $p(l)$ and the vapour pressure is $p(g)$. Therefore the integration is

$$\int_{p*}^{p(g)} dp(g)/p(g) = (1/RT) \int_{p*}^{p(l)} V_m(l) \, dp(l).$$

By assuming that the volume of the liquid is the same throughout the small range of pressures involved this becomes

$$\ln\{ p(g)/p^* \} = \{V_m(l)/RT\}\{ p(l) - p^* \},$$

or

(7.7.4)° $$p(g) = p^* \exp[\{V_m(l)/RT\}\{ p(l) - p^* \}].$$

This formula shows that the vapour pressure increases when the pressure on the liquid increases: molecules are squeezed out into the vapour.

One way of increasing the pressure on the liquid is to disperse it as droplets. The pressure differential across a curved surface is $2\gamma/r$, and so substituting this in the last expression gives the

(7.7.5)° | **Kelvin equation**: $p(\text{mist}) = p(\text{bulk}) \exp\{2\gamma V_m(l)/rRT\}$

where $p(\text{mist})$ is the vapour pressure of a sample of mist with droplet radius r, and $p(\text{bulk})$ is the vapour pressure above a plane surface of the same material.

The analogous expression for the vapour pressure of a liquid on the inside of bubbles can be written at once. The pressure of the liquid surrounding a cavity is less than the pressure of the vapour in the bubble. It is quite easy to see that the only change necessary in the last equation is in the sign of the exponent:

(7.7.6)° $$p(\text{bubbles}) = p(\text{bulk}) \exp\{-2\gamma V_m(l)/rRT\}.$$

Neither effect is very great, but both have important consequences. In the case of droplets of water of radius 10^{-3} mm and 10^{-6} mm the ratios

$p(\text{mist})/p(\text{bulk})$ at room temperature are about 1.001 and 2.95 respectively. The second figure is deceptive because at that radius the droplet is only about 40 molecules in diameter, and so the whole basis of the calculation is suspect.

Example (Objective 12). Calculate the vapour pressure inside a bubble of water vapour, and outside a drop of water, in each case taking the radius as 10 nm. At 298 K the surface tension of water is $72.0 \, \text{mN m}^{-1}$ and the vapour pressure across a flat surface is $3.167 \times 10^3 \, \text{N m}^{-2}$.

● *Method.* In order to calculate the vapour pressure across the curved surfaces we use eqns (7.7.5) and (7.7.6). The molar volume of liquid water is $17.97 \times 10^{-6} \, \text{m}^3 \text{mol}^{-1}$. ($M_m = 18.02 \, \text{g mol}^{-1}$, $\rho = 0.9970 \, \text{g cm}^{-3}$ at 25 °C.)

● *Answer.* Both equations depend on the quantity

$$2\gamma V_m(\text{l})/rRT = \frac{2 \times (7.20 \times 10^{-2} \, \text{N m}^{-1}) \times (1.797 \times 10^{-5} \, \text{m}^3 \text{mol}^{-1})}{(1.00 \times 10^{-8} \, \text{m}) \times (8.314 \, \text{J K}^{-1} \text{mol}^{-1}) \times (298.15 \, \text{K})} = 0.104.$$

A pressure of $3.167 \times 10^3 \, \text{N m}^{-2}$ is equivalent to $3.126 \times 10^{-2} \, \text{atm}$. Therefore, from eqn (7.7.5):

$$p(\text{mist}) = (3.126 \times 10^{-2} \, \text{atm}) \exp(0.104) = 3.47 \times 10^{-2} \, \text{atm},$$

and from eqn (7.7.6):

$$p(\text{bubbles}) = (3.126 \times 10^{-2} \, \text{atm}) \exp(-0.104) = 2.82 \times 10^{-2} \, \text{atm}.$$

● *Comment.* It is questionable whether droplets of this size can be treated in this way because they contain so few molecules. Nevertheless the orders of magnitude of the vapour pressures are approximately correct.

The important consequence of the increased vapour pressure of droplets is the stabilizing influence it has on the condensation of gas into liquid. Consider the formation of a cloud. Warm, moist air rises into cooler regions higher in the atmosphere. At some altitude the temperature becomes so low that the vapour becomes thermodynamically unstable with respect to the liquid, and a cloud should form as the water vapour liquefies. The initial process in condensation can be imagined as a sticking together of a swarm of water molecules into a microscopic droplet. But it is unlikely that the swarm will be very big, and from the preceding discussion we know that it will have an exaggerated vapour pressure. Therefore, instead of growing it evaporates. This effect stabilizes the vapour because an initial tendency to condense is eliminated by the built-in evaporation mechanism.

Clouds do form, and so there must be a mechanism. Two processes overcome this difficulty. The first is that a sufficiently large number of molecules might stick into a droplet so big that the evaporative effect does not have time to operate before the droplet has picked up an even bigger collection of molecules and has increased its radius to an extent that ensures it will not re-evaporate. The chance of one of these spontaneous *nucleation* centres forming is low, and so in rain formation it is not a dominant mechanism. The other process depends on the presence of minute dust particles or other kinds of foreign matter. These act as nuclei for condensation, and the water vapour can condense on them. The observation that rain is more likely on weekdays than at weekends may reflect the abundance of industrial dust and its activity in nucleation. The *cloud chamber* works on the principle of nucleation. In a very pure environment a supersaturated mixture of air and water vapour does not condense, but when ionizing radiation, in the form of some charged elementary particle, passes through, the ions formed in its path act as nuclei for condensation, and the trajectory is mapped out as a streak of condensed water. The hydrogen *bubble chamber* works on a similar principle, but depends on the nucleation of superheated hydrogen.

The cooling of water vapour may lead to a system in which the water exists as vapour even though its chemical potential lies above the chemical potential of the liquid. This may also happen for other liquids, and the thermodynamically unstable vapour phase is then said to be *supersaturated*. Liquids may also be supercooled below their freezing points, for crystallization also requires nucleation, or superheated above their boiling points. Superheating occurs because the vapour pressure of a liquid enclosing a bubble is lower than the bulk, and so any small bubble of vapour tends to condense back into the bulk. For example, heating an unstirred beaker of water can raise its temperature above its boiling point, and it is then thermodynamically unstable with respect to its vapour. Violent bumping often ensues as spontaneous fluctuation leads to bubbles big enough to survive. In order to ensure smooth boiling at the true boiling point, nucleation centres, such as small pieces of glass, or bubbles of air, should be introduced.

Capillary action. The phenomenon of liquids rising up capillary tubes is a consequence of surface tension and the thermodynamic tendency of a liquid to minimize its surface area.

Consider what happens when a glass capillary tube is first immersed in water or any liquid that has a tendency to stick to the walls. The energy is lowest when a thin film covers as much of the glass as possible. As this film creeps up the inside wall it has the effect of curving the surface of the liquid inside the tube, Fig. 7.17a. This implies that the pressure just underneath the curving meniscus is less than the atmospheric pressure by an amount approximately equal to $2\gamma/r$, where r is the radius of the tube (we assume a hemispherical surface). The pressure just under the surface outside the tube is p (the atmospheric pressure), but inside the tube the pressure in only $p - 2\gamma/r$. The excess external pressure presses the liquid up the tube until hydrostatic equilibrium (equal pressures at equal depths) has been reached, Fig. 7.17b. This equilibrium is established when a column of liquid of density ρ has reached a height h such that

$$\text{pressure} = \frac{\rho\pi r^2 hg}{\pi r^2} = 2\gamma/r.$$

($\pi r^2 h$ is the volume in the tube, $\rho\pi r^2 h$ its mass, $\rho\pi r^2 hg$ the downwards

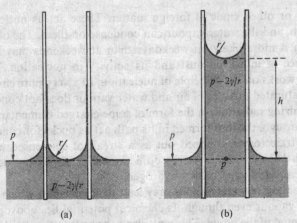

Fig. 7.17. Capillary action: the curvature of the surface in (a) leads to a pressure difference, but this is balanced in (b) by the liquid rising in the tube.

force, and πr^2 the area over which the force is exerted.) Therefore

(7.7.7) $h = 2\gamma/\rho g r.$

This simple expression provides a very accurate way of measuring the surface tension of liquids.

Example (Objective 14). In an experiment to measure the surface tension of water over a range of temperatures, a capillary tube of internal diameter 0.4 mm was supported vertically in the sample. The density of the sample was measured in a separate experiment. The following results were obtained:

$t/^\circ C$	10	15	20	25	30
h/cm	7.56	7.46	7.43	7.36	7.29
$\rho/g\,cm^{-3}$	0.9997	0.9991	0.9982	0.9971	0.9957

Find the temperature variation of the surface tension.

● *Answer*. This is simply an application of eqn (7.7.7) in the form $\gamma = \frac{1}{2}\rho h g r$, with $g = 9.81\,m\,s^{-2}$ and $r = 0.2\,mm$. The following results are obtained:

$t/^\circ C$	10	15	20	25	30
$\gamma/N\,m^{-1}$	7.41×10^{-2}	7.31×10^{-2}	7.28×10^{-2}	7.20×10^{-2}	7.12×10^{-2}

● *Comment*. The surface tension decreases with increasing temperature because thermal motion disrupts the intermolecular bonds and tends to make the environment of the molecules in the bulk similar to those at the surface; hence there is less change of energy in moving from surface to bulk.

When the liquid and the material of the capillary attract each other less strongly than the liquid molecules attract each other (mercury and glass is a common example) the liquid in the tube retracts from the walls. This curves the surface with the convex, low pressure side upwards. Just beneath the meniscus the pressure must be greater than atmospheric, and in order to equalize hydrostatic pressures at the same depth throughout the fluid the surface of the liquid falls, which gives capillary depression. In many cases there is a non-zero angle between the edge of the meniscus and the side of the wall. If this *contact angle* is θ, eqn (7.7.7) is modified by multiplying the r.h.s. by $\cos \theta$. This modification is examined in the Problems.

Further reading

The quest for absolute zero. K. Mendelssohn; McGraw-Hill, New York, 1966.

Thermodynamics (Schaum Outline Series). M. M. Abbott, H. C. Van Ness; McGraw-Hill, New York, 1976.

Determination of melting and freezing temperatures. E. L. Skau and J. C. Arthur; In *Techniques of chemistry* (A. Weissberger and B. W. Rossiter, eds.) V, 105, Wiley-Interscience, New York, 1971.

Determination of boiling and condensation temperatures. J. R. Anderson; In *Techniques of chemistry* (A. Weissberger and B. W. Rossiter, eds.) V, 199, Wiley-Interscience, New York, 1971.

Determination of surface and interfacial tension. A. E. Alexander and J. B. Hayter; in *Techniques in chemistry* (A. Weissberger and B. W. Rossiter, eds.) V, 501, Wiley-Interscience, New York, 1971.

Introduction to phase transitions and critical phenomena. H. E. Stanley; Clarendon Press, Oxford, 1971.

Physico-chemical constants of pure organic compounds. J. Timmermans; Elsevier, Amsterdam, 1956.

Vapor pressure of organic compounds. T. E. Jordan; Interscience, New York, 1954.

Tables of physical and chemical constants. G. W. C. Kaye and T. H. Laby; Longmans, London, 1973.

Handbook of chemistry and physics. Chemical Rubber Co., Cleveland, 1980.

Problems

7.1. At 25 °C the enthalpy of the graphite → diamond phase transition is 1.8961 $kJ\,mol^{-1}$, and the entropy is $-3.2552\,J\,K^{-1}\,mol^{-1}$. What is the molar Gibbs function for the transition at 298 K? Is the spontaneous direction favoured by a rise of temperature?

7.2. The densities of graphite and diamond are $2.25\,g\,cm^{-3}$ and $3.52\,g\,cm^{-3}$ respectively. What is the value of ΔA_m at 298 K for the transition when the sample is subjected to a pressure of (a) 1 atm, (b) 500 kbar?

7.3. The rate of change of the chemical potential with temperature, as expressed by the coefficient $(\partial \mu / \partial T)_p$, is an important quantity in the discussion of phase transitions and equilibria. What is the difference in slope on either side of (a) the freezing point of water, (b) the conventional boiling point of water? Use the data in Table 4.7.

7.4. Another important gradient is the rate of change of chemical potential with pressure, $(\partial \mu / \partial p)_T$. Calculate the change in this gradient on either side of the two transitions referred to in the last Problem. You need to know the densities of ice and water at 0 °C ($0.917\,g\,cm^{-3}$ and $1.000\,g\,cm^{-3}$) and of water and steam at 100 °C ($0.958\,g\,cm^{-3}$ and $0.598\,kg\,m^{-3}$).

7.5. By how much does the chemical potential of water supercooled to -5 °C exceed that of ice at that temperature?

7.6. A cloud chamber operates on the principle that vapour may be supercooled, and the passage of an ionizing particle can generate ions which induce condensation. By how much does the chemical potential of the water vapour exceed that of water at that temperature? Suppose, instead, that the pressure was increased to 1.2 atm at 100 °C. What then is the difference of chemical potentials?

7.7. We have just calculated the relative rise in chemical potential when the temperature of liquid water is decreased. When the pressure is increased the chemical potential of ice increases more rapidly than that of water, and so the chemical potentials of the two phases can be brought back into equality, and therefore the phases into an equilibrium. Use this approach to estimate the freezing point of water under a pressure of 1000 atm.

7.8. When benzene freezes at 5.5 °C its density changes from $0.879 \, \text{g cm}^{-3}$ to $0.891 \, \text{g cm}^{-3}$. Its enthalpy of fusion is $10.59 \, \text{kJ mol}^{-1}$. Use the same technique as in the last Problem to estimate the freezing point of benzene under a pressure of 1000 atm.

7.9. The Clapeyron equation, eqn (7.2.1) is sometimes more useful when it is upside down, for then dT/dp gives the dependence of the transition temperature on the pressure. What change in the boiling point of water at 1 atm is brought about by a 10 mmHg change of pressure?

7.10. The change in enthalpy of a species is given by $dH = C_p dT + V dp$. The Clausius equation relates dp and dT for equilibria, and so the last equation can be used to obtain an expression for the change in ΔH along the phase boundary when the temperature changes and the phases remain in equilibrium. Find this expression, and solve it for $\Delta H(T)$ in the case when C_p of both phases may be assumed independent of temperature. Use this temperature dependence of the transition enthalpy in the Clausius–Clapeyron equation to show that the vapour pressure depends on temperature according to the expression $p/p^* = (T/T^*)^a$, where $a = \Delta H_m(T^*)/RT^* + (\Delta C_{p,m}/2R) \ln(T/T^*)$, and plot the phase boundary for the vaporization of water using p^*, T^* for the normal boiling point.

7.11. The enthalpy of fusion of mercury is $2.292 \, \text{kJ mol}^{-1}$, and it freezes at 234.3 K under 1 atm pressure, the change in volume being $0.517 \, \text{cm}^3 \, \text{mol}^{-1}$. At what temperature will the bottom of a 10 m high column of mercury (of density $13.6 \, \text{g cm}^{-3}$) freeze?

7.12. Several methods exist for the determination of vapour pressure. When it lies between 10^{-3}–10^3 mmHg a pressure gauge can be used directly. When it is very low (as for an almost involatile solid) an *effusion method* can be used (p. 877). The *gas saturation method* is a straightforward technique for liquids, and we concentrate on it here. Let a volume $V(\text{g})$ of some gas, measured at temperature T and pressure $p(\text{g})$, be bubbled slowly through the sample maintained at the temperature T. The loss of mass by the sample is determined: let this be $m(l)$, its R.M.M. being $M_r(l)$. Show that the vapour pressure of the liquid at that temperature, $p(l)$, is related to the mass loss by $p(l) = Am(l)p(\text{g})/[Am(l)+1]$, where $A = RT/M_r(l)V(\text{g})p(\text{g})$. ($m(l)$ should strictly be written $m(l)/\text{g}$.)

7.13. The vapour pressure of geraniol, which is a component of oil of roses and other perfumes, with $M_r = 148.4$, was measured at 110 °C. When 5 dm³ of nitrogen at 1 atm was passed through the heated liquid the loss of weight was 0.32 g. What is the vapour pressure of geraniol at this temperature?

7.14. The last experiment was repeated at 140 °C, the mass loss then being 243 mg for 1 dm³ of nitrogen passed. What is the molar enthalpy of vaporization of geraniol? Estimate its conventional boiling point.

7.15. 50 dm³ of dry air were slowly bubbled through a thermally insulated beaker containing 250 g of water. If the initial temperature was 25 °C, what is the final temperature? The vapour pressure of water at 25 °C is 23.8 mmHg and its molar heat capacity is $75.5 \, \text{J} \, \text{K}^{-1} \, \text{mol}^{-1}$.

7.16. In July in Los Angeles the incident sunlight at ground level has a power density of $1.2 \, \text{kW} \, \text{m}^{-2}$ at noon. A swimming pool of area $10 \, \text{m} \times 5 \, \text{m}$ is directly exposed to the radiation. What is the rate of loss of water?

7.17. An open vessel containing (a) water, (b) benzene, (c) mercury stands in a laboratory measuring $5 \, \text{m} \times 5 \, \text{m} \times 3 \, \text{m}$. The temperature is 25 °C. What quantity (in g) of each material will be found in the air if there is no ventilation? (The vapour pressures are 24 mmHg, 98 mmHg, and 1.7×10^{-3} mmHg respectively.)

7.18. The *relative humidity* of air is the ratio of the partial pressure of water vapour to its vapour pressure at that temperature. What is (a) the partial pressure, (b) the amount present in grams, when the relative humidity is 70 per cent in the laboratory in the last Problem?

7.19. The vapour pressure of nitric acid depends on the temperature as follows:

$t/°C$	0	20	40	50	70	80	90	100
p/mmHg	14.4	47.9	133	208	467	670	937	1282

What is (a) the conventional boiling point, (b) the molar enthalpy of vaporization?

7.20. The ketone carvone is a component of the oil that gives spearmint its characteristic odour. Its vapour pressure depends on the temperature as follows:

$t/°C$	57.4	100.4	133.0	157.3	203.5	227.5
p/mmHg	1.00	10.0	40.0	100	400	760

The R.M.M. is 150.2. What is the molar enthalpy of vaporization and the conventional boiling point?

7.21. Combine the barometric formula for the dependence of pressure on altitude, p. 51, with the Clausius–Clapeyron equation, and predict how the boiling point of a liquid depends on the altitude and the ambient temperature (be careful to distinguish the various temperatures in the expressions). Taking the mean ambient temperature as 20 °C, predict the boiling point of water at 10 000 ft.

7.22. Vapour pressure is often reported in the form $\lg(p/\text{mmHg}) = b - 0.052\,23a/(T/K)$ (see for example, *Handbook of chemistry and physics*, The Chemical Rubber Co.). How is the molar enthalpy of vaporization related to the parameters a and b? The values for white phosphorus are $a = 63\,123$ and $b = 9.6511$ in the range 20–44 °C. What is the vapour pressure of white phosphorus at 25 °C, and what is the molar enthalpy of sublimation of the solid?

7.23. On a cold, dry morning after a frost the temperature was -5 °C. The partial pressure of water vapour in the atmosphere dropped to 2 mmHg. Will the frost sublime? What partial pressure of water would ensure that the frost remained? Use data from the *Example* on p. 189.

7.24. Construct the phase diagram for benzene in the vicinity of its triple point (36 mmHg, 5.50 °C) on the basis of the following data:

$\Delta H^{\ominus}_{\text{melt,m}} = 10.6 \, \text{kJ} \, \text{mol}^{-1}$ $\Delta H^{\ominus}_{\text{vap,m}} = 30.8 \, \text{kJ} \, \text{mol}^{-1}$,

$\rho(\text{s}) = 0.91 \, \text{g} \, \text{cm}^{-3}$, $\rho(\text{l}) = 0.899 \, \text{g} \, \text{cm}^{-3}$.

7.25. Base your answer to this Problem on Fig. 7.7. What changes would be observed when water vapour at 1 atm and 400 K is cooled at constant pressure to 260 K? Suppose you observed the rate of cooling of the sample when it was in contact with a cool bath. What would you notice?

7.26. What modification to the observation would appear when the cooling was carried out at 0.006 atm?

7.27. On the basis of the phase diagram in Fig. 7.8, state what you would observe when a sample of carbon dioxide, initially at 1 atm and 298 K, is subjected to the following cycle. (a) Isobaric heating to 320 K, (b) isothermal compression to 100 atm, (c) isobaric cooling to 298 K, (d) isothermal decompression to 80 atm, (e) isobaric cooling to 210 K, (f) isothermal decompression to 1 atm, (g) isobaric heating to 298 K.

7.28. What is the coefficient of performance of a refrigerator working with thermodynamic efficiency in a room of temperature 20 °C when the interior is at a temperature of (a) 0 °C, (b) −10 °C?

7.29. How much work is needed to freeze 250 g of water originally at 0 °C in a refrigerator standing in a room at 20 °C? What is the minimum time for bringing about total freezing when the power consumption of the refrigerator is 100 W?

7.30. Quite often we have to find out how much work has to be done in order to lower the temperature of a body; this is slightly more involved than the last Problem because the temperature of the object, and therefore the coefficient of performance, is changing. Nevertheless it is quite a simple job to find the work by writing $dw = dq/c_0(T)$, and relating dq to dT through the heat capacity C_p. Then the total work is the integral of the resulting expression. First we assume that the heat capacity is independent of temperature in the range of interest. Find an expression for the amount of work needed to cool an object from T_i to T_f when the refrigerator is in a room with a temperature T_h.

7.31. Apply the calculation in the last Problem to find the amount of work needed to freeze 250 g of water put into the refrigerator at 20 °C. How long will it take when the refrigerator operates at 100 W? Suppose the water was put in at 25 °C: what amount of work is then required?

7.32. Calculate the minimum amount of work needed to lower the temperature of a 1 g block of copper from 1.10 K to 0.10 K, the surroundings being at 1.20 K. In the first place, make a crude estimate by assuming that the heat capacity remains constant at 3.9×10^{-5} J K^{-1} mol^{-1} and that the coefficient of performance can be evaluated at the mean temperature of the block.

7.33. Repeat the last calculation in a more realistic way by noting that the molar heat capacity varies as $AT^3 + BT$, where $A = 4.82 \times 10^{-5}$ J K^{-4} mol^{-1} and $B = 6.88 \times 10^{-4}$ J K^{-2} mol^{-1}, and taking into account the variation of the coefficient of performance with temperature.

7.34. In a more ambitious experiment the aim was to cool the copper block to 10^{-6} K. How much work would be needed, and for how long, at least, must a 1 μW refrigerator operate?

7.35. How much work must be done in order to cool the air in an otherwise empty room of size 3 m × 5 m × 5 m from 30 °C to 22 °C when the ambient temperature is (a) 20 °C, (b) 30 °C? Assume c_0 is constant over the range, and take $C_{p,m} = 29$ J K^{-1} mol^{-1} for the air, which has a mean density of 1.2 mg cm^{-3} and $M_r \approx 29$.

7.36. An electric heater rated at 1 kW is left on in the same room. An air conditioner is also left running. What is the minimum power rating of the air conditioner that is needed in order to maintain the room at 25 °C when the ambient temperature is 30 °C?

7.37. The assumption that the number of molecules in the surface of a sample is much less than the total number fails when the sample is dispersed as very small droplets. But how small do the droplets have to be before the effect is significant? Estimate the ratio of the number of water molecules ($r \approx 120\,\text{pm}$) on the surface of a spherical droplet to the total number in the entire droplet, when the radius of the droplets is (a) $10^{-5}\,\text{mm}$, (b) $10^{-2}\,\text{mm}$, (c) $1.0\,\text{mm}$.

7.38. A sample of benzene of mass $100\,\text{g}$ is dispersed as droplets of radius $10^{-3}\,\text{mm}$. The surface tension of benzene is $2.8 \times 10^{-2}\,\text{N m}^{-1}$ and its density is $0.88\,\text{g cm}^{-3}$. What is the change of Gibbs function? What is the minimum amount of work necessary to bring about the dispersal?

7.39. By how much is the vapour pressure of benzene changed when it is dispersed in the form of small droplets of radius (a) $10^{-2}\,\text{mm}$, (b) $10^{-4}\,\text{mm}$, both at $25\,°\text{C}$.

7.40. A carburettor jet generates a fine spray of fuel in an air stream. Suppose the operating temperature is $60\,°\text{C}$ and the fuel has a vapour pressure of $100\,\text{mmHg}$ (like heptane). First suppose that the jet is badly adjusted and simply provides puddles of fuel. If the air, at a pressure of $1\,\text{atm}$, passes through sufficiently slowly to become saturated, what mass of fuel will be carried into the engine for the passage of $10\,\text{dm}^3$ of air? Now adjust the jet so that it provides a spray of droplets of radius $10^{-4}\,\text{mm}$. How much fuel is transported for the same flow of air? (In practice, the flow would transport the droplets themselves; here we assume that only their vapour is transported.) Take $\gamma = 2.8 \times 10^{-2}\,\text{N m}^{-1}$ and $\rho = 0.879\,\text{g cm}^{-3}$.

7.41. The surface tension of water is $7.28 \times 10^{-2}\,\text{N m}^{-1}$ at $20\,°\text{C}$ and $5.80 \times 10^{-2}\,\text{N m}^{-1}$ at $100\,°\text{C}$. The densities are respectively $0.998\,\text{g cm}^{-3}$ and $0.958\,\text{g cm}^{-3}$. To what height will water rise in tubes of internal radius (a) $1\,\text{mm}$, (b) $0.1\,\text{mm}$ at these two temperatures?

7.42. At $30\,°\text{C}$ the surface tension of ethanol in contact with its vapour is $2.189 \times 10^{-2}\,\text{N m}^{-1}$, and its density is $0.780\,\text{g cm}^{-3}$. How far up a tube of internal diameter $0.2\,\text{mm}$ will it rise? What pressure is needed to push the meniscus back level with the surrounding liquid? From tables of density, surface tension, and vapour pressure, find the temperature at which, if the upper end of the capillary is sealed, the vapour pressure is itself sufficient to depress the meniscus.

7.43. A glass tube of internal diameter $1.00\,\text{cm}$ surrounds a glass rod of diameter $0.98\,\text{cm}$. How high will water rise in the space between them at $25\,°\text{C}$.

7.44. The capillary rise when a fluid meets the material of the tube tangentially is given by eqn (7.7.7). Some liquids make a non-zero *contact angle* θ with the wall. Only the vertical component of the force due to surface tension draws the liquid up the tube; on this basis deduce an expression for capillary rise for a general contact angle.

7.45. In Chapter 6 we saw how to deduce the Maxwell relations (p. 163). Now regard μ as a function of σ as well as T and p, and deduce the relation $(\partial V/\partial \sigma)_{p,T} = (\partial \gamma/\partial p)_{\sigma,T}$ on the basis that $d\mu$ is an exact differential. The derivative $\partial V/\partial \sigma$ can be evaluated very easily for spherical droplets of radius r. Derive this coefficient, and show that the new Maxwell type of relation leads at once to the Laplace equation, eqn (7.7.3).

7.46. Show that the velocity of formation of a hole in the skin of a bubble is of the order of $\sqrt{(2\gamma/\rho\delta)}$, where γ is the surface tension of the fluid, ρ its density, and δ the thickness of the film. Estimate the velocity in a film of (a) water, $\gamma = 7.2 \times 10^{-2}\,\text{N m}^{-1}$, (b) soapy water, $\gamma = 2.6 \times 10^{-2}\,\text{N m}^{-1}$.

8 Changes of state: physical transformations of simple mixtures

Learning objectives

After careful study of this chapter you should be able to

(1)   Define *partial molar quantity* (p. 216) and describe how any thermodynamic quantity depends on the composition of the system (eqn (8.1.6)).

(2)   Derive the *Gibbs–Duhem equation* (eqn (8.1.7)) connecting changes of the chemical potentials of the components of a system.

(3)   Deduce and use an expression for the *Gibbs function of mixing* of perfect gases (eqn (8.2.1)) and use the result to derive expressions for the *entropy and enthalpy of mixing* (p. 221) and the *excess functions* (p. 227).

(4)   State *Raoult's Law* for the partial pressure of a gas above a mixture (eqn (8.2.7)).

(5)   Explain the meaning of *ideal* (p. 224), *ideal dilute* (p. 224), and *regular solutions* (p. 227).

(6)   State *Henry's Law* (eqn (8.2.8)) and use it to derive the solubility of gases in liquids (p. 224).

(7)   Explain what is meant by *colligative property* and give examples (p. 228).

(8)   Derive an expression relating the *elevation of boiling point* of an ideal solution to its composition (eqn (8.3.1)) and use it to determine the relative molar mass of an involatile solute (p. 231).

(9)   Derive an expression relating the *depression of freezing point* of an ideal solution to its composition (eqn (8.3.2)) and use it to determine the relative molar mass of the solute (p. 233).

(10)   Derive and use an expression for the *solubility* of an ideal solute (eqn (8.3.4)).

(11)   Define *osmotic pressure* (p. 235), derive the *van't Hoff equation* (eqn (8.3.9)) relating osmotic pressure to composition.

(12)   Construct and interpret *vapour-pressure diagrams* for a mixture of two volatile liquids (p. 236).

(13)   Relate the compositions of a liquid and its vapour to the total pressure and to the partial vapour pressures (p. 237).

(14)   Use the *lever rule* to determine from the phase diagram the relative amounts of the phases present (p. 239).

(15)   Interpret a *temperature-composition diagram* and use it to determine the course of distillation of a mixture (p. 240).

(16)   Explain the significance of the term *azeotrope* (p. 242).

(17) Define the *activity* and *activity coefficient* of a component in a real mixture (p. 244), and the *standard states* of the solvent and the solute (p. 244 and Box 8.1).

(18) Explain and use the *method of intercepts* for measuring partial molar quantities (p. 249).

Introduction

This chapter moves away from pure materials and the limited but important changes they can undergo, and begins the study of mixtures. Unreactive mixtures are considered in this chapter, and reactions are discussed in the next. The development is based on the chemical potential, and we shall see how this simple and natural idea unifies all the equilibrium properties of simple mixtures.

Almost everything we do will be confined to systems with only two components: these are *binary mixtures*. Few new concepts are introduced by considering more complex mixtures, and so these will not be discussed until Chapter 10. Since we are dealing with binary mixtures we shall constantly be able to make use of the relation $x_A + x_B = 1$ for the mole fractions of the components.

8.1 Preparing to study mixtures: partial molar quantities

Imagine an indefinitely large volume of water. When a further 1 mol of H_2O is added the volume increases by 18 cm^3. The quantity 18 cm^3 mol^{-1} is the *molar volume* of pure water. Now suppose that 1 mol of water is —added to a large volume of pure ethanol. It is found that the volume increases by only 14 cm^3. The reason for this is that water and ethanol molecules stick together in different ways, and the volume 1 mol occupies depends on the environment of its molecules. In the present example there is so much ethanol and so little water that every water molecule is surrounded by pure ethanol. 14 cm^3 is the volume 1 mol of water molecules occupies when every water molecule is surrounded by an indefinitely extensive region of ethanol. The quantity 14 cm^3 mol^{-1} is the *partial molar volume* of water in pure ethanol, Fig. 8.1.

The partial molar volume of a substance in a mixture of some general composition can be defined in terms of the increase of volume that occurs when the material is added to an indefinitely large sample of the solution. If n_A is the amount of substance added and ΔV the volume increase observed, then

$$\Delta V = V_{A,m}(x_A, x_B)n_A$$

where $V_{A,m}(x_A, x_B)$ is the partial molar volume of A when the solution has a composition described by the mole fractions x_A and x_B.

The definition of partial molar volume has depended on the constancy of the original composition of the solution. The system is taken to be so large that the addition of A does not change the mole fractions except to

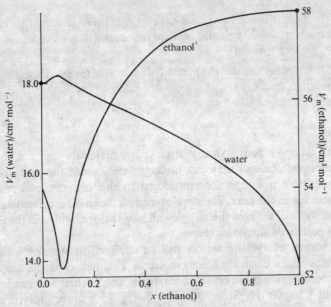

Fig. 8.1. The partial molar volumes of water and ethanol in aqueous ethanol at 25 °C.

an infinitesimal extent. The same constancy can be assured if the sample is finite, but the addition of A is limited to an infinitesimal amount. Then the last equation becomes

$$dV = V_{A,m}(x_A, x_B)\,dn_A.$$

This is a more powerful way of defining a partial molar quantity, as the following development will illustrate.

When an amount dn_A is added to a solution of composition x_A, x_B the volume changes by $V_{A,m}(x_A, x_B)\,dn_A$. Likewise when dn_B of B is added the volume changes by an amount $V_{B,m}\,dn_B$ (for simplicity we omit the composition labels on the partial molar volumes). The overall change when infinitesimal amounts of both A and B are added to the solution with the pressure and temperature constant is therefore

$$dV = V_{A,m}\,dn_A + V_{B,m}\,dn_B.$$

V is a state function that depends on the amounts of A and B present. Therefore dV is an exact differential, and can be written

$$dV = (\partial V/\partial n_A)_{p,T,n_B}\,dn_A + (\partial V/\partial n_B)_{p,T,n_A}\,dn_B.$$

It follows by comparing the two expressions for dV that the partial molar volumes may be identified with the partial derivatives:

(8.1.1) $V_{A,m} = (\partial V/\partial n_A)_{p,T,n_B} \qquad V_{B,m} = (\partial V/\partial n_B)_{p,T,n_A}.$

The concept of partial molar quantity can be extended to any of the thermodynamic state functions. One already encountered is the *partial molar Gibbs function*, or the chemical potential. In eqn (6.3.2) it was written

(with small change of notation)

(8.1.2) $\mu_A = (\partial G/\partial n_A)_{p,T,n_B,\ldots}, \qquad \mu_B = (\partial G/\partial n_B)_{p,T,n_A,\ldots}$.

We now see that μ_A and μ_B can be interpreted either as the change in the Gibbs function of a large sample of the system when unit amount of A or B is added, or as the coefficients which give the change in the Gibbs function for a system containing amounts n_A of A and n_B of B, etc. when infinitesimal amounts of A and B are added and the other intensive variables (pressure, temperature) are constant:

(8.1.3) $dG = \mu_A \, dn_A + \mu_B \, dn_B.$

Partial molar quantities permit one to state the *total* volume, *total* Gibbs function, etc. of a mixture of arbitrary composition. For instance, if the partial molar volumes at a composition x_A, x_B are $V_{A,m}$ and $V_{B,m}$, then the total volume of a sample containing an amount n_A of A and n_B of B (with $x_A = n_A/n$, $x_B = n_B/n$; $n = n_A + n_B$) is

(8.1.4) $V = n_A V_{A,m}(x_A, x_B) + n_B V_{B,m}(x_A, x_B)$

or, for short,

(8.1.5) $V = n_A V_{A,m} + n_B V_{B,m}$

where $V_{A,m}$ and $V_{B,m}$ are the partial molar volumes at the specified composition. This equation is obvious in the case of only one component because it reduces to $V = n_A V_{A,m}$, but the fact that it works for a mixture may seem mysterious. It can be proved formally, but the following argument is a justification.

Consider a volume of the solution of the specified composition, and let it be so large that however much A and B are added their mole fractions do not change. Now add an amount n_A of A: the volume changes by $n_A V_{A,m}(x_A, x_B)$; then add an amount n_B of B and the volume changes by $n_B V_{B,m}(x_A, x_B)$. The total change is given by $n_A V_{A,m} + n_B V_{B,m}$. The sample has an enlarged volume, but the proportions of A and B are still the same. Now scoop out of this enlarged volume a sample containing an amount n_A of A and n_B of B. The volume of this sample is simply $V = n_A V_{A,m} + n_B V_{B,m}$. Because V is a state function, the same sample could have been prepared simply by mixing the appropriate amounts of A and B; and so we have demonstrated the validity of eqn (8.1.5).

The same argument applies to any other partial molar quantity. Therefore the Gibbs function for a sample containing an amount n_A of A, n_B of B, etc., is given by

(8.1.6) $G = n_A \mu_A + n_B \mu_B + \ldots = \sum_J n_J \mu_J.$

Example (Objective 1). A corrupt barman attempts to prepare 100 cm³ of some drink by mixing 30 cm³ of ethanol with 70 cm³ of water. Does he succeed? If not, what volumes should have been mixed in order to arrive at a mixture of the same strength but of the required volume?

● *Method.* The bàrman had understood neither the concept nor the importance of partial molar quantities. In order to find the total volume of the mixture we need to know the molar composition, and the partial molar volumes at that composition. The former we get from the densities and molar masses of the liquids; the latter we get from Fig. 8.1. In order to find the required recipe we find the solution to $100 \text{ cm}^3 = n_A V_{A,m} + n_B V_{B,m}$ with n_A/n_B the same as in the corrupt attempt.

● *Answer.* $n(H_2O) = (70.0 \text{ cm}^3) \times (1.00 \text{ g cm}^{-3})/(18.02 \text{ g mol}^{-1}) = 3.80 \text{ mol}.$

$\qquad n(EtOH) = (30.0 \text{ cm}^3) \times (0.785 \text{ g cm}^{-3})/(46.07 \text{ g mol}^{-1}) = 0.511 \text{ mol}.$

$\qquad x(H_2O) = (3.80 \text{ mol})/[(3.80 \text{ mol}) + (0.511 \text{ mol})] = 0.884.$

$\qquad x(EtOH) = 1 - 0.884 = 0.116.$

From Fig. 8.1 the partial molar volumes at this molar composition are $V_m(H_2O) = 18.0 \text{ cm}^3 \text{ mol}^{-1}$ and $V_m(EtOH) = 53.6 \text{ cm}^3 \text{ mol}^{-1}$. Therefore the total volume of the mixture is

$$V = (3.80 \text{ mol}) \times (18.0 \text{ cm}^3 \text{ mol}^{-1}) + (0.511 \text{ mol}) \times (53.6 \text{ cm}^3 \text{ mol}^{-1})$$
$$= 97.2 \text{ cm}^3.$$

A mixture of the same relative composition but of total volume 100 cm^3 will have the same mole fractions of components but a different overall amount n. Therefore

$$100 \text{ cm}^3 = n\{(0.884) \times (18.0 \text{ cm}^3 \text{ mol}^{-1}) + (0.116) \times (53.6 \text{ cm}^3 \text{ mol}^{-1})\}$$
$$= n(22.13 \text{ cm}^3 \text{ mol}^{-1})$$

so that

$$n = 4.52 \text{ mol}.$$

The mixture should therefore contain an amount $(4.52 \text{ mol}) \times (0.884) = 4.00 \text{ mol}$ of water, and $(4.52 \text{ mol}) \times (0.116) = 0.524 \text{ mol}$ of ethanol. The initial volumes corresponding to these amounts are respectively 72.0 cm^3 and 30.8 cm^3.

● *Comment.* It would probably be unwise to attempt to explain this to the barman.

Partial molar quantities can be determined in a variety of ways. One method consists of measuring the property Q and the gradient $\partial Q/\partial n$ at the composition of interest. This is not very accurate. A better way is by the *method of intercepts*: this is outlined in the Appendix (p. 248).

One further property of partial molar quantities has important implications. Suppose the concentrations are varied by small amounts, then according to eqn (8.1.5) we might expect the volume of the system to change both because the n_J change and because the $V_{J,m}$ change (they also depend on the concentration). Expressed algebraically, a general change in V (at constant temperature and pressure) is given by

$$dV = n_A \, dV_{A,m} + V_{A,m} \, dn_A + n_B \, dV_{B,m} + V_{B,m} \, dn_B,$$

as may be seen by differentiating eqn (8.1.5). But we have seen already that

$$dV = V_{A,m} \, dn_A + V_{B,m} \, dn_B.$$

If these two equations are to be consistent (they must ᴜe equal because V is a state function) it follows that

$$n_A \, dV_{A,m} = -n_B \, dV_{B,m}.$$

In other words, the partial molar volumes cannot change independently of one another. If, for some reason, $V_{A,m}$ increases, then $V_{B,m}$ must decrease in accord with the last equation. This can be seen in Fig. 8.1: increases in the partial molar volume of one component are faithfully mirrored by decreases in the other, and vice versa.

The same line of reasoning applies to the partial molar Gibbs functions. Equation (8.1.6) therefore implies that

(8.1.7)

$$\sum_J n_J \, d\mu_J = 0.$$

This is the *Gibbs–Duhem equation*, which relates the changes in chemical potential of the components of a system at constant temperature and pressure. Some of the consequences of this important relation are explored in the Problems.

One final word of warning. Molar volumes and entropies are definitely positive, but partial quantities need not be. This is not as strange as it might seem. For instance, when 1 mol of $MgSO_4$ is added to a large volume of water the total volume *decreases* by 1.4 cm^3. According to the definition of the partial molar volume it follows that the partial molar volume of $MgSO_4$ in pure water is -1.4 cm^3 mol^{-1}. The contraction occurs because the salt breaks up the structure of water, which collapses by a small amount.

8.2 The thermodynamics of mixing

If two gases at a pressure p are confined in separate vessels, and then the vessels are joined, the gases will mix and in time the mixture will be uniform. The mixing process is spontaneous, and must correspond to rolling down a Gibbs hill. We now see how to make this idea precise.

Why gases mix, but sometimes don't. Consider an amount n_A of gas A and n_B of gas B, in separate containers but both at a temperature T and a pressure p, Fig. 8.2a. The Gibbs function for the total system is $n_A\mu_A + n_B\mu_B$, where μ_J is the chemical potential of gas J at the pressure p and temperature T. For simplicity we assume the gases to be perfect, in which case the initial Gibbs function is

$$G_i = n_A\{\mu_A^\ominus(T) + RT\ln(p/p^\ominus)\} + n_B\{\mu_B^\ominus(T) + RT\ln(p/p^\ominus)\}.$$

After mixing (Fig. 8.2b) each gas exerts a partial pressure and their sum $p_A + p_B$ is p; the Gibbs function is then

$$G_f = n_A\{\mu_A^\ominus(T) + RT\ln(p_A/p^\ominus)\} + n_B\{\mu_B^\ominus(T) + RT\ln(p_B/p^\ominus)\}.$$

The change of the Gibbs function on mixing, $\Delta G_{mix} = G_f - G_i$, is

$$\Delta G_{mix} = n_A RT\ln(p_A/p) + n_B RT\ln(p_B/p).$$

Dalton's Law (p. 34) allows p_A/p to be replaced by the mole fraction x_A,

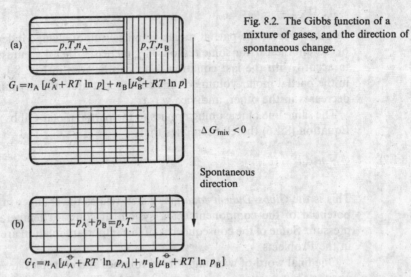

Fig. 8.2. The Gibbs function of a mixture of gases, and the direction of spontaneous change.

(a) $G_i = n_A [\mu_A^\ominus + RT \ln p] + n_B [\mu_B^\ominus + RT \ln p]$

$\Delta G_{mix} < 0$

Spontaneous direction

(b) $G_f = n_A [\mu_A^\ominus + RT \ln p_A] + n_B [\mu_B^\ominus + RT \ln p_B]$

where $x_A = n_A/n$ and n is the total amount of gas $n_A + n_B$. The same can be done for p_B/p. Then the Gibbs function of mixing becomes

(8.2.1)° $\Delta G_{mix} = nRT\{x_A \ln x_A + x_B \ln x_B\}.$

The last expression confirms that the mixing of perfect gases is a natural process. (Both x_A and x_B are less than 1, and so their logarithms are negative.) This conclusion does more than confirm the common-sense view that gases do mix if they are allowed to, for it also enables the drop in the Gibbs function and its pressure and temperature dependence to be expressed quantitatively. For example, we see that ΔG is independent of the pressure, but depends linearly on the temperature.

Example (Objective 3). A container is divided into two compartments. One contains 3 mol of hydrogen, the other 1 mol of nitrogen. The temperature is 25 °C, and the pressure in each compartment is 1 atm. Calculate the Gibbs function of mixing of the two gases when the partition is withdrawn.

- *Method.* Use eqn (8.2.1); we therefore require the mole fractions of the two gases.

- *Answer.* $x(H_2) = (3 \text{ mol})/(3 \text{ mol} + 1 \text{ mol}) = \frac{3}{4}$

  $x(NH_3) = (1 \text{ mol})/(3 \text{ mol} + 1 \text{ mol}) = \frac{1}{4}.$

  From eqn (8.2.1):

  $\Delta G_{mix} = (4 \text{ mol}) \times (8.314 \text{ J K}^{-1} \text{ mol}^{-1}) \times (298.15 \text{ K}) \times (\frac{3}{4} \ln \frac{3}{4} + \frac{1}{4} \ln \frac{1}{4})$

  $= (9.92 \text{ kJ}) \times (-0.562)$

  $= -5.58 \text{ kJ}.$

- *Comment.* This assumes perfect behaviour. The same result would be obtained for any overall pressure. Note that if the two gases were 3 mol of ammonia and 1 mol of ammonia, withdrawing the partition would give $\Delta G_{mix} = 0$ although the present calculation would seem to predict that $\Delta G_{mix} = -5.58 \text{ kJ}$. This is a point of some subtlety, and its resolution depends on the recognition that the ammonia molecules in one part of the container are indistinguishable from those initially in the other part. This point is dealt with in Part 2.

The quantitative expression for ΔG_{mix} gives the *entropy of mixing*. Since $(\partial G/\partial T)_{p,n}$ is equal to $-S$, it follows immediately that

$(8.2.2)°$ $$\Delta S_{mix} = -(\partial \Delta G_{mix}/\partial T)_{p,n_A,n_B} = -nR\{x_A \ln x_A + x_B \ln x_B\}.$$

Since $\ln x$ is negative, the entropy of mixing is a positive quantity. This conforms to the interpretation of mixing as a dispersal (both gases are now free to occupy the total volume), and of an increasing entropy as a signpost of natural change. The entropy of mixing in the last *Example* is readily found to be $18.7 \text{ J K}^{-1} \text{ mol}^{-1}$.

From ΔG_{mix} and ΔS_{mix} we may calculate the enthalpy of mixing of two perfect gases. Using $\Delta G = \Delta H - T\Delta S$, because the mixing is occurring at constant temperature ($\Delta T = 0$), leads to

$(8.2.3)°$ $$\Delta H_{mix} = 0 \qquad (\text{const. } p, T).$$

The enthalpy change is zero, as should be expected for a system in which there are no interactions between the molecules. Therefore the whole of the driving force for mixing comes from the increase in entropy of the system. In other words, perfect gases mix solely on account of their tendency to spread into the total volume made available by opening the partition between the two containers.

A knowledge of ΔG_{mix} is enough to determine the volume change on mixing; since $(\partial G/\partial p)_{T,n} = V$, we can get ΔV_{mix} by differentiating ΔG_{mix} with respect to pressure. But ΔG_{mix} is independent of p, and so the volume of mixing is zero:

$(8.2.4)$ $$\Delta V_{mix} = 0 \qquad (\text{const. } p, T).$$

This is also to be expected for a system without interactions.

Knowing ΔG and ΔS led to ΔH. Knowing ΔH and ΔV now leads to ΔU. We have seen that ΔH_{mix} and ΔV_{mix} are zero, and as the mixing is proceeding at constant total pressure we find.

$(8.2.5)°$ $$\Delta U_{mix} = \Delta H_{mix} - p\Delta V_{mix} = 0 \qquad (\text{const. } p, T).$$

There is no change of internal energy on mixing.

All these conclusions are modified when the gases depart from ideality. Nevertheless it normally remains true that the Gibbs function decreases when real gases are free to mix, but it has a more complex pressure, temperature, and composition dependence than that given by eqn (8.2.1). The entropy of the system normally increases, but not by the amount $-\Delta G_{mix}/T$, which means that there is also an enthalpy of mixing. In the real case ΔG_{mix} depends on the pressure because it is affected by intermolecular forces, and so the volume of the sample will in general change (for T and p constant).

Note that we say that ΔG_{mix} will *normally* decrease when gases are free to mix. There are exceptions in special circumstances. When two gases are confined to the same vessel, and the conditions are arranged so that they are above their critical temperatures (and so by definition are not liquids), and the pressure is raised to large values, some gases do not mix but behave like immiscible liquids. In these cases the enthalpy of mixing is positive and dominates the favourable $T \Delta S$ term; therefore ΔG_{mix} is negative for the unmixing process.

The chemical potential of liquids. Consider a container in which there is a liquid in equilibrium with its vapour, Fig. 8.3. From the uniformity of the chemical potential throughout the system it is known that $\mu(l) = \mu(g)$. Since the chemical potential of a gas can be expressed in terms of its pressure, $\mu = \mu^\ominus + RT \ln(p/p^\ominus)$, this equality lets us relate the chemical potential of a liquid to its vapour pressure.

If only one component is present (the component A), the chemical potential of the pure liquid is related to that of its pure vapour (which is assumed to behave as a perfect gas) by

$$\mu_A^*(l) = \mu_A^*(g) = \mu_A^\ominus + RT \ln(p_A^*/p^\ominus)$$

where p_A^* is the vapour pressure of the pure liquid at that temperature (the star * is used to indicate a pure component).

If several components are present the chemical potential of A in the liquid is still equal to its chemical potential in the vapour (otherwise A would not be in equilibrium). If this partial vapour pressure is p_A we can write

$$\mu_A(l) = \mu_A(g) = \mu_A^\ominus + RT \ln(p_A/p^\ominus).$$

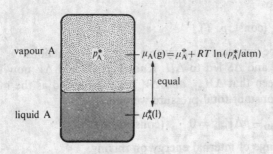

vapour A p_A^* $\mu_A(g) = \mu_A^\ominus + RT \ln(p_A^*/atm)$

equal

liquid A $\mu_A^*(l)$

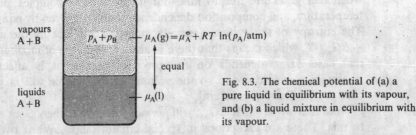

vapours A+B $p_A + p_B$ $\mu_A(g) = \mu_A^\ominus + RT \ln(p_A/atm)$

equal

liquids A+B $\mu_A(l)$

Fig. 8.3. The chemical potential of (a) a pure liquid in equilibrium with its vapour, and (b) a liquid mixture in equilibrium with its vapour.

The last two equations can be combined to eliminate the standard potential, giving

$(8.2.6)°$ $$\mu_A(l) = \mu_A^*(l) + RT \ln(p_A/p_A^*).$$

This equation expresses the chemical potential μ_A of a liquid in a mixture in terms of the chemical potential of the pure liquid, μ_A^*, and the vapour pressures of the two cases.

The final step involves doing something about p_A/p_A^*. When only A is present this ratio is unity. When no A is present it is zero. A number of measurements on the vapour pressures of mixtures of closely related liquids (such as benzene and toluene) support the conclusion that p_A/p_A^* varies approximately linearly with the mole fraction of A in the liquid. This conclusion is embodied in the idealization

$(8.2.7)°$ **Raoult's Law: $p_A = x_A p_A^*$,**

where x_A is the mole fraction of A in the liquid mixture. The pressure dependence predicted by Raoult's Law is illustrated in Fig. 8.4a.

Some solutions obey Raoult's Law very well, especially when the components are chemically similar. Then a molecule of one species interacts with its neighbours as though they were its own kind, and the vapour pressure of that component is almost the same as when it is actually in

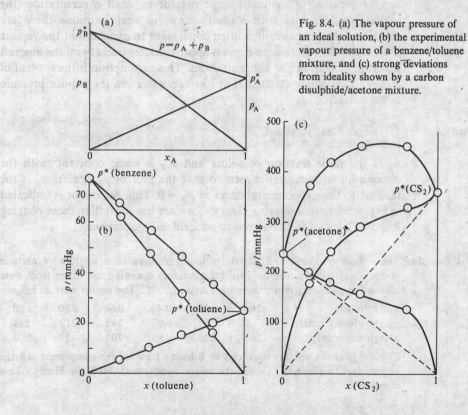

Fig. 8.4. (a) The vapour pressure of an ideal solution, (b) the experimental vapour pressure of a benzene/toluene mixture, and (c) strong deviations from ideality shown by a carbon disulphide/acetone mixture.

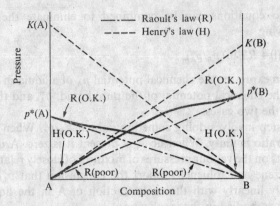

Fig. 8.5. The regions of applicability of Raoult's (R) and Henry's (H) law behaviour.

a pure environment of its own kind. This is illustrated in Fig. 8.4b which shows the behaviour of benzene and toluene (methylbenzene). Solutions that obey Raoult's Law throughout their composition range are called *ideal*. Equations relating to ideal solutions will be distinguished by ° attached to the equation number.

Dissimilar species deviate strongly from Raoult's Law, but even they conform when the system is nearly pure. This means that for the component in excess, the 'solvent', the vapour pressure is given quite well by eqn (8.2.7) in the vicinity of $x \approx 1$, Fig. 8.4c. Whereas in an ideal solution the vapour pressure of the component present in small concentration (the solute) also accords with Raoult's Law, in real solutions there are deviations. Nevertheless it is often convenient to assume that the vapour pressure of the solute does depend approximately linearly on the amount of solute present at low concentrations. This assumption is the content of *Henry's Law* which states that at low concentration the vapour pressure of the solute obeys

(8.2.8)   *Henry's Law:* $p_B = x_B K_B$

x_B is the mole fraction of solute and K_B is some constant (with the dimensions of pressure) chosen so that the plot of p_B against x_B is the tangent to the experimental curve at $x_B = 0$. This behaviour is indicated in Fig. 8.5. Systems obeying Henry's Law are less ideal than those obeying Raoult's Law and are referred to as *ideal dilute solutions*.

Example. The vapour pressure and partial vapour pressures of a mixture of acetone (propanone) and chloroform (trichloromethane) spanning the range from pure acetone to pure chloroform were measured at 35 °C. The results were as follows:

x (chloroform)	0.00	0.20	0.40	0.60	0.80	1.00
p (chloroform)/mmHg	0	35	82	142	219	293
p (acetone)/mmHg	347	270	185	102	37	0

Confirm that the mixture conforms to Raoult's Law for the component in large excess, and to Henry's Law for the minor component. Find the Henry's Law constants.

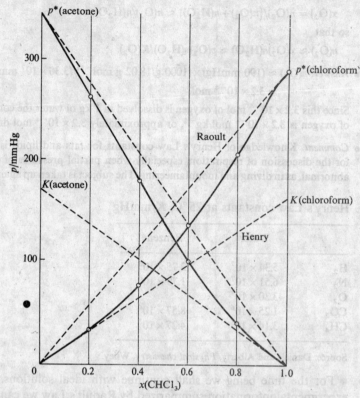

Fig. 8.6. Experimental vapour pressures of acetone/chloroform mixtures.

- *Method.* Plot the partial vapour pressures and the total vapour pressure against mole fraction. Raoult's Law can be tested by comparing the data with straight lines drawn according to $p_A = x_A p_A^*$. Henry's Law can be tested by finding some value of K_A such that the line $p_A = x_A K_A$ fits the data for small x_A.

- *Answer.* The data are plotted in Fig. 8.6 together with the Raoult's Law lines. Henry's Law requires K(acetone) = 175 mmHg and K(chloroform) = 165 mmHg.

- *Comment.* Notice how the data deviate from both Raoult's and Henry's Laws for quite small deviations from $x_A = 1$ and $x_A = 0$ respectively. A way of accommodating such real systems into the limiting law behaviour suggested by Raoult's and Henry's Laws, is dealt with on p. 243.

Some Henry's Law data are listed in Table 8.1, and their application to gas solubility estimations is illustrated in the following *Example*.

Example (Objective 6). Use the Henry's Law data to estimate the solubility of oxygen in water at 25 °C and a partial pressure of 190 mmHg.

- *Method.* The mole fraction of solute (oxygen) is given by Henry's Law as $x = p/K$, p being its partial pressure. Calculate the amount of oxygen dissolved in 1000 g water; since the amount of dissolved gas is small approximate the mole fraction as follows:

$$x(O_2) = n(O_2)/\{n(O_2)+n(H_2O)\} \approx n(O_2)/n(H_2O),$$

so that

$$n(O_2) \approx x(O_2)n(H_2O) \approx p(O_2)n(H_2O)/K(O_2).$$

- *Answer.* $n(O_2) \approx (190 \text{ mmHg}) \times (1000 \text{ g}/18.02 \text{ g mol}^{-1})/(3.30 \times 10^7 \text{ mmHg})$

$$\approx 3.2 \times 10^{-4} \text{ mol.}$$

Since this 3.2×10^{-4} mol of oxygen is dissolved in 1 kg of water the concentration of oxygen is 3.2×10^{-4} mol kg^{-1}, or approximately 3.2×10^{-4} mol dm^{-3}.

- *Comment.* Knowledge of Henry's Law constants for fats and lipids is important for the discussion of respiration, especially when partial pressures of oxygen are abnormal, as in diving and mountaineering. The subject is taken up again on p. 279.

Table 8.1. Henry's Law constants at 25 °C, K/mmHg

	Water	Benzene
H_2	5.34×10^7	2.75×10^6
N_2	6.51×10^7	1.79×10^6
O_2	3.30×10^7	
CO_2	1.25×10^6	8.57×10^4
CH_4	3.14×10^5	4.27×10^5

Source: Daniels and Alberty, *Physical chemistry*, Wiley.

For the time being we shall continue with ideal solutions. With the experimental information summarized by Raoult's Law we can write eqn (8.2.6) as

$(8.2.9)°$ $$\mu_A(l) = \mu_A^*(l) + RT \ln x_A.$$

This important equation, which can also be taken as an alternative definition of an ideal solution, enables the chemical potential of a component of a liquid mixture to be expressed in terms of the relative amount of the species present.

Why some liquids mix. Consider the Gibbs function of mixing of two liquids A and B. If they form an ideal solution, then at constant temperature and pressure the Gibbs function changes from its initial value

$$G_i = n_A \mu_A^*(l) + n_B \mu_B^*(l)$$

for the separate, pure liquids to its final value

$$G_f = n_A \{\mu_A^*(l) + RT \ln x_A\} + n_B \{\mu_B^*(l) + RT \ln x_B\}$$

for the mixture. Consequently, the change in Gibbs function is

$(8.2.10)°$ $$\Delta G = G_f - G_i = nRT \{x_A \ln x_A + x_B \ln x_B\},$$

where $n = n_A + n_B$ is the total amount of liquid. This function, and the corresponding expressions for the entropy and enthalpy of mixing, are plotted in Fig. 8.7.

The expression for the Gibbs function of mixing of two liquids to give an ideal solution is the same as that for the mixing of two perfect gases. All the conclusions drawn there are valid here: the entropy of mixing is positive (and is the driving force for mixing: the liquid molecules spread into the total available volume), the enthalpy of mixing is zero (the average molecular interactions in the mixture are the same as in the pure materials), and there is no change of total volume. Note however, that ideality implies something different from gas perfection: in a perfect gas there are no interactions between the molecules. In an ideal solution there are interactions, but the average A–B interactions in the mixture are the same as the mean of the A–A and B–B interactions in the pure liquids.

Real solutions are composed of molecules for which A–A, A–B, and B–B interactions are all different. An enthalpy change may therefore occur when A is mixed with B. Furthermore, the molecules of one type might tend slightly to congregate together rather than to be uniformly distributed throughout the mixture. If the enthalpy change is large it may overcome the entropy change and, as a consequence, ΔG may be positive for the mixing process. Unmixing is then spontaneous, and the liquids are immiscible.

Real solutions are often discussed in terms of the *excess functions*, G^E, S^E, etc. An excess function is defined as the difference between the observed thermodynamic function of mixing and the function of mixing for the ideal mixture. In the case of the excess entropy, for instance,

$$S^E = \Delta S - nR\{x_A \ln x_A + x_B \ln x_B\}.$$

Deviations of the excess functions from zero then indicate the extent to which the solution is non-ideal. A useful model system is the *regular solution*: in this the excess enthalpy is taken to be non-zero but the excess

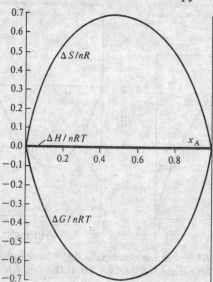

Fig. 8.7. Gibbs function, enthalpy, and entropy of mixing of an ideal binary mixture.

entropy is taken as zero. A regular solution can be visualized as a mixture in which the two kinds of molecules are distributed randomly even though the A–A, B–B, and A–B interactions differ. Two examples of how excess functions depend on composition are illustrated in Fig. 8.8.

8.3 Solutions of non-volatile solutes: colligative properties

We are now sufficiently well-equipped to make light work of explaining the effect of dissolved matter on the boiling and freezing points of liquids, and some related phenomena. In this section we see how to calculate the elevation of boiling point, the depression of freezing point, and the osmotic pressure of ideal solutions. These properties depend on the amount of solute present but not on its nature, and for that reason they are called *colligative* (denoting 'depending on the collection'). It is also helpful to think of 'colligative' as denoting properties bound together in a common explanation.

We make two assumptions. The first is that *the solute is non-volatile*. This implies that it does not appear in the vapour phase, and so the solvent vapour is the only gas present. The second assumption is that *the non-volatile solute does not dissolve in the solid solvent*. This is quite a drastic simplification although it is true of many mixtures. It can be avoided at the expense of more algebra, but that introduces no new principles. Both these restrictions will be removed in the qualitative discussion of Chapter 10.

The common cause behind the colligative properties is the modification of the chemical potential of the liquid solvent by the presence of the solute. The discussion of the last section, and in particular eqn (8.2.9), $\mu_A = \mu_A^* + RT\ln x_A$, shows that the chemical potential of a liquid is reduced by an amount $RT\ln x_A$ when a solute is present (x_A being the mole fraction

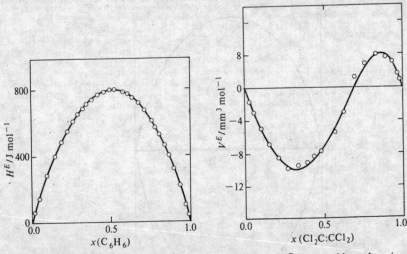

Fig. 8.8. Excess functions. (a) H^E for benzene/cyclohexane, (b) V^E for tetrachloroethane/cyclopentane at 25 °C. (K. N. Marsh and I. A. McLure, *Int. DATA Ser., Selec. Data Mixtures, Ser. A*. 1973, 2 and 9.)

of the solvent, $\ln x_A$ is negative). The chemical potentials of the vapour and solid are unchanged by the presence of the non-volatile solute. As can be seen from Fig. 8.9, this implies that the vapour–liquid equilibrium occurs at a higher temperature while the solid–liquid equilibrium occurs at a lower one. Hence the boiling point is raised and the freezing point is lowered.

The physical basis of the lowering of the chemical potential cannot lie in the modification of the intermolecular forces, because it happens even in the case of ideal solutions. If it is not an enthalpy effect it must be an entropy effect. In the absence of a solute the pure liquid solvent has some entropy (reflecting its disorder) and some enthalpy. The vapour pressure arises from the tendency of the universe to maximum entropy, and one positive contribution to this comes from the evaporation of the liquid to form an even more random gas. When a solute is present there is an extra randomness present in the solution that was not present in the pure solvent. Therefore the tendency for the solvent to escape and acquire the higher entropy of its vapour is not so compelling. When there is already additional randomness in the solution not so much vapour has to be formed in order to equalize the chemical potentials of the two phases, and so the vapour pressure is lowered. A similar consideration applies to the tendency to melt, but now the enhanced randomness of the solution increases the tendency of the solid to break up. It is able to do this at a lower temperature than for the pure solvent (for a given temperature $-T\,\Delta S$ is more negative because ΔS is more positive), and so the melting point is depressed.

The guiding principle for the quantitative discussion is to look for the temperature at which one phase (the vapour or the solid) has the same chemical potential as the solvent in the solution. This is the equilibrium temperature, and therefore is the new boiling or freezing point. The next few paragraphs show this idea in action.

How the boiling point is raised. The equilibrium of interest is between the solvent vapour and the solvent in the solution, Fig. 8.10a. We shall denote the solvent by A and the solute by B. The equilibrium is attained at a temperature given by

$$\mu_A^*(g) = \mu_A(l) = \mu_A^*(l) + RT\ln x_A.$$

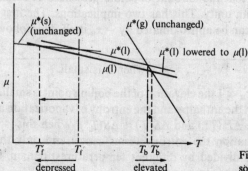

Fig. 8.9. The chemical potential of a solvent in the presence of a solute.

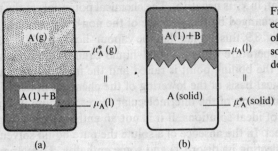

Fig. 8.10. (a) Liquid–vapour equilibrium and the elevation of boiling point. (b) Liquid–solid equilibrium and the depression of freezing point.

This can be rearranged into

$$\ln(1-x_B) = \{\mu_A^*(g) - \mu_A^*(l)\}/RT = \Delta G_{vap,m}(T)/RT.$$

where $\Delta G_{vap,m}$ is the molar Gibbs function of vaporization of the pure solvent and x_B is the mole fraction of the solute ($x_A + x_B = 1$).

There are various ways of solving this equation for T in terms of x_B. One uses the Gibbs–Helmholtz equation (p. 166) and takes into account the temperature dependence of ΔG exactly. That is left as an exercise (Problem 8.25). A simpler procedure is as follows. When $x_B = 0$ the boiling point is T^*, and so in this case

$$\ln 1 = \Delta G_{vap,m}(T^*)/RT^*.$$

Combining this equation with the last, and using $\Delta G = \Delta H - T\,\Delta S$ leads to

$$\ln(1-x_B) - \ln 1 = \left\{\frac{\Delta G_{vap,m}(T)}{RT}\right\} - \left\{\frac{\Delta G_{vap,m}(T^*)}{RT^*}\right\}$$

$$= \left\{\frac{\Delta H_{vap,m}(T)}{RT} - \frac{\Delta S_{vap,m}(T)}{R}\right\}$$

$$- \left\{\frac{\Delta H_{vap,m}(T^*)}{RT^*} - \frac{\Delta S_{vap,m}(T^*)}{R}\right\}.$$

Notice that we have been careful to allow for the temperature dependence of the Gibbs function.

Now suppose that the amount of solute present is so small that x_B is much smaller than unity. This has two implications. The first is that the $\ln(1-x_B)$ terms can be approximated by $-x_B$ itself. This follows from the series expansion

$$\ln(1-x) = -x - \tfrac{1}{2}x^2 - \ldots \approx -x \text{ when } x \text{ is small.}$$

Furthermore, if $x_B \ll 1$ the elevation of the boiling point is small. Therefore, we may take both the enthalpy and the entropy of vaporization as constant, so that $\Delta H(T) \approx \Delta H(T^*)$ and $\Delta S(T) \approx \Delta S(T^*)$. When this is introduced into the last equation the entropy terms cancel and only the enthalpy terms (which are divided by different temperatures) remain. Combining these approximations gives

$$x_B = -\left\{\frac{\Delta H_{vap,m}}{R}\right\}\left(\frac{1}{T}-\frac{1}{T^*}\right).$$

This equation can be rearranged into an expression for the boiling point T when an amount x_B of solute is present. The final expression can be made even simpler by continuing to make the approximation that T is close to T^*. The two reciprocals combine to give $(T^*-T)/TT^*$, but this is almost the same as $(T^*-T)/(T^*)^2$. On writing $\delta T = T - T^*$ for the elevation of the boiling point, the final expression is

(8.3.1)° $\qquad \delta T \approx (RT^{*2}/\Delta H_{vap,m})x_B.$

This formula makes no reference to the nature of the solute, only to its amount (x_B). That is why the term 'colligative' applies in this case. The boiling point elevation does depend on the properties of the solvent, and the biggest change occurs for solvents with high boiling points but low molar enthalpies of vaporization. For example, in the case of benzene $\Delta H_{vap,m}$ is 30.8 kJ mol^{-1} and $T^* = 353$ K; therefore $\delta T \approx (33.6$ K$)x_B$, and so a mole fraction of 0.01 of any non-volatile solute raises the boiling point by 0.34 K.

Boiling point elevation can be used to determine the relative molecular mass of soluble, non-volatile materials. The method is outlined in the following *Example*.

Example (Objective 8). Show that the elevation of boiling point can be written $\delta T = K_b m_B$ when the molality, m_B, of the solute is small. Calculate the value of the *ebullioscopic constant* K_b for benzene. In an experiment 10 g of a solid were dissolved in 100 g of benzene and the boiling point was found to rise from 80.10 °C to 80.90 °C. What is the relative molar mass of the solute?

• *Method.* The calculation hinges on eqn (8.3.1) and the smallness of the mole fraction of the solute. From the latter it follows that

$\qquad x_B = n_B/(n_A+n_B) \approx n_B/n_A.$

Then use $m_B = n_B/1$ kg, or $n_B = m_B \times 1$ kg, and $n_A = 1$ kg/$M_{A,m}$ so that $x_B \approx m_B M_{A,m}$. The manipulation of eqn (8.3.1) is then straightforward. In order to evaluate K_b we need the following data for the solvent benzene: boiling point (80.15 °C, 353.25 K), enthalpy of vaporization (30.8 kJ mol^{-1}), molar mass ($M_{A,m}$, 78.11 g mol^{-1}).

• *Answer.* From eqn (8.3.1) and the smallness of x_s,

$\qquad \delta T \approx (RT_b^{*2}/\Delta H_{vap,m})m_B M_{A,m} = K_b m_B,$

where $K_b = RT_b^{*2}M/\Delta H_{vap,m}$. Substituting values of the quantities gives

$$K_b = \frac{(8.314 \text{ J K}^{-1} \text{ mol}^{-1}) \times (353.25 \text{ K})^2 \times (78.11 \text{ g mol}^{-1})}{(30.8 \text{ kJ mol}^{-1})}$$

$\qquad = 2.63$ K mol^{-1} kg $\quad$ or $\quad$ 2.63 K/(mol kg^{-1}).

This means that a 1 mol kg^{-1} solute concentration would raise the boiling point 2.63 K (if the solution remained ideal at that concentration).

From the data $\delta T = 0.80$ K, and so the substance is present at a molality

$\qquad m_B = (0.80 \text{ K})/(2.63 \text{ K mol}^{-1} \text{ kg}) = 0.304$ mol kg^{-1}.

If the molar mass is $M_{B,m}$, a solution made from 10 g of solute in 100 g of solvent has a molality of $(10 \text{ g}/M_{B,m}) \times 10 \text{ kg}^{-1}$. Therefore the molar mass of the solute can be found from

$$M_{B,m} = (10 \text{ g} \times 10 \text{ kg}^{-1})/(0.304 \text{ mol kg}^{-1}) = 329 \text{ g mol}^{-1}.$$

Check that x_B is small: $x_B \approx m_B M_{A,m} = (0.304 \text{ mol kg}^{-1}) \times (78.11 \times 10^{-3} \text{ kg mol}^{-1}) = 0.02$.

• *Comment.* Remember that the derivation assumes that the solute is involatile at the boiling point of the solvent. Note that the ebullioscopic constant is large for solvents with high boiling points, high molar masses, but low enthalpies of vaporization. Note too that K_b depends only on the solvent and is independent of the solute. Other values are listed in Table 8.2.

Table 8.2. Cryoscopic and ebullioscopic constants

Solvent	$K_f/\text{K kg mol}^{-1}$	$K_b/\text{K kg mol}^{-1}$
Acetic acid	3.90	3.07
Benzene	5.12	2.53
Camphor	40	—
CS_2	3.8	2.37
CCl_4	30	4.95
Naphthalene	6.94	5.8
Phenol	7.27	3.04
Water	1.86	0.51

Source: G. W. C. Kaye and T. H. Laby, *Tables of physical and chemical constants*, Longmans.

How the freezing point is depressed. The equilibrium of interest is between pure solid solvent (the solute is assumed to be insoluble in the solid solvent) and the solvent in the solution containing a mole fraction x_B of the solute B_8, see Fig. 8.10b. At the freezing point the chemical potentials of the pure solid and the contaminated solvent are equal:

$$\mu_A^*(s) = \mu_A(l) = \mu_A^*(l) + RT\ln x_A.$$

The only difference between this and the last calculation is the appearance of the solid's chemical potential rather than the vapour's. Therefore, we can write the result directly from eqn (8.3.1):

$(8.3.2)°$ $\delta T \approx (RT^{*2}/\Delta H_{\text{melt,m}})x_B.$

Here $\delta T = T^* - T$ is the freezing point depression. T^* is the freezing point of the pure solid, and $\Delta H_{\text{melt,m}}$ its molar enthalpy of fusion (melting).

The depression is a colligative property because it depends only on the amount and not the nature of the solute. Big depressions are favoured by high-melting solvents with low enthalpies of fusion. In the case of benzene $\Delta H_{\text{melt,m}} = 9.84 \text{ kJ mol}^{-1}$ and $T^* = 278.7 \text{ K}$; this gives a depression of $T = (65.6 \text{ K})x_B$. This is bigger than the corresponding elevation of boiling point, as we would expect from the general discussion of the effect of lowering the chemical potential (see Fig. 8.9). Measurement of the freezing-point depression is also a method of determining relative molar mass, and the procedure is illustrated in the following *Example*.

Example (Objective 9). Show that the freezing point depression can be expressed in the form $\Delta T = K_f m_B$, where the *cryoscopic constant* depends only on the solvent. Calculate the value of K_f for benzene. A 10 g amount of a solute in 100 g of benzene lowered its freezing point from 5.50 °C to −0.74 °C. What is the molar mass of the solute?

- *Method*. The calculation is virtually identical to that in the last *Example*, being based on eqn (8.3.2) and the smallness of the mole fraction of the solute. For numerical values of K_f we need to know the freezing point of benzene (5.5 °C), the enthalpy of fusion (9.84 kJ mol^{-1}), and its molar mass (78.11 g mol^{-1}).

- *Answer*. $\Delta T = K_f m_B$,

  where

  $$K_f = RT_f^{*2}M_{A,m}\Delta H_{\text{melt,m}}.$$

  For benzene,

  $$K_f = \frac{(8.314 \text{ J K}^{-1}\text{ mol}^{-1})\times(278.6 \text{ K})^2\times(78.11 \text{ g mol}^{-1})}{(9.84 \text{ kJ mol}^{-1})}$$

  $$= 5.12 \text{ K/(mol kg}^{-1}).$$

  From the data

  $$m_B = (6.24 \text{ K})/(5.12 \text{ K/mol kg}^{-1}) = 1.22 \text{ mol kg}^{-1},$$

  and by the same process as before this corresponds to a molar mass

  $$M_{B,m} = (10 \text{ g} \times 10 \text{ kg}^{-1})/(1.22 \text{ mol kg}^{-1}) = 82 \text{ g mol}^{-1}.$$

- *Comment*. The accuracy of the result depends on the validity of the assumption of ideal behaviour and lack of solute in the solid solvent. The cryoscopic constant is large for solvents with high molar masses, and for this reason camphor ($K_f = 40 \text{ K/mol kg}^{-1}$) is often used. Other typical values are given in Table 8.2.

How to estimate solubility. Although it is not strictly a colligative property, the present technique may be used to estimate the solubility of a solid in a solvent. If a lump of solid is left in contact with a solvent it will dissolve until the solvent has become saturated. The saturated solution corresponds to the case in which the chemical potential of the pure solid is equal to the chemical potential of the *solute* in the saturated solution, Fig. 8.11. The former is μ^* (solid) and the latter is $\mu_B^*(l) + RT \ln x_B$, where $\mu_B^*(l)$ is the chemical potential of the pure liquid *solute* (ideal, Raoult's Law behaviour is assumed). At saturation

$$\mu_B^*(s) = \mu_B(l) = \mu_B^*(l) + RT \ln x_B.$$

This is the same as the starting equation in the last section, except that the quantities refer to the solute, B, not the solvent, A.

The starting point is the same, but the aim is different. In the present case we wish to find the solubility (the mole fraction x_B in solution) at some temperature T given information about the melting point and

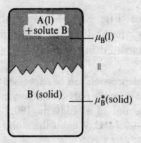

Fig. 8.11. Liquid solute/solid equilibrium and solubility equilibrium.

enthalpy of fusion of the solute. At the melting point T^* of the pure liquid solute $\Delta G_{\text{melt,m}}$ is zero, and so $\Delta G_{\text{melt,m}}(T^*)/RT^*$ is also zero. Since the last equation tells us that

$$\ln x_B = -\{\mu_B^*(l) - \mu_B^*(\text{solid})\}/RT = -\Delta G_{\text{melt,m}}(T)/RT,$$

it is also true that

$$\ln x_B = -\left\{\frac{\Delta G_{\text{melt,m}}(T)}{RT} - \frac{\Delta G_{\text{melt,m}}(T^*)}{RT^*}\right\}.$$

The assumption that both ΔS and ΔH change hardly at all over the temperature range of interest turns this equation into

(8.3.3)° $$\ln x_B = -\left\{\frac{\Delta H_{\text{melt,m}}}{R}\right\}\left(\frac{1}{T} - \frac{1}{T^*}\right).$$

Close to the melting point $T \approx T^*$, and so

(8.3.4)° $$x_B \approx \exp\left\{\frac{\Delta H_{\text{melt,m}}}{RT^{*2}}(T^* - T)\right\}.$$

Therefore the mole fraction of solute present in the saturated solution decreases exponentially as the temperature is lowered from its melting point. This equation also expresses quantitatively the common-sense view that, at a given temperature, solutes with high melting points and large enthalpies of fusion have low solubilities.

Example (Objective 10). Estimate the solubility of naphthalene in benzene at room temperature (25 °C).

- *Method.* If ideal behaviour is assumed we can use eqn (8.3.3). In order to do so we need the enthalpy of fusion of naphthalene (19.0 kJ mol^{-1}) and its melting point (80.2 °C). The mole fraction in solution can then be converted into molality.

- *Answer.* From eqn (8.3.3)

$$x(\text{naphthalene}) = \exp\left\{\frac{-(19.0 \text{ kJ mol}^{-1})}{(8.314 \text{ J K}^{-1} \text{ mol}^{-1})}\left(\frac{1}{298.15 \text{ K}} - \frac{1}{378.4 \text{ K}}\right)\right\}$$

$$= \exp(-1.62) = 0.198.$$

Since the mole fraction is related to the amounts of substance present by $x_B = n_B/(n_A + n_B)$, we have $n_B = x_B n_A/(1 - x_B)$. Therefore the amount of naphthalene in 1000 g of benzene is

$$n_B = (0.198) \times (1000 \text{ g}/78.1 \text{ g mol}^{-1})/(1 - 0.198) = 3.16 \text{ mol}.$$

Therefore the molality of the solution at 25 °C is about 3.16 mol kg^{-1}.

● *Comment.* The answer assumes ideal behaviour: the experimental result is 4.6 mol kg^{-1}. Note that the mole fraction x(naph.) is independent of the solvent (so long as it is ideal), but the molality does depend on the solvent's molar mass.

Osmosis. The phenomenon of osmosis is the tendency of a pure solvent to enter a solution separated from it by a membrane permeable to the solvent but not to the solute (such a membrane is called *semipermeable*), Fig. 8.12. One of the important examples of this process is transport through membranes into the interior of cells. It can be demonstrated by separating an aqueous sugar solution from pure water by a membrane made of cellulose. The membrane permits the passage of water molecules but not of the big sugar molecules. Osmosis causes the water to penetrate into the solution, and so the bulk of solution is increased (Fig. 8.12). The passage can be stopped by applying a pressure to the solution, and the pressure required to inhibit all flow is called the *osmotic pressure* of the solution.

The thermodynamic description of the process is straightforward. The chemical potentials of the *solvent* on both sides of the semipermeable membrane must be equal. On one side we have the pure solvent with chemical potential $\mu_A^*(l)$. On the other is the solvent containing solute B, but it experiences an additional pressure when the system is at equilibrium. Because of the random distribution of solute the chemical potential of a solvent in a mixture is less than that of the pure solvent (an entropy effect), and so the pure solvent has a thermodynamic tendency to flow into the solution. However, the chemical potential increases with pressure, and so a balance is reached when

(8.3.5) $$\mu_A^*(l; \text{ pressure } p) = \mu_A(l; \text{ amount } x_A; \text{ pressure } p + \Pi).$$

Π is the additional, osmotic pressure required for equilibrium. The presence

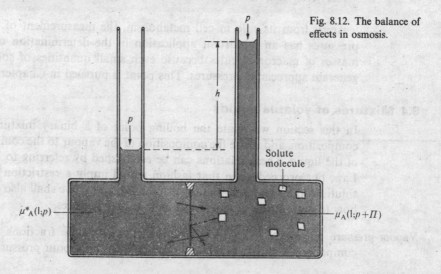

Fig. 8.12. The balance of effects in osmosis.

p

h

p

Solute molecule

$\mu_A^*(l;p)$ $\mu_A(l;p+\Pi)$

of solute can be taken into account in the normal way:

(8.3.6)　　　$\mu_A(l;\, x_A;\, p+\Pi) = \mu_A^*(l;\, p+\Pi) + RT\ln x_A.$

We saw how to correct for change of pressure on p. 167:

(8.3.7)　　　$\mu_S^*(p+\Pi) = \mu_S^*(p) + \int_p^{p+\Pi} V_m^*\, dp.$

When eqn (8.3.7) is inserted into eqn (8.3.6), and eqn (8.3.6) used on the r.h.s. of eqn (8.3.5), the equilibrium condition becomes

$$\mu_A^*(l;\, p) = \mu_A^*(l;\, p) + RT\ln x_A + \int_p^{p+\Pi} V_m^*\, dp$$

or

(8.3.8)°　　　$-RT\ln x_A = \int_p^{p+\Pi} V_m^*\, dp.$

The last equation relates the osmotic pressure to the mole fraction of the solute present (through $x_A = 1 - x_B$). It demonstrates that osmosis is a colligative property because Π depends only on the amount of solute and not on its nature.

The equation can be simplified for dilute solutions. First $\ln x_A$ is replaced by $\ln(1 - x_B) \approx -x_B$. Then it is assumed that the molar volume of the solvent does not change over the pressure range of interest. That being so, V_m^* may be taken outside the integral. This gives

$$RTx_B = \Pi V_m^*.$$

The mole fraction of the solute is $n_B/(n_A + n_B)$, which is virtually equal to n_B/n_A when the solution is dilute. Since $n_A V_m^*$ is the actual volume of the solvent, V, the equation simplifies into the

(8.3.9)°　　　*van't Hoff equation*: $\Pi V = n_B RT.$

Apart from its role in cell metabolism, the measurement of osmotic pressures has an important application in the determination of molar masses of macromolecules because even small quantities of solute can generate appreciable pressures. This point is pursued in Chapter 24.

8.4 Mixtures of volatile liquids

In this section we relate the boiling point of a binary mixture to its composition, and relate the composition of the vapour to the composition of the liquid. These relations can be established by referring to Raoult's Law, but proceeding in that fashion would imply a restriction to ideal solutions. This restriction must be removed, and so we shall also examine how to enlarge the discussion to include real mixtures.

Vapour-pressure diagrams. If an ideal solution contains mole fractions x_A of a component A and x_B of a component B, then the vapour pressures of the two components are

$$p_A = x_A p_A^* \quad \text{and} \quad p_B = x_B p_B^*$$

where p_A^* and p_B^* are the vapour pressures of the pure liquids at that temperature. The total vapour pressure is therefore

$$
\begin{aligned}
p = p_A + p_B &= x_A p_A^* + x_B p_B^* \\
&= p_B^* + (p_A^* - p_B^*) x_A.
\end{aligned}
$$

(8.4.1)°

This shows that the total vapour pressure at some fixed temperature changes linearly with the composition, being p_B^* when A is absent, and p_A^* when B is absent. This line is plotted in Fig. 8.13a.

Figure 8.13a is also a phase diagram. This is because the straight line represents the pressure where the two phases, liquid and gas, are in equilibrium at different compositions of the liquid mixture. If the mixture of some composition a is subjected to a pressure p' which is greater than the total vapour pressure, the whole sample is present as liquid. All points above the vapour-pressure line correspond to a liquid being the stable phase, and so the whole of that region has been labelled 'liquid'. Beneath the vapour-pressure line, corresponding to lower pressures, the vapour is the stable phase. If the liquid mixture of composition a is subjected to the pressure p'' the formation of vapour is favoured, and so the whole sample evaporates. All the points in the region below the vapour-pressure line correspond to the vapour as the stable phase, and it has been labelled accordingly.

When liquid and vapour are in equilibrium their compositions are not necessarily the same. Common sense suggests that the vapour should be richer in the more volatile component. This can be confirmed as follows. We know the partial pressures p_A and p_B, and so the mole fractions of A and B in the gas, y_A and y_B, are given by Dalton's Law:

$$y_A = p_A/p \quad \text{and} \quad y_B = p_B/p.$$

But Raoult's Law relates the partial pressures to the mole fractions of A and B *in the liquid*, x_A and x_B. From eqn (8.4.1) we get

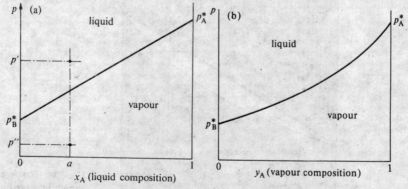

Fig. 8.13. The dependence of the vapour pressure of a mixture on (a) the composition x_A of the liquid, (b) the composition of the vapour.

$$y_A = \frac{x_A p_A^*}{p_B^* + (p_A^* - p_B^*)x_A}, \qquad y_B = 1 - y_A.$$

These two equations relate the vapour composition (y_A, y_B) to the liquid composition (x_A, x_B) when the two phases are in equilibrium.

The last equations conform to the expectation that the vapour is richer in the more volatile component (the component with the greater vapour pressure). For instance, suppose the liquid has equal amounts of A and B, then $x_A = \frac{1}{2}$ and $y_A = p_A^*/(p_A^* + p_B^*)$ and $y_B = p_B^*/(p_A^* + p_B^*)$, which means $y_A > y_B$ if $p_A^* > p_B^*$. In the limit of a non-volatile solute, one for which the pure material has zero vapour pressure at the temperature of interest, none of it will appear in the gas: $y_B = 0$ if $p_B^* = 0$. This case reduces to the work of the last section.

Equation (8.4.1) shows how the total pressure of the vapour in equilibrium with a mixture depends on the liquid's composition. The dependence of the total pressure on the *vapour's* composition can be obtained because we know how to express y_A in terms of x_A. This leads to

$(8.4.2)°$
$$p = \frac{p_A^* p_B^*}{p_A^* + (p_B^* - p_A^*)y_A}.$$

This curve is plotted in Fig. 8.13b. This diagram is also a phase diagram because it represents the boundary between the stable regions of two phases.

Since y_A and y_B are related unambiguously to x_A and x_B it does not matter which pair we choose as composition variables: if we are interested more in the liquid we choose x, but if we are more interested in the vapour we choose y. When discussing distillation both the gas and the liquid compositions are of interest, and then it is sensible to combine both diagrams into one. This has been done in Fig. 8.14 where the composition axis is labelled with z_A, the mole fraction of A corresponding to the *total* composition of the system. Above the upper line we can be confident that only liquid is present, because any point there falls into the liquid region of

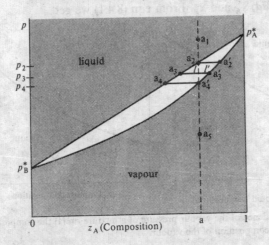

Fig. 8.14. The dependence of vapour pressure on the overall composition.

the phase diagram irrespective of whether z is identified with x or with y: the liquid composition is also the total composition. Likewise, below the lower line the system is entirely gas and the composition of the vapour is now also the total composition. Suppose, though, that we encounter a point that falls between the two lines: what does it signify?

A point in the unshaded sector of the diagram indicates not only *qualitatively* that both liquid and vapour are present simultaneously, but it also indicates *quantitatively* the relative amounts of each. The relative amounts of the two phases are given by the *lever rule*.

Refer to Fig. 8.14. Suppose the overall composition is a and the pressure is p_3: since the point (a, p_3) falls in the unshaded sector we know at once that both phases are present in equilibrium. In order to find the relative amounts the distances l and l' are measured along the *tie-line*, and the proportions are given by the

> *Lever rule*: $n(\text{liquid})/n(\text{vapour}) = l'/l$.

In the present example, since $l' \approx \frac{2}{3} l$, the amount of liquid is about $\frac{2}{3}$ the amount of vapour.

The basis of the lever rule can be demonstrated as follows. Let the amounts of liquid and vapour be $n(l)$ and $n(g)$ and the total amount be n. The overall amount of A is nz_A; but the overall amount is also the sum of the amounts in the two phases, $n(l)x_A$ and $n(g)y_A$:

$$nz_A = n(l)x_A + n(g)y_A.$$

Furthermore, as $n = n(l) + n(g)$ we also have

$$nz_A = n(l)z_A + n(g)z_A.$$

On equating these two expressions it follows that

$$n(l)(x_A - z_A) = n(g)(z_A - y_A),$$

or

$$n(l)/n(g) = (y_A - z_A)/(z_A - x_A) = l'/l$$

as was to be proved.

In order to see in more detail how the rule is used, consider what happens when a mixture of composition a_1 in Fig. 8.14 is subjected to decreasing pressure. The system remains entirely liquid until the pressure is reduced to p_2, at which point both liquid and vapour can coexist. At p_2 the liquid has composition a_2 and the vapour present has the composition a_2'. Since $l'/l \approx \infty$ along the tie line at p_2, there is only a trace of vapour present. When the pressure is lowered further to p_3, the composition of the liquid shifts to a_3 and that of the vapour to a_3', but the overall composition remains at a. The relative amounts of liquid and vapour are given by the ratio l'/l measured along the tie-line at p_3, and now the amount of liquid is about $\frac{2}{3}$ the amount of vapour; furthermore, the vapour of composition a_3' is richer than the liquid in A, the more volatile component. When the pressure is reduced still further to p_4 the sample becomes virtually entirely gaseous: the vapour's composition is a'_4 and the trace of liquid (because $l'/l \approx 0$) that remains has the composition a_4 (rich in B). Only an infinitesimal reduction of pressure is now needed to eliminate the stability of the last trace of liquid, and from then on the sample is wholly gaseous with composition a.

How to think about distillation. Reducing the pressure at constant temperature is one way of doing distillation but it is more common to distil at constant pressure by raising the temperature. In order to discuss this process we require a *temperature–composition diagram*. Figure 8.15 shows such a diagram for an ideal binary mixture. The interpretation is almost the same as in the case of the pressure–composition diagram, but there are a few points of difference. The principal one is that, because the vapour phase is stable at high temperatures, the new diagram is upside-down in relation to the pressure diagram.

Using the diagram involves the same kind of thinking as in the case of the vapour pressure diagram. For instance, take a liquid mixture of composition a and heat it from its initial state a_1. It boils at the temperature T_2. At that temperature the liquid has composition a_2 and the vapour (which is present only as a trace) has a composition a'_2. The vapour is richer in the more volatile component (A), just as common sense would suggest. The diagram enables one to state both the vapour's composition at the boiling point (a'_2), and the boiling point itself (the temperature corresponding to the tie line $a_2 a'_2$). The diagram corresponds to a single

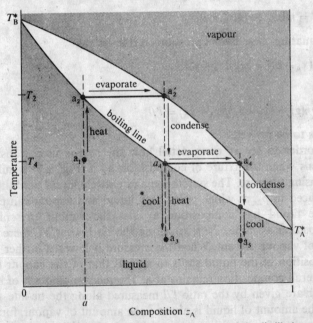

Fig. 8.15. A temperature–composition diagram, and the distillation process.

overall pressure, and is modified if the pressure is changed.

In a distillation experiment the vapour is withdrawn and condensed. If the vapour in the last example is drawn off and completely condensed, then the first drop gives a sample of liquid of composition a_3 which is richer in the more volatile component than the original mixture. Furthermore, the liquid remaining is richer in the less volatile component, and so the boiling point shifts to higher values.

Fractional distillation repeats the boiling and condensation cycle several times. If the condensate from the last example (of composition a_2', but now cooled to the point a_3) is reheated it boils at T_4, and yields a vapour of composition a_4' which is even richer in the more volatile component. That vapour is drawn off and the first drop condenses to a liquid of composition a_5, and the process can be repeated. Repeating the process leads to a condensate which in the end will be virtually pure volatile component.

The discussion above applies to many real mixtures. The actual boiling point–composition curves often resemble those drawn in Fig. 8.15, especially when the two liquids are of such similar character that they form an almost ideal solution. Nevertheless, in a number of very important cases the deviations from ideality are marked and sometimes completely upset the distillation process. The deviations can lead to a maximum in the boiling point curve, Fig 8.16a, or to a minimum, Fig. 8.16b. The former corresponds to a reduction of the vapour pressure from the ideal value, and suggests that the liquid phase is stabilized by the molecular interactions. For such mixtures, the excess Gibbs function is negative (more favourable to mixing than ideal). Examples include chloroform/acetone and nitric acid/water mixtures. Boiling point curves showing minima indicate that the mixture is destabilized relative to the ideal, and so the molecular interactions are unfavourable. For such systems the excess Gibbs function is positive (less favourable to mixing than ideal), and there may be contributions from the enthalpy and the entropy changes. Examples

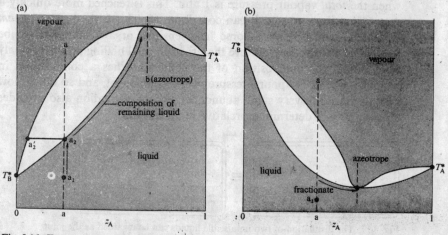

Fig. 8.16. Two types of azeotropic mixture: (a) high boiling azeotrope, (b) low boiling azeotrope.

include CS_2/acetone, dioxane/water, and ethanol/water mixtures.

The deviations from ideality are not always so strong as to lead to a hump or a trough in the boiling point curves, but when they do there are important consequences for distillation. Consider a liquid of composition a on the left of the maximum in Fig. 8.16a. The vapour (at a_2') of the boiling mixture (at a_2) is richer in the component B, and as evaporation proceeds the composition of the remaining liquid moves from left to right in the diagram as the vapour is drawn off. The boiling point of the mixture rises, and the vapour gets richer in A. When so much B has been evaporated that the liquid has reached the composition b, the vapour has the same composition as the liquid, and so evaporation proceeds without any further change of composition. The mixture is then called an *azeotrope* (which comes from the Greek for 'boiling without changing'). When this point has been reached, distillation cannot separate the two liquids, for the condensate retains the composition of the liquid. One example of this behaviour is HCl/water, which is azeotropic at 80 per cent water (by mass) and then boils (unchanged) at 108.6 °C. The type of system in Fig. 8.16b also shows azeotropic properties, as can be seen by starting with the composition a and attempting to fractionate the mixture. In the fractionation cycle the vapour composition shifts towards the azeotropic composition but cannot be concentrated further in the direction of A. A mixture of ethanol and water shows this behaviour, and boils unchanged when the water content is 4 per cent and the temperature 78 °C.

The final type of system to consider is the *distillation of two immiscible components*, like oil and water. As they are immiscible we can regard their mixture as unscrambled with each placed in a separate vessel, Fig. 8.17. If one component has a vapour pressure p_A^* and the other p_B^*, then the total pressure exerted is $p = p_A^* + p_B^*$ and the mixture boils when $p_A^* + p_B^* = 1$ atm. The composition of the vapour is $y_A = p_A^*/(p_A^* + p_B^*) = p_A^*/\text{atm}$. The presence of the other component means that the pair boil at a lower temperature than either would alone because boiling begins when the *total* vapour pressure is 1 atm. This is reached more quickly if A can contribute p_A^* and B can contribute p_B^*. This is the basis of *steam distillation*, which enables some heat-sensitive organic materials to be distilled at a lower temperature than their normal boiling point. The only snag is that the composition of the condensate has a composition in proportion to the vapour pressure of the component, and so oils of low-volatility distil in very small abundance. Steam distillation also provides a messy way of determining relative molar masses.

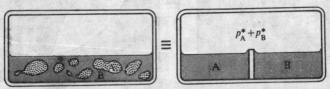

Fig. 8.17. The distillation of two immiscible liquids can be regarded as the joint distillation of the separated components.

8.5 Real solutions and activities

In this section we examine some of the complications that enter when the thermodynamics of real solutions are discussed. So far we have considered systems that obey Henry's Law or Raoult's Law, and the relevant information is summarized in Fig. 8.18. The general form of the chemical potential of a solute or solvent is

(8.5.1)
$$\mu_A(l) = \mu_A^*(l) + RT \ln(p_A/p_A^*)$$

where p_A^* is the vapour pressure of pure A and p_A its vapour pressure for the mixture of interest. In the case of an ideal solution both solvent and solute obey Raoult's Law at all concentrations, and the last equation becomes

$$\mu_A(l) = \mu_A^*(l) + RT \ln x_A.$$

The standard state is the pure liquid, and is attained when $x_A = 1$.

When the solution deviates from Raoult's Law the form of the last equation may be preserved if the mole fraction x_A is replaced by the *activity* a_A:

(8.5.2)
$$\mu_A(l) = \mu_A^*(l) + RT \ln a_A.$$

The activity is *defined* by this equation; but if it is to be useful it must be related to the composition of the system. Since eqn (8.5.1) is always valid (the only approximation being that the vapour behaves perfectly, but this can be removed by introducing the additional complication of dealing with fugacities) comparison with eqn (8.5.2) gives

(8.5.3)
$$a_A = p_A/p_A^*.$$

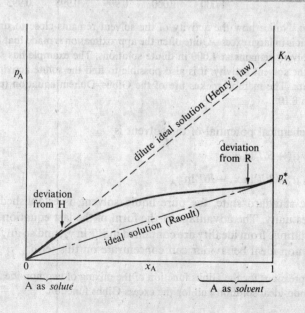

Fig. 8.18. A summary of Raoult's Law and Henry's Law behaviour.

Therefore one way of determining the activity of a component in a solution where its mole fraction is x_A is to measure its vapour pressure.

In the case of the component present in excess (the solvent) the definition of a standard state is straightforward. All solvents obey Raoult's Law increasingly closely as they approach purity, and so the activity of the solvent, a_A, becomes equal to the mole fraction x_A as $x_A \rightarrow 1$. One useful way of expressing this is to introduce the *activity coefficient* γ_A by the definition

(8.5.4) $a_A = \gamma_A x_A.$

Then as $x_A \rightarrow 1$ (pure solvent), $\gamma_A \rightarrow 1$ also.

Example (Objective 17). From the data in the following table for the vapour pressure of water against the molality of sugar (sucrose) at 25 °C calculate the activity of the solvent and its activity coefficient.

m(sucrose)/mol kg^{-1}	0.000	0.200	0.500	1.000	2.000
p(water)/mmHg	23.75	23.66	23.52	23.28	22.75

• *Method.* The activity is given by $a = p/p^*$, where p^* is the vapour pressure of the pure material. The activity coefficient is given by eqn (8.5.4). The molality can be converted into a mole fraction so long as we know the molar mass of the solvent (18.02 g mol^{-1}).

• *Answer.* The mole fraction of water is

$$x(\text{water}) = \frac{(1 \text{ kg}/18.02 \text{ g mol}^{-1})}{(1 \text{ kg}/18.02 \text{ g mol}^{-1}) + m(\text{sucrose}) \times (1 \text{ kg})}$$

Therefore we can construct the following table:

m(sucrose)/mol kg^{-1}	0.000	0.200	0.500	1.000	2.000
x(water)	1.000	0.996	0.991	0.982	0.965
a(water)	1.0000	0.9962	0.9903	0.9802	0.9579
γ(water)	1.000	1.000	0.999	0.998	0.993

• *Comment.* Notice how the activity of the solvent remains close to unity: this is because it is in large excess. Quite often the approximation is made that the activity of the solvent remains at 1.000 in dilute solutions. The example has shown how to find the solvent activity: it is also possible to find the solute activity from the same data. The method makes use of the Gibbs–Duhem equation (p. 219). See Problem 8.10.

The chemical potential of the solvent is

$$\mu_A = \mu_A^* + RT \ln a_A$$

(8.5.5) $$= \mu_A^* + RT \ln x_A + RT \ln \gamma_A$$

and the standard state, the pure liquid solvent, is established when x_A becomes unity. The advantage of the form of the last equation is that all the deviations from ideality are contained in $RT \ln \gamma_A$, and so any investigation of non-ideal behaviour can concentrate on this term.

Example. Find expressions for the Gibbs function of the mixing of two non-ideal liquids to form a non-ideal solution and for the excess Gibbs function.

- *Method.* Use the reasoning that led to the Gibbs function of mixing for an ideal solution (p. 219) but use eqn (8.5.5) for the chemical potential of the components in the mixture. G^E is the difference between this ΔG and that in eqn (8.2.10).

- *Answer.* Repeating the arguments on p. 219 gives

  before mixing: $\qquad G = n_A\mu_A^*(l) + n_B\mu_B^*(l)$

  after mixing: $\qquad G = n_A\{\mu_A^*(l) + RT\ln a_A\} + n_B\{\mu_B^*(l) + RT\ln a_B\}$

  change: $\qquad \Delta G = n_A RT\ln a_A + n_B RT\ln a_B$

  or

  $$\Delta G_m = RT(x_A\ln a_A + x_B\ln a_B)$$
  $$= RT(x_A\ln x_A + x_B\ln x_B + x_A\ln\gamma_A + x_B\ln\gamma_B).$$

  The molar excess Gibbs function is therefore

  $$G_m^E = RT(x_A\ln\gamma_A + x_B\ln\gamma_B).$$

- *Comment.* Since the activity coefficients might depend on the pressure, the enthalpy of mixing is non-zero and the entropy of mixing might no longer have the ideal value (eqn (8.2.2) on p. 221).

The activity and standard state of the component in low abundance (the solute, denoted B) need more careful treatment. The major problem is that the solution approaches ideal dilute behaviour at low concentrations of solute, concentrations far away from being pure liquid solute. First we establish the meaning of a standard state of a solute that does obey Henry's Law, and then we admit deviations from ideal dilute behaviour.

At sufficiently low concentrations solutes obey Henry's Law, $p_B = K_B x_B$. In this expression K_B is some constant, which is equal to the vapour pressure of the pure solute only if the solute also obeys Raoult's Law (see Fig. 8.18). The equality of the chemical potentials in the liquid and vapour phases gives

$$\mu_B(l) = \mu_B(g) = \mu_B^\ominus + RT\ln(p_B/p^\ominus)$$
$$= \mu_B^\ominus + RT\ln(K_B/p^\ominus) + RT\ln x_B.$$

Now pretend that the solute obeys Henry's Law over the whole concentration range and not just in the limit of low concentrations. If this were so the chemical potential of the hypothetical pure solute (which we denote $\mu_B^\dagger$) would be obtained by setting $x_B = 1$ in the last expression:

$$\mu_B^\dagger = \mu_B^\ominus + RT\ln(K_B/p^\ominus).$$

Of course, Henry's Law is not obeyed over this range, but we shall see that it is valid to take this extrapolated *hypothetical* state as the standard state. Then the chemical potential at the mole fraction x_B is

(8.5.6) $\qquad \mu_B = \mu_B^\dagger + RT\ln x_B.$

Now we can take the important step of permitting deviations from ideal dilute (Henry's Law) behaviour. Since the solute obeys Henry's Law at sufficiently low concentrations, we may use the same device as in the case of the standard states of real gases (p. 173). We assert that the chemical potential of the non-ideal solution is given by

(8.5.7) $\qquad \mu_B = \mu_B^\dagger + RT \ln a_B.$

The standard state is established when the solute is present at unit mole fraction *and* is behaving in accord with Henry's Law. This definition, although peculiar, has a well-defined significance which we explore by referring to Fig. 8.19 which is like Fig. 6.2 for real gases. In order to relate the chemical potential of some composition x_B where the activity is a_B, eqn (8.5.7) is used to trace its value down the actual curve, until the composition is so low that the curve coincides with that for the hypothetical ideal Henry's Law dependence. Then eqn (8.5.6) is used to move to the standard state value of the chemical potential.

It is important to be quite clear about the significance of the definition of the standard state in relation to eqn (8.5.7). The chemical potential of a real system certainly has the *value* $\mu_B^\dagger$ when the solute is present at unit activity; nevertheless, the system is not then in its standard state because this is a hypothetical, unattainable state. This hypothetical standard state can be pictured as the pure solute, but one in which the environment of each molecule is the same as at infinite dilution. The advantage of this cumbersome definition is the same as in the case of real gases: the deviations from ideality (which now means deviations from Henry's Law) occur only in the logarithmic terms because the standard state is defined to refer to ideal behaviour.

The last remark suggests that we should introduce an activity coefficient γ_B through

(8.5.8) $\qquad a_B = \gamma_B x_B.$

The solute behaves ideally (in the Henry's Law sense) at infinite dilution, and so we can be sure that

(8.5.9) $\qquad a_B \to x_B \quad \text{or} \quad \gamma_B \to 1 \quad \text{as} \quad x_B \to 0.$

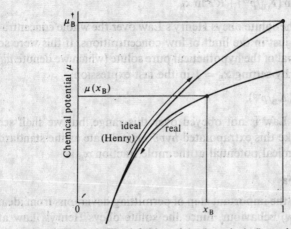

Fig. 8.19. The definition of the (hypothetical) standard state of a solute on the basis of Henry's Law.

Notice that deviations from ideality disappear at infinite dilution in this Henry's Law convention of standard states, but at the pure material in the Raoult's Law convention for solvents.

Compositions are often expressed as molarities or molalities. In order to deal with these instead of mole fractions we are led to another standard state, but it follows quite naturally from what has been done so far. Now the standard state is defined as *a state at unit molality but in which the solution is behaving ideally*. It may be visualized as a hypothetical state in which the molality is 1 mol kg^{-1}, but in which the environment of every solute molecule is the same as at infinite dilution. If the chemical potential of this state is denoted $\mu_B^\ominus$ and $m^\ominus$ is used to denote 1 mol kg^{-1} the chemical potential at a general molality m_B is

$$\mu_B(m_B) = \mu_B^\ominus + RT \ln \gamma_B + RT \ln (m_B/m^\ominus)$$

where the new activity coefficient has the property

$$\gamma_B \to 1 \quad \text{as} \quad m_B \to 0.$$

A solute activity may be introduced by

(8.5.10)
$$a_B = \gamma_B m_B/m^\ominus, \qquad m^\ominus \overset{\text{def}}{=} 1 \text{ mol kg}^{-1}; \qquad \gamma_B \to 1 \quad \text{as } m_B \to 0.$$

(According to this definition, a_B is dimensionless.) Then we arrive at a succinct expression for the chemical potential of the solute at any molality:

(8.5.11)
$$\mu(m) = \mu^\ominus + RT \ln a.$$

It is important to be aware of the different definitions of standard states and activities, and these are summarized in Box 8.1. We shall put activities and standard states to work in the next few chapters, and we shall see that their applications are much less confusing than their definitions.

Box 8.1.   **Standard states**

For the *solvent* (the major component) use Raoult's Law and take the standard state as the pure solvent. Then

$$\mu_A = \mu_A^* + RT \ln a_A$$

with

$$a_A = \gamma_A x_A \qquad \gamma_A \to 1 \quad \text{as} \quad x_A \to 1 \text{ (pure solvent)}.$$

For the *solute* (the minor component) use Henry's Law and take the standard state as the *hypothetical* pure solute in which the environment of each molecule is the same as at infinite dilution. Then

$$\mu_B = \mu_B^\dagger + RT \ln a_B$$

with

$$a_B = \gamma_B x_B, \qquad \gamma_B \to 1 \quad \text{as} \quad x_B \to 0 \text{ (infinite dilution)}.$$

Alternatively, use Henry's Law but take the standard state as the hypothetical state at *unit molality* (or, alternatively, at unit molarity) but in which the environment of each molecule is the same as at infinite dilution. Then

$$\mu_B = \mu_B^\ominus + RT \ln a_B$$

with

$$a_B = \gamma_B m_B/m^\ominus, \qquad \gamma_B \to 1 \quad \text{as } m_B \to 0, \qquad m^\ominus \stackrel{\text{def}}{=} 1 \text{ mol kg}^{-1}.$$

Appendix: the method of intercepts

Consider some extensive thermodynamic property X (such as the volume, or the Gibbs function). Let the binary solution contain an amount n_A of A and n_B of B, and write the total amount as $n = n_A + n_B$ and the mole fractions as $x_A = n_A/n$ and $x_B = n_B/n$. Define the *mean molar property* as $X_m = X/n$. The partial molar properties are $X_{A,m} = (\partial X/\partial n_A)_{n_B}$, and $X_{B,m} = (\partial X/\partial n_B)_{n_A}$.

Concentrate on $X_{A,m}$. This can be written (with constant pressure and temperature understood) as

$$X_{A,m} = (\partial X/\partial n_A)_{n_B} = (\partial n X_m/\partial n_A)_{n_B}$$
$$= X_m + n(\partial X_m/\partial n_A)_{n_B}.$$

Now, since $x_B = n_B/n$, it follows that

$$(\partial X_m/\partial n_A)_{n_B} = (\partial x_B/\partial n_A)_{n_B}(dX_m/dx_B) = (-n_B/n^2)(dX_m/dx_B),$$

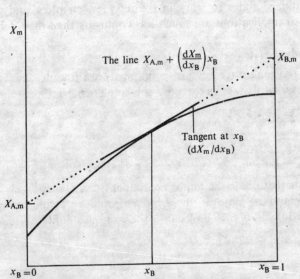

Fig. 8A.1. The extrapolation needed to find the partial molar quantities $X_{A,m}$ and $X_{B,m}$ at the composition x_B.

and so

$$X_{A,m} = X_m + n(-n_B/n^2)(dX_m/dx_B),$$

or

$$X_m = X_{A,m} + (dX_m/dx_B)x_B.$$

The last equation is the key result. It is the equation of a straight line of slope (dX_m/dx_B) and intercept $X_{A,m}$. Therefore, plot X_m against x_B, and at the chosen value of x_B draw a tangent to the experimental curve. This is the line referred to, and it intercepts the $x_B = 0$ axis at $X_{A,m}$, Fig. 8A.1. The same argument, but with B and A interchanged, shows that the intercept at $x_B = 1$ gives the value of $X_{B,m}$ at the selected composition.

Further Reading

Physical chemistry of surfaces. A. W. Adamson; Wiley-Interscience, New York, 1967.

Physical surfaces. J. R. Bikerman; Academic Press, New York, 1970.

Interfacial phenomena. J. T. Davies and E. K. Rideal; Academic Press, New York, 1973.

The physical chemistry of surface films. W. D. Harkins; Reinhold, New York, 1952.

Liquids and liquid mixtures. J. S. Rowlinson; Butterworths, London, 1969.

Regular and related solutions. J. H. Hildebrand, J. M. Prausnitz, and R. L. Scott; Van Nostrand Reinhold, New York, 1970.

Solubilities of non-electrolytes. J. H. Hildebrand and R. L. Scott; Reinhold, New York, 1950.

Determination of solubility. W. J. Mader and L. T. Grady; In *Techniques of chemistry* (A. Weissberger and B. W. Rossiter, eds.) V, 257, Wiley-Interscience, New York, 1971.

Determination of osmotic pressure. J. R. Overton; In *Techniques of chemistry* (A. Weissberger and B. W. Rossiter, eds.) V, 309, Wiley-Interscience, New York, 1971.

International critical tables, IV; McGraw-Hill, 1927.

Landolt-Börnstein Tables, IV; McGraw-Hill, 1927.

Biological membranes. D. S. Parsons (ed.); Clarendon Press, Oxford, 1975.

Problems

8.1.   The volume of an aqueous solution of sodium chloride at 25 °C was measured at a series of molalities m, and it was found that the data could be fitted to the expression $V/cm^3 = 1003 + 16.62m + 1.77m^{3/2} + 0.12m^2$ (where $m \equiv m/mol\ kg^{-1}$ and V refers to the volume of a solution formed from 1 kg of water). Find the partial molar volume of the components at $m = 0.1\ mol\ kg^{-1}$ by explicit differentiation.

8.2.   Different behaviour is shown by an aqueous solution of magnesium sulphate. At 18 °C the total volume of a solution formed from 1 kg of water is given approximately by $V/cm^3 = 1001.21 + 34.69\ (m - 0.07)^2$, the expression applying up to about 0.1 mol kg^{-1}. What is the partial molar volume of (a) salt, (b) solvent at 0.05 mol kg^{-1}?

8.3.   The *method of intercepts* is described in the Appendix. Use it to plot the partial molar volume of the HNO$_3$ in aqueous nitric acid at 20 °C on the basis of the following data, where w is the weight fraction of HNO_3.

100 w	2.162	10.98	20.80	30.00	39.2	51.68
$\rho/g\ cm^{-3}$	1.01	1.06	1.12	1.18	1.24	1.32

100 w	62.64	71.57	82.33	93.40	99.60
$\rho/g\ cm^{-3}$	1.38	1.42	1.46	1.49	1.51

8.4. The densities of aqueous solutions of copper sulphate at $20\,°C$ were determined and are reported below. Determine and plot the partial molar volume of anhydrous copper sulphate in the range given.

%	5	10	15	20
$\rho/g\ cm^{-3}$	1.051	1.107	1.167	1.230

(% means the number of grams of the salt per 100 g of solution).

8.5. The partial molar volumes of acetone and chloroform in a solution containing a mole fraction 0.4693 of chloroform are $74.166\ cm^3\ mol^{-1}$ and $80.235\ cm^3\ mol^{-1}$ respectively. What is the volume of a solution of mass 1 kg? What is the volume of the unmixed components? (For the second part you need to know that the molar volumes are $73.993\ cm^3\ mol^{-1}$ and $80.665\ cm^3\ mol^{-1}$ respectively.)

8.6. What proportions of ethanol and water should be mixed in order to form $100\ cm^3$ of mixture containing 50 per cent by mass of ethanol? What change of volume is brought about by adding $1\ cm^3$ of ethanol to $100\ cm^3$ of a 50 per cent (by weight) ethanol/water mixture? Base your answers on the information in Fig. 8.1.

8.7. When chloroform is added to acetone at $25\,°C$ the volume of the mixture varies with composition as follows:

$x(c)$	0	0.194	0.385	0.559	0.788	0.889	1.000
$V_m/cm^3\ mol^{-1}$	73.99	75.29	76.50	77.55	79.08	79.82	80.67

$x(c)$ is the mole fraction of the chloroform. Determine the partial molar volumes of the two components at these compositions, and plot the result.

8.8. The *Gibbs-Duhem equation* for the chemical potentials of the components of a mixture was quoted on p. 219, eqn (8.1.7). Show that in the case of partial molar volumes the corresponding equation is $n_A\,dV_{A,m} + n_B\,dV_{B,m} = 0$. Equations of this type have a lot of important implications, and save a lot of effort. We shall illustrate this in the next few Problems. As a first step, show that the equation implies that the slopes of the graphs of partial molar volumes (or any such quantity) of the two components of a binary mixture must be equal and opposite when $x_A = x_B$. Check the consistency of your results for the last three Problems by seeing whether this is so.

8.9. The way that the Gibbs-Duhem equation links the properties of the components of a mixture can be illustrated by the following question. First derive the *Gibbs-Duhem-Margules equation*, that $(\partial \ln f_A/\partial \ln x_A)_{p,T} = (\partial \ln f_B/\partial \ln x_B)_{p,T}$. This can be obtained as a straightforward development of the Gibbs-Duhem equation, and in the limit of gases that behave perfectly, reduces to a connection between vapour pressures and composition of the components of a mixture. Use it to show that if Raoult's Law applies to one component of a binary mixture, then it must also apply to the other.

8.10. As a third example of how the Gibbs-Duhem equation can be used, show that the partial molar volume (or any partial molar quantity) of a component B can be obtained if the partial molar volume (etc.) of component A is known for all compositions up to the composition of interest. You should find an integral involving x_A, x_B, and $V_{A,m}$. Show that the partial molar volume of acetone in chloroform at $x = \frac{1}{2}$ can be obtained in this way from the data in Problem 8.7.

8.11. We now pass on to look at the way that the thermodynamic properties of substances change when mixtures are prepared. In the first place, consider the case of a container of volume 5 dm³ which is divided into two compartments of equal size. In the left compartment there is nitrogen at 1 atm and 25 °C; in the right compartment there is hydrogen at the same pressure and temperature. What are the changes in the entropy and the Gibbs function of the system when the partition is removed and the gases are able to mix isothermally?

8.12. In a similar experiment, the nitrogen was initially at 3 atm and the hydrogen at 1 atm. Find the entropy and the Gibbs function changes in this case. Do gases always mix spontaneously? What would the changes of entropy and Gibbs function be if the hydrogen were replaced by nitrogen in this sample?

8.13. Air is a mixture with the composition listed on p. 35. What is its change of entropy on being mixed from its constituents?

8.14. The same formulas for the thermodynamic functions of mixtures apply to ideal liquid mixtures as to perfect gas mixtures. Calculate the Gibbs function, entropy, and enthalpy of mixing when 500 g of n-hexane is mixed with 500 g of n-heptane at 25 °C.

8.15. What proportions of n-hexane and n-heptane (a) by mole fraction, (b) by weight should be mixed in order to achieve the greatest change in entropy?

8.16. Henry's Law is useful for several things, and provides a simple way of finding the solubility of gases in liquids. For instance, what is the solubility of carbon dioxide in water at 25 °C when its partial pressure is (a) 0.1 atm, (b) 1.0 atm? Henry's Law data will be found in Table 8.1.

8.17. The mole fractions of nitrogen and oxygen in air at room temperature at sea level are approximately 0.782 and 0.209. What are the molalities in a vessel of water left open to the atmosphere at 25 °C?

8.18. A water carbonating plant is available for use in the home. It operates by providing carbon dioxide at 10 atm. Estimate the composition of the soda water it produces.

8.19. But how do we determine the value of Henry's Law constant? That can be found quite accurately by plotting vapour pressure data for various compositions and extrapolating the dilute solution data as in Fig. 8.5. The table below lists the vapour pressure of methyl chloride above its mixture with water at 25 °C. Find Henry's Law constant for methyl chloride.

$m(CH_3Cl)/mol\ kg^{-1}$	0.029	0.051	0.106	0.131
$p/mmHg$	205.2	363.2	756.1	945.9

Don't forget to convert to mole fractions.

8.20. At 90 °C the vapour pressure of toluene is 400 mmHg, and that of o-xylene is 150 mmHg. What is the composition of a liquid mixture that will boil at 90 °C when the pressure is 0.5 atm? What is the composition of the vapour produced?

8.21. In order to determine relative molar mass by the methods of depression of freezing point or elevation of boiling point it is necessary to know the cryoscopic or ebullioscopic constants K_f and K_b. In most cases these are available in the literature, or can be determined experimentally by dissolving a known amount of a material with known M_r. Sometimes, though, it is useful to be able to make a rapid assessment of their magnitudes. Calculate K_f and K_b for carbon tetrachloride on the basis that $\Delta H_{melt,m} = 2.5 \text{ kJ mol}^{-1}$, $T_f = 250.3 \text{ K}$, $\Delta H_{vap,m} = 30.0 \text{ kJ mol}^{-1}$, $T_b = 350 \text{ K}$, $M_r = 153.8$.

8.22. Colligative properties depend on the total number of particles dissolved in a solvent: sometimes this means keeping alert. What is the freezing point of 100 g of water containing 2 g of sodium chloride?

8.23. Instead of measuring boiling points we could measure the vapour pressure depression directly, and relate that to the R.M.M. of the solute. The vapour pressure of a sample of 500 g of benzene is 400 mmHg at 60.6 °C, but it fell to 386 mmHg when 19 g of an involatile organic compound was dissolved in it. What is the R.M.M. of the compound?

8.24. The value of an M_r obtained by depression of freezing point may be in error for a variety of reasons, such as the solution being too concentrated for the approximations involved to be valid, or because it behaves non-ideally. But what is the 'intrinsic' sensitivity of the method? A fluorocarbon is believed to be either $CF_3(CF_2)_3CF_3$ or $CF_3(CF_2)_4CF_3$; the cryoscopic constant of camphor is 40 K/mol kg^{-1}. To what precision must the temperature be measured if the freezing point depression of a solution of 1 g of the fluorocarbon in 100 g of camphor is to be used to distinguish the possibilities?

8.25. When we deduced the form of the expressions for the depression of freezing point and the elevation of boiling point we used a simple approximation of assuming that both ΔH and ΔS for the transitions are independent of temperature. This is quite a good approximation for small concentrations and some materials, but we ought to be able to deal with the more general case. We shall see, for example, that something rather odd happens in the case of water. As a first step return to $\ln x_A = \Delta G_{melt,m}(T)/RT$, p. 230, and use the Gibbs–Helmholtz equation to find an expression for $d \ln x_A$ in terms of dT. Then integrate the left-hand side from $x_A = 1$ to the x_A of interest, and integrate the right-hand side from the transition temperature of interest for the pure material T_t^* to the transition point of the solution. In order to make the integration, assume that $\Delta H_{t,m}(T) \approx \Delta H_{t,m}(T_t^*)$. This leads in a more formal way to the equation already derived on p. 232.

8.26. Differentiating and then integrating the differential might seem to be a waste of time. It is not really, because we need not have made the assumption that $\Delta H(T)$ is independent of temperature. By putting in the appropriate dependence we can obtain a more accurate expression for the change of the transition temperatures. Assume now that the heat capacities of the solid and liquid pure solvent are independent of temperature. Derive an expression for the freezing point depression on this basis, and calculate the cryoscopic constant for water. Compare the result with that obtained by assuming ΔH to be independent of temperature. ($C_{p,m} = 75.3 \text{ J K}^{-1} \text{ mol}^{-1}$ for water and 24.3 J K^{-1} mol^{-1} for ice.)

8.27. What is the freezing point of a 250 cm^3 glass of water sweetened with five cubes of sugar (7.5 g of sucrose)?

8.28. Common salt is spread on roads to prevent the formation of ice. The cost of the salt is £0·80/100 kg; $1.92/100 kg. A rich new source of $CaCl_2$ was discovered and mined at £0·62/100 kg; $1.49/100 kg. Which salt is more cost-effective?

8.29. Potassium fluoride is very soluble in glacial acetic acid, and the solutions have a number of peculiar properties. In an attempt to discover something about their structure, freezing point depression data were obtained by taking a solution of known molality and then diluting it several times. (J. Emsley, *J. chem. Soc. A*,

2702 (1971).) The following data were obtained:

$m(KF)/mol\ kg^{-1}$	0.015	0.037	0.077	0.295	0.602
$\Delta T_f/K$	0.115	0.295	0.470	1.381	2.67

Find the apparent R.M.M. of the solute, and suggest an interpretation. (This system was also the basis of Problem 4.28: the result there has a bearing on the data here.) Use $\Delta H_{melt,m} = 11.4\ kJ\ mol^{-1}$ and $T_f^* = 290\ K$.

8.30. In a study of the properties of aqueous solutions of thorium nitrate $Th(NO_3)_4$ (A. Apelblat, D. Azoulay, and A. Sahar, *J. chem. Soc. Faraday Trans.* 1, 1618 (1973)) a freezing point depression of 0.0703 K was observed for a $0.0096\ mol\ kg^{-1}$ aqueous solution. What is the apparent number of ions per solute molecule? Take $K_f = 1.86\ K/mol\ kg^{-1}$.

8.31. Predict the ideal solubility of lead in bismuth at 280 °C on the basis that for lead $T_f^* = 327\ °C$ and $\Delta H_{melt,m} = 5.2\ kJ\ mol^{-1}$.

8.32. Anthracene has an enthalpy of fusion of $28.8\ kJ\ mol^{-1}$ and melts at 217 °C. What is its ideal solubility in benzene at 25 °C?

8.33. The osmotic pressures of solutions of polystyrene in toluene were measured with the intention of measuring the mean R.M.M. of the polymer. The pressure was expressed in terms of the height of the solvent h, its density being $1.004\ g\ cm^{-3}$. The following results were obtained at 25 °C:

$c/g\ dm^{-3}$	2.042	6.613	9.521	12.602
h/cm toluene	0.592	1.910	2.750	3.600

What is the mean R.M.M. of the polymer?

8.34. The R.M.M. of a newly isolated enzyme was determined by dissolving it in water, measuring the osmotic pressure of various solutions, and then extrapolating the data to zero concentration. The following data were obtained at 20 °C.

$c/mg\ cm^{-3}$	3.221	4.618	5.112	6.722
h/cm water	5.746	8.238	9.119	11.990

What is the R.M.M. of the enzyme? More elaborate Problems of this kind, and better ways of handling the data, will be found in Chapter 24.

8.35. An industrial process led to the production of a mixture of toluene (T) and n-octane (O), and the vapour pressure of the mixture was investigated with a view to designing a separation plant. The following temperature–composition data were obtained at 760 mmHg: x is the mole fraction in the liquid and y the mole fraction in the vapour at equilibrium.

$t/°C$	110.9	112.0	114.0	115.8	117.3	119.0	120.0	123.0
x_T	0.908	0.795	0.615	0.527	0.408	0.300	0.203	0.097
y_T	0.923	0.836	0.698	0.624	0.527	0.410	0.297	0.164

The boiling points are T: 110.6 °C, O: 125.6 °C. Plot the temperature–composition diagram for T and O. What is the composition of the vapour in equilibrium with liquid of composition (a) $x_T = 0.250$, (b) $x_O = 0.250$?

8.36. The table below lists the vapour pressures of mixtures of ethyl iodide (I) and ethyl acetate (A) at 50 °C. Find the activity coefficients of both components on (a) the Raoult's Law basis, (b) the Henry's Law basis with I regarded as solute.

x_A	0	0.0579	0.1095	0.1918	0.2353	0.3718
$p_I/mmHg$	0	28.0	52.7	87.7	105.4	155.4
$p_A/mmHg$	280.4	266.1	252.3	231.4	220.8	187.9

x_1	0.5478	0.6349	0.8253	0.9093	1.0000
p_1/mmHg	213.3	239.1	296.9	322.5	353.4
p_A/mmHg	144.2	122.9	66.6	38.2	0

(The data are from *International critical tables*, Vol. 3, p. 288; McGraw-Hill, New York (1928).)

8.37. The industrial process referred to in Problem 8.35 produced a mixture containing a mole fraction 0.300 of toluene. At what temperature will it boil when the pressure is 760 mmHg? What is the composition of the first drop of distillate? What proportions of liquid and vapour are present at the boiling point? What is the composition of the liquid and the vapour at 1 °C above the boiling point? How much of each phase is present?

8.38. The next step in the analysis of the design problems for the separation plant was to set up a fractional distillation column. First we need some terminology. When the fractionation process was described on p. 240 we saw that it could be regarded as occurring in a sequence of zig-zag steps through the phase diagram: the horizontal lines in these steps are called *theoretical plates*. In a column set up in a pilot plant it was found that when a mixture of 0.300 toluene and 0.700 *n*-octane was heated, the distillate had a composition $x_T = 0.700$. How many theoretical plates had that column?

8.39. That degree of separation was not adequate for the subsequent syntheses involved in the plant. You needed a mixture with not less than $x_T = 0.900$. What is the minimum number of theoretical plates in the column that has to be designed?

8.40. Tabulated data can be analysed for their thermodynamic content. The next few Problems invite you to analyse some data taken directly from *International critical tables*, a principal source for this kind of material (see Vol. 3, p. 287). As a first step, take the vapour pressure data for benzene (B) in acetic acid (A) in the table below, and plot the vapour pressure composition curve for the mixture at 50 °C. Then confirm that Raoult's and Henry's Laws are obeyed in the appropriate regions.

x_A	0.0160	0.0439	0.0835	0.1138	0.1714	0.2973
p_A/mmHg	3.63	7.25	11.51	14.2	18.4	24.8
p_B/mmHg	262.9	257.2	249.6	244.8	231.8	211.2

x_A	0.3696	0.5834	0.6604	0.8437	0.9931
p_A/mmHg	28.7	36.3	40.2	50.7	54.7
p_B/mmHg	195.6	153.2	135.1	75.3	3.5

8.41. Deduce the activities and activity coefficients of the components on the basis of Raoult's Law. Regarding B as solute, express its activity and activity coefficients on the basis of Henry's Law.

8.42. What is the excess Gibbs function of mixing of benzene and acetic acid over the composition range $x_A = 0$ to $x_A = 1$ at 25 °C.

9   Changes of state: chemical reactions

Learning objectives

After careful study of this chapter you should be able to

(1)   Determine the *direction of spontaneous change* of a chemical reaction on the basis of the Gibbs function (p. 258).

(2)   Define *extent of reaction* (p. 260) and *stoichiometric coefficient* (p. 261) and write general expressions for chemical reactions (eqn (9.1.6)).

(3)   Define the *Gibbs function of reaction* (eqn (9.1.16)) and calculate it from tables of Gibbs functions of formation (p. 264).

(4)   Express the *equilibrium constant* of a reaction in terms of the standard Gibbs functions of the reactants and products (eqn (9.1.14)).

(5)   Distinguish the *thermodynamic equilibrium constant* from the pressure and composition equilibrium constants (p. 265).

(6)   State how a *catalyst* affects the equilibrium constant (p. 266).

(7)   Express mathematically how the equilibrium constant depends on the temperature, and derive and use the *van't Hoff equation* (eqn (9.2.2)).

(8)   State how the temperature affects the position of equilibrium of exothermic and endothermic reactions (p. 269).

(9)   Define and use the *Giauque function* to predict the equilibrium constants of reactions at different temperatures (p. 270).

(10)   State how the extent of reaction depends on the pressure (p. 271).

(11)   Discuss the thermodynamic basis of the *extraction of metals* (p. 274), *biological processes* (p. 276), and *respiration* (p. 279).

(12)   Define the *biological standard state*, (p. 277).

Introduction

Chemical reactions proceed towards an equilibrium in which, in general, both products and reactants are present but show no further tendency to change their relative amounts. Sometimes the proportion of products greatly overwhelms the remnant of reactants in the equilibrium mixture, and for all practical purposes the reaction is complete. In a wide variety of important cases the reaction mixture at equilibrium has significant amounts of both reactants and products. In this chapter we see how thermodynamics can be used to predict the equilibrium amounts under any conditions of reaction.

All this is of vital importance. In industry it is worse than useless to build a sophisticated plant if the overall reaction has a thermodynamic tendency to run in the wrong direction. If a plant is to be run economically, we must know how to maximize yields. Does that mean raising or lowering the temperature or the pressure? Thermodynamics provides a very simple recipe for predicting these effects for any reaction. We may be interested in the way that food is used in complicated sequences of reactions which ultimately are used to warm the body, power muscular contraction, and energize the nervous system. Some reactions have a natural tendency to occur (like the oxidation of carbohydrates), and these may be coupled to other reactions in order to drive them in unnatural but necessary directions. With thermodynamics we can sort out the reactions that need to be driven, and calculate the spare driving force of reactions that occur spontaneously.

We have seen that the natural direction of all change at constant temperature and pressure is towards minimum Gibbs function. The idea is entirely general, and in this chapter we use it as a basis for the discussion of chemical transformations.

Consider the reaction

$$A + B \rightleftharpoons C + D.$$

Is the natural tendency for the mixture $A + B$ to go to the mixture $C + D$, or the opposite? This can be answered by calculating the Gibbs function of each mixture. If the sum of the Gibbs functions of $A + B$ is higher than that of $C + D$, then the reaction has a tendency to roll from left to right. If $C + D$ lies higher up the Gibbs hill than $A + B$, then the reaction has a natural tendency to roll from right to left: see Fig. 9.1.

ΔG was introduced on the basis of arguments about the natural tendency of the entropy of the universe to increase during a spontaneous change. ΔG concentrates our attention on the system, and lets us avoid having to think explicitly about the entropy of the surroundings. In an isothermal change ΔG can be related to the enthalpy and entropy changes of the system by

$$\Delta G = \Delta H - T\Delta S.$$

The entropy change of the surroundings is hidden in the term ΔH, because if the enthalpy of the system changes by ΔH, the amount of heat

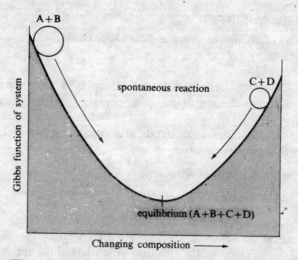

Fig. 9.1. The direction of spontaneous change is down to the valley of the Gibbs function.

transferred to the surroundings at constant pressure is $-\Delta H$, and so their entropy changes by $-\Delta H/T$.

In order to determine whether the change in the Gibbs function will be negative we have to know both the enthalpy and the entropy of the reaction. At low temperatures $T\Delta S$ is often small, and so $\Delta G \approx \Delta H$, which is negative for an exothermic reaction. This approximation, which is often quite good, led Berthelot to his erroneous view that reactions occurred because the reaction tended to occur in the direction that lowered the energy (the enthalpy) of the mixture. This view rules out all endothermic reactions, a point which Berthelot refused to acknowledge. At higher temperatures, or when ΔH is small, the $T\Delta S$ term may dominate, and $\Delta G \approx -T\Delta S$. This is negative if ΔS is positive, and so the reaction then tends to occur in the direction that maximizes the entropy of the reaction mixture.

It is important to be clear that it is the *overall* entropy that determines the direction of natural change. When $T\Delta S$ dominates, it is the change of entropy of the reaction system itself which is mainly determining the entropy change of the universe, and therefore the natural direction of the reaction. When ΔH dominates, the entropy change of the surroundings is dominant. Sometimes ΔH and $T\Delta S$ are of similar importance in determining the sign of ΔG, sometimes they are in opposition, and sometimes they reinforce each other. The contribution of the randomness of the system gets more important as the reaction temperature is raised because $T\Delta S$ gains in importance.

Sometimes it is argued that $\Delta G = \Delta H - T\Delta S$ implies a struggle between the tendency of the enthalpy of the system to seek low values, and the entropy to seek high values. This is a very good way of remembering the expression $\Delta G = \Delta H - T\Delta S$, but it is wholly misleading. The fallacy was discussed in Chapter 5.

9.1 Which way is down-hill?

We begin by studying the simplest possible reaction:

$$A \rightleftharpoons B.$$

Even though this looks trivial, there are examples of it, e.g. the isomerization of a molecule like n-pentane to iso-pentane, or the racemization of L-alanine.

Suppose an infinitesimal amount $d\xi$ of A (where ξ is the Greek letter xi) turns into B, then we can write:

change in the amount of A present $= -d\xi$,

change in the amount of B present $= +d\xi$.

By how much does the Gibbs function for the system change when this change occurs? At constant temperature and pressure the change in G can be written in terms of the chemical potentials of the species involved and the change in the amounts of the species present:

$$dG = \mu_A \, dn_A + \mu_B \, dn_B \qquad (\text{const } p, T).$$

Therefore in the present case

(9.1.1) $$dG = -\mu_A \, d\xi + \mu_B \, d\xi = (\mu_B - \mu_A) \, d\xi \qquad (\text{const } p, T).$$

This equation can be reorganized to

(9.1.2) $$(\partial G/\partial \xi)_{p, T} = \mu_B - \mu_A.$$

It should be understood that μ_A and μ_B depend on the composition of the system, and so their values change, and so does the slope of G, as the reaction proceeds. If at some stage of the reaction the chemical potential of A is greater than that of B, the slope is negative, but it changes sign when μ_B exceeds μ_A. The chemical reaction proceeds in the direction of decreasing G; thus

so long as $\mu_A > \mu_B$ the reaction proceeds from A to B

so long as $\mu_A < \mu_B$ the reaction proceeds from B to A.

This is illustrated in Fig. 9.2. We have switched attention from the Gibbs function to the chemical potential. Now μ is fulfilling the role its name suggests.

When $\mu_A = \mu_B$ the slope of G with respect to the extent of reaction, ξ, is zero. This occurs at the minimum of the curve in Fig. 9.2, and corresponds to the position of chemical equilibrium:

when $\mu_A = \mu_B$ the reaction is at equilibrium.

Both μ_A and μ_B depend on the conditions, and in particular on the concentrations of A and B. The simplest assumption is to suppose that the species are perfect gases, in which case their chemical potentials are determined by their partial pressures:

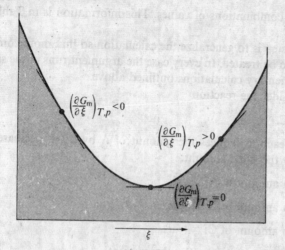

Fig. 9.2. The slope at various points on the Gibbs hill for a reaction.

$$\mu_J = \mu_J^{\ominus} + RT \ln(p_J/p^{\ominus}), \qquad J = A \text{ or } B.$$

The condition of equilibrium then reads

$$\mu_A^{\ominus} + RT \ln(p_A/p^{\ominus})_e = \mu_B^{\ominus} + RT \ln(p_B/p^{\ominus})_e$$

or

$$RT \ln(p_B/p_A)_e = -(\mu_B^{\ominus} - \mu_A^{\ominus}),$$

where the subscript e denotes the equilibrium values of the partial pressures. This can be made tidier by writing the *molar standard Gibbs function* for the reaction as follows:

$$\Delta G_m^{\ominus} = \mu_B^{\ominus} - \mu_A^{\ominus},$$

then the equation becomes

(9.1.3a)°　　$$\ln(p_B/p_A)_e = -\Delta G_m^{\ominus}/RT$$

or

(9.1.3b)°　　$$(p_B/p_A)_e = \exp(-\Delta G_m^{\ominus}/RT).$$

When written like this it can be seen very clearly that as $\Delta G_m^{\ominus}$ gets more negative the partial pressure of B present in the equilibrium mixture increases exponentially. In other words, as the standard chemical potential of pure B drops beneath that of pure A, the equilibrium lies more strongly in favour of B.

When $\Delta G_m^{\ominus}$ is positive at the temperature of the experiment, the exponential is less than 1 and so at equilibrium the partial pressure of A exceeds that of B. When $\Delta G_m^{\ominus}$ is strongly positive the equilibrium lies hard over to the left in favour of A, and the yield of B will be very small. This we anticipated, but the equation expresses the relation quantitatively. In order to use eqn (9.1.3) we need to know the tabulated standard molar Gibbs functions of formation for the reactants and products at the temperature of interest, because $\Delta G_m^{\ominus}$ can then be found by taking the appropriate combinations of values. This information is in Table 5.3 on p. 155.

The next step is to generalize the calculation so that more complicated reactions can be treated. In every case the argument runs in the same way as the rudimentary calculations outlined above.

First consider the reaction

$$A + B \rightleftharpoons C.$$

When the reaction reduces the amount of A by $d\xi$, B decreases and C increases by the same amount:

change in amount of A: $-d\xi$

change in amount of B: $-d\xi$

change in amount of C: $+d\xi$.

The Gibbs function changes by the amount

$$dG = \mu_A \, dn_A + \mu_B \, dn_B + \mu_C \, dn_C = -\mu_A \, d\xi - \mu_B \, d\xi + \mu_C \, d\xi$$

$$= (-\mu_A - \mu_B + \mu_C) \, d\xi,$$

and so, since the reaction is occurring at constant temperature and pressure,

$$(\partial G/\partial \xi)_{p,T} = \mu_C - \mu_A - \mu_B = 0 \text{ at equilibrium.}$$

If A, B, and C are all perfect gases their chemical potentials are all of the form used in the first example. Therefore at equilibrium the partial pressures are such that

$$\{\mu_C^\ominus + RT\ln(p_C/p^\ominus)_e\} - \{\mu_A^\ominus + RT\ln(p_A/p^\ominus)_e\} - \{\mu_B^\ominus + RT\ln(p_B/p^\ominus)_e\}$$
$$= 0$$

or

$$RT\ln\{(p_C/p^\ominus)/(p_A/p^\ominus)(p_B/p^\ominus)\}_e = -(\mu_C^\ominus - \mu_A^\ominus - \mu_B^\ominus).$$

Two operations on this will make it clearer. First we define the standard molar Gibbs function for the reaction in the same way as before:

$$\Delta G_m^\ominus = G_m^\ominus(\text{products}) - G_m^\ominus(\text{reactants}) = \mu_C^\ominus - \mu_A^\ominus - \mu_B^\ominus.$$

Then we introduce the equilibrium constant

$$K = \{(p_C/p^\ominus)/(p_A/p^\ominus)(p_B/p^\ominus)\}_e.$$

(Notice that it is a dimensionless quantity.) Then

$(9.1.4)^\circ$ $\qquad RT\ln K = -\Delta G_m^\ominus,$

or

$(9.1.5)^\circ$ $\qquad K = \exp(-\Delta G_m^\ominus/RT).$

We have arrived at an equation which is virtually the same as before, the only difference being that the equilibrium constant and the $\Delta G_{m}^{\ominus}$ have a slightly more complicated structure.

Example (Objective 4). Calculate the equilibrium constant for the dimerization of nitrogen(IV) oxide (NO_2) at 25 °C.

- *Method.* The reaction is
$$NO_2 + NO_2 \rightleftharpoons N_2O_4,$$
and so it has the form of the reaction treated above. The equilibrium constant is
$$K = (p_{N_2O_4}/p^{\ominus})/(p_{NO_2}/p^{\ominus})^2$$
and the molar Gibbs function of the reaction is
$$\Delta G_{m}^{\ominus} = \mu^{\ominus}(N_2O_4) - 2\mu^{\ominus}(NO_2),$$
or, in terms of the Gibbs functions of formation,
$$\Delta G_{m}^{\ominus} = \Delta G_{f}^{\ominus}(N_2O_4) - 2\Delta G_{f}^{\ominus}(NO_2).$$
The values can be obtained from Table 5.3, and eqn (9.1.5) used to find K.

- *Answer.* $\Delta G_{m}^{\ominus}(298\,K) = 97.8\,kJ\,mol^{-1} - 2(51.3\,kJ\,mol^{-1}) = -4.8\,kJ\,mol^{-1}$

From eqn (9.1.5)
$$K = \exp[-(-4.8\,kJ\,mol^{-1})/(8.314\,J\,K^{-1}\,mol^{-1}) \times (298.15\,K)]$$
$$= \exp(1.94) = 6.93.$$

- *Comment.* The value of K just calculated gives the equilibrium constant in terms of the partial pressures. Often we are interested in the extent of dissociation under a given total pressure. We shall see how to make the conversion on p. 271.

The case of a general reaction can be treated in the same way. Consider one that can be written
$$2A + 3B = C + 2D.$$

An alternative way of writing it is
$$0 = -2A - 3B + C + 2D.$$

The numbers $-2, -3, 1, 2$ are the *stoichiometric coefficients*, the negative signs going with the reactants and the positive with the products. (The reason for this choice will become apparent in a moment.) An entirely general form for a reaction is therefore

(9.1.6) $$0 = \sum_{J} v_J J,$$

where the stoichiometric coefficient of species J is v_J.

When the reaction proceeds by an amount $d\xi$ the amounts of reactants and products change as follows:

change in amount of A: $-2d\xi$, or $v_A d\zeta$ in general

change in amount of B: $-3d\xi$, or $v_B d\zeta$ in general

change in amount of C: $+d\xi$, or $v_C d\zeta$ in general

change in amount of D: $+2d\xi$, or $v_D d\zeta$ in general.

(The amounts of reactants decrease when the reaction advances, hence the negative signs for the reactant coefficients.) At constant temperature and pressure it follows that

$$(\partial G/\partial\xi)_{T,p} = -2\mu_A - 3\mu_B + \mu_C + 2\mu_D$$

$$= \sum_J \nu_J \mu_J \text{ in general.}$$

At equilibrium $(\partial G/\partial\xi)_{p,T}$ is zero. By the same arguments as before we get

(9.1.7) $$RT\ln K = -\Delta G_m^\ominus,$$

where now

$$\Delta G_m^\ominus = -2\mu_A^\ominus - 3\mu_B^\ominus + \mu_C^\ominus + 2\mu_D^\ominus$$

(9.1.8) $$= \sum_J \nu_J \mu_J^\ominus \text{ in general.}$$

The equilibrium constant is

$$K = \{(p_C/p^\ominus)(p_D/p^\ominus)^2/(p_A/p^\ominus)^2(p_B/p^\ominus)^3\}_e$$

$$= \{(p_C/p^\ominus)(p_D/p^\ominus)^2(p_A/p^\ominus)^{-2}(p_B/p^\ominus)^{-3}\}_e$$

for the particular case, and in general

(9.1.9) $$K = \{(p_A/p^\ominus)^{\nu_A}(p_B/p^\ominus)^{\nu_B}(p_C/p^\ominus)^{\nu_C}(p_D/p^\ominus)^{\nu_D}\}_e.$$

$$= \prod_J (p_J/p^\ominus)_e^{\nu_J}.$$

Example (Objective 4). Calculate the equilibrium constants for the reactions

(a) $\frac{1}{2}N_2(g) + \frac{3}{2}H_2(g) \rightleftharpoons NH_3(g)$

(b) $\frac{3}{2}N_2(g) + \frac{1}{2}H_2(g) \rightleftharpoons N_3H(g)$

at 25 °C.

● *Method.* Each reaction is written in the form of eqn (9.1.6) and the stoichiometric coefficients ν_J identified. Then eqns (9.1.7) and (9.1.8) are implemented, with the use of the data in Table 5.3. Note that the Gibbs functions of reaction in these two cases are simply the Gibbs functions of formation of NH_3 or N_3H.

● *Answer.* For the ammonia synthesis, reaction (a), we have

$$\Delta G_m^\ominus = \mu^\ominus(NH_3) - \frac{1}{2}\mu^\ominus(N_2) - \frac{3}{2}\mu^\ominus(H_2) = \Delta G_f^\ominus(NH_3)$$

$$= -16.5\,kJ\,mol^{-1}.$$

$$K = \{(p_{NH_3}/p^\ominus)/(p_{N_2}/p^\ominus)^{1/2}(p_{H_2}/p^\ominus)^{3/2}\}_e$$

$$= \exp\{-(-16.5\,kJ\,mol^{-1})/(8.314\,J\,K^{-1}\,mol^{-1}) \times (298.15\,K)\}$$

$$= \exp(+6.66) = 778.$$

For the hydrogen azide synthesis, reaction (b), we have

$$\Delta G_m^\ominus = \mu^\ominus(N_3H) - \frac{3}{2}\mu^\ominus(N_2) - \frac{1}{2}\mu^\ominus(H_2)$$

$$= 328.0\,kJ\,mol^{-1}.$$

$$K = \{(p_{N_3H}/p^\ominus)/(p_{N_2}/p^\ominus)^{3/2}(p_{H_2}/p^\ominus)^{1/2}\}_e$$

$$= \exp\{-(328.0\,kJ\,mol^{-1})/(8.314\,J\,K^{-1}) \times (298.15\,K)\}$$

$$= \exp(-132.3) = 3.5 \times 10^{-58}.$$

- *Comment*. The hydrogen azide formation reaction lies very strongly in favour of the elements, and even if $p_{H_2} \approx p_{N_2} \approx 1000\,\text{atm}$, $p_{N_3H} \approx 10^{-52}\,\text{atm}$ at equilibrium.

Equations (9.1.7) to (9.1.9) constitute an explicit recipe for calculating equilibrium constants for reactions of perfect gases. The only snag in the development so far is that in neither industry nor biology is there much interest in perfect gases. We must turn what we have done into something useful.

A recipe for equilibrium constants: real gases. In the case of real gases, everything that went before applies except that instead of replacing μ_J by $\mu_J^{\ominus} + RT \ln(p_J/p^{\ominus})$ the fugacity f_J is used in the logarithm. If in the first example A and B are real gases, this leads to

$$RT \ln(f_B/f_A)_e = -\Delta G_m^{\ominus}.$$

The standard states involved in $\Delta G^{\ominus}$ were described on p. 173.

In order to obtain the amounts of gases present in the equilibrium mixture it is necessary to relate the fugacities to the pressures. On writing $f_J = \gamma_J p_J$, where p_J is the pressure of component J, and γ_J its fugacity coefficient, the last equation becomes

$$RT \ln(\gamma_B p_B/\gamma_A p_A)_e = -\Delta G_m^{\ominus},$$

or

$$RT\{\ln(p_B/p_A)_e + \ln(\gamma_B/\gamma_A)_e\} = -\Delta G_m^{\ominus}.$$

The equilibrium constant $K = (p_B/p_A)_e$ can be extracted from this if the fugacity coefficients are known.

The form of the equilibrium equation remains the same as in the earlier examples when real gases are involved, the only change being that the equilibrium constant K is replaced by an analogous expression in terms of the fugacities. K is then called the *thermodynamic equilibrium constant* to underline that when it is used in the equilibrium equation the result is exact. In order to use K we need to relate the fugacities to pressures, mole fractions, or molalities. That is normally done by referring to published tables, or making some estimate of the fugacity coefficients.

Another recipe: real reactions. Only a short step is needed to obtain the equilibrium constant for any type of reaction. The general expression for the chemical potential of some substance J is

$$\mu_J = \mu_J^{\ominus} + RT \ln a_J,$$

where a_J is the activity of the substance when it is present at the concentration of interest. Activities and standard states were discussed on p. 244; remember that a is defined as a dimensionless quantity.

All the previous analysis now follows through, the only change being that the thermodynamic equilibrium constant is expressed in terms of the activities. For instance, for the reaction

$$A + B \rightleftharpoons C + D$$

we get

(9.1.10) $RT \ln K = -\Delta G_m^{\ominus}$

where

(9.1.11) $\Delta G_m^{\ominus} = \mu_C^{\ominus} + \mu_D^{\ominus} - \mu_A^{\ominus} - \mu_B^{\ominus}$

and now

(9.1.12) $K = (a_C a_D / a_A a_B)_e$

where the subscript e denotes the activities at the equilibrium concentrations. In the general case of a reaction

(9.1.13) $0 = \sum_J \nu_J J,$

the equilibrium equation is

(9.1.14) $\Delta G_m^{\ominus} = -RT \ln K,$

where the thermodynamic equilibrium constant is

(9.1.15) $K = \{a_A^{\nu_A} a_B^{\nu_B} a_C^{\nu_C} a_D^{\nu_D} \ldots \}_e = \prod_J (a_J^{\nu_J})_e,$

and the standard Gibbs function for the reaction is

(9.1.16) $\Delta G_m^{\ominus} = \sum_J \nu_J \mu_J^{\ominus}.$

It should be noticed that eqn (9.1.14) provides a way of measuring the standard Gibbs functions of reactions: a determination of the equilibrium constants (by determining the amounts of the components present at equilibrium) is all that is necessary. This is a *non-calorimetric* determination of $\Delta G_m^{\ominus}$. Notice, however, that $\Delta G_m^{\ominus}$ is defined in terms of the standard states of the *pure* components. A $\Delta G_m^{\ominus}$ obtained in this way therefore includes a contribution from the Gibbs function of mixing of the reactants (and of unmixing of the products) as well as the Gibbs function for the truly chemical part of the reaction. This is significant when $\Delta G_m^{\ominus}$ is interpreted in terms of enthalpies and entropies.

Example (Objective 4). Calculate the equilibrium constants for the following reactions on the basis of the Gibbs function information given in Table 5.3.

    (a) $Ag(s) + \frac{1}{2}Cl_2(g) \rightleftharpoons AgCl(s)$

    (b) $2Al(s) + \frac{3}{2}O_2(g) \rightleftharpoons Al_2O_3(s)$

    (c) $NH_3(g) + HCl(g) \rightleftharpoons NH_4Cl(s)$

(d) $H_2(g) + CO_2(g) \rightleftharpoons H_2O(l) + CO(g)$.

- *Method.* In each case the reaction is written in the form of eqn (9.1.13) and the equilibrium constant is calculated on the basis of eqns (9.1.14) and (9.1.16).

- *Answer.*

(a) $K = a_{AgCl(s)}/\{a_{Ag(s)}(f_{Cl_2}/p^{\ominus})^{1/2}\}$

$\quad = 1/(f_{Cl_2}/p^{\ominus})^{1/2} \approx 1/(p_{Cl_2}/p^{\ominus})^{1/2}$.

$\Delta G_m^{\ominus} = -109.8 \,\text{kJ mol}^{-1}, \qquad K = 1.7 \times 10^{19}$.

(b) $K = a_{Al_2O_3(s)}/\{a_{Al(s)}^2(f_{O_2}/p^{\ominus})^{3/2}\}$

$\quad = 1/(f_{O_2}/p^{\ominus})^{3/2} \approx 1/(p_{O_2}/p^{\ominus})^{3/2}$.

$\Delta G_m^{\ominus} = -1582.4 \,\text{kJ mol}^{-1}, \qquad K = 1.7 \times 10^{277}$

(c) $K = a_{NH_4Cl(s)}/\{(f_{NH_3}/p^{\ominus})(f_{HCl}/p^{\ominus})\}$

$\quad = 1/\{(f_{NH_3}/p^{\ominus})(f_{HCl}/p^{\ominus})\} \approx 1/\{(p_{NH_3}/p^{\ominus})(p_{HCl}/p^{\ominus})\}$

$\Delta G_m^{\ominus} = (-203.0 \,\text{kJ mol}^{-1}) - (-16.5 \,\text{kJ mol}^{-1}) - (-95.3 \,\text{kJ mol}^{-1})$

$\quad = -92.2 \,\text{kJ mol}^{-1}, \qquad K = 1.4 \times 10^{16}$.

(d) $K = a_{H_2O(l)}(f_{CO}/p^{\ominus})/\{(f_{H_2}/p^{\ominus})(f_{CO_2}/p^{\ominus})\}$

$\quad = (f_{CO}/p^{\ominus})/\{(f_{H_2}/p^{\ominus})(f_{CO_2}/p^{\ominus})\} \approx (p_{CO}/p^{\ominus})/\{(p_{H_2}/p^{\ominus})(p_{CO_2}/p^{\ominus})\}$

$\Delta G_m^{\ominus} = (-237.2 \,\text{kJ mol}^{-1}) + (-137.2 \,\text{kJ mol}^{-1}) - (-394.4 \,\text{kJ mol}^{-1})$

$\quad = 20.0 \,\text{kJ mol}^{-1}, \qquad K = 3.1 \times 10^{-4}$.

- *Comment.* All the equilibria relate to 298 K: we shall shortly see how to generalize the calculation to other temperatures (p. 266): this could be done already if we knew the values of the Gibbs functions at those temperatures of interest.

Expressing the thermodynamic equilibrium constant in terms of mole fractions or concentrations is a simple matter once the activity coefficients are known. For instance, on writing $a_J = \gamma_J x_J$ where x_J is the mole fraction of species J, the equilibrium constant in eqn (9.1.12) becomes

$$K = (\gamma_C x_C \gamma_D x_D / \gamma_A x_A \gamma_B x_B)_e$$

(9.1.17)
$$= (\gamma_C \gamma_D / \gamma_A \gamma_B)_e (x_C x_D / x_A x_B)_e$$

$$= K_\gamma K_x.$$

In this expression K_γ and K_x have the same form as K itself, but K_γ is expressed in terms of the activity coefficients, and K_x in terms of the mole fractions. It follows that if we can obtain the activity coefficients from published data or calculate them, we can get K_x from K/K_γ. The same remark applies to the general reaction, and its thermodynamic equilibrium constant can also be written in the form $K_\gamma K_x$.

9.2 The response of reactions to the conditions

In this section we investigate how the equilibrium concentrations and pressures depend on the conditions of the reaction. We shall be able to say whether a rise of temperature will shift the reaction in favour of the products, or drag it back towards the reactants. The technological importance of this kind of calculation is obvious: it would be futile to raise the temperature of the reaction in order to make it go faster if by so doing it were made to go faster in the wrong direction.

How equilibrium responds to a catalyst. It doesn't. Any claims that an equilibrium can be shifted by an enzyme or a catalyst (which are materials that change the rate of the reaction without suffering any net change) can be dismissed by appealing to the laws of thermodynamics.

Consider an exothermic reaction, and the hypothetical catalyst which, it is claimed, shifts the equilibrium towards the products. Take an uncatalysed reaction at equilibrium, and add the catalyst. If the claim were true, the reaction would proceed until the new equilibrium is reached, and as it formed products the system would evolve heat. This heat can be turned into work. Now remove the catalyst. The equilibrium shifts back, and heat is drawn in from the surroundings. In the end everything is in its original condition, except that some of the heat of the surroundings has been turned into work. This is contrary to the Second Law in the form stated on p. 136. Therefore, in all probability, the claim is fraudulent.

The advantage of catalysts is that they increase the speeds of reactions, but they do so without affecting equilibrium concentrations. Therefore, if the equilibrium lies hard to the right (towards products) at low temperatures, but to the left at high temperatures, then a catalyst might be found which can operate at low temperatures so that the favourable equilibrium can be established speedily. The details of the operations of catalysts are described in Part 3.

How equilibrium responds to temperature. If the equilibrium constant is known at a temperature T^*, its value at another temperature T can be found as follows. In each case

$$\ln K(T^*) = -\Delta G_m^{\ominus}(T^*)/RT^*$$
$$\ln K(T) = -\Delta G_m^{\ominus}(T)/RT.$$

Subtracting these relates the two equilibrium constants:

$$\ln K(T) = \ln K(T^*) - \left\{ \frac{\Delta G_m^{\ominus}(T)}{RT} - \frac{\Delta G_m^{\ominus}(T^*)}{RT^*} \right\}$$

Now write $\Delta G = \Delta H - T\Delta S$, and make the approximation that neither ΔH nor ΔS changes significantly over the temperature range of interest:

$$\ln K(T) = \ln K(T^*) - \left\{ \frac{\Delta H_m^{\ominus}(T)}{RT} - \frac{\Delta H_m^{\ominus}(T^*)}{RT^*} - \frac{\Delta S_m^{\ominus}(T)}{R} + \frac{\Delta S_m^{\ominus}(T^*)}{R} \right\}$$

(9.2.1)
$$\approx \ln K(T^*) - \left\{ \frac{\Delta H_m^{\ominus}(T^*)}{R} \right\} \left\{ \frac{1}{T} - \frac{1}{T^*} \right\}.$$

The last equation is very remarkable because it predicts the shift in equilibrium when the temperature is changed; all we need to know is the standard molar enthalpy of reaction.

Example (Objective 7). The equilibrium constants for the ammonia and hydrogen azide syntheses were calculated for a temperature of 298.15 K in the *Example* on p. 262. Use the enthalpy information in Table 4.1 to find the equilibrium constant when the temperature is changed by $+100$ K and -100 K in each case, assuming the conditions are such that all the compounds remain gaseous.

- *Method.* Assume that the enthalpies are constant over the range of temperatures under consideration, and then use eqn (9.2.1). For the ammonia synthesis we have $\Delta H_m^{\ominus}(298\,K) = -46.1\,kJ\,mol^{-1}$, and for the azide synthesis $\Delta H_m^{\ominus}(298\,K) = +264.0\,kJ\,mol^{-1}$.

- *Answer.* The equilibrium constants are 778 and 3.5×10^{-58} respectively at $T^* = 298.15$ K. Taking $T = 398.15$ K leads to

$$\ln K(398.15\,K) = \ln K(298.15\,K) + (\Delta H_m^{\ominus}/R)[(100\,K)/(298.15\,K)(398.15\,K)].$$

For ammonia:

$$\ln K(398.15\,K) = 6.66 + (-46.11\,kJ\,mol^{-1}/8.314\,J\,K^{-1}\,mol^{-1}) \times$$
$$\times (8.424 \times 10^{-4}\,K^{-1}) = 1.99.$$

For hydrogen azide:

$$\ln K(398.15\,K) = -132 + (264.0\,kJ\,mol^{-1}/8.314\,J\,K^{-1}\,mol^{-1}) \times$$
$$\times (8.424 \times 10^{-4}\,K^{-1})$$
$$= -132 + 26.75 = -106.$$

For a drop of temperature of 100 K the appropriate equation is

$$\ln K(198.15\,K) = \ln K(298.15\,K)$$
$$+ (\Delta H_m^{\ominus}/R)\{(-100\,K)/(198.15\,K)(298.15\,K)\},$$

and we find for ammonia:

$$\ln K(198.15\,K) = 6.66 + 9.39 = 16.05$$

for hydrogen azide:

$$\ln K(198.15\,K) = -132 - 53.7 = -186.$$

These results can be collected in the following table:

K	198 K	298 K	398 K
ammonia	9.32×10^6	778	7.28
hydrogen azide	2.1×10^{-81}	3.5×10^{-58}	9.2×10^{-47}

The entries are the equilibrium constants for the syntheses.

- *Comment.* Notice how the exothermic reaction (the ammonia synthesis) is favoured by a drop in temperature (the equilibrium shifts towards the products) but disfavoured by a rise in temperature. The opposite holds true for the endothermic azide synthesis.

Some severe approximations were made in the interests of simplicity, but it is a simple matter to obtain an exact expression for $d(\ln K)/dT$ by using the Gibbs–Helmholtz equation (p. 166). This equation states that

$$(\partial(G/T)/\partial T)_p = -H/T^2.$$

It immediately implies the

(9.2.2)

> *van't Hoff isochore:* $\mathrm{d}\ln K/\mathrm{d}T = \Delta H_{\mathrm{m}}^{\ominus}(T)/RT^2.$

Another form of this exact equation can be obtained by recognizing that

$$(\mathrm{d}/\mathrm{d}T)(1/T) = -1/T^2 \quad \text{or} \quad \mathrm{d}T/T^2 = -\mathrm{d}(1/T),$$

for then

(9.2.3)

> $\mathrm{d}\ln K/\mathrm{d}(1/T) = -\Delta H_{\mathrm{m}}^{\ominus}(T)/R.$

This too is exact.

Example (Objective 7). If we knew how the enthalpy of reaction depended on the temperature it would be possible to integrate eqn (9.2.2) and so find an exact connection between the equilibrium constant and the temperature. In Chapter 4 we saw how to relate the temperature dependence of the reaction enthalpy to the heat capacities of the reactants and products. Deduce an expression for the temperature dependence of the equilibrium constant on the basis that the heat capacity of each component of the reaction can be written in the form $C_{p,\mathrm{m}} = a + bT + cT^{-2}$.

- *Method.* The integration of eqn (9.2.2) from a temperature T^* to T gives

$$\ln K(T) - \ln K(T^*) = \int_{T^*}^{T} \frac{\Delta H_{\mathrm{m}}^{\ominus}(T)\,\mathrm{d}T}{RT^2}.$$

From Chapter 4 (p. 110) we also know that

$$\Delta H_{\mathrm{m}}(T) = \Delta H_{\mathrm{m}}^{\ominus}(T^*) + \int_{T^*}^{T} \Delta C_{p,\mathrm{m}}(T)\,\mathrm{d}T$$

where $\Delta C_{p,\mathrm{m}}$ is the change in heat capacity between reactants and products, eqn (4.1.2) on p. 110. All that it is necessary to do is to express ΔC in terms of the changes Δa, Δb, and Δc in the coefficients in the expression for the temperature dependence of $C_{p,\mathrm{m}}$ for each component, evaluate the integral to find $\Delta H(T)$, and then insert that result in the first integral to find $\ln K(T)$.

- *Answer.* All the manipulations are straightforward. They give:

$$\ln K(T) = \ln K(T^*) + (\Delta a/R)\ln(T/T^*) + A(T,T^*)\,\delta T + B(T,T^*)\,\delta T^2,$$

where $\delta T = T - T^*$ and

$$A(T,T) = \Delta H_{\mathrm{m}}^{\ominus}(T^*)/RTT^* - \Delta a/RT$$
$$B(T,T^*) = \Delta b/2RT + \Delta c/2RT^2T^{*2}.$$

- *Comment.* If we took $T^* = 298\,\mathrm{K}$, the standard tables could be used to give K at that temperature and the expression just deduced could be used to find the equilibrium constant at any other temperature. In order to use the last equation we need to know the reaction enthalpy at $T^* = 298\,\mathrm{K}$ (this is given in standard tables, for example Tables 4.1 and 4.2) and the coefficients a, b, and c. The latter are known for a variety of materials, and a list is given in Table 4.3.

The van't Hoff equation provides a way of measuring the enthalpy of a reaction without using a calorimeter, but in practice it is inaccurate. Equilibrium compositions are measured over a range of temperatures (at constant pressure), and $\ln K$ is plotted against $1/T$. It follows from eqn

(9.2.3) that the slope is $-\Delta H_m^{\ominus}/R$ (Fig. 9.3). Note that on account of the temperature dependence of the reaction enthalpy, this slope depends on the temperature, and so the line is not expected to be perfectly straight. In practice, however, $\Delta H^{\ominus}$ normally depends only weakly on the temperature.

Eqn (9.2.2) shows that the slope $d \ln K/dT$ is negative for an exothermic reaction ($\Delta H < 0$), but positive for an endothermic reaction ($\Delta H > 0$). A negative slope (downhill from left to right) means that $\ln K$, and therefore K itself, gets smaller as the temperature rises. Therefore, the equilibrium shifts away from products (which reduces the numerator in K) and towards reactants (and so increases the denominator). The opposite occurs in the case of an endothermic reaction. We can summarize this important conclusion as follows:

exothermic reactions: a rise in temperature favours the reactants.

endothermic reactions: a rise in temperature favours the products.

In order to find the reason for this behaviour we need look only as far as the expression $\Delta G = \Delta H - T\Delta S$, but written in the form $-\Delta G/T = -\Delta H/T + \Delta S$. When the reaction is exothermic $-\Delta H/T$ represents a positive change in the entropy of the surroundings, and is a driving force for the reaction going from left to right. But if the temperature is raised $-\Delta H/T$ gets smaller, and so the increasing entropy of the surroundings is a less potent driving force, and the equilibrium lies less far to the right. When the reaction is endothermic the principal driving force is the increasing entropy of the system, because the entropy of the surroundings decreases as enthalpy is sucked into the system. The effect of the unfavourable change in their entropy is reduced if the temperature is raised ($\Delta H/T$ is reduced) and so the reaction can proceed more strongly towards products.

The conclusions in the box above are a special case of *Le Châtelier's Principle*, which states that *a system at equilibrium, when subjected to a perturbation, responds in a way that tends to eliminate its effect.*

Tables of standard Gibbs functions of formation often refer to 25 °C, and there might be insufficient data on the enthalpy of formation to apply

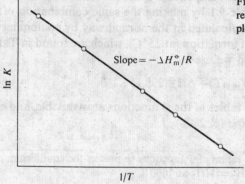

Fig. 9.3. Determination of the reaction enthalpy from the van't Hoff plot.

Slope $= -\Delta H_m^{\ominus}/R$

$\ln K$

$1/T$

van't Hoff's isochore reliably in order to find K at another temperature. Furthermore, $\Delta G^\ominus$ commonly varies quite strongly with temperature, and so interpolation between values quoted at two temperatures in order to obtain $\Delta G^\ominus$ for the particular temperature of interest might lead to significant inaccuracies. For this reason it is common to tabulate a quantity Φ which is related to $\Delta G^\ominus$ but which varies much more slowly with temperature. It is sometimes called the *Giauque function*. There are actually two such Φ functions, $\Phi_0(T) = \{G_m^\ominus(T) - H_m^\ominus(0)\}/T$ and $\Phi_{298}(T) = \{G_m^\ominus(T) - H_m^\ominus(298.15\,\text{K})\}/T$. Converting from one to the other is simple because

(9.2.4) $$\Phi_{298}(T) = \Phi_0(T) - \{H_m^\ominus(298.15\,\text{K}) - H_m^\ominus(0)\}/T,$$

and the difference of the enthalpies is often listed. An example is given in Table 9.1. Note that G and H in the definition of Φ refer to the single substance, and so are not Gibbs functions and enthalpies of formation: they are the *actual* Gibbs function and enthalpy at the stated temperature (the appearance of $-H^\ominus(0)$ in the definition literally does away with the problem of not knowing absolute enthalpies). The values of Φ may be compiled either by calculation using the techniques of statistical thermodynamics (an example is given in Part 2) or calorimetrically from heat capacity measurements down to low temperatures, as explained in Chapter 5.

The form of the Φ functions makes them quite easy to use for calculating equilibrium constants at any temperature. The crucial quantity controlling the magnitude of K is $\Delta G_m^\ominus(T)/T$, which can be found from the tables using

(9.2.5) $$G_m^\ominus(T)/T = \Phi_{298}(T) + H_m^\ominus(298.15\,\text{K})/T$$

$$= \Phi_0(T) - \{H_m^\ominus(298.15\,\text{K}) - H_m^\ominus(0)\}/T + H_m^\ominus(298.15\,\text{K})/T,$$

so that for the reaction,

(9.2.6) $$\Delta G_m^\ominus(T)/T = \Delta\Phi_0(T) - \Delta\{H_m^\ominus(298.15\,\text{K}) - H_m^\ominus(0)\}/T$$
$$+ \Delta H_m^\ominus(298.15\,\text{K})/T.$$

Everything in this apparently cumbersome expression is normally available. $\Delta\Phi_0$ can be obtained from Table 9.1 by taking the differences of the values for the various components of the reaction; the second term is also obtained from Table 9.1 by making the same combination of terms; the third term can be calculated in the normal way by combining the values of the enthalpies of formation (at 25 °C), which are found in Table 4.1 and 4.2. If Φ_{298} is listed we use

(9.2.7) $$\Delta G_m^\ominus(T)/T = \Delta\Phi_{298}(T) + \Delta H_m^\ominus(298.15\,\text{K})/T$$

directly. Extensive tables of the Φ functions are available, and are referred to in *Further reading* (p. 280).

Example (Objective 9). Use the tables of Φ functions to predict the equilibrium constant for the reaction $\frac{1}{2}N_2 + \frac{3}{2}H_2 \rightleftharpoons NH_3$ at 1000 K.

Table 9.1. Giauque functions, $-\Phi_0(T)/\text{J K}^{-1}\,\text{mol}^{-1}$

	298 K	500 K	1000 K	1500 K	2000 K	$\dfrac{[H_m^{\ominus}(298) - H_m^{\ominus}(0)]}{\text{kJ mol}^{-1}}$
$H_2(g)$	102.2	116.9	137.0	148.9	157.6	8.468
$N_2(g)$	162.4	177.5	197.9	210.4	219.6	8.669
$O_2(g)$	176.0	191.1	212.1	225.1	234.7	8.682
$Cl_2(g)$	192.2	208.6	231.9	246.2	256.6	9.180
$Br_2(g)$	212.8	230.1	254.4	269.1	279.6	9.728
$C(s)$	2.2	4.85	11.6	17.5	22.5	1.050
$CO(g)$	168.4	183.5	204.1	216.6	225.9	8.673
$CO_2(g)$	182.3	199.5	226.4	244.7	258.8	9.364
$CH_4(g)$	152.5	170.5	199.4	221.1	239	10.029
$NH_3(g)$	159.0	176.9	203.5	221.9	236.6	9.92
$H_2O(g)$	155.5	172.8	196.7	211.7	223.1	9.908
$HCl(g)$	157.8	172.8	193.1	205.4	214.3	8.640

Source: G. N. Lewis and M. Randall, *Thermodynamics* (revised by K. S. Pitzer and L. Brewer), McGraw-Hill.

- *Method.* We need the value of $\Delta G_m^{\ominus}(1000\,\text{K})$ for the reaction; then use $RT\ln K = -\Delta G_m^{\ominus}$. Use eqn (5.4.2) to find $\Delta G_m^{\ominus}$, and Table 4.1 for $\Delta H_m^{\ominus}$.

- *Answer.* For the ammonia synthesis $\Delta H_m^{\ominus}(298\,\text{K}) = -46.1\,\text{kJ mol}^{-1}$. From Table 9.1 we have

$$\Delta \Phi_0(1000\,\text{K}) = (-203.5\,\text{J K}^{-1}\,\text{mol}^{-1}) - \tfrac{1}{2}(-197.9\,\text{J K}^{-1}\,\text{mol}^{-1})$$
$$- \tfrac{3}{2}(-137.0\,\text{J K}^{-1}\,\text{mol}^{-1}) = 101.0\,\text{J K}^{-1}\,\text{mol}^{-1}$$
$$\Delta\{H_m^{\ominus}(298.15\,\text{K}) - H_m^{\ominus}(0)\} = 9.92\,\text{kJ mol}^{-1} - \tfrac{1}{2}(8.669\,\text{kJ mol}^{-1})$$
$$- \tfrac{3}{2}(8.468\,\text{kJ mol}^{-1}) = -7.12\,\text{kJ mol}^{-1}.$$

Then, from eqn (9.2.6)

$$\Delta G_m^{\ominus}(1000\,\text{K})/(1000\,\text{K}) = 101.0\,\text{J K}^{-1}\,\text{mol}^{-1} - (-7.12\,\text{kJ mol}^{-1})/(1000\,\text{K})$$
$$+ (-46.1\,\text{kJ mol}^{-1})/(1000\,\text{K}) = 62.0\,\text{J K}^{-1}\,\text{mol}^{-1}.$$

It follows that

$$K(1000\,\text{K}) = \exp\{-(62.0\,\text{J K}^{-1}\,\text{mol}^{-1})/(8.314\,\text{J K}^{-1}\,\text{mol}^{-1})\}$$
$$= e^{-7.46} = 5.76 \times 10^{-4}.$$

- *Comment.* Note that $\Phi(T)$ is in a useful form for equilibrium calculations because K depends on $\Delta G^{\ominus}/T$.

How equilibrium responds to pressure. The equilibrium constant K depends on $\Delta G^{\ominus}$, but the standard Gibbs function is a property that is defined for species at a specific pressure. Therefore $\Delta G^{\ominus}$ does not vary when the pressure of the experiment is changed, and so K is independent of pressure. At constant temperature it follows that

$$(\partial K/\partial p)_T = 0.$$

This does not imply that the amounts of the species at equilibrium do not change. For **example**, consider the perfect gas equilibrium

$$A \rightleftharpoons 2B.$$

If an amount n of A was present initially, then at equilibrium its abundance will have fallen to an amount we shall write $n(1 - \alpha_e)$ and the amount of B will have risen from zero to $2n\alpha_e$. It follows that at equilibrium the mole fractions of A and B are

$$x_{A,e} = n(1 - \alpha_e)/\{n(1 - \alpha_e) + 2n\alpha_e\} = (1 - \alpha_e)/(1 + \alpha_e)$$

$$x_{B,e} = 2\alpha_e/(1 + \alpha_e).$$

The equilibrium constant may be expressed in terms of these mole fractions as follows:

$$K = \{(p_B/p^{\ominus})^2/(p_A/p^{\ominus})\}_e = \{(x_B p/p^{\ominus})^2/(x_A p/p^{\ominus})\}_e$$

$$= (x_B^2/x_A)_e(p/p^{\ominus})$$

$$= \{4\alpha_e^2/(1 - \alpha_e^2)\}(p/p^{\ominus}).$$

The equilibrium constant K is independent of pressure, and therefore α_e must depend on pressure in such a way that the r.h.s. of this expression does not depend on p. Rearranging the expression leads to

(9.2.8) $$\alpha_e = \{K/(K + 4p/p^{\ominus})\}^{1/2}$$

and so the amount of A and B at equilibrium can be predicted for any pressure, see Fig. 9.4. Conversely, their amounts can be used to find K and then $\Delta G^{\ominus}$.

The last result is a special case of Le Châtelier's Principle. In the reaction $A \rightleftharpoons 2B$ a shift to the right increases the pressure exerted by the reaction mixture, and a shift to the left decreases it. When pressure is applied to a system at equilibrium the principle asserts that it will adjust to absorb the effect of the pressure increase: in this case the equilibrium should shift

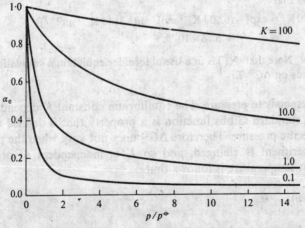

Fig. 9.4. The pressure dependence of the extent of the reaction $A \rightleftharpoons 2B$ at equilibrium.

in the direction of the reactant A (to smaller values of α_e). This shift is predicted by the last equation.

The effect of pressure can be expressed in a slightly more general way by considering the perfect gas reaction

$$0 = \sum_J \nu_J J$$

for which the equilibrium constant is

$$K = \{(p_A/p^\ominus)^{\nu_A}(p_B/p^\ominus)^{\nu_B}(p_C/p^\ominus)^{\nu_C}(p_D/p^\ominus)^{\nu_D}\}_e$$

$$= \{x_A^{\nu_A}x_B^{\nu_B}x_C^{\nu_C}x_D^{\nu_D}\}_e(p/p^\ominus)^{\nu_A+\nu_B+\nu_C+\nu_D}$$

$$= (p/p^\ominus)^\nu K_x,$$

where K_x is the equilibrium constant expressed in terms of the mole fractions and $\nu = \nu_C + \nu_D + \nu_A + \nu_B$, the difference between the numbers of gas molecules in the reaction equation in the products and the reactants (because ν_A, ν_B are negative). We have seen that $(\partial K/\partial p)_T$ is zero, and so on differentiating both sides of the last equation we find

$$0 = (\partial K/\partial p)_T = (\partial K_x/\partial p)_T(p/p^\ominus)^\nu + (\nu K_x/p^\ominus)(p/p^\ominus)^{\nu-1},$$

or

$(9.2.9)^\circ$

$$(\partial \ln K_x/\partial \ln p)_T = -\nu.$$

This result indicates that when the pressure is raised the equilibrium constant K_x increases if ν is negative but decreases if it is positive. ν is positive if the number of gas molecules in the reaction equation is greater in the products than in the reactants, and so this is in accord with the previous discussion because a greater K_x indicates a larger amount of products. If $\nu = 0$ changing the pressure will not affect the value of K_x.

Example (Objective 10). Predict the effect of pressure on the composition of the equilibrium mixture in the synthesis of ammonia.

• *Method*. The reaction is

$$\tfrac{1}{2}N_2(g) + \tfrac{3}{2}H_2(g) \rightleftharpoons NH_3(g)$$

and so we can state the change in terms of the stoichiometric coefficients of the gases. If the gases are assumed to behave perfectly we also know that K is independent of pressure. Hence we can state how K_x must depend on the pressure.

• *Answer*. Since $\nu = 1 - \tfrac{1}{2} - \tfrac{3}{2} = -1$ the connection between K and K_x is $K = (p/p^\ominus)^{-1}K_x$. Since K is independent of pressure it follows that $K_x \propto p/p^\ominus$. Therefore an increase in pressure increases the value of K_x and favours the formation of ammonia.

• *Comment*. The Haber synthesis of ammonia is run at high pressure in order to make use of this result. In precise work it is necessary to base arguments on the pressure independence of the thermodynamic equilibrium constant, and therefore to take note of the pressure dependence of the fugacity coefficients in order to draw conclusions about the behaviour of K_x.

9.3 Applications to selected systems

In this section we examine some of the conclusions that can be drawn from the equilibrium equation $\Delta G_m^\ominus = -RT \ln K$, or $K = \exp(-\Delta G_m^\ominus/RT)$. For convenience Fig. 9.5 shows how K depends on $\Delta G_m^\ominus$ for several different temperatures. If $\Delta G_m^\ominus$ for a reaction is known at the temperature of interest, then a glance at the diagram establishes the value of the equilibrium constant.

In many biological reactions, and often in others, equilibria are established at about room temperature. At 25 °C $RT = 2.48 \, \text{kJ mol}^{-1}$. Suppose $\Delta G_m^\ominus = g \, \text{kJ mol}^{-1}$, then $\ln K = -g/2.48$. But $2.303 \lg x \approx \ln x$, where $\lg$ means $\log_{10}$ and so $\lg K \approx -g/5.71$, or

(9.3.1) $K \approx 10^{-g/5.71}.$

This gives a quick way of estimating an equilibrium constant at room temperature. Notice that $K > 1$ (so that products dominate reactants) if g is negative (so that $\Delta G_m^\ominus < 0$).

Extraction of metals from their oxides. Metals can be obtained from their oxides by reduction with carbon if the equilibria

$$MO + C \rightleftharpoons M + CO$$

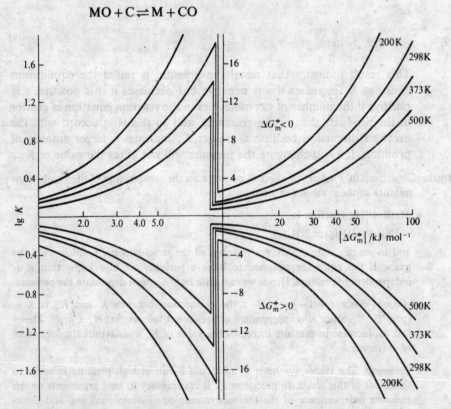

Fig. 9.5. The dependence of the equilibrium constant on the standard Gibbs function of reaction at various temperatures. Note the changes of scale.

or

$$MO + \tfrac{1}{2}C \rightleftharpoons M + \tfrac{1}{2}CO_2$$

lie sufficiently far to the right. These equilibria can be discussed in terms of the thermodynamic functions for the reactions

(i) $M + \tfrac{1}{2}O_2 \rightleftharpoons MO$,

(ii) $\tfrac{1}{2}C + \tfrac{1}{2}O_2 \rightleftharpoons \tfrac{1}{2}CO_2$,

(iii) $C + \tfrac{1}{2}O_2 \rightleftharpoons CO$,

(iv) $CO + \tfrac{1}{2}O_2 \rightleftharpoons CO_2$.

The temperature dependence of these reactions depends on the entropy change through $(\partial \Delta G_m / \partial T)_p = -\Delta S_m$. Since in reaction (iii) there is an increase in the amount of gas, $\Delta S_m^{\ominus}$ is large and positive, and so $\Delta G_m^{\ominus}$ decreases sharply with increasing temperature. In reaction (iv) there is a similar decrease in the amount of gas, and so $\Delta G_m^{\ominus}$ increases sharply. In reaction (ii) the amount of gas is constant, there is therefore little entropy change, and so $\Delta G_m^{\ominus}$ hardly changes with temperature. These remarks are illustrated in Fig. 9.6 (note that ΔG decreases upwards!).

The Gibbs function for reaction (i) indicates the metal's affinity for oxygen. At room temperature $\Delta H_m^{\ominus}$ dominates $\Delta S_m^{\ominus}$ and so $\Delta G_m^{\ominus}$ is governed by the enthalpy of formation of the oxide. This is indicated on the left of Fig. 9.6. The entropy of reaction is approximately the same for all metals because the reactions correspond to the elimination of gaseous oxygen to form a compact, solid oxide. Therefore, the temperature dependence of ΔG_m is approximately the same for all metals; this is indicated in Fig. 9.6 by the similar slopes of the lines at low temperatures. The kinks at higher temperature correspond to evaporation of the metal (less pronounced kinks occur at the melting points of the metal and the oxide).

Reduction of the oxide depends on the competition of the carbon for the oxygen bound to the metal. The Gibbs function for the relevant processes can be expressed in terms of the Gibbs function for the oxidation reactions:

$$MO + C \rightleftharpoons M + CO \qquad \Delta G_m^{\ominus} = \Delta G_m^{\ominus}(iii) - \Delta G_m^{\ominus}(i)$$

$$MO + \tfrac{1}{2}C \rightleftharpoons M + \tfrac{1}{2}CO_2 \qquad \Delta G_m^{\ominus} = \Delta G_m^{\ominus}(ii) - \Delta G_m^{\ominus}(i)$$

$$MO + CO \rightleftharpoons M + CO_2 \qquad \Delta G_m^{\ominus} = \Delta G_m^{\ominus}(iv) - \Delta G_m^{\ominus}(i)$$

and the equilibrium lies to the right if $\Delta G_m^{\ominus} < 0$, which is the case when the line for $\Delta G_m^{\ominus}(i)$ lies below (is more positive than) the line for one of the carbon reactions (ii)–(iv). At any temperature we are able to predict the success of the reduction simply by examining the diagram: reactions move spontaneously downwards in the diagram. For example, CuO can be reduced to copper at any temperature above room temperature. Ag_2O has $\Delta G^{\ominus}(i) > 0$ when the temperature is above about $200\,°C$, and so reduction may be caused even in the absence of carbon simply by heating (reaction (i) is spontaneous in the backwards direction above 200 °C). Al_2O_3, however, cannot be reduced until the temperature has been raised to above 2000 °C and so it is a stable refractory material until this temperature has been reached.

The position of equilibrium at any temperature can be obtained from the vertical separation of the appropriate lines in the diagram. Similar diagrams may be constructed on the basis of sulphides, nitrides, phosphides, and so on, and their interpretation is basically the same as described here.

Biological activity: thermodynamics of ATP. An important component of living systems is adenosine triphosphate, ATP (Fig. 9.7). The function of ATP is to act as a store of the energy made available when food is consumed, and to be able to provide the energy for a wide variety of biochemical processes, including muscular contraction, vision, and reproduction. The basis of ATP's activity is its ability to lose the terminal phosphate group and to form the diphosphate ADP, making $\Delta G \approx -70 \, kJ \, mol^{-1}$ available

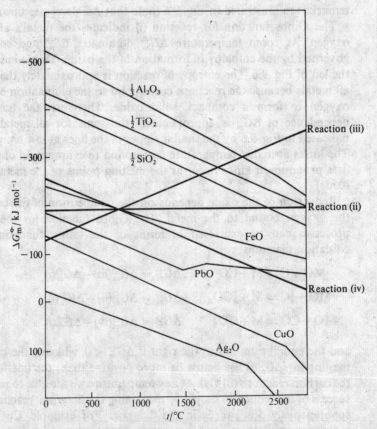

Fig. 9.6. The standard Gibbs functions for the oxidation reactions involved in the discussion of metal extraction. This is called an *Ellingham diagram*.

for driving other reactions. In a typical cell cycle, ADP is converted to ATP and then ATP provides the energy at the demand of some activity. In order to assess this thermodynamically we have to know the Gibbs function, enthalpy, and entropy of the hydrolysis reaction

$$ATP^{4-} + H_2O \rightleftharpoons ADP^{3-} + HPO_3^{2-} + H^+.$$

The standard state of this reaction corresponds to unit activity of H^+ (approximately $1\,mol\,dm^{-3}$ and $pH = 0$), but this is not appropriate to most biological conditions. Therefore, the standard thermodynamic functions of biological reactions are normally defined to relate to systems in which $pH = 7.0$ (a hydrogen ion activity of 10^{-7}). We shall adopt this convention in this section, and label the standard functions $G^{\oplus}$, $H^{\oplus}$, and $S^{\oplus}$.

At $37\,°C$ ($310\,K$, blood heat) the standard functions for ATP hydrolysis have the following values: $\Delta G_m^{\oplus} = -30\,kJ\,mol^{-1}$, $\Delta H_m^{\oplus} = -20\,kJ\,mol^{-1}$, and $\Delta S_m^{\oplus} = +34\,J\,K^{-1}\,mol^{-1}$. In a solution in which $pH = 7.0$ the ATP exists as ATP^{4-}, and a large part of the value of $\Delta G_m^{\oplus}$ (the sign of which indicates that the hydrolysis is thermodynamically favourable) can be ascribed to the decrease in electrostatic repulsions that occurs when it is hydrolysed. The entropy of the reaction is also favourable because of the release of a phosphate group, and because $\Delta S_m^{\oplus}$ is large the hydrolysis of ATP is sensitive to temperature changes.

ATP has a strong thermodynamic tendency to hydrolyse to ADP, and for this reason the ADP—PO_3 bond has been called a *high-energy phosphate bond*. This name is intended to signify a high tendency to break apart, and so the name is not consistent with the normal meaning of strong bonds. Nor, even, is it of very 'high energy' in the biological sense. The function of ATP depends upon it being intermediate in activity. Thus it acts as a phosphate donor to a number of acceptors (e.g. glucose), but the ADP can be recharged to ATP by some even more powerful phosphate donors that appear in the respiration cycle.

The efficiency of some biological processes can be assessed in the light of the value of $\Delta G_m^{\oplus}$ given above. The energy source of anaerobic cells is the glycolysis reaction (p. 113), in which a glucose molecule is changed

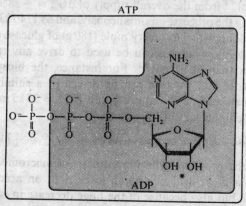

Fig. 9.7. The molecules adenosine triphosphate (ATP) and adenosine diphosphate (ADP).

into ethanol, butanol, acetone, or lactic acid. The glycolysis leading to lactic acid is moderately well understood, and overall the reaction is

$$= 2CH_3CH(OH)COOH,$$
$$\Delta G_m^{\ominus} = -218 \text{ kJ mol}^{-1}.$$

If the glucose were decomposed directly into the two lactic acid molecules, the reaction enthalpy $(-120 \text{ kJ mol}^{-1})$ would appear as heat and the cell would be unable to pursue any of its functions. The glycolysis reaction is coupled to a reaction in which two ADP molecules are converted into ATP. This process occurs in a complex sequence of reactions involving about 11 enzymes, but overall the net effect is

$$\text{glucose} + 2HPO_4^{2-} + 2ADP^{3-} = 2 \text{ lactate}^- + 2ATP^{4-} + 2H_2O.$$

The overall standard Gibbs function for this reaction is $-218 \text{ kJ mol}^{-1} - 2(-30 \text{ kJ mol}^{-1})$, or -158 kJ mol^{-1}. In this way, part of the standard Gibbs function of the glucose has been preserved for application to the other cell processes.

Aerobic respiration is a much more efficient process for providing power for cells. The standard Gibbs function for the combustion of glucose is $\Delta G_m^{\ominus} = -2880 \text{ kJ mol}^{-1}$, and so terminating the consumption of fuel at the stage of lactic acid is a poor use of resources. In aerobic respiration the oxidation of glucose is carried to completion, and the extremely complex system of reactions is adapted to preserving as much of the energy of the molecule as possible. The respiration cycle has the following overall result:

$$\text{glucose} + 6O_2 + 38P_i + 38ADP = 38ATP + 6CO_2 + 44H_2O$$

(P_i is an abbreviation for a phosphate group) and the important feature is the generation of 38 molecules of ATP in the process. Each ATP extracts $\Delta G_m^{\ominus} = -30 \text{ kJ mol}^{-1}$ from the overall supply of $\Delta G_m^{\ominus} = -2880 \text{ kJ mol}^{-1}$ available for driving the processes spontaneously, and so 1140 kJ are stored and available for doing work for every mole (180 g) of glucose consumed.

Every ATP molecule available can be used to drive any reaction for which $\Delta G_m^{\ominus}$ is less than 30 kJ mol^{-1}. For instance, the biosynthesis of sucrose from glucose and fructose can be driven (if a suitable enzyme system is available) because it requires only $\Delta G_m^{\ominus} = 23 \text{ kJ mol}^{-1}$. This indicates that the equilibrium of the reaction

$$\text{glucose} + \text{fructose} + \text{ATP} = \text{sucrose} + \text{ADP} \qquad \Delta G_m^{\ominus} = -7 \text{ kJ mol}^{-1}$$

lies in favour of the products. The biosynthesis of macromolecules like proteins requires a great deal of driving force both on account of the enthalpy required, but also because of the huge decrease in entropy that occurs when so many small molecules are assembled in a precisely determined sequence. For instance, the formation of a peptide link in a protein corresponds to an increase in the standard Gibbs function of about $17 \, \text{kJ mol}^{-1}$, but the biosynthesis occurs indirectly, and the sequence of reactions employed is equivalent to the consumption of three ATP molecules for each link. In a moderately small protein like myoglobin, where there are about 150 peptide linkages, the construction alone requires the consumption of 450 molecules of ATP, or about 12 mol of glucose for every mole of protein formed.

Biological activity: thermodynamics of respiration. Myoglobin and haemoglobin (abbreviated to Mb and Hb respectively) both play a special role in respiration and can be used to illustrate some thermodynamic points.

Myoglobin is a protein found in muscle, and it appears in abundance in the flesh of diving animals (e.g. the whale). The amount of oxygen it can absorb is related to the partial pressure of the oxygen. The *oxygen saturation curve* is given by

$$s = n(\text{MbO}_2)/\{n(\text{Mb}) + n(\text{MbO}_2)\}$$

where $n(\text{Mb})$ is the amount of unoxygenated Mb and $n(\text{MbO}_2)$ the amount of oxygen-bearing MbO_2. When the protein is fully saturated $s = 1$. The experimental curve is shown in Fig. 9.8a.

Before we discuss it we can give a simple thermodynamic argument to account for its form. The equilibrium is

$$\text{Mb} + \text{O}_2 = \text{MbO}_2$$

and so the equilibrium constant is given by

$$K = \{n(\text{MbO}_2)/n(\text{Mb})(p(\text{O}_2)/p^{\ominus})\}_e = \exp(-\Delta G_m^{\ominus}/RT)$$

where $\Delta G_m^{\ominus} = \mu^{\ominus}(\text{MbO}_2) - \mu^{\ominus}(\text{Mb}) - \mu^{\ominus}(\text{O}_2)$. The oxygen saturation curve can be written

$$s = \frac{n(\text{MbO}_2)/n(\text{Mb})}{1 + n(\text{MbO}_2)/n(\text{Mb})} = \frac{p(\text{O}_2)}{p^* + p(\text{O}_2)}$$

where $p^*/p^{\ominus} = \exp(\Delta G_m^{\ominus}/RT)$. This curve is proportional to $p(\text{O}_2)$ at low partial pressures ($p(\text{O}_2) \ll p^*$) but approaches a constant ($s = 1$) when the pressure is high. This is the experimental behaviour, and so oxygen uptake by myoglobin appears to be a simple equilibrium process.

The experimental value of p^* is about 5 mmHg. The partial pressure of oxygen in the lung is about 20–40 mmHg (3 to 6 kPa) and so myoglobin is virtually fully saturated with oxygen under normal conditions. Only when the supply of oxygen is curtailed, when the supply itself is cut off, or when metabolic processes are so rapid that they reduce the local oxygen density in the body fluids, does the myoglobin unload its oxygen. Myoglobin is a major reserve supply, not a primary supply, and that accounts for its high abundance in diving species.

The oxygen saturation curve for haemoglobin is shown in Fig. 9.8b. It

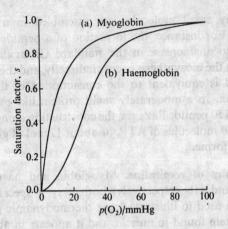

Fig. 9.8. Oxygen saturation curves for (a) myoglobin and (b) haemoglobin.

differs in two major respects from the one for myoglobin. In the first place it is sigmoidal (S-shaped). This is good design because the environment of the blood in resting muscle is equivalent to a local partial pressure of oxygen of bout 40 mmHg: at this pressure the haemoglobin is still at $s \approx 0.75$ and so has lost only a little of the oxygen it picked up by equilibration in the lungs. Furthermore this pressure corresponds to the uppermost point on the steepest part of the curve, and so as soon as the muscle needs to be activated the oxygen can be supplied to a metabolic process with great efficiency.

The thermodynamic calculation of the saturation curve in the present case is more difficult than before. In the first place each Hb molecule consists of four folded protein chains each bearing a Fe^{2+} ion to which the O_2 molecule will attach. Therefore each Hb may exist in one of five states: Hb, $Hb(O_2)_n$, $n = 1, 2, 3, 4$. Simply proceeding in the fashion of myoglobin does not lead to a sigmoidal curve. In order to get a curve of the form shown in Fig. 9.8b it is necessary to suppose that the oxygen uptake is a co-operative effect, and that the presence of one or more O_2 molecules already bound to the Hb assists the attachment of more. In this way the low-pressure curves, where O_2 molecules are attaching to Hb molecules, have a low value of s; but as the pressure rises, the ability of the HbO_2 molecules to take up more oxygen increases, and there is a rapid shift of equilibrium in favour of the saturated species.

Further reading

Elementary chemical thermodynamics. B. Mahan; Benjamin, New York, 1963

Chemical thermodynamics. I. M. Klotz and R. M. Rosenberg; Benjamin, New York, 1972.

The principles of chemical equilibrium. K. G. Denbigh; Cambridge University Press, 1971.

Thermodynamics. G. N. Lewis and M. Randall, revised by K. S. Pitzer and L. Brewer; McGraw-Hill, New York, 1961.

Energy changes in biochemical reactions. I. Klotz; Academic Press, New York, 1967.

Bioenergetics. A. L. Lehninger; Benjamin, New York, 1965.

Inorganic chemistry. C. S. G. Phillips and R. J. P. Williams; Clarendon Press, Oxford, 1965.

Some thermodynamic aspects of inorganic chemistry. D. A. Johnson; Cambridge University Press, 1968.

Elementary thermodynamics for geologists. B. J. Wood and D. G. Fraser; Oxford University Press, 1976.

Thermodynamic data. See references at the end of Chapter 4, p. 123.

Problems

9.1. Use the thermodynamic data in Table 9.1 to decide which of the following reactions are spontaneous at 25 °C in the direction written, the species being present in their standard states:

(a) $HCl(g) + NH_3 \rightarrow NH_4Cl$
(b) $2Al_2O_3 + 3Si \rightarrow 3SiO_2 + 4Al$
(c) $Fe + H_2S(g) \rightarrow FeS + H_2$
(d) $FeS_2 + 2H_2 \rightarrow Fe + 2H_2S(g)$
(e) $2H_2O_2 + H_2S(aq) \rightarrow H_2SO_4(aq) + 2H_2$.

9.2. Which of the reactions in Problem 9.1 are favoured by a rise in temperature at constant pressure?

9.3. What is the enthalpy of a reaction for which the equilibrium constant (a) is doubled, (b) is halved when the temperature is increased by 10 K from 25 °C.

9.4. Suppose you make an error of 10 per cent in the determination of an equilibrium constant at 25 °C. What error does that imply for the value of $\Delta G_m^{\ominus}$? Conversely, suppose you make an error of 10 per cent in the determination of $\Delta G_m^{\ominus}$: what error does that imply for the equilibrium constant at that temperature?

9.5. On the basis that $\Delta G_m^{\ominus}(298 \, K)$ for the reaction $\frac{1}{2}N_2 + \frac{3}{2}H_2 \rightarrow NH_3$ is $-16.5 \, kJ \, mol^{-1}$, find the equilibrium constant at 25 °C for (a) the reaction as written, (b) the reaction $N_2 + 3H_2 \rightarrow 2NH_3$, (c) the reaction $NH_3 \rightarrow \frac{1}{2}N_2 + \frac{3}{2}H_2$.

9.6. The standard Gibbs function for the synthesis of ammonia was given in the last Problem. What is the value of ΔG_m when the pressures of the (perfect gas) components are $p(N_2) = 3 \, atm$, $p(H_2) = 1 \, atm$, $p(NH_3) = 4 \, atm$?

9.7. Calculate the equilibrium constant at 298 K for the reaction $CO(g) + H_2(g) \rightarrow H_2CO(g)$ given that $\Delta G_m^{\ominus}$ for the formation of a liquid product is $28.95 \, kJ \, mol^{-1}$ at 25 °C and that the vapour pressure of formaldehyde at 298 K is 1500 mmHg.

9.8. When ammonium chloride is heated the vapour pressure at 427 °C is 4560 mmHg. At 459 °C the vapour pressure has risen to 8360 mmHg. What is (a) the equilibrium constant for the dissociation, (b) the Gibbs function, (c) the enthalpy, (d) the entropy of dissociation at 427 °C? Assume the vapour behaves as a perfect gas.

9.9. In the *Dumas vapour density method* a liquid is allowed to evaporate in a glass bulb heated to some selected temperature; then the bulb is sealed and cooled, the amount of sample being determined by weighing or by titration. In this way a known volume is filled with vapour at a known temperature and pressure, and by a known amount of material. In one such experiment, acetic acid was evaporated at 437 K and the amount of acid in the bulb of volume $21.45 \, cm^3$ was 0.0519 g when the external pressure was 764.3 mmHg. In a second experiment in the same vessel heated to 471 K the amount present was found to be 0.038 g when the external pressure was the same. What is the equilibrium constant for the dimerization of the acid in the vapour? What proportion of the vapour is dimeric at each temperature? What is the enthalpy of dimerization?

9.10. The techniques described in the chapter can be applied to all kinds of reactions, including ionic reactions in solution. The only difficulty is that ionic solutions are strongly non-ideal, and thermodynamic equilibrium constants may have to be strongly corrected in order to get concentration equilibrium constants. We shall see how to do this in Chapter 11, but even now we can arrive at solubility products and estimates of solubilities of ionic salts. For instance, use the information in Table 9.1 and 12.2 to find the *solubility product* $K_{sp} = a(Ag^+)a(Cl^-)$ and the solubility (as a molality) of silver chloride in water at 25 °C. Ignore all effects of non-ideality in the latter part.

9.11. The *ionization constant* of water, $K_w = a(H^+)a(OH^-)$ is another special type of equilibrium constant which plays an important role in governing the behaviour of acids and bases. At 20 °C its value is 0.67×10^{-14}, at 25 °C its value is 1.00×10^{-14}, and at 30 °C its value is 1.45×10^{-14}. This is enough information to be able to deduce the standard enthalpy of the ionization of water, the reaction $H_2O(l) \rightarrow H^+(aq) + OH^-(aq)$. Deduce its value at 25 °C. What would the error bounds be if K_w were known to within ± 0.01?

9.12. Hydrogen and carbon monoxide have been investigated with a view to using them in high temperature fuel cells, and so their solubility in various molten nitrates is of some technological interest. The solubility in a $NaNO_3/KNO_3$ mixture was examined (E. Desimoni, F. Paniccia, and P. G. Zambonin, *J. chem. Soc. Faraday Trans.* 1, 2014 (1973)) with the following results:

For H_2: $\lg s(H_2) = -5.39 - 768\,(T/K)^{-1}$;

For CO: $\lg s(CO) = -5.98 - 980\,(T/K)^{-1}$.

$s(A)$ is the solubility expressed in units of $mol\,cm^{-3}\,bar^{-1}$. Find the standard molar enthalpies of solution of the two gases at 570 K.

9.13. Everyone knows that when blue crystals of copper sulphate are heated they crumble and lose their colour as a result of the dehydration reaction $CuSO_4.5H_2O \rightarrow CuSO_4 + 5H_2O$. Given the following data for 298 K, discuss the process from the viewpoint of thermodynamics, and predict the temperature at which the vapour pressure of water reaches (a) 10 mmHg, (b) 1 atm.

	$\Delta H_f^{\ominus}/kJ\,mol^{-1}$	$\Delta G_f^{\ominus}/kJ\,mol^{-1}$
$CuSO_4$	-769.86	-661.9
$CuSO_4.5H_2O$	-2277.98	-1879.9
$H_2O(g)$	-241.83	-228.59

9.14. Calculations on the positions of chemical equilibrium get much more involved when the components are mixed in arbitrary proportions initially, and the equilibrium composition is required. A good way of working through problems of this kind is to draw up a table with columns headed by the names of all the species involved, and then successive rows labelled (1) the initial amounts, (2) the change in one of the components that is known to occur; sometimes this is in fact the unknown in the problem, and so it is just left as x, (3) the changes in the amounts of all components that this implies, basing the argument on the stoichiometry of the reaction, (4) the final composition (perhaps still in terms of the unknown x), (5) the mole fractions. Then the equilibrium constant can be expressed in terms of this last line, and then solved for the unknown x. See how this works in practice with the following problems. First consider a flask which was filled with 0.30 mol hydrogen, 0.4 mol iodine vapour, and 0.2 mol hydrogen iodide, the total pressure being 1 atm. What is the composition when the mixture has reached thermal equilibrium at 25 °C, the equilibrium constant being $K = 870$.

9.15. You will have noticed that the set of rules given in the last Problem is virtually in the form of a computer algorithm. If you have access to a small computer, write a program that will predict the final composition of the reaction mixture for arbitrary initial compositions. You may find it instructive to do the same for some of the other examples that follow.

9.16. Now consider the general reaction $2A + B \rightarrow 3C + 2D$, all components gases. It was found that when 1 mol of A, 2 mol of B, and 1 mol of D were mixed and the whole allowed to come to equilibrium, the mixture contained 0.9 mol of C. Is this enough information to find the equilibrium constant?

9.17. The equilibrium constant for the reaction $N_2 + 3H_2 \rightarrow 2NH_3$ was treated in Problem 9.5. Express K in terms of an extent of the reaction α defined so that $\alpha = 0$ corresponds to the initial state and $\alpha = 1$ to pure $NH_3(g)$, given that the nitrogen and hydrogen were initially present in stoichiometric proportions when the catalyst was added and behave as perfect gases.

9.18. We have seen that the thermodynamic equilibrium constant is independent of pressure, yet the extent of reaction at equilibrium does depend on pressure (recall the discussion on p. 271). Find α_e at a pressure p. In fact, there are two ways of solving this type of problem. One is to attempt to solve the equilibrium constant expression for α_e in terms of p; but in the present case that means solving a quartic equation, which is tiresome. The other method involves noticing that if $K \propto 1/p^2$, then the constant of proportionality must be proportional to p^2 for K to be independent of pressure. Go on from there.

9.19. Plot α_e for the ammonia synthesis as a function of total pressure from 0.1 atm to 1000 atm; use a logarithmic scale. What is the composition of the equilibrium mixture when the pressure is 500 atm?

9.20. Triethylamine (TEA) and 2,4-dinitrophenol (DNP) form a molecular complex in chlorobenzene, the binding depending on proton transfer and electrostatic attraction. The equilibrium constant for the formation of the complex has been measured at various temperatures by monitoring the optical absorption of the solution (K. J. Ivin, J. J. McGarvey, E. L. Simmons, and R. Small, *J. chem. Soc. Faraday Trans.* 1, 1016 (1973)) with the following results:

$t/°C$	17.5	25.2	30.0	35.5	39.5	45.0
K	$29\,670 \pm 1230$	$14\,450 \pm 560$	9270 ± 70	5870 ± 120	3580 ± 30	2670 ± 70

where $K = c(\text{complex})/c(\text{TEA})c(\text{DNP})$ and $c(A) \equiv c(A)/\text{mol dm}^{-3}$. Calculate the standard enthalpy and entropy of formation of the complex from its components at 20 °C.

9.21. At high temperatures gaseous iodine dissociates and the vapour phase contains I_2 molecules and I atoms. The extent of dissociation can be measured by monitoring the pressure change, and two sets of results are as follows:

T/K	973	1073	1173
p/mmHg	0.06244	0.07500	0.09181
$10^4 n(\text{I})/\text{mol } I_2$	2.4709	2.4555	2.4366

Find the equilibrium constant for the reaction and the molar enthalpy of dissociation as the mean temperature. Use $V = 342.68\,\text{cm}^3$.

9.22. Nitrogen(IV) oxide NO_2 is in equilibrium with its dimer at room temperature. Since NO_2 is a brown gas and N_2O_4 is colourless, the amounts present at equilibrium may be studied either spectroscopically or by monitoring the pressure. The following data were obtained at two temperatures:

	$p(NO_2)/\text{mmHg}$	$p(N_2O_4)/\text{mmHg}$
298 K	46	23
305 K	68	30

Calculate the equilibrium constant for the reaction, the standard molar Gibbs function, enthalpy, and entropy of dimerization at 298 K.

9.23. When light passes through a cell of length l containing an absorbing gas at a pressure p the amount of absorption is proportional to pl. In the case of a cell in which there is the equilibrium $2NO_2 \rightleftharpoons N_2O_4$, with NO_2 the absorbing species, show that when two cells of lengths l_1 and l_2 are used, and the pressures needed to obtain equal amounts of absorption are p_1 and p_2 respectively, that the equilibrium constant is given by

$$K = (r^2 p_1 - p_2)^2 / r(r-1)(p_2 - r p_1), \qquad r = l_1/l_2.$$

The following data were obtained (R. J. Nordstrum and W. H. Chan, *J. phys. Chem.* **80**, 847 (1976)):

Absorbance	p_1/mmHg	p_2/mmHg
0.05	1.00	5.47
0.10	2.10	12.00
0.15	3.15	18.65

$l_1 = 395\,\text{mm}, l_2 = 75\,\text{mm}$

Find the equilibrium constant for the reaction.

9.24. At normal temperatures BF_3 acts as a catalyst for the equilibrium between acetaldehyde (ethanal, CH_3CHO) and paraldehyde (a trimer of CH_3CHO). The partial pressures of the components in the reaction $3CH_3CHO(g) \rightleftharpoons (CH_3CHO)_3(g)$ are too low for accurate determination of the equilibrium constant by direct measurement, but this can be overcome by ensuring that liquid forms of the two components are always present (W. K. Busfield, R. M. Lee, and D. Merigold, *J. chem. Soc. Faraday Trans.* 1, 936 (1973)). On the basis of perfect behaviour, show that the equilibrium constant for the trimerization can be written $K = p_P^0 (p_A^0 - p)(p_A^0 - p_P^0)^2 / p_A^{03} (p - p_P^0)^3$, where p_A^0 and p_P^0 are the saturated vapour pressures of acetaldehyde and paraldehyde, and p is the total pressure. Use the enthalpies of vaporization $\Delta H_{\text{vap,m}} \approx 25.6\,\text{kJ mol}^{-1}$ (acetaldehyde) and $41.5\,\text{kJ mol}^{-1}$ (paraldehyde) and the following data to find the enthalpy of trimerization in the gas phase.

$t/^\circ C$	20.0	22.0	26.0	28.0	30.0	32.0	34.0	36.0	38.0	40.0
$p/\text{kN m}^{-2}$	23.9	27.3	36.5	42.6	49.9	56.9	65.1	74.3	85.0	96.2

9.25. The enthalpy and entropy of trimerization of acetaldehyde in the gas phase are $-133.5\,\text{kJ mol}^{-1}$ and $-457.5\,\text{J K}^{-1}\text{mol}^{-1}$ respectively (see the last Problem). Given the enthalpies of vaporization quoted there, and their boiling points of 294 K(acetaldehyde), 398 K(paraldehyde), find the values of the enthalpy and entropy of the trimerization in the liquid phase. What is the ratio of the equilibrium constants for the trimerization in the gas and liquid phases at 25 °C. Use $\ln(p/\text{kN m}^{-2}) = \text{const.} - \Delta H_{\text{vap,m}}/RT$ with const. $(A) = 15.1$ and const. $(B) = 17.2$.

9.26. In Chapter 6 we saw that if the attractive part of the van der Waals interaction dominates, the fugacity of a gas is related to the pressure by $f(p) = p\exp(-ap/R^2T^2)$. First show that $(\partial \ln K_p/\partial p)_T = -(\partial \ln K_\gamma/\partial p)_T$. Then find an expression for the pressure dependence of K_p, the first of these two coefficients, in the case of the reaction $I_2 + H_2 \rightarrow 2HI$, and estimate its value at 25 °C and 500 atm. By how much will K_p change when the pressure is increased by 50 atm?

9.27. We have also seen that ΔG tells us the maximum amount of non-pV work that may be available from a system. In an *electrochemical cell* this work may be tapped electrically, and we shall explore this in Chapter 12. At this point, though, we have enough information to be able to assess how much electrical work can be obtained from cells that depend on any type of reaction. Suppose that a cell is designed with copper and zinc electrodes and is driven by the reaction $Zn + CuSO_4(aq) \rightarrow ZnSO_4(aq) + Cu$. What is the maximum amount of electrical work available from the consumption of 1 mol of zinc at 25 °C when all the materials are in their standard states? Use the data in Table 12.2 for the ions. What is the maximum amount of heat? Why do the two figures differ?

9.28. Suppose an electrical cell could be designed that converted coal directly into electrical work; what is the amount of work available from the consumption of 100 kg of coal (regarded as graphite)? What is the maximum amount of work available if the coal is burnt and the thermal output is converted into work in a plant operating between 150 °C (hot source) and 30 °C (cold sink)?

9.29. The next few Problems take a closer look at the way of dealing with the temperature dependence of the Gibbs function of a reaction, and hence with the temperature dependence of the equilibrium constants of reactions. First we look at a way of getting explicit expressions for G at different temperatures. The entropy and enthalpy at a temperature T_2 can be related to their values at T_1 if the temperature dependence of the heat capacity is known. Find $\Delta G(T_2)$ for a reaction in terms of $\Delta G(T_1)$ and the parameters a, b, c in the expression $C_{p,m} = a + bT + c/T^2$.

9.30. Use the heat capacity data in Table 4.3 to find $\Delta G_m^\ominus(372\,K)$ for the synthesis of water, given $\Delta G_m^\ominus(298\,K) = -237.2\,kJ\,mol^{-1}$ and $\Delta H_m^\ominus(298\,K) = -285.8\,kJ\,mol^{-1}$.

9.31. We saw that the Giauque function $\Phi_0(T) = [G_m^\ominus(T) - H_m^\ominus(0)]/T$ or $\Phi_{298}(T) = [G_m^\ominus(T) - H_m^\ominus(298\,K)]/T$ is useful for discussing equilibria at different temperatures because it varies only slowly with temperature. One method of finding its value was described in Problem 5.35. Here we shall look at ways of using it. As a first step, show how to relate the entropy of a substance, $S^\ominus(T)$, to $\Phi_0(T)$ and the enthalpy difference $H_m^\ominus(T) - H_m^\ominus(0)$.

9.32. The Giauque functions Φ_0 and Φ_{298} are tabulated in Table 9.1. Use the information given there to calculate $\Delta G_m^\ominus$ for the following reactions at the stated temperatures.

(a) $N_2 + 3H_2 \rightarrow 2NH_3$ 1000 K

(b) $H_2O + CO_2 \rightarrow H_2 + CO$ 500 K, 2000 K.

9.33. Find the equilibrium constants for the reactions listed in the preceding Problem.

9.34. Finally, link the two methods of arriving at the temperature dependence of equilibrium constants. Express $\Phi_{298}(T)$ in terms of its value at 298 K and the parameters a, b, c in the heat capacity expression.

10 Equilibria: the general situation

Learning objectives

After careful study of this chapter you should be able to

(1) State and derive the *phase rule* (eqn (10.1.1)).

(2) Define the terms *phase* (p. 288), *component* (p. 289), and *degree of freedom* (p. 288).

(3) Apply the phase rule to one-component systems (p. 290).

(4) Explain the term *triple point* (p. 291).

(5) Account for the form of *cooling curves* and explain how they are used to construct phase diagrams (p. 291).

(6) Use the phase rule to interpret *liquid–vapour composition diagrams* of two-component systems (p. 291).

(7) Construct and interpret *liquid–liquid phase diagrams* (p. 292) and explain the term *consolute temperature*.

(8) Describe the distillation of *partially miscible liquids* in terms of the phase rule and the lever rule (p. 294).

(9) Construct and interpret *liquid–solid phase diagrams* (p. 296) and define the term *eutectic* and *eutectic halt* (p. 298).

(10) Interpret phase diagrams in which the components take part in a reaction (p. 299), and define the terms *phase reaction, peritectic reaction,* and *incongruent melting* (p. 301).

(11) Interpret the phase diagram for *steel* (p. 301).

(12) Describe the principles and applications of *zone-refining* and *zone-levelling* (p. 302).

(13) Construct *three-component phase diagrams* using triangular co-ordinates (p. 304).

(14) Interpret triangular-coordinate phase diagrams for three partially miscible liquids and for solutions of two salts (p. 306).

(15) Explain the term *common-ion effect* (p. 308).

ntroduction

In this chapter we establish a systematic way of discussing the changes mixtures undergo when they are heated and cooled, and when their compositions are changed. The scheme enables one to know at a glance whether two or three substances are mutually miscible, whether a particular equilibrium can exist over a range of conditions, and whether the system has to be tuned to a definite pressure, temperature, and composition before an equilibrium can be established. We do this by developing the phase diagrams first encountered in Chapter 8, and begin by deriving the *phase rule*. This simple piece of analysis, which is due to Gibbs†, is probably the most elegant result of the whole of chemical thermodynamics.

10.1 The phase rule

We have already met special cases of the phase rule. On p. 191, we saw that the number of intensive degrees of freedom, or variance F, of a one-component system is given by $F = 3 - P$, where P is the number of phases in equilibrium. If only one phase is present in the system the result $F = 2$ means that the temperature and pressure can be varied independently. If two phases are in equilibrium so that $F = 1$, then only one intensive variable is independent, and the vapour pressure is determined by the temperature.

The derivation of the one-component phase rule (p. 191) was based on the argument that the chemical potentials of the phases in equilibrium must be equal, and so expressions like $\mu(l; p, T) = \mu(g; p, T)$ amount to equations relating the pressure and temperature. When one such equation exists p is fixed by the temperature (or vice versa), and so the number of degrees of freedom is reduced from 2 to 1. When there are two equations (when three phases are in equilibrium) the pair of simultaneous equations can be satisfied only for specific values of temperature and pressure, and the variance drops to zero. We shall now apply the same kind of reasoning to a system of C components and P phases.

Suppose first the constraints are ignored; how many intensive variables are available? We count two for the pressure and temperature. (In less conventional chemistry this number would increase to 3 or more if we were interested in the effect of a magnetic or a gravitational field.) The composition of every phase can be specified if the amount of each component is stated. But it is not necessary to specify the overall bulk of the phase; all we need specify is the mole fraction of every component.

† J. Willard Gibbs spent most of his working life at Yale, and may justly be regarded as, among other things, the originator of chemical thermodynamics. He reflected for years before publishing his conclusions, and then did so in precisely expressed papers in an obscure journal (*The Transactions of the Connecticut Academy of Arts and Sciences*). He needed interpreters before the power of his work was recognized, and before it was applied to industrial processes. It may be judged from this that he was of a retiring disposition and not of a practical turn; but that is certainly not so, for it is recorded that he prescribed and made his own spectacles and held a number of patents.

Since $x_1 + x_2 \ldots + x_C = 1$, one of the mole fractions is fixed if all the others have been specified, and so specifying $C-1$ mole fractions specifies the composition of a phase. There are P phases, and so the total number of composition variables available is $P(C-1)$.

At this point the total number of variables at our disposal is $P(C-1)+2$. The presence of equilibria between phases reduces the freedom to vary these arbitrarily, and these constraints must now be counted.

Let the chemical potential of one of the components be μ_J. At equilibrium the chemical potential of a species must be the same in every phase. (The argument for this was given on p. 182.) If there are P phases it follows that at equilibrium

$$\mu_J(\text{phase } 1) = \mu_J(\text{phase } 2) = \ldots = \mu_J(\text{phase } P).$$

That is, $P-1$ equations have to be satisfied by the component J. There are C components, and so the total number of equations that have to be satisfied is $C(P-1)$.

Every equation reduces the freedom to vary one of the $P(C-1)+2$ variables. It follows that the total number of degrees of freedom is

$$F = P(C-1) + 2 - C(P-1),$$

or

(10.1.1) $$F = C - P + 2.$$

This is the celebrated *phase rule*.

When the phase rule is applied, special attention has to be paid to what is meant by the terms 'component' and 'phase'.

What, if anything, is meant by 'phase'? The formal definition of a phase is *a state of matter that is 'uniform throughout, not only in chemical composition but also in physical state'*. (The words are Gibbs's.)

This definition is consistent with the common-sense notion of the term. It means that a gas or a gaseous mixture is a single phase, a crystal form a single phase, and a liquid mixture of two totally miscible liquids a single phase. Ice is a single phase even though it might be chipped into small fragments. A slurry of ice and water is a 2-phase system even though it is difficult to map the boundary between the two phases.

A dispersion of one metal in another is a 2-phase system if the metals are immiscible, but it is a 1-phase system if they are miscible. This example shows that it is not always easy to decide whether a system consists of one phase or two. A *solution* of a solid A in solid B is uniform on a molecular scale, which means that molecules of A are surrounded by molecules of both A and B, and any sample cut from it, however small, is representative of the composition of the whole. A *dispersion* is uniform only on a macroscopic scale, for close inspection shows that it consists of grains or droplets of one component in a matrix of the other. A lump cut on a molecular scale might come entirely from one of the minute grains of pure A, and so it would not be representative of the whole, Fig. 10.1.

The situation is worse when very fine liquid dispersions are involved because vapour pressure depends on the droplet size. What now constitutes a phase? What also constitutes a phase in a gravitational field? Fortunately we need not analyse these tricky problems, but we must be aware that they exist in case, by some mischance, it is necessary to apply the phase rule in one of these grey areas.

What is meant by 'component'? The number of components is *the minimum number of independent species necessary to define the composition of all the phases present in the system.*

The definition is easy to apply when the species do not react, for then we simply count their number. For instance, water is a 1-component system, and a mixture of ethanol and water is a 2-component system. If the species do react, and are at equilibrium, we have to take into account the significance of the expression 'all the phases' in the definition above. Thus, in the case of NH_4Cl in equilibrium with its vapour (NH_3 and HCl), since both phases have the formal composition 'NH_4Cl' it is a 1-component system. In contrast, $CaCO_3$ in equilibrium with its vapour (CO_2) is a 2-component system because we have to specify 'CO_2' for the gas phase and '$CaCO_3$' for the solid phase (since the three species are connected by an equilibrium, the concentration of CaO is not independent). There are two components whether we start from $CaCO_3$, or equal amounts of CaO and CO_2, or arbitrary amounts of them.

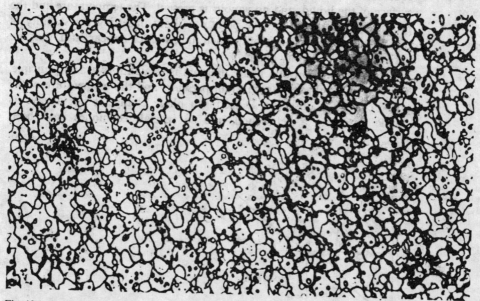

Fig. 10.1. A photomicrograph of a superplastic nickel–chromium iron alloy (H. W. Hayden, R. C. Gibson, and J. H. Brophy, *Scientific American*, March 1969).

10.2 One-component systems

One-component systems were dealt with in detail in Chapter 7, and here we summarize the information in the light of the phase rule.

Since $C = 1$ the variance of a one-component system is $3 - P$. When only one phase is present $F = 2$ and p and T can be varied independently. A single phase is therefore represented by an area on a p, T-graph.

When *two phases* are in equilibrium $F = 1$, which means that the pressure is not at our disposal if we have decided on a temperature. Alternatively, we can select a pressure, but having done so the liquid and vapour (or solid and liquid) come into equilibrium at a single definite temperature. Therefore, the boiling point (or any other transition point) occurs at a definite and well-defined temperature at a given pressure.

When *three phases* are in equilibrium (solid, liquid, and vapour; or solid, solid, and liquid) $F = 0$, and so this special condition, the *triple point*, can be attained only at a definite temperature and pressure, which is an intrinsic property of the system and not at our disposal.

When *four phases* are in equilibrium . . . but four phases cannot be in equilibrium in a one-component system because it is not possible for F to be negative.

This has been quite abstract. Now turn to the phase diagram for water shown in Fig. 10.2. Consider what happens when a system at a is cooled at constant pressure (of 1 atm). The system remains entirely gaseous until the temperature reaches b, when liquid appears. The 2-phase region has been struck, and the two phases coexist at a definite temperature (for fixed pressure). If the temperature is taken below b to c, the system moves squarely into the 1-phase, liquid region. The piston now rests on the surface of the liquid, and no vapour is present. The temperature and the pressure can rove around the point c without upsetting the stability of that single phase, and it is only when lumps of ice appear, at d, that the variance drops to unity again.

The determination of a phase diagram is straightforward in principle. The system is subjected to constant pressure, and the temperature of the phase change is observed. Detecting a phase change is not always as simple as seeing a kettle boil, and so special techniques have been

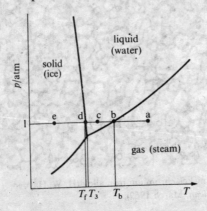

Fig. 10.2. The phase diagram for water.

developed. An important method is *thermal analysis*, which takes advantage of the effect of the enthalpy change during first-order transitions. In the method a sample is allowed to cool, and its temperature is monitored as a function of time. At a first-order phase transition heat is evolved, and so the rate of cooling is retarded. In fact, the phase rule shows that the cooling must stop altogether until the transition is complete. This is because at a fixed pressure the temperature can take only a single value while the two phases are both present. Therefore the cooling curve along the isobar cde in Fig. 10.2 has the shape shown in Fig. 10.3. The transition temperature is obvious, and can be used to mark the point d on Fig. 10.2. This method is particularly useful for solid–solid transitions where simple visual inspection of the sample is inadequate, and also in the study of binary systems.

10.3 Two-component systems

The first part of this section summarizes the information in Chapter 8 in terms of the phase rule.

If two components are present, $F = 4 - P$. For simplicity we can agree to keep the pressure constant (at 1 atm), and so one of the degrees of freedom has been discarded. Now $F' = 3 - P$ (F' denotes that one degree of freedom has been discarded; F'' that two have been discarded). The maximum number of degrees of freedom is available when $P = 1$, for then $F' = 2$. One of these is the temperature, and the other is the composition (the mole fraction of a species).

How to interpret liquid–vapour composition diagrams. A typical diagram for totally miscible liquids is reproduced as Fig. 10.4: this shows a low boiling azeotrope. The phase diagram is labelled '1-phase' and '2-phase'. The former region is bivariant ($F' = 2$, the pressure is taken to be constant), and the latter univariant ($F' = 1$). In order to see what these remarks signify in practice, consider what happens when a sample of composition a is heated.

Initially, at T_1, the entire sample is liquid, and as the two components are miscible, there is only one phase, therefore $F' = 2$. This phase remains

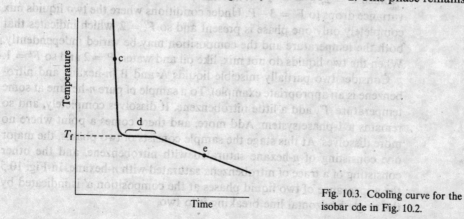

Fig. 10.3. Cooling curve for the isobar cde in Fig. 10.2.

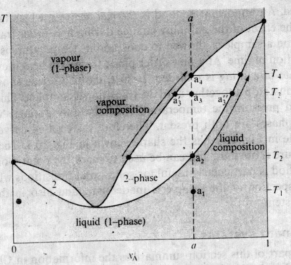

Fig. 10.4. Typical temperature-composition diagram for a two-component liquid mixture forming a low-boiling azeotrope.

stable if the temperature is changed by an arbitrary amount within this 1-phase region. When the temperature has been raised to T_2 the vapour phase exists in equilibrium with the liquid. Now $P = 2$ and so $F' = 1$. Although the temperature may be varied arbitrarily (between T_2 and T_4), the two phases are in equilibrium only if the system has a specific composition (e.g. at T_3 the vapour a'_3, the liquid a''_3) and the relative amounts in accord with the lever rule. When the temperature is raised to just above T_4, the whole sample is a gas; then $P = 1$ so $F' = 2$, and the gas can exist with arbitrary composition and temperature.

How to interpret liquid–liquid phase diagrams. We deal in this section with binary mixtures ($C = 2$) at temperatures and pressures where the vapour is absent. In particular we consider pairs of liquids that are only *partially miscible*, which means they do not mix in all proportions at all temperatures.

Since $C = 2$, $F = 4 - P$. When the pressure is restricted to 1 atm, the variance drops to $F' = 3 - P$. Under conditions where the two liquids mix completely only one phase is present and so $F' = 2$, which indicates that both the temperature and the composition may be varied independently. When the two liquids do not mix, like oil and water, $P = 2$ and so $F' = 1$.

Consider two partially miscible liquids A and B (*n*-hexane and nitrobenzene is an appropriate example). To a sample of pure *n*-hexane at some temperature T', add a little nitrobenzene. It dissolves completely, and so remains a 1-phase system. Add more, and there comes a point where no more dissolves. At this stage the sample consists of two phases, the major one consisting of *n*-hexane saturated with nitrobenzene, and the other consisting of a trace of nitrobenzene saturated with *n*-hexane. In Fig. 10.5 the appearance of two liquid phases at the composition a' is indicated by the single horizontal line breaking into two.

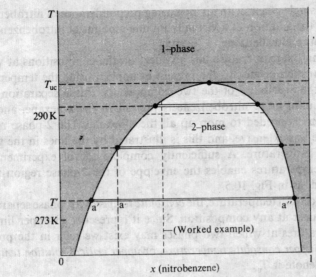

Fig. 10.5. Temperature–composition diagram for nitrobenzene and n-hexane.

Example (Objective 6). A mixture of 50 g n-hexane and 50 g nitrobenzene is prepared at 290 K. What is the composition of the two phases, and in what proportions do they occur? To what temperature must the mixture be raised in order to obtain a single phase?

● *Method.* The question can be discussed on the basis of Fig. 10.5 and the lever rule (p. 239). We need to convert the masses of the components to mole fractions. Relative molar masses are 86.2 for n-hexane and 123 for nitrobenzene.

● *Answer.* $n(n\text{-hexane}) = (50\ \text{g})/(86.2\ \text{g mol}^{-1}) = 0.58\ \text{mol}$

$n(\text{nitrobenzene}) = (50\ \text{g})/(123\ \text{g mol}^{-1}) = 0.41\ \text{mol}$

Therefore $x_{\text{Hex}} = 0.59$ and $x_{\text{NB}} = 0.41$.

The composition $x_{\text{NB}} = 0.41$ occurs in the 2-phase region when the temperature is 290 K, and the composition of the phases is given by the intersection of the tie-line with the curve. One phase has a composition $x_{\text{NB}} = 0.37$, the other has $x_{\text{NB}} = 0.83$. The amount of each phase is given by the lever rule. In the present case the ratio of the amounts is $0.03/0.20 = 0.15/1.0$. Heating the sample to 292 K will take it to the single phase region.

● *Comment.* Since the phase diagram has been constructed experimentally these conclusions are 'exact'. They would be modified if the system were subjected to a different pressure.

When more nitrobenzene is added n-hexane shifts out of the n-hexane-rich layer into the nitrobenzene-rich layer. This implies that the proportion of nitrobenzene-rich phase grows at the expense of the other. A point is reached where so much nitrobenzene is present that it can act as solvent for all the n-hexane present, and so the system reverts to one phase. This is indicated by the fusion of the two lines in Fig. 10.5 at the concentration a″. Thereafter more nitrobenzene simply dilutes the solution, and it remains a single phase, with an increasing preponderance of nitrobenzene, until after the addition of a virtually infinite amount of nitrobenzene it is virtually pure nitrobenzene.

Changing the temperature has an effect on the compositions at which phase separation occurs. In the present example, raising the temperature enhances the miscibility of the two components. Phase separation does not occur until more nitrobenzene is present in the *n*-hexane, and less nitrobenzene is needed to mop up all the *n*-hexane. The 2-phase region is therefore less extensive, and this is illustrated by the lines in the figure at higher temperature. A sufficiently complete set of experiments at different temperatures enables the envelope of the 2-phase region to be constructed, as in Fig. 10.5.

Above a certain temperature, the *consolute temperature*, phase separation does not occur at any composition. Since it represents an upper limit to the temperatures at which two phases may exist we refer in the present case to the *upper consolute temperature*, or *upper critical solution temperature*, and denote it T_{uc}.

Although it may seem quite natural that there should be upper consolute temperatures, where the more violent molecular motion overcomes the tendency of molecules of the same species to stick together in swarms and therefore to form two phases, some systems show a *lower consolute temperature* beneath which they mix in all proportions and above which they may form two layers. An example is the system water/triethylamine, Fig. 10.6a. This behaviour suggests that at low temperatures the two kinds of molecule form a weak complex so that solubility is enhanced. At higher temperatures the complexes are broken up and the two types of molecule cluster in swarms of their own kind.

Some systems show both lower and upper consolute temperatures. This is because after the 'complexes' have been disrupted, the thermal motion at higher temperatures homogenizes the mixture just as in the case of conventional partially miscible liquids. The most famous example of this is the system nicotine/water, which shows partial miscibility between 61 °C and 210 °C, Fig. 10.6b.

The distillation of partially miscible liquids. Boiling a 1-phase liquid mixture gives a vapour which generally has a different composition. Removing and then condensing this vapour might yield a mixture that falls in the 2-phase region. That is, a homogeneous solution might distil as a milky, 2-phase mixture. This section will show how to represent, and anticipate, such behaviour.

Consider a system forming a low-boiling azeotrope. The combination of azeotrope formation and partial miscibility is quite common, because both properties indicate a tendency for the two kinds of molecules to get out of each other's way.

The liquid–vapour phase diagram can be constructed by combining Fig. 10.4 with Fig. 10.5. This is easy when the azeotrope boils well above the consolute temperature, because then a diagram of the kind shown in

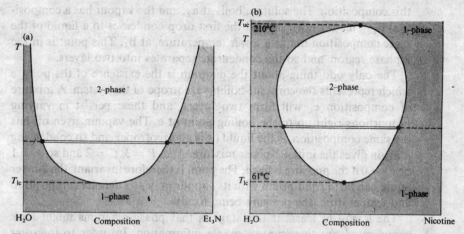

Fig. 10.6. Two-component temperature-composition phase diagrams. (a) Water and triethylamine, showing a lower consolute temperature, (b) Water and nicotine showing both upper and lower consolute temperatures.

Fig. 10.7a is obtained. Distillation of a mixture of composition a leads to a vapour of composition b_1 which, when withdrawn, condenses to the completely miscible solution b_2. When this distillate is cooled to b_3, phase separation occurs. (This applies to the first drop of distillate. If distillation continues, the composition of the remaining liquid changes, and so the composition of the distillate also changes. In the end, when the whole sample has been evaporated and condensed, the composition of the liquid is the same as at the beginning.)

A slightly more involved case is when the upper consolute temperature does not exist; then the distillate drips out of the still as a 2-phase mixture (so long as the composition is appropriate). Fig. 10.7b summarizes this kind of behaviour.

Consider the point a_1, and proceed to distil a homogeneous liquid of

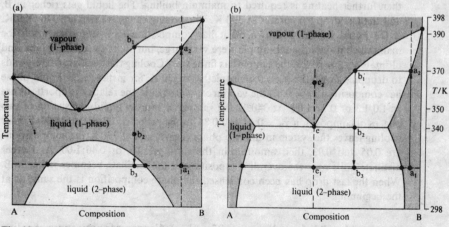

Fig. 10.7. Temperature composition phase diagrams for the discussion of the evaporation of two partially miscible liquids.

this composition. The solution boils at a_2, and the vapour has a composition b_1. This is drawn off and the first drop condenses to a liquid of the same composition but at a lower temperature, at b_3. This point is in the 2-phase region, and so the condensate separates into two layers.

The only odd thing about the diagram is the existence of the point e which represents the constant-boiling azeotrope of the system. A mixture of composition e_1 will form two layers, and these persist in varying proportions right up to the boiling point at e. The vapour given off has the same composition as the liquid (it is an azeotrope), and so condensing it again gives the initial 2-phase mixture. At e, $P = 3$, $C = 2$ and so $F = 1$ or $F' = 0$ if the pressure is fixed. The point is therefore invariant: the 2-layer liquid can be in equilibrium with its vapour only at a definite composition and temperature (the pressure being fixed).

The lesson to learn at this stage is that phase diagrams might look complicated but they convey simple information. In order to interpret them it is helpful to think *operationally*. This means that definite situations should be imagined, and the diagram should be proceeded through as we progressed through Fig. 10.7b. Phases switch on and off, systems boil and freeze, relative amounts change: just think of the reality of the information the diagrams provide. Another useful rule is to concentrate on the significance of the *lines* rather than the areas. We see the importance of this remark in the next few paragraphs.

Example (Objective 7). State the changes that occur when a mixture with composition x_B $= 0.95$ is boiled and the vapour condensed. Base the analysis on Fig. 10.7b, using the temperature scale on the right.

Method. The numbers and compositions of the phases can be read off the diagram in the manner already stated. The proportions of the phases are determined by the lever rule.

Answer. The initial point is in the single-phase region. When heated it boils at 370 K, giving a vapour of composition $x_B = 0.66$. If the vapour is not drawn off then further heating is required to maintain boiling. The liquid gets richer in B, and the last drop evaporates at 390 K. The boiling range is therefore 370–390 K. If the initial vapour is drawn off it has composition $x_B = 0.66$ (this would be maintained if the original sample were very large, but shifts to higher values, and ultimately $x_B = 0.95$, if the sample has finite size). Cooling the distillate corresponds to dropping down the $x_B = 0.66$ isopleth. At 350 K, for instance, the liquid phase has composition $x_B = 0.87$, the vapour $x_B = 0.49$, and the relative proportions are as 1.0/1.3 or 0.77:1.00. At 340 K the sample is entirely liquid, consisting of two phases of composition $x_B = 0.44$ and 0.84, in the ratio 1.1:1.3 or 0.85:1.00. Futher cooling leaves the system in the two phase region, and at 298 K the compositions are 0.05 and 0.93, their amounts in the ratio 1.7:3.7, or 0.46:1.00. As further distillate boils over the overall composition in the distillate becomes richer in B. When the last drop has been condensed the phase composition is the same as at the beginning.

How to interpret liquid–solid phase diagrams. The phase diagrams for mixtures of solids are often like those for mixtures of liquids, but in·place of the regions representing liquids and vapours they now represent solids and liquids.

As an example, consider a pair of metals that are almost wholly immiscible right up to their melting points (e.g., antimony and bismuth). Figure 10.8 summarizes what happens when a liquid mixture is cooled. (Note how closely it resembles Fig. 10.7b.)

Consider a liquid mixture of composition a at the point a_1. When it is cooled to a_2 it enters the 2-phase region labelled liquid + A. At this temperature almost pure solid A begins to come out of solution, and so the solution gets richer in B. At a_2 the solid A and the liquid are in equilibrium, and the lever rule shows that virtually the whole of the sample is still liquid.

On cooling to a_3 more of the slightly impure A precipitates, and the relative amounts of solid and liquid (which are still in equilibrium) are given by the lever rule: there is much more solid A than before, and the supernatant liquid, with composition b_3, is richer in B (follow the line marked 'liquid composition').

At a_4 there is less liquid (now of composition e) and more impure solid A. The liquid now solidifies to give a two-phase mixture of almost pure A and almost pure B. At a_5, for instance, the compositions of the two phases are a_5'' and a_5' respectively.

One function of this example has been to show how important it is to concentrate on the lines. The overall composition of the system has remained constant (on the *isopleth* a–a; isopleth comes from the Greek for 'same abundance') but the points where the horizontal lines cut the phase boundary have given us the information we have described: *the points at the ends of the tie lines give the composition of each phase*, and *the distances from the isopleth give their relative abundances.*

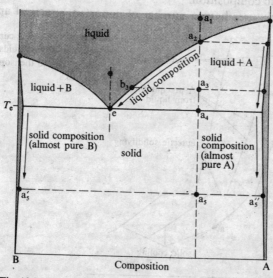

Fig. 10.8. Phase diagram for two almost immiscible solids and their liquids.

The mixture of composition e plays a role in solid–liquid transitions similar to the one played by the azeotrope in liquid–vapour transitions. If a liquid mixture is prepared with this *eutectic* composition (the word comes from the Greek for 'easily melted'), then it solidifies at the lowest temperature of all the mixtures (or melts at the lowest temperature). It is also the mixture that solidifies (and melts) without change of composition. Mixtures to the right of e precipitate A as they cool, mixtures to the left precipitate B, but only the eutectic freezes at a definite temperature (but one that depends on the pressure, $F = 1$, or $F' = 0$) without gradually unloading one or other of the components from the liquid.

The *thermal analysis* procedure described on p. 291 is a very useful practical way of determining phase diagrams. Consider the rate of cooling at the compositions a and e, Fig. 10.9. The liquid cools without change until it reaches a_2, when pure A begins to appear. The rate of cooling then changes because the progressive solidification of A delivers some enthalpy of melting which retards the cooling. When the eutectic composition is reached in the remaining liquid the temperature remains constant until the whole sample has solidified (because under constant pressure conditions it is an invariant point, $F' = 0$). This is the *eutectic halt*, Fig. 10.9. If the sample has the eutectic composition, the cooling proceeds at an approximately constant rate until the eutectic temperature is reached, then there is a long eutectic halt while the whole of the sample solidifies. Once the sample is solid, the cooling continues.

Observation of the cooling curves gives a very clear indication of the structure of the phase diagram. The liquid–solid 2-phase boundary can be determined from the points where the rate of cooling changes, and from the position of the eutectic halt we can determine the eutectic temperature T_e. The sample giving the longest eutectic halt corresponds to the eutectic composition.

Eutectic mixtures have a number of important practical applications,

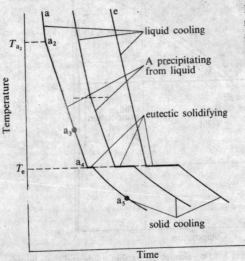

Fig. 10.9. Cooling curve for a mixture of two liquids. The labels a and e refer to the isopleths in Fig. 10.8.

and their study is essential to an understanding of the formation and properties of alloys. One technologically important eutectic mixture is used for solder. The eutectic composition is 67 per cent by weight of tin and 33 per cent lead: this melts at 183 °C and is used in electrical solder.

Another eutectic of interest is a salt/water mixture. Common salt in water has a phase diagram like that shown in Fig. 10.8, but the pure salt melting point lies at a very high temperature. The eutectic composition is 23.3 per cent by weight of salt, and it melts at -21.1 °C. Two applications of this eutectic may be mentioned. When salt is added to ice under *isothermal conditions* (e.g., salt spread on an icy road) the system melts if the temperature is greater than -21.2 °C and the eutectic composition has been achieved. When salt is added to ice under *adiabatic conditions* (e.g., to ice held in a vacuum flask) the ice melts, but in doing so absorbs heat from the rest of the mixture. The temperature falls, and if enough salt has been added, cooling continues down to the eutectic temperature.

How to take reactions into account. In some systems a mixture of A and B reacts to give a compound such as AB. An example is provided by aniline and phenol. At equilibrium this is a 2-component system because the amount of the third species, AB, is determined by the amounts of A and B present, and by the appropriate equilibrium constant.

Consider a sample of equal amounts of A and B, and for the present suppose that the equilibrium lies strongly in favour of AB, then at that composition we have the pure compound AB. (It must be emphasized that we do mean compound, and not just a mixture.) We heat A and B together, and form the liquid AB. Cooling the liquid leads to solidification at the melting point of the compound AB.

Suppose now more A is added to the system (or the amount of B is decreased), then a mixture of A and AB will be present. This was considered in the last section, the only difference being that the whole of Fig. 10.8 has to be squashed into the range of compositions between pure A and an equimolar mixture of A and B, Fig. 10.10. To the left of the diagram is another phase diagram resembling the first, but in this region B is in excess. Interpreting the information in the diagram proceeds as for Fig. 10.8, but the solid deposited down the isopleth a is the compound AB, slightly contaminated with A, and the two-phase solid beneath a_4 is a mixture of AB and A (each one slightly contaminated by the other).

One important modification occurs when the compound is stable only in the solid phase, and falls apart in the liquid. The phase diagram indicating this behaviour is shown in Fig. 10.11. A typical example is the alloy Na_2K, which is a definite compound formed on mixing sodium and potassium metals in the appropriate proportions. By considering what changes occur as samples of composition a and b are cooled you can see that the solid Na_2K is never in equilibrium with its liquid, but with liquid Na and K instead.

First consider what happens when a liquid of composition a and temperature T_1 (Fig. 10.11) is cooled. At T_2 some solid Na (slightly

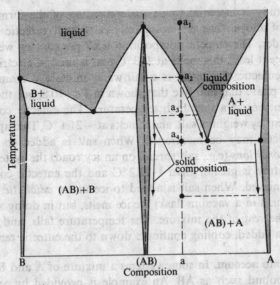

Fig. 10.10. Phase diagram for a system in which A and B react to form the compound AB.

contaminated with K) is deposited and the remaining liquid becomes
richer in K. When the temperature has fallen to T_3 the system enters a
2-phase solid region, the phases being solid Na and solid compound Na_2K
(each slightly contaminated by the other). Now consider isopleth b. The
initial precipitation (at T'_2) is of dirty solid Na, until at T_3 a reaction
occurs with the result that solid Na_2K is formed. At this stage there is
present liquid Na + K and a little solid Na_2K (slightly contaminated by
some K), but there is still no liquid Na_2K. As cooling proceeds the amount
of solid Na_2K increases, until at T_4, opposite b_4, the eutectic composition
of the liquid is attained; the liquid then changes, to give a 2-phase solid

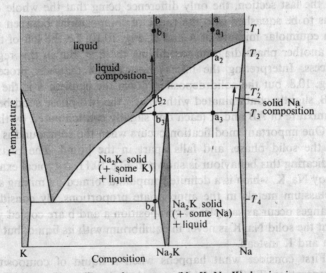

Fig. 10.11. Phase diagram for a system (Na, K, Na_2K) showing incongruent melting.

mixture of slightly dirty K and slightly dirty Na$_2$K. No liquid Na$_2$K has appeared at any stage. The underlying reason is that Na$_2$K is too feeble to survive melting, and falls apart instead.

The behaviour just outlined is called *incongruent melting*, a *phase reaction*, or a *peritectic reaction*. The last name comes from the Greek for 'around' and 'melting' and reflects the appearance of the solid formed on cooling down an isopleth, for it is observed that it consists of a distribution of minute sodium crystals, each surrounded by a shell of the compound.

Steel. A good example of a binary solid system is provided by the iron/carbon phase diagram. Modern steels contain a number of ingredients designed to make them resistant to corrosion under a variety of conditions, or to give them particular properties under conditions of stress. Primitive steel, however, which is composed of iron and its carbides, provides a straight-forward example of the application of a binary phase diagram.

Steel is mostly iron, and rarely contains more than a few per cent of carbon. Therefore, we need concentrate on only the iron-rich end of the phase diagram and ignore the regions close to pure carbon. The fragment of the phase diagram of interest is shown in Fig. 10.12.

First consider a fairly crude steel with about 2 per cent carbon: this composition is represented by the vertical line a. When the liquid is cooled to a$_1$ some solid solution of carbon in iron, called austenite, appears, and the liquid becomes correspondingly richer in carbon. When the temperature has fallen to a$_2$ the liquid has reached a eutectic composition (e$_1$), and the temperature remains constant until all of it has disappeared. The solid formed is a mixture of austenite and *cementite* (Fe$_3$C). (The line for the latter would appear well to the right on this diagram.) Once this solid mixture has been formed the temperature drops, and at a$_3$ there is another eutectic halt (e$_2$) while solid iron, contaminated by a little carbon,

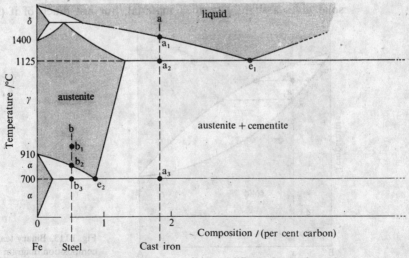

Fig. 10.12. Phase diagram for steel. (α, γ, δ denote different forms of pure iron.)

precipitates. The material formed is *cast iron*. From the phase diagram it can be seen that cast iron never contains just one phase until the melting point is reached. One of the consequences of this is brittleness. Nevertheless, cast iron is resistant to corrosion, and irons with compositions close to the eutectic e_1 (and melting at 1125 °C) may be cast quite readily.

When more of the carbon is eliminated a higher grade of steel is obtained. The process is represented by a material of composition b. When a mixture of iron and carbon is at the point b_1 only a single phase, austenite, is present. In this state the steel can be rolled very easily. On cooling to b_2 some iron (contaminated with a little carbon) is formed, and the solid austenite phase gets richer in carbon. This continues until point b_3, when the iron, the austenite, and the cementite (whose line, remember, is far to the right) are in equilibrium. The temperature halts, all the austenite disappears, and beneath 700 °C the system consists of the two phases of cementite and iron. This is steel, and its rigidity can be traced to the presence of the carbide Fe_3C.

Ultrapurity and controlled impurity: zone refining. Advances in technology have called for materials of extreme purity. For instance, semiconductor devices consist of almost pure silicon or germanium, but doped to a precisely controlled extent. If these materials are to operate successfully, the impurity level must be kept down to no more than about 1 in 10^9 (which is about one grain of salt in 5 tons of sugar).

The technique of *zone refining* was developed during the early 1950s in response to this need. The liquid–solid binary system shown in Fig. 10.13 indicates the basis of the technique.

Consider a liquid of composition a, largely composed of the component B, but contaminated by some A. When the liquid is cooled to a_1 a solid of composition b_1 appears, and according to the diagram the solid is a solution of A and B, but it is richer in B than the liquid. Removing the solid gives a slightly purified material, but not much of it (lever rule).

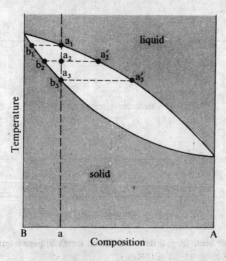

Fig. 10.13. Binary temperature–composition diagram and the steps leading to zone refining.

Further conventional cooling of the original system delivers more B-rich solid (b_2), and leaves B-depleted liquid (a'_2). Finally the whole sample freezes just beneath a_3 to give a sample of the same composition (a) as the original, dirty liquid. Conventional freezing has left us with a material no purer than the initial sample.

We could strain off the early crystals, and then repeat the procedure on them. This *fractional crystallization* would, in each cycle, step the system towards pure B, in the manner of fractional distillation. But the procedure is very slow and wasteful.

The modern procedure is based on the recognition that our interpretation of Fig. 10.13 is idealistic. It assumes that the freezing is so slow that the composition of the solid phase is uniform, and has throughout the equilibrium composition demanded by the phase diagram. For example, it assumes that the composition of the solid shifts uniformly from b_2 to b_3 when the system is cooled from a_2 to a_3. In any real process this will not be the case because the solute A does not have the time to disperse in and out of the solid at the demand of the slope of the Gibbs function.

Consider a dirty liquid of composition a, and let it be cooled to a_1. Solid of composition b_1 is deposited and dirtier liquid remains. Now let this liquid be cooled without permitting the system to come to overall equilibrium. The first solid deposited remains at its composition b_1 and the new deposit, if the temperature is dropped to a_2, is of composition b_2 leaving an even dirtier liquid of composition a'_2. Cooling along the isopleth a'_2 without equilibrating with the solids b_1, b_2 deposits solid b_3, and leaves dirtier a'_3. The process continues until the last liquid to solidify is heavily contaminated with A.

There is some everyday evidence that the freezing of impure liquids proceeds as we have described. An ice cube is clear near the surface, but misty in the core. This is because the water used to make ice normally contains dissolved air. Freezing normally proceeds from the outside, and the air is accumulated in the retreating liquid phase. It cannot escape from the interior of the cube, and so when that finally freezes it occludes the air in a mist of tiny bubbles.

In the technique of zone refining the sample is normally in the form of a narrow cylinder. This is heated in a narrow, disc-like zone which is swept from one end of the sample to the other. The advancing liquid zone accumulates the contaminants as it passes, for they are preferentially partitioned into the liquid phase. One pass may have the effect of reducing the impurity content only slightly, because a really dirty liquid phase cannot accumulate much more dirt. For this reason it is normal in zone refining to devise a multiple passage system, and a train of hot and cold zones are swept repeatedly from one end of the sample to the other. The zone right at the end of the sample is the impurity dump: when the heater has passed its position, cooling occurs, and the dirty liquid simply cools to dirty solid, which can be discarded.

A modification of the technique, *zone levelling*, is used to introduce controlled amounts of impurity (for example, indium into germanium). A rich sample of the desired impurity is put at the head of the sample, and made molten. This zone is then dragged through the length of the sample. If the solute is very insoluble in the major component, the liquid zone deposits a uniform distribution of impurity into the pure material, and the final sample is the uniformly doped material.

Zone refining has found application in fields other than the semiconductor industry. Ultrapure materials often have markedly different properties from conventionally pure samples. For instance, bismuth is normally regarded as a hard, brittle metal, yet when it has been zone refined it forms rods which can be bent without fracture. We have already seen how a few per cent of carbon produces brittle cast iron: when all traces of impurity are removed from iron it retains its ductility down almost to absolute zero. Organic chemicals may be refined, and substances thought to have a foul odour have been rendered odourless by purification.

10.4 Three-component systems

Three-component systems have a variance that may reach 4. Even imposing constant temperature and pressure constraints still leaves two intensive variables, and the representation of phase equilibria is on the verge of being absurdly complicated. We shall examine only the very simplest type of three-component systems in order to acquire some familiarity with their representation in terms of triangular diagrams.

Sorting out triangular coordinates. If the system is composed of the components A, B, and C, their mole fractions must sum to unity: $x_A + x_B + x_C = 1$. We need a diagram that automatically ensures that this condition is fulfilled. The equilateral triangle has the required property. It follows from elementary geometry that the sum of the distances to a point measured parallel to each of the three edges is equal to the side of the triangle. Figure 10.14 illustrates this for an equilateral triangle of unit side. If the mole fractions x_A, x_B, and x_C are represented by the three distances, then a mixture of any composition can be represented by a point on the interior of the triangle.

Figure 10.14 shows how this works in practice. Imagine the triangle marked off by lines parallel to each side, and let the distances along these lines represent the mole fraction of each component. In the figure the point P indicates a mole fraction $x_A = 0.50$. A similar grid can be imagined for the B mole fraction, with zero lying along the AC line, and unity at the B apex. Point P corresponds to $x_B = 0.10$. The concentration of C, $x_C = 0.4$, is ensured by the geometry of the triangle.

The edge AB corresponds to $x_C = 0$, and likewise for the other two edges. Therefore the three edges refer to the three binary systems (A, B), (B, C), and (C, A).

An important property of a triangular diagram is the significance of a straight line joining an apex to a point on the opposite edge, Fig. 10.14. Any point along the indicated line represents a composition that is

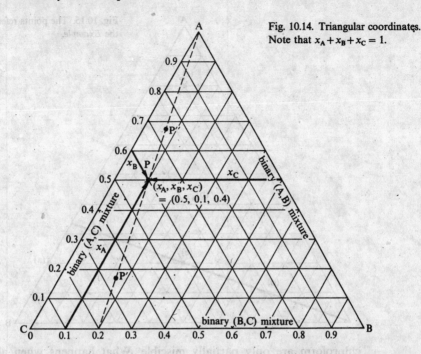

Fig. 10.14. Triangular coordinates. Note that $x_A + x_B + x_C = 1$.

progressively richer in A as it passes from P′, through P to P″, but the significant point is that B and C remain present in the same initial proportion. The validity of this observation depends on the properties of similar triangles, and on showing that $x'_B/x'_C = x''_B/x''_C$. Therefore, if we wish to represent the changing composition of a system as A is added, all that it is necessary to do is to draw the line from the apex A to the point on BC representing the initial binary system. Any ternary system formed by adding A lies at some point on this line.

Example (Objective 13). Mark the following points on a triangular coordinate system:

(a) $x_A = 0.20$, $x_B = 0.80$, $x_C = 0.00$
(b) $x_A = 0.42$, $x_B = 0.26$, $x_C = 0.32$
(c) $x_A = 0.80$, $x_B = 0.10$, $x_C = 0.10$
(d) $x_A = 0.10$, $x_B = 0.20$, $x_C = 0.70$
(e) $x_A = 0.20$, $x_B = 0.40$, $x_C = 0.40$
(f) $x_A = 0.30$, $x_B = 0.60$, $x_C = 0.10$

• *Method.* x_A is measured along either edge leading to the apex A, likewise for B. The mole fraction of C always takes care of itself.

• *Answer.* The points are plotted in Fig. 10.15.

• *Comment.* Note that points (d), (e), and (f) have x_A/x_B in constant ratio and fall on a straight line, as stated in the text.

Partially miscible liquids. A good example of a 3-component system is water/ chloroform/acetic acid. Water and acetic acid are completely miscible in all proportions, and so are chloroform and acetic acid. Water and

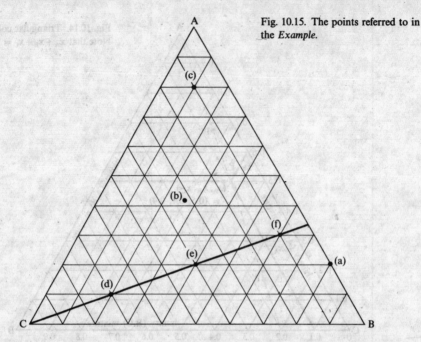

Fig. 10.15. The points referred to in the *Example.*

chloroform are only partially miscible. What happens when all three are present together?

The 3-component phase diagram for room temperature (and pressure) is shown in Fig. 10.16. This shows that $CHCl_3/CH_3CO_2H$ and CH_3CO_2H/H_2O binary mixtures form 1-phase systems, but $CHCl_3/H_2O$ (the bottom edge of the triangle) has a 1-phase region at the ends and a 2-phase region at intermediate concentrations. The latter behaviour is analogous to what we saw in more detail in Fig. 10.5.

We might expect that a 1-phase system will be formed from a 2-phase $CHCl_3/H_2O$ mixture if acetic acid is added. Figure 10.16 confirms that this is so, because at high enough acid concentration the system enters a 1-phase region whatever the initial proportions of chloroform and water.

Following the line $a_1a_2a_3a_4$ shows this in more detail. We begin with a binary mixture of composition a_1. The relative amounts and compositions of the two phases can be read off in the normal way (use the lever rule for the former). Adding acetic acid takes the system along the line joining a_1 to the apex. At some point a_2 the solution still has two phases, but there is more water in the chloroform phase (the phase at a_2') and more chloroform in the water (the phase at a_2'') because the acetic acid assists both to dissolve. Note, however, that there is more acetic acid in the water-rich phase than in the chloroform-rich phase (a_2'' is nearer the acetic acid apex than is a_2') and also note that the tie lines have to be drawn in if the diagram is to be interpreted quantitatively (we have to know which points on the phase boundary, representing the compositions of the two phases, relate to the overall composition a_2). At a_3 two phases are present, but the chloroform-rich layer is there only as a trace. Further

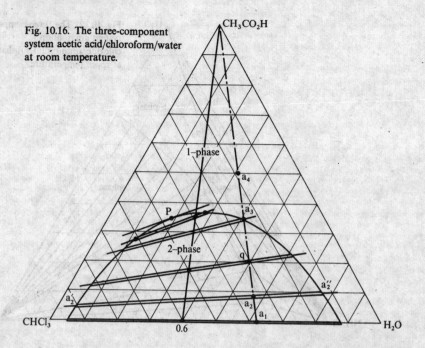

Fig. 10.16. The three-component system acetic acid/chloroform/water at room temperature.

addition of acid (e.g., to a_4) makes the whole system a single phase.

Example (Objective 14). A mixture is prepared consisting of a mole fraction 0.60 of chloroform and 0.40 of water. Describe the changes that occur when acetic acid (ethanoic acid) is added to the mixture.

● *Method*. Base the discussion on Fig. 10.16. The relative proportions of chloroform and water remain constant, and so the addition of acetic acid corresponds to motion along the line from the acetic acid apex to the point $x_c = 0.60$ on the opposite base line. The tie lines give the phase compositions, the lever rule their proportions.

● *Answer*. Initially the composition is $x_c = 0.60$, $x_w = 0.40$, $x_a = 0.00$: we shall denote this (0.60, 0.40, 0.00). This composition lies in the two-phase region, the phase compositions being (0.95, 0.05, 0.00) and (0.12, 0.88, 0.00) with relative proportions 4.4/3.3 = 1.3 : 1.0. Addition of acetic acid takes the system along the line mentioned. When sufficient acetic acid has been added such that its mole fraction is 0.18 the overall composition is (0.49, 0.33, 0.18) and the system consists of two phases, one of composition (0.82, 0.06, 0.12) and the other of composition (0.17, 0.60, 0.23) in relative abundance 2.9/2.8 = 1.0 : 1.0. When enough acid has been added to bring x_a to 0.365 the system consists of a trace of a phase with composition (0.64, 0.11, 0.25) and a dominating phase of composition (0.35, 0.28, 0.37). Further addition of acid takes the system into the single-phase region, where it remains right to the point of pure acetic acid.

The role of added salts. Ammonium chloride is soluble in water. Ammonium sulphate is soluble in water. What is the solubility of a mixture of the two salts? Figure 10.17 illustrates the room-temperature phase diagram. The tie

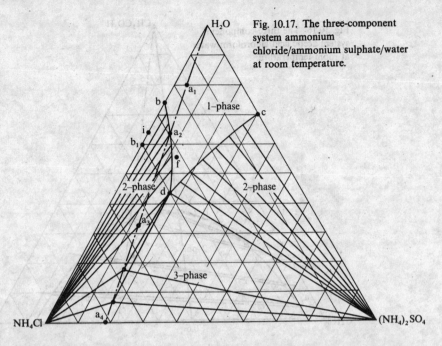

Fig. 10.17. The three-component system ammonium chloride/ammonium sulphate/water at room temperature.

lines have been indicated to guide the interpretation. (In fact, there should be a solid mass of tie lines, but that would defeat the object of the illustration.) The point b indicates the solubility of ammonium chloride in water: a binary mixture of composition b_1 consists of the undissolved solid and the supernatant, saturated solution of composition b. The point c indicates the solubility of the sulphate.

Consider a ternary solution of composition a_1. This is unsaturated, and is a single phase. Now evaporate the water: the composition moves along line a_1–a_4. At a_2 it strikes the 2-phase region, and according to the tie lines (all of which have an apex in common) some solid chloride will crystallize out of the solution. The liquid gets relatively richer in sulphate, and its composition moves towards d. When enough water has been removed the overall composition a_3 is reached and the saturated solution of composition d, solid chloride, and solid sulphate (see where the two tie lines go) are in equilibrium. There are three phases, and three components, and so $F = 2$. Both pressure and temperature are fixed, and so this is an invariant point ($F'' = 0$). From this point on the evaporation of the water diminishes the amount of saturated solution, and deposits solid chloride and sulphate. When point a_4 has been reached, the whole of the solvent has disappeared, and we have the binary mixture of two solids.

The composition of the saturated ternary solution (d) corresponds to a smaller mole fraction of water than either of the two binary solutions (b and c). This indicates that the two salts form a more concentrated solution overall than either does alone: this is the *common-ion effect*, where the presence of one salt alters the solubility of the other.

Example (Objective 14). A solution of 50 g ammonium chloride in 30 g water is prepared at room temperature. 45 g of ammonium sulphate is then added. Describe the states of the initial and final systems.

- *Method*. Figure 10.17 can be used if the masses are converted to mole fractions. Compositions of phases and their abundances can be read off the phase diagram, with use of the lever rule.

- *Answer*. The molar masses of the three components are water: 18.02 g mol^{-1}; ammonium chloride: 53.49 g mol^{-1}; ammonium sulphate: 132.1 g mol^{-1}. It follows that 30 g water is 1.66 mol, 50 g ammonium chloride is 0.93 mol, and 45 g ammonium sulphate is 0.34 mol. The initial mole fraction composition is therefore (0.64, 0.36, 0.00) and the final (0.57, 0.21, 0.12) for (water, ammonium chloride, ammonium sulphate). From Fig. 10.17 we see that the initial system consists of ammonium chloride with a saturated solution of composition (0.64, 0.36, 0.00) in relative proportion 0.9/5.8 = 0.16 : 1.00. After addition of sulphate there is only one phase.

Further reading

Freezing points, triple points, and phase equilibria. R. C. Parker and D. S. Kristol; *J. chem. educ.*, 658, 51 (1974).

Phase transitions. R. Brout; Benjamin, Reading, 1965.

Phase diagrams. A. Alper; Academic Press, New York, 1970.

The phase rule and its applications. A. Findlay, A. N. Campbell, and N. O. Smith; Dover, New York, 1951.

The phase rule and heterogeneous equilibria. J. E. Ricci; Van Nostrand, New York, 1951.

High pressure chemistry. R. S. Bradley and D. C. Munro; Pergamon, Oxford, 1965.

Geochemistry. W. S. Fyfe; Clarendon Press, Oxford, 1974.

Problems

10.1.   The compound *p*-azoxyanisole forms a liquid crystal. 5 g of the solid was put into a tube which was then evacuated and sealed. Use the phase rule to prove that the solid will melt at a definite temperature, and that the liquid crystal phase will make a transition to a normal liquid phase at a definite temperature.

10.2.   State how many *components* there are in the following systems. (a) NaH_2PO_4 in water in equilibrium with water vapour, but disregarding the possibility that the salt ionizes in solution; (b) the same, but taking into account the possibility of complete ionization into all possible ions; (c) $AlCl_3$ in water, noting that hydrolysis and precipitation of $Al(OH)_3$ occur.

10.3.   Two phases are in equilibrium if the chemical potentials of the species present are the same in each phase. Show that two phases are in thermal equilibrium only if their temperatures are the same, and that they are in mechanical equilibrium only if they are at the same pressure.

10.4.   Blue copper sulphate crystals decompose and release their water of hydration when heated. How many phases and components are present in an otherwise empty heated vessel?

10.5.   Ammonium chloride also dissociates when it is heated. How many phases and components are present when the salt is heated in an otherwise empty vessel?

How many phases and components are present when ammonia is added before the salt is heated?

10.6.   A saturated solution of sodium sulphate, with excess of the salt, is at its boiling point in a closed vessel. How many phases and components are present? How many degrees of freedom are there, and what are they?

10.7.   Now suppose that the solution referred to in the last Problem was not saturated. How many components, phases, and degrees of freedom are there at the boiling point, and what are the degrees of freedom?

10.8.   Both MgO and NiO are highly refractory materials. Nevertheless, when the temperature is high enough they do melt, and the temperature at which the mixture melts is of considerable interest in the ceramics industry. Draw the temperature-composition diagram for the MgO/NiO system on the basis of the data below, where x is the composition of the solid, and y that of the liquid (as a mole fraction).

$t/°C$	1960	2200	2400	2600	2800
$x(MgO)$	0	0.35	0.60	0.83	1.00
$y(MgO)$	0	0.18	0.38	0.65	1.00

10.9.   On the basis of the MgO/NiO phase diagram just constructed, state (a) the melting point of composition $x(MgO) = 0.30$; (b) the composition of the system, in terms of the nature, composition, and proportions of the phases, when a solid containing $x(MgO) = 0.30$ is heated to $2200\,°C$; (c) the temperature at which a liquid of composition $y(MgO) = 0.70$ will begin to solidify.

10.10.   The bismuth–cadmium system is of interest in metallurgy, and we can use it to illustrate several points made in the chapter. In fact the solid–liquid phase diagram calculated on the basis of the Clausius–Clapeyron equation is very close to the experimental phase diagram throughout the composition range, and so we begin by constructing it. Use the information that $T_f(Bi) = 544.5$ K, $T_f(Cd) = 594$ K, $\Delta H_{f,m}(Bi) = 10.88$ kJ mol^{-1}, $\Delta H_{f,m}(Cd) = 6.07$ kJ mol^{-1}, and the knowledge that they are insoluble in each other in the solid state.

10.11.   On the basis of the phase diagram constructed in the last Problem, state what you would observe when a liquid containing $x(Bi) = 0.70$ is cooled slowly from 550 K. What relative amounts of solid and liquid are present at (a) 460 K, (b) 350 K? What would the solid be if the cooling took place very rapidly?

10.12.   Plot a graph of the number of phases in equilibrium and, on the same diagram, the variance of the system, as the $x(Bi) = 0.70$ mixture is cooled from 550 K to 450 K.

10.13.   Sketch the cooling curve for the mixture treated in the last Problem.

10.14.   Now we move on to more conventional liquid mixtures, and begin by constructing the phase diagram for m-toluidine and glycerol. In the experiments on which data below were obtained (at $p = 1$ atm), solutions of m-toluidine were made up in glycerol, and then warmed from room temperature. The mixture was observed to lose its turbidity at the temperature t_1, and then on further heating to become turbid again at t_2. Plot the phase diagram on the basis of the data, and find the upper and lower consolute temperatures.

$100w$	18	20	40	60	80	85
$t_1/°C$	48	18	8	10	19	25
$t_2/°C$	53	90	120	118	83	53

$100w$ is the percentage by weight of m-toluidine in the mixture.

10.15. Now use the phase diagram drawn in the last Problem to state what happens as *m*-toluidine is added dropwise to glycerol at 60 °C. State the number of phases present at each concentration, their composition, and their relative amounts.

10.16. In Fig. 10.7a is shown the phase diagram for two partially miscible liquids, and the water/isobutanol (2-methylpropan-1-ol) system resembles it quite closely. State what happens when a mixture of water (A) and isobutanol (B) of composition b_3 is heated, at each stage stating the composition, number, and amounts of the phases in equilibrium.

10.17. The phase diagram for the silver/tin system is shown in Fig. 10.18. Label the regions and state what will be observed when liquids of compositions a and b are cooled to 200 K.

10.18. Indicate on the phase diagram the feature that denotes an incongruent melting point. What is the composition of the eutectic mixture? At what temperature does the eutectic melt? Sketch the cooling curves for the isopleths a and b.

10.19. On the basis of the silver/tin phase diagram, state (a) the solubility of silver in tin at 800 °C, (b) the solubility of the compound Ag_3Sn in silver at 460 °C, (c) the solubility of the compound in silver at 300 °C, (d) the solubility of tin in silver at 400 °C.

10.20. Sketch the phase diagram for the magnesium/copper system on the basis of the following information. $T_f(Mg) = 648$ °C, $T_f(Cu) = 1085$ °C; two intermetallic compounds are formed with melting points $T_f(MgCu_2) = 800$ °C, $T_f(Mg_2Cu) = 580$ °C; eutectics of composition and melting points 10 per cent, 33 per cent, 65 per cent by weight magnesium, 690 °C, 560 °C, and 380 °C, respectively.

10.21. A sample of Mg/Cu alloy containing 25 per cent magnesium by weight was prepared in a crucible heated to 800 °C in an inert atmosphere. On the basis of the phase diagram drawn in the last Problem, state what will be observed if the melt is cooled slowly to room temperature. State what phases are in equilibrium at each temperature, their compositions, and their relative abundances.

10.22. Sketch the temperature–time cooling curve for the molten alloy referred to in the last Problem.

10.23. At first sight triangular coordinates may appear quite confusing, but the next few Problems will give practice in thinking about the information they contain. As a first step, mark the following features on triangular coordinates. (a) The point

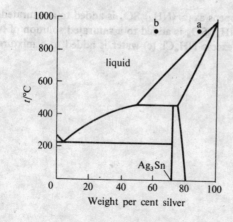

Fig. 10.18. The silver/tin phase diagram.

denoting the composition $x_A = 0.2$, $x_B = 0.2$, $x_C = 0.6$; (b) the same for $x_A = 0$, $x_B = 0.2$, $x_C = 0.8$; (c) the same for $x_A = x_B = x_C$; (d) the point denoting a mixture formed with 25 per cent NaCl, 25 per cent $Na_2SO_4 \cdot 10H_2O$, the rest water; (e) the line denoting the composition where NaCl and $Na_2SO_4 \cdot 10H_2O$ retain the same relative abundances as water is added to the solution.

10.24.   One of the properties of triangular phase diagrams is that a straight line from an apex (A) to an opposite edge (BC) represents the mixtures where x_B and x_C are in the same relative proportions however much A is present. Prove this property on the basis of the properties of similar triangles.

10.25.   Now we turn to the actual construction of a triangular phase diagram (you will find similar data in Vol. 3 of *International critical tables*). Methanol (M), diethyl ether (E), and water (W) form a partially miscible ternary system. The phase diagram at 20 °C was determined by adding the methanol to various binary ether–water mixtures, and noting the mole fractions of methanol x_M^* at which complete miscibility occurred. Plot the phase diagram on the basis of the following data:

x_E(E, W):	0.10	0.20	0.30	0.40	0.50	0.60	0.70	0.80	0.90
x_M^*(M, E, W):	0.20	0.27	0.30	0.28	0.26	0.22	0.17	0.12	0.07

x_E(E, W) is the mole fraction of ether in the original binary mixture, and x_M (M, E, W) is the mole fraction of methanol in the ternary mixture. How many phases will a mixture of 5 g methanol, 30 g diethyl ether, and 50 g water show at 20 °C? How many grams of water would have to be removed or added to change the number of phases?

10.26.   Refer to the ternary mixture with the phase diagram shown in Fig. 10.16. How many phases are there, what are their compositions, and in what relative proportions do they occur, in a mixture formed from 2.3 g water, 9.2 g chloroform, and 3.1 g acetic acid?

10.27.   State what would be observed if (a) water, (b) acetic acid is added to the mixture referred to in the last Problem.

10.28.   The phase diagram for the ternary system $NH_4Cl(A)/(NH_4)_2SO_4(B)/$ water (C) at 25 °C is illustrated in Fig. 10.17. What is the nature of the system that contains (a) $x_A = 0.2$, $x_B = 0.4$, $x_C = 0.4$; (b) $x_A = 0.4$, $x_B = 0.4$, $x_C = 0.2$; (c) $x_A = 0.2$, $x_B = 0.1$, $x_C = 0.7$; (d) $x_A = 0.40$, $x_B = 0.16$, $x_C = 0.44$?

10.29.   What is the solubility of (a) NH_4Cl in water at 25 °C, (b) $(NH_4)_2SO_4$ in water at 25 °C?

10.30.   Explain what happens as (a) $(NH_4)_2SO_4$ is added to a saturated solution of NH_4Cl in water; (b) $(NH_4)_2SO_4$ is added to a saturated solution of NH_4Cl in water in the presence of excess NH_4Cl; (c) water is added to a mixture of 25 g NH_4Cl, 75 g $(NH_4)_2SO_4$.

11 Equilibrium electrochemistry: ions and electrodes

Learning objectives

After careful study of this chapter you should be able to

(1) Define the *activity*, the *activity coefficient*, and the *mean activity coefficient* of ions in solution (p. 315).

(2) Describe the physical basis of the *Debye–Hückel theory* of ionic solutions and the formation and role of the *ionic atmosphere* in determining the mean activity coefficients of ions (p. 316).

(3) State the form of a *shielded Coulomb potential* (eqn (11.2.4)).

(4) Define *ionic strength* (eqn (11.2.6)).

(5) Derive and use the *Debye–Hückel Limiting Law* for the mean activity coefficient (eqn (11.2.11)) and indicate how it may be extended to more concentrated solutions (p. 326).

(6) Describe and calculate how the *Debye length* depends on the ionic strength, temperature, and relative permittivity of the medium (p. 323).

(7) Define the terms *anode* and *cathode* when used in galvanic cells and in electrolytic cells (p. 330).

(8) Define the *electrochemical potential* of an ion (eqn (11.3.2)).

(9) Derive an expression for the *potential difference across an interface* in terms of the *standard potential difference* and the activity of ions (eqn (11.3.6)).

(10) Describe the construction of, and derive an expression for the potential difference across, a *gas|inert metal electrode* (eqn (11.4.2)).

(11) Describe the construction of, and derive an expression for the potential difference across, an *ion|insoluble salt|metal electrode* (eqn (11.4.3)).

(12) Derive and use an expression for the potential difference at a *redox electrode* (eqn (11.4.4)).

(13) Describe the formation of a *liquid junction potential* (p. 339).

(14) Derive an expression for the potential difference across a *membrane* (p. 339).

Introduction

The principal objective of modern electrochemistry is to study the behaviour and reactions of ions and molecules in environments where they can take part in the transfer of electrons. The information obtained has important applications to a wide variety of processes, including power cells, fuel cells, catalysis, corrosion, and the function of biological membranes. None of these, though, is strictly an equilibrium process. Nevertheless, a knowledge of equilibrium electrochemistry forms a basis for understanding non-equilibrium processes, and it has many important applications in its own right.

The formulation of thermodynamics presented in the preceding chapters is directly applicable to the discussion of ions in solution. Only two changes of detail are necessary in order to accommodate the presence of charge. In the first place, ions interact electrostatically over long distances and so they deviate strongly from ideal behaviour even at very low concentrations. In the second place, when several phases are present (e.g., a metal electrode immersed in an electrolyte solution) each phase may be at a different electric potential. The ions respond to these potentials in the sense that a favourable potential decreases their chemical activity and an unfavourable one enhances it. This remark is the heart of electrochemistry.

In this chapter we examine these two related points by seeing how ions respond to electric potentials arising from the presence of other ions (which leads to an expression for the activity coefficient) and from the existence of different phases in contact (which leads to expressions for electrode potentials).

11.1 Activities of ions in solution

Abundant evidence confirms that many materials exist as ions in solution. Much of this evidence is based on the measurement of conductivities and on the investigation of colligative properties.

The conduction of electricity can be explained very easily on the basis that charged ions exist and transport current through the solution. The details of this motion, and the properties that determine the mobilities of the ions, need not concern us until dynamic processes are examined in Part 3.

The evidence from colligative properties (properties that depend on the amount rather than the nature of the solute) depends on the observation of greater vapour pressure depressions and greater osmotic pressures than would be predicted if molecules dissolved but remained associated. For example, when sodium chloride dissolves in water the osmotic pressure is about twice that expected on the basis of the number of NaCl units present. This is explained if NaCl is present as Na^+ and Cl^-, for then there are twice as many dissolved particles.

Activities and standard states. The thermodynamic properties of ions in solution depend on their chemical potentials μ_i. If the electrolyte behaved as an ideal dilute solute (in the sense of conforming to Henry's Law) the chemical potential would be related to the molality m_i by (p. 248)

$(11.1.1)°$ $\qquad \mu_i = \mu_i^{\ominus} + RT \ln (m_i/m^{\ominus})$

and the standard state of the ionic solution would be established at $m_i = m^{\ominus} \equiv 1 \, \text{mol kg}^{-1}$. The chemical potential of a real solution can be written in a similar way if we introduce the activity a_i through

$(11.1.2)$ $\qquad \mu_i = \mu_i^{\ominus} + RT \ln a_i.$

This expression is nothing more than a definition of activity, and becomes useful only when we have found a way of relating a_i to m_i. Note that according to this definition the activity is dimensionless. The activity can be expressed in terms of the molality by introducing the *activity coefficient* γ_i through

$(11.1.3)$ $\qquad a_i = \gamma_i m_i / m^{\ominus},$

where γ_i is an as yet unknown function of the nature, and in particular of the molality, of the solution. The main aim of the first part of this Chapter is to find the dependence of γ_i on m_i.

The standard state of a real solution of an electrolyte is defined in the same way as for non-electrolytes, Chapter 6 and 8. There we saw that the standard state is a *hypothetical* state of molality $m^{\ominus}$ where all the interactions leading to departure from ideality have been eliminated. The advantage of this subtle definition was stressed on p. 246: in this way all the deviations from ideality in the expression for μ are incorporated in the activity, and hence in the activity coefficient. As in the case of non-electrolytes, although the chemical potential takes its standard value $\mu_i^{\ominus}$ when $a_i = 1$, the system is not then actually in its standard state, for the standard state is a purely hypothetical condition. At low molalities the solution approaches ideality (in the sense of conforming to Henry's Law), and so the activity becomes equal to $m/m^{\ominus}$ as the latter approaches zero. Therefore the ionic activity coefficient γ_i approaches unity at low molalities: $\gamma_i \to 1$ and $a_i \to m_i/m^{\ominus}$ as $m_i \to 0$.

If the chemical potential of a monovalent cation is denoted μ_+ and that of a monovalent anion by μ_-, the total chemical potential of the electrically neutral solute is

$$\mu_+ + \mu_- = \mu_+^{\ominus} + \mu_-^{\ominus} + RT \ln a_+ + RT \ln a_-$$
$$= \mu_+^{\ominus} + \mu_-^{\ominus} + RT \ln (m_+/m^{\ominus}) + RT \ln (m_-/m^{\ominus}) + RT \ln \gamma_+ \gamma_-$$

The electrostatic interactions leading to the deviations from ideality are contained in the last term, and although we shall see that individual ion activity coefficients can be calculated, there is no way of disentangling the product $\gamma_+ \gamma_-$ experimentally and of ascribing a part to the cations and a part to the anions. The best the experimenter can do is to ascribe the deviations from ideality equally to both kinds of ion. Therefore for practical applications we introduce the *mean ionic activity coefficient*

(11.1.4) $\gamma_\pm = (\gamma_+\gamma_-)^{\frac{1}{2}}$

and write chemical potentials as

$$\mu = \mu_+ + \mu_-$$

$$\mu_+ = \mu_+^\ominus + RT\ln(m_+/m^\ominus) + RT\ln\gamma_\pm$$

$$\mu_- = \mu_-^\ominus + RT\ln(m_-/m^\ominus) + RT\ln\gamma_\pm$$

so that the non-ideality term is distributed equally.

When the salt is M_pX_q, which dissolves to give a solution containing the ions $M^{+|z_+|}$ and $X^{-|z_-|}$ in the ratio $p{:}q$, the mean activity coefficient is given by

(11.1.5) $\gamma_\pm = (\gamma_+^p\gamma_-^q)^{1/n}, \qquad n = p+q,$

and the chemical potential of either species is given by

(11.1.6) $\mu_i = \mu_i^\ominus + RT\ln(m_i/m^\ominus) + RT\ln\gamma_\pm.$

Once more the non-ideality term is distributed equally between the types of ion.

The principal problem remaining is the determination of the $\gamma_\pm$. Once its dependence on the molality and other properties of the solution (e.g., the temperature or the nature of the solvent) has been established, we shall be able to investigate the thermodynamic consequences of the chemical potential (such as equilibrium constants and solubilities).

11.2 A model of ions in solution: the Debye–Hückel theory

Ionic solutions depart sharply from ideality on account of the long range of their electrostatic interactions. The energy of interaction of two neutral molecules falls off as approximately R^{-6}, where R is their separation, but the Coulomb interaction between two charged ions falls off only as R^{-1}. This indicates not only that ions interact over long distances, but also that the departures from ideality are likely to be dominated by the direct electrostatic, Coulombic interactions. This point of view is taken here, and is used to build a model of the structure of an ionic solution. As in the case of the van der Waals model of a real gas, the Debye–Hückel theory is an excellent example of the way in which essential physical features are discerned and then incorporated into a quantitative model. In the present case the problem is more intricate, but every step is guided by a clear appreciation of the underlying physical ideas.

Oppositely charged ions attract each other. This suggests that cations and anions are not uniformly distributed, but that anions tend to be found in the vicinity of cations, and vice versa, Fig. 11.1. Overall the solution is neutral, but in the vicinity of any given ion there is a predominance of ions of opposite charge (these are called *counter-ions*). On the average more counter-ions than like-ions pass by any given ion, and they come and go in all directions. This time-averaged, spherical haze of opposite charge is called the *ionic atmosphere* of the ion.

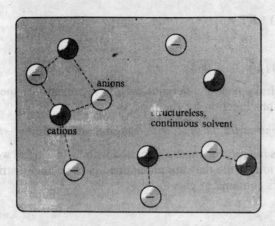

Fig. 11.1. The general picture underlying the Debye–Hückel theory.

anions

cations

structureless, continuous solvent

The energy, and therefore the chemical potential, of the central ion is lowered by its favourable electrostatic interaction with its ionic atmosphere. The main task is to find a way of formulating the effect quantitatively.

The whole of the departure from ideality is assumed to be due to these electrostatic interactions, and so we can imagine a solution in which all the ions have their actual positions, but in which all the ion–atmosphere electrostatic interactions have been turned off, Fig. 11.2. The determination of the activity coefficient then reduces to finding the change in chemical potential when the charges on the ions are returned to their true values while their average distribution is held constant.

The difference in chemical potentials is a difference of molar Gibbs function between the hypothetical uncharged state and the true, charged state. Under conditions of constant pressure and temperature ΔG is equal to $w_{e,max}$, the work other than the work of expansion done on charging the system. This suggests the following strategy:

(1) Denote the chemical potential of an ideal dilute solution $\mu^{\circ}(m_i)$ (it depends on m_i in accord with eqn (11.1.1)); then it follows from eqn (11.1.2) that the activity coefficient for the true solution is given by

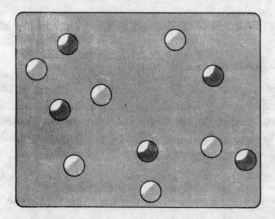

Fig. 11.2. The ions in their final distribution, but with their charges quenched.

$$\mu(m_i) = \mu^\circ(m_i) + RT \ln \gamma_i$$

or

$$\ln \gamma_i = (1/RT)\{\mu(m_i) - \mu^\circ(m_i)\}.$$

(2) The chemical potential difference $\mu - \mu^\circ$ is identified with the change in the molar Gibbs function for the process

hypothetical uncharged, ideal state $\rightarrow$ *true state*, $\Delta G_m = \mu(m_i) - \mu^\circ(m_i)$.

(3) The Gibbs function change is identified with the electrical work, w_e, of charging the ions while they are maintained in their final distribution. These remarks combine to give

(11.2.1) $\ln \gamma_i = w_e/RT.$

The problem has been reduced to finding the final distribution of the ions, and the work of charging unit amount in that distribution.

Example Measurements of the kind to be described later show that the activity coefficient of potassium chloride in water at 25 °C is 0.901 when the concentration is 0.01 mol dm^{-3}, 0.769 at 0.1 mol dm^{-3}, and 0.606 at 1.0 mol dm^{-3}. Estimate the amount of work involved in charging the hypothetical uncharged solution.

- *Method.* The electrical work is calculated from eqn (11.2.1) written in the form

  $w_e = RT \ln \gamma.$

- *Answer.* At a concentration 0.01 mol dm^{-3} the work is

  $w_e = (8.314 \text{ J K}^{-1} \text{ mol}^{-1}) \times (298.15 \text{ K}) \ln(0.901) = -0.258 \text{ kJ mol}^{-1}.$

  At 0.1 mol dm^{-3} we find $w_e = -0.651$ kJ mol^{-1}, and at 1.0 mol dm^{-3} $w_e = -1.242$ kJ mol^{-1}.

- *Comment.* The calculation indicates the size of energies we are dealing with. Notice that the work is negative in each case. This means that the energy of the system drops when interionic interactions are taken into account (the ions attract each other). The effect is greater at higher concentrations, because the ions are then closer together on average.

The ionic atmosphere. The Coulomb potential at a distance r from an ion of charge $z_i e$ is

(11.2.2) $\phi_i(r) = (z_i e/4\pi\varepsilon_0)(1/r).$

(A brief review of electrostatics is given in the Appendix on p. 340.) This is the potential due to an isolated ion in a vacuum. In solution two modifications are needed. In the first place the solvent decreases the strength of the field (Fig. 11.3), and if its permittivity is ε (epsilon) the potential is

(11.2.3) $\phi_i(r) = (z_i e/4\pi\varepsilon)(1/r).$

The permittivity is often expressed as $\varepsilon = \varepsilon_r \varepsilon_0$ where ε_r is the *relative*

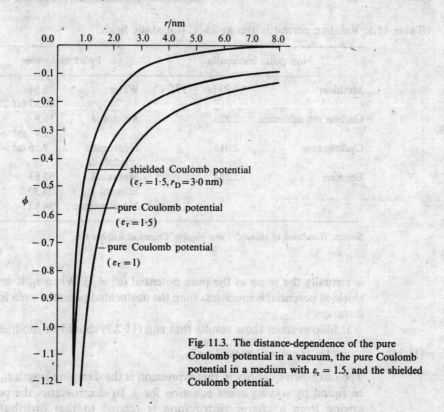

Fig. 11.3. The distance-dependence of the pure Coulomb potential in a vacuum, the pure Coulomb potential in a medium with $\varepsilon_r = 1.5$, and the shielded Coulomb potential.

permittivity (dielectric constant) of the medium. Since $\varepsilon_r > 1$ the potential is reduced from its vacuum value. This reduction of strength is very important in many solvents. For example, the relative permittivity of water is 78.5, and so at a given distance the Coulomb field is reduced from the vacuum value by almost two orders of magnitude. This is one reason why water is such a successful solvent: the Coulombic interactions are so strongly reduced by the solvent that the ions interact only weakly and do not conglomerate into a crystal. A list of relative permittivities is given in Table 11.1.

The second modification arises from the presence of the ionic atmosphere. A probe measuring the electrical potential in the vicinity of an ion enters the weak, oppositely charged ionic atmosphere as it is withdrawn from the position of the central ion, and so the potential falls off more rapidly than is predicted by eqn (11.2.3). The central charge is said to be *shielded* by the atmosphere and the appropriate potential is the shielded Coulomb potential in which $1/r$ is replaced by

$$1/r \rightarrow (1/r)\exp(-r/r_D).$$

The parameter r_D, which is called the *shielding length* or the *Debye length*, determines how strongly the potential is damped from its pure Coulomb value, Fig. 11.3. When r_D is very large $r/r_D \rightarrow 0$ and the shielded potential

Table 11.1. Relative permittivities at 25 °C (for static fields)

Non-polar molecules		Polar molecules	
Methane	1.70 (at −173 °C)	Water	78.54
			80.37 (at 20 °C)
Carbon tetrachloride	2.228	Ammonia	16.9
			22.4 (at −33 °C)
Cyclohexane	2.015	Hydrogen sulphide	9.26 (at −85 °C)
Benzene	2.274	Methanol	32.63
		Ethanol	24.30
		Nitrobenzene	34.82

Source: *Handbook of chemistry and physics,* Chemical Rubber Co.

is virtually the same as the pure potential ($e^0 = 1$). When r_D is small the shielded potential is much less than the unshielded potential even for short distances.

It follows from these results that eqn (11.2.3) should be modified to

(11.2.4) $\phi_i(r) = (z_i e / 4\pi\varepsilon)(1/r) \exp(-r/r_D).$

The unknown quantity in this expression is the damping length r_D. It can be found by solving some equation for ϕ. In electrostatics the potential arising from a charge distribution is related to that distribution by Poisson's equation (Appendix, p. 340). In the case of a spherically symmetrical charge distribution, where the *charge density* at a distance r from the central ion is $\rho_i(r)$ in any direction, the potential depends only on the radius and so it is permissible to use the following simplified form of the equation:

$$(1/r^2)(d/dr)(r^2 d\phi_i/dr) = -\rho_i(r)/\varepsilon.$$

Substituting the expression for $\phi_i(r)$ into this equation gives

(11.2.5) $\phi_i(r)/r_D^2 = -\rho_i(r)/\varepsilon.$

This equation might give an expression for r_D, but first it is necessary to know the charge density $\rho_i(r)$.

In order to find the charge density another equation is needed. Debye and Hückel proposed that the charge density at any point should be regarded as arising from the competition between the electrostatic attraction of the central ion for its counter-ions and the disruptive effects of thermal agitation. The energy of interaction of an ion of charge $z_j e$ with the central ion of charge $z_i e$ when they are separated by a distance r relative to their energy of interaction when they are infinitely far apart, is $\Delta E = z_j e \phi_i(r)$, where ϕ_i is the potential generated by the ion i. The Boltzmann distribution (p. 10) gives the proportion of ions with this energy when the temperature of the system is T:

$$\frac{\mathcal{N}_j(r)}{\mathcal{N}_j^0} = \frac{\text{number of ions of type j per unit volume where the potential is } \phi_i(r)}{\text{number of ions of type j per unit volume where the potential is zero}}$$

$$= e^{-\Delta E/kT}.$$

This means that

$$\mathcal{N}_j(r)/\mathcal{N}_j^0 = \exp\{-z_j e\phi_i(r)/kT\}.$$

In this expression $\mathcal{N}_j^0$ is the average number of ions of type j in unit volume of solution.

The *charge density* at a distance r from the ion i is the concentration there of ions of each type multiplied by the charge each one carries:

$$\rho_i(r) = \mathcal{N}_+(r)z_+ e + \mathcal{N}_-(r)z_- e$$

$$= \mathcal{N}_+^0 z_+ e \exp\{-z_+ e\phi_i(r)/kT\} + \mathcal{N}_-^0 z_- e \exp\{-z_- e\phi_i(r)/kT\}.$$

(It should be remembered that $z_j e$ is the charge of the ion j, so that z_+ is a positive number but z_- is negative.)

At this point a major simplification is to suppose that the average electrostatic interaction energy is small compared with kT. If it were large the attraction of ions would overcome the scattering influence of the thermal motion and the ions would condense into a crystal lattice. If we accept this simplification we may use the expansion $e^x = 1 + x + \dots$ to write the last relation as

$$\rho_i(r) = (\mathcal{N}_+^0 z_+ + \mathcal{N}_-^0 z_-)e - \{(\mathcal{N}_+^0 z_+^2 + \mathcal{N}_-^0 z_-^2)e^2 \phi_i(r)\}/kT + \dots$$

The first term on the right is zero because it is the concentration of charge in a uniform solution, and the solution is electrically neutral. The unwritten terms (represented by the dots) are assumed to be too small to be significant.

Example Estimate the size of the exponent in the expression for $\rho(r)$ and confirm that it is small for dilute aqueous solutions at room temperature.

- *Method.* If we ignore damping, the potential ϕ is equal to $(z_+ e/4\pi\varepsilon_0\varepsilon_r)(1/r)$. It depends upon r, and clearly gets very large for small r. Guess the average value of r (later on we shall be able to do this more accurately) as $r \approx 10$ nm (later we shall see that this is about right for a concentration of 0.001 mol dm^{-3}). For water at 25 °C, $\varepsilon_r = 78.5$. Take $z_+ = 1$ and $z_- = -1$.

- *Answer.* The quantity to be calculated is

$$-z_- e\phi_+/kT \approx (e^2/4\pi\varepsilon_0\varepsilon_r)(1/r)(1/kT)$$

$$\approx \left(\frac{(1.602 \times 10^{-19} \text{ C})^2}{4\pi \times (8.854 \times 10^{-12}) \text{ J}^{-1} \text{ C}^2 \text{ m}^{-1} \times 78.5)}\right)$$

$$\times \left(\frac{1}{1.0 \times 10^{-8} \text{ m}}\right)\left(\frac{1}{(1.381 \times 10^{-23} \text{ J K}^{-1}) \times (298.15 \text{ K})}\right)$$

$$\approx 0.071.$$

● *Comment*. In this case the quantity is significantly smaller than unity. Note that it becomes comparable to unity when $r \approx 3$ nm. We shall see that that is an appropriate value when the concentration is about 0.01 mol dm^{-3}, and so the theory will then fail.

The factor $\mathcal{N}^0_+ z^2_+ + \mathcal{N}^0_- z^2_-$ can be simplified as follows. First the number concentrations can be expressed in terms of molalities through

$$\mathcal{N}^0_j = m_j L \rho$$

where ρ is the solvent density, since we ignore the small change of volume caused by the presence of the salt, and L is Avogadro's constant. (m_j is the amount per unit mass, $m_j \rho$ is the amount per unit volume, and multiplying by L converts amounts into numbers.) Then we introduce a quantity, the *ionic strength*, I, which occurs widely whenever ionic solutions are discussed:

(11.2.6)
$$I = \tfrac{1}{2} \sum_j m_j z^2_j.$$

In the case of two types of ion at molalities m_+ and m_- this takes the form

(11.2.6a)
$$I = \tfrac{1}{2}(m_+ z^2_+ + m_- z^2_-).$$

Notice how I emphasizes the charges of the ions, because z_+ and z_- appear as their squares. For instance, in the case of a (1,1)-electrolyte (in which $z_+ = 1$ and $|z_-| = 1$) the ionic strength is $I = \tfrac{1}{2}(m_+ + m_-) = m$, where m is the solute molality (note that $m_+ = m_-$ because of electric neutrality). For a (1,2)-electrolyte which gives the ions $2M^+ + X^{2-}$, the ionic strength of a solution of molality m is $I = \tfrac{1}{2}(m_+ + 4m_-) = 3m$ because $m_- = m$ and $m_+ = 2m$. Table 11.2 summarizes the relation of ionic strength and molality in a readily useable form.

Table 11.2. Ionic strength and molality

The ionic strength I is related to the molality m by the relation $I = km$, where k is a number that depends on the valencies of the ions in the solution. The table below gives the value of k for various charge types. All have been derived on the basis of the definition of ionic strength in eqn (11.2.6).

	X^-	X^{2-}	X^{3-}	X^{4-}
M^+	1	3	6	10
M^{2+}	3	4	15	12
M^{3+}	6	15	9	42
M^{4+}	10	12	42	16

For example, the ionic strength of the salt M_2X_3, which is understood to be $M_2^{3+}X_3^{2-}$, at a molality m is $15\,m$.

The expression for the charge density in the vicinity of i now becomes

(11.2.7) $\rho_i(r) = -(2\rho e^2 LI/kT)\phi_i(r)$

which means that the charge density and the potential are directly proportional (but opposite in sign, because counter-ions predominate). Equation (11.2.5) can now be solved for r_D with no difficulty:

(11.2.8)
$$r_D^2 = \varepsilon RT/2\rho L^2 e^2 I.$$

This expression for r_D determines the potential (through eqn (11.2.4)) and the charge density (through eqn (11.2.7)) at any point near an ion.

Example (Objective 6). Calculate the shielding length r_D for a $1:1$ electrolyte in aqueous solution at $25\,°C$ when the concentration is (a) 0.001 mol kg^{-1}, (b) 0.01 mol kg^{-1}, (c) 0.10 mol kg^{-1}, (d) 1.0 mol kg^{-1}.

- *Method.* Simply substitute appropriate values into eqn (11.2.8). For a $1:1$ electrolyte (which means $z_+ = 1, z_- = -1$) the ionic strength is equal to the molality because eqn (11.2.6) reduces to $I = m$ since $m_+ = m_-$.

- *Answer.* The density of the solvent is $\rho = 0.997$ g cm^{-3}, or 0.997×10^3 kg m^{-3} and its relative permittivity is 78.5. Hence

$$r_D^2 = \frac{(8.854 \times 10^{-12}\ J^{-1}\ C^2\ m^{-1}) \times 78.5 \times (8.314\ J\ K^{-1}\ mol^{-1}) \times (298.15\ K)}{2 \times (0.997 \times 10^3\ kg\ m^{-3}) \times (6.022 \times 10^{23}\ mol^{-1})^2 \times (1.602 \times 10^{-19}\ C)^2 \times I}$$

$$= 9.284 \times 10^{-20}(I/mol\ kg^{-1})^{-1}\ m^2.$$

Therefore we have the following results:

$m/mol\ kg^{-1}$	0.001	0.01	0.1	1.0
r_D/nm:	9.64	3.05	0.96	0.30

- *Comment.* If we regard r_D as an indication of the mean value of r used in the last *Example* (p. 321), we see that the truncation of the exponential used in the Debye theory is valid only for molalities significantly less than 0.01 mol kg^{-1}. The discussion on p. 324 will confirm that r_D can be interpreted as an average charge separation: see the remarks following eqn (11.2.9) on p. 324. Notice what an important role the high permittivity of water plays in extending the range of validity of the theory.

Before putting these results to use we should investigate whether the expression for the Debye length, r_D, is plausible. It predicts that if we ignore the effect of density and relative permittivity, r_D increases with temperature. This is expected because thermal agitation disrupts the ionic atmosphere and weakens its shielding effect (in practice, though, the quantity $\varepsilon_r T/\rho$, which gives the overall temperature dependence and not just that stemming from the Boltzmann equation, actually decreases for water as the temperature is raised). The Debye length decreases with increasing ionic strength: the higher the concentration of ions the more effective the shielding. Furthermore, since the ionic strength emphasizes the charge of the ions, even a low concentration of highly charged ions may form an effective shield. This is illustrated in Fig. 11.4. The Debye length increases with increasing permittivity: when ε_r is large the marshalling effect of the central ion is weak, and the ionic atmosphere is diffuse.

The activity coefficient. The strategy for calculating the activity coefficient involves determining the electrical work of charging the central ion when it is

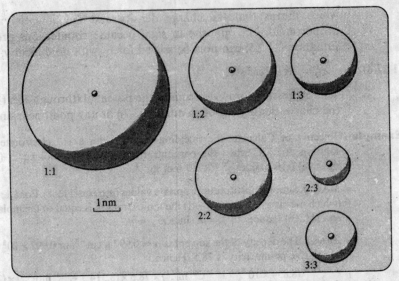

Fig. 11.4. The thickness of the ionic atmosphere (r_D) for ions of various charge types and $m = 0.01$ mol kg^{-1}.

already surrounded by its ionic atmosphere. It follows that we need to know the potential at the ion arising from the atmosphere, ϕ_{atmos}. This is the difference between the total potential (as given by eqn (11.2.4)) and the potential arising from the central ion itself:

$$\phi_{atmos}(r) = \phi_i(r) - \phi_{central\ ion}(r) = (z_i e/4\pi\varepsilon) \left\{ \frac{e^{-r/r_D}}{r} - \frac{1}{r} \right\}.$$

The magnitude of the potential at the ion (at $r = 0$) is then obtained by taking the limit as $r \to 0$:

$$\phi_{atmos}(0) = (z_i e/4\pi\varepsilon)\{[1 - r/r_D + \tfrac{1}{2}(r/r_D)^2 - \ldots]/r - (1/r)\}_{r \to 0}$$

(11.2.9) $$\qquad\qquad = -(z_i e/4\pi\varepsilon)(1/r_D).$$

This result indicates another interpretation of the Debye length: it shows that the potential at the central ion of charge $z_i e$ due to its atmosphere is equivalent to the potential arising from a single charge $-z_i e$ at a distance r_D.

If the charge of the central ion were q and not $z_i e$ the potential would be

$$\phi_{atmos}(0) = -(q/4\pi\varepsilon)(1/r_D).$$

The work involved in adding a charge dq to a region where the electrical potential is $\phi_{atmos}(0)$ is $\phi_{atmos}(0)\,dq$ (Appendix, p. 340). This is dw_e. Therefore the total work of fully charging unit amount of ions is

(11.2.10) $$w_e = L \int_0^{z_i e} \phi_{atmos}(0)\,dq = L \int_0^{z_i e} (-q/4\pi\varepsilon r_D)\,dq = -z_i^2 e^2 L/8\pi\varepsilon r_D.$$

It follows from eqn (11.2.1) that the activity coefficient of ions of type i is

$$\ln \gamma_i = -z_i^2 e^2 L / 8\pi\varepsilon r_D R T.$$

Example In the *Example* on p. 318 we saw that the electrical work involved in charging an electrolyte solution from a hypothetical uncharged state was -0.26 kJ mol^{-1} in the case of potassium chloride at a molality of 0.01 mol kg^{-1}. Check that the Debye–Hückel theory predicts the right order of magnitude for this quantity.

- *Method.* The work is given by eqn (11.2.10). The value of r_D for an 0.01 mol kg^{-1} aqueous solution of a 1:1 electrolyte was calculated on p. 323: $r_D = 3.05$ nm.

- *Answer.* From eqn (11.2.10) we have

$$w_e = \frac{-(1.602 \times 10^{-19}\ \text{C})^2 \times (6.022 \times 10^{23}\ \text{mol}^{-1})}{8\pi(8.854 \times 10^{-12}\ \text{J}^{-1}\ \text{C}^2\ \text{m}^{-1}) \times 78.5 \times (3.05 \times 10^{-9}\ \text{m})}$$

$$= -0.290\ \text{kJ mol}^{-1}.$$

- *Comment.* The quantities are in good agreement, suggesting that the Debye–Hückel theory is reliable and valid at this concentration.

The mean activity coefficient for an ionic solution formed from M_pX_q can now be calculated:

$$\ln \gamma_\pm = (1/n)(p \ln \gamma_+ + q \ln \gamma_-)$$

$$= -(1/n)(e^2 L / 8\pi\varepsilon r_D R T)(pz_+^2 + qz_-^2)$$

where $n = p + q$. But $pz_+ + qz_- = 0$ for neutrality, and so

$$\ln \gamma_\pm = -|z_+ z_-|(L e^2 / 8\pi\varepsilon r_D R T).$$

The expression for r_D given by eqn (11.2.8) may now be inserted into the last equation. If the logarithms are converted to base 10 this gives

$$\lg \gamma_\pm = -1.825 \times 10^6 |z_+ z_-| \left\{ \frac{(I/\text{mol kg}^{-1}) \times (\rho/\text{g cm}^{-3})}{\varepsilon_r^3 (T/\text{K})^3} \right\}^{1/2}$$

$$= -A |z_+ z_-| (I/\text{mol kg}^{-1})^{1/2}.$$

For water at 25 °C, where the density is 0.997 g cm^{-3} and the relative permittivity is 78.54, the value of A is 0.509. This gives the *Debye–Hückel Limiting Law* for water at 25 °C as

(11.2.11)
$$\lg \gamma_\pm = -0.509 |z_+ z_-| (I/\text{mol kg}^{-1})^{1/2}.$$

The name 'limiting law' is applied for the same reason as in the case of gases. Ionic solutions of arbitrary molality may have activity coefficients that differ from eqn (11.2.11), yet all solutions are expected to conform to this expression in the limit of sufficiently low molalities. They should conform if the basic model is correct and the approximations are tenable. In order to test it we turn to experiment.

Experimental tests, improvements, and extensions. In this section we examine the results of measuring activity coefficients. The measurement techniques will be described in the next chapter and some will be found in the Problems at the end of this chapter. Figure 11.5 shows how the observed activities of electrolytes of different valence type depend on the square-root of the ionic strength, and compares them with the theoretical curves based on eqn (11.2.11). The agreement at low molalities (less than 10^{-2} or 10^{-3} mol kg^{-1} depending on the valence type) is remarkable, and convincing evidence for the validity of the model. Nevertheless the departures from the theoretical curves above these molalities are large, and show that the approximations made in the Debye–Hückel calculation are valid only at very low concentrations. Some experimental values are listed in Table 11.3.

What are the approximations, and how may they be removed? Four principal approximations were made: (i) the bulk relative permittivity was used; (ii) the sizes of the ions were ignored and they were treated as points; (iii) the spherically symmetrical Poisson equation was combined with the Boltzmann equation, and the expression was linearized; and (iv) the departures from ideality were ascribed solely to Coulombic interactions.

The simplest approximation to remove is the assumption of negligible ion size. In the case of a (1,1)-electrolyte $r_D \approx 100$ ionic radii at molalities of 10^{-4} mol kg^{-1} but $r_D \approx 10$ radii at 10^{-1} mol kg^{-1}. This indicates that the assumption of point-ions is untenable even at moderate molalities (above 10^{-2} mol kg^{-1}). When the ions are considered as having a radius r_I we might expect that the shielded Coulomb potential falls off as

$$\phi(r) = A'(1/r)\exp\{-(r-r_1)/r_D\} \qquad (r \geqslant r_1).$$

(A' is some constant.) This form allows the potential its full, Coulombic

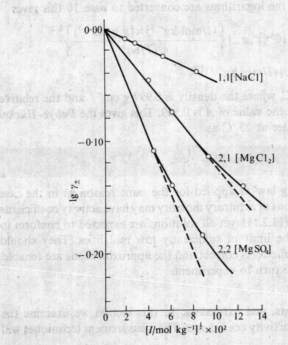

Fig. 11.5. The dependence of the activity coefficient on concentration for various valence types, and the Debye–Hückel limiting law predictions.

Table 11.3. Mean activity coefficients in aqueous solution at 25 °C

$m/\text{mol kg}^{-1}$	HCl	KCl	CaCl$_2$	H$_2$SO$_4$	LaCl$_3$	In$_2$(SO$_4$)$_3$
0.001	0.966	0.966	0.888	—	0.853	—
0.005	0.930	0.927	0.789	0.643	0.716	0.16
0.01	0.906	0.902	0.732	0.545	0.637	0.11
0.05	0.833	0.816	0.584	0.341	0.417	0.035
0.10	0.798	0.770	0.524	0.266	0.356	0.025
0.50	0.769	0.652	0.510	0.155	0.303	0.014
1.00	0.811	0.607	0.725	0.131	0.387	—
2.00	1.011	0.577	1.554	0.125	0.954	—

Source: S. Glasstone, *Introduction to electrochemistry*, Van Nostrand.

strength on the surface of the ion, when $r = r_I$, and the shielding operates only outside this distance.

This choice of this potential modifies the expression for the work of charging the central ion because now the charge is added to the surface of the ion where the potential is $\phi_{\text{atmos}}(r_I)$. The same sequence of manipulations that led to eqn (11.2.11) now leads to

$$(11.2.12) \qquad \lg \gamma_\pm = -\left\{ \frac{A}{(1+r_I/r_D)} \right\} |z_+ z_-| (I/\text{mol kg}^{-1})^{1/2}.$$

The Debye length which occurs in the denominator also depends inversely on the square root of the ionic strength, eqn (11.2.8), and so this expression has the form

$$(11.2.13) \qquad \lg \gamma_\pm = -\left\{ \frac{A}{1+A^*(I/\text{mol kg}^{-1})^{1/2}} \right\} |z_+ z_-| (I/\text{mol kg}^{-1})^{1/2},$$

where A^* is another constant.

When the solution is very dilute the denominator in eqn (11.2.12) is almost unity, and it reverts to the limiting law. The criterion for this is $r_I \ll r_D$, which is equivalent to assuming that the ions are of negligible size. When the concentration rises to the point where r_I is small but no longer negligible in comparison with $r_D (r_I \approx r_D/10)$, the approximation $(1+x)^{-1} \approx 1-x$ applied to the last equation gives the *extended Debye-Hückel Law*.

$$(11.2.14) \qquad \lg \gamma_\pm \approx -A |z_+ z_-| (I/\text{mol kg}^{-1})^{1/2} + AA^* |z_+ z_-| (I/\text{mol kg}^{-1}).$$

This predicts that the deviations from the limiting law should correspond to an increase in the activity coefficient (the second, new term is positive), and it accounts for the positive departures from the limiting law shown by the experimental lines in Fig. 11.5. The equation can be fitted to the observed curves by choosing an appropriate value of the constant A^*, which is best regarded as a variable parameter. A curve drawn in this way is shown in Fig. 11.6. It is clear that eqn (11.2.14) accounts for the

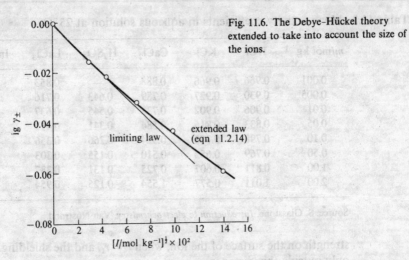

Fig. 11.6. The Debye-Hückel theory extended to take into account the size of the ions.

activity coefficient over a moderate range of dilute solutions; nevertheless it is still very poor for molalities in the vicinity of 1 mol kg^{-1}.

Several attempts have been made to improve the limiting law, but the real problem lies deep in the structure of the model. The fundamental flaw is the combination of the Boltzmann distribution with the Poisson equation. The Poisson-Boltzmann equation fails to allow for the inter-actions between the ions that constitute the atmosphere: each of them is also attempting to construct its own atmosphere at the same time as they are being collected around the ion of interest. The linearization step effectively ignores this competition, and becomes valid only in the limit of infinite dilution.

A self-consistent theory of ionic solutions is still awaited. The theory will probably be based on a dynamical view of ions in solution, and it may draw on the techniques used to account for the properties of plasma (ionized gases) which are central to the understanding of controlled nuclear fusion and rocket exhausts. Until that theory is available we shall have to use the Debye-Hückel Limiting Law, eqn (11.2.11) or the extended theory, eqn (11.2.14) and accept that the predictions are valid only for low molalities.

11.3 The role of electrodes

A large number of reactions of the form $A + B \rightleftharpoons C + D$ can be regarded as composed of a *reduction*, or *electronation*, step

reduction: $A + e^- \rightarrow C$

and an *oxidation*, or *de-electronation*, step

oxidation: $B \rightarrow D + e^-$.

One way of bringing about the reaction is to mix the reagents A and B, then the electron released by B in the oxidation step is transferred to a neighbouring A species. In the bulk of the mixture the spatial direction of the electron transfers is random, Fig. 11.7a, and they occur without doing any useful work (apart from pV-work as the reaction system expands or contracts). Another way of running the reaction is to separate the species A and B, but to connect them by an electric circuit making contact with the solutions via two electrodes, one of which acts as a source of electrons and the other as a sink, Fig. 11.7b. Then B sheds its electron into one electrode, and A collects its from another. The two electrodes of this cell need to be connected by a wire so that the electrons dumped by B can be transported to A. As the reaction proceeds an electric current is established in the external circuit, and may be used to do work.

The electrode where *oxidation* occurs is called the *anode*; that where *reduction* occurs is called the *cathode*. In the case of an electrochemical cell acting as a source of electricity the anode is at a low potential relative to the cathode. The reason for this can be understood as follows. In order for the reaction to proceed, the species undergoing reduction (A) withdraws electrons from its electrode (the cathode), so leaving a net positive charge on it. The opposite occurs at the other electrode, where oxidation is

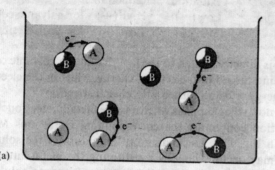

(a)

Fig. 11.7. (a) Electron transfers in the bulk of an oxidation–reduction system (b) The directional nature of electron transfers in an electrochemical cell.

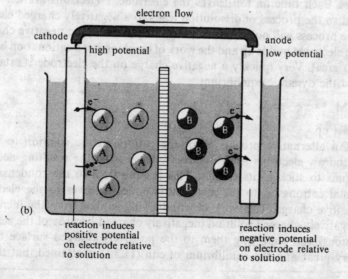

electron flow

cathode
high potential

anode
low potential

(b)

reaction induces
positive potential
on electrode relative
to solution

reaction induces
negative potential
on electrode relative
to solution

occurring: species B dumps electrons on its electrode (the anode), and gives to it a net negative charge. A test positive charge brought up to the cathode therefore has a higher energy than one brought up to the anode. Therefore the electric potential (which is defined in terms of positive test charges) at the cathode is higher than at the anode.

Confusion is sometimes encountered when discussing cathodes and anodes because their relative potentials are different when a cell is used in electrolysis. In this case, the anode is still the place where oxidation occurs, and the cathode the place where reduction occurs, but electrical work is now being used to produce chemical change. Now anions (negative species) have to be induced to move to the electrode in order to be oxidized, and cations have to be drawn to the other electrode in order to be reduced. Therefore the anode must now be made relatively positive with respect to the cathode so that the anions are induced to go there, and to donate their electrons.

The conventions may be summarized as follows:

Process:	Anode Oxidation	Cathode Reduction	
Potential:	Low ($-$)	High ($+$)	for galvanic cell
	High ($+$)	Low ($-$)	for electrolytic cell

The change is directly related to the fact that in the galvanic cell chemical energy is being turned into electrical energy, while in the electrolytic cell, electrical energy is being turned into chemical energy.

Consider now what happens when a single metallic electrode M is dipped into a solution containing the corresponding metal ions M^{+z}. This is illustrated in Fig. 11.8. Depending on the nature of the solution one of two things may occur. The ions constituting the metal lattice may have a tendency to break away from the metal and go into solution as M^{+z} ions. Each time an ion leaves the electrode, z electrons are left behind, and so the process of dissolution leads to a negatively charged electrode. The process will not continue indefinitely because the negative charge on the electrode builds up and the work of breaking out a cation soon becomes too great. Very quickly a negative charge on the electrode is established and the dynamic equilibrium

(11.3.1) $$M^{+z} + z e^- \rightleftharpoons M$$

is set up.

An alternative process is for the cations in the solution to tend to withdraw electrons from the electrode, and for the resulting neutral M atoms to stick to its surface. This is equivalent to the condensation of metal cations on to the metal surface, and it gives to the electrode a positive charge. The work of withdrawing cations from the increasingly negatively charged solution (negatively charged because of the remaining anions) and supplying them to the positively charged surface becomes very great and the equilibrium of eqn (11.3.1) is attained, but this time

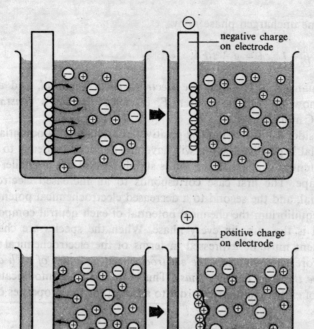

negative charge
on electrode

positive charge
on electrode

Fig. 11.8. Charge changes at the surface of an electrode.

with a positive charge on the electrode.

The modification of the charge on the electrode appears to an outside observer as a change of its electric potential. If an observer compares the work of bringing up a positive test charge from infinity he finds that more work is needed to bring it up to an electron-deficient piece of metal than to an uncharged piece. We can say that the electric potential of the electron-deficient electrode is greater than the potential of a neutral electrode. Likewise the addition of a positive test charge to an electron-excess electrode needs less work than the addition to the neutral electrode, and so its electric potential is less than that of the neutral electrode.

The electrochemical potential. In the section on ion activities we saw that the electric potential of the ionic atmosphere in the vicinity of an ion gives a contribution to the latter's Gibbs function. When a bulk phase is maintained at some electric potential we expect a similar modification of the Gibbs function. This can be looked at in another way. The Gibbs function measures the non-pV work available. When a charged species is added to a system in which the electric potential is ϕ the work done is different from the case when the potential is zero. Bringing up unit amount of charges of magnitude $z_i e$ from infinity into a region where the potential is ϕ involves an extra amount of work of magnitude $z_i e \phi L$. Therefore the partial molar Gibbs function in the charged phase $\bar{\mu}_i$ is related to that in the same uncharged phase μ_i by

(11.3.2) $$\bar{\mu}_i = \mu_i + L z_i e \phi = \mu_i + z_i F \phi.$$

The quantity $\bar{\mu}_i$ is called the *electrochemical potential*, and eL, which from now on will be written F, is called *Faraday's constant* ($1F = 96485\,C\,mol^{-1}$).

Consider a positive ion ($z_i e$ positive) in a phase with potential ϕ. If the potential is positive the ion will have an enhanced tendency to escape. If the potential is negative the ion is stabilized and has a smaller tendency to escape. The first case corresponds to an increased electrochemical potential, and the second to a decreased electrochemical potential.

At equilibrium the chemical potential of each neutral component of a system is the same in every phase. When the species are charged this statement must be expressed in terms of the electrochemical potential. Therefore, *at equilibrium the electrochemical potential of each component must be the same in every phase*. This is the principal modification of the work of earlier chapters in order to account for the properties of charged species.

The interfacial potential difference. In order to see how the electrochemical potential is employed consider the simplest type of electrode. The metal ion | metal electrode is illustrated in Fig. 11.9 and is denoted $M^{+z} \,|\, M$. Numerous examples of this type can be given, e.g., a silver electrode in contact with a silver nitrate solution, or copper in contact with copper sulphate. The equilibrium is between the M^{+z} ions in the solution and the metal atoms and electrons in the solid electrode:

$$M^{+z} + z e^- \rightleftharpoons M.$$

At equilibrium appropriately weighted sums of the electrochemical potentials of the 'reactants' and the 'products' are equal, and so by the same arguments as in Chapter 9

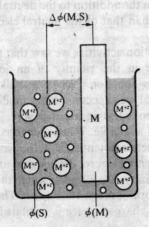

Fig. 11.9. Electric potentials at an electrode–electrolyte interface.

(11.3.3) $\qquad \bar{\mu}_{M^{+z}}(S) + z\bar{\mu}_{e^-}(M) = \bar{\mu}_M(M),$

where $\bar{\mu}(S)$ denotes an electrochemical potential in the solution and $\bar{\mu}(M)$ denotes an electrochemical potential in the metal electrode.

In the present case the M^{+z} ions are in the solution where the electric potential is $\phi(S)$ and the electrons and metal atoms are in the metal electrode where the electric potential is $\phi(M)$. The condition of equilibrium, eqn (11.3.3), is expressed in terms of the electric potentials as follows:

$$\{\mu_{M^{+z}}(S) + zF\phi(S)\} + z\{\mu_{e^-}(M) - F\phi(M)\} = \mu_M(M).$$

(The electrochemical potential of the atoms M is the same as their chemical potential because they are uncharged.) Therefore at equilibrium the *electric potential difference* between the solution and the metal is

(11.3.4) $\qquad \Delta\phi = \phi(M) - \phi(S) = (1/zF)\{\mu_{M^{+z}}(S) + z\mu_{e^-}(M) - \mu_M(M)\}.$

The first thing to do with this expression is to ascertain that it represents the expected behaviour. If the chemical potential of the cations in solution is large the above expression is positive. A high chemical potential implies a strong escaping tendency, and in this case the escape is of the ions from the solution into the electrode. That direction of migration tends to build up a positive charge on the electrode, and to make its potential positive relative to the solution. This is in fact the prediction of the last equation. The opposite charge is expected to appear on the electrode when the electrode tends to dissolve and to form cations. This tends to happen when the chemical potential of the atoms of the electrode is high, and eqn (11.3.4) does indeed predict that the potential of the electrode should drop below that of the solution when that is so.

The potential of the electrode depends on the concentration of M^{+z} ions present in the solution. The dependence can be deduced from eqn (11.3.4) by expressing the chemical potential of the ions in terms of their activity, for then

(11.3.5) $\qquad \Delta\phi = (1/zF)\{\mu_M^{\ominus}{}_{+z} + z\mu_{e^-}(M) - \mu_M(M)\} + (RT/zF)\ln a_{M^{+z}}.$

The appearance of this expression can be simplified. Let the molality of the ions in solution be arranged so that they are at unit activity; then the *standard potential difference* of the interface is

$$\Delta\phi^{\ominus} = (1/zF)\{\mu_M^{\ominus}{}_{+z} + z\mu_{e^-}(M) - \mu_M(M)\}.$$

Once this potential difference is known, the potential difference for any other ionic activity can be predicted by combining it with eqn (11.3.5):

(11.3.6) $\qquad \Delta\phi = \Delta\phi^{\ominus} + (RT/zF)\ln a_{M^{+z}}.$

We can check that this equation gives physically plausible predictions. For instance, when the activity of the ions in the solution rises (for example, when their molalities are increased) the ions have an increased tendency to leave the solution. That leads to a positive charge on the electrode and to an increase in its electric potential relative to the solution. The last equation conforms to this picture because it predicts that $\Delta\phi$ increases as a_{M^+} increases.

It is often convenient to convert potential difference expressions to logarithms to the base 10 using $\ln x = 2.3026 \lg x$. At room temperature $2.3026RT/F = 0.059\,15$ V. It follows that changing the activity by a factor of 10 changes the potential difference $\Delta\phi$ by 59.15 mV if $z = 1$.

11.4 The electric potential at interfaces

The potential difference across an interface at equilibrium can be related to the electrochemical potentials of the species involved. We saw on p. 332 how to do this in the case of the metal ion | metal electrode, and now we examine other cases.

The gas | inert metal electrode. The construction of the gas | inert metal electrode is illustrated in Fig. 11.10. The inert metal electrode acts as a source or sink of electrons but takes no other part in the reaction. The gas is bubbled over the surface of the electrode while it is bathed in a solution of ions related to the gas. For example, if the gas is chlorine the solution should contain Cl^- ions, and if it is hydrogen the solution should contain H^+ ions. The electrode is represented by $G^\pm | G_2 | Pt$ or $G^\pm | G_2, Pt$ (if the inert metal is platinum) and in each case the equilibrium can be expressed as

Case (a) $\tfrac{1}{2}G_2 + e^- \rightleftharpoons G^-$

Case (b) $G^+ + e^- \rightleftharpoons \tfrac{1}{2}G_2$.

The conditions for equilibrium are

Case (a): $\tfrac{1}{2}\bar{\mu}_{G_2}(g) + \bar{\mu}_{e^-}(M) = \bar{\mu}_{G^-}(S)$

Case (b): $\bar{\mu}_{G^+}(S) + \bar{\mu}_{e^-}(M) = \tfrac{1}{2}\bar{\mu}_{G_2}(g)$.

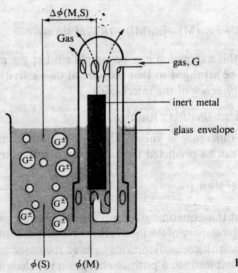

Fig. 11.10. A gas/inert metal electrode.

The gases, being uncharged, have electrochemical potentials equal to their chemical potentials. The ions are in the solution where the electrical potential is $\phi(S)$ and the electrons are in the inert metal where the potential is $\phi(M)$. Expressing the electrochemical potentials in terms of the potentials and the chemical potentials gives

(11.4.1a) Case (a): $\frac{1}{2}\mu_{G_2}(g) + \{\mu_{e^-}(M) - F\phi(M)\} = \{\mu_{G^-}(S) - F\phi(S)\}$,

(11.4.1b) Case (b): $\{\mu_{G^+}(S) + F\phi(S)\} + \{\mu_{e^-}(M) - F\phi(M)\} = \frac{1}{2}\mu_{G_2}(g)$.

Introducing

$$\mu_{G_2}(g) = \mu_{G_2}^\ominus + RT\ln(f_{G_2}/p^\ominus)$$

$$\mu_{G^+}(S) = \mu_{G^+}^\ominus + RT\ln a_{G^+}$$

$$\mu_{G^-}(S) = \mu_{G^-}^\ominus + RT\ln a_{G^-}$$

and rearranging in order to obtain the potential difference as $\Delta\phi = \phi(M) - \phi(S)$ gives

Case (a):

$$\Delta\phi = (1/F)\{\frac{1}{2}\mu_{G_2}^\ominus + \mu_{e^-}(M) - \mu_G^\ominus(S)\} + (RT/F)\ln(f_{G_2}^{1/2}/a_{G^-})$$

Case (b):

$$\Delta\phi = -(1/F)\{\frac{1}{2}\mu_{G_2}^\ominus - \mu_{e^-}(M) - \mu_{G^+}^\ominus(S)\} - (RT/F)\ln(f_{G_2}^{1/2}/a_{G^+}).$$

(For clarity the $f/p^\ominus$ has been abbreviated to f.)

When the gas is at unit fugacity and the ions are at unit activity, the metal–solution interface potential difference has its standard value $\Delta\phi^\ominus$. Then the potential difference at any fugacity and activity can be calculated from

(11.4.2a) Case (a): $\Delta\phi = \Delta\phi^\ominus + (RT/F)\ln(f_{G_2}^{1/2}/a_{G^-})$

(11.4.2b) Case (b): $\Delta\phi = \Delta\phi^\ominus - (RT/F)\ln(f_{G_2}^{1/2}/a_{G^+}).$

We can check that these equations make sense. Consider case (a), where the gas is in equilibrium with a solution of negative ions. Raising the activity of the negative ions G^- increases their tendency to discharge at the electrode. This indicates that we should expect the potential of the electrode to shift to more negative values. Since a_{G^-} occurs in the denominator of the logarithmic term, eqn (11.4.2a) does predict that $\Delta\phi$ decreases (gets more negative) as a_{G^-} is raised. The opposite effect is expected to occur when the equilibrium involves positive ions, and this is confirmed by the change of sign between eqn (11.4.2a) and (11.4.2b).

Example (Objective 10). Chlorine is bubbled over a platinum electrode dipping into a sodium chloride solution at 25 °C. Find the change in metal–solution potential difference when the chlorine pressure is increased from 1 atm to 2 atm.

● *Method.* The change is given by eqn (11.4.2a). The chloride activity remains constant; the chlorine may be assumed to behave perfectly at the pressure of interest, and so fugacities may be replaced by pressures. Use $RT/F = 25.69$ mV.

- **Answer.** $\Delta\phi(2\,\text{atm}) - \Delta\phi(1\,\text{atm}) = (RT/F)\ln(2\,\text{atm}/1\,\text{atm})^{1/2}$

$$= (25.69\,\text{mV}) \times \tfrac{1}{2}\ln 2 = 8.90\,\text{mV}.$$

- **Comment.** The potential difference increases when the pressure of the gas is increased because the $\tfrac{1}{2}Cl_2 + e^- \rightleftharpoons Cl^-$ equilibrium is driven to the right (in favour of Cl^-) and the withdrawal of electrons from the electrode makes it more positive.

The ion | insoluble salt | metal electrode. The construction of this electrode is illustrated in Fig. 11.11. It consists of a metal M covered by a layer of insoluble salt MX, the whole immersed in a solution containing the anion X^-. The electrode is denoted $X^- | MX | M$ or $X^- | MX,M$. A common example is the chloride | silver chloride | silver electrode which we look at in more detail below. The importance of this type of electrode is the dependence of its interfacial potential difference on the activity of the *anion* X^- in the solution.

We may picture the electrode as a system of two interfaces, one between the metal electrode and the metal ions in the insoluble salt:

$$M^+ + e^- \rightleftharpoons M,$$

and the other between the X^- anions in the solution and the X^- anions in the coating of insoluble salt:

$$MX \rightleftharpoons M^+ + X^-.$$

The overall equilibrium is

$$MX + e^- \rightleftharpoons M + X^-,$$

and this is established when

$$\bar{\mu}_{MX} + \bar{\mu}_{e^-}(M) = \bar{\mu}_M(M) + \bar{\mu}_{X^-}(S).$$

The electrochemical potential of the salt MX is the same as its chemical potential because it is electrically neutral, and the potentials of the metal

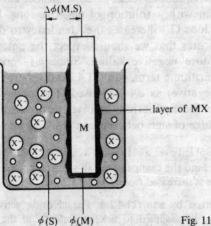

$\phi(S)$ $\phi(M)$ Fig. 11.11. An ion | insoluble salt | metal electrode.

and solution phases are $\phi(M)$ and $\phi(S)$ respectively. Therefore the last equation becomes

$$\mu_{MX} + \mu_{e^-}(M) - F\phi(M) = \mu_M(M) + \mu_{X^-}(S) - F\phi(S),$$

or, writing $\Delta\phi = \phi(M) - \phi(S)$,

$$\Delta\phi = (1/F)\{\mu_{MX} + \mu_{e^-}(M) - \mu_{X^-}(S) - \mu_M(M)\}$$

The chemical potential of the ion X^- can be expressed in terms of its activity in the usual way, and the insoluble salt, being a pure solid, is in its standard state. If the standard potential difference is identified as $(1/F)\{\mu_{MX}^\ominus + \mu_{e^-}(M) - \mu_{X^-}^\ominus(S) - \mu_M(M)\}$, the potential difference at any anion activity can be expressed as

(11.4.3) $$\Delta\phi = \Delta\phi^\ominus - (RT/F)\ln a_{X^-}.$$

As anticipated, the potential difference depends on the activity of the anions in the solution. For example, the $Cl^- | AgCl | Ag$ electrode, in which the overall reaction is

$$AgCl + e^- \rightleftharpoons Ag + Cl^-,$$

has a potential difference that depends on the chloride activity. The *calomel electrode* $Cl^- | Hg_2Cl_2 | Hg$ behaves similarly; its overall reaction equilibrium is

$$\tfrac{1}{2}Hg_2Cl_2 + e^- \rightleftharpoons Hg + Cl^-$$

and its potential depends on the Cl^- activity.

Example (Objective 11). Calculate the change of potential difference between a silver | silver chloride electrode and a solution saturated with silver chloride when an excess of 0.01 mol kg^{-1} solution of potassium chloride solution is added.

- *Method.* The potential difference is determined by the chloride ion activity according to eqn (11.4.3). When silver chloride is present alone the chloride molality is very low (10^{-5} mol kg^{-1}) but when the potassium chloride is added it rises sharply to 0.01 mol kg^{-1}. The activity coefficient is 0.906 at this concentration (Table 11.3) and 0.996 at 10^{-5} mol kg^{-1} (use eqn (11.2.11)). Use $RT/F = 25.69$ mV.

- *Answer.* $\Delta\phi(10^{-2}$ mol $kg^{-1}) - \Delta\phi(10^{-5}$ mol $kg^{-1})$
    $= -(25.69$ mV$)\ln(0.01 \times 0.906/0.996 \times 10^{-5})$
    $= -0.175$ V.

- *Comment.* The solubility of silver chloride can be measured by an adaptation of this result. We shall see how in the next chapter (p. 364).

Oxidation–reduction electrodes. In the general sense of the term all electrodes involve reduction (electronation) and oxidation (de-electronation), but the term oxidation–reduction electrode, or *redox electrode*, is normally confined to the case where a species exists in the solution in two oxidation states. The equilibrium is

$$Ox + ve^- \rightleftharpoons Red^{v-}$$

where Ox is the oxidized form and Red the reduced, and we have allowed for the possibility that v electrons are involved in the reduction. This type of redox reaction includes

$$Fe^{3+} + e^- \rightleftharpoons Fe^{2+}$$

(where Red is Fe^{2+} and Ox is Fe^{3+}) and organic equilibria such as the hydroquinone equilibrium

which is a model of important biological reactions, and the reduction of anthracene:

The redox electrode is denoted Red, Ox | M, where M is the inert metal making electrical contact with the solution.

The equilibrium can be expressed in terms of the electrochemical potentials of the species in the appropriate phases. Both the oxidized and the reduced forms are in the solution where the electric potential is $\phi(S)$. The electrons are confined to the metal electrode which is dipping into the solution and acting as a source or sink of electrons, depending on the demands of the reaction.

The equilibrium condition is

$$\bar{\mu}_{Ox}(S) + v\bar{\mu}_{e^-}(M) = \bar{\mu}_{Red}(S).$$

Bearing in mind that Red has v more negative charges than Ox, this rearranges to

$$\Delta\phi = (1/vF)\{\mu_{Ox}(S) - \mu_{Red}(S) + \mu_{e^-}(M)\}.$$

In terms of activities this reads

$$\Delta\phi = (1/vF)\{\mu_{Ox}^\ominus - \mu_{Red}^\ominus + \mu_{e^-}(M)\} + (RT/vF)\ln(a_{Ox}/a_{Red}).$$

In a solution in which the activities of both Ox and Red are unity the interfacial difference is $\Delta\phi^\ominus$. Therefore the potential difference at a general molality is

(11.4.4) $$\Delta\phi = \Delta\phi^\ominus + (RT/vF)\ln(a_{Ox}/a_{Red}).$$

You should notice that this equation also applies to all the previous cases we have discussed. That is, it is an easy way of remembering eqns (11.3.6), (11.4.2), and (11.4.3).

The importance of this equation sometimes arises from the following alternative interpretation. Instead of treating the potential as arising from the redox equilibrium, we can imagine controlling the position of the equilibrium by modifying the potential difference across the solution | electrode interface. If in some way we can ensure that the potential difference is $\Delta\phi$, then the concentration of the reduced and oxidized components will adjust so that eqn (11.4.4) is satisfied.

Liquid junctions. Electric potential differences occur across the junction of ionic solutions. A simple case is the junction between different concentrations of hydrochloric acid. At the junction the mobile protons diffuse into the more dilute solution. The bulkier chloride ions follow, but initially do so more slowly. These different rates of diffusion of oppositely charged species set up a potential difference, the junction potential. Since the interface blurs unless it is constantly renewed, this potential is a non-equilibrium phenomenon, and so, even though it plays an important role in later discussions, it cannot be analysed using equilibrium thermodynamics.

Membrane potentials. Consider two solutions containing different concentrations of the salt MX, and let them be separated by a membrane permeable only to M^+. This ion tends to diffuse into the more dilute solution, but the anion X^- cannot follow and so a potential difference is set up across the membrane, Fig. 11.12. The process will not proceed indefinitely because the separation of charges retards the further migration of cations. At equilibrium the electrochemical potentials of M^+ are the same on either side of the barrier. If the solutions are denoted α and β the condition of equilibrium is $\bar{\mu}_{M^+}(\alpha) = \bar{\mu}_{M^+}(\beta)$, or in terms of the potentials

$$\mu_{M^+}(\alpha) + F\phi(\alpha) = \mu_{M^+}(\beta) + F\phi(\beta).$$

Expressing this in terms of the activities of M^+ gives the equilibrium potential difference $\Delta\phi = \phi(\alpha) - \phi(\beta)$ as

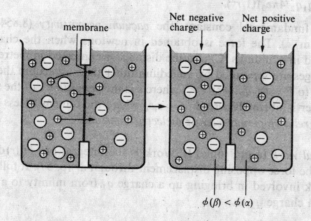

$$\phi(\beta) < \phi(\alpha)$$

Fig. 11.12. Formation of a membrane potential.

(11.4.5) $\Delta\phi = (RT/F)\ln\{a_{M^+}(\beta)/a_{M^+}(\alpha)\}.$

If β is the more concentrated phase this potential difference is positive, $\phi(\alpha) > \phi(\beta)$, because positive ions have passed into the α solution and have discarded anions in the β solution, Fig. 11.12.

One important example of a membrane system resembling this description is the biological cell wall, which is more permeable to K^+ than either Cl^- or Na^+. Inside the cell the K^+ concentration is about 10–30 times that on the outside, and is maintained at that level by a specific pumping operation fuelled by ATP (p. 277) and governed by enzymes. If the system is approximately at equilibrium the potential difference, $\phi(\text{inside}) - \phi(\text{outside})$, across the cell wall is predicted to be

$$\Delta\phi \approx (0.059\,\text{V})\lg(1/20) \approx -70\,\text{mV},$$

which accords quite well with the measured value. This potential difference plays a particularly interesting role in the transmission of nervous impulses. K^+ and Na^+ pumps occur in high abundance in the cells of the nervous system and when the nerve is inactive there is a high K^+ concentration inside the cells and a high Na^+ concentration outside. The potential difference across the cell wall is about -70 mV. When the cells are subjected to a pulse of about 20 mV the structure of the membrane adjusts and becomes permeable to Na^+. This causes a drop in the interfacial potential difference as the Na^+ ions flood in. This change of potential difference triggers the neighbouring cells and the pulse of collapsing potential is passed along the nerve. Behind the pulse the sodium and potassium pumps restore the concentration differential ready for the next pulse.

Appendix: electrostatics

Electrostatic force (units: newton, N). The basic equation of electrostatics is that for the magnitude of the Coulomb force between two charges q_1, q_2 separated by a distance r in a vacuum:

$$F = (q_1 q_2/4\pi\varepsilon_0)(1/r^2).$$

ε_0 is a fundamental constant, the *vacuum permittivity* (8.854×10^{-12} $J^{-1}\,C^2\,m^{-1}$). The force is obtained in newtons when the charges are expressed in coulombs (C) and the distance is expressed in metres. When the charges are separated by a medium other than a vacuum the force is reduced to $F = (q_1 q_2/4\pi\varepsilon)(1/r^2)$ where ε is the *permittivity* of the medium. The latter is normally written $\varepsilon = \varepsilon_r\varepsilon_0$ where the dimensionless quantity ε_r is the *relative permittivity* or *dielectric constant*.

Electrical work (units: joule, J). Work is the integral of $-F(r)\,dr$, where $F(r)$ is the force opposing displacement through dr (p. 57). It follows that the work involved in bringing up a charge q_1 from infinity to a distance r from a charge q_2 is

$$w_e = -\int_\infty^r F(r)\,dr = -(q_1 q_2/4\pi\varepsilon_0)\int_\infty^r (1/r^2)\,dr = (q_1 q_2/4\pi\varepsilon_0)(1/r).$$

Electric potential (units: volt, V; $1\ \text{V} = 1\ \text{J C}^{-1}\,\text{s}^{-1}$). Doing work of the amount just calculated results in raising the potential energy, $\mathscr{V}$, of the system by the same amount. This potential energy can be written

$$\mathscr{V} = q_1 \phi(r), \qquad \phi(r) = (q_2/4\pi\varepsilon_0)(1/r),$$

where $\phi(r)$ is called the *potential* at r arising from the charge q_2. (Distinguish between potential and potential energy.) The particular form quoted is the *Coulomb potential*: in a medium it is reduced by a factor ε_r. When the charge distribution is more complex than a single point it is described in terms of a *charge density* ρ (units: C m^{-3}): the electric potential arising from a given charge distribution is the solution to *Poisson's equation*:

$$\nabla^2 \phi = -\rho/\varepsilon_0$$

where ∇^2 is $(\partial^2/\partial x^2)+(\partial^2/\partial y^2)+(\partial^2/\partial z^2)$. If the distribution is spherically symmetrical, so too is $\phi(r)$, and the equation then simplifies to the form given on p. 320.

Electric field strength (units: V m^{-1}). Just as the energy of a charge q_1 can be written $\mathscr{V} = q_1 \phi$, so the force on q_1 can be written $F = q_1 E$, where E is the magnitude of the electric field strength arising from q_2 (or from some more general charge distribution). The electric field strength (which, like the force, is actually a vector quantity) is the negative gradient of the electric potential:

$$\mathbf{E} = -\nabla\phi.$$

(∇ is the vector derivative, or *gradient*, $\mathbf{i}(\partial/\partial x)+\mathbf{j}(\partial/\partial y)+\mathbf{k}(\partial/\partial z)$.)

Further reading

Modern electrochemistry. J. O'M. Bockris and A. K. N. Reddy; Plenum, New York, 1970.

Electrolyte solutions. R. A. Robinson and R. H. Stokes; Academic Press, New York, 1959.

The physical chemistry of electrolytic solutions. H. S. Harned and B. B. Owen; Reinhold, New York, 1958.

Ionic solution theory. H. L. Friedman; Wiley–Interscience, New York, 1962.

Treatise on electrochemistry. G. Kortum; Elsevier, Amsterdam, 1952.

Problems

11.1. The *ionic strength* was defined on p. 322. It is a crucial quantity in the discussion of ionic solutions, and the first few Problems give practice in manipulating it. As a first step, find an expression for the ionic strength in terms of the concentration c expressed in mol dm^{-3}.

11.2. Show that the ionic strengths of solutions of KCl, $MgCl_2$, $FeCl_3$, $Al_2(SO_4)_3$, $CuSO_4$, are related to their molalities as follows: $I(KCl) = m$, $I(MgCl_2) = 3m$,

$I(FeCl_3) = 6m, I[Al_2(SO_4)_3] = 15m, I(CuSO_4) = 4m.$

11.3. Calculate the ionic strength of a solution which is 0.1 mol kg^{-1} in KCl and 0.2 mol kg^{-1} in $CuSO_4$.

11.4. 5 g of KCl is added to a solution prepared by adding 5 g of $FeCl_3$ to 100 g water. What is the ionic strength of the solution?

11.5. What molality of $CuSO_4$ has the same ionic strength as a 1 mol kg^{-1} solution of KCl?

11.6. The other set of quantities which appeared through the chapter were the mean activities and mean activity coefficients. Demonstrate, starting from $\mu(M_pX_q) = p\mu(M) + q\mu(X)$, that the activity of a solution of a salt M_pX_q can be written $a(M_pX_q) = p^pq^q\gamma_\pm^{p+q}m^{p+q}$, where m is its molality (actually $m/m^\ominus$) and $\gamma_\pm$ is the mean activity coefficient $(\gamma_+^p\gamma_-^q)^{1/(p+q)}$.

11.7. Express the activities of the salts KCl, $MgCl_2$, $FeCl_3$, $CuSO_4$, and $Al_2(SO_4)_3$ in terms of the molality m and the mean activity coefficients.

11.8. Estimate the amount of electrical work involved at 25 °C in charging 2 dm^3 of a hypothetical uncharged 0.50 mol kg^{-1} solution of (a) NaCl, with $\gamma_\pm = 0.679$ at 25 °C, (b) $In_2(SO_4)_3$, with $\gamma_\pm = 0.014$.

11.9. The Coulomb potential has a very long range. Consider a spherical shell of electrons around a single proton, there being one electron in every square centimetre of the surface of the shell. What force does the proton exert on the *entire* shell of electrons, when the latter's radius is (a) 10 cm, (b) 1 m, (c) 10^6 km?

11.10. The shielded Coulomb potential has a much shorter range. Repeat the calculation on the basis that the potential is of the form of eqn (11.2.4) with $\varepsilon_r = 1$ but $r_D = 10$ cm.

11.11. The Debye–Hückel theory lets us calculate the shielding length in dilute ionic solutions. Find its value in a 0.001 mol kg^{-1} solution of magnesium iodide at (a) 25 °C, (b) 0 °C.

11.12. The Debye–Hückel theory is by no means limited to water as solvent, but the latter's high relative permittivity and consequent suppression of the Coulombic forces between ions means that it is applicable at higher concentrations. What is the shielding length in the case of a 0.001 mol kg^{-1} solution of magnesium iodide in liquid ammonia at −33 °C. The solvent density is 0.69 g cm^{-3}, and $\varepsilon_r = 22$.

11.13. Calculate the value of the Debye–Hückel A-coefficient for solutions in liquid ammonia at −33 °C. $\rho(NH_3) = 0.69$ g cm^{-3} and $\varepsilon_r = 22$ at this temperature.

11.14. Calculate the mean activity coefficient for aqueous solutions of NaCl at 25 °C at the molalities 0.001, 0.002, 0.005, 0.010, 0.020 mol kg^{-1}. The experimental values are 0.9649, 0.9519, 0.9275, 0.9024, and 0.8712. By plotting lg $\gamma_\pm$ against $I^{1/2}$ confirm that the Debye–Hückel law gives the correct limiting behaviour.

11.15. But how are the activity coefficients actually measured? One of the most important methods makes use of the electrochemical cell, and we shall meet that in the next chapter. In the next few Problems, however, we shall see some alternative methods. Several methods hinge on the Gibbs–Duhem equation (p. 219) and so we begin by doing some manipulations involving it. The Gibbs–Duhem equation links the chemical potentials, and therefore the activities, of solvent and solute. It follows that the vapour pressure (and hence the colligative properties) of the solvent should depend on the activities of the ions it contains. As a first step in finding the connection, repeat the freezing point depression calculation in terms of the Gibbs–Helmholtz equation, Problem 8.25, but do not assume that the solution is ideal. Derive the expression $d \ln a_A = (M_{A,m}/K_f) d\delta T$, where δT is the depression, $M_{A,r}$ the R.M.M. of the solvent, and a_A its activity. Use the Gibbs–Duhem equation to deduce that $d \ln a_B = (-1/m_B K_f) d\delta T$ where a_B is the activity of the solute and m_B its molality.

11.16. The *osmotic coefficient* for a $1:1$ electrolyte may be defined as $\varphi = \Delta T/2mK_f$, where m is the solute molality. It is important to remember in what follows that T depends on m, and that $\varphi \to 1$ in infinitely dilute solutions. There are two ways of dealing with the equation $d \ln a = (-1/mK_f) d\delta T$ derived above. The first assumes that the activity coefficient for the solute is given by the Debye–Hückel Limiting Law. Assume that this is so, and hence show that the osmotic coefficient at some molality m is given by $\varphi = 1 - \frac{1}{3} A m^{1/2}$.

11.17. The cryoscopic constant of water is 1.858 K/(mol kg^{-1}). When sodium chloride was dissolved in water the following freezing point depressions were observed. Confirm that the correct limiting law behaviour is obeyed at low concentrations.

$m/\text{mol kg}^{-1}$	0.001	0.002	0.005	0.010	0.020
$10^3 \, \delta T/K$	3.696	7.376	18.36	36.43	72.5

11.18. The other way of using the result of Problem 11.15 involves setting it up in such a way that it can be used to *measure* the activity coefficients, and not merely to test some theory about them. Begin by finding $d\varphi/dm$, and then deducing that

$$-\ln \gamma_\pm = 1 - \varphi(m) + \int_0^m (1/m)[1 - \varphi(m)] \, dm.$$

11.19. The following freezing point depressions were observed when KCl was dissolved in water. What is the mean activity coefficient for KCl in 0.05 mol kg^{-1} solution? Use the expression just derived, but take the limiting law value of φ (Problem 11.16) for the integration from $m = 0$ to 0.01 mol kg^{-1}.

$m/\text{mol kg}^{-1}$	0.01	0.02	0.03	0.04	0.05
$\delta T/K$	0.0355	0.0697	0.1031	0.137	0.172

11.20. The *dissociation constant* of an acid electrolyte MA is defined as $K = a(M^+)a(A^-)/a(MA)$. The degree of dissociation α of acetic acid was measured at $25\,^{\circ}$C over a range of concentrations, the number of ions present being determined by measuring the conductivity of the solution. Use the data reported below to confirm that the Debye–Hückel Limiting Law correctly predicts the limiting behaviour of γ at low concentrations by demonstrating that $\lg K_c$ plotted against $\sqrt{(\alpha c)}$, where K_c is the dissociation constant in terms of concentrations, should be a straight line.

$10^3 c(MA)/\text{mol dm}^{-3}$	0.0280	0.1114	0.2184	1.0283	2.414	5.9115
α	0.5393	0.3277	0.2477	0.1238	0.0829	0.0540

11.21. The pK_a value of an acid is defined as $-\lg K_a$, where K_a is the dissociation constant (in terms of activities). Use the Debye–Hückel Limiting Law to estimate the value of pK_a', where the prime indicates the dissociation constant in terms of the concentration, for 0.1 mol dm^{-3} acetic acid at $25\,^{\circ}$C given that its value at zero ionic strength is 4.756.

11.22. One of the properties of a salt that is modified by ionic interactions of various kinds is its solubility. The next few Problems explore some of these effects. We begin by defining the *solubility product*, K_{sp}, because this summarizes the equilibrium between the amounts of ions in solution in a saturated solution. K_{sp}

for a salt M_pX_q which dissolves in some solvent to give the ions $M^{+|z_+|}$ and $X^{-|z_-|}$ is defined as $K_{sp} = a_+^p a_-^q$, the activities being those of the ions in the saturated solution. As a first step, express K_{sp} in terms of the molality m and the mean activity coefficient of the ions in the saturated solution. In very dilute solutions (as of sparingly soluble salts) it may be possible to set the activity coefficients equal to unity. Assume that this is the case for AgCl and BaSO$_4$ in water, and calculate the values of their solubility products on the basis that their saturated solutions are found to have concentrations 1.34×10^{-5} mol dm^{-3} and 9.51×10^{-4} mol dm^{-3} respectively.

11.23. The solubility product is an equilibrium constant, and so it ought to be possible to find its value from tables of thermodynamic data. Calculate the solubility of AgBr in water at 25 °C on the basis of the following information:

$$\Delta G^\ominus_{f,m}(\text{AgBr, s}) = -95.9 \text{ kJ mol}^{-1},$$

$$\Delta G^\ominus_{f,m}(\text{Ag}^+, \text{aq}) = 77.1 \text{ kJ mol}^{-1},$$

$$\Delta G^\ominus_{f,m}(\text{Br}^-, \text{aq}) = -104.0 \text{ kJ mol}^{-1}.$$

11.24. The solubility of a salt is modified if the solution contains another salt with one ion in common: this is the *common-ion effect*, and was met at the end of Chapter 10. Now we can deal with it quantitatively. Suppose that a 1:1 salt MX has a solubility product K_{sp} and that the saturated solution is so dilute that $\gamma_\pm \approx 1$; let the saturated concentration be $c_0(\text{MX})$. Now consider the saturated solubility in a solution containing a concentration $c(\text{NX})$ of some freely soluble and fully dissociated 1:1 salt NX. Show that the solubility of MX shifts to $c_0'(\text{MX}) = \frac{1}{2}[c(\text{NX})^2 + 4K_{sp}]^{1/2} - \frac{1}{2}c(\text{NX})$.

11.25. Investigate the effect of taking into account the deviations from ideality. Suppose that the ionic strength is dominated by the added freely soluble salt, but that it is still possible to use the Debye–Hückel Limiting Law. Show that $c(\text{MX}) \approx m(\text{MX}) \approx K_{sp}^{1/2} + A K_{sp}^{1/4}$ when $A K_{sp}^{1/4} \ll 1$.

11.26. Calculate the solubility of AgBr in water containing KBr at a concentration of 0.01 mol dm^{-3}; take account of activity coefficients.

11.27. The solubility of a salt may also be modified by the presence of another salt even though there are no ions in common. This arises because the addition of ions changes the ionic strength of the solution, and therefore the activity coefficients are changed too. The solubility product itself remains unchanged, but its relation to concentrations is modified as a result of the change in the activity coefficients. Find an expression for the solubility of a sparingly soluble 1:1 salt MX in the presence of a freely soluble salt at ionic strength I in the region where the Debye–Hückel Limiting Law is applicable.

11.28. What is the solubility of AgCl in the following solutions at 25 °C: (a) 0.1 mol dm^{-3} KCl, (b) 0.01 mol dm^{-3} KCl, (c) 0.01 mol dm^{-3} KNO$_3$? Take account of activity coefficients (Problem 11.25). Evaluate K_{sp} from Tables 5.3 and 12.2.

11.29. Another way of determining activity coefficients is to write $K_{sp} = K'_{sp}K_\gamma$, where K'_{sp} is the solubility product expressed in molalities and K_γ the appropriate combination of activity coefficients, to measure the solubility at some concentration, and then, knowing K_{sp}, to get K_γ. The solubility of AgCl in aqueous magnesium sulphate was measured at 25 °C with the following results:

$m/(\text{mol MgSO}_4/\text{kg})$	0.001	0.002	0.003	0.004	0.006	0.010
$10^5 m/(\text{mol AgCl/kg})$	1.437	1.482	1.547	1.575	1.598	1.650

where m(mol AgCl/kg) is the saturation molality. Find the thermodynamic solubility product (by extrapolation to zero ionic strength) and the mean ionic activity coefficient when the $MgSO_4$ concentration is 0.004 mol kg^{-1}.

11.30.   What is the difference in electrochemical potential when a Cu^{+2} ion moves from one region of a solution to another, the electrical potential difference being 2 V? Show that negative ions tend to move to regions of relatively positive potential.

11.31.   The pH of a solution is defined as pH $= -\lg a(H^+)$, and we shall explore this quantity in more detail in the next chapter. Show that the potential difference across a Pt, $H_2(g)$ electrode is proportional to the pH of the solution. By how much does the potential difference change when the electrode is transferred from a strongly acid solution (pH ≈ 0) to a strongly alkaline solution (pH ≈ 14).

11.32.   In Section 11.4 we examined the potential difference across a gas | metal electrode at which the reaction was $\frac{1}{2}G_2 + e^- \rightleftharpoons G^-$. Suppose we were interested in a reduction that required two electrons, as in $\frac{1}{2}O_2 + 2e^- \rightleftharpoons O^{2-}$. How does the potential difference across an electrode operating by virtue of this reaction depend on the pressure of the oxygen? Does the potential difference increase or decrease when the pressure is raised? Account physically for your answer.

11.33.   The antimony | antimony oxide electrode Sb, Sb_2O_3 | OH^- is reversible with respect to hydroxide ions and so it has played an important role in electrochemical measurements. Set up the expression for the dependence of the potential difference across the electrode solution interface, in terms of the OH^- activity in the solution. By how much does the interfacial potential difference change when the NaOH concentration is changed from 0.01 mol kg^{-1} to 0.05 mol kg^{-1}. Use the Debye–Hückel Limiting Law to estimate the activity coefficients you require.

11.34.   Consider an electrode that responds to the equilibrium between chromium (III) ions and chromium (VI) oxide ions, $Cr_2O_7^{2-}$, according to the reaction

$$Cr_2O_7^{2-}(aq) + 14H^+(aq) + 6e^- \rightleftharpoons 2Cr^{3+}(aq) + 7H_2O.$$

Find an expression for the potential difference across the electrode interface.

12 Equilibrium electrochemistry: electrochemical cells

•

Learning objectives

After careful study of this chapter you should be able to

(1)   Describe the construction of a *cell with a liquid junction* and a *cell without a liquid junction* (p. 347), and explain the notation for them (p. 348).

(2)   Describe how the equilibrium *e.m.f.* is measured (p. 349).

(3)   Explain how to recognize if a cell is *thermodynamically reversible*, and indicate the sources of irreversibility.

(4)   Define the *electromotive force* (e.m.f.) of a cell (p. 350).

(5)   Define *electrode potential* and describe the *sign convention* (p. 350).

(6)   Relate the e.m.f. of a cell to the *spontaneous direction of change* of the cell reaction (p. 352).

(7)   Derive and use the *Nernst equation* for the concentration dependence of the e.m.f. of a cell, and define the term *standard e.m.f.* (eqn (12.1.2)).

(8)   Relate the standard e.m.f. to the *equilibrium constant* of the cell reaction (eqn (12.1.3)).

(9)   Describe the methods of measuring *standard electrode potentials* (p. 358), and explain the significance of their order in tables (p. 358).

(10)   Describe the measurement of *activity coefficients* (p. 359).

(11)   Use e.m.f. data to deduce *Gibbs functions* of reactions (p. 360).

(12)   Explain the convention for tabulating *Gibbs functions of formation* of ions (p. 361) and state and use the *Born equation* (eqn (12.3.3)) to account for their values.

(13)   Relate the *temperature coefficient of e.m.f.* to the entropy of reaction (eqn (12.3.4)) and the enthalpy of reaction (eqn (12.3.5)), and explain the convention for listing *standard ion entropies* (p. 363).

(14)   Define *solubility product* and deduce its value from e.m.f. measurements (p. 364).

(15)   Describe the electrochemical basis of *potentiometric titrations* (p. 365).

(16)   Define *pH* and *pK*, and the terms *conjugate base* and *conjugate acid*, (p. 367).

(17)   Describe how to use e.m.f. measurements to determine pH and pK (p. 370).

(18)   Explain the electrochemical basis of *indicator detection* of the endpoints of titrations (p. 370).

(19)   Describe the action of a *buffer solution* (p. 374).

Introduction

In this chapter we examine the information that can be obtained by studying the potential difference between two electrodes combined to form an electrochemical cell. We shall see that it is possible to draw up a scheme for discussing the equilibrium position of any reaction in solution in terms of the electrode potentials associated with the oxidation and reduction processes into which it may be analysed. We shall also see how electro-chemical measurements are used to obtain values of the thermodynamic quantities associated with reactions, such as their Gibbs functions, entropies, and enthalpies. Then we shall turn to the ways in which electro-chemical arguments are used in the discussion of acids and bases and in analytical techniques.

12.1 Electrochemical cells

An electrochemical cell consists of two electrodes dipping into an electrolyte. When it acts as a source of electrical energy it is called a *galvanic cell*; when it is connected to an external source of electric current which drives a chemical reaction within it, it is called an *electrolytic cell*. In this chapter we are concerned with the cell as a producer of electricity.

The simplest type of cell, a *cell without a liquid junction*, has a single electrolyte common to both electrodes: Fig. 12.1. In this case the potential difference across the terminals is the sum of the potential differences across the two electrode–electrolyte interfaces (but we shall have to be more precise shortly). In some cases it is necessary to immerse the electrodes in different electrolytes, and the additional source of potential difference, the *liquid junction*, must be taken into account. An example of such a *cell with a liquid junction* is shown in Fig. 12.2. The liquid junction contributes to the overall potential difference, and so it has to be estimated if the latter is to be interpreted in terms of the properties of the individual

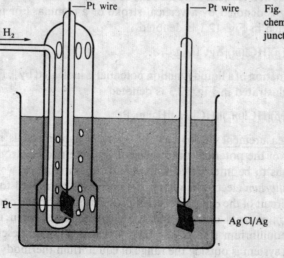

Fig. 12.1. A simple electro-chemical cell without a liquid junction.

(a) Static junction

Fig. 12.2. A cell with a liquid junction.

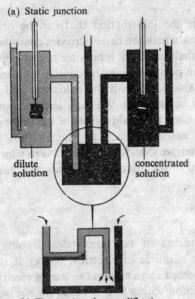

dilute
solution

concentrated
solution

(b) Flowing junction modification

electrodes. Alternatively it can be minimized by joining the two half-cells through a third electrolyte. In practice this is done by means of a salt bridge, a saturated solution of potassium chloride, Fig. 12.3. The reason for the success of the salt bridge is not wholly clear, but it is related to the fact that the high concentration of K^+ and Cl^- ions ensures that they carry most of the current in the region of the junctions irrespective of its direction, and also, being of similar size, they carry equal shares. Furthermore, the bridge introduces two junction potentials, and it is both hoped and thought that these tend to cancel.

The notation for cells is based on that for individual electrodes. A phase boundary is denoted by a vertical stroke or a comma. For instance, the cell depicted in Fig. 12.1 is denoted

$$Pt, H_2(p)|HCl(m)|AgCl, Ag.$$

The elimination of a liquid junction potential is indicated by $\|$. For instance, the cell illustrated in Fig. 12.3 is denoted

$$Pt, H_2(p)|HCl(m_L)\|HCl(m_R)|H_2(p), Pt.$$

The measurement underlying the study of galvanic cells is the determination of the potential difference between the two electrodes; but this remark has to be interpreted with care. In the first place, we are dealing with *equilibrium* electrochemistry. If a current is permitted to flow in the outside circuit of the cell the reaction proceeds inside it, the concentrations of the components change, and the current ceases when the reaction reaches equilibrium (the cell is then 'exhausted'). The description of a changing system is outside the range of equilibrium thermodynamics, and

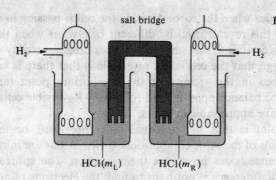

salt bridge

Fig. 12.3. Cell with a salt bridge.

H₂

H₂

$HCl(m_L)$ $HCl(m_R)$

so we have to measure the potential difference when the cell is held at a particular constant composition. Therefore it is necessary to measure it without allowing any current to flow. The classical procedure is to balance the cell against an opposing source of potential difference by using a *potentiometer*, Fig. 12.4. The cell's potential difference is compared with that of a *Weston standard cell*. This is formed from $Cd^{2+}|Cd(Hg)$ in combination with $SO_4^{2-}|Hg_2SO_4, Hg$, the electrolyte being a saturated aqueous solution of $CdSO_4$. The Weston cell's potential, $1.01807\,V$ at $25\,°C$, is only weakly dependent on the temperature. The modern method of measuring potential differences without drawing current is to use an electronic voltmeter.

The next point to establish is that the potential difference measured refers to a *thermodynamically reversible* process in the cell. Sources of irreversibility include the different diffusional characteristics of ions of

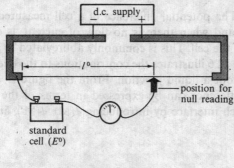

d.c. supply

− +

l^0

position for
null reading

standard
cell (E^0)

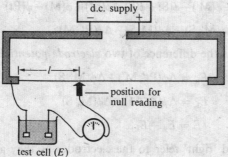

d.c. supply

− +

l

position for
null reading

test cell (E)

Fig. 12.4. Measuring the cell e.m.f. with a potentiometer. The cell e.m.f. is given by $(l/l^0)E^0$, where E^0 is the e.m.f. of the standard cell and l, l^0 are the positions at which the cells are balanced against the constant d.c. supply.

different mobilities when the current within the cell is passing in different directions, and this reveals itself by different behaviour when the cell is off-balance. Ascertaining that the cell is operating reversibly is often difficult. One sign that the cell is not reversible is that there is very little current produced in the vicinity of the equilibrium point (and as a consequence, the balance point is hard to locate). Reversible cells just off balance often give appreciable currents.

The final point is that the potential difference must be measured between materials of the same composition. If two metals are in contact a potential difference is set up across their junction. The source of this *contact potential difference* is not hard to find: the electrons of one metal may find it energetically favourable to spill across into the other. This provides an excess of negative charge on the second metal and a deficiency on the first, Fig. 12.5. The transfer occurs until the positive charge on the first exerts a compensating force, but by then there will be an interfacial potential difference. In general, the metals of the electrodes of a cell are different, and the wires from the measuring instruments set up a contact potential at the connections. This must be avoided by ensuring that both terminals of the cell are of the same material. For example, in the cell shown in Fig. 12.1 we shall agree to fix a strip of platinum to the silver electrode and to measure the potential difference between the two pieces of platinum.

In fact we shall go even farther than this. We shall invariably quote the potential differences of cells as measurements between terminals composed of platinum. In this way a contact potential of the form $\phi(Pt) - \phi(M)$ will be common to all measurements of cell potential differences.

E.m.f. and electrode potentials. The potential difference of a cell measured between platinum terminals, and when there is no flow of current, is called the *electromotive force* of the cell. This is commonly abbreviated to e.m.f. and denoted by E. Figure 12.6 illustrates the contributions to the overall e.m.f. of a typical cell without a liquid junction. From the figure $E = \phi(Pt') - \phi(Pt)$. The overall difference can be expressed in terms of the potential differences across each interface by introducing $\phi(M)$, $\phi(M')$, and $\phi(S)$ as follows:

$$E = \phi(Pt') - \phi(Pt)$$
$$= \phi(Pt') - \phi(M') + \phi(M') - \phi(S) + \phi(S) - \phi(M) + \phi(M) - \phi(Pt)$$
$$= \Delta\phi(Pt', M') + \Delta\phi(M', S) - \Delta\phi(M, S) - \Delta\phi(Pt, M).$$

This can be written as the difference of two *electrode potentials*:

right-hand electrode: $\quad E_R = \Delta\phi(Pt', M') + \Delta\phi(M', S)$

left-hand electrode: $\quad\quad E_L = \Delta\phi(Pt, M) + \Delta\phi(M, S)$

(12.1.1) $\quad$ cell e.m.f.: $\quad\quad\quad\quad\quad E = E_R - E_L.$

The labels 'left' and 'right' refer to the electrodes as they are written

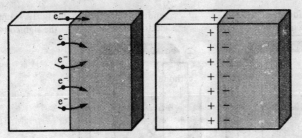

Fig. 12.5. Formation of a contact potential difference between two metals.

in the cell description $M|M^+(aq)||M'^+(aq)|M'$. Then reporting that the e.m.f. of the cell $Pt, H_2(p)|HCl(m)|AgCl, Ag|Pt$, or more briefly $H_2(p)|HCl(m)|AgCl, Ag$, is $+0.2\,V$ indicates that the $Cl^-|AgCl, Ag$ electrode has a positive potential relative to the other. If the cell is denoted $Ag, AgCl|HCl(m)|H_2$, the e.m.f. would be reported as $-0.2\,V$. The point to appreciate is that *the sign always refers to the potential across the electrodes $E_R - E_L$, where R and L refer to the cell as written.*

The sign convention. In a cell for which E is positive, the right-hand electrode has a higher electrical potential than the left ($E_R > E_L$). This indicates that it is positively charged relative to the left. This, in turn, indicates that if a circuit were connected between the terminals, electrons would flow from left to right, Fig. 12.7. (Electric current is conventionally regarded as the flow of a positively charged substance, and so the current flow is opposite to the flow of the electrons. This convention is confusing, and so we shall always express processes in terms of the flow of electrons.)

A positive e.m.f. signifies a deficiency of electrons on the right-hand electrode. This deficiency is caused by the tendency of the reaction in the cell to extract electrons from that electrode. Therefore, if $E > 0$ there is a tendency for reduction to take place at the right-hand electrode:

$$M'^+ + e^- \rightarrow M'.$$

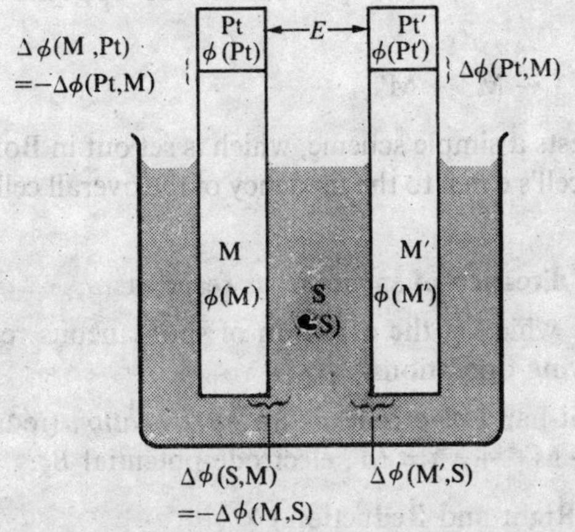

$\Delta\phi(M, Pt)$
$= -\Delta\phi(Pt, M)$

$\Delta\phi(Pt', M)$

$\Delta\phi(S, M)$
$= -\Delta\phi(M, S)$

$\Delta\phi(M', S)$

Fig. 12.6. Potential differences contributing to the overall e.m.f. of a cell.

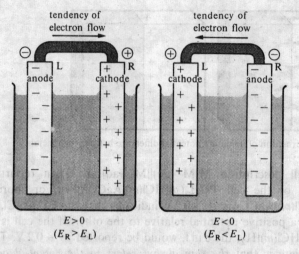

tendency of electron flow   tendency of electron flow   Fig. 12.7. Implications of the sign convention.

$$E > 0$$
$$(E_R > E_L)$$

$$E < 0$$
$$(E_R < E_L)$$

That electrode is therefore the *cathode* of the cell (defined, p. 330, as the site of reduction). The relative excess of electrons in the left-hand electrode (when $E > 0$) indicates that in its vicinity the reaction tends towards oxidation (de-electronation), and so it is the *anode*:

$$M \rightarrow M^+ + e^-.$$

An overall positive e.m.f. of the cell therefore arises from the tendency of the overall cell reaction to proceed as follows:

$$E > 0: \qquad M + M'^+ \rightarrow M^+ + M'.$$

The observation of a positive e.m.f. indicates that the reaction has a *tendency* to occur in the direction indicated; but only if the outside circuit is completed, and the electrons that are dumped at one electrode (the left) are delivered to the other, will the reaction actually proceed in the direction indicated.

Conversely, if the cell registered a negative e.m.f. we would know that the tendency of the reaction is to proceed in the opposite direction, namely

$$E < 0: \qquad M + M'^+ \leftarrow M^+ + M'.$$

This analysis suggests a simple scheme, which is set out in Box 12.1, for relating the sign of a cell's e.m.f. to the tendency of the overall cell reaction.

Box 12.1   **Cell e.m.f. and the direction of spontaneous reaction**

In order to decide which is the direction of spontaneous reaction, perform the following operations.

1. Write the right-hand electrode as an *electronation* (reduction) $\frac{1}{2}$-*cell reaction*: $M'^+ + e^- \rightleftharpoons M'$, electrode potential E_R.

(Remember R for Right and Reduction.)

2. Write the left-hand electrode reaction as an *electronation* $\frac{1}{2}$*-cell reaction*: $M^+ + e^- \rightleftharpoons M$, electrode potential E_L.

3. Subtract Right − Left to get both the overall cell reaction and the overall cell e.m.f.:

$R - L$, *cell reaction*: $M'^+ - M^+ + e^- - e^- \rightleftharpoons M' - M$,

or

$M'^+ + M \rightleftharpoons M' + M^+$.

$R - L$, *cell e.m.f.*: $E = E_R - E_L$.

4. Then,
    if $E > 0$ the cell reaction has a tendency to go Left → Right (that is, in the direction indicated by $>$).
    if $E < 0$ the cell reaction has a tendency to go Left ← Right (that is, in the direction indicated by $<$).

Example (Objective 6). One of the reactions important in the corrosion of iron in an acidic environment is $Fe + 2HCl(aq) + \frac{1}{2}O_2 \rightarrow FeCl_2(aq) + H_2O$. Which is the spontaneous direction of this reaction when the activity of the Fe^{2+} is unity and $a(H^+) = 1$?

● *Method.* Use the scheme set out in Box 12.1. Begin by finding the half-cell reactions that combine to give the overall reaction. For future convenience, write all reactions as though they involved single electron transfers. Electrode potentials are in Table 12.1; ions are present at unit activity, so use the standard values.

● *Answer.* The cell reaction can be expressed as

$Fe + 2H^+ + 2e^- + \frac{1}{2}O_2 \rightleftharpoons Fe^{2+} + 2e^- + H_2O$

or

$\frac{1}{2}Fe + H^+ + e^- + \frac{1}{4}O_2 \rightleftharpoons \frac{1}{2}Fe^{2+} + e^- + \frac{1}{2}H_2O$.

This can be obtained by writing

R: $H^+ + e^- + \frac{1}{4}O_2 \rightleftharpoons \frac{1}{2}H_2O$, $E_R^\ominus = 1.229\ V$
L: $\frac{1}{2}Fe^{2+} + e^- \rightleftharpoons \frac{1}{2}Fe$, $E_L^\ominus = -0.440\ V$

as $R - L$ is the overall reaction. The overall e.m.f. is $E^\ominus = E_R^\ominus - E_L^\ominus = 1.669\ V$. Since $E^\ominus > 0$, the reaction has a spontaneous tendency to run to the right when the conditions are those specified.

● *Comment.* The conclusion applies when $a(H^+) = 1$ and $a(Fe^{2+}) = 1$; when the reaction has progressed there will come a point where the activities of the ions have changed to the point where the thermodynamic tendency to the right no longer holds ($E = 0$), and the reaction ceases. Read on.

Suppose the molalities of the components of the cell are adjusted so that $E = 0$. At these molalities the reaction in the cell has no tendency to go to the right or the left, and so the cell reaction is at equilibrium. Even if a wire connected the two electrodes there would be no flow of current because the reaction has no tendency to proceed in either direction. The cell is exhausted when the reaction is at equilibrium. This is a very important conclusion, because if we can predict the concentrations which guarantee $E = 0$ we shall have found the equilibrium constant for the reaction in the cell. The next section explores this point.

Table 12.1. Standard electrode potentials at 25 °C, $E^{\ominus}/V$

$Li^+ + e^- \rightleftharpoons Li$	-3.045	$AgBr + e^- \rightleftharpoons Ag + Br^-$	0.071
$K^+ + e^- \rightleftharpoons K$	-2.924	$\frac{1}{2}Sn^{4+} + e^- \rightleftharpoons \frac{1}{2}Sn^{2+}$	0.139
$Rb^+ + e^- \rightleftharpoons Rb$	-2.925	$Cu^{2+} + e^- \rightleftharpoons Cu^+$	0.158
$\frac{1}{2}Ba^{2+} + e^- \rightleftharpoons \frac{1}{2}Ba$	-2.90	$AgCl + e^- \rightleftharpoons Ag + Cl^-$	0.2223
$\frac{1}{2}Sr^{2+} + e^- \rightleftharpoons \frac{1}{2}Sr$	-2.89	$\frac{1}{2}Cu^{2+} + e^- \rightleftharpoons \frac{1}{2}Cu$	0.340
$\frac{1}{2}Ca^{2+} + e^- \rightleftharpoons \frac{1}{2}Ca$	-2.76	$Cu^+ + e^- \rightleftharpoons Cu$	0.522
$Na^+ + e^- \rightleftharpoons Na$	-2.712	$\frac{1}{3}I_3^- + e^- \rightleftharpoons \frac{3}{2}I^-$	0.534
$\frac{1}{2}Mg^{2+} + e^- \rightleftharpoons \frac{1}{2}Mg$	-2.375	$\frac{1}{2}I_2 + e^- \rightleftharpoons I^-$	0.535
$\frac{1}{2}Be^{2+} + e^- \rightleftharpoons \frac{1}{2}Be$	-1.85	$Fe^{3+} + e^- \rightleftharpoons Fe^{2+}$	0.770
$\frac{1}{3}Al^{3+} + e^- \rightleftharpoons \frac{1}{3}Al$	-1.706	$\frac{1}{2}Hg_2^{2+} + e^- \rightleftharpoons Hg(l)$	0.799
$\frac{1}{2}Zn^{2+} + e^- \rightleftharpoons \frac{1}{2}Zn$	-0.763	$Ag^+ + e^- \rightleftharpoons Ag$	0.7996
$\frac{1}{2}Fe^{2+} + e^- \rightleftharpoons \frac{1}{2}Fe$	-0.409	$Hg^{2+} + e^- \rightleftharpoons \frac{1}{2}Hg_2^{2+}$	0.905
$\frac{1}{2}Cd^{2+} + e^- \rightleftharpoons \frac{1}{2}Cd$	-0.403	$\frac{1}{2}Br_2(l) + e^- \rightleftharpoons Br^-$	1.065
$Tl^+ + e^- \rightleftharpoons Tl$	-0.37	$H^+ + \frac{1}{4}O_2(g) + e^-$	
$\frac{1}{2}Ni^{2+} + e^- \rightleftharpoons \frac{1}{2}Ni$	-0.23	$\rightleftharpoons \frac{1}{2}H_2O(l)$	1.229
$AgI + e^- \rightleftharpoons Ag + I^-$	-0.152	$\frac{7}{3}H^+ + \frac{1}{6}Cr_2O_7^{2-} + e^-$	
$\frac{1}{2}Sn^{2+} + e^- \rightleftharpoons \frac{1}{2}Sn$	-0.136	$\rightleftharpoons \frac{7}{6}H_2O(l) + \frac{1}{3}Cr^{3+}$	1.33
$\frac{1}{2}Pb^{2+} + e^- \rightleftharpoons \frac{1}{2}Pb$	-0.126	$\frac{1}{2}Cl_2(g) + e^- \rightleftharpoons Cl^-$	1.3583
$H^+ + e^- \rightleftharpoons H_2(g)$	0	$\frac{1}{3}Au^{3+} + e^- \rightleftharpoons \frac{1}{3}Au$	1.42
		$\frac{8}{5}H^+ + \frac{1}{5}MnO_4^- + e^-$	
		$\rightleftharpoons \frac{4}{5}H_2O + \frac{1}{5}Mn^{2+}$	1.491
		$Ce^{4+} + e^- \rightleftharpoons Ce^{3+}$	1.443
		$\frac{1}{2}S_2O_8^{2-} + e^- \rightleftharpoons SO_4^{2-}$	2.05
Basic solution:			
$\frac{1}{2}Fe(OH)_2 + e^-$		$O_2 + e^- \rightleftharpoons O_2^-$	-0.56
$\rightleftharpoons \frac{1}{2}Fe + OH^-$	-0.877	$\frac{1}{2}S + e^- \rightleftharpoons \frac{1}{2}S^{2-}$	-0.48
$H_2O + e^- \rightleftharpoons \frac{1}{2}H_2(g) + OH^-$	-0.828	$\frac{1}{2}O_2 + \frac{1}{2}H_2O + e^-$	
		$\rightleftharpoons \frac{1}{2}HO_2^- + \frac{1}{2}OH^-$	-0.076

Source: *Handbook of chemistry and physics*, Chemical Rubber Co.

Concentration-dependence of the e.m.f. Consider a cell $A|A^{\nu+}||B^{\nu+}|B$ for which the reaction is

$$A + BX \rightleftharpoons AX + B, \quad \text{or} \quad A + B^{\nu+} + X^{\nu-} \rightleftharpoons A^{\nu+} + X^{\nu-} + B$$

and which may be expressed in terms of the half-cell reactions

$$R: \quad B^{\nu+} + \nu e^- \rightleftharpoons B \qquad E_R$$
$$L: \quad A^{\nu+} + \nu e^- \rightleftharpoons A \qquad E_L.$$

ν is the number of electrons transferred in the reduction step. Note that while $A + BX \rightleftharpoons AX + B$ may be the actual form of the reaction (as in

$Zn + CuSO_4(aq) \rightleftharpoons ZnSO_4(aq) + Cu$, when $v = 2$) it should also be regarded as standing for a general reaction in which v electrons are transferred. Thus even something as complicated as

$$Pb(s) + PbO_2(s) + 2H_2SO_4(aq) \rightleftharpoons 2PbSO_4(s) + 2H_2O(l),$$

which can be broken down into

R: $PbO_2(s) + SO_4^{2-}(aq) + 4H^+ + 2e^- \rightleftharpoons PbSO_4(s) + 2H_2O$

L: $PbSO_4(s) + 2e^- \rightleftharpoons Pb(s) + SO_4^{2-}(aq)$

is captured by it (and incidentally has $v = 2$). Examples of the general way of interpreting $A + BX \rightleftharpoons AX + B$ will be given later.

The potential differences across electrode–electrolyte interfaces are given by the expression derived on p. 338:

$$\Delta\phi = \Delta\phi^\ominus + (RT/vF)\ln\{a(M^{v+})/a(M)\}.$$

Therefore the overall e.m.f. of the cell is

(12.1.2)
$$E = E_R - E_L = E^\ominus + (RT/vF)\ln\left\{\frac{a(A)a(B^{v+})}{a(A^{v+})a(B)}\right\}.$$

This is the *Nernst equation*; $E^\ominus$ is the *standard e.m.f.* of the cell, its e.m.f. when all the species are at unit activity.

Example (Objective 7). At what activity of Fe^{2+} does iron cease dissolving in hydrochloric acid of $a(H^+) = 1$?

- *Method.* This develops the last *Example*. Look for the value of $a(Fe^{2+})$ for which $E_L = E_R$. Use the Nernst equation, and note that $\ln x = 2.303 \lg x$, and $2.303\,RT/F = 0.0592\,V$.

- *Answer.* Use the two half-cell reactions in the last *Example*; $v = 1$ for both. The Nernst equation then reads

$E_R = E_R^\ominus + (RT/F)\ln a(H^+) = E_R^\ominus$ at $a(H^+) = 1$,

$E_L = E_L^\ominus + (RT/F)\ln a(Fe^{2+})^{1/2} = E_L^\ominus + (2.303\,RT/2F)\lg a(Fe^{2+})$.

$E = E_R - E_L = 0$ when $E_L^\ominus + (2.303\,RT/2F)\lg a(Fe^{2+}) = E_R^\ominus$,

or when

$\lg a(Fe^{2+}) = (E_R^\ominus - E_L^\ominus)/\frac{1}{2}(2.303\,RT/F) = (1.669\,V)/\frac{1}{2}(0.0592\,V) = 56.4$.

Consequently, dissolution ceases only when the activity of Fe^{2+} rises to 2.5×10^{56}.

- *Comment.* This result shows that iron will continue to dissolve in strong hydrochloric acid virtually indefinitely.

If the reagents are all at their equilibrium concentrations the e.m.f. of the cell is zero: then

$$0 = E^\ominus + (RT/vF)\ln\left\{\frac{a(A)a(B^{v+})}{a(A^{v+})a(B)}\right\}_e.$$

The thermodynamic equilibrium constant for the cell reaction is

$$K = \left\{ \frac{a(A^{v+})a(X^{v-})a(B)}{a(A)a(B^{v+})a(X^{v-})} \right\}_e = \left\{ \frac{a(A^{v+})a(B)}{a(A)a(B^{v+})} \right\}_e.$$

Therefore

(12.1.3) $E^{\ominus} = (RT/vF)\ln K.$

Thus, if we know the standard e.m.f. of the cell corresponding to the reaction of interest, it is a simple matter to find the equilibrium constant. The procedure is summarized in Box 12.2. Clearly, a great deal depends on a knowledge of standard e.m.f.s, and so to these we now turn.

Box 12.2   **How to obtain equilibrium constants from standard electrode potentials**

Consider the reaction $A + BX \rightleftharpoons AX + B$, for which the equilibrium constant is

$$K = \left\{ \frac{a(A^{v+})a(X^{v-})a(B)}{a(A)a(X^{v-})a(B^{v+})} \right\}_e.$$

The reaction may be expressed as the following $\frac{1}{2}$-cell reactions:

R: $B^{v+} + ve^{-} \rightleftharpoons B$

L: $A^{v+} + ve^{-} \rightleftharpoons A.$

Then

$$E^{\ominus} = E^{\ominus}_{R} - E^{\ominus}_{L} = E^{\ominus}/(RT/vF)\ln K$$

and so

$$\lg K = (2.303\,RT/vF).$$

The practical form of this relation at 298.15 K is

$$\lg K = vE^{\ominus}/(0.059\,15\,\text{V}).$$

Example (Objective 8). Calculate the equilibrium constant for the reaction $Cu(s) + Cl_2(g) \rightleftharpoons CuCl_2(aq)$.

- *Method.* Use the technique set out in Box 12.2 with $A = Cu$, $BX = Cl_2$. The reaction involves a 2-electron transfer, and so $v = 2$.

- *Answer.* The half-reactions are

     R: $Cl_2(g) + 2e^{-} \rightleftharpoons 2Cl^{-}(aq),$ $E^{\ominus}_{R} = E^{\ominus}(Cl_2, Cl^{-}) = 1.3595\,\text{V}$

     L: $Cu^{2+}(aq) + 2e^{-} \rightleftharpoons Cu(s),$ $E^{\ominus}_{L} = E^{\ominus}(Cu^{2+}, Cu) = 0.337\,\text{V}.$

Then $E^{\ominus} = E^{\ominus}_{R} - E^{\ominus}_{L} = 1.360\,\text{V} - 0.337\,\text{V} = 1.023\,\text{V}$. The equilibrium constant for the reaction $Cu + Cl_2 \rightleftharpoons CuCl_2$ is

$$K = \{a(Cu^{2+})a(Cl^{-})^2/a(Cu)a(Cl_2)\}_e = \{a(Cu^{2+})a(Cl^{-})^2/(p/p^{\ominus})\}_e$$

where p is the pressure of chlorine (fugacity actually). From the Box,

$$\lg K = 2E^{\ominus}/(2.303RT/F) = 2(1.023\ \text{V})/(0.05915\ \text{V}) = 34.59.$$

Therefore, $K = 3.9 \times 10^{34}$

- **Comment.** In this case the equilibrium lies strongly towards the salt solution.

12.2 Standard electrode potentials

The standard e.m.f. of a cell, $E^{\ominus}$, may be expressed as the difference $E_R^{\ominus} - E_L^{\ominus}$ of two *standard electrode potentials*. The potential of a single electrode never appears in the expressions for measurable quantities (such as E and K) and so the standard potential of one electrode may be arbitrarily assigned the value zero and all others tabulated on that basis. The one selected to have zero potential is the *standard hydrogen electrode* (S.H.E.). This electrode is constructed as in Fig. 11.10 on p. 334: in its standard form the temperature is 298.15 K, the hydrogen ion activity is unity, and the hydrogen fugacity is 1 atm (for all but the most exact experiments this corresponds to a pressure of 1 atm):

$$\text{S.H.E.:}\quad H^+(aq, a_{H^+} = 1)|H_2(g, f = 1\ \text{atm}), Pt \qquad E^{\ominus}(H^+, H_2) \overset{\text{def}}{=} 0.$$

The standard electrode potential of any other system (or *couple*) is then defined as the e.m.f. of a cell in which the couple forms the right-hand electrode and the S.H.E. forms the left, and in which all the species are at unit activity:

$$E_R^{\ominus} = E_R^{\ominus} - E^{\ominus}(H^+, H_2) = E^{\ominus}(\text{cell}).$$

Once the standard potential of an electrode has been measured against the S.H.E. it may be used in combination with other electrodes in order to determine their standard potentials, and with suitable choices of combinations of electrodes a table of values may be compiled. A list is given in Table 12.1.

Sometimes tables of electrode potentials are encountered, especially in the older literature, in which all the signs are opposite to those given here. This indicates the adoption of a different convention. We have listed *standard reduction potentials*; the alternative tables list standard oxidation potentials. It is easy, and important, to check which convention is being used by noting the sign of the Cu^{2+}, Cu couple: in our convention this has a positive value. Therefore, if another table lists it as negative, reverse all the signs and proceed as in this book.

The significance of the sign of the standard electrode potential should always be kept in mind. It can be arrived at as follows. The standard e.m.f. is obtained from the Table by forming the difference $E^{\ominus} = E_R^{\ominus} - E_L^{\ominus}$; then the equilibrium constant for the cell reaction can be calculated from eqn (12.1.3) in the form

$$K = \exp(\nu F E^{\ominus}/RT).$$

K is large (products favoured, right hand couple is reduced) if $E^{\ominus}$ is positive. $E^{\ominus}$ is positive when $E_R^{\ominus} > E_L^{\ominus}$. Therefore, couples with relatively high standard electrode potentials are reduced by couples with relatively low standard electrode potentials. In other words: *high is reduced by low*. For instance, the Zn^{2+}, Zn couple $(-0.763\,V)$ is lower than the Cu^{2+}, Cu couple $(+0.337\,V)$ and so in the reaction

$$Zn + CuSO_4(aq) \rightleftharpoons ZnSO_4(aq) + Cu$$

the equilibrium lies to the right. The value of the equilibrium constant is 1.5×10^{27} at $25\,°C$, which is strongly in favour of the reduction of Cu(II) to copper (i.e. to the displacement of copper from the solution by zinc).

The layout of the Table of standard electrode potentials is designed to show the order of reducing power: higher up the Table (more negative; that is, lower in value) displaces lower down (more positive): *higher up reduces lower down*. The quantitative value of the equilibrium constant then comes from the set of rules summarized in Box 12.1.

Example (Objective 9). Will zinc displace copper and magnesium from aqueous solutions of their sulphates?

- *Method.* Decide whether $E^{\ominus}(M^{2+}, M)$, $M = Cu$, Mg, is positive relative to $E^{\ominus}(Zn^{2+}, Zn)$.

- *Answer.* From Table 12.1, $E^{\ominus}(Zn^{2+}, Zn) = -0.763\,V$, $E^{\ominus}(Cu^{2+}, Cu) = 0.337\,V$, and $E^{\ominus}(Mg^{2+}, Mg) = -2.37\,V$. Since $E^{\ominus}(Cu^{2+}, Cu) > E^{\ominus}(Zn^{2+}, Zn)$, zinc will displace copper. $E^{\ominus}(Mg^{2+}, Mg) < E^{\ominus}(Zn^{2+}, Zn)$, and so zinc will not displace magnesium.

- *Comment.* These conclusions can be arrived at with the minimum of fuss simply by determining the relative order of the couples in the table of reduction potentials. The quantitative value of $\lg K$ at $25\,°C$ is given by $2E^{\ominus}/59.15\,mV$, which is 37 for Cu. The displacement is therefore also quantitatively extensive ($K \approx 10^{37}$ at $25\,°C$).

The determination of standard electrode potentials. The procedure for measuring $E^{\ominus}$ of an electrode can be illustrated by a specific case, and we choose the important $Cl^-|AgCl, Ag$ electrode.

The basic measurement is the determination of the e.m.f. of the *Harned cell*

$$Pt, H_2(1\ atm)|HCl(m)|AgCl, Ag.$$

The observed e.m.f. is related to the standard electrode potentials by

$$E = E^{\ominus}(Cl^-|AgCl, Ag) - (RT/F)\ln(a_{H^+}a_{Cl^-}).$$

The activities can be expressed in terms of the molality m and the mean activity coefficient $\gamma_{\pm}$ through eqn (11.1.5), $\gamma_{\pm}^2 = \gamma_+\gamma_-$:

(12.2.1)
$$E = E^{\ominus}(Cl^-|AgCl, Ag) - (RT/F)\ln(m/m^{\ominus})^2 - (RT/F)\ln\gamma_{\pm}^2,$$

or

$$E + 2(RT/F)\ln(m/m^{\ominus}) = E^{\ominus}(Cl^-|AgCl, Ag) - 2(RT/F)\ln\gamma_{\pm}.$$

From the Debye-Hückel theory, $\lg\gamma_{\pm} = -0.509(m/mol\ kg^{-1})^{1/2}$ for a dilute 1:1 electrolyte. The last expression may therefore be written

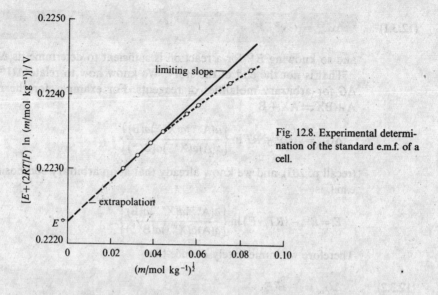

Fig. 12.8. Experimental determination of the standard e.m.f. of a cell.

$$E + 2(RT/F)\ln(m/m^{\ominus}) = E^{\ominus}(\text{Cl}^-|\text{AgCl},\text{Ag}) + 2.34\,(RT/F)(m/m^{\ominus})^{1/2}.$$

(In precise work, the $\sqrt{m}$ term is brought to the left, and a higher order correction term from the extended Debye–Hückel theory is employed on the right.) The expression on the left may be determined for a range of molalities of HCl in the cell, plotted against the square-root of the molality, and then extrapolated to $m = 0$. The intercept is the value of $E^{\ominus}(\text{Cl}^-|\text{AgCl},\text{Ag})$, Fig. 12.8.

Measuring activity coefficients. Once the standard electrode potential of an electrode in a cell is known, the activity coefficients of the relevant ions can be determined for any ionic strength simply by measuring the e.m.f. of the cell with the ions at the concentration of interest. For instance, if the mean activity of the ions in aqueous HCl of molality m is required, eqn (12.2.1) can be used in the form

$$\ln\gamma_{\pm} = (F/2RT)(E^{\ominus} - E) - \ln(m/m^{\ominus})$$

once E has been measured. Knowledge of activity coefficients is essential for the precise interpretation of e.m.f. determinations of concentrations, of pH, and in titrations. We shall see examples below.

12.3 Thermodynamic data from cell e.m.f.s

Measurements of e.m.f.s are a convenient source of data on the ΔG, ΔH, and ΔS of a reaction. In practice the standard values $\Delta G^{\ominus}$, $\Delta H^{\ominus}$, and $\Delta S^{\ominus}$ are what are usually determined.

Measurement of $\Delta G^{\ominus}$. In order to relate $\Delta G^{\ominus}$ to $E^{\ominus}$, combine $E^{\ominus} = (RT/vF)\ln K^{\ominus}$ with the expression $\Delta G_m^{\ominus} = -RT\ln K$, derived in Chapter 9 (p. 264). Then

(12.3.1)
$$\Delta G_m^{\ominus} = -\nu F E^{\ominus},$$

and so knowing $E^{\ominus}$ for a reaction is sufficient to determine its $\Delta G_m^{\ominus}$.

That is not the end of the story. We know how to relate $\Delta G^{\ominus}$ to the ΔG for arbitrary molalities of reagents. For example, for the reaction $A + BX \rightleftharpoons AX + B$

$$\Delta G_m = \Delta G_m^{\ominus} + RT \ln \left\{ \frac{a(A^{\nu+})a(X^{\nu-})a(B)}{a(A)a(X^{\nu-})a(B^{\nu+})} \right\}$$

(recall p. 263), and we know already that at an arbitrary composition the e.m.f. is

$$E = E^{\ominus} - (RT/\nu F) \ln \left\{ \frac{a(A^{\nu+})a(X^{\nu-})a(B)}{a(A)a(X^{\nu-})a(B^{\nu+})} \right\}.$$

Therefore we immediately deduce that

(12.3.2)
$$\Delta G_m = -\nu F E.$$

The method of using these two important results is illustrated in the following *Example*.

Example (Objective 11). Find the values of $\Delta G_m^{\ominus}$ and of ΔG_m when pH = 4, for the reaction $Pb + PbO_2(s) + 2H_2SO_4(aq) \rightleftharpoons 2PbSO_4(s) + 2H_2O(l)$ at $25\,^{\circ}C$.

- *Method.* For the first part, use eqn (12.3.1) with $\nu = 2$. For the second part convert from $E^{\ominus}$ to E using the Nernst equation, eqn (12.1.2). It is necessary to know $a(SO_4^{2-})$. Use $a(SO_4^{2-}) = \gamma_{\pm} m(SO_4^{2-})$, $a(H^+) = \gamma_{\pm} m(H^+)$, and $m(SO_4^{2-}) \approx \frac{1}{2} m(H^+)$ to obtain $a(SO_4^{2-}) \approx \frac{1}{2}\gamma_{\pm} m(H^+) = \frac{1}{2}a(H^+)$. Use $RT/F = 25.69\,mV$.

- *Answer.* The half-reactions and their potentials are

  R: $PbO_2(s) + SO_4^{2-}(aq) + 4H^+(aq) + 2e^- = PbSO_4(s) + 2H_2O(l)$,
       $E_R^{\ominus} = 1.68\,V$

  L: $PbSO_4(s) + 2e^- = Pb(s) + SO_4^{2-}(aq)$, $E_L^{\ominus} = -0.41\,V$

  R−L: $PbO_2 + Pb + 2H_2SO_4(aq) = 2PbSO_4(s) + 2H_2O(l)$, $E^{\ominus} = 2.09\,V$

  $\Delta G_m^{\ominus} = -2FE^{\ominus} = -2 \times (9.648 \times 10^4\,C\,mol^{-1}) \times (2.09\,V)$
          $= -403\,kJ\,mol^{-1}$.

  In the case of general ion activities, eqn (12.1.2) is used by making the identifications ($\triangleq$ means 'corresponds to')

  R reaction:

     $B \triangleq PbSO_4(s) + 2H_2O(l)$; $a_B \triangleq a(PbSO_4)a^2(H_2O)$

     $B^{\nu+} \triangleq PbO_2(s) + SO_4^{2-}(aq) + 4H^+(aq)$; $a_{B^{\nu+}} \triangleq a(PbO_2)a(SO_4^{2-})a(H^+)^4$

  L reaction:

     $A \triangleq Pb + SO_4^{2-}$; $a_A \triangleq a(Pb)a(SO_4^{2-})$

     $A^{\nu+} \triangleq PbSO_4$; $a_{A^{\nu+}} \triangleq a(PbSO_4)$

  and using $a(PbSO_4) = a(H_2O) = a(PbO_2) = 1$,

$$E = E^{\ominus} + (RT/2F)\ln\left\{\frac{a(\text{Pb})a(\text{SO}_4^{2-})a(\text{PbO}_2)a(\text{SO}_4^{2-})a(\text{H}^+)^4}{a(\text{PbSO}_4)a(\text{PbSO}_4)a^2(\text{H}_2\text{O})}\right\}$$

$$= E^{\ominus} + (RT/2F)\ln\{a(\text{H}^+)^4 a(\text{SO}_4^{2-})^2\}$$

$$= E^{\ominus} + (RT/2F)\ln\{a(\text{H}^+)^4[\tfrac{1}{2}a(\text{H}^+)]^2\}, \text{ using } a(\text{SO}_4^{2-}) \approx \tfrac{1}{2}a(\text{H}^+),$$

$$= E^{\ominus} + (3RT/F)\ln\{a(\text{H}^+)/4\} = E^{\ominus} - (3RT/F)\ln 4 + 2.303(3RT/F)\lg a(\text{H}^+)$$

$$= E^{\ominus} - (3RT/F)\ln 4 - 2.303(3RT/F)\text{pH}$$

$$= 2.09\,\text{V} - 3 \times (0.02569\,\text{V})\ln 4 - 2.303 \times 3 \times (0.02569\,\text{V}) \times 4$$

$$= 1.27\,\text{V}.$$

It follows that $\Delta G_m = -2FE = -246\,\text{kJ}\,\text{mol}^{-1}$.

- *Comment.* The reaction is the basis of the action of the lead accumulator (such as a car battery). The calculation shows that the maximum amount of electrical work available from the consumption of 1 mol (207 g) of lead in a reaction with all components in their standard states is 403 kJ. Less work is available in a weaker acid.

Values of the standard Gibbs function for ionic reactions in solution can be obtained from tables of the Gibbs functions of formation of the ions involved. These tables are set up by defining the standard Gibbs function of formation of one species of ion to be zero, and then quoting the values for all other ions on that basis. The convention is adopted that *the standard Gibbs function of formation of the proton in aqueous solution is zero* (at 298.15 K):

$$\tfrac{1}{2}\text{H}_2(\text{g}) \rightarrow \text{H}^+(\text{aq}) \qquad \Delta G_f^{\ominus}(\text{H}^+,\text{aq}) \overset{\text{def}}{=} 0.$$

The Gibbs functions of formation of other ions can be found from measurements of $E^{\ominus}$. For example, the standard Gibbs function of formation of the Ag^+ ion in water can be found from the standard Gibbs function for the reaction

$$\text{Ag}^+(\text{aq}) + \tfrac{1}{2}\text{H}_2(\text{g}) \rightleftharpoons \text{H}^+(\text{aq}) + \text{Ag}(\text{s})$$

since

$$\Delta G_m^{\ominus} = \Delta G_f^{\ominus}(\text{H}^+,\text{aq}) - \Delta G_f^{\ominus}(\text{Ag}^+,\text{aq}) = -\Delta G_f^{\ominus}(\text{Ag}^+,\text{aq}).$$

This reaction occurs in the cell $\text{H}_2|\text{H}^+(\text{aq})\|\text{Ag}^+(\text{aq})|\text{Ag}$, which has the standard e.m.f. $E^{\ominus}(\text{Ag}^+,\text{Ag}) - E^{\ominus}(\text{H}^+,\text{H}_2) = E^{\ominus}(\text{Ag}^+,\text{Ag})$. Therefore on the basis that $-\nu FE^{\ominus} = \Delta G_m^{\ominus}$ and $\nu = 1$ we find

$$\Delta G_f^{\ominus}(\text{Ag}^+,\text{aq}) = FE^{\ominus}(\text{Ag}^+,\text{Ag}) = 77.10\,\text{kJ}\,\text{mol}^{-1},$$

as in Table 12.2.

The standard Gibbs function of formation of an ion can be estimated theoretically using a result due to Born. The *Born equation* identifies the *Gibbs function of solvation* of an ion (that is, the Gibbs function for the process $\text{M}^+(\text{g}) \rightarrow \text{M}^+(\text{aq})$) with the electrical work involved in transferring an ion from a vacuum to the solvent, the latter being regarded as a continuous dielectric of relative permittivity ε_r:

(12.3.3) $\Delta G^{\ominus}_{\mathrm{m,solvation}} = -\frac{1}{2}(z_i^2 e^2 L/4\pi\varepsilon_0 r_i)\{1-(1/\varepsilon_r)\},$

where z_i is the charge of the ion and r_i its radius. You should note that
the solvation standard Gibbs function is negative and that its magnitude
is greatest for small ions. The standard Gibbs function of formation of
the ion in solution is the sum of this quantity and the standard Gibbs
function of formation of the gaseous ions.

Table 12.2. Standard entropies and standard Gibbs functions and enthalpies of
formation of ions in water at 298.15 K

Ion	$\Delta H^{\ominus}_f/\mathrm{kJ\,mol^{-1}}$	$S^{\ominus}/\mathrm{J\,K^{-1}\,mol^{-1}}$	$\Delta G^{\ominus}_f/\mathrm{kJ\,mol^{-1}}$
H^+	0	0	0
Li^+	-278.5	14.2	-293.8
Na^+	-239.7	60.2	-261.9
K^+	-251.2	102.5	-282.3
NH_4^+	-132.5	113.4	-79.4
Ag^+	105.6	72.7	77.1
Ca^{2+}	-543.0	-55.2	-553.0
Cu^{2+}	64.8	-99.6	65.5
Zn^{2+}	-153.9	-112.1	-147.0
Fe^{2+}	-89.1	-137.7	-78.9
Fe^{3+}	-48.5	-315.9	-4.6
Al^{3+}	-531.4	-321.7	-485.3
OH^-	-229.94	-10.8	-157.30
F^-	-332.6	-13.8	-278.8
Cl^-	-167.2	56.5	-131.3
Br^-	-121.5	82.4	-104.0
I^-	-55.2	111.3	-51.6
SO_4^{2-}	-909.3	20.1	-741.6
PO_4^{3-}	-2277.4	-221.8	-1018.8
HPO_4^{2-}	-1292.1	-33.5	-1089.3
$H_2PO_4^-$	-1296.3	-90.4	-1130.4
$CH_3CO_2^-$	-486.0	86.6	-369.4

Estimated absolute values for $H^+(g) \rightarrow H^+(aq)$ are $\Delta H^{\ominus}_m = -1090\,\mathrm{kJ\,mol^{-1}}, \Delta S^{\ominus}_m = -21\,\mathrm{J\,K^{-1}}$
$\mathrm{mol^{-1}}$, and $\Delta G^{\ominus}_m = -1104\,\mathrm{kJ\,mol^{-1}}$.
Source: G. W. C. Kaye and T. H. Laby, *Tables of physical and chemical constants*, Longmans.

The temperature-dependence of the e.m.f. The e.m.f. of a cell is related to the change
in the Gibbs function for the cell reaction by eqn (12.3.2). The variation
of ΔG with temperature is given by eqn (6.2.2):

$(\partial\Delta G/\partial T)_p = -\Delta S.$

Therefore the variation of the cell e.m.f. with temperature is

(12.3.4) $(\partial E/\partial T)_p = \Delta S_m/\nu F.$

Simply measuring the variation of the e.m.f. of a cell with temperature gives the value of the entropy change of the cell reaction. The same technique applied to $E^\ominus$ gives the standard molar entropy change $\Delta S_m^\ominus$, and this provides a non-calorimetric method of determining ion entropies. Since $\Delta G = \Delta H - T\Delta S$ the two results can be combined to give the enthalpy change of the reaction:

(12.3.5) $$\Delta H_m = \Delta G_m + T\Delta S_m = -vF[E - T(\partial E/\partial T)_p].$$

This is a purely non-calorimetric method of determining the enthalpy of a reaction; the enthalpies of formation (and solvation) of ions were discussed in Chapter 4 (p. 118).

Example (Objective 13). At 20 °C the standard e.m.f. of the cell Pt, $H_2|HCl(aq), Hg_2Cl_2(s)|Hg$ is 0.2699 V and at 30 °C it is 0.2669 V. Find the values of $\Delta G_m^\ominus$, $\Delta H_m^\ominus$, and $\Delta S_m^\ominus$ at 25 °C.

- *Method.* The cell reaction is $\frac{1}{2}Hg_2Cl_2(s) + \frac{1}{2}H_2(g) \rightarrow Hg(l) + HCl(aq)$, and written in this way it corresponds to a 1-electron transfer. Use eqn (12.3.1) for $\Delta G_m^\ominus$, and find $\Delta S_m^\ominus$ from the temperature coefficient of $\Delta G_m^\ominus$. Find $\Delta H_m^\ominus$ from $\Delta G_m^\ominus = \Delta H_m^\ominus - T\Delta S_m^\ominus$.

- *Answer.* From eqn (12.3.1), $\Delta G_m^\ominus = -FE^\ominus$. Therefore

$$\Delta G_m^\ominus(293.15\,\text{K}) = -(9.6485 \times 10^4\,\text{C mol}^{-1}) \times (0.2699\,\text{V}) = -26.04\,\text{kJ mol}^{-1}$$

and

$$\Delta G_m^\ominus(303.15\,\text{K}) = -(9.6485 \times 10^4\,\text{C mol}^{-1}) \times (0.2669\,\text{V}) = -25.75\,\text{kJ mol}^{-1}.$$

Hence, by linear interpolation (in this case by taking the mean)

$$\Delta G_m^\ominus(298.15\,\text{K}) = -25.90\,\text{kJ mol}^{-1}$$

and

$$\left(\frac{\partial \Delta G_m^\ominus}{\partial T}\right)_{298\,\text{K}} \approx \frac{[-25.75 - (-26.04)]\,\text{kJ mol}^{-1}}{(303.15 - 293.15)\,\text{K}} = 29.00\,\text{J K}^{-1}\,\text{mol}^{-1}.$$

It follows that

$$\Delta S_m^\ominus = -(\partial \Delta G_m^\ominus/\partial T)_{298\,\text{K}} = -29.00\,\text{J K}^{-1}\,\text{mol}^{-1}$$
$$\Delta H_m^\ominus = \Delta G_m^\ominus + T\Delta S_m^\ominus$$
$$= -25.75\,\text{kJ mol}^{-1} + (298.15\,\text{K}) \times (-29.00\,\text{J K}^{-1}\,\text{mol}^{-1})$$
$$= -34.40\,\text{kJ mol}^{-1}.$$

- *Comment.* In order to obtain the changes for the reaction written as $Hg_2Cl_2(s) + H_2(g) \rightarrow 2Hg + 2HCl(aq)$, simply multiply these values by 2. Then at 298 K, $\Delta G_m^\ominus = -51.80\,\text{kJ mol}^{-1}$, $\Delta S_m^\ominus = -58.00\,\text{J K}^{-1}\,\text{mol}^{-1}$, and $\Delta H_m^\ominus = -68.80\,\text{kJ mol}^{-1}$.

Yet again we encounter the feature of electrochemical measurements that they always relate to electrically neutral collections of ions. Therefore we are free to ascribe to one type of ion an arbitrary entropy of zero. The normal convention is to select *the standard entropy of the aquated proton as zero*: $S^\ominus(H^+, aq) \overset{\text{def}}{=} 0$. A list of standard ion entropies is given in Table 12.2. They vary in the manner one might expect when one considers the ability of ions to order the water molecules in their vicinity. For example, small, highly-charged ions induce local structure in the organization of the surrounding water molecules, and this decreases the entropy of the system more than in the case of large, singly-charged ions. Remember that these remarks about order refer to the effect relative to that induced by the proton: they are entropies relative to $S^{\ominus}(H^+, aq) = 0$.

The actual Third Law entropy of the proton in aqueous solution can be estimated on the basis of a model of the structure it induces in the water, and there is some agreement on the value $-21\,J\,K^{-1}\,mol^{-1}$, the negative value representing the ordering of the solvent induced by the proton.

12.4 Simple applications of e.m.f. measurements

Electrochemical techniques are the basis of a wide variety of calculations and measurements in chemistry. The next few paragraphs indicate some of them.

Solubility products. The solubility of a sparingly soluble salt MX can be discussed in terms of the equilibrium

(12.4.1) $MX(s) \rightleftharpoons M^+(aq) + X^-(aq)$.

The activity of the pure solid MX is unity, and so the equilibrium constant for this reaction is $K_s = a_{M^+} a_{X^-}$. K_s is called the *solubility product* for the salt. When the solubility is so low that the activity coefficients are unity, the concentration of ions can be estimated from $K_s \approx (m_{M^+}/m^{\ominus})(m_{X^-}/m^{\ominus})$. Since in the absence of other species $m_{M^+} = m_{X^-}$, at equilibrium (i.e., in the saturated solution)

(12.4.2) $m_{M^+} \approx K_s^{1/2} m^{\ominus}$.

The magnitude of K_s can be predicted from a knowledge of standard electrode potentials chosen so that the overall cell reaction is the solubility equilibrium reaction eqn (12.4.1).

Consider the case of silver chloride. The cell $Ag|Ag^+||Cl^-|AgCl,Ag$ can be analysed as follows:

right-hand electrode:

$AgCl + e^- \rightleftharpoons Ag + Cl^-(aq),\ E_R = E^{\ominus}(Cl^-|AgCl,Ag) - (RT/F)\ln a_{Cl^-}$

left-hand electrode:

$Ag^+(aq) + e^- \rightleftharpoons Ag,\qquad E_L = E^{\ominus}(Ag^+|Ag) + (RT/F)\ln a_{Ag^+}$

overall (R − L):

$AgCl \rightleftharpoons Ag^+(aq) + Cl^-(aq),\ E = E^{\ominus}(Cl^-|AgCl,Ag) - E^{\ominus}(Ag^+|Ag)$
$\qquad\qquad - (RT/F)\ln a_{Ag^+} a_{Cl^-}$.

$E = 0$ at the equilibrium of the cell reaction (which is the dissolution equilibrium of interest), and so (as $v = 1$),

$(RT/F)\ln K_s = E^{\ominus}(Cl^-|AgCl,Ag) - E^{\ominus}(Ag^+|Ag)$.

At 25 °C the two standard potentials are respectively $+0.2223$ V and $+0.7991$ V (Table 12.1); therefore

$$\lg K_s = \{(0.2223 \text{ V}) - (0.7991 \text{ V})\}/(0.05915 \text{ V}) = -9.75.$$

This shows that the solubility product is very small ($K_s \approx 10^{-10}$), and that the concentration in a saturated solution is approximately 10^{-5} mol kg^{-1}

Potentiometric titrations. In a redox titration a reduced form of some ion (e.g., Fe^{2+}) is oxidized by the addition of some oxidizer (e.g., Ce^{4+}), and the overall reaction

$$Fe^{2+} + Ce^{4+} \rightleftharpoons Fe^{3+} + Ce^{3+}$$

may be discussed in terms of the two half-cell reactions (with $v = 1$):

$$Ce^{4+} + e^- \rightleftharpoons Ce^{3+} \qquad E^{\ominus}(Ce^{4+}, Ce^{3+}) = 1.61 \text{ V}$$

$$Fe^{3+} + e^- \rightleftharpoons Fe^{2+} \qquad E^{\ominus}(Fe^{3+}, Fe^{2+}) = 0.771 \text{ V}.$$

The equilibrium constant for the reaction at 25 °C is therefore 1.52×10^{14}, so cerium is a good choice for the titration because the equilibrium lies so strongly towards products (Fe^{3+}). At the *end-point* of the titration just enough Ce(IV) has been added to eliminate the Fe(II), and our problem is to see how monitoring the e.m.f. of a cell can be used to detect this point. Since the equilibrium constant is 10^{14} and not infinity, there will be a trace of Fe(II) present even at (and after) the end point; but it will be extremely small.

A cell is constructed by dipping a platinum electrode into the titration mixture, and having another electrode in electrical contact with the solution (perhaps through a salt bridge). During the course of the reaction the e.m.f. of the cell is monitored. The presence of the iron ions gives rise to an electrode potential $E(Fe^{3+}, Fe^{2+})$ and the cerium ions give rise to a potential $E(Ce^{4+}, Ce^{3+})$; but as these share the same electrode, and one electrode can be at only one potential, they must be equal. Therefore at any stage of the titration the electrode potential may be written

$$E = E^{\ominus}(Ce^{4+}, Ce^{3+}) + (RT/F) \ln(a_{Ce^{4+}}/a_{Ce^{3+}})$$

$$= E^{\ominus}(Fe^{3+}, Fe^{2+}) + (RT/F) \ln(a_{Fe^{3+}}/a_{Fe^{2+}}),$$

and we may use whichever form is convenient. We shall simplify the discussion by making the approximation that activities can be replaced by molalities.

Suppose that initially the amount of Fe(II) is f; as Ce(IV) is added this is reduced to $f(1-x)$ and the amount of Fe(III) rises to fx, where x depends on the amount of Ce(IV) added (if K were infinite, fx would be equal to the amount of oxidant added). At some intermediate stage in the titration it follows that the electrode has a potential

$$E = E^{\ominus}(Fe^{3+}, Fe^{2+}) + (RT/F) \ln\{x/(1-x)\}.$$

The function $\ln\{x/(1-x)\}$ occurs widely in electrochemistry and is plotted in Fig. 12.9: when x is very small $\ln\{x/(1-x)\} \approx \ln x$, which is a large negative number. In the vicinity of $x = \frac{1}{2}$, $\ln\{x/(1-x)\} \approx \ln 1 = 0$; when x approaches unity, $\ln\{x/(1-x)\}$ becomes large and positive. A rapid rise, slow variation around $x = \frac{1}{2}$, and then a further rapid rise is characteristic of a variety of electrochemical measurements, and it can often be traced back to the properties of the $\ln\{x/(1-x)\}$ function.

As far as the present problem is concerned, the $\ln\{x/(1-x)\}$ behaviour indicates that initially the electrode potential is strongly negative, rises rapidly to the vicinity of $E^{\ominus}(Fe^{3+}, Fe^{2+})$ as Ce(IV) is added, and is exactly equal to that value when enough has been added to equalize the molalities (actually activities) of Fe(II) and Fe(III). As more cerium is added the electrode potential rises quite slowly at first, but then the rapid, characteristic second rise of the $\ln\{x/(1-x)\}$ function takes over, and the potential of the electrode moves rapidly away from the vicinity of $E^{\ominus}(Fe^{3+}, Fe^{2+})$. The line in Fig. 12.9 barely moves from the zero axis (on the scale drawn) until x has reached about 0.98, and so the end-point is signalled by the sudden abrupt rise in the potential away from $E^{\ominus}(Fe^{3+}, Fe^{2+})$.

How high does the potential rise? When enough Ce(IV) has been added to eliminate all except a trace of Fe(II) it becomes easier to discuss the electrode potential in terms of the cerium molality (activity). If an amount c of Ce(IV) has been added, the amount present in the solution is about $c-f$, because all Fe(II) has been oxidized (this is exact in the case of an infinite equilibrium constant; here it is a good approximation). The amount of Ce(III) left is virtually f, however much Ce(IV) is added once the end-point has been passed. Therefore the potential of the electrode is

$$E = E^{\ominus}(Ce^{4+}, Ce^{3+}) + (RT/F)\ln\{(c-f)/f\}.$$

This function rises to $E^{\ominus}(Ce^{4+}, Ce^{3+})$ by the time $c = 2f$ (about twice as much cerium as was needed to reach the end-point), and then climbs only slowly as more cerium is added (because $\ln\{(c-f)/f\} \approx \ln c$ when $c \gg f$, and $\ln c$ is a function that increases only slowly with increasing c).

The complete behaviour is depicted in Fig. 12.10: all we need to be

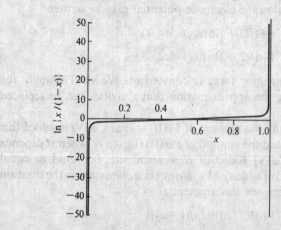

Fig. 12.9. The function $\ln\{x/(1-x)\}$.

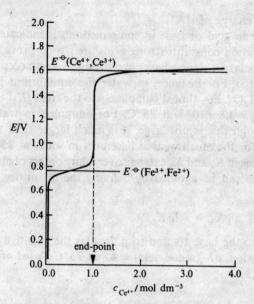

Fig. 12.10. Changes of e.m.f. during the titration of Fe(II) and Ce(IV)

able to do in order to detect the end-point is to note the rapid change of cell e.m.f. as the potential of the platinum electrode shifts from the vicinity of $E^{\ominus}(Fe^{3+}, Fe^{2+})$ to the vicinity of $E^{\ominus}(Ce^{4+}. Ce^{3+})$.

pK and pH. The Brønsted–Lowry classification defines an acid as a *proton donor* and a base as a *proton acceptor*. These definitions can be generalized, but they are adequate for the present discussion. An acid HA and a base B take part in the following equilibria in water:

$$HA + H_2O \rightleftharpoons A^- + H_3O^+ \qquad K_a = a_{H^+}a_{A^-}/a_{HA}$$

$$B + H_2O \rightleftharpoons BH^+ + OH^- \qquad K_b = a_{BH^+}a_{OH^-}/a_B.$$

In each case it has been assumed that the activity of water is constant and absorbed into the definition of K_a and K_b. These quantities are called the acid and base *dissociation constants*. Similar definitions apply when solvents other than water are involved.

The species A^- acts as a proton acceptor in the first equilibrium; it is therefore a base, according to the Brønsted–Lowry definition, and is called the *conjugate base* of the acid HA. Similarly BH^+ is a proton donor, and is the conjugate acid of the base B.

In the case of a polyacid (which is an acid such as H_2SO_4 capable of donating more than one proton to an acceptor) it is necessary to distinguish between the first, second, etc. dissociation constants K_{a1}, K_{a2}, etc. For instance,

$$H_2A + H_2O \rightleftharpoons HA^- + H_3O^+ \qquad K_{a1} = a_{H^+}a_{HA^-}/a_{H_2A}$$

$$HA^- + H_2O \rightleftharpoons A^{2-} + H_3O^+ \qquad K_{a2} = a_{H^+}a_{A^{2-}}/a_{HA}$$

Generally $K_{a2} < K_{a1}$ because the second proton is more difficult to remove on account of the charge of HA^-.

The strength of an acid or base in some medium is indicated by the value of its dissociation constant: strong acids are strong proton donors, and for them K_a is large; strong bases are strong proton acceptors, and for them K_b is large. For instance, in water the equilibrium $H_2SO_4 + H_2O \rightleftharpoons HSO_4^- + H_3O^+$ lies almost completely to the right ($K_{a1} \sim \infty$) but for acetic acid $K_a = 1.8 \times 10^{-5}$ at 25 °C. For ammonia in water $K_b = 1.77 \times 10^{-5}$; although this is not large, it is much larger than the value $K_b = 1.5 \times 10^{-14}$ for the much weaker base urea in water at 25 °C.

The span of values of K_a and K_b extends over many orders of magnitude. This inconvenience can be avoided by defining the quantities pK_a and pK_b through

(12.4.3) $$pK_a = -\lg K_a, \qquad pK_b = -\lg K_b.$$

The logarithm is to the base 10, and so p denotes the negative power of 10. With this notation, pK_a(acetic acid) = 4.8 at 25 °C. A list of values is given in Table 12.3.

Table 12.3. Dissociation constants at 25 °C, pK in aqueous solution

	pK_{a1}	pK_{a2}		pK_b
Acetic acid	4.756		Aniline	9.42
Butyric acid	4.820		Ethylamine	3.25
Lactic acid	3.860		Diethylamine	3.02
Succinic acid	4.207	5.638	Triethylamine	3.24
Phosphoric acid	2.148	7.198	Glycine	11.65
Ammonium ion	9.245		Hydrazine	5.52
Methylammonium ion	10.624		Ammonium hydroxide	4.75
Glycine	2.350	9.780		

Note that K_a (water) $= K_w/a_{H_2O}$ and $pK_w = 14.00$, $K_w = 1.008 \times 10^{-14}$ at 25 °C. Also $pK_w = 14.94$ at 0 °C, 13.68 at 35 °C, and 13.02 at 60 °C.
Source: G. W. C. Kaye and T. H. Laby, *Tables of physical and chemical constants*, Longmans; and *Handbook of chemistry and physics*, Chemical Rubber Co.

Water, like many other substances, can act as both an acid and a base. Therefore even in pure water there is the *autoprotolysis equilibrium*

$$H_2O + H_2O \rightleftharpoons H_3O^+ + OH^- \qquad K_a = a_{H^+}a_{OH^-}/a_{H_2O}.$$

It is obviously absurd to absorb one water activity into K_a and not the second, and so it is normal to express this equilibrium in terms of the constant K_w:

$$K_w = a_{H^+}a_{OH^-}.$$

At 25 °C $K_w = 1.008 \times 10^{-14}$, which indicates that only a very small proportion of the water molecules are dissociated. Since the concentration of ions is so low, the activities may be replaced by molalities:

$$K_w \approx (m_{H^+}/m^{\ominus}) \times (m_{OH^-}/m^{\ominus}).$$

Furthermore, since the liquid is electrically neutral, $m_{H^+} = m_{OH^-}$; consequently the molality of protons (as H_3O^+) in pure water at 25 °C is

$$m_{H^+} \approx K_w^{\frac{1}{2}} m^{\ominus} = 1.004 \times 10^{-7} \, \text{mol kg}^{-1}.$$

The molality of protons (or, more precisely, their activity) plays a central role in many chemical processes. Its magnitude can vary over a wide range. For instance, when a strong acid is present at $1 \, \text{mol kg}^{-1}$ the proton molality is about $1 \, \text{mol kg}^{-1}$; in pure water it is about $10^{-7} \, \text{mol kg}^{-1}$; when a very strong base is present at $1 \, \text{mol kg}^{-1}$, so that $m_{OH^-} \approx 1 \, \text{mol kg}^{-1}$, the proton molality drops to around $K_w/(m_{OH^-}/\text{mol kg}^{-1}) \approx 10^{-14} \, \text{mol kg}^{-1}$. The *pH scale* reduces this wide span into the range of about 0 to 14, which is much more manageable. Initially pH was actually defined as $-\lg(m_{H^+}/\text{mol kg}^{-1})$, but now it is expressed in terms of the activity through

(12.4.4)

$$pH = -\lg a_{H^+}.$$

In water at 25 °C since $a_{H^+} = K_w^{\frac{1}{2}} = 1.004 \times 10^{-7}$, the pH is 7.00, and so this value corresponds to neutrality at 25 °C. (At 45 °C $K_w = 4.02 \times 10^{-14}$, and so then neutrality corresponds to pH = 6.70.) In an acidic aqueous solution a_{H^+} is greater than in pure water, and so the pH is then less than 7; in basic solutions the pH is greater than 7.

The autoprotolysis equilibrium is maintained in the presence of acids or bases, and so if an acid is present, so that a_{H^+} is enhanced, a_{OH^-} must be such as to maintain $a_{H^+}a_{OH^-} = K_w$. Hence in a basic solution (where we might know a_{OH^-}) the pH can be obtained from $a_{H^+} = K_w/a_{OH^-}$. Note that it is legitimate to replace a_{H^+} (and a_{OH^-}) by molalities only if *all* the ions are present in low concentration, not merely H_3O^+ or OH^- species (all ions contribute to activity coefficients via the total ionic strength). In elementary applications it is normal to simplify the presentation by replacing activities by molalities even though the results then lack thermodynamic precision. The approximation can be avoided by drawing on known values of activity coefficients. Thus, at low ionic concentrations $a_{H^+} \approx m_{H^+}/m^{\ominus}$, and pH can then be interpreted as the negative power of 10 of the proton molality. In an aqueous solution of HCl of unit molality, $m_{H^+} \approx 1 \, \text{mol kg}^{-1}$ because there is virtually complete dissociation; at that molality the mean activity coefficient $\gamma_{\pm}$ has the value 0.811 (Table 11.3, p. 327), and so $a_{H^+} = 0.811$, implying pH = 0.09. Note that there is nothing mysterious about a negative pH, for it merely indicates an activity exceeding unity. For instance, in a $2 \, \text{mol kg}^{-1}$ HCl solution, so that $m_{H^+} \approx 2 \, \text{mol kg}^{-1}$, the activity coefficient is 1.011 (Table 11.3) and so the activity is 2.022, implying pH = -0.31. In an aqueous solution of a base in which $m_{OH^-} \approx 1.0 \, \text{mol kg}^{-1}$, the proton activity is of the order of $K_w/1.0 \approx 10^{-14}$, and so the pH is about 14. There is no fundamental reason why the pH should not exceed 14 in a strong basic solution, but for most purposes it is sufficient to confine attention to the range 0 to 14.

The measurement of pH and pK. The measurement of pH is of fundamental importance in chemistry. It is the basis of monitoring a wide variety of biological and industrial processes, and is of crucial assistance in monitoring laboratory experiments, such as acid–base titrations, rates of reactions, and so on. The measurement of pH is also the basis of the determination of the strengths of acids and bases.

The determination of pH is very simple in principle, for it is based on the measurement of the potential of a hydrogen electrode immersed in the solution. The left-hand electrode is generally a calomel electrode with potential $E(\text{cal})$; the right-hand electrode is the hydrogen electrode with potential $E^{\ominus}+(RT/F)\ln a_{H^+}$; the e.m.f. of the composite cell is

$$E = \{E^{\ominus}(H^+, H_2)+(RT/F)\ln a_{H^+}\} - E(\text{cal})$$

$$= (-59.15\,\text{mV})\text{pH} - E(\text{cal}).$$

It follows that the pH is given as $(E-E(\text{cal}))/(59.15\,\text{mV})$ once the e.m.f. has been determined. In practice indirect methods prove much more convenient, and the hydrogen gas electrode is replaced by the *glass electrode*, which is also sensitive to H^+ activity, has a potential that depends linearly on pH, and is much more readily handled. This electrode is calibrated against solutions of standard pH values.

The procedure for determining pK_a involves the determination of the pH of a series of mixtures of the acid and its salt, MA. If the solution is prepared with an acid molality m and a salt molality m' we can write

$$m_{HA} = m - m_{H^+} \quad \text{(because every } H^+ \text{ dissociated depletes HA)}$$

$$m_{A^-} = m_{H^+} + m' \quad \text{(because every HA that dissociates delivers an } A^- \text{ for every } H^+ \text{ produced, and every salt molecule is fully dissociated and gives an } A^- \text{ ion).}$$

It follows that the acid dissociation constant is

$$K_a = \left\{ \frac{(m_{H^+}/m^{\ominus}) \times (m_{A^-}/m^{\ominus})}{(m_{HA}/m^{\ominus})} \right\} \left\{ \frac{\gamma_{H^+}\gamma_{A^-}}{\gamma_{HA}} \right\}$$

(12.4.5)
$$= \left\{ \frac{m_{H^+}(m_{H^+} + m')}{(m - m_{H^+})m^{\ominus}} \right\} \gamma_{\pm}^2.$$

We have set $\gamma_{HA} = 1$. $\gamma_{\pm}$ is the mean activity coefficient of H^+ and A^- at the molalities involved. The procedure is therefore to measure pH, convert it to proton molality, and then to find K_a and pK_a by estimating (or measuring) the relevant activity coefficients.

Acid–base titrations. Measurements of pH may be used to monitor the course of acid–base titrations. In the course of the titration of a strong base by a strong acid the end-point is at pH = 7.0 (at 25 °C), but in other cases (e.g., weak acid–strong base titrations) the end-point is at other pH values on account of the presence of *solvolysis reactions* (*hydrolysis* in the case of water). The next few paragraphs show how solvolysis may be taken into account, how the pH of the end-point may be predicted, and how the pH changes during the course of the titration. They also show how to predict the pH of an arbitrary mixture of a weak acid or base and its salt, and the pH of an aqueous solution of a salt of a weak acid and strong base, or vice versa.

Consider the case of a weak acid HA being titrated with a strong base MOH. At any stage in the titration the acid molality is m_{HA}^0 and the molality of the salt (which is fully dissociated) is m_{MA}^0. At the beginning of the titration no base has been added, and so $m_{MA}^0 = 0$. At the end-point just enough base has been added to turn all the acid into salt, and so then $m_{HA}^0 = 0$. The problem is to find the pH of the end-point, bearing in mind the existence of the equilibria

$$HA + H_2O \rightleftharpoons H_3O^+ + A^- \qquad K_a = m_{H^+} m_{A^-}/m_{HA}$$
$$2H_2O \rightleftharpoons H_3O^+ + OH^- \qquad K_w = m_{H^+} m_{OH^-}.$$

All molalities will be denoted simply as m (that is $m \equiv m/m^\ominus$) and activity coefficients will be ignored (that is, all $\gamma = 1$). m_{HA} is the molality of the undissociated acid HA.

At every stage in the titration there are two constraints on the composition of the solution: it is electrically neutral, and the number of A groups is constant (although distributed as A^-, HA in various proportions). That is,

$$m_{H^+} + m_{M^+} = m_{A^-} + m_{OH^-} \quad \text{(charge neutrality)}$$
$$m_{HA} + m_{A^-} = m_{HA}^0 + m_{MA}^0 \quad \text{(A group conservation)}.$$

The four relations above can be combined to give an equation for the proton molality by eliminating m_{OH^-}, m_{A^-}, and m_{HA}:

(12.4.6)
$$m_{H^+} = \frac{m_{HA}^0}{1 + (m_{H^+}/K_a)} + \frac{K_w}{m_{H^+}} - \frac{m_{MA}^0}{1 + (K_a/m_{H^+})}.$$

This can be rearranged into a quartic equation for m_{H^+}, and the explicit solution obtained. The form of the solution is very obscure, and it is more instructive to examine approximations. (This also gives very good practice at making approximations by assessing magnitudes.) We therefore consider a case in which $K_a \approx 10^{-4}$ and $m_{HA}^0 \approx 0.1 \, \text{mol kg}^{-1}$ initially. Since $K_w \approx 10^{-14}$ it follows that initially $K_w/m_{H^+} \approx 10^{-13}$, $m_{H^+}/K_a \approx 10^3$, and $K_a/m_{H^+} \approx 10^{-3}$; the wide range of these values enables us to neglect various terms, as we shall now see.

At the start of the titration no salt is present ($m_{MA}^0 = 0$) and so the pH is that of an aqueous solution of the weak acid of molality m_{HA}^0. The third term on the right of the last expression is zero, the second term is of the order of 10^{-13}, and the first term is about $10^{-1}/(1 + 10^3) \approx 10^{-4}$. This therefore dominates the second, and so

$$m_{H^+} \approx m_{HA}^0/(1 + m_{H^+}/K_a) \approx m_{HA}^0 K_a/m_{H^+}$$

because it is also the case that $m_{H^+}/K_a \gg 1$. Therefore

$$m_{H^+} \approx (m_{HA}^0 K_a)^{\frac{1}{2}},$$

so that *the pH of a solution of a weak acid is given by*

(12.4.7) $$pH \approx \tfrac{1}{2}pK_a - \tfrac{1}{2}\lg(m_{HA}^0/m^\ominus).$$

In the case of an $0.1\,mol\,kg^{-1}$ aqueous solution of acetic acid we deduce that $pH \approx \tfrac{1}{2}(4.756) - \tfrac{1}{2}\lg 0.1 \approx 2.88$; the approximations are valid for this solution.

At the half-way point $m_{MA}^0 = m_{HA}^0$ because enough base has been added to convert half the acid present to salt. The ratio m_{H^+}/K_a is still large (still about 10^3 for the conditions we are assuming) and furthermore $m_{H^+}/m_{MA}^0 \ll 1$ because the acid is weak. K_w/m_{H^+} is still of the order of 10^{-13}, and therefore negligible. Therefore eqn (12.4.6) rearranges to

$$m_{H^+}/m_{MA}^0 \approx 1/(1 + m_{H^+}/K_a) - 1/(1 + K_a/m_{H^+}),$$

or

$$0 \approx 1/(m_{H^+}/K_a) - 1$$

when the appropriate terms are neglected. This implies that $m_{H^+} \approx K_a$, or

(12.4.8) $$pH \approx pK_a.$$

(Once again the magnitudes have to be checked for consistency. For the acetic acid titration, $m_{MA}^0 \approx 0.05\,mol\,kg^{-1}$ at this stage, and so $m_{H^+}/m_{MA}^0 \approx 10^{-4}/5 \times 10^{-2} \approx 2 \times 10^{-3}$, which is much smaller than unity.)

At the end-point all the acid has been replaced by salt, and so $m_{HA}^0 = 0$. The first term on the right of eqn (12.4.6) is therefore zero. Furthermore, if m_{H^+} is less than about $10^{-8}\,mol\,kg^{-1}$ (that is, $pH > 8$), the ratio K_w/m_{H^+} dominates m_{H^+} (at least 10^{-6} compared with less than 10^{-8}). Therefore, since $K_a/m_{H^+} \gg 1$ (so long as K_a exceeds about 10^{-6}), eqn (12.4.6) can be approximated by

$$0 \approx K_w/m_{H^+} - m_{MA}^0 m_{H^+}/K_a.$$

The solution is $m_{H^+} \approx (K_w K_a/m_{MA}^0)^{\frac{1}{2}}$, which implies that at the end-point

(12.4.9) $$pH \approx \tfrac{1}{2}pK_a + \tfrac{1}{2}pK_w + \tfrac{1}{2}\lg(m_{MA}^0/m^\ominus).$$

Since all the acid is now present as salt this pH is also what one obtains when the salt of a weak acid and strong base is simply dissolved in water at a molality m_{MA}^0. The corresponding expression for the pH of a solution of a salt of a strong acid and weak base is

(12.4.10) $$pH \approx -\tfrac{1}{2}pK_b + \tfrac{1}{2}pK_w - \tfrac{1}{2}\lg(m_B^0/m^\ominus).$$

Example   (Objective 18). What is the pH at the equivalence point of a titration of $0.01\,mol\,kg^{-1}$ butyric acid ($pK_a = 4.82$) with a concentrated solution of a strong base?

- *Method.* The *equivalence point* of a monobasic, monoacid system is the point where the amount of added base is equal to the amount of acid initially present: all the acid is then present, formally, as the salt. It follows that the pH of the equivalence point is the same as the pH of an aqueous solution of the salt (sodium butyrate). Therefore use eqn (12.4.9). Assume a temperature of 25 °C.

- *Answer.* Use $pK_a = 4.820$, $pK_w = 14.00$, $m_{MA} = 0.01 \text{ mol kg}^{-1}$; then

  $$pH = \tfrac{1}{2}(14.00) + \tfrac{1}{2}(4.820) + \tfrac{1}{2} \lg 0.01 = 7.00 + 2.41 - 1.00 = 8.41.$$

- *Comment.* An indicator for the titration would therefore have to be one with a colour change at about pH = 8.4. We stated 'concentrated solution of strong base' because that avoids questions of volume change of the sample.

When so much strong base has been added that the titration has been carried past the end-point, the pH is dominated by the amount of base present, and since $m_{H^+} m_{OH^-} \approx K_w$ at every stage (except for activity coefficient corrections) the final pH is determined by $m_{H^+} \approx K_w/m_{OH}^0$ where $m_{OH^-}^0$ is the excess base present:

(12.4.11) $$pH \approx pK_w + \lg(m_{OH^-}^0/m^{\ominus}).$$

The general form of the pH curve throughout the calculation can be drawn on the basis of these results and is illustrated in Fig. 12.11. The pH rises slowly from the value given by eqn (12.4.7) until the neutralization point has been approached, then it changes rapidly, rising to the value given by eqn (12.4.9) at equivalence, passing through it, and then flattening out towards its final value as given by eqn (12.4.11). The end-point of the titration can be detected very easily by observing where the pH changes most rapidly. The slow-quick-slow variation of pH is another example of the form of the logarithmic dependence, like that in

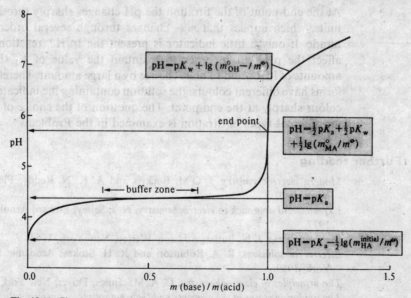

Fig. 12.11. Changes in pH during the titration of a weak acid with a strong base.

Fig. 12.9.

The slow rise of pH in the vicinity of $m_{HA}^0 = m_{MA}^0$ (eqn 12.4.8) is the basis of the function of *buffer solutions*. Such solutions maintain the approximate pH of a solution when small amounts of acids or bases are added and play a vital role in biological processes. For example, blood has to be maintained in the region pH = 7.0–7.9, and saliva at pH = 6.8. Digestion takes place in an acidic environment, and in the stomach the gastric juices are buffered to lie in the range 1.6–1.8 (the cells of the stomach are adapted to generate a litre or so of approximately 1 mol kg^{-1} HCl each day). Failure of the buffering process may be fatal, and its function depends on ionic equilibria.

The mathematical basis of buffering is the logarithmic pH dependence given by eqn (12.4.8) which is quite flat in the vicinity of pH $\approx$ pK_a. The physical reason is that when acid is added to a buffered solution the equilibrium $HA + H_2O \rightleftharpoons H_3O^+ + A^-$ is driven to the left in order to maintain the K_a equilibrium, and it can occur because A^- is provided by the salt MA. When a base is added the K_w equilibrium is maintained by a reduction of proton activity, but the protons suppressed can be recovered by the dissociation of HA, which proceeds in order to maintain the K_a equilibrium.

The rapid change of pH in the vicinity of the end-point is the basis of *indicator* detection. An acid–base indicator is normally some large organic molecule which can exist as either a protonated or an unprotonated form, one form being coloured. If the two forms are denoted InH^+ and In, then in solution there is an equilibrium

$$In + H^+ (as\ H_3O^+) \rightleftharpoons InH^+, \qquad K_I = a_{InH^+}/a_{In}a_{H^+}.$$

At the end-point of the titration the pH changes sharply through several units, which implies that m_{H^+} changes through several orders of magnitude. If only a little indicator is present the In/H^+ reaction does not affect the pH, but in order to maintain the value of K_I the relative amounts of InH^+ and In must change by a large amount. Therefore, as the forms have different colours, the solution containing the indicator changes colour sharply at the end-point. The question of the choice of indicators for different types of titration is examined in the Problems.

Further reading

Modern electrochemistry. J. O'M. Bockris and A. K. N. Reddy; Plenum, New York, 1970.

Experimental approach to electrochemistry. N. J. Selley; Edward Arnold, London, 1977.

Ionic equilibria. J. N. Butler; Addison–Wesley, New York, 1964.

Electrolyte solutions. R. A. Robinson and R. H. Stokes; Academic Press, New York, 1959.

The principles of electrochemistry. D. A. MacInnes; Dover, New York, 1961.

The oxidation states of the elements and their potentials in aqueous solutions. W. M. Latimer; Prentice-Hall, Englewood Cliffs, N.J., 1952.

The study of ionic equilibria. H. S. Rossotti; Longmans, London, 1978.
Chemical applications of e.m.f. H. S. Rossotti; Longmans, London, 1978.
Acid–base equilibria. E. J. King; Pergamon, Oxford, 1965.
Treatise on electrochemistry. G. Kortum; Elsevier, Amsterdam, 1951.
The proton in chemistry. R. P. Bell; Cornell University Press, Ithaca, 1959.
The glass electrode. G. Eisenman, R. Bates, G. Matlock, and S. M. Friedman; Interscience, New York, 1965.
Determination of pH: theory and practice. R. G. Bates; Wiley, New York, 1973.
The initiation of the heartbeat (2nd ed.). D. Noble; Clarendon Press, Oxford, 1979.
Electrochemical data. B. E. Conway; Elsevier, Amsterdam, 1952.

Problems

12.1. Write the cell reactions and the left and right half-cell reactions for the following cells: (a) $Pt,H_2(g)|HCl(aq)|AgCl,Ag$; (b) $Pt|FeCl_2,FeCl_3| \times |SnCl_4,SnCl_2|Pt$; (c) $Cu|CuCl_2||MnCl_2,HCl|MnO_2,Pt$; (d) $Ag,AgCl|HCl(aq)| \times |HBr(aq)|AgBr,Ag$.

12.2. Devise cells in which the following are the reactions: (a) $Zn(s)+CuSO_4(aq) \rightleftharpoons ZnSO_4(aq)+Cu(s)$; (b) $AgCl(s)+\frac{1}{2}H_2(g) \rightleftharpoons HCl(aq)+Ag(s)$; (c) $H_2(g)+\frac{1}{2}O_2(g) \rightleftharpoons H_2O(l)$; (d) $Na(s)+H_2O(l) \rightleftharpoons NaOH(aq)+\frac{1}{2}H_2(g)$; (e) $H_2(g)+I_2(g) \rightleftharpoons 2HI(aq)$.

12.3. What are the standard e.m.f.s at $25\,°C$ of the cells in the preceding two problems? Which electrode is the positive electrode?

12.4. Calculate $\Delta G^{\ominus}_m(298\,K)$ for the following reactions on the basis of the electrode potential data in Table 12.1: (a) $2Na(s)+2H_2O(l) \rightarrow 2NaOH(aq)+H_2(g)$; (b) $K(s)+H_2O(l) \rightarrow KOH(aq)+\frac{1}{2}H_2(g)$; (c) $K_2S_2O_8(aq)+2KI(aq) \rightarrow I_2(s)+2K_2SO_4(aq)$; (d) $Pb(s)+ZnCO_3(aq) \rightarrow PbCO_3(aq)+Zn(s)$.

12.5. Calculate the equilibrium constants for the following reactions at $25\,°C$: aqueous solutions where appropriate. (a) $Sn(s)+CuSO_4(aq) \rightleftharpoons Cu(s)+SnSO_4(aq)$, (b) $2H_2(g)+O_2(g) \rightleftharpoons 2H_2O(l)$; (c) $Cu^{2+}(aq)+Cu(s) \rightleftharpoons 2Cu^{+}(aq)$.

12.6. The cell $Zn|ZnSO_4(a_+ = 1)||CuSO_4(a_+ = 1)|Cu$ was set up in a laboratory experiment. Use the information in Table 12.1 to state (a) its e.m.f. at $25\,°C$, (b) the value of $\Delta G^{\ominus}_m$ for the cell reaction, (c) the equilibrium constant for the cell reaction. What is the ratio of the activities of the two electrolytes when the cell is 'exhausted'?

12.7. Another cell that was investigated involved a redox reaction that depended on the transfer of two electrons in each act of ionic reduction. This was the cell $Al|Al^{3+}(aq)||Sn^{2+},Sn^{4+}(aq)|Pt$. Write the cell reaction. State (a) the cell e.m.f. when the activities are all 0.1, 1.0, (b) the value of $\Delta G^{\ominus}$ for the reaction, (c) the equilibrium constant for the reaction. Which is the positive electrode? Which way do electrons tend to flow?

12.8. The same cell as in the last Problem was set up, the only difference being that the concentrations of Al^{3+}, Sn^{2+}, and Sn^{4+} were each $0.01\,mol\,kg^{-1}$. What is the cell e.m.f.?

12.9. Devise a cell in which the overall reaction is $Pb(s)+Hg_2SO_4(s) \rightleftharpoons PbSO_4(s)+2Hg(l)$. What is its e.m.f. when the electrolyte it contains is saturated with the two salts? $[K_{sp}(PbSO_4) = 2.43 \times 10^{-8}, K_{sp}(Hg_2SO_4) = 1.46 \times 10^{-6}.]$

12.10. What is the e.m.f. at $25\,°C$ of the cell $Pt,H_2(1\,atm)|HCl(m_1)| \times |HCl(m_2)|H_2(1\,atm),Pt$ when $m_1 = 0.1\,mol\,kg^{-1}$ and $m_2 = 0.2\,mol\,kg^{-1}$? What would be the e.m.f. of the cell if the pressure of gas at the right-hand electrode were increased to 10 atm? The mean activity coefficients of HCl(aq) are $\gamma_1 = 0.798$ and $\gamma_2 = 0.790$.

12.11. Hydrogen behaves imperfectly at high pressure, and its equation of state may be summarized by the virial expansion $pV_m/RT = 1 + 5.37 \times 10^{-4}(p/\text{atm}) + 3.5 \times 10^{-8}(p/\text{atm})^2$. The e.m.f. of the cell Pt,H_2|HCl $(0.1 \text{ mol kg}^{-1})$| Hg_2Cl_2,Hg has been measured up to pressures of 1000 atm (W. R. Hainsworth, H. J. Rowley, and D. A. McInnes, *J. Amer. chem. Soc.* **46**, 1437 (1924)), and good agreement between experiment and a theoretical expression was obtained up to 600 atm on the assumption that the only significant volume change was that associated with the hydrogen. Above that pressure the volume change of the other components became significant. What is the e.m.f. of the cell when the pressure is 500 atm? Take $\gamma_\pm(\text{HCl}) = 0.798$, Table 11.3.

12.12. The fugacity of a gas can be determined electrochemically on account of the dependence of the electrode potential on it. In the cell Pt,H_2(1 atm)|HCl(0.01 mol kg^{-1})|Cl_2,Pt the following values of the e.m.f. were obtained at 25 °C for various pressures of chlorine. What is the fugacity and the fugacity coefficient of chlorine at the pressures quoted? Table 11.3 gives $\gamma_\pm$.

p/atm	1	50	100
E/V	1.5962	1.6419	1.6451

12.13. A leak developed in a cell which had been designed to operate as Pt,H_2(1 atm)|HCl(m)|AgCl,Ag. As a result water was able to dribble in and dilute the hydrochloric acid so that its molality changed with time according to $m/\text{mol kg}^{-1} = 1.00/(1 + \kappa t)$. What will be the time-dependence of the cell potential? First neglect activity coefficients, and then include them by using the Debye–Hückel Limiting Law. Sketch the result of the calculations, and calculate $(\partial E/\partial t)$.

12.14. A fuel cell converts chemical energy directly into electrical energy without involving the wasteful process of combustion, heating of steam for use in a turbine, and the conversion of thermal energy into electrical energy. What is the maximum e.m.f. obtainable from a cell in which hydrogen and oxygen, both at 1 atm pressure and 25 °C, combine on a catalyst surface? What is the maximum amount of electrical work available per mole of hydrogen?

12.15. Much modern research goes into investigating the employment of hydrocarbon gases in fuel cells. Calculate the maximum amount of heat, the maximum amount of work, and the maximum amount of electrical work available from the oxidation of butane at 25 °C and 1 atm.

12.16. Electrochemical cells are very important for finding the values of activities and activity coefficients of electrolytes in solutions (not just aqueous solutions). For a 1:1 electrolyte in aqueous solution at 25 °C the extended Debye–Hückel Law is that $\lg \gamma_\pm = -0.509(m/\text{mol kg}^{-1})^{1/2} + B(m/\text{mol kg}^{-1})$. As a first step in seeing how activity coefficients are measured, show that a plot of $E + 0.1183 \lg(m/\text{mol kg}^{-1}) - 0.0602(m/\text{mol kg}^{-1})^{1/2}$ against m, where E is the e.m.f. of the cell Pt,H_2(1 atm)|HCl(m)|AgCl,Ag, should give a straight line with intercept $E^\ominus$ and slope $-0.1183B$. Once $E^\ominus$ has been determined it is a simple matter to measure E and relate it to the activity coefficient.

12.17. The e.m.f. of the cell referred to in the last Problem has been measured at a series of concentrations of HCl. The data for 25 °C are below. What is (a)

$E^\ominus$ of the cell, (b) $E^\ominus$ of the AgCl,Ag electrode?

$m/\text{mol kg}^{-1}$	0.1238	0.02563	0.009138	0.005619	0.003215
E/mV	341.99	418.24	468.60	492.57	520.53

12.18. The e.m.f. of the same cell as in the last Problem is 0.3524 V at 25 °C when the HCl is at a concentration 0.100 mol kg^{-1}. What is (a) the activity of the HCl, (b) the mean activity coefficient of the HCl at this concentration, (c) the pH of the acid?

12.19. If the temperature dependence of a cell e.m.f. is known the information can be used to determine the enthalpy and entropy of the cell reaction: this is a straightforward way of determining these quantities without using a calorimeter, but one source of inaccuracy comes from the step where a small quantity is obtained as the difference of two large numbers. The e.m.f. of the AgCl,Ag electrode has been measured very carefully at a series of temperatures (R. G. Bates and V. E. Bower, *J. Res. Nat. Bur. Stand.* **53**, 283 (1954)), and the results were found to fit the expression

$$E^\ominus/\text{V} = 0.23659 - 4.8564 \times 10^{-4}(t/°C) - 3.4205 \times 10^{-6}(t/°C)^2 + 5.869 \times 10^{-9}(t/°C)^3.$$

Find the Gibbs function, enthalpy, and entropy of the electrode reaction, and those of the Cl$^-$ ion in water at 25 °C.

12.20. The cell Pt,H$_2$(1 atm)|HCl(aq)|AgCl,Ag had an e.m.f. of 0.332 V at 25 °C: what is the pH of the solution?

12.21. The solubility of AgBr is 2.1×10^{-6} mol kg^{-1} at 25 °C. What is the e.m.f. of the cell Ag|AgBr(aq)|AgBr,Ag at that temperature?

12.22. The standard e.m.f. of the cell Ag|AgI(aq)|AgI,Ag is -0.9509 V at 25 °C. What is (a) the solubility of AgI in water at this temperature, (b) the solubility product?

12.23. Measurements of the e.m.f. of cells of the type Ag,AgX|MX(m_1)|M$_x$Hg| |MX(m_2)|AgX,Ag, where M$_x$Hg denotes an amalgam and the electrolyte is an alkali metal halide dissolved in ethylene glycol, have been reported (U. Sen, *J. chem. Soc. Faraday Trans. I* **69**, 2006 (1973)) and a selection for LiCl at 25 °C is given below. Estimate the activity coefficient at the concentration marked * and then use this value to calculate the activity coefficients from the measured e.m.f. as at the other concentrations.

$m/\text{mol kg}^{-1}$	0.0555	0.0914*	0.1652	0.2171	1.040	1.350
E/V	-0.0220	0.0000	0.0263	0.0379	0.1156	0.1336

Base your answer on the use of the extended Debye–Hückel expression $\lg \gamma_\pm = -Am^{\frac{1}{2}}/(1 + A^*m^{\frac{1}{2}}) + Bm$, $m \equiv m/\text{mol kg}^{-1}$, using $A = 1.461$, $A^* = 1.7$, $B = 0.2$.

12.24. The e.m.f. of the cell Pt,H$_2$(1 atm)|HCl(m)|Hg$_2$Cl$_2$,Hg has been measured with high precision (G. J. Hills and D. J. G. Ives, *J. chem. Soc.*, 311 (1951)) with the following results at 25 °C.

$100m/\text{mol kg}^{-1}$	0.16077	0.30769	0.50403	0.76938	1.09474
E/mV	600.80	568.25	543.66	522.67	505.32

Find the standard e.m.f. of the cell and the experimental values of the HCl activity coefficients at these molalities. (Make a least-squares fit of the data to the best straight line, and quote the coefficient of determination.)

12.25. In view of the importance of knowing the enthalpy of ionization of water, and because there were discrepancies between the published values, careful measurements of the e.m.f. of the cell $Pt,H_2(1\,atm)|NaOH,NaCl|AgCl,Ag$ have been reported (C. P. Bezboruah, M. F. G. F. C. Camoes, A. K. Covington, and J. V. Dobson, *J. chem. Soc. Faraday Trans. I* **69**, 949 (1973)). Among the data is the following information: $m(NaOH) = 0.0100\,mol\,kg^{-1}$; $m(NaCl) = 0.011\,25$ $mol\,kg^{-1}$;

$t/°C$	20	25	30
E/V	1.047\,74	1.048\,64	1.049\,42

Calculate the values of pK_w at these temperatures, and the enthalpy and entropy of ionization of water at 25 °C.

12.26. There is some interest in the structure of water/urea mixtures on account of the protein denaturation that occurs in them. The e.m.f. of the cell $Pt,H_2(1\,atm)|HCl(m)$, water, urea$|AgCl,Ag$ has been measured for a variety of different mixtures and temperatures (K. K. Kundu and K. Mazumdar, *J. chem. Soc. Faraday Trans. I* **69**, 806 (1973)). The following data were obtained at 25 °C for a 29.64 weight per cent urea/water mixture (relative permittivity 91.76, density $1.0790\,g\,cm^{-3}$). Find the standard e.m.f. of the $AgCl,Ag$ electrode in this medium.

$m(HCl)/mol\,kg^{-1}$	0.005\,58	0.013\,10	0.019\,2	0.024\,6
E/V	0.561\,6	0.518\,7	0.499\,9	0.487\,8

$m(HCl)/mol\,kg^{-1}$	0.034\,9	0.041\,1
E/V	0.470\,8	0.462\,9

12.27. Calculate the A-coefficient from the Debye–Hückel Limiting Law for the urea/water mixture referred to in the last Problem, and compare its value with experiment.

12.28. When MA, the salt formed from a strong base and a weak acid, is dissolved in water, the anions A^- tend to form HA from the water molecules, so leaving an excess of OH^-. In such a case we expect the pH to exceed 7.0. On the basis of the acid dissociation constant of HA, K_a, and the water dissociation constant, K_w, find an expression which, when solved, gives the pH in terms of the concentration of added salt c. Show, when c greatly exceeds K_a, c_{H^+}, and K_w/c_{H^+}, that $pH \approx \frac{1}{2}(pK_a + pK_w - \lg c)$.

12.29. When MA is the salt of a weak base and a strong acid the pH of the solution is less than 7.0 on account of the reaction of M^+ with the water to remove OH^- and to leave an excess of H^+. Derive an expression for the pH of a solution formed from a concentration c of salt, and choose a set of conditions that lead to a simplification as in the last Problem.

12.30. What is the pH of (a) a $0.1\,mol\,dm^{-3}$ solution of ammonium chloride, (b) a $0.1\,mol\,dm^{-3}$ solution of sodium acetate, (c) a $1.0\,mol\,dm^{-3}$ solution of sodium acetate, (d) a $0.1\,mol\,kg^{-1}$ solution of sodium acetate, (e) a $0.1\,mol\,dm^{-3}$ solution of acetic acid. Data in Table 12.3; note that $K_b = K_w/K_a$.

12.31. An aqueous $0.1\,mol\,dm^{-3}$ solution of sodium lactate was prepared at 25 °C. What is the pH of the solution? Are the concentrations such that the simplified expression for the pH may be used? What is the pH of the end point of the titration of lactic acid with NaOH?

12.32. One very important aspect of salts of a weak acid (or base) and a strong base (or acid) is their ability to stabilize the pH of liquids (e.g., fluids in the body, like blood plasma). As the first step in deriving an expression for this action, find an expression for the pH of a solution formed from a weak acid (nominal concentration c_a) and a salt formed from a strong base and the same weak acid (nominal concentration c_s). You should be able to solve this equation for the pH by assuming that $c_s \gg c_{H^+}$ and $c_s \gg K_w/c_{H^+}$.

12.33. Plot the pH of a solution containing $0.1 \, mol \, dm^{-3}$ of sodium acetate and a variable amount of acetic acid. Notice that the pH changes least quickly in the region of pH $\approx pK_a$. This region of stability is the *buffering region*, and the stability is greatest for equimolar mixtures of acid and salt. In what region of pH will a mixture of boric acid, $pK_a = 9.14$, and sodium borate (Na_2HBO_3), molalities $0.1 \, mol \, kg^{-1}$ for both, be an effective buffer?

12.34. From the data in Table 12.3, select combinations of acids and salts that will have their maximum buffering ability at (a) pH ≈ 2.2, (b) pH ≈ 7.0.

12.35. Quartic equations can be solved in a closed form, just like quadratic equations, but the expression for the four roots is very clumsy: see, for instance, M. Abramowitz and I. A. Stegun, *Handbook of mathematical functions*, Dover (1965), Section 3.8.3. If you have access to a computer or a programmable calculator, solve eqn (12.4.6) for $[H^+]$, and hence pH, for a wide range of concentrations of added weak acid. Take $K_a = 2.69 \times 10^{-5}$ for the whole range of acid concentrations.

12.36. In order to deal with corrosion processes we have to treat them as dynamical processes. Nevertheless, the *tendency* of a system to corrode may be assessed thermodynamically, and we look into that here (the full subject is taken up in Part 3). Metals corrode by being oxidized, the driving reactions being the reductions

$$2H_3O^+ + 2e^- \rightleftharpoons 2H_2O + H_2 \qquad E^\ominus = 0$$
$$O_2 + 4H^+ + 4e^- \rightleftharpoons 2H_2O \qquad E^\ominus = 1.23 \, V \left.\right\} \text{acid conditions}$$

$$O_2 + 2H_2O + 4e^- \rightleftharpoons 4OH^- \qquad E^\ominus = 0.401 \, V, \text{alkaline conditions}$$

Which of the following metals will corrode in water made slightly acid (pH ≈ 6.0)? Which will corrode in water made slightly alkaline (pH ≈ 8.0)? Which corrode only in strong acid (pH ≈ 1) and which corrode only in strong alkali (pH ≈ 14)? Take the criterion of corrosion to be the thermodynamic tendency to form *at least* a $10^{-6} \, mol \, dm^{-3}$ solution of ions in the vicinity of the metal surface. The metals to consider are (a) Al, (b) Cu, (c) Fe, (d) Ag, (e) Pb, (f) Au.

12.37. An industrial process produces a $1 \, mol \, dm^{-3}$ solution of sodium acetate. Can it be collected and stored in mild steel vessels?

12.38. Superheavy elements are now of considerable interest. Shortly before it was believed that the first had been discovered an attempt was made to predict the chemistry of element 115 (O. L. Keller, C. W. Nestor, and B. Fricke, *J. phys. Chem.* **78**, 1945 (1974)). In one part of the paper the enthalpy and entropy of the reaction $(115)^+(aq) + \frac{1}{2}H_2(g) \rightleftharpoons (115)(s) + H^+(aq)$ were estimated from the following data: $\Delta H_{sub,m} = 1.5 \, eV \, mol^{-1}$, $I = 5.2 \, eV$, $\Delta H(115^+, aq)_{solv,m} = -3.22 \, eV$, $S^\ominus(115^+, aq) = 1.34 \times 10^{-3} \, eV \, K^{-1} \, mol^{-1}$, $S^\ominus(115, s) = 0.69 \times 10^{-3} \, eV \, K^{-1} \, mol^{-1}$. Estimate the expected standard electrode potential of the $115^+, 115$ couple.

Part 2 · Structure

13 The microscopic world: quantum theory

Learning objectives

After careful study of this chapter you should be able to

(1) List the characteristics of *black-body radiation* and explain how Planck's hypothesis of the quantization of energy accounts for them (p. 390).

(2) Explain how the characteristics of the *heat capacities* of solids conflict with classical physics but can be accounted for on the basis of the quantum theory (p. 392).

(3) Describe the experimental observations of the *photoelectric effect* and show that they imply the existence of photons (p. 394).

(4) Describe the observations of the *Compton effect* and show that they are evidence for the quantum theory (p. 394).

(5) Summarize the evidence for the *wave nature* of matter (p. 395) and quote *de Broglie's relation* (eqn (13.2.6)).

(6) Explain why *atomic* and *molecular spectra* are evidence for the quantum theory (p. 396).

(7) Write the *Schrödinger equation* (eqn (13.3.1)).

(8) Demonstrate the connection between the *curvature* of the wavefunction and the *kinetic energy* of the particle it describes (p. 398).

(9) Describe the *Born interpretation* of the wavefunction (p. 399).

(10) State and interpret the *Heisenberg uncertainty relation* (eqn (13.3.6)) and explain the significance of *complementary observables* (p. 401).

(11) List the necessary *characteristics of the wavefunction* (p. 404) and indicate why they imply quantization (p. 404).

(12) Solve the Schrödinger equation for a *particle in a square well* (p. 404), describe the properties of the solutions (p. 407), and draw the form of the wavefunctions (p. 407).

(13) Explain the term *zero-point energy* (p. 407).

(14) Explain the quantum mechanical basis of *tunnelling* (p. 408) and state when it is likely to be important (p. 409).

(15) Describe the procedure of *separation of variables* (p. 411) and apply it to the case of a particle in a two-dimensional well.

(16) Define the term *degeneracy* (p. 413).

(17) Write the expression for the energy levels of a *harmonic oscillator* (eqn (13.5.2)) and draw and interpret the form of the wavefunctions (p. 414).

(18) Deduce the energy levels of a *rotating body* in two and three dimensions (eqn (13.6.4) and eqn (13.6.7)).

(19) Explain the role of *cyclic boundary conditions* (p. 419) in the quantization of angular momentum and rotational energy.

(20) List the properties of the *angular momentum* of a system and state the significance of the quantum numbers l and m_l (p. 421).

(21) Describe *space quantization* (p. 422) and interpret the *Stern–Gerlach experiment* (p. 422).

(22) List the properties of the *spin* of an electron (p. 425).

(23) Describe the *vector model* of angular momentum (p. 427) and use the *Clebsch–Gordon* series to combine angular momenta (p. 426).

(24) Explain the significance of an *expectation value* (p. 429), and show how it may be calculated from a wavefunction (p. 429).

(25) Write down the *Hamiltonian operator* for a system (eqn (13.7.4)), and express the Schrödinger equation in terms of it (eqn (13.7.10)).

(26) Write down the form of the operators for *linear momentum* (eqn (13.7.8)) and *kinetic energy* (eqn (13.7.6)).

(27) Explain the terms *eigenvalue*, *eigenstate*, *eigenfunction*, and *eigenvalue equation* (p. 431).

Introduction

In Part 1 the properties of bulk matter were examined from the point of view of thermodynamics. In Part 2 the viewpoint is shifted to the study of the behaviour of individual atoms and molecules.

We are familiar with the way everyday objects move along trajectories in response to forces, how they can be brought to rest and examined, and how they can be accelerated into definite states of motion. This behaviour can be discussed quantitatively by solving the equations of *classical mechanics*, which are based on the laws of motion introduced by Newton. Until the present century it was assumed that classical concepts and laws could also be applied to species as small as atoms. Experimental evidence was accumulated, however, which indicated that classical mechanics failed to account for the behaviour of very small objects, and it was not until 1926 that a technique of calculation for such systems was devised. This new mechanics is called *quantum mechanics*.

It should not be thought that quantum mechanics applies only to microscopic, atomic, phenomena and not to macroscopic, everyday, processes. Quantum theory is more fundamental than classical mechanics, and in some way it must also 'explain' classical theory. In this chapter we encounter some of the aspects of quantum theory, and see how to think about processes on an atomic scale and how the familiar concepts of classical mechanics have their basis in quantum mechanics.

13.1 Classical mechanics: some central ideas

The classical mechanical approach to the description of the behaviour of a system can be illustrated by two equations. One equation expresses the total energy of the particle in terms of its kinetic energy $\frac{1}{2}mv^2$, where v is its speed at the time of interest and m its mass, and the potential energy $V(x)$ at the point in space occupied by the particle at that instant:

$$E = \tfrac{1}{2}mv^2 + V(x), \quad \text{where } x \text{ and } v \text{ are functions of } t.$$

In terms of the *linear momentum*, $p = mv$, this becomes

(13.1.1) $$E = p^2/2m + V(x).$$

The last equation can be used and interpreted in various ways. For example, if we know both the momentum and position of the particle we can calculate its total energy. Alternatively, since $p = m(\mathrm{d}x/\mathrm{d}t)$, eqn (13.1.1) is a differential equation for x as a function of time. If the particle's energy is fixed it might be possible to solve the differential equation and hence obtain the position and momentum as functions of time. A statement of both $x(t)$ and $p(t)$ is called a *trajectory* of the particle. The implication of this is that the whole future behaviour of the particle can be predicted if its present position and velocity are known. There appeared to be no limit to the accuracy with which these initial values might be known or controlled, and so it appeared that the trajectory of a particle could be predicted with arbitrary precision.

The simplest example of this procedure is the case of a constant potential, so that V is independent of x. A gas particle in an otherwise perfect vacuum, and free from a gravitational field, is an example in which $V = 0$ everywhere. If the particle is known to have an energy E, then

$$E = p^2/2m \quad \text{or} \quad (2E/m)^{1/2} = \mathrm{d}x/\mathrm{d}t,$$

and the solution of the differential equation is

$$x(t) = x(0) + (2E/m)^{1/2}t.$$

The constant energy E can be written in terms of the initial momentum $p(0)$, and so the trajectory is

(13.1.2) $$x(t) = x(0) + p(0)t/m, \qquad p(t) = p(0).$$

It follows that knowing the position and momentum of the particle at some initial time gives the trajectory at all later times.

The other basic equation of classical physics is Newton's Second Law. This law expresses the relation between the acceleration, $\ddot{x} = \mathrm{d}^2x/\mathrm{d}t^2$, of a particle and the force $F(x)$ it experiences. Using $\ddot{x} = \dot{p}/m$, the law is

(13.1.3) $$\dot{p} = F(x).$$

It follows that if we know the force acting on the particle in every region of space we might be able to solve this equation and find its momentum at all times, and from that its position. This calculation of the trajectory is equivalent to the method based on E, but is more suitable in some applications. Furthermore, it may be used to illustrate another feature of classical mechanics by asking what energy a system may acquire when it is accelerated by some force. As an example consider a particle that experiences a constant force F for a time τ (the Greek letter tau), and is then allowed to travel freely. Newton's equation becomes

$dp/dt = F$ for times between $t = 0$ and $t = \tau$

$dp/dt = 0$ for times later than $t = \tau$.

The first equation has the solution

$$p(t) = p(0) + Ft \qquad (0 \leqslant t \leqslant \tau)$$

and at the end of the action of the force the particle's momentum is

$$p(\tau) = p(0) + F\tau.$$

The second equation has the solution $p = $ constant, and so at all times later than $t = \tau$ the particle has a momentum

$$p(t) = p(\tau) = p(0) + F\tau \qquad (t \geqslant \tau).$$

For simplicity we suppose that the particle is initially at rest, and henceforth set $p(0) = 0$. The kinetic energy is $p(t)^2/2m$, and so in the present case its value is $F^2\tau^2/2m$ at all times after the force ceases to act. Therefore, the total energy of the accelerated particle has been increased to $F^2\tau^2/2m$ by the force. The important point is that, since F and τ may be varied in an arbitrary fashion, *the energy acquired by the particle may take any value.*

The same type of calculation may be applied to more complex cases. For example, we can investigate how much energy may be imparted to a rotating body. In the case of translation the linear momentum p is related to the linear velocity v by $p = mv$; in the case of rotation the *angular momentum* J is related to the *angular velocity* ω, omega, by $J = I\omega$ where I is the moment of inertia of the body. (The analogous roles of m and I, of v and ω, and of p and J in the translational and rotational cases should be remembered because they provide a ready way of constructing or remembering equations.) In order to accelerate the rotation it is necessary to apply a torque T (a twisting force), and Newton's equation of motion is:

$$\dot{J} = T.$$

If the torque is applied for a time τ the energy of a body is increased by $T^2\tau^2/2I$, where I is its moment of inertia. This equation predicts that an appropriate control over the magnitude of the torque and the length of time for which it is applied should allow the experimenter to excite the body to any energy of rotation.

The final example is the harmonic oscillator. When a particle is subject to a restoring force linearly proportional to its displacement from some point it undergoes *simple harmonic motion* about the point (like a pendulum). The force is $F(x) = -kx$, where k, the *force constant*, determines its strength: a strong spring has a greater force constant than a weak one. The negative sign appears because the force is directed towards the origin (when x is positive, representing displacement to the right, F is negative, representing pushing to the left; when x is negative, representing displacement to the left, F is positive, representing pushing to the right). Newton's Second Law now reads

$$m(d^2x/dt^2) = -kx$$

and a solution is

(13.1.4) $$x(t) = A \sin \omega t, \quad \text{with } \omega = (k/m)^{1/2}.$$

The momentum at any instant is $m\dot{x}$, and so

(13.1.5) $$p(t) = m\omega A \cos \omega t.$$

The properties of this motion will be familiar: the position of the oscillating particle varies harmonically (as $\sin \omega t$) with a frequency $\omega/2\pi$; the momentum is least when the particle is at its maximum displacement A, and vice versa. The total energy is $\frac{1}{2}kA^2$. This comes from the result of classical physics that the force is related to the gradient of the potential by

(13.1.6) $$F(x) = -dV/dx,$$

and because $F = -kx$, it follows that $V = \frac{1}{2}kx^2$; therefore the total energy is

$$E = p^2/2m + V = p^2/2m + \frac{1}{2}kx^2 = \frac{1}{2}kA^2$$

by substitution of equations (13.1.4) and (13.1.5) and use of $\cos^2\theta + \sin^2\theta = 1$.

The preceding discussion shows that the energy of an oscillating particle can be raised to any value by a suitably controlled impulse. Once excited, the oscillation occurs at a frequency $\omega/2\pi$, which is determined by the strength of the restoring force (as measured by k) and the inertia of the oscillating particle (as measured by m), and is *independent* of the initial impulse. It is important to emphasize that the *frequency* is a function of the structure (as represented by k, m) of the oscillator and independent of the energy; the *amplitude* is determined by the energy through $A^2 = 2E/k$.

The lessons of these examples are that classical physics (1) predicts a precise trajectory for particles, and (2) allows the translational, rotational, and vibrational modes of motion to be excited to any extent simply by controlling the impressed forces and torques. These conclusions accord with everyday experience.

Everyday experience, however, does not encompass individual atoms, and careful laboratory experiments have shown that the laws of classical physics are wholly untenable when the responses of particles to small transfers of energy are examined. Both conclusions mentioned in the last paragraph are overturned by quantum theory. In the first place eqn (13.1.2) fails because it turns out that it is impossible to know both $x(0)$ and $p(0)$ to arbitrary precision. In the second place we shall see that it is not possible even to think in terms of a definite trajectory. In the third place it is not possible to inject arbitrary amounts of energy into systems; therefore the discussion based on eqn (13.1.3) fails, and so Newton's equation itself must be inadequate. The results of 'classical physics are in fact only an approximation to the true behaviour of particles, and the approximation fails when small masses, small moments of inertia, and small transfers of energy are involved.

13.2 The failures of classical physics

A number of experiments done late in the nineteenth century and early in this century, gave results totally at variance with the predictions of classical physics. All, however, could be explained on the basis that *classical physics is wrong in allowing systems to possess arbitrary amounts of energy*. When this key idea was pursued, quantum mechanics emerged.

Black-body radiation. A hot object emits electromagnetic radiation. At high temperatures an appreciable proportion of the radiation appears in the visible region of the spectrum, and a higher proportion of short-wavelength (blue) light appears as the temperature is raised. This is seen when an object glowing red hot becomes white hot when heated further. This behaviour is illustrated in Fig. 13.1, which shows the energy output at various wavelengths for several temperatures. The curve depicted relates to the radiation emitted by an ideal emitter, which is one capable of emitting and absorbing all frequencies of radiation without favour. Such a material is called a *black body*, and a very good approximation to it is a pin-hole in a container, because the radiation leaking out of the hole has been reflected around inside so many times that it has come to thermal equilibrium with the walls.

Two principal features are apparent in the results depicted in the figure. The first is that the wavelength of the peak shifts to smaller values as the temperature is raised and the short wavelength tail increasingly spreads through the whole of the visible region. This corresponds to a shift of

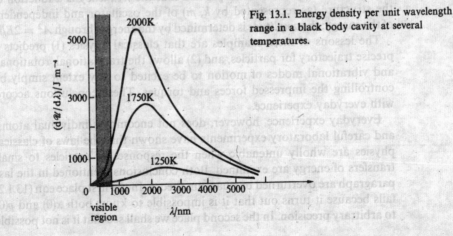

Fig. 13.1. Energy density per unit wavelength range in a black body cavity at several temperatures.

colour towards the blue, as already mentioned. Analysis of the data confirms that there is a quantitative relation between λ_{max}, the wavelength at maximum emission, and the absolute temperature:

Wien's Displacement Law: $T\lambda_{max} = \text{constant}$.

The experimental value of the constant is 2.9×10^{-3} m K; and so at 1000 K, $\lambda_{max} \approx 2900$ nm. The second feature was noted by Stefan: he considered the total amount of energy emitted over all wavelengths at a particular temperature. More energy is expected to be present in the electromagnetic field in the cavity at high temperatures than at low, and if the energy is summed over all wavelengths the total amount per unit volume, the *energy density* $\mathcal{U}$, obeys

Stefan's Law: $\mathcal{U} = aT^4$,

where a is a constant independent of the material. An alternative form of this observation relates to the *excitance M* (which used to be called the *emittance*) of a black body. The excitance is the power emitted per unit area. Since the power emitted is proportional to the energy density, the excitance is also proportional to T^4:

$$M = \sigma T^4.$$

σ is the *Stefan-Boltzmann constant*; its experimental value is 5.67×10^{-8} W m^{-2} K^{-4}, so that each 1 cm^2 of the surface of a black body heated to 1000 K radiates about 5.7 W.

Rayleigh, with minor help from Jeans, took a classical viewpoint and thought of the electromagnetic field as a collection of oscillators, one for each possible frequency of light. The appearance of light of some frequency was interpreted as the excitation of one oscillator, and the *intensity* of light of that frequency was related to the *amplitude* of oscillation of the corresponding oscillator. According to the equipartition theorem (p. 14), an oscillator in equilibrium with a source at a temperature T possesses a mean energy kT (k is Boltzmann's constant). Therefore the energy density in the range of wavelengths λ to $\lambda + d\lambda$ is the number of oscillators per unit volume in that range, $d\mathcal{N}(\lambda)$, multiplied by their average energy, kT; hence $d\mathcal{U}(\lambda) = kT\,d\mathcal{N}(\lambda)$. An explicit calculation of $\mathcal{N}$ then gives the

(13.2.1)   *Rayleigh–Jeans Law:* $d\mathcal{U}(\lambda) = (8\pi kT/\lambda^4)\,d\lambda$.

Unfortunately (for Rayleigh and Jeans), as λ decreases this expression increases without going through a maximum, Fig. 13.2. It suggests, for example, that oscillators of extremely short wavelength (corresponding to ultraviolet light, X-rays, and even γ-rays) are strongly excited even at room temperature. (According to classical physics, objects would be visible in the dark: there would in fact be no darkness.) This absurd result is the *ultraviolet catastrophe*; but the prediction is unavoidable if classical mechanics is used.

Planck studied this problem from his position as an expert in thermo-

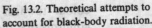

Fig. 13.2. Theoretical attempts to account for black-body radiation.

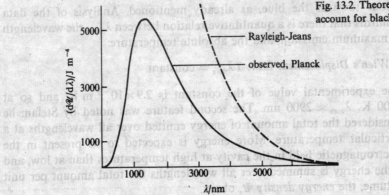

dynamics, and found that he could account for the experimental observations by requiring that a radiation mode of a given frequency could be excited to only a discrete set of energies. To be explicit, the experimental results appeared to demand that a radiation mode of frequency ν (and wavelength $\lambda = c/\nu$) can possess an energy which must be some integral multiple of $h\nu$, where h is a constant and the same for all frequencies of vibration. *Planck's constant h* has been found by experiment to have the value 6.626×10^{-34} J s. Since a radiation mode of frequency ν can possess only the energies 0, $h\nu$, $2h\nu$, and so on, it strongly suggests that it should be thought of as consisting of the appropriate numbers of particles, each one carrying the energy $h\nu$. These particles are called *photons*. Then if the ray carries an energy E through a region, the number of photons passing through is $E/h\nu$. For instance, a 100 W yellow lamp (100 W = 100 J s^{-1}, yellow light has $\lambda = 550$ nm, or $\nu = 5.5 \times 10^{14}$ s^{-1}) is generating 100 J s^{-1}/$(6.626 \times 10^{-34}$ J s $\times 5.5 \times 10^{14}$ s$^{-1}) = 2.77 \times 10^{20}$ s^{-1}, or 2.77×10^{20} yellow photons each second.

The implications of Planck's hypothesis for black-body radiation are as follows. The walls of the black body possess an amount of thermal energy, depending on their temperature. This energy excites the electromagnetic field inside the cavity, which acquires some of the energy of the walls. At equilibrium the field is excited to a point where there is no net flow of energy between the walls and the electromagnetic field (except in so far as a small amount of flow has to be maintained in order to replenish the radiation lost through the pinhole to the detector). In classical theory electromagnetic oscillators of all frequencies are set in motion under the influence of the excitation caused by the material composing the cavity, and so even very high frequencies are stimulated. In quantum theory, however, the oscillators are excited only if they can acquire at least an energy $h\nu$. Put another way, a photon of light of frequency ν can be created only if an energy $h\nu$ is available. *The effect of Planck's hypothesis is to damp out the contribution from the high frequency oscillators*, for they cannot pick up enough energy from the walls. Detailed calculation shows that the energy density in the range λ to $\lambda + d\lambda$ is given by the

> **Planck distribution:**
>
> (13.2.2)
> $$d\mathcal{U}(\lambda) = \left(\frac{8\pi hc}{\lambda^5}\right)\left\{\frac{\exp(-hc/\lambda kT)}{1-\exp(-hc/\lambda kT)}\right\}d\lambda.$$

The exponential term, which is absent in the Rayleigh–Jeans formula, predicts the quenching of the energy density when λ is small, since $\exp(-hc/\lambda kT) \approx 0$ when $hc/\lambda kT$ is large, which it is when λ is small. This expression fits the experimental curve very well at all wavelengths, Fig. 13.2, and h may be determined by adjusting it to get the best fit. Notice that if the constant h is erroneously set equal to zero, Planck's distribution reduces to the Rayleigh–Jeans Law (expand $\exp(-hc/\lambda kT) \approx 1 - hc/\lambda kT$ in the numerator and denominator, then let h go to zero). Therefore the classical result is recovered in the limit $h \to 0$.

Example. A spherical cavity of volume 1 cm³ is heated to 1500 K. What is the amount of energy inside carried by radiation of wavelengths in the range (a) 550–575 nm (yellow), (b) 1000–1025 nm (infrared)?

- *Method.* Use eqn (13.2.2), regarding the wavelength ranges as virtually infinitesimal ($\delta\lambda \approx d\lambda$) in each case.

- *Answer.* The middle of the first wavelength range is 562.5 nm; in the range $\delta\lambda \approx 25$ nm centred on this wavelength we have

$$\delta\mathcal{U} \approx \{8\pi \times (6.626 \times 10^{-34}\ \text{J s}) \times (2.998 \times 10^8\ \text{m s}^{-1})/(562.5 \times 10^{-9}\ \text{m})^5\}$$

$$\times \left\{\exp\left[\frac{-(6.626 \times 10^{-34}\ \text{J s}) \times (2.998 \times 10^8\ \text{m s}^{-1})}{(562.5 \times 10^{-9}\ \text{m}) \times (1.381 \times 10^{-23}\ \text{J K}^{-1}) \times (1500\ \text{K})}\right]\right\}\delta\lambda$$

$$= [(8.866 \times 10^7\ \text{J m}^{-4})\exp(-17.05)](25 \times 10^{-9}\ \text{m})$$

$$= 8.75 \times 10^{-8}\ \text{J m}^{-3}.$$

Therefore, in a cavity of volume 1 cm³ the energy in the wavelength range 550–575 nm is 8.75×10^{-14} J.

Repeating the calculation but with $\lambda = 1012.15$ nm and $\delta\lambda = 25$ nm leads to $\delta\mathcal{U} = 9.02 \times 10^{-12}$ J m⁻³, a factor of 103 larger.

- *Comment.* The infrared oscillations of the electromagnetic field are over 100 times more strongly activated than the visible, yellow oscillators at this temperature. Since every photon has an energy $h\nu = hc/\lambda$ it follows that there are about 2.5×10^5 yellow photons and 4.6×10^7 infrared photons inside the cavity at 1500 K.

The Planck distribution may also be used to account for the Stefan and Wien Laws. The Stefan Law can be deduced by summing the energy density in every wavelength range $d\lambda$ from $\lambda = 0$ to $\lambda = \infty$ at a set temperature; this gives $\mathcal{U} = aT^4$ with $a = 4\sigma/c$, and $\sigma = \pi^2 k^4/60c^2h^3$. Substitution of the fundamental constants then gives $\sigma = 5.670\,32 \times 10^{-8}$ W m⁻² K⁻⁴, in accord with the experimental value. The Wien Law can be deduced by looking for the wavelength at which $d\mathcal{U}/d\lambda$ is zero: for high temperatures or short wavelengths this gives the condition $\lambda_{max}T = hc/5k = 2.878 \times 10^{-3}$ m K, also in accord with experiment.

The key result of this section is the realization that oscillators cannot possess arbitrary amounts of energy. This *quantization* (from the Latin *quantum*, amount) is wholly alien to classical theory. It meant that the whole structure of physics had to be revised.

Heat capacities. Accounting for black-body radiation involves examining how energy is taken up by the electromagnetic field. Accounting for the heat capacities of materials involves examining the way that heat may be taken up by the internal modes of motion of material objects. Therefore a study of heat capacities might be expected to show up quantum behaviour.

If classical physics were valid the mean energy of vibration of an atom oscillating in one dimension in a metal would be given by the equipartition principle (p. 14) as kT. Each of the N atoms in a block is free to vibrate in three directions, and so the total energy of the block is expected to be $3NkT$. For unit amount of substance N is replaced by L, Avogadro's constant, and the molar vibrational energy is then $3RT$. The heat capacity C_V relates the rise in energy of a sample to its rise in temperature at constant volume:

$$dU = C_V \, dT.$$

(This definition is described in detail on p. 65.) Therefore to calculate C_V it is necessary to calculate dU/dT. In the present case $U_m = 3RT$, and so classical physics predicts that $C_{V,m} = 3R$, independent of temperature. This result is known as *Dulong and Petit's Law*, for it had been proposed by them on the basis of experimental evidence.

Einstein was stimulated to study heat capacities when technological advances made it possible to test Dulong and Petit's Law at low temperatures. Significant deviations were observed: every metal was found to have a molar heat capacity lower than $3R$ at low temperatures, and the value appeared to approach zero at absolute zero. Einstein incorporated Planck's hypothesis into the calculation of heat capacities by assuming that every atom in the metal could vibrate about its equilibrium position with a single frequency v, and then allowing the atomic oscillators to possess only the energies nhv, n being an integer. It is important to notice the analogies between this discussion and that of black-body radiation. There Planck dealt with electromagnetic oscillations of a range of frequencies from zero to infinity; Einstein dealt with a material oscillator confined to a single frequency.

Einstein's calculation first gave an expression for the total vibrational energy of unit amount of metal (the details are given in Chapter 21):

$$U_m = \frac{3Lhv \exp(-hv/kT)}{[1 - \exp(-hv/kT)]}.$$

Notice the similarity of this to the Planck distribution (p. 391). Extracting C_V now involves differentiating with respect to T. This gives, using $Lk = R$, the

(13.2.3)   *Einstein formula:* $C_{V,m} = 3R\left(\dfrac{h\nu}{kT}\right)^2 \left\{\dfrac{\exp(-h\nu/kT)}{[1-\exp(-h\nu/kT)]^2}\right\}.$

At very high temperatures kT becomes much larger than $h\nu$ and so $\exp(-h\nu/kT)$ can be expanded using $e^{-x} = 1-x+\ldots$ and keeping only the lowest-order terms in $h\nu/kT$. The result is

$$C_{V,m} = 3R\left(\frac{h\nu}{kT}\right)^2 \left\{\frac{1-(h\nu/kT)+\ldots}{[1-1+(h\nu/kT)+\ldots]^2}\right\} = 3R,$$

in accord with the classical calculation. At low temperatures $\exp(-h\nu/kT)$ drops to zero, and so Einstein's formula accounts qualitatively for the lowering of the heat capacity at low temperatures, Fig. 13.3.

The physical reasons for the success of Einstein's and Planck's theories are similar. In contrast to the classical viewpoint, the atomic oscillators can pick up only discrete amounts of energy. Therefore at moderately low temperatures not all the oscillators may collect enough energy to oscillate. At higher temperatures, where there is enough internal energy for all the oscillators to be active, all $3N$ oscillations are stimulated and the sample approaches its classical heat capacity.

Einstein's model is a simplification because he required all the atoms in the solid to oscillate with the same, single frequency whereas they vibrate with a range of frequencies. This complication can be accommodated by averaging the Einstein model over the frequencies present, and the expression obtained is known as the *Debye formula* for the heat capacity.

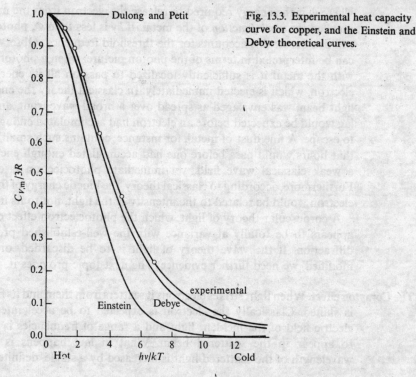

Fig. 13.3. Experimental heat capacity curve for copper, and the Einstein and Debye theoretical curves.

The details of this modification, which, as Fig. 13.3 shows, gives improved agreement with experiment, need not deflect us from the main conclusion that quantization must be introduced in order to explain the low-temperature thermal properties of solids.

The photoelectric effect. Evidence of a different kind comes from the study of the energies of electrons ejected from metals when they are irradiated with light of different frequencies. The properties of this photoelectric effect are quite different from what would be expected classically. Three observations are important for the present argument:

(1) No electrons are ejected, regardless of the intensity of the light, unless its frequency exceeds a threshold value characteristic of each metal.

(2) The kinetic energy of the ejected electrons is linearly proportional to the frequency of the incident light.

(3) Even at low intensities of light, electrons are ejected immediately if the frequency is above threshold.

These observations strongly suggest that the photoelectric effect depends on an electron suffering a collision with a localized projectile carrying enough energy to knock it out of the metal. If we suppose that light carries energy in packets of $h\nu$, then conservation of energy requires that the kinetic energy of the ejected electron should obey

(13.2.4)
$$\tfrac{1}{2}mv^2 = h\nu - \Phi$$

where Φ is the energy required to release the electron from the metal: this is called the *work-function* of the metal. If $h\nu$ is less than Φ, photoejection will not occur: this accounts for the threshold frequency. Observation (3) can be interpreted in terms of the photon picture: when a photon collides with the metal it is sufficiently localized to pass on all its energy to an electron, which is ejected immediately. In classical theory the energy of a light beam was envisaged as spread over a broad wave front, and a time lag would be expected before an electron had accumulated enough energy to escape. A fine dust of metal, for instance, contains such small particles that hours would pass before one had accumulated enough energy from a weak classical wave field, yet immediate photoejection is observed. Furthermore, according to classical theory, the kinetic energy of the ejected electron would be related to the intensity of the light, not to its frequency.

A corpuscular theory of light, which the photoelectron effect demands, appears to be totally at variance with the well-established property of diffraction. If the wave theory of light is to be discarded, or at least modified, we need further evidence. The next topic provides it.

The Compton effect. When light strikes electrons it scatters from them and its frequency is shifted. Classically the electron is expected to be accelerated by the electric field of the incident light and a range of frequencies is expected to appear in the scattered beam. What in fact happens is that the wavelength of the scattered light is increased by a single, definite amount which depends only on the angle through which the light is scattered. Furthermore, the shift of frequency is independent of the wavelength of the incident light. This behaviour is called the *Compton effect*. We have already seen that the photon has an energy hv; if we also suppose that it is a particle of *momentum* hv/c (or h/λ, where λ is its wavelength, $\lambda = c/v$), then by regarding the scattering as a collision between a photon and an electron (of mass m_e), Fig. 13.4, and requiring both linear momentum and energy to be conserved, the following expression is obtained for the wavelength shift accompanying scattering through an angle θ:

(13.2.5) $\delta\lambda = (h/m_e c)(1 - \cos\theta),$

which is verified experimentally. The quantity $h/m_e c$ is called the *Compton wavelength* of the electron: its magnitude is 2.426×10^{-12} m (2.426 pm), and so the maximum shift of wavelength, which occurs for $\theta = 180°$, is only 4.832 pm, whatever the initial wavelength.

The diffraction of electrons. Both the photoelectric effect and the Compton effect indicate that light has attributes characteristic of particles. Although contrary to the long-established wave theory of light, a similar view had been held before, but dismissed. No significant physicist, however, had taken the view that matter was wavelike. Nevertheless, experiments in 1925 forced people even to that conclusion. The crucial experiment was performed by Davisson and Germer, who observed the diffraction of electrons from a crystal lattice, Fig. 13.5. Their experiment was a lucky accident because a chance rise in temperature caused their polycrystalline sample to crystallize, and the ordered planes of atoms then acted as a diffraction grating.

This diffraction experiment, which has since been repeated with other particles (including molecular hydrogen), shows quite clearly that particles have wave properties. We have also seen that waves have particle properties, and thus we are brought to the heart of the revolution of modern physics: *when examined on an atomic scale the concepts of particle and wave melt together, particles take on characteristics of waves, and waves the characteristics of particles.*

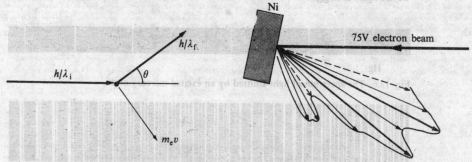

Fig. 13.4. Directions and momenta involved in the Compton effect.

Fig. 13.5. Davisson and Germer's observation of electron diffraction.

Some progress towards coordinating these properties was taken by de Broglie when, in 1924, he suggested that any particle, not only photons, travelling with a momentum p should have in some sense a wavelength given by the

(13.2.6)   *de Broglie relation:* $\lambda = h/p$.

This expression was confirmed for particles by the Davisson–Germer experiment, and for photons by the Compton effect. We shall build on it in the next section.

Atomic and molecular spectra. The most directly compelling evidence for quantization comes from the observation of the frequencies of light absorbed and emitted by atoms and molecules. A typical atomic spectrum is shown in Fig. 13.6, and another type of spectrum, which later we shall see can be ascribed to changes in the energy of molecular vibration and rotation as well as to electronic movements, is shown in Fig. 13.7. The obvious feature of both types of spectrum is the appearance of radiation at a series of discrete frequencies. This can be understood if the energy of the atoms and molecules is confined to certain discrete values, and if the atom or molecule can shed its energy only in discrete steps, Fig. 13.8. As it drops from one energy state to a lower one its change of energy is radiated as light. If the light carries away a packet of energy ΔE, the frequency should be $\nu = \Delta E/h$ if the previous discussion is applicable.

The spectra shown exhibit lines in a complicated pattern. This suggests that the energy levels of atoms and molecules are not arranged in a simple fashion. We must construct a form of mechanics—*quantum mechanics*—that allows us to account for, and to predict, the permitted energies of such systems.

13.3 The dynamics of microscopic systems

We take the de Broglie relation $p = h/\lambda$ as the starting point for the construction of a theory of mechanics of particles. The central feature of de Broglie's suggestion, and Davisson and Germer's experiment, was the

Hg

Fig. 13.6. The spectrum of light emitted by an excited mercury atom.

Fig. 13.7. The spectrum of light absorbed as a molecule changes its state (spectrum of ScF provided by Dr R. F. Barrow).

association of wave character with a particle. Therefore, we abandon the classical concept of localized particles and replace it with the idea that the position of a particle is distributed like the amplitude of a wave. The concept of *wavefunction* is introduced to replace the classical concept of trajectory, and the mechanics consists of setting up a scheme for calculating how the wavefunction is distributed in a system. One of our tasks will be to see how to extract all the properties of a system on the basis of a knowledge of its wavefunction.

The Schrödinger equation. In 1926 Schrödinger proposed an equation which, when solved, gives the wavefunction for any system. Its position is as central to quantum mechanics as Newton's equations are to classical mechanics. Just as Newton's equations were an inspired postulate which enable us to calculate the trajectory of particles, so Schrödinger's equation can be regarded as an inspired postulate that enables us to calculate the wavefunction. For a particle free to move in one dimension the equation reads

(13.3.1)
$$(-\hbar^2/2m)(d^2/dx^2)\psi(x) + V(x)\psi(x) = E\psi(x).$$

ψ (the Greek letter psi) is the wavefunction, $V(x)$ is the potential energy of the particle when it is at x, E is its total energy (kinetic + potential), and $\hbar$ (read h cross or h bar) is a convenient modification of Planck's constant: $\hbar = h/2\pi$. Various ways of expressing this equation, of incorporating the time-dependence of the wavefunction, and of extending it to three dimensions, will be encountered, but they are left until Box 13.2 at the end of the chapter.

The form of this equation can be justified to a certain extent by the following remarks. Consider first the case when the particle is free to move in a region where its potential energy is zero. The Schrödinger equation

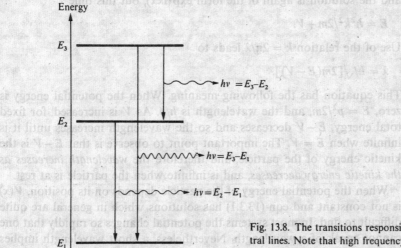

Fig. 13.8. The transitions responsible for spectral lines. Note that high frequency radiation is emitted when the energy change is large.

is then simply

(13.3.2) $-(\hbar^2/2m)\,d^2\psi/dx^2 = E\psi$

and a solution is

(13.3.3) $\psi(x) = \exp(ikx) = \cos kx + i \sin kx.$

with

(13.3.4) $k = \sqrt{(2mE)}/\hbar$ or $E = k^2\hbar^2/2m.$

These are the bare bones of the solution, but it is the interpretation that is significant. In the first place, since the potential energy is zero, the total energy of the particle is also its kinetic energy. The kinetic energy is related to the momentum p by $E = p^2/2m$, and so comparison with eqn (13.3.4) shows that the particle's momentum is $p = k\hbar$. The next step is to recognize that $\cos kx$ (or $\sin kx$) is a wave with wavelength $2\pi/k$. This can be seen by comparing $\cos kx$ with the standard form of a harmonic wave, $\cos(2\pi x/\lambda)$, implying $k = 2\pi/\lambda$. Combining these two pieces of argument leads to

$$p = k\hbar = (2\pi/\lambda)(h/2\pi) = h/\lambda,$$

which is de Broglie's relation. Therefore, in this case at least, Schrödinger's equation has led to an experimentally verified conclusion. Even so, we have not yet *interpreted* the wave nature of matter: all we have done so far is to deduce a connection between the momentum of a particle and the wavelength of the associated wave; we have not yet said how that wave is 'associated'.

If the particle moves in a region of constant but non-zero potential the Schrödinger equation reads

$$(-\hbar^2/2m)\,d^2\psi/dx^2 = (E - V)\psi$$

and the solution is again of the form $\exp(ikx)$, but this time

$$E = \hbar^2 k^2/2m + V.$$

Use of the relation $k = 2\pi/\lambda$ leads to

(13.3.5) $\lambda = h/\sqrt{[2m(E - V)]}.$

This equation has the following meaning. When the potential energy is zero, $E = p^2/2m$, and the wavelength is h/p. As V is increased, for fixed total energy, $E - V$ decreases and so the wavelength increases until it is infinite when $E = V$. The important point to observe is that $E - V$ is the kinetic energy of the particle, and therefore the *wavelength increases as the kinetic energy decreases*, and is infinite when the particle is at rest.

When the potential energy of the particle depends on its position, $V(x)$ is not constant and eqn (13.3.1) has solutions which in general are quite difficult to find. In most systems the potential changes so rapidly that one cannot speak of a wavelength. Nevertheless, a short wavelength implies a *high curvature* of the wavefunction, and so where the concept of wavelength is undefined we may still think in terms of curvature. The *curvature* of a function ψ is simply $d^2\psi/dx^2$, and so a high kinetic energy is associated with a wavefunction with a large second-derivative. This can be summarized as follows:

high curvature implies *high kinetic energy* implies *high momentum*.

At this point we can begin to see the emergence of Newton's mechanics out of quantum mechanics. Consider the system depicted in Fig. 13.9. The potential is changing and therefore its gradient is not zero. In classical physics the force is proportional to the gradient of the potential ($F = -dV/dx$). Therefore, the depicted system corresponds to a particle subjected to a constant force directed to the right. In the figure the wavelength of ψ is seen to decrease from left to right. But decreasing wavelength implies increasing momentum. Therefore, in some sense, the quantum-mechanical picture represents the acceleration of a particle under the influence of a constant force. This is the content of Newton's Second Law.

Schrödinger's equation is a second-order differential equation and therefore it has an infinite number of solutions. For example, in the case of a free particle $\exp(ikx)$ is a solution, so is $a\exp(ikx)$, and k, and hence E; may take any value. The next step in the argument involves putting an interpretation on ψ. We shall see that the implication of this interpretation is to discard some of the solutions. Discarding solutions means forbidding some values of E. In this way we are getting very close to the basis of the quantization of energy.

The interpretation of the wavefunction. The interpretation of ψ is based on a suggestion made by Born. The *Born interpretation* draws an analogy with the wave theory of optics in which the *square of the amplitude* of an electromagnetic wave is interpreted as the *intensity* of radiation. We have seen evidence for a corpuscular theory of light, and this suggests that intensity could also be interpreted in terms of the number of photons present: high-intensity monochromatic radiation (high incident energy) implies a large number of photons (each carrying the same energy $h\nu$). If that is the correct

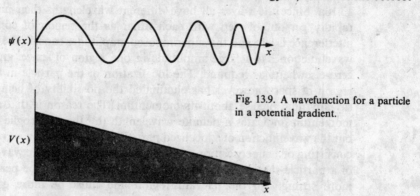

Fig. 13.9. A wavefunction for a particle in a potential gradient.

interpretation for light, the analogy for particles is that the wavefunction is an amplitude whose square (or square modulus $\psi^*\psi$ if ψ is complex) indicates the probability of finding the particle at each point of space. The Born interpretation of ψ is, therefore, that $\psi^*(x)\psi(x)\,dx$ *is proportional to the probability of finding the particle in an infinitesimal region between x and x + dx.*

Example (Objective 9). The wavefuntion of an electron in the lowest state of the hydrogen atom is the function $\psi(\mathbf{r}) = (1/\pi a_0^3)^{1/2} \exp(-r/a_0)$, where $a_0 = 53$ pm (0.53 Å). Notice that $\psi(\mathbf{r})$ is independent of θ and φ (latitude and longitude). What is the probability of finding the electron inside a small sphere of volume 1 pm³ (about 1/100 the volume of the atom) centred (a) at the nucleus, (b) at a point 50 pm (0.50 Å) from the nucleus?

- *Method.* The probability is given by $\psi^2(\mathbf{r})\,d\tau$. The volume is so small that it may be regarded as infinitesimal, and we write $d\tau \approx 1$ pm³.

- *Answer.* At the nucleus $r = 0$ and so

$$[\psi^2(\mathbf{r})\,d\tau]_{r=0} = (1/\pi a_0^3)\,d\tau = (1/\pi)(1/53\,\text{pm})^3\,(1\,\text{pm}^3) = 1/53^3\,\pi = 2.14 \times 10^{-6}.$$

At a point 50 pm from the nucleus in an arbitrary (but definite) direction,

$$[\psi^2(\mathbf{r})\,d\tau]_{r=50\,\text{pm}} = \{(1/\pi)(1/53\,\text{pm})^3 \exp[2 \times (-50\,\text{pm}/53\,\text{pm})]\}\,(1\,\text{pm}^3)$$
$$= (2.14 \times 10^{-6})\exp(-1.89) = 3.24 \times 10^{-7}.$$

- *Comment.* The probability of finding the electron at a point 50 pm from the nucleus in some definite but arbitrary direction is only 15 per cent of the probability of finding it at the nucleus itself. The probability of finding the electron in the same volume at a distance of 1 mm is strictly non-zero, but it is so small (≈ 10 raised to the power -2×10^7) that it is wholly negligible. Notice that the dimensions of ψ are $(1/\text{length})^{3/2}$.

How is this discussion to be reconciled with the manifest truth that particles can be located in well-defined positions? If a particle is known to occupy a well-defined position, and known to be nowhere else, its wavefunction must have a significant amplitude in that region, and be zero elsewhere. A wavefunction of this form can be constructed by superimposing a large number of functions of different wavelengths with amplitudes that match (are in phase) at the location of the particle, Fig. 13.10a. Since the waves all have different wavelengths their amplitudes rapidly get out of step with each other as the region of constructive interference is left, and the superimposed amplitudes cancel. The resultant wavefunction has a large amplitude in one region of space, and almost zero elsewhere, as required. The localization of the particle in a definite region of space however has eliminated the possibility of being able to say anything precise about its momentum. The reason is that a definite momentum indicates a definite wavelength (by the de Broglie relation), but the wavefunction of a localized particle is described by a wavefunction consisting of a superposition of a range of wavelengths. The wavefunction of a particle that is localized even more definitely must be peaked even more strongly in the relevant region and must be more completely

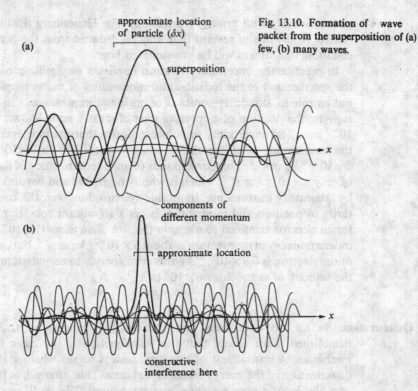

approximate location of particle (δx)

(a)

superposition

components of different momentum

(b)

approximate location

constructive interference here

Fig. 13.10. Formation of wave packet from the superposition of (a) few, (b) many waves.

annihilated elsewhere. But to achieve this, an even greater range of wavelengths must be superimposed, and so the particle has an even more indefinite momentum, Fig. 13.10b. In the limit of total localization at a precisely specified point, all possible wavelengths must be superimposed, and so the momentum is wholly indefinite. If the position is precisely specified the momentum is completely uncertain.

The outcome of this discussion is that position and momentum are *complementary properties* of a particle. This means that at a given instant only one of them may be specified with arbitrary precision. The more definite the position, the less definite the momentum, and vice versa. An essential feature of nature is that if one of these properties is known, the other must remain unknown. Classical physics supposed, falsely, that both could be known simultaneously.

The quantitative connection between the uncertainty in the position (δq) and the uncertainty in the momentum (δp) is the

(13.3.6)   *Heisenberg uncertainty relation:* $\delta p \delta q \geqslant \frac{1}{2}\hbar$.

The p and q refer to the same direction in space: therefore whereas momentum parallel to x and position along x are complementary (and therefore cannot be specified simultaneously with a precision greater than allowed by $\delta p_x \delta x \geqslant \frac{1}{2}\hbar$), momentum along some perpendicular axis (p_y) and position along x are not complementary (and can be specified simultaneously with arbitrary precision). The Heisenberg relation in eqn (13.3.6) is just one of several that may be deduced from the Schrödinger equation, and others will be encountered later.

In practice the uncertainty relation implies a negligible constraint on the specification of the location and momentum of macroscopic objects, but its role in atomic dynamics is of profound importance. For example, suppose the position of a speck of dust of mass 1 μg is known to within 10^{-3} mm. The uncertainty principle indicates that the indeterminancy of the momentum is of the order of $\delta p \approx (1 \times 10^{-34}$ J s$/2 \times 10^{-6}$ m$)$, or 5×10^{-29} kg m s^{-1}. This corresponds to an indeterminancy of the velocity of only 5×10^{-20} m s^{-1}, which is wholly negligible and beyond detection by laboratory instruments. In atomic systems, however, the complementarity of position and momentum plays a significant role. For example, for an electron confined to a region the size of an atom $(1 \times 10^{-10}$ m$)$ the indeterminancy in momentum is about 5×10^{-25} kg m s^{-1}, but as the mass of the electron is 0.9×10^{-30} kg this corresponds to an indeterminancy in the velocity of as much as 6×10^5 m s^{-1}.

Quantization. So far we have seen that the classical concept of trajectory is demolished if we accept that the wavefunction is the basic feature of mechanics. A crucial test of this approach is to see whether it leads to quantization of the energy levels of a system. This, after all, is the reason why the Schrödinger equation was introduced. We shall now examine how the Schrödinger equation, together with the Born interpretation of the wavefunction, does indeed succeed in accounting for quantized energy levels.

If the Born interpretation of the wavefunction is to be consistent and plausible, the properties of the wavefunction are restricted. For example, if $\psi^*(x)\psi(x)\,\mathrm{d}x$ is the probability of finding the particle in the region $\mathrm{d}x$, the sum of such probabilities over the whole of space must be unity because the particle is certainly somewhere if it exists at all. This can be expressed mathematically by requiring

$$(13.3.7) \qquad \int_{-\infty}^{\infty} \psi^*(x)\psi(x)\,\mathrm{d}x = 1.$$

Wavefunctions which satisfy this condition are said to be *normalized* (or normalized to unity). If we are dealing with a three-dimensional problem each volume element is $\mathrm{d}x\mathrm{d}y\mathrm{d}z$, and the probability of the particle appearing in such a volume at the point $\mathbf{r}$ is $\psi^*(\mathbf{r})\psi(\mathbf{r})\mathrm{d}x\mathrm{d}y\mathrm{d}z$. The requirement that the particle must be somewhere in the universe is the sum over the probabilities that it appears in every element $\mathrm{d}x\mathrm{d}y\mathrm{d}z$ into which the universe can be divided, Fig. 13.11. The total probability is therefore

$$\int_{-\infty}^{\infty} \mathrm{d}x \int_{-\infty}^{\infty} \mathrm{d}y \int_{-\infty}^{\infty} \mathrm{d}z \, \psi^*(\mathbf{r})\psi(\mathbf{r}) = 1.$$

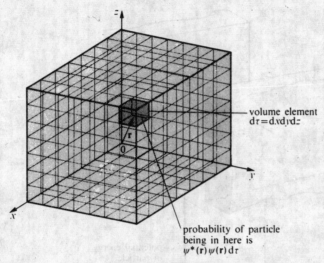

volume element
$d\tau = dx\,dy\,dz$

probability of particle
being in here is
$\psi^*(\mathbf{r})\,\psi(\mathbf{r})\,d\tau$

Fig. 13.11. The interpretation of the wavefunction.

Such an integral looks ugly and complicated, and so is usually abbreviated to

(13.3.8)
$$\int \psi^*\psi\,d\tau = 1.$$

$d\tau$ is called the *infinitesimal volume element*.

Example. The wavefunction for the ground state of the hydrogen atom is proportional to $\exp(-r/a_0)$, a_0 a constant. Find the *normalization constant N* such that $\psi(\mathbf{r}) = N\exp(-r/a_0)$ is normalized to unity.

- *Method.* Find N such that eqn (13.3.8) is satisfied. The volume element in three-dimensional space is $dx\,dy\,dz$, but it is more convenient to work in polar coordinates and to use $d\tau = r^2\,dr\sin\theta\,d\theta\,d\varphi$, allowing r to range from 0 to ∞, θ from 0 to π, and φ from 0 to 2π.

- *Answer.* From eqn (13.3.8):
$$\int \psi^*\psi\,d\tau = \int_0^\infty dr\,r^2 \int_0^\pi d\theta\sin\theta \int_0^{2\pi} d\varphi\,[N^2\exp(-2r/a_0)]$$
$$= N^2[-\cos\theta]\Big|_0^\pi\,[\varphi]\Big|_0^{2\pi} \int_0^\infty dr\,[r^2\exp(-2r/a_0)]$$
$$= N^2[2]\,[2\pi]\,[a_0^3/4] = N^2\pi a_0^3.$$
Therefore, in order for the integral to be equal to unity, $N = (1/\pi a_0^3)^{1/2}$.

- *Comment.* This value of N was used in the *Example* on p. 400. Do not forget the factor of r^2 in the expression for the volume element in polar coordinates.

This *total probability criterion* (or *normalization condition*) puts stringent requirements on the wavefunction, because it is not satisfied by every

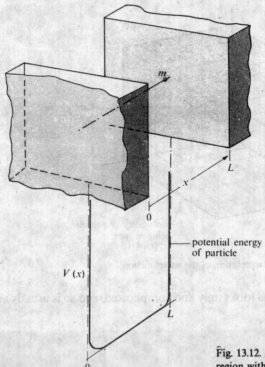

Fig. 13.12. A particle in a one-dimensional
region with impenetrable walls.

possible solution of the Schrödinger equation.* If a solution does not
satisfy it, it must be discarded, and if it is discarded it follows that the
corresponding value of E, the energy in eqn (13.3.1), cannot occur. Thus
we arrive at quantization: *only some energies are possible, because other
energies correspond to untenable properties of the distribution of the particle.*

These remarks have been quite abstract, but we now look at the way
they apply to simple systems.

13.4 Translational motion

We begin by considering the quantum mechanics of a particle of mass m
free to move in a straight line between two walls, Fig. 13.12. The potential
energy $V(x)$ is constant (and taken to be zero) at all points of the particle's
path except where it hits the walls, and there $V(x)$ rises to a very high value
as the particle squashes into them. The simplest case is the *infinite square
well*, where the particle's potential energy shoots up to infinity immediately
it touches the walls.

The Schrödinger equation and its solution. The Schrödinger equation for the region
between $x = 0$ and $x = L$, where the potential energy is zero, is

---

* Although the total probability condition is the main criterion it is also necessary to impose
the constraints that the *wavefunction be continuous* everywhere, and (except in a few cases)
that *its derivative $d\psi/dx$ be continuous*. These constraints arise essentially in order that the
Schrödinger equation, with its second derivative $d^2\psi/dx^2$, is defined everywhere.

(13.4.1) $(-\hbar^2/2m)\,d^2\psi/dx^2 = E\psi(x)$

and its solution is

(13.4.2) $\psi(x) = A\sin kx + B\cos kx, \qquad k = (2mE)^{1/2}/\hbar.$

A and B are constants with values to be determined.

The particle cannot penetrate outside the region of zero potential because the potential energy is infinite. The clearest way of seeing this is to lower the potential energy from infinity to some very large, constant value V for values of x inside the material of the walls. Then the Schrödinger equation for the region $x > L$ (or $x < 0$) is

$$(-\hbar^2/2m)\,d^2\psi/dx^2 + V\psi = E\psi,$$

or

$$d^2\psi/dx^2 = (2m/\hbar^2)(V-E)\psi.$$

The second derivative $d^2\psi/dx^2$ is the *curvature* of ψ. It is necessary to remember that the curvature of a curve ⌣ is positive while that of ⌢ is negative. Suppose that the wavefunction has a positive value just inside the material of the walls; then according to the last equation the curvature is positive (because V is so large that it exceeds E, and ψ is positive) and so the wavefunction curls rapidly up to infinite values as x increases, and the total probability criterion, eqn (13.3.7) cannot be satisfied. If ψ has a negative value just inside the material of the walls the curvature is negative, and so the wavefunction droops down to negatively infinite values. This too is unacceptable if the Born interpretation is to be tenable. Therefore, the wavefunction must be zero inside the material of the walls, and this requirement becomes increasingly stringent as the potential energy V is allowed to approach infinity.

At this stage we know that the wavefunction has the form given in eqn (13.4.2), with the additional requirement that at the walls it must drop to zero amplitude. Consider the wall at $x = 0$. According to eqn (13.4.2) the amplitude of the wavefunction is $\psi(0) = B$ (because $\sin 0 = 0$ and $\cos 0 = 1$). But for the solution to be acceptable $\psi(0)$ must vanish, and therefore B must be zero. This implies that the wavefunction is

$\psi(x) = A\sin kx.$

The amplitude at the other wall is $\psi(L) = A\sin kL$. This too must be zero. Taking $A = 0$ would imply that $\psi(x) = 0$ for all x, which is absurd (and in conflict with the Born interpretation). Therefore kL must be chosen so that $\sin kL = 0$. This demands that kL is some integral multiple of π because $\sin\theta = 0$ when $\theta = 0, \pi, 2\pi, \ldots$. Therefore the boundary condition forces us to conclude that the only values of k permitted are those for which

$kL = n\pi, \qquad n = 1, 2, \ldots$

($n = 0$ is eliminated, because it implies $k = 0$, or $\psi(x) = 0$ everywhere.) Since k is related to E, it follows that *the energy of the system is confined to the values*

(13.4.3) $E = n^2\hbar^2\pi^2/2mL^2 = n^2h^2/8mL^2$, $n = 1, 2, \ldots$

The energy of the particle is *quantized*, and, as anticipated, it is by virtue of the presence of boundaries that this quantization occurs.

Before discussing this result in more detail we shall find the complete wavefunctions. The remaining task is the determination of the value of the *normalization constant* A. This can be found by substituting ψ into the total probability expression, eqn (13.3.7), and remembering that ψ is zero outside the range $x = 0$ to $x = L$:

$$1 = \int_{-\infty}^{\infty} \psi^*(x)\psi(x)\,dx = A^2 \int_{0}^{L} \sin^2 kx\,dx = A^2L/2, \quad \text{or} \quad A = (2/L)^{1/2}.$$

Therefore the complete solution to the problem is

(13.4.4)
$$\left.\begin{array}{ll}\textit{energies:} & E_n = n^2h^2/8mL^2 \\ \textit{wavefunctions:} & \psi_n(x) = (2/L)^{1/2}\sin(n\pi x/L)\end{array}\right\} n = 1, 2, \ldots$$

The energies and wavefunctions have been labelled with the *quantum number* n. A statement of the value of a quantum number enables us to state the form of the wavefunction and the energy of a particle whose distribution it describes.

Example (Objective 12). An electron is confined to a molecule of length 1.0 nm (10 Å, about 5 atoms long). What is its minimum energy? What is the minimum excitation energy from this state? What is the probability of finding it in the region of the molecule lying between $x = 0.49$ nm and $x = 0.51$ nm? What is the probability of finding it between $x = 0$ and $x = 0.2$ nm?

• *Method.* Use eqn (13.4.4) for the energy with $m = m_e$ and calculate E_1, the ground state (zero-point) energy. The minimum excitation energy is $E_2 - E_1$. The distribution of the electron is given by $\psi_1(x)$. For the first part the region $0.49 \leqslant x \leqslant 0.51$ nm is almost infinitesimal and so the probability is approximately $\psi^2(x = 0.50 \text{ nm})\,\delta x$ with $\delta x = 0.02$ nm. For the final part the region $0 \leqslant x \leqslant 0.2$ nm is not infinitesimal (on the scale of the molecule) and so integrate $\psi_n^2(x)\,dx$ from $x = 0$ to $x = l$ to get the total probability of being in that region, and then set $l = 0.2$ nm and $n = 1$.

• *Answer.* $E_1 = h^2/8m_eL^2 = \dfrac{(6.6262 \times 10^{-34}\,\text{J s})^2}{8 \times (9.1095 \times 10^{-31}\,\text{kg}) \times (1 \times 10^{-9}\,\text{m})^2}$

$\qquad = 6.025 \times 10^{-20}\,\text{J (or 36.28 kJ mol}^{-1}\text{)}.$

$\qquad E_2 = 4h^2/8m_eL^2 = 24.099 \times 10^{-20}\,\text{J (or 145.1 kJ mol}^{-1}\text{)}$

$\qquad E_2 - E_1 = 108.8\,\text{kJ mol}^{-1}\text{(or 1.13 eV)}.$

For the distribution we require

$$\psi_1^2(x)_{x = 0.50\,\text{nm}} = (2/1.0 \times 10^{-9}\,\text{m})\sin^2(0.5\pi/1.0) = 2 \times 10^9\,\text{m}^{-1}.$$

Therefore, the probability of being in the range 0.49 nm to 0.51 nm is $(2 \times 10^9\,\text{m}^{-1}) \times (0.02\,\text{nm}) = 0.04$, or 1 in 25.

The probability of being in the range $x = 0$ to $x = l$ is

$$\int_{0}^{l} \psi_n^2(x)\,dx = (2/L)\int_{0}^{l} \sin^2(n\pi x/1.2)\,dx = (1/L) - (1/2n\pi)\sin(2n\pi l/L)$$

$$= (0.2 \text{ nm}/1.0 \text{ nm}) - (1/2\pi)\sin(0.4\pi) = 0.0486,$$

or 1 in 20.6.

● *Comment.* The electron in a box is a model of molecular structure: it can be used for obtaining rough orders of magnitude of transition energies.

The properties of the solutions. The shapes of some of the wavefunctions are illustrated in Fig. 13.13. It is easy to see the source of the energy quantization in pictorial terms. Each wavefunction is a standing wave and, in order to fit into the cavity, successive functions must possess one more half-wavelength. Increasing the number of half-wavelengths in a cavity of constant length implies a sharpening of the curvature of the wavefunction. This in turn implies that the kinetic energy of the particle is increased in each successive quantum level.

Because n cannot be zero the lowest energy that a particle may possess is not zero (as classical mechanics would allow) but $E_1 = h^2/8mL^2$. This lowest, irremovable energy is called the *zero-point energy*. Its source can be understood in either of two ways. First, the uncertainty principle requires that if a particle is confined within a finite region, then it must possess kinetic energy. For example, since the position of the particle is uncertain to no more than $\delta x \approx L$, the condition $\delta x \delta p \geqslant \hbar/2$ implies that there must be an uncertainty in the momentum of order $\delta p \approx \hbar/2L$, at least. This suggests the presence of kinetic energy of the order of at least $\delta p^2/2m \approx \hbar^2/8mL^2$, which is about the magnitude of the zero-point energy. The second way of understanding the occurrence of the zero-point energy

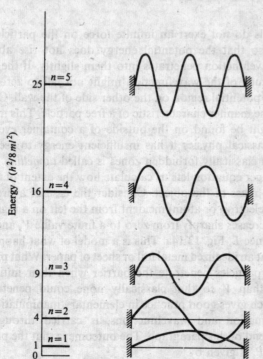

Fig. 13.13. The energies and wavefunctions of a particle in a one-dimensional infinite square well.

is to note that if the wavefunction is to be zero at the walls, but smooth, continuous, yet not zero everywhere, then it must possess curvature; but if it is curved the particle possesses kinetic energy. Therefore, if the particle is in the cavity, it must possess kinetic energy.

The separation between neighbouring quantum levels is

$$\Delta E = E_{n+1} - E_n = (2n+1)h^2/8mL^2$$

and this decreases as the length L of the container increases. The separation ΔE is extremely small when the container is large. It follows that a *free particle*, moving in an unbounded region of space, *has unquantized translational energy levels*. For this reason atoms and molecules free to move in laboratory-sized vessels may be treated as though their translational energy were unquantized.

The distribution of the position of a particle in a box is not uniform at low energies. Since the probability of discovering the particle in the region dx at the point x is $(2/L)\sin^2(n\pi x/L)\,dx$ the probability varies with x. The effect is pronounced when n is small, and Fig. 13.13 shows that in the lowest energy level there is an apparent repulsion from the vicinity of the walls. At high quantum numbers the distribution becomes more uniform. Classically a particle bouncing backwards and forwards between reflecting walls spends equal amounts of time, on the average, at all points. Therefore, the classical distribution is attained in the limit of very high quantum numbers. This is an aspect of the *correspondence principle*, which states that classical behaviour emerges when high quantum numbers are reached.

Quantum leaks. If the walls do not exert an infinite force on the particle when it touches them (so that the potential energy does not rise abruptly to infinity), the wavefunction penetrates into them slightly. If the walls are thin, the amplitude of the wavefunction might not fall to zero before it reaches the low potential region on the other side of the wall. Once there it oscillates in the manner characteristic of a free particle. This means that the particle might be found on the outside of a container even though according to classical physics it has insufficient energy to escape. This leaking through classically forbidden zones is called *tunnelling*.

The Schrödinger equation lets us calculate how the extent of tunnelling depends on the mass of the object. Consider the case of a projectile of mass m (e.g. an electron or atom) incident from the left on a region where the potential increases sharply from zero to a finite value V, and remains there for a distance L, Fig. 13.14a. This is a model of what happens when atoms are fired at an idealized metal foil or sheet of paper. What proportion of the incident particles penetrate the barrier when their initial kinetic energy is less than V so that classically none could penetrate? The calculation, which gives good practice in elementary manipulations of the Schrödinger equation and wavefunctions, is carried through in the *Example* at the end of this paragraph. The outcome is that the probability of penetration, P, is given by

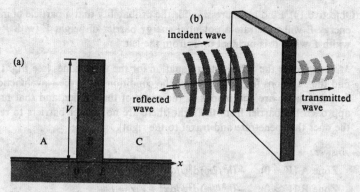

Fig. 13.14. (a) A potential barrier, (b) the incident, reflected, and transmitted waves.

(13.4.5) $P = 1/(1+G)$,

where

$$G = \frac{\{\exp([2m(V-E)/\hbar^2]^{\frac{1}{2}}L) - \exp(-[2m(V-E)/\hbar^2]^{\frac{1}{2}}L)\}^2}{4(E/V)[1-(E/V)]}.$$

When the barrier is high and long so that $\{2m(V-E)/\hbar^2\}^{\frac{1}{2}}L \gg 1$, the first exponential dominates the second, and since then $G \gg 1$, $P \approx 1/G$. In this case

(13.4.6) $P \approx 4(E/V)\{1-(E/V)\}\exp\{-2[2m(V-E)\hbar^2]^{\frac{1}{2}}L\}$.

This shows that P depends exponentially on the mass of the projectile and the length of the barrier, and only low mass projectiles (electrons, protons) are likely to be able to tunnel significantly. Figure 13.15 shows how a proton's and a deuteron's tunnelling abilities depend on the energy for a barrier of height 5 eV and width 0.1 nm (1 Å).

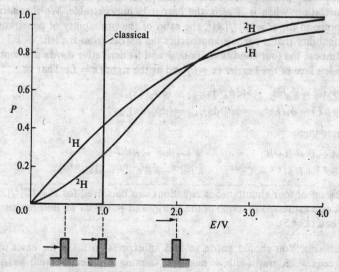

Fig. 13.15. Probability of transmission of a proton and a deuteron through a potential barrier for $E < V$.

Example (Objective 14). Find an expression for the probability that a particle of mass m and energy E will penetrate a potential energy barrier of height V (with $V > E$) and width L when it is incident on it from the left.

- *Method.* Write the Schrödinger equation for the three zones (see Fig. 13.14), and find the solutions for each. Impose the conditions that the wavefunction and its first derivative are continuous at the edges of the barrier, and that no particles approach the barrier from the right (there are no further barriers to reflect back the ones that penetrate and travel to the right).

- *Answer.*

   Zone A $(V = 0)$: $-(\hbar^2/2m)\mathrm{d}^2\psi/\mathrm{d}x^2 = E\psi$

   Zone B $(V > 0)$: $-(\hbar^2/2m)\mathrm{d}^2\psi/\mathrm{d}x^2 + V\psi = E\psi$

   Zone C $(V = 0)$: $-(\hbar^2/2m)\mathrm{d}^2\psi/\mathrm{d}x^2 = E\psi$.

The energy E is variable (we are free to fire the projectiles with any kinetic energy) and we take $E < V$ (although the case $E > V$ can be explored in the same way). The three equations have the solutions:

   Zone A: $\psi_A(x) = A\mathrm{e}^{ikx} + A'\mathrm{e}^{-ikx}$ $k = (2mE/\hbar^2)^{\frac{1}{2}}$

   Zone B: $\psi_B(x) = B\mathrm{e}^{k'x} + B'\mathrm{e}^{-k'x}$ $k' = [2m(V-E)/\hbar^2]^{\frac{1}{2}}$

   Zone C: $\psi_C(x) = C\mathrm{e}^{ikx} + C'\mathrm{e}^{-ikx}$ $k = (2mE/\hbar^2)^{\frac{1}{2}}$.

Note that, because $E < V$, the wavefunction inside the barrier is not an oscillating wave: it is a superposition of an increasing and a decreasing exponential. We have seen already that e^{ikx} is a wave of momentum $k\hbar$ in the direction of x; similarly e^{-ikx} is a wave of momentum $k\hbar$ towards $-x$, Fig. 13.14b. Since the projectiles are being fired from the left, the component $A'\mathrm{e}^{-ikx}$ is the part of the wavefunction corresponding to them being reflected from the barrier. On the right there can be no projectiles travelling towards $-x$, and so we also know that $C' = 0$. This term $C\mathrm{e}^{ikx}$ then represents the amplitude of the wavefunction for particles that successfully penetrate the barrier: when $|C|$ is large, there is a high probability of penetration; when it is zero, the barrier is impenetrable. We therefore seek an expression for $P = |C|^2/|A|^2$, the ratio of the probability of penetration to the probability that a particle approaches the barrier from the left.

Impose the four conditions: both ψ and its derivative $\mathrm{d}\psi/\mathrm{d}x$ are continuous at the left face of the barrier $(x = 0)$ and at the right $(x = L)$. That is:

$\psi_A(0) = \psi_B(0)$; $(\mathrm{d}\psi_A/\mathrm{d}x)_{x=0} = (\mathrm{d}\psi_B/\mathrm{d}x)_{x=0}$

$\psi_B(L) = \psi_C(L)$; $(\mathrm{d}\psi_B/\mathrm{d}x)_{x=L} = (\mathrm{d}\psi_C/\mathrm{d}x)_{x=L}$.

These imply

$A + A' = B + B'$; $ikA - ikA' = Bk' - B'k'$

$B\mathrm{e}^{k'L} + B'\mathrm{e}^{-k'L} = C\mathrm{e}^{ikL}$; $k'B\mathrm{e}^{k'L} - k'B'\mathrm{e}^{-k'L} = ikC\mathrm{e}^{ikL}$.

This set of four simultaneous equations can be solved for $P = |C|^2/|A|^2$. When k, k' are expressed in terms of the energy and potential we find the result quoted as eqn (13.4.5).

- *Comment.* You should go on to find an expression for P for cases in which E exceeds V. A trap (with V negative, forming a potential well) has interesting properties, as it becomes transparent at some values of the incident energies. Solve that problem (Problem 13.25).

Motion in more dimensions. When the particle is allowed to move over a rectangular surface and is not confined only to a line, we have to consider motion in two dimensions. This introduces one new technique, the *separation of variables* (which lets us express the solutions in terms of wavefunctions in one dimension), and a new concept, *degeneracy*, the possession of the same energy by different wavefunctions.

Consider a particle confined to the xy-plane, where the potential is zero everywhere except at the walls (at $x = 0$, L and $y = 0$, L') where it rises abruptly to infinity. The Schrödinger equation is

(13.4.7) $$-(\hbar^2/2m)(\partial^2/\partial x^2 + \partial^2/\partial y^2)\psi(x,y) = E\psi(x,y)$$

because the particle has freedom to move in two dimensions. We attempt a solution of the form $\psi(x,y) = X(x)Y(y)$, where X is a function of x alone, and Y of y alone. Then

$$-(\hbar^2/2m)\{Y(d^2X/dx^2) + X(d^2Y/dy^2)\} = EXY,$$

and so, dividing by XY,

$$-(\hbar^2/2m)(1/X)(d^2X/dx^2) - (\hbar^2/2m)(1/Y)(d^2Y/dy^2) = E.$$

Now for the cunning part. The first term appears to depend on x; the second certainly does not; but their sum is a constant (E). Therefore even the first term cannot actually depend on x, but must be equal to some constant which we write E_X. Likewise the second term must also be a constant E_Y, and $E_X + E_Y = E$. It follows that the single equation can be written as two equations, one for each variable

$$-(\hbar^2/2m)(1/X)(d^2X/dx^2) = E_X$$

$$-(\hbar^2/2m)(1/Y)(d^2Y/dy^2) = E_Y.$$

This achieves the separation of variables. (You should note that it does not always work for more complicated systems, but we shall see it in action again when we get to the hydrogen atom.)

The two new equations are each the same as the equations for 1-dimensional translation. It follows that

(13.4.8) $$X_n(x) = (2/L)^{\frac{1}{2}}\sin(n\pi x/L), \quad E_{X,n} = n^2h^2/8mL^2$$

$$Y_{n'}(y) = (2/L')^{\frac{1}{2}}\sin(n'\pi y/L'), \quad E_{Y,n'} = n'^2h^2/8mL'^2.$$

The total energy is therefore

(13.4.9) $$E_{nn'} = E_{X,n} + E_{Y,n'} = (h^2/8m)\{(n/L)^2 + (n'/L')^2\} \quad n,n' = 1,2,3\ldots$$

and the complete wavefunctions are the products $\psi_{nn'}(x,y) = X_n(x)Y_{n'}(y)$. Some of them are illustrated in Fig. 13.16.

A particularly interesting case occurs when the surface is square, for then $L = L'$, and

(13.4.10) $$E_{nn'} = (h^2/8mL^2)(n^2 + n'^2).$$

This is interesting because whereas, for instance, the functions $\psi_{1,2}$ and

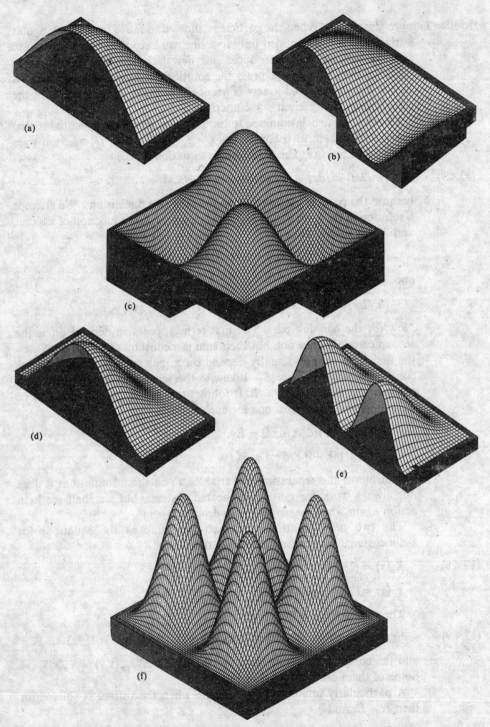

Fig. 13.16. Wavefunctions and probability densities for a particle confined to a rectangular surface. (a) ψ_{11} section, (b) ψ_{21} section (rotate by 90° for ψ_{12}), (c) ψ_{22}; (d) ψ_{11}^2 section, (e) ψ_{21}^2 section, (f) ψ_{22}^2. Each section is half the entire function.

$\psi_{2,1}$ correspond to different distributions of the particle (one can be generated from the other by rotating the square by 90°) they both have the same energy ($5h^2/8mL^2$). Such functions are called *degenerate*. The level with energy $5h^2/8mL^2$ is said to be *doubly degenerate*.

It is not hard to see that a particle in a three dimensional rectangular box (an actual box) has energies

(13.4.11) $E_{nn'n''} = (h^2/8m)\{(n/L)^2 + (n'/L')^2 + (n''/L'')^2\},$

and if the box is a cube, so that $L = L' = L''$,

(13.4.12) $E_{nn'n''} = (h^2/8mL^2)(n^2 + n'^2 + n''^2).$

Multiply degenerate energy levels are now possible (e.g. the six states corresponding to an energy $14h^2/8mL^2$).

13.5 Vibrational motion

A particle undergoing simple harmonic motion about the point $x = 0$ experiences a restoring force $-kx$, and, as was seen on p. 387, possesses a potential energy $\frac{1}{2}kx^2$ when the displacement is x. The Schrödinger equation for a particle of mass m is

(13.5.1) $(-\hbar^2/2m)\mathrm{d}^2\psi/\mathrm{d}x^2 + \frac{1}{2}kx^2\psi = E\psi.$

This differential equation can be solved, but it is helpful to notice the similarities to the square well potential. As in that case, the particle is trapped in a symmetrical well in which the potential energy rises to very large (in fact, infinite) values for sufficiently large displacements, Fig. 13.17. Therefore, we should expect the boundary conditions to require the wavefunction to sink to zero at large displacements, but less sharply than in the abrupt, square well case. As boundary conditions have to be satisfied we should expect the energy of the particle to be quantized, as in the square well. The wavefunctions should resemble those of the square well with two minor differences. First, their amplitude can drop towards zero more slowly at large displacements because the potential climbs to infinity as x^2, and not abruptly. Secondly, the potential energy increases as x changes from zero, the kinetic energy therefore decreases for a given total energy, and so the curvature varies in a more complex way across the range of permitted displacements.

Solution of eqn (13.5.1), together with the boundary condition that ψ sinks towards zero at large displacements, confirms these anticipations. The *energy* of an oscillator is found to be restricted to the values

(13.5.2) $E_v = (v + \frac{1}{2})\hbar\omega, \qquad v = 0, 1, 2, \dots$

where $\omega = (k/m)^{1/2}$. The wavefunctions have the shapes shown in Fig. 13.17 and the mathematical forms listed in Table 13.1. The energy of each level arises from two sources. There is a contribution from the kinetic energy, which increases successively up the ladder of v values because the

Table 13.1. Harmonic oscillator wavefunctions

Write $y = \alpha x$, where x is the displacement from equilibrium, and $\alpha^2 = m\omega/\hbar$, $\omega^2 = k/m$. Then the wavefunctions are

$$\psi_v(x) = (\alpha/2^v v! \pi^{1/2})^{1/2} H_v(y) \exp(-\tfrac{1}{2}y^2)$$

where the functions $H_v(y)$ are the following polynomials (the *Hermite polynomials*):

$H_0(y) = 1$	$H_4(y) = 16y^4 - 48y^2 + 12$
$H_1(y) = 2y$	$H_5(y) = 32y^5 - 160y^3 + 120y$
$H_2(y) = 4y^2 - 2$	$H_6(y) = 64y^6 - 480y^4 + 720y^2 - 120$
$H_3(y) = 8y^3 - 12y$	$H_7(y) = 128y^7 - 1344y^5 + 3360y^3 - 1680y$

General expressions for these polynomials and their properties can be written for any value of n.

Source: M. Abramowitz and I. A. Stegun; *Handbook of mathematical functions*, Dover.

wavefunction gets increasingly buckled, Fig. 13.17. There is also a contribution from the potential energy. This increases with v because the wavefunction spreads over a greater range of displacements and so the particle samples regions of higher potential energy, Fig. 13.18.

Example (Objective 17). The wavefunction for the ground state of a harmonic oscillator of mass m and force-constant k is proportional to $\exp(-\tfrac{1}{2}\alpha^2 x^2)$, where $\alpha^2 = m\omega/\hbar$ and $\omega^2 = k/m$. Confirm that this is a solution and corresponds to an energy

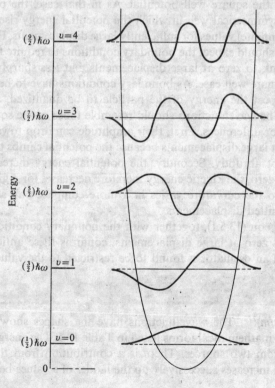

Fig. 13.17. The potential energy of a simple harmonic oscillator and some allowed solutions.

$E_0 = \frac{1}{2}\hbar\omega$. What is the magnitude of the zero point energy in a $^{35}Cl_2$ molecule?

- *Method.* Substitute the function in the l.h.s. of eqn (13.5.1) and show that the equation is satisfied with $E_0 = \frac{1}{2}\hbar\omega$. For the second part, use $m = \frac{1}{2}m(^{35}Cl)$ and $k = 328.6 \, N \, m^{-1}$. (The factor $\frac{1}{2}$ arises because both chlorine atoms move: it will be explained later, p. 584.)

- *Answer.* $d^2\psi_0/dx^2 = (d/dx)[-\alpha^2 x \exp(-\frac{1}{2}\alpha^2 x^2)]$

$$= -\alpha^2 \exp(-\frac{1}{2}\alpha^2 x^2) + \alpha^4 x^2 \exp(-\frac{1}{2}\alpha^2 x^2)$$

$$= -\alpha^2\psi_0 + \alpha^4 x^2\psi_0.$$

Therefore the l.h.s. of eqn (13.5.1) is equal to

$$(-\hbar^2/2m)(d^2\psi_0/dx^2) + \frac{1}{2}kx^2\psi_0 = (\hbar^2\alpha^2/2m)\psi_0 - (\hbar^2\alpha^4/2m)x^2\psi_0 + \frac{1}{2}kx^2\psi_0.$$

But as $\hbar\alpha^2/2m = \frac{1}{2}\hbar\omega$, $\hbar^2\alpha^4/2m = \frac{1}{2}\omega^2 m = \frac{1}{2}k$, the l.h.s. is equal to

$$\frac{1}{2}\hbar\omega\psi_0 - \frac{1}{2}kx^2\psi_0 + \frac{1}{2}kx^2\psi_0 = \frac{1}{2}\hbar\omega\psi_0,$$

The r.h.s. of eqn (13.5.1) is equal to

$$(v + \frac{1}{2})\hbar\omega\psi_0 = \frac{1}{2}\hbar\omega\psi_0 \quad \text{for} \quad v = 0.$$

The two sides are equal. This confirms that the function is a solution with the stated energy. For $^{35}Cl_2$, from the table on the last endpaper,

$$m(^{35}Cl) = (34.969) \times (1.6605 \times 10^{-27} \, kg) = 5.807 \times 10^{-26} \, kg.$$

Consequently,

$$\omega = [(328.6 \, N \, m^{-1})/\frac{1}{2}(5.807 \times 10^{-26} \, kg)]^{1/2} = 1.064 \times 10^{14} \, s^{-1}.$$

and so

$$E_0 = \frac{1}{2}(1.0546 \times 10^{-34} \, J \, s) \times (1.064 \times 10^{14} \, s^{-1}) = 5.610 \times 10^{-21} \, J.$$

- *Comment.* The zero-point energy may seem small, but it is equivalent to $3.38 \, kJ \, mol^{-1}$ and so it is a significant chemical quantity. Note that ω is a circular

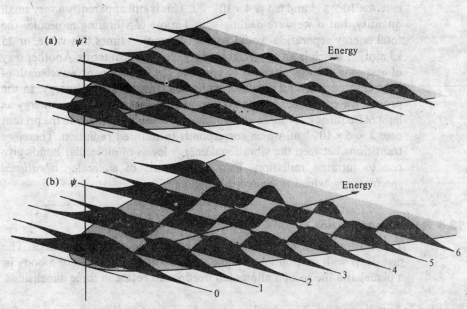

Fig. 13.18. (a) Probabilities and (b) wavefunctions for the harmonic oscillator.

frequency (radians per second); convert to cycles per second v with $\omega = 2\pi v$. Therefore $v(^{35}Cl_2) = 1.7 \times 10^{13}$ Hz.

An important feature of the harmonic oscillator is that the separation between neighbouring energy levels is the same for all v:

$$\Delta E = E_{v+1} - E_v = \hbar\omega.$$

There is a zero-point energy of $\frac{1}{2}\hbar\omega$ which arises because the particle is confined by the potential. Whereas the classical picture of an oscillator allows the particle to be quite still (for example, a pendulum hanging quietly in its equilibrium position), quantum theory requires it always to have some energy: the lowest state can be visualized as one in which the particle is incessantly fluctuating around its equilibrium position. Note that the wavefunction for the ground (lowest) state is peaked closest to the zero-displacement positions, representing this clustering fluctuation.

At high quantum levels the wavefunction has its largest amplitude near the limits of its range, and possesses only a small amplitude near the equilibrium position. This is approaching the classically predicted behaviour, where the particle moves most slowly close to its turning points and is therefore most probably found there, but travels with maximum velocity through its minimum displacements. Once again we see the attainment of classical behaviour in the correspondence limit of large quantum numbers.

The energy separations $\hbar\omega$ are negligibly small for macroscopic objects, but are of considerable significance for objects the size of atoms. For example, a force constant in a chemical bond is typically of the order of 300 N m^{-1}, and the mass of a proton is 1.7×10^{-27} kg; therefore $\omega \approx 4 \times 10^{14}$ s^{-1} and $\hbar\omega \approx 4 \times 10^{-20}$ J. This is still apparently a very small quantity, but if we were dealing with 1 mole of vibrating molecules the total energy separation would be 6×10^{23} mol^{-1} times this value, or 25 kJ mol^{-1}, which is an appreciable energy in chemical terms. Another way of assessing the magnitude is to consider the frequency or wavelength of light emitted when an escillator emits one quantum of its energy. In the process an energy $\hbar\omega$ is emitted: the photon carries away this energy as light of frequency $v = \omega/2\pi$ and wavelength $\lambda = c/v = 2\pi c/\omega$. In the present case $\lambda \approx 5 \times 10^{-6}$ m, which corresponds to infrared radiation. Therefore transitions between the vibrational energy levels of molecular bonds give rise to infrared radiation. This is the basis of *molecular vibrational spectroscopy*, which is treated in Chapter 17.

13.6 Rotational motion

The full treatment of rotation can be broken down into two steps. The first deals with motion on a ring (corresponding to rotation of a body in a plane) and the second allows full freedom to rotate in three dimensions.

Rotation in two dimensions. The simplest example of rotational motion is that of a particle of mass m moving around a circular path of radius r. The particle possesses momentum as it circulates, and therefore its wavefunction is of wavelength $\lambda = h/p$. The *angular momentum* of the particle is of magnitude $J = pr$, and the kinetic energy $p^2/2m$ can be re-expressed in terms of the angular momentum as $J^2/2mr^2$. The product mr^2 is called the *moment of inertia* I of the system. Thus we may write

(13.6.1) $\qquad E = J^2/2I.$

Not all values of the momentum are permitted, and therefore the energy of the rotating particle is quantized. This may be seen by considering what wavelengths are permitted; if discrete values of λ are permitted, discrete values of p, J, and therefore E, are implied.

Suppose for the moment that λ takes an arbitrary value. The wavefunction varies round the ring as the angle φ changes from 0 to 2π in the manner depicted in the Fig. 13.19a. When φ increases beyond 2π the wavefunction continues to vary, but if λ is arbitrary the wave will not match that from the first circuit, φ continues to circulate around the ring, and each time the wavefunction varies in the manner shown. The *total amplitude* of the wavefunction at any angle φ' determines the total probability of finding the particle there. But because of the superposition of positive and negative values of the amplitudes of successive circuits this total amplitude is zero for any point φ'. Therefore the total probability of finding the particle on the ring is zero if the wavelength is arbitrary. This is absurd, conflicts with the Born interpretation, and is therefore untenable.

A physically satisfactory solution can be recovered if λ is matched to the circumference of the ring. For example, if the wavelength is infinite (meaning that the amplitude is constant), the amplitudes on successive

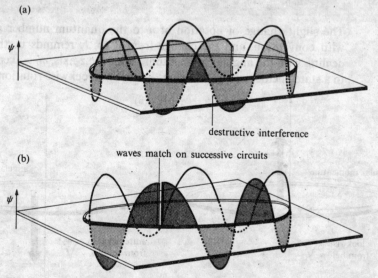

Fig. 13.19. (a) Unacceptable and (b) acceptable wavefunctions for a particle on a ring.

circuits do not interfere destructively. If λ is equal to one circumference, the *nodes* (points of zero amplitude) and *antinodes* (points of maximum amplitude) match on successive circuits and the wave is not annihilated, Fig. 13.19b. The same is true for any integral fraction of the circumference ($\lambda = \infty$, $C/1$, $C/2$, ..., C being the circumference $2\pi r$). Therefore the wavelength is limited to the values $\lambda = 2\pi r/n$, $n = 0, 1, 2, \ldots$, the magnitude of the linear momentum is therefore limited to $p = h/\lambda = nh/2\pi r = n\hbar/r$. It follows that the magnitude of the angular momentum is limited to $J = pr = n\hbar$, and the energy is limited to the values

(13.6.2) $E = n^2\hbar^2/2I, \qquad n = 0, 1, 2, \ldots$

A moment's thought shows that this cannot be complete. The linear momentum may arise from motion in either direction, clockwise or anticlockwise. Therefore angular momentum corresponds either to clockwise or to anticlockwise rotation. We shall adopt the convention of depicting the angular momentum by a vector perpendicular to the plane of rotation, as shown in Fig. 13.20. A clockwise rotation viewed from below is represented by a vector sticking up out of the plane, Fig. 13.20a, and an anticlockwise rotation by a vector penetrating into the plane, Fig. 13.20b. The former vector has a positive component on the z-axis (Fig. 13.20a), and the latter has a negative component. The preceding remarks about the quantization of the angular momentum restricted only the *magnitude* of the angular momentum: either sense of rotation is permitted. Therefore although the magnitude of J is limited to integral multiples of $\hbar$ its absolute value may take both positive and negative integral multiples.

This discussion leads to the conclusion that the angular momentum of a particle confined to a plane is restricted to the values

(13.6.3) $J_z = m_l\hbar, \qquad m_l = 0, \pm 1, \pm 2, \ldots$

The slight change of notation of n to the quantum number m_l accords with convention, and the change from J to J_z reminds us that we are dealing with the angular momentum about the z-axis, $m_l > 0$ corresponds to a state of clockwise rotation, $m_l < 0$ to anticlockwise rotation as viewed

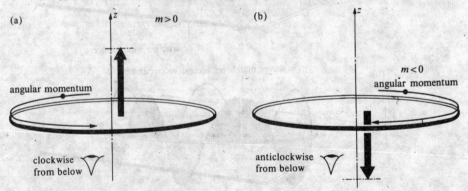

Fig. 13.20. The vector representation of angular momentum.

from below. The energies that the particle may possess are accordingly

(13.6.4)
$$E_{m_l} = m_l^2(\hbar^2/2I), \qquad m_l = 0, \pm 1, \pm 2, \ldots$$

The appearance of m_l as its square means that the energy of rotation is independent of the sense of rotation, as is desirable intuitively. Although these conclusions have been based on the model of a single particle moving on a circle, the results apply to any body of moment of inertia I constrained to rotate about only one axis.

These conclusions are confirmed by solving the Schrödinger equation explicitly. Instead of writing the equation in x, y coordinates, it is sensible in the present case to use the polar coordinates r, φ, where $x = r\cos\varphi$, $y = r\sin\varphi$. Since r is a constant, $\partial^2/\partial x^2 + \partial^2/\partial y^2$ turns into $(1/r^2)(\partial^2/\partial\varphi^2)$ by standard manipulations of the differentials. Since $V = 0$ at all points on the ring, the Schrödinger equation in 2-dimensions (eqn 13.4.7) becomes:

$$-(\hbar^2/2m)(1/r^2)(\mathrm{d}^2/\mathrm{d}\varphi^2)\psi(\varphi) = E\psi(\varphi).$$

The solutions are:

(13.6.5)
$$\psi_{m_l}(\varphi) = (1/2\pi)^{\frac{1}{2}}e^{im_l\varphi}, \quad E_{m_l} = m_l^2\hbar^2/2I$$

where $I = mr^2$. The factor $(1/2\pi)^{\frac{1}{2}}$ ensures that the function is normalized to unity. Since ψ must match at points separated by a complete revolution, $\psi_{m_l}(\varphi) = \psi_{m_l}(\varphi + 2\pi)$. This is called a *cyclic boundary condition*. But

$$\psi_{m_l}(\varphi + 2\pi) = (1/2\pi)^{\frac{1}{2}}e^{im_l\varphi}e^{2\pi im_l} = (e^{2\pi im_l})\psi_m(\varphi).$$

As $e^{i\pi} = -1$, it follows that $2m_l$ must be an even number, and therefore that m_l must be an integer $(0, \pm 1, \pm 2, \ldots)$ in order to satisfy the cyclic boundary condition.

The essential conclusions to be drawn are that the energy is quantized, and so too is the angular momentum. Notice also the connection between the number of nodes and the angular momentum: as the number of nodal points increases the angular momentum also increases because the wavefunction becomes more buckled, Fig. 13.21.

Example (Objective 18). What is the minimum rotational energy and the minimum angular momentum of a disc the size of a benzene molecule (moment of inertia 2.93×10^{-45} kg m^2) when it is rotating in a plane. What is the approximate quantum number for the angular momentum of a gramophone record rotating at 33 r.p.m.?

- *Method.* Use eqn (13.6.4) with $m_l = 1$. For the second part you need to know that $I = \frac{1}{2}MR^2$, M the mass of the disc (150 g) and R its radius (10 cm). Then use $m_l\hbar$ = angular momentum.

- *Answer.* $E_l = m_l^2(\hbar^2/2I)$, $m_l = 1$

$$= (1.0546 \times 10^{-34}\,\mathrm{J\,s})^2/2(2.93 \times 10^{-45}\,\mathrm{kg\,m^2})$$

$$= 1.898 \times 10^{-24}\,\mathrm{J} \quad \text{or} \quad 1.14\,\mathrm{J\,mol^{-1}}\,(\text{not kJ mol}^{-1}).$$

$(m_l\hbar)_{m_l=1} = 1.055 \times 10^{-34}\,\mathrm{J\,s}$ is the minimum angular momentum.

For the gramophone record rotating at a frequency v (in cycles per second, Hz)

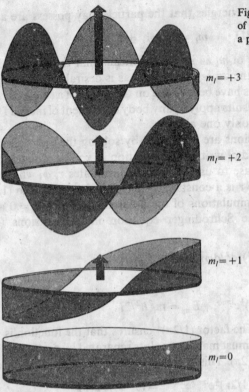

Fig. 13.21. Wavefunctions and some
of the permitted angular momenta of
a particle on a ring.

$m_l = +3$

$m_l = +2$

$m_l = +1$

$m_l = 0$

the angular momentum is $2\pi\nu I$. Therefore at 33 r.p.m., which is 33/60 Hz,

$$|m_l| = (2\pi\nu)(\tfrac{1}{2}MR^2)/\hbar$$
$$= \pi \times (33/60)\,\text{Hz} \times (0.15\,\text{kg}) \times (10 \times 10^{-2}\,\text{m})^2/(1.055 \times 10^{-34}\,\text{J s})$$
$$= (2.59 \times 10^{-3}\,\text{kg m}^2\,\text{s}^{-1})/(1.055 \times 10^{-34}\,\text{J s}) = 2.5 \times 10^{31}.$$

• *Comment.* The minimum rotational energy is really zero because m_l may be zero: we have calculated the minimum energy of benzene when it is actually rotating. Note that the rotational energy of $1.14\,\text{J mol}^{-1}$ is just large enough to be chemically significant. The gramophone record can be treated as a classical rotor because the quantum number is very high; its m_l is in fact negative.

Rotation in three dimensions. When we turn to rotation in three dimensions the extra complication of being able to move over the whole surface of a sphere instead of only around the equator introduces the further requirement that the wavefunctions must match as a path is traced over the poles, Fig. 13.22a. This additional boundary condition introduces another quantum number. The total kinetic energy and angular momentum of the particle now arise from rotation about all three axes. This apparently artificial model of a mass point moving over the surface of a sphere is used in the discussion of electrons in atoms and of molecules free to rotate about their centres of mass. The basis of the latter application is illustrated in Fig. 13.22b: a rotating molecule of moment of inertia I is equivalent to a

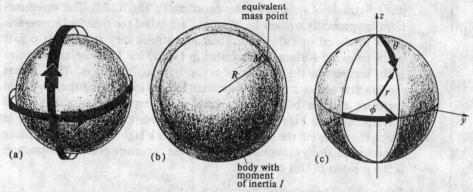

Fig. 13.22. (a) A particle on the surface of a sphere has wavefunctions that have to satisfy two continuity conditions. (b) A solid body of moment of inertia can be represented by the motion of a single point, mass M, moving on a spherical surface of radius R such that $I = MR^2$. (c) Polar coordinates; θ is the *colatitude* and φ the *azimuth*.

point mass M rotating about a centre at a distance R, the *radius of gyration* of the molecule: $I = MR^2$.

Explicit solution of the Schrödinger equation for the full problem (using the separation of variables technique) gives the following results.

(1) The *angular momentum about the z-axis* is quantized, and confined to the values

(13.6.6)
$$J_z = m_l \hbar, \qquad m_l = 0, \pm 1, \pm 2, \ldots \pm l.$$

This is the same as before, except for the upper limit of l, which for a given value of l confines m_l to $2l+1$ possible values: more on this below.

(2) *The magnitude of the angular momentum*, allowing for rotation about all directions in space, is quantized and confined to the values

(13.6.7)
$$J = [l(l+1)]^{1/2}\hbar, \qquad l = 0, 1, 2, \ldots ..$$

The new quantum number l, the *angular momentum quantum number*, determines the magnitude of the angular momentum through the formula just quoted, and the amount of this angular momentum that can be ascribed to motion about the z-axis is determined by m_l through the preceding equation.

(3) The *energy of rotation is quantized*, it is related to the angular momentum by $E = J^2/2I$, and so

(13.6.8)
$$E_l = l(l+1)\hbar^2/2I.$$

Notice that E_l is independent of m_l: the energy is independent of the orientation of the motion.

(4) The wavefunctions are products of a φ-dependent part $\exp(im_l\varphi)$ (of the form characteristic of a particle on a ring: you can regard a spherical surface as a collection of rings) and a part that depends on the angle θ

(the colatitude of spherical polar coordinates, Fig. 13.22). The wavefunctions are generally written $Y_{l,m_l}(\theta,\varphi)$ and are called the *spherical harmonics*. A few are listed in Table 13.2. Their amplitudes vary at different points of the spherical surface as illustrated in Fig. 13.23. The dominant feature is the increase in the number of nodal lines as l increases. This reflects the fact that higher angular momentum implies higher kinetic energy, and therefore a more buckled wavefunction. Observe too how the states corresponding to high angular momentum about the z-axis are those in which the nodes cut the equator: this indicates a high kinetic energy arising from motion along the equator, because the curvature of the wavefunction is the greatest in this direction.

Table 13.2. The spherical harmonics $Y_{lm_l}(\theta,\varphi)$

l	m_l	$Y_{lm_l}(\theta,\varphi)$
0	0	$1/2\pi^{\frac{1}{2}}$
1	0	$\frac{1}{2}(3/\pi)^{\frac{1}{2}}\cos\theta$
	± 1	$\mp\frac{1}{2}(3/2\pi)^{\frac{1}{2}}\sin\theta\, e^{\pm i\varphi}$
2	0	$\frac{1}{4}(5/\pi)^{\frac{1}{2}}(3\cos^2\theta-1)$
	± 1	$\mp\frac{1}{2}(15/2\pi)^{\frac{1}{2}}\cos\theta\sin\theta\, e^{\pm i\varphi}$
	± 2	$\frac{1}{4}(15/2\pi)^{\frac{1}{2}}\sin^2\theta\, e^{\pm 2i\varphi}$
3	0	$\frac{1}{4}(7/\pi)^{\frac{1}{2}}(2\cos\theta-3\cos\theta\sin\theta)$
	± 1	$\mp\frac{1}{8}(21/\pi)^{\frac{1}{2}}(4\cos^2\theta\sin\theta-\sin^3\theta)\, e^{\pm i\varphi}$
	± 2	$\frac{1}{4}(105/2\pi)^{\frac{1}{2}}\cos\theta\sin^2\theta\, e^{\pm 2i\varphi}$
	± 3	$\mp\frac{1}{8}(35/\pi)^{\frac{1}{2}}\sin^3\theta\, e^{\pm 3i\varphi}$

Source: L. Pauling and E. B. Wilson; *Introduction to quantum mechanics*, McGraw-Hill.

Example (Objective 20). Calculate the first five rotational energy levels of a hydrogen molecule ($I = 4.603 \times 10^{-48}$ kg m^2), the corresponding values of its angular momentum, and the number of spatial orientations the molecule may adopt in each case.

● *Method.* Use eqn (13.6.8) with $l = 0, 1, 2, 3, 4$. For the magnitude of the angular momentum use eqn (13.6.7). For a state with quantum number l there are $2l+1$ permitted orientations, eqn (13.6.6). For rotating molecules the quantum number J is used in place of l. The magnitude of the angular momentum is denoted $|J|$.

● *Answer.* We need $\hbar^2/2I = 1.2081 \times 10^{-21}$ J. Then draw up the following table

J	0	1	2	3	4
$E_J/10^{-21}$ J	0.	2.42	7.25	14.5	24.2
$E_J/$kJ mol^{-1}	0	1.46	4.37	8.73	14.6
$|J|/$J s 10^{-34}	0	1.49	2.58	3.65	4.72
Orientations	1	3	5	7	9

● *Comment.* We see from the third row that these rotational energies are even more significant in chemical applications than the rotations of benzene: the smaller the molecule the greater the rotational energy separations.

Space quantization. We have indicated that m_l is confined to the $2l+1$ discrete values $l, l-1, \ldots -l$ for any value of l. This means that the component of

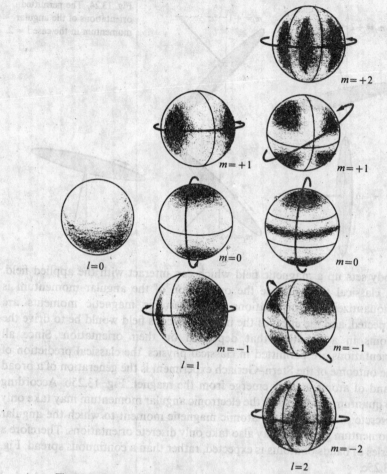

Fig. 13.23. The patterns of wavefunctions for a particle on the surface of a sphere.

angular momentum about the z-axis may take only $2l+1$ values. If the angular momentum of the particle is represented by a vector with a length proportional to the magnitude of the angular momentum and an orientation perpendicular to the plane of rotation, as in Fig. 13.20, the z-component of the vector will have a magnitude proportional to $m_l \hbar$. It follows that the orientation of the plane of rotation may take only a discrete range of values, as shown in Fig. 13.24 for $l = 2$. The remarkable implication of this result is that even *orientation in space* is quantized for a rotating body.

This astonishing result, which means that a rotating body may not take up an arbitrary orientation with respect to some axis specified by some externally applied field, is called *space quantization*. It was confirmed by an experiment first performed by Stern and Gerlach in 1921. They shot a beam of silver atoms through an inhomogeneous magnetic field, Fig. 13.25a. The idea behind the experiment is that a rotating charged

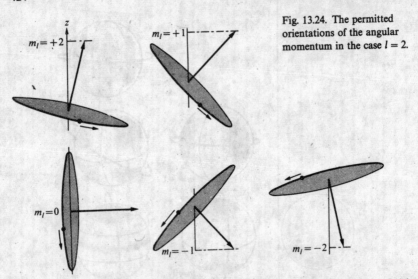

Fig. 13.24. The permitted orientations of the angular momentum in the case $l = 2$.

body sets up a magnetic field which can interact with the applied field. In classical theory, since the orientation of the angular momentum is unquantized, all orientations of the atomic magnetic moments are expected, and the effect of the inhomogeneous field would be to drive the atoms in a direction that depended on their orientation. Since all orientations are permitted in classical physics, the classical prediction of the outcome of the Stern–Gerlach experiment is the generation of a broad band of atoms as they emerge from the magnet, Fig. 13.25b. According to quantum theory, since the electronic angular momentum may take only discrete orientations, the atomic magnetic moment to which the angular momentum gives rise may also take only discrete orientations. Therefore a series of bands of atoms is expected, rather than a continuous spread, Fig. 13.25c.

In their first experiment Stern and Gerlach confirmed the classical prediction. Nevertheless the experiment was technically difficult because

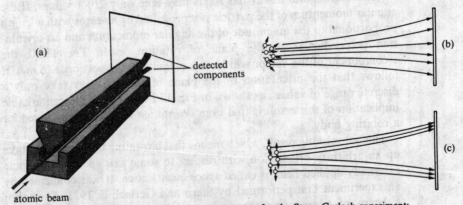

Fig. 13.25. (a) The experimental arrangement for the Stern–Gerlach experiment; (b) the classically expected result; and (c) the quantum prediction.

the atoms in the beam collided with each other, scattered, and blurred the result. When the experiment was repeated they used a beam of very low intensity in order to cut down the smearing effect of collisions, and after exposing the detection plate for hours they finally obtained the quantum result of a set of blurred but definitely discrete bands.

Spin. Using silver atoms as the projectiles Stern and Gerlach observed *two* bands. This seems to conflict with one of the quantum predictions because an angular momentum l gives rise to $2l+1$ orientations, which is equal to 2 only if $l = \frac{1}{2}$, contrary to the conclusion that l must be an integer. The resolution of this conflict came with the suggestion that the angular momentum observed in the case of the silver atoms was not that due to the motion of the electron round the nucleus, its *orbital angular momentum*, but could be ascribed to the angular momentum of an electron spinning about its own axis, the *spin angular momentum*. The wavefunction for an electron spinning at a single point of space does not have to satisfy the same type of boundary conditions as those for a particle migrating on the surface of a sphere, and so the values of l might have different restrictions on their possible values. In order to distinguish this spin angular momentum from orbital angular momentum we use s to denote the spin angular momentum quantum number (in place of l), and m_s for the projection on the z-axis. The magnitude of the spin is still $[s(s+1)]^{1/2}\hbar$ and the projection $m_s\hbar$ is restricted to the $2s+1$ values given by $m_s = s, s-1, \ldots, -s$.

The detailed analysis of the spin of a particle is quite sophisticated, but the results are simple: for an *electron*, only one value of s is allowed, and this is $s = \frac{1}{2}$, corresponding to an angular momentum of magnitude $[\frac{1}{2}(\frac{1}{2}+1)]^{1/2}\hbar$. This spin angular momentum is an intrinsic property of an electron like its rest mass or charge, and every electron has exactly the same amount of spin angular momentum. The spin of an electron can no more be decreased or increased than its rest mass or charge: it is a fixed property, characteristic of the species. Different electrons may have their spin angular momentum oriented in different directions, and the amount of angular momentum about the z axis is $m_s\hbar$, where m_s is either $+\frac{1}{2}$ or $-\frac{1}{2}$, representing two different directions of spin, Fig. 13.26.

This description of the properties of the electron is confirmed by the result of the Stern–Gerlach experiment applied to silver atoms if we

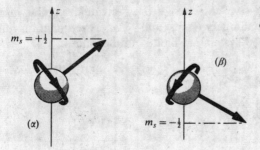

Fig. 13.26. The permitted states of electron spin.

suppose that each atom possesses an angular momentum that can be ascribed to an electron having spin but no orbital motion. The reason why the atom behaves in this way will be dealt with in the next chapter.

Different elementary particles are characterized by different but characteristic amounts of spin angular momentum. For example, a proton and a neutron both have a spin quantum number of $\frac{1}{2}$, and so electrons, neutrons, and protons all have the same, unchangeable amount of spin angular momentum. Since the masses of the proton and neutron are so much greater than that of the electron, the same momentum implies in classical terms that they rotate much more slowly than an electron. Some elementary particles have a spin quantum number of 1, which means that they have an irremoveable spin angular momentum of $2^{1/2}\hbar$. Some types

---

Box 13.1   Angular momentum

The quantum numbers for the *magnitude* of the angular momentum are written l for orbital angular momentum ($l = 0, 1, 2, \ldots$), s for spin angular momentum ($s = \frac{1}{2}$ for an electron), and j in general. These quantum numbers are always *non-negative*. The magnitude of the angular momentum is related to the quantum number through the expression $[j(j+1)]^{1/2}\hbar$.

The quantum numbers denoting the *orientation* of the angular momentum are written m_l for orbital momenta, m_s for spin momenta, and m_j for momenta in general. The projection of the angular momentum on the z-axis (an arbitrarily selected axis) is limited to the $2j+1$ values of $m_j\hbar$ with $m_j = j, j-1, j-2, \ldots -j$. Note that m_j (which means m_l or m_s or any other type of angular momentum) may take either positive or negative values: if j is integral so is m_j; if j is half-integral so is m_j.

When there are several sources of angular momentum the total angular momentum is denoted by the corresponding capital letter. Thus L denotes the total orbital angular momentum, S the total spin angular momentum, and J the total angular momentum of some general kind. The letter j is also used to denote the combined orbital and spin momentum of a single electron, and J the angular momentum of a rotating molecule.

The permitted values of the total angular momentum are given by the *Clebsch–Gordan series*. If the individual angular momenta have quantum numbers j_1, m_{j1} and j_2, m_{j2}, the quantum numbers for the combined system are limited to the values

$$J = j_1 + j_2 \qquad j_1 + j_2 - 1, \qquad j_1 + j_2 - 2, \ldots |j_1 - j_2|,$$

and

$$M_J = m_{j1} + m_{j2}.$$

of mesons have this spin, but for our purposes the most significant particle with unit spin quantum number is the photon. The importance of the angular momentum of the photon will become apparent when spectra are discussed.

The properties of angular momentum have been collected in Box 13.1.

The vector model. Throughout the preceding discussion we have referred to the magnitude of the angular momentum, and of the z-component; no mention was made of the x- and y-components. The reason for this omission is that quantum theory requires that these components are indeterminate if the z-component is known. This follows from the uncertainty principle, for if the x-, y-, and z-components were known, the orientation as well as the angular momentum of the particle would be specified, Fig. 13.27a, but this is forbidden by the principle. *If J_z is known it is impossible to say anything about the values of J_x and J_y.* It follows that Fig. 13.27a gives a false impression of the information knowable about the system, because it suggests definite values for J_x and J_y; a better picture must imply the indeterminance of J_x and J_y.

The *vector model of angular momentum* makes use of the picture in Fig. 13.27b. The cones represent possible, but indeterminate, x- and y-orientations of the angular momentum: a given cone has a definite z-projection, but the other projections are indefinite. The vector representing the angular momentum may be envisaged as lying with its tip on any point on the mouth of the cone.

The range of positions of a vector on the surface of the cone is sometimes referred to as *precession*. This name, however, is used in classical physics to suggest that the vector's tip is moving in a circle around the cone under the influence of a force. At this stage of our discussion that is an impression to be avoided; the cone should be regarded simply as a depiction of the range of possible, but unknowable orientations of the angular momentum vector.

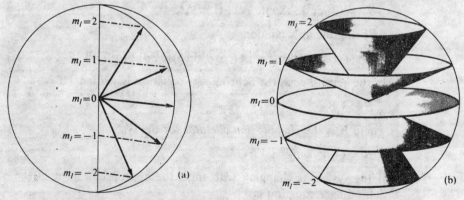

Fig. 13.27. Constructing the vector model of angular momentum. The uncertainty principle does not permit as complete a specification as in (a), and we have to accept that the vectors lie at indeterminate positions as represented by the cones drawn in (b).

In the *vector model* a given magnitude of angular momentum is depicted by a cone of the appropriate side. For example, if the angular momentum quantum number is j (where j is either l or s) the magnitude of momentum is $[j(j+1)]^{1/2}\hbar$ and the side of the cone is $[j(j+1)]^{1/2}$ units long. The orientation of the motion is depicted by drawing the cone so that its circular mouth cuts the z-axis at m_j units. For example, in the case of $j = 2$, five cones are permitted, as drawn in Fig. 13.27b.

Box 13.2

The Schrödinger equation

For *one-dimensional* problems

$$(-\hbar^2/2m)(d^2\psi/dx^2) + V(x)\psi = E\psi, \qquad \psi = \psi(x)$$

or

$$(d^2\psi/dx^2) + (2m/\hbar^2)[E - V(x)]\psi = 0.$$

In these expressions $V(x)$ is the potential energy of the particle. For example, for a free particle $V(x) = 0$ (or some constant), and for a harmonic oscillator $V(x) = \frac{1}{2}kx^2$.

For *three-dimensional* problems

$$(-\hbar^2/2m)\nabla^2\psi + V(x, y, z)\psi = E\psi, \qquad \psi = \psi(x, y, z) = \psi(\mathbf{r})$$

where ∇^2 is read 'del squared' and is the abbreviation

$$\nabla^2 = (\partial^2/\partial x^2) + (\partial^2/\partial y^2) + (\partial^2/\partial z^2).$$

The equation may also be encountered in the form

$$\nabla^2\psi + (2m/\hbar^2)[E - V(\mathbf{r})]\psi = 0.$$

For systems with spherical symmetry it is more appropriate to regard ψ as a function of the polar coordinates r, θ, φ. Then ∇^2 takes the form

$$\nabla^2 = (\partial^2/\partial r^2) + (2/r)(\partial/\partial r) + (1/r^2)\Lambda^2,$$

Λ^2 being the abbreviation

$$\Lambda^2 = (1/\sin\theta)^2(\partial^2/\partial\varphi^2) + (\cos\theta/\sin\theta)(\partial/\partial\theta) + (\partial^2/\partial\theta^2).$$

In the *general case*, the Schrödinger equation is written

$$H\psi = E\psi$$

where H is the *Hamiltonian operator* for the system,

$$H = (-\hbar^2/2m)\nabla^2 + V(\mathbf{r}).$$

If the system is changing with time, the appropriate form of the Schrödinger equation is

$$H\psi = i\hbar\,\partial\psi/\partial t.$$

13.7 Some mathematical aspects of quantum theory

Various ways of expressing the Schrödinger equation are collected in Box 13.2. The central feature of any system is the wavefunction ψ, and all the information about the properties of the system is embedded in it. In order to extract this information we construct a quantity called the *expectation value* for the property of interest.

The procedure can be illustrated as follows. Suppose we wish to know the average position of a particle along the x-axis. Since the probability of finding the particle in the region dx at the position x is $\psi^*(x)\psi(x)\,dx$, the average value of x receives a contribution x weighted by this factor. The average value is the sum (integral) of all the positions appropriately weighted:

$$\langle x \rangle = \int_{-\infty}^{\infty} x\psi^*(x)\psi(x)\,dx.$$

It is convenient to write this in the more symmetrical form

(13.7.1) $$\langle x \rangle = \int_{-\infty}^{\infty} \psi^*(x)x\psi(x)\,dx$$

which is referred to as the *expectation value* of the x-component of position. The three-dimensional analogue, the mean position in three-dimensions, is

(13.7.2) $$\langle \mathbf{r} \rangle = \int \psi^*(\mathbf{r})\mathbf{r}\psi(\mathbf{r})\,d\tau.$$

The interpretation of $\langle \mathbf{r} \rangle$ is as follows: in a large number of determinations of the position of the particle, a range of results might be obtained, and their average is $\langle \mathbf{r} \rangle$.

The expectation values of other observables—energy, momentum, etc.—may be calculated in the same way. For example, since the one-dimensional Schrödinger equation is

(13.7.3) $$-(\hbar^2/2m)(d^2/dx^2)\psi + V(x)\psi = E\psi$$

multiplying by ψ^* and integrating over all space gives

$$-(\hbar^2/2m)\int_{-\infty}^{\infty} \psi^*(d^2/dx^2)\psi\,dx + \int_{-\infty}^{\infty} \psi^*V(x)\psi\,dx = E\int_{-\infty}^{\infty} \psi^*\psi\,dx = E.$$

The two integrals on the left are of the form of expectation values, and on writing

(13.7.4) $$H = -(\hbar^2/2m)(d^2/dx^2) + V(x)$$

we find

(13.7.5) $$E = \int_{-\infty}^{\infty} \psi^*H\psi\,dx = \langle H \rangle,$$

observable

and *the energy is the expectation value of the quantity H.* This H plays a very special role in quantum mechanics, and is called the *Hamiltonian* or *Hamiltonian operator* for the system. The name 'Hamiltonian' reflects the contribution made by Hamilton to the formulation of mechanics during the nineteenth century (his formalism proved to be ideally suited for adaptation to quantum theory). The name 'operator' reflects the appearance of derivatives in the definition of H: the operator d^2/dx^2 *operates* on a function ψ and gives as a result the second derivative $d^2\psi/dx^2$. (Note that x, if it appears in H, should be regarded as the operator 'multiply by x'.)

Operators are central to quantum theory. We can see this as follows. The energy of the system can be divided into two parts, one corresponding to the kinetic energy and the other to the potential energy. The expectation value may also be divided in the same way, and in one dimension we write

$$E = T + V = \langle H \rangle = \langle (-\hbar^2/2m)(d^2/dx^2) \rangle + \langle V(x) \rangle.$$

This indicates that in order to evaluate the mean kinetic energy of the system it is necessary to evaluate the expectation value of the operator $(-\hbar^2/2m)(d^2/dx^2)$. This shows that *the operator corresponding to kinetic energy is* $(-\hbar^2/2m)(d^2/dx^2)$. We write

(13.7.6) $\qquad T_{\text{operator}} = -(\hbar^2/2m)(d^2/dx^2).$

In three dimensions this becomes

$$T_{\text{operator}} = -(\hbar^2/2m)[(\partial^2/\partial x^2)+(\partial^2/\partial y^2)+(\partial^2/\partial z^2)]$$

(13.7.7) $\qquad\qquad\quad = -(\hbar^2/2m)\nabla^2,$

where ∇^2 ('del-squared') is a convenient abbreviation for the sum of the three second-derivatives. These expressions for T are the basis of the discussion on p. 398 relating curvature to kinetic energy.

Example (Objective 24). What is the average kinetic energy of a harmonic oscillator in its ground state?

- *Method.* Calculate the expectation value of the kinetic energy operator, eqn (13.7.6) using the wavefunction given in Table 13.1. Two standard integrals are required.

- *Answer.* From eqn (13.7.6).

$$\langle T \rangle = \int_{-\infty}^{\infty} \psi_0(x) T_{\text{op}} \psi_0(x)\, dx$$

$$= (\alpha/\pi^{1/2}) \int_{-\infty}^{\infty} [e^{-(\alpha^2 x^2)/2}(-\hbar^2/2m)(d^2/dx^2)e^{-(\alpha^2 x^2)/2}]\, dx$$

$$= (-\alpha\hbar^2/2m\pi^{1/2}) \int_{-\infty}^{\infty} e^{-(\alpha^2 x^2)/2}(-\alpha^2 e^{-(\alpha^2 x^2)/2}+\alpha^4 x^2 e^{-(\alpha^2 x^2)/2})\, dx$$

$$= (-\alpha\hbar^2/2m\pi^{1/2})[-\alpha^2 \int_{-\infty}^{\infty} e^{-\alpha^2 x^2}\, dx + \alpha^4 \int_{-\infty}^{\infty} x^2 e^{-\alpha^2 x^2}\, dx]$$

$$= (-\alpha\hbar^2/2m\pi^{1/2})[-(\pi\alpha^2)^{1/2}+\tfrac{1}{2}(\pi\alpha^2)^{1/2}]$$

$$= \tfrac{1}{4}\alpha^2\hbar^2/m = \tfrac{1}{4}\hbar\omega.$$

• *Comment*. We see that the mean kinetic energy is half the zero-point energy; therefore in the present case $\langle T \rangle = \langle V \rangle$. This is in fact true for all states of the harmonic oscillator (i.e., that $\langle T \rangle = \langle V \rangle$), and is a special case of the *virial theorem*. This asserts that if the potential is of the form $V \propto x^s$, then $\langle T \rangle = \frac{1}{2}s\langle V \rangle$. In the hydrogen atom, for example, $V \propto 1/r$, so that $s = -1$ and $\langle T \rangle = -\frac{1}{2}\langle V \rangle$.

We know from classical theory that T may be expressed as $(1/2m)p^2$. This suggests that we may also find *the operator corresponding to the linear momentum*. For instance, on writing the *classical expression*

$$T = (1/2m)p^2,$$

and then setting

(13.7.8)
$$p_{\text{operator},x} = (\hbar/i)\frac{\mathrm{d}}{\mathrm{d}x}$$

we get the *quantum expression*:

$$T_{\text{operator}} = (1/2m)p_{\text{operator},x}^2 = -(\hbar^2/2m)\mathrm{d}^2/\mathrm{d}x^2$$

which agrees with the earlier analysis, eqn (13.7.6).

Now we are in a position to see how to calculate the mean momentum of any system. The average momentum of a particle described by a wavefunction ψ is the expectation value of the momentum operator. In one dimension this means

(13.7.9)
$$\langle p_x \rangle = \int_{-\infty}^{\infty} \psi^*(x)p_{\text{operator},x}\psi(x)\,\mathrm{d}x = (\hbar/i)\int_{-\infty}^{\infty}\psi^*(x)\left(\frac{\mathrm{d}\psi}{\mathrm{d}x}\right)\mathrm{d}x$$

The y- and z-components can be found analogously.

Having introduced the Hamiltonian for the system it is possible to write the Schrödinger equation in a very succinct form. Comparing eqns (13.7.3) and (13.7.4) we see that the Schrödinger equation is

(13.7.10)
$$H\psi = E\psi.$$

This form of equation, where an operator (H) operates on a function (ψ) and gives a number (E) multiplying the original function is called an *eigenvalue equation*. The recognition that the Schrödinger equation is of this form has led to the introduction of various different names for E and ψ. E is also called the *energy eigenvalue*, and ψ is called the *eigenfunction* or *eigenstate* of the system for this energy. All these words stem from *eigen*, which is the German word for 'characteristic'.

This development of quantum theory, the discussion of properties in terms of the expectation values of appropriately selected operators, is the basis of the formal theory. All the conclusions of quantum theory can be deduced by examining the properties of the operators representing the various observable properties. In particular the Heisenberg uncertainty relation can be derived, and the way that systems change with time can be expressed in a powerful and general way.

Further reading

The strange story of the quantum. B. Hoffman; Dover, New York, 1959.

Black-body theory and the quantum discontinuity, 1894–1912. T. S. Kuhn; Clarendon Press, Oxford, 1978.

The conceptual development of quantum mechanics. M. Jammer; McGraw-Hill, New York, 1966.

Quanta: a handbook of concepts. P. W. Atkins; Clarendon Press, Oxford, 1974.

Quantization (lecture cassette and workbook). P. W. Atkins; Royal Society of Chemistry, London, 1981.

The physical principles of the quantum theory. W. Heisenberg; Dover, New York, 1930.

Lectures in physics. R. P. Feynman, R. B. Leighton, and M. Sands; Freeman, San Francisco, 1963.

Molecular quantum mechanics. P. W. Atkins; Clarendon Press, Oxford, 1970.

Introduction to quantum mechanics. L. Pauling and E. B. Wilson; McGraw-Hill, New York, 1935.

Contemporary quantum chemistry. J. Goodisman; Plenum, New York, 1977.

Quantum theory of matter. J. C. Slater; McGraw-Hill, New York, 1968.

Quantum mechanics. L. I. Schiff; McGraw-Hill, New York, 1968.

Quantum mechanics. A. Messiah; Wiley, New York, 1961.

Quantum mechanics. A. S. Davydov; Pergamon, Oxford, 1976.

The principles of quantum mechanics. P. A. M. Dirac; Clarendon Press, Oxford, 1958.

The tunnel effect in chemistry. R. P. Bell; Chapman and Hall, London, 1980.

Atoms and molecules. M. Karplus and R. N. Porter; Benjamin, Menlo Park, 1970.

Problems

13.1. One of the key ideas in quantum theory is the concept of the *photon*. The first few Problems give some practice in dealing with its properties. First we assess the energy carried by photons of different wavelengths. Calculate the energy per photon, and the energy per mole of photons, when their wavelength is (a) 600 nm (red), (b) 550 nm (yellow), (c) 400 nm (blue), (d) 200 nm (ultraviolet), (e) 150 pm (X-ray), (f) 1 cm (microwave).

13.2. What are the momenta of the photons in the last Problem? What speed would a stationary hydrogen atom attain if a photon collided with it and was absorbed?

13.3. A glow-worm of mass 5 g emits red light (650 nm) with a power of 0.1 W entirely in the backward direction. What velocity will it have accelerated to after 10 years if released in free space (and assumed to live)?

13.4. A sodium light emits yellow light (550 nm). How many photons does it emit each second if its power is (a) 1 W, (b) 100 W?

13.5. The Planck distribution, eqn (13.2.2), gives the energy in a wavelength range $d\lambda$ at the wavelength λ. Estimate the energy density lying between the wavelengths 650 nm and 655 nm (regarding the range $\Delta\lambda$ as virtually infinitesimal) at equilibrium inside a cavity of volume 100 cm^3 when its temperature is (a) 25 °C, (b) 3000 °C.

13.6. One way of deducing the value of Planck's constant h is to fit his distribution law to the measured radiation energy emitted by a heated black body. Less accurate, but simpler, is to derive an expression for the wavelength corresponding to the emission maximum, and to fit that to experiment. Derive the expression

$\lambda_{max}T = constant$ in the region of short wavelengths, and find an expression for the constant in terms of h, c, and k.

13.7. The wavelength of the emission maximum from a small pinhole in an electrically heated container was determined at a series of temperatures. From the data below, deduce a value of Planck's constant.

$t/°C$	1000	1500	2000	2500	3000	3500
λ_{max}/nm	2181	1600	1240	1035	878	763

13.8. The peak in the sun's emitted energy occurs at about 480 nm. Assuming it to behave as a black-body emitter, what is the temperature of the surface?

13.9. The Einstein theory of heat capacities leads to the expression quoted in eqn (13.2.3). The Einstein frequency for copper is 7.1×10^{12} Hz. What is its molar heat capacity at (a) 200 K, (b) 298 K, (c) 700 K? What are the classical values at these temperatures?

13.10. The Einstein frequency is often expressed in terms of an equivalent temperature θ_E. When the actual temperature is much greater than θ_E the heat capacity is virtually the classical value. To what temperature does 7.1×10^{12} Hz correspond?

13.11. The basic equation for the photoelectric effect is eqn (13.2.4). The work function for caesium is 2.14 eV. What is the kinetic energy and the speed of the electrons emitted when the metal is irradiated with light of wavelength (a) 700 nm, (b) 300 nm?

13.12. The photoelectric effect is the basis of the spectroscopic technique known as *photoelectron spectroscopy* (Chapter 18). An X-ray photon of wavelength 150 pm plunges into the inner part of an atom and ejects an electron. The speed of the latter was measured and found to be 2.14×10^7 m s^{-1}. How tightly was it bound in the atom?

13.13. By how much does the wavelength of radiation change when it scatters from (a) a free electron, (b) a free proton, and is detected at 90° to the initial line of flight.

13.14. But from where does the expression for the Compton effect come? It can be derived quite simply by setting up the conditions for the conservation of energy and of linear momentum before and after the collision of the photon and the particle. The only tricky part is the need to take into account the relativistic effects arising from the high speeds involved. When the electron is at rest it possesses the energy $m_e c^2$. When it is in motion with a momentum of magnitude p its energy is $\sqrt{(p^2 c^2 + m_e^2 c^4)}$. Let the photon, wavelength λ_i, strike the electron and be scattered with a new wavelength λ_f through an angle θ, and the electron, initially stationary, move off with a momentum of magnitude p at an angle θ' to the incoming photon. Set up the three conservation expressions (energy, momentum along line of approach, momentum perpendicular to the line of approach), eliminate θ', then eliminate p, and hence arrive at an expression for $\delta\lambda$.

13.15. Calculate the size of the quantum involved in the excitation of (a) an electronic motion of period 10^{-15} s, (b) a molecular vibration of period 10^{-14} s, (c) a pendulum of period 1 s. Express your results in kJ mol^{-1}.

13.16. What is the de Broglie wavelength of (a) a mass of 1 g travelling at 1 cm s^{-1}, (b) the same, travelling at 100 km s^{-1}, (c) a helium atom in a container at room temperature, (d) an electron accelerated through a potential difference of 100 V, 1 kV, 100 kV?

13.17. The uncertainty principle limits our ability to determine simultaneously the momentum and position of a particle. But why were the classical physicists unaware of the restraints it implies? Calculate the minimum uncertainty in the speed of a ball of mass 500 g that is known to be within 10^{-6} m of a point on a bat. What is the minimum uncertainty in the position of a bullet of mass 5 g that is known to have a speed somewhere between 350.00001 m s^{-1} and 350.00000 m s^{-1}?

13.18. An electron is confined to a linear region with a length of the order of the diameter of an atom (≈ 0.1 nm). What are the minimum uncertainties in its linear momentum and speed?

13.19. The Born interpretation of the wavefunction is described on p. 399. Suppose a wavefunction has the form $(2/L)^{1/2} \sin(\pi x/L)$, as for a particle in a box of length L. Let the box be 10 nm long. What is the probability of finding the particle (a) between $x = 4.95$ nm and 5.05 nm, (b) between $x = 1.95$ nm and 2.05 nm, (c) between $x = 9.90$ and 10.00 nm, (d) in the right half of the box, (e) in the central third of the box?

13.20. The wavefunction for the electron in the ground state of the hydrogen atom is $\psi(r,\theta,\varphi) = (1/\pi a_0^3)^{1/2} \exp(-r/a_0)$, where $a_0 = 53$ pm. What is the probability of finding the electron somewhere inside a small sphere of radius 1.0 pm centred on the nucleus? Now suppose the same tiny sphere is moved to surround a point at a distance 53 pm from the nucleus: what is the probability that the electron is inside it?

13.21. In order to use the Born interpretation directly it is necessary that the wavefunction is normalized to unity. Normalize to unity the following wavefunctions: (a) $\sin(n\pi x/L)$ for the range $0 \leqslant x \leqslant L$, (b) c, a constant in the range $-L \leqslant x \leqslant L$, (c) $\exp(-r/a_0)$ in three dimensions, (d) $x\exp(-r/2a_0)$ in three-dimensional space. In order to integrate over three dimensions you need to know that the volume element is $d\tau = r^2 dr \sin\theta d\theta d\varphi$, with $0 \leqslant r \leqslant \infty$, $0 \leqslant \theta \leqslant \pi$, $0 \leqslant \varphi \leqslant 2\pi$. Use $\int_0^\infty x^n e^{-ax} dx = n!/a^{n+1}$.

13.22. The energy levels of a particle of mass m in a box of length L are given by eqn (13.4.4). Suppose that the particle is an electron, and that the box represents a long conjugated molecule. What are the energy separations in J, kJ mol^{-1}, eV, and cm^{-1} between the levels (a) $n = 2$ and $n = 1$, (b) $n = 6$ and $n = 5$, in both cases taking $L = 1$ nm (10 Å).

13.23. A gas molecule in a flask has quantized translational energy levels, but how important are the effects of quantization? Calculate the separation between the lowest two energy levels for an oxygen molecule in a container of length 5 cm. At what value of the quantum number n does the energy of the molecule equal $\frac{1}{2}kT$, when $T = 300$ K? What is the separation of this level from its nearest neighbour?

13.24. Set up the Schrödinger equation for a particle of mass m in a three-dimensional square well with sides L_X, L_Y, and L_Z (and volume $V = L_X L_Y L_Z$). Show that the wavefunction requires three quantum numbers for its specification, and that $\psi(x,y,z)$ can be written as the product of three wavefunctions for one-dimensional square wells. Deduce an expression for the energy levels, and specialize it to the case of a cubic box of side L.

13.25. Employ the same technique as in the *Example* on p. 410 to find an expression for the transmission P when E exceeds the barrier height V.

13.26. Calculate the probability that a particle will be reflected back from a dip in an otherwise uniform potential. Take the depth of the well as V and its width as L. Plot the probability as a function of E/V with $V = 5$ eV, $L = 0.1$ nm, for a proton and a deuteron.

13.27. Particles can penetrate into regions where classical mechanics forbids them to go. This is because the wavefunction is continuous at the boundaries of containers, and so it seeps into regions of high potential before dropping to zero. Take the wavefunction for the ground state of a harmonic oscillator formed from a mass m on a spring of force-constant k. Plot the probability density on graph paper and estimate by graphical integration the probability that the mass will be found in regions where for the same total energy it is forbidden by classical mechanics. Take the mass to be that of (a) a proton, (b) a deuteron, and the force constant to be that of a typical bond, 500 N m^{-1}.

13.28. The wavefunction for the lowest state of a harmonic oscillator has the form of a *Gaussian function* $\exp(-gx^2)$, where x is the displacement from equilibrium. Show that this function satisfies the Schrödinger equation for a harmonic oscillator, and find g in terms of the mass m and the force-constant k. What is the (zero-point) energy of the oscillator with this wavefunction? What is its minimum excitation energy?

13.29. Several types of motion are simple harmonic, and their energy levels are all given by eqn (13.5.2). Find the minimum excitation energies of the following oscillators: (a) a pendulum of length 1 m in a gravitational field; (b) the balance wheel of a watch; (c) the 33 kHz quartz crystal of a watch; (d) the atomic oscillators that appear in the Einstein theory of heat capacity of copper; (e) the bond between two oxygen atoms ($k = 1177$ N m^{-1}).

13.30. The following molecules were found to absorb radiation at the wave-numbers quoted (*Spectra of diatomic molecules*. G. Herzberg; van Nostrand (1950)), the energy going into the excitation of the vibration of the bond. Since both atoms of the diatomic molecules move when the bond vibrates, the mass to use in the expression $\omega = (k/\mu)^{1/2}$ is $\mu = m_1 m_2/(m_1 + m_2)$. Find the force constants of the bonds in the molecules, and arrange them in order of increasing stiffness, HCl, 2989.74 cm^{-1}; HBr, 2649.72 cm^{-1}; HI, 2309.5$_3$ cm^{-1}; CO, 2170.21 cm^{-1}; NO, 1904.03 cm^{-1}.

13.31. The Schrödinger equation for a particle of mass m confined to a circle of radius R is $-(\hbar^2/2mR^2)(\partial^2\psi/\partial\varphi^2) = E\psi$, where $\psi = \psi(\varphi)$. Solve this equation, impose the appropriate conditions on the solutions, and confirm that the particle's energy is confined to the values given in eqn (13.6.4).

13.32. The rotation of an HI molecule can be visualized as the orbiting of the hydrogen atom at a distance of 160 pm from a stationary iodine atom (this is quite a good approximation, but to be precise we would have to take into account the motion of both atoms around their joint centre of mass). Suppose that the molecule rotates only in a plane. How much energy (in kJ mol^{-1} and cm^{-1}) is needed to excite the stationary molecule into rotation? What, apart from zero, is the minimum angular momentum of the molecule?

13.33. Take the same model of the HI molecule as in the last Problem and suppose that it is known to possess (a) zero angular momentum, (b) one unit of angular momentum. Use the form of the wavefunction for the two states to decide on the location of the hydrogen atom.

13.34. Calculate the energies of the first four rotational levels of HI, allowing it to rotate in three dimensions about its centre of mass. Express your answer in kJ mol^{-1} and cm^{-1}.

13.35. The three-dimensional form of the Schrödinger equation, expressed in polar coordinates, is set out in Box 13.2. Show that the following wavefunctions $\psi_{l,m}(\theta, \varphi)$ for a particle on a spherical surface satisfy the equation, and that in each case the energy and angular momentum of the particle are given by eqns (13.6.8) and (13.6.7). (a) $\psi_{0,0} = 1/2\pi^{1/2}$, (b) $\psi_{1,0} = \frac{1}{2}(3/\pi)^{1/2} \cos\theta$, (c) $\psi_{2,-1} = \frac{1}{2}(15/2\pi)^{1/2} \cos\theta \sin\theta \exp(-i\varphi)$, (d) $\psi_{3,3} = -\frac{1}{8}(35/\pi)^{1/2} \sin^3\theta \exp(3i\varphi)$.

13.36. Confirm that $\psi_{3,3}(\theta, \varphi)$ is normalized to unity. Take the integration over the surface of a sphere.

13.37. Discuss the quantum mechanics of bicycling; suppose that $h = 6.6$ J s instead of 6.6×10^{-34} J s.

13.38. In the vector model of angular momentum a state with quantum numbers l, m_l (or s, m_s) is represented by a vector of length $\sqrt{[l(l+1)]}$ units and of z-component m units. Draw scale diagrams of the state of an electron with (a) $s = \frac{1}{2}$, $m_s = \frac{1}{2}$, (b) $l = 1$, $m_l = +1$, (c) $l = 2$, $m_l = 0$.

13.39. Derive an expression for the half-angle of the apex of the cone of precession in terms of the quantum numbers l, m_l (or s, m_s). What is its value for the α-state of an electron spin? Show that the minimum possible angle approaches zero as l approaches infinity.

13.40. Draw the cones of precession of an electron with $l = 6$. In what region of the spherical surface is an electron most likely to be found when $l = 6$ and (a) $m_l = 0$, (b) $m_l = +6$, (c) $m_l = -6$?

13.41. In the last few Problems of this chapter we look into some of the simpler mathematical aspects of quantum theory, at the level of Section 13.7. As a first step, find the expectation value of the position of a particle which is described by the following wavefunctions (a) the level $n = 1$ of a particle in square well of length L, (b) the unnormalized function $\cos\varphi$ for a particle on a ring of radius R.

13.42. Of more interest is the *root mean square deviation* of the particle from its mean position. This is defined as $\delta x = \sqrt{(\langle x^2 \rangle - \langle x \rangle^2)}$, and indicates the spread of the distribution of the particle. Calculate the value of δx and $\delta\varphi$ for the states referred to in the last Problem.

13.43. The distribution of a harmonically oscillating particle spreads over a greater range of displacements as its excitation energy is increased (just as a pendulum swings with increasing amplitude). Calculate the mean displacement and the root mean square displacement of a particle in the states $n = 0$, $n = 1$, $n = 2$ of a harmonic oscillator, using the wavefunctions set out in Table 13.1 on p. 414.

13.44. The expectation value of momentum is evaluated by using eqn (13.7.9). What is the average momentum of a particle described by the following wavefunctions: (a) $\exp(ikx)$, (b) $\exp(-k|x|)$, (c) $\cos kx$, (d) $\exp(-\alpha x^2)$, each one in the range $-\infty \leqslant x \leqslant \infty$?

13.45. Which of the following functions are eigenfunctions of the operator d/dx: (a) $\exp(ikx)$, (b) $\cos kx$, (c) k, (d) kx, (e) $\exp(-\alpha x^2)$? Give the eigenvalue where appropriate.

13.46. Which of the functions in the last Problem are also eigenfunctions of d^2/dx^2, and which are eigenfunctions only of d^2/dx^2? Give the eigenvalues where appropriate.

13.47. The *commutator* of two operators A_{op} and B_{op}: is written $[A_{op}, B_{op}]$ and is defined as the difference $A_{op}B_{op} - B_{op}A_{op}$. It can be evaluated by taking some convenient function ψ (which can be left unspecified) and evaluating both $A_{op}B_{op}\psi$ and $B_{op}A_{op}\psi$, and finding the difference in the form $C_{op}\psi$. Then C_{op} is identified as the commutator $[A_{op}, B_{op}]$. Quite often C_{op} turns out to be a simple numerical factor. An extremely important commutator is that of the operator for the components of momentum and the components of position. Find $[x_{op}, y_{op}]$, $[x_{op}, x_{op}]$, $[p_{x,op}, p_{y,op}]$, $[x_{op}, p_{x,op}]$, $[x_{op}, p_{y,op}]$.

13.48. One of the reasons why the commutator is so important is that it lets us identify at a glance the observables that are restricted by the uncertainty relation. Thus, if A_{op} and B_{op} have a non-zero commutator, the observables A and B cannot in general be determined simultaneously. Can p_x and x be determined simultaneously? Can p_x and y? Can the three components of position be specified simultaneously?

13.49. Another important commutator is that for the components of angular momentum. From classical theory $l_x = yp_z - zp_y$, $l_y = zp_z - xp_z$, $l_z = xp_y - yp_z$; hence write the corresponding operators. Show that $[l_{x,op}, l_{y,op}] = i\hbar l_{z,op}$. Can l_x and l_y be determined in general simultaneously? This is the restraint that underlies the vector model of angular momentum. Why?

14 Atomic structure and atomic spectra

Learning objectives

After careful study of this chapter you should be able to

(1) Describe the main features of the *spectrum of atomic hydrogen* (p. 439) and state the *Ritz combination principle* (p. 440).

(2) State the dependence of the *energy levels* of the hydrogen atom on the *principal quantum number* (p. 441).

(3) Describe the shapes and significance of the *atomic orbitals* of the hydrogen atom (p. 442).

(4) Define *radial distribution function* and explain its significance (p. 448).

(5) Use the *Bohr frequency condition* and the *selection rules* to predict the form of the spectrum of the hydrogen atom (p. 448).

(6) State the *Aufbau principle* (p. 450) and the *Pauli exclusion principle* (p. 451) and use them to account for the structure of *manyelectron atoms*.

(7) Explain the terms *penetration*, *shielding*, and *effective nuclear charge* (p. 451).

(8) Define *ionization energy* of an atom and explain how it varies in the periodic table (p. 452).

(9) Describe the strategy of *self-consistent field* calculations of the structure of atoms (p. 454).

(10) Describe the *spin-orbit interaction* (p. 456) and show how it is responsible for the *fine structure* of spectra (p. 458).

(11) Write and use the *Clebsch–Gordan series* for the addition of angular momenta (eqn (14.3.1)).

(12) Describe the construction and significance of *term symbols* (p. 459).

(13) Describe *Russell–Saunders coupling* and *jj-coupling* and state the *selection rules* for many-electron atoms (p. 461).

(14) Write expressions for the *magnetic moment* of an electron in terms of its orbital and spin angular momenta (p. 462), and for its energy in a magnetic field (eqns (14.3.5) and (14.3.6)).

(15) Describe the *normal Zeeman effect* (p. 462).

(16) Explain the source of the *Landé g-factor* and show how it accounts for the *anomalous Zeeman effect* (p. 465).

Introduction

The rudiments of the quantum-mechanical description of the basic types of motion were established in the preceding chapter. In this chapter we move on to their application to atoms. The experimental information at our disposal comes from atomic spectra. We must be able to interpret this spectral information and relate it to the energies of electrons in atoms. Information about atomic energy levels is of central importance in a rational discussion of the structures and reactions of molecules, and the information we shall describe has extensive chemical applications.

14.1 The structure and spectrum of atomic hydrogen

When an electric discharge is passed through hydrogen the molecules are dissociated and the atoms emit light at a series of frequencies, Fig. 14.1. The first important contribution to the understanding of this observation was made by Balmer, who pointed out that the wavelength of the light in the visible region (the series of lines now called the *Balmer series*) fitted an expression which written in modern notation has the form,

$$\frac{1}{\lambda} = R_H\left(\frac{1}{4} - \frac{1}{n^2}\right), \qquad n = 3, 4, \ldots$$

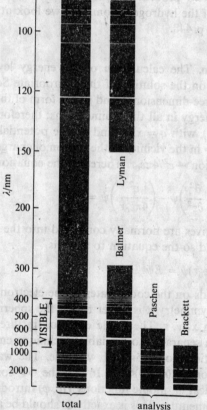

Fig. 14.1. The spectrum of atomic hydrogen.

where R_H is a constant now called the *Rydberg constant*, and having the value $109\,677\,cm^{-1}$, or $3.2898 \times 10^{15}\,Hz$. As further lines were discovered in the ultraviolet (the *Lyman series*) and the infrared (the *Paschen series*) it was noted that all could be fitted to the expression

(14.1.1) $$\frac{1}{\lambda} = R_H\left(\frac{1}{n_1^2} - \frac{1}{n_2^2}\right),$$

with $n_1 = 1$ (Lyman), $n_1 = 2$ (Balmer), $n_1 = 3$ (Paschen), and in each case $n_2 = n_1 + 1, n_1 + 2,\ldots$ The form of this expression strongly suggests that the inverse wavelength (and through $v = c/\lambda$, the frequency) of every line in the spectrum can be written as the difference of two terms, each of the form R_H/n^2. This statement amounts to the *Ritz combination principle*: the position of any spectral line (of any atom) can be expressed as the difference of two *terms*. Two terms T_1 and T_2 *combine* to produce a spectral line at a wavelength given by

$$1/\lambda = T_2 - T_1.$$

There are three points to examine. The first is to show why the terms in hydrogen have the form R_H/n^2, the second is to calculate the value of R_H, and the third is to explain why not all combinations of terms appear in the spectrum. The first two problems depend on a knowledge of the permitted energy levels of the hydrogen atom, and we look at that now. The third is taken up on p. 448.

The structure of the hydrogen atom. The calculation of the energy levels of the hydrogen atom depends on the solution of the appropriate Schrödinger equation. The atom is three-dimensional, and so the form of the equation must allow for kinetic energy in all three dimensions: therefore we must include $(-\hbar^2/2m)(d^2/dq^2)$, with $q = x, y$, and z. The potential energy of the electron of charge $-e$ in the vicinity of the proton of charge e is given by the Coulomb expression $-e^2/4\pi\varepsilon_0 r$. Therefore the equation is

$$(-\hbar^2/2m)\left(\frac{\partial^2}{\partial x^2} + \frac{\partial^2}{\partial y^2} + \frac{\partial^2}{\partial z^2}\right)\psi - \left(\frac{e^2}{4\pi\varepsilon_0 r}\right)\psi = E\psi.$$

The three second-derivatives are normally combined into the symbol ∇^2 introduced on p. 425, and so the equation to solve is

(14.1.2) $$(-\hbar^2/2m)\nabla^2\psi - (e^2/4\pi\varepsilon_0 r)\psi = E\psi.$$

The wavefunction depends on the coordinates of the electron, and since the atom has spherical symmetry it is better to use the spherical coordinates r, θ, φ rather than x, y, z. The single equation in three variables can be separated (by the separation of variables technique employed on p. 411) into three equations, one for each variable. The wavefunction $\psi(r, \theta, \varphi)$ is then of the form $R(r)\Theta(\theta)\Phi(\varphi)$. In fact the $\Theta\Phi$ combinations turn out to be the spherical harmonic functions $Y(\theta, \varphi)$ introduced in the discussion of angular momentum. This is exactly as should be anticipated, because the electron is expected to possess orbital angular momentum about the central nucleus.

Solving eqn (14.1.2) and imposing boundary conditions that ensure the Born interpretation (p. 399) is tenable leads to the following conclusions, Three quantum numbers emerge: two come from the spherical nature of the problem, and are simply the angular momentum l and m_l quantum numbers (p. 421); the third, n, arises from the freedom of the electron to vary its distance from the nucleus. The wavefunctions are therefore labelled ψ_{nlm_l}, and the restrictions on the quantum numbers that emerge from the boundary conditions are

(14.1.3)

$$n = 1, 2, 3, \ldots$$
$$l = 0, 1, 2, \ldots n-1; \ m_l = l, l-1, l-2, \ldots, -l.$$

The *energies* might be expected to depend on all three quantum numbers, or at least the first two, but it is a peculiarity of the hydrogen atom that the energy depends only on n. For this reason n is called the *principal quantum number*. The permitted energies are

(14.1.4)

$$E_n = -\left(\frac{me^4}{32\pi^2\varepsilon_0^2\hbar^2}\right)\frac{1}{n^2},$$

which are of the form $-hcR_H/n^2$, R_H being a constant.

The spectrum of atomic hydrogen can now be accounted for by supposing that as its electron drops from a state with principal quantum number n_2 (and energy $-hcR_H/n_2^2$) to one with quantum number n_1 (and energy $-hcR_H/n_1^2$), it emits this difference in energy in a packet, a photon of energy $h\nu$ and frequency ν. From the conservation of energy the frequency is given by

$$h\nu = hcR_H/n_1^2 - hcR_H/n_2^2, \quad \text{or} \quad 1/\lambda = \nu/c = R_H/n_1^2 - R_H/n_2^2$$

precisely as the experimental observations require. The value of the Rydberg constant R_H can be calculated by inserting the values of the fundamental constants into eqn (14.1.4), and on doing so there is almost exact agreement with experiment.

Example Calculate the value of the Rydberg constant for the hydrogen atom, predict the wavelengths of the first four transitions in the Lyman series, and find the *ionization energy* of the atom.

● *Method.* The Rydberg constant R_H can be identified in eqn (14.1.4), and in order to be strictly accurate, we should interpret m as the *reduced mass* of the electron, $m = m_e m_p/(m_e + m_p)$. For the Lyman series set $n_1 = 1$ and $n_2 = 2, 3, \ldots$. For the ionization energy I (the energy required to remove an electron from the ground state atom) take $n_1 = 1$ and $n_2 = \infty$ (corresponding to zero binding energy).

● *Answer.* $hcR_H = me^4/32\pi^2\varepsilon_0^2\hbar^2 = me^4/8h^2\varepsilon_0^2$.

$$m/\text{kg} = \frac{(9.109\,53 \times 10^{-31}) \times (1.672\,65 \times 10^{-27})}{(9.109\,53 \times 10^{-31}) + (1.672\,65 \times 10^{-27})} = 9.104\,57 \times 10^{-31}.$$

Therefore, using the data on the end-papers

$$hcR_H = \frac{(9.104\,57 \times 10^{-31}\,\text{kg}) \times (1.602\,19 \times 10^{-19}\,\text{C})^4}{8 \times (6.626\,18 \times 10^{-34}\,\text{J s})^2 \times (8.854\,188 \times 10^{-12}\,\text{J}^{-1}\,\text{C}^2\,\text{m}^{-1})^2}$$

$$= 2.178\,72 \times 10^{-18}\,\text{J}.$$

If R_H is required in cm^{-1}, divide by hc:

$$R_H = (2.178\,72 \times 10^{-18}\,\text{J})/(6.626\,18 \times 2.997\,925 \times 10^{-34} \times 10^{10}\,\text{J cm})$$

$$= 1.096\,78 \times 10^5\,\text{cm}^{-1}.$$

For the Lyman series, form the following table:

n_2	2	3	4	5	...	∞
$[(1/1)^2 - (1/n_2)^2]$	3/4	8/9	15/16	24/25	...	1
$R_H[,,]/10^5\,\text{cm}^{-1}$	0.8226	0.9749	1.0282	1.0529	...	1.096 78
λ/nm	121.57	102.57	97.255	94.975	...	91.176

- *Comment.* The ionization energy is $1.0968 \times 10^5\,\text{cm}^{-1}$, which converts to $13.60\,\text{eV}$. Notice that light of very short wavelength (91 nm) is needed to ionize the ground state atom. Using $m = m_e$ in the expression for the Rydberg constant gives the quantity denoted R_∞ as listed on the endpapers.

Atomic orbitals. The wavefunctions of the hydrogen atom are important because they provide a basis for the discussion of the structure of all atoms and molecules. Wavefunctions for electrons in atoms and molecules are called *orbitals* and we shall use this name from now on.

The amplitude $\psi_{nlm_l}(\mathbf{r})$ varies from place to place, and the probability of finding the electron in some infinitesimal region $d\tau$ at the point $\mathbf{r}$ is $|\psi_{nlm_l}(\mathbf{r})|^2\,d\tau$. The spatial variation of the distribution of the electron can be represented by indicating the value of $|\psi_{nlm_l}(\mathbf{r})|^2$ by the density of shading in a diagram, and some of the total wavefunctions and probabilities are represented in Fig. 14.2. Their *radial* variation is depicted in Fig. 14.3 and tabulated in Table 14.1. In order to simplify their presentation, atomic orbitals are normally represented by diagrams of the shape of the surface that captures about 90 per cent of the electron probability. This results in diagrams typified by the illustrations on the right of Fig. 14.2.

The *ground state orbital*, the orbital corresponding to the lowest energy, is very simple: it is spherically symmetrical, and its amplitude decays exponentially with the distance from the nucleus:

lowest energy orbital $(n = 1, l = 0, m_l = 0)$:

(14.1.5) $$\psi = (1/\pi a_0^3)^{1/2} \exp(-r/a_0).$$

a_0 is a collection of fundamental constants, $4\pi\varepsilon_0\hbar^2/m_e e^2$, which has the dimensions of length and magnitude 53 pm, and is called the *Bohr radius*. The exponential form of this orbital shows that the most probable place to find the electron is at the nucleus (where ψ, and therefore ψ^2, is a maximum). This is consistent with the view that the electron should drift on to the nucleus in order to achieve its lowest potential energy. What we

Electron densities

Boundary surfaces

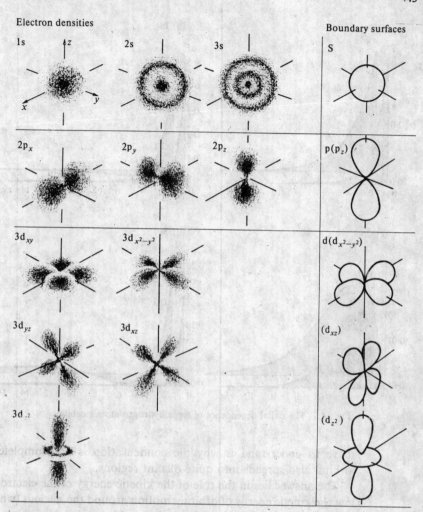

Fig. 14.2. The density distribution in some hydrogen atomic orbitals. A selection of boundary surfaces is shown on the right.

Table 14.1. Hydrogen atomic orbitals

Hydrogen atomic orbitals are of the form $\psi_{nlm_l}(r, \theta, \phi) = R_{nl}(r)Y_{lm_l}(\theta, \phi)$. The *radial* dependence, R_{nl}, is as follows; the Y_{lm_l} are given in Table 13.2.

1s $\quad R_{10}(r) = (Z/a_0)^{3/2}2\exp(-\rho/2)$

2s $\quad R_{20}(r) = (Z/a_0)^{3/2}(1/2\sqrt{2})(2-\rho)\exp(-\rho/2)$

2p $\quad R_{21}(r) = (Z/a_0)^{3/2}(1/2\sqrt{6})\rho\exp(-\rho/2)$

3s $\quad R_{30}(r) = (Z/a_0)^{3/2}(1/9\sqrt{3})(6-6\rho+\rho^2)\exp(-\rho/2)$

3p $\quad R_{31}(r) = (Z/a_0)^{3/2}(1/9\sqrt{6})(4-\rho)\rho\exp(-\rho/2)$

3d $\quad R_{32}(r) = (Z/a_0)^{3/2}(1/9\sqrt{30})\rho^2\exp(-\rho/2)$

where $\rho = 2Zr/na_0$, $a_0 = 52.92\,\text{pm}$, and Z is the atomic number.

Source: L. Pauling and E. B. Wilson, *Introduction to quantum mechanics*, McGraw-Hill.

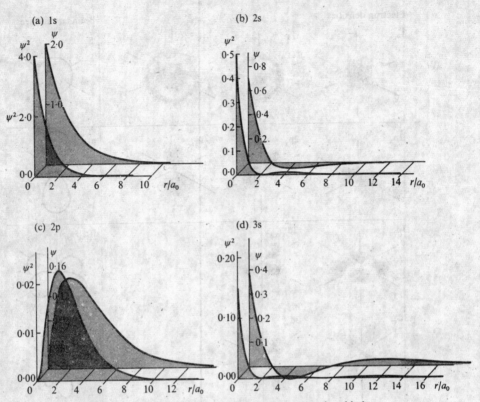

Fig. 14.3. The radial dependence of some hydrogen atomic orbitals.

have to understand is why the condensation is not complete, for the orbital also spreads into quite distant regions.

The answer lies in the role of the kinetic energy of the electron. This is *not* the kinetic energy of orbiting motion around the nucleus (which could result in a centrifugal force holding the electron off the nucleus) because the angular momentum of the electron in the ground state is zero. This can be seen from Fig. 14.2 because a spherically symmetrical orbital has no nodes, and therefore no angular momentum. More explicitly, since $l = 0$, the magnitude of the angular momentum, $[l(l+1)]^{1/2}\hbar$, is zero. The relevant kinetic energy is that associated with the curvature of the orbital in the *radial* direction. Classically this represents motion of the electron swinging in and out along a radius. In order to confine the electron more closely to the nucleus the radial part of its wavefunction must be more sharply kinked; but curvature raises its kinetic energy. The observed ground state, with the electron clustering close to the nucleus, but also spread significantly into regions distant from it, is the result of the compromise between moderately low potential energy and only moderately high kinetic energy, Fig. 14.4.

All spherically symmetric and therefore zero angular momentum orbitals are called *s-orbitals*. The lowest energy s-orbital is that with $n = 1$, and

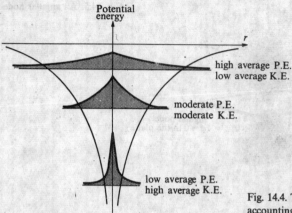

Fig. 14.4. The balance of energies accounting for the structure of the ground state of atomic hydrogen.

is called 1s: When $n = 2$ and $l = 0$ we have the 2s-orbital. This is higher in energy than 1s for two reasons. Inspection of Fig. 14.2 or 14.3 shows that it possesses a radial node, and so there is more curvature, and therefore more kinetic energy, associated with the wavefunction in the radial direction. Furthermore the 2s-orbital spreads further from the nucleus; therefore the potential energy of an electron in it is higher than when it is in a 1s-orbital. Similar remarks may be made for the higher s-orbitals, 3s, 4s, and so on.

When $n = 1$ the only value permitted to l is 0, but when $n = 2$ the orbital angular momentum quantum number may take the values 0 (giving the 2s-orbital) or 1. When $l = 1$ the atomic orbitals are called *p-orbitals*. When $n = 2$, $l = 1$ we have a 2p-orbital. This differs from the 2s-orbital in so far as an electron occupying it possesses orbital angular momentum (of magnitude $\sqrt{2}\hbar$). This angular momentum is a consequence of the presence of the *angular node*, Fig. 14.2, which introduces curvature into the *angular* variation of the wavefunction. The presence of this orbital angular momentum has a profound effect on the radial form of the orbital. Whereas the s-orbitals all have non-zero values at the nucleus, the p-orbitals vanish there. This can be understood as a result of the orbital angular momentum flinging the electron away from the vicinity of the nucleus. The Coulombic force on the electron is proportional to $1/r^2$, but the centrifugal force is proportional to J^2/r^3, and so for sufficiently small distances the latter always overcomes the former so long as the angular momentum is not identically zero. This centrifugal effect also appears in the atomic orbitals with $l = 2$ (which are called *d-orbitals*), $l = 3$ (*f-orbitals*), and the higher orbitals ($g, h, i, \ldots$-*orbitals*). All these orbitals with $l \neq 0$ have zero amplitude at the nucleus, and consequently there is zero probability that the electron will be found there.

The 2p-orbital has no radial node but the 3p-orbital has one. As in the case of the 2s-orbital, this has the effect of increasing the energy of the electron in that state because the increased curvature raises the kinetic energy and the increased spread of the wavefunction samples regions of

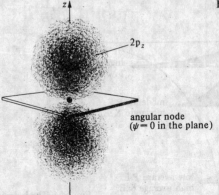

Fig. 14.5. An angular node in a p-orbital.

higher potential energy.

Since $l = 1$ for the p-orbitals, m_l can take the three values $+1, 0, -1$. Different values of m_l correspond to orbitals with different orientations of the orbital angular momentum. The p-orbital with $m_l = 0$ has zero component of angular momentum about the z-axis, and for this reason it is called a p_z-orbital: its form is illustrated in Figs. 14.2 and 14.5. This indicates that the electron collects in pools along the z-axis, as illustrated by the density of shading in the diagram. The xy-plane is a *nodal plane* cutting through the nucleus, and in this plane there is zero probability of finding the electron. The other two p-orbitals may be represented by similar shapes oriented along the x- and the y-axes, Fig. 14.2: they are therefore referred to as p_x- and p_y-orbitals.*

When $n = 3$, l may take the values 0, 1, 2: this gives one 3s-orbital, three 3p-orbitals, and the 3d-orbitals. Of the last there are five distinct orbitals ($m_l = 2, 1, 0, -1, -2$) corresponding to different orientations of the orbital angular momentum with respect to the z-axis. These five orbitals are depicted in Fig. 14.2. All have zero amplitude at the nucleus (because of the centrifugal effect of the orbital angular momentum of magnitude $\sqrt{6}\hbar$). They have no radial nodes, but each possesses two nodal planes which buckle the orbital in an angular direction, and give rise to the orbital angular momentum. The distribution of the electron that occupies these orbitals is depicted by the density of shading in Fig. 14.2.

It will be recalled (p. 441) that the energy of an electron in the hydrogen atom depends on the principal quantum number of the orbital it occupies and is independent of its orbital angular momentum state. This implies that an electron in a 2s-orbital has the same energy as an electron in any of the 2p-orbitals, and an electron in the 3s-orbital has the same energy as one in any of the 3p-orbitals or any of the five 3d-orbitals, and so on.

---

* There is a subtlety here. The orbitals with $m_l = +1$ and $m_l = -1$ correspond to running waves circulating in opposite senses around the z-axis. The p_x- and p_y-orbitals are standing waves, formed by superimposing the running waves. This makes them easier to draw, but to visualize the orbitals with $m_l = \pm 1$, the appropriate combination of the p_x and p_y orbitals has to be taken. (These combinations are the complex functions $p_x \pm ip_y$.)

This degeneracy is unique to hydrogen-like atoms and due to the peculiar nature of the Coulomb potential.

The wavefunction of the ground state can be used to give some estimate of the size of the hydrogen atom. Since the orbital decays exponentially, one view could be that the atom is infinitely big because the amplitude drops to zero only in the limit of r reaching infinity. This point of view is not particularly enlightening, and so we seek some other criterion of size. One assessment of the size could be based on discovering the radius at which the electron is most likely to be found. This can be calculated by the following argument.

The wavefunction tells us the probability of finding the electron in an infinitesimal volume element at each point in space. We can imagine a probe with a volume $d\tau$ and sensitive to electrons moved around inside the atom: since the probability density in the ground state varies as

$$\psi^*\psi = (1/\pi a_0^3)\exp(-2r/a_0),$$

the reading from the detector declines exponentially as the probe is moved outwards along any radius; but, because $\psi^*\psi$ is independent of the latitude and longitude, the detector gives a constant reading as it is moved around at constant radius, Fig. 14.6a. Now consider the probability that the electron can be found anywhere on a spherical shell of thickness dr at the radius r. The sensitive volume of the probe is the whole volume of the shell, which is $4\pi r^2 dr$, and the sensitive volume of this probe gets bigger as r increases (Fig. 14.6b). The probe gets bigger, but

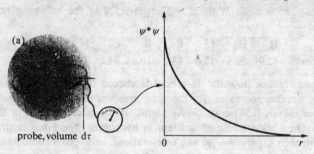

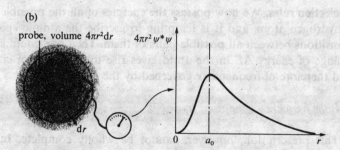

Fig. 14.6. The variation of (a) density and (b) the radial distribution function in the ground state of atomic hydrogen.

the probability density $\psi^*\psi$ gets smaller, and the reading follows the curve shown in Fig. 14.6b. Such a curve is known as the *radial distribution function* (r.d.f.), because it indicates the probability of finding the electron at a selected radius irrespective of the orientation of the point. Thus an electron at r, whatever its latitude or longitude, gives a reading on the detector. The most probable radius at which the electron will be found is therefore the maximum value of the r.d.f.

$$4\pi r^2 \psi^*\psi = (4r^2/a_0^3)\exp(-2r/a_0).$$

This maximum occurs at $r = a_0$, Fig. 14.6b, and so the most probable radius at which the electron will be found in the 1s-orbital is the Bohr radius $a_0 = 53$ pm. When the same technique is applied to the 2s-orbital the radius of greatest probability moves out to $5.2a_0 = 276$ pm, which reflects the puffing up of the atom as its energy increases.

Example (Objective 4). Find the *most probable radius* at which an electron will be found when it occupies a 1s-orbital of an atom with atomic number Z, and tabulate the values for the atoms from H to Ne in the periodic table.

- *Method.* The wavefunction for a 1s-orbital on a nucleus of charge Ze is $(Z^3/\pi a_0^3)^{1/2}\exp(-Zr/a_0)$, Table 14.1. Form the radial distribution function $4\pi r^2\psi^2$ and find by differentiation the value of r for which it is a maximum.

- *Answer.* $4\pi r^2\psi^2 = 4\pi(Z^3/\pi a_0^3)r^2\exp(-2Zr/a_0)$.

  $(d/dr)4\pi r^2\psi^2 = (4\pi Z^3/\pi a_0^3)[2r-(2Z/a_0)r^2]\exp(-2Zr/a_0) = 0$ at $r = r^*$.

  Therefore, $r^* = a_0/Z$. With $a_0 = 52.9$ pm $(0.529$ Å$)$ the following table may be formed:

	H	He	Li	Be	B	C	N	O	F	Ne
r^*/pm	52.9	26.5	17.6	13.2	10.6	8.82	7.56	6.61	5.88	5.29

- *Comment.* Notice how the 1s-orbital is sucked in towards the nucleus as the atomic number increases. By the time uranium is reached the most probable distance is only 0.58 pm (0.058 Å), about 100 times closer than in the hydrogen atom. (On a scale where $r^* \approx 10$ cm in hydrogen, $r^* \approx 1$ mm in uranium.) The electron then experiences strong accelerations, and relativistic effects are important.

The spectral selection rules. We now possess the energies of all the possible states of a hydrogen atom, and it is tempting to ascribe the atomic spectrum to transitions between all possible pairs of them. Then one could say that a change of energy ΔE in the atom gives rise to a photon of energy ΔE, and therefore of frequency ν governed by the

(14.1.6)   *Bohr frequency condition:* $h\nu = \Delta E$.

This prescription, however, cannot be wholly complete. In the last chapter it was stated that a photon possesses angular momentum, its spin being unity. If a photon leaves an atom, for the overall angular momen-

tum to be conserved the electronic angular momentum must change by
an amount that compensates for what is carried away by the spinning
photon. This means, for instance, that an electron in a d-orbital ($l = 2$)
cannot drop down into an s-orbital ($l = 0$) with an emission of a photon
because the emitted photon cannot carry away enough angular momentum. An electron in a d-orbital (henceforth we shall call this a *d-electron*)
can drop into a p-orbital (become a *p-electron*) because l changes by
unity and the balance of the angular momentum can be carried off by the
photon. From this discussion it follows that some transitions are *allowed*
and others are *forbidden*. The statements of which transitions are allowed
are the *selection rules* for the system. In the present case they are

$$\Delta l = \pm 1$$

$$\Delta n = \text{any integer.}$$

We see that the principal quantum number may change by any amount
in the transition because it governs the energy and not the angular
momentum.

These remarks enable Fig. 14.7 to be constructed showing the allowed
transitions and the lines they give rise to in the spectrum. Such a diagram

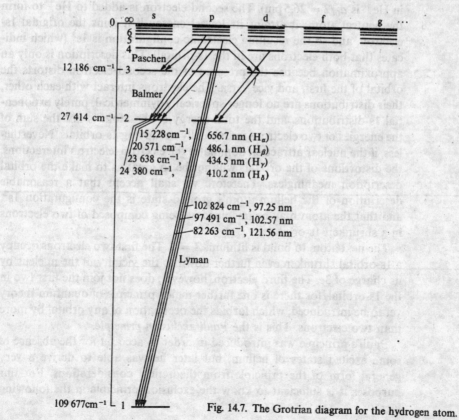

Fig. 14.7. The Grotrian diagram for the hydrogen atom.

is called a *Grotrian diagram*: it summarizes the information we possess on the energy levels, the structure, and the allowed transitions in the hydrogen atom.

14.2 The structure of many-electron atoms

The Schrödinger equations for atoms other than hydrogen are extremely complicated because all the electrons interact with each other. Even in the case of helium no analytical expression for the orbitals and energies can be given, and resort has to be made to numerical solutions obtained with computers. Later in this section we shall indicate the procedure adopted, but for the purpose of understanding the structure of these atoms it is sufficient to adopt a much simpler, qualitative approach which is based on the orbitals already found for the hydrogen atom.

The *Aufbau* principle. The ground state of the hydrogen atom consists of one electron occupying the 1s-orbital: we say that its *electronic configuration* is 1s. The helium atom has two electrons. We may imagine forming the atom by adding first one electron to the bare nucleus (of charge 2e). This electron will occupy a 1s-orbital that differs from the hydrogen atom 1s-orbital only in as much as it is less diffuse because the stronger nuclear charge sucks the electron closer to the nucleus (the most probable radius in He$^+$ is $a_0/2 = 26.5$ pm). The second electron is added to He$^+$ to form the neutral atom: it attains its lowest energy if it joins the original 1s-electron, and so the resulting electronic configuration is 1s^2 (which indicates that both electrons are in the 1s-orbital). This description is only an approximation because the presence of the second electron distorts the orbital of the first, and vice versa. The electrons interact with each other, their distributions are no longer spherically symmetrical, purely exponential 1s-distributions, and the total energy of the atom is not the sum of the energies of two electrons individually occupying 1s-orbitals. Nevertheless, if the nuclear attraction dominates the electron–electron interactions, the distortions of the orbitals will not be so great as to make the orbital description meaningless. Therefore we shall accept that a reasonable description of the helium atom's ground state is the configuration 1s^2, and that the atom may be imagined as being composed of two electrons in a shrunken 1s-orbital.

The next atom to build is lithium, $Z = 3$. The first two electrons occupy a 1s-orbital shrunken even further down to the vicinity of the nucleus by its charge of 3e. The third electron, however, does not join the first two in the 1s-orbital, for there is one further major principle of quantum theory yet to be introduced, which forbids the occupation of any orbital by more than two electrons. This is the *Pauli exclusion principle*.

Pauli's principle was introduced in order to account for the absence of some excited states of helium, but later he was able to derive a very general form of the principle from theoretical considerations. For our purposes it is sufficient to know the exclusion principle in the following form:

No more than two electrons may occupy any orbital, and if two do occupy it their spin directions must be opposed.

This remarkable feature of nature is the key to the structure of complex atoms, to chemical periodicity, and to molecular structure. Opposed electron spins are said to be *paired* or *antiparallel*.

Returning to lithium, the Pauli principle forbids the third electron to enter the 1s-orbital, therefore it enters the next lowest energy orbital. This is one of the orbitals with $n = 2$; but does it enter the 2s-orbital or one of the three 2p-orbitals? In the case of the hydrogen atom we saw that these orbitals are degenerate (have the same energy); but this is not the case in lithium, or any other atom, because the presence of the other electrons (in the 1s-orbitals) destroys the simple Coulombic $1/r$ form of the interaction. Of course all the individual electrostatic interactions remain of the Coulombic $1/r$ form: what is meant by this remark is that the electrostatic potential experienced by any one electron is *non-central* because the species it interacts with are not all at the geometrical centre of the atom (in contract to the situation in the hydrogen atom). We have seen that an s-electron has non-zero probability of being found at the nucleus, whereas a p-electron cannot occur there. This suggests that, because the 2s-electron penetrates through the shielding 1s-electrons, it experiences a more favourable electrostatic potential than the 2p-electron, which does not penetrate through the negatively charged shield, Fig. 14.8. The combined effects of *penetration* and *shielding* result in the separation of the energies of 2s- and 2p-electrons, and the 2s-electrons lie lower in energy (are more strongly bound) than the 2p-electrons. The same remarks apply to electrons of higher principal quantum number, and we can expect the energies of the $n = 3$ shell to lie in the order 3s < 3p < 3d because of the effects of their penetration (or lack of it) through the inner, shielding $n = 1$ and $n = 2$ shells of electrons, Fig. 14.8. Another way of expressing these conclusions is to ascribe different *effective nuclear charges* depending on the orbital occupied by the electron of interest. Thus a sodium 3p-electron would, on account of shielding, experience a lower

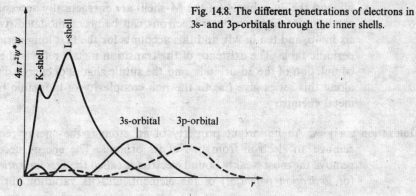

Fig. 14.8. The different penetrations of electrons in 3s- and 3p-orbitals through the inner shells.

effective nuclear charge than a 3s-electron.

We can now state the electronic configuration of the neutral lithium atom. The first two electrons enter the 1s-orbital and do so with opposed spins. The third electron occupies the next lowest orbital, which is 2s. Therefore the configuration is $1s^2 2s$, and the structure can be envisaged as a central nucleus surrounded by a spherical shell of two electrons, and that surrounded by another spherical shell of a single electron.

The generalization of this scheme is known as the *Aufbau principle* (*Aufbau* is German for 'building-up'). In order to construct the electronic configuration of an atom of atomic number Z we envisage the atomic orbitals arranged in the energy sequence $1s < 2s < 2p < 3s < 3p < 3d < \ldots$, and then feed Z electrons into the lowest-energy arrangement compatible with the Pauli principle. The only point to remember is that, although there is only one 1s-orbital, or 2s-orbital, etc., there are three orbitals of type 2p, or 3p, etc., and five of type 3d, 4d, etc., and seven ($m_l = 3, 2, 1, 0, -1, -2, -3$) of type 4f. For example, the carbon atom has $Z = 6$, and so six electrons must be fed into the available orbitals. Two enter 1s, and complete the innermost shell (with $n = 1$; this is known as the *K-shell*). Two enter the 2s-orbitals, and the remaining two enter the slightly higher-energy 2p-orbitals. Its configuration is therefore $1s^2 2s^2 2p^2$. The fluorine atom has $Z = 9$. This is three greater than carbon, and so three more electrons must be injected into the array of atomic orbitals. The 2p-orbitals can accommodate up to six electrons, and so the overall configuration is $1s^2 2s^2 2p^5$. Neon, with $Z = 10$, has one more electron, and therefore its configuration is $1s^2 2s^2 2p^6$. This completes the $n = 2$ shell (also called the *L-shell*) and so the next electron has to enter the 3s-orbital. Sodium, with $Z = 11$, therefore has the structure $1s^2 2s^2 2p^6 3s$, and, like lithium with $1s^2 2s$, consists of a single s-electron outside a completed core of electron shells.

This analysis has brought us to the basis of chemical periodicity. The L-shell is completed by eight electrons, and so the element with $Z = 3$ (Li) should have chemical properties similar to the element with $Z = 11$ (Na). Likewise Be ($Z = 4$) and Mg ($Z = 12$) should be similar, and so on up to the rare gas atoms He ($Z = 2$), Ne ($Z = 10$), Ar ($Z = 18$) which have complete shells of electrons. Above $Z = 18$ the N-shell ($n = 4$) begins to fill, and the 3d-orbitals (of the M-shell) are energetically accessible. As a result, up to eighteen outer electrons can be accommodated (two in 4s, six in 4p, and ten in 3d), and this accounts for the first long period of the periodic table. The existence of the transition metals reflects the gradual completion of the 3d-orbitals, and the subtle shades of energy differences along this series give rise to the rich complexity of inorganic transition metal chemistry.

Ionization energies. An important property of an atom is the energy required to remove an electron from one of its orbitals. The energy necessary to remove the most weakly bound electron is called the *first ionization energy* (or *ionization potential*) of the element, and its variation through the

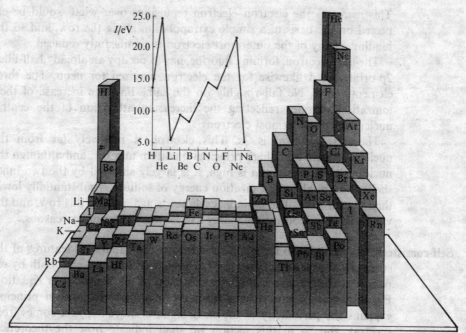

Fig. 14.9. Ionization energies of neutral atoms.

periodic table is illustrated in Fig. 14.9 (numerical values are given in Table 4.9). We can interpret this pattern as follows.

Lithium has a low ionization energy, reflecting the ease with which its outermost electron can be removed. This electron is in a 2s-orbital, which is on the average quite distant from the nucleus. Its binding energy is small even though the nuclear charge is three times that of hydrogen: this is because the 1s-electrons almost succeed in shielding two units of charge and the effective nuclear charge is only $Z_{eff} \approx 1.3$. The next element (beryllium) has two 2s-electrons, but the nuclear charge has risen to $4e$. This suggests they are more strongly bound than in the case of lithium, and this is indeed the case. The ionization energy drops from beryllium to boron, because the outermost electron occupies the less strongly bound 2p-orbital. Carbon has a higher ionization energy because the next electron is also in a 2p-orbital, but the nuclear charge has increased by one unit and the binding is correspondingly greater. Nitrogen has a still higher ionization energy because of the further increase in the nuclear charge.

There follows a kink in the curve which reduces the ionization energy of oxygen below what would be expected by simple extrapolation. The reason is interesting, and it raises a further detail about the *Aufbau* principle. In the case of boron, carbon, and nitrogen there are respectively one, two, and three 2p-electrons. There are three 2p-orbitals, and so in all cases only one electron need occupy each p-orbital. Electrons occupying $2p_x$ and $2p_y$ orbitals are on the average further apart than if both were crammed into one $2p_x$ orbital. When oxygen is reached, its configuration is $2p^4$, and so one 2p-orbital must be doubly occupied.

This increases the electron–electron repulsions over what would be expected on the basis of a simple extrapolation along the row, and so the binding energy of the outermost electron is significantly reduced.

The next electron, forming fluorine, has to occupy an already half-filled 2p-orbital, and likewise for the electron required for neon. The three elements O, F, Ne fall roughly on the same line, the increase of their ionization energies reflecting the increasing attraction of the central nucleus for its outermost electrons.

The next electron is 3s. This electron is relatively far from the nucleus because of its high principal quantum number, and although the nuclear charge of sodium is high, it is largely shielded by the two inner shells. This is why the ionization energy of sodium is substantially lower than that of neon. The periodic cycle starts again along this row, and the variation of the ionization energy can be traced to similar reasons.

Self-consistent field orbitals. So far we have indicated the qualitative features of the structure of many-electron atoms. These ideas can be tested both by experiment and by direct numerical solution of the Schrödinger equation. Experimental support comes from the way that the *Aufbau* principle explains the periodicity of the elements (and more support comes from spectra), and so in this section we shall indicate how the theoretical calculations are performed.

The essential difficulty of solving the Schrödinger equation for a many-electron atom is the complexity arising from all the electron–electron interactions. It is hopeless to expect to find analytical solutions, and so a scheme has been developed for finding numerical solutions on electronic computers. The scheme was introduced first by Hartree and then modified by Fock in order to take into account the impossibility of distinguishing between individual electrons in an atom. In broad outline their method is as follows.

Imagine that we have some rough idea of the structure of the atom. In the sodium atom, for example, the *orbital approximation* suggests that the configuration is $1s^2 2s^2 2p^6 3s$, and the orbitals may be approximated by hydrogen-like atomic orbitals, but sucked in by the higher charge. Now consider the 3s-electron. A Schrödinger equation can be written for this electron by giving it a potential energy due to both the nuclear attraction and the electronic repulsions from the other electrons distributed in their approximate orbitals. This equation is of the form

$$(-\hbar^2/2m_e)\nabla^2\psi_{3s} - (Ze^2/4\pi\varepsilon_0 r)\psi_{3s} + V_{ee}\psi_{3s} = E\psi_{3s},$$

where V_{ee}, which depends on the distribution of all the other electrons, is the repulsion term. This equation may be solved for ψ_{3s}, and the solution obtained will normally be different from the solution guessed originally. The procedure may now be repeated for another orbital; 2p, for example. First the Schrödinger equation for the 2p-electron is written in terms of the kinetic energy, the nuclear attraction, and the electron–electron repulsions, but the last term arises from the electrons distributed in the guessed orbitals, except for the 3s-electron which is distributed in the improved orbital found from the first step. Numerical solution of this equation gives an improved 2p-orbital. The process is repeated for the 2s-orbitals and the 1s-orbitals, each time using the improved orbitals to calculate an expression for the repulsive potential. At the end of the exercise one possesses a set of orbitals which differ from the original guessed set. Then the whole procedure is repeated using the improved orbitals, and a second improved set of orbitals obtained. The cycling continues until the orbitals obtained are insignificantly different from those used at the start of the latest cycle. The orbitals are then self-consistent and are accepted as solutions of the problem.

Some of the self-consistent field (SCF) Hartree–Fock (HF) atomic orbitals (AO) for sodium are illustrated in Fig. 14.10. They differ significantly from the orbitals used for hydrogen, but the general features are similar. These SCF–HF calculations show the grouping of electron density into shells, as was anticipated by the early chemists. Furthermore, the s- and p-orbitals show properties of penetration of the kind mentioned previously, and so the numerical calculations have confirmed the qualitative discussion that lay under the explanation of chemical periodicity.

14.3 The spectra of complex atoms

The spectra of atoms rapidly become very complicated as the atomic number increases, but there are some important features which are moderately simple. The general idea is, as we have seen, very straightforward: lines in the spectrum—either in emission or in absorption—occur when the atom changes its energy by ΔE and emits or absorbs a photon of frequency $v = \Delta E/h$. The change in energy arises because an electron shifts from one orbital to another. It follows that an inspection of the

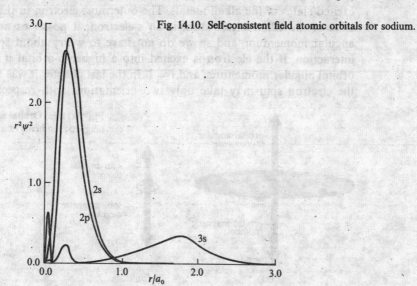

Fig. 14.10. Self-consistent field atomic orbitals for sodium.

spectrum will give information about the energies of electrons in atoms. That, however, is only the primitive approach and several details help (and sometimes hinder) the analysis of a spectrum.

Spin–orbit interaction. An electron has spin, and by virtue of its spin it possesses a magnetic moment. An electron with orbital angular momentum is effectively a circulating current, and so it gives rise to a magnetic field with a strength proportional to its angular momentum, Fig. 14.11. The magnetic moment due to the electron's spin interacts with the magnetic field due to its orbital motion: this is the *spin–orbit interaction*. The energy of the interaction depends on the relative orientation of the spin magnetic moment and the orbital magnetic field. The orientation of the spin magnetic moment is determined by the orientation of the spin angular momentum, and the orientation of the orbital field is determined by the orientation of the orbital angular momentum. The energy of the atom therefore depends on the relative orientation of the spin and orbital angular momenta.

The property that determines the strength of the interaction is the magnitude of the nuclear charge. This can be understood most readily by imagining an observer fixed to the orbiting electron. He will see the nucleus swinging around him; the nucleus is charged, and so he is encompassed by a ring of current. The greater the nuclear charge the stronger the current and the magnetic field to which it gives rise. The effect depends strongly on the atomic number of the nucleus, and while spin–orbit coupling is almost negligible in the hydrogen atom it is large in heavy atoms like sulphur and chlorine (about $380\,\mathrm{cm}^{-1}$ and $580\,\mathrm{cm}^{-1}$ respectively for their 3p-electrons).

It remains to determine the way in which spin–orbit coupling appears in the spectrum, and for simplicity the following remarks are confined to the excited states of the alkali metals. The outermost electron in the ground state of the alkali metal atom is an s-electron; it possesses no orbital angular momentum, and so we do not have to worry about spin–orbit interaction. If the electron is excited into a higher p-orbital it acquires orbital angular momentum, and $l = 1$. In the last chapter it was seen that the electron spin may take only two orientations with respect to any

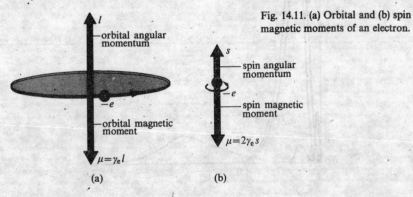

Fig. 14.11. (a) Orbital and (b) spin magnetic moments of an electron.

orbital angular momentum

orbital magnetic moment

$\mu = \gamma_e l$

(a)

spin angular momentum

$-e$

spin magnetic moment

$\mu = 2\gamma_e s$

(b)

chosen axis, and so in this case the spin of the p-electron may be aligned parallel to the orbital momentum, Fig. 14.12a, or opposite (antiparallel) to it, Fig. 14.12b. The former case gives the greater *total* angular momentum for the atom, for the two sources of momentum augment each other.

A detailed study of the properties of angular momentum shows that if an angular momentum with quantum number l (and magnitude $[l(l+1)]^{1/2}\hbar$) is combined with an angular momentum with quantum number s (and magnitude $[s(s+1)]^{1/2}\hbar$) the resulting angular momentum has the magnitude $[j(j+1)]^{1/2}\hbar$, where the *total angular momentum quantum number* may take the values given by the *Clebsch–Gordan* series (Box 13.1):

(14.3.1)

$$j = l + s, l + s - 1, l + s - 2, \ldots |l - s|.$$

The Clebsch–Gordan series comes from an analysis of the way that quantum angular momenta can be combined to give a total angular momentum that is also quantized. If the orbital angular momentum is represented by a vector **l** and the spin by a vector **s**, then their resultant **j** depends on their relative orientations. There is a maximum value (when **l** and **s** lie parallel), a minimum value (when they are antiparallel), and a discrete series of intermediate orientations (when s is greater than $\frac{1}{2}$) which give some intermediate overall momentum, Fig. 14.13. The Clebsch–Gordan series is a rule of thumb for deciding which intermediate values are possible. A simple way of remembering it is to decide whether a triangle can be constructed with sides of length l, s, and j: if it can, that value of j is permitted. This is the *triangle condition*.

Example (Objective 11). Find the values of the total angular momentum that may arise from (a) a d-electron with spin, (b) an f-electron with spin, (c) an s-electron with spin.

- *Method.* Use the Clebsch–Gordan series, eqn (14.3.1), with $s = \frac{1}{2}$ and (a) $l = 2$, (b) $l = 3$, (c) $l = 0$. Decide on the minimum value, $|l - s|$, first.

- *Answer.* (a) $|l - s| = |2 - \frac{1}{2}| = \frac{3}{2}$, hence $j = 2 + \frac{1}{2}$, $2 + \frac{1}{2} - 1$, or $\frac{5}{2}$, $\frac{3}{2}$.
  (b) $|l - s| = |3 - \frac{1}{2}| = \frac{5}{2}$, hence $j = 3 + \frac{1}{2}$, $3 + \frac{1}{2} - 1$, or $\frac{7}{2}$, $\frac{5}{2}$.

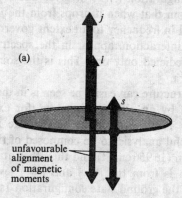

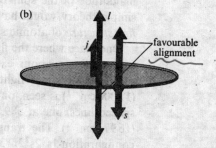

Fig. 14.12. The two levels of different energy for an electron with spin and orbital momentum.

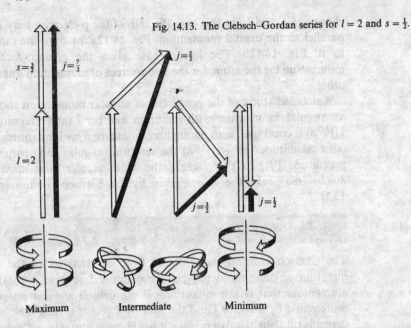

Fig. 14.13. The Clebsch–Gordan series for $l = 2$ and $s = \frac{1}{2}$.

(c) $|l - s| = |0 - \frac{1}{2}| = \frac{1}{2}$, hence $j = 0 + \frac{1}{2}$, or just $\frac{1}{2}$.

- *Comment.* Note that j is always a *positive* number. There is only one value of j when $l = 0$ because the total angular momentum is the same as the spin angular momentum and no question arises as to various relative alignments of orbital and spin momenta.

In the case of a single p-electron, $l = 1$, $s = \frac{1}{2}$, and so the only permitted values of j are $\frac{3}{2}$ and $\frac{1}{2}$. The former corresponds to the previously mentioned state of high momentum, and the latter to the state of low momentum.

In the $j = \frac{3}{2}$ level the spin magnetic moment is in an energetically less favourable orientation with respect to the orbital field than when $j = \frac{1}{2}$. Therefore the energy of the ...np configuration is split into two, Fig. 14.12. When the excited atom drops from the $j = \frac{3}{2}$ level of the ...2p configuration to the ground state configuration (...2s) it gives out a photon of slightly different frequency from that when it drops from the $j = \frac{1}{2}$ level. Therefore two lines, separated in frequency to an extent governed by the magnitude of the spin–orbit interaction, appear in the spectrum where simple theory would have predicted only one. This is the source of the *fine structure* of atomic spectra.

One place where the fine structure can easily be seen is in the emission from sodium vapour excited by an electric discharge (for example, in one kind of street lighting). The characteristic yellow line at 589 nm $(17\,000\ \text{cm}^{-1})$ is seen, on careful analysis, to be composed of two closely spaced lines, one at 589.76 nm $(16\,956\ \text{cm}^{-1})$ and the other at 589.16 nm $(16\,973\ \text{cm}^{-1})$. The transition is from the $j = \frac{3}{2}$ and $j = \frac{1}{2}$ levels of the configuration $1s^2 2s^2 2p^6 3p$ to the ground state configuration $1s^2 2s^2 2p^6 3s$.

Therefore the magnitude of the spin–orbit coupling of a 3p-electron is of the order of $17\,\text{cm}^{-1}$ in sodium. A similar analysis for the other alkali metals gives values that can range from $0.3\,\text{cm}^{-1}$ for Li to $554\,\text{cm}^{-1}$ for Cs, and the increase in magnitude accords with the earlier discussion (p. 456).

Term symbols and selection rules. So far we have used expressions such as 'the $j = \frac{3}{2}$ level of the configuration $1s^2 2s^2 2p^6 3p$'. It would be convenient to have a more succinct way of conveying this information. This is the role of a *term symbol*, which is a symbol looking like $^2P_{3/2}$ or $^3D_2$. The symbol gives three pieces of information:

(1) The *letter* (e.g., P, D in the above examples) indicates the <u>total orbital angular momentum</u> of the atom. If the atom contains only one electron, the orbital momentum is given by the value of l, and just as the orbital nomenclature uses the identification $l = 0, 1, 2, 3, 4, \ldots \leftrightarrow$ s, p, d, f, g, ..., the term symbols use S, P, D, F, G, ... When there is more than one electron the total orbital angular momentum is obtained by combining the orbital angular momenta of the individual electrons by using the Clebsch–Gordan series. Thus if one electron has l_1 and the other has l_2 their combined orbital angular momentum has a magnitude $[L(L+1)]^{1/2}\hbar$, where L is one of the values

(14.3.2)
$$L = l_1 + l_2, l_1 + l_2 - 1, l_1 + l_2 - 2, \ldots |l_1 - l_2|.$$

Thus two p-electrons ($l_1 = 1, l_2 = 1$) can combine to give $L = 2, 1, 0$, and these states are denoted D, P, S respectively. In the term $^2P_{3/2}$, for example, we see that the total orbital angular momentum of the atom corresponds to $L = 1$, and in $^3D_2$ it corresponds to $L = 2$. In the case of a *single electron outside a closed shell* the orbital angular momentum of the whole atom is the same as the orbital momentum of the single outermost electron. This is because the closed shell possesses no net orbital motion. Therefore the configuration $1s^2 2s^2 2p^6 3p$ can give rise only to a P term, and $1s^2 2s^2 2p^6 3s$ only to an S term.

Example (Objective 11). Find the states of total orbital angular momentum that may arise from the combination of the orbital momenta of (a) two d-electrons, (b) one d-electron and an f-electron, (c) three p-electrons. Write the term letter in each case.

● *Method.* Use the Clebsch–Gordan series in each case; start by finding the minimum value of the series. In (c) couple two electrons, and then combine the third with each combined state.

● *Answer.*

(a) Minimum value: $|l_1 - l_2| = |2 - 2| = 0$; then $L = 2 + 2, 2 + 2 - 1, \ldots 0 = 4, 3, 2, 1, 0$ (G, F, D, P, S respectively).

(b) Minimum value: $|l_1 - l_2| = |2 - 3| = 1$; then $L = 3 + 2, 3 + 2 - 1, \ldots 1 = 5, 4, 3, 2, 1$ (H, G, F, D, P, respectively).

p^3

(c) First coupling: minimum value $|l_1 - l_2| = |1 - 1| = 0$, then $L' = 1 + 1$, $1 + 1 - 1, \ldots 0 = 2, 1, 0$. Now couple $l_3 = 1$ with $L' = 2$, to give $L = 3, 2, 1$; couple $l_3 = 1$ with $L' = 1$, to give $L = 2, 1, 0$; couple $l_3 = 1$ with $L' = 0$, to give $L = 1$. The overall result is therefore $L = 3, 2, 2, 1, 1, 1, 0 \,(\text{F}, 2\text{D}, 3\text{P}, \text{S})$.

● *Comment.* For several momenta, the overall coupling can always be broken down into a chain of pairwise couplings.

(2) The *left superscript* in the term symbol (e.g., 2 in $^2\text{P}_{3/2}$) gives *the number of possible orientations of the total spin* of the atom: it is often called the *multiplicity* of the term, but this name has to be used with care, as we see below. For example, if the atom contains a single unpaired electron its total spin is $\frac{1}{2}$ and there are two permitted orientations. Therefore 2 is used in the left-superscript position. If the atom contains several electrons outside a completed shell each one contributes its angular momentum to a total spin. If the total spin is S the number of permitted orientations is $2S + 1$, and so that number appears as the left superscript. If there are two electrons S may be either 1 or 0 (use the Clebsch–Gordan series for the coupling of two momenta $s_1 = \frac{1}{2}$ and $s_2 = \frac{1}{2}$). In the former case $2S + 1 = 3$ (there are three orientations $M_S = 1, 0, -1$) and so the term symbol carries the superscript 3. This is the significance of the 3 in ^3D, which is called a *triplet term*. In the case of $S = 0$ there can be only one value of M_S ($2S + 1 = 1$, corresponding to $M_S = 0$), and so the term is a *singlet*, such as ^1D. A singlet term has zero net spin angular momentum.

(3) The *right subscript* on the term symbol (e.g., the $\frac{3}{2}$ in $^2\text{P}_{3/2}$) indicates the value of the total angular momentum quantum number J, and therefore it also indicates the relative orientation of the spin and orbital momenta (J is large when the momenta are parallel). If there is a single electron outside the closed shell the value of J is the value of j given by the Clebsch–Gordan series, eqn (14.3.1). If there are several electrons things are more complicated, but when the spin–orbit interaction is weak (as in atoms of low atomic number) the *Russell–Saunders coupling scheme* can be used. This is based on the view that if the coupling is weak it is effective only when all the orbital momenta are operating in concert. Therefore the orbital momenta are supposed first to couple into some total L (as described above) and the spins into some overall total S, and only then do these two total momenta interact by the magnetic spin–orbit coupling. The permitted values of J are then given by

(14.3.3)
$$J = L + S, L + S - 1, L + S - 2, \ldots |L - S|.$$

If $S < L$ the number of values of J is $2S + 1$, which is the same as the number of orientations of the spin. Therefore, in this case, the number of values of J is given by the left superscript on the term symbol. In the case of ^3D the three levels are $^3\text{D}_3$, $^3\text{D}_2$, $^3\text{D}_1$. When $S > L$ the Clebsch–Gordan series terminates after only $2L + 1$ values (because J cannot be negative), and so only $2L + 1$ levels are permitted. In the case of ^2S, for instance, there is only the one level $^2S_{1/2}$ even though it is a doublet term.

Example (Objective 12). Write the term symbols for the ground state configurations of Na and F atoms, and the excited configuration $1s^2 2s^2 2p3p$ of carbon.

- *Method.* Write the configurations; then ignore the closed inner cores. Couple the orbital momenta and find L. Couple the spins and find S. Couple L and S and find J. Express the term as $^{2S+1}\{L\}_J$, where $\{L\}$ is the appropriate letter. For fluorine, treat the single gap in $2p^6$ as a single particle.

- *Answer.* For Na the configuration is $1s^2 2s^2 2p^6 3s$, or $(1s^2 2s^2 2p^6)3s$ if the core is marked off. In this configuration $S = s = \frac{1}{2}, L = l = 0$, and so $J = j = \frac{1}{2}$. The term symbol is therefore $^2S_{1/2}$.

  The configuration of F is $1s^2 2s^2 2p^5$. Use the trick of writing this as a closed core plus a hole, $(1s^2 2s^2 2p^6)2p^{-1}$. Then it is also in effect a one-particle problem with $S = s = \frac{1}{2}, L = l = 1$, and so $J = j = \frac{3}{2}, \frac{1}{2}$. The term symbols are therefore $^2P_{1/2}$ and $^2P_{3/2}$.

  The carbon configuration of interest is $(1s^2 2s^2)2p3p$. This is a two-electron problem, with $l_1 = l_2 = 1$, so that $L = 2, 1, 0$; and $s_1 = s_2 = \frac{1}{2}$, so that $S = 1, 0$. Terms are of the form ^3D, ^3P, ^3S when $S = 1$, and ^1D, ^1P, and ^1S when $S = 0$. For ^3D, since $L = 2$, $S = 1$, we have $J = 3, 2, 1$; for ^3P, $J = 2, 1, 0$; for ^3S, $J = 1$ only because $L = 0$. For the singlet terms (with $S = 0$) $J = L$ only, and so only $^1P_1$, $^1D_2$, $^1S_0$ occur. The complete list is therefore $^3D_3$, $^3D_2$, $^3D_1$ (the triplet of *levels* of the ^3D term), $^3P_2$, $^3P_1$, $^3P_0$, $^3S_1$, $^1D_2$, $^1P_1$, $^1S_0$.

- *Comment.* The reason why we treated an *excited* configuration of carbon is that in the ground configuration, $1s^2 2s^2 2p^2$, the Pauli principle forbids some of the states, and deciding which survive (^1D, ^3P, ^1S) is a little complicated.

Any state of the atom and any spectral transition can be specified by writing the complete term symbol. For example, the transitions giving rise to the yellow doublet in sodium can be written $1s^2 2s^2 2p^6 3p\ ^2P_{3/2} \rightarrow 1s^2 2s^2 2p^6 3s\ ^2S_{1/2}$ and $^2P_{1/2} \rightarrow ^2S_{1/2}$. Note that the configuration is not always specified, and that *the upper energy term precedes the lower*. The corresponding absorptions would therefore be ascribed to the transitions $^2P_{3/2} \leftarrow ^2S_{1/2}$ and $^2P_{1/2} \leftarrow ^2S_{1/2}$.

The selection rules for complex atomic spectra can be expressed in terms of the term symbols because they convey information about the angular momentum of the species, and on p. 448 we saw that the conservation of angular momentum governed what transitions the spectrum could show. An emitted or incident photon causes a change of angular momentum corresponding to the unit spin it carries. An analysis results in the following selection rules when Russell–Saunders coupling is valid (for light atoms):

$\Delta S = 0$ (the light does not affect the internal motion of the electrons).

$\Delta L = \pm 1, 0$, with $\Delta l = \pm 1$ (l is the angular momentum of an individual electron; L that of the overall atom).

$\Delta J = \pm 1, 0$ but $J = 0$ cannot go to $J = 0$.

Russell–Saunders coupling fails when the spins and orbital angular momenta of individual electrons couple strongly (giving *jj-coupling*). The quantum numbers S and L then lose their relevance, and so the selection rules fail. For this reason transitions between singlet and triplet states $(S = 0 \leftrightarrow S = 1)$ are permitted in heavy atoms even though they are forbidden in light atoms.

The effect of magnetic fields. Since orbital and spin angular momenta are accompanied by magnetic moments it might be expected that the application of a magnetic field should lead to a modification of an atom's spectrum. This effect, the *Zeeman effect*, has been observed.

The orbital angular momentum of an electron about some z-axis takes the values $m_l \hbar$. The magnetic moment is proportional to the angular momentum, and so the component of magnetic moment on the z-axis may be written $\gamma_e m_l \hbar$, where γ_e is a constant called the *magnetogyric ratio* of an electron. Treating the magnetic moment as arising from the circulation of an electron of charge $-e$ moving with angular momentum $m_l \hbar$ gives the value of γ_e as $-e/2m_e$: the negative sign, reflecting the negative charge of the electron, indicates that the magnetic moment is antiparallel to the angular momentum. Sometimes $e/2m_e$ is combined with $\hbar$, then the positive quantity $\mu_B = e\hbar/2m_e$ is called the *Bohr magneton*. With this notation the possible components of the magnetic moment along the z-axis are

(14.3.4) $$\mu_z = \gamma_e m_l \hbar = -(e\hbar/2m_e)m_l = -\mu_B m_l.$$

If a magnetic field B is applied along the z-axis the energy of the magnetic moment is $-\mu_z B$ (this is a result from standard magnetic theory). Therefore an electron with quantum number m_l is given an extra energy of magnitude

(14.3.5) $$E = -\mu_z B = \mu_B m_l B.$$

If we consider a p-electron, since $l = 1$, m_l may take the three values $+1$, 0, -1. In the absence of a magnetic field these states are degenerate.

The degeneracy is removed by a field and an electron with $m_l = 1$ moves up in energy by an amount $\mu_B B$, that with $m_l = 0$ remains unshifted, and that with $m_l = -1$ moves down by an amount $-\mu_B B$. The Zeeman effect consists of seeing three lines where there was only one in the absence of the field, Fig. 14.14.

The splitting brought about by a magnetic field is very small. For instance, putting in the values of the fundamental constants gives $\mu_B = 9.274 \times 10^{-24} \, \mathrm{J\,T^{-1}}$, or $0.4669 \, \mathrm{cm^{-1}\,T^{-1}}$. Therefore a field of $2\,\mathrm{T}$ ($20\,000$ Gauss) is required to give a splitting of $1\,\mathrm{cm^{-1}}$, which is very small in comparison with the basic optical transitions occurring in the region of $20\,000\,\mathrm{cm^{-1}}$. Confirmation that one is observing the Zeeman effect, however, also comes from a study of the *polarization* of the three lines. Since $\Delta m_l = 1$ corresponds to an opposite change of orbital momentum from $\Delta m_l = -1$ it should not be surprising that the transitions are circularly polarized in opposite senses, Fig. 14.14. The light from the $\Delta m_l = 0$ transition, however, is linearly polarized, and observable only in a different direction, Fig. 14.14.

What has been described so far is the *normal* Zeeman effect. Much more common is the *anomalous* Zeeman effect, which, instead of giving three lines, gives a more complex arrangement. The source of this complexity can be traced to the spin of the electron, and the unusual property that its magnetic moment is *not* given by $\gamma_e m_s \hbar$, as one might expect by analogy with orbital motion, but by $2\gamma_e m_s \hbar$. The extra factor of 2 emerges from a correct relativistic treatment of the problem. The factor is not strictly 2; its exact value is 2.002 319..., and we shall write this as g_e, the *g-factor* of the electron. It follows that an electron with spin orientation m_s has an energy in a magnetic field B given by

(14.3.6)
$$E = -g_e(-e/2m_e)m_s \hbar B = g_e \mu_B m_s B.$$

Now we can see the source of the anomalous Zeeman effect. When there is spin present in an atom we speak in terms of the quantum numbers s, l, and j (for one electron): the total angular momentum is obtained by combining the spin and orbital momenta, as in Fig. 14.15. If the magnetic moments bore the same relation to the angular momentum irrespective of whether they were orbital or spin, the resultant magnetic

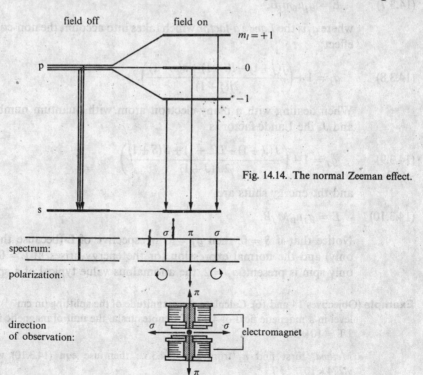

Fig. 14.14. The normal Zeeman effect.

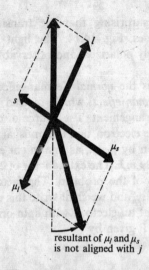

Fig. 14.15. The total magnetic moment of an electron is not aligned along the total angular momentum.

resultant of μ_l and μ_s is not aligned with j

moment would lie along the resultant total angular momentum. Since, however, the spin magnetic moment is anomalous, the total magnetic moment is not collinear with the total angular momentum. Therefore an atom with total angular momentum quantum number j, with an orientation given by m_j, does not have an energy given simply by $\mu_B m_j B$. In fact the energy can be expressed as

(14.3.7) $E = g_j \mu_B m_j B,$

where g_j is the *Landé g-factor* which takes into account the non-collinearity effect:

(14.3.8) $g_j = 1 + \left(\dfrac{j(j+1) - l(l+1) + s(s+1)}{2j(j+1)} \right).$

When dealing with a many-electron atom with quantum numbers L, S, and J, the Landé factor is

(14.3.9) $g_J = 1 + \left(\dfrac{J(J+1) - L(L+1) + S(S+1)}{2J(J+1)} \right).$

and the energy shifts are

(14.3.10) $E = g_J \mu_B M_J B.$

Notice that if $S = 0$, then $g_J = 1$, irrespective of L (because then $J = L$ only) and the normal expression for the energy arises. If $L = 0$, so that only spin is present, $g_J = 2$, the anomalous value typical of a spin.

Example (Objectives 14 and 16). Calculate the magnitude of the splitting (in cm^{-1}) of a $^2P_{3/2}$ level in a magnetic field of 4.0 T. T denotes tesla, the unit of magnetic induction; $1\,T = 10\,kG$.

• *Method.* First find g_J from eqn (14.3.9); then use eqn (14.3.10) with $\mu_B = 9.274 \times 10^{-24}\,J\,T^{-1}$.

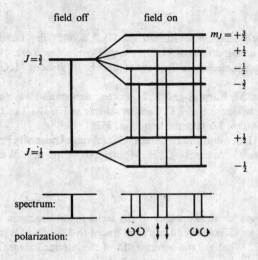

field off   field on

$m_J = +\frac{3}{2}$
$+\frac{1}{2}$
$-\frac{1}{2}$
$-\frac{3}{2}$

$J=\frac{3}{2}$

$+\frac{1}{2}$

$J=\frac{1}{2}$

$-\frac{1}{2}$

spectrum:

polarization:

Fig. 14.16. The anomalous Zeeman effect for the transition $^2D_{3/2} \rightarrow {}^2P_{1/2}$.

- **Answer.** Since $J = \frac{3}{2}$, $L = 1$, $S = \frac{1}{2}$, we have

$$g_{3/2} = 1 + \frac{\frac{3}{2}(\frac{3}{2}+1) - 1(1+1) + \frac{1}{2}(\frac{1}{2}+1)}{2(\frac{3}{2})(\frac{3}{2}+1)} = 1 + \frac{1}{3} = \frac{4}{3}.$$

$$E = g_{3/2}\mu_B M_J B = (\tfrac{4}{3}) \times (9.274 \times 10^{-24}\,\text{J T}^{-1}) \times (4.0\,\text{T})M_J = (4.946 \times 10^{-23}\,\text{J})M_J.$$

Convert to cm^{-1} by dividing by $hc = 1.986\,48 \times 10^{-23}\,\text{J cm}$

$$E/hc = [(4.946 \times 10^{-23}\,\text{J})/(1.986\,48 \times 10^{-23}\,\text{J cm})]M_J = (2.490\,\text{cm}^{-1})M_J.$$

- **Comment.** The $^2P_{3/2}$ level splits up into four states ($M_J = \frac{3}{2}, \frac{1}{2}, -\frac{1}{2}, -\frac{3}{2}$) with a separation $2.490\,\text{cm}^{-1}$. The separation of the two states of the $^2P_{1/2}$ level of the term is only half this value because the spin and orbital moments nearly oppose each other ($g_{1/2} = \frac{2}{3}$).

In order to see how to explain the appearance of anomalous Zeeman spectra, consider the transition $^2D_{3/2} \rightarrow {}^2P_{1/2}$. Since in the upper state $L = 2$, $S = \frac{1}{2}$, $J = \frac{3}{2}$ the Landé factor is $\frac{4}{3}$ and the energies of the four M_J states are $\frac{4}{3}\mu_B M_J B$, $M_J = \frac{3}{2}, \frac{1}{2}, -\frac{1}{2}, -\frac{3}{2}$. In the lower state the quantum numbers are $S = \frac{1}{2}$, $L = 1$, $J = \frac{1}{2}$ and the Landé factor is $\frac{2}{3}$, implying energy shifts of magnitude $\frac{2}{3}\mu_B M_J B$. This is illustrated in Fig. 14.16. Also shown there are the six possible transitions allowed by the selection rule $\Delta M_J = 0, \pm 1$. These lie at six different frequencies, which accounts for the anomalous spectrum. The lines in the spectrum can be disentangled because the polarization of the emitted light differs: with $\Delta M_J = 1$ the light is right-circularly polarized, with $\Delta M_J = -1$ the light is left-circularly polarized, and with $\Delta M_J = 0$ the light is linearly polarized.

When the field is very strong the orbital and spin magnetic moments are uncoupled and align separately with respect to its direction. If a transition then occurs only the orbital angular momentum is affected (because light in the optical region does not affect the spin motion directly). Therefore the spectrum reverts to the normal Zeeman effect with the spin playing no role. This reversion is called the *Paschen–Back effect*.

Further reading

Structure and spectra of atoms. W. G. Richards and P. R. Scott; Wiley, London, 1976.

Atomic spectra and atomic structure. G. Herzberg; Dover, New York, 1944.

Introduction to quantum mechanics. L. Pauling and E. B. Wilson; McGraw-Hill, New York, 1935.

Atoms and molecules. M. Karplus and R. N. Porter; Benjamin, New York, 1970.

Atomic spectra and the vector model. C. Candler; Hilger and Watts, London, 1964.

Atomic spectra. H. Kuhn; Longmans, London, 1962.

Atomic structure. E. U. Condon and H. Odabasi; Cambridge University Press, 1980.

Atomic energy levels. (3 Volumes). C. E. Moore; NBS-Circ. 467, Washington, 1949, 1952, 1958.

Problems

14.1.   The *Humphreys series* is another of the series in the spectrum of atomic hydrogen. It begins at 12 368 nm and has been traced to 3281.4 nm. What are the transitions involved? What are the wavelengths of the intermediate transitions?

14.2.   A series of lines in the spectrum of atomic hydrogen lies at the wavelengths 656.46 nm, 486.27 nm, 434.17 nm, and 410.29 nm. What is the wavelength of the next line in the series? What energy is required to ionize the hydrogen atom when it is in the lower state involved in these transitions?

14.3.   The doubly-ionized lithium ion, Li^{2+}, has only one electron, and its spectrum is expected to resemble hydrogen's. From the data below (which is taken from C. E. Moore, *Atomic energy levels*, Natl. Bur. Stds., Circ. 467 (1949), a rich source of information of this kind) show that the energy levels do have the form K/n^2, and find the value of the constant K. Data: Lyman series at 740 747 cm^{-1}, 877 924 cm^{-1}, 925 933 cm^{-1}.

14.4.   There is enough information in the last Problem for you to predict the wavenumbers of the Balmer series of Li^{2+}. What are the wavenumbers of the two lowest energy transitions of the series?

14.5.   The Rydberg constant for an atom in which the nucleus is assumed to be infinitely heavy (and therefore immobile) is given by eqn (14.1.4) with $m = m_e$. In fact it is necessary to allow for the slight motion of the nucleus. This can be done quite simply, for the Rydberg constant for a nucleus of mass m_N, R_N, is related to the infinite mass constant, R_∞, by $R_N = R_\infty/(1 + m_e/m_N)$. The correction term lets us use spectroscopy to 'weigh' the nucleus. Calculate the mass of the deuteron on the basis that the first line of the Lyman series lies at 82 259.098 cm^{-1} for H, and at 82 281.476 cm^{-1} for D.

14.6.   Positronium consists of an electron and a positron (same mass, opposite charge) orbiting around their common centre of mass. The broad features of the spectrum are therefore expected to resemble those of hydrogen, the differences arising largely from the mass relations. Where will the first three lines of the Balmer series of positronium lie? What is the binding energy in the ground state?

14.7.   One of the most famous of the obsolete theories of the hydrogen atom was proposed by Bohr. It has been displaced by quantum mechanics, but by a remarkable coincidence (not the only one where the Coulomb potential is concerned) the energies it predicts agree exactly with those obtained from the Schrödinger equation. The *Bohr atom* is imagined as an electron circulating about a central nucleus. The Coulombic force of attraction, $Ze^2/4\pi\varepsilon_0 r^2$, is balanced by the centrifugal effect of the circular orbiting motion of the electron. Bohr proposed that the angular momentum was limited to some integral multiple of $\hbar = h/2\pi$. When the two forces are balanced, the atom remains in a 'stationary state' until it makes a spectral transition. Find the energies of the hydrogen-like atom on the basis of this model.

14.8. What features of the Bohr model are untenable in the light of quantum mechanics? How does the Bohr ground state differ from the actual ground state? If numerical agreement is exact, is there no experimental way of eliminating the Bohr model in favour of the quantum mechanical model?

14.9. A hydrogen-like 1s-orbital in an atom of atomic number Z is the exponential function $\psi_{1s}(r) = (Z^3/\pi a_0^3)^{1/2} \exp(-Zr/a_0)$. Form the radial distribution function and derive an expression for the most probable distance of the electron from the nucleus. What is its value in the case of (a) helium, (b) fluorine?

14.10. In 1976 it was mistakenly believed that the first of the 'superheavy' atoms had been discovered in some mica: the atomic number was believed to be 126. What is the most probable distance of the innermost electrons from the nucleus in an atom of this element? (In such elements the Coulombic forces are so strong that relativistic corrections are very significant: ignore them here.)

14.11. What is the magnitude of the angular momentum of an electron that occupies the following orbitals: (a) 1s, (b) 3s, (c) 3d, (d) 2p, (e) 3p? Give the number of radial and angular nodes in each case.

14.12. Is an electron on average further away from the nucleus when it occupies a 2p-orbital in hydrogen than when it occupies a 2s-orbital? What is the most probable distance of an electron in a 3s-orbital from the nucleus? Orbitals are given in Table 14.1.

14.13. Take the exponential 1s-orbital for the ground state of hydrogen and confirm that it satisfies the Schrödinger equation for the atom (consult Box 13.2), and that its energy is $-R_H$. Now modify the nuclear charge from e to Ze. What is the binding energy of the electron in the ion F^{8+}?

14.14. What is the most probable *point* (not radius) at which an electron will be found if it occupies a $2p_z$-orbital on hydrogen?

14.15. What is the *degeneracy* of the level of the hydrogen atom that has the energy (a) $-R_H$, (b) $-R_H/9$, (c) $-R_H/25$?

14.16. Which of the following transitions are allowed in the normal electronic spectrum of an atom: (a) $2s \rightarrow 1s$, (b) $2p \rightarrow 1s$, (c) $3d \rightarrow 2p$, (d) $5d \rightarrow 3s$, (e) $5p \rightarrow 3s$?

14.17. How many electrons can enter the following sets of atomic orbitals: (a) 1s, (b) 3p, (c) 3d, (d) 6g?

14.18. Write the configurations of the first 18 elements of the periodic table ($Z = 1$ to 18). What problems arise with the next 18?

14.19. Ionization energies I may be determined in a variety of ways. Spectroscopy may be used by looking for the energy of excitation at which the line structure is replaced by a continuum. In favourable cases, such as the alkali metals and some ions, I can be determined by extrapolating the gradually converging series of lines. Use the data in Problem 14.3 to find the ionization energy of Li^{2+}.

14.20. A series of lines in the spectrum of neutral lithium (the spectrum known as Li(I)) arise from combinations of the state $1s^2 2p\ ^2P$ with $1s^2 nd\ ^2D$, and occur at 610.36 nm, 460.29 nm, and 413.23 nm. The d-orbitals involved are hydrogen-like. It is known that the $1s^2 2p\ ^2P$ term lies at 670.78 nm above the ground term $1s^2 2s\ ^2S$. What is the ionization potential of the neutral atom in its ground state?

14.21. An alternative method for measuring the ionization energy is to expose the atom to high energy monochromatic radiation, and to measure the kinetic energy or the speed of the electrons it ejects (recall Problem 13.12). When 58.4 nm light from a helium discharge lamp is directed into a sample of krypton, electrons are ejected with a velocity of $1.59 \times 10^6 \, \mathrm{m \, s^{-1}}$. The same radiation releases electrons from rubidium vapour with a speed of $2.45 \times 10^6 \, \mathrm{m \, s^{-1}}$. What are the ionization energies of the two species?

14.22. By how much does the ionization potential of deuterium differ from that of ordinary hydrogen atoms?

14.23. The Clebsch–Gordan series crops up quite often in discussions of atoms and molecules, and it gives a quick way of arriving at the allowed angular momenta of a composite system. As a first example, use it to decide what values of the total angular momentum quantum number j, and the magnitude of the total angular momentum, which a single electron with $l = 3$ may possess.

14.24. Suppose that an electron is part of a molecule which is rotating with an angular momentum corresponding to a quantum number $J_{mol} = 20$. What are the permitted angular momenta of the entire system?

14.25. An object rotates in space with an angular momentum corresponding to the quantum number j_1. Inside the rotating object there is another object rotating with an angular momentum corresponding to the quantum number j_2. The objects may be macroscopic, in which case j_1 and j_2 could be extremely large (as for a tornado on earth) but we shall confine attention to microscopic objects and small momenta. Suppose (a) $j_1 = 5$ and $j_2 = 3$, (b) $j_1 = 3$, $j_2 = 5$: what values may the total angular momentum quantum number take, and what values are permitted to the total angular momenta of the composite system?

14.26. The characteristic emission from potassium atoms when heated is purple and lies at 770 nm. On close inspection the line is seen to be composed of two closely spaced components, one at 766.70 nm and the other at 770.11 nm. Account for this observation, and deduce what quantitative information you can.

14.27. Suppose an atom has (a) two, (b) three, (c) four electrons in different orbitals. What values of the total spin quantum number S may the atom possess? What would be the multiplicity in each case?

14.28. What values of J may arise in the following terms: ^1S, ^2P, ^3P, ^3D, ^2D, ^1D, ^4D? How many states (distinguished by different values of M_J) occur in each level?

14.29. What (electric dipole) transitions are allowed between the terms encountered in the last Problem?

14.30. Write the possible term symbols for the following atomic configurations: (a) Li $(1s^2)2s$, (b) Na $(1s^2 2s^2 2p^6)3p$, (c) Sc $(1s^2 ...)3d$, (d) Br $(1s^2 ...)4p^5$. In each case the configuration in brackets denotes closed inner shells and subshells.

14.31. Calculate the magnetic flux density that is required in the Zeeman experiment in order to produce a splitting of $1 \, \mathrm{cm^{-1}}$ between the states of a ^1P term.

14.32. Calculate the value of the Landé g-factor for (a) a singlet term, (b) a term in which J has its maximum value for $L = 4$ and $S = 2$, (c) a term in which J has its minimum value for $L = 4$ and $S = 2$.

14.33. Calculate the form of the (anomalous) Zeeman effect which occurs when a magnetic field of 4.0 T (40 kG) is applied to an atom in which the transition $^3S \rightarrow {}^3P$ is being monitored.

15 Molecular structure

Learning objectives

After careful study of this chapter you should be able to

(1) State and justify the basis of the *Born–Oppenheimer approximation* (p. 470).

(2) Describe the basis of the *molecular orbital theory* of the hydrogen molecule-ion, and the basis of the *linear combination of atomic orbitals* procedure (p. 471).

(3) Describe the construction of *bonding* and *antibonding orbitals* and explain the source of their effects (p. 473).

(4) Account for the form of a *molecular potential energy curve* (p. 474).

(5) Explain the importance of the *electron pair* in chemical bonding (p. 476).

(6) Construct molecular orbital *energy level diagrams* for diatomic molecules (p. 476).

(7) Describe the structures of the *homonuclear diatomics* on the basis of molecular orbital theory (p. 480).

(8) Define and interpret the *overlap integral* (eqn (15.2.1)) and explain the significance of *orthogonal orbitals* (p. 490).

(9) Explain the notation g, u and Σ, Π, Δ in the *term symbols* for linear molecules (p. 482).

(10) Outline the structure of the *heteronuclear diatomics* (p. 483) and explain the basis of *covalent*, *polar*, and *ionic bonds* (p. 485).

(11) Explain the significance of *electronegativities* and use them to discuss the polarities of bonds (p. 486).

(12) Describe the *hybridization* of orbitals (p. 487) and analyse bond formation in terms of *overlap* and *promotion* (p. 490).

(13) Relate hybridization to *bond angle* (eqn (15.3.1)).

(14) Account for the structures of *polyatomic molecules* in terms of *hybridization, promotion,* and *electron–electron repulsions* (p. 496).

(15) Describe the formation and appearance of sp-, sp^2-, and sp^3-*hybrid orbitals* (p. 494).

(16) Explain the structures of *water, ammonia,* and *methane* on the basis of molecular orbital theory (p. 496).

(17) Describe the structures of *double* and *triple bonds* (p. 499).

(18) Explain the basis of *aromatic stability* (p. 500).

(19) Describe the basis of the *band structure* of metals (p. 502) and account for electrical conductivity (p. 504).

(20) Outline the *crystal field theory* of transition metal complexes and explain the terms *low spin* and *high spin* complexes (p. 504).

(21) Outline the *molecular orbital theory* of transition metal complexes (p. 507).

(22) State the basis of the *valence bond theory* of the chemical bond (p. 509) and write the wavefunctions for the hydrogen molecule and benzene (p. 510).

(23) Describe *ionic–covalent resonance* and *resonance* between Kekulé structures (p. 512).

Introduction

A principal aim of physical chemistry is to understand the shapes of molecules and the strengths of chemical bonds. *Valence* is the study of what governs the strengths of bonds, why atoms form definite numbers of bonds and why these bonds are disposed in a three-dimensional array to give the molecule a definite geometry. An adequate theory of valence has to explain these features both qualitatively and quantitatively. It must also explain why the electron pair plays such an important role in chemistry. We shall see that quantum theory provides a basis for a theory of valence which has become so highly developed that with the use of computers it is now possible to predict with success and accuracy the structure of moderately complex molecules.

In the last chapter the hydrogen atom was taken as the primitive species, and the description of complex atoms was based on the lessons learnt by studying it. The same procedure is adopted here. We use the simplest molecule of all, the hydrogen molecule-ion, H_2^+, to discover the essentials of the structure of molecules, and then go on to discuss more complex and chemically more interesting species.

15.1 The structure of the hydrogen molecule-ion

The hydrogen molecule-ion consists of three particles: two protons and one electron. The protons repel each other but attract the electron. The balance between the kinetic energy and these repulsions and attractions must account for the stability of the species. The exact Schrödinger equation deals with all three particles simultaneously, but the nature of the species suggests a simplification which is known as the *Born–Oppenheimer approximation*.

The Born–Oppenheimer approximation uses the fact that both protons are very much more massive than the single electron (by a factor of nearly 2000). The nuclei therefore move much more sluggishly than the electron, and may be regarded as fixed while the electron moves through the whole volume of the molecule. Explicit calculation shows that the nuclei move only about 1 mm while the electron speeds through a distance of about 1 m, and the error of assuming that the nuclei are pinned down to a fixed distance is very small. In molecules other than H_2^+ the nuclei are even heavier, and the approximation is usually even more reasonable. The Born–Oppenheimer approximation therefore implies that the motion of the nuclei may be neglected, and that it is sufficient to consider the motion and distribution of the electrons for static nuclear positions. This has the effect of reducing the H_2^+ problem to a single-particle Schrödinger equation for the electron in the electrostatic field of two stationary protons.

The potential energy of the electron in the field of two protons is

$$(15.1.1) \qquad V = -(e^2/4\pi\varepsilon_0)\left(\frac{1}{r_A} + \frac{1}{r_B}\right)$$

where r_A and r_B are its distances from nuclei A and B. This expression may be inserted in the Schrödinger equation

$$(-\hbar^2/2m_e)\nabla^2\psi + V\psi = E\psi$$

and exact solutions obtained. Such a procedure, however, does not give a great deal of insight into the form of the orbitals and energies, and so we shall adopt a more approximate approach.

Molecular orbital theory of H_2^+. When the electron is very close to nucleus A the term $1/r_A$ is much bigger than $1/r_B$ and so the potential energy of the electron reduces from the V given in eqn (15.1.1) to $-(e^2/4\pi\varepsilon_0)(1/r_A)$. The Schrödinger equation then becomes that for a single hydrogen atom, and its ground state is described by the orbital ψ_{1s} centred on nucleus A: we denote this $1s_A$. On the other hand, if the electron is close to nucleus B the potential energy reduces from the full V to a term proportional to $1/r_B$, and the Schrödinger equation is that for hydrogen atom B, and its ground state is an electron in $1s_B$. On this basis we assume that the overall distribution of the electron can be described by the wavefunction

$$(15.1.2) \qquad \psi \approx 1s_A + 1s_B.$$

This fits the preceding discussion, because when the electron is close to A its distance from B is large, the amplitude of the function $1s_B$ is small, and therefore the wavefunction is just $1s_A$ (see Fig. 15.1). Conversely, when the electron is close to B it is far from A, the amplitude of $1s_A$ at that point is small, and the wavefunction there is almost pure $1s_B$.

The wavefunction in eqn (15.1.2) is like an atomic orbital, but spreads over the whole molecule, and so it is called a *molecular orbital*. Since the molecular orbital (m.o.) is formed by the addition of two atomic orbitals according to eqn (15.1.2) the approximation is known as the *linear combination of atomic orbitals* or *l.c.a.o.* procedure. Note that the exact solution also gives molecular orbitals spreading through the whole nuclear framework of the molecule, and writing them as an l.c.a.o. is only an approximation.

The m.o. given in eqn (15.1.2) determines the distribution of the electron. According to the Born interpretation (p. 470) the probability of discovering the electron in a volume $d\tau$ at the point $\mathbf{r}$ is proportional to $\psi^*(\mathbf{r})\psi(\mathbf{r})d\tau$, and so in the present case the probability is proportional to

$$(1s_A + 1s_B)^2 \, d\tau = (1s_A)^2 \, d\tau + (1s_B)^2 \, d\tau + 2(1s_A)(1s_B) \, d\tau.$$

(Remember that $1s_A$ is an abbreviation for $\psi_{1s}(\mathbf{r})$, with $\mathbf{r}$ measured from A; we are not troubling about normalization factors at this stage.) This distribution can be analysed as follows.

In the region close to A the amplitude of $1s_B$ is small and the probability of finding the electron is dominated by $(1s_A)^2$: it then resembles an isolated hydrogen atom, and the probability falls off exponentially with distance. In the region close to B the probability is similarly dominated by $(1s_B)^2$. The very important feature of the molecular orbital appears when we investigate the probability of finding the electron in the region between the nuclei, where both atomic orbitals have about the same amplitude. According to the expression quoted above, the probability of finding the electron there is the sum of the probability that it would be there if it were on the hydrogen atom A (this is given by the value of $(1s_A)^2$ in that region), plus the probability that it would be there if it were on the hydrogen atom B (given by $(1s_B)^2$), plus—and this is the important point—an *extra* contribution to the probability proportional to $2(1s_A)(1s_B) \, d\tau$ which arises from the third term in $\psi^2 \, d\tau$. Therefore, the probability of finding the electron in the internuclear region is enhanced above what one would expect if one simply had a hydrogen atom at the same distance from that point, Fig. 15.1.

The reason for this behaviour is the constructive interference of the two wavefunctions. Each has a positive amplitude in the internuclear region, and so the total amplitude is augmented there. We shall constantly discover that the *electrons accumulate in the regions where atomic orbitals overlap and interfere constructively*.

The following explanation of the strength of bonds is often given. The

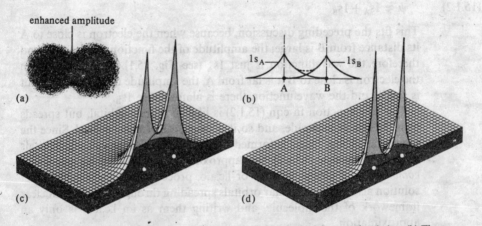

Fig. 15.1. (a) The orbital overlap responsible for bonding in the hydrogen molecule ion. (b) The constructive interference in the internuclear region. (c) The orbital amplitude in a plane containing the two nuclei. (d) The probability density. Note the internuclear electron density. ((b) is a side view of (c).)

accumulation of electron density in the internuclear region puts the electron predominantly into a position where it can interact favourably with both positive nuclei. This lowers the energy of the molecule and accounts for its stability.

Unfortunately this neat explanation is probably incorrect in the case of H_2^+ (at least). This is because shifting the electrons away from the nuclei and depositing them in the internuclear region *raises* their potential energy. The modern explanation is more subtle and does not emerge clearly from the simple l.c.a.o. treatment given here. The broad feature of the explanation is that at the same time as electrons are shifted into the internuclear region the atomic orbitals shrink closer to their respective nuclei. This improves the electron–nuclear attraction to an extent greater than its loss in filling up the internuclear region, and so there is a significant lowering of potential energy. The kinetic energy of the electron is also modified by these changes, but the net effect is that the improved electron–nuclear attraction dominates and is the main reason for the lowering of the energy of the electron in the molecule, and therefore for the strength of the bond in H_2^+.

We have to get this explanation into perspective. Many accounts of chemical bonding ascribe the stabilization to the improvement of the electron–nuclear attraction because the electrons are accumulated in the 'bonding region' between the nuclei. Explicit calculations on H_2^+ suggest that this is untenable, but it may still be true of more complex molecules for which the same, detailed calculations have not yet been made. When electrons are accumulated between nuclei, they are generally found to be bonding electrons. Therefore, throughout the following discussion *we ascribe the strength of chemical bonds to the accumulation of electron density in the bonding region*, and leave open the question whether in molecules more complex than H_2^+ the strength arises from the shrinking of the orbitals that this allows, or from a simple change in the nucleus–electron interaction because of the density in the internuclear region.

Bonding and antibonding orbitals. The orbital $1s_A + 1s_B$ is formed by the constructive interference of the two atomic orbitals. An electron that occupies it helps to bind the two nuclei together, but there are two reasons why it cannot draw them into contact. First, at very small internuclear separations there is not enough room to accommodate the electron between the nuclei, and so its binding effect is reduced. Second, the nucleus–nucleus repulsion increases as their separation is reduced, and at small separations it dominates any binding influence of the electron. The total energy of the molecule can be calculated at all internuclear separations, and the resulting *molecular potential energy curve* has the form shown in Fig. 15.2. The energy decreases initially with R, on account of the redistribution of the electrons into the internuclear region, and then increases for the reasons just mentioned. The stable conformation of the molecule corresponds to the minimum of the curve: R_e is called the *equilibrium bond length*. Since the molecular orbital $1s_A + 1s_B$ is responsible for the bonding it is termed

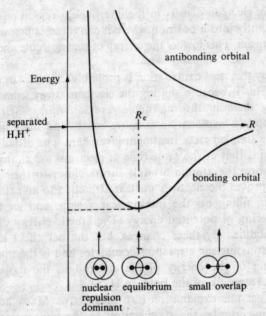

Fig. 15.2. Molecular potential energy curve for the hydrogen molecule ion. (The antibonding curve is discussed later.)

a *bonding orbital*. (In general, a bonding m.o. is one which, if occupied, contributes to a lowering of the energy of the molecule.) Calculations based on this model give $R_e = 130$ pm (1.30 Å; note that 100 pm or 0.10 nm corresponds to 1.00 Å). The lowering of the molecule's energy below that of the infinitely separated H plus H^+ species (which is approximately the dissociation energy of H_2^+) is calculated as 0.9 eV, or 87 kJ mol^{-1}. The experimental values are 106 pm and 2.6 eV respectively, and so this crude calculation, while inaccurate, is not absurdly wrong.

Another molecular orbital of H_2, the next-higher exact solution of the Schrödinger equation, can be modelled by combining the two 1s-orbitals with opposite sign, $1s_A - 1s_B$. In this case the atomic orbitals interfere destructively where they overlap. As a consequence, an electron that occupies $1s_A - 1s_B$ is excluded from the internuclear region. This can be understood on the basis of the probability distribution, which is proportional to

$$(1s_A - 1s_B)^2 \, d\tau = (1s_A)^2 \, d\tau + (1s_B)^2 \, d\tau - 2(1s_A)(1s_B) \, d\tau.$$

The crucial difference between this and the bonding distribution arises from the third term, which affects the electron density in the internuclear region most strongly (because $1s_A$ and $1s_B$ both have appreciable amplitudes there, and their product is significant). It *subtracts* electron density, and therefore reduces the electron density in the internuclear region, Fig. 15.3. In fact, on the plane passing through the mid-point of the bond (where $1s_A$ and $1s_B$ have equal amplitudes) destructive interference is complete and a node appears in the orbital.

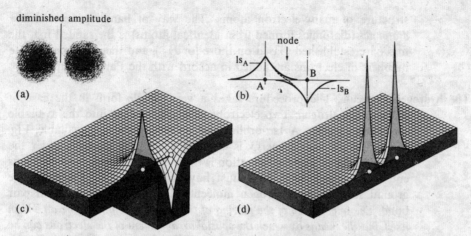

Fig. 15.3. (a) The orbital overlap responsible for antibonding in the hydrogen molecule ion. (b) The destructive interference in the internuclear region. (c) The orbital amplitude in a plane containing the two nuclei. (d) The probability density. Note the internuclear node. ((c) is a side view of (b).)

An electron that occupies $1s_A - 1s_B$ tends to draw the nuclei apart. This is partly due to the fact that if it is excluded from the internuclear region then it is located more strongly outside, and so exerts a force that tends to stretch the molecule (there are also kinetic energy contributions). Because of its disrupting contribution, the m.o. $1s_A - 1s_B$ is called an *antibonding orbital*. (In general, an antibonding orbital is one which, if occupied, contributes to an increase of the total molecular energy). The dependence of its energy on the internuclear separation is shown by the upper curve in Fig. 15.2.

A representation of the energies of the bonding and antibonding orbitals at the equilibrium separation R_e is shown in Fig. 15.4. This is termed a *molecular orbital energy level diagram*. The label σ denotes an orbital of cylindrical symmetry about the internuclear axis (it is the analogue of s in atoms); the star on σ^* denotes an antibonding orbital.

15.2 The structure of diatomics

In the last chapter we saw that the hydrogen atomic orbitals together with the *aufbau* principle provided a basis for the discussion of the

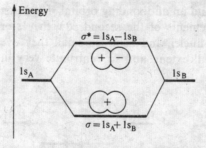

Fig. 15.4. Molecular orbital energy level diagram for orbitals formed from 1s-orbitals.

structure of many-electron atoms. The way of handling *homonuclear diatomics* (diatomics formed from identical atoms) is the same. First, the m.o.s are established (based on those for H_2^+), and then the appropriate numbers of electrons are added in accord with the Pauli principle.

The hydrogen molecule. The procedure takes a very simple form in the case of molecular hydrogen. Two electrons must be injected into the available molecular orbitals. A $1s\sigma$-orbital and a $1s\sigma^*$-orbital are available, Fig. 15.5a, and the lowest energy is attained when both electrons enter the bonding orbital. The configuration is therefore $1s\sigma^2$. Since both electrons enter the same orbital they must have opposed spins. Therefore, the ground state of the hydrogen molecule is formed by an electron pair bond. *The importance of the pairing of the spins is seen not to be an end in itself, but the means by which the spatial arrangement of the electrons can be made most favourable.* This is the basis of the importance of electron pairs throughout bonding theory.

The same argument can be extended to show why diatomic helium He_2 is unstable. The molecular orbitals are formed from the overlap of the two helium 1s-orbitals, and are $1s\sigma$ and $1s\sigma^*$. The molecule possesses four electrons; the first two enter $1s\sigma$, and must do so with paired spins. The next electron cannot join them because of the Pauli principle, and so enters the antibonding $1s\sigma^*$-orbital, Fig. 15.5b. This tends to weaken the bonding due to the first pair. The fourth electron joins it with opposed spin. At this stage the molecule has one complete bond, and one complete antibond. Their effects cancel (in fact the antibond wins slightly over the bond) and the molecule He_2 is less stable than the two separate atoms. When one of the helium atoms has been excited into an upper state (for example, when a 1s-electron has been promoted to 2s) the argument just presented fails because the molecule formed has a configuration $1s\sigma^2 1s\sigma^* 2s\sigma$, where $2s\sigma$ is an orbital formed from the 2s-orbitals. The antibonding effect of $1s\sigma^*$ is insufficient to overcome the bonding effect of $1s\sigma^2$, and so the molecule He...He* survives until it discards its energy by radiation or collision. These weakly bonded excited state dimers (excimers) of the noble gases have been detected.

The homonuclear molecular orbitals. We have seen that two 1s-orbitals can be combined to give a bonding and an antibonding orbital. The same is true of two 2s-orbitals. In the elements of the second row the 1s-orbitals are sucked very close to the nuclei, and being so small they overlap their neighbours to only a small extent, and so contribute very little to the

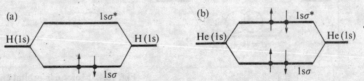

Fig. 15.5. Electronic structures of (a) H_2 and (b) He_2.

bonding energy. The 2s-orbitals are large and can overlap their neighbours significantly. Therefore the molecular orbitals they form play a significant role in the energies of the second row diatomics. (The 3s-orbitals play this role in the diatomics of the third row.)

In the second row diatomics we also have to allow for the overlap of the 2p-orbitals. This introduces a new feature.

Consider first the overlap of two $2p_z$-orbitals, which are the orbitals directed along the internuclear axis. These overlap very strongly, and may do so in a way that leads to constructive interference and a strong accumulation of electron density between the nuclei, Fig. 15.6a. They may also interfere destructively, giving rise to a nodal plane bisecting the internuclear distance. The first gives a strongly bonding m.o.; the second a strongly antibonding m.o. Both have cylindrical symmetry about the

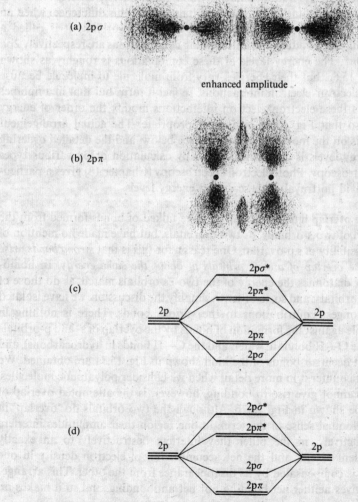

Fig. 15.6. Formation of (a) a 2pσ-orbital and (b) a 2pπ-orbital; (c) approximate order of energies; (d) order of energies often found.

internuclear axis, and so both are σ-orbitals. In order to distinguish them from the $2s\sigma$-orbitals they are labelled $2p\sigma$ and $2p\sigma^*$.

The $2p_x$- and $2p_y$-atomic orbitals have their regions of maximum amplitude off the axis; nevertheless, two $2p_x$-orbitals on different atoms may overlap to give constructive interference and enhanced electron density in the general region of the internuclear axis; but it is off-axis, and away from the prime bonding region. They may also overlap in a way that leads to destructive interference, and if they do so the electron density is eliminated from these off-axis positions. The former arrangement gives a bonding m.o., and the latter an antibonding m.o., but both can be expected to be weaker than in the σ case because the accumulation of electron density lies away from the optimum positions. Furthermore, the electron density they represent is not cylindrically symmetrical, Fig. 15.6b, and so the distribution is both qualitatively and quantitatively different from the σ-orbital. The notation for these orbitals reflects this difference: when an m.o. is formed from the broadside overlap of p-orbitals it is called a *π-orbital*. The bonding and antibonding combinations are respectively $2p\pi$ and $2p\pi^*$. The energy levels of these combinations is roughly as shown in Fig. 15.6c, but their energies vary from molecule to molecule because of the electron–electron interactions. In fact it turns out that in a number of cases these electron–electron interactions modify the order of energy levels so that Fig. 15.6d is more appropriate. The actual arrangement depends on the molecule, as we shall see below, and the detailed ordering of energy levels is ascertained either by calculation or by various types of spectroscopy. Photoelectron spectroscopy (Chapter 18) gives a particularly vivid portrayal of these deeper energy levels.

s,p-overlap and the overlap integral. So far we have talked of bonds formed from the overlap of two s-orbitals, or two p-orbitals, but have made no mention of the possibility of s,p-overlap. One reason for this is that *strong bonds arise from the overlap of atomic orbitals of about the same energy*. In homonuclear diatomics the energy of the two 2s-orbitals match, as do those of the 2p-orbitals; and so, in order to simplify the discussion, we have isolated the strongest contributions to the various bonds. There is nothing in principle against the formation of bonds by the overlap of s- and p-orbitals (e.g., the O—H bonds in water, and the C—H bonds in hydrocarbons), and when it occurs σ-bonds of the kind shown in Fig. 15.7a are obtained. We shall encounter it in more detail when we consider polyatomic molecules. What cannot give rise to bonding, however, is the attempted overlap of the type shown in Fig. 15.7b. Although the two orbitals do 'overlap' in the colloquial sense of the term, in one region their amplitudes interfere constructively, in the other they interfere destructively to an exactly equivalent extent, and the net accumulation of electron density in one region is compensated by its disappearance from the other. This arrangement gives neither net bonding nor net antibonding, and so it makes no contribution to the bond strengths.

The strength of any bond depends on the degree to which the orbitals

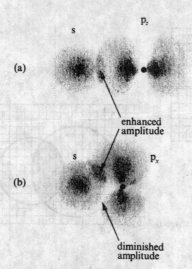

Fig. 15.7. Overlapping of s- and p-orbitals.

overlap each other. A measure of this is given by the *overlap integral S*, which is defined as follows:

(15.2.1)
$$S = \int \psi_A(\mathbf{r})\psi_B(\mathbf{r})\,d\tau.$$

If the atomic orbital on A, ψ_A, is small wherever that on B, ψ_B, is large, and vice versa, the product of their amplitudes is everywhere small and the integral—the sum of these products—is small. If ψ_A and ψ_B are large in the same region of space their product is large at points in this region, and the integral S may be large. This is illustrated for the overlap of two 1s-orbitals in Fig. 15.8a and b.

Consider now the overlap integral for the arrangement in Fig. 15.8c where a 1s-orbital overlaps a 2p-orbital broadside on. At some point $\mathbf{r}_1$ the product $\psi_{1s}(\mathbf{r}_1)\psi_{2p}(\mathbf{r}_1)$ may be appreciable. There exists a symmetrical point $\mathbf{r}'_1$, however, where the product has exactly the same magnitude, but is opposite in sign. When the integral is taken these two contributions are added together, and sum to zero. For every point in the upper half of the diagram there exists a point that cancels in the lower half. Therefore the integral S is zero. The vanishing of the overlap in this case fits our discussion about the vanishing of any net bonding effect for this arrangement of orbitals. Orbitals that have $S = 0$ are said to be *orthogonal*.

The magnitude of the overlap integral is a good guide to the strength of bonds formed from various orbitals, and typical values for effective bonds lie in the range $S = 0.2$–0.3.

Example (Objective 8). Calculate the magnitude of the overlap between the two hydrogen 1s-orbitals in the hydrogen molecule.

● *Method.* Tables of overlap integrals give $S = [1 + (R/a_0) + \frac{1}{3}(R^2/a_0^2)]\exp(-R/a_0)$: simply substitute the equilibrium bond length $R = 74.16$ pm and use $a_0 = 52.92$ pm.

● *Answer.* $S = (1 + 1.401 + 0.655) \times (0.246) = 0.753$.

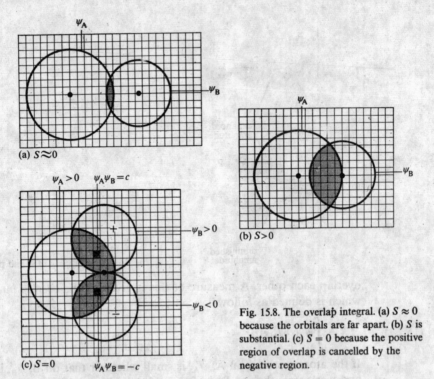

Fig. 15.8. The overlap integral. (a) $S \approx 0$ because the orbitals are far apart. (b) S is substantial. (c) $S = 0$ because the positive region of overlap is cancelled by the negative region.

● *Comment.* This overlap integral is larger than typical ones between bonded atoms: two hydrogen atoms are very close together in H_2.

The configurations of the diatomics. The energies of the atomic orbitals are arrayed as depicted on the left and right of Fig. 15.9. With these energies established it is a straightforward task to estimate the positions of all the molecular orbitals that may be formed, and these are denoted by the horizontal lines at the centre of the diagram. It must be emphasized that the energy levels are entirely qualitative, and in accurate work they must be calculated by a modification of the self-consistent field calculation described in the preceding chapter. Nevertheless the orbitals give a basis for a discussion of diatomics, as we shall now see.

In order to arrive at the electronic configuration of any diatomic, the appropriate number of electrons are fed into the molecular orbitals established in Fig. 15.9. For neutral diatomics composed of atoms of atomic number Z, $2Z$ electrons are to be inserted.

The procedure can be illustrated by N_2, which has 14 electrons. The first electron enters the lowest m.o.; this is $1s\sigma$. It can be joined by the second if the spins pair. The next two cannot join them, and so enter the next lowest orbital: this is $1s\sigma^*$. The next pair enter $2s\sigma$, the next enter $2s\sigma^*$, and the next enter $2p\pi$. The $2p\pi$-level is really a pair of orbitals, one formed from $2p_x$-overlap, and the other from $2p_y$-overlap. Four electrons may enter this set, a pair in each, and so 12 electrons have been accommodated to this point. The last two enter $2p\sigma$. The configuration of

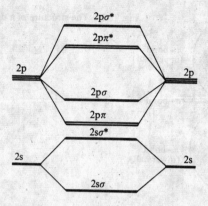

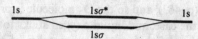

Fig. 15.9. Molecular orbital energies for the second row homonuclear diatomics.

the nitrogen molecule is therefore $1s\sigma^2 1s\sigma^{*2} 2s\sigma^2 2s\sigma^{*2} 2p\pi^4 2p\sigma^2$.

The implication of this configuration is as follows. We can count the number of bonds holding the atoms together. The $1s\sigma$-orbital is full and counts as one complete (but very weak) bond; but its effect is cancelled by the full $1s\sigma^*$ antibond. Likewise the complete $2s\sigma$ bond and $2s\sigma^*$ antibond cancel. The $2p\sigma$ bond, which is complete, is not cancelled because the corresponding antibond is unoccupied; likewise for the two $2p\pi$-bonds. Therefore the net number of bonds is three. This accords very well with the chemical view of the bonding in molecular nitrogen, that it is a triply bonded species, which in the older literature was denoted N≡N.

The O_2 molecule ($2Z = 16$) has two more electrons, which must occupy the antibonding $2p\pi^*$-orbital. Its configuration is therefore $\dots 2p\pi^4 2p\sigma^2 2p\pi^{*2}$, and the presence of one more complete antibond means that the net number of bonds is two. This accords with the classical view that oxygen is doubly bonded and written O=O. Note that this double bond is a single σ-bond plus a single π-bond: this is a common structure of double bonds, Fig. 15.10.

The molecular orbital description of O_2 has a further success. This emerges when we ask which of the $2p\pi^*$-orbitals the electrons occupy. Energetic considerations suggest that $2p_x\pi^* 2p_y\pi^*$ is a better arrangement than $(2p_x\pi^*)^2$ because the electrons are then further apart and repel each other less. But since the two electrons then occupy different space orbitals, the Pauli principle no longer constrains them to have opposed spins. It turns out that *electrons with parallel spins have a lower energy than a corresponding pair with opposed spins* (this is *Hund's rule**), and so O_2 is

---

* The explanation of Hund's rule has not yet been fully settled, but the underlying reason is that electrons with the same spin orientation have an intrinsic tendency to avoid each other. This can be demonstrated by letting electron 1 occupy an orbital ψ_a and electron 2 occupy ψ_b; the wave function has then to be written $\psi_a(r_1)\psi_b(r_2) - \psi_a(r_2)\psi_b(r_1)$ if it is to be

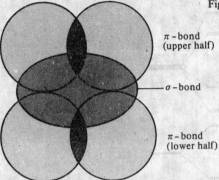

Fig. 15.10. The structure of a double bond.

π-bond
(upper half)

σ-bond

π-bond
(lower half)

predicted to have a pair of electrons with the same spin, and therefore with a resultant spin angular momentum. We have seen that spin gives rise to a magnetic moment (p. 463), and so the O_2 molecule is predicted to be magnetic. This is in fact the case, and liquid oxygen sticks to a magnet. This is excellent confirmation of the general correctness of the simple molecular orbital description of its structure.

Molecular fluorine, $2Z = 18$, has a configuration $\ldots 2p\pi^{*4}$ and the extra two electrons, entering the antibonding orbital, reduce to one the net number of bonds. Therefore F_2 is a singly-bonded species, in accord with classical chemical views, and its bond strength is weak (its dissociation energy is about $154\,\mathrm{kJ\,mol^{-1}}$; that of N_2 is $942\,\mathrm{kJ\,mol^{-1}}$). Molecular neon $2Z = 20$, has two further electrons: these enter the $2p\sigma^*$-orbital and destroy the remaining bonding character. This fully accords with the monatomic nature of ground state neon and its companions in Group 0 of the periodic table.

Another piece of information that comes directly from the bonding diagram is whether ionic species (e.g., N_2^+ and N_2^-) are more or less strongly bonded than the neutral molecules. This is taken up in the Problems.

More about notation. In the case of homonuclear diatomics it is common to see the subscripts g or u attached to the labels σ and π. These denote the *parity* of the orbital in the following way. Consider any point in a homonuclear molecule, and note whether the amplitude of an orbital there is positive or negative. Now travel through the centre of the molecule and note the orbital's amplitude at the point an equal distance on the other side, Fig. 15.11. If the amplitude is the same we use the symbol g (from *gerade*, the German for even). If the amplitude has an opposite sign we apply the symbol u (from *ungerade*, odd). Inspection of Fig. 15.11 shows that a σ-orbital is g if it is bonding, but u if it is antibonding. A π-orbital, however, is u if it is bonding, but g if it is antibonding. This is more than an alternative to adding * to denote antibonding, because the transitions that give rise

---

quantum mechanically acceptable. If the electrons were to approach each other the point r_1 would approach the point r_2, and when $r_1 = r_2$ the wavefunction vanishes because the two parts cancel. This avoiding tendency is called *spin correlation*.

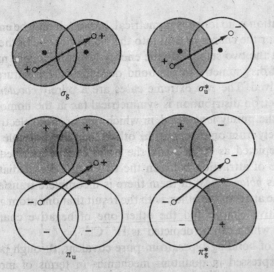

Fig. 15.11. Inversion symmetry and the g,u classification.

σ_g σ_u^* π_u π_g^*

to the optical absorption spectra of molecules have selection rules that may be expressed in terms of g and u (see p. 608). In the case of a many electron molecule, note the g or u character of every electron, and form the product of the parities using $g \times g = g$, $g \times u = u$, $u \times u = g$. Thus any closed shell species (such as H_2 and N_2) is of g parity overall.

Another classification is the analogue of S, P, D,... for many-electron atoms. In the case of linear molecules, the electronic orbital angular momentum about the internuclear axis is well-defined and forms the basis of the scheme. A single electron in a σ-orbital (or a σ^*-orbital) has no orbital angular momentum: the σ-orbital is cylindrically symmetrical and has no angular nodes. The *term symbol* for the H_2^+ and the H_2 molecules is therefore Σ (the analogue of S). The same term symbol applies to all closed-shell diatomics, because a closed shell has no net orbital angular momentum. The ground states of the N_2 and F_2 molecules are therefore also denoted Σ (or, more fully, $^1\Sigma_g$). In contrast to σ-electrons, a π-electron does have an angular node, and so it has one unit of angular momentum around the internuclear axis. If a molecule has a single π-electron outside a closed shell, its state is classified as Π (the analogue of P). If there are two π-electrons (as in O_2) the term symbol may be either Σ, if the electrons are orbiting in opposite directions, or Δ (corresponding to a net orbital momentum of 2 units) if the orbital momenta are aligned. In the case of O_2 it is known that Σ has the lower energy. Since we have already seen that the two π-electrons are unpaired, their spin is $S = 1$, indicating that the ground state is a triplet. Hence it is denoted $^3\Sigma$. Since the parity of π^* is g, and there are two such electrons outside a close shell, the overall parity is also g. Consequently the oxygen ground state is $^3\Sigma_g$.

The structure of heteronuclear diatomics. A *heteronuclear diatomic* is a molecule of the form AB: chemically interesting species include CO and HCl. There are two consequences of the presence of different atoms: one is that the

electron distribution is no longer symmetrical, because it may be energetically favourable for the charge to drift into the vicinity of one of the atoms;
the other is that the two sets of atomic energy levels no longer match.

The degree of dissymmetry in the bond depends on the nature of the
two atoms involved. The two extreme cases are a purely *covalent bond*,
in which the electron distribution is symmetrical (as in the homonuclear
diatomics) and the purely *ionic bond*, in which one or more electrons are
transferred entirely from one atom to the other, then the molecule is more
appropriately depicted as A^+B^-, and the bond can be ascribed to the
Coulombic force of attraction between the two ions. Intermediate cases
are referred to as *polar bonds*: in them there is incomplete transfer of an
electron from one atom to the other, with the result that one atom acquires
a shade of positive charge and the other one of negative charge. An
example is HCl, which can be depicted as $H^{\delta+}Cl^{\delta-}$.

The sequence of bonding types, from pure covalent, through polar, to
pure ionic, is expressed in quantum mechanics in terms of molecular
orbitals constructed from various amounts of atomic orbitals on each
nucleus. That is, in general an m.o. has the form

(15.2.2) $$\psi = c_A\psi_A + c_B\psi_B,$$

where ψ_A and ψ_B are the atomic orbitals contributing to the bond, and
c_A and c_B are their coefficients. The proportion of ψ_A in the bond is c_A^2,
and the proportion of ψ_B is c_B^2. A pure covalent bond has $c_A^2 = c_B^2$,
and a pure ionic bond has $c_A^2 = 0$ and $c_B^2 = 1$ (for the species A^+B^-).
An actual example of a polar species is HCl, for which the bonding m.o.
has the form $\psi = 0.57(1s_H) + 0.73(2p_{Cl})$. (The normalization of this orbital
takes account of overlap, with $S \approx \frac{1}{3}$.) An electron occupying ψ therefore
spends about 62 per cent (from $0.73^2/(0.57^2 + 0.73^2)$) of its time on the
chlorine atom. In HF the orbital has the form $0.45(1s_H) + 0.82(2p_F)$, and
this molecule's greater polarity is reflected by the greater time spent by
the electron in the fluorine orbital (about 77 per cent).

Example (Objective 10). What is the form of the wavefunction that describes a molecular
orbital in which an electron spends 90 per cent of its time in an orbital ψ_A on A and
10 per cent in ψ_B on B in the polar molecule AB?

- *Method.* The wavefunction is written in the form $\psi = c_A\psi_A + c_B\psi_B$ such that
$|c_A|^2 = 0.90$ and $|c_B|^2 = 0.10$.

- *Answer.* Since $|c_A|^2 = 0.90$, $c_A = \pm 0.95$, and since $|c_B|^2 = 0.10$, $c_B = \pm 0.32$. For
the bonding orbital we take the in-phase overlap, so that $\psi \approx 0.95\psi_A + 0.32\psi_B$.

- *Comment.* Overlap has been ignored. Note that the *squares* of the coefficients
determine the probabilities.

The actual electron distribution, and hence an assessment of the polarity
of the bond, can be obtained either from experiment (for example, the
measurement of dipole moments, Chapter 23) or from detailed m.o.

calculations. For simple applications, though, it is possible to predict approximate polarities by setting up a table of *electronegativities*, χ (chi), which are numbers denoting the ability of the atom to attract electrons. There are several electronegativity scales, of various levels of sophistication. The earliest were due to Pauling and to Mulliken. The *Mulliken scale* equates the electronegativity to the mean of the electron affinity and the ionization energy of the atom (expressed in electronvolts): $\chi^M = \frac{1}{2}(E_a + I)/\mathrm{eV}$. The *Pauling scale* depends on a more involved definition, but it turns out that the two are roughly proportional, being related by $\chi^P = 0.336\chi^M - 0.207$. A list of values is given in Table 15.1. Once the electronegativities have been established they can be used to estimate the polarity of the bond (expressed as percentage ionic character, or *ionicity*), dipole moments, and even the bond strengths themselves. The relevant expressions are given in the footnote to the table.

Table 15.1. Pauling electronegativities, χ^P

H 2.20						
Li 0.98	Be 1.57	B 2.04	C 2.55	N 3.04	O 3.44	F 3.98
Na 0.93	Mg 1.31	Al 1.61	Si 1.90	P 2.19	S 2.58	Cl 3.16
K 0.82	Ca 1.00	Ga 2.01	Ge 2.01	As 2.18	Se 2.55	Br 2.96
Rb 0.82	Sr 0.95	In 1.78	Sn 1.96	Sb 2.05		I 2.66

Dipole moment: $p/\mathrm{debye} = \chi_A^P - \chi_B^P$
Ionicity/per cent $= 16|\chi_A^P - \chi_B^P| + 3.5|\chi_A^P - \chi_B^P|^2$
Covalent–ionic resonance energy/eV $= (\chi_B^P - \chi_B^P)^2$

Source: A. L. Allred, *J. Inorg. Nucl. Chem.* **17**, 215 (1961).

The other important structural feature of heteronuclear diatomics is the consequence of the mismatch of the atomic energy levels: we run into the problem that there is no clear-cut way of selecting which orbitals should be combined. This is illustrated for the case of LiH in Fig. 15.12. Although Li2s lies closer to H1s, the Li2p, orbitals are not far away. It is clearly unjust to allow H1s to form a bond with only Li2s, and a better description of the bond is to express it as a linear combination of all three orbitals:

(15.2.3) $\psi = c_1(\mathrm{Li2s}) + c_2(\mathrm{Li2p}) + c_3(\mathrm{H1s})$,

where the coefficients c_1, c_2, and c_3 determine (through their squares) the relative contributions of each. An approximate calculation gives

(15.2.4) $\psi = 0.323(\mathrm{Li2s}) + 0.231(\mathrm{Li2p}) + 0.685(\mathrm{H1s})$.

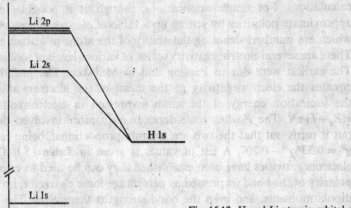

Fig. 15.12. H and Li atomic orbital energy levels.

The method of calculation of the coefficients is based on the variation theorem. This states that *no approximate wavefunction can give an energy lower than the exact ground state energy of the system (or molecule).* This means that if the energy of the molecule is calculated using eqn (15.2.3), it will always be higher than the true ground state energy whatever the values of the coefficients c_n. The best coefficients (e.g. those in eqn (15.2.4)) are then determined by varying their values until the energy reaches a minimum. At this stage, though, we appear to have lost an attractive feature of m.o. theory, for it appears to be impossible to regard the Li–H bond as being formed from the overlap of *two* orbitals, one on Li and the other on H. This is in fact not so, for it is possible to resort to the device of *hybridization*, and to group the two lithium orbitals together and regard the bond as being formed from the overlap of H1s with a *hybrid orbital* on Li. Then the m.o. is written

(15.2.5) $\psi = 0.397(\text{Li, hybrid}) + 0.685(\text{H1s})$
             with $(\text{Li, hybrid}) = 0.813(\text{Li2s}) + 0.582(\text{Li2p})$.

(We shall discuss the actual figures below.) Now we have retained the simplicity of the basic m.o. picture, but at the expense of having made the atomic contribution more complicated.

In a sense the use of hybridization is unnecessary, for it arises from a wish to find a formally simple description of the bond: there is no mathematical reason why it has to be introduced, and no compelling physical reason. Nevertheless, its introduction does help in the discussion of the factors contributing to the strengths of bonds and the shapes of molecules. This can be seen in the case of LiH. The shape of the Li-hybrid orbital is shown in Fig. 15.13. It is easy to see that the bulk of its amplitude lies in the internuclear region. The quantum mechanical source of this 'distortion' is the interference between the contributing Li2s and Li2p orbitals: there is constructive interference on one side of the nucleus (where the amplitude of Li2s and Li2p have the same sign) and destructive on the other (where the signs are opposite). As a consequence there is a

(a)

(b)

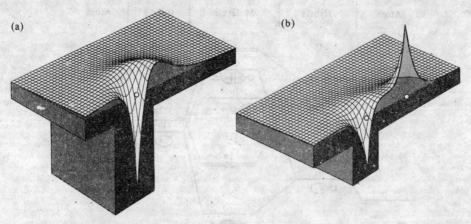

Fig. 15.13. (a) Li hybrid orbital, (b) LiH bonding molecular orbital.

stronger bond: because the hybrid is directed into the internuclear region, it has a greater overlap with H1s than either Li2s or Li2p would have had alone, Fig. 15.13b. Exactly the same overlap enhancement is present in the 'three-orbital' description of the bond, eqn (15.2.4), but thinking of the bond in terms of a hybrid orbital focuses attention on it.

The question that immediately arises is why the hybrid orbital takes the form 0.813(Li2s) + 0.582(Li2p) and not some other composition (such as 0.71(Li2s) + 0.71(Li2p), which is more strongly directed and therefore has a better overlap). The reason lies in an opposing energy contribution. If the hybrid orbital is 0.813(Li2s) + 0.582(Li2p) it is 66 per cent Li2s and 34 per cent Li2p (take squares of the coefficients), and so an electron in it can be regarded as spending 34 per cent of its time in an excited state of the atom. In a hybrid like 0.71(Li2s) + 0.71(Li2p) the proportion is higher, at 50 per cent. Therefore, in order to achieve better overlap, it is necessary to invest more energy in the *promotion* of the atom. The two effects are in opposition, and the outcome, the observed mixture, is the resulting compromise.

As we have emphasized, the concept of hybridization is introduced in order to focus attention on aspects of molecular structure, in particular the competition between various contributions to the energy. It also provides a moderately simple scheme for predicting the configuration and bonding characteristics of heteronuclear molecules (as well as the shapes of polyatomics: we shall see that shortly). As an example consider the molecule CO. The atomic energy level arrays of the two atoms are shown in Fig. 15.14. The C2s, C2p, O2p, and (to some extent) O2s orbitals all have similar energies and are expected to participate in bonding more or less equally. Instead of thinking of the C—O σ-bond as arising from superposition of four orbitals (2s and $2p_z$ on each atom), we can think of both atoms as hybridizing and then forming hybrid–hybrid orbitals.

First, consider the form of the hybrids. The carbon hybrid Ch1, Fig. 15.14, is a C2s-orbital contaminated with some $C2p_z$. It lies slightly above the C2s energy level on account of the 2p admixture; it is directed strongly

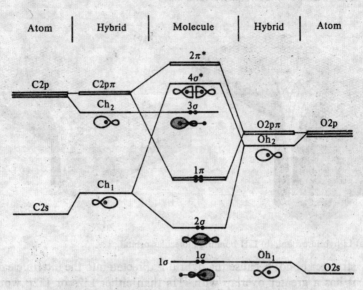

Fig. 15.14. Energy levels of atomic, hybrid, and molecular orbitals in CO.

into the internuclear region. Just as C2s is contaminated by $C2p_z$, so $C2p_z$ is contaminated by C2s. The resulting hybrid, Ch2, lies lower in energy than $2p_z$ on account of the favourable 2s admixture, and is directed out of the internuclear region. The oxygen hybridizes less extensively because of the greater 2s, 2p separation; nevertheless it forms Oh1, which is O2s contaminated by a little $O2p_z$, and is directed out of the internuclear region, and Oh2, which is $O2p_z$ contaminated by a little O2s, and is directed into the internuclear region.

Next, form the m.o.s from the array of hybrids in the normal way. The main feature is the similarity in energy of Ch1 and Oh2. Since they are also directed towards each other they can be expected to give a strong bond (and a strong antibond). Neither Ch2 nor Oh1 is directed into the internuclear region, and they remain *non-bonding orbitals*. (Take care to distinguish between non-bonding and anti-bonding!) There are also two bonding and two antibonding π-orbitals formed from the broadside overlap of $O2p_x$–$C2p_x$ and $O2p_y$–$C2p_y$. They are expected to be weak on account of their poor overlap and energy mismatch.

Finally, feed in the electrons according to the *aufbau* principle. There are 10 electrons to accommodate (the 4 1s-core electrons are being ignored). The resulting configuration is therefore $1\sigma^2 2\sigma^2 1\pi^4 3\sigma^2$ (σ and π are often labelled sequentially, as 1σ, 2σ, etc., especially when their parentage is obscure). There is a non-bonding lone pair on the oxygen (1σ, which is Oh1), a strong σ-bond (2σ), a pair of π-bonds, and a non-bonding lone pair on the carbon (3σ, which is Ch2). The carbon lone pair is directed away from the carbon atom, and accords with the ability of CO to attach itself strongly to transition metal ions. The overall structure is therefore a triply bonded structure (one strong σ-bond and two weak π-bonds) with two lone pairs, one on each atom.

15.3 The structure of polyatomics

As soon as we turn to polyatomic molecules, which consist of more than two atoms and include most of the really important molecules of chemistry, we have to explain why they have their characteristic shapes. Why is water triangular, ammonia pyramidal, methane tetrahedral, benzene a planar hexagon, and so on?

The example of water can be used to illustrate the general approach to the problem. The configuration of the oxygen atom is $1s^2 2s^2 2p_z^2 2p_y 2p_x$, Fig. 15.15. The two perpendicular $2p_x$ and $2p_y$ orbitals are half full. We can imagine forming a σ-orbital by bringing up one hydrogen atom's 1s-orbital along the x-axis, and the other's 1s-orbital along the y-axis. Two perpendicular σ-bonds are formed in this way, Fig. 15.15, and each may accommodate the electrons provided by the oxygen atom and the hydrogen atom. The model predicts that the water molecule is bent, the HOH angle being 90°. This picture is remarkably good, for the water molecule is bent; but its angle is 104.45°, and the discrepancy must be explained.

The description given so far ignores the possibility that the oxygen's 2s-orbital may participate. We should construct m.o.s from all six available atomic orbitals (2s, $2p_x$, $2p_y$, $2p_z$ on O and 1s on each H), and then use the aufbau principle to arrive at the configuration. The physical basis of the role of the 2s-orbital, however, is brought out more clearly by adopting the language of hybridization introduced in the last section, and by regarding each O—H bond as being formed from the overlap of a H1s-orbital and an oxygen hybrid. The observed structure can then be explained in terms of the competition between the lowering of energy arising from enhanced overlap and the raising of the energy as a result of the promotion of an O2s-electron. In order to see how to use hybridization to account for a change in the bond angle from 90° to 104° it is first necessary to go into the way that hybrid orbitals are composed.

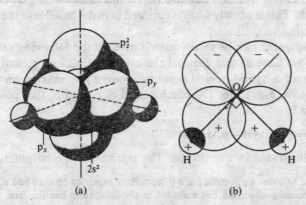

(a)   (b)

Fig. 15.15. Primitive description of the structure of water. (a) Shows the valence orbitals; (b) is a two-dimensional drawing showing the formation of the s–p bonds.

Orthogonality and hybridization. In quantum chemistry *orthogonality* has the technical meaning that one orbital has zero overlap with another. (The overlap integral was defined in eqn. (15.2.1).) In this sense, a p_x-orbital is orthogonal to a p_y-orbital on the same nucleus, to a p_z-orbital, and also to an s-orbital (on the same nucleus). However, instead of regarding orthogonality as only an abstract, mathematical property of orbitals it can be regarded as a signal that two wavefunctions are distinct. The overlap integral measures, in fact, the extent to which one orbital resembles another. For instance, the overlap of two identical orbitals based on the same nucleus is unity: there is perfect resemblance. If you refer to Fig. 15.8, p. 480 again, then (a) and (b) show that S increases as ψ_B is pushed into the region occupied by ψ_A, for in (b) ψ_B is beginning to resemble ψ_A more than it does in (a). There is zero resemblance, strict orthogonality, between the two orbitals illustrated in Fig. 15.8c whatever their separation, and zero resemblance between three p-orbitals (and five d-orbitals, and so on). *Orbitals that are orthogonal are distinct.*

When we turn to H_2O, we want to construct a description in which the two bonds are distinct yet equivalent. Chemical data tells us that the two bonds are *equivalent*. The point of using hybridization is to set up a description in terms of two *distinct* orbitals, one O—H_a and the other O—H_b. In other words, the two hybrids on the oxygen atom must be mutually orthogonal. This has both qualitative and quantitative implications.

Consider the qualitative implication first. When only pure O2p-orbitals are used in the m.o. description, we have two bonds that are both equivalent (because the O2p_x-H1s bond is the same as the O2p_y-H1s) and distinct (because the p_x- and p_y-orbitals are orthogonal). When s-orbital character is allowed both orbitals acquire some. Therefore they cease to be orthogonal on account of their common s-character. Only if they bend away from each other, so reducing their resemblance, can they be regarded as being distinct. Consequently we can predict that the admission of s-character not only increases the bond strength (on account of the enhanced O—H overlap) but also that it increases the angle between the two bonds. This is exactly what is required in order to achieve the observed bond angle.

The argument can be made quantitative by the following calculation. Suppose we have at our disposal the orbitals s, p_x, and p_y and we want to construct a hybrid orbital pointing along a line making an angle $\frac{1}{2}\Theta$ to the x-axis, Fig. 15.16. The combination of p_x and p_y that points along the line is*

$$p(\tfrac{1}{2}\Theta) = p_x \cos\tfrac{1}{2}\Theta + p_y \sin\tfrac{1}{2}\Theta$$

This is illustrated in Fig. 15.16a. The equivalent orbital pointing along

---

* Recall (p. 443) that a p_x-orbital has a θ-dependence expressed by $\cos\theta$ and p_y by $\sin\theta$. A p-orbital pointing along the line indicated in the illustration has the form $\cos(\theta - \tfrac{1}{2}\Theta)$ because it is like a p_x orbital but rotated through an angle $\tfrac{1}{2}\Theta$. But $\cos(\theta - \tfrac{1}{2}\Theta) = \cos\theta\cos\tfrac{1}{2}\Theta + \sin\theta\sin\tfrac{1}{2}\Theta$, which is proportional to $p_x\cos\tfrac{1}{2}\Theta + p_y\sin\tfrac{1}{2}\Theta$.

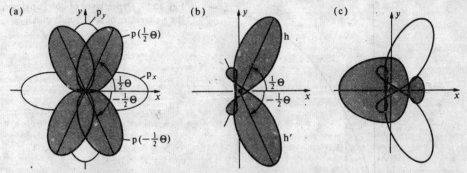

Fig. 15.16. (a) Formation of $p(\frac{1}{2}\Theta)$ and $p(-\frac{1}{2}\Theta)$; (b) formation of sp-hybrids making an angle Θ; (c) the remaining (lone pair) orbital.

a line at $-\frac{1}{2}\Theta$ is

$$p(-\tfrac{1}{2}\Theta) = p_x \cos\tfrac{1}{2}\Theta - p_y \sin\tfrac{1}{2}\Theta.$$

The next step is to add s-orbital character. The same amount has to be added to both in order to keep them equivalent. This gives rise to the two equivalent hybrid orbitals illustrated in Fig. 15.16b:

$$h = as + bp(\tfrac{1}{2}\Theta), \qquad h' = as + bp(-\tfrac{1}{2}\Theta)$$

where a and b are the two mixing coefficients. These coefficients can be found by requiring the hybrid orbitals to satisfy two conditions. The first is the normalization condition, that eqn (13.3.8), p 403, must be satisfied. This leads to

$$\int h^2 \, d\tau = a^2 \int s^2 \, d\tau + b^2 \int p(\tfrac{1}{2}\Theta)^2 \, d\tau + 2ab \int sp(\tfrac{1}{2}\Theta) \, d\tau$$

$$= a^2 + b^2 = 1$$

because p and s are individually normalized (so that the integrals over p^2 and s^2 are unity) and are mutually orthogonal (so that the integral over sp is zero). The same condition applied to h' also leads to $a^2 + b^2 = 1$. The second condition is the requirement that the hybrids be distinct, that is, orthogonal. Then

$$\int hh' \, d\tau = \int (as + bp_x \cos\tfrac{1}{2}\Theta + bp_y \sin\tfrac{1}{2}\Theta)$$

$$\times (as + bp_x \cos\tfrac{1}{2}\Theta - bp_y \sin\tfrac{1}{2}\Theta) \, d\tau$$

$$= a^2 + b^2 \cos^2\tfrac{1}{2}\Theta - b^2 \sin^2\tfrac{1}{2}\Theta = a^2 + b^2 \cos\Theta = 0.$$

(All the sp_x, sp_y, and $p_x p_y$ integrals are zero, and the s^2, p_x^2, and p_y^2 integrals are unity.) On combining these two conclusions we arrive at

(15.3.1) $$\cos\Theta = -a^2/(1-a^2), \qquad \text{or} \qquad a^2 = \cos\Theta/(\cos\Theta - 1).$$

This expression is plotted in Fig. 15.17. The immediate conclusion is that as the s-orbital character, a^2, increases, the angle between the hybrids

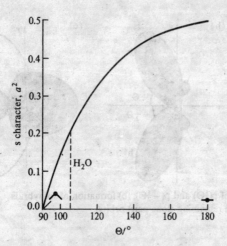

Fig. 15.17. Dependence of bond angle on the s-character of A–B bonds in AB_2.

changes from $90°$ (pure p, $a^2 = 0$) and reaches $180°$ when its contribution is 50 per cent ($a^2 = \frac{1}{2}$). In the case of H_2O, an angle of $104.45°$ indicates that $a^2 = 0.20$, $b^2 = 1 - a^2 = 0.80$ and hence that $a = 0.45$ and $\underline{b} = 0.89$; the hybrids are therefore of the form $0.45s + 0.89p$.

Before going on to make use of these remarks, it should be noted that linear combination of three atomic orbitals leads to the formation of *three* orthogonal hybrids (and, in general, N a.o.s lead to N hybrids). So far we have found only two. The third will have the form $a's + b'p$ where p is an orbital pointing along the $-x$ direction in Fig. 15.16. The coefficients a' and b' can be obtained by imposing the same two conditions as before. Normalization gives $a'^2 + b'^2 = 1$; orthogonality to the two other hybrids gives $aa' + bb' \cos \frac{1}{2}\Theta = 0$. After a little trigonometric manipulation these conditions yield $a'^2 = (1 + \cos \Theta)/(1 - \cos \Theta)$ for the proportion of s-orbital as a function of the bond angle. In the case of H_2O it follows that $a' = 0.77$, implying a hybrid of the form $0.77s + 0.63p$, and therefore having 60 per cent s-character. Its shape is indicated in Fig. 15.16c.

Now we can return to the factors that govern the structure of H_2O. In accord with the m.o. procedure, we first establish the orbitals and then feed in the appropriate number of electrons in accord with the *aufbau* principle.

Suppose first that 2s-orbital involvement is neglected. The orbital diagram then resembles Fig. 15.18a. The H1s-orbitals overlap the $O2p_x$- and $O2p_y$-orbitals in the manner already described. The bonds are quite strong; but the bond angle corresponding to the lowest energy is $90°$, not the observed value of $104°$. There is no promotion energy. This is obvious, but the following more pedantic analysis will prove useful shortly. The configuration of the oxygen atom in the molecule can be arrived at by adding up the number of electrons in each type of orbital. There is the 2s lone pair ($2s^2$), one electron in $2p_x(2p^1)$ one in $2p_y(2p^1)$ and a $2p_z$ lone pair ($2p^2$): this gives $2s^2 2p^{1+1+2}$ or $2s^2 2p^4$ overall. The configuration of the ground state free atom is also $2s^2 2p^4$, hence the conclusion that there is no promotion energy.

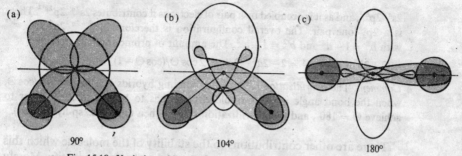

90°   104°   180°

Fig. 15.18. Variation of bond angle, hybridization, and overlap in H_2O.

Next, consider 2s-orbital involvement. Strong overlap is obtained with a hybrid of 50 per cent s-character and 50 per cent p-character. The bond angle is 180°, Fig. 15.17. Although the bonds are strong (on account of the excellent overlap characteristics of the hybrids, Fig. 15.18c) there is a large promotional energy. The argument runs as before. Since each hybrid is 50 per cent s, 50 per cent p the configuration of the oxygen is $2s^{0.5}2p^{0.5}$ for each bond, which is $2s^1 2p^1$ for the two. There are now two 2p lone pairs, and so the overall configuration of the oxygen atom is $2s^1 2p^{1+4}$, or $2s^1 2p^5$. This should be compared with the ground state configuration $2s^2 2p^4$. We conclude that in order to achieve the linear conformation an entire s-electron has to be promoted.

Consider now an intermediate case, Fig. 15.18b. We choose the bond angle corresponding to the observed value, and assess the extent of promotion of the oxygen. The hybrids are 20 per cent s- and 80 per cent p-character. We can therefore ascribe a configuration $2s^{0.2}2p^{0.8}$ to the oxygen atom when it is in the molecule. The two bonds together therefore contribute $2s^{0.4}2p^{1.6}$. There is one pure 2p lone pair (the one perpendicular to the molecular plane). The other lone pair is no longer pure p: we have already calculated that it has 60 per cent s-character. It follows that each electron in it contributes $2s^{0.6}2p^{0.4}$ to the oxygen configuration and hence that the pair contributes $2s^{1.2}2p^{0.8}$. The overall configuration of the atom in the molecule is therefore $1s^2 2s^{1.6}2p^{4.4}$. In this case 0.4 of an electron has had to be promoted in order to achieve the specified hybridization. This is the compromise we sought: the molecule adopts a bond angle of 104° because although promotional energy has to be invested, it is not so great that it overcomes the extra bonding power of the resulting hybrids. The total energy is lower at 104° than at any other angle.

Example (Objectives 13 and 14). Find an expression for the extent of promotion of a $2s^2 2p^4$ atom needed in order to achieve two equivalent bonds making an angle Θ.

- *Method.* Find an expression of the form $2s^m 2p^n$ on the basis of the hybridization expressions $a^2 = \cos\Theta/(\cos\Theta - 1)$ and $a'^2 = (1 + \cos\Theta)/(1 - \cos\Theta)$ deduced above. The fraction of electron to be promoted is $n - 4$.

- *Answer.* One bond is formed from a $2s^{a^2}2p^{b^2}$ hybrid, so is the other. Together they contribute $2s^{2a^2}2p^{2b^2}$ to the final configuration. The lone pair is hybridized into

$2s^{a'2}2p^{b'2}$, and as it is occupied by a pair of electrons it contributes $2s^{2a'2}2p^{2b'2}$. There is a $2p_x$ lone pair. The overall configuration is therefore $2s^{2a^2+2a'2}2p^{2+2b^2+2b'2}$ with $b^2 = 1 - a^2$ and $b'^2 = 1 - a'^2$. The amount of promotion is therefore

$$(2 + 2b^2 + 2b'^2) - 4 = 2 - 2a^2 - 2a'^2 = 2\cos\Theta/(\cos\Theta - 1).$$

- *Comment.* The *hybridization ratio* of the bonding hybrids is b^2/a^2, or $-1/\cos\Theta$, when the bond angle is Θ. A whole electron has to be promoted in order to achieve $\Theta = 180°$, and the hybridization ratio is then unity (i.e., sp-hybridized).

There are other contributions to the stability of the molecule which this simple analysis does not bring out, but which are easy to identify. Merely by looking at the diagram of the molecule in Fig. 15.18 it can be seen that an increase of the bond angle has the energetically advantageous effect of reducing both the proton–proton repulsion and the repulsion between the pairs of electrons in the two O–H bonds. This is a further reward for investing promotional energy. We shall see the same effect contributing to the stability of other polyatomic molecules.

Hybridization: some special cases. In a variety of important molecules, such as methane, there are several equivalent bonds. An important question is therefore the nature and composition of hybrid orbitals that contribute to such symmetrical cases.

Suppose the atomic configuration is $2s2p_x$: what is the form of the two equivalent hybrid orbitals that may be formed? We write one as $a2s + b2p_x$ and the other as $a'2s + b'2p_x$. Since they are equivalent we must insist that $a^2 = a'^2$ and $b^2 = b'^2$. Since they are distinct they are orthogonal. Hence $aa' + bb' = 0$. Since they are both normalized $a^2 + b^2 = 1$ and $a'^2 + b'^2 = 1$. These conditions are sufficient to give the two hybrids as $(2s + 2p_x)/\sqrt{2}$ and $(2s - 2p_x)/\sqrt{2}$. They each have equal amounts of s- and p-character, and so are referred to as *sp-hybrids*. One is directed strongly towards $+x$, the other strongly towards $-x$ (Fig. 15.19a).

Fig. 15.19. Three important types of hybrid orbitals.

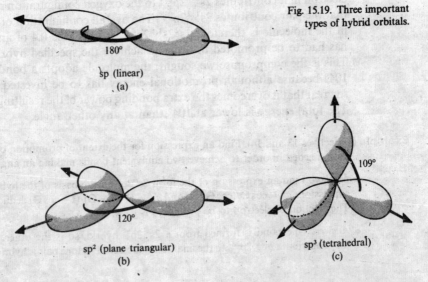

180°

sp (linear)

(a)

120°

sp² (plane triangular)

(b)

109°

sp³ (tetrahedral)

(c)

Example (Objective 15). Form a hybrid orbital from equal proportions of a hydrogen 2s-orbital and a hydrogen $2p_z$-orbital, and plot the probability amplitude and the probability density along the axis.

- *Method.* The two orbitals, taken from Table 14.1, are

$$\psi_{2s} = \tfrac{1}{4}(1/2\pi a_0^3)^{1/2}[(r/a_0) - 2]\exp(-r/2a_0)$$

$$\psi_{2p_z} = \tfrac{1}{4}(1/2\pi a_0^3)^{1/2}(z/a_0)\exp(-r/2a_0).$$

The hybrid is $\psi_h = (\psi_{2s} + \psi_{2p_z})/\sqrt{2}$. The common factor of $(1/4\sqrt{2})(1/2\pi a_0^3)^{1/2} = 1.83 \times 10^{-4}\,\text{pm}^{-3/2}$ may be ignored for the present purpose.

- *Answer.* Plot ψ_h and ψ_h^2 for $z/a_0 = -10,\ldots,10, 12, 14, 16$, finding the following values:

z/a_0	-10	-8	-6	-4	-2	-1	0
ψ_h	-0.01	-0.04	-0.10	-0.27	-0.74	-1.21	-2.00
ψ_h^2	0.0001	0.0016	0.010	0.073	0.55	1.46	4.00

z/a_0	0	0.5	1	1.5	2	4	6	8	10	12	14	16
ψ_h	-2.0	-0.78	0	0.47	0.74	0.81	0.50	0.26	0.12	0.05	0.02	0.01
ψ_h^2	4.00	0.61	0	0.22	0.55	0.66	0.25	0.07	0.014	0.003	0.004	0.0001

These are shown in Fig. 15.20.

- *Comment.* Notice how the amplitude and probability density are greater at positive values of z than at negative values: the p_z- and s-orbitals interfere constructively in this region.

Three equivalent orbitals can be formed if the orbitals available are $2s2p_x2p_y$. This is just a special case of the H_2O calculation, for we can set $\Theta = 120°$ and find three hybrids each of the form $(s + \sqrt{2}p)/\sqrt{3}$ directed towards the apices of an equilateral triangle. The s, p character is in the proportion 1:2, and so each one is termed an sp^2-hybrid. They are illustrated in Fig. 15.19b.

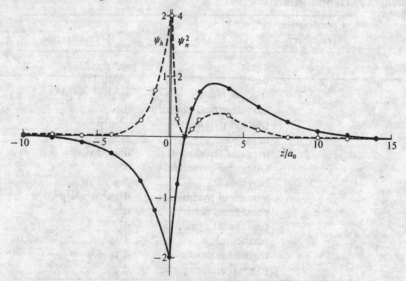

Fig. 15.20. Amplitude and probability distribution for the sp-hybrid constructed in the *Example*.

Four equivalent orbitals can be formed if the orbitals available are $2s2p_x2p_y2p_z$. The presence of $2p_z$, which is perpendicular to the other two p-orbitals, implies that the four orbitals form a three-dimensional array. Straightforward geometry, based on the requirement that the four hybrids are equivalent and distinct, leads to the form $(s+\sqrt{3}p)/2$ for each one, and each one is directed towards the apex of a regular tetrahedron, Fig. 15.19c. Since the s, p proportions are in the ratio 1:3 these four orbitals are termed sp^3-*hybrids*.

Numerous other forms of hybridization may occur, and those involving d-orbitals are particularly important. Some possibilities are listed in Table 15.2. A more detailed depiction of the amplitudes of hybridized orbitals is given in Fig. 15.21. This shows the amplitudes in one plane of an unhybridized 2p-orbital and the effect of mixing in a 2s-orbital to form an sp-hybrid. (In thinking about this diagram you should recall that a 2s-orbital has a radial node, and is quite strongly negative at the nucleus, but positive out beyond the node.)

The shapes of typical polyatomics. In this section we outline the main features of some polyatomics to illustrate how hybridization and energy considerations account for their shapes.

(1) *Water, H_2O.* The unpromoted configuration of oxygen is $1s^22s^22p^4$: from this a 90° structure is predicted. The bonds are simply H1s/O2p overlaps, and are not particularly strong. They are also close together, and so they repel each other electrostatically. Promotion of oxygen to its *valence state*, $1s^22s^{1.6}2p^{4.4}$ requires energy but the hybrids then formed give strong bonds with the hydrogen atoms, and *non-bonded interactions* are minimized.

Table 15.2. Hybrid orbitals

Coordination number	Shape	Hybridization
2	linear	**sp**,* dp
	bent	$\mathbf{p^2}$, ds, d^2
3	trigonal planar	$\mathbf{sp^2}$, dp^2, ds^2, d^3
	unsymmetrical planar	dsp
	trigonal pyramidal	$\mathbf{p^3}$, d^2p
4	tetrahedral	$\mathbf{sp^3}$, d^3s
	irregular tetrahedral	d^2sp, dp^3, d^3p
	tetragonal pyramidal	d^4
5	bipyramidal	$\mathbf{dsp^3}$, d^3sp
	tetragonal pyramidal	d^2sp^2, d^4s, d^2p^3, d^4p
	pentagonal planar	d^3p^2
	pentagonal pyramidal	d^5
6	octahedral	$\mathbf{d^2sp^3}$
	trigonal prismatic	d^4sp, d^5p
	trigonal antiprismatic	d^3p^3

Source: H. Eyring, J. Walter, and G. E. Kimball, *Quantum chemistry*, Wiley.
* Important forms are in bold.

Example (Objective 14). Account for the shape of the water molecule on the basis of the electrostatic repulsions between electron clouds representing the bonds and the lone pairs.

- *Method.* Consider two bonding clouds and two lone-pair clouds. The latter are more diffuse, and repulsions involving them either with each other or with the bonding clouds are stronger than repulsions between bonding clouds, Fig. 15.22a.

- *Answer.* Four identical charge clouds arrange themselves tetrahedrally around the central oxygen. Repulsions between the lone pairs will move two of the clouds apart. Moving them apart is expressed by changing them from sp³ towards sp-hybrids (so that they lie more directly opposite, on different sides of the nucleus). That rehybridization requires losing a proportion of p-character, and so the bonding clouds acquire it and change from sp³ towards pure p. That brings them closer together; but their repulsions are weaker than that between the lone pairs, and so the rehybridization is successful. The bond angle is therefore expected to decrease from pure tetrahedral (109°).

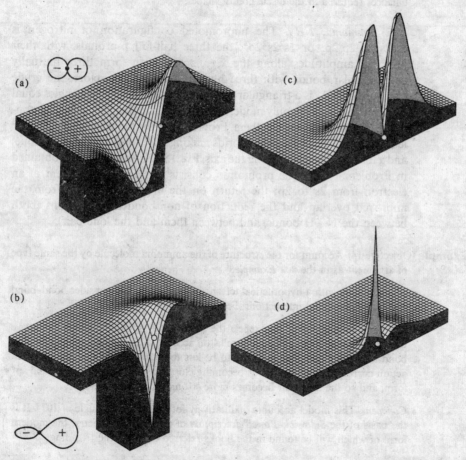

Fig. 15.21. Probability amplitude in a plane through (a) p-orbital, (b) sp-hybrid; probability density in an a plane through (c) p-orbital, (d) sp-hybrid. Note the distortion caused by admixture of s-character.

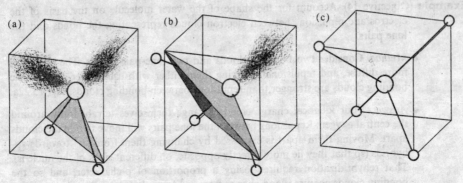

Fig. 15.22. Non-bonding interactions in (a) H_2O, (b) NH_3, and (c) CH_4.

- *Comment.* This model accounts qualitatively for the 104.5° angle in the water molecule. The technique used here is a common way of looking at the energy balances required in the discussion of shapes.

(2) *Ammonia, NH_3.* The unpromoted configuration of nitrogen is $1s^2 2s^2 2p_x 2p_y 2p_z$ or $1s^2 2s^2 2p^3$: the three half-full p-orbitals, with their principal amplitudes along the *x, y, z* axes, can form three mutually perpendicular bonds with three hydrogen atoms. Therefore, the crude structure of NH_3 is a triangular pyramid with all three HNH angles equal to 90°. The bonds are only moderately strong, and there are powerful non-bonded repulsions between them. Promotion of the atom to $1s^2 2s^{1.22} 2p^{3.78}$ lets us form three equivalent hybrids making 107° (the experimental value) and a lone pair pointing along the axis, Fig. 15.22b (the figures are obtained in Problem 15.23). The promotion consists of raising 78 per cent of an electron from 2s to 2p; the return on the investment comes from the improved overlap and the reduction of non-bonded interactions (both between the N—H bonds, and between them and the lone pair).

Example (Objective 14). Account for the structure of the ammonia molecule by the same type of argument as in the last *Example*.

- *Method.* Start with a hypothetical tetrahedral molecule, and consider bond–bond and bond–non-bond interactions, Fig. 15.22b.

- *Answer.* The single lone-pair repels the three bonding clouds, which can move together, like the closing of an umbrella, on account of their lower repulsions. The bond angles are therefore expected to be less than 109°. Note that the umbrella action corresponds to the bonds becoming closer to pure p from their initial sp^3 form, and so the lone-pair becomes more strongly s in character.

- *Comment.* This model accounts qualitatively for the observed angle of 107°. It is the basis of the *Sidgwick–Powell description* of molecular geometry, the modern form of which will be found in the books referred to in Further Reading.

(3) *Methane, CH_4.* The unpromoted configuration of carbon is $1s^2 2s^2 2p_x 2p_y$: on this basis *two* bonds may be formed when hydrogen atoms approach, and the molecule CH_2 may be expected to be the typical

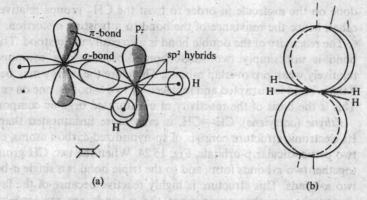

Fig. 15.23. (a) The structure of a double bond, and (b) the source of its torsional rigidity.

hydrocarbon. If, however, the atom is promoted to $1s^2 2s 2p_x 2p_y 2p_z$ there are *four* orbitals available for bonding. These four atomic orbitals may form sp³-hybrids directed towards the corners of a regular tetrahedron. Therefore the structure of methane formed from this valence state of carbon is a regular tetrahedron. It is important to observe how the promotional energy of an entire electron is recovered: four bonds can be formed instead of only two, the bonds are strong because the hybridization emphasizes the amplitude in the bonding regions, and the non-bonded interactions are minimized in a tetrahedral arrangement of charges, Fig. 15.22c. Furthermore, the promotional energy is small because the 2s-electron is shifted into an empty 2p-orbital, where its electron-repulsion terms are minimal. (In N and O promotion of 2s was to already half-occupied 2p-orbitals). It appears that carbon is ideally suited to the formation of tetravalent, tetrahedral compounds, and the variety and number of organic compounds is a consequence of this property.

15.4 Unsaturated and aromatic hydrocarbons

The structure of ethene (ethylene), $CH_2{=}CH_2$, is easy to understand in the light of what has already been described. The two carbon atoms are both approximately sp²-hybridized, giving three almost equivalent hybrid orbitals in a plane, Fig. 15.23a, and a single 2p-orbital perpendicular to it. Hydrogen atoms bond to two sp²-hybrids on each atom, and the remaining sp²-hybrids overlap to form a σ-bond between the two CH_2 groups. The formation of this bond draws the remaining 2p-orbitals into a position where they overlap and form a π-bond. The *double bond* of unsaturated compounds is this $\sigma^2 \pi^2$ bonding arrangement.

One characteristic feature of carbon–carbon double bonds is immediately explained by this description. The *torsional rigidity* of the double bond (the absence of free rotation) arises from the overlap of the two p-orbitals. The view along the bond is shown in Fig. 15.23b: as one CH_2 group is rotated relative to the other the overlap decreases, the bond weakens, and the energy of the molecule rises. Therefore work has to be done on the molecule in order to twist the CH_2 groups relative to each other, hence the resistance of the bond to a twisting distortion.

The *reactivity* of the double bond is also easily understood. The double bond is not simply two single bonds, because one component is a relatively weak p/p overlap π-bond. Therefore the energy may be lowered if the π-bond is disrupted and replaced by two σ-bonds, one on each atom. This is the basis of the reactivity of unsaturated organic compounds.

Ethyne (acetylene), CH≡CH, is even more unsaturated than ethene. Its electronic structure consists of sp-hybridized carbon atoms, each with two perpendicular p-orbitals, Fig. 15.24. When the two CH groups come together two π-bonds form, and so the triple bond is a single σ-bond plus two π-bonds. This structure is highly reactive because of the favourable energy changes that occur when the π-bonds are replaced by strong σ-bonds.

Conjugated double bonds (alternating single and double bonds) confer a special stability on organic molecules, especially in cyclic systems. The stability of these *aromatic* compounds is best discussed on the basis of their most familiar representative, benzene.

Benzene and aromatic stability. The molecular orbitals of benzene may be constructed as follows. The six carbon atoms are sp²-hybridized. This gives three trigonally (triangularly) disposed orbitals at 120°, and a single, perpendicular p-orbital. One hydrogen atom is attached to one sp²-hybrid of each atom, and then the atoms are arranged into a regular hexagon, each bound to its neighbours by two σ-bonds, Fig. 15.25. The internal angle of a regular hexagon is 120°, and so the arrangement of sp²-orbitals is ideal for forming strong σ-bonds. This is the first feature of the stability of the ring: a hexagon permits strain-free σ-bonding.

The formation of the six σ-bonds draws the perpendicular p-orbitals into positions where they overlap to form π-bonds. From six atomic orbitals it is possible to form six molecular orbitals: they are shown in Fig. 15.26 together with their energies. The strongest bonding orbital is

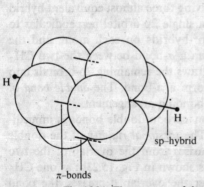

Fig. 15.24. The structure of the triple bond in HC≡CH.

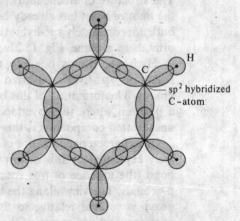

Fig. 15.25. The σ-framework in benzene.

formed when all the atomic orbitals overlap constructively and lead to an accumulation of electron density in all six bonding regions. The higher orbitals are successively less strongly bonding because they contain regions of destructive interference, and therefore reduced electron density, in some bonding regions—they contain successively more antibonding character. The highest energy orbital of all is antibonding between all neighbouring atoms.

The *aufbau* principle is now used to determine the electronic configuration of benzene. Each atom provides one electron for the π-system. (6 electrons are provided by each carbon; 2 remain as $1s^2$ and contribute insignificantly to the bonding; 3 have been used in the σ-framework, one in each σ-bond; that leaves 1 in each $2p\pi$-orbital.) Two enter the lowest orbital (this is usually labelled a). Four can enter the two degenerate orbitals labelled e. Therefore the configuration is a^2e^4. The essential point is that *the only molecular orbitals occupied are those with net bonding characteristics*. This is the second feature underlying the stability of the benzene ring.

The stability of an aromatic compound can therefore be traced to two features. First, the geometry of the regular hexagon is ideal for the formation of strong σ-bonds: the σ-structure of benzene is relaxed and without strain. Second, the π-orbitals are such as to be able to accommodate all the electrons in *bonding* orbitals.

15.5 Metals

The extreme example of conjugated systems is provided by metals, where atom after atom lies in a three-dimensional array and takes part in

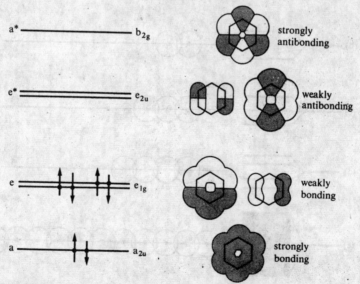

Fig. 15.26. The π-orbitals of benzene. (The labels on the right of the energy levels are the symmetry labels for D_{6h} symmetry: see the next chapter.)

bonding spreading throughout the sample. The easiest entry into an
understanding of the structure of metals is to consider a single, indefinitely long line of atoms, each one carrying a single valence electron in
an s-orbital. The molecular orbitals available to these electrons can be
constructed by imagining what happens when each atom is slid into
position on the line. In the spirit of the *aufbau* principle we first construct
all possible molecular orbitals and then inject into them the appropriate
number of electrons.

Metal molecular orbitals: band structure. One atom provides one atomic s-orbital at
some energy, Fig. 15.27a. When the second atom is brought up it overlaps the first and forms a bonding orbital and an antibonding orbital, Fig.
15.27b. The third is brought up and overlaps its nearest neighbour (and
only slightly its next-nearest) and from these three orbitals three molecular
orbitals are formed, Fig. 15.27c. The fourth atom leads to the formation
of a fourth molecular orbital, and at this stage we can see that the general
effect of bringing up successive atoms is slightly to spread the range of
energies covered by the orbitals, and also to fill in the range with orbital
energies. When N atoms have been slotted on to the line there are N
molecular orbitals covering a band of finite width. When N is indefinitely
large, the orbital energies are indefinitely close, and form a virtually continuous *band*. Nevertheless this virtually continuous band consists of N
different molecular orbitals, the lowest-energy orbitals in the band being
predominantly bonding, and the highest-energy predominantly antibonding.

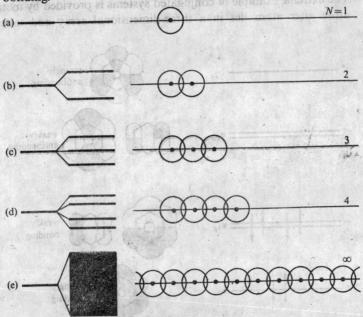

Fig. 15.27. The formation of a band of N orbitals by the successive addition of
atoms to a line.

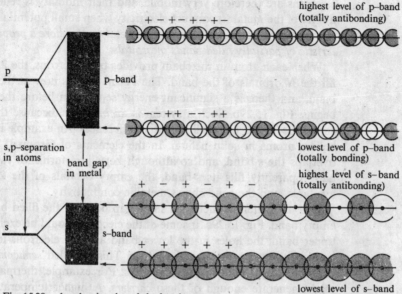

Fig. 15.28. s-band, p-band, and the band gap.

The band formed from s-orbitals is called the *s-band*. If the atoms carry p-orbitals the same procedure may be followed, and the band of molecular orbitals is called the *p-band*, Fig. 15.28. If the atomic p-levels lie higher than the atomic s-levels, the p-band lies higher than the s-band, unless it is so broad (strong overlap) that the bands overlap.

Now we consider the structure of a metal formed from atoms able to contribute one electron (e.g., the alkali metals). Applying the *aufbau* principle, we have N orbitals and N electrons: the latter may occupy the lowest $\frac{1}{2}N$ orbitals spreading through the metal, Fig. 15.29a. The important feature of this structure is that there are unfilled orbitals lying very close to the uppermost filled level, the *Fermi level*, and so it requires exceptionally little energy to excite the uppermost-energy electrons. The

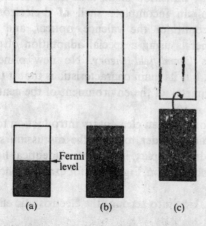

Fig. 15.29. Band filling leads to (a) metal, (b) insulator, (c) semiconductor.

electrons are therefore very mobile, and their mobility is reflected in the ability of the metal to conduct electricity when small potential differences are applied. *Electrical conductivity of metals is therefore a property characteristic of partially filled bands of orbitals.*

When each atom in the chain provides two electrons, the $2N$ electrons fill the N orbitals of the band. The Fermi level is now at the top of the band, and there is a significant energy separation before the next band begins, Fig. 15.29b. Such a system is an *insulator*, because the electrons cannot be shifted by small potential differences. An example is a chain of helium atoms in solid helium. In the elements of Group II the p-band overlaps the s-band, and so although beryllium furnishes two electrons and apparently fills its s-band, the empty orbitals of the 2p-band are available, and the solid shows metallic conductivity.

In some materials there is a small gap between the filled band and an empty band, Fig. 15.29c. If some of the electrons could be excited into the upper band, the holes in the lower band and the electrons in the upper band could move through the lattice. This is called *semiconductivity*. It can be brought about in several ways. For example, thermal excitation might generate enough of these *carriers*. A higher temperature implies more carriers, and so the conductivity increases with temperature. Another way of forming carriers is to introduce impurities in an otherwise ultrapure material. If these impurities can trap electrons they withdraw electrons from the full band, leaving holes which permit conduction: this is *p-type semiconductivity* (*p* indicating the introduction of holes which are positive relative to the negatively charged electrons filling the band). Alternatively, the impurity might carry excess electrons (e.g., phosphorus atoms introduced into germanium) and these electrons swim in the otherwise empty bands, giving *n-type semiconductivity* (where *n* denotes negatively charged carriers). The formation of ultrapure materials and the controlled addition of impurities is discussed in Chapter 10.

15.6 Transition metal complexes: ligand-field theory

The feature that unifies the transition metal ions into a distinguishable group is the existence of an incomplete shell of d-electrons. These electrons can largely account for the valence, optical, and magnetic properties of transition metals using a special adaptation of molecular orbital theory known as *ligand-field theory*. No new principles are involved in this approach, but its main characteristic is that it takes fully into account the high symmetry of the environment of the central metal ion.

This section gives no more than an elementary introduction to the type of arguments involved, and in order to keep the discussion simple we concentrate on an octahedral complex, in which six identical ligands are at the vertices of a regular octahedron, Fig. 15.30; for example, the molecule $[Co(NH_3)_6]^{3+}$

The first step involves taking into account the electrostatic effect of the

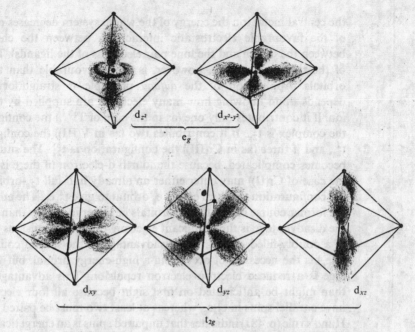

Fig. 15.30. The classification of d-orbitals in an octahedral environment.

ligands on the electrons that occupy the d-orbitals of the central atom. The ligands are regarded as point centres of electron repulsion (for example, they might carry lone-pairs of electrons which are directed towards the central atom, and then an electron in the vicinity of the ligand will be repelled). This approximation is called *crystal-field theory*.

From Fig. 15.30 it is clear that the five d-orbitals fall into two sets: $d_{x^2-y^2}$ and d_{z^2} point directly towards the ligand positions whereas d_{xy}, d_{yz}, and d_{zx} point between them. An electron occupying one of the former pair has a less favourable potential energy than when it occupies any of the other three, and so the energies of the d-orbitals are split as illustrated in Fig. 15.31. For ease of reference the $d_{x^2-y^2}$ and d_{z^2} orbitals are referred to as e_g-orbitals, and the other three are called t_{2g}-orbitals. (This nomenclature will be dealt with in the next chapter.)

The crystal-field theory gives a general but crude explanation of the structure of a transition-metal complex. When the six ligands approach

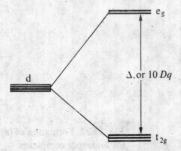

Fig. 15.31. The energy splitting according to crystal-field theory.

the central metal ion the energy of the whole system decreases on account of the favourable electrostatic interactions between the charges (i.e., between the cation and the lone pair electrons of the ligands). The energy of the three t_{2g} orbitals, however, is more favourable than the two e_g orbitals. Application of the *aufbau* principle is straightforward, and depends upon knowing how many electrons are supplied by the central ion. If it contributes only one (as in the case of Ti^{3+}) the configuration of the complex is t_{2g}^1. If it contributes two (as in V(III)) the configuration is t_{2g}^2, and if three (as in Cr(III)) the configuration is t_{2g}^3. The situation now becomes complicated, because the fourth d-electron (if there is one, as in the case of Cr(II)) may enter either an already half-full t_{2g}-orbital to give the configuration t_{2g}^4 or an empty e_g orbital to give $t_{2g}^3 e_g$. The advantage of the former course is that the t_{2g} orbitals lie lower in energy than the e_g, but the disadvantage is the significant electron–electron repulsions that occur in a doubly-filled orbital. The disadvantage of the second course (giving $t_{2g}^3 e_g$) is the necessity of occupying a high-energy orbital, but the advantage is a reduced electron–electron repulsion. This advantage is better than might be anticipated on first sight because all four electrons may have parallel spins in $t_{2g}^3 e_g$, whereas at least two must be paired in t_{2g}^4, and Hund's rule (p. 481) indicates that unpaired spins is an energetically favourable arrangement.

Which structure actually occurs depends on a variety of factors, but an important one is the separation of the e_g and t_{2g} orbitals (this is written Δ or $10Dq$). If $10Dq$ is large, the t_{2g}^4 configuration is favoured, and the spin-paired arrangement occurs: such molecules are called *low-spin complexes*, Fig. 15.32a. If $10Dq$ is small, the advantage of small electron–electron repulsions outweighs the disadvantage of occupying a high-energy orbital, and the $t_{2g}^3 e_g$ configuration is expected. This gives a *high-spin complex*, Fig. 15.32b.

Low-spin and high-spin complexes differ in their optical and magnetic properties. Their magnetic differences can be understood by recalling that a spin angular momentum gives rise to a magnetic moment (p. 463). A high-spin complex has a greater spin magnetic moment than the corresponding low-spin complex, and it is more strongly paramagnetic (see p. 801).

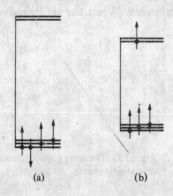

(a)   (b)

Fig. 15.32. Formation of (a) low-spin and (b) high-spin complexes.

Example (Objective 20). Find the number of unpaired spins in a d^6 complex in the presence of weak and strong octahedral ligand fields.

- *Method.* Use the *aufbau* principle, allowing maximum multiplicity to be the dominant factor in weak fields, but not in strong.

- *Answer.* In a weak crystal field the t_{2g}/e_g separation is small, the former lying below the latter; the first five electrons enter t_{2g} and e_g with parallel spins (maximum multiplicity), the sixth must enter t_{2g} and must pair. The configuration is therefore $t_{2g}^4 e_g^2$, and there are 4 unpaired electrons. In a strong crystal field the maximum multiplicity configuration requires too great an expenditure of energy because the e_g-orbitals are far above the t_{2g}. All six electrons enter t_{2g}, and to do so they must have paired spins. The configuration is therefore t_{2g}^6, and there are no unpaired electrons.

- *Comment.* The strong field, *low-spin complex*, having no unpaired electron spins, is diamagnetic. The weak field, *high-spin complex* is strongly paramagnetic. This calculation is quite simple minded: spin–orbit coupling and intermediate field cases generally have to be taken into account.

An indication of the magnitude of $10Dq$ can be obtained from the optical spectra of transition metal complexes. For example, $[Ti(H_2O)_6]^{3+}$ shows an absorption at $20\,000\,cm^{-1}$ which can be ascribed to the excitation of the t_{2g} electron up into an e_g-orbital. This value, which is equivalent to $2.5\,eV$, is therefore the magnitude of $10Dq$ in the complex.

There is one major deficiency of crystal-field theory. It attempts to ascribe the bonding of the complex to point-charge electrostatic interactions between d-electrons localized on a central metal ion and electron pairs localized in orbitals confined to the ligands. The discussion in earlier sections of this chapter suggests that this cannot be wholly true, and that in the actual molecule there are molecular orbitals spreading over both cation and ligands. The molecular orbital approach develops this point of view, and sets up orbitals of octahedral symmetry spreading over the whole complex.

The m.o. approach proceeds as follows. Suppose the ligand orbitals of interest are represented by six spheres (these spheres could represent the lone-pair orbitals on the NH_3). From six atomic orbitals may be constructed six molecular orbitals spreading over the six ligands (at this stage we neglect the central atom).

Inspection of the diagram shows that two of the six m.o.s have a shape that gives non-zero overlap with the two e_g-orbitals of the central ion, and four have the wrong shape for any net overlap with either the e_g- or the t_{2g}-orbitals. With this in mind the transition metal ion is introduced into the centre of the ligand octahedron, when its e_g-orbitals overlap with the appropriate ligand orbital combinations. This gives rise to two e_g bonding orbitals and two e_g^* antibonding orbitals. The full array of orbitals is illustrated in Fig. 15.33 and their energies are depicted in Fig. 15.34.

Into this array of orbitals must be injected the appropriate number of electrons in accord with the building-up principle. Each ligand provides

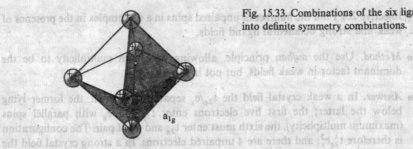

a_{1g}

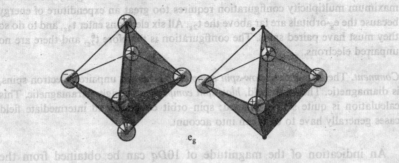

e_g

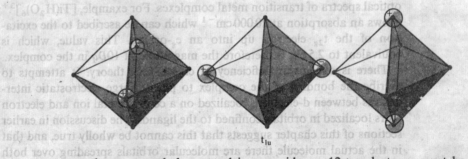

t_{1u}

two electrons, and the central ion provides n: $12 + n$ electrons must be accommodated. Of these, 4 will enter the two e_g-bonding orbitals, 8 will enter the non-bonding orbitals (the four orbitals confined to the ligands), and the remaining n have to be distributed between the t_{2g}-orbitals (confined to the metal, because they have no net overlap with the ligands) and the e_g^*-orbitals spread over ion and ligands. The problem at this stage is just the same as in the crystal-field account, because n electrons have to be distributed among the (t_{2g}, e_g^*) orbitals. The difference lies both in the source of the $10Dq$ splitting and in the spread of the e_g^*-orbital on to the ligands. The occurrence of low-spin and high-spin complexes is accounted for not simply in terms of the electrostatic point-charge model but also in terms of the energy splittings that occur on bond and antibond formation.

There is plenty of evidence to support the spread of electrons from the central ion on to the ligands: evidence coming from magnetism, spectroscopy, and magnetic resonance (Chapter 19). The modern approach to the description of bonding in complexes of various symmetries takes into account both the electrostatic interactions characteristic of crystal-field

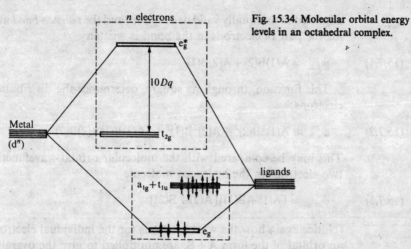

Fig. 15.34. Molecular orbital energy levels in an octahedral complex.

theory and the overlap interactions of the molecular orbital approach. This composite theory is called *ligand-field theory*.

15.7 Valence-bond theory

The whole of the discussion until now has centred on the molecular orbital theory of the chemical bond, and this was presented as a natural extension to molecules of the ideas developed for atoms. The theory was not the first, but it has been developed much more extensively than its rival because it is more easily adapted for computers. Valence-bond theory is still encountered (especially through the language it introduced, including concepts such as 'resonance'), and the following sections indicate its main content.

Valence-bond theory accepts the guidance that came from the early chemists, and concentrates on the electron pair right from the beginning rather than seeing its importance through the building-up principle. Thus it considers the bond in molecular hydrogen as the primitive structure (rather than H_2^+ as in the m.o. theory), and constructs its wavefunction by the following argument.

When two hydrogen atoms are very far apart they are each described as either $1s_A$ or $1s_B$, and if electron 1 occupies atom A and electron 2 occupies atom B the wavefunction for the pair is

$$\psi = \psi_{1s_A}(\mathbf{r}_1)\psi_{1s_B}(\mathbf{r}_2).$$

(Notice that the wavefunctions of separate systems are multiplied to produce the overall function.) Henceforth this is abbreviated to A(1)B(2). This is an accurate description when the atoms are infinitely separated: the valence-bond theory assumes that it is also an adequate description when the atoms are at their equilibrium separation in the molecule. It admits, however, one important modification. When the atoms are together in a molecule it is impossible to say whether it is electron 1 that is in $1s_A$ and electron 2 in $1s_B$, or vice versa. Therefore the wavefunction A(2)B(1) is admitted as an equally valid description, and the *valence-bond wavefunction* for the pair of electrons in the bond is written

(15.7.1) $$\psi_{v.b.} \approx A(1)B(2) + A(2)B(1).$$

This function, through its square, determines the distribution of both electrons:

(15.7.2) $$\psi^2_{v.b.} \approx A(1)^2B(2)^2 + A(2)^2B(1)^2 + 2A(1)B(1)A(2)B(2).$$

This may be compared with the molecular orbital wavefunction for the two electrons in the hydrogen molecule

(15.7.3) $$\psi_{m.o.} \approx [A(1) + B(1)][A(2) + B(2)].$$

(Notice again how the wavefunctions for the individual electrons, each in an orbital of the form $A + B$, are multiplied to give the overall function.) The distribution of the electrons is determined by the square of this function:

(15.7.4)
$$\psi^2_{m.o.} \approx A(1)^2B(2)^2 + A(2)^2B(1)^2 + 4A(1)A(2)B(1)B(2)$$
$$+ A(1)^2A(2)^2 + B(1)^2B(2)^2 + \ldots.$$

We have not written all the terms, but the four unwritten ones may easily be added.

Comparison of eqns (15.7.2) and (15.7.4) underlines both the similarities and the differences between the two theories. The two expressions are different, therefore the electron distributions are also different. Nevertheless they both have in common the terms involving $A(1)B(1)$ and $A(2)B(2)$, which, as we saw on p. 472, represent electrons distibuted with an enhanced amplitude in the internuclear region. Just as in the m.o. theory, the strength of the bond according to valence-bond theory can be traced in large part to the effects of the accumulation of electron density in the bonding region between the two nuclei.

The differences between the two theories are represented by the extra terms in $\psi^2_{m.o.}$, and in particular the term $A(1)^2A(2)^2$ and its analogues. This term represents a distribution in which both electrons 1 and 2 occupy the same *atomic* orbital. Therefore its presence represents the contribution of the *ionic terms* H^-H^+ in the molecular orbital description of H_2. The corresponding terms do not appear in $\psi^2_{v.b.}$, and so the v.b. theory forbids the appearance of these ionic terms.

The exclusion of ionic terms is physically implausible, and therefore we can anticipate that the v.b. description of H_2 (and any other diatomic molecule) could be improved by adding terms that permit both electrons to appear in the same atom to some extent. A wavefunction that puts both electrons into $1s_A$ is $\psi_{1s_A}(r_1)\psi_{1s_A}(r_2)$, or $A(1)A(2)$ in the abbreviated notation. If both are in $1s_B$ the wavefunction contains a term $B(1)B(2)$. Since both H^-H^+ and H^+H^- are equally probable the original valence-bond wavefunction ought to be augmented by an ionic contribution

$$\psi^{ion} \approx A(1)A(2) + B(1)B(2).$$

The new, improved wavefunction is therefore

(15.7.5) $$\psi \approx \psi_{v.b.} + c\psi^{ion},$$

where the value of c governs the extent of ionic character mixed in to the original purely covalent v.b. structure. The improvement of the wavefunction by this admixture is called *ionic–covalent resonance*, and the improvement in energy can be traced to the extra freedom it gives to the distribution of the electrons.

If $c = 1$ the wavefunction just given becomes identical to the m.o. wavefunction in eqn (15.7.3). This choice of c gives equal weight to the ionic and covalent structures, which is unreasonable; therefore the m.o. treatment is wrong in so far as it *over-emphasizes* the role of ionic distributions. The actual value of c which gives the best description of this type (the best value being found by applying the variation theorem, p. 486), is about 0.25, which indicates that the ionic structures contribute about $c^2 \approx 0.06$, or 6 per cent, to the overall structure of H_2. In heteronuclear molecules the ionic terms are much more important, and in species like NaCl they wholly dominate the vestigial covalent character.

Resonance and aromaticity. In broad outline the v.b. description of benzene examines a single Kekulé structure and identifies three bonds, Fig. 15.35. If the atoms are labelled $A, B, \ldots, F$ and the electrons $1, 2, \ldots, 6$, one bond is $A(1)B(2) + A(2)B(1)$ because formally it is like an isolated hydrogen molecule, the difference being that A and B now refer to carbon $2p\pi$-orbitals. Another bond is $C(3)D(4) + C(4)D(3)$, and the third is $E(5)F(6) + E(6)F(5)$. Therefore the overall wavefunction for the Kekulé structure is

$$\psi_{v.b.}^{Kek.1} \approx [A(1)B(2) + A(2)B(1)][C(3)D(4) + C(4)D(3)]$$

$$\times [E(5)F(6) + E(6)F(5)]$$

because the independent bond wavefunctions are just multiplied together. Although this wavefunction might look complicated it can be analysed as in the case of H_2 by concentrating on each of the bonding regions individually. The π-bonds in benzene, in the Kekulé form quoted, receive their strength from the overlap accumulations, just as in H_2. Notice, too, how this way of writing the wavefunction concentrates on the individual electron pairs forming each bond. This is the characteristic feature of v.b. theory already alluded to.

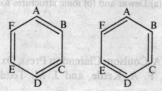

Fig. 15.35. The Kekulé structures for benzene.

The Kekulé structure represented by the function just quoted is plainly too restrictive on the electron distribution because it forbids accumulation of electron density between the unbonded carbon atoms (e.g., between B and C). This can be repaired by admitting the other Kekulé structure, with its wavefunction based on terms such as $B(1)C(2) + B(2)C(1)$, and so on. Therefore, the actual structure of the molecule is better described by the sum of the two Kekulé wavefunctions:

$$\psi = \psi_{v.b.}^{Kek.\,1} + \psi_{v.b.}^{Kek.\,2}.$$

The admission of both Kekulé forms gives a more realistic description of the electron distribution and allows the electrostatic interactions to be relaxed by distributing the electrons to all parts of the ring. The energy is lowered and the overall structure is more stable than either of the contributory structures alone. This mixing of states of equal (or almost equal) energy is called *resonance*.

Bonding may also occur between non-nearest-neighbour atoms, and the Dewar structures of benzene, Fig. 15.36, with bonds described by wavefunctions including the factors $A(1)D(2) + A(2)D(1)$, must also be allowed to contribute. These structures are less stable than the Kekulé forms, because the A–D bonds, etc., are long and weak; therefore, although they must be allowed to take part in the resonance, they contribute more weakly. Calculation shows that the best description is obtained by allowing each Dewar structure to contribute about 6 per cent and each Kekulé structure about 40 per cent.

The admission of resonance among covalent structures is not the complete story. No mention has been made so far of ionic structures of the kind shown in Fig. 15.36. The number of these is very great, especially for molecules more complex than benzene, and the difficulty of incorporating them all is one of the reasons why valence-bond theory has undergone less development than molecular orbital theory.

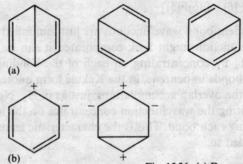

Fig. 15.36. (a) Dewar and (b) ionic structures for benzene.

Further reading

The shape and structure of molecules. C. A. Coulson; Clarendon Press, Oxford, 1973. ●
The chemical bond. J. N. Murrell, S. F. A. Kettle, and J. M. Tedder; Wiley, Chichester, 1978.

Valence theory. J. N. Murrell, S. F. A. Kettle, and J. M. Tedder; Wiley, New York, 1965.

Quantum chemistry. J. P. Lowe; Academic Press, New York, 1978.

Coulson's Valence. R. McWeeny; Oxford University Press, 1979.

Electron densities in molecules and molecular orbitals. J. R. Van Wazer and I. Absar; Academic Press, New York, 1975.

The nature of the chemical bond. L. Pauling; Cornell University Press, Ithaca, 1960.

Molecular orbital theory. C. J. Ballhausen and H. B. Gray; Benjamin, New York, 1965.

Quantum theory of matter. J. C. Slater; McGraw-Hill, New York, 1968.

The organic chemist's book of orbitals. W. L. Jorgensen and L. Salem; Academic Press, New York, 1973.

Molecular geometry. R. J. Gillespie; Van Nostrand, New York, 1972.

Atoms and molecules. M. Karplus and R. N. Porter; Benjamin, New York, 1970.

Molecular wavefunctions. E. Steiner; Cambridge University Press, Cambridge, 1976.

An introduction to transition metal chemistry. L. E. Orgel; Methuen, London, 1966.

Introduction to ligand fields. B. N. Figgis; Wiley, New York, 1966.

Introduction to ligand field theory. C. J. Ballhausen; McGraw-Hill, New York, 1962.

Band theory of metals. S. L. Altmann; Pergamon, Oxford, 1970.

Problems

15.1. Wavefunctions are just like any other mathematical functions, and they can be manipulated in the same way. One of their aspects that was encountered throughout this chapter was their interference (just like waves in any kind of medium). The first few Problems in this chapter give some practice with calculations of this kind. First, show that a wave $\cos k_1 x$ centred on A (so that, for this wave, x is measured from A) interferes with a similar wave $\cos k_2 x$ centred on B (so that, for this wave, x is measured from B) to give an enhanced amplitude half-way between A and B when $k_1 = k_2 = \pi/2R$, but that the waves interfere destructively if $k_1 R = \frac{1}{2}\pi$, $k_2 R = \frac{3}{2}\pi$. (R is the separation of A and B.)

15.2. In molecules the atomic wavefunctions are not extended waves (like $\cos kx$) but are localized around the nuclei. Their amplitudes spread into common regions of space, and interference occurs that is essentially of the same kind as in the last Problem. The expression in eqn (15.1.2) indicates the form of the superposition involved in the H_2 molecule. Take $1s_A = \exp(-r/a_0)$ with r measured from A (and questions of normalization ignored at this stage), and $1s_B = \exp(-r/a_0)$ with r measured from B. Plot the amplitude of the bonding molecular orbital along the internuclear axis for an internuclear distance of 106 pm.

15.3. Repeat the calculation for the antibonding combination $1s_A - 1s_B$, and notice how the node appears in the internuclear region.

15.4. Sketch the forms of the interference patterns that arise with the waves in (a) $\cos kx$ form with $k = \pi/2R$, (b) the hydrogen 1s-orbital form, superimpose in the form $\psi_A + \beta\psi_B$ with (i) $\beta = 0.5$, (ii) $\beta = -0.5$. This models what happens in heteronuclear molecules.

15.5. Plot the electron density along the internuclear axis (both between and beyond the nuclei) for the H_2^+ molecule using the orbitals constructed above. Make sure that the orbitals are normalized to unity: this is easily done by dividing the value of $1s_A + 1s_B$ by $1218\,pm^{3/2}$ and that of $1s_A - 1s_B$ by $622\,pm^{3/2}$. Superimpose on this diagram the simple sum of electron densities $(1s_A^2 + 1s_B^2)/9.35 \times 10^5 \text{ pm}^3$ that would be obtained if we disregarded interference effects (the divisor ensures that the normalization is correct).

15.6. The diagram just constructed shows the densities along the line joining the nuclei. In order to clarify the shifts of electron density that take place on molecule formation it is usual to plot the *difference density*. Plot the change in electron density that occurs on the formation of a bond and an antibond, by subtracting the appropriate curves.

15.7. Imagine a small electron-sensitive probe of volume 1 pm^3 being inserted into a H_2^+ molecule in its ground state. What is the probability that it will register the presence of an electron when it is inserted at the following positions: (a) at nucleus A, (b) at nucleus B, (c) half-way between A and B, (d) at a point arrived at by moving 20 pm from A along the axis towards B and then 10 pm perpendicularly?

15.8. What probabilities would occur if the same probe were inserted into a H_2^+ molecule in which the electron had just been excited into the antibonding orbital (and the molecule had not yet dissociated)?

15.9. The energy of an H_2^+ molecule with an internuclear distance R is given by the expression $E = E_H - [V_1(R) + V_2(R)]/[1 + S(R)] + e^2/4\pi\varepsilon_0 R$, where E_H is the energy of an isolated hydrogen atom, V_1 is the attraction between the electron in the orbital centred on one nucleus and the other nucleus, V_2 is the attraction between the overlap density and a nucleus, and S is the overlap integral between the two atomic orbitals. All the terms depend on the internuclear distance (except E_H) and their values are given below. Plot the potential energy curve for the molecule and find (a) the bond dissociation energy (in eV), (b) the equilibrium bond length.

R/a_0	0	1	2	3	4
V_1/R_H	1.000	0.729	0.473	0.330	0.250
V_2/R_H	1.000	0.736	0.406	0.199	0.092
S	1.000	0.858	0.587	0.349	0.189

You will need to evaluate the nuclear repulsion term. In this table $E_H = -\frac{1}{2}\bar{R}_H$, $\bar{R}_H = 27.3 \text{ eV}$ and $a_0 = 53 \text{ pm}$. Why does V_2 drop off so rapidly?

15.10. From the same data, plot the energy of an H_2^+ molecule when the electron occupies the antibonding orbital. (The energy expression changes by $V_2 \to -V_2$ and $S \to -S$.)

15.11. But where does the energy expression come from? The search can be broken into two steps, one easy, the other moderately involved. Try this part at least. Normalize the orbital $1s_A + 1s_B$ to unity (assume that $1s_A$ and $1s_B$ are both individually normalized). Finding the normalization factor introduces the overlap integral S into the discussion.

15.12. This is the involved step. Write the Schrödinger equation for H_2^+. We know, however, that $1s_A$ and $1s_B$ both individually satisfy the hydrogen atom Schrödinger equation. You should recognize the occurrence of the atom equation inside the H_2^+ equation, and therefore you will be able to replace parts of the left-hand side by E_H. Proceed from that point, and deduce the bonding and antibonding energy expressions. The nuclear repulsion term is then tacked on at the end.

15.13. The *overlap integral* can be calculated by graphical, numerical, or analytical integration. The simplest case is when we are interested in the overlap of two hydrogen 1s-orbitals on nuclei separated by a distance R, for then $S = [1 + (R/a_0) + (R^2/3a_0^2)] \exp(-R/a_0)$. Plot this function for $0 \leqslant R \leqslant \infty$, and confirm the entries in the table of data for Problem 15.9.

15.14. When an s-orbital approaches a p-orbital along the latter's axis, the overlap increases and then drops to zero when the centres of the orbitals coincide. Why? The analytical expression for 1s, 2p overlap of this kind is $(R/2a_0)$ $[1 + (R/a_0) + (R^2/3a_0^2)] \exp(-R/a_0)$. Plot this function, confirm the above remark, and find the separation for which the overlap is a maximum.

15.15. Use Fig. 15.9 to give the configurations of the following species: H_2^-, N_2, O_2, CO, NO, CN.

15.16. Which of the species N_2, NO, O_2, C_2, F_2, CN would you expect to be stabilized (a) by the addition of an electron to form AB^-, (b) by ionization to AB^+?

15.17. A transition metal was vaporized in a furnace, and spectroscopic analysis showed the presence of diatomic molecules and ions. But what types of molecular orbitals may be formed when atoms stick together by using their d-orbitals? Give the configurations of the diatomics formed when each atom brings up (a) 2, (b) 5, (c) 8 d-electrons.

15.18. Draw the molecular orbital diagram for (a) CO, (b) XeF, and use the *aufbau* principle to put in the appropriate number of electrons. Is XeF^+ likely to be more stable than XeF?

15.19. Give the g and u character of the following types of molecular orbital: (a) π^* in F_2, (b) σ^* in NO, (c) δ in Tl_2, (d) δ^* in Fe_2, (e) the six orbitals of the benzene molecule (Fig. 15.26).

15.20. What is the energy required to remove a potassium ion from its equilibrium distance of 294 pm in the K^+Br^- molecule?

15.21. The ionization energies of some alkali metal atoms and the electron affinities of some halogen atoms are listed below. Draw up a table showing the separation of the metal and halogen atoms at which it becomes energetically favourable for the M^+X^- species to form.

$I/kJ\,mol^{-1}$	520.3 (Li)	495.8 (Na)	418.9 (K)
$E_a/kJ\,mol^{-1}$	332.7 (F)	348.6 (Cl)	328 (Br)

15.22. Find an expression for the extent of promotion of a $2s^2 2p^3$ atom needed in order to achieve three equivalent and coplanar bonds making an angle $120°$ to each other.

15.23. Find an expression for the extent of promotion of a $2s^2 2p^3$ atom in order to achieve pyramidal hybridization with three equivalent bonds making angles Θ to each other and Φ to a lone pair. Express your answer in terms of Θ alone, and evaluate the expression for NH_3 ($\Theta = 106.7°$).

15.24. Which of the following species do you expect to be linear: CO_2, NO_2, NO_2^+, NO_2^-, SO_2, H_2O, H_2O^{2+}, H_2O^+? Give reasons.

15.25. Which of the following species do you expect to be planar: NH_3, NH_3^{2+}, CH_3, NO_3^-, CO_3^{2-}? Give reasons.

15.26. Construct the molecular orbital diagram of (a) ethene, (b) ethyne (acetylene) on the basis that the molecules are formed from the appropriately hybridized CH_2 or CH fragments.

15.27. Consider the molecule $CH_2:CH.CH:CH_2$ as the prototype of a conjugated polyene, and concentrate on the two sets of π-orbitals. Since they are not at great separations we may expect them to overlap weakly. Draw the molecular orbital diagram for butadiene allowing for weak $\pi-\pi$ overlap, and sketch the π-orbitals for the complete system.

15.28. A simple model of conjugated polyenes allows their electrons to roam freely along the chain of atoms. The molecule is then regarded as a collection of independent particles confined to a box, and the molecular orbitals are taken to be the square-well wave functions. This is the *free-electron molecular orbital* (f.e.m.o.) picture of structure. As a first example of employing it, take the butadiene molecule with its four π-electrons and show that for its f.e.m.o. description we require the lowest two particle-in-a-box wavefunctions. Sketch the form of the orbitals required, and compare them with the orbitals obtained in the last Problem.

15.29. An advantage of the f.e.m.o. approach is that it lets one arrive at some quantitative conclusions with very little effort. What is the minimum excitation energy of butadiene?

15.30. Consider the f.e.m.o. description of the molecule $CH_2:CH.CH:CH.CH:CH.CH:CH_2$ and regard the electrons as being in a box of length $8R_{CC}$ (as in this case, an extra $\frac{1}{2}R_{CC}$ is often added at each end of the molecule). What is the minimum excitation energy of this molecule? What colour does the molecule absorb from white light? What colour does it then appear? Sketch the form of the uppermost filled orbital. Take $R_{CC} = 140$ pm.

15.31. Write the d-electron configurations for d^n species ($n = 1-10$) in (a) strong, (b) weak octahedral crystal fields. Give the number of unpaired electron spins in each case.

15.32. In the discussion of crystal and ligand field theory in the text we concentrated on octahedral complexes. The same type of argument may be applied to complexes of different symmetry, and tetrahedral and planar square complexes are often encountered. Sketch the splitting of the five d-orbitals that would be expected to take place when these two types of complex are formed.

15.33. Why is d^4 the most easily spin-paired configuration in tetrahedral complexes?

15.34. Just as we constructed six molecular orbitals from the ligand orbitals in order to arrive at an m.o. description of octahedral complexes, do the same for the description of tetrahedral complexes. Write the m.o. configuration of the tetrahedral complex $[Zn(NH_3)_4]^{2+}$. Would you expect $[Zn(NH_3)_4]^{3+}$ to be more stable?

15.35. Write the valence-bond wavefunction for the HF molecule (regarding it as being formed from an H1s-orbital and an F2p$_z$-orbital) (a) supposing it to be purely covalent, (b) supposing it to be purely ionic, (c) supposing it to be 80 per cent covalent and 20 per cent ionic.

15.36. Draw the covalent and ionic structures that should be included in a complete description of the π-electron structure of cyclobutadiene.

15.37. Draw the covalent and singly polar (i.e., one positive and one negative charge) valence-bond structures of the π-electrons in naphthalene. (Go as far as covalent structures with one long bond.)

16 Symmetry: its description and consequences

Learning objectives

After careful study of this chapter you should be able to

(1) Define *symmetry operation* and *symmetry element*, and detect the presence of an *axis of symmetry*, a *plane of symmetry*, a *centre of symmetry*, and an *axis of improper rotation* in a body (p. 518).

(2) Classify molecules according to their *point-group symmetry* (p. 521).

(3) From a knowledge of the symmetry group of the molecule deduce whether it can be polar or optically active (p. 527).

(4) State the *group property* (p. 530) and find the *matrix representative* of a symmetry operation (p. 531).

(5) Define *character* (p. 533), *irreducible representation* (p. 534), *direct sum* (p. 534), and *class* (p. 533).

(6) Deduce the transformation properties of the functions x, y, z, x^2, y^2, xy, xz, etc. (p. 537).

(7) Use *character tables* to decide when an integral must vanish (p. 541).

(8) Use character tables to select orbitals having *non-zero overlap* (p. 543).

(9) Use character tables to deduce *selection rules* for spectral transitions (p. 545).

(10) Use character tables to construct *symmetry-adapted orbitals* (p. 546).

(11) Define *crystal system* and *crystal class* (p. 548) and state the distinguishing characteristic of the seven systems (p. 549).

(12) Show that the unit cell of a crystal can have only 1,2,3,4,6-fold rotation symmetry (p. 552).

(13) Define and draw the fourteen *Bravais lattices* and distinguish *primitive*, *body-centred*, and *face-centred* unit cells (p. 554).

(14) Explain the meaning of *space group*, *screw axis*, and *glide plane* (p. 553).

Introduction

In the preceding chapter we saw the underlying quantum-mechanical reasons for the structures and shapes adopted by molecules, and in the chapters following this we shall see how these structures are determined experimentally. When molecules and ions conglomerate into crystals they do so to give extensive structures with well-defined symmetry. In this chapter we sharpen the common-sense appreciation of shape and symmetry and show that it is possible to discuss the symmetries of molecules and crystals in a quantitative manner. One result of this approach is a system of classifying any molecule or crystal according to its symmetry. Classification, however useful it may be, is not an end in itself, and we shall discover that the same quantitative approach also enables us to discuss a range of molecular properties without becoming involved in detailed calculations.

The quantitative discussion of symmetry is called *group theory*. Much of group theory is a systematic summary of common sense about the symmetry of objects, and this common-sense viewpoint should never be forgotten. Nevertheless, because group theory is systematic, its rules can be applied in a straightforward, mechanical way, and in some cases unexpected results are obtained. In most cases the theory provides a simple, direct technique for arriving at useful conclusions with the minimum of labour, and this is the aspect stressed here.

16.1 Symmetry elements of objects

Some objects are 'more symmetrical' than others. For example, a sphere is more symmetrical than a cube because it looks the same after it has been rotated through any angle about any diameter, whereas the cube looks the same if it is rotated through 90°, 180°, or 270° about axes emerging from the centre of its faces, Fig. 16.1, or by 120° and 240° about any of the four axes passing through opposite corners. Similarly the NH_3 molecule is 'more symmetrical' than H_2O because it looks the same after rotation by 120° or 240° about the axis shown in Fig. 16.2, whereas the water molecule looks the same only after a rotation of 180°. Operations that leave objects looking the same are called *symmetry operations*. There is a corresponding *symmetry element* for each operation. These are the points, lines, and planes with respect to which the operation is carried out. Objects can be classified into groups according to their symmetry by listing all their symmetry elements. This puts the sphere into a different group from the cube, and the NH_3 molecule into a different group from the H_2O molecule.

Symmetry operations are not confined to rotations. For instance, a sphere has a *plane of symmetry* which may be imagined to be a mirror cutting the sphere into hemispheres. Reflection (the symmetry operation) of the sphere in this mirror plane (the corresponding symmetry element) gives an image that is indistinguishable from the original sphere, and so it too is a symmetry operation of the sphere. The mirror plane may lie

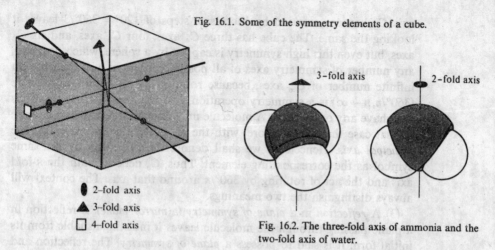

Fig. 16.1. Some of the symmetry elements of a cube.

■ 2–fold axis

▲ 3–fold axis

□ 4–fold axis

3–fold axis   2–fold axis

Fig. 16.2. The three-fold axis of ammonia and the two-fold axis of water.

in any orientation that passes through the centre of the sphere. In H_2O, however, there are only two mirror planes, Fig. 16.3, which once again makes slightly more explicit the expression the 'high symmetry' of a sphere.

For the purpose of discussing the symmetries of molecules it is sufficient to consider five categories of symmetry operation and element.

(1) The *identity*. The identity might seem a trivial operation for it consists of doing nothing to the object. Since every object is indistinguishable from itself if nothing is done to it, every object possesses at least this symmetry. The reason for including it is partly due to the desire to classify every molecule into symmetry classes, and some (e.g. CHClBrF) might have no other symmetry. Another reason is technical and connected with the formulation of group theory. We shall denote the identity by E.

(2) A *rotation* (the operation) about an *axis of symmetry* (the corresponding element). If rotation through an angle of $360°/n$ leaves the molecule in an indistinguishable condition it is said to have an *n-fold axis of symmetry*, and the presence of that element is denoted C_n. Thus the water molecule has a 2-fold axis of symmetry C_2 because rotation by $180° = 360°/2$ is a symmetry operation. The NH_3 molecule has a 3-fold

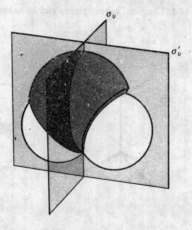

σ_v

σ'_v

Fig. 16.3. The two mirror planes of water.

axis C_3 because rotation by successive steps of $120° = 360°/3$ leaves it
looking the same. The cube has three C_4 axes, four C_3 axes, and six C_2
axes, but even this high symmetry is capped by a sphere, which possesses
any number of symmetry axes of all possible orders of n: it even has an
infinite number of C_∞ axes, because rotation by an infinitesimal angle
$(360°/n, n \rightarrow \infty)$ is a symmetry operation, and the axes of these rotations
may have any orientation. A molecule may possess several rotation axes:
in that case the one (or more) with the greatest value of n is called the
principal axis. Henceforth we shall denote the operation by the same
symbol as the corresponding element. Thus, C_n denotes both the n-fold
axis and the act of rotating by $360°/n$ around that axis. The context will
always distinguish the two meanings.

(3) A *reflection* in a *plane of symmetry* (a *mirror plane*). If reflection in
a plane passing through the molecule leaves it indistinguishable from its
initial form it is said to possess a *plane of symmetry*. The reflection and
the plane are denoted σ. The mirror planes may take various orientations
with respect to axes of symmetry that happen to be present. For example,
in H_2O there are two mirror planes, and both pass through the C_2 axis,
Fig. 16.3. When a mirror plane contains the principal axis of symmetry
it is denoted σ_v (v for vertical). Both planes in H_2O are σ_v; the three planes
in NH_3 are all σ_v. When the mirror plane is perpendicular to the principal
axis, so that σ is a horizontal plane if C_n is a vertical axis, it is denoted
σ_h. The benzene molecule, for example, has a C_6 axis and a σ_h plane (as
well as several other elements). Another possibility arises when the mirror
plane is vertical and contains the principal axis, but has the additional
feature of bisecting the angle between two C_2 axes that are themselves
perpendicular to the principal axis: this arrangement, which is illustrated
in Fig. 16.4, is denoted σ_d (d for diagonal or dihedral).

(4) An *inversion* through a *centre of symmetry*. Imagine taking every
point of an object, moving it to the centre of the molecule, and then taking
it out an equal distance to the other side. If the appearance of the
molecule is left unchanged, it is said to possess a *centre of inversion*, i.
Neither H_2O nor NH_3 possesses an inversion centre, but both the sphere

Fig. 16.4. Diagonal mirror planes and an example (ethane).

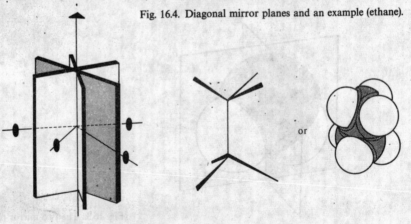

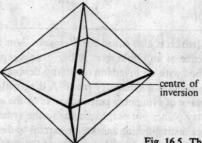

Fig. 16.5. The centre of inversion of a regular octahedron.

and the cube do; so do benzene and the regular octahedron, Fig. 16.5, but
the regular tetrahedron and the CH_4 molecule do not.

(5) An *improper rotation*, or a *rotary-reflection* about an *axis of
improper rotation* or a *rotatory-reflection axis*. An improper rotation is
composite: an object possesses such an axis if it is indistinguishable after
an *n*-fold rotation followed by a horizontal reflection. This composite
operation and element are denoted S_n. Neither H_2O nor NH_3 has an S_n
axis, but CH_4 has three S_4 axes, Fig. 16.6.

When we come to the symmetry of crystals a further aspect of symmetry has to be taken into account: the indistinguishability of crystalline
arrays under a translational motion. This topic is taken up on p. 550.

Symmetry classification of molecules. In order to classify molecules according to
their symmetry, all that is necessary is to list all their symmetry elements,
and to group together molecules that give the same list. This puts CH_4 and
CCl_4, which are both regular tetrahedrons, into the same group, and H_2O
into another. The name of the group to which a particular molecule
belongs is determined by the symmetry elements that characterize it.

Two systems of notation are in use, the *Schoenflies system* and the
Hermann–Mauguin system. The former is more common for the discussion
of individual molecules, but the latter is used almost exclusively in the
discussion of crystal symmetry. The following paragraphs explain the
Schoenflies system; the Hermann–Mauguin system is described in Table
16.1.

Fig. 16.6. An S_4 rotatory-reflection axis of methane.

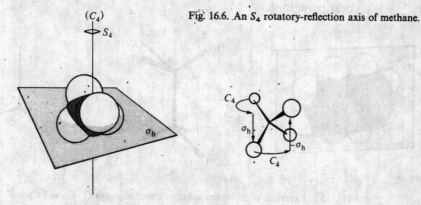

Table 16.1. Notation for point groups

In the *International System* (which is also called the *Hermann–Mauguin System*) a number n denotes the presence of an n-fold axis. A bar over a symbol indicates that the operation is combined with an inversion (so that $\bar{6}$ denotes C_3 combined with i, or C_{3h}). The letter m denotes a symmetry plane. A diagonal line indicates that the group contains a plane of symmetry perpendicular to the symmetry axis (so that $2/m$ is C_{2h}). Symmetry planes not perpendicular to the symmetry axis are denoted by m without further distinguishing marks. Be careful to distinguish symmetry operations of the same type but in different classes (so that D_{4h} becomes $4/mmm$, there being one symmetry plane perpendicular to the axis, the remaining reflections falling into two different classes).

The table below translates the Schoenflies System into the International, Hermann–Mauguin System. The only groups translated are the 32 crystallographic point groups (see p. 548).

$C_i:\bar{1}$	$C_s:m$			
$C_1:1$	$C_2:2$	$C_3:3$	$C_4:4$	$C_6:6$
	$C_{2v}:2mm$	$C_{3v}:3m$	$C_{4v}:4mm$	$C_{6v}:6mm$
	$C_{2h}:2/m$	$C_{3h}:\bar{6}$	$C_{4h}:4/m$	$C_{6h}:6/m$
	$D_2:222$	$D_3:32$	$D_4:422$	$D_6:622$
	$D_{2h}:mmm$	$D_{3h}:\bar{6}2m$	$D_{4h}:4/mmm$	$D_{6h}:6/mmm$
	$D_{2d}:\bar{4}2m$	$D_{3d}:\bar{3}m$	$S_4:\bar{4}$	$S_6:\bar{3}$
$T:23$	$T_d:\bar{4}3m$	$T_h:m3$	$O:43$	$O_h:m3m$

The group D_2 is also sometimes denoted V and called the *Vierer group* (group of four).

Source: M. Hamermesh, *Group theory*, Addison–Wesley.

(1) *The groups* C_s, C_i, C_1. If a molecule possesses a plane of symmetry as the only element apart from the identity, it is classified as belonging to the group C_s. An example is the quinoline molecule (Fig. 16.7a). If a molecule possesses the identity and the inversion as the only elements, like *meso*-tartaric acid (Fig. 16.7b) it belongs to the group C_i. If it possesses no symmetry element other than the identity (like CHFClBr) it belongs to C_1 (Fig. 16.7c): the basis of this notation is described in the next paragraph.

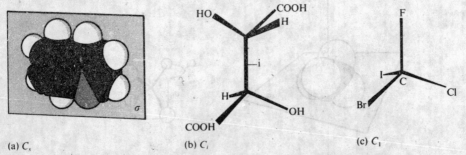

(a) C_s (b) C_i (c) C_1

Fig. 16.7. Examples of molecules belonging to the groups C_s, C_i, and C_1.

(2) *The groups C_n.* If a molecule (or any object) possesses the identity element and an n-fold axis of symmetry, it belongs to the group C_n. (Note that C_n is now playing a triple role: it is a label for one of the symmetry elements present, the label of the corresponding operation, and a name of the group.) The object of least symmetry is one possessing only the identity, E, as its sole element of symmetry. But such a molecule may also be regarded as possessing a C_1 axis, because rotating it by 360° also leaves it apparently unchanged: consequently C_1 (in this context) is simply another symbol for E. Therefore such molecules belong to the group C_1, which is characterized by the element E alone. Another example of a C_n group member is illustrated in Fig. 16.8.

(3) *The groups C_{nv}.* Objects in these groups possess a C_n axis and n vertical reflection planes σ_v. Water, for example, has the symmetry elements $(E, C_2, \sigma_v, \sigma'_v)$ and so belongs to C_{2v}. The NH_3 molecule, which possesses $(E, C_3, \sigma_v, \sigma'_v, \sigma''_v)$, or $(E, C_3, 3\sigma_v)$ for short (but see below, p. 528), belongs to C_{3v}. All groups include the identity E, and we shall no longer mention it as a requirement, except in the listing of their elements.

(4) *The groups C_{nh}.* Objects possessing a C_n axis and a perpendicular, horizontal mirror plane belong to the group C_{nh}. Note that the group C_{2h} automatically possesses the inversion element i. An example of C_{2h} is *trans* CHCl=CHCl, which, together with another example, is shown in Fig. 16.9.

(5) *The groups D_n.* Objects possessing a C_n axis and n 2-fold axes perpendicular to C_n belong to D_n. An example is shown in Fig. 16.10.

(6) *The groups D_{nh}.* Objects belong to D_{nh} if they belong to D_n and possess a horizontal mirror plane σ_h, Fig. 16.11. The flat, triangular molecule BF_3 belongs to D_{3h} because it has a C_3 axis, three 2-fold rotation axes (one along each B–F bond), and, being planar, a mirror plane perpendicular to the C_3 axis. Note that if the molecule were bent (as NH_3) the horizontal σ_h plane would disappear, and the three C_2 rotations would be replaced by three σ_v planes: this would lower the symmetry to C_{3v}. An important example of D_{6h} is the benzene molecule: it possesses the elements $(E, C_6, 6C_2, \sigma_h)$ among some others which the presence of these elements imply. A uniform cylinder belongs to $D_{\infty h}$, but a cone to $C_{\infty v}$. This implies that homonuclear diatomics belong to $D_{\infty h}$, and heteronuclear diatomics to $C_{\infty v}$.

(7) *The groups D_{nd}.* The classification D_{nd} is also based on D_n, but

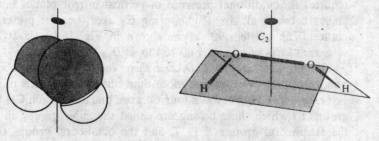

Fig. 16.8. Example of a molecule belonging to the group C_2.

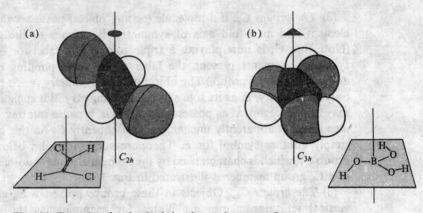

Fig. 16.9. Examples of molecules belonging to the groups C_{nh}

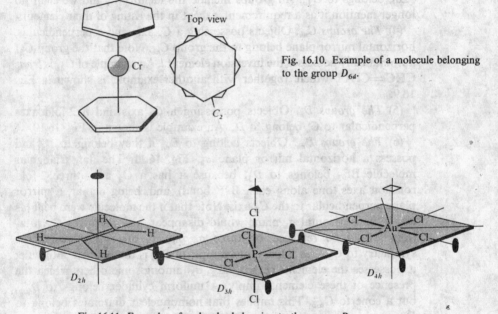

Fig. 16.10. Example of a molecule belonging to the group D_{6d}.

Fig. 16.11. Examples of molecules belonging to the groups D_{nh}.

requires the additional presence of vertical mirror planes bisecting the angles between all the neighbouring C_2 axes (i.e., the presence of $n\sigma_d$ planes). The twisted, 90° allene shown in Fig. 16.12a is D_{2d}, and the staggered form of ethane, Fig. 16.12b, is D_{3d}.

(8) *The cubic groups T, O, and their derivatives.* A number of very important molecules possess more than one principal axis of symmetry. For example, CH_4 possesses four C_3 axes, one along each CH bond. The groups to which these belong are called the *cubic groups*, in particular the tetrahedral groups T, T_d, T_h and the octahedral groups O, O_h, Fig. 16.13. The *group T_d* is the group of the regular tetrahedron (e.g., CH_4)

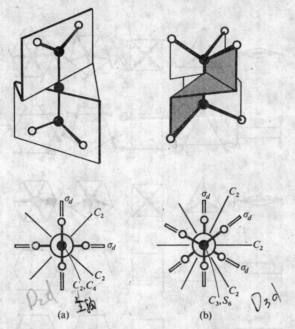

Fig. 16.12. Examples of molecules belonging to the groups D_{nd}.

(a)   (b)

and the *group* O_h is the group of the regular octahedron. If the object possesses the *rotational* symmetry of the tetrahedron or the octahedron, but none of its planes of reflection, it belongs to the simpler groups T or O. The group T_h is slightly peculiar because it is based on T, but also includes the centre of inversion.

(9) The *full rotation group*, R_3, is the group of operations shown by a spherical object. An atom belongs to R_3, but no molecule does. Exploring the consequences of R_3 symmetry is a very important way of applying group-theoretical arguments to atoms.

The identification of the symmetry elements possessed by a molecule enables it to be classified into its group. These groups are usually called *point groups* in order to distinguish them from the *space groups* we meet when dealing with translational symmetry, as in crystals. In many cases this is facilitated by comparing the structure with the shapes shown in Fig. 16.14.

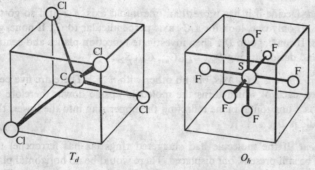

T_d O_h

Fig. 16.13. Examples of molecules belonging to the groups T_d and O_h.

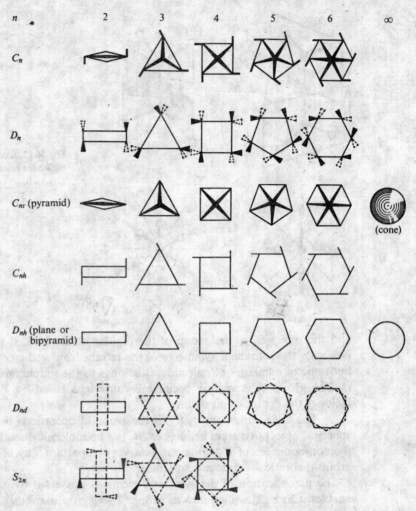

Fig. 16.14. Examples of objects belonging to a variety of point groups.

Example (Objective 2). Identify the point group to which the sandwich molecule ruthenocene (two eclipsed cyclopentadiene rings) belongs.

- *Method.* Decide if it has more than one n-fold axis, $n \geqslant 3$: if so go to the cubic groups. If only one, look for a C_2 axis perpendicular to C_n. If none, go to the C_n groups. If some, go to D_n. Then investigate reflection planes and centres of inversion, and decide among D_{nh}, D_{nd}, C_{nv}, C_{nh}, S_{2n}.

- *Answer.* There is a C_5 axis, but no other with $n \geqslant 3$. There are five perpendicular C_2 axes, each of which turns the molecule upside down: therefore turn to D_5. There is a horizontal plane reflecting the upper ring into the lower: therefore the molecule is D_{5h}.

- *Comment.* If the molecule had staggered rings (as has ferrocene) the C_2 axes would be still present, but displaced. There would be no horizontal plane, but five dihedral planes: therefore the molecule would be D_{5d}.

Some immediate consequences of symmetry. As soon as the point group of a molecule is known, it is possible to state some consequences relating to its properties. For instance, only molecules belonging to the groups C_n, C_{nv}, and C_s may have an *electric dipole moment*, and in the case of C_n and C_{nv} it must lie along the rotation axis. This can be understood as follows. If the molecule is C_n it cannot possess a charge distribution corresponding to a dipole moment perpendicular to the axis, Fig. 16.15a; but as the group makes no reference to the symmetry of the molecule between the 'top' and 'bottom' of the molecule, a charge distribution giving a dipole along the symmetry axis may exist, Fig. 16.15b. The same remarks apply to C_{nv}, which makes no reference to 'up–down' symmetry. In all the other groups, such as C_{3h}, D, etc., there are symmetry operations corresponding to turning the molecule upside down. Therefore, as well as having no dipole moment perpendicular to the axis, such a molecule can have none along the axis, otherwise turning it upside down, by reflection or rotation, would not be a symmetry operation.

The other property we can comment on is the *optical activity* of a molecule (this is discussed in Chapter 23). A molecule can rotate the plane of polarized light only if it cannot be superimposed on its mirror image. The symmetry element to look for is an *axis of improper rotation*, S_n: if one is present the object can be superimposed on its mirror image, and therefore cannot be optically active. If S_n is absent, the superposition is impossible, and optical activity may occur. Note that one has to be careful to include any improper rotation axes that may be implied by the group rather than just written explicitly. For example, the groups C_{nh} all include S_n in a concealed form because they include C_n and σ_h. Any group containing the inversion as an element also possesses at least the element S_2, because an inversion can be envisaged as a 180° rotation followed by a σ_h reflection, Fig. 16.16. It follows that molecules with centres of inversion cannot be optically active. An optically active

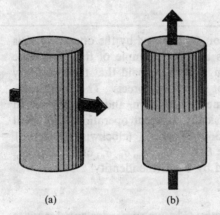

(a)   (b)

Fig. 16.15. A molecule with a C_n axis cannot have a dipole moment perpendicular to its axis (a), but one may exist parallel to the axis (b).

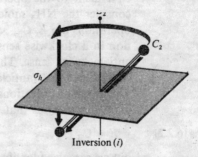

Inversion (i)

Fig. 16.16. Any group containing an inversion also possesses at least an S_2 element.

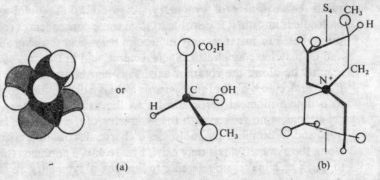

Fig. 16.17. Optically active (a) and inactive (b) molecules.

molecule is illustrated in Fig. 16.17a. Not all molecules without a centre of inversion are active: for instance if their symmetry is S_4 they lack an i element, but possess S_4, which implies inactivity. An example is given in Fig. 16.17b.

Other molecular properties may be analysed once the group structure is known, but the information is buried in the group with greater subtlety. In order to get the information we have to proceed to a more rigorous analysis of group theory, but even so we shall merely skim the surface of this very subtle and powerful subject. At this point we turn from the qualitative to the quantitative aspects of symmetry.

16.2 Groups, representations, and characters

Consider the symmetry operations of the H_2O molecule (C_{2v}). We may imagine the performance of one symmetry operation (e.g., the C_2 rotation) followed by another (e.g., σ_v'). The final arrangement of the molecule is identical to the initial, and so the sequence of two operations has the same effect as the identity operation E. We could write this symbolically as

$$(16.2.1) \qquad E = \sigma_v' C_2$$

which is shorthand for 'the operation C_2 *followed* by the operation σ_v' is equivalent to the operation E'. As another example of this procedure consider the NH_3 molecule. Although we have said that the symmetry operations are E, C_3, $3\sigma_v$, this is not sufficiently precise. The 3-fold rotation in a clockwise sense is a symmetry operation, and so is that in an anticlockwise sense. These are physically different operations. We shall denote them C_3^+ (anticlockwise by $360°/3$) and C_3^- (clockwise by $360°/3$). The group of operations is therefore E, C_3^+, C_3^-, σ_v, σ_v', σ_v'', or E, $2C_3$, $3\sigma_v$. Now it is obvious that C_3^+ followed by C_3^- is the identity:

$$(16.2.2) \qquad E = C_3^- C_3^+,$$

but we can also identify C_3^+ followed by C_3^+ (two successive anticlockwise rotations by $120°$, an overall rotation of $240°$) with a single clockwise rotation of $120°$. Symbolically this is written

(16.2.3) $C_3^- = C_3^+ C_3^+.$

Furthermore, suppose C_3^+ is followed by σ_v. Inspection of Fig. 16.18 confirms that this operation could have been achieved by a single σ_v'' operation. Symbolically

(16.2.4) $\sigma_v'' = \sigma_v C_3^+.$

(Note that the operations all refer to some fixed background of elements, and the symmetry planes, for example, are not interchanged by rotations. Furthermore, the *second* acting operator is written to the *left* of the first.) The table of all these combinations is called the *group multiplication table*, and that for C_{3v} is shown below:

first → second ↓	E	C_3^+	C_3^-	σ_v	σ_v'	σ_v''
E	E	C_3^+	C_3^-	σ_v	σ_v'	σ_v''
C_3^+	C_3^+	C_3^-	E	σ_v'	σ_v''	σ_v
C_3^-	C_3^-	E	C_3^+	σ_v''	σ_v	σ_v'
σ_v	σ_v	σ_v''	σ_v'	E	C_3^-	C_3^+
σ_v'	σ_v'	σ_v	σ_v''	C_3^+	E	C_3^-
σ_v''	σ_v''	σ_v'	σ_v	C_3^-	C_3^+	E

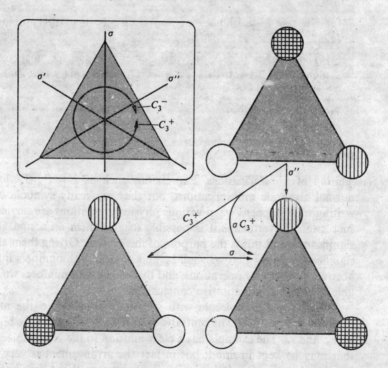

Fig. 16.18. The symmetry elements of the group C_{3v} and the composition $\sigma'' = \sigma C_3^+$.

The very important feature of all these results is that *the successive operations by a series of symmetry operations can always be expressed as a single one of the symmetry operations of the group*. This property is called the *group property*, and is the main feature of the structure of group theory. A set of operations forms a *group* if they satisfy the group property together with some other mild conditions: these are set out in Box 16.1. All symmetry operations of molecules satisfy these conditions, and that is why the theory of the symmetry of molecules is called *group theory*.

Box 16.1

Properties of a group

A group is a set of objects, or *elements* (such as symmetry elements) $G = \{g_1, g_2 \ldots g_N\}$, together with a rule of combination so that the symbol $g_i g_j$ has a well-defined meaning (such as the symmetry operation corresponding to the element g_j followed by the symmetry operation corresponding to g_i), and which satisfy the following criteria:

(1) The set includes the *identity* element; this is an element normally denoted E such that $Eg_i = g_i E = g_i$ for all the elements in the set.

(2) The set includes the inverse of every element in the set; the inverse of g_i (written g_i^{-1}) being that element for which $g_i g_i^{-1} = g_i^{-1} g_i = E$.

(3) The rule of combination is *associative*, so that the combination $(g_i g_j)g_k$ is the same as $g_i(g_j g_k)$.

(4) The combination of any two elements of the set must itself be a member of the set; that is $g_i g_j = g_k$, where g_k is a member of G. This is called the *group property*.

Note that the definition does not require $g_i g_j = g_j g_i$ (except in the special cases defining E and g_i^{-1}). Groups for which $g_i g_j = g_j g_i$ are called *commutative groups* or *Abelian groups*.

The representation of transformations. Expressions such as $E = C_3^- C_3^+$ look like normal algebraic multiplications, but they are really symbolic ways of writing what happens when various physical operations are carried out in succession. Nevertheless it is possible to give them an actual algebraic significance, and this is the purpose of this section. Giving them algebraic significance means that we shall be able to deal with numbers instead of abstract symbols for operations, and by dealing with numbers we shall be able to arrive at quantitative conclusions.

Consider a C_{3v} molecule with s-orbitals attached to the atoms as depicted in Fig. 16.19. The central s-orbital we label s_N and the remainder s_A, s_B, and s_C. The example has a clear allusion to the NH_3 molecule, and that may be kept in mind; but in fact the arrangement is very general because it applies to anything with C_{3v} symmetry. What have been called s-orbitals could be a wide variety of other functions or objects, such as p_z

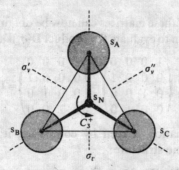

Fig. 16.19. An orbital basis for a C_{3v} molecule.

orbitals (with z parallel to the symmetry axis) or even real rubber balls.

Now consider what happens to these functions when a symmetry operation is applied to the molecule. Under the operation σ_v the following change occurs: $(s_N, s_A, s_C, s_B) \leftarrow (s_N, s_A, s_B, s_C)$. This transformation can be expressed using matrix multiplication (this subject is reviewed in the Appendix). We can find some matrix, denoted $\mathbf{D}(\sigma_v)$, such that the last relation is reproduced:

$$(16.2.5) \qquad (s_N, s_A, s_C, s_B) = (s_N, s_A, s_B, s_C) \begin{pmatrix} 1 & 0 & 0 & 0 \\ 0 & 1 & 0 & 0 \\ 0 & 0 & 0 & 1 \\ 0 & 0 & 1 & 0 \end{pmatrix}.$$

This matrix, $\mathbf{D}(\sigma_v)$, is called a *representative* of the operation σ_v. The same technique may be used to find matrices that reproduce the other symmetry operations. For instance, the C_3^+ operation has the effect $(s_N, s_B, s_C, s_A) \leftarrow (s_N, s_A, s_B, s_C)$; that is, A goes to the position initially occupied by B, B goes to the position of C, and C goes to A. This can be expressed as

$$(16.2.6) \qquad (s_N, s_B, s_C, s_A) = (s_N, s_A, s_B, s_C) \begin{pmatrix} 1 & 0 & 0 & 0 \\ 0 & 0 & 0 & 1 \\ 0 & 1 & 0 & 0 \\ 0 & 0 & 1 & 0 \end{pmatrix}.$$

The matrix here is denoted $\mathbf{D}(C_3^+)$. The operation σ_v'', which has the effect $(s_N, s_C, s_B, s_A) \leftarrow (s_N, s_A, s_B, s_C)$ can be represented by the matrix multiplication:

$$(16.2.7) \qquad (s_N, s_C, s_B, s_A) = (s_N, s_A, s_B, s_C) \begin{pmatrix} 1 & 0 & 0 & 0 \\ 0 & 0 & 0 & 1 \\ 0 & 0 & 1 & 0 \\ 0 & 1 & 0 & 0 \end{pmatrix}.$$

This matrix is denoted $\mathbf{D}(\sigma_v'')$. The matrix representative of the identity E leaves the functions (s_N, s_A, s_B, s_C) unchanged, and so it has the form

$$(16.2.8) \qquad \mathbf{D}(E) = \begin{pmatrix} 1 & 0 & 0 & 0 \\ 0 & 1 & 0 & 0 \\ 0 & 0 & 1 & 0 \\ 0 & 0 & 0 & 1 \end{pmatrix},$$

as may be verified by inspection.

A very important property of these matrices can now be obtained. Using the rules of matrix multiplication to evaluate the product $\mathbf{D}(\sigma_v)\mathbf{D}(C_3^+)$ gives

$$\mathbf{D}(\sigma_v)\mathbf{D}(C_3^+) = \begin{pmatrix} 1 & 0 & 0 & 0 \\ 0 & 1 & 0 & 0 \\ 0 & 0 & 0 & 1 \\ 0 & 0 & 1 & 0 \end{pmatrix}\begin{pmatrix} 1 & 0 & 0 & 0 \\ 0 & 0 & 0 & 1 \\ 0 & 1 & 0 & 0 \\ 0 & 0 & 1 & 0 \end{pmatrix} = \begin{pmatrix} 1 & 0 & 0 & 0 \\ 0 & 0 & 0 & 1 \\ 0 & 0 & 1 & 0 \\ 0 & 1 & 0 & 0 \end{pmatrix} = \mathbf{D}(\sigma_v'').$$

The importance of this equation is seen by comparing it with eqn (16.2.4): it has precisely the same structure:

$$\sigma_v C_3^+ = \sigma_v'' \text{ compared with } \mathbf{D}(\sigma_v)\mathbf{D}(C_3^+) = \mathbf{D}(\sigma_v'').$$

Whichever group elements are chosen, the matrix representatives multiply together analogously. Therefore *the whole of the group multiplication table is reproduced by the algebraic multiplication of the matrix representatives*. The set of six matrices is called a *matrix representation* of the C_{3v} group, and its discovery means that the link has been established between the symbolic manipulations of the group and algebraic manipulations involving numbers.

Example (Objective 4). Consider the four hydrogen 1s-orbitals of methane. Find matrix representatives for the operations C_3^+ and S_4^+ (clockwise from below) and confirm that they satisfy the group multiplication property.

- *Method.* Methane belongs to T_d. C_3^+ runs along a C–H bond, (e.g., C–H$_1$) and so it rotates the three other orbitals into each other. S_4^+ rotates clockwise by 90° about a bisector of a CH$_2$ angle (e.g., H$_1$CH$_2$) and then reflects across a horizontal plane. Find the 4×4 matrices that reproduce these changes.

- *Answer.* Under one C_3^+ operation $(H_1, H_2, H_3, H_4) \rightarrow (H_1, H_3, H_4, H_2)$ and under S_4^+, $(H_1, H_2, H_3, H_4) \rightarrow (H_3, H_4, H_2, H_1)$. These changes are also brought about by the following matrices:

$$\mathbf{D}(C_3^+) = \begin{pmatrix} 1 & 0 & 0 & 0 \\ 0 & 0 & 0 & 1 \\ 0 & 1 & 0 & 0 \\ 0 & 0 & 1 & 0 \end{pmatrix} \qquad \mathbf{D}(S_4^+) = \begin{pmatrix} 0 & 0 & 0 & 1 \\ 0 & 0 & 1 & 0 \\ 1 & 0 & 0 & 0 \\ 0 & 1 & 0 & 0 \end{pmatrix}$$

because $(H_1, H_2, H_3, H_4)\mathbf{D}(C_3^+) = (H_1, H_3, H_4, H_2)$ and $(H_1, H_2, H_3, H_4)\mathbf{D}(S_4^+) = (H_3, H_4, H_2, H_1)$, as required. To check the group property, form

$$\mathbf{D}(C_3^+)\mathbf{D}(S_4^+) = \begin{pmatrix} 1 & 0 & 0 & 0 \\ 0 & 0 & 0 & 1 \\ 0 & 1 & 0 & 0 \\ 0 & 0 & 1 & 0 \end{pmatrix}\begin{pmatrix} 0 & 0 & 0 & 1 \\ 0 & 0 & 1 & 0 \\ 1 & 0 & 0 & 0 \\ 0 & 1 & 0 & 0 \end{pmatrix} = \begin{pmatrix} 0 & 0 & 0 & 1 \\ 0 & 1 & 0 & 0 \\ 0 & 0 & 1 & 0 \\ 1 & 0 & 0 & 0 \end{pmatrix}$$

The new matrix turns the original basis into (H_4, H_2, H_3, H_1), which is the effect on the original basis of reflection in the mirror plane containing H_2 and H_3. But $C_3^+ S_4^+ = \sigma_d$ in the group, and so the group multiplication is reproduced.

- *Comment.* The matrices obtained depend on the *basis* selected (in this case the four hydrogen 1s-orbitals). This four-dimensional basis leads to a four-dimensional representation. You should sketch the effects of C_3^+, S_4^+, and their product. Does $C_3^+ S_4^+ = S_4^+ C_3^+$?

The character of symmetry operations. In common parlance the rotations C_3^+ and C_3^- of the group C_{3v} have the same character: they merely differ in direction. Likewise the three reflections have the same character, but are different from the rotations. This notion can be given precise quantitative significance.

Inspection of the matrix representation of C_{3v} using the s-orbitals as a *basis* for the discussion shows a remarkable fact. If we sum the diagonal elements of each matrix representative we get the following numbers:

	$\mathbf{D}(E)$	$\mathbf{D}(C_3^+)$	$\mathbf{D}(C_3^-)$	$\mathbf{D}(\sigma_v)$	$\mathbf{D}(\sigma_v')$	$\mathbf{D}(\sigma_v'')$
$\chi =$	4	1	1	2	2	2

Matrices representing operations of the same type are seen to have identical diagonal sums. This is a far-reaching result. We shall call the sum of the diagonal elements of the matrix representative of an operation the *character* of the operation and denote it χ (chi). Symmetry operations in the same class have the same character. Thus the two rotations constitute one class and the three reflections another.

Example (Objective 5). What are the characters of the operations C_3^+, S_4^+ and σ_d in the basis used in the last exercise for methane?

- *Method.* Refer to the D-matrices calculated in the last *Example* and sum the diagonal elements.

- *Answer.* For C_3^+: $\chi(C_3) = 1+0+0+0 = 1$

   For S_4^+: $\chi(S_4) = 0+0+0+0 = 0$

   For σ_d: $\chi(\sigma_d) = \chi(C_3 S_4) = 0+1+1+0 = 2$.

- *Comment.* A quick rule for determining the character is to count 1 every time an atom is left unchanged by the symmetry operations because only these atoms give an entry on the diagonal. In some bases there might be a sign change: if $(...f...) \rightarrow (...-f...)$ then -1 appears on the diagonal of the matrix, so count -1. The character of the identity is clearly $\chi(E) = 4$ in the present basis. The rule also gives $\chi(C_2) = 0$ for this basis.

The character of an operation depends on the basis used to set up the matrix representation. For example, if instead of considering the set of four s-orbitals, attention were confined to s_N, none of the symmetry operations change it and so every operation brings about the trivial transformation $s_N \leftarrow s_N$. Therefore every operation can be represented by the same rule $s_N = s_N 1$ where 1 may still be regarded as a matrix, but only in a trivial sense. The character of every operation is 1, because every matrix representative is 1, and so the table of characters is

	$\mathbf{D}(E)$	$\mathbf{D}(C_3^+)$	$\mathbf{D}(C_3^-)$	$\mathbf{D}(\sigma_v)$	$\mathbf{D}(\sigma_v')$	$\mathbf{D}(\sigma_v'')$
$\chi =$	1	1	1	1	1	1

It remains true that the characters of the same class of operation are equal, but the example serves to emphasize that the characters of different classes *may* be the same. Furthermore, it is obvious that, because $1 \times 1 = 1$ the matrices for this basis do reproduce all the group multiplication table, but they do so in a trivial and not particularly informative way. For this reason the representation with 1 representing every element is called an *unfaithful representation* of the group.

Irreducible representations. Although s_N alone is a basis for an unfaithful representation of the group, the representation *is* a representation, and should not be discarded as being of no interest. In fact the next few sections will show that the representation with 1 for each element is the most significant representation for many chemical applications.

When working with the basis (s_N, s_A, s_B, s_C) the matrices were 4×4, and the representation *four-dimensional*. Nevertheless, inspection of the representation shows that every matrix has the form

$$\begin{pmatrix} 1 & 0 & 0 & 0 \\ 0 & & & \\ 0 & & & \\ 0 & & & \end{pmatrix}$$

and the symmetry operations never mix s_N with the other three functions of the basis. This suggests that the basis can be cut into two parts, one consisting of s_N alone, and the other (s_A, s_B, s_C). The s_N is a basis for the unfaithful representation, as we have seen, and the other three are a basis for a three-dimensional representation consisting of the following matrices:

$$\begin{array}{cccccc} \mathbf{D}(E) & \mathbf{D}(C_3^+) & \mathbf{D}(C_3^-) & \mathbf{D}(\sigma_v) & \mathbf{D}(\sigma_v') & \mathbf{D}(\sigma_v'') \\ \begin{pmatrix} 1 & 0 & 0 \\ 0 & 1 & 0 \\ 0 & 0 & 1 \end{pmatrix} & \begin{pmatrix} 0 & 0 & 1 \\ 1 & 0 & 0 \\ 0 & 1 & 0 \end{pmatrix} & \begin{pmatrix} 0 & 1 & 0 \\ 0 & 0 & 1 \\ 1 & 0 & 0 \end{pmatrix} & \begin{pmatrix} 1 & 0 & 0 \\ 0 & 0 & 1 \\ 0 & 1 & 0 \end{pmatrix} & \begin{pmatrix} 0 & 1 & 0 \\ 1 & 0 & 0 \\ 0 & 0 & 1 \end{pmatrix} & \begin{pmatrix} 0 & 0 & 1 \\ 0 & 1 & 0 \\ 1 & 0 & 0 \end{pmatrix} \\ \chi = \quad 3 & 0 & 0 & 1 & 1 & 1 \end{array}$$

Notice that the characters still satisfy the rule about symmetry operations of the same class. These matrices are the same as those of the 4-dimensional representation except for the elimination of the first row and the first column. We say that the original 4-dimensional representation has been **reduced** to the sum (more precisely the *direct sum*) of a 1-dimensional representation spanned by s_N plus a 3-dimensional representation spanned by (s_A, s_B, s_C). This fits the common-sense view that the central s_N orbital plays a role different from the other three.

At this point the results can be expressed symbolically by letting $D^{(1)}$ denote the 1-dimensional representation, $D^{(3)}$ the 3-dimensional, and $D^{(4)}$ the 4-dimensional. Then the reduction can be symbolized by

$$D^{(4)} = D^{(1)} + D^{(3)}.$$

The fact that this is a symbolic notation should not be lost sight of: it does *not* say that the 4-dimensional matrices are the sum, in the normal sense, of the 1- and 3-dimensional matrices. But, it *does* say that the 4-member basis set can be divided into two independent bases, one with one member, and the other with three, and that the corresponding representations are respectively 1- and 3-dimensional.

The representation $D^{(1)}$, the set of matrices 1, 1, 1, 1, 1, 1, obviously cannot be reduced any further by a suitable selection from the basis (which now only has one member). Therefore $D^{(1)}$ is called an *irreducible representation* of the group. The question now arises, however, as to whether the 3-dimensional representation $D^{(3)}$ can be cut into representations of lower order.

We now show that $D^{(3)}$ is reducible, and can be expressed in terms of another 1-dimensional representation and a 2-dimensional representation, both these representations being irreducible. In order to do this we switch attention from the orbitals s_A, s_B, s_C and concentrate on three different linear combinations:

$$\begin{aligned} s_1 &= s_A + s_B + s_C \\ s_2 &= 2s_A - s_B - s_C \\ s_3 &= s_B - s_C. \end{aligned}$$

(16.2.9)

These are sketched in Fig. 16.20, and even at this stage it can be seen that, because of the presence of the node in the second and third combinations, they have a symmetry different from the first. The decomposition $D^{(3)} = D^{(1)} + D^{(2)}$ is beginning to emerge.

We can construct matrix representatives for the new basis very simply because we know how the components transform. For example, under the reflection σ_v we know that $s_A \to s_A$, $s_B \to s_C$, $s_C \to s_B$; therefore $(s_1, s_2, -s_3) \leftarrow (s_1, s_2, s_3)$, and this can be achieved by the matrix multiplication

$$(s_1, s_2, -s_3) = (s_1, s_2, s_3) \begin{pmatrix} 1 & 0 & 0 \\ 0 & 1 & 0 \\ 0 & 0 & -1 \end{pmatrix},$$

and so we find the form of the representative $\mathbf{D}(\sigma_v)$ in the new basis. The matrix representative of C_3^+ takes a little more calculation, but depends on the transformations $s_A \to s_B$, $s_B \to s_C$, $s_C \to s_A$. Substitution in the

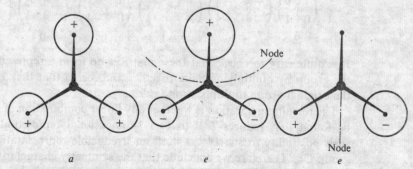

Fig. 16.20. Three (symmetry adapted) linear combinations of the orbitals shown in Fig. 16.19.

expressions for s_1, s_2, s_3 gives $(s_i, -\frac{1}{2}s_2 + \frac{3}{2}s_3, -\frac{1}{2}s_2, -\frac{1}{2}s_3) \leftarrow (s_1, s_2, s_3)$, and so the matrix representative can be obtained from

$$(s_1, -\frac{1}{2}s_2 + \frac{3}{2}s_3, -\frac{1}{2}s_2 - \frac{1}{2}s_3) = (s_1, s_2, s_3)\begin{pmatrix} 1 & 0 & 0 \\ 0 & -\frac{1}{2} & -\frac{1}{2} \\ 0 & \frac{3}{2} & -\frac{1}{2} \end{pmatrix}.$$

The complete representation and its characters may be found in this way: we obtain

$D(E)$	$D(C_3^+)$	$D(C_3^-)$	$D(\sigma_v)$	$D(\sigma_v')$	$D(\sigma_v'')$
$\begin{pmatrix} 1 & 0 & 0 \\ 0 & 1 & 0 \\ 0 & 0 & 1 \end{pmatrix}$	$\begin{pmatrix} 1 & 0 & 0 \\ 0 & -\frac{1}{2} & -\frac{1}{2} \\ 0 & \frac{3}{2} & -\frac{1}{2} \end{pmatrix}$	$\begin{pmatrix} 1 & 0 & 0 \\ 0 & -\frac{1}{2} & \frac{1}{2} \\ 0 & -\frac{3}{2} & -\frac{1}{2} \end{pmatrix}$	$\begin{pmatrix} 1 & 0 & 0 \\ 0 & 1 & 0 \\ 0 & 0 & -1 \end{pmatrix}$	$\begin{pmatrix} 1 & 0 & 0 \\ 0 & -\frac{1}{2} & \frac{1}{2} \\ 0 & \frac{3}{2} & \frac{1}{2} \end{pmatrix}$	$\begin{pmatrix} 1 & 0 & 0 \\ 0 & -\frac{1}{2} & -\frac{1}{2} \\ 0 & -\frac{3}{2} & \frac{1}{2} \end{pmatrix}$

$\chi = \quad 3 \qquad\qquad 0 \qquad\qquad 0 \qquad\qquad 1 \qquad\qquad 1 \qquad\qquad 1$

A number of important features emerge. First the characters conform to the principle concerning operations of the same class. Second, the characters are the same as for the original 3-dimensional basis. This illustrates a point that is generally valid: *taking linear combinations of a basis set leaves the characters unchanged*. Thirdly, and most important of all, the representative 3×3 matrices are all in the block diagonal form

$$\begin{pmatrix} 1 & 0 & 0 \\ 0 & & \\ 0 & & \end{pmatrix}$$

and the s_1 combination is not mixed with the other two by any symmetry transformation of the group. Therefore we have succeeded in making the reduction

$$D^{(3)} = D^{(1)} + D^{(2)},$$

where s_1 forms a basis for the same $1,1,1,1,1,1$ representation as before, and $D^{(2)}$ is a 2-dimensional representation on the basis (s_2, s_3). The matrix representation of this is the set of 2×2 matrices formed from the 3×3 representation, but with the first row and column sliced off:

$D(E)$	$D(C_3^+)$	$D(C_3^-)$	$D(\sigma_v)$	$D(\sigma_v')$	$D(\sigma_v'')$
$\begin{pmatrix} 1 & 0 \\ 0 & 1 \end{pmatrix}$	$\begin{pmatrix} -\frac{1}{2} & -\frac{1}{2} \\ \frac{3}{2} & -\frac{1}{2} \end{pmatrix}$	$\begin{pmatrix} -\frac{1}{2} & \frac{1}{2} \\ -\frac{3}{2} & -\frac{1}{2} \end{pmatrix}$	$\begin{pmatrix} 1 & 0 \\ 0 & -1 \end{pmatrix}$	$\begin{pmatrix} -\frac{1}{2} & \frac{1}{2} \\ \frac{3}{2} & \frac{1}{2} \end{pmatrix}$	$\begin{pmatrix} -\frac{1}{2} & -\frac{1}{2} \\ -\frac{3}{2} & \frac{1}{2} \end{pmatrix}$

$\chi = \quad 2 \qquad\qquad -1 \qquad\qquad -1 \qquad\qquad 0 \qquad\qquad 0 \qquad\qquad 0.$

It is quite easy to check that these matrices do form a representation of the group by multiplying pairs together and seeing that they reproduce the original group multiplication table.

The remaining question is whether any linear combination of s_2, s_3 can be found that reduces $D^{(2)}$ to two 1-dimensional representations. No such possibility exists: $D^{(2)}$ is itself an irreducible representation of the group C_{3v}. Therefore we conclude that the symmetry characteristics of s_N and s_1 are the same—they are both a basis for the same 1-dimensional irreducible representation, and the pair s_2, s_3 are of a different symmetry, but must be treated together as a pair. These features accord entirely with the diagrams of these functions and their linear combinations.

The lists of the characters of *all possible* irreducible representations of various groups are called *character tables*. The character table for the group C_{3v} is as follows:

C_{3v}	E	$2C_3$	$3\sigma_v$
A_1	1	1	1
A_2	1	1	-1
E	2	-1	0

The columns are headed by the operations that characterize the group. It is not necessary to display the character for every element because those in the same class are the same, and so the columns refer to the classes of operation, but the number 2 in $2C_3$ indicates that there are two C_3 rotations in the group. The column on the left is the name of the *symmetry species* of the irreducible representation. Where we have used $D^{(1)}$ to label a 1-dimensional representation, convention suggests that we use A. There are two species of 1-dimensional irreducible representation (distinguished by different sets of characters) in this group, and so they are labelled A_1 and A_2. 2-dimensional irreducible representations are labelled E, and this label should not be confused with the label for the identity operation.

Perhaps the most surprising thing about this table is that there are so few species of irreducible representations, yet these three exhaust all possibilities. A very elegant theorem in group theory gives the result that

> the number of species of = the number of
> irreducible representations classes.

In C_{3v} there are three classes (three columns in the character table), and so the three irreducible representations are the *only* species of irreducible representations of the group.

The whole of the discussion has centred on the group C_{3v}, but the remarks are entirely general, and the characters of the possible irreducible representations of any group may be listed. These character tables are of such usefulness and importance that a selection is given in Table 16.3 at the end of this chapter (p. 555). There is only one more task to do, and then we shall demonstrate how to use character tables in typical problems, and show why they are so important.

Transformations of other bases. The development so far has concentrated on a set of orbitals, and we have seen that s_N and s_1 behave differently from s_2 and s_3. Instead of these objects consider now the *functions* x, y, z with x, y, z measured along three orthogonal molecular axes based on the central atom. Any point in the molecule is then characterized by some set of numbers (x, y, z). This amounts to using a three-dimensional position vector $\mathbf{r} = (x, y, z)$ set in the molecule as a basis for finding a representation. When the molecule undergoes a σ_v reflection the point (x, y, z) is carried into the point $(-x, y, z)$, Fig. 16.21. (Note the fixed coordinate system is an unchanging background to the molecule.) The transformation $(-x, y, z) \leftarrow (x, y, z)$ can be written in matrix notation as

$$(-x, y, z) = (x, y, z)\begin{pmatrix} -1 & 0 & 0 \\ 0 & 1 & 0 \\ 0 & 0 & 1 \end{pmatrix}.$$

Similarly, under a $120°$ rotation C_3^+ we have $(-\tfrac{1}{2}x - \tfrac{1}{2}\sqrt{3}y, +\tfrac{1}{2}\sqrt{3}x - \tfrac{1}{2}y, z) \leftarrow (x, y, z)$, and so the representative of this operation in this basis is the matrix in the expression

$$(-\tfrac{1}{2}x - \tfrac{1}{2}\sqrt{3}y, \tfrac{1}{2}\sqrt{3}x - \tfrac{1}{2}y, z) = (x, y, z)\begin{pmatrix} -\tfrac{1}{2} & \tfrac{1}{2}\sqrt{3} & 0 \\ -\tfrac{1}{2}\sqrt{3} & -\tfrac{1}{2} & 0 \\ 0 & 0 & 1 \end{pmatrix}.$$

The complete collection of representatives gives the following representation:

$$\mathbf{D}(E) \qquad\qquad \mathbf{D}(C_3^+) \qquad\qquad \mathbf{D}(C_3^-)$$

$$\begin{pmatrix} 1 & 0 & 0 \\ 0 & 1 & 0 \\ 0 & 0 & 1 \end{pmatrix} \quad \begin{pmatrix} -\tfrac{1}{2} & \tfrac{1}{2}\sqrt{3} & 0 \\ -\tfrac{1}{2}\sqrt{3} & -\tfrac{1}{2} & 0 \\ 0 & 0 & 1 \end{pmatrix} \quad \begin{pmatrix} -\tfrac{1}{2} & -\tfrac{1}{2}\sqrt{3} & 0 \\ \tfrac{1}{2}\sqrt{3} & -\tfrac{1}{2} & 0 \\ 0 & 0 & 1 \end{pmatrix}$$

$$\chi = 3 \qquad\qquad\qquad 0 \qquad\qquad\qquad 0$$

$$\mathbf{D}(\sigma_v) \qquad\qquad \mathbf{D}(\sigma_v') \qquad\qquad \mathbf{D}(\sigma_v'')$$

$$\begin{pmatrix} -1 & 0 & 0 \\ 0 & 1 & 0 \\ 0 & 0 & 1 \end{pmatrix} \quad \begin{pmatrix} \tfrac{1}{2} & -\tfrac{1}{2}\sqrt{3} & 0 \\ -\tfrac{1}{2}\sqrt{3} & -\tfrac{1}{2} & 0 \\ 0 & 0 & 1 \end{pmatrix} \quad \begin{pmatrix} \tfrac{1}{2} & \tfrac{1}{2}\sqrt{3} & 0 \\ \tfrac{1}{2}\sqrt{3} & -\tfrac{1}{2} & 0 \\ 0 & 0 & 1 \end{pmatrix}$$

$$1 \qquad\qquad\qquad 1 \qquad\qquad\qquad 1$$

This shows that the 3-dimensional representation is reducible, because all matrices have a block-diagonal form from which the parts affecting z may be sliced off. The characters of the remaining 2-dimensional representation are

$$2 \qquad -1 \qquad -1 \qquad 0 \qquad 0 \qquad 0$$

and by comparing this with the C_{3v} character table it is clear that (x, y) constitutes a basis for the irreducible representation of symmetry species E. The conclusion of this important piece of analysis is that z behaves in one fashion, and x, y form a pair behaving in another.

Example (Objective 6). Find how the functions x, y, z transform in the group C_{2v}.

- *Method.* The group C_{2v} has the elements $E, C_2, \sigma_v, \sigma_v'$. Find the effect of each on the three functions, and then write the matrix representation. Identify the block-diagonal form.

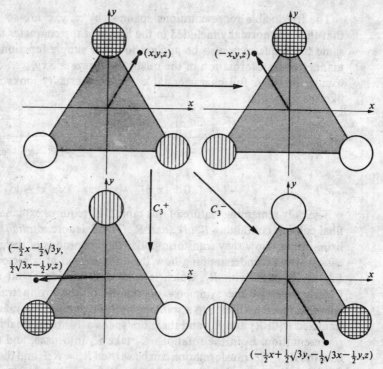

Fig. 16.21. Transformations of the functions (x, y, z) in a C_{3v} molecule.

● *Answer.* Under E, $(x, y, z) \to (x, y, z)$; under C_2, $(x, y, z) \to (-x, -y, z)$; under σ_v, $(x, y, z) \to (-x, y, z)$; under σ'_v, $(x, y, z) \to (x, -y, z)$. The matrix representation is therefore:

$$
\begin{array}{ccccc}
\text{basis} & E & C_2 & \sigma_v & \sigma'_v \\[4pt]
(x, y, z) &
\begin{pmatrix} 1 & 0 & 0 \\ 0 & 1 & 0 \\ 0 & 0 & 1 \end{pmatrix} &
\begin{pmatrix} -1 & 0 & 0 \\ 0 & -1 & 0 \\ 0 & 0 & 1 \end{pmatrix} &
\begin{pmatrix} -1 & 0 & 0 \\ 0 & 1 & 0 \\ 0 & 0 & 1 \end{pmatrix} &
\begin{pmatrix} 1 & 0 & 0 \\ 0 & -1 & 0 \\ 0 & 0 & 1 \end{pmatrix}
\end{array}
$$

This is already in block-diagonal form, and may be broken down into the following 1-dimensional representations:

$$
\begin{array}{ccccc}
x: & 1 & -1 & -1 & 1 \\
y: & 1 & -1 & 1 & -1 \\
z: & 1 & 1 & 1 & 1
\end{array}
$$

The characters of the representatives are the numbers themselves (because the matrices are 1×1), and so the irreducible representations spanned by x, y, and z are B_1, B_2, and A_1 respectively.

● *Comment.* This sequence of operations can always be used. Fortunately, for most groups the representations spanned by x, y, and z are tabulated, and the transformation properties of more complicated functions can always be expressed in terms of them.

The irreducible representations spanned by x, y, z are so important that they are normally included in the listing of the character tables. The same technique may also be applied to other simple functions, and the manner of transformation of the quadratic forms x^2, xy, xz,...z^2 is also usually listed. A full character table therefore usually looks something like the following:

C_{3v}	E	$2C_3$	$3\sigma_v$			
A_1	1	1	1	z,	$x^2+y^2+z^2$, $2z^2-x^2-y^2$,	
A_2	1	1	-1			R_z
E	2	-1	0	(x,y)	(xz, yz) (xy, x^2-y^2)	(R_x, R_y)

The only remaining feature of this table that requires explanation is the final column containing R_x, R_y, and R_z. These denote *rotations*, and their listing shows how they transform under the operations of the group. The easiest way of understanding how they are arrived at is pictorially, and Fig. 16.22 deals with R_z, which is rotational motion about the z-axis. Clearly none of the symmetry operations of the group transforms a spinning motion about z to a spinning motion about x and y, and so R_z is unmixed with R_x and R_y, and therefore spans a 1-dimensional irreducible representation. Both the rotations C_3 take R_z into itself, and so in each case the $R_z \leftarrow R_z$ transformation can be written $R_z = R_z 1$, and the character of this 1-dimensional representation is $\chi(C_3) = 1$. A reflection, however, reverses the sense of rotation, Fig. 16.22, and so under each of the σ_v operations $-R_z \leftarrow R_z$. This is represented by $-R_z = R_z(-1)$, and so the matrix representative is -1, and its character likewise. It follows that R_z spans a basis having the characters of the irreducible representation A_2. The R_x and R_y rotations are slightly more difficult to visualize, but it should be intuitively obvious that they are intermingled by the symmetry transformations, and therefore jointly span E.

All the character tables at the end of this chapter are in the form of the C_{3v} table shown above. They show at a glance how various functions, and the rotations, transform. Why this is so useful is explained in the next section.

16.3 Using character tables

Although the characters, the sums of the diagonal elements of the matrices of a representation, do not contain all the information contained in the

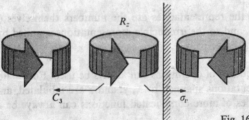

Fig. 16.22. Transformations of rotations.

matrices, they do contain enough to make them of central importance in chemistry. One of the reasons is that the character tables let one say at a glance whether an integral is zero without any need to evaluate it in detail. This saves a great deal of time, and permits a rapid qualitative assessment of the properties of molecules.

Vanishing integrals. Suppose we had to evaluate the following integral:

$$I = \int f_1(\mathbf{r}) f_2(\mathbf{r}) \, d\tau$$

where f_1 and f_2 are wavefunctions spreading through a molecule. For example, f_1 might be one atomic orbital and f_2 another; then I would be the overlap integral between them (p. 478), and if we knew that it vanished we could say without hesitation that f_1 and f_2 do not overlap to contribute to the bonding in the molecule. The key point is that *the value of the integral is independent of the orientation of the molecule*, Fig. 16.23. In terms of group theory we would say that I is unchanged by any symmetry transformation of the molecule. Therefore every symmetry transformation brings about the trivial change $I \leftarrow I$. *Therefore the integral must be a basis for the totally symmetric, 1-dimensional irreducible representation A_1 of the molecular symmetry group.*

Now suppose we know that f_1 and f_2 are each members of bases for irreducible representations. Let f_1 be a member of the basis for an irreducible representation D_1 and f_2 that for another irreducible representation D_2. How do we find out how their product $f_1 f_2$ transforms? The importance of this is that if $f_1 f_2$ does *not* transform as A_1, the integral must disappear, because I is a basis for A_1. The character tables enable us to say by inspection what irreducible representations $f_1 f_2$ spans. The rule is as follows:

(1) *Ascertain the characters for the irreducible representations spanned by the basis of which f_1 and f_2 are members, and write them in two rows in the order of the operations.*

For example: Let f_1 be the s-orbital s_N in the NH_3 molecule, and f_2 be the combination s_3, Fig. 16.20. In C_{3v} the former spans A_1 and the latter is a member of the basis for E. Therefore from the C_{3v} character table we write

$$f_1: \quad 1 \qquad 1 \qquad 1.$$
$$f_2: \quad 2 \qquad -1 \qquad 0.$$

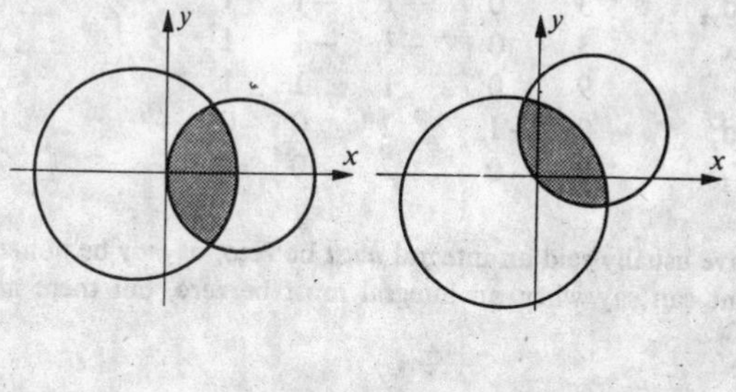

Fig. 16.23. The value of an integral (e.g. an area) is independent of the coordinate system used to evaluate it.

(2) *Multiply the numbers in each column, and write them in the same order.*

For example: The present example gives

$$f_1 f_2: \quad 2 \quad -1 \quad 0.$$

(3) *Inspect the row so produced, and see if it can be decomposed into a sum of characters of the irreducible representations of the group. If this sum does not include A_1 the integral must be zero.*

For example: In C_{3v} the numbers produced are always expressible in the form $c_{1\chi}(A_1) + c_{2\chi}(A_2) + c_{3\chi}(E)$ and the integral must disappear if $c_1 = 0$. In the present example the characters $2, -1, 0$ are those of E alone, and so the integral must be zero. Inspection of the form of these functions shows why this is so: s_3 has a node running through the s_N orbital, Fig. 16.20.

If we had taken f_2 to be the combination s_1, and f_1 to be s_N, Fig. 16.20, since each spans A_1, their characters are both 1, 1, 1 and the product is 1, 1, 1, which is A_1 itself. Therefore s_N and s_1 may have a non-vanishing overlap.†

The same technique may be used to say whether integrals of the form

$$I = \int f_1(\mathbf{r}) f_2(\mathbf{r}) f_3(\mathbf{r}) \, d\tau$$

necessarily disappear. This integral must be unchanged by any symmetry transformation of the molecule, and so it must span A_1 of the appropriate group. Therefore the triple product $f_1 f_2 f_3$ must contain a component that spans A_1. This can be found by extending the procedure already given by multiplying three sets of characters together and then seeing whether the product contains the characters for A_1. An example of this procedure is given below.

Example (Objective 7). Does the integral $\int d_{z^2} x d_{yz} d\tau$ vanish in a tetrahedral molecule?

- *Method.* Refer to the T_d character table. Find the characters of the representations spanned by d_{z^2}, x, and d_{yx}; then form $x d_{yz}$; then $d_{z^2} x d_{yz}$. Check whether the last includes A_1.

- *Answer.* From the table, x and d_{yx} are members of bases that span T_2 and d_{z^2} a member of one that spans E (the last is recognized by noting that $3z^2 - r^2 = 2z^2 - x^2 - y^2$). Draw up the following table:

	E	$8C_3$	$3C_2$	$6S_4$	$6\sigma_d$
$f_3 = d_{yz}$	3	0	-1	-1	1
$f_2 = x$	3	0	-1	-1	1
$f_2 f_3$	9	0	1	1	1
$f_1 = d_z^2$	2	-1	2	0	0
$f_1 f_2 f_3$	18	0	2	0	0

---

† Notice that we have usually said an integral *must* be zero, or *may* be non-zero. A group-theoretical argument can say when an integral must be zero, but there might be other

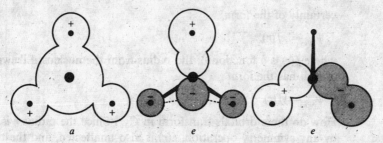

Fig. 16.24. Non-vanishing overlap integrals, and acceptable bonding orbitals in NH_3.

The characters are the sum of $A_1 + A_2 + 2E + 2T_1 + 2T_2$. Therefore the integral is not necessarily zero.

- *Comment.* Three remarks can be made. (a) This integral occurs in the theory of electronic spectra, and we encounter it again in Chapter 18. (b) The same conclusion would have been arrived at by considering d_{xy} in place of d_{yz} because d_{xy} also belongs to T_2. Closer inspection of the symmetry of the problem shows that the integral vanishes. *Be warned that arguments based on character tables show only when an integral is necessarily zero.* (c) Finding the sum of characters to see if A_1 is included could be a long job. Here is a simple rule: multiply the characters (18, 0,...) by the number of elements at the head of each column (1, 8,...) and add together the numbers so produced $(18 + 0 + ... = 24)$. Divide by the order of the group (24). That gives the number of times (1) that A_1 appears in the reducible representation.

Orbitals with non-vanishing overlap. The rules just given enable one to say immediately which atomic orbitals may have a net overlap in a molecule. We have seen that the overlap integral between s_N and the linear combination s_3 is zero in a C_{3v} molecule, and therefore the central 2s-orbital has no net interaction with that combination of orbitals. It may, however, have a non-zero overlap with the combination s_1, and so one of the molecular orbitals in NH_3 may be envisaged as arising from the overlap of these two sets of orbitals, Fig. 16.24. Because this orbital (and the antibonding partner) transform as A_1 it is called an a_1-orbital. The general rule is that *only orbitals of the same symmetry type may have non-vanishing net overlap,* and so only these orbitals may be combined into bonding and antibonding combinations.

The combinations s_2 and s_3 have E symmetry. Does the nitrogen atom carry orbitals which may have a net overlap with them? Intuition suggests that the p_x- and p_y-orbitals on nitrogen ought to be suitable. That this is the case can be checked immediately. In the hydrogen atom the $2p_x$-orbital has the algebraic form

$$\psi_{2p_x} = Nx \exp(-r/2a_0)$$

and although the nitrogen atom is not exactly the same its $2p_x$-orbital is

reasons why it is zero even though symmetry arguments permit it to be non-zero. In the present example, for instance, the s_N orbital may be so small that it does not spread into the regions where s_1 is non-zero.

certainly of the form

$$\psi_{2p_x} = xf(r),$$

where $f(r)$ is a function of the radius from the nucleus. Likewise the $2p_y$-orbital has the form

$$\psi_{2p_y} = yf(r).$$

How do these orbitals transform in C_{3v}? Since the radius r is unaffected by any symmetry operation, $f(r)$ is also unaffected, and the transformation behaviour is governed by the factors x and y. But in C_{3v} the functions x and y span E (see the table on p. 555). Therefore the p_x- and p_y-orbitals on the central nitrogen atom have E symmetry, and may have a non-vanishing overlap with the s_2 and s_3 combinations on the protons. The *e-orbitals* that result are depicted in Fig. 16.24.

The power of the method can be illustrated by asking whether any of the d-orbitals on the central atom can take part in bonding with the hydrogen atoms. It is sufficient to know that d-orbitals have the following form, whatever the atom involved:

$$d_{z^2} \propto (3z^2 - r^2)f(r) \qquad d_{x^2-y^2} \propto (x^2 - y^2)f(r)$$

$$d_{xy} \propto xyf(r) \qquad d_{yz} \propto yzf(r) \qquad d_{xz} \propto xzf(r).$$

Their symmetries can be read off the character tables simply by seeing how the quadratic forms xy, xz, etc., transform. Reference to the C_{3v} table shows that d_{z^2} has A_1 symmetry, the pair $(d_{x^2-y^2}, d_{xy})$ has E symmetry, and the pair (d_{xz}, d_{yz}) also has E symmetry. It follows immediately that a molecular orbital may be formed by overlap of the nitrogen (or phosphorus, etc.) d_{z^2}-orbital with the s_1 combination of the hydrogens, and either or both the other pairs of d-orbitals may overlap the two combinations s_2 and s_3.

Although the technique has been illustrated with the group C_{3v} it is entirely general, and the importance of knowing how s-, p-, and d-orbitals overlap is one of the reasons why the functions x, xz, etc., are listed in the character tables.

Example (Objective 8). The four hydrogen (1s-orbitals in methane span $A_1 + T_2$. With which orbitals of the carbon atom can they overlap? What if the carbon atom had d-orbitals available?

- *Method.* Look for s-, p-, and d-orbitals spanning A_1 and T_2.

- *Answer.* The A_1 combination of hydrogen orbitals can overlap with the carbon's s-orbital (which spans A_1) and the T_2 combination with the three p-orbitals (which span T_2). If d-orbitals are available, the d_{xy}, d_{yz}, and d_{zx} orbitals span T_2, and so they may form bonds and antibonds with the T_2 combination of hydrogen orbitals. No d-orbital transforms as A_1 in the group, and so both d_{z^2} and $d_{x^2-y^2}$ are non-bonding.

- *Comment.* Molecular orbitals formed from A_1-A_1 overlap are labelled a_1, and those from T_2-T_2 overlap are labelled t_2. The pair of d-non-bonding orbitals (which span E) would be called e-orbitals.

Selection rules. In Chapter 18 we shall see that the intensity of spectral lines arising from a molecular transition between some initial state with a wavefunction ψ_i and a final state with a wavefunction ψ_f depends on a quantity called the *transition (electric) dipole moment*. The z-component of this quantity is defined through

$$p_z \propto \int \psi_f z \psi_i \, d\tau,$$

and in stating when this is necessarily zero we are stating a *selection rule* for the molecule. The transition dipole moment has precisely the form of the integral already encountered, $I = \int f_1 f_2 f_3 \, d\tau$, and so we can say immediately whether particular transitions are allowed once we know the symmetry of the states involved. The transition dipole moment also has x- and y-components, defined in a similar way, and so to be sure that a certain transition cannot occur we have to investigate whether all three components p_x, p_y, and p_z are zero.

The procedure can be illustrated by considering whether a water molecule (C_{2v}) can give out a spectral line when an electron drops from an a_1-orbital (formed as illustrated in Fig. 16.25) to a b_1-orbital (also illustrated there). We have to try three possibilities, one for each of the components of the transition dipole moment. The function f_2 in $\int f_1 f_2 f_3 \, d\tau$ is x, y, or z, and reference to the table shows that these transform as B_1, B_2, and A_1 respectively. The three calculations run as follows:

	x-component				y-component				z-component			
f_1(which is A_1)	1	1	1	1	1	1	1	1	1	1	1	1
f_2(x, y, or z)	1	−1	1	−1	1	−1	−1	1	1	1	1	1
f_3(which is B_1)	1	−1	1	−1	1	−1	1	−1	1	−1	1	−1
$f_1 f_2 f_3$	1	1	1	1	1	1	−1	−1	1	−1	1	−1

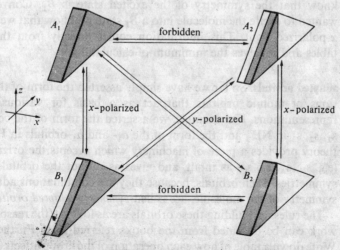

Fig. 16.26. The polarization of optical transitions in a C_{2v} molecule.

These composite characters are respectively A_1, A_2, and B_1; only the first contains A_1, and *so only the x-component of the transition dipole moment may be non-vanishing.* Therefore we can conclude that the electric dipole transition from the b_1-orbital to the a_1-orbital is allowed. We can even state that the radiation is thrown off as a photon plane-polarized in the x-direction. This is because the x-component of the transition dipole is responsible for emitting (or absorbing) light polarized in the molecule-fixed x-direction, and analogously for the y-, and z-components, Fig. 16.25.

Example (Objective 9). Is $p_x \rightarrow p_y$ an allowed (electric dipole) transition in a tetrahedral molecule?

- *Method.* See whether $\int p_y q p_x \, d\tau$, with $q = x, y, z$, spans A_1. Find whether A_1 is in $f_1 f_2 f_3$ using the technique in the *Example* on p. 542.

- *Answer.* Using the same technique as above, write

$f_1 = p_x(T_2)$	3	0	-1	-1	1
$f_2 = z(T_2)$	3	0	-1	-1	1
$f_3 = p_y(T_2)$	3	0	-1	-1	1
$f_1 f_2 f_3$	27	0	-1	-1	1

The method described in the Comment in the last *Example* on p. 543 gives the number of times A_1 occurs as 1. The same conclusion is reached with $q = x$ and $q = y$; hence $p_x \rightarrow p_y$ is symmetry-allowed in a tetrahedral molecule.

- *Comment.* Closer analysis (looking at the representatives themselves rather than at the characters, the sums of their diagonal elements) shows that only z is effective. Hence the light will be z-polarized.

This type of analysis is very useful for analysing molecular spectra, because if we know that the ground state of a C_{2v} molecule has A_1 symmetry, and in an observation of its spectrum we detect x-polarized light emitted as an excited state decays into the ground state, then we know that the symmetry of the excited state is B_1. Conversely if we wanted to excite the molecule into a B_1 state we know that we have to use x-polarized light. This information comes directly from the character tables and involves the minimum of calculation.

Symmetry-adapted orbitals. So far we have simply asserted the form of the combinations of atomic orbitals that act as a basis for various irreducible representations. For example, we asserted the form of the combinations s_1, s_2, s_3 in NH_3 and the form of the a_1- and b_1-orbitals in H_2O. Group theory provides a piece of machinery which accepts the original orbitals (e.g. s_N, s_A, s_B, s_C) as input, and gives as output the orbitals of various symmetries. Such orbitals, because they are combinations adapted to the symmetry of the molecule, are called *symmetry-adapted orbitals*.

The rules for building these orbitals are as follows (the reasons why they work can be obtained from the books referred to in Further Reading). Write down a table of how each operation of the group affects each orbital. For example, in NH_3 the original four orbitals behave as follows:

original set	s_N	s_A	s_B	s_C
under E	s_N	s_A	s_B	s_C
C_3^+	s_N	s_B	s_C	s_A
C_3^-	s_N	s_C	s_A	s_B
σ_v	s_N	s_A	s_C	s_B
σ_v'	s_N	s_B	s_A	s_C
σ_v''	s_N	s_C	s_B	s_A

In order to generate the orbital of some symmetry take each column in turn and
(i) multiply each member of the column by the character of the corresponding operation.
(ii) Add together all the orbitals in each column with the factors as determined in (i).
(iii) Divide the sum by the order of the group (the number of elements).

In A_1 all the characters are 1, and so the first column gives $s_N + s_N + \ldots = 6s_N$. The order of the group is 6, and so the orbital of A_1 symmetry that can be constructed from the nitrogen orbital is simply s_N. Applying the same technique to the column under s_A gives $\frac{1}{6}(s_A + s_B + s_C + s_A + s_C + s_B) = \frac{1}{3}(s_A + s_B + s_C)$ as the orbital of A_1 symmetry that can be built from the three hydrogen 1s-orbitals. The same A_1 orbital is generated from the remaining two columns, and so they give no further information. One of the molecular orbitals of overall A_1 symmetry (the a_1-orbital) is then just the appropriate linear combination

$$\psi_{a_1} = c_N s_N + c_H(s_A + s_B + s_C),$$

where the coefficients c_N, c_H have to be obtained by solving the appropriate Schrödinger equation and do not come directly from the symmetry of the problem.

Suppose now we make a mistake by trying to generate the molecular orbital of A_2 symmetry. The previous analysis shows that the present basis gives only a_1- and e-orbitals. What shall we get when the rules are applied? The characters for the six operations are respectively 1, 1, 1, -1, -1, -1. Therefore the first column sums to zero, but so also do the other three. Therefore the appropriate linear combinations of A_2 symmetry vanish for both the nitrogen and the hydrogen sets of orbitals. The rules even eliminate mistakes.

When we try to generate the two e-orbitals the rules leave some extra work to do. This always happens for representations of dimension greater than 1 because they have been constructed on the properties of the characters, and as these do not contain all the information in the matrices themselves, they do not always give unambiguous results. In the one-dimensional representations the characters are also the representatives, and so the ambiguity is absent. In fact, since the characters are sums of elements, the rules give the sum of the orbitals forming a basis for E.

Table 16.2. The 7 crystal systems and the 32 crystal classes

The systems	The classes	
	Schoenflies notation	Hermann–Mauguin notation
1. Triclinic	C_1, C_i	$1, \bar{1}$
2. Monoclinic	C_s, C_2, C_{2h}	$m, 2, 2/m$
3. Orthorhombic	C_{2v}, D_2, D_{2h}	$2mm, 222, mmm$
4. Rhombohedral	$C_3, C_{3v}, D_3, D_{3h}, S_6$	$3, 3m, 32, \bar{6}2m, \bar{3}$
5. Tetragonal	$C_4, C_{4v}, C_{4h}, D_{2d}, D_4,$ D_{4h}, S_4	$4, 4mm, 4/m, 422, \bar{4}2m,$ $4/mmm, \bar{4}$
6. Hexagonal	$C_6, C_{6v}, C_{6h}, C_{3h}, D_{3d},$ D_6, D_{6h}	$6, 6mm, 6/m, \bar{6}, \bar{3}m,$ $622, 6/mmm$
7. Cubic, or Regular	T, T_d, T_h, O, O_h	$23, \bar{4}3m, m3, 43, m3m$

Source: M. Hamermesh, *Group theory*, Addison–Wesley.
Trigonal is also called rhombohedral.

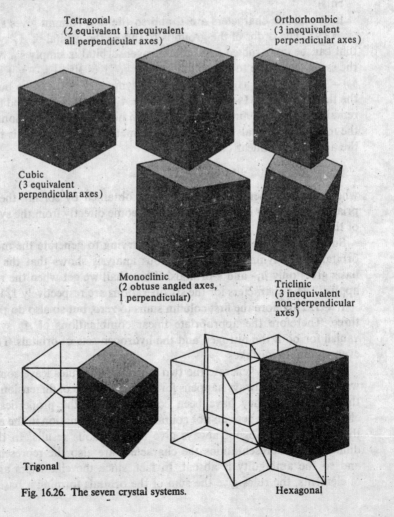

Tetragonal
(2 equivalent 1 inequivalent
all perpendicular axes)

Orthorhombic
(3 inequivalent
perpendicular axes)

Cubic
(3 equivalent
perpendicular axes)

Monoclinic
(2 obtuse angled axes,
1 perpendicular)

Triclinic
(3 inequivalent
non-perpendicular
axes)

Trigonal

Hexagonal

Fig. 16.26. The seven crystal systems.

This can be seen happening as follows. The characters of the operations in the representation E are respectively 2, -1, -1, 0, 0, 0. Therefore the column under s_N gives $2s_N - s_N - s_N + 0 + 0 + 0 = 0$. The other columns give $\frac{1}{6}(2s_A - s_C - s_B)$, $\frac{1}{6}(2s_B - s_A - s_C)$, and $\frac{1}{6}(2s_C - s_B - s_A)$. But we cannot have a basis of three orbitals for a two-dimensional representation, and in fact these three orbitals are not linearly independent (any one can be expressed as the sum of the other two). If we take the difference of the second pair we obtain $\frac{1}{2}(s_B - s_C)$, and so that and the first of the three are taken to be the pair of e-orbitals for the molecule.

The technique illustrated here for C_{3v} can be applied to all the molecular point groups, and so the rules will construct any required symmetry-adapted combination.

16.4 The symmetry of crystals

At an early stage it was suggested that the regular external form of crystals implied internal regularity. That crystals are regular symmetrical bodies suggests that the techniques described in this chapter should be a basis both for their classification and for a rapid assessment of their physical properties. Furthermore, since crystals are composed of ions, atoms, or molecules stacked together in a manner that is responsible for the external appearance or *morphology* (*morphos* is the Greek for 'form') of the crystal, it is also reasonable to expect group theory to provide a way of discussing how the local symmetry of molecules in the crystal can account for their overall symmetry.

Crystals from the outside: symmetry and classification. Inspection of a wide variety of crystals leads to the conclusion that all can be regarded as conforming to one of seven regular figures. These basic regular figures are called the seven *crystal systems*, Fig. 16.26 and Table 16.2. Which system a given crystal belongs to is determined by measuring the angles between its faces and deciding how many axes are needed to define the principal features of its shape. For example, if three equivalent and mutually perpendicular axes are required, the crystal belongs to the cubic system, Fig. 16.27a. If one principal axis (b in Fig. 16.27b) perpendicular to two that make an obtuse angle are required, the crystal belongs to the *monoclinic system*.

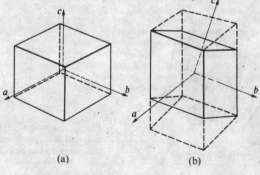

(a)   (b)

Fig. 16.27. Crystals belonging to (a) the cubic system and (b) the monoclinic system.

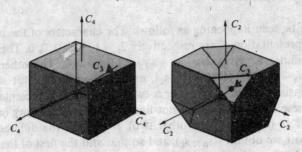

Fig. 16.28. Classification taking rotation into account.

The basic systems can also be discussed in terms of their symmetry. The cubic system, for instance, has four 3-fold axes (and three 4-fold axes), Fig. 16.28 while the monoclinic system has one 2-fold axis. The *essential* symmetries are also listed in Fig. 16.26. Crystals must possess these in order to belong to the particular system, but they may also possess others, in which case they belong to different *classes* of a particular system. For instance, the crystals shown in Fig. 16.28 both belong to the cubic system, but one has a 4-fold axis while the other has only 2-fold axes (and has the symmetry of a tetrahedron). When the additional symmetries are taken into account it turns out that there are 32 possible crystal classes. The way they are distributed over the crystal systems is shown in Table 16.2. Crystallographers favour the Hermann–Mauguin system of nomenclature over the Schoenflies (p. 522) and so the table contains a dictionary for conversion.

It is important to emphasize that at this stage we have simply reported on the empirical results from the morphological analysis of crystals. The system and class of a crystal may be determined by examining its appearance and measuring the angles between its faces. The only difficulty in the procedure is that crystal faces may grow at different rates, and so the appearance of the crystal may be distorted from the ideal forms illustrated in Fig. 16.26, see Fig. 16.29, and deciding on the symmetry class is clouded by the presence of this kind of distortion. That, however, is a technical problem which need not delay us at this stage.

Crystals from the inside: lattices and unit cells. A crystal is composed of an array of ions, atoms, or molecules. The regular external morphology suggests that the crystal is formed from small units which are themselves symmetrical. These basic units, which consist of a few atoms or molecules, are called

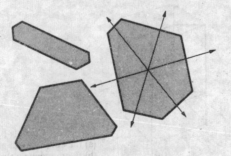

Fig. 16.29. Several crystals showing different face developments, but all belonging to the hexagonal system.

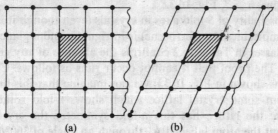

(a)   (b)

Fig. 16.30. Two unit cells of two-dimensional lattices.

the *unit cells* of the lattice. If we imagine a 2-dimensional lattice of points, like those shown in Fig. 16.30, it is easy to see that the unit cells must possess the overall symmetry of the crystal. Take for example the cubic lattice in Fig. 16.30a: the overall crystal has a C_4 axis perpendicular to the plane, and this implies that the unit cell must also have a C_4 axis, for otherwise the crystal itself would not possess one. It follows that we should expect to be able to account for the crystal morphology by inspecting the symmetry of the unit cells. Some of the ways of stacking together unit cells to produce crystal faces are illustrated in Fig. 16.31: the morphology depends on the rate of growth of the different types of face, but the underlying unit cell structure is uniform.

A unit cell cannot have a symmetry corresponding to any arbitrarily chosen point group. The reason for this lies in the requirement that the unit cells have a shape that allows them to be stacked together and fill all space (or at least the space occupied by the relatively huge macroscopic crystal). This requirement has a dramatic effect on the symmetries that are permissible. The reason can be seen very clearly in the case of a 2-dimensional lattice, where there are only five shapes of unit cell capable of filling a plane, and these have the rotational symmetry C_1, C_2, C_3, C_4, or C_6. No other rotational symmetries are possible: this corresponds to the impossibility of covering a floor with regular pentagons (C_5) or

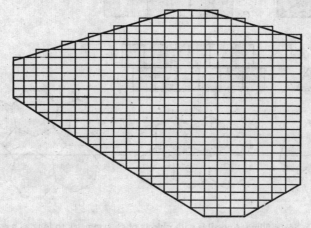

Fig. 16.31. Crystal form and unit-cell stacking.

regular *n*-gons, $n \geqslant 7$, Fig. 16.32.

The impossibility of 5-fold axes in crystals (even though it is permitted in isolated molecules, e.g., ferrocene) is intriguing, and a glance at the list of crystal classes in Table 16.2 confirms the absence of any group with a 5-fold axis. The proof that it cannot occur runs as follows. Consider the line of atoms shown in Fig. 16.33: we can imagine that this line has been isolated from some crystal lattice which shows *n*-fold rotational symmetry. Since the lattice has this *n*-fold symmetry the crystal may be rotated about the atom labelled A_3 through an angle of $360°/n$, and after this rotation the lattice must be indistinguishable from its initial form. That implies that in the full lattice there must also be an atom at the lattice point A_4' etc. The same rotation may be applied about any other atom on the lattice and all the points of the crystal will be generated (we ignore the presence of edges in the crystal, and consider only an infinite lattice). Rotations through $-360°/n$ also generate lattice points in the same way. For example, a rotation of $-360°/n$ about atom A_2 carries A_1 into the lattice point A_1''.

Now we come to the crux of the argument. Consider the distance between the lattice points on the original line; let it be a. The lattice points A_1'' and A_4' are separated by $a + 2a\cos(360°/n)$, and lie parallel to the original line. If, however, the atoms are to give a regular, periodic array, their separation must be an integral multiple of a. Therefore $\cos(360°/n)$, which cannot exceed unity, is limited to the values $1(n = 1, 2)$, $\frac{1}{2}(n = 3, 6)$, and $0(n = 4)$. This implies that only C_n, $n = 1, 2, 3, 4, 6$, are allowed.

When we turn to 3-dimensional arrays there are found to be 14 types of unit cell that can stack together and fill all space: these are the *Bravais lattices*. They are illustrated in Fig. 16.34. The lattices with points only at the corners are called *primitive*; when they contain a point at the centre

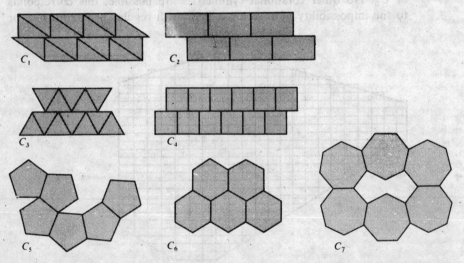

Fig. 16.32. Space filling is possible with objects of C_n symmetry so long as $n = 1,2,3,4,6$.

they are *body-centred*; and when they have atoms in their faces they are called *face-centred*. Note that the 14 Bravais lattices fall into seven groups (as represented by each primitive cell) and these regular figures have precisely the same symmetries as the crystal systems. Thus we arrive at the first underlying feature of the structure of crystals: the seven crystal systems reflect the existence of the seven regular shapes that may be packed together to fill space.

This observation can be extended from the crystal systems to the crystal classes. *The occurrence of the crystal classes indicates the presence of the corresponding symmetry in the individual unit cells.* Therefore a tetrahedral crystal morphology indicates the presence of a unit cell with tetrahedral symmetry. The value of studying crystal morphology is now apparent: from external observations we are able to state the symmetry of the unit cells, and therefore the way the molecules pack together to give the crystal. The actual positions of the atoms, however, cannot yet be stated, only the symmetry of their arrangement; but even that is valuable information. In Chapter 22 we shall examine how the detailed structure is determined.

Space groups: arranging unit cells in space. How do unit cells of a given symmetry stack together? For example, a brick wall can be built in different patterns even though the unit cells (the bricks) are all the same. The two complications are that in a crystal the problem is 3-dimensional and the unit cells are not necessarily rectangular. This problem involves thinking about *translation symmetry* as well as the local symmetry. This points to an examination of *space groups*, the symmetry of objects arranged in infinite arrays, filling all space. As well as the rotations, etc. of the point groups we now have to consider three other symmetry operations connected with displacement in space. The first of these is the *simple translation*, Fig. 16.35a. This just moves an object through some distance in a straight line. The second is the *screw axis* which twists by some fraction of 360° as well as translating, Fig. 16.35b. The third is a *glide plane*, which is a translation followed by a reflection across a plane containing the translation axis, Fig. 16.35c. The combination of these symmetry elements with the 32 classes of point groups of the unit cells cannot be done in an arbitrary fashion, and so only a limited number of space groups can be constructed. In fact there are only 230 space groups, and these account for the structure

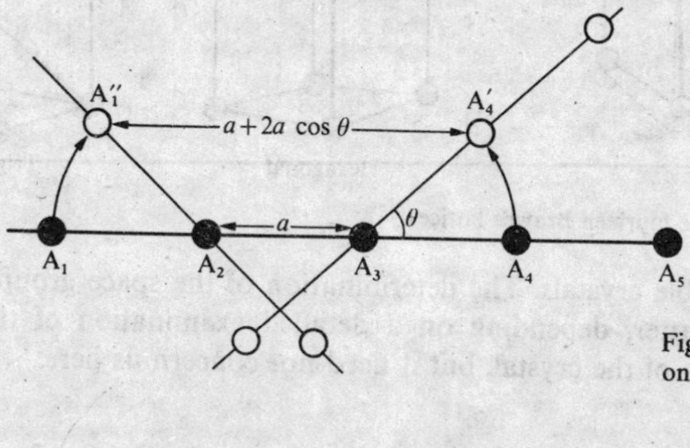

Fig. 16.33. Why *n* is permitted only the values 1,2,3,4,6.

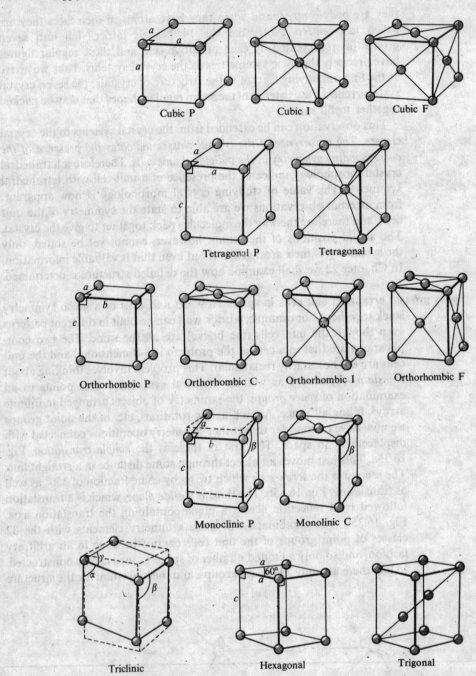

Fig. 16.34. The fourteen Bravais lattices.

of all possible crystals. The determination of the space group is a complicated matter, depending on a detailed examination of the internal constitution of the crystal, but it need not concern us here.

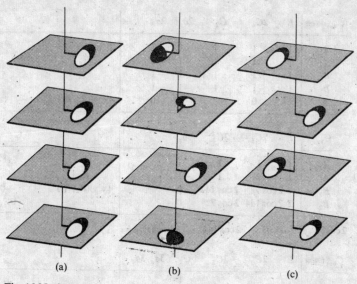

Fig. 16.35. A translation, a screw axis, and a glide plane.

Properties of crystals. Just as a knowledge of the symmetry group of molecules enables one to make immediate statements about its physical properties, the same may be done for crystals. For example, although the individual units composing a crystal might not themselves be optically active, because individually they are superimposable on their mirror image (p. 527), the same might not be true of the overall crystal that they constitute. Thus in order to determine whether or not a crystal is optically active its symmetry should be inspected to see if it lacks an improper-rotation axis (recall p. 527). Quartz, for example, has symmetry 32 (which translates into T in the Schoenflies system), and is optically active. On the other hand, calcite has a centre of symmetry (its symmetry is $\bar{3}m$, or D_{3d}) and so we know at once that it is not optically active.

Table 16.3. Character tables

$C_{2v}, 2mm$	E	C_2	$\sigma_v(xz)$	$\sigma_v'(yz)$		
A_1	1	1	1	1	z	x^2, y^2, z^2
A_2	1	1	-1	-1	R_z	xy
B_1	1	-1	1	-1	x, R_y	xz
B_2	1	-1	-1	1	y, R_x	yz

$C_{3v}, 3m$	E	$2C_3$	$3\sigma_v$		
A_1	1	1	1	z	x^2+y^2, z^2
A_2	1	1	-1	R_z	
E	2	-1	0	$(x, y)(R_x, R_y)$	$(x^2-y^2, xy)(xz, yz)$

$C_{4v}, 4mm$	E	$2C_4$	C_2	$2\sigma_v$	$2\sigma_d$		
A_1	1	1	1	1	1	z	x^2+y^2, z^2
A_2	1	1	1	-1	-1	R_z	
B_1	1	-1	1	1	-1		x^2-y^2
B_2	1	-1	1	-1	1		xy
E	2	0	-2	0	0	$(x,y)(R_x,R_y)$	(xz, yz)

C_{5v}	E	$2C_5$	$2C_5^2$	$5\sigma_v$		
A_1	1	1	1	1	z	x^2+y^2, z^2
A_2	1	1	1	-1	R_z	
E_1	2	$2\cos 72°$	$2\cos 144°$	0	$(x,y)(R_x,R_y)$	(xz, yz)
E_2	2	$2\cos 144°$	$2\cos 72°$	0		(x^2-y^2, xy)

$2\cos 72° = 0.61803$ $2\cos 144° = -1.61803$

$C_{6v}, 6mm$	E	$2C_6$	$2C_3$	C_2	$3\sigma_v$	$3\sigma_d$		
A_1	1	1	1	1	1	1	z	x^2+y^2, z^2
A_2	1	1	1	1	-1	-1	R_z	
B_1	1	-1	1	-1	1	-1		
B_2	1	-1	1	-1	-1	1		
E_1	2	1	-1	-2	0	0	$(x,y)(R_x,R_y)$	(xz, yz)
E_2	2	-1	-1	2	0	0		(x^2-y^2, xy)

$C_{\infty v}$	E	$2C_\infty^\varphi$	$\cdots$	$\infty\sigma_v$		
$A_1(\Sigma^+)$	1	1	$\cdots$	1	z	x^2+y^2, z^2
$A_2(\Sigma^-)$	1	1	$\cdots$	-1	R_z	
$E_1(\Pi)$	2	$2\cos\varphi$	$\cdots$	0	$(x,y)(R_x,R_y)$	(xz, yz)
$E_2(\Delta)$	2	$2\cos 2\varphi$	$\cdots$	0		(x^2-y^2, xy)
$E_3(\Phi)$	2	$2\cos 3\varphi$	$\cdots$	0		
$\vdots$	$\vdots$	$\vdots$		$\vdots$		

$T_d, \overline{4}3m$	E	$8C_3$	$3C_2$	$6S_4$	$6\sigma_d$		
A_1	1	1	1	1	1		$x^2+y^2+z^2$
A_2	1	1	1	-1	-1		
E	2	-1	2	0	0		$(2z^2-x^2-y^2, x^2-y^2)$
T_1	3	0	-1	1	-1	(R_x, R_y, R_z)	
T_2	3	0	-1	-1	1	(x,y,z)	(xy, xz, yz)

$O, 43$	E	$8C_3$	$3C_2$	$6C_4$	$6C_2'$		
A_1	1	1	1	1	1		$x^2+y^2+z^2$
A_2	1	1	1	-1	-1		
E	2	-1	2	0	0		$(2z^2-x^2-y^2, x^2-y^2)$
T_1	3	0	-1	1	-1	$(x,y,z)(R_x,R_y,R_z)$	
T_2	3	0	-1	-1	1		(xy, xz, yz)

Source: P. W. Atkins, M. S. Child and C. S. G. Phillips, *Tables for Group theory*, Oxford University Press.

Appendix: matrices

Matrices are simply arrays of numbers with special rules for combining them together. In general a matrix may have any shape, but we shall consider only $n \times n$ square matrices.

A square *n-dimensional* matrix is the array of n^2 *elements*

$$\mathbf{M} = \begin{pmatrix} M_{11} & M_{12} & \cdots & M_{1n} \\ M_{21} & M_{22} & \cdots & M_{2n} \\ \vdots & \vdots & \ddots & \vdots \\ M_{n1} & M_{n2} & \cdots & M_{nn} \end{pmatrix}.$$

The element in the *row r* and *column c* is denoted M_{rc} (the labels r and c are a kind of map reference). Two matrices are equal, and written $\mathbf{M} = \mathbf{N}$ only if all corresponding elements are equal: $M_{rc} = N_{rc}$ for all r and c.

The *addition* of two matrices is written $\mathbf{M} + \mathbf{N} = \mathbf{P}$ and defined through $P_{rc} = M_{rc} + N_{rc}$ (that is, corresponding elements are added). For example, if

$$\mathbf{M} = \begin{pmatrix} 1 & 2 \\ 3 & 4 \end{pmatrix} \quad \text{and} \quad \mathbf{N} = \begin{pmatrix} 5 & 6 \\ 7 & 8 \end{pmatrix}$$

then

$$\mathbf{M} + \mathbf{N} = \begin{pmatrix} 1 & 2 \\ 3 & 4 \end{pmatrix} + \begin{pmatrix} 5 & 6 \\ 7 & 8 \end{pmatrix} = \begin{pmatrix} 6 & 8 \\ 10 & 12 \end{pmatrix}.$$

The *multiplication* of two matrices is written as $\mathbf{MN} = \mathbf{P}$ and defined through $P_{rc} = \Sigma_q M_{rq} N_{qc}$. This rule can be memorized on the basis of the following diagram:

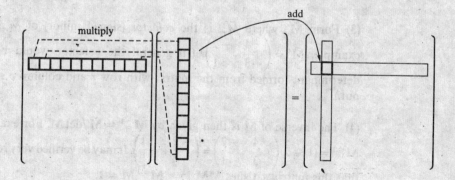

For example, using the same matrices as above,

$$\mathbf{MN} = \begin{pmatrix} 1 & 2 \\ 3 & 4 \end{pmatrix} \begin{pmatrix} 5 & 6 \\ 7 & 8 \end{pmatrix} = \begin{pmatrix} 1 \times 5 + 2 \times 7 & 1 \times 6 + 2 \times 8 \\ 3 \times 5 + 4 \times 7 & 3 \times 6 + 4 \times 8 \end{pmatrix} = \begin{pmatrix} 19 & 22 \\ 43 & 50 \end{pmatrix}.$$

Note that, in general, $NM \neq MN$. In the present case, for example, the product $NM = \begin{pmatrix} 23 & 34 \\ 31 & 46 \end{pmatrix}$. This means that matrix multiplication is *non-commutative*.

There are several types of matrix having special names or properties. Among them are the following:

A *diagonal matrix* is one in which all $M_{rc} = 0$ except those for which $r = c$.

For example, $M = \begin{pmatrix} 1 & 0 \\ 0 & 2 \end{pmatrix}$ is diagonal but $\begin{pmatrix} 0 & 1 \\ 2 & 0 \end{pmatrix}$ is not.

The *unit matrix* is written 1: it is diagonal with all non-zero elements equal to 1. Thus $1 = \begin{pmatrix} 1 & 0 \\ 0 & 1 \end{pmatrix}$ is a 2-dimensional unit matrix.

The *transpose* of a matrix M is written $\tilde{M}$; their elements are related by $\tilde{M}_{rc} = M_{cr}$. Thus if $M = \begin{pmatrix} 1 & 2 \\ 3 & 4 \end{pmatrix}$, $\tilde{M} = \begin{pmatrix} 1 & 3 \\ 2 & 4 \end{pmatrix}$ (interchange rows and columns).

The *inverse of a matrix* M is written M^{-1} and is the matrix which satisfies $MM^{-1} = M^{-1}M = 1$. It can be constructed by following the set of rules below:

(1) Find the *determinant* of the matrix M, $\det M$. For example, if $M = \begin{pmatrix} 1 & 2 \\ 3 & 4 \end{pmatrix}$, $\det M = \begin{vmatrix} 1 & 2 \\ 3 & 4 \end{vmatrix} = 1 \times 4 - 2 \times 3 = -2$. If $\det M$ is zero the matrix M is *singular* and M^{-1} does not exist (just as 0^{-1} is not defined in ordinary arithmetic). If $\det M \neq 0$ the inverse does exist, so read on.

(2) Form the transpose $\tilde{M}$ of M. For example, $\tilde{M} = \begin{pmatrix} 1 & 3 \\ 2 & 4 \end{pmatrix}$.

(3) Form $\tilde{M}'$, where $\tilde{M}'_{rc}$ is the *cofactor* (signed minor) of $\tilde{M}_{rc}$. For example, $\tilde{M}' = \begin{pmatrix} 4 & -2 \\ -3 & 1 \end{pmatrix}$. (In general the cofactor would be the determinant formed from the matrix with row r and column c struck out.)

(4) The inverse of M is then given by $M^{-1} = \tilde{M}'/\det M$. For example, $M^{-1} = (1/-2)\begin{pmatrix} 4 & -2 \\ -3 & 1 \end{pmatrix} = \begin{pmatrix} -2 & 1 \\ \frac{3}{2} & -\frac{1}{2} \end{pmatrix}$. It may be verified very readily that this matrix satisfies $MM^{-1} = M^{-1}M = 1$.

One important application of matrices (apart from their role as representatives of symmetry operations, as we saw in the chapter) is in the solution of *simultaneous equations*. Suppose you encounter a set of n simultaneous linear equations of the form

$$M_{11}x_1 + M_{12}x_2 + \dots M_{1n}x_n = c_1$$
$$M_{21}x_1 + M_{22}x_2 + \dots M_{2n}x_n = c_2$$
$$\vdots \qquad \vdots \qquad \vdots \qquad \vdots$$
$$M_{n1}x_1 + M_{n2}x_2 + \dots M_{nn}x_n = c_n,$$

and you wish to find the x. Express the equations in terms of the matrix $\mathbf{M}$ and the $1 \times n$ matrices (or *column vectors*)

$$\mathbf{x} = \begin{pmatrix} x_1 \\ x_2 \\ \vdots \\ x_n \end{pmatrix} \qquad \mathbf{c} = \begin{pmatrix} c_1 \\ c_2 \\ \vdots \\ c_n \end{pmatrix}$$

in the form

$$\mathbf{Mx} = \mathbf{c}$$

(check this by applying the matrix multiplication rule). Since $\mathbf{M}^{-1}\mathbf{M} = 1$, multiply both sides from the left by $\mathbf{M}^{-1}$ and obtain

$$\mathbf{x} = \mathbf{M}^{-1}\mathbf{c}$$

(because $1\mathbf{x} = \mathbf{x}$). This is a solution of the set of equations: all that is necessary to do is to invert the matrix, and the set of rules above tells how this is to be done.

Further reading

Symmetry. H. Weyl; Princeton University Press, Princeton, 1952.

Geometric symmetry. E. H. Lockwood and R. H. Macmillan; Cambridge University Press, Cambridge, 1978.

Symmetry in chemistry. H. H. Jaffe and M. Orchin; Wiley, New York, 1965.

Chemical applications of group theory. F. A. Cotton; Wiley, New York, 1971.

Group theory and symmetry in chemistry. L. H. Hall; McGraw-Hill, New York, 1969.

Symmetry and spectroscopy. D. C. Harris and M. D. Bertolucci; Oxford University Press, New York, 1978.

Molecular quantum mechanics. P. W. Atkins; Clarendon Press, Oxford, 1970.

Group theory and chemistry. D. M. Bishop; Clarendon Press, Oxford, 1973.

Introduction to crystal geometry. M. J. Buerger; McGraw-Hill, New York, 1971.

Tensors and group theory for the physical properties of crystals. W. A. Wooster; Clarendon Press, Oxford, 1973.

Group theory and quantum mechanics. M. Tinkham; McGraw-Hill, New York, 1964.

Tables for group theory. P. W. Atkins, M. S. Child, and C. S. G. Phillips; Clarendon Press, Oxford, 1970.

Problems

16.1. Name the point groups to which the following objects belong: a sphere, an isosceles triangle, an equilateral triangle, an unsharpened pencil, a sharpened pencil, a three-bladed propellor, a snow flake, a table, yourself.

16.2. List the symmetry elements of the following molecules and name the point groups to which they belong: NO_2(bent), CH_3Cl, CCl_3H, $CH_2:CH_2$, *cis* $CHCl:CHCl$, *trans* $CHCl:CHCl$, naphthalene, anthracene, chlorobenzene.

16.3.   Do the same for the following molecules: CH_3CH_3(staggered), cyclohexane (chair), B_2H_6, CO_2, $Co(en)_3^{3+}$ ('en' means ethylenediamine: disregard its detailed structure), S_8(crown).

16.4.   The group C_{2h} consists of the elements E, C_2, σ_h, i. Construct the group multiplication table. Find an example of a molecule possessing the symmetry of this group.

16.5.   The group D_{2h} has a C_2 axis perpendicular to the principal 2-fold axis, and it also has a horizontal mirror plane. Show that the group must also have a centre of inversion.

16.6.   Which of the molecules in Problems 16.2 and 16.3 can have a permanent electric dipole moment?

16.7.   Which of the molecules in Problems 16.2 and 16.3 may be optically active?

16.8.   Consider a water molecule (which has C_{2v} symmetry), and take as a basis for constructing molecular orbitals the two hydrogen 1s-orbitals, and the oxygen 2s- and 2p-orbitals. Set up the 6×6 matrices that reproduce the effects of the symmetry operations of the group in this basis.

16.9.   Use the matrices derived in the last Problem to confirm that they represent the group multiplications (a) $C_2\sigma_v = \sigma_v'$, (b) $\sigma_v\sigma_v' = C_2$ correctly.

16.10.   The (one-dimensional) matrices $D(C_3) = 1$, $D(C_2) = 1$ and $D(C_3) = 1$, $D(C_2) = -1$ are different matrix representatives for the group C_{6v} in the sense that they reproduce correctly the multiplication $C_3C_2 = C_6$, with $D(C_6) = 1$ for the first case and -1 for the second. Use the character table to confirm these remarks. What are the representatives of σ_v and σ_d in each case?

16.11.   One of the advantages of character tables is that one can use them to arrive at conclusions very quickly with the minimum of work. As the first of a set of exercises in their use, consider the NO_2 molecule (C_{2v}). The combination $p_{x1} - p_{x2}$, where p_{x1} and p_{x2} are the orbitals on the two oxygens (x perpendicular to the plane), spans A_2. Are there any orbitals on the central nitrogen atom with which that combination can have a net overlap? What if the central atom were sulphur?

16.12.   The ground state of NO_2 is of symmetry A_1. To what excited states may it be excited when it absorbs light (by electric dipole transition), and what is the polarization of the light that it is necessary to use?

16.13.   The ClO_2 molecule (which has C_{2v} symmetry) was trapped in a solid. Its ground state is known to be of B_1 symmetry (it has a single electron in a p_x-orbital outside a closed core, and p_x has B_1 symmetry). Light polarized parallel to the y-axis excited the molecule to an upper electronic state. What is the symmetry of that state?

16.14.   What states of (a) benzene, (b) naphthalene may be reached by the absorption of light from their A_{1g} ground states, and what are the polarizations of the transitions?

16.15.   What irreducible representations of the group T_d do the hydrogen 1s-orbitals span in methane? Are there both s- and p-orbitals on the central atom that may form bonds with them? Could d-orbitals, even if they were available on carbon, play a role in bonding in methane?

16.16. Suppose methane were distorted to (a) C_{3v} symmetry, by lengthening one bond, (b) to C_{2v} symmetry, by opening it out with a scissors action: would more d-orbitals become available for bonding?

16.17. Find the appropriate symmetry adapted combinations of the basis used in the discussion of H_2O in Problem 16.8. Deal with the oxygen and the hydrogen pair separately, and state which oxygen orbitals may have net overlap with which hydrogen orbital combinations.

16.18. Return to Problem 16.8 and confirm, by determining the trace of all the matrices, (a) that symmetry elements of the same class have the same character, (b) that the representation is reducible, (c) that the basis spans $3A_1 + B_1 + 2B_2$.

16.19. The f-orbitals are less familiar than the s-, p-, and d-orbitals, but they play an important role in lanthanide chemistry. Even though their shapes are unfamiliar, we can still draw conclusions about their properties from the viewpoint of their symmetry. But what are their symmetries? Their algebraic forms are $z(5z^2 - 3r^2)f(r)$, $y(5y^2 - 3r^2)f(r)$, $x(5x^2 - 3r^2)f(r)$, $z(x^2 - y^2)f(r)$, $y(x^2 - z^2)f(r)$, $x(z^2 - y^2)f(r)$, and $xyzf(r)$. What irreducible representations do these orbitals span in molecules with (a) C_{2v}, (b) C_{3v}, (c) T_d, (d) O_h symmetry?

16.20. Consider a lanthanide ion at the centre of (a) a tetrahedral, (b) an octahedral complex. On the basis of crystal field theory, what sets of orbitals will the seven f-orbitals break up into?

16.21. Which of the following transitions is allowed in (a) a tetrahedral complex, (b) an octahedral complex: (i) $d_{xy} \rightarrow d_{z^2}$, (ii) $d_{xy} \rightarrow f_{xyz}$?

16.22. In the text we arrived at the symmetry properties of rotations by a simple pictorial argument. Now we do it slightly more formally. This approach recognizes that the angular momentum about the z-direction is defined as $xp_y - yp_x$, where the p_x and p_y are components of linear momentum. Confirm that this z-component of angular momentum is a basis for A_2 in C_{3v}.

16.23. As what irreducible representations do translations and rotations transform in the following groups: (a) C_{2v}, (b) C_{3v}, (c) C_{6v}, (d) O_h?

16.24. The response of a molecule to an electric field is governed by the latter's ability to stretch the electronic structure (this is a translational distortion). The result of this stretching is a tendency to mix excited states into the ground state of the molecule (thus p-orbitals may be mixed into s-orbitals, and the molecule bulges on the side where the two orbitals interfere constructively). Is it necessary to consider p-, d-, and f-orbitals being mixed into the s-orbitals of the central atom of (a) octahedral complexes, (b) tetrahedral complexes?

16.25. A magnetic field tends to distort a molecule by twisting it around the direction along which it is applied. A twisting distortion has the symmetry of a rotation. Which states does the magnetic field tend to mix into the ground state in (a) NO_2, (b) an octahedral Ti^{3+} complex, (c) benzene, the field perpendicular to the plane?

16.26. Group theory gives a rapid way of detecting when an integral is zero. Here are two examples. The integral of $\sin \theta \cos \theta$ over a range that is symmetrical about $\theta = 0$ is zero. Prove this by a symmetry argument, beginning at $f_1 = \sin \theta$, $f_2 = \cos \theta$, showing that these span different irreducible representations of the group C_h (or C_s or C_i), and using the argument described in Section 16.3. Why might the integral not disappear when the range is not symmetrical about the origin?

16.27. Does the product *xyz* vanish when integrated over (a) a cube, (b) a tetrahedron, (c) a hexagonal prism, each centred on the origin?

16.28. Three crystals of iron pyrites, or fool's gold, FeS_2, are shown in the diagram. They have been quite carefully drawn from the accepted viewpoint, which K. Lonsdale, *Crystals and X-rays*, Bell (1948) describes 'as if the eye were looking down from a little above and slightly to the right, like a benevolent Conservative government'. Check that they all have all the features that make them members of the cubic system. But of what class are they?

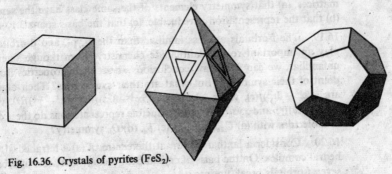

Fig. 16.36. Crystals of pyrites (FeS_2).

16.29. A crystal of sphaelerite (ZnS) is shown. To what crystal system and class does it belong?

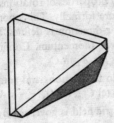

Fig. 16.37. A crystal of sphaelerite (ZnS).

16.30. The only element of symmetry possessed by blue crystals of copper sulphate pentahydrate is a centre of inversion. To what system and class do they belong?

16.31. Crystals of sucrose are found to possess one 2-fold axis as their only symmetry element. To what system do they belong?

16.32. When a crystal of rock salt is examined externally it is found to possess three 4-fold axes, four 3-fold axes, six 2-fold axes, three mirror planes, and a centre of inversion. To what system and what class do such crystals belong?

16.33. The morphology of crystals tells us a great deal about the inner symmetry, but even simple unit cells may give rise to a wide variety of external forms (the drawings of pyrites illustrate this). In order to experiment with the variety of forms that may stem from even a cubic unit cell, procure 1 lb or so of cubical sugar lumps and see what external forms may be constructed by stacking them together.

16.34. Show, using sugar cubes, that an equilateral triangular face may be formed at a corner of an otherwise perfect cube, and calculate the angle that it makes to the uppermost flat surface.

17 Determination of molecular structure: rotational and vibrational spectra

Learning objectives

After careful study of this chapter you should be able to

(1) List the basic features of *absorption, emission,* and *Raman spectroscopy* (p. 564).

(2) Relate the intensity of a spectral line to the population of the initial state (eqn (17.1.2)).

(3) State the *gross selection rules* for *rotational, vibrational,* and *electronic* transitions (p. 566).

(4) Account for the *width* of a *spectral line* in terms of the *Doppler effect* and *lifetime broadening* (p. 568).

(5) Explain the processes of *spontaneous* and *stimulated emission,* and calculate their relative rates (p. 571 and eqn (17.1.5)).

(6) Deduce the *rotational energy levels* of molecules in terms of their moments of inertia (eqns (17.2.3) and (17.2.5)) and explain the significance of the quantum numbers J, K (p. 576).

(7) Apply the *rotational selection rules* to account for pure rotational spectra, and interpret rotational spectra in terms of molecular geometry (p. 577).

(8) State the gross and specific selection rules for *rotational Raman spectra* and account for the formation of *Stokes* and *anti-Stokes lines* (p. 580).

(9) Express the *vibrational energy levels* of a diatomic molecule in terms of the force constant and masses of the atoms, with and without allowing for the effects of anharmonicity (eqn (17.3.2) and eqn (17.3.5)).

(10) Apply the selection rules to account for the vibrational spectrum of a diatomic molecule (p. 591).

(11) Interpret the vibrational spectrum in terms of the *Morse potential* (p. 585) and find the *dissociation energy* (p. 584).

(12) Explain the source of the *P, Q, R-branches* in a vibration–rotation spectrum (p. 589).

(13) List the selection rules and account for the *vibrational Raman spectra* of diatomic molecules (p. 591).

(14) Describe the significance of and count the number of *normal modes* of a polyatomic molecule (p. 592).

(15) Define the terms *infrared active and inactive,* and *Raman active and inactive,* and state the *exclusion rule* (p. 596).

Introduction

The basic reason for the appearance of spectral lines from molecules is the same as for atomic spectroscopy: when a molecule's energy drops from one value to another the excess is emitted as a photon. The difference between molecular spectroscopy and atomic spectroscopy lies in the greater variety of ways in which a molecule may change its energy, for not only may the electrons make transitions between orbitals, but the molecule may also change its energy of vibration and rotation. By examining a molecular spectrum we can obtain a great deal of information on the strength of the molecule's bonds, and about its size and shape.

The richness of information obtainable from molecular spectra as compared with atomic spectra has to be paid for by greater complexity in their appearance and interpretation. We shall see that whereas a pure rotational spectrum may be obtained, a molecular vibrational spectrum does not normally consist of lines due solely to vibrational changes but in addition shows a pattern of lines resulting from simultaneous rotational energy changes, and a molecular electronic spectrum (which is treated in the next chapter) shows a structure due to both vibrational and rotational changes. The simplest way of dealing with this complexity is to tackle each type of energy change in turn, and then to see how simultaneous excitations of different types of motion affect the appearance of spectra. All types of spectra are, however, linked by some common features, and we examine these first.

17.1 General features of spectroscopy

Spectra can be obtained experimentally in three ways: by emission spectroscopy, absorption spectroscopy, and Raman spectroscopy.

In *emission spectroscopy* a molecule drops from a high energy state into a lower one. The excess energy is emitted as a photon, and the experimenter observes the frequencies corresponding to each transition energy. If the transition is from a state of energy E_1 to a state of energy E_2, the spectrum shows a line at a frequency v given by

(17.1.1)
$$hv = E_1 - E_2.$$

Sometimes this relation is expressed in terms of the vacuum wavelength $\lambda = c/v$ or the *wavenumber* $\tilde{v} = v/c$, and the units for the latter are usually chosen as cm^{-1}. The conversions between these different quantities and their relation to different parts of the electromagnetic spectrum are summarized in Fig. 17.1.

In *absorption spectroscopy* the experimenter observes what frequencies of radiation are absorbed from incident radiation as it passes through the sample. If light of frequency v is absorbed, it signifies that the molecule can be excited by a photon of energy hv. This is possible if the molecule, initially in a state of energy E_2, can be excited to a state of energy E_1, with v, E_2, and E_1 satisfying the relation quoted in eqn (17.1.1). For

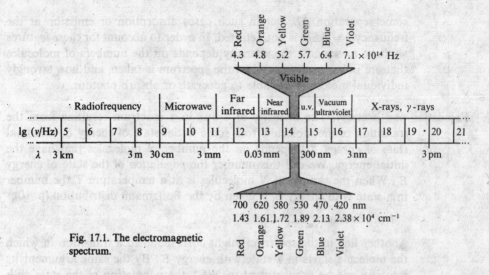

Fig. 17.1. The electromagnetic spectrum.

example, if the molecule is in its lowest energy level (its *ground state*) and absorbs blue light of wavelength 470 nm (and therefore appears red when looked at in white light) we can deduce that there is an accessible excited state at an energy corresponding to 21 000 cm^{-1} (2.6 eV) above the ground state. Absorption and emission spectroscopy both give the same information about energy levels of molecules, but practical considerations generally determine which technique is employed.

Raman spectroscopy is a technique which explores the energy levels of molecules by examining the light scattered by them. The incident beam of a single, definite frequency, consists of a stream of photons, all of the same energy. When the beam passes through the sample some of the photons collide with the molecules. When they do so, a number give up some of their energy, and emerge with a lower energy. These less energetic photons contribute lower frequencies, the *Stokes lines*, to the emergent beam. Others of the photons that collide may collect energy from the molecules (if they are already excited), and these contribute higher frequencies, the *anti-Stokes lines*, to the emergent beam. The experimenter analyses the frequencies present in the light scattered perpendicular to the incoming beam, and then interprets them in terms of the molecular energy levels. The shifts in frequency can be quite small, and unless the incident beam has a sharply defined wavelength (is almost *monochromatic*) they may be obscured. Raman spectroscopy has recently undergone considerable development because the availability of lasers has provided light sources which fit this requirement, and are so intense that smaller samples and shorter exposures can be used.

The intensity of spectral lines. A glance at any of the spectra illustrated in this chapter and the next two shows that their lines occur with a variety of intensities: some are strong, others weak. Some lines which might be expected do not appear at all; a molecule may be known to possess energy levels with some separation ΔE, but in such cases absorption or emission at the frequency $v = \Delta E/h$ is not detected. In order to account for these features we have to see how the intensity depends on the numbers of molecules that are in various states when the spectrum is taken, and how strongly individual molecules are able to generate or absorb photons.

Population and intensity. The intensity of a line in a spectrum resulting from the transition of the molecule from some initial state of energy E_i to a final state of energy E_f depends on the number of molecules that have the initial energy. We call this number the *population* of the state of energy E_i. When the sample of N molecules is at a temperature T the number in a state with energy E_i is given by the Boltzmann distribution (p. 10):

$$N(E_i) \propto N \exp(-E_i/kT).$$

Another line in the spectrum might correspond to a transition in which the molecule starts in a state with energy E_i'. By the same argument its intensity will be proportional to $N(E_i')$, the population of the state with energy E_i'. It follows that the intensities of the two lines are in the ratio

(17.1.2) $I(E_i)/I(E_i') = N(E_i)/N(E_i') = \exp[-(E_i - E_i')/kT].$

(We are ignoring for the present the possibility that factors other than populations of initial levels may affect the intensities.)

If $E_i - E_i'$ is very much bigger than kT the number of molecules with energy E_i is very much less than the number with energy E_i'. In this case the intensity of the line starting at E_i is much less than that of the line starting at E_i'. This can be illustrated by considering a sample at room temperature, when kT corresponds to 200 cm^{-1}. The first electronically excited state of a molecule is often several tens of thousands of cm^{-1} above the ground state, and so an absorption spectrum normally shows lines originating from molecules in their electronic ground states because the others are not significantly populated. Vibrational spectra of samples at room temperature can also usually be accounted for entirely in terms of transitions from the lowest vibrational state because the first vibrationally excited state lies a few hundred cm^{-1} above the ground state and is only slightly populated. In contrast, molecular rotational levels usually lie only fractions of cm^{-1} apart, and many are populated, even at room temperature; therefore a rotational absorption spectrum shows lines due to transitions of molecules from a range of populated rotational states.

Arrangements can be made to excite molecules to an artificial, non-equilibrium population distribution before their spectra are determined. For example, photolysis, an electrical discharge, or a chemical reaction may lead to molecules in a variety of electronically and vibrationally excited states, and the spectrum then shows lines due to transitions starting from these excited states.

Selection rules and intensity. In Chapter 14 we met the idea of a *selection rule* which governs whether an atomic transition was *allowed* or *forbidden*: allowed transitions appeared as lines in the spectrum; forbidden lines were absent even though energy levels of the appropriate separation were present. In this section we have to develop the idea and apply it to the various types of transitions that occur in molecules.

The basic idea is that an electromagnetic wave of frequency v is generated by an electric dipole oscillating at the frequency v. We shall now see how this picture applies to the different types of emission or absorption transitions (Raman transitions will be treated later).

(1) *Rotational transitions.* A molecule possessing a permanent dipole moment appears to be a fluctuating dipole when it rotates and the observer is in the plane of rotation, Fig. 17.2. The observer sees no fluctuating dipole if the molecule carries no permanent dipole. Therefore, *only molecules with permanent electric dipole moments can emit or absorb radiation by making a transition between different states of rotation.* The permanent dipole moment can be regarded as a handle with which the molecule can stir the electromagnetic field into oscillation (and vice versa for absorption).

Example (Objective 3). State which of the following molecules can show a pure rotational microwave spectrum: CO_2, OCS, N_2, $CH_2 : CH_2$, benzene, water. Which can show a vibrational spectrum?

- *Method.* Select the ones with permanent electric dipole moments. For the vibrational part, judge which have vibrational modes that change the dipole moment (including modes that change it from zero).

- *Answer.* OCS and H_2O are the only ones with a permanent electric dipole moment, and so only those can show a pure rotational spectrum. All the others, except N_2, possess at least one vibrational mode that changes the dipole moment, and so all except N_2 can show a vibrational spectrum.

- *Comment.* Not all the vibrational modes of complex molecules are active in the infrared. For example, the breathing mode of benzene (in which the ring swells and contracts) is *inactive*.

(2) *Vibrational transitions.* When bonds stretch or bend in a vibrating molecule the dipole moment may vary. This oscillating dipole shakes the electromagnetic field into oscillation, and so radiation may be emitted or absorbed. Some vibrations do not change the molecular dipole moment (e.g., the stretching motion of a homonuclear diatomic, Fig. 17.3) and so neither interact with nor give rise to radiation. Therefore *only vibrations that are accompanied by a changing dipole moment can be excited by, or generate, electromagnetic radiation.*

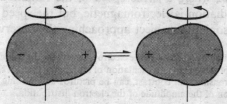

Fig. 17.2. A rotating polar molecule looks like an oscillating dipole.

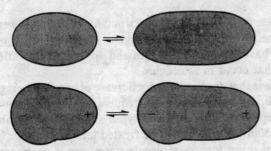

Fig. 17.3. An oscillating non-polar molecule does not look like an oscillating dipole, but a polar molecule does.

(3) *Electronic transitions*. The dipole moment of a molecule may change during an electronic redistribution. If the electron redistribution occurs symmetrically, however, there is no net change in the dipole, and the transition is forbidden. Notice the importance of focusing attention on the *change* of the distribution. For example, in an atom, which can have no permanent dipole moment, an $s \rightarrow p_z$ transition is allowed because it corresponds to a movement of charge along the z-axis.* In contrast, a $2s - 1s$ transition is spherically symmetrical. It therefore involves no transient dipole moment, and so is forbidden. In molecules we have to look for the same kind of transient dipole moment (it is called a *transition dipole moment*) associated with the redistribution of charge. An example is given in Fig. 17.4; one transition illustrated for the SO_4^{2-} molecule is allowed (it has a transition moment qualitatively similar to that for the $s \rightarrow p_z$ transition in an atom) but the second (resembling the s-s transition in an atom) has zero transition dipole, and is forbidden.

So far we have discussed the general features that a molecule must possess if it is to show an emission or an absorption spectrum. The statement of the general features is called a *gross selection rule*. In Chapter 14 we also saw that the permitted transitions could be expressed in terms of changes of a quantum number (e.g., $\Delta l = \pm 1$). The same kind of *specific selection rules* can be listed for molecular transitions. We shall see some examples of these rules when we discuss the different kinds of transitions.

Line widths. Different lines in a spectrum have different widths. In this section we shall see some of the reasons why absorption and emission lines spread over a range of frequencies.

One important mechanism that gives rise to width in a spectral line is the classical *Doppler shift*. When an object emitting radiation of wavelength λ recedes from an observer with a speed v the observer detects radiation of wavelength $(1 + v/c)\lambda$, where c is the speed of the radiation (c is the speed of light if the radiation is electromagnetic, but the speed of sound if the wave motion is sonic). An object approaching the observer with a

---

* A detailed analysis of the time-dependence of the electron density for the transitions $s \rightarrow p_z$ can be interpreted in terms of there being an oscillation of amplitude along $+z$, then along $-z$, and so on. The 1s-2s transition, in contrast, can be pictured as a pulsing, spherically symmetrical breathing motion of the amplitude of the electron distribution.

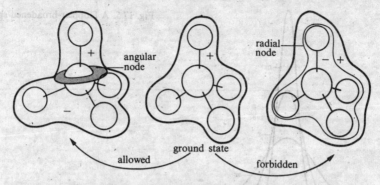

Fig. 17.4. A representation of two types of electronic transition in SO_4^{2-}.

speed v and emitting radiation of wavelength λ appears to the static observer to be emitting at a wavelength $(1-v/c)\lambda$. Molecules may attain very high speeds in a gas (p. 14), and a static observer of radiation from a collection of moving molecules sees a corresponding range of wavelengths in the light he detects. Some molecules in the sample are moving towards the observer, some away; some move quickly, others slowly. The detected line shape is the absorption or emission profile arising from all the Doppler shifts in the sample.

In order to predict the shape of the spectral line it is necessary to know the proportion of molecules having each velocity. Since this is given by the Maxwell distribution (p. 14 and in more detail on p. 865), the line shape can be calculated for any temperature. The Maxwell distribution is a bell-shaped Gaussian curve (depending on e^{-x^2}), and so the Doppler line shape is also a Gaussian curve, Fig. 17.5. The width of the line (the *width at half-height*, Fig. 17.5) is given by

(17.1.3)
$$\Delta\lambda = 2(\lambda/c)(2kT\ln 2/m)^{1/2}.$$

This increases with increasing temperature because the molecules then possess a wider range of speeds. In order to obtain lines of maximum sharpness (which is one step towards obtaining spectra in which the lines have their maximum resolution) it follows that one should attempt to work with samples at low temperatures. Another practical aspect of Doppler broadening is its application to the determination of the temperature of the surfaces of stars from the measurement of their spectral line widths.

Example (Objective 4). The sun emits a spectral line at 677.4 nm which has been identified as arising from a transition in highly ionized ^{57}Fe. Its width is 0.0053 nm. What is the temperature of the sun's surface?

● *Method.* Invert eqn (17.1.3). Use $m(^{57}\text{Fe}) = 56.94 \times 1.6605 \times 10^{-27}$ kg.

● *Answer.*

$$T = (\Delta\lambda/2\lambda)^2(mc^2/2k\ln 2)$$

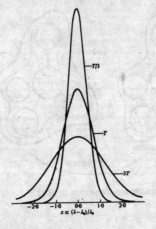

Fig. 17.5. A Doppler-broadened spectral line.

$$= \frac{(0.0053 \text{ nm}/2 \times 677.4 \text{ nm})^2 \times (9.455 \times 10^{-26} \text{ kg}) \times (2.998 \times 10^8 \text{ m s}^{-1})^2}{2 \times (1.381 \times 10^{-23} \text{ J K}^{-1}) \ln 2}$$

$$= 6.8 \times 10^3 \text{ K}.$$

● *Comment.* Notice that, for a given temperature, the width is greatest for lines of long wavelength.

Even at very low temperatures, where Doppler broadening is negligible, spectral lines do not have zero width (that is, infinite sharpness) because a mechanism called *lifetime broadening* is always in operation. When the Schrödinger equation is solved for a system that changes with time, it turns out that it is impossible to specify the energy of its quantum levels exactly. If a system changes its state at a rate $1/\tau$ (for example, if it lives for a time τ) its energy levels are blurred to an extent of the order of δE, where

(17.1.4) $\delta E \approx \hbar/\tau.$

It follows that if an excited molecule lives for about a time τ, then its energy, instead of being exactly E, may be anywhere in a range δE around the energy E. Only if the state has an infinitely long lifetime can its energy be specified exactly. This relation between energy spread and lifetime is reminiscent of the Heisenberg uncertainty relation (p. 401), and is a consequence of it.

No excited state has an infinite lifetime; therefore no excited state has a precise energy. It follows that spectral lines always spread over a range of frequencies. *Short-lived excited states are characterized by broad lines; long-lived excited states are characterized by narrower lines.* The extent of broadening can be estimated from eqn (17.1.4). Expressing δE in wavenumbers through $\delta E = hc\delta\tilde{\nu}$ and τ in seconds leads to

$$\delta\tilde{\nu}/\text{cm}^{-1} \approx 5 \times 10^{-12}(\tau/\text{s})^{-1}.$$

Therefore a lifetime of 10^{-10} s gives rise to a width of about 0.05 cm^{-1}, and whether or not that is significant depends on the resolution aimed at. We shall meet examples of lifetime broadening in subsequent sections.

Two principal mechanisms affect the lifetime of excited states. One is the *stimulated emission process* and the other the *spontaneous emission process*.

Stimulated emission. The presence of light bathing the sample acts as a stimulant for emission processes just as it is able to stimulate absorptions (the oscillating electric field shakes the molecule into an upper state, or into a lower one). It follows that the lifetime of an excited state is reduced if the molecules are bathed with light of a matching frequency. The rate of deactivation is therefore expected to obey

rate of stimulated emission $= B\mathscr{U}(v)$,

where $\mathscr{U}(v)$ is the energy density of radiation already present at the frequency of the transition. B is known as *Einstein's coefficient of stimulated emission*.

Spontaneous emission. The name spontaneous emission reflects the tendency of all excited states to discard energy and sink to a lower energy level, even though light is not present initially. The process is not really 'spontaneous' in the sense that it does not have a cause, and it is reasonable to look for a mechanism.

An explanation of the process can be found by returning to the discussion of black-body radiation (p. 388). There we saw that a fruitful model of the electromagnetic field is as a collection of oscillators, one for each possible frequency of light. When no light is present no oscillators are excited. We also saw in Chapter 13 that an oscillator does not lose all its energy even when it is in its lowest energy state (p. 413). Therefore, even when there are no excitations present (that is, no photons) the electromagnetic oscillators are not completely still. These stray, zero-point electric fields can shake the molecule and stimulate it to generate a photon. Since initially no photons are present we regard the process as spontaneous.

The total rate of deactivation of the excited state is

rate of emission $= A + B\mathscr{U}(v)$,

where A is *Einstein's coefficient of spontaneous emission*. The relative importance of the spontaneous and stimulated processes depends on the intensity and the ratio of A to B. Theoretical considerations show that

(17.1.5) $\qquad A/B = 8\pi h(v/c)^3$,

and so the spontaneous process is expected to be very important at high frequencies. For example, it takes only 10^{-7} s for an excited atom to emit a photon of visible light, but about 10^{+19} s (which is about the age of the universe) for a nucleus to reverse its magnetic moment in a magnetic field of about 1 T (the energy separation then being only about 10^{-4} cm^{-1}).

Example (Objective 5). A mercury lamp has its light concentrated into a 1 cm^2 cross-section beam, and is passed into a sample where there are excited mercury atoms. What are the relative rates of spontaneous and stimulated emission at 254 nm when the lamp is operating at 100 W?

- *Method.* Calculate A/B. The relative rates are then given by $A/B\mathcal{U}$ where $\mathcal{U}$ is to be understood as the energy per unit volume per unit frequency range. Calculate $\mathcal{U}$ by thinking in terms of photons: the lamp emits N photons in 1 second across an area 1 cm^2. Each photon moves at a speed c and so sweeps through a distance $c \times 1 \text{ s}$ in 1 second. Therefore the number of photons per unit volume is $N/(1 \text{ cm}^2) \times c \times 1 \text{ s}$. Each carries an energy $h\nu$. Therefore the energy density is $Nh\nu/c(1 \text{ cm}^2 \text{ s})$. If the width of the spectral line is $\delta\nu$, the energy per unit volume per unit frequency range is $Nh\nu/c(1 \text{ cm}^2 \text{ s})\delta\nu$. Take $\delta\nu$ equivalent to 1 cm^{-1} (i.e., $\delta\nu \approx 3 \times 10^{10} \text{ Hz}$). Note that $Nh\nu/1 \text{ s}$ is the energy per second, or power, of the lamp. Assume all the power goes into a single frequency emission line.

- *Answer.*

$$\mathcal{U} = \frac{Nh\nu}{c(1 \text{ cm}^2 \times 1 \text{ s})\delta\nu} = \frac{(100 \text{ W})}{(2.998 \times 10^{10} \text{ cm s}^{-1}) \times (1 \text{ cm}^2) \times (3 \times 10^{10} \text{ Hz})}$$

$$= 1.11 \times 10^{-19} \text{ J cm}^{-3} \text{ s}.$$

Therefore, since

$$A/B = 8\pi h(\nu/c)^3 = 8\pi h(1/\lambda)^3 = 8\pi \times (6.626 \times 10^{-34} \text{ J s}) \times (1/254 \times 10^{-7} \text{ cm})^3$$

$$= 1.02 \times 10^{-18} \text{ J cm}^{-3} \text{ s},$$

it follows that

$$A/B\mathcal{U} = (1.02 \times 10^{-18} \text{ J cm}^{-3} \text{ s})/(1.11 \times 10^{-19} \text{ J cm}^{-3} \text{ s}) = 9.16$$

- *Comment.* This shows that the rate of spontaneous emission is 9 times that of the stimulated emission. Another approach, applicable when the excited atoms are in a hot environment, would be to interpret $\mathcal{U}$ as the energy density per unit frequency range for a black-body radiator, and to use Planck's expression.

Other processes can also stimulate an excited state to drop into a lower one. One of the most important is *collisional deactivation* or *collisional broadening* in which the energy of the excited state is not radiated but is transferred into the motion of the molecule that collides with it. This mode of deactivation is very important in liquids, and for vibrational and rotational motion in gases. Then it always exceeds the natural linewidth and, except at the lowest pressures, the Doppler width too.

All these processes have the same effect. Irrespective of the method of deactivation, if the lifetime of the excited state is reduced the line in the spectrum acquires a breadth given by eqn (17.1.4). If several deactivation processes act, and each one causes the excited state to decay at a rate $1/\tau_1$, $1/\tau_2$, etc., the overall width of the line is the sum of the individual widths:

(17.1.6) $\delta E \approx \hbar(1/\tau_1 + 1/\tau_2 + \ldots).$

17.2 Pure rotation spectra

First we find expressions for the rotational energy levels of molecules in terms of their sizes and shapes, and then calculate the frequencies of the allowed transitions by applying the selection rules. In the final step we predict the appearance of the spectrum by taking into account the populations of the rotational levels when the sample is maintained at some specified temperature.

The rotational energy levels. The rotational energy levels may be obtained by solving the Schrödinger equation for the molecule. There is, fortunately, a very effective short-cut. This depends on noting that the classical expression for the energy of a body rotating about some axis x is

$$E = \tfrac{1}{2}I_{xx}\omega_x^2,$$

where ω_x is the angular velocity about that axis and I_{xx} is the moment of inertia (the two subscripts on I appear for a technical reason—moment of inertia relates a response, angular momentum, to a driving stimulus, torque, and both response and stimulus have direction; this is a technical matter which need not concern us here: just regard the subscripts as overblown labels). A body free to rotate about all three axes has an energy

$$E = \tfrac{1}{2}I_{xx}\omega_x^2 + \tfrac{1}{2}I_{yy}\omega_y^2 + \tfrac{1}{2}I_{zz}\omega_z^2.$$

The classical angular momentum of a body of moment of inertia I_{xx} and angular velocity ω_x is $J_x = I_{xx}\omega_x$; therefore, in terms of the angular momentum about each molecular axis, the energy is

$$(17.2.1) \qquad E = J_x^2/2I_{xx} + J_y^2/2I_{yy} + J_z^2/2I_{zz}.$$

This is the key equation for the rest of the section. The quantum-mechanical properties of angular momentum were outlined in Chapter 13, and can be applied to this equation in order to obtain the quantum-mechanical properties of the rotational energy.

It is convenient to break the discussion down according to the type of molecule being considered. We shall describe *spherical tops*, which are molecules with all three moments of inertia equal (such as methane), *symmetric tops*, in which two moments of inertia are the same, but different from a third (such as ammonia or methyl chloride), and *linear molecules* (such as carbon dioxide and any diatomic). The energy levels and rotational spectra of *asymmetric tops*, which have all three moments of inertia different (such as water), are very complicated and we shall not consider them any further. The energies all depend on the moments of inertia of the molecules; expressions for these are collected in Table 17.1.

Spherical top molecules. When I_{xx}, I_{yy}, and I_{zz} are identical the energy expression reduces to

$$E = (1/2I)(J_x^2 + J_y^2 + J_z^2) = J^2/2I.$$

Table 17.1. Moments of inertia

The expressions below give the moments of inertia for molecules of the type shown.

1. Diatomics

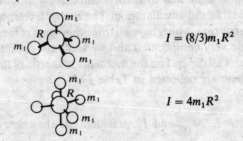

$$I = (m_1 m_2/m)R^2, \quad m = m_1 + m_2$$

2. Linear triatomics

$$I = (m_1 m_3/m)(R + R')^2 \\ + (m_2/m)(m_1 R^2 + m_3 R'^2)$$

$$I = 2m_1 R^2, \quad m = m_1 + m_2 + m_3$$

3. Symmetric tops

$$I_\parallel = 2m_1 R^2(1 - \cos\theta),$$
$$I_\perp = m_1 R^2(1 - \cos\theta)$$
$$+ (m_1/m)(m_2 + m_3)R^2(1 + 2\cos\theta)$$
$$+ (m_3/m)R'\{(3m_1 + m_2)R'$$
$$+ 6m_1 R\sqrt{[\tfrac{1}{3}(1 + 2\cos\theta)]}\}$$
$$m = 3m_1 + m_2 + m_3$$

$$I_\parallel = 2m_1 R^2(1 - \cos\theta), \quad m = 3m_1 + m_2$$
$$I_\perp = m_1 R^2(1 - \cos\theta)$$
$$+ (m_1 m_2/m)R^2(1 + 2\cos\theta)$$

$$I_\parallel = 4m_1 R^2$$
$$I_\perp = 2m_1 R^2 + 2m_3 R'^2$$

4. Spherical tops

$$I = (8/3)m_1 R^2$$

$$I = 4m_1 R^2$$

Source: C. H. Townes and A. Schawlow; *Microwave spectroscopy*, McGraw-Hill.

This is the classical expression, and J^2 is the square of the magnitude of the classical angular momentum. It can be turned into the quantum expression by introducing the quantum result that the magnitude of an angular momentum is confined to the values $[J(J+1)]^{1/2}\hbar$, where $J = 0, 1, 2, \ldots$ (p. 421). Therefore the energy of a rotating spherical top molecule is limited to the values

(17.2.2) $$E_J = J(J+1)\hbar^2/2I, \qquad J = 0, 1, 2, \ldots.$$

The factor $\hbar^2/2I$ is usually written hcB, so that B, the *rotational constant*, is expressed in wavenumbers, usually cm^{-1}. Then we have the simple expression

(17.2.3) $$E_J = hcBJ(J+1); \qquad J = 0, 1, 2, \ldots.$$

The separation in energy between neighbouring rotational levels is

(17.2.4) $$E_J - E_{J-1} = 2hcBJ, \quad or \quad \tilde{v}_J - \tilde{v}_{J-1} = 2BJ.$$

This decreases as the moment of inertia increases; large molecules have closely spaced rotational energy levels. The magnitude of the separation can be estimated by considering CCl_4. From the bond lengths and masses of the atoms we find a moment of inertia of 4.85×10^{-45} kg m², and so $hcB = 1.04 \times 10^{-22}$ J (J is joule, J is the quantum number!) or $B = 5.24$ cm^{-1}.

Symmetric top molecules. In symmetric tops $I_{xx} = I_{yy}$ but I_{zz} is different: a cylinder is an example. We shall write $I_\parallel$ for the moment of inertia parallel to the axis (I_{zz}), and $I_\perp$ ($= I_{xx} = I_{yy}$) for the moment of inertia perpendicular to it. (In group-theoretical language a symmetric top molecule is one with at least a 3-fold axis of symmetry.)

The basic classical expression gives for the energy

$$E = (1/2I_\perp)(J_x^2 + J_y^2) + (1/2I_\parallel)J_z^2.$$

This can be expressed in terms of the magnitude of the angular momentum $J^2 = J_x^2 + J_y^2 + J_z^2$ by adding and subtracting $J_z^2/2I_\perp$:

$$E = (1/2I_\perp)(J_x^2 + J_y^2 + J_z^2) - (1/2I_\perp)J_z^2 + (1/2I_\parallel)J_z^2$$

$$= (1/2I_\perp)J^2 + [(1/2I_\parallel) - (1/2I_\perp)]J_z^2.$$

In order to get the quantum expression, the square of the classical magnitude J^2 is replaced by the quantum value $J(J+1)\hbar^2$. Quantum theory also shows (p. 421) that the component of angular momentum about any axis is restricted to the values $K\hbar$, where $K = 0, \pm 1, \ldots, \pm J$. ($K$ is the quantum number denoting projection on to a molecular axis; M denotes a projection on to a laboratory axis.) Therefore the component J_z is restricted to these values. Consequently, the energy of a rotating symmetric top molecule may take the values

(17.2.5) $$E_{JK}/hc = BJ(J+1)+(A-B)K^2 \quad \begin{cases} J = 0,1,2,\ldots \\ K = 0,\pm 1,\pm 2,\ldots,\pm J \end{cases}$$

where $B = \hbar/4\pi cI_\perp$ and $A = \hbar/4\pi cI_\parallel$ (since $\hbar = h/2\pi$).

The quantum number K plays a role because the energy of rotation depends on how the total angular momentum is distributed. When K is large (near $+J$ or $-J$) most of the molecular rotation is about the symmetry axis, Fig. 17.6a, but when it is zero the whole of the motion is end-over-end rotation, Fig. 17.6b. In the former case the energy depends mainly on $I_\parallel$, and in the second entirely on $I_\perp$. Note also that the energy depends on K^2, so that the states with the same value of $|K|$, but opposite signs, have the same energy. This reflects the physical requirement that molecules with opposite signs of K are rotating about their axes at the same rate but in different directions; but the sense of rotation cannot affect the energy.

Example (Objective 6). The ammonia molecule is a symmetric top (it has a C_3 axis) with bond length 101.2 pm and bond angle 106.7°. Find its rotational energy levels.

• *Method.* Calculate the rotational constants A and B. The moments of inertia can be found from the expressions in Table 17.1. Use data for $^{14}N^1H_3$ (endpapers).

• *Answer.* In the notation of Table 17.1, $m_1 = 1.0078 \times (1.6605 \times 10^{-27}$ kg), $m_2 = 14.0031 \times (1.6605 \times 10^{-27}$ kg), $R = 101.2$ pm, $\theta = 106.7°$. Therefore

$$I_\parallel = 2 \times (1.6735 \times 10^{-27}\text{ kg}) \times (101.2 \times 10^{-12}\text{ m})^2 \times (1-\cos 107°)$$
$$= 4.4128 \times 10^{-47}\text{ kg m}^2.$$

$$I_\perp = 2.2064 \times 10^{-47}\text{ kg m}^2$$
$$+ \left\{ \frac{(1.6735 \times 10^{-27}\text{ kg}) \times (2.3252 \times 10^{-26}\text{ kg})}{3 \times (1.6735 \times 10^{-27}\text{ kg}) + (2.3252 \times 10^{-26}\text{ kg})} \times (101.2 \times 10^{-12}\text{ m})^2 \times (1+2\cos 106.7°) \right\}$$

$$= 2.8059 \times 10^{-47}\text{ kg m}^2.$$

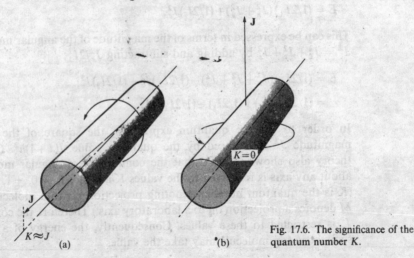

Fig. 17.6. The significance of the quantum number K.

Hence

$$A = h/4\pi cI_\parallel = (1.054\,59 \times 10^{-34} \text{ J s})/4\pi \times (4.4128 \times 10^{-47} \text{ kg m}^2) \times$$
$$(2.997\,925 \times 10^{10} \text{ cm s}^{-1})$$

$$= 6.344 \text{ cm}^{-1}.$$

$$B = h/4\pi cI_\perp = (1.054\,59 \times 10^{-34} \text{ J s})/4\pi \times (2.8003 \times 10^{-47} \text{ kg m}^2) \times$$
$$(2.997\,925 \times 10^{10} \text{ cm s}^{-1})$$

$$= 9.977 \text{ cm}^{-1}.$$

It follows that

$$E_{JKM}/\text{cm}^{-1} = 9.997J(J+1) + (6.344 - 9.977)K^2$$
$$= 9.997J(J+1) - 3.633 K^2,$$

with $J = 0, 1, 2, \ldots$ and $K = J, J-1, \ldots, -J$.

- *Comment.* In the case of $J = 1$, the energy for the molecule to rotate about its axis ($K = J$) is 16.32 cm^{-1}, but end-over-end rotation ($K = 0$) with the same total angular momentum ($J = 1$) requires an energy equivalent to 19.95 cm^{-1}.

Linear molecules. In linear molecules (such as the diatomics, CO_2, and C_2H_2) the whole of the angular momentum is about an axis perpendicular to the line of atoms. This indicates that we may use the last equation with the imposed restriction that $K \equiv 0$. Therefore the energy levels are

(17.2.6)
$$E_J = hcBJ(J+1), \qquad J = 0, 1, 2, \ldots$$

where $B = \hbar/4\pi cI$, I being the single moment of inertia ($I_\perp$).

The reason why the angular momentum must be about the axis perpendicular to the line of atoms can be seen most clearly in the case of a diatomic molecule. In order to specify the positions of the two atoms we could specify the three coordinates of each. Since the atoms can change their position by changing any one of these six coordinates we say there are six *degrees of freedom*. There is, however, an alternative way of expressing their positions. We could state the position of the centre of mass of the molecule (which takes three coordinates), the distance between the atoms (one coordinate), and the orientation of the molecule (two more). On this basis, six degrees of freedom are available, but their physical nature is now clearer. Three of the degrees of freedom (the motion of the centre of mass) are translations of the molecule as a whole; one (changing the atomic separation) is a vibration. Only two degrees of freedom remain, and these correspond to a rotation of the molecule as it changes the orientation of its axis: the rotation around the axis itself does not enter into the specification of the orientation of the molecule.

Rotational transitions. Since typical values of B for small molecules are in the region of 10 cm^{-1} (for HCl, $B = 10.59$ cm^{-1}), transitions of molecules in the region of $J \approx 10$ occur at $2BJ \approx 200$ cm^{-1}. This lies in the *microwave region* of the electromagnetic spectrum, and so pure rotational spectroscopy is a microwave technique. In order to use it, though, we have to know exactly which transitions are allowed.

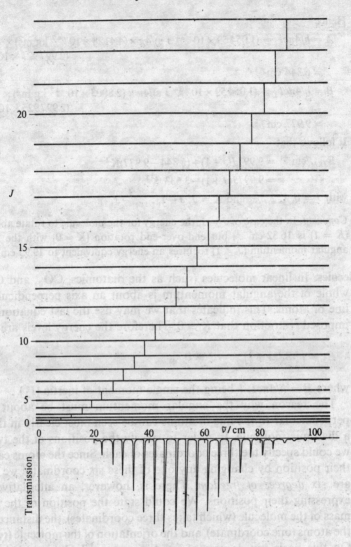

Fig. 17.7. The transitions giving rise to a pure rotation spectrum.

The *gross selection rule* was introduced on p. 564: a molecule must possess a permanent electric dipole moment if it is to show an emission or absorption rotational spectrum. Therefore, neither a spherical top molecule nor a homonuclear diatomic can show a pure rotational spectrum. A symmetric top molecule must have its dipole moment (if it has one at all) along its symmetry axis (otherwise it would not be axially symmetric). Therefore it cannot be accelerated into different states of rotation about this axis by emission or absorption of light. It follows that we can expect to observe rotational spectra only from heteronuclear diatomics, other polar linear molecules, and from end-over-end rotation of polar symmetric tops.

The *specific selection rule* (which tells us which changes may occur in J and K) can be obtained in the same way as for atoms (p. 449). Since a photon has a spin angular momentum, when it is ejected from a molecule during a rotational transition the angular momentum of the molecule must change by a compensating amount. The photon carries one unit of angular momentum, and so in a transition the molecule's angular momentum must change by one unit. Therefore we arrive at the selection rule

$$\Delta J = \pm 1.$$

We have already seen that the light cannot accelerate rotation around the symmetry axis of a symmetric top, and so another selection rule is

$$\Delta K = 0.$$

Applying these rules to the energy expressions gives the allowed absorption wavenumbers $(\tilde{\nu})$ for the transitions $J + 1 \leftarrow J$ as

(17.2.7) $$\tilde{\nu} = 2B(J+1); \qquad J = 0, 1, 2, \ldots.$$

The spectrum that results is shown in Fig. 17.7. The most significant feature is that it consists of a series of lines with wavenumbers $2B, 4B, 6B, \ldots$, all with separation $2B$. Measuring the spacing gives B, and hence the moment of inertia perpendicular to the symmetry axis of the molecule. Since the masses of the atoms in the molecule are known it is a simple matter to deduce information about the shape and size of the molecule. Unfortunately, though, in a molecule like A–B–C the analysis leads to a single quantity $I_\perp$, but there are *two* bond lengths. This deficiency can be overcome by using an isotopically substituted molecule, such as A′–B–C, and determining its moment of inertia too. Then, by assuming that $R(A\text{–}B) = R(A'\text{–}B)$, both $R(A\text{–}B)$ and $R(B\text{–}C)$ can be extracted from these two measurements. A famous example of this procedure is the study of OCS, and the actual calculation is set in Problems 17.23–4.

Example (Objective 7). Predict the form of the rotational spectrum of ammonia.

- *Method.* Energy levels were calculated in the last *Example*. The molecule has a permanent electric dipole (along C_3), and the selection rules are $\Delta J = \pm 1$, $\Delta K = 0$.

- *Answer.* The wavenumbers of the transitions $J + 1 \leftarrow J$, $K \leftarrow K$ are $\tilde{\nu} = 2B(J+1)$, $J = 0, 1, 2, \ldots$. Therefore the spectrum is a series of lines at 19.95 cm^{-1}, 39.91 cm^{-1}, 59.86 cm^{-1}, 79.82 cm^{-1}, etc. The separation is $2B = 19.95 \text{ cm}^{-1}$.

- *Comment.* Not all the lines have the same intensity. One reason is that the rotational levels are differently populated: the maximum intensity transition occurs in the vicinity of $J \approx 6$ when the sample is at room temperature. Another reason is that the spin orientations of the protons modify the appearance of the spectrum because the Pauli principle allows only some of the rotational states to be occupied.

Rotational Raman spectra. The gross selection rule for rotational Raman spectra is that the molecule must possess an *anisotropic polarizability*. The meaning of this is as follows. When a molecule is put in an electric field it is distorted because of the forces that act on the electrons and the nuclei. The extent of distortion is determined by the molecule's *polarizability*: a high polarizability indicates that a large distortion can be caused even by a moderate field. For example, a xenon atom is much more easily distorted, and is therefore more *polarizable* than a helium atom because its outer electrons are less tightly bound to the central nucleus. In the case of an atom it doesn't matter in which direction the field is applied because the same distortion is obtained in any direction; we say the polarizability is *isotropic*. The same is true of a spherical top molecule. In general, however, the polarizability of a molecule depends on the direction in which the applied field lies. For example, it is easier to distort a hydrogen molecule when the field is applied along the bond direction than when it is applied perpendicular to it. In these cases we say that the molecular polarizability is *anisotropic*.

According to the gross selection rule, methane cannot give a rotational Raman spectrum, whereas the hydrogen molecule, and any other diatomic (either homonuclear or heteronuclear) can. This is a major reason for the importance of rotational Raman spectra: they enable us to examine many of the molecules inaccessible to pure rotation, microwave spectra.

The *specific selection rule* in Raman spectra is

Raman: $\Delta J = 0, \pm 2.$

The $\Delta J = 0$ rule plays a role in rotation–vibration transitions, and we shall ignore it for the present. The appearance of the 2 in the rule can be justified in a number of ways. The physical reason can be traced to the fact that the polarizability of a molecule returns to its initial value twice on every revolution. Therefore half a full rotation corresponds to a whole oscillation of the polarizability, Fig. 17.8. The selection rule does not conflict with the earlier argument based on unit spin of the photon and overall conservation of angular momentum: two photons are involved in the Raman process, one coming in, the other going out.

The classical reason for the need for an anisotropic polarizability and the appearance of the factor of 2 is as follows. In an electric field E a molecule acquires a dipole moment p of magnitude αE, where α is the polarizability. If the molecule is irradiated with light of frequency ω_0 there is a time-dependent induced dipole moment of magnitude

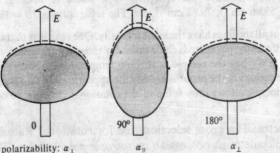

polarizability: $\alpha_\perp$ | $\alpha_\parallel$ | $\alpha_\perp$

Fig. 17.8. The polarization of a molecule returns to its initial value after a rotation of only 180°.

$$p(t) = \alpha E(t) = \alpha E_0 \cos \omega_0 t.$$

If the molecule is rotating its polarizability will be changing (if it is anisotropic), and we could write

$$\alpha = \alpha_0 + \Delta\alpha \cos 2\omega_R t,$$

where α_0 is the average polarizability and ω_R is the rotational frequency. Substituting this into the last equation, and using some trigonometric relations, gives

$$p(t) = \alpha_0 E_0 \cos \omega_0 t + \Delta\alpha E_0 \cos 2\omega_R t \cos \omega_0 t$$

$$= \alpha_0 E_0 \cos \omega_0 t + \tfrac{1}{2}\Delta\alpha E_0 [\cos (\omega_0 + 2\omega_R)t + \cos (\omega_0 - 2\omega_R)t].$$

This shows that the molecular dipole induced by the field has a component oscillating at ω_0, and also components at $\omega_0 \pm 2\omega_R$. This oscillating dipole radiates, and hence lines of these frequencies appear in the scattered light. We see that the Raman lines occur only if $\Delta\alpha \neq 0$, hence the need for anisotropy.

When we turn to the quantum mechanical version of the Raman effect, all we need to do is to find what energies may be transferred to or from the photons by applying the selection rule $\Delta J = \pm 2$ to the energy level expressions derived on pp. 573–4. When the molecule makes a transition with $\Delta J = +2$ the scattered light leaves the molecule in a higher rotational state and emerges with a lower frequency. Such transitions account for the *Stokes lines* of the spectrum:

Stokes lines $(\Delta J = +2)$: $\tilde{v} = \tilde{v}_0 - 2B(2J+3)$ for $J+2 \leftarrow J$.

These lines appear on the low-frequency side of the incident light of wavenumber $\tilde{v}_0$ and frequency $v_0 = c\tilde{v}_0$, and at displacements $6B$, $10B$, $14B$, ... The intensity falls off at the lines corresponding to high values of J because these states are only sparsely populated, Fig. 17.9.

When the molecule makes a transition with $\Delta J = -2$ during the collision the photon picks up the excess energy and emerges with a higher frequency (greater wavenumber). These transitions account for the *anti-Stokes lines* in the spectrum:

anti-Stokes lines $(\Delta J = -2)$: $\tilde{v} = \tilde{v}_0 + 2B(2J-1)$ for $J \rightarrow J-2$.

The lines occur at $6B$, $10B$, $14B$, ... to the high-frequency side of the incident light.

Analysing the structure of a Raman spectrum consists of measuring the separations between neighbouring lines in the Stokes or anti-Stokes series, and equating them to $4B$. Once B has been found the moment of inertia of the molecule perpendicular to its symmetry axis can be found. As in the case of microwave rotational spectra, these moments of inertia can then be interpreted in terms of the bond lengths and angles. The technique can be applied to non-polar as well as to polar molecules so long as they have anisotropic polarizabilities. This excludes spherical tops, but includes all symmetric tops and linear molecules.

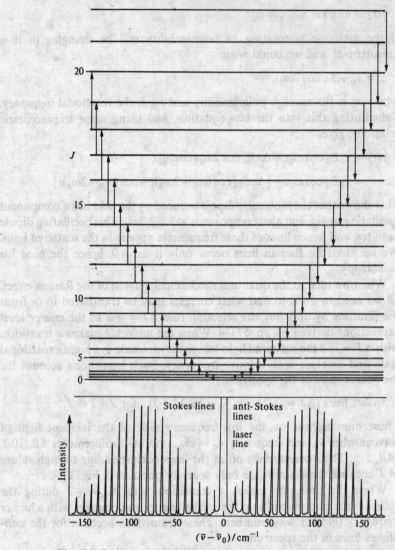

Fig. 17.9. Stokes and anti-Stokes rotational Raman lines and the corresponding transitions.

Example (Objective 8). Predict the form of the rotational Raman spectrum of ammonia gas when it is irradiated with sharply defined laser light of wavelength 336.732 nm.

- *Method.* The molecule is active in the Raman because the end-over-end rotation modulates its polarizability as viewed by a stationary observer. The value of B was calculated on p. 576. The Stokes and anti-Stokes lines are at frequencies given by the expressions above. Convert to wavelengths.

- *Answer.* $B = 9.977 \text{ cm}^{-1}$, $\lambda_0 = 336.732 \text{ nm}$, which is equivalent to $29\,697.2 \text{ cm}^{-1}$. Hence Stokes lines will appear at

    $$\tilde{\nu}/\text{cm}^{-1} = 29\,697.2 - 19.95(2J+3) = 29\,637.4, \ 29\,597.5, \ 29\,557.6, \ 29\,517.7, \ldots$$

    or

on the right of the minimum the curve rises and then flattens out as the bond is made ...... that the ...... repel...

In regions close to ...... the potential energy is a ...... , as shown in Fig. 17.10. Therefore a reasonable approximation to the potential energy curve at small displac...... is

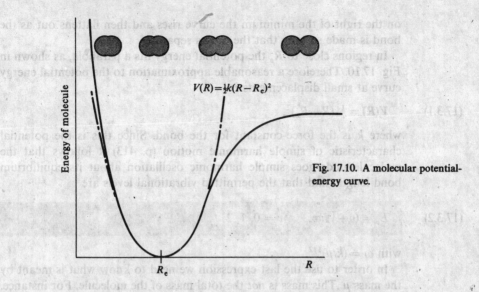

$$V(R) = \tfrac{1}{2}k(R - R_e)^2$$

Fig. 17.10. A molecular potential-energy curve.

where k is the force constant for the bond. Since this is the potential characteristic of simple harmonic motion (p. 413) it follows that the bond undergoes simple harmonic oscillation about its equilibrium ...... that the permitted vibrational levels are

with $\omega = (k/\mu)$...

In order to use the last expression we need to know what is meant by the mass μ. This mass is not the total mass of the molecule. For instance, if one of the masses is extremely large so that the molecule is like a single atom attached by ...... deform...... by the smaller mass. When the mass ...... contribute to .... properties and is...

$\lambda/\mathrm{nm} = 337.412, 337.867, 338.323, 338.780, \ldots$

and anti-Stokes lines at

$$\tilde{v}/\mathrm{cm}^{-1} = 29\,697.2 + 19.95(2J - 1) = 29\,757.1, 29\,797.0, \ldots$$

or

$\lambda/\mathrm{nm} = 336.055, 335.605, \ldots$

- *Comment.* The appearance of the spectrum will be as follows: a strong central line at 336.732 nm surrounded on either side by lines of increasing, then decreasing intensity (the intensity distribution being due to thermal population effects). The spread of the entire spectrum is very small (about 500 cm^{-1} at room temperature), and so the wavelength of the incident light must be sharply defined.

17.3 Vibrations of diatomics

We begin this section by finding expressions for the vibrational energy levels of a molecule, and then apply the appropriate selection rules in order to account for the spectrum. First we consider the vibrations of a diatomic molecule, where the only mode of oscillation is the stretching and contracting of the bond, and then go on to see how the spectrum is affected by simultaneous excitations of rotational transitions. In the next section we see how a straightforward extension of the discussion enables us to discuss the vibrations of polyatomic molecules.

Molecular vibrations. A typical molecular potential-energy curve of a diatomic molecule is illustrated in Fig. 17.10. This curve shows how the energy of the molecule changes as the distance between its nuclei—its bond length—changes. There is a minimum in the curve, corresponding to the equilibrium bond length of the molecule (R_e). On the left of the minimum the energy rises sharply because it is difficult to press two atoms together;

on the right of the minimum the curve rises and then flattens out as the bond is made so long that the atoms separate.

In regions close to R_e the potential energy fits a parabola, as shown in Fig. 17.10. Therefore a reasonable approximation to the potential energy curve at small displacements is

(17.3.1) $$V(R) = \tfrac{1}{2}k(R - R_e)^2,$$

where k is the force-constant for the bond. Since this is the potential characteristic of simple harmonic motion (p. 413) it follows that the molecule undergoes simple harmonic oscillation about its equilibrium bond length, and that the permitted vibrational levels are

(17.3.2) $$E_v = (v + \tfrac{1}{2})\hbar\omega, \qquad v = 0, 1, 2, \ldots$$

with $\omega = (k/\mu)^{1/2}$.

In order to use the last expression we need to know what is meant by the mass μ. This mass is *not* the total mass of the molecule. For instance, if one of the masses is extremely large (so that the molecule is like a single atom attached by a spring to a massive wall), the frequency is determined by the smaller mass. When the masses are similar they should contribute to μ in about equal proportions. The expression for μ has these properties and is

(17.3.3) $$1/\mu = 1/m_1 + 1/m_2.$$

For example, if object 1 is a brick wall, $m_1 \approx \infty$ and $1/\mu \approx 1/m_2$ implying $\mu \approx m_2$, the mass of the lighter object. The mass μ as defined by the last equation is called the *reduced mass* of the molecule.

So much for the physical justification of the form of μ. We now show that its form can be deduced rigorously. Consider two masses m_1 and m_2. They oscillate around their mutual centre of mass, which remains stationary. The centre of mass is determined by the relation $m_1 R_1 = m_2 R_2$, R_1 and R_2 being the distances of 1 and 2 from the centre of mass at any instant. Since $R_1 + R_2 = R$, where R is the bond length at that instant, it follows that

$$R_1 = [m_2/(m_1 + m_2)]R, \qquad R_2 = [m_1/(m_1 + m_2)]R.$$

Both atoms experience a restoring force proportional to the amount the bond has been stretched, $R - R_e$. Therefore the accelerations of the atoms obey

$$m_1(d^2 R_1/dt^2) = -k(R - R_e), \qquad m_2(d^2 R_2/dt^2) = -k(R - R_e).$$

Substituting for R_1 in the first or for R_2 in the second gives the single equation

$$[m_1 m_2/(m_1 + m_2)](d^2 R/dt^2) = -k(R - R_e),$$

which is the classical equation of motion for a particle of mass

$$\mu = m_1 m_2/(m_1 + m_2)$$

undergoing simple harmonic motion along R with a frequency $\omega = (k/\mu)^{1/2}$:

$$\mu \ddot{R} = -k(R - R_e).$$

The expression for μ can be rearranged into the form of eqn (17.3.3).

The energy levels given by eqn (17.3.2) are only an approximation to the true vibrational energies because they are based on the parabolic approximation to the true potential-energy curve. When the vibrations of the molecule are so energetic that the swing of the atoms takes them to large displacements from the equilibrium bond length, the potential energy differs from the parabolic approximation. The true potential is less confining than the parabola (Fig. 17.10), and so we should expect the energy levels to be less widely spaced at high energies than at low.

One approach to calculating the energy levels at moderate excitations is to suggest a mathematical form for the potential energy that resembles the one in Fig. 17.10 more closely than the simple parabola. A particularly fruitful suggestion was made by Morse when he proposed the expression

(17.3.4) $$V(R) = D_e\{1 - \exp[-a(R - R_e)]\}^2,$$

where D_e is the depth of the potential energy minimum and $a = (\mu/2D_e)^{1/2}\omega$. The shape of this curve is shown in Fig. 17.11, and even though there are only two adjustable parameters (k, which appears in ω, and D_e) the shape is reasonable and a number of molecules can be fitted moderately well. The Schrödinger equation can be expressed in terms of the Morse potential energy, and solved for the energy levels. These may be written

(17.3.5) $$E_v = (v + \tfrac{1}{2})\hbar\omega - (v + \tfrac{1}{2})^2 x_e \hbar\omega,$$

where the *anharmonicity constant* $x_e = \hbar a^2/2\mu\omega$. The second term subtracts from the first, confirming the expected convergence of the levels at high quantum numbers. This is illustrated in Fig. 17.11. More general expressions for the potential energy lead to further terms in this expression, and in general

$$E_v = (v + \tfrac{1}{2})\hbar\omega - (v + \tfrac{1}{2})^2 x_e \hbar\omega + \ldots.$$

Vibrational spectra of diatomics. As explained earlier in the chapter, the *gross selection rule* is that the dipole moment must change during a vibration if it is to be spectroscopically active. Stretching the bond between two atoms of a homonuclear diatomic does not change its dipole moment from zero; therefore homonuclear diatomics show no vibrational spectra. The dipole moment of a heteronuclear diatomic does change when its bond length

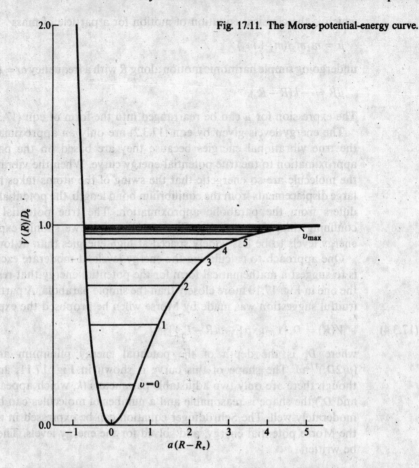

Fig. 17.11. The Morse potential-energy curve.

changes, and so these molecules can show a vibrational absorption or emission spectrum.

The specific selection rule for a harmonic oscillator is

(17.3.6) $\Delta v = \pm 1$.

The basis of this rule can be traced to the conservation of angular momentum of the combined photon + molecule system. Its application to the energy level expressions gives the change of energy

(17.3.7a) $\Delta E_v = \hbar\omega, \qquad v+1 \leftarrow v$

in the harmonic expression, or

(17.3.7b) $\Delta E_v = \hbar\omega - 2(v+1)x_e\hbar\omega + \ldots$

if anharmonicity is taken into account. The energy appears (or disappears) as a photon of energy $h\nu$, and so the light emitted or absorbed has the wavenumber $\tilde{\nu} = \Delta E_v/hc$, or

(17.3.8) $\tilde{\nu} = \tilde{\nu}_0 - 2(v+1)x_e\tilde{\nu}_0 + \ldots$

with $\tilde{\nu}_0 = \omega/2\pi c$. In the harmonic approximation $\tilde{\nu} = \tilde{\nu}_0$.

A typical molecular force constant is 500 N m^{-1} (516 N m^{-1} for HCl), and a typical reduced mass is about 10^{-26} kg (the value for $^1H^{35}Cl$ is 1.63×10^{-27} kg). Therefore $\omega \approx 5.6 \times 10^{14}$ rad s^{-1}, which corresponds to a wavelength of 3.3×10^{-6} m or a wavenumber of 2990 cm^{-1}. This radiation lies in the *infrared* region of the spectrum. Similar magnitudes apply to other molecules, and so vibrational spectroscopy is an infrared technique.

At room temperature kT corresponds to 200 cm^{-1}, and from the discussion on p. 10 on the role of the Boltzmann factor, it follows that the great majority of molecules are in their vibrational ground states at room temperature. Consequently the dominant transition in infrared absorption spectroscopy is from $v = 0$ to $v = 1$.

If the molecules are formed in some vibrationally excited state, e.g., in the reaction

$$H_2 + F_2 \rightarrow 2HF^*,$$

where HF* is a 'hot', vibrationally excited species, transitions downwards from v to $v-1$, $v-1$ to $v-2$, etc. may be observed in the infrared emission spectrum. In the harmonic approximation all these transitions generate photons of the same frequency, and the spectrum is predicted to be a single line. When the fact that the vibrations may be slightly anharmonic is taken into account, two changes are observed. In the first place the transition energies are given by eqn (17.3.7b), and now depend weakly on v. Therefore the transitions starting in different levels generate photons of slightly different frequencies, and several lines appear in the spectrum. The second consequence of anharmonicity is that the selection rule (which assumes harmonic motion) is infringed, and transitions with $\Delta v = \pm 2$ (these are called the *second harmonics*) are allowed as well as the $\Delta v = \pm 1$ transitions (the *first harmonic*, or *fundamental*). The second harmonics give photons with wavenumbers

$$\tilde{v} = 2\tilde{v}_0 - 2(2v + 3)x_e\tilde{v}_0 + \dots \qquad (v+2 \leftarrow v).$$

The appearance of lines originating from different levels is useful because x_e, and hence D_e, may be determined.

Example (Objective 11). The infrared emission from a 'hot' HF molecule appeared as a series of lines at 3958.38 cm^{-1}(1 → 0), 3778.25 cm^{-1}(2 → 1), 3598.10 cm^{-1}(3 → 2), and a weak second harmonic at 7736.63 cm^{-1}(2 → 0). What is the force constant of the HF bond, and the molecule's dissociation energy?

- *Method.* Fit the first harmonics ($v + 1 \rightarrow v$) to eqn (17.3.8). The second harmonic gives no further information, but can be used as a check. Find $\tilde{v}_0$ and $\tilde{v}_0 x_e$. From the Morse curve $\tilde{v}_0 x_e = hc\tilde{v}_0^2/4D_e = \tilde{v}_0^2/4(D_e/hc)$. Hence find $D_e/hc = \tilde{v}_0/4x_e$.

- *Answer.* Write $\tilde{v}(v) = \tilde{v}_0 - 2(v+1)x_e\tilde{v}_0$. Then $\tilde{v}_0 = \tilde{v}(0) + \tilde{v}(1) - \tilde{v}(2)$ and $\tilde{v}_0 x_e = \frac{1}{2}\{\tilde{v}(0) - \tilde{v}(1)\} = \frac{1}{2}\{\tilde{v}(1) - \tilde{v}(2)\}$. Therefore $\tilde{v}_0 = (3958.38$ cm^{-1} + 3778.25 cm^{-1} − 3598.10 cm^{-1}) = 4138.53 cm^{-1} and $\tilde{v}_0 x_e = \frac{1}{2}(3958.38$ cm^{-1} − 3778.25 cm^{-1}) or $\frac{1}{2}(3778.25$ cm^{-1} − 3598.10 cm^{-1}), giving a mean of 90.07 cm^{-1}. Hence $x_e = (90.07$ cm$^{-1})/(4138.53$ cm^{-1}) = 0.02176 and $D_e/hc = (4138.53$ cm$^{-1})/4 \times 0.02176 = 47\,540$ cm^{-1} (5.89 eV).

● *Comment.* When more frequencies are available a graphical technique ($\bar{v}(v)$ plotted against v) is more appropriate and gives a result closer to the true value ($\leqslant 6.4\ \text{eV}$). The connection between D_e and the bond dissociation energy, $D^{\ominus}(\text{H-F})$, is made below.

The analysis of a vibrational spectrum gives two pieces of data: the circular frequency ω and the anharmonicity constant x_e. The first gives the force constant of the bond, and this is a measure of its rigidity. The second gives the value of D_e for the molecule, and therefore the energy required to break the bond. Some care, however, is needed in interpreting the significance of D_e. As defined, D_e is the depth of the molecular potential energy curve, but the molecule must possess at least a zero-point energy $\frac{1}{2}\hbar\omega(1-\frac{1}{2}x_e)$ (see Fig. 17.11 and eqn (17.3.5)). The bond dissociation energy, is therefore related to D_e by

(17.3.9) $$D^{\ominus}(\text{A-B}) = D_e - \tfrac{1}{2}\hbar\omega(1-\tfrac{1}{2}x_e).$$

This is a valuable method of determining strengths of chemical bonds, and the method has been refined to overcome the approximations inherent in the Morse curve analysis. Many of the molecular bond strengths given in Table 17.2 were obtained by this spectroscopic method. (Note that the molar bond dissociation *enthalpy*, which was discussed from a thermo-chemical point of view in Chapter 4 (p. 114), is $DH^{\ominus}_m = LD^{\ominus} + RT$.)

Vibration–rotation spectra. When the vibrational spectrum of a gaseous heteronuclear diatomic is analysed under high resolution, each line is found to consist of a large number of closely spaced components, Fig. 17.12. For this reason molecular spectra are often called *band spectra* in contrast to atomic spectra, which are called *line spectra*. The separation between the lines in the bands are of the order $1\ \text{cm}^{-1}$, and this points to the interpretation that the structure is due to the excitation of rotational motion during the vibrational transition.

The stimulation of a rotational transition during a vibrational change can be explained in a number of ways. In the first place the amount of energy involved in the vibrational transition is so great that plenty is

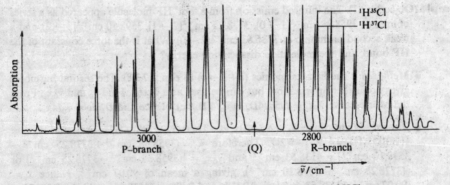

Fig. 17.12. A high-resolution vibration–rotation spectrum of HCl.

available for use in exciting a rotational transition. In the second place a rotating body accelerates when its moment of inertia is reduced, and decelerates when it is increased (a familiar example is the way that an ice-skater makes himself rotate faster by drawing in his arms). This effect is a consequence of the conservation of angular momentum $J = I\omega$ in the absence of any external torques: if I changes, ω must change in the opposite direction in order to preserve the value of the product $I\omega$. In the present case the vibrational change is analogous to a transient change of the moment of inertia, and the molecule responds by changing its rotational state.

Table 17.2. Properties of diatomic molecules

Molecule	Vibrational wavenumber $\tilde{v}_0/\text{cm}^{-1}$	Rotational constant B/cm^{-1}	Bond length R_e/pm	Force Constant $k/\text{N m}^{-1}$	Dissociation energy $D^{\ominus}/\text{eV}$	Dissociation energy $D^{\ominus}/\text{kJ mol}^{-1}$
$^1\text{H}_2^+$	2321.8	29.8	106	160.0	2.651	255.8
$^1\text{H}_2$	4400.39	60.864	74.138	574.9	4.478	432.1
$^2\text{H}_2$	3118.46	30.442	74.154	577.0	4.556	439.6
$^1\text{H}^{19}\text{F}$	4138.32	20.9557	91.680	965.7	5.85	564.4
$^1\text{H}^{35}\text{Cl}$	2990.95	10.5934	127.45	516.3	4.433	427.7
$^1\text{H}^{81}\text{Br}$	2648.98	8.46488	141.44	411.5	3.759	362.7
$^1\text{H}^{127}\text{I}$	2308.09	6.5108	160.92	313.8	3.0	294.9
$^{14}\text{N}_2$	2358.07	1.9987	109.76	2293.8	9.760	941.7
$^{16}\text{O}_2$	1580.361	1.44567	120.75	1176.8	5.115	493.5
$^{19}\text{F}_2$	[891.8]	[0.8828]	[141.78]	445.1	[1.60]	[154.4]
$^{35}\text{Cl}_2$	559.71	0.2441	198.75	322.7	2.480	239.3

Source: *American Institute of Physics handbook*, McGraw-Hill.

Detailed analysis shows that the selection rule for the rotational change that accompanies a vibrational change is the by now familiar $\Delta J = \pm 1$, but in exceptional circumstances (when there is electronic angular momentum about the axis of the linear molecule, as in NO), the overall conservation of angular momentum can also be satisfied with $\Delta J = 0$.

The form of the vibration–rotation spectrum can be predicted as follows. The energy levels of a vibrating, rotating molecule are

(17.3.10) $$E_{v,J} = (v+\tfrac{1}{2})\hbar\omega + hcBJ(J+1).$$

In more detailed treatments the anharmonicity corrections are included and the rotational constant B is allowed to depend on the vibrational state of the molecule (because in higher vibrational states the swing of the atoms swells the molecule slightly, but significantly). When a vibrational transition occurs not only does v change to $v+1$ (for absorption) but J changes by ± 1 and sometimes 0. The vibrational absorption is therefore divided into three sets of lines called *branches*, and these are labelled P, Q, and R, according to the value of ΔJ.

P-branch. These lines correspond to transitions with $\Delta v = 1$ and $\Delta J = -1$. If the molecule is initially in a rotational state J the transition is $J \to J - 1$ and the overall energy change is

$$\Delta E = \hbar\omega - 2BhcJ.$$

This accounts for the lines at the wavenumbers $\tilde{v}_0 - 2B, \tilde{v}_0 - 4B, \ldots$, Figs. 17.12 and 17.13.

Q-branch. This is the set of lines arising from $\Delta v = 1$, $\Delta J = 0$. Their wavenumbers are given by

$$\tilde{v} = \Delta E/hc = \hbar\omega/hc = \tilde{v}_0,$$

for all values of J. Thus the Q-branch, when it is allowed, is a single line at the wavenumber of the vibrational transition frequency. In real molecules the rotational constants of the vibrational states v and $v + 1$ are slightly different, and so the Q-branch appears as a cluster of very closely spaced lines. Figure 17.12 shows a gap where the Q-branch is expected: this is because it is forbidden for the molecule used in the experiment (HCl).

R-branch. Transitions with $\Delta v = 1$ and $\Delta J = +1$ constitute the R-branch, and their wavenumbers are given by

$$\tilde{v} = \tilde{v}_0 + 2B(J + 1).$$

These lines are displaced by $2B, 4B, \ldots$ to high frequency of the fundamental, as illustrated in Fig. 17.13.

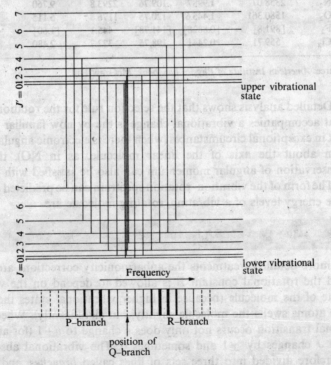

Fig. 17.13. The formation of P, Q, and R branches.

The separation between the lines in the P- and R-branches of a vibrational transition gives the value of the rotational constant, and so the bond length of the molecule can be obtained from the vibrational spectrum without needing to take a pure rotation, microwave spectrum (though the latter gives more precise results).

Vibrational Raman spectra of diatomics. The *gross selection rule* for the appearance of a vibrational Raman spectrum depends on the polarizability of the molecule. In this case the important criterion is whether the polarizability changes when the molecule vibrates. If it does, the vibration is *Raman active* and can exchange energy with photons during a collision. Both homonuclear and heteronuclear diatomics swell and contract during a vibration, and the control of the nuclei over the electrons, and hence the molecular polarizability, varies accordingly. Both types of molecule may therefore give vibrational Raman spectra.

The *specific selection rule* is $\Delta v = \pm 1$. The lines to high frequency of the incident light, the anti-Stokes lines, correspond to $\Delta v = -1$. These are usually very weak because very few molecules of any sample are in an excited vibrational state initially, and therefore only very few collisions lead to a transfer of energy from the molecules to the photons. The lines to low frequency, the Stokes lines, correspond to $\Delta v = +1$. Superimposed on the lines is a branch structure which arises from the simultaneous

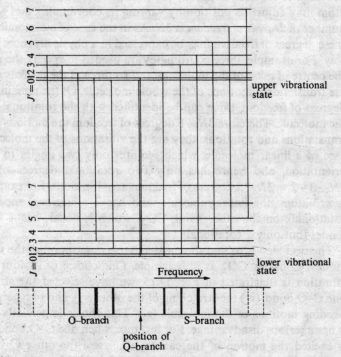

Fig. 17.14. The formation of O, Q, and S branches.

rotational transitions excited by the collision. The selection rules for these rotational transitions are $\Delta J = 0, \pm 2$ (as in pure rotational Raman spectra, p. 580), and the branches are called the *O-branch* (for $\Delta J = -2$), the *Q-branch* (for $\Delta J = 0$, if it is allowed—see the comment about Q-branches above), and the *S-branch* (for $\Delta J = +2$), Fig. 17.14.

The information available from vibrational Raman spectra augments that available from ordinary spectroscopy. In particular, Raman spectroscopy provides a way of studying homonuclear diatomic molecules: from the spacing of the Stokes lines their force-constants can be deduced, and this measures the rigidity of the bonds. The rotational structure of the spectra can be interpreted in terms of the rotational constants of the molecules, using the relations in Table 17.1.

17.4 Vibrations of polyatomics

In a diatomic molecule there is only one mode of vibration, the stretching of the bond, but in polyatomics we have to allow for bonds to stretch and angles to change as the molecule distorts. A linear molecule (like CO_2) built from N atoms may vibrate in $3N-5$ different ways, and a non-linear polyatomic (like H_2O) can vibrate in $3N-6$ ways. The method of arriving at these figures is the same as that used on p. 577 to count the number of degrees of freedom of a diatomic molecule, and the argument runs as follows.

Consider a non-linear molecule of N atoms. The total number of coordinates required to specify the positions of all the atoms is $3N$. Each atom may adjust its position by varying its coordinates, and so the total number of *degrees of freedom* is $3N$. As in the case of a diatomic molecule, these degrees of freedom may be expressed in a physically more significant way. For example, three coordinates are needed to specify the position of the centre of mass of the molecule, and so three of the degrees of freedom are translational motions of the whole molecule. Of the remaining $3N-3$ degrees of freedom, three can be identified with the rotational motion of the molecule. Therefore $3N-6$ degrees of freedom are motions other than translations and rotations: they are the vibrations of the molecule. In the case of a linear molecule, which requires only two angles to specify its orientation, and hence has only two rotational degrees of freedom, $3N-3-2 = 3N-5$ degrees of freedom are vibrations. For example, H_2O is a 3-atom non-linear molecule, and has 3 vibrational modes (and 3 rotational); on the other hand, CO_2, which is linear, has 4 vibrational modes (but only 2 rotational).

The next step is to find the most suitable description of the vibrational modes. Consider CO_2 as an example. One choice of its four modes of vibration is illustrated in Fig. 17.15: we have selected the stretching of one C–O bond, (v_L) the stretching of the other (v_R), and two perpendicular bending motions of the molecule (v_2). This choice is perfectly valid, but it has a serious disadvantage. For instance, when one C–O bond vibration is excited the motion of the carbon atom sets the other C–O bond in motion, and so the energy of vibration flows backwards and forwards

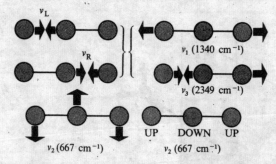

Fig. 17.15. Alternative choices for describing the vibrations of carbon dioxide.

v_1 (1340 cm^{-1})

v_3 (2349 cm^{-1})

v_2 (667 cm^{-1})   UP   DOWN   UP v_2 (667 cm^{-1})

between the two bonds.

The description is much simpler if one mode of vibration is taken to be the *symmetric stretch* of both bonds simultaneously, v_1 in Fig. 17.15. In this mode the carbon atom is buffetted simultaneously from both sides and the same motion can continue indefinitely. Another mode is the *anti-symmetric stretch* v_3, in which one bond is stretched and the other simultaneously compressed. These modes of motion are stable in the sense that if one is excited it does not excite the other. They are two of the *normal modes* of the molecule. The two other normal modes are the two perpendicular bending motions v_2. Normal modes are mutually independent synchronous motions of groups of atoms, and any one mode may be excited without leading to excitation of the others.

The four normal modes of carbon dioxide (and the $3N-6$ of non-linear polyatomics) are the key to the description of the vibrations of molecules. Each independent normal mode behaves like an independent harmonic oscillator (if anharmonicities are neglected), and so each may be excited into an energy $(v_Q + \frac{1}{2})\hbar\omega_Q$, where ω_Q is the frequency of the normal mode Q. This frequency depends both on the force-constant k_Q for the mode (for instance, a bond is often more resistant to stretching than to bending) and on the reduced mass μ_Q, the relation being $\omega_Q = (k_Q/\mu_Q)^{1/2}$. The reduced mass of the mode depends on how much mass is swung about by the vibration. For example, in the symmetric stretch in carbon dioxide the carbon atom is static and the reduced mass of that mode is determined by the masses of the two oxygen atoms; in the antisymmetric stretch all three atoms are moving, and so all three contribute to the reduced mass of this mode.

Example (Objective 14). Determine the symmetries of the normal vibrations of methane.

• *Method.* There are $3 \times 5 = 15$ possible displacements, of which $15-6 = 9$ are vibrations. They can be classified by group theory. The simplest way of doing this is to notice that under a symmetry operation only the displacement coordinates that remain unchanged contribute to the diagonal of the transformation matrix, and so they contribute 1 to the character of the operation. If the coordinate is reversed, -1 appears on the diagonal. Refer to Fig. 17.16.

• *Answer.* Under E no displacement coordinates are changed, and so the transformation matrix is diagonal with 1 for each non-zero element: the basis is 15-dimensional (15 possible displacements), and so the character of E is $\chi(E) = 15$. Under C_3 no

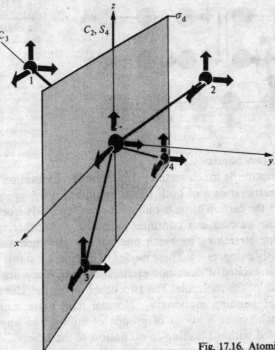

Fig. 17.16. Atomic displacements in methane.

displacements are left unchanged, and so $\chi(C_3) = 0$. Under C_2 the z-displacement on the central atom is left unchanged, while the x- and y-displacements both change sign and so $\chi(C_2) = -1$. Under S_4 the z-displacement on the central atom is reversed, and so $\chi(S_4) = -1$. Under σ_d the z-displacement on C, H_3, H_4 are left unchanged, and so $\chi(\sigma_d) = 3$. The characters are therefore 15 0 -1 -1 3, corresponding to $A_1 + E + T_1 + 3T_2$. Three of the displacements are overall translations of the molecule: these transform as x, y, z; that is, T_2. Three displacements are rotations, and transform as T_1. Hence the vibrations span $A_1 + E + 2T_2$.

- *Comment.* We shall soon see that this type of symmetry analysis gives a quick way of deciding which modes are active.

Vibrational spectra of polyatomics. The gross selection rule is that the normal mode of vibration must result in a change of dipole moment if it is to be spectroscopically active. The symmetric stretch of carbon dioxide leaves the dipole moment unchanged (at zero), and so this mode is *infrared inactive*. The antisymmetric stretch and the two bending modes all change the dipole moment, and so these three modes are *infrared active*. The active modes obey the specific selection rule $\Delta v_Q = \pm 1$, and so the energy of the fundamental transition of each active mode is $\hbar \omega_Q$. Therefore absorptions at the frequencies $v_Q = \omega_Q/2\pi$ (and wavenumbers $\tilde{v}_Q = v_Q/c$) are expected to appear in the spectrum. From an analysis of the spectrum we can establish the force constant for all the active normal modes, and therefore build up a picture of the stiffness of various parts of the molecule. The three vibrational modes of water are illustrated in Fig. 17.17. Notice

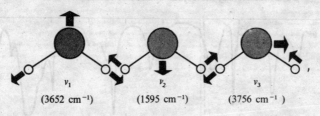

Fig. 17.17.
Vibrational modes
of water.

v_1
(3652 cm^{-1})

v_2
(1595 cm^{-1})

v_3
(3756 cm^{-1})

that the predominantly bending mode (v_2) is at a significantly lower frequency than the other modes, both of which involve the stretching and squashing of bonds. Bonds are often easier to bend than to stretch.

Superimposed on this simple scheme are the complications arising from anharmonicities and the effects of molecular rotation. Very often the sample is liquid or solid, and the molecules are not able to rotate freely. In a liquid, for example, a molecule may be able to rotate for only a very short time before it is struck by a neighbour, and so it changes its rotational state very frequently. Since the lifetimes of rotational states of molecules in liquids are very short, the rotational energy levels are ill-defined. Collisions between molecules in a liquid occur at the rate of about 10^{13} s^{-1}, and even allowing for only a 10 per cent success rate in knocking the molecule into a different rotational state a lifetime broadening (p. 570) of more than 1 cm^{-1} can easily be achieved. The rotational structure of the vibrational transition is blurred by this effect, and so the vibrational spectra of molecules in condensed phases usually show broad lines with no rotational structure.

One very important application of infrared spectroscopy of polyatomic molecules in condensed phases is in chemical analysis. The vibrational spectra of different groups in a molecule act as fingerprints because the contribution of the normal modes of a group is often quite similar when it occurs in different molecules, and the identity of a molecule can often be established by examining its infrared spectrum and accounting for the absorption bands by referring to a table of characteristic vibrational frequencies, Table 17.3. This is illustrated in Fig. 17.18.

Table 17.3. Typical vibration wavenumbers, $\tilde{v}/\text{cm}^{-1}$

C–H	stretch	2850–2960	O–H	stretch	3590–3650
C–H	bend	1340–1465	H-bonds		3200–3570
C–C	stretch, bend	700–1250	C=O	stretch	1640–1780
C=C	stretch	1620–1680	C≡N	stretch	2215–2275
C≡C	stretch	2100–2260	N–H	stretch	3200–3500
CO_3^{2-}		1410–1450	C–F	stretch	1000–1400
NO_3^-		1350–1420	C–Cl	stretch	600–800
NO_2^-		1230–1250	C–Br	stretch	500–600
SO_4^{2-}		1080–1130	C–I	stretch	500
silicates		900–1100			

Source: L. J. Bellamy, *The infrared spectra of complex molecules* and *Advances in infrared group frequencies*, Chapman and Hall.

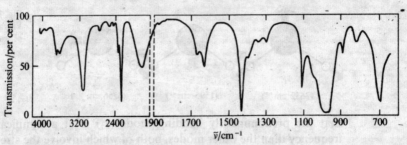

Fig. 17.18. An infrared vibrational fingerprint.

Vibrational Raman spectra of polyatomics. The normal modes of vibration of polyatomics are *Raman active* if they are accompanied by a changing
polarizability. For instance, the symmetric stretch of carbon dioxide
alternately swells and contracts the molecule; this changes the polarizability, and so this mode is Raman active. The other modes of vibration
of carbon dioxide leave the polarizability unchanged. For instance, in the
antisymmetric stretch the effect of the expansion of one bond is compensated by the contraction of the other. None of these other modes is Raman
active, and so the Raman spectrum consists of a single line corresponding
to the frequency of the symmetric stretch.

There is a useful way of distinguishing the active and inactive modes.
The *exclusion rule* states that, *if the molecule has a centre of symmetry,
then no modes can be both infrared and Raman active.* In cases of high
symmetry there may be modes inactive to both. The rule applies to carbon
dioxide, but not to water (because it has no centre of symmetry). Note
that the exclusion rule implies that both infrared and Raman spectra must
be used to obtain all the information about the rigidity (the *force field*)
of a molecule that possesses a centre of symmetry.

Example (Objective 15). State which of the normal modes of water are infrared and Raman
active.

- *Method.* Use group theory as follows. (a) A normal mode is *infrared active* if its
symmetry is that of a translation (x, y, z). (b) A normal mode is *Raman active* if
its symmetry is that of a quadratic form (x^2, xy,...). Determine the vibrational
symmetries as in the last *Example*.

- *Answer.* H_2O belongs to C_{2v} and its vibrational symmetries are $2A_1 + B_2$. In C_{2v}
translations transform as $A_1(z)$, $B_2(y)$, and $B_1(x)$; therefore all modes are infrared
active. Quadratic forms transform as A_1, A_2, B_1, B_2, and so the vibrational modes
are all Raman active.

- *Comment.* The A_1 vibrations are z-polarized and the B_2 vibration is y-polarized.
Sketch the form of the vibrations and confirm the physical basis of this conclusion.

A major application of Raman spectroscopy is to the identification of
organic and inorganic species in solution, and to the elucidation of their
structure. In order to be successful it is important to use both aspects of
Raman 'fingerprinting': not only the position of the vibrational Raman

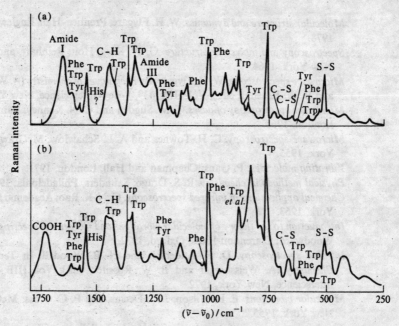

Fig. 17.19. The vibrational Raman spectrum of lysozyme in water and the superposition of the Raman spectra of the constituent amino acids. (From *Raman spectroscopy* by D. A. Long. Copyright © 1977, McGraw-Hill Inc. Used with the permission of the McGraw-Hill Book Company.)

lines, but also their polarization (the orientation of the electric field of the scattered light). The change in the polarization of the light when it is scattered can be interpreted in terms of the symmetry of the vibration and hence of the molecular framework. In order to be used most effectively, Raman vibrational studies should be used in conjunction with infrared absorption spectra. As an example of the technique, Fig. 17.19 shows the vibrational Raman spectrum of a solution of lysozyme in water, and for comparison a superposition of the Raman spectra of the constituent amino acids. The differences are an indication of the effects of conformation, environment, and specific interactions (such as S–S linking) in the enzyme molecule.

Further reading

Introduction to molecular spectroscopy. G. M. Barrow; McGraw-Hill, New York, 1962.

Fundamentals of molecular spectroscopy. C. N. Banwell; McGraw-Hill, New York, 1972.

The determination of molecular structure (2nd edn). P. J. Wheatley; Clarendon Press, Oxford, 1968.

Molecular structure: the physical approach. J. C. D. Brand and J. C. Speakman; Arnold, London, 1960.

Spectroscopy. D. H. Whiffen; Longman, London, 1972.

Molecular quantum mechanics. P. W. Atkins; Clarendon Press, Oxford, 1970.

Molecular structure and dynamics. W. H. Flygare; Prentice-Hall, Englewood Cliffs, 1978.

Spectroscopy and molecular structure. G. W. King; Holt, Reinhart, and Winston, New York, 1964.

Microwave spectroscopy. W. H. Flygare; In *Techniques of chemistry* (A. Weissberger and B. W. Rossiter, eds.), Vol. IIIA, 439, Wiley-Interscience, New York, 1972.

Microwave spectroscopy of gases. T. M. Sugden and C. N. Kenney; Van Nostrand, London, 1965.

Microwave spectroscopy. C. H. Townes and A. L. Schawlow; McGraw-Hill, New York, 1955.

Vibrating molecules. P. Gans; Chapman and Hall, London, 1971.

Physical methods in chemistry. R. S. Drago; Saunders, Philadelphia, 1977.

Chemical applications of infrared spectroscopy. C. N. R. Rao; Academic Press, New York, 1963.

Introduction to the theory of molecular vibrations and vibrational spectroscopy. L. A. Woodward; Clarendon Press, Oxford, 1972.

Infrared spectroscopy. D. H. Anderson and N. B. Woodall; In *Techniques of Chemistry* (A. Weissberger and B. W. Rossiter, eds.), Vol. IIIB, 1, Wiley-Interscience, New York, 1972.

Molecular vibrations. E. B. Wilson, J. C. Decius, and P. C. Cross; McGraw-Hill, New York, 1955.

The infrared spectra of complex molecules. L. J. Bellamy; Chapman and Hall, London, 1975.

Infrared and Raman spectra of polyatomic molecules. G. Herzberg; Van Nostrand, 1945.

Raman spectroscopy. H. A. Szymanski; Plenum, New York, 1967.

Raman spectroscopy. D. A. Long; McGraw-Hill, New York, 1977.

Raman spectroscopy. J. R. Durig and W. C. Harris; In *Techniques of chemistry* (A. Weissberger and B. W. Rossiter, eds.), Vol. IIIB, 85, Wiley-Interscience, New York, 1972.

Laser Raman spectroscopy. T. R. Gilson and P. J. Hendra; Wiley, New York, 1970.

Problems

17.1.   Which of the following molecules may show a pure rotational microwave spectrum: H_2, HCl, CH_4, CH_3Cl, CH_2Cl_2, H_2O, H_2O_2, NH_3, NH_4Cl?

17.2.   Which of the following molecules may show absorption in the infrared: H_2, HCl, CO_2, H_2O, CH_3CH_3, CH_4, CH_3Cl, N_2, N_3^-?

17.3.   Which of the following molecules may show a pure rotational Raman spectrum: H_2, HCl, CH_4, CH_3Cl, CH_2Cl_2, CH_3CH_3, H_2O, SF_6?

17.4.   Which of the following molecules can show a vibrational Raman spectrum: H_2, HCl, CH_4, H_2O, CO_2, CH_3CH_3, SF_6?

17.5.   What is the Doppler-shifted wavelength of a red ($\lambda = 660$ nm) traffic light when approaching it at 50 m.p.h.? At what speed of approach would it appear to be green ($\lambda = 520$ nm)?

17.6.   A spectral line of $^{48}Ti^{8+}$ in a distant star was found to be shifted from 654.2 nm to 706.5 nm and to be broadened to 61.8 pm. What is the speed of recession and the surface temperature of the star?

17.7.   Microwave spectroscopy is a very precise technique, but the breadth of lines may obscure details. What is the Doppler width (expressed as a fraction of the wavelength of the transition) for any kind of transition in (a) HCl, (b) ICl at 298 K? What would be the widths of rotational (in MHz) and vibrational (in cm^{-1}) transitions of these species? ($B(ICl) = 0.114$ cm^{-1}.)

17.8. From the relation $\tau\delta E \approx \hbar$ deduce the lifetime of a state that gives rise to a line of width (a) 0.1 cm^{-1}, (b) 1 cm^{-1}, (c) 100 MHz.

17.9. A molecule in a liquid undergoes about 10^{13} collisions each second. Suppose (a) that every collision is effective in deactivating the molecule vibrationally, (b) that one in a hundred collisions is effective. What is the width (in cm^{-1}) of vibrational transitions to this excited state?

17.10. The number of collisions a molecule undergoes in unit time in a gas where the pressure is p is $4\sigma(kT/\pi m)^{1/2}Lp/RT$. What is the collision-limited lifetime of an excited molecular state, assuming that every collision quenches? What is the width of the rotational transitions in HCl ($\sigma = 0.30$ nm^2) when the pressure of the gas is 1 atm? What must be the pressure in order that collision broadening is less important than Doppler broadening?

17.11. What proportions of chlorine molecules ($\tilde{\nu} = 559.7$ cm^{-1}) are in the ground and first excited vibrational states at (a) 273 K, (b) 298 K, (c) 500 K?

17.12. What is the value of J in the most highly populated rotational level of ICl at room temperature? Do not forget that there are $2J+1$ states of the same energy for each value of J. ($B = 0.114$ cm^{-1}.)

17.13. What is the value of J in the most highly populated rotational level of CH_4 at room temperature? There is a mild trick here: although the energy levels are given by the same expression as for a linear molecule we have to remember that there are $(2J+1)^2$ states of the same energy for every value of J. Why? Use $B = 5.24$ cm^{-1}.

17.14. The *reduced mass* was a quantity that appeared at several points in the chapter, and it is important to know how to calculate it. Calculate (in kg) the reduced masses and the moments of inertia of (a) $^{1}H^{35}Cl$, (b) $^{2}H^{35}Cl$, (c) $^{1}H^{37}Cl$. Take $R = 127.45$ pm.

17.15. The rotational constant of NH_3 has the value $cB = 298$ GHz. Compute the separation of the lines in GHz, cm^{-1}, and mm, and show that the value of B is consistent with an N–H bond length of 101.4 pm and a bond angle of 106° 47'. (Refer to Table 17.1 for information about moments of inertia.)

17.16. A space probe was designed to look for signs of CO in the atmosphere of Saturn. It was decided to employ a microwave technique from an orbiting satellite. Given the bond length of the molecule as 112.82 pm, where will the first four transitions of $^{12}C^{16}O$ lie? What resolution is needed if it was desired to distinguish the 1–0 line in the $^{12}C^{16}O$ spectrum from that of $^{13}C^{16}O$ in order to examine the relative abundances of the two carbon isotopes?

17.17. Rotational absorption lines from HCl gas were found at the following positions (R. L. Hausler and R. A. Oetjen, *J. chem. Phys.* **21**, 1340 (1953)): 83.32, 104.13, 124.73, 145.37, 165.89, 186.23, 206.60, 226.86 cm^{-1}. Find the moment of inertia and bond length of the molecule. (The species used was $^{1}H^{35}Cl$.)

17.18. Predict the positions of the microwave absorption lines in DCl ($^{2}H^{35}Cl$) on the basis of the data in the preceding Problem.

17.19. Is HCl the same size as DCl? The data from the rotational structure of the infrared spectra of the species are as follows (I. M. Mills, H. W. Thompson, and R. L. Williams, *Proc. R. Soc. A* **218**, 29 (1953); J. Pickworth and H. W. Thompson, *Proc. R. Soc. A* **218**, 37 (1953)).

J:	0	1	2	3	4	5	6
$^1H^{35}Cl$ { R-branch	2906.25	2925.92	2944.99	2963.35	2981.05	2998.05	$3014.50\,cm^{-1}$
P-branch	—	2865.14	2843.63	2821.59	2799.00	2775.77	2752.01
$^2H^{35}Cl$ { R-branch	2101.60	2111.94	2122.05	2131.91	2141.53	2150.93	2160.06
P-branch	—	2080.26	2069.24	2058.02	2046.58	2034.95	2023.12

17.20. Show that the moment of inertia of a diatomic molecule composed of two atoms of masses m_A and m_B and of bond length R is given by $I = [m_A m_B/(m_A + m_B)]R^2$. What is the moment of inertia of (a) $^1H_2$, (b) $^{127}I_2$?

17.21. The pure rotation spectrum of HI consists of a series of lines separated by $13.10\ cm^{-1}$. What is the bond length of the molecule?

17.22. From a thermodynamic point of view the copper monohalides CuX are expected to exist mainly as polymers in the gas phase, and so it proved difficult to get enough intensity to measure the spectra of the monomers. This was overcome by flowing the halogen gas over chips of copper heated to 1000–1100 K (E. L. Manson, F. C. de Lucia, and W. Gordy, *J. chem. Phys.* **63**, 2724 (1975)). For CuBr the $J = 13$–14, 14–15, 15–16 transitions occurred at 84 421.34, 90 449.25, 96 476.72 MHz. Find the rotational constant and the bond length of CuBr.

17.23. The microwave spectrum of $^{16}O^{12}C^{32}S$ shows absorption lines at 24.325 92, 36.488 82, 48.651 64, 60.814 08 GHz. What is the moment of inertia of the molecule? Note that the individual bond lengths cannot be obtained from these data; but read on.

17.24. In a study of the microwave spectra of various isotopically substituted OCS molecules (C. H. Townes, A. N. Holden, and F. R. Merritt, *Phys. Rev.* **74**, 1113 (1948)) the following transitions were observed:

$J \to J+1$	$1 \to 2$	$2 \to 3$	$3 \to 4$	$4 \to 5$
$^{16}O^{12}C^{32}S$	24.325 92	36.488 82	48.651 64	60.814 08 GHz
$^{16}O^{12}C^{34}S$	23.732 33		47.462 40	

Using the expression for the moment of inertia given in Table 17.1, and assuming that the bond lengths are unchanged in isotopic substitution, find the lengths of the C–O and C–S bonds in OCS.

17.25. The set of equations known as *Kraitchman's equations* relate the change of rotational constant (or moments of inertia) to the positions at which isotopic substitution is made. Show that when isotopic substitution is made at a distance z along the axis from the original centre of mass of a symmetric top molecule, z and ΔB, the change of the perpendicular rotational constant, are related by $(z/pm)^2 = 1.685\,90 \times 10^5\ (\Delta B/cm^{-1})/[(B/cm^{-1})(B'/cm^{-1})(\Delta M_r)]$, where B is the rotational constant before substitution, B' that after, and $\Delta M_r = M_r(M_r' - M_r)/M_r'$.

17.26. The microwave spectra of various isotopic species of $ClTeF_5$ show rigid symmetric top behaviour (A. C. Legon, *J. chem. Soc. Faraday Trans. II*, 29 (1973)). Four fluorines lie in a square, the tellurium atom lies just above the plane they form, and the other flourine lies beneath this plane and the chlorine above. Use Kraitchman's equation derived in the last Problem to deduce the Te–Cl bond length on the assumption that all four TeF bond lengths have the same length. Data: 11–10 transition in $^{35}Cl^{126}TeF_5$: 30 711.18 MHz; the same in $^{35}Cl^{125}TeF_5$: 30 713.24 MHz; the same in $^{37}Cl^{126}TeF_5$: 29 990.54 MHz.

17.27. The moments of inertia of the linear mercury(II) halides are very large, and as a consequence the O- and S-branches of the vibrational Raman spectra show little structure. Nevertheless the peaks of both branches can be distinguished and have been used to measure the rotational constants (R. J. H. Clark and

D. M. Rippon, *J. chem. Soc. Faraday Trans. II* **69**, 1496 (1973)). Show, from a knowledge of the value of J corresponding to the population maximum, that the separation of the peaks of the O- and S-branches is given by the *Placzek–Teller relation* $\delta\tilde{v} = \{32BkT/hc\}^{\frac{1}{2}}$. The following widths were obtained at the temperatures stated: $HgCl_2(282\,°C)$: $23.8\,cm^{-1}$; $HgBr_2(292\,°C)$: $15.2\,cm^{-1}$; $HgI_2(292\,°C)$: $11.4\,cm^{-1}$. Calculate the rotational constants of these molecules.

17.28. The hydrogen halides have the following fundamental vibration frequencies: HF $(4141.3\,cm^{-1})$, $H^{35}Cl$ $(2988.9\,cm^{-1})$, $H^{81}Br$ $(2649.7\,cm^{-1})$, $H^{127}I$ $(2309.5\,cm^{-1})$. Find the force constants of the hydrogen–halogen bonds.

17.29. From the data in the last Problem, predict the fundamental vibration frequencies of the deuterium halides.

17.30. The Morse curve, eqn (17.3.4), is very useful as a simple representation of the molecular potential energy curve. When RbH was studied it was found that $\tilde{v}_0 = 936.8\,cm^{-1}$ and $x\tilde{v}_0 = 14.15\,cm^{-1}$. Plot the potential energy curve from 50 pm to 800 pm around $R_e = 236.7\,pm$.

17.31. The rotation of a molecule may weaken its bonds. This can be illustrated by plotting the potential energy curve allowing for the kinetic energy of rotation of the molecule. Thus $V^*(R) = V(R) + B(R)J(J+1)$ is plotted, where V is the Morse potential and $B = \hbar/4\pi cmR^2$, m being the reduced mass. Plot these curves for RbH on the same graph as in the last Problem, taking $J = 40$, 80, and 100. Notice how the dissociation energy is affected by the rotation. (Taking $hcB(R_e) = 3.020\,cm^{-1}$ will simplify the calculation of $B(R)$.)

17.32. The first five vibrational energy levels of HCl were found to lie at 1331.84, 3917.44, 6398.94, 8776.34, 11 049.6 cm^{-1}. What is the dissociation energy of the molecule?

17.33. The vibrational energy levels of NaI lie at the following wavenumbers: 142.81, 427.31, 710.31, 991.81 cm^{-1}. Show that they fit an expression $(v+\frac{1}{2})\tilde{v} - (v+\frac{1}{2})^2 x\tilde{v}$, and deduce the force-constant, zero-point energy, and dissociation energy of the bond.

17.34. The HCl molecule has a potential energy that is quite well described by a Morse curve with $D_e = 5.33\,eV$ and $x\tilde{v} = 52.05\,cm^{-1}$, $\tilde{v} = 2689.7\,cm^{-1}$. Assuming that the potential remains unchanged on deuteration, predict the dissociation energies $D^{\ominus}$ of (a) HCl, (b) DCl.

17.35. The expressions for the P-, Q-, and R-branches (p. 590) assume that the moment of inertia of the molecule is the same in the excited vibrational states as in the ground state. Derive expressions for the positions of the lines without making these assumptions.

17.36. Lines in the P-branch of $^1H^{35}Cl$ were observed at 2865.1, 2843.6, and 2821.6 cm^{-1} for $J = 1$, 2, and 3, and in the R-branch at 2096.2, 2925.9, 2945.0, 2963.3 cm^{-1} for $J = 0$, 1, 2, 3 (I. M. Mills, H. W. Thompson, and R. L. Williams, *Proc. R. Soc. A* **218**, 29 (1953)). Find the force-constant of the bond and the bond lengths of the upper and lower vibrational states.

17.37. The vibrational Raman spectrum of $^{35}Cl_2$ shows a series of Stokes lines separated by 0.9752 cm^{-1} and a similar series of anti-Stokes lines. What is the bond length of Cl_2?

17.38. Which of the three vibrations of a general AB_2 molecule are infrared or Raman active when the molecule is (a) bent, (b) linear?

17.39. Consider the vibrational mode that corresponds to the uniform expansion of the benzene ring. Is it (a) Raman active, (b) infrared active?

18 Determination of molecular structure: electronic spectroscopy

Learning objectives

After careful study of this chapter you should be able to

(1) Define *absorption coefficient, molar absorption coefficient, extinction coefficient, optical density*, and *absorbance* (p. 604), and use the *Beer–Lambert Law* to calculate the intensity of light after passage through a sample (eqn (18.1.1)).

(2) Relate the intensity of a transition to the *oscillator strength* (eqn (18.1.4)).

(3) Account for the colour of *transition metal complexes* (p. 608).

(4) Describe the transitions responsible for absorption by *carbonyl groups* and the $C\!=\!C$ *double bond* (p. 609).

(5) Explain the basis of the *Franck–Condon principle* and use it to account for the vibrational structure of electronic transitions (p. 611).

(6) Describe the mechanisms of *fluorescence* (p. 614) and *phosphorescence* (p. 615).

(7) Describe *laser action* (p. 616) and the terms *population inversion* and *super-radiant mode* (p. 619).

(8) Describe the action of the *ruby laser*, the *neodymium laser*, the *He–Ne laser*, and the CO_2 *laser* (p. 618).

(9) Explain the terms *dissociation limit, predissociation*, and *internal conversion* (p. 620).

(10) Describe the basis of *photoelectron spectroscopy* (p. 622) and relate a photoelectron spectrum to the electronic structure of a molecule (p. 624).

(11) Explain how *X-ray photoelectron spectroscopy* can be used in chemical analysis and for the study of the inner electron shells in molecules (p. 626).

Introduction

The energies involved when molecules make transitions by changing their electronic distributions are of the order of several electron volts (1 eV corresponds to about 8000 cm^{-1}) and so the photons emitted or absorbed lie in the visible and ultraviolet regions of the spectrum (which spreads from about $14\,000 \text{ cm}^{-1}$ for red light to $21\,000 \text{ cm}^{-1}$ for blue and on to $50\,000 \text{ cm}^{-1}$ for the far ultraviolet), Table 18.1. Studying the electronic absorption and emission spectra of molecules enables us to account for the colour of substances. For instance, if some object contains molecules that absorb blue light, it will look red in reflected white light. The green of vegetation is due to the reflection of white sunlight from leaves, etc., that contain chlorophyll, a molecule able to absorb in both the red and the blue regions of the spectrum, Fig. 18.1. Electronic spectra give information about the electronic structures of molecules, and by analysing them we are able to deduce the energies of electrons in different molecular orbitals and therefore to test theories of molecular structure. Another way of obtaining similar information is to see how much energy it takes to expel an electron from a molecule. This branch of electronic spectroscopy, called *photoelectron spectroscopy*, is dealt with in the last part of the chapter.

Table 18.1. Colour, frequency, and energy of light

Colour	Wavelength	Frequency	Wavenumber	Energy		
	λ/nm	$\nu/10^{14}\,\text{Hz}$	$\tilde{\nu}/10^4\,\text{cm}^{-1}$	eV	kJ mol^{-1}	kcal mol^{-1}
Infrared	1000	3.00	1.00	1.24	120	28.6
Red	700	4.28	1.43	1.77	171	40.8
Orange	620	4.84	1.61	2.00	193	46.1
Yellow	580	5.17	1.72	2.14	206	49.3
Green	530	5.66	1.89	2.34	226	53.9
Blue	470	6.38	2.13	2.64	254	60.8
Violet	420	7.14	2.38	2.95	285	68.1
Near ultraviolet	300	10.0	3.33	4.15	400	95.7
Far ultraviolet	200	15.0	5.00	6.20	598	142.9

Source: J. G. Calvert and J. N. Pitts, *Photochemistry*, Wiley.

When an electronic transition occurs in a molecule a large change of energy is involved, and the redistribution of the electron changes the electrostatic forces operating on the nuclei. The molecule responds to the change of forces by bursting into vibration. In other words, an electronic transition is usually accompanied by a vibrational transition. The extra vibrational structure in a spectrum can be resolved if the sample is gaseous, but in a liquid or solid sample the lines merge together and result in a broad, almost featureless line, Fig. 18.2. The vibrational transitions that accompany electronic transitions are themselves accompanied by

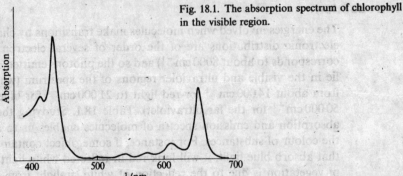

Fig. 18.1. The absorption spectrum of chlorophyll in the visible region.

rotational transitions. It follows that the electronic spectra of gaseous samples can be very complicated. For that reason we shall begin by discussing the electronic spectra of liquids, where the fine details of the spectra are blurred into broad bands.

Another factor that complicates electronic spectra is that the molecules might absorb enough energy to fall apart. This is important because dissociation is one of the fundamental processes of photochemistry. We shall have to see how dissociation affects the spectrum, and what photo-chemical information can be obtained by studying it.

18.1 Measures of intensity

In this section we establish an experimental measure of the intensities of transitions. Consider the reduction of intensity that occurs when light of intensity I passes through a slab of material of thickness dx which contains absorbing species at a concentration C. The loss of intensity is proportional to the thickness, the concentration, and the intensity of the incident light, and so we can write

$$dI = -\alpha CI\,dx$$

where the coefficient α is called the *absorption coefficient*, and depends

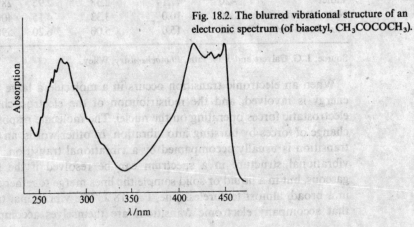

Fig. 18.2. The blurred vibrational structure of an electronic spectrum (of biacetyl, $CH_3COCOCH_3$).

both on the molecule under study and the frequency of the light. This equation, which can be rewritten as

$$d \ln I = -\alpha C \, dx,$$

applies to every layer into which the sample can be divided, and to obtain the intensity (I_f) that emerges from a sample of thickness l when the incident intensity is I_i we sum all the successive changes:

$$\int_{I_i}^{I_f} d \ln I = -\int_0^l \alpha C \, dx.$$

If the concentration is uniform through the sample, C is independent of x and the expression integrates to the

(18.1.1)   Beer–Lambert Law: $I_f = I_i \exp(-\alpha Cl)$.

We see that the intensity of the radiation falls off exponentially with the thickness of the sample, and depends both on the concentration of the absorbing species and its ability to absorb light at the frequency being used. Most work in this field uses logarithms to the base 10 in the expression of the Beer–Lambert Law (since they are easier to use in practical applications), and so the last equation is normally written

(18.1.2a) $I_f = I_i 10^{-\varepsilon Cl}$

or

(18.1.2b) $\lg(I_f/I_i) = -\varepsilon Cl$.

In these alternative forms ε, epsilon, (which is equal to $\alpha/2.303$) is called the *molar absorption coefficient* (formerly the *extinction coefficient*) of the species at the relevant frequency. The product εCl is called the *absorbance*, A, (formerly the *optical density*) of the sample.

Example (Objective 1). Light of wavelength 256 nm passes through a 1 mm path length cell containing a 0.05 mol dm^{-3} solution of benzene. The light intensity is reduced to 16 per cent of its initial value. Calculate the absorbance and molar absorption coefficient of the sample. What would be the transmittance through a cell of path length 2 mm?

- *Method.* The absorbance and molar absorption coefficient can be obtained from eqn (18.1.2b); the absorbance is $-\lg(I_f/I_i)$, and the molar absorption coefficient is $= -(1/Cl)\lg(I_f/I_i)$. Use $I_f = 0.16 \, I_i$.

- *Answer.* $\lg(I_f/I_i) = \lg 0.16 = -0.80$. The absorbance of the sample is therefore 0.80. The molar absorption coefficient is

$$\varepsilon = -(1/Cl)\lg(I_f/I_i) = 0.80/(0.05 \text{ mol dm}^{-3}) \times (0.1 \text{ cm}) = 160 \text{ mol}^{-1} \text{ dm}^3 \text{ cm}^{-1}.$$

When the path length is 2 mm,

$$\lg(I_f/I_i) = -(160 \text{ mol}^{-1} \text{ dm}^3 \text{ cm}^{-1}) \times (0.05 \text{ mol dm}^{-3}) \times (0.2 \text{ cm}) = -1.60.$$

Therefore $I_f/I_i = 0.025$, and so the emergent light will have 3 per cent of its incident intensity.

● *Comment.* The molar absorption coefficient is sometimes expressed in $mol^{-1} cm^2$; in the present example that would give $\varepsilon = 160\,000\, mol^{-1} cm^2$ because $dm^3 = 1000\, cm^3$. This choice of units ($cm^2\, mol^{-1}$) suggests that ε can be interpreted as a (molar) cross-section for photon absorption.

Many absorptions spread over a range of frequencies, and so quoting the absorption coefficient for a single frequency might not give a true indication of the intensity of absorption of a molecular transition. The *integrated absorption coefficient* is the sum of the absorption coefficients for all the frequencies in the band:

(18.1.3) $$\mathscr{A} = \int \alpha(v)\, dv,$$

and this does measure the total strength of a broad spectral band. $\mathscr{A}$ is sometimes expressed in terms of the dimensionless quantity f, the *oscillator strength* of the transition:

(18.1.4) $$f = (4m_e c\varepsilon_0/Le^2)\mathscr{A} = 6.257\,30 \times 10^{-19} \times (\mathscr{A}/m^2\, mol^{-1}\, s),$$

where ε_0 is a fundamental constant (the vacuum permittivity, see p. 340). It turns out that intense transitions have oscillator strengths of about unity. Another indication of the intensity of absorption is the *maximum* value of the molar absorption coefficient of a band. This is written ε_{max}. Typical values for strong transitions are of the order of 10^4–$10^5\, cm^{-1}\, dm^3\, mol^{-1}$, indicating that in a $1\, mol\, dm^{-3}$ solution the intensity of light (of the appropriate frequency) drops to $\frac{1}{10}$ of its initial value by the time it has passed through about $0.1\, mm$ of solution. Some values are listed in Table 18.2.

Table 18.2. Absorption characteristics of some groups and molecules

Group	$\tilde{v}_{max}/cm^{-1}$	λ_{max}/nm	$\varepsilon_{max}/dm^3\, mol^{-1}\, cm^{-1}$
C=C	55 000	183	250
	57 300	174	16 500
	58 600	170	16 500
	62 000	162	10 000
C=O	34 000	295	10
	54 000	185	Strong
—NO$_2$	36 000	278	10
	47 500	210	10 000
—N=N—	28 800	347	15
	>38 500	<260	Strong
Benzene ring	39 000	255	200
	50 000	200	6 300
	55 500	180	100 000
Cu^{2+}(aq)	12 400	810	10
$Cu(NH_3)_4^{2+}$	16 700	600	50

Source: Principally J. G. Calvert and J. N. Pitts, *Photochemistry*, Wiley.

Example (Objective 2). A transition in benzene has a molar absorption coefficient of $160\,mol^{-1}$ $dm^3\,cm^{-1}$ (last *Example*) and spreads over about $4000\,cm^{-1}$. Estimate its oscillator strength.

- *Method.* Use eqn (18.1.4), approximating the area under the absorption curve by height × width. Use $d\nu = c\,d\tilde{\nu}$ (from $\nu = c\tilde{\nu}$) in eqn (18.1.3).

- *Answer.* $\alpha = 2.303\varepsilon = 2.303 \times (160\,000\,mol^{-1}\,cm^2) = 3.68 \times 10^5\,mol^{-1}\,cm^2$.

$$\mathscr{A} = \int \alpha(\nu)\,d\nu = c \int \alpha(\tilde{\nu})\,d\tilde{\nu} \approx c\alpha_{max} \times (\text{width})$$

$$= (2.998 \times 10^{10}\,cm\,s^{-1}) \times (3.68 \times 10^5\,mol^{-1}\,cm^2) \times (4000\,cm^{-1})$$

$$= 4.42 \times 10^{19}\,mol^{-1}\,cm^2\,s^{-1} = 4.42 \times 10^{15}\,mol^{-1}\,m^2\,s^{-1}.$$

In order to find f we need the factor

$$4m_e c\varepsilon_0/Le^2$$

$$= \frac{4 \times (9.109\,53 \times 10^{-31}\,kg) \times (2.997\,925 \times 10^8\,ms^{-1}) \times (8.854\,88 \times 10^{-12}\,J^{-1}C^2m^{-1})}{(6.022\,05 \times 10^{23}\,mol^{-1}) \times (1.602\,19 \times 10^{-19}\,C)^2}$$

$$= 6.257\,30 \times 10^{-19}\,mol\,m^{-2}\,s.$$

It follows that

$$f = (4.42 \times 10^{15}\,mol^{-1}\,m^2\,s^{-1}) \times (6.257 \times 10^{-19}\,mol\,m^{-2}\,s) = 2.77 \times 10^{-3}.$$

- *Comment.* Note that f is a dimensionless quantity. In the present example it is small: this indicates that the transition is 'forbidden'.

The advantage of introducing the oscillator strength is that it can be interpreted theoretically. We have seen already that the intensity of any spectroscopic transition depends on the magnitude of a transition dipole moment (p. 564). This quantity, which measures the shift of charge (and its direction) during a transition, can be given a precise definition in terms of the wavefunctions ψ_i and ψ_f for the initial and final states of the molecule:

(18.1.5) $$\mathbf{p} = -e \int \psi_f^* \mathbf{r} \psi_i \, d\tau.$$

If the wavefunctions are known, $\mathbf{p}$ may be calculated by evaluating this integral. The connection between the transition moment and the observed intensity is

(18.1.6) $$f = (8\pi^2/3)(m_e\nu/he^2)|\mathbf{p}|^2.$$

Therefore we may estimate the intensity of the line if we know the wavefunctions for the molecule. This is an important link between theory and experiment.

Example. Find the value of the transition dipole in the last *Example*.

- *Method.* We have seen that $f = 2.77 \times 10^{-3}$. Substitute in eqn (18.1.6) and find $|\mathbf{p}|$. The transition is at 256 nm.

- *Answer.* We need the factor

$$3he^2/8\pi^2 m_e = (3/8\pi^2)\left\{\frac{(6.626\,18 \times 10^{-34}\,\text{J s}) \times (1.602\,19 \times 10^{-19}\,\text{C})^2}{(9.109\,53 \times 10^{-31}\,\text{kg})}\right\}$$

$$= 7.094\,58 \times 10^{-43}\,\text{m}^2\,\text{s}^{-1}\,\text{C}^2.$$

$$v = c/\lambda = (2.9979 \times 10^8\,\text{m s}^{-1})/(256 \times 10^{-9}\,\text{m}) = 1.17 \times 10^{15}\,\text{Hz}.$$

Then

$$|\mathbf{p}|^2 = (7.094\,58 \times 10^{-43}\,\text{m}^2\,\text{s}^{-1}\,\text{C}^2) \times (2.77 \times 10^{-3})/(1.17 \times 10^{15}\,\text{Hz})$$

$$= 1.68 \times 10^{-60}\,\text{C}^2\,\text{m}^2,$$

so that $|\mathbf{p}| = 1.30 \times 10^{-30}\,\text{C m}$.

• *Comment.* Convert the transition dipole moment to debye (D) by $1\,\text{D} = 3.34 \times 10^{-30}\,\text{C m}$: this gives p = 0.39 D. The transition can be visualized as corresponding to moving about 40 per cent of an electron through a distance about equal to the radius of an atom.

18.2 Chromophores

In a number of molecules the absorption of a photon can be traced to the excitation of the electrons of a small group of atoms. For example, when a carbonyl group is present in a molecule, an absorption at about 290 nm is normally observed, although its precise position depends on the nature of the rest of the molecule. Groups of this kind are called *chromophores* (from the Greek for 'colour bringer'), and their presence often accounts for the colour of materials. There are several types of chromophore, but we shall confine our attention to three: transition metal ions, double bonds (such as C=C), and lone-pairs of electrons.

The colour of *transition metal complexes* can normally be ascribed to the presence of low-lying d-orbitals. In Chapter 15 we saw that the presence of the ligands splits the d-orbitals of the central atom into two sets, Fig. 18.3. The separations are usually not very large, and the excitation of an electron from a t_{2g}-orbital into an e_g-orbital accounts for absorptions in the visible region of the spectrum. The spectrum of $[\text{Ti}(\text{H}_2\text{O})_6]^{3+}\text{Cl}_3^-$ is shown in Fig. 18.4, and the transition at $20\,000\,\text{cm}^{-1}$ (500 nm) can be ascribed to the $e_g \leftarrow t_{2g}$ transition of the single d-electron.

The only drawback to this explanation is that d–d transitions are forbidden in octahedral complexes. The selection rule for electronic transitions in complexes with a centre of symmetry is called the *Laporte selection rule*: it states that *the only allowed transitions are those involving a change of parity*. That is, g → u and u → g are allowed, but g → g and

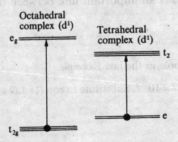

Fig. 18.3. Electronic transitions in transition metal complexes.

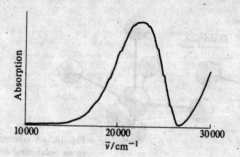

Fig. 18.4. The electronic absorption of $[Ti(H_2O)_6]^{3+}Cl_3^-$ in aqueous solution.

u → u are forbidden. This can be traced back to the forbidding of d–d transitions in free atoms (p. 448), or it can be explained by a group-theoretical argument.*

The forbidden nature of the $e_g \leftarrow t_{2g}$ transition can be eliminated if the symmetry of the molecule is destroyed by a vibration. For example, the vibration shown in Fig. 18.5 destroys the centre of symmetry of the complex, and therefore destroys the g, u classification of the orbitals. In the absence of a centre of symmetry a d–d transition is allowed, and so the complex is able to absorb. Since the electronic d–d transitions depend on vibrations of the complex they are called *vibronic*. In practice the transition becomes only weakly allowed by this vibronic mechanism, and so d–d transitions in octahedral complexes are usually weak ($f \approx 10^{-4}$).

Another way that a transition metal complex may absorb light is by the transfer of an electron from the ligands into the d-orbitals of the central ion, or vice versa. In such a *charge-transfer transition* the electron moves through a considerable distance, which means the transition moment may be large and the absorption intensity correspondingly great. This mode of chromophore activity is shown, for example, by the MnO_4^- ion, and accounts for its intense violet colour ($\lambda = 420–700$ nm, $f \approx 3 \times 10^{-2}$).

In the case of the $C{=}C$ *double bond*, absorption lifts a π-electron into the empty antibonding π^*-orbital, Fig. 18.6. We therefore speak of a (π^*, π)-*transition* (which is normally read 'π-to-π-star transition'). The energy of the transition is about 7 eV for an unconjugated double bond, and so it absorbs at about 180 nm, which is in the ultraviolet. When the double bond is part of a conjugated chain the energies of the molecular orbitals lie closer together and the (π^*, π)-transitions shift into the visible region of the spectrum. We shall see an important example of this in a moment.

The transition responsible for absorption in *carbonyl compounds* can be

---

* In accord with the general discussion on p. 545, the integral that defines the transition dipole moment, p. 607, must have A_{1g} symmetry in the octahedral group, or else it disappears. A glance at the character table for the group O_h shows that the functions x, y, z transform as T_{1u}. Therefore the integral

$$\mathbf{p} = -e \int \psi^*_{t_{2g}} \mathbf{r} \psi_{e_g} \, d\tau$$

cannot transform as A_{1g} because the characters of T_{2g}, T_{1u}, E_g combine to give irreducible representations of u symmetry, not g. Hence $e_g \leftarrow t_{2g}$ transitions are symmetry forbidden.

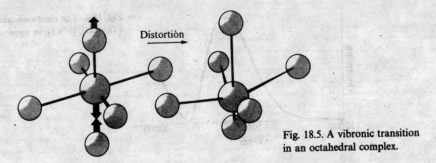

Distortion

Fig. 18.5. A vibronic transition
in an octahedral complex.

traced to the non-bonding lone-pairs of electrons located on the oxygen. One of these electrons may be excited into the empty π^*-orbital of the group, Fig. 18.7, and the transition is then called a (π^*, n)-*transition* (read '*n*-to-π-star transition'). Typical absorption energies are in the region of 4 eV (290 nm), and because (π^*, n)-transitions are often symmetry forbidden the absorptions are weak (f lies in the range 2×10^{-4} to 6×10^{-4}).

•An important example of (π^*, π)- and (π^*, n)-transitions is provided by the mechanism responsible for vision. The retina of the eye contains 'visual purple', which is a protein in combination with 11-*cis*-retinal:

The 11-*cis*-retinal acts as a chromophore, and is the primary receptor of photons entering the eye. A solution of 11-*cis*-retinal absorbs at about 380 nm, but in combination with the protein (a link which might involve elimination of the terminal carbonyl) the absorption maximum shifts to about 500 nm, which tails into the blue. The conjugated double bonds are a major reason why the molecule is able to absorb over the whole of the visible spectrum, but they also play another role. In its electronically excited state the conjugated double bond chain can isomerize, and one half of the chain can twist about an excited C=C bond and form all-*trans*-retinal:

This has a different shape from the original isomer, and is unable to fit into the protein. The primary step in vision therefore seems to be a photon absorption which is followed by isomerization of the carbon chain; this uncoiling of the molecule triggers a nerve impulse to the brain.

Fig. 18.6. The transition in the carbon–carbon double bond.

18.3 Vibrational structure and the Franck–Condon principle

The breadth of electronic absorption bands in liquid samples is due to the unresolved vibrational structure. This structure arises because electronic transitions are accompanied by changes of the vibrational state of the molecules, and in some liquid samples, and in gases, the vibrational structure can be resolved and used to obtain information about the vibrational characteristics of excited-state molecules. The appearance of the vibrational structure of an electronic spectrum can be accounted for in terms of the *Franck–Condon principle*.

The Franck–Condon principle is based on the view that, because nuclei are so much more massive than electrons, the electronic redistribution occurs while the nuclei stay at their initial separations. The electronic transition rapidly builds up charge density in new regions, and this charge density exerts a force on the nuclei. This new force drives them into oscillation, and they swing backwards and forwards from their original separation. The stationary equilibrium separation of the nuclei in the initial state of the molecule therefore becomes the *turning point* of the vibration of the excited electronic state.

The quantum-mechanical description presents this model in a slightly disguised form, and refines it to the point where it is possible to calculate the intensity of the transitions to different vibrational levels of the electronically excited molecule.

Consider the molecular potential-energy curves drawn in Fig. 18.8. They relate to different electronic states of the same molecule. Before the absorption occurs the molecule is in the ground vibrational state of its lower electronic state. The form of the lowest vibrational wavefunction shows that the most probable location of the nuclei is at their equilibrium separation R_e. It follows that most electronic transitions will occur while the nuclei are at this separation.

When an electronic transition occurs, the molecule is excited to the state represented by the upper curve. According to the Franck–Condon

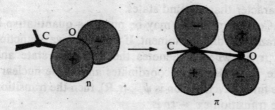

Fig. 18.7. The transition in the carbonyl group.

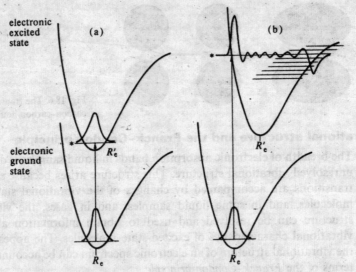

Fig. 18.8. The quantum-mechanical basis of the Franck–Condon principle.

principle the nuclear framework remains constant during the actual transition, and so we may imagine the energy of the molecule as rising up the vertical line marked in Fig. 18.8. This is the origin of the expression *vertical transition*, which is used to denote an electronic transition that occurs without change of nuclear geometry.

The vertical transition cuts through several vibrational levels of the upper electronic state. The level marked * is the one in which the nuclei are most probably at the initial separation R_e (because the wavefunction has maximum amplitude there), and so this is the most probable level for the termination of the excitation. It is not, however, the only vibrational level at which the excitation may end, because several of its neighbours also have a high probability of having nuclei at the separation R_e. Therefore transitions take place to all the vibrational levels in this region, but most intensely to the level having a wavefunction that overlaps the initial vibrational state wavefunction most favourably. The vibrational structure of the spectrum depends on the relative displacement of the two molecular potential-energy curves, and a long *progression* of vibrations (a lot of vibrational structure in the spectrum) is stimulated if the two states are significantly displaced. The upper curve is usually, but not always, displaced to longer bond lengths, because excited states usually have more antibonding character than ground states.

The Franck–Condon principle may be put on a quantitative basis by analysing the transition dipole moment. If the initial wavefunction of the molecule is $\psi_{\varepsilon,v}(\mathbf{r}, \mathbf{R})$, where ε denotes the electronic state and v the vibrational state, $\mathbf{r}$ the electronic coordinates and $\mathbf{R}$ the nuclear coordinates, and if the final wavefunction is $\psi_{\varepsilon',v'}(\mathbf{r}, \mathbf{R})$, then the transition dipole moment for the transition $\varepsilon'v' \leftarrow \varepsilon v$ is

$$p = -e \int \psi_{\varepsilon',v'}(\mathbf{r},\mathbf{R}) \mathbf{r} \psi_{\varepsilon,v}(\mathbf{r},\mathbf{R}) \, d\tau_{elec} \, d\tau_{nuc},$$

where the integrations are over all the electronic and nuclear coordinates. The combined electronic and nuclear (vibrational) wavefunctions can be expressed as the products $\psi_e(\mathbf{r})\psi_v(\mathbf{R})$, where $\psi_v(\mathbf{R})$ is a vibrational wavefunction (this separation is an approximation depending on the Born-Oppenheimer approximation, p. 470). The last integral then factorizes into two parts:

$$p = -e \int \psi_{\varepsilon'}(\mathbf{r}) \mathbf{r} \psi_\varepsilon(\mathbf{r}) \, d\tau_{elec} \int \psi_{v'}(\mathbf{R}) \psi_v(\mathbf{R}) \, d\tau_{nuc}.$$

The second factor is the overlap between the vibrational wavefunctions of the initial and final states. Since the intensity of the transition depends on the square of the transition dipole, it follows that the greatest intensity is for a transition between vibrational states having the greatest overlap. This is the formal basis of the preceding discussion. It is possible to estimate the overlap integrals that appear in the transition moment, and so quantitative estimates of the intensity distribution can be made.

Example (Objective 5). Consider the transition from one excited electronic state, with bond length R_e, to another with bond length R'_e. The intensity is determined by the overlap of the vibrational wavefunctions for the two states. Calculate the overlap between the vibrational ground states of each of the two electronic states, and show that the intensity of the 0–0 transition is greatest when the two equilibrium bond lengths are the same. For simplicity, let the force constants in the two electronic states be the same.

- *Method.* The intensity is proportional to the square of the overlap between the two vibrational states: this is the *Franck-Condon factor*. Calculate $\int \psi'_0(R)\psi_0(R)\,dR$, where ψ_0 is the ground vibrational state of the lower electronic state, and ψ'_0 is the ground vibrational state of the upper electronic state. Use the wavefunctions in Table 13.1.

- *Answer.* $\psi_0(R) = (\alpha/\pi^{1/2})^{1/2} \exp[-\tfrac{1}{2}\alpha^2(R-R_e)^2]$

  $\psi'_0(R) = (\alpha/\pi^{1/2})^{1/2} \exp[-\tfrac{1}{2}\alpha^2(R-R'_e)^2],$

  with $\alpha^2 = m\omega_0/\hbar$. (The functions are both Gaussians, one is centred on R_e, the other on R'_e.) The overlap is

  $$S = \int \psi'_0(R)\psi_0(R)\,dR = (\alpha/\pi^{1/2}) \int_{-\infty}^{\infty} \exp[-\tfrac{1}{2}\alpha^2(R-R_e)^2 - \tfrac{1}{2}\alpha^2(R-R'_e)^2]\,dR.$$

  It is not difficult to manipulate this into an expression proportional to an integral over e^{-x^2}, the value of which is $\pi^{1/2}$. Therefore the overlap integral is

  $S = \exp[-\tfrac{1}{4}\alpha^2(R_e-R'_e)^2],$

  and the intensity is proportional to $S^2 = \exp[-\tfrac{1}{2}\alpha^2(R_e-R'_e)^2]$. This is unity when $R_e = R'_e$, and decreases as the equilibrium bond lengths of the two states diverge.

- *Comment.* In the case of Br_2, $R_e = 228$ pm and there is an upper state with $R'_e = 266$ pm. Taking the vibrational wavenumber as $250\,cm^{-1}$ gives $S^2 = 5 \times 10^{-10}$; therefore the intensity of the 0–0 transition is only 5×10^{-10} of what it would have been if the potential curves were directly above each other.

It should be noticed that the splitting between the vibrational lines of an electronic absorption spectrum depends on the vibrational energies of the *excited* electronic state: it follows that electronic spectra may be used to assess the rigidities of electronically excited molecules.

18.4 The fate of electronically excited states

We now turn to the processes that take place after the absorption of light and the formation of an electronically excited species. A common fate for the excitation energy is for it to be transferred into vibration, rotation, and translation of the neighbouring molecules. This *thermal degradation* of the energy transforms it into thermal energy of the environment. Much more interesting is the possibility that the excited molecule may be able to take part in a chemical reaction. The isomerization of the chromophore in visual purple is an example of this, and the general subject of *photochemistry* will be taken up in Part 3. In this section we shall be concerned mainly with *radiative decay*, which is the ability of a molecule to discard its excitation energy by emitting a photon.

Two mechanisms for radiative deactivation have been identified: *fluorescence* and *phosphorescence*. In fluorescence the emitted radiation ceases immediately the exciting source is removed, but in phosphorescence it may persist for long periods (even hours, but characteristically seconds, or fractions of seconds). This difference of behaviour suggests that fluorescence is an immediate conversion of absorbed light into re-emitted energy, but that in phosphorescence the energy is stored in some kind of reservoir, from which it slowly leaks.

Fluorescence. The sequence of steps involved in fluorescence is depicted in Fig. 18.9. Initial absorption of light carries the molecule into the vibrational energy levels of the upper molecular state. If the absorption spectrum were monitored it would look like that shown in Fig. 18.9a. In the excited state the molecule is subjected to collisions with its environment (e.g., the solvent molecules) and its thermal energy is discarded as thermal motion of the surroundings. The collisions succeed in lowering the molecule down its ladder of vibrational energy levels, but they may be unable to withdraw the larger electronic energy difference and quench the electronic excitation energy. The molecule might therefore live long enough to undergo a spontaneous emission, and to emit the excess energy as radiation as it drops to the lower electronic state.

The downward step occurs vertically, in accord with the Franck–Condon principle, and a series of lines appear as the *fluorescence spectrum*, Fig. 18.9b. The vibrational structure of this spectrum is characteristic of the lower electronic state (in contrast to the structure of the absorption spectrum), and so this is a valuable method of studying the vibrational characteristics of the ground state.

The mechanism accounts for the observations that fluorescence radia-

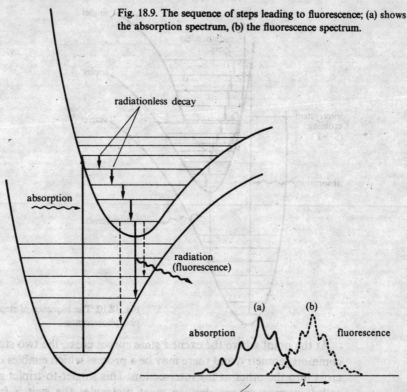

Fig. 18.9. The sequence of steps leading to fluorescence; (a) shows the absorption spectrum, (b) the fluorescence spectrum.

radiationless decay

absorption

radiation (fluorescence)

(a)   (b)

absorption   fluorescence

tion has a lower frequency than the incident radiation: the fluorescence occurs after some energy has been discarded into the solvent. The vivid oranges and greens of fluorescent dyes are an everyday manifestation of the energy degradation process: they absorb in the blue and ultraviolet, and fluoresce in the visible. The mechanism also suggests that the intensity of fluorescence should depend on the ability of the solvent to withdraw the larger quantum of energy required to lower the molecule from one electronic state to the other. It is indeed found that fluorescence can be eliminated by selecting a different solvent. For example, a solvent with widely spaced vibrational energy levels (such as water) may be able to accept the large quantum of electronic energy, but one with more closely spaced levels (such as heavy, flabby $SeOCl_2$) might not.

Phosphorescence. The sequence of events leading to phosphorescence is depicted in Fig. 18.10. The first steps are the same as in fluorescence, but the presence of a second excited state of the molecule plays a decisive role. This other excited state is called a *triplet state*, and it differs from the first (a *singlet state*) because the spins of two of its electrons are arranged so that they are parallel rather than opposed, Fig. 18.10.*

* The state is called a *triplet* for the reason discussed on p. 460. When two electrons have parallel spins the total spin angular momentum is $S = 1$ (each individual electron having $s = \frac{1}{2}$). This angular momentum can take *three* orientations with respect to some direction in space ($M_S = +1, 0, -1$).

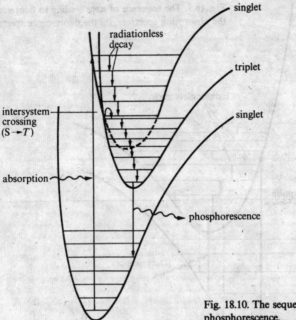

Fig. 18.10. The sequence of steps leading to phosphorescence.

At the point where the excited state curves cross, the two states share a common geometry, and there may be a process which enables one state to cross into the other at the intersection. This singlet-to-triplet switching is called *intersystem crossing*. In most molecules the switch from singlet (paired spins) to the triplet (two unpaired spins) is forbidden, but in some molecules it is weakly allowed. This happens when the molecule contains a heavy atom, because its strong spin–orbit interaction can reverse the relative orientations of pairs of electrons.

If the molecule crosses over into the triplet it continues to deposit energy into the environment, and as it does so it drops down the triplet's ladder of vibrational energies. The ladder terminates at the ground vibrational state of the excited triplet, and now the energy of the molecule is trapped. The solvent cannot extract the final, large quantum of electronic excitation energy, and the molecule cannot radiate its energy because a return to the ground state involves a forbidden singlet–triplet transition. The radiative transition, however, is not completely forbidden, because spin–orbit coupling is present and breaks the selection rule. The molecules are therefore able to emit slowly, and the emission may last long after the original excited state is formed.

This mechanism accords with the experimental observation that the energy seems to be trapped in a slowly leaking reservoir. It also suggests (and this is also confirmed experimentally) that the phosphorescence should be most intense from solid samples. This is because the environment then collides less effectively with the molecule, and the intersystem crossing step has time to occur as the singlet excited state slowly loses vibrational energy and falls past the intersection point. The mechanism

equilibrium population

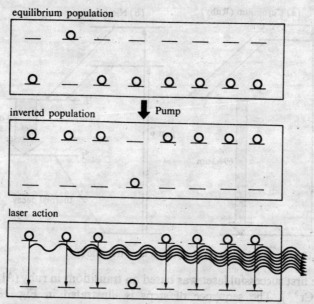

inverted population

Pump

laser action

Fig. 18.11. The sequence of steps leading to laser action.

also suggests that the amount of phosphorescence depends on the presence of a heavy atom: this also is in accord with experiment. Finally, the mechanism predicts that on account of the presence of the unpaired electron spins, the excited reservoir state should be magnetic. This has been confirmed by observing magnetism in phosphorescent, excited molecules using the sensitive resonance techniques described in the next chapter.

Laser action. Both fluorescence and phosphorescence are modes of return to the ground state by spontaneous emission. Laser action, as the acronym *laser* (Light Amplification by Stimulated Emission of Radiation) suggests, depends on emission by stimulated processes.

The laser process is shown schematically in Fig. 18.11. By some means a majority of the molecules in the sample are excited, or pumped, into the upper state. The sample is contained in a cavity between two mirrors, and when a molecule emits spontaneously the photon it generates ricochets backwards and forwards. Its presence stimulates other molecules to emit: they add more photons of the same frequency to the cavity, and these photons stimulate more molecules to emit. The cascade of energy builds up very rapidly, and if one of the mirrors is half-silvered the radiation may be tapped.

The characteristics of laser radiation follow from its manner of generation: it is monochromatic (because photons stimulate photons of the same frequency), coherent (because the phases of the electric field of the photons are in step—another feature of the stimulating process), and non-divergent (because photons travelling at angles to the cavity axis are not trapped, and do not stimulate others).

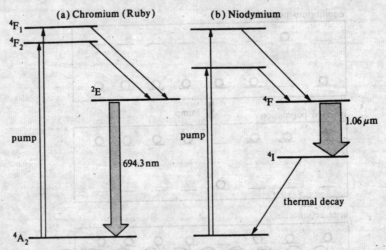

Fig. 18.12. The (a) ruby and (b) niodymium laser transitions.

The first successful laser was based on transitions in ruby (Al_2O_3 doped with Cr^{3+}). The sequence of steps is illustrated in Fig. 18.12a. The *population inversion* (pumping a majority of ions into an excited state) is achieved with a powerful flash from another source (such as a xenon lamp or another laser). These excited states switch their electronic distributions and change into the state labelled ^2E: the transition from ^2E to the ground state is the laser transition, and gives rise to red 694 nm radiation. The neodymium laser works on a slightly different principle, for the laser transition takes place between two excited states (labelled ^4F and ^4I in Fig. 18.12b). The advantage of this type of operation is that as the population inversion is between the two excited states, it is much easier to achieve than between an excited state and the heavily populated ground state.

Laser action occurs in liquids and gases as well as in solids. *Dye lasers* have intensely absorbing organic dyes (such as Rhodamine 6G dissolved in methanol) as the active medium. They have the advantage of operating over a wide range of wavelengths, and by a modification of the design of the cavity (for instance, by incorporating a diffraction grating in place of one of the mirrors) the radiation can be tuned to a selected wavelength. The disadvantage of dye lasers, which resulted in them being the last of the types to be developed, is that the excited state is very rapidly relaxed back to the ground state by the surrounding fluid medium. Gas lasers are widely used, and since cooling can be effected by ensuring that the gas flows rapidly through the cavity, they can be used to generate high powers. The pumping step normally involves a species other than the laser material itself. For instance, in the helium–neon laser the initial step is the excitation of helium atoms to the state $1s2s\ ^1S$, Fig. 18.13a. This spectroscopically forbidden transition ($2s \leftarrow 1s$) is brought about in an electric discharge (for example, by the collision of rapidly moving electrons with the atoms). By chance the energy of the excited state happens to match an excitation

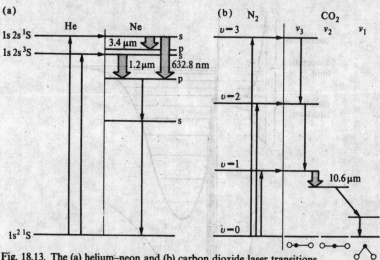

Fig. 18.13. The (a) helium–neon and (b) carbon dioxide laser transitions.

energy of neon, and during a He–Ne collison an efficient transfer of energy may occur, leading to the formation of neon in a highly excited configuration with less highly excited states remaining largely unpopulated. Laser action generating 632.8 nm radiation (among about 100 other lines) can then be induced. The CO_2 laser works on a similar principle, Fig. 18.13b. Most of the working gas is nitrogen: in the course of a radiofrequency discharge it is vibrationally excited by electron collisions. The vibrational excitation levels happen to coincide with the ladder of energy levels of the antisymmetric stretching mode of CO_2, which picks up energies by resonant transfer during collisions, Fig. 18.13b. Laser action then occurs from the lowest ($v = 1$) level of this mode to the $v = 1$ level of the symmetric stretch, which has remained unpopulated during the collisions. This results in the generation of 10.6 μm radiation (in the infrared). In some cases (the N_2 gas laser at 337.1 nm) the efficiency of the stimulated transition process is so great that a single passage is sufficient to generate radiation, and the cavity uses only one mirror. Such lasers are then said to be in a *super-radiant mode*.

Chemical reactions may also be used to generate species with non-equilibrium, inverted populations. For instance, the photolysis of Cl_2 leads to the formation of Cl atoms, which attack H_2 molecules present in the mixture. This produces HCl and H atoms. The latter go on to attack Cl_2 molecules, and lead to the formation of 'hot' (vibrationally excited) HCl. Since these molecules have non-equilibrium populations of their vibrational states, laser action can result as they emit radiation on their return to the lower states. Such processes are remarkable examples of the direct conversion of chemical energy into coherent electromagnetic radiation.

Dissociation and predissociation. Another fate for an electronically excited molecule is fragmentation into two or more components. This *dissociation* is illustrated in Fig. 18.14. The onset of dissociation can be detected by observing

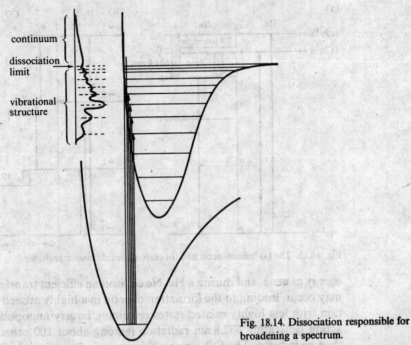

continuum

dissociation
limit

vibrational
structure

Fig. 18.14. Dissociation responsible for
broadening a spectrum.

where the vibrational structure of the electronic spectrum ends: this is the
dissociation limit of a spectrum. Finding the dissociation limit is a valuable
method for determining the bond dissociation energy. Transitions to states
of even higher energy result in disruption, and since the particles move
apart and do not undergo relative oscillation, the vibrational structure is
lost and the line structure is replaced by a continuous band.

In some cases, however, the vibrational structure disappears, but returns
at higher energies. This behaviour, which is called *predissociation*, can be
interpreted in terms of the molecular energy curves shown in Fig. 18.15.
When the molecule makes a transition to the vibrational energy levels in
the region where one molecular potential-energy curve crosses the other,
there may be a reorganization of its electrons, switching from one state
to another (the nuclear geometry of the two states is the same at the
crossing point). This process is called an *internal conversion*. If the molecule
converts into the new state it may have enough energy to dissociate.
Therefore the energy levels in the vicinity of the intersection of the curves
have unbound character and the vibrational structure of the spectrum
blurs into a continuum. As soon as the excitation source supplies enough
energy to raise the molecule into the vibrational states above the predissociation zone the molecule is unable to convert into the dissociative
state, the levels resume their well-defined, vibrational character, and the
spectrum reverts to a line structure.

Another way that predissociation shows up is in a difference between
absorption and emission spectra. This is shown in Fig. 18.16 for the case
of the molecule S_2. The absorption spectrum shows a series of lines

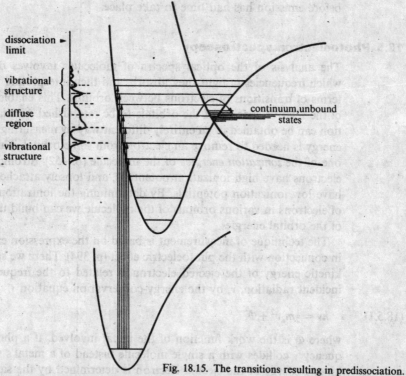

dissociation
limit

vibrational
structure

diffuse
region

vibrational
structure

continuum, unbound
states

Fig. 18.15. The transitions resulting in predissociation.

throughout the range illustrated, and there is some broadening on the right. The emission spectrum for the same range shows a series of lines only at lower frequencies (on the left) because the molecules in the upper vibrational states, which lie near the crossing of the type shown in Fig. 18.15, have made the transition into the continuum and have dissociated

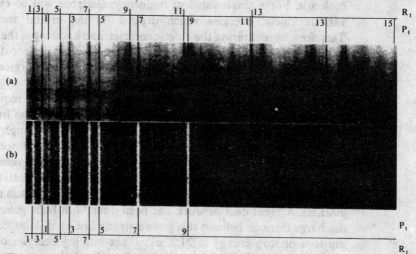

Fig. 18.16. A spectrum showing predissociation (J. M. Ricks and R. F. Barrow, *Canad. J. Phys.* **47**, 2423 (1969), with permission); (a) absorption, (b) emission.

before emission has had time to take place.

18.5 Photoelectron spectroscopy

The analysis of the optical spectra of molecules involves determining which frequencies of light they absorb, and then interpreting the data in terms of transitions of electrons between orbitals. This enables a picture of the relative spacing of the orbitals to be established. Similar information can be obtained in an entirely different way by measuring how much energy is needed to remove an electron from some orbital. This energy is one of the *ionization energies* of the molecule (p. 452). Strongly attached electrons have high ionization potentials, and loosely attached electrons have low ionization potentials. By determining the ionization potentials of electrons in various orbitals of the molecule we can build up a picture of the orbital energies.

The technique of measurement is based on the expression encountered in connection with the photoelectric effect (p. 394). There we saw that the kinetic energy of the ejected electron is related to the frequency of the incident radiation, v, by the energy-conservation equation

(18.5.1) $$hv = \tfrac{1}{2}m_e v^2 + \Phi,$$

where Φ is the work function of the metal involved. If a photon of frequency v collides with a single molecule instead of a metal's surface, the kinetic energy of the ejected electron is determined by the same kind of expression:

(18.5.2) $$hv = \tfrac{1}{2}m_e v^2 + I_i,$$

where Φ has been replaced by I_i, the ionization energy for the electron. The index i labels the orbital occupied by the electron in the un-ionized molecule. For a single incident frequency, electrons will be ejected with various kinetic energies which depend on the orbital they occupy. Therefore, by measuring the kinetic energies, and knowing v, the ionization potentials can be established. The ejected electrons are called *photoelectrons*, and the determination of their spectrum of energies gives rise to the name *photoelectron spectroscopy* (p.e.s.).

The experimental technique is straightforward. The first requirement is a source of radiation which is monochromatic (so that the frequency is well defined), energetic (so that each photon carries enough energy to ionize the molecule), and intense (so that a good flow of photoelectrons is obtained). Since the ionization energies of molecules are several electron-volts even for the outermost, valence electrons, it is essential to work in the ultraviolet region of the spectrum (with wavelengths less than about 200 nm). A great deal of work has been done with light generated by a discharge through helium: this gives a strong line at 58.4 nm, corresponding to a photon energy of 21.22 eV. If the binding energies of electrons much deeper in the molecule (that is, the strongly bound core electrons) are being studied, much higher photon energies have to be used in order to prise them out. It is then necessary to use X-rays, and the technique is then called *X-p.e.s.* Typical X-ray sources are magnesium (1253.6 eV) and aluminium (1486.6 eV).

Example (Objective 10). Using 58.43 nm light generated by a transition in helium that had been stimulated by an electric discharge, it was found that electrons were ejected from N_2 molecules with kinetic energies 5.63 eV and 4.53 eV. What were their binding energies in the molecule?

• *Method.* Use eqn (18.5.2). A useful conversion is $1\,eV \triangleq 8065.5\,cm^{-1}$.

• *Answer.* 58.43 nm light corresponds to a wavenumber of $(1/58.43 \times 10^{-7}\,cm) = 1.711 \times 10^5\,cm^{-1}$, and therefore to an energy of 21.22 eV. Then from eqn (18.5.2),

    (a) $21.22\,eV = 5.63\,eV + I_i$ or $I_i = 15.59\,eV$.

    (b) $21.22\,eV = 4.53\,eV + I_i$ or $I_i = 16.69\,eV$.

The binding energies are therefore $-15.59\,eV$ and $-16.69\,eV$.

• *Comment.* The more tightly bound electrons emerge with the lower kinetic energy (and velocity). This result is analysed below.

As well as an ionization source it is necessary to have some method for determining the kinetic energy of the photoelectrons. This is normally done by passing them between the poles of an electromagnetic or electrostatic analyser, for the extent of deflection depends on their velocity. At some field strength electrons of a certain velocity, and therefore kinetic energy, will be deflected through an arc that brings them to the position of a detector. As the field is changed, electrons of different energies reach the detector and the electron flux can be measured for each field strength. The field strength can then be related to the electron kinetic energy. The photoelectron spectrum consists of a plot of the electron current incident on the detector against the kinetic energy, or against the difference $h\nu - \frac{1}{2}mv^2$ (which gives I_i directly). The apparatus is illustrated schematically in Fig. 18.17.

Typical photoelectron spectra are shown in Fig. 18.18. For the top spectrum the sample is nitrogen gas, and the scale of ionization energy increases from right to left. The main features of this spectrum can be

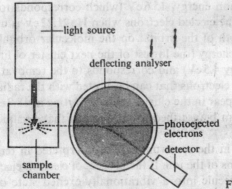

light source

deflecting analyser

photoejected electrons

detector

sample chamber

Fig. 18.17. A photoelectron spectrometer.

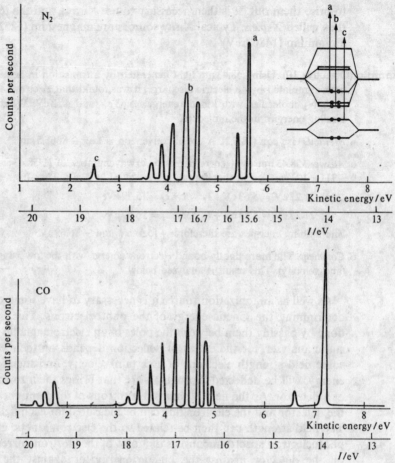

Fig. 18.18. Photoelectron spectra of N_2 and CO.

interpreted in terms of the molecular-orbital diagram described in Chapter 15 (p. 481) and shown in the figure. If we disregard the detailed structure of the spectrum we see that the lines fall into three main groups. The least tightly bound electrons are those in the uppermost molecular orbital, $2p\sigma_g$, and its ionization energy, 15.6 eV (which corresponds to a kinetic energy of 5.6 eV for the ejected electrons when He 21.22 eV is used), gives a measure of the depth of the orbital on the molecular orbital diagram. The next ionization energy (the lowest of the next cluster of lines) lies at 16.7 eV (kinetic energy 4.5 eV) and corresponds to the removal of a $2p\pi_u$ electron. The deepest electrons that can be ejected with the radiation used in this experiment appear in the cluster of lines starting at 18.8 eV (kinetic energy 2.4 eV) and can be ascribed to an electron that occupies $2s\sigma_u^*$ in the original molecule.

The fine structure in the spectrum can be interpreted in terms of the excitation of vibrations of the ion. If the process of electron ejection takes the ground-state molecule into a vibrationally excited state of the ion, some of the photon's energy is not available for the kinetic energy of the electron, and the energy conservation equation becomes

(18.5.3) $$h\nu = \tfrac{1}{2}m_e v^2 + I_i + \Delta E_{\text{vib}}^+,$$

where ΔE_{vib}^+ is the amount of energy used to excite the ion into vibration.

Returning to the nitrogen spectrum we see that the ejection of the $2p\sigma_g$ and $2s\sigma_u^*$ electrons stimulates very little vibration, but ejection of a $2p\pi_u$ electron is accompanied by a large amount. Little vibrational excitation is expected if the bond lengths of the molecule and the ion are similar, but a great deal of vibration is expected if they are different. This is because the sudden removal of an electron from the molecule may leave the ion in a compressed or stretched state relative to its new equilibrium geometry. The observed spectrum of N_2 therefore indicates that neither the $2p\sigma_g$ nor the $2s\sigma_u^*$ orbitals play much role in the bonding in N_2, but that the $2p\pi_u$ orbital has a strong effect on the size of the molecule and therefore plays a significant role in its bonding. This is confirmed by measuring the splitting between the vibrational lines and comparing them with the splitting in the un-ionized molecule. A conventional spectroscopic investigation of N_2 shows that it has a vibrational frequency of $2345\,\text{cm}^{-1}$ (this is the splitting between its two lowest vibrational energy levels). When a $2p\sigma_g$ electron is removed the structure in the photoelectron spectrum gives a vibrational splitting of the ion of $2150\,\text{cm}^{-1}$, showing that the $2p\sigma_g$ electron had little effect on the stiffness of the bond. In contrast, removal of the $2p\pi_u$ electron drops the vibrational frequency from $2345\,\text{cm}^{-1}$ to $1810\,\text{cm}^{-1}$, showing that it had a marked contribution to the bond stiffness.

In X-p.e.s. the energy of the incident photons is so great that electrons are ejected from the tightly-bound atomic cores. As a first approximation we would not expect the core ionization energies to be sensitive to the bonds between the atoms: they are too tightly bound to be affected greatly by the changes in distribution of the valence electrons that occur when bonds are formed. This turns out to be largely true, and since the inner-shell ionization energies are characteristic of the individual atom rather than the overall molecule, the determination of the X-p.e.s. spectrum gives lines characteristic of the elements present in a compound or alloy. For instance, the 1s-orbital ionization energies of the first row elements are Li($50\,\text{eV}$), Be($110\,\text{eV}$), B($190\,\text{eV}$), C($280\,\text{eV}$), N($400\,\text{eV}$), O($530\,\text{eV}$), and F($690\,\text{eV}$). Detection of one of these ionization energies in an X-p.e.s. spectrum indicates the presence of the corresponding element. Figure 18.19, for example, shows the result of an X-p.e.s. analysis of lunar rock brought back from an Apollo mission. This chemical analysis application is responsible for the alternative name *E.S.C.A.*, which stands for Electron Spectroscopy for Chemical Analysis. Note, however, that the technique is mainly limited to the study of surface layers, because even though X-rays may penetrate into the bulk, the ejected electrons cannot escape except from within a few nanometers from the surface. Nevertheless, like most limitations, the restriction can be turned to advantage: X-p.e.s.

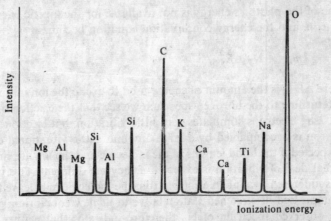

Fig. 18.19. The X-ray photoelectron spectrum of a sample of moon dust (adapted from *Physical methods and molecular structure*, The Open University Press, 1977).

is particularly valuable for the study of the surface structure of catalysts (Chapter 29).

While it is largely true that inner-shell ionization energies are unaffected by the nature of the molecule, it is not wholly true, and small but significant shifts can be detected which depend on the local environment of the atoms. For example, the azide ion, N_3^-, gives an X-p.e.s. spectrum which, although in the region of 400 eV typical of nitrogen 1s-electrons, is split into two lines separated by about 6 eV, Fig. 18.20. The structure of the ion is $N{=}N{=}N$, with charge distribution $(-, +, -)$ along the line of atoms. The presence of the negative charges on the terminal atoms lowers their inner-shell ionization energies, and the positive charge on the central atom raises its: hence the two lines in the spectrum. The observation of chemical shifts can give limited but valuable information about the presence of chemically inequivalent but otherwise identical atoms.

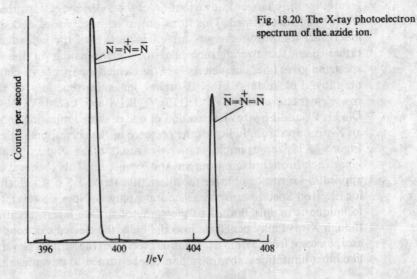

Fig. 18.20. The X-ray photoelectron spectrum of the azide ion.

One of the most appealing features of p.e.s. is its ability to portray the energies of the molecular orbitals in molecules, and to confirm—or refute—their order as calculated by molecular orbital theories. Since it can also be used to study negative ions it is a rich source of data on the electron affinities of the corresponding neutral species. It is important, however, to realize its limitations. The identification of the ionization energy of a molecule with the energy of an electron in a molecular orbital is an approximation (called *Koopmans' theorem*), and in a detailed analysis one has to be very careful to take into account a number of complicating features. Moreover, although we have discussed the ionization process in terms of single electrons, we should really discuss it in terms of the overall states of the molecule and the ion. These, however, are technical problems, and they should not be allowed to obscure the main power of the method. A photoelectron spectrum comes very close to the direct portrayal of molecular orbital energy levels, and to the confirmation of the general features of the discussions of molecular bonding.

Further reading

Quanta: a handbook of concepts. P. W. Atkins; Clarendon Press, Oxford, 1974.

Theory and applications of ultraviolet spectroscopy. H. H. Jaffe and M. Orchin; Wiley, New York, 1962.

Visible and ultraviolet spectroscopy. F. Grum; in *Techniques of chemistry* (A. Weissberger and B. W. Rossiter, eds.), Vol. IIIB, 207, Wiley-Interscience, New York, 1972.

Ultraviolet and visible spectroscopy. C. N. R. Rao; Butterworths, London, 1967.

Spectroscopy. D. H. Whiffen; Longman, London, 1972.

The theory of the electronic spectra of organic molecules. J. N. Murrell; Chapman and Hall, London, 1971.

Spectra of diatomic molecules. G. Herzberg; Van Nostrand, New York, 1950.

Electronic spectra and electronic structure of polyatomic molecules. G. Herzberg; Van Nostrand, New York, 1966.

Inorganic electronic spectroscopy. A. B. P. Lever; Elsevier, Amsterdam, 1968.

Molecular quantum mechanics. P. W. Atkins; Clarendon Press, Oxford, 1970.

Dissociation energies. A. G. Gaydon; Chapman and Hall, London, 1952.

Spectroscopy and molecular structure. G. W. King; Holt, Reinhart, and Winston, New York, 1964.

Lasers. B. A. Lengyel; Wiley-Interscience, New York, 1971.

Handbook of lasers. R. J. Pressley; Chemical Rubber Co., Cleveland, Ohio, 1971.

Lasers and their application to physical chemistry. A. F. Haught; *Ann. Rev. Phys. Chem.* **19**, 343 (1968).

The determination of fluorescence and phosphorescence. N. Wotherspoon, G. K. Oster, and G. Oster; in *Techniques of chemistry* (A. Weissberger and B. W. Rossiter, eds.), Vol. IIIB, 429, Wiley-Interscience, New York, 1972.

Photochemistry. J. G. Calvert and J. N. Pitts; Wiley, New York, 1966.

Photophysics of aromatic molecules. J. B. Birks; Wiley, London, 1970.

Molecular photoelectron spectroscopy. D. W. Turner, C. Baker, A. D. Baker, and C. R. Brundle; Wiley-Interscience, New York, 1970.

Physical methods and molecular structure (2). Open University Press, 1977.

Photoelectron spectroscopy. J. H. Eland; Open University Press, 1977.

Problems

18.1. The following data were obtained for the absorption of light by a 1 cm sample of bromine dissolved in carbon tetrachloride using a 2 mm path-length cell. What is the extinction coefficient of bromine at the wavelength employed?

$[Br_2]/mol\,dm^{-3}$	0.001	0.005	0.010	0.050
Transmission/per cent	81.4	35.6	12.7	3×10^{-3}

18.2. In another experiment the same cell was filled with $0.01\,mol\,dm^{-3}$ benzene, and the wavelength of the spectrometer changed to 256 nm, where there is a maximum in the absorption. At this wavelength the transmitted intensity was 48 per cent of the incident intensity. What is the molar absorption coefficient (extinction coefficient) of benzene at this wavelength?

18.3. What are the absorbances (optical densities) of the two samples in the last two Problems?

18.4. What would be the percentage of transmission by (a) $0.01\,mol\,dm^{-3}$ benzene, (b) an $0.001\,mol\,dm^{-3}$ bromine solution in carbon tetrachloride when the sample cell is (i) 0.1 cm, (ii) 10 cm thick, the wavelengths being the same as above.

18.5. A swimmer enters a gloomier world (in one sense) as he dives to greater depths. Given that the mean molar absorption coefficient for sea water in the visible region is $6.2 \times 10^{-5}\,dm^3\,mol^{-1}\,cm^{-1}$, calculate the depth at which a diver will experience (a) half the surface intensity of light, (b) one tenth that intensity.

18.6. The absorption bands of many molecules in solution have widths at half-height of about $5000\,cm^{-1}$. In such cases, estimate (by assuming a triangular line shape) the integrated absorption coefficient for a band in which (a) $\varepsilon_{max} \approx 1 \times 10^4\,dm^3\,mol^{-1}\,cm^{-1}$, (b) $\varepsilon_{max} \approx 5 \times 10^2\,dm^3\,mol^{-1}\,cm^{-1}$. (The width at half-height means the width of the band where $\varepsilon = \frac{1}{2}\varepsilon_{max}$.)

18.7. What are the oscillator strengths of the transitions in the last Problem?

18.8. A glance at any optical spectrum of a molecule in solution shows that the line shape is only poorly approximated by a triangle. In many cases it is better to assume that it is a Gaussian (proportional to e^{-x^2}) centred on the wave number or frequency corresponding to ε_{max}. Assume such a line shape, and show how to express the integrated absorption coefficient in terms of the values of ε_{max} and $\Delta\tilde{\nu}_{1/2}$, the width at half-height; deduce the relation $\mathscr{A} \approx 2.4515\,c\varepsilon_{max}\,\Delta\tilde{\nu}_{1/2}$. Both these quantities may be measured quite readily, and so we have a quick method of determining both $\mathscr{A}$ and the oscillator strength of the band.

18.9. Quick it may be, but how reliable is the assumption of a Gaussian line shape? The absorption spectrum of azoethane ($CH_3CH_2N_2$) between $24000\,cm^{-1}$ and $34000\,cm^{-1}$ is shown in Fig. 18.21. First estimate $\mathscr{A}$ and f for the band by assuming that its shape is Gaussian. Then integrate the absorption band graphically. This can be done either by counting squares on graph paper (tedious) or by tracing on to paper, cutting out the shape, and weighing (a general procedure for integrating tiresome shapes). Is the transition forbidden or allowed?

18.10. A more complex absorption spectrum, that of biacetyl in the range 250–470 nm, is shown in Fig. 18.2. Estimate the oscillator strengths of both principal transitions, and decide whether they are forbidden or allowed. You will notice that the spectrum is plotted on a wavelength scale: this means that extra work has to be done before arriving at the f-values of the transitions. There are three approaches. One is to estimate $\Delta\tilde{\nu}_{1/2}$ from the wavelength data and to use the Gaussian line shape approximation (Gaussian in wavenumber, that is). The

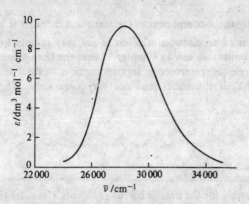

Fig. 18.21. The absorption spectrum of azoethane ($CH_3CH_2N_2$).

second is to transfer the spectrum to wavenumber scale, and to proceed by graphical integration. The third is to find an expression for f in terms of wavelength integration, and to use the spectrum directly. Use all three methods. ε_{max} (280 nm peak) = 11 dm^3 mol^{-1} cm^{-1}, ε_{max} (430 nm) = 18 dm^3 mol^{-1} cm^{-1}.

18.11. A lot of information about the energy levels and wavefunctions of small inorganic molecules can be obtained from their ultraviolet spectra. An example with considerable vibrational structure is shown in Fig. 18.22: it is the spectrum of gaseous SO_2 at 25 °C. Find the oscillator strength of the transition. Is it allowed or forbidden? What electronic states are accessible from the A_1 ground state of this C_{2v} molecule?

18.12. The oscillator strength is a quantity that, as we have seen, can be measured from experimental spectra. It is also a quantity that can be calculated from knowledge of the structure of the molecule, and in particular, from a knowledge of the molecular orbitals of the ground and excited states involved in the transition. We can get some practice in this kind of calculation by dealing with some simple models of molecules. First, consider an electron in a 1-dimensional square well of length L. Calculate the oscillator strength for the transitions $n+1 \leftarrow n$ and $n+2 \leftarrow n$. You should begin by calculating the transition dipole moment, eqn (18.1.5), in the form $\mu_x = -e \int_0^L \psi_n(x) x \psi_n(x) \, dx$, and then using eqn

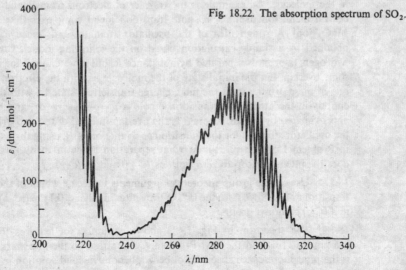

Fig. 18.22. The absorption spectrum of SO_2.

(18.1.6) to find f. (Only the x-component of the transition moment is non-zero.)

18.13. The behaviour of an electron in a box is not very remote from that of electrons in real molecules: we saw in Chapter 15 that the f.e.m.o. model gave a rough idea of the energies of spectroscopic transitions; now we shall see that it can be used to estimate their intensities (but not very accurately). The molecule

(β-carotene) is responsible for the orange colour of carrots. Consider the polyene chain up to the mark as the chromophore. Estimate the excitation energy (in cm^{-1}), suggest what colour that implies the carrots will appear in white light, and then use the result obtained in the last Problem to estimate the extinction coefficient of the molecule. Estimate the thickness of carrot soup (supposed clear and molar in the above molecule) that gives 50 per cent absorption of incident light.

18.14. Suppose you were a colour chemist and had been asked to intensify the colour of a dye without changing its nature, and the colour of the dye of interest depended largely on a transition involving a long polyene chain. Would you choose to lengthen or to shorten the chain? Would that shift the apparent colour of the dye towards the red or the blue?

18.15. Consider an electron in an atom to be oscillating with a simple harmonic motion (that was an early model of atomic structure). The wavefunctions for the electron are given on p. 414. Show that the oscillator strength for the transition of this electron from its ground vibrational state to its first vibrational state is exactly $\frac{1}{3}$. If the electron were able to vibrate in three dimensions the oscillator strength for all possible transitions out of the ground state would be unity: this is the source of the name *oscillator strength* for the quantity f.

18.16. In the text we saw that the class of transitions that give rise to *charge-transfer spectra* involves the migration of an electron from one region to another in the molecule. An example is the transfer of electrons from ligand orbitals to central metal ion orbitals in some transition metal complexes (like the purple MnO_4^- ion). A rough idea of the oscillator strength, and hence ε_{max}, can be obtained by a simple calculation based on the following model. Consider two hydrogen 1s-orbitals separated by a distance R, but close enough for there to be some overlap. Let the initial state of the molecule be with the electron entirely in one of these orbitals, and the final, charge-transferred, state be with the electron entirely in the other. The transition dipole is $p = -e \int \psi_f r \psi_i d\tau$, and this can be approximated by $-eRS$, where S is the overlap integral of the two orbitals. Plot the oscillator strength for the transition as a function of R, using the value of $S(R)$ in Problem 15.13 on p. 514. At what separation is the intensity greatest? Why does the intensity drop to zero both as $R \to 0$ and as $R \to \infty$?

18.17. Use simple group-theoretical arguments to decide which of the following transitions are allowed: (a) the (π^*, π)-transition in ethene, (b) the (π^*, n)-transition in a (C_{2v}) carbonyl group.

18.18. The spectrum in Fig. 18.23 shows a number of features that will be examined in this and the next Problem. Concentrate on the line marked A. This is the phosphorescence spectrum of benzophenone in solid solution in ethanol at

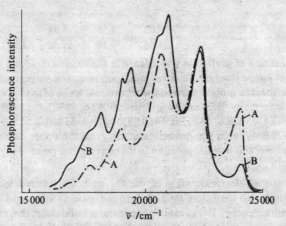

Fig. 18.23. Phosphorescence of naphthalene and benzophenone.

low temperatures during irradiation with 360 nm light. What can be said about the vibrational energy levels of the carbonyl chromophore (a) in its ground electronic state, (b) in its excited electronic state?

18.19. When naphthalene is irradiated with 360 nm light it does not absorb; but the line marked B in the diagram is the phosphorescence spectrum of a solid solution of a mixture of naphthalene and benzophenone, and a component due to the naphthalene can be seen. Account for this observation.

18.20. The fluorescence spectrum of anthracene vapour shows a series of peaks of increasing intensity with individual maxima at 440 nm, 410 nm, 390 nm, and 370 nm, and followed by a sharp cut-off at shorter wavelengths. The absorption spectrum rises sharply from zero to a maximum at 360 nm with a trail of peaks of lessening intensity at 345 nm, 330 nm, and 305 nm. Account for these observations.

18.21. Consider an indicator, such as bromophenol blue, which follows the equilibrium $InH \rightleftharpoons H^+ + In^-$ (HIn might itself be charged) and where the InH and In^- forms absorb in different parts of the spectrum (so that there is a colour change in response to the pH). Suppose that $A(InH)$ and $A(In^-)$ are the absorbances of a sample of the indicator when it is in the acid (InH) and base (In^-) forms. When a mixture of the forms is present let the absorbance at the same wavelength be $A(mix)$. Show that the degree of dissociation α of the indicator is given by $[A(mix) - A(InH)]/[A(In^-) - A(InH)]$. Hence show that the pH of the solution can be expressed in terms of the dissociation constant K_{In} of the indicator and the optical density $A(mix)$. Sketch the form of the pH dependence of $A(mix)/A(InH)$ in the case of a colourless In^- form in order to appreciate the rapidity with which the colour change occurs. Take $pK_{In} \approx 4$.

18.22. When pyridine is added to a solution of iodine in carbon tetrachloride the 520 nm band of absorption shifts towards 450 nm. The absorbance of the solution at 490 nm, however, remains constant: this is called an *isobestic point*. Use the technique of the last Problem, and especially the relation of α to $A(mix)$, to show that an isobestic point should occur when two absorbing species are in equilibrium.

18.23. The data below are for the molar absorption coefficient ($10^{-3} \varepsilon/mol^{-1}$ $dm^3 cm^{-1}$) of a solution of *p*-nitrophenol in water at different pH (A. I. Biggs, *Trans. Faraday Soc.* **50**, 800 (1954)). Find the pK value for the ionization of the nitrophenol.

pH	4	5	6	7	8	9	10
317 nm	9.72	9.72	9.03	5.55	1.81	1.39	1.39
407 nm	—	—	1.66	9.16	17.50	18.33	18.33

18.24. A transition of particular importance in the oxygen molecule gives rise to the *Schumann–Runge band* in the ultraviolet spectrum. The wavenumbers (cm^{-1}) of transitions from the ground state to the vibrational levels of the upper electronic state (which is formally $^3\Sigma_u^-$) are as follows: 50 062.6, 50 725.4, 51 369.0, 51 988.6, 52 579.0, 53 143.4, 53 679.6, 54 177.0, 54 641.8, 55 078.2, 55 460.0, 55 803.1, 56 107.3, 56 360.3, 56 570.6. What is the dissociation energy of the upper electronic state? You should answer this by extrapolation to the convergence point of the vibrational sequence.

18.25. The excited electronic state of the O_2 molecule referred to in the last Problem is known to dissociate into one ground state O atom and one excited state atom with an energy $190 \, kJ \, mol^{-1}$ above the ground state; this excited atom is responsible for a great deal of photochemical mischief in the atmosphere. Ground state oxygen dissociates into two ground state atoms. Combine this information with that in the last Problem to find the dissociation energy of ground state oxygen.

18.26. Spin angular momentum is conserved when a molecule breaks up into atoms. What atomic multiplicities are permitted when (a) an oxygen molecule, (b) a nitrogen molecule breaks up into two atoms?

18.27. The photoelectron spectra of N_2 and CO are shown in Fig. 18.18. Ascribe the lines to the ionization processes involved, and classify the orbitals into bonding (or antibonding) and non-bonding in the light of the amount of vibrational structure. Analyse the bands near 4 eV in terms of the vibrational spacing of the ions.

18.28. The photoelectron spectrum of NO can be described as follows (D. W. Turner, *Physical Methods in Advanced Inorganic Chemistry* (eds. H. A. O. Hill and P. Day), Wiley-Interscience (1968)). Using He 584 pm (21.21 eV) radiation a single strong peak is observed at 4.69 eV and there is a long series of 24 lines starting at 5.56 eV and ending at 2.2 eV. A shorter series of 6 lines begins at 12.0 eV and ends at 10.7 eV. Account for this spectrum.

18.29. The highest energy electrons in the photoelectron spectrum of water (using 21.21 eV He radiation) are at about 9 eV and show a large vibrational spacing of 0.41 eV. The symmetric stretching mode of un-ionized water lies at $3652 \, cm^{-1}$. What conclusions can you draw about the nature of the orbital from which the electron is ejected?

18.30. In the same spectrum of water the band at about 7 eV shows a long vibrational series with an interval of 0.125 eV. The bending mode of water lies at $1596 \, cm^{-1}$. What conclusions can you draw about the bonding characteristics of the orbital occupied by the ejected electron?

19 Determination of molecular structure: resonance techniques

Learning objectives

After careful study of this chapter you should be able to

(1) Explain the basis of the *electron spin resonance* experiment (p. 634).

(2) Explain the significance of the *g-value* and describe how it is measured (p. 636).

(3) Describe the source of the *hyperfine interaction* and the *hyperfine structure* of an e.s.r. spectrum (p. 638).

(4) Distinguish between *dipolar (anisotropic)* and *contact (isotropic)* hyperfine interactions (p. 640).

(5) Map the *unpaired electron density* around the rings of aromatic radicals (p. 641).

(6) Calculate *population differences* of electron (p. 642) and nuclear (p. 646) spin states.

(7) Indicate the mechanism of *spin-lattice* and *transverse relaxation processes* (p. 643).

(8) Describe the basis of the *nuclear magnetic resonance* experiment (p. 644).

(9) Explain the significance of the *shielding constant* and the *chemical shift* (p. 646) and express the latter on the δ- and τ-scales.

(10) Account for the appearance of the *fine-structure* of an n.m.r. spectrum and explain the term *proton decoupling* (p. 651).

(11) Deduce the *rates of processes* from the widths of spectral lines and the appearance of the spectrum (p. 652).

(12) Explain the essential features of *Fourier transform n.m.r.* and indicate why the *free induction decay* signal gives spectral information (p. 653).

(13) Describe the *Mössbauer effect* and the *Mössbauer experiment* (p. 655).

(14) Explain the terms *isomer shift* and *electric quadrupole* splitting in a Mössbauer spectrum (p. 658), and indicate how they are used.

Introduction

Resonance occurs when two pendulums are joined by the same slightly flexible support. If one is set in motion the other is forced into oscillation by the motion of the common axle, and the energy of oscillation flows backwards and forwards between the two. This process occurs most efficiently when the frequencies of the two oscillators are identical. The condition of strong effective coupling on account of the possession of identical frequencies is called *resonance*, and the excitation energy is said to *resonate* between the two coupled oscillators.

Resonance is the basis of a number of everyday phenomena, including tuning radios to the weak oscillations of the electromagnetic field caused by a distant transmitter. In this chapter we examine some spectroscopic applications. These have the common characteristic of depending on the matching of a system of energy levels to a source of radiation of well-defined frequency, and observing the strong absorption when a resonance condition is attained.

19.1 Electron spin resonance

An electron possesses spin angular momentum and, because of this momentum, a spin magnetic moment, p. 463. The spin may take two orientations denoted α, β with respect to some selected direction, and these orientations correspond to the angular momentum projections $m_s \hbar$, $m_s = \pm\frac{1}{2}$. The spin magnetic moment can therefore take two orientations with respect to an applied magnetic field, and the energies of the two states are given by eqn (14.3.6) on p. 463:

(19.1.1) $$E_{m_s} = g_e \mu_B m_s B, \qquad m_s = \pm\tfrac{1}{2};$$

g_e is 2 (or, to be precise, 2.0023), B is the applied magnetic field, and μ_B is the Bohr magneton, $eh/2m_e$. This shows that the energy of electrons with α-spin ($m_s = +\frac{1}{2}$) rises and that of β-spin ($m_s = -\frac{1}{2}$) electrons falls as the magnetic field is increased, Fig. 19.1. The energy separation of the two spin states is

(19.1.2) $$\Delta E = E_{1/2} - E_{-1/2} = g_e \mu_B B.$$

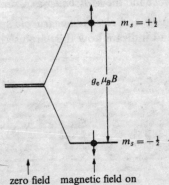

Fig. 19.1. Electron spin energy levels in a magnetic field.

If the sample is bathed in radiation of a frequency v the unpaired electron spins of the sample have energy levels that come into resonance with the radiation when the magnetic field has been adjusted so that

(19.1.3) $hv = g_e \mu_B B.$

When this condition is satisfied the energy levels are in resonance with the surrounding radiation, and the spins may absorb its energy strongly. The establishment of this *resonance condition* (eqn (19.1.3)) is detected by observing a strong absorption of the incident radiation as the spins are stimulated to flip up from β-states to α-states. The *electron spin resonance* (e.s.r.) technique (which is also called *electron paramagnetic resonance*, (e.p.r.)) is the study of the properties of molecules containing unpaired electrons by observing the magnetic fields at which they come into resonance with an applied radiation field of definite frequency.

The technique. Magnetic fields of the order of 0.1–1.0 T (1–10 kG) can readily be established without resorting to elaborate techniques. A field of 0.3 T corresponds to resonance with an electromagnetic field of frequency 10 GHz (10^{10} c/s) and wavelength 3 cm. This is the magnetic field used in most commercial e.s.r. spectrometers; 3 cm radiation falls into the *X-band*

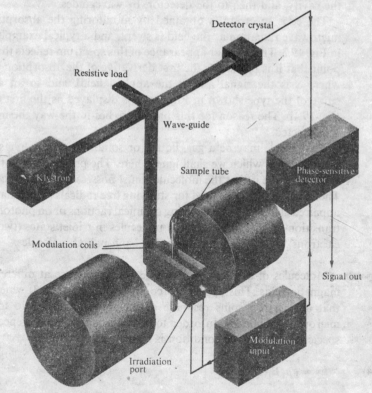

Fig. 19.2. An electron spin resonance spectrometer (K. A. McLauchlan, *Magnetic Resonance*, Clarendon Press, Oxford (1972)).

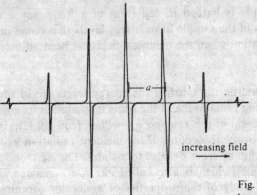

Fig. 19.3. The e.s.r. spectrum of the benzene radical anion in solution.

of microwaves, and so e.s.r. is a microwave technique.

The basic spectrometer arrangement is illustrated in Fig. 19.2. It consists of a source of microwaves (the generator is called a *klystron*), a cavity into which the sample is inserted in a glass or quartz container, a detector for the radiation, and an electromagnet giving a field which can be varied in the region of 0.3 T. The microwaves are brought from the klystron to the cavity, and then to the detector, by waveguides.

The e.s.r. spectrum is obtained by monitoring the absorption of the microwave radiation as the field is swept, and a typical example is shown in Fig. 19.3. The peculiar appearance of this spectrum reflects the technical point that it is plotted as the first derivative of the absorption (as dS/dB, where S is the signal and B the applied field), and so an absorption curve of the type shown in Fig. 19.4a is displayed as the first derivative, Fig. 19.4b. The reason for this procedure lies in the way the microwaves are detected.

The sample may be a gas, liquid, or solid, but the gas phase involves complications which we shall ignore here. The principal limitation on the type of sample is that the molecules must possess unpaired electron spins. Therefore e.s.r. may be used for studying free radicals (which have a single unpaired electron) formed during chemical ractions or on photolysis, many transition-metal complexes, and molecules in triplet states (two unpaired electrons). It is insensitive to normal, spin-paired molecules.

The g-value. Molecules of different kinds come into resonance at different applied magnetic fields. This appears to conflict with eqn (19.1.3) which suggests that all electron spins resonate at the same applied field for a given microwave frequency. In order to accommodate the differences between species the resonance condition is rewritten in the form

(19.1.4) $h\nu = g\mu_B B,$

where g, the *g-value* of the electron, is an experimental parameter which depends on the molecule under study. For a free electron $g = g_e = 2.0023$

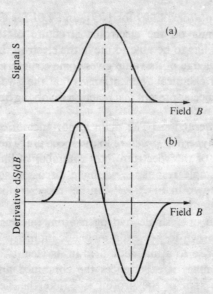

Fig. 19.4. An absorption curve and its derivative.

(see p. 463), and for many radicals g lies in the range 1.9 to 2.1. Its value for the species being studied can be obtained from the magnetic field corresponding to the centre of the e.s.r. spectrum.

Example (Objective 2). The centre of the e.s.r. spectrum of the methyl radical occurs at 329.4 mT when the spectrometer is using 9.233 GHz microwaves. What is its g-value?

- *Method.* Use eqn (19.1.4) in the form $g = h\nu/\mu_B B$.
- *Answer.* $g = \dfrac{(6.626\,18 \times 10^{-34}\,\text{J s}) \times (9.233 \times 10^9\,\text{Hz})}{(9.274\,08 \times 10^{-24}\,\text{J T}^{-1}) \times (0.3294\,\text{T})} = 2.0027.$

- *Comment.* Many organic free radicals have a g of about this value. Inorganic radicals have g in the general area of 1.97–2.02. Transition metal ions often have g-values over a wider range (e.g. 0–4).

The reason for the deviation of g from g_e is that the electron spin magnetic moment interacts with the *local* magnetic field, and this local field can differ from the applied field. This is because the applied field can force the electron to circulate through the molecular framework, and the orbital motion sets up a small additional magnetic field δB. The magnitude of δB is proportional to B itself, and so we could write the total local field $B + \delta B$ as $B - \sigma B$, where σ is some constant (the negative sign in $B - \sigma B$ is a matter of definition). The resonance condition is satisfied when

$$h\nu = g_e \mu_B B_{\text{local}} = g_e \mu_B (B + \delta B) = g_e \mu_B (1 - \sigma) B.$$

When this is compared with the previous equation, g can be identified with $g_e(1 - \sigma)$. The constant σ may be positive or negative, depending on the species, and so g may be less than g_e or more. Typical values are

2.002 26 for the hydrogen atom, 1.999 for NO_2, and 2.01 for ClO_2.

The value of g depends on the electronic structure of the species because the applied field has to be able to move the electron through the molecule, and so a knowledge of its value gives some structural information. More important for chemical applications is that the g-value can be used to help identify the species present in a reaction involving free radicals.

Hyperfine structure and hyperfine interactions. The main importance of e.s.r. comes from the observation of *hyperfine structure* in the spectrum. The hyperfine structure is the splitting of the spectrum into a number of lines centred on the position of the single resonance we have discussed so far. This structure (of the kind shown in Fig. 19.3) can be traced to the presence of *nuclear* magnetic moments.

Consider first the role of a single hydrogen atom nucleus in a free radical. The proton is known to possess a magnetic moment. This small magnetic moment gives rise to a magnetic field in its vicinity, and when the electron spin approaches it experiences the combined applied and nuclear fields. Since the spin of the proton is $\frac{1}{2}$ (like the electron), its magnetic moment can take two orientations with respect to the applied field. One orientation adds to the local field at the electron, the other subtracts from it. If the z-component of the nuclear angular momentum is denoted $m_I h$, $m_I = \pm\frac{1}{2}$, the total local field can be expressed as

$$B_{loc} = B + am_I,$$

where a is some constant called the *hyperfine coupling constant* (we ignore the effects giving rise to the g shift). If the proton has one orientation the local field is changed to $B + \frac{1}{2}a$, and if it has the other it is changed to $B - \frac{1}{2}a$.

Half the molecules in the sample have protons with $m_I = \frac{1}{2}$; the other half have protons with $m_I = -\frac{1}{2}$. Therefore, half the molecules will resonate when the applied field satisfies the condition

$$hv = g_e\mu_B(B + \tfrac{1}{2}a) \quad \text{or} \quad B = (hv/g_e\mu_B) - \tfrac{1}{2}a$$

and the other half will resonate when the applied field has a value given by

$$hv = g_e\mu_B(B - \tfrac{1}{2}a), \quad \text{or} \quad B = (hv/g_e\mu_B) + \tfrac{1}{2}a.$$

Therefore, instead of a single line, the spectrum shows two lines separated by a magnetic field of magnitude a, equally disposed about the original centre of the spectrum and each with half the total intensity. This is illustrated in different ways in Fig. 19.5.

If the unpaired electron comes close to a nitrogen nucleus the spectrum is split into three lines of equal intensity. This is because the ^{14}N nucleus has unit spin, and therefore three possible orientations ($m_I = -1, 0, +1$), and each spin orientation is possessed by one third of all the radicals in the sample. In general a nucleus of spin quantum number I splits the spectrum into $2I + 1$ lines of equal intensity, and their positions are given by

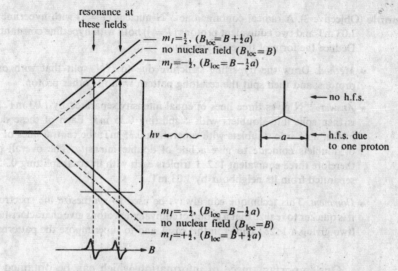

Fig. 19.5. The hyperfine splitting due to a single proton.

(19.1.5) $hv = g\mu_B B + am_I, \qquad m_I = I, I-1, \ldots -I.$

When there are several magnetic nuclei present in the same radical each contributes to the splitting of the spectrum. For example, if there are two protons the spectrum is split into four lines which are given by the four values at which B satisfies the resonance condition

$$hv = g\mu_B B + a_1 m_{I_2} + a_2 m_{I_2}; \qquad m_{I_1} = \pm\tfrac{1}{2}, \quad m_{I_2} = \pm\tfrac{1}{2}.$$

The hyperfine coupling constant a_1 may differ from a_2 if the hydrogen atoms are at inequivalent sites in the radical, because the unpaired electron may occur with a greater density in the vicinity of one nucleus. When the protons are *equivalent* (so that $a_1 = a_2$, as in the case of the two methylene protons in the ethyl radical $CH_3\dot{C}H_2$) the line with $m_{I_1} = \tfrac{1}{2}$, $m_{I_2} = -\tfrac{1}{2}$ coincides with the line with $m_{I_1} = -\tfrac{1}{2}$, $m_{I_2} = \tfrac{1}{2}$, and so the spectrum is a triplet of lines with relative intensities $1:2:1$.

It is not hard to show that if the radical contains N equivalent protons, the spectrum splits into $N+1$ lines in the *binomial intensity distribution* given by Pascal's triangle:

N	intensity distribution
0	1
1	1 1
2	1 2 1
3	1 3 3 1
4	1 4 6 4 1

and so on. The spectrum from the negative ion of benzene was shown in Fig. 19.3: its 7 lines are in the ratio $1:6:15:20:6:15:1$, confirming the presence of six equivalent protons in this radical.

Example (Objective 3). A radical contains one ^{14}N nucleus ($I = 1$) with hyperfine constant 1.03 mT and two equivalent protons ($I = \frac{1}{2}$) both with hyperfine constant 0.35 mT. Deduce the form of the e.s.r. spectrum.

- *Method.* Draw the hyperfine structure due to ^{14}N, split that with one of the protons, and then split the resulting pattern with the other proton.

- *Answer.* ^{14}N gives three lines of equal intensity separated by 1.03 mT. Each line is then split into doublets with a splitting 0.35 mT. Each of these doublets is further split into doublets with separation 0.35 mT: the central lines of each pair of doublets coincide to give a line of double intensity. The overall pattern is therefore three equivalent 1:2:1 triplets each with internal splitting 0.35 mT and separated from its neighbour by 1.03 mT.

- *Comment.* This technique can always be used to synthesize the spectrum. Often it is quicker to realize that a group of equivalent protons give characteristic patterns (two giving a 1:2:1 triplet in this case), and to superimpose the patterns directly.

One important piece of information which can be obtained from an analysis of the hyperfine structure of a spectrum is the identity of the radical (the hyperfine structure is a type of fingerprint). Moreover, since the magnitude of the splitting depends on the distribution of the unpaired electron in the vicinities of the magnetic nuclei present, one can also use the measured splittings to map the molecular orbital it occupies. But before we can do this mapping we have to know what governs the magnitude of the hyperfine coupling.

The hyperfine interaction is a magnetic interaction between the magnetic moment of the unpaired electron and the magnetic nucleus. There are two types of contribution. An electron in a p-orbital does not approach the nucleus very closely, and so it experiences a magnetic field that appears to arise from a point magnetic dipole. This is the *dipole–dipole interaction*. A characteristic of this type of interaction is that it is *anisotropic*; that is, its magnitude depends on the orientation of the radical with respect to the applied magnetic field. Furthermore, in fluid samples the rotation of the radicals results in the interaction averaging to zero. An s-electron is distributed spherically about its nucleus irrespective of whether the radical is rotating, and so it shows no dipolar interaction. Nevertheless, an s-orbital has the property of not vanishing at the nucleus, and so an s-electron may approach the nucleus so closely that it is no longer correct to treat the latter as a point magnetic dipole. Another magnetic interaction then comes into play: this is the *Fermi contact interaction*. It is *isotropic*; that is, independent of the radical's orientation, and consequently it is shown even by rapidly tumbling radicals in fluids.

The magnetic fields generated in the two cases can be quite large. For example, an electron in the 2p-orbital of a nitrogen atom experiences an average field of about 3.4 mT (34 G) from the ^{14}N nucleus. The 1s-electron in the hydrogen atom experiences a nuclear field of about 50 mT (500 G) as a result of the contact interaction, a 2s-electron in nitrogen experiences 55.2 mT, and a 2s-electron in the fluorine atom a massive 1.7 T (17 kG).

The magnitude of the contact interaction can be interpreted in terms of the s-character of the unpaired spin, and the dipole–dipole interaction in terms of the p-character. This gives valuable information on the structure of the molecular orbital and the hybridization of the atomic orbitals occupied by the unpaired electron, see Problem 19.17. Magnitudes for atoms are given in Table 19.1.

Table 19.1. Hyperfine coupling constants for atoms

Isotope	Spin	Isotropic coupling/mT	Anisotropic coupling/mT
^{1}H	$\frac{1}{2}$	50.8 (1s)	
^{2}H	1	7.8 (1s)	
^{13}C	$\frac{1}{2}$	113.0 (2s)	6.6 (2p)
^{14}N	1	55.2 (2s)	4.8 (2p)
^{19}F	$\frac{1}{2}$	1720 (2s)	108.4 (2p)
^{31}P	$\frac{1}{2}$	364 (3s)	20.6 (3p)
^{35}Cl	$\frac{3}{2}$	168 (3s)	10.0 (3p)
^{37}Cl	$\frac{3}{2}$	140 (3s)	8.4 (3p)

Source: P. W. Atkins and M. C. R. Symons, *The structure of inorganic radicals*, Elsevier.

Unfortunately there is a further problem. Figure 19.3 shows the spectrum of the benzene negative ion, in which the unpaired electron occupies a π-orbital. The sample is fluid; but how can such an electron have a contact interaction with protons lying in the nodal plane of its orbital? This interaction in aromatic radical ions occurs widely, and is the characteristic feature of the e.s.r. spectra of organic radicals.

The *polarization mechanism* of hyperfine interaction accounts for the coupling. This mechanism recognizes that the paired electrons in the C–H bond are slightly influenced by the magnetic field arising from the proton in the sense that an electron of the same spin orientation of the proton (i.e., opposite magnetic moment) has a tendency to be found more often in the vicinity of the proton than the other electron (which has opposite spin). The latter electron is therefore more likely to be found at the other end of the C–H bond, and therefore predominates in the vicinity of the unpaired π-electron. This π-electron therefore detects the magnetic orientation of the proton, even though it is not coupled to it directly. As a result the resonance line is split into two (because the proton spin may have two orientations), and both calculation and experiment show that the splitting is as though the proton spin is giving rise to a field of about 2.8 mT (28 G) at the neighbouring π-electron.

When we return to the spectrum of (benzene)$^-$ we see that the hyperfine splitting is 0.47 mT, which is $\frac{1}{6} \times 2.8$ mT. This can now be interpreted in the light of the polarization mechanism. The unpaired electron is spread around the ring, and occupies a π-orbital on each carbon atom with equal probability. Therefore, we expect it to have a polarization interaction with each neighbouring proton of $\frac{1}{6}$th the strength of the interaction of a whole electron. This type of analysis is a powerful way of mapping the distribution of an electron spin around the rings of organic aromatic radicals (and therefore of mapping the molecular orbital it occupies). It may also be applied to heterocyclic molecules to investigate the change of the electron distribution under the influence of the heteroatom.

Example (Objective 5). The e.s.r. spectrum of (naphthalene)⁻ can be interpreted as arising from two groups of four equivalent protons. Those at the α-positions have a splitting of 0.490 mT, and those at the β-positions have one of 0.183 mT. Map the unpaired spin density around the ring.

- *Method.* The *McConnell equation*, $a_H = Q\rho$, summarizes the discussion in the last paragraph: ρ is the unpaired electron density on a carbon atom, Q is a constant, and a_H the hyperfine splitting constant for the proton attached to the carbon. For (benzene)⁻, $0.47\ mT = Q \times (\frac{1}{6})$, and so $Q = 2.8\ mT$. Use this value for (naphthalene)⁻.

- *Answer.* For the α-positions, $\rho = (0.490\ mT/2.8\ mT) = 0.18$. For the β-positions, $\rho = (0.183\ mT/2.8\ mT) = 0.07$.

- *Comment.* Notice how the unpaired electron accumulates at the α-positions. The same value of Q may be used for approximate mappings in heterocyclics.

Line-shapes and relaxation. The remaining unexplained features of e.s.r. spectra are the intensities and shapes of the resonance lines. The observed intensity is proportional to the difference in population between the α- and β-spin levels. If the populations were equal the radiation field would induce as many spins to flip up from β to α (and absorb energy) as it would stimulate spins to flop down from α to β (and emit energy), and the net absorption would be zero. The size of the population difference in a sample at a temperature *T* can be calculated from the Boltzmann distribution (p. 10). The α level lies at an energy $g\mu_B B$ above the β level, and so the ratio of the populations is

$$N_\alpha/N_\beta = \exp(-g\mu_B B/kT).$$

It follows that the difference is given by

$$(19.1.6) \qquad \frac{N_\beta - N_\alpha}{N_\beta + N_\alpha} = \frac{1 - \exp(-g\mu_B B/kT)}{1 + \exp(-g\mu_B B/kT)}$$

$$= \frac{1 - (1 - g\mu_B B/kT + \ldots)}{1 + (1 + g\mu_B B/kT + \ldots)} \approx g\mu_B B/2kT,$$

because $g\mu_B B/kT$ is small. At room temperature kT corresponds to $200\ cm^{-1}$ and in a typical experiment ($B = 0.3$ T) $g\mu_B B$ is equivalent to $0.3\ cm^{-1}$. Therefore the ratio just calculated is only 7×10^{-4}, indicating an imbalance of populations of only 0.07 per cent. In fact it may be surprising that any absorption is seen at all. This is because when absorption occurs spins are pumped into the upper state, the population difference is decreased, and may be expected to disappear. Therefore, one might expect the small initial absorption to vanish shortly after the start of the experiment.

The problem of weak absorption is overcome by building a spectro-meter with sufficient sensitivity, and commercial instruments can now detect as few as 10^{11} spins. Such a number might seem large, but not when expressed in molar terms: 10^{11} spins is about 10^{-12} mol. These figures show that e.s.r. is an extremely sensitive technique for the detection of radicals, and this makes it suitable for detecting intermediates in chemical reactions, and the weak magnetism of phosphorescent triplet states (p. 615).

The second of the problems, the disappearance of the absorption, is eliminated in most systems by the radicals themselves. This is because mechanisms exist for returning spins to the lower energy state. These mechanisms are called *relaxation processes*. If they are fast enough the original population distribution is maintained even though the sample is absorbing radiation, because the absorbed energy is given up by the spins (which drop back to β states in the process) and dissipated as thermal energy in the rest of the sample. Only when the microwave power is increased, so that the rate of absorption is so great that the relaxation processes are unable to cope with the flux of incoming energy, does the population difference behave in the way first described: this condition is called *saturation* of the line.

The presence of relaxation processes affects the widths of the absorption lines. This is because the lifetimes of the spin levels are shortened if relaxation occurs, and so their energy is made imprecise: this is the *lifetime broadening* effect discussed on p. 570. Transitions then come into resonance over a range of applied fields, and the line broadens. This broadening can be analysed into two contributions, one arising from *spin–lattice relaxation*, also called *longitudinal relaxation*, and the other from *transverse relaxation*. The rate of spin–lattice relaxation is expressed in terms of the *spin–lattice relaxation time* T_1, and the rate of transverse relaxation by the *transverse relaxation time* T_2. Each mechanism contributes about $\hbar/T$ to the energy spread of the lines. In liquid samples $T_2 \approx 10^{-8}$ s and $T_1 \approx 10^{-6}$ s, and so the former gives spreads of about 0.3 mT (3 G) and the latter of about 0.003 mT (30 mG).

The longitudinal, T_1, process is a true energy relaxation process. It occurs because the motion of the radical causes fluctuations in the local magnetic fields, and these fluctuations can stimulate the electron to change its spin state from β to α (Fig. 19.6a and b) and transfer its energy to the lattice. The rate of the fluctuations determines the efficiency of the process (because the energy transfer is most efficient if the motion of the molecules causes fluctuations at the resonance frequency), and so a study of line widths is a source of valuable information on how molecules move in fluids. Linewidth studies are able to show how rapidly molecules rotate, and whether the rotation occurs preferentially about one axis of the radical. They can also show that some types of molecule rotate in a series of small steps ($\approx 5°$), but that others jump through about 1 rad ($\approx 57°$) in each step.

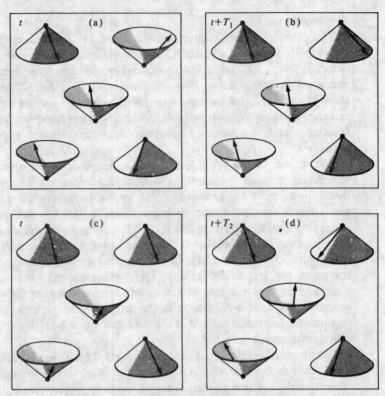

Fig. 19.6. Spin relaxation. A predominance of α-spins (a) relaxes back to a Boltzmann distribution of populations (b) by a longitudinal (T_1) process. An initial set of phase relations (c) relaxes back to random relative phases without changing α and β populations (d) by a transverse (T_2) process because of the different precession frequencies.

The transverse, T_2, process is different, Fig. 19.6c and d. It can be understood on the basis that different radicals in the sample are in slightly different magnetic environments. This can be seen as follows. Although the anisotropic interactions (e.g., the dipole–dipole hyperfine interaction) average to zero, if the radicals are rotating slowly some remain in one orientation for a significant time, and resonate at the appropriate value of the applied field; others linger in other orientations and resonate at different fields. This suggests that the radicals of the sample resonate over a spread of fields. The spread of resonance fields appears as a broadening of the line, which can be expressed by analogy with the lifetime broadening relation as an *effective* lifetime T_2 defined as $T_2 = \hbar/\delta E$. A study of T_2 yields information on molecular motion because rotation averages the anisotropic interactions to zero more effectively as it quickens.

19.2 Nuclear magnetic resonance

The basic principles of nuclear magnetic resonance (n.m.r.) are the same as for e.s.r., the fundamental difference merely being that the experiment monitors the reversal of nuclear magnetic moments. Every nucleus with spin possesses a magnetic moment, and by analogy with the discussion in the last section the energy of a nucleus of spin quantum number I, projection m_I, in a magnetic field B is

(19.2.1) $$E_{m_I} = -g_I \mu_N m_I B$$

where g_I is the nuclear g-factor (see Table 19.2) and μ_N is the *nuclear magneton*, $e\hbar/2m_p$ (the sign change in the equation reflects the opposite charges of electrons and nuclei). We shall concentrate on the n.m.r. of protons, which have $g_I = 2.79270$, but nuclear resonance of other nuclei (especially ^{13}C) are of increasing importance in structural studies. ^{13}C gives rise to special problems of detection on account of its low natural abundance, and the special techniques employed are treated later (p. 653). Note that the nuclear magneton is about 2000 times smaller than the Bohr magneton, and so nuclear moments are about that much weaker than the electron moment. This can be understood on the basis that the spin angular momenta of nuclei are about the same as that of the electron (when $I = \frac{1}{2}$ the momenta are identical) but their masses are much greater; therefore, on the classical picture of spin, the nuclei may be imagined as rotating much more slowly than electrons. That implies a lower circulating current, and therefore a lower magnetic moment.

Table 19.2. Nuclear spin properties

Isotope (*: radioactive)	Natural abundance (per cent)	Spin I	Magnetic moment† μ/μ_N	Nuclear g-value	N.m.r. frequency at 1 T (v/MHz)
^{1}n*	—	$\frac{1}{2}$	-1.9130	-3.8260	29.167
^{1}H	99.9844	$\frac{1}{2}$	2.79285	5.5857	42.576
^{2}H	0.0156	1	0.85745	0.85745	6.536
^{3}H*	—	$\frac{1}{2}$	-2.12765	-4.2553	32.4338
^{13}C	1.108	$\frac{1}{2}$	0.7023	1.4046	10.705
^{14}N	99.635	1	0.40356	0.40356	3.076
^{17}O	0.037	$\frac{5}{2}$	-1.893	-0.75720	5.772
^{19}F	100	$\frac{1}{2}$	2.62835	5.2567	40.054
^{31}P	100	$\frac{1}{2}$	1.1317	2.2634	17.238
^{33}S	0.74	$\frac{3}{2}$	0.6434	0.4289	3.266
^{35}Cl	75.4	$\frac{3}{2}$	0.8218	0.5479	4.171
^{37}Cl	24.6	$\frac{3}{2}$	0.6841	0.4561	3.472

† The nuclear magneton has the value $\mu_N = e\hbar/2m_p = 5.05082 \times 10^{-27}$ J T^{-1}; μ is the magnetic moment of the spin state with the largest values of m_I: $\mu = g_I \mu_N I$.
Source: G. W. C. Kaye and T. H. Laby, *Tables of physical and chemical constants*, Longmans.

The resonance condition for the transition of the proton from its lower (α) to its upper (β) spin state, Fig. 19.7, is

(19.2.2) $$hv = \tfrac{1}{2}g_I \mu_N B - (-\tfrac{1}{2}g_I \mu_N B) = g_I \mu_N B$$

and insertion of the appropriate values indicates that at a field of 1.5 T

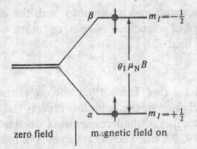

Fig. 19.7. Nuclear spin energy levels in a magnetic field (for $I = \frac{1}{2}$).

(15 kG) the resonance frequency is 60 MHz (60×10^6 c/s). At that field n.m.r. is a *radio-frequency* technique. Much higher fields are available from superconducting magnets, and modern spectrometers work at fields of about 7 T (70 kG), which corresponds to a frequency of 300 MHz. One advantage of working at high fields is that intensities are enhanced. The population difference at a field B and temperature T is given by a straightforward modification of eqn (19.1.6) as

(19.2.3)
$$(N_\alpha - N_\beta)/(N_\alpha + N_\beta) \approx g_I \mu_N B/2kT$$

which for 1.5 T at room temperature gives only 2.6×10^{-6} for protons (and less for other common nuclei), so that the excess of α spin over β spins is very small. When the applied field is increased to 7 T the population difference is greater in proportion, and so the signal intensity is correspondingly greater.

The technique. The experimental arrangement is much the same as·in e.s.r., the principal difference being the substitution of a radio-frequency source and detector for the microwave components. Furthermore the magnet operates at higher fields, and because the spectrum has components split by very small energies (as described below) the magnetic field has to be exceptionally well controlled and homogeneous over the bulk of the sample. One way of ensuring that the sample experiences a homogeneous field is to rotate it rapidly so that small inhomogeneities are averaged out. The experiment is performed either by sweeping the magnetic field over a small range and monitoring the absorption of radio-frequency power at constant frequency, or sweeping the frequency at constant applied field.

A typical experimental result (for ethanol) is shown in Fig. 19.8. Two main features require explanation: why the spectrum is split into several blocks, and why these blocks consist of several closely-spaced narrow lines.

The chemical shift. As in the case of e.s.r. the resonant magnetic moment interacts with the *local* field, and this may differ from the applied field because of the currents the latter might stir up in the electrons of the molecule. In n.m.r. it is conventional to write the local field in the form

$$B_{loc} = (1 - \sigma)B$$

where σ is called the *shielding constant*. This form relies on the supposition

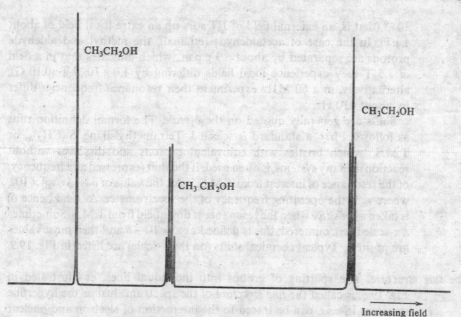

Fig. 19.8. The n.m.r. spectrum of ethanol. The bold letters denote the protons giving rise to the resonance.

that the currents stirred up are proportional to the strength of the fields applied externally, so that there is an extra contribution of the form $-\sigma B$ to the overall field experienced by the protons. (The same argument was used for g-values, p. 637.) The extra field σB is called the *chemical shift* of the group of protons. The ability of the applied field to induce currents depends on the molecule's electronic structure in the vicinity of the proton of interest. Therefore, protons in different chemical groups have different shielding constants, and the resonance condition

$$h\nu = g_I \mu_N (1 - \sigma) B$$

is satisfied at different values of B for protons in different chemical environments. This lets us understand the general structure of the spectrum of ethanol shown in Fig. 19.8. The methyl protons form one group and come into resonance at a position governed by their chemical shift. The two methylene protons are in a different part of the molecule; they therefore have a different chemical shift, and come into resonance at another magnetic field. Finally, the proton on the hydroxyl group is in another magnetic environment characterized by yet another chemical shift, and comes into resonance at yet another value of the applied field. Note that we can distinguish which group of lines corresponds to which group of protons by their relative intensities (the areas under the absorption curves). The group intensities are in the ratio 3:2:1 because there are 3 methyl protons, 2 methylene protons, and 1 hydroxyl proton.

The magnitude of the chemical shift is normally expressed in parts per million (p.p.m.) because the shielding constant is of the order of 10^{-5}–

10^{-6} (that is, an external field of 1 T stirs up an extra local field of about 1 μT). In the case of acetaldehyde (ethanal), the methyl and aldehyde protons are separated by about 6.9 p.p.m., which indicates that in a field of 1.5 T they experience local fields differing by 1.0×10^{-5} T (0.10 G); alternatively, in a 60 MHz experiment their resonance frequencies differ by about 410 Hz.

Shifts are generally quoted on the δ-*scale*. The formal definition runs as follows. First, a standard is selected. Tetramethylsilane, $Si(CH_3)_4$, or T.M.S., which bristles with equivalent protons and dissolves without reaction in many systems, is often used. If the shift (expressed as a frequency) of the resonance of interest from T.M.S. is $\Delta\nu$, the value of δ is $(\Delta\nu/\nu_0) \times 10^6$, where ν_0 is the operating frequency of the spectrometer. $\Delta\nu$ (and hence δ) is taken as *positive* when the resonance is downfield from T.M.S. Sometimes a τ-scale is encountered: this is defined as $\tau = 10 - \delta$ and then most values are positive. Typical chemical shifts (on the δ-scale) are listed in Fig. 19.9.

The fine structure. The splitting of groups into individual lines, as illustrated in Fig. 19.8, is called the *fine structure* of the spectrum. Just as the hyperfine structure in e.s.r. can be traced to the interaction of electron and nuclear spins, so the fine structure in n.m.r. can be ascribed to the magnetic interactions between the nuclei in the molecule.

First we can examine whether two protons can have a direct, dipole-dipole interaction. At a point a distance R from a proton with spin projection m_I there is a magnetic field given by

(19.2.4) $B^{\text{nucl}} = -g_I\mu_N(\mu_0/4\pi)(1/R^3)(1 - 3\cos^2\theta)m_I.$

The quantities R and θ are defined in Fig. 19.10. The magnitude of this field is of the order of 0.1 mT when $R = 0.3$ nm, and so it is of the order of magnitude of the observed splittings from the solid sample in Fig. 19.11. In liquids the angle θ drifts rapidly over all values; therefore, although the direct dipole–dipole interaction makes an important contribution to the spectra of solid samples (and can be used to determine the inter-proton distances R), it averages to zero in fluid samples.*

Another mechanism for interaction between magnetic nuclei depends on the polarization interaction (p. 641). The magnetic field arising from one nucleus affects the distribution of the electrons in its bond, and the effect can be transmitted through neighbouring bonds to other nuclei in the molecule. Detailed calculation confirms that this mechanism can account for the splittings observed in fluid samples.

We can now move on to see how the appearance of the spectrum can be explained. The ethanol spectrum will form the basis of the discussion.

Consider first the methyl protons. It turns out that *equivalent protons do not show any fine structure* (they *do* interact, but the selection rules for

---

* The volume element is proportional to $\sin\theta\,d\theta (d\tau = r^2\,dr\sin\theta\,d\theta\,d\varphi)$, and as θ can range from 0 to π the average value is proportional to the integral $\int_0^\pi (1 - 3\cos^2\theta)\sin\theta\,d\theta = 0$.

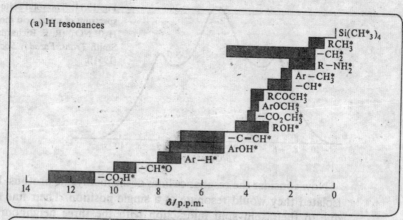

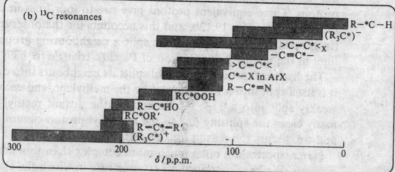

Fig. 19.9. The range of typical chemical shifts: (a) ^{1}H resonances, (b) ^{13}C resonances.

the spectral transitions forbid any transition that would give an indication of the interaction energy). Therefore an isolated methyl group, as in CH_3Cl, gives a single unsplit line. Next to the methyl group in ethanol, however, are the two methylene protons. The methyl line is first split into two by virtue of the two orientations of one of the methylene protons, and then each line is split again as a result of the interaction with the second proton. Therefore the methyl resonance is split into three lines with intensity ratio 1:2:1, Fig. 19.12a. The splitting is normally denoted J.

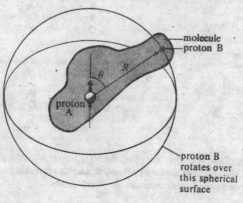

Fig. 19.10. The angle involved in the dipole–dipole interaction, and the effect of molecular rotation.

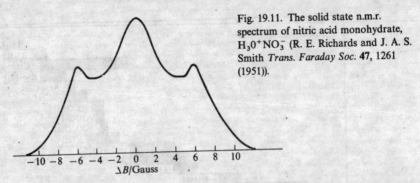

Fig. 19.11. The solid state n.m.r.
spectrum of nitric acid monohydrate,
$H_3O^+NO_3^-$ (R. E. Richards and J. A. S.
Smith *Trans. Faraday Soc.* **47**, 1261
(1951)).

Now consider the resonance lines of the methylene protons. If they were isolated they would resonate at a single position. Their fine structure is due to their spin–spin interaction with the three neighbouring methyl protons. Three equivalent protons give rise to four lines in the intensity ratio $1:3:3:1$, Fig. 19.12b, and this accounts for the observed structure. (In general N equivalent protons split a neighbouring group into $N+1$ lines in the intensity distribution of Pascal's triangle (p. 639).)

The hydroxyl proton in ethanol splits its neighbours into doublets, and it is itself split into a $1:2:1$ triplet by the methylene, and each line is very weakly split into a $1:3:3:1$ quartet by the distant methyl protons. In many cases the splitting due to the hydroxyl proton cannot be detected, and we discuss the reason for this in the next section.

N.m.r. spectra are often much more complex than this simple analysis suggests. We have described an extreme case in which the chemical shifts are all much greater than the spin–spin couplings, and it is simple to say which protons are equivalent, and which are not. When the chemical shift

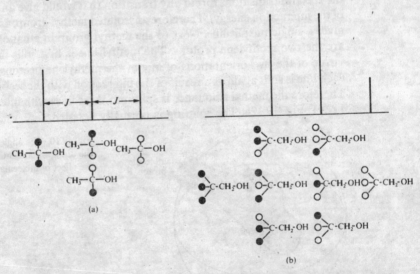

Fig. 19.12. The source of the fine structure of (a) the CH_3 resonances and (b) the CH_2 resonances in ethanol. Bold-face **H** denotes protons. ● and ○ represent protons of opposite spin giving rise to the fine structure in the H resonance.

becomes comparable to the coupling constants it becomes necessary to make use of a computer in order to evaluate all the energy levels and the transition probabilities. The spectra are then referred to as *second-order spectra*. This introduces another advantage of working at very high fields. The chemical shift is σB, and increases with field; on the other hand the spin–spin interaction is independent of the applied field. Therefore when the field is increased, the separation between chemically shifted groups of lines increases, but the spin–spin interactions remain the same. It follows that a complicated spectrum can be simplified considerably by working at high fields: the groups of lines are resolved, and the muddling of energy levels and transition probabilities is reduced.

Another way of simplifying the appearance of a spectrum is to use a second, strong source of radiofrequency radiation to stimulate transitions of a group of nuclei. For instance, if the methyl protons in ethanol are made to undergo many transitions, the methylene protons sense that their neighbours have rapidly alternating spin states, and so their fine-structure is averaged away. Therefore the quartet of lines in the methylene resonance collapses to a single line. This is a *proton decoupling* technique.

Example (Objective 9, 10). Figure 19.13 shows an experimental 60 MHz spectrum. Suggest an interpretation. What changes would occur if the spectrum were obtained at 300 MHz?

- *Method.* Look for groups with characteristic chemical shifts: refer to Fig. 19.9. Account for the fine-structure by the same argument as used for ethanol. Report δ-values, J-values, and identify the compound.

- *Answer.* The resonance at $\delta = 3.4$ corresponds to $-CH_2-$ protons in an ether; that at $\delta = 1.2$ corresponds to $-CH_3-$ protons in CH_3CH_2. The spin–spin splitting of the $-CH_2-$ resonance (a quartet) is characteristic of a neighbouring $-CH_3$ group, and the spin–spin splitting of the $-CH_3$ resonance is characteristic of a nearby

Fig. 19.13. An n.m.r. spectrum.

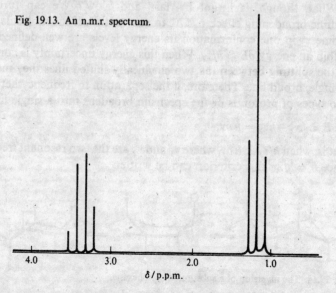

–CH$_2$–. The splitting constant is $J = 6.0$ Hz (the same for each group). The compound is CH$_3$CH$_2$.O.CH$_2$CH$_3$. At 300 MHz the groups would be 5 times further apart, but the splitting within them would remain the same.

• *Comment.* At a lower field (or in a different molecule) the chemical shifts may be comparable to the fine-structure splittings. Then the appearance of the spectrum would be much more complicated and a computer would normally be used to analyse it.

Line-shapes and rate processes. The motion of nuclei can affect the shapes and widths of lines in n.m.r. just as it does in e.s.r., and the spin–lattice and transverse relaxation times can be discussed in precisely the same way. The important difference is that, because the nuclear magnetic moments are so weak, the relaxation times are long, and are often of the order of seconds.

N.m.r. is particularly well suited to studying several types of rapid molecular change, including interconversion of isomeric forms and exchange of protons with other molecules. The occurrence of these processes can be detected from the appearance of the spectrum, and we shall examine two typical examples.

Consider a molecule that is able to jump between two conformations, for example, the chair–boat inversion of the substituted cyclohexane in Fig. 19.14. In one conformation the methyl protons are in an axial position (a) and in the other they are equatorial (e). Their chemical shifts are different in the two cases, and when the inversion rate is slow the spectrum shows two sets of lines, one from molecules with an axial methyl, and one from molecules with an equatorial methyl. When the inversion rate is fast the molecules change from one conformation to the other so rapidly that the spectrum shows only a single line corresponding to the mean of the two chemical shifts.

What, though, is meant by 'fast' and 'slow'? We can turn to the lifetime-broadening effect, p. 570, to answer this. If the molecule lives for a time τ_J in one conformation its energy levels are well-defined only to within an energy $\delta E \approx \hbar/\tau_J$. When this energy uncertainty is comparable to the splitting between the two chemically shifted lines they merge into a single, broad line. Therefore, if the separation in frequency between the two types of proton is $\delta\nu$ the spectrum broadens into a single line when

$$\delta E \gtrsim h\nu_1 - h\nu_2 = h\,\delta\nu;$$

that is, when $\hbar/\tau_J \gtrsim h\,\delta\nu$, where ν_1 and ν_2 are the two resonant frequencies. Since $\hbar = h/2\pi$ this criterion can be written

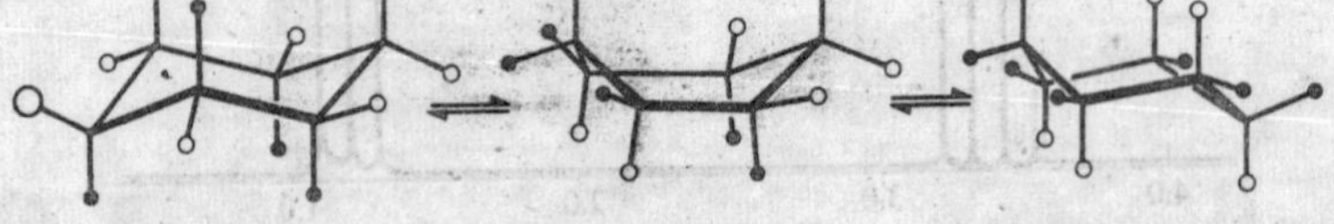

Fig. 19.14. The inversion of a substituted cyclohexane.

(19.2.5) *Collapse of structure when* $\tau_J \lesssim 1/2\pi \, \delta\nu$.

For example, if the chemical shift of the two groups is 100 Hz the two lines merge into one if interconversion lifetimes are less than about 2 ms.

Example (Objective 11). The NO group in N,N-dimethylnitrosamine, $(CH_3)_2N{-}NO$, rotates and as a result the magnetic environments of the two methyl groups are interchanged. At 60 MHz the two methyl resonances are separated by 39 Hz. At what rate of tautomerization will the resonance collapse into a single line in a 60 MHz spectrometer?

- *Method.* Use eqn (19.2.5).

- *Answer.* $\tau_J \leqslant 1/2\pi \, (39 \text{ Hz}) = 4.1$ ms.

- *Comment.* In a 300 MHz spectrometer the collapse will occur when $\tau_J \leqslant 0.82$ ms. This more rapid jumping may be stimulated by heating the sample: the change of temperature required can be used to determine the *activation energy* of the tautomerization (see Problem 19.34).

A similar explanation accounts for the loss of structure in solvents able to exchange protons with the sample. For example, alcoholic hydroxyl protons are able to exchange with water protons. When this *chemical exchange* occurs, a molecule ROH with an α-spin proton (we write this $ROH^{(\alpha)}$) rapidly turns into $ROH^{(\beta)}$ and then into $ROH^{(\alpha)}$, because the protons provided by the solvent molecules in successive exchanges have random spin orientations. Therefore, instead of seeing a spectrum composed of contributions from both $ROH^{(\alpha)}$ and $ROH^{(\beta)}$ molecules (that is, a spectrum with lines split into doublets as a result of spin–spin coupling with the hydroxylic proton) all that is seen is a single, unsplit line at the mean position. The effect is observed when the rate of chemical exchange $1/\tau_{ex}$ is so great that the energy uncertainty is greater than the doublet splitting, and so collapse of the fine structure is expected when

$$\tau_{ex} \lesssim 1/2\pi \, \delta\nu,$$

where $\delta\nu$ is the frequency separation of the doublets being modulated by the exchanging protons. Since this splitting is often very small (about 1 Hz), a proton must stay attached to the same molecule for longer than about $\tau_{ex} \approx (1/2\pi \text{ Hz}) \approx 0.1$ s if splitting in the spectrum is to survive. In water the chemical exchange rate is much faster than that, and so alcohols in aqueous solution show no splitting from the hydroxyl protons. When very dry alcohol is used the exchange rate is much slower and the splitting can be detected.

The importance of chemical exchange effects is that they may be used to study rapid chemical processes, and to see how their rates vary as the conditions (e.g., the temperature and solvent) are changed.

Fourier transform spectroscopy. One of the main problems with n.m.r. is the low intensity of the transitions on account of the extremely small population differences between the nuclear spin states. This is particularly severe in the case of ^{13}C, a nucleus of prime interest in organic and biochemical studies. The sensitivity problem is exacerbated by the low natural abundance of the isotope (1.1 per cent) and its small magnetic moment ($0.702 \, \mu_N$, only one quarter that of the proton) which not only results in a population difference one quarter of that for protons but also reduces the strength of coupling with the detector being used to observe the transitions.

The modern version of n.m.r. relies on computer-based data acquisition and transformation techniques. The essential feature of *Fourier transform (F.T.) spectroscopy* is the observation of the time evolution of the nuclear spin state of the sample from a specially prepared initial state, and then the transformation of this time-resolved behaviour into information about the energy levels present. The initial state is prepared by exposing the sample to a brief burst of high power radiofrequency radiation; typically 1 μs bursts of 1 kW are used. This radiation distorts the equilibrium population of the nuclear energy levels. After the radiation ceases, the system emits its excess energy as the populations revert towards equilibrium. This emission takes the form of radiation at all the allowed transition frequencies, and it appears as a decaying, oscillating signal called a *free induction decay*, the oscillations arising from the beats between all the contributing frequencies, Fig. 19.15. All the spectral information is contained in the signal (just as dropping a piano generates a signal that contains all the information about its frequency spectrum), and it can be analysed by finding the superposition of frequencies that gives the observed shape. This analysis step is the Fourier transformation that gives the technique its name, and it is a well-defined mathematical procedure which can be performed on a computer (which is normally an integral part of the spectrometer). The transform of Fig. 19.15 is in fact the proton spectrum of ethanol already depicted in Fig. 19.8.

When ^{13}C is the nucleus being examined it is not normally necessary to take into account ^{13}C-^{13}C spin-spin coupling because the low

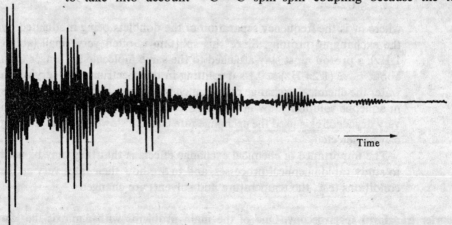

Time

Fig. 19.15. Free induction decay from ethanol.

abundance of the nucleus makes it very unlikely that two such nuclei will occur in the same molecule (unless it has been enriched). If the molecule contains many protons the spectrum (and the time evolution) may be excessively complex. In order to avoid this difficulty, ^{13}C-F.T. n.m.r. spectra are normally taken with protons decoupled either totally or selectively in groups (achieved, for instance, by a proton decoupling technique as described in the last section). The decoupling has the additional advantage of enhancing sensitivity, because transition intensity is concentrated into a single spectral line instead of being spread over several.

19.3 Mössbauer spectroscopy

Nuclear magnetic resonance spectroscopy is based on the absorption of photons of very low energy—their frequencies of the order of 60 MHz (and wavelengths 5 m) are about the lowest used in spectroscopy. In contrast, Mössbauer spectroscopy lies at the other extreme. It is based on γ-ray photons with frequencies of the order of 10^{19} Hz and wavelengths of about 100 pm (10^{-10} m). But even though the photon energies are so different in the two forms of spectroscopy, the Mössbauer lines are so sharply resolved that they reveal splittings of the same order of magnitude as those in n.m.r.

The emission and absorption of γ-rays. The basis of Mössbauer spectroscopy is the resonant absorption of a γ-ray photon by a nucleus. A typical experiment uses a sample of ^{57}Co as the primary source of an isotope of iron, ^{57}Fe. The isotope ^{57}Co decays slowly (its half-life is 270 days) and forms an excited nuclear state of ^{57}Fe denoted $^{57}Fe^*$. This state decays very rapidly to ground state ^{57}Fe (its half-life is about 2×10^{-7} s) and emits a γ-ray, Fig. 19.16, which is the photon used in the experiment. When the proton strikes a ground-state ^{57}Fe nucleus in another part of the spectrometer it is absorbed if its energy matches the $^{57}Fe^*$-^{57}Fe energy separation (this is the resonant absorption step), but will not be absorbed if the energies do not match.

The reason why the resonance is so sharp can be traced to a number

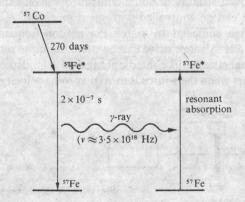

Fig. 19.16. Transitions involved in the Mössbauer effect using ^{57}Fe.

of factors. In the first place the life of the $^{57}Fe^*$ is long enough for its energy to be very well defined. A lifetime of about 2×10^{-7} s corresponds to a frequency uncertainty of only about 2 MHz. The γ-ray photon emitted by $^{57}Fe^*$ (and absorbed by ^{57}Fe) has a frequency of about 3.5×10^{18} Hz (corresponding to an energy of 14.4 keV), and so the line-width is only about 10^{-12} of the photon frequency. That, however, is not the end of the story, because a photon of such high energy also possesses a significant momentum (given by $h\nu/c$). When $^{57}Fe^*$ decays and emits the photon, there is a recoil which gives the de-excited nucleus a velocity of about 10^2 m s^{-1}. This recoil has a damaging effect on the purity of the photon's frequency because the Doppler effect has to be taken into account. If the source of a photon moves at some velocity v with respect to an observer, its frequency ν is shifted by an amount $\nu v/c$ (recall p. 569). Even for $v \approx 1$ cm s^{-1} the photon from a $^{57}Fe^*$ would suffer a frequency shift of about 100 MHz, which is enough to obliterate the natural line-width. One crucial aspect of the Mössbauer experiment overcomes this recoil effect. If the $^{57}Fe^*$ and ^{57}Fe nuclei are held rigidly in a crystal lattice, the momentum is taken up by the whole bulky sample. This implies an extremely small recoil velocity, and therefore an extremely small Doppler shift and broadening of the lines. The *Mössbauer effect* is this recoil-free emission and resonant absorption of γ-rays.

The Mössbauer spectrometer. The practical aspects of the technique are as follows. The γ-ray source is a thin foil of ^{57}Co, Fig. 19.17. This isotope is suitable because its reasonably long life implies a constant supply of $^{57}Fe^*$ over a period of hours or days during which the experiment is in progress. The sample being studied is carried in a similar foil, which is placed in front of a counter. Many isotopes may be used in the equipment, but ^{57}Fe and ^{119}Sn are frequently used, and we shall concentrate on them, and especially on the former.

The resonant absorption occurs when the nuclear energy levels of source and absorber match. The energy of the emitted photon can be modified by making use of the Doppler effect. (Now it plays a useful role, and is not a hindrance.) This is achieved by mounting the source on a movable support: if it is moved steadily at some controlled velocity the Doppler shift of the photon can be controlled with great precision. A range of a few mm s^{-1} is often sufficient to match the photon frequency to the absorber energy levels. A screw drive can be used, but quite often the foil is driven electromagnetically, like the diaphragm of a loudspeaker.

Absorption occurs when the source is moved at some velocity; but quite

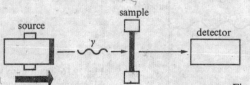

Fig. 19.17. A Mössbauer spectrometer.

often absorption is observed for several relative velocities. The raw data of Mössbauer spectroscopy is a plot of the number of photon counts against the velocity of the source. A typical spectrum is shown in Fig. 19.18: positive velocities indicate motion of the source towards the absorber; negative velocities indicate motion away. Positive velocities indicate that the photon frequency must be increased in order that the absorber species can resonate, and therefore that the $^{57}Fe^*$-^{57}Fe separation is greater in the absorber than in the source.

Example (Objective 13). In a $^{57}Fe^*$ Mössbauer experiment it was found that the source had to be moved towards the sample at 2.2 mm s^{-1} in order to obtain resonant absorption of the γ-ray. What is the shift in energy (in MHz) between sample and source?

- *Method.* The Doppler shift is of magnitude vv/c. The $^{57}Fe^*$ γ-ray has an energy of 14.4 keV: convert this to a frequency (via 1 eV $\triangleq$ 8065.5 cm^{-1} and multiplication by c).

- *Answer.* $v/c = (14.4 \text{ keV}) \times (8065 \text{ cm}^{-1}/\text{eV})$. Therefore
$$\delta v = (14.4 \text{ keV}) \times (8065.5 \text{ cm}^{-1}/\text{eV}) \times (2.2 \text{ mm s}^{-1})$$
$$= 25.5 \times 10^6 \text{ s}^{-1}, \text{ or } 25.5 \text{ MHz}.$$

- *Comment.* This frequency corresponds to the wavenumber 8.52×10^{-4} cm^{-1}. A remarkable feature of the Mössbauer experiment is that such small shifts are measured using photons of such enormous energy.

Information from the spectra. We have to understand two features of the spectra. The first is why the resonance is shifted relative to the source frequency. The second is why the spectra are often split into several lines.

The change in position of the resonance is called the *isomer shift*. The excited and ground state nuclei differ in radius to a small but significant extent (^{57}Fe shrinks by about 0.2 per cent when it is excited to $^{57}Fe^*$), and so its electrostatic interaction with the surrounding electrons changes on excitation. The effect is significant only for electrons in s-orbitals (which can approach the nucleus) and so the energy of the nucleus changes by an amount proportional to the change in nuclear radius and the s-electron density at the nucleus. If the former is written δR and the latter $\psi_s^2(0)$,

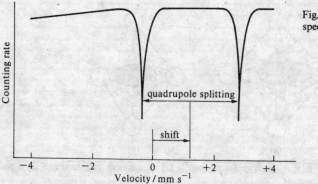

Fig. 19.18. Mössbauer spectrum of $FeSO_4$.

the energy shift is

$$\delta E \propto \psi_s^2(0)\,\delta R.$$

Although δR is the same in source and absorber, the s-electron density might be different for the two cases (because the ^{57}Fe nuclei are in different chemical environments) and the isomer shift is given by

$$\delta E \propto \{[\psi_s^2(0)]_{\text{absorber}} - [\psi_s^2(0)]_{\text{source}}\}\,\delta R.$$

By measuring the isomer shift it is possible to assess the extent to which s-orbitals are involved in the bonding.

As an example, the Mössbauer spectra of ^{119}Sn compounds show an isomer shift of about 20 MHz relative to ^{119}Sn as white tin in the case of Sn(II) covalent molecules (such as $SnCl_2$) but of about -50 MHz for Sn(IV) covalent molecules (such as $SnCl_4$). The presence of the two extra s-electrons in Sn(II) leads to a more positive shift than in Sn(IV). Evidently this technique provides a very direct way of determining the valence of an element in a compound.

The structure of a Mössbauer spectrum arises from two principal interactions. In a number of cases (in ^{57}Fe, in particular) the nuclear spins of the unexcited and excited isotopes are different. For instance $I(^{57}\text{Fe}) = \frac{1}{2}$ but $I(^{57}\text{Fe*}) = \frac{3}{2}$. This change of spin is accompanied by a change of the charge distribution in the nucleus, so that whereas ^{57}Fe has a spherical distribution of charge, that in ^{57}Fe* is concentrated at the poles. In an octahedral environment (such as in $Fe(CN)_6^{4-}$) the energy of the nucleus is independent of its orientation, Fig. 19.19a; but if the environment has only axial symmetry (as in $Fe(CN)_5NO^{2-}$) the energy does depend on the nuclear orientation, Fig. 19.19b. This accounts for the *electric quadrupole splitting* of the Mössbauer spectrum in some compounds and illustrated in Fig. 19.18. Its analysis lets one build up a picture of the symmetry of the electron distribution close to the nucleus.

The importance of ^{57}Fe Mössbauer studies reflects more than the fact that it is a convenient material to work with. Iron occurs in a variety of biologically and technologically important compounds (such as haemoglobin and rust) and Mössbauer spectroscopy has been used to analyse them and to explore, for example, the mechanism of corrosion.

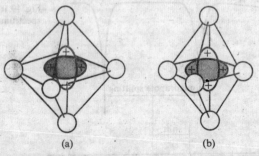

Fig. 19.19. A quadrupolar nucleus in (a) a regular environment and (b) a distorted environment.

Further reading

Magnetic resonance. K. A. McLauchlan; Clarendon Press, Oxford, 1972.

Spectroscopy. D. H. Whiffen; Longman, London, 1972.

Electron spin resonance in chemistry. P. B. Ayscough; Methuen, London, 1967.

Electron spin resonance, elementary theory and practical applications. J. E. Wertz and J. R. Bolton; McGraw-Hill, New York, 1972.

Introduction to magnetic resonance. A. Carrington and A. D. McLachlan; Harper and Row, New York, 1967.

Electron spin resonance. P. H. Rieger; in *Techniques of chemistry* (A. Weissberger and B. W. Rossiter, eds.), Vol. IIIA, 499, Wiley–Interscience, New York, 1972.

Chemical and biochemical aspects of electron spin resonance spectroscopy. M. Symons; Van Nostrand Reinhold, New York, 1978.

Physical methods in chemistry. R. S. Drago; Saunders, Philadelphia, 1977.

Nuclear magnetic resonance spectroscopy. R. M. Lynden-Bell and R. K. Harris; Nelson, London, 1969.

Nuclear magnetic resonance. J. D. Roberts; McGraw-Hill, New York, 1959.

Nuclear magnetic resonance. N. Muller; in *Techniques of chemistry* (A. Weissberger and B. W. Rossiter, eds.), Vol IIIA, 599, Wiley–Interscience, New York, 1972.

High-resolution nuclear magnetic resonance. J. A. Pople, W. G. Schneider, and H. J. Bernstein; McGraw-Hill, New York, 1959.

The principles of nuclear magnetism. A. Abragam; Clarendon Press, Oxford, 1961.

Fourier transform n.m.r. spectroscopy. D. Shaw; Elsevier, Amsterdam, 1976.

Fourier transform n.m.r. techniques: a practical approach. K. Mullen and P. S. Pregosin; Academic Press, New York, 1976.

Experimental aspects of Mössbauer spectroscopy. R. H. Herber and Y. Hazony; in *Techniques of chemistry* (A. Weissberger and B. W. Rossiter, eds.), Vol. IIID, 215, Wiley–Interscience, New York, 1972.

Physical methods in advanced inorganic chemistry. H. A. O. Hill and P. Day (eds.); Interscience, London, 1968.

Problems

19.1. Some commercial e.s.r. spectrometers use 8 mm microwave radiation. What magnetic field is then needed to bring an electron spin into resonance?

19.2. Commercial e.s.r. spectrometers can detect resonance from a sample containing as few as 10^{10} spins per cm^3. What is the molar concentration corresponding to this detection limit? Suppose a sample is known to contain 2.5×10^{14} spins. How many of these spins would be in the α-state at 25 °C, and how many in the β-state?

19.3. Calculate the relative population difference, $\delta N/N = (N_\beta - N_\alpha)/(N_\beta + N_\alpha)$, for electron spins in a magnetic field of 0.3 T (3 kG) at (a) 4 K, (b) 300 K.

19.4. The centre of the e.s.r. spectrum of atomic hydrogen lies at 329.12 mT in a spectrometer operating at 9.2231 GHz (1 GHz = 10^9 Hz). What is the g-value of the electron in a hydrogen atom?

19.5. The benzene radical anion, $C_6H_6^-$, has a g-value of 2.0025. At what field would you search for resonance in a spectrometer operating at (a) 9.302 GHz, (b) 33.67 GHz?

19.6. The mean g-value of the stable ClO_2 radical is 2.0102. What is the value of the additional local field that this implies when the applied field is (a) 3400 G, (b) 12 300 G?

19.7. The triangular molecule NO_2 has a single unpaired electron and it can be trapped in a solid matrix or prepared inside a crystal by radiation damage of nitrite ions. With the radical held rigidly in a definite orientation the magnetic field may be applied at selected orientations. When it is parallel to the O–O direction, the centre of the spectrum lies at 333.64 mT in a spectrometer operating at 9.302 GHz, but when the field lies along the bisector of the ONO angle the resonance shifts to 331.94 mT. What are the g-values in the two directions?

19.8. The e.s.r. spectrum of atomic hydrogen consists of two lines. In a spectrometer working at 9.302 GHz one line appeared at 357.3 mT and the other at 306.6 mT. What is the hyperfine coupling constant for the atom?

19.9. A radical containing two equivalent protons shows a three-line e.s.r. spectrum with an intensity distribution in the ratio $1:2:1$. The lines occurred at 330.2 mT, 332.5 mT, and 334.8 mT. What is the hyperfine coupling constant for each proton?

19.10. The spectrometer used to obtain the spectrum described in the last Problem used a klystron generating microwaves of frequency 9.319 GHz. What is the g-value of the radical, and what are the coupling constants in MHz? (You should derive the useful rule that 0.1 mT $\approx$ 2.8 MHz.)

19.11. A radical containing two inequivalent protons with splitting constants 2.0 mT and 2.6 mT gives a spectrum centred on 332.5 mT. At what fields do the hyperfine lines lie, and what are their relative intensities?

19.12. The radical referred to in the last Problem isomerizes, and the motion interchanges the two protons. At low temperatures the spectrum is as described above. At room temperature the central pair of lines merge into a single line of double the intensity, and the spectrum is then as described in Problem 19.9. What is the rate of isomerization at which the spectrum switches from one form to the other?

19.13. Predict the intensity distribution in the hyperfine lines of the e.s.r. spectrum of the radicals $\cdot CH_3$ and $\cdot CD_3$. ($I = 1$ for ^2H.)

19.14. The hyperfine coupling constant in the methyl radical is 0.23 mT. Use the information in Table 19.2 to predict the splitting between the hyperfine lines of the $\cdot CD_3$ radical's spectrum. What is the overall width of the spectra in each case?

19.15. Predict the appearance of the e.s.r. spectrum of the ethyl radical, CH_3CH_2, using $a(CH_2) = 0.224$ mT and $a(CH_3) = 0.268$ mT for the respective protons. What changes will occur in the spectrum when deuterons are substituted for methylene protons?

19.16. The p-dinitrobenzene radical anion can be prepared by reduction of the nitrobenzene. The radical-anion has two equivalent nitrogen nuclei ($I = 1$) and four equivalent protons. Predict the form of the e.s.r. spectrum using $a(N) = 0.148$ mT, $a(H) = 0.112$ mT.

19.17. When an electron occupies a 2s-orbital on nitrogen it has a hyperfine interaction of 55.2 mT with the nucleus (this is the magnetic field generated by the nucleus and averaged over the region of the s-orbital where its presence is experienced). The spectrum of NO_2, on the other hand, shows an isotropic contact hyperfine interaction of 5.7 mT. For what proportion of its time can the unpaired electron of NO_2 be regarded as occupying the nitrogen 2s-orbital?

19.18. The nuclear field experienced by an electron when it occupies a 2p-orbital in the nitrogen atom has a magnitude of 3.4 mT. In NO_2 the anisotropic part of the hyperfine coupling has a magnitude of 1.3 mT. What proportion of its time does the unpaired electron spend in the nitrogen 2p-orbital in NO_2?

19.19. The information in the last two Problems is sufficient to map the unpaired electron distribution over the molecule. What is the total probability that the unpaired electron will be found on the nitrogen? What is the probability that it will be found on either oxygen atom? What is the hybridization of the central nitrogen atom? Does this hybridization support the suggestion that NO_2 is a bent molecule? Use the discussion of hybridization ratio of p. 492 to interpret p/s in terms of the bond angle by showing that $\cos \Theta = -\lambda/(\lambda+2)$, where $\lambda = b'^2/a'^2$. Estimate Θ; the experimental value is 134°.

19.20. Two different radicals were created when an inorganic solid was damaged with γ-rays. One had a g-value of 2.0101 and the other had a g-value of 2.0060. What is the separation of their resonances (in mT) when they are examined with an e.s.r. spectrometer working at (a) 9.218 GHz, (b) 34.22 GHz. The former consisted of three lines split by 0.60 mT, and the latter was a $1:3:3:1$ quartet with a spacing of 0.50 mT. Sketch the appearance of the spectra at the two operating frequencies.

19.21. The hyperfine coupling constants observed in various radical anions are listed below (in mT). Use the benzene value, and symmetry, to map the probability of finding the unpaired electron in the $2p\pi$-orbital on each carbon atom.

19.22. The pyridyl radical (I) has the proton hyperfine coupling constants shown (in mT), and the carboxy derivative (II) has the coupling constants also shown (H. Zemel and R. W. Fessenden, *J. phys. Chem.* **79**, 1419 (1975)):

Calculate the unpaired electron spin density around the rings.

19.23. The motion of nitroxide radicals trapped in a solid clathrate cage was examined at low temperatures and the hyperfine interaction of the electron with the nitrogen nucleus was found to depend on the orientation of the radical in the applied field (A. A. McConnell, D. D. MacNicol, and A. L. Porte, *J. Chem. Soc. A*, 3516 (1971)). When the field was perpendicular to the bond the coupling constant varied between 113.1 MHz and 11.2 MHz, and when it was parallel to the bond the coupling constant was 14.1 MHz. When the temperature was raised to 115 K motion about the parallel axis began to influence the appearance of the spectrum. How rapidly does the molecule rotate in the clathrate cage at that temperature? (This Problem is taken further in Problem 28.24 on p. 1000.)

19.24. Use the table of nuclear properties. Table 19.2, to predict the magnetic fields at which ^{1}H, ^{2}H, ^{13}C, ^{14}N, ^{19}F, and ^{31}P nuclei come into resonance with a radio-frequency field of (a) 60 MHz, (b) 300 MHz.

19.25. Calculate the relative population difference $\delta N/N$ for a proton in fields of 0.3 T, 1.5 T, and 7.0 T at (a) 4 K, (b) 300 K.

19.26. What magnetic field would be required in order to use an e.s.r. X-band (9 GHz) spectrometer to observe proton n.m.r. spectra, and a 60 MHz n.m.r. spectrometer to observe e.s.r.?

19.27. A scientist investigates the possibility of *neutron* spin resonance. He has available a commercial n.m.r. spectrometer operating at 60 MHz. Use the information in Table 19.2 to calculate the field required for resonance, and estimate the relative population difference at room temperature. Which is the lower spin state?

19.28. The chemical shift of the methyl protons in acetaldehyde (ethanal) is $\delta = 2.20$ p.p.m.; that of the aldehydic proton is $\delta = 9.80$ p.p.m. What is the difference in local magnetic field between the two regions of the molecule when the applied field is (a) 1.5 T, (b) 7.0 T?

19.29. What is the splitting (in Hz) between the methyl and aldehyde proton resonances when the spectrometer is operating at (a) 60 MHz, (b) 300 MHz?

19.30. Sketch the form of the ethanal n.m.r. spectrum using the δ-values quoted above and a spin–spin coupling constant of 2.90 Hz. Indicate the relative intensities of the lines. How does the appearance of the spectrum change when it is recorded at 300 MHz instead of 60 MHz?

19.31. The solid-state spectrum of gypsum shows the presence of a direct dipole–dipole magnetic interaction which varies with angle in accord with the expression quoted on p. 650. The spectrum can be interpreted in terms of an extra magnetic field of 0.715 mT generated by one proton and experienced by the other. What is the distance separating the protons in the solid?

19.32. In a normal liquid, molecular rotation averages the direct dipolar interaction to zero (p. 650). A molecule dissolved in a kind of liquid called a *liquid crystal* (p. 799) might not rotate freely in all directions, and so the dipolar interaction might not average to zero. Suppose a molecule is trapped so that, although the vector separating two protons may rotate freely around the direction of an applied field, the colatitude varies only between 0 and θ_{max}. Average the expression for the dipolar interaction over this restricted range of orientations, and confirm that the average disappears when θ_{max} is allowed to be π (free rotation over a sphere). What is the average value of the local dipolar field for the molecule in the last Problem if it were dissolved in a liquid crystal that enabled it to rotate up to $\theta_{max} = 30°$?

19.33. N.m.r. is used moderately often to obtain information about rate processes. Here is an example of the type of application. Two groups of protons are made equivalent by the isomerization of a molecule. At low temperatures, where the interconversion is slow, one group resonates with $\delta = 4.0$ p.p.m. and the other with $\delta = 5.2$ p.p.m. At what rate of interconversion will the two signals merge into a single line when the spectrometer is operating at 60 MHz?

19.34. The same compound as in the last Problem was investigated at different temperatures and in different spectrometers. It was found that the 60 MHz spectrum collapsed into a single line at 280 K, but at 300 MHz the temperature had to be raised to 300 K for the same effect. Estimate the activation energy (p. 23) for the interconversion.

19.35. The Mössbauer effect depends upon the recoil-less emission of a γ-ray. How important is it to embed the emitter nucleus in a bulky, rigid, crystal lattice?

Calculate the velocity of recoil of (a) a free ^{57}Fe atom, (b) a ^{57}Fe atom that is part of a rigid 100 mg crystal. Find the Doppler shift of the 14.4 keV γ-ray in each case.

19.36.   The Mössbauer spectrum of $Na_4Fe(CN)_6$ consists of a single line, while that of $Na_2Fe(CN)_5NO$ shows a pair of lines, each of about half the intensity and approximately symmetrically displaced about the original resonance. Account for these observations.

19.37.   The isomer shifts relative to Au in Pt in the Mössbauer spectra of AuI, AuBr, and AuCl were found to be 0.125 cm s^{-1}, 0.143 cm s^{-1}, and 0.161 cm s^{-1} respectively (V. G. Bhide, G. K. Shencry, and M. S. Multani, *Solid State Commun.* **2**, 221 (1964)). What does this imply about the ionic nature of these species?

20 Statistical thermodynamics: the concepts

Learning objectives

After careful study of this chapter you should be able to

(1) Define the terms *population*, *configuration*, and *weight* (p. 666).

(2) Explain the basis of finding the *most probable configuration*, and deduce the form of the *Boltzmann distribution* (eqn (20.1.8)).

(3) Define and calculate the *molecular partition function* (eqn (20.1.9)) and explain its significance (p. 671).

(4) Relate the thermodynamic *internal energy* to the partition function (eqn (20.1.12)).

(5) Calculate the internal energy of a collection of independent structureless particles (eqn (20.1.16)).

(6) Define the term *canonical ensemble* (p. 675).

(7) Explain the terms *thermodynamic limit* and *ergodic hypothesis* (p. 676).

(8) Define the *canonical partition function* (eqn (20.2.3)), and relate it to the *internal energy* of a system (eqn (20.2.5)).

(9) Relate the molecular partition function to the canonical partition function for *distinguishable* (eqn (20.2.6)) and *indistinguishable* (eqn (20.2.7)) molecules.

(10) Explain how the *thermal reservoir* controls the thermodynamic properties of the system (p. 679).

(11) State the molecular significance of *heat* and *work* (p. 682).

(12) Define the *statistical mechanical entropy* (eqn (20.3.7)).

(13) Relate the entropy to the canonical and molecular partition functions (eqns (20.3.10) and (20.3.11)).

(14) Show that the parameter β is equal to $1/kT$ (p. 683).

(15) Derive and use the *Sackur–Tetrode equation* for the entropy of a monatomic gas (eqn (20.3.14)).

(16) Explain the terms *Bose–Einstein* and *Fermi–Dirac* statistics (p. 690).

Introduction

The energies of atoms and molecules are confined to discrete values, and the last few chapters have shown how these energies may be calculated, determined spectroscopically, and related to the sizes and shapes of molecules. The next major step is to show how a knowledge of these permitted energy levels can be used to account for the behaviour of matter in bulk. In this chapter we set up the link between quantum theory and thermodynamics.

The crucial step in going from quantum mechanics to thermodynamics is the recognition that, whereas quantum mechanics deals with the *detailed* arrangement and motion of molecules, thermodynamics deals with their *average* behaviour. For example, the pressure exerted by a gas is interpreted as the average force per unit area exerted by the molecules, and in order to specify the pressure it is not necessary to know which molecule is colliding with the wall at some instant. Nor is it necessary to follow the fluctuations in the pressure as different numbers of molecules happen to collide with the wall at different instants, because the chance that these fluctuations are appreciable is very small—it is highly improbable that there will be a sudden lull in the number of collisions, or a sudden storm. *Statistical thermodynamics* is based on the principle that thermodynamic observables are averages of molecular properties, and it sets up a scheme for calculating these averages.

This chapter introduces statistical thermodynamics in three stages. The first, the derivation of the Boltzmann distribution for independent molecules, is of restricted applicability, but it has the advantage of taking us directly to a result of central importance in a straightforward and elementary way: statistical thermodynamics can be *used* once the Boltzmann distribution has been deduced. Then the arguments are elaborated a little (but not much) in order to accommodate systems composed of interacting molecules. Finally, the presentation is turned round and the basis of the subject is examined from an entirely different perspective.

The aim of these two chapters should not be lost from sight: it is to show how all the functions of thermodynamics—both those arising from the First Law, such as the internal energy, the enthalpy, and the pressure, and those based on the Second Law, such as the entropy, the Gibbs function, and equilibrium constants—can be calculated on the basis of structural information about the molecules involved.

20.1 Molecular energy levels and the Boltzmann distribution

Consider a system composed of N molecules. Although the total energy can be specified it is not possible to be definite about how that energy is distributed. Molecular collisions take place all the time, and result in the ceaseless redistribution of energy not only between the molecules but also among their different modes of motion. The closest we can come to a description of the distribution of energy is the specification of the numbers of molecules having particular energies. Thus at any instant there may be

n_i molecules in a state of energy ε_i, and in an equilibrium system this *population* remains virtually constant even though the identities of the molecules in that state may change with every collision. The problem we address in this section is the calculation of these populations for any type of molecule and at any temperature. The only restriction is that the molecules should be independent in the sense that the total energy of the system is the sum of the energies of the individual molecules. We are discounting (at this stage) the possibility that in a real system a contribution to the total energy may arise from interactions between molecules.

Any individual molecule may exist in states with the energies $\varepsilon_1, \varepsilon_2, \ldots$. We shall almost always take ε_1, the lowest state, as the zero of energy, $\varepsilon_1 = 0$, and measure all the other energies relative to it. This entails that we shall have to add some constant to the calculated energy of the system in order to arrive at the internal energy based on some other energy origin, U (for example, the zero point energy of oscillators may have to be added). In the sample as a whole there will be n_1 molecules in the state with energy ε_1, n_2 with ε_2, and so on. The specification of the set of populations n_1, $n_2, \ldots$ amounts to the statement of the *configuration* of the system.

The configuration of the system fluctuates with time because the populations change. We can envisage a large number of different configurations. One, for instance, might be $n_1 = N$, $n_2 = 0$, $n_3 = 0, \ldots$, or $\{N,0,0,\ldots\}$, corresponding to every molecule being in its lowest energy state. Another might be $n_1 = N-2$, $n_2 = 2$, $n_3 = 0, \ldots$, or $\{N-2,2,0,\ldots\}$. This is intrinsically more likely to occur than the first because it can be achieved in more ways. The first configuration can be achieved only in one way, because there is only one way of putting all the molecules into the state of energy ε_1. In contrast, the two molecules to go into ε_2 in the second configuration can be chosen in $\frac{1}{2}N(N-1)$ different ways, and so there are $\frac{1}{2}N(N-1)$ equally likely ways of forming it.* If the system were free to fluctuate from the first configuration to the second (for instance, under the influence of collisions), inspection would almost always result in it being found in the second. In other words, a system free to switch between the two configurations would show properties characteristic almost exclusively of the second.

A general configuration $\{n_1, n_2, \ldots\}$ can be achieved in W different ways, called its *weight*, where

(20.1.1) $W(n_1, n_2, \ldots) = N!/n_1!n_2!n_3!\ldots$

(Bear in mind that $0! = 1$.) This expression is a generalization of $\frac{1}{2}N(N-1)$, to which it reduces in the case $\{N-2,2,0,\ldots\}$. It is encountered in elementary statistics when calculating the number of ways in which N objects can be sorted into bins with n_i in bin i: now the objects are the molecules and the bins are their states.

---

* The number $\frac{1}{2}N(N-1)$ arises from simple combinatorial arguments: one candidate for entering ε_2 can be selected in N ways; there remain $N-1$ candidates for the second choice, leading to a total of $N(N-1)$ choices. But we should not distinguish the choice (Jack,Jill) from (Jill,Jack), and so only half the choices lead to distinguishable configurations.

We have seen that one configuration $\{N-2,2,0,\ldots\}$ dominates another, $\{N,0,0,\ldots\}$, and it is easy to believe that there may be others that greatly dominate both. Is there, in fact, a configuration with so great a weight that it so overwhelms all the rest in importance that there is negligible error in supposing that the system will always be found in it, and consequently that the system will display properties characteristic of that configuration to the virtual exclusion of all others? If that configuration exists, we ought to be able to find it by the usual rules of looking for the maximum value of a function (by examining where derivatives disappear). In this case the relevant function is the weight W, eqn (20.1.1).

Many conceivable configurations are inadmissible. If this were not the case, the maximum of W could be written down without any further work, for $\{1,1,1,\ldots\}$ has $W = N!$, the maximum value of the r.h.s. of eqn (20.1.1). Other analogous configurations such as $\{0,0,1,0,1,1,\ldots\}$ have the same value of W, and there are, it would appear, an infinite number of different configurations all having the same, maximum weight $N!$, and achieved by putting each molecule into a different, but arbitrarily selected state. The feature missing from the discussion so far is the fact that many configurations conflict with the specified total energy of the system. For instance, the configurations such as $\{N,0,0,\ldots\}$, $\{N-2,2,0,0,\ldots\}$, and $\{1,1,1,\ldots\}$, among numerous others, may be inadmissible because they cannot be achieved with the specified total energy. It follows that in looking for the configuration with the maximum weight, we have to ensure that it also satisfies the *total energy criterion*:

$$(20.1.2) \qquad \sum_i n_i \varepsilon_i = E,$$

where E is the total energy of the system.

The maximum value of W can be found by varying the n_i and looking for the values for which $\mathrm{d}W = 0$. But this immediately introduces another constraint, because although the individual populations may vary, the total number of molecules present in the system is fixed at N. In other words, we have to find a maximum of W subject not only to the energy constraint but also to the requirement that any variation of the n_i must not contravene the *total number criterion*:

$$(20.1.3) \qquad \sum_i n_i = N.$$

We are now equipped to start the calculation. We require the set of numbers n_1^*, n_2^*,... for which W has its maximum value, W^*. It is equivalent, and it turns out to be simpler, to find the conditions for which $\ln W$ is a maximum. Since $\ln W$ depends on all the n_i, when a configuration changes so that the n_i change to $n_i + \mathrm{d}n_i$, $\ln W$ changes to $\ln W + \mathrm{d}(\ln W)$, where

$$(20.1.4) \qquad \mathrm{d}(\ln W) = \sum_i (\partial \ln W / \partial n_i)\, \mathrm{d}n_i.$$

At a maximum the change $\mathrm{d}(\ln W)$ vanishes, but the presence of the two equations involving E and N means that when the n_i change they do so subject to the *constraints*

(20.1.5) $\sum_i \varepsilon_i \, dn_i = 0$ and $\sum_i dn_i = 0.$

These two equations prevent us from solving the equation for $d(\ln W)$ simply by setting all $(\partial \ln W/\partial n_i) = 0$ because the dn_i in eqn (20.1.4) are not all independent. The way to take constraints into account was devised by Lagrange and is called the *method of undetermined multipliers*. The basis of the method is described in Appendix A1 at the end of this chapter; all we need here is the very simple rule that *a constraint should be multiplied by some constant and then added to the main variation equation*, eqn (20.1.4). The variations are then treated as though they were independent, and the constants are evaluated at the end of the calculation.

The technique can be seen in operation as follows. The two constraints are multiplied by the constants $-\beta$ and α respectively and then added to eqn (20.1.4):

$$d(\ln W) = \sum_i (\partial \ln W/\partial n_i) \, dn_i + \alpha \sum_i dn_i - \beta \sum_i \varepsilon_i \, dn_i$$

$$= \sum_i \{(\partial \ln W/\partial n_i) + \alpha - \beta \varepsilon_i\} \, dn_i.$$

Following the rules, all the dn_i are now treated as independent. In that case, the only way of satisfying $d(\ln W) = 0$ is to require that for every i

(20.1.6) $(\partial \ln W/\partial n_i) + \alpha - \beta \varepsilon_i = 0$

for the n_i taking their most probable values n_i^*.

In order to solve the last equation we make an approximation which is very good indeed for large numbers. A simplified form of *Stirling's approximation* is that

(20.1.7)   for x large: $\ln x! \approx x \ln x - x.$

use of this approximation in the expression for $\ln W$ gives

$$\ln W = \ln(N!/n_1! n_2! \ldots) = \ln N! - \ln(n_1! n_2! \ldots)$$

$$= \ln N! - \sum_j \ln n_j! \approx (N \ln N - N) - \sum_j (n_j \ln n_j - n_j)$$

$$\approx N \ln N - \sum_j n_j \ln n_j$$

because the sum over the n_j is equal to N. Since N is a constant, differentiation with respect to n_i gives

$$(\partial \ln W/\partial n_i) \approx -\sum_j \{\partial(n_j \ln n_j)/\partial n_i\}$$

$$\approx -\sum_j \{(\partial n_j/\partial n_i) \ln n_j + n_j(\partial \ln n_j/\partial n_i)\} \approx -\{\ln n_i + 1\}$$

because, for $j \neq i$, n_j is independent of n_i so that $(\partial n_j/\partial n_i) = 0$; and for $j = i$

the differential coefficient is unity; furthermore, $(\partial \ln n_j / \partial n_i) = (1/n_j)(\partial n_j / \partial n_i)$. The 1 in the last line of the last equation can be neglected in comparison with $\ln n_i$, and so eqn (20.1.6) becomes

$$-\ln n_i + \alpha - \beta \varepsilon_i = 0$$

when $n_i = n_i^*$, its most probable value. The most probable population of the state of energy ε_i is therefore

$$n_i^* = \exp(\alpha - \beta \varepsilon_i).$$

The final step is to find the values of α and β. Since

$$N = \sum_i n_i^* = \sum_i e^\alpha e^{-\beta \varepsilon_i},$$

it follows that

$$e^\alpha = N \Big/ \sum_i e^{-\beta \varepsilon_i}.$$

Hence we arrive at the form of the *Boltzmann distribution*:

(20.1.8)
$$n_i^*/N = e^{-\beta \varepsilon_i} \Big/ \sum_i e^{-\beta \varepsilon_i}.$$

The identification of the parameter β with $1/kT$, where T is the absolute temperature, can be made on the basis of a variety of arguments. The best is in terms of the statistical definition of the entropy, and so we shall set it aside until that has been introduced (p. 681).

The molecular partition function. In eqn (20.1.8) we have arrived at a result that is central to statistical thermodynamics, for it enables the most probable populations of the molecular states in a system to be specified at any temperature. The sum over $\exp(-\beta \varepsilon_i)$ in the denominator is so important that it carries a special name, the *molecular partition function*, q:

(20.1.9)
$$\textit{molecular partition function: } q = \sum_i \exp(-\beta \varepsilon_i).$$

This function is sometimes expressed in a slightly different form. Note that the sum in eqn (20.1.9) is over the *states* of the individual molecules. It may happen that several states have the same energy, and so give the same contribution to the sum. If, for instance, the energy ε_j arises from g_j different states (that is, the energy level is g_j-*fold degenerate*), there are g_j terms in the sum all with the value $\exp(-\beta \varepsilon_j)$. We could therefore write

(20.1.10)
$$q = \sum_j g_j \exp(-\beta \varepsilon_j),$$

where the sum is now over the different *sets* of energy levels rather than the individual states. We shall normally employ the former notation, and sum over states.

Example (Objective 3). Write an expression for the rotational partition function of a linear, heteronuclear molecule.

- *Method.* The energy of a rotating linear molecule is $\varepsilon_{JM} = hcBJ(J+1)$, $J = 0, 1, 2, \ldots$. Every J-level has $2J+1$ states, corresponding to $2J+1$ different orientations of the rotating molecule: these states are distinguished by the quantum number M_J. The energies are independent of M_J, and so every J-level is $(2J+1)$-fold degenerate. Therefore, sum either over J and M_J, or only over J, but in the latter case multiplying each term in the sum by the degeneracy $g_J = (2J+1)$.

- *Answer.* $q = \sum_{J,M_J} \exp[-\beta hcBJ(J+1)] = \sum_{J=0}^{\infty} \sum_{M_J=-J}^{J} \exp[-\beta hcBJ(J+1)]$

$$= \sum_{J=0}^{\infty} (2J+1)\exp[-\beta hcBJ(J+1)].$$

The last line can be obtained directly by noting that $g_J = 2J+1$.

- *Comment.* Be careful with degeneracies: in a rotating spherical top molecule the degeneracy of a level J is $(2J+1)^2$.

The importance of the molecular partition function is that it contains all the thermodynamic information about a system of independent particles at thermal equilibrium. We have already seen that it appears in the Boltzmann expression for the most probable population of states. In terms of q we have

(20.1.11) $\qquad n_i^*/N = (1/q)e^{-\beta\varepsilon_i}.$

With the relative populations of energy levels established it is a simple matter to obtain an expression for the total energy of the system. The total energy for any specified configuration is

$$E = \sum_i n_i \varepsilon_i.$$

If we concentrate on the most probable configuration of the system, this becomes

$$E = (N/q)\sum_i \varepsilon_i \exp(-\beta\varepsilon_i).$$

The sum can be manipulated into a form involving only q. First notice that

$$(d/d\beta)\exp(-\beta\varepsilon_i) = -\varepsilon_i \exp(-\beta\varepsilon_i).$$

Then

$$E = (N/q)\sum_i (-d/d\beta)\exp(-\beta\varepsilon_i). = -(N/q)(d/d\beta)\sum_i \exp(-\beta\varepsilon_i).$$

$$= -(N/q)(dq/d\beta).$$

There are several points to be made about this result. In the first place, since $\varepsilon_1 = 0$ by definition, the value of E should be interpreted as the difference between the actual internal energy and its value at absolute zero (when only the lowest energy state is occupied). That is, $U = U(0) + E$. Furthermore, the partition function depends on quantities like the volume of the system (as we shall see explicitly soon), and so the derivative with respect to β is really a partial derivative with the volume (and perhaps some other parameters, such as external fields) held constant. The final expression relating the molecular partition function to the thermodynamic internal energy of a collection of independent particles is therefore

(20.1.12) $U - U(0) = -(N/q)(\partial q/\partial \beta)_V = -N(\partial \ln q/\partial \beta)_V.$

The last equation confirms that we need know only the partition function in order to calculate the internal energy. The next chapter will expand this by going on to show that *all* thermodynamic functions can be calculated once we know q (and its generalization for interacting molecules). In that respect q plays a role very similar to the wavefunction of quantum mechanics, which contains all the dynamical information about an individual molecule. The analogy can be made even sharper by writing the expression for the mean energy per molecule, $\varepsilon = E/N$.

$\varepsilon q = -(\partial q/\partial \beta)_V$

and noticing its striking resemblance to the time-dependent Schrödinger equation (p. 425):

$H\psi = -(\hbar/i)(\partial \psi/\partial t),$

where ψ is the wavefunction and H is the energy operator for the molecule. You should take pleasure from noting that temperature (essentially β) is imaginary time.

An interpretation of the partition function. Some understanding of the significance of the molecular partition function can be obtained first by considering its range of values, and then by calculating its value in a simple case.

When the temperature of the system is close to absolute zero the parameter $\beta = 1/kT$ is close to infinity. Then every term except one in the sum

$q = \sum_i e^{-\beta \varepsilon_i}$

is zero because each one has the form e^{-x} with $x \to \infty$. The exception is the term with $\varepsilon_1 \equiv 0$, because then $\varepsilon_1/kT = 0$ for all values of T, including the limit of $T \to 0$. As there is only one surviving term when $T = 0$, and its value is 1, it follows that

at $T = 0$, $q = 1$.

At the other extreme, consider the case when T is so large that for every term in the sum $\varepsilon_i/kT \to 0$. Since $e^{-x} = 1$ when $x = 0$, every term in the sum then contributes the value 1. It follows that the sum is equal to the number of molecular states, which in general is infinite:

as $T \to \infty$, $q \to \infty$.

(In some cases the system has only a finite number of energy levels; then the partition function has an upper limit equal to the number of states.)

We see that the partition function gives an indication of *the number of states that are thermally accessible to the system at the temperature of interest*. At absolute zero, only the ground state is accessible, and $q = 1$. At the highest temperatures virtually all the states are thermally accessible, and so q approaches infinity.

As an explicit example of the calculation of q and its role as an indicator of the range of thermally accessible states, consider a simple energy spectrum where the levels form a uniform ladder with separation ε, Fig. 20.1 (we shall see soon that this is less artificial than it might seem). The partition function takes the form

$$q = 1 + e^{-\beta\varepsilon} + e^{-2\beta\varepsilon} + \ldots$$

$$= 1 + (e^{-\beta\varepsilon}) + (e^{-\beta\varepsilon})^2 + \ldots$$

This geometrical progression can be summed explicitly (or alternatively we can recognize the expansion $(1-x)^{-1} = 1 + x + x^2 + \ldots$), and so

(20.1.13) $q = (1 - e^{-\beta\varepsilon})^{-1}$.

It follows that the proportion, P_j, of molecules in the state with energy ε_j is given by

(20.1.14) $P_j = n_j^*/N = (1 - e^{-\beta\varepsilon})e^{-\beta\varepsilon_j}$

and we can arrive at numerical values of both q and P_j by substituting the appropriate value of the temperature. This has been done in Fig. 20.2, where the populations of the states at various temperatures are shown and the distributions of molecular states have been labelled with the numerical value of q calculated from eqn (20.1.13). At very low temperatures q is close to unity, as already anticipated, and the only significant population is in the lowest state. As the temperature is raised, the population breaks out of the lowest state, and the upper states become progressively populated. As this happens the value of the partition function rises from unity, and so its value gives a general idea of the dispersal of populations. At very high temperatures the population is widely distributed over numerous states (in the diagram, several are significantly populated) and the partition function has a correspondingly larger value.

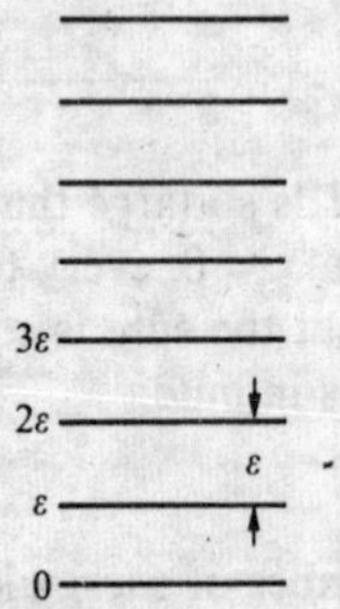

Fig. 20.1. The array of molecular energy levels used to calculate the molecular partition function in eqn (20.1.13).

Example (Objective 3). Find the proportion of iodine molecules in the ground, first excited, and second excited vibrational states at 25 °C.

- *Method.* The vibrational energy levels have a constant separation between neighbours of 214.6 cm^{-1}, and so they correspond to the type of system just described. Take $\varepsilon \triangleq 214.6 \text{ cm}^{-1}$ and identify j with the vibrational quantum number v. The zero of energy can be defined as the state with $v = 0$.

- *Answer.* For future problems, as well as this, it will be convenient to have kT in cm^{-1}:

$$(kT/hc)_{298.15\text{K}} = \frac{(1.380\,66 \times 10^{-23} \text{ J K}^{-1}) \times (298.15 \text{ K})}{(6.626\,18 \times 10^{-34} \text{ J s}) \times (2.997\,925 \times 10^{10} \text{ cm s}^{-1})} = 207.223 \text{ cm}^{-1}.$$

For the present problem we require

$$\varepsilon\beta = (214.6 \text{ cm}^{-1})/(207.223 \text{ cm}^{-1}) = 1.036.$$

The relative populations are then given by

$$P_v = (1 - e^{-\beta\varepsilon}) e^{-v\beta\varepsilon} = 0.645 \exp(-1.036v).$$

Hence, $P_0 \doteq 0.645$, $P_1 = 0.229$, $P_2 = 0.081$, etc.

- *Comment.* The iodine bond is not strong and the atoms are heavy: as a result the vibrational energy separations are small, and even at room temperature several vibrational levels are significantly populated. Note how the value of the partition function (1.550) reflects this small but significant population spread.

The translational partition function. As an example of the use of the formula for q we apply it to a perfect monatomic gas in a container of volume V. The length of the container in the x-direction is X (and similarly Y, Z for the other two directions, with $XYZ = V$). The partition function for any one of the molecules is

$$q = \sum_j \exp(-\beta\varepsilon_j),$$

where ε_j is a translational energy of the molecule in the container and j the quantum number labelling the state. Since ε_j is the sum of the kinetic energies in three dimensions, $\varepsilon_j = \varepsilon_{j(X)} + \varepsilon_{j(Y)} + \varepsilon_{j(Z)}$, the partition function factorizes as follows:

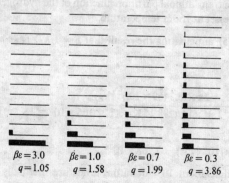

$\beta\varepsilon = 3.0$	$\beta\varepsilon = 1.0$	$\beta\varepsilon = 0.7$	$\beta\varepsilon = 0.3$
$q = 1.05$	$q = 1.58$	$q = 1.99$	$q = 3.86$

Fig. 20.2. Populations of molecular states at various temperatures.

$$q = \sum_{\text{all } j} \exp\{-\beta\varepsilon_{j(X)} - \beta\varepsilon_{j(Y)} - \beta\varepsilon_{j(Z)}\}$$

$$= \left\{\sum_{j(X)} \exp(-\beta\varepsilon_{j(X)})\right\}\left\{\sum_{j(Y)} \exp(-\beta\varepsilon_{j(Y)})\right\}\left\{\sum_{j(Z)} \exp(-\beta\varepsilon_{j(Z)})\right\}$$

$$= q_X q_Y q_Z.$$

We use the discussion of the energy levels of a particle in a box, p. 404, to introduce

$$\varepsilon_{j(X)} = j^2(h^2/8mX^2), \quad j = 1, 2, 3, \ldots,$$

where m is the molecule's mass. Then the sum to evaluate is

$$q_X = \sum_{j=1}^{\infty} \exp(-j^2 h^2 \beta/8mX^2).$$

In a container the size of a typical laboratory vessel the translational energy levels are very close, and so the sum can be replaced by an integral:

$$q_X = \int_1^{\infty} \exp(-j^2 h^2 \beta/8mX^2)\,dj.$$

Extending the lower limit to $j = 0$ makes a negligible error but turns the integral into a standard form. This is achieved by making the substitution $x^2 = j^2 h^2 \beta/8mX^2$, implying $dj = (8mX^2/h^2\beta)^{1/2}\,dx$:

$$q_X = (8mX^2/h^2\beta)^{1/2} \int_0^{\infty} \exp(-x^2)\,dx = (8mX^2/h^2\beta)^{1/2}(\pi^{1/2}/2).$$

This is the molecular partition function for translational motion along the x-direction. The only change for the other two degrees of freedom is the replacement of X by Y or Z, and so the partition function for motion in three dimensions is

$(20.1.15)°$ $q = (2\pi mX^2/h^2\beta)^{1/2}(2\pi mY^2/h^2\beta)^{1/2}(2\pi mZ^2/h^2\beta)^{1/2} = (2\pi m/h^2\beta)^{3/2}V.$

Example (Objective 3). Calculate the translational partition function of the hydrogen molecule confined to a $100\,cm^3$ vessel at room temperature.

- *Method.* Use eqn (20.1.15), anticipating the result that $\beta = 1/kT$. Take $m(H_2)$ from the final endpaper.

- *Answer.*

$$q = \left\{\frac{2\pi \times (3.3469 \times 10^{-27}\,kg)}{(6.626\,18 \times 10^{-34}\,Js)^2 \times [1/(298.15\,K) \times (1.380\,66 \times 10^{-23}\,JK^{-1})]}\right\}^{3/2}$$
$$\times (100\,cm^3)$$
$$= (1.976 \times 10^{20})^{3/2} \times (10^{-4}\,m^3) = 2.768 \times 10^{26}.$$

- *Comment.* This shows that about 10^{26} quantum levels are thermally accessible, even at room temperature, for this light molecule.

Several applications of this simple expression will be met in the next chapter. For the present we illustrate how to calculate the internal energy of the gas using eqn (20.1.13). First find $(\partial q/\partial \beta)_V$:

$$(\partial q/\partial \beta)_V = \frac{\partial}{\partial \beta}\left\{\left(\frac{2\pi m}{h^2\beta}\right)^{3/2} V\right\} = (2\pi m/h^2)^{3/2}V(-\tfrac{3}{2}\beta^{-5/2}).$$

Consequently, by (20.1.12),

(20.1.16)° $\qquad U - U(0) = -\dfrac{N(2\pi m/h^2)^{3/2}V(-\tfrac{3}{2}\beta^{-5/2})}{(2\pi m/h^2)^{3/2}V(\beta^{-3/2})} = 3N/2\beta.$

Therefore, if we knew β, we could immediately give the value of the internal energy of a sample of a perfect monatomic gas of N particles. If we anticipate the result that $\beta = 1/kT$, then the internal energy is $3NkT/2$, or $(3/2)nRT$, the equipartition value quoted on p. 15.

Example What is the constant-volume heat capacity of a monatomic gas?

- *Method.* By definition, $C_V = (\partial U/\partial T)_V$. Anticipate that $\beta = 1/kT$. The last equation gives an expression for U.

Answer. $C_V = (3N/2)\,\mathrm{d}(1/\beta)/\mathrm{d}T = (3N/2)\,\mathrm{d}(kT)/\mathrm{d}T = (3/2)Nk = (3/2)nR.$

The molar heat capacity is therefore $C_{V,\mathrm{m}} = 3R/2 = 12.47\,\mathrm{J\,K^{-1}\,mol^{-1}}$.

- *Comment.* This agrees almost exactly with experimental values for monatomic gases. In more complex molecules other modes of motion contribute to U and therefore also to C_V. This is treated in the next chapter.

20.2 Refinements: ensembles and reservoirs

In this section we show how the conclusions of the last section may be generalized in order to accommodate systems composed of interacting particles. The crucial new concept is the *ensemble*. Like so many scientific terms this has basically its everyday meaning, but sharpened and refined into a precise significance.

In order to set up the ensemble we first think of the system, and the much larger reservoir with which it is in thermal contact, as constituting an isolated unit, and then form the ensemble by reproducing this unit many times. Every reproduction has exactly the same temperature, and the system within each unit has exactly the same volume and constitution. This imaginary collection of replications is called the *canonical ensemble*. The point about the ensemble is that it is a set of *imaginary* replications, and so we are free to let the number of members, $\mathbb{N}$, be as large as we like, and when appropriate we let the number become infinite.* (Note that the

---

*There are two other important types of ensemble used in statistical thermodynamics. In the *microcanonical ensemble* the condition of constant temperature is replaced by the requirement that every system in each member of the ensemble has exactly the same energy: every system is itself isolated. In the *grand canonical ensemble* the volume of every system in each unit is the same, each system is in contact with a thermal reservoir of temperature T (as in the canonical ensemble), but matter can be imagined as able to pass between each unit so that the composition of each system can fluctuate.

number $\mathbb{N}$ is quite unrelated to N. N is determined by the number of molecules in the actual system; $\mathbb{N}$ is the number of imaginary repetitions of that system.) The point of introducing the canonical ensemble is as follows. In the first place we know very little about the actual state of the system at any time. If it were totally isolated, it would be in a definite quantum state, and it would remain in that state. But it is not isolated; it is in contact with a reservoir, and the exchange of energy that this allows corresponds to the system being jostled among its quantum states. Not only do we not know the state of the system when it was first prepared, we are also unable to calculate its detailed evolution in time.

The ensemble is a way round this mountain of difficulty. First we identify the *time average* of properties of the system as the measurable, bulk thermodynamic properties. Thus the time average of the system's energy is identified as its thermodynamic internal energy U. Then we assume that we get exactly the same value for the bulk property if, instead of evaluating the time average of a single system, we *average the property over all members of the ensemble at some single instant*, taking the limit $\mathbb{N} \to \infty$. The introduction of the ensemble has the effect of turning a tricky time-dependent problem into one that is time-independent, and time-independent calculations are much easier to handle. The assumption that the two averaging procedures are equivalent is called the *ergodic hypothesis*, and its validity has been the subject of a great deal of erudite discussion.

Dominating configurations. Once the ensemble has been conceived, the argument runs along the same lines as in the last section, but instead of having to think of some overall energy E of the system distributed over molecules we have to think of some overall ensemble energy $\mathbb{E}$ distributed over the systems. If this total energy is $\mathbb{E}$ and there are $\mathbb{N}$ members of the ensemble, the average energy of any member is $\mathbb{E}/\mathbb{N}$, and it is this quantity which is to be identified with the thermodynamic internal energy in the thermodynamic limit of $\mathbb{N}$ (and $\mathbb{E}$) approaching infinity.

We think of the $\mathbb{N}$ systems of the ensemble as being in thermal contact with each other, Fig. 20.3, so that all the other systems are, in effect, the thermal reservoir for any particular system. This thermal contact allows the energies of individual members of the ensemble to fluctuate, but as

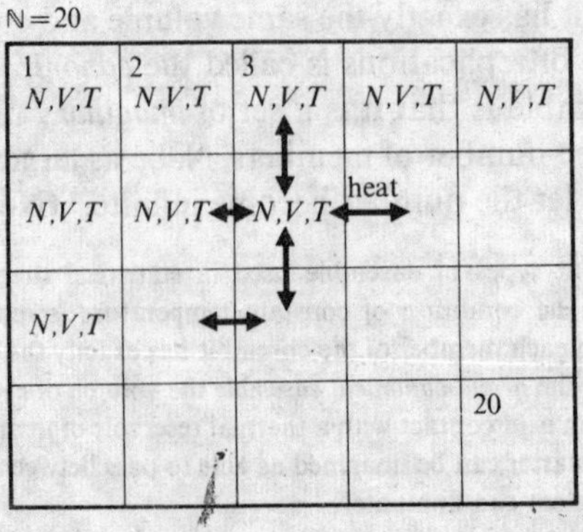

$\mathbb{N} = 20$

1 N,V,T	2 N,V,T	3 N,V,T	N,V,T	N,V,T
N,V,T	$N,V,T \leftrightarrow N,V,T \leftrightarrow$	heat		
N,V,T	$\leftrightarrow$			
			20	

Fig. 20.3. The canonical ensemble ($\mathbb{N} = 20$).

the entire ensemble is isolated from the rest of the world, its total energy has the constant value $\mathbb{E}$. If the number of members of the ensemble being in a state i with energy E_i is denoted $\mathfrak{m}_i$, we can speak of the configuration of the ensemble and the weight $\mathbb{W}$ of each configuration. Just as in the single molecule case, some of these configurations are much more probable than others. For instance, it is very unlikely that the whole of the energy $\mathbb{E}$ will happen to accumulate in only a few members. By analogy with the earlier case, we find the most probable configuration (with weight $\mathbb{W}^*$) of the ensemble, and evaluate all the thermodynamic properties by taking an average over the ensemble having that complexion. Then since the most probable configuration is much more probable than the complexions deviating significantly from it, we can be confident that the results are reliable. Since $\mathbb{N}$ is the number of *imaginary* replications of the system, we are free to let it become infinite. In this *thermodynamic limit* the most probable configuration is overwhelmingly the most probable.

The quantitative discussion proceeds exactly as in the earlier case, with the modification that in place of N, n_i we now have $\mathbb{N}, \mathfrak{m}_i$; if we disregard the question of the total amount of energy, the weight of a configuration in which $\mathfrak{m}_1$ of the members of the ensemble are in state 1 with energy E_1, $\mathfrak{m}_2$ are in state 2 with energy E_2, etc., is

(20.2.1) $$\mathbb{W}(\mathfrak{m}_1, \mathfrak{m}_2, \ldots) = \mathbb{N}!/\mathfrak{m}_1!\mathfrak{m}_2!\ldots$$

We then seek the configuration $\mathfrak{m}_1^*, \mathfrak{m}_2^*, \ldots$ corresponding to the maximum value of this weight, $\mathbb{W}^*$, subject to the two conditions that the total energy of the ensemble is the constant $\mathbb{E}$ and the total number of members is the constant $\mathbb{N}$. The outcome of the calculation has the same form as before: the most probable configuration of the ensemble is the one in which the populations are given by

(20.2.2) $$\mathfrak{m}_i^*/\mathbb{N} = \exp(-\beta E_i)\Big/\sum_i \exp(-\beta E_i).$$

This is called the *canonical distribution*.

The shape of the canonical distribution, apparently an exponentially decreasing function of energy, is not entirely what casual inspection of eqn (20.2.2) might suggest. It has to be noted that the expression applies to the probability of occurrence of a *single state i* (of the entire system) of energy E_i. There may in fact be numerous states with almost identical energies, and in fact the number of such states (which is normally expressed in terms of the *density of states*, the number of states per unit energy range) is a very sharply rising function of the energy. It follows that the probability of a member of the ensemble having a specified *energy* (as distinct from being in a specified state) is $\mathfrak{m}_i^*/\mathbb{N}$, a sharply decreasing function, multiplied by a sharply increasing function. It follows that the distribution is a sharply peaked function at some energy: most members of the ensemble have an energy very close to the mean value.

The canonical partition function. The sum of $\exp(-\beta E_i)$ over the states of the members of the ensemble is called the *canonical partition function*, Q:

(20.2.3)

> *canonical partition function,* $Q = \sum_i \exp(-\beta E_i)$.

In terms of this notation, the proportion, $\mathbb{p}_i$, of members that are in the state i with energy E_i in an ensemble having the most probable configuration is

(20.2.4) $\mathbb{p}_i = \mathbb{n}_i^*/\mathbb{N} = (1/Q)\exp(-\beta E_i)$.

Like the molecular partition function, the canonical partition function carries all the thermodynamic information about the system, but it is more general because nowhere have we had to regard the system as a collection of independent molecules. As in the molecular case the mean energy of the ensemble, which is to be identified with the thermodynamic internal energy of the system, is related to Q by

$$U - U(0) = \sum_i \mathbb{p}_i E_i = (1/Q)\sum_i E_i \exp(-\beta E_i).$$

The sum can be expressed in terms of the derivative of Q using the same argument as was used for q on p. 670:

(20.2.5) $U - U(0) = -(1/Q)(\partial Q/\partial \beta)_V = -(\partial \ln Q/\partial \beta)_V$.

The canonical partition function for independent molecules. We now show how the canonical partition function is related to the molecular partition function when the system is composed of independent molecules. The energy of any member of the ensemble is then the sum of the energies of all its N individual molecules:

$$E_i = \varepsilon_1^{(i)} + \varepsilon_2^{(i)} + \ldots + \varepsilon_N^{(i)}$$

where $\varepsilon_1^{(i)}$ is the energy of molecule 1 in the member of the ensemble happening to be in the state i of energy E_i. The canonical partition function is then

$$Q = \sum_i \exp\{-\beta(\varepsilon_1^{(i)} + \varepsilon_2^{(i)} + \ldots + \varepsilon_N^{(i)})\},$$

where the sum is over all the states available to the system. All these states can be covered by letting every molecule enter every one of its own individual states (although we meet an important proviso below). Therefore, instead of summing over the collective states i we sum over all the individual *molecular* states, ε_{1j} for molecule 1, ε_{2j} for molecule 2, and so on. This leads to

$$Q = \left(\sum_j e^{-\beta \varepsilon_{1j}}\right)\left(\sum_j e^{-\beta \varepsilon_{2j}}\right)\ldots\left(\sum_j e^{-\beta \varepsilon_{Nj}}\right).$$

If all the molecules are the same it is unnecessary to label them, and the last line reduces to

$$Q = \left(\sum_j e^{-\beta \varepsilon_j}\right)\left(\sum_j e^{-\beta \varepsilon_j}\right) \cdots \left(\sum_j e^{-\beta \varepsilon_j}\right) = \left(\sum_j e^{-\beta \varepsilon_j}\right)^N.$$

We immediately recognize the appearance of the molecular partition function q, p. 649; consequently

(20.2.6)

$$Q = q^N.$$

Unfortunately the derivation of eqn (20.2.6) breaks down in a very important case. This can be seen as follows. If all the molecules are the same and free to move, we cannot distinguish one from another. Suppose molecule 1 is in some state a, molecule 2 in a state b, and molecule 3 in a state c, then one member of the ensemble would have the energy $\varepsilon_a + \varepsilon_b + \varepsilon_c$. This member, however, is indistinguishable from one formed by putting molecule 1 in state b, molecule 2 in state c, and molecule 3 in state a, or some other permutation. In the case of indistinguishable molecules, it follows that we have counted too many states in going from the sum over members of the ensemble to a sum over molecular states, and so writing $Q = q^N$ overestimates Q. The detailed argument is quite involved, Appendix A2 at the end of this chapter, but at temperatures well above absolute zero it turns out that the correction factor is $1/N!$, and so, *for indistinguishable molecules,*

(20.2.7)

$$Q = q^N/N!.$$

When are molecules distinguishable, so that we use eqn (20.2.6), and when are they indistinguishable, so that we use eqn (20.2.7)? In the first place they must be of the same species. An argon atom is never indistinguishable from a methane molecule or a neon atom. Their identity, though, is not the only criterion. If the identical atoms are arranged in a crystal lattice we can label every one of them with a set of coordinates. Identical molecules held stationary in a lattice are therefore distinguishable, and we use eqn (20.2.6). On the other hand, identical molecules free to move in a gas are indistinguishable because there is no way of keeping track of the identity of a given molecule. In that case we use eqn (20.2.7). We shall see the consequence of this rule in due course.

Another viewpoint: the role of the reservoir. In this section we present another viewpoint which emphasizes the controlling role played by the reservoir with which the system of interest is in thermal contact.

Consider a system with energy E_i in contact with a reservoir such that the total energy of the entire isolated unit is E_{total}. The energy of the reservoir is $E_{total} - E_i$. We also conceive of an ensemble of replications of this unit, each replication having the same energy E_{total}, but not necessarily having the same partition of the energy between the system and the reservoir. We seek the probability that if any member of the ensemble is selected, it will be found to be in the energy state E_i.

It is plausible to suppose that the probability $P(E_i)$ of the *system* being in a given state with energy E_i is proportional to the number of ways that the *reservoir* can adjust itself to accommodate the remainder of the energy. If the reservoir can accommodate this energy in $W'(E_{total} - E_i)$ different ways, that is, it has $W'(E_{total} - E_i)$ states of energy $E_{total} - E_i$, it follows that

$$P(E_i) = CW'(E_{total} - E_i),$$

where C is some constant. The statement just made is one that underlies much of statistical thermodynamics and is called the *principle of equal a priori probabilities*. It asserts that, *unless there is any information to the contrary, all possibilities are assumed equally probable*. In the present case, if the states are there, they can all be occupied without favouring one particular type or another (e.g., without favouring states that correspond to a lot of vibrational motion, or states of the same energy but which correspond to less vibration and more translation).

As the reservoir is exceedingly large its energy is very much larger than the energy E_i of the system. Since E_i is much smaller than $E_{total} - E_i$, and therefore much smaller than E_{total}, the number of states of the reservoir at $E_{total} - E_i$ can be related to the number at E_{total} by a Taylor expansion. In fact it is better to work with logarithms (which change less sharply than the quantities themselves), and so we write

$$\ln W'(E_{total} - E_i) = \ln W'(E_{total}) - \left(\frac{\partial \ln W'}{\partial E}\right)_{E_{total}} E_i + \dots.$$

The other terms in the series are higher powers of E_i and are negligible because E_i is so small. Furthermore, the differential coefficient is independent of the state of the system, and depends only on the total energy of system plus reservoir. Therefore it can be written as a constant β,

$$(20.2.8) \qquad \beta = \left(\frac{\partial \ln W'}{\partial E}\right)_{E_{total}}.$$

This leads to

$$\ln W'(E_{total} - E_i) = \ln W'(E_{total}) - \beta E_i$$

or

$$W'(E_{total} - E_i) = W'(E_{total}) e^{-\beta E_i}$$

so that

$$P(E_i) = CW'(E_{total}) e^{-\beta E_i}.$$

The number $W'(E_{total})$, a property of the reservoir, is independent of the state of the system, and so may be combined with C to give a new constant C'. This constant can be determined from the condition that the total probability of finding the system in some state must be unity:

$$\sum_i P(E_i) = C' \sum_i e^{-\beta \varepsilon_i} = 1,$$

so that

$$C' = 1 \Big/ \sum_i e^{-\beta E_i}.$$

The final result for the probability of selecting a system with energy E_i from the ensemble is therefore

(20.2.9) $$P(E_i) = e^{-\beta \varepsilon_i} \Big/ \sum_i e^{-\beta \varepsilon_i},$$

which is precisely the canonical distribution again.

Note the central role played by the reservoir: the parameter β, which has the dimensions 1/energy, eqn (20.2.8), is defined solely in terms of the thermal reservoir, which in turn is characterized by a single quantity, its temperature T. We shall see later that β can be identified as $1/kT$. Since the distribution is defined in terms of the property of the reservoir we can expect it to be of very wide significance: *any* system, of whatever composition and of any smallness, in contact with a reservoir of temperature T will be distributed in accord with eqn (20.2.9).

We shall shortly turn to the statistical definition of entropy, and eqn (20.2.8) points the way. Recall from Chapter 6, eqn (6.1.4), that

$$(\partial S/\partial U)_V = 1/T.$$

Equation (20.2.8) is exactly of this form if β is identified as $1/kT$ and $\ln W$ identified as S/k. In the next section we pursue this connection, and show that the entropy can be calculated from $k \ln W$. The whole of chemical thermodynamics is then open to calculation on the basis of a knowledge of molecular energy levels.

20.3 Statistical thermodynamics and the Second Law

Up to this point we have concentrated on the connection between the partition function and the First Law concept of internal energy. We have seen that U may be calculated from the partition function and its temperature dependence. The claim that the partition function contains *all* thermodynamic information can be substantiated only if it can be used to discuss the entropy, the principal Second Law concept.

Heat, work, and the statistical entropy. When entropy was introduced in Chapter 5, it was presented as a measure of the *distribution* of energy. Since the partition function is also a measure of the distribution of energy, it is reasonable to suspect a relation between them. The nature of the relation can be discovered by thinking about the connection between the internal energy U and the energy levels and populations.

In order to proceed we write the general expression for a change in the internal energy of the system when it interacts with its surroundings. Since

$$U - U(0) = (1/\mathbb{N})\sum_i \mathfrak{m}_i^* E_i, \qquad \mathbb{N} \to \infty,$$

a change dU may arise either from a modification of the energy levels of the system (so that E_i, the energy of system i in the ensemble, changes by dE_i) or from changing the number of members of the ensemble in the state i (so that $\mathfrak{m}_i^*$ changes by $d\mathfrak{m}_i$). The most general change is therefore

(20.3.1) $\qquad dU = dU(0) + (1/\mathbb{N})\sum_i \mathfrak{m}_i^*\, dE_i + (1/\mathbb{N})\sum_i E_i\, d\mathfrak{m}_i.$

From thermodynamics we know that for a reversible change

(20.3.2) $\qquad dU = dq_{rev} + dw_{rev},$

and so there must be some connection between the two quantities. The connection is established by realizing that *energy levels of a system change when its size is changed*: they do not change when the system is heated, Fig. 20.4. On the other hand, *when the system is heated* the populations of the states change (that is, the most probable configuration is modified), but the energy levels themselves remain the same. This points to the identifications

(20.3.3) $\qquad dq_{rev} = (1/\mathbb{N})\sum_i E_i\, d\mathfrak{m}_i$

(20.3.4) $\qquad dw_{rev} = dU(0) + (1/\mathbb{N})\sum_i \mathfrak{m}_i^*\, dE_i.$

We see that *the reversible transfer of heat corresponds to the redistribution of populations among fixed energy levels*, and *reversible work of expansion or compression corresponds to the change of the energy levels themselves while the populations remain the same*. Since

(20.3.5) $\qquad dS = dq_{rev}/T,$

by equating the two expressions for dq_{rev} a relation can be established between the *change of entropy* and the change of the configuration of the ensemble:

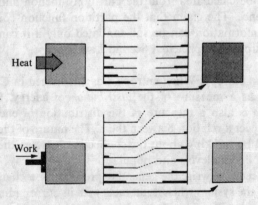

Fig. 20.4. The difference between the effects of heat and work.

(20.3.6) $T\,dS = (1/\mathbb{N})\sum_i E_i\,d\mathfrak{m}_i.$

We now show that the same expression for dS can be derived if an entropy is *defined* by the expression

(20.3.7)   *statistical entropy:* $S = (k/\mathbb{N})\ln \mathbb{W}^*$

taking, as always, the thermodynamic limit of $\mathbb{N} \to \infty$. In this expression $\mathbb{W}^*$ is the weight of the most probable configuration of the canonical ensemble.

The calculation runs as follows. A change in S arises from a change in $\ln \mathbb{W}^*$, and a change in $\ln \mathbb{W}^*$ arises from a change in the configuration of the ensemble. This means that we can write

$$dS = (k/\mathbb{N})\,d(\ln \mathbb{W}^*) = (k/\mathbb{N})\sum_i (\partial \ln \mathbb{W}^*/\partial \mathfrak{m}_i)\,d\mathfrak{m}_i.$$

Since $\mathbb{W}^*$ is the weight of the most probable distribution it satisfies eqn (20.1.6):

$$(\partial \ln \mathbb{W}^*/\partial \mathfrak{m}_i) + \alpha - \beta E_i = 0.$$

The equation for dS therefore becomes

$$dS = (k/\mathbb{N})\sum_i (-\alpha + \beta E_i)\,d\mathfrak{m}_i = (k\beta/\mathbb{N})\sum_i E_i\,d\mathfrak{m}_i.$$

We have used $\sum_i d\mathfrak{m}_i = 0$ (because the number of members of the ensemble remains constant at $\mathbb{N}$); the sum over $E_i\,d\mathfrak{m}_i$ is still a constraint, but it is no longer zero in this calculation because now the system is not isolated. The last equation is exactly the same as the expression we sought, eqn (20.3.6), so long as the parameter β is identified as $1/kT$.

The relation $dS = dq_{\mathrm{rev}}/T$ clinches the identity of the thermodynamic and statistical entropies, and the thermodynamic criterion of spontaneous change, $dS > 0$, can be interpreted as an ensemble shifting towards a more probable configuration. Moreover, as the temperature of the system is lowered, the value of $\mathbb{W}^*$, and hence S, declines because fewer configurations are compatible with the total energy. In the limit of T attaining absolute zero, $\mathbb{W}^* = 1$, so that $\ln \mathbb{W}^* = 0$, because only *one* configuration (every member of the ensemble in the lowest level) is compatible with a total energy of zero (above the zero point value). It follows that S *approaches zero as T approaches zero*. This conforms to the Third Law of thermodynamics, that *the entropies of all perfect crystals tend to the same value at absolute zero*, discussed on p. 153, although we shall have to refine it to take into account imperfect crystals (this is done on p. 714).

The expression for the entropy can be turned into a very simple and useful form by making a few modifications and introducing the partition function. Since we are interested in the thermodynamic limit, the number of members in the ensemble is large enough to permit the use of Stirling's approximation. This leads to

$$S = (k/N)\ln(N!/m_1^*!\,m_2^*!\ldots) = (k/N)(\ln N! - \sum_i \ln m_i^*!)$$

$$= (k/N)(N\ln N - N - \sum_i m_i^* \ln m_i^* + \sum_i m_i^*)$$

$$= (k/N)(N\ln N - \sum_i m_i^* \ln m_i^*).$$

The two parts of this expression can be combined by replacing the N multiplying $\ln N$ by $\sum_i m_i^*$; then

$$S = k\sum_i \{(m_i^*/N)\ln N - (m_i^*/N)\ln m_i^*\} = -k\sum_i (m_i^*/N)\ln(m_i^*/N).$$

The ratio m_i^*/N is the proportion of members of the ensemble having the energy E_i (in its most probable configuration), and we write it p_i. Then the entropy is given by the succinct expression

(20.3.8)
$$S = -k\sum_i p_i \ln p_i.$$

The last expression can be developed even further, and we can achieve the goal of expressing the entropy in terms of the partition function. Since

(20.3.9)
$$p_i = m_i^*/N = (1/Q)\exp(-\beta E_i), \quad \text{or} \quad \ln p_i = -\beta E_i - \ln Q.$$

It follows that

$$S = -k(-\beta\sum_i p_i E_i - \sum_i p_i \ln Q) = k\beta[U - U(0)] + k\ln Q,$$

because the sum over the p_i is unity, and the sum over $p_i E_i$ is the mean energy, or the internal energy $U - U(0)$ in the thermodynamic limit. It has already been established that $\beta = 1/kT$, and so the following immensely important expression for the statistical mechanical entropy is obtained:

(20.3.10)
$$S = [U - U(0)]/T + k\ln Q.$$

In the case of a gas of non-interacting particles the canonical partition function can be replaced by $q^N/N!$, with the result that

$$S = [U - U(0)]/T + kN\ln q - k\ln N!$$

and since the number of molecules in the sample, $N = nL$, is almost always large enough for Stirling's approximation to be valid, this turns into

$$S = [U - U(0)]/T + nR\ln q - k(N\ln N - N)$$

or

(20.3.11)°
$$S = [U - U(0)]/T + nR(\ln q - \ln N + 1).$$

We have already seen how to calculate the internal energy from the partition function, and so the last expression amounts to an explicit formula for the entropy in terms of molecular properties.

Example (Objective 13). Calculate the vibrational contribution to the entropy of iodine at 25 °C.

- *Method.* The vibrational levels are evenly spaced with a separation of 214.6 cm^{-1}. We therefore use q given by eqn (20.1.13) and as calculated in the *Example* on p. 673. Use $S_m = [U - U_m(0)]/T + R \ln q$ from eqn (20.3.10). Use 1 cm$^{-1} \triangleq 11.96$ J mol^{-1} (endpaper 1), giving $\varepsilon \triangleq 214.6$ cm$^{-1} \triangleq 2.567$ kJ mol^{-1}.

- *Answer.* Since $q = (1 - e^{-\beta\varepsilon})^{-1}$ and $U - U(0) = -N(1/q)(\partial q/\partial \beta)_V$, $U - U(0) = nL\varepsilon[e^{-\beta\varepsilon}/(1 - e^{-\beta\varepsilon})]$. From the last *Example* $\beta\varepsilon = 1.036$ and $q = 1.550$.

  $U_m - U_m(0) = (2.567 \text{ kJ mol}^{-1}) \times (0.3549) \times (1.550) = 1.412 \text{ kJ mol}^{-1}$

  $[U_m - U_m(0)]/T = 4.736 \text{ J K}^{-1} \text{mol}^{-1}$.

  $R \ln q = (8.31441 \text{ J K}^{-1} \text{mol}^{-1}) \ln 1.550 = 3.644 \text{ J K}^{-1} \text{mol}^{-1}$.

  Hence

  $S_m = 4.736 \text{ J K}^{-1} \text{mol}^{-1} + 3.644 \text{ J K}^{-1} \text{mol}^{-1} = 8.380 \text{ J K}^{-1} \text{mol}^{-1}$.

- *Comment.* Why don't we use $N!$ in S? Because we are not dealing with translations: we count it once, not everywhere. This is taken further in the next chapter.

The entropy of a monatomic gas. A very simple example of the expressions just derived is the calculation of the absolute entropy of a gas of non-interacting atoms. Since they are not interacting, the molecular partition function q can be used to find U and S. (The canonical partition function would have to be retained if we were dealing with interacting atoms. That is a much more involved problem and the conclusions are sketched in Chapter 23.)

Since the only degree of freedom for the atoms is translation, the molecular partition function is

$$q = (2\pi m/h^2 \beta)^{3/2} V$$

and the internal energy is given by eqn (20.1.16) as

(20.3.12)° $\qquad U - U(0) = -(N/q)(\partial q/\partial \beta)_V = 3N/2\beta = 3nRT/2.$

It follows that the entropy is

$$S = 3nR/2 + nR\{\ln(2\pi mkT/h^2)^{3/2}V - \ln nL + 1\}$$

$$= nR\{\ln e^{3/2} + \ln(2\pi mkT/h^2)^{3/2}V - \ln nL + \ln e\}$$

(20.3.13)° $\qquad = nR \ln\{e^{5/2}(2\pi mkT/h^2)^{3/2}V/nL\},$

which is the *Sackur–Tetrode equation* for the entropy of a monatomic perfect gas. Since the gas is perfect, V can be replaced by $nLkT/p$, which leads to an expression for the entropy as a function of the temperature and pressure:

(20.3.14)° $\qquad S = nR \ln\{e^{5/2}(2\pi mkT/h^2)^{3/2}(kT/p)\}.$

Connection with the material of Part 1 is made by noting that the change of entropy when a perfect gas expands isothermally from V_i to V_f is given by the difference of two expressions of the form of eqn (20.3.13)

$$\Delta S = S_f - S_i = nR \ln aV_f - nR \ln aV_i = nR \ln (V_f/V_i)$$

where aV is the collection of quantities inside the logarithm. This expression is exactly the same as deduced on the basis of thermodynamic arguments where only *changes* of thermodynamic quantities could be treated (p. 134).

Example (Objective 13). What is the value of the translational entropy of gaseous iodine at 25 °C and 1 atm?

• *Method.* The translational entropy can be obtained by regarding I_2 as a monatomic gas (the 'internal' modes are taken care of by adding the rotational and vibrational contributions). Use eqn (20.3.14) and data from the final endpaper.

• *Answer.* $m(I_2) = 2 \times 126.9 \times (1.6605 \times 10^{-27} \text{kg}) = 4.214 \times 10^{-25} \text{kg}$

so that

$$2\pi mkT/h^2 = \frac{2\pi \times (4.214 \times 10^{-25} \text{kg}) \times (1.380\,66 \times 10^{-23} \text{J K}^{-1}) \times (298.15 \text{K})}{(6.626\,18 \times 10^{-34} \text{J s})^2}$$

$$= 2.482 \times 10^{22} \text{m}^{-2}.$$

It follows from eqn (20.3.14) that

$$S_m = (8.3144 \text{J K}^{-1} \text{mol}^{-1}) \ln [e^{5/2} \times (2.482 \times 10^{22} \text{m}^{-2})^{3/2}$$
$$\times (1.3866 \times 10^{-23} \text{J K}^{-1}) \times (298.15 \text{K})/(1.013\,25 \times 10^5 \text{N m}^{-2})]$$

$$= (8.3144 \text{J K}^{-1} \text{mol}^{-1}) \ln (1.9357 \times 10^9) = 177.8 \text{J K}^{-1} \text{mol}^{-1}.$$

• *Comment.* The rotational entropy at room temperature can be calculated as described in the next chapter. It is $65.9 \text{J K}^{-1} \text{mol}^{-1}$. The total entropy of the molecule as a gas at 1 atm and 298 K is the sum of translational ($177.8 \text{J K}^{-1} \text{mol}^{-1}$), vibrational ($8.4 \text{J K}^{-1} \text{mol}^{-1}$), and rotational ($65.9 \text{J K}^{-1} \text{mol}^{-1}$) contributions, or $252.1 \text{J K}^{-1} \text{mol}^{-1}$.

Appendix A1: undetermined multipliers

Suppose we want to find the maximum (or minimum) value of some function f that depends on several variables, $x_1, x_2, \ldots, x_n$. When the variables undergo a small change $x_i \to x_i + \delta x_i$ the function changes from f to $f + \delta f$, where

$$\delta f = \sum_{i=1}^{n} (\partial f/\partial x_i) \delta x_i.$$

At a stationary point (a maximum or minimum) a small variation of the x_i leads to no change in f, and so $\delta f = 0$:

(20A1.1)   *at a stationary point:* $\sum_{i=1}^{n} (\partial f/\partial x_i) \delta x_i = 0.$

If the δx_i are all independent, all the δx_i are arbitrary, and so the way of satisfying the last condition is to let all $(\partial f/\partial x_i) = 0$. That is the conventional way of looking for the maximum or minimum of a function.

In some cases of interest the variations δx_i are not all independent. (We encountered an example of this in the chapter: the populations n_i were varied, but we had to ensure that the sum $n_1 + n_2 + \ldots$ remained equal to a fixed number N.) Since the δx_i are no longer independent, the simple solution to eqn (20A1.1) is no longer valid. We treat it in the following way.

Let the *constraint* connecting the variables be written

$$g(x_1, x_2, \ldots, x_n) = 0.$$

For example, in the chapter we had $n_1 + n_2 + \ldots = N$ which becomes $g(n_1, n_2, \ldots) = -N + n_1 + n_2 + \ldots = 0$ when written in this form. This constraint is always valid, and so g remains unchanged when the x_i are varied:

(20A1.2)
$$\delta g = \sum_{i=1}^{n} (\partial g/\partial x_i)\delta x_i = 0.$$

Multiply this equation by some parameter λ and add it to eqn (20A1.1):

(20A1.3)
$$\sum_{i=1}^{n} [(\partial f/\partial x_i) + \lambda(\partial g/\partial x_i)]\,\delta x_i = 0.$$

This is an equation for one of the δx, δx_n for instance, in terms of all the other δx_i. All those other δx_i $(i = 1, 2, \ldots, n-1)$ are independent and arbitrary because there is only one constraint on the system. (For example, requiring $\delta n_n = -\delta n_1 - \delta n_2 - \ldots - \delta n_{n-1}$ would guarantee that the constraint on the example in the chapter was fulfilled even though the δn_1, $\delta n_2, \ldots \delta n_{n-1}$ were regarded as arbitrary.) But here is the trick: the parameter λ is arbitrary; and so choose it so that the coefficient of δx_n in eqn (20A1.3) vanishes; that is, choose it so that

(20A1.4)
$$(\partial f/\partial x_n) + \lambda(\partial g/\partial x_n) = 0.$$

Then eqn (20A1.3) becomes

$$\sum_{i=1}^{n-1} [(\partial f/\partial x_i) + \lambda(\partial g/\partial x_i)]\,\delta x_i = 0.$$

Now the $n-1$ variations δx_i are independent, and so the solution is

$$(\partial f/\partial x_i) + \lambda(\partial g/\partial x_i) = 0, \qquad i = 1, 2, \ldots, n-1.$$

But eqn (20A1.4) has exactly the same form as this, and so the stationary values are found by solving

(20A1.5)
$$(\partial f/\partial x_i) + \lambda(\partial g/\partial x_i) = 0, \qquad \text{all } x_i.$$

This was exemplified in the text where the case of two constraints (and two undetermined multipliers λ_1 and λ_2, or α and $-\beta$) was treated.

The multipliers cannot always remain undetermined. One approach is to solve eqn (20A1.4) instead of incorporating it into the general minimization scheme. In the chapter we used the alternative procedure of keeping λ undetermined until a molecular property was calculated for which the answer was already known. Thus α was found by making sure that

$n_1 + n_2 + \ldots = N$ (and not any other number), and β was found at a later stage by calculating dS and fitting it to a known result.

Appendix A2: quantum statistics

The reduction of Q to q^N, eqn (20.2.6) depends on the *distinguishability* of the particles. The result can be obtained by considering the molecules to have available the energy levels $\varepsilon_1, \varepsilon_2, \ldots$, and for a general state of the system to have n_1 particles in ε_1, n_2 in ε_2, etc. Then summing over the states of the system is equivalent to summing over the occupation numbers of each molecular level, subject to the condition $n_1 + n_2 + \ldots = N$. The total number of ways of arriving at a specified distribution in which N particles are distributed so that n_1 occupy level ε_1, n_2 occupy level ε_2, etc., is $N!/n_1!n_2!\ldots$. Therefore the canonical partition function for a system of N distinguishable molecules is

$$(20A2.1) \qquad Q = \sum_{\text{all } n}' (N!/n_1!n_2!\ldots) \exp\{-\beta(n_1\varepsilon_1 + n_2\varepsilon_2 + \ldots)\}$$

where the sum is over all $n_1, n_2, \ldots$, the prime indicating that it is subject to $\sum_j n_j = N$. This expression is precisely the expansion of $(e^{-\beta\varepsilon_1} + e^{-\beta\varepsilon_2} + \ldots)^N$, as may be verified by using the binomial expansion

$$(x+y)^N = \sum_r \left\{ \frac{N!}{r!(N-r)!} \right\} x^r y^{N-r}$$

repeatedly; and so, for distinguishable particles,

$$(20A2.2) \qquad Q = (e^{-\beta\varepsilon_1} + e^{-\beta\varepsilon_2} + \ldots)^N = q^N.$$

Identical particles, however, are *indistinguishable* if they are free to exchange places, and the expression just developed then fails. When the particles are ordinary molecules they obey a type of statistics known as *Bose–Einstein statistics*, which can be developed as follows.

When n_1 molecules occupy the energy state ε_1, it does not matter which of the N molecules are there: since the molecules are indistinguishable there is only *one* distinct way of having n_1 molecules in ε_1, n_2 in ε_2, etc. Therefore the canonical partition function is

$$(20A2.3) \qquad Q_{\text{BE}}(N) = \sum_{\text{all } n}' \exp\{-\beta(n_1\varepsilon_1 + n_2\varepsilon_2 + \ldots)\}$$

where BE stands for Bose–Einstein, and the sum is restricted so that $\sum_j n_j = N$. This is the same as eqn (20A2.1), except for the absence of the weighting factor.

The value of Q depends very strongly on the value of N because increasing N by a small amount permits many more terms to occur in the sum. This gives a clue to a clever way of evaluating it. If $Q(N)$ is multiplied by a sharply decreasing function of N, the product will be a sharply peaked function. We therefore form $Q(N')e^{-\alpha N'}$ where $e^{-\alpha N'}$ is the sharply decreasing function. If α, a function of N, is chosen adroitly, this product peaks sharply at $N' = N$. If it is summed over all N', the only significant contribution to the sum comes from terms near $N' = N$:

$$Z = \sum_{N'} Q(N')e^{-\alpha N'} \approx Q(N)e^{-\alpha N} \delta N.$$

δN is the range of N' values that contribute significantly. It follows that

(20A2.4) $$\ln Z = \ln Q(N) - \alpha N + \ln \delta N$$

and the final term can be neglected in comparison with the other much larger terms. The value of Z (the *grand partition function*) is

$$Z = \sum_N \sum_{\text{all } n}' e^{-\alpha N} \exp\{-\beta(n_1\varepsilon_1 + n_2\varepsilon_2 + \ldots)\}$$

$$= \sum_{\text{all } n} e^{-\alpha N} \exp\{-\beta(n_1\varepsilon_1 + n_2\varepsilon_2 + \ldots)\}$$

(20A2.5) $$= \sum_{\text{all } n} \exp\{-[(\alpha + \beta\varepsilon_1)n_1 + (\alpha + \beta\varepsilon_2)n_2 + \ldots]\}.$$

The first line comes from the substitution of eqn (20A2.3). The second line recognizes that, since the sum is now over *all* values of N, the restraint $\sum_j n_j = N$ can be discarded: summing over unrestricted values of $n_1, n_2, \ldots$ covers all the values reached by summing over the restricted values and then summing over the constraint. The last line is simply

(20A2.6) $$Z = \left(\sum_{n_1} e^{-(\alpha + \beta\varepsilon_1)n_1}\right)\left(\sum_{n_2} e^{-(\alpha + \beta\varepsilon_2)n_2}\right) \ldots = \left(\frac{1}{1 - e^{(-\alpha + \beta\varepsilon_1)}}\right)\left(\frac{1}{1 - e^{-(\alpha + \beta\varepsilon_2)}}\right) \ldots$$

so that

(20A2.7) $$\ln Z = -\sum_j \ln\{1 - e^{-(\alpha + \beta\varepsilon_j)}\}.$$

Hence

(20A2.8) $$\ln Q_{\text{BE}} = \alpha N - \sum_j \ln\{1 - e^{-(\alpha + \beta\varepsilon_j)}\}.$$

Now we have to relate α to some property. It was chosen so that $Q(N')e^{-\alpha N'}$ is sharply peaked at $N' = N$. Taking logarithms, and looking for a maximum

$$(\partial/\partial N')\{\ln Q(N') - \alpha N'\} = 0 \qquad \text{at } N' = N.$$

Hence (noting that α is a function of N),

$$\{\partial \ln Q(N)/\partial N\} - \alpha(N) = 0.$$

Now, since $\ln Q(N) = \ln Z(\alpha) + \alpha N$, it follows that

$$\partial \ln Q/\partial N = \alpha + (\partial\alpha/\partial N)\{\partial[\ln Z + \alpha N]/\partial\alpha\}$$

$$= \alpha + (\partial\alpha/\partial N)\{(\partial \ln Z/\partial\alpha) + N\}.$$

Combining this with the condition in the previous equation gives

$$(\partial \ln Z/\partial\alpha) + N = 0.$$

But

$$\ln Z = -\sum_j \ln \left\{ 1 - e^{-(\alpha + \beta \varepsilon_j)} \right\}$$

and so only a little manipulation is needed to show that

(20A2.9) $$N = \sum_j \left\{ \frac{1}{e^{\alpha + \beta \varepsilon_j} - 1} \right\},$$

a condition that may be used to determine α.

Although these expressions may be developed further we have space to examine only their high-temperature, classical limits. When many states are accessible there are many terms in eqn (20A2.9); therefore every term must be small. This implies that $e^{\alpha + \beta \varepsilon_j} \gg 1$. When this is so

$$N \approx \sum_j \left\{ \frac{1}{e^{\alpha + \beta \varepsilon_j}} \right\} = \sum_j e^{-\alpha - \beta \varepsilon_j} = q e^{-\alpha}.$$

Therefore, in this limit,

(20A2.10) $$\alpha = \ln(q/N).$$

Furthermore, in the same limit, making use of Stirling's approximation,

$$\ln Q \approx \alpha N + \sum_j e^{-\alpha - \beta \varepsilon_j} \approx N \ln(q/N) + N \approx N \ln q - N \ln N + N$$

(20A2.11) $$\approx \ln q^N - \ln N!.$$

In this way we arrive at the identification $Q \approx q^N/N!$ used in the text.

Bose–Einstein statistics are not the only kind of statistics used in quantum theory. We know that particles such as electrons obey the Pauli principle (p. 451) which forbids more than one to occupy any state. Bose–Einstein statistics are then clearly inappropriate, because they permit any number to occupy a given state. In *Fermi–Dirac statistics*, a level may be empty, or it may contain one particle (e.g., one electron). The canonical partition function is the same as in eqn (20A2.3) because electrons are indistinguishable, but there is the further restriction that $n_1 = 0$ or 1 only, likewise for all other n_i. The same technique as above may be employed, and eqn (20A2.6) is simply replaced by

$$Z = \left(\sum_{n_1 = 0,1} e^{-(\alpha + \beta \varepsilon_1)n_1} \right) \left(\sum_{n_2 = 0,1} e^{-(\alpha + \beta \varepsilon_2)n_2} \right) \cdots = (1 + e^{-(\alpha + \beta \varepsilon_1)})(1 + e^{-(\alpha + \beta \varepsilon_2)}) \cdots$$

so that

(20A2.12) $$\ln Q_{FD} = \alpha N + \sum_j \ln \left\{ 1 + e^{-(\alpha + \beta \varepsilon_j)} \right\}.$$

It may readily be verified that in the classical limit $Q_{FD} = q^N/N!$, yielding once again the factor of $1/N!$ for indistinguishable particles.

Quantum statistics are essential for the discussion of low-temperature phenomena such as superconductivity and superfluidity; the discussion of the electrical and thermal properties of electrons in metals is based on Fermi–Dirac statistics.

Further reading

Entropy and energy levels. R. P. H. Gasser and W. G. Richards; Clarendon Press, Oxford, 1974.

The second law. H. A. Bent; Oxford University Press, New York, 1965.

Statistical thermodynamics. B. J. McClelland; Wiley, New York, 1973.

Statistical mechanics. F. C. Andrews; Wiley, New York, 1975.

Elementary statistical thermodynamics. L. K. Nash; Addison-Wesley, Reading, Mass., 1968.

Fundamentals of statistical and thermal physics. F. Reif; McGraw-Hill, New York, 1965.

Thermodynamics and statistical mechanics. P. T. Landsberg; Oxford University Press, 1978.

An introduction to statistical mechanics. T. L. Hill; Addison-Wesley, Reading, Mass., 1960.

The third law of thermodynamics. J. Wilks; Clarendon Press, Oxford, 1961.

Boltzmann's distribution law. E. A. Guggenheim; Interscience, New York, 1955.

Statistical mechanics. J. E. Mayer and M. G. Mayer; Wiley, New York, 1940.

The principles of statistical mechanics. R. C. Tolman; Clarendon Press, Oxford, 1938.

Statistical mechanics. A. Münster; Springer, Berlin, 1974.

Problems

20.1. A sample of 5 molecules has a total energy $U = 5(\varepsilon_0 + \varepsilon)$. Each molecule is able to occupy states of energy $\varepsilon_0 + j\varepsilon$, j an integer $(0, 1, \ldots)$. What is the weight of the configuration corresponding to the energy being equally shared among the molecules?

20.2. In the first Problem we calculated the weight of one configuration, but there are several others, all of greater importance. Draw up a table with columns headed by the energy of a molecule (from ε_0 up to $\varepsilon_0 + 5\varepsilon$) and write beneath them all the configurations of the system compatible with the total energy $5(\varepsilon_0 + \varepsilon)$. Start, for example, with $4, 0, 0, 0, 0, 1$ (only one member can have the energy $\varepsilon_0 + 5\varepsilon$ and the other 4 must then all have ε_0). Find the weight of each configuration (use eqn (20.1.1)). What is the most probable configuration?

20.3. In the case of a sample composed of 9 molecules we are approaching the region of numbers where the averages are just beginning to have thermodynamic significance, but where averages can still be dealt with exactly and numerically. Draw up a table of configurations for $N = 9$, total energy $9(\varepsilon + \varepsilon_0)$, and therefore with each molecule, on the average, having the same energy as in the last problem.

20.4. Before working out the weights of the configuration in the $N = 9$ system, look at the ones you have drawn up in the table and guess (by looking for 'exponential' form) which of the configurations will turn out to be the most probable (with the greatest weight). Now calculate the weights of all the configurations (not nearly such a long task as it might seem at first sight) and identify the most probable one.

20.5. The most probable configuration is characterized by the *temperature* of the system. But what is the temperature of the system we have been considering? It must be such as to give a mean value $\varepsilon_0 + \varepsilon$ for the energy of each molecule and a total energy $N(\varepsilon + \varepsilon_0)$, the molecules being allowed only the energies $\varepsilon_0 + j\varepsilon$. Show that the temperature of a system can be obtained by plotting $\ln(n_j/n_0)$ against j, the n_j being the populations for the most probable configuration (in the thermodynamic limit). Apply this method to the 9 molecule system in order to see how closely it corresponds to the thermodynamic limit (by assessing the quality of the straight line). Let the spacing of the molecular energy levels be $\varepsilon = 50 \, \text{cm}^{-1}$. What is the temperature of this system?

20.6.   When N is large, show that if the mean energy per molecule is $\varepsilon_0 + a\varepsilon$ when molecules can occupy the ladder of energy levels $\varepsilon_0 + j\varepsilon$, then the temperature of the sample is given by $\beta = (1/\varepsilon) \ln(1 + 1/a)$. Find the temperature for the case treated in the last set of problems ($a = 1$, $\varepsilon \triangleq 50 \, \text{cm}^{-1}$). What is the value of q at the temperature for which the mean molecular energy is $\varepsilon_0 + a\varepsilon$?

20.7.   The 'temperature' is a parameter that has significance only for the most probable configuration, and we should not expect to get a good straight line for others. Choose two other configurations for the 9 molecule system, one close to most probable, one far from it, and plot $\ln(n_j/n_0)$ against j. The straight lines should be very poor.

20.8.   Calculate the molecular partition function for the 9 molecule system at 104 K (the result of Problem 20.6), 100 K, and 108 K. Then find the gradient $dq/d\beta$ at 160 K and so confirm explicitly that the total energy (which we know to be $9(\varepsilon_0 + \varepsilon)$ at that temperature) is given by $-N(d \ln q/d\beta)$. Any discrepancy comes from the estimation of the temperature of the ensemble.

20.9.   We now begin to explore the calculation and manipulation of the molecular partition function. As a first step, state for which systems it is essential to include a factor of $1/N!$ in going from Q to q: (a) a sample of helium gas; (b) a sample of carbon monoxide gas; (c) the same sample of carbon monoxide, but now frozen to a solid; (d) water vapour; (e) ice; (f) an electron gas; (g) an electron gas in a metal.

20.10.   An argon atom is trapped in a cubical box of volume V. What is its translational partition function at (a) 100 K, (b) 298 K, (c) 10000 K, (d) 0 K, when the box is of side 1 cm?

20.11.   The form of the translational partition function, as normally used, is valid when a huge number of energy levels are accessible. When does the normal expression become invalid and when do we have to resort to the explicit summation, eqn (20.1.9)? Estimate what the temperature of the argon in the last Problem would have to be in order for the partition function to drop to about 10. What is the exact value of the partition function at that temperature?

20.12.   There are several types of partition function that can be calculated by direct summation of exponentials, using spectroscopic data for the molecular energy levels involved. First we deal with the tellurium atom, which has several low-lying excited states. Find the electronic partition function for atoms at (a) 298 K, (b) 5000 K on the basis of the following data from atomic spectroscopy: ground state (5-fold degenerate), $4751 \, \text{cm}^{-1}$ (3-fold degenerate), $4707 \, \text{cm}^{-1}$ (singly degenerate), $10559 \, \text{cm}^{-1}$ (5-fold degenerate).

20.13.   What proportion of the tellurium atoms are in (a) their ground level, (b) the level at $4751 \, \text{cm}^{-1}$ at the two temperatures of the last Problem?

20.14.   Most of the molecules we shall encounter have electronic states that are so high in energy above the ground state that only the latter need be considered for its thermodynamic properties. There are several exceptions, one interesting case being NO, for this has an electronically excited state at only $121.1 \, \text{cm}^{-1}$ above the ground state. Both this and the ground state are doubly degenerate.

Calculate and plot the electronic partition function of NO from zero to 1000 K. What is the distribution of populations at room temperature?

20.15. What is the mean electronic internal energy of the NO molecule at room temperature? Answer this first by calculating an expression for U at a general temperature (use the partition function found above), and then substitute the value $T = 298$ K.

20.16. The *Example* on p. 725 assumed that the vibration of the I_2 molecules were harmonic. In fact it has vibrational energy levels at the following wavenumbers above the zero-point level: $213.30 \, \text{cm}^{-1}$, $425.39 \, \text{cm}^{-1}$, $636.27 \, \text{cm}^{-1}$, $845.93 \, \text{cm}^{-1}$, $1054.38 \, \text{cm}^{-1}$. Calculate the vibrational partition function by explicit summation of the expression for q at (a) 100 K, (b) 298 K.

20.17. What proportion of the iodine molecules are in the ground and first two excited vibrational states at the two temperatures in the last Problem?

20.18. What is the average vibrational energy of molecular iodine at those two temperatures? Evaluate the sum for U, eqn (20.1.19), explicitly using the data in Problem 20.15.

20.19. An electron spin can adopt two orientations in a magnetic field, the energies being $\pm \mu_B B$. Find an expression for the electron spin partition function and mean energy, and plot these as a function of applied field at 298 K and 4 K. What are the relative populations of the spin levels at these two temperatures?

20.20. A three-level system, such as a nitrogen nucleus in a magnetic field, can also be expressed in a simple closed form. Derive an expression for the partition function and mean energy of a nitrogen nucleus ($I = 1$) in a magnetic field at 4 K. Take the $M_I = 0$ state as the zero of energy.

20.21. A peculiarity of 2-level systems can now be explored. Suppose by some artificial means we contrive to invert the populations of the spin levels (methods do exist for this and lasers make use of them). More electrons will then be in the upper energy state than in the lower. This is a non-equilibrium condition, and the distribution of populations is far from being the most probable. Nevertheless, we can still *formally* express the ratio of populations in terms of a single parameter, which we are free to call 'temperature'. Demonstrate that this temperature must be *less* than absolute zero. What 'temperature' corresponds to (a) an inversion of the equilibrium population at 298 K? (b) an inversion of the equilibrium populations at 10 K, (c) total inversion, with all electrons in the upper spin state?

20.22. Under what circumstances is it permissible or meaningful to speak of negative temperatures in connection with 3-level systems?

20.23. What is the average entropy per molecule when N is large enough for eqn (20.3.10) to be applied to the system treated in Problem 20.6; that is, when the average energy per molecule is $\varepsilon_0 + a\varepsilon$? Show that $S/Nk = (1 + a)\ln(1 + a) - a\ln a$, and hence that $S/Nk = 2\ln 2$ when the mean molecular energy is $\varepsilon_0 + \varepsilon$.

20.24. Calculate the electronic contribution to the molar entropy of the tellurium atom at (a) 298 K, (b) 5000 K; use the data in Problem 20.12.

20.25. Calculate the electronic contribution to the molar entropy of the NO molecule at (a) 298 K, (b) 500 K; use the data in Problem 20.14.

20.26. Calculate the vibrational contribution to the molar entropy of the iodine molecule at (a) 100 K, (b) 298 K; use the data in Problem 20.15.

20.27. Calculate the dependence of the molar entropy of a collection of independent, localized electron spins on the strength of an applied magnetic field. What do you *expect* the entropy of the spins to be at (a) $B = 0$, (b) $B = \infty$, and what do you *calculate* it to be?

20.28. Suppose 1 mol of argon atoms are first held rigidly in position, and then allowed to move freely: what is the change of entropy?

20.29. What is the entropy of a dilute electron gas at (a) 298 K, (b) 5000 K? Take $V = 10\,dm^3$.

20.30. Confirm that the statistical thermodynamic expression for the entropy of a monatomic ideal gas accounts correctly for the dependence of the entropy on (a) its pressure, (b) its temperature.

21 Statistical thermodynamics: the machinery

Learning objectives

After careful study of this chapter you should be able to

(1) Calculate the *translational contribution* to the molecular partition function (eqn (21.1.4)).

(2) Define and calculate the *symmetry number* of a molecule (p. 698).

(3) Calculate the *rotational contribution* to the molecular partition function (eqn (21.1.5)) and estimate it at high temperatures (eqns (21.1.7) and (21.1.8)).

(4) Calculate the *vibrational contribution* to the partition function (eqn (21.1.10)).

(5) Calculate the *electronic contribution* to the partition function (eqn (21.1.12)).

(6) Relate the *internal energy* (eqn (21.2.1)), *entropy* (eqn (21.2.2)), *enthalpy* (eqn (21.2.5)), *Helmholtz function* (eqn (21.2.3)), and *Gibbs function* (eqn (21.2.7)) to the partition function.

(7) Define the *molar partition function* (p. 705).

(8) Relate the *pressure* to the partition function (eqn (21.2.4)) and use the result to show that $\beta = 1/kT$.

(9) Calculate the *mean energy* of modes of molecular motion (p. 706).

(10) State the *equipartition theorem* and explain its limitations (p. 708).

(11) Use the equipartition principle to calculate *heat capacities* (eqn (21.3.18)).

(12) Calculate *vibrational heat capacities* at low temperatures (eqn (21.3.17)).

(13) Define *residual entropy* and calculate it for disordered systems (p. 714).

(14) Relate the *equilibrium constant* of a reaction to the molecular partition functions of the reactants and products (eqn (21.3.25)).

(15) Use *spectroscopic data* to calculate the equilibrium constant of a chemical reaction (p. 719).

Introduction

The previous chapter set out the framework of statistical thermodynamics. We saw that the central concept is the partition function Q, and that once it is known the thermodynamic functions U and S can be calculated. The partition function is the crucial link between thermodynamics, spectroscopy, and quantum theory, because spectroscopic data, or molecular parameters such as masses and force-constants, can be used to calculate its value at some temperature, and then U and S can be calculated from Q.

In this chapter we explore the ways of using statistical thermodynamics to discuss problems of chemical significance. This involves finding out how to calculate the partition function given a set of spectroscopic data or molecular constants. The problems we shall look at include the calculation of heat capacities, entropies, and chemical equilibrium constants, and in the process we shall discover new ways of thinking about their significance.

21.1 How to calculate the partition function

In this section we confine attention to a system of non-interacting molecules, such as a perfect gas. The molecules, however, are allowed to have internal structure, and so the energy of the system is divided among the translational, rotational, vibrational, and electronic modes of motion. Since the molecules are indistinguishable and independent of each other $Q = q^N/N!$, and so we can concentrate on the calculation of the molecular partition function q,

(21.1.1)
$$q = \sum_j \exp(-\beta\varepsilon_j), \qquad \beta = 1/kT.$$

The first step depends on the separation of the energy of an individual molecule into contributions from its various modes of motion. If the total molecular energy can be written as the sum

$$\varepsilon_j = \varepsilon^{\text{translation}} + \varepsilon^{\text{rotation}} + \varepsilon^{\text{vibration}} + \varepsilon^{\text{electronic}},$$

the molecular partition function reduces to a product of the partition functions for each mode:

$$q = \sum_{\substack{(\text{trans., rot.,} \\ \text{vib., elec., states})}} \exp(-\beta\varepsilon^t - \beta\varepsilon^r - \beta\varepsilon^v - \beta\varepsilon^e)$$

$$= \sum_{\text{trans.}} \exp(-\beta\varepsilon^t) \sum_{\text{rot.}} \exp(-\beta\varepsilon^r) \sum_{\text{vib.}} \exp(-\beta\varepsilon^v) \sum_{\text{elec.}} \exp(-\beta\varepsilon^e)$$

(21.1.2)
$$= q^t q^r q^v q^e.$$

This factorization is only an approximation (except for the separation of translation) because the modes of motion are not quite independent of each other, but for most purposes it is satisfactory. Its great advantage, though, is that the various contributions to the partition function can be investigated separately.

The translational contribution. The translational partition function was derived on p. 674, eqn (20.1.15):

$(21.1.3)°$ $\qquad q^t = (2\pi m/h^2\beta)^{3/2}V = (2\pi mkT/h^2)^{3/2}V.$

This applies to a molecule of mass m in a container of volume V. The appearance of this expression, which will occur frequently, can be simplified by writing

$(21.1.4)°$ $\qquad q^t = \tau V, \qquad \tau = (2\pi mkT/h^2)^{3/2}$

(τ is the Greek letter tau). Notice that the value of q^t rises to infinity as the temperature is increased, because an extremely large number of quantum states become accessible. Even at room temperature $q^t \approx 2 \times 10^{28}$ for a molecule of oxygen in a 100 cm^3 vessel.

The rotational contribution. One method of calculating q^r is to substitute the experimental values of the rotational energy levels into the expression for the rotational partition function and then to sum the exponentials numerically. One point to remember is that more than one rotational state may correspond to the same energy. For instance, in the case of a diatomic molecule the states are labelled with the quantum numbers J, M but the energy depends only on J (p. 577). Since there are $2J+1$ values of M for any given J, there are $2J+1$ states of the same energy for each value of J. The rotational partition function is then

$(21.1.5)$ $\qquad q^r = \sum_{J,M} \exp(-\beta E_J) = \sum_J (2J+1)\exp(-\beta E_J).$

Example (Objective 3). Calculate the rotational partition function of HCl at room temperature (25 °C).

- *Method.* Use eqn (21.1.5), evaluated term by term. The rotational energy levels are given by $E_J = hcBJ(J+1)$, and $B = 10.591$ cm^{-1}.

- *Answer.* We shall require $hcB/kT = (10.591$ cm$^{-1})/(207.22$ cm$^{-1}) = 0.051\,11$. Draw up the following table:

J	0	1	2	3	4	5	6	7	8	9	10
$J(J+1)$	0	2	6	12	20	30	42	56	72	90	110
$e^{-J(J+1)B/kT}$	1	0.903	0.736	0.542	0.360	0.216	0.117	0.057	0.025	0.010	0.004
$2J+1$	1	3	5	7	9	11	13	15	17	19	21

Then form the sum required by eqn (21.1.5). This is equal to 19.9, hence $q^r = 19.9$ at this temperature. Taking J up to 16 gives $q = 19.902$.

- *Comment.* Notice that about ten J-levels are significantly populated. Later we shall encounter the approximation that $q^r \approx kT/hcB$. In the present case this gives $q^r \approx 19.6$, in very good agreement with the exact value, and with much less work.

Another point that has to be guarded against is including too many states in this sum. This is a subtle problem and depends on the observation that although a heteronuclear diatomic (like HCl) can occur with any value of J, an individual homonuclear diatomic molecule can occur only with even values of J or only with odd values. Fortunately this complication can be taken into account in a simple way when the temperature is so high that many rotational levels are populated, and so we concentrate on that case.

The rotational energy levels of many molecules lie close enough together for a large number to be populated at room temperature and above. At room temperature $kT \cong 200$ cm^{-1} whereas the rotational constants of the molecules HCl, I_2, CH_4, and CO_2 are 10.6 cm^{-1}, 0.04 cm^{-1}, 5.2 cm^{-1}, and 0.39 cm^{-1} respectively. If we confine our attention to molecules no lighter than these, and to temperatures no lower than room temperature, then to a good approximation so many closely spaced rotational states will be populated that the sum in eqn (21.1.5) can be replaced by an integral.

At this stage we have to take into account the fact that not all values of J can be taken by every molecule. When large numbers of quantum levels are occupied the problem can be approached classically, and this leads to a neat and simple solution. In the case of a homonuclear diatomic, rotation by 180° interchanges two equivalent nuclei, and as the new orientation is indistinguishable from the original, we have to divide the integral by 2 to avoid counting indistinguishable orientations twice:

homonuclear diatomic: $q^r \approx \frac{1}{2} \int_0^\infty (2J+1) \exp\{-\beta E_J\} \, dJ$.

In the case of a heteronuclear diatomic, the rotation by 180° leads to a distinguishable orientation (HCl $\rightarrow$ ClH) and so the factor $\frac{1}{2}$ does not appear. These conclusions can be combined by writing

(21.1.6) $q^r \approx (1/\sigma) \int_0^\infty (2J+1) \exp\{-\beta E_J\} \, dJ$,

where σ, the *symmetry number*, takes the values 1 for heteronuclear diatomics and 2 for homonuclear diatomics.

The same problem arises for other types of symmetrical molecule, and a symmetry number can be used to correct the integral over J to eliminate false counting of energy levels. In the case of linear molecules the last expression is appropriate because their rotational properties are the same as those of a diatomic molecule: for CO_2 $\sigma = 2$, but for OCS $\sigma = 1$. In more complex molecules more than two orientations might be indistinguishable. For instance, whereas for H_2O $\sigma = 2$, for NH_3 $\sigma = 3$ and for $CH_2:CH_2$ $\sigma = 4$. Sometimes the symmetry number may be quite large: in benzene, for example, there are 12 equivalent orientations, and $\sigma = 12$ (see the Problems for the way σ is determined).

The integral in eqn (21.1.6) can be evaluated and explicit expressions obtained for the rotational partition functions of linear, including diatomic, molecules. As found on p. 577, $E_J = hcBJ(J+1)$, where B is the rotational constant of the molecule and is related to its moment of inertia by $B = \hbar/4\pi cI$. It follows that

$$q^r \approx (1/\sigma) \int_0^\infty (2J+1) \exp\{-\beta hcBJ(J+1)\} \, dJ.$$

Although this integral looks complicated it can be evaluated without much effort by noticing that it can also be written as

$$q^r \approx \int_0^\infty (-1/\sigma\beta hcB)[(d/dJ) \exp\{-\beta hcBJ(J+1)\}] \, dJ.$$

The integral of a derivative of a function is the function itself, and so

$$q^r \approx (-1/\sigma\beta hcB) \exp\{-\beta hcBJ(J+1)\}|_0^\infty = 1/\sigma\beta hcB.$$

The approximate form of the rotational partition function for a linear molecule is therefore

(21.1.7)
$$q^r \approx kT/hcB\sigma = 2IkT/\hbar^2\sigma.$$

Approximate rotational partition functions for other types of molecule can be found in the same way, leading to

(21.1.8)
$$q^r \approx (\pi^{1/2}/\sigma)\{(2I_AkT/\hbar^2)(2I_BkT/\hbar^2)(2I_CkT/\hbar^2)\}^{1/2}$$

where I_A, I_B, and I_C are their three moments of inertia.

Example (Objective 3). Estimate the rotational partition function for ethene at room temperature (25 °C).

- *Method.* Use eqn (21.1.8) with $\sigma = 4$. The rotational constants are $A = 4.828 \text{ cm}^{-1}$, $B = 1.0012 \text{ cm}^{-1}$, and $C = 0.8282 \text{ cm}^{-1}$. Begin by expressing q^r in terms of A, B, C through $A = \hbar/4\pi cI_A$ (for A expressed in wave numbers), and likewise for B and C.

- *Answer.* $q^r = (\pi^{1/2}/\sigma)[(kT/Ahc)(kT/Bhc)(kT/Chc)]^{1/2} = (1/\sigma)(kT/hc)^{3/2}(\pi/ABC)^{\frac{1}{2}}$. At 298.15 K, $kT/hc = 207.22 \text{ cm}^{-1}$, and so

$$q^r = \tfrac{1}{4}(207.22 \text{ cm}^{-1})^{3/2}(\pi/4.828 \times 1.0012 \times 0.8282 \text{ cm}^{-3})^{1/2}$$

$$= 660.6.$$

- *Comment.* Ethene is quite a big molecule, the energy levels are close together; as a consequence many rotational levels are populated even at room temperature.

The general conclusion at this stage is that *molecules with large moments of inertia have large rotational partition functions*. This reflects the closeness of the rotational energy levels in large, heavy molecules, corresponding to many states being populated at normal temperatures.

The vibrational contribution. In the case of a diatomic molecule the vibrational partition function can be evaluated by substituting the measured vibrational energy levels into the expression

(21.1.9)
$$q^v = \sum_j \exp(-\beta\varepsilon_j^v).$$

In a polyatomic molecule every normal mode of vibration (p. 593) has its own independent set of energy levels, and so

$$q^v = q^v(1)q^v(2)\dots$$

where $q^v(K)$ is the partition function for the Kth normal mode. This is a direct method for calculating the partition function from the information given by vibrational spectroscopy, and avoids having to make assumptions about the absence of anharmonicity.

So long as the vibrational excitation is not too great the approximation can be made that the vibrations are simple harmonic. Then the vibrational energies are given by the expression $E_v = (v+\frac{1}{2})\hbar\omega$, $v = 0, 1, 2, \dots$, where $\omega = (k/\mu)^{1/2}$, k is the force-constant for the vibration and μ the reduced mass; this result was discussed on p. 584. If we measure energies from the zero-point level, the permitted values are $E' = \hbar\omega$. Then using the harmonic approximation leads to a simple expression for the vibrational partition function:

$$q^v = \sum_v \exp(-v\hbar\omega\beta) = \sum_v (e^{-\hbar\omega\beta})^v.$$

This series was encountered on p. 672 (which is no accident: the ladder-like array of levels in Fig. 20.1 is exactly the same as that of a simple harmonic oscillator). The series can be summed in the same way, and gives

(21.1.10)
$$q^v = \frac{1}{1-\exp(-\hbar\omega\beta)} = \frac{1}{1-\exp(-hc\tilde{v}/kT)}$$

where $\tilde{v}$ is the wavenumber of the mode. When the molecule has several normal modes the total vibrational partition function is formed by multiplying expressions like this, but with the appropriate values of the vibrational frequencies of each mode.

Example (Objective 4). The wavenumbers of the three normal modes of the water molecule are 3656.7 cm^{-1}, 1594.8 cm^{-1}, and 3755.8 cm^{-1}. What is the vibrational partition function at (a) 298.15 K, (b) 1500 K?

● *Method.* Use eqn (21.1.10) for each mode. At 298.15 K, kT/hc is equal to 207.22 cm^{-1}, and at 1500 K it equals 1042.5 cm^{-1}.

● *Answer.* Draw up the following table:

mode:	1	2	3
$\tilde{v}/\text{cm}^{-1}$	3656.7	1594.8	3755.8
$(hc\tilde{v}/kT)_{298\text{ K}}$	17.646	7.696	18.125
$(hc\tilde{v}/kT)_{1500\text{ K}}$	3.508	1.530	3.603
$q_i^v(298\text{ K})$	1.0000	1.0005	1.0000
$q_i^v(1500\text{ K})$	1.0309	1.2764	1.0280

The overall vibrational partition function is $q = q_1^v q_2^v q_3^v$. At 298 K, $q^v = 1.0005$; at 1500 K, $q^v = 1.3527$.

Comment. The vibrations of the water molecule are of high frequency, and at room temperature virtually all molecules are in the ground vibrational state. There is only a small departure from this even at 1500 K.

In many molecules the vibrational frequencies are so high that $\hbar\omega\beta \gg 1$ (and $hc\tilde{v}\beta \gg 1$). For example, the lowest vibrational wavenumber of methane is 1306 cm^{-1}, and so $hc\tilde{v}\beta \approx 6$ at room temperature. C–H stretches normally lie in the range 2850–2960 cm^{-1} and so $hc\tilde{v}\beta \approx 14$ at room temperature. In these cases $\exp(-hc\tilde{v}\beta)$ may be neglected in the denominator of q^v (for example, $e^{-6} \approx 0.002$) and the vibrational partition function for a single mode is simply $q^v \approx 1$ (implying that only the lowest state is occupied).

The electronic contribution. Electronic energy separations are usually very large, and so the exponentials $\exp(-\beta\varepsilon^e)$ are all very small, except for the ground state, for which $\varepsilon^e = 0$. Therefore, in most cases,

(21.1.11) $q^e = 1.$

An important exception to this result occurs in the case of atoms having a degenerate ground state (that is, when several states of the atom have the same energy). If there are g states corresponding to the same energy, so that the ground state is *g-fold degenerate*, the partition function has the value

(21.1.12) $q^e = g.$

The alkali metals, for example, have doublet ground states, and so in the absence of a magnetic field $q^e = 2$ (corresponding to the two, equal energy, orientations of the unpaired electron spin).

Some molecules also have electronically degenerate ground states, and a few have excited electronic states that lie very close to the ground state. One case is nitrogen(II) oxide, NO, which has the electronic configuration $\dots \pi^1$ (see the diagram on p. 481, and add to the orbitals 15 electrons). The π electron can orbit around the internuclear axis in two senses, clockwise and anticlockwise, and its spin may be aligned in the same sense as its orbital momentum, or opposite to it. This gives the four states illustrated in Fig. 21.1. These four states fall into two groups: when the orbital and spin momenta are parallel the energy of the molecule is slightly greater than when they are opposed (because of the spin–orbit coupling, p. 458). The first two states constitute the level denoted $^2\Pi_{3/2}$ and the second two states constitute the level $^2\Pi_{1/2}$. These designations need not trouble us: the important point is that the electronic energy levels of NO are as shown in Fig. 21.1, the lowest state is doubly degenerate, and an excited level lies only 121 cm^{-1} above it, and this level is also doubly degenerate.

The electronic partition function for NO can be obtained as follows. Denoting the energies of the two levels by $\varepsilon_{1/2}$ and $\varepsilon_{3/2}$ and writing $\varepsilon_{1/2} = 0$

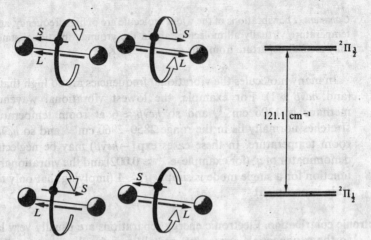

121.1 cm^{-1}

$^2\Pi_{\frac{3}{2}}$

$^2\Pi_{\frac{1}{2}}$

Fig. 21.1. Electronic states of nitrogen(II) oxide.

and $\varepsilon_{3/2} = \delta$, leads to

$$q^e = \sum_j g_j \exp(-\beta\varepsilon_j) = 2\exp(-\beta\varepsilon_{1/2}) + 2\exp(-\beta\varepsilon_{3/2})$$

(21.1.13)
$$= 2(1 + e^{-\beta\delta}).$$

At absolute zero $q^e = 2$ because only the two states of the lower level are accessible; at very high temperatures q^e approaches 4. At room temperature $1/\beta$ correspond to 200 cm^{-1}, and so $q^e \approx 2.8$. Some of the consequences of this behaviour are examined in Problem 21.31.

The overall partition function. We now have the partition functions for all the separate modes of motion: for convenience they are collected in a practical form in Box 21.1. The overall partition function is their product. For a diatomic molecule this is

(21.1.14)
$$q \approx (2\pi mkT/h^2)^{3/2} V(2IkT/\hbar^2\sigma)\left\{\frac{1}{1-\exp(-\hbar\omega/kT)}\right\}.$$

The corresponding expression for more complicated molecules can be written down just as readily by multiplying the parts listed in the Box. The overall partition function is approximate because it assumes that the rotational levels are close together and the vibrations are harmonic. This approximation could be avoided by evaluating the sums explicitly on the basis of the spectroscopic data.

21.2 How to calculate the thermodynamic functions

We have at our command methods of calculating the two principal thermodynamic functions, the internal energy U and the entropy S:

Box 21.1 **Contributions to the molecular partition function**

1. *Translation.*

$$q^t = \tau V, \qquad \tau = (2\pi m kT/h^2)^{3/2}$$

$$\tau/\text{m}^{-3} = 1.8792 \times 10^{26} \, (T/K)^{3/2} M_r^{3/2}$$

$$q_m^{t\ominus}/L = \tau(kT/p^{\ominus}) = 0.025\,61 \, (T/K)^{5/2} M_r^{3/2}$$

2. *Rotation.*
(a) Linear molecule:

$$q^t = (1/\sigma)(2IkT/\hbar^2) = kT/\sigma hcB$$

$$= 0.6950(1/\sigma)(T/K)/(B/\text{cm}^{-1}).$$

(b) Non-linear molecule:

$$q^r = (\pi^{1/2}/\sigma)[(2I_AkT/\hbar^2)(2I_BkT/\hbar^2)(2I_CkT/\hbar^2)]^{1/2}$$

$$= (\pi^{1/2}/\sigma)(kT/hc)^{3/2}(1/ABC)^{1/2}$$

$$= 1.0270(1/\sigma)(T/K)^{3/2}/[(A/\text{cm}^{-1})(B/\text{cm}^{-1})(C/\text{cm}^{-1})]^{1/2}$$

3. *Vibration.*

$$q^v = 1/[1 - \exp(-\hbar\omega/kT)] = 1/[1 - \exp(-hc\tilde{v}/kT)].$$

$$= 1/\{1 - \exp[-1.4388(\tilde{v}/\text{cm}^{-1})/(T/K)]\}$$

4. *Electronic.* $q^e = g$, where g is the degeneracy of the ground state (normally the only electronic state thermally accessible). At high temperatures, evaluate explicitly.

Values of kT expressed in cm^{-1} are listed on the first endpaper.

(21.2.1) $\quad U - U(0) = -(1/Q)(\partial Q/\partial \beta)_V = -(\partial \ln Q/\partial \beta)_V$

(21.2.2) $\quad S = [U - U(0)]/T + k \ln Q.$

In the case of independent molecules these simplify by making the substitution $Q = q^N$ (distinguishable particles) or $Q = q^N/N!$ (indistinguishable particles). All the other thermodynamic functions can be based on U and S, and so we also have a route to the calculation of every thermodynamic function. The actual relations can be established by drawing on a few of the results obtained in Part 1.

As a first step we express the *Helmholtz function, A,* in terms of the partition function. Since

$$A = U - TS$$

and at $T = 0$, $A = U(0)$, substitution for U and S leads at once to

(21.2.3)
$$A - A(0) = -kT \ln Q.$$

With this result established, the *pressure* exerted by the system can be related to its canonical partition function by using a result obtained by the same argument as on p. 165:

(21.2.4)
$$p = -(\partial A/\partial V)_T = kT(\partial \ln Q/\partial V)_T.$$

This result is entirely general, and may be used for a real gas as well as a perfect gas (or for a liquid). Since it relates the pressure to the volume and temperature, it is a very important route to the determination of equations of state of real gases and relating them to the intermolecular interactions (which have to be built into Q). That it gives the correct equation of state of a perfect gas can be checked by substituting $Q = q^N/N!$:

$$p = kT(1/Q)(\partial Q/\partial V)_T = NkT(1/q)(\partial q/\partial V)_T.$$

Since only q^t depends on the volume ($q^t = \tau V$) this becomes

$$p = nLkT(1/q^t q^r q^v q^e)(\partial q^t q^r q^v q^e/\partial V)_T = nRT(1/q^t)(\partial q^t/\partial V)_T$$

$$= nRT(1/\tau V)\tau = nRT/V,$$

as required. (This can be regarded as yet another way of deducing the relation $\beta = 1/kT$.)

Since both U and p can be related to Q, the *enthalpy H* can also be related to it, and so a way has been found of calculating this important thermodynamic quantity. From the definition $H = U + pV$, and noting that at $T = 0$, $H(0) = U(0)$,

(21.2.5)
$$H - H(0) = -(\partial \ln Q/\partial \beta)_V + kTV(\partial \ln Q/\partial V)_T.$$

The set of thermodynamic quantities can be completed by finding a way of calculating the *Gibbs function G*, p. 146. Since

$$G = H - TS = U + pV - TS,$$

we find

(21.2.6)
$$G - G(0) = -kT \ln Q + kTV(\partial \ln Q/\partial V)_T.$$

In the case of a perfect gas (including one composed of molecules with structure) this expression simplifies considerably. In the first place pV in the definition of G can be replaced by nRT; then inserting the expressions for U and S leads to

$$G - G(0) = -kT \ln Q + nRT.$$

Replacing Q by $q^N/N!$ gives

$$G - G(0) = -NkT \ln q + kT \ln N! + nRT$$

$$= -nRT \ln q + kT(N \ln N - N) + nRT,$$

so that

(21.2.7)° $$G - G(0) = -nRT \ln (q/N)$$

where $N = nL$. It follows that the molecular partition function can be used directly to determine the Gibbs function, the central function of chemical thermodynamics. In later work it will prove convenient to write q/N as q/nL and to refer to q/n as q_m, the *molar partition function* (with dimensions mol^{-1}). Then

(21.2.8)° $$G - G(0) = -nRT \ln (q_m/L).$$

These important relations are collected in Box 21.2.

Box 21.2 **Statistical thermodynamic relations**

In terms of the canonical partition function Q:

$$U - U(0) = -(\partial \ln Q/\partial \beta)_V$$

$$S = [U - U(0)]/T + k \ln Q$$

$$p = kT(\partial \ln Q/\partial V)_T$$

$$H - H(0) = -(\partial \ln Q/\partial \beta)_V + kTV(\partial \ln Q/\partial V)_T$$

$$A - A(0) = -kT \ln Q$$

$$G - G(0) = -kT \ln Q + kTV(\partial \ln Q/\partial V)_T.$$

In the case $Q = q^N/N!$ (free, indistinguishable particles, external modes, e.g., translation)

$$U - U(0) = -N(\partial \ln q/\partial \beta)_V$$

$$S = [U - U(0)]/T + nR(\ln q - \ln N + 1)$$

$$G - G(0) = -nRT \ln (q/N) = -nRT \ln (q_m/L).$$

In the case $Q = q^N$ (internal modes, e.g., rotation)

$$U^i - U^i(0) = -N(\partial \ln q^i/\partial \beta)_V$$

$$S^i = [U - U(0)]/T + nR \ln q^i$$

$$G^i - G^i(0) = -nRT \ln q^i.$$

In general, for indistinguishable non-interacting particles,

$$Q = (q^{ex}q^i)^N/N! = [(q^{ex})^N/N!](q^i)^N.$$

Hence

$$S^{total} = S + S^i, \quad U^{total} = U + U^i, \quad \text{etc.}$$

Example (Objective 6). Calculate the value of the function $\Phi_0(T) = [G_m^\ominus(T) - H_m^\ominus(0)]/T$ for water at 1500 K.

- *Method.* $G_m^\ominus(T)$ may be calculated from eqn (21.2.8); then $\Phi_0(T) = -R\ln(q_m^\ominus/L)$. Use the expression in Box 21.1 for $q_m^\ominus/L$. The vibrational partition function was calculated in the *Example* on p. 700. Use the equations in Box 21.1 for the other contributions. The rotational constants are 27.8778 cm^{-1}, 14.5092 cm^{-1}, and 9.2869 cm^{-1}; $\sigma = 2$.

- *Answer.* For the translational contribution, Box 21.1 gives $q_m^{t\ominus}/L = (0.025\,61) \times (18.015)^{3/2} \times (1500)^{5/2} = 1.706 \times 10^8$. For the rotational contribution,

$$q^r = 1.0270 \times \tfrac{1}{2} \times (1500)^{3/2} \times (1/27.8778 \times 14.5092 \times 9.2869)^{1/2}$$
$$= 486.7.$$

For the vibrational contribution, we have already seen that

$$q^v = 1.353.$$

The complete molar partition function is therefore

$$q_m^\ominus/L = (1.706 \times 10^8) \times (486.7) \times (1.353) = 1.123 \times 10^{11}.$$

It follows that

$$\Phi_0(1500) = -R\ln(q_m^\ominus/L) = -(8.3144 \text{ J K}^{-1} \text{ mol}^{-1})\ln(1.123 \times 10^{11})$$
$$= -211.5 \text{ J K}^{-1} \text{ mol}^{-1}.$$

- *Comment.* This accords with the value quoted in Table 9.1, 211.7 J K^{-1} mol^{-1}, the discrepancy arising from the use of slightly different data. Other values in Table 9.1 can be constructed from spectroscopic data in a similar way.

21.3 Using statistical thermodynamics

Any thermodynamic property can now be deduced from a knowledge of the energy levels of molecules; thermodynamics and spectroscopy have been combined. In this section we indicate how to do the calculation for four important problems: the equipartition principle, heat capacities, residual entropies, and equilibrium constants. These are by no means the only applications of statistical thermodynamics, and more will be encountered in Part 3.

Average energies and the equipartition principle. Quite often it is useful to know the average energy locked up in various molecular modes when the molecule forms part of a system at a temperature T. This can be obtained by using the expression for the thermodynamic internal energy, and the factorization of the overall molecular partition function:

$$U - U(0) = -N(1/q)(\partial q/\partial \beta) = -N(1/q^t q^r q^v q^e)(\partial q^t q^r q^v q^e/\partial \beta)$$
$$= N\{-(1/q^t)(\partial q^t/\partial \beta) - (1/q^r)(\partial q^r/\partial \beta) - \ldots\}.$$

Since $[U - U(0)]/N$ is the sum of the mean energies of all the modes,

$$(21.3.1)^\circ \qquad U - U(0) = N\{\langle\varepsilon^t\rangle + \langle\varepsilon^r\rangle + \langle\varepsilon^v\rangle + \langle\varepsilon^e\rangle\},$$

we arrive at

$(21.3.2)°$ $\langle \varepsilon^m \rangle = -(1/q^m)(\partial q^m/\partial \beta).$

where m = t, r, v, or e.

The *mean translational energy* of a molecule can be found from the translational partition function. In order to show a pattern emerging, we consider a one-dimensional system first and then pass on to three dimensions. From p. 674 the translational partition function for a molecule of mass m is $q^t = (2\pi m/\beta h^2)^{1/2}X$, so the mean translational energy is

$$\langle \varepsilon^t \rangle = -(\beta h^2/2\pi m)^{1/2}(2\pi m/h^2)^{1/2}(-\tfrac{1}{2}\beta^{-3/2})$$
$$= 1/2\beta = \tfrac{1}{2}kT.$$

In three dimensions the same calculation leads to

$(21.3.3)°$ $\langle \varepsilon^t \rangle = \tfrac{3}{2}kT.$

In both cases the mean energy is independent of the mass of the molecule and the size of the container. (This fits in with the statement, p. 81, that the internal energy of a perfect gas is independent of the volume it occupies: $(\partial U/\partial V)_T = 0$.)

In classical mechanics the kinetic energy T of a particle of mass m is related to the components of its velocity by

$$T = \tfrac{1}{2}mv_x^2 + \tfrac{1}{2}mv_y^2 + \tfrac{1}{2}mv_z^2.$$

We see that the mean energy can be arrived at by setting the mean value of each of the quadratic terms in this expression equal to $\tfrac{1}{2}kT$. This is a general result, as we shall see by considering the rotational energy and the vibrational energy of a molecule.

The *mean rotational energy* of a linear molecule is

$$\langle \varepsilon^r \rangle = -(1/q^r)(\partial q^r/\partial \beta)$$

with

$$q^r = \sum_J (2J+1)\exp\{-\beta hcBJ(J+1)\}.$$

When the temperature is low we have to deal with this expression term by term. In the case of a heteronuclear diatomic all J values contribute to the sum; then, writing $b = \beta hcB$,

$$q^r = 1 + 3e^{-2\beta hcB} + 5e^{-6\beta hcB} + \ldots = 1 + 3e^{-2b} + 5e^{-6b} + \ldots,$$

and so

$(21.3.4)$ $$\langle \varepsilon^r \rangle = \frac{hcB\{6e^{-2b} + 30e^{-6b} + \ldots\}}{\{1 + 3e^{-2b} + 5e^{-6b} + \ldots\}}$$

and the mean energy drops to zero as $T \to 0$ (or $b \to \infty$). When the temperature is so high that many rotational levels are occupied we can use the approximate form of the partition function, p. 699:

$$\langle \varepsilon^r \rangle = -(\sigma \hbar^2 \beta/2I)(2I/\sigma \hbar^2)(-1/\beta^2)$$

$(21.3.5)$ $$= 1/\beta = kT.$$

The classical expression for the rotational energy of a linear molecule is

$$T = \tfrac{1}{2}I\omega_x^2 + \tfrac{1}{2}I\omega_y^2$$

(there is no z-component because it has zero moment of inertia about the line of atoms), where I is the moment of inertia and ω_x and ω_y the components of angular velocity. We see that, *in the classical limit*, the mean energy of kT can be arrived at by apportioning $\tfrac{1}{2}kT$ to each of the two quadratic terms. Extrapolating this result to non-linear molecules, where three modes of rotation are allowed, suggests that the mean rotational energy ought to be $\tfrac{3}{2}kT$. This can be confirmed by using eqn (21.1.8).

The *mean vibrational energy* in the classical limit (which means high temperatures) can now be predicted by writing the classical expression for the energy:

$$E = T + V = \tfrac{1}{2}mv_x^2 + \tfrac{1}{2}kx^2,$$

and noticing the presence of *two* quadratic terms. This suggests a mean energy of $2(\tfrac{1}{2}kT)$. But are we correct in ascribing equal amounts of energy to the potential and kinetic energies? This can be answered by an explicit calculation. The exact partition function for a harmonic oscillator is

(21.3.6)
$$q^v = \frac{1}{1 - \exp(-\hbar\omega\beta)}$$

and so, as

$$(\partial q^v/\partial\beta) = \frac{-\hbar\omega \exp(-\hbar\omega\beta)}{\{1 - \exp(-\hbar\omega\beta)\}^2},$$

we find

(21.3.7)
$$\langle \varepsilon^v \rangle = \hbar\omega \left\{ \frac{\exp(-\hbar\omega\beta)}{1 - \exp(-\hbar\omega\beta)} \right\}.$$

This is the exact result (apart from a zero point energy contribution of $\tfrac{1}{2}\hbar\omega$). When the temperature is so high that $\hbar\omega\beta \ll 1$ it simplifies as follows:

$$\langle \varepsilon^v \rangle \approx \hbar\omega \left\{ \frac{1 - \hbar\omega\beta}{1 - 1 + \hbar\omega\beta} \right\} \approx 1/\beta - \hbar\omega \approx 1/\beta,$$

which confirms that the mean energy is kT in the classical limit.

These conclusions may be summarized by the statement of the

> **equipartition principle**: when quantum effects can be ignored, the average energy of every quadratic term in the energy expression has the same value, $\tfrac{1}{2}kT$.

Note the importance of the limitation of this simple rule to the cases where quantum effects are negligible. When this limitation is unimportant the equipartition principle gives a very simple rule for totting up the total average energy of a molecule, and therefore of estimating U. We shall see an application in the following section.

Another question that can be answered is what is the chance of finding a molecule that has an energy much greater than the mean energy? This can be answered explicitly by using the Boltzmann distribution (p. 669), but sometimes it is helpful to have a rough guide to the spread of energies. In statistics the spread of a distribution is often measured in terms of the *root mean square deviation* of the property. In the present case we are interested in the energy, and the *root mean square deviation of the energy* from its mean value is defined as

(21.3.8)
$$\delta\varepsilon = \sqrt{\{\langle\varepsilon^2\rangle - \langle\varepsilon\rangle^2\}}.$$

For brevity, we shall refer to $\delta\varepsilon$ as the *fluctuation* in energy of the molecules in the sample. If all the molecules happened to be in the same state of energy ε^*, the average energy would be ε^*, and the mean of the squares of the energy would be ε^{*2}: in such a case $\delta\varepsilon = 0$. When the molecules are distributed over many states of different energy, the mean of the squared energy will no longer equal the square of the mean energy, Fig. 21.2, and so $\delta\varepsilon$ is no longer zero. A large value of $\delta\varepsilon$ corresponds to a wide spread of populations, and a small value to a narrow spread.

The square of the mean energy is $(1/q)^2(\partial q/\partial\beta)^2_V$. The mean of the square can be calculated as follows:

$$\langle\varepsilon^2\rangle = \sum_j p_j\varepsilon_j^2 = (1/q)\sum_j \varepsilon_j^2\exp(-\beta\varepsilon_j)$$
$$= (1/q)\sum_j(\partial^2/\partial\beta^2)\exp(-\beta\varepsilon_j) = (1/q)(\partial^2 q/\partial\beta^2)_V.$$

Consequently the energy fluctuation is related to the partition function by

(21.3.9)
$$\delta\varepsilon^2 = (1/q)(\partial^2 q/\partial\beta^2)_V - (1/q)^2(\partial q/\partial\beta)^2_V.$$

The application of the last expression can be illustrated by considering a harmonic oscillator. Since the exact partition function is given in eqn (21.3.6), straightforward differentiation leads to

(21.3.10)
$$\delta\varepsilon = \frac{\hbar\omega\exp(-\tfrac{1}{2}\hbar\omega\beta)}{1-\exp(-\hbar\omega\beta)}$$

and the dependence of this function on the temperature is shown in Fig. 21.3: notice that $\delta\varepsilon = 0$ at absolute zero because all the molecules are then

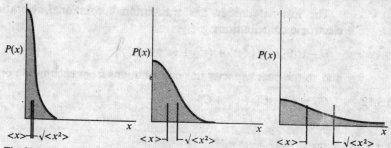

Fig. 21.2. Mean and root mean square energies of a distribution.

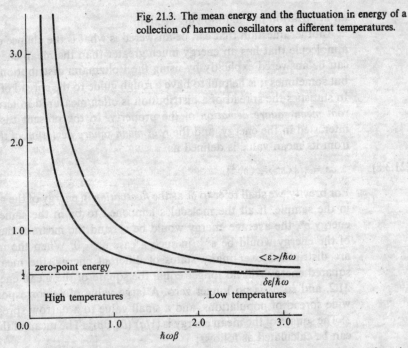

Fig. 21.3. The mean energy and the fluctuation in energy of a collection of harmonic oscillators at different temperatures.

in their ground vibrational state and there is no spread of energies. At high temperatures the exponentials may be expanded, and the last expression reduces to

$$\delta\varepsilon \sim 1/\beta = kT,$$

and the fluctuation of energies is seen to be equal to the mean energy: the fluctuation grows as the mean energy of excitation grows.

Heat capacities. The heat capacity at constant volume C_V is

(21.3.11a) $C_V = (\partial U/\partial T)_V.$

This expression was first encountered on p. 66. Sometimes it is convenient to write C_V as a derivative with respect to β: as $\beta = 1/kT$ implies that $d\beta = -dT/kT^2$, the last equation can also be written

(21.3.11b) $C_V = -(1/kT^2)(\partial U/\partial \beta)_V.$

The internal energy has translational, rotational, vibrational, and electronic contributions:

$$U - U(0) = N\{\langle \varepsilon^t \rangle + \langle \varepsilon^r \rangle + \langle \varepsilon^v \rangle + \langle \varepsilon^e \rangle\},$$

and so the heat capacity has contributions from each mode of motion:

(21.3.12) $C_V = C_V^t + C_V^r + C_V^v + C_V^e$

where

(21.3.13a) $C_V^m = N(\partial \langle \varepsilon^m \rangle / \partial T)_V$

(21.3.13b) $\qquad = -(N/kT^2)(\partial\langle\varepsilon^m\rangle/\partial\beta)_V$

with m denoting the type of mode (m = t, r, v, or e). From now on we treat a gas of non-interacting molecules, and so there is no contribution to the translational energy from intermolecular interactions. This means that we are considering a perfect gas, but we are allowing the molecules to have internal structure.

The temperature is always high enough for the mean translational energy to be given by the equipartition principle. It follows that

(21.3.14)° $\qquad C_V^t = N(\partial\tfrac{3}{2}kT/\partial T)_V = \tfrac{3}{2}Nk = \tfrac{3}{2}nR.$

In the case of monatomic gases this is the only contribution, and so molar heat capacities equal to $\tfrac{3}{2}R = 12.47$ J K^{-1} mol^{-1} should be observed. This is in excellent agreement with the experimental values for the rare gases (He, for example, has the theoretical value over a range of 2000 K).

When the temperature is high enough for the rotations of the molecules to be regarded as unquantized, the equipartition principle may be used to estimate their contribution to the heat capacity. In the case of a linear molecule $\langle\varepsilon^r\rangle = kT$, and so

(21.3.15) $\qquad C_V^r = Nk = nR.$

When the temperature is so low that only the lowest rotational state is occupied, $\langle\varepsilon^r\rangle = 0$ and there is no contribution to the heat capacity from the rotations. At intermediate values the value of C_V^r can be obtained by differentiating eqn (21.3.4). Even without doing that we see that the rotational contribution to the heat capacity rises from zero (when $kT \ll hcB$) to nR (when $kT \gg hcB$). Since the translational contribution is always present, we should expect the overall heat capacity of a diatomic molecule ($C_{V,m}^t + C_{V,m}^r$) to rise from 12.47 J K^{-1} mol^{-1} to 20.8 J K^{-1} mol^{-1} as the temperature rises past $T = hcB/k$. The heat capacity of hydrogen is shown in Fig. 21.4, and this behaviour can be seen very clearly.

In the case of non-linear molecules the mean rotational energies in the high-temperature region approach $3kT/2$, and so their contribution to the molar heat capacities amounts to a maximum of $3R/2$.

Molecular vibrations contribute to the heat capacity, but only at temperatures high enough for their excitation. When the temperature is so high, or the bonds so feeble, that equipartition may be used to estimate their mean energy, every normal mode contributes $\langle\varepsilon^v\rangle = kT$, and so every one contributes an amount R to the molar heat capacity. In the case of a diatomic molecule, only one normal mode exists (the stretching and compressing of the bond), and so at high temperatures

(21.3.16) $\qquad C_{V,m}^v = R.$

It should be noticed that the contribution from a single vibrational mode is R whereas other modes contribute only $\tfrac{1}{2}R$: this is a reflection of the two contributions to the energy, kinetic and potential, in the case of a harmonic oscillator. Another consequence of this greater contribution

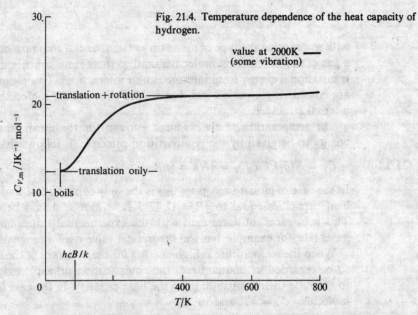

Fig. 21.4. Temperature dependence of the heat capacity of hydrogen.

value at 2000K ——
(some vibration)

to the heat capacity is encountered when the rotation of a methyl group attached to some molecule is hindered, for example when the temperature is lowered so that it is trapped in a potential well, Fig. 21.5. When it is hot enough to rotate there is a contribution $\frac{1}{2}R$ to the molar heat capacity of the molecule. At lower temperatures, where the rotation is quenched into an oscillation, the same mode contributes R; and so lowering the temperature leads to an increase in the heat capacity above $\frac{1}{2}R$, even though the full R might not be attained.

When equipartition cannot be used to estimate the mean excitation energy of vibrations we have to use the exact expression in eqn (21.3.7). Differentiation with respect to T leads to

(21.3.17) $$C_V = nR\left\{\frac{(\hbar\omega/kT)\exp(-\frac{1}{2}\hbar\omega\beta)}{1-\exp(-\hbar\omega\beta)}\right\}^2$$

and the shape of this function, Fig. 21.6, shows how the vibrational contribution rises from zero at low temperatures and attains its limiting value of nR when $kT \gg \hbar\omega$.

The total heat capacity of a gas can be calculated by adding the

High temperature   Low temperature

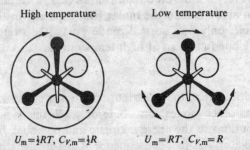

$U_m = \frac{1}{2}RT,\ C_{V,m} = \frac{1}{2}R$ $U_m = RT,\ C_{V,m} = R$

Fig. 21.5. Heat capacity of a hindered rotor at high and low temperatures.

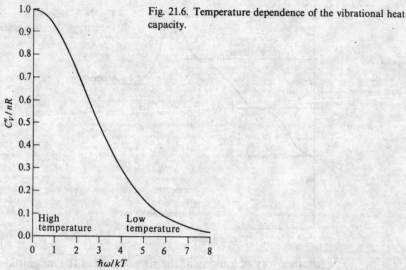

Fig. 21.6. Temperature dependence of the vibrational heat capacity.

contributions from translation, rotation, and vibration. In a few cases there is also an electronic contribution to the heat capacity, but the energy levels are normally sufficiently far apart for it to be negligible. (The case of nitrogen(II) oxide is one exception and the calculation of its electronic heat capacity is set as a Problem.) Often it is sufficient to estimate the approximate temperature dependence of the heat capacity by counting the number of modes of motion that actively contribute to the storing of energy. In all gases the translational modes are fully active and contribute $3nR/2$; in many cases rotations are fully active and contribute $2(\frac{1}{2}nR)$ in linear molecules and $3(\frac{1}{2}nR)$ in non-linear molecules: the number of active rotations we shall denote v_r^*. If v_v^* vibrations are active and the temperature is high enough for equipartition to be valid, there is a further contribution of $v_v^* nR$. It follows that the total heat capacity is

(21.3.18)
$$C_V = \tfrac{1}{2}nR(3 + v_r^* + 2v_v^*)$$

and such an expression gives a straightforward way of accounting for (or predicting) the temperature dependence of C_V of the type shown in Fig. 21.7, the different modes becoming active at various temperatures.

Example (Objective 11). Estimate the heat capacity of steam at 100 °C.

- *Method.* Use eqn (21.3.18) after deciding which modes of motion are active at 100 °C.

- *Answer.* Translations and rotations are active. Vibrations will not contribute because they lie in the range 1500–3500 cm^{-1}, which is much greater than kT ($kT \approx 260$ cm^{-1} at 100 °C). Therefore, from eqn (21.3.18):
  $$C_{V,m} = \tfrac{1}{2}R(3 + 3 + 0) = 3R = 24.9 \text{ J K}^{-1} \text{ mol}^{-1}.$$

- *Comment.* The experimental value is 26.1 J K^{-1} mol^{-1}. Part of the discrepancy can be ascribed to the approximation for the rotational partition function.

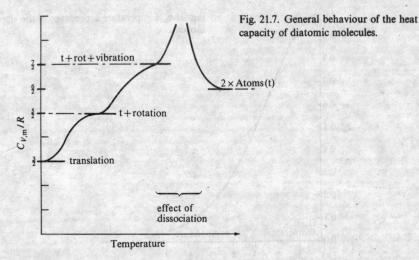

Fig. 21.7. General behaviour of the heat capacity of diatomic molecules.

Another way of looking at the significance of the magnitude of the heat capacity starts from its definition in terms of the molecular partition function:

$$C_V = -(1/kT^2)(\partial U/\partial \beta)_V = (N/kT^2)\left\{\frac{\partial}{\partial \beta}\frac{1}{q}\left(\frac{\partial q}{\partial \beta}\right)\right\}_V$$

$$= (N/kT^2)\{(1/q)(\partial^2 q/\partial \beta^2)_V - (1/q^2)(\partial q/\partial \beta)_V(\partial q/\partial \beta)_V\}.$$

The bracketed term is the square of the energy fluctuation, p. 709, and so

(21.3.19) $C_V = nR(\delta \varepsilon/kT)^2.$

This neat expression is exact, apart from the assumption that the molecules are independent. From the discussion of the temperature dependence of the fluctuations in molecular energy, p. 709, we know that $\delta \varepsilon$ drops to zero at absolute zero; hence the heat capacity is also predicted to drop to zero, in accord with experiment. At higher temperatures the molecular energy fluctuates over a wide range, especially if many closely spaced energy levels are accessible, and so the heat capacity rises above zero. This analysis relates the heat capacity to the fluctuations present in an assembly of molecules: the magnitude of the fluctuations determines the readiness of the system to accommodate energy.

Residual entropies. Entropy may be calculated from spectroscopic data. Entropy may also be measured calorimetrically, as described in Chapter 5, p. 141. In many cases there is very good agreement, but in some the experimental entropy is less than the calculated values. One possible explanation is that the experimenter failed to detect a phase change, and so omitted to take into account a term of magnitude $\Delta H/T$, where ΔH is the transition's enthalpy and T its temperature. Another possibility is that some disorder is present in the solid even at absolute zero. The entropy of the solid at absolute zero is then greater than zero, and is called the *residual entropy*. The source and magnitude of the residual entropy can be understood by considering a crystal composed of molecules AB, where A and B are of similar size. There may be so little energy difference between the energy of the arrays ... AB AB AB AB ... and ... AB BA AB BA ... that the molecules adopt either orientation at random in the solid. Since solidification is not an infinitely slow process, the random array may be frozen in and survive even at absolute zero.

The entropy arising from residual disorder can be calculated on the basis of the fundamental expression for the entropy

$$S = (k/N) \ln W$$

in the limit of $N \to \infty$. Suppose that two molecular orientations are equally probable, and that the sample contains N molecules. In one member of the ensemble the same energy can be achieved in 2^N different ways (because every molecule can take two orientations, and there are N molecules). There are N members of the ensemble, and every one can be formed in 2^N different ways. Therefore the weight of the configuration, the overall number of ways of achieving the same energy, is $(2^N)^N$. It follows that the entropy is

(21.3.20)
$$S = (k/N) \ln (2^N)^N = k \ln 2^N = Nk \ln 2 = nR \ln 2,$$

with $nL = N$. This predicts a residual molar entropy of $R \ln 2 = 5.76 \text{ J K}^{-1} \text{ mol}^{-1}$ for molecules that can adopt either of two orientations at absolute zero. If three orientations are energetically equivalent (or very nearly so) a straightforward modification of the derivation predicts a residual molar entropy of $R \ln 3$, and so on for other possibilities.

An example where the residual molar entropy is approximately $6 \text{ J K}^{-1} \text{ mol}^{-1}$ is solid nitrogen(II) oxide. X-ray diffraction studies show

that the NO molecules stick together into a rectangular dimer
$$\begin{matrix} \text{N--O} \\ \vdots \quad \vdots \\ \text{O--N} \end{matrix}$$

This rectangle can adopt two orientations in the crystal with approximately equal energy, and so a residual molar entropy of about $R \ln 2$ is expected. The $FClO_3$ molecule can adopt four orientations with approximately equal energy, and so a residual molar entropy of about $R \ln 4 = 11.5 \text{ J K}^{-1} \text{ mol}^{-1}$ can be expected: the experimental value is $10.1 \text{ J K}^{-1} \text{ mol}^{-1}$. An instructive example, which has often been used to illustrate the striking agreement between theory and experiment, is carbon monoxide. The calorimetric entropy was measured as $193 \text{ J K}^{-1} \text{ mol}^{-1}$ whereas the spectroscopic value is $197.9 \text{ J K}^{-1} \text{ mol}^{-1}$ at the same temperature. The difference, $5 \text{ J K}^{-1} \text{ mol}^{-1}$, is not far off $R \ln 2$, and this was explained by supposing that the CO molecules could take up either of two orientations in the crystal, which would then have a structure such as ... CO CO OC CO CO OC ... Recent work, however, has questioned whether the calorimetric experiments missed a phase change at low temperature, and so it may be that there is in fact no residual entropy. This example emphasizes the care that has to be taken in obtaining and interpreting experimental data. It is possible that other examples of residual entropy might eventually be eliminated in the same way.

A final example of residual entropy is provided by ice. The discrepancy between calorimetric and spectroscopic determinations of the entropy of water vapour leads to the conclusion that there is a randomness present in the structure of ice corresponding to a residual entropy of magnitude $3.4 \, J \, K^{-1} \, mol^{-1}$. This can be traced to the hydrogen-bonded nature of the ice crystal, and the tetrahedral arrangement of hydrogen atoms around every oxygen atom. Two of the hydrogen atoms are attached by normal, short, σ-bonds, and the two others attached by long hydrogen bonds. There is a randomness in which two of the four bonds are short, and an approximate analysis of the problem leads to the prediction of a residual molar entropy of about $R \ln \frac{3}{2} = 3.37 \, J \, K^{-1} \, mol^{-1}$, in good agreement with the experimental value.

Equilibrium constants. The basis of the calculation of the equilibrium constant for the reaction $A \rightleftharpoons B$ can be understood by referring to Fig. 21.8a. One set of lines corresponds to the energy levels of A, and the other set to the energy levels of B. The population of the levels is governed by a Boltzmann distribution, and is independent of the nature of the energy levels and, in particular, independent of whether a level belongs to A or to B. We can therefore imagine a single Boltzmann distribution spreading, without distinction, over the two sets of energy levels. If the spacings of the A and B levels are the same, and A lies beneath B, the diagram indicates that A will dominate the reaction mixture at equilibrium. If, however, B has a higher density of states, as in Fig. 21.8b, then even though it lies higher in energy it may predominate in the reaction mixture because we are

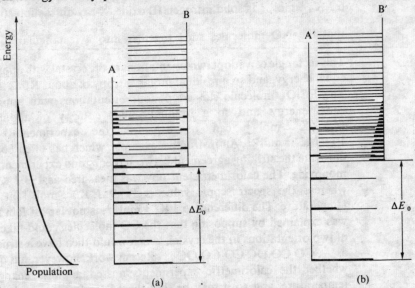

Fig. 21.8. The array of energy levels for A and B molecules in the equilibrium $A \rightleftharpoons B$, and the dependence of the equilibrium on the density of states. The Boltzmann distribution is shown on the left.

interested in the *total* numbers of A and B molecules present at equilibrium.

At some temperature T the number of molecules in some state i of the joint system is

$$n_i = N(1/q)\exp(-\beta\varepsilon_i)$$

where N is the total number of molecules. The total number of molecules of type A is the sum of these n_i over the states of molecule A with energies ε_a:

$$N_A = \sum_{i\,of\,A} n_i = (N/q)\sum_a \exp(-\beta\varepsilon_a).$$

Likewise the number of B molecules in the mixture is the sum over the states of B with energies ε_b':

$$N_B = \sum_{i\,of\,B} n_i = (N/q)\sum_b \exp(-\beta\varepsilon_b').$$

The sum over the states of A is nothing other than the molecular partition function for molecule A, q_A:

$$N_A = Nq_A/q.$$

The sum over the states of B is also a partition function, but it is not quite the same as the ones met so far because the energies ε_b' are measured from the ground state of the total system, which in the present example happens to be the ground state of A. The recovery of the conventional energy zero is easy to bring about, because $\varepsilon_b' = \varepsilon_b + \Delta E_0$, where ΔE_0 is the separation between the lowest levels of A and B, Fig. 21.8. Then

$$N_B = (N/q)\exp(-\beta\Delta E_0)\sum_b \exp(-\beta\varepsilon_b)$$

$$= (Nq_B/q)\exp(-\beta\Delta E_0).$$

The equilibrium constant of the reaction is therefore

(21.3.21a) $$K = N_B/N_A = (q_B/q_A)\exp(-\beta\Delta E_0).$$

This gives a way of calculating the value of K simply from a knowledge of the partition function, and therefore from spectroscopic data or molecular properties. The form of the expression can be modified so that K is expressed in terms of concentrations or pressures. For instance, since $q_B = (\tau q^i)_B V$, where q^i is the partition function for the internal modes, we find

(21.3.21b) $$K_c = (N_B/V)/(N_A/V) = [(\tau q^i)_B/(\tau q^i)_A]\exp(-\beta\Delta E_0).$$

Then, since $N_A/V = n_A L/V = p_A/kT$,

(21.3.21c) $$K_p = p_B/p_A = [(\tau q^i)_B/(\tau q^i)_A]\exp(-\beta\Delta E_0).$$

The content of eqn (21.3.21a) can be appreciated by considering a simple example in which A has only one energy level accessible at the temperature of interest (so that $q_A = 1$) and B has a large number of equally spaced,

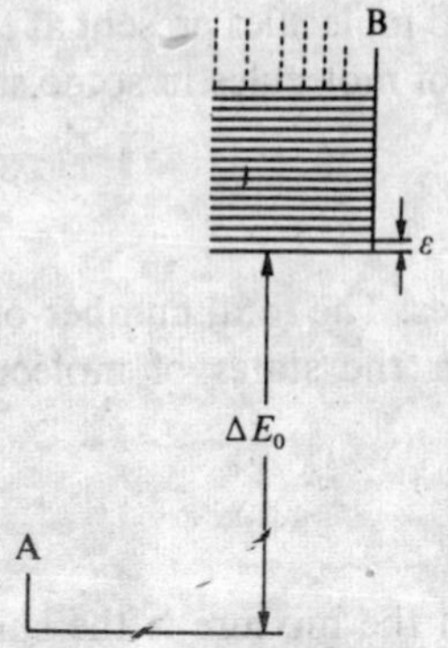

Fig. 21.9. A model for an equilibrium calculation.

close levels, Fig. 21.9. The partition function of B is the same as that met on p. 672 (and the same as that of a harmonic oscillator), and so

$$q_B = (1 - e^{-\beta\varepsilon})^{-1}.$$

If the separation ε is small compared to kT this simplifies to $q_B = kT/\varepsilon$. It follows that when the energy separation of A and B is ΔE_0, the equilibrium constant is

(21.3.22) $K \approx (kT/\varepsilon)\exp(-\Delta E_0/kT).$

This result can be used to investigate the shift in populations as the separation ΔE_0 is reduced, but ε held constant. When ΔE_0 is very large the exponential term dominates and $K \approx 0$, implying the presence of very little B at equilibrium. When ΔE_0 is small but still positive, the value of K might still exceed unity, reflecting the predominance of B by virtue of its large number of accessible states. The same kind of discussion can be made for the case in which the energy separations are held constant, but the temperature is changed. At low temperatures the exponential is about zero and so $K \approx 0$, and the system consists almost entirely of A. At high temperatures the exponential is approximately unity, and the pre-exponential factor is greater than one, indicating a predominance of B. We see that a rise of temperature favours B in the equilibrium, simply because so many of its states become accessible.

At this point no contact has been made with the normal, chemical discussion of equilibrium in terms of the Gibbs functions and chemical potentials of the participants, yet this was the basis of so much of the discussion of Part 1. This crucial link between spectroscopy and thermodynamics can be formed on the basis of the expressions for the Gibbs function presented in the preceding section, and the main reason for digressing in the opening part of this section is to give some insight into the underlying connections.

We derived an expression for the Gibbs function of a collection of independent molecules on p. 705,

$$G - G(0) = -nRT\ln(q_m/L),$$

and we know the connection between K and the *standard* Gibbs function of the reaction $\Delta G_m^{\ominus}$, p. 264:

$$\Delta G_m^{\ominus} = -RT \ln K.$$

All we have to do is to combine these components.

In the first place we require the molar Gibbs function, and so form G/n. Furthermore, we want its standard value, and so the molar partition function must be evaluated at a pressure of 1 atm: we indicate this by $q_m^{\ominus}$. Since only the translational factor in the partition function depends on the pressure (through the presence of V_m), we evaluate q_m with $V_m = V_m^{\ominus} = RT/p^{\ominus}$, taking $p^{\ominus} = 1$ atm (that is, $p^{\ominus} = 1.013\,25 \times 10^5$ N m^{-2}). For a component A, it follows that

(21.3.23) $$G_{A,m}^{\ominus} - G_{A,m}^{\ominus}(0) = -RT \ln(q_{A,m}^{\ominus}/L).$$

The equilibrium constant for the $A \rightleftharpoons B$ reaction is then

$$-RT \ln K_p = G_{B,m}^{\ominus} - G_{A,m}^{\ominus} = -RT \ln(q_{B,m}^{\ominus}/q_{A,m}^{\ominus}) + G_{B,m}^{\ominus}(0) - G_{A,m}^{\ominus}(0),$$

and so

(21.3.24) $$K_p = (q_{B,m}^{\ominus}/q_{A,m}^{\ominus}) \exp\{-[G_{B,m}^{\ominus}(0) - G_{A,m}^{\ominus}(0)]/RT\},$$

and when it is realized that $G_{B,m}^{\ominus}(0) - G_{A,m}^{\ominus}(0)$ is the same as $U_{B,m}^{\ominus}(0) - U_{A,m}^{\ominus}(0)$, and written ΔE_0, we obtain the same expression as before.

The equilibrium constant for a general reaction, such as

$$A + B \rightleftharpoons C + D + F$$

can now be written down immediately in terms of the molecular partition functions for all five components:

(21.3.25) $$K_p = (q_{C,m}^{\ominus} q_{D,m}^{\ominus} q_{F,m}^{\ominus}/q_{A,m}^{\ominus} q_{B,m}^{\ominus} L) \exp(-\Delta E_0/kT),$$

with ΔE_0 the energy separation between the lowest levels of $A + B$ and $C + D + F$.

An illustration of the application of this formula is the gaseous equilibrium between a diatomic molecule X_2 and its atoms:

$$X_2(g) \rightleftharpoons 2X(g).$$

The equilibrium constant is given by

$$K_p = (q_{X,m}^{\ominus 2}/q_{X_2,m}^{\ominus} L) \exp(-D_m^{\ominus}/RT),$$

where $D_m^{\ominus}$ is the molar dissociation energy of the molecule. The atomic partition functions relate only to translational motion and any electronic degeneracy g (p. 701):

$$q_{X,m}^{\ominus} = g\tau_X V_m^{\ominus}, \qquad V_m^{\ominus} = RT/p^{\ominus}.$$

On the other hand, the diatomic molecule has rotational and vibrational degrees of freedom as well as translational motion. Therefore

$$q_{X_2,m}^{\ominus} = \tau_{X_2} V_m^{\ominus} q_{X_2}^{r} q_{X_2}^{v}.$$

It follows that the equilibrium constant is

(21.3.26)
$$K_p = \frac{g^2 \tau_X^2 V_m^\ominus \exp(-D_m^\ominus/RT)}{\tau_{X_2} q_{X_2}^r q_{X_2}^v L}.$$

In the case of $Na_2 \rightleftharpoons 2Na$ the values of all the parameters are known, and at $T = 1000$ K we find $K_p \approx 2.4$.

Example (Objective 15). Calculate the equilibrium constant $K^\ominus$ for the $Na_2(g) \rightleftharpoons 2Na(g)$ equilibrium at 1000 K using the following spectroscopic data: $B = 0.1547$ cm^{-1}, $\tilde{v} = 159.2$ cm^{-1}, $D_m^\ominus = 70.4$ kJ mol^{-1} (0.73 eV). Use $M_r = 22.99$ for Na. The atoms are in doublet states.

- *Method.* Use the expressions in Box 21.1 and then substitute into eqn (21.3.26).

- *Answer.*
  Translation of Na_2: $\tau = (1.8792 \times 10^{26}) \times (1000 \times 45.98)^{3/2}$ m^{-3}
  $\qquad\qquad\qquad\qquad = 1.853 \times 10^{33}$ m^{-3}.
  Rotation of Na_2: $q^r = \frac{1}{2}(0.6950) \times (1000/0.1547) = 2246$.
  Vibration of Na_2: $q^v = 1/[1 - \exp(-1.4388 \times 159.2/1000)] = 4.885$
  Electronic of Na_2: $q^e = g = 1$.
  Translation of Na: $\tau = (1.8772 \times 10^{26}) \times (1000 \times 22.99)^{3/2}$ m^{-3}
  $\qquad\qquad\qquad\quad = 6.551 \times 10^{32}$ m^{-3}.
  Electronic of Na: $q^e = g = 2$.
  $\qquad\qquad V_m^\ominus = (8.3144$ J K^{-1} mol$^{-1}) \times (1000$ K$)/(1.013\,25 \times 10^5$ N m$^{-2})$
  $\qquad\qquad\qquad = 8.2057 \times 10^{-2}$ m^3 mol^{-1}.

  From eqn (21.3.26),
  $$K_p = \frac{g^2 \tau_{Na}^2 V_m^\ominus \exp(-D_m^\ominus/RT)}{\tau_{Na_2} q_{Na_2}^r q_{Na_2}^v L}$$

  $$= \frac{4 \times (6.551 \times 10^{32} \text{ m}^{-3})^2 \times (8.2057 \times 10^{-2} \text{ m}^3 \text{ mol}^{-1}) \exp(-70.4/8.3144)}{(1.853 \times 10^{33} \text{ m}^{-1}) \times (2246) \times (4.885) \times (6.022 \times 10^{23} \text{ mol}^{-1})}$$

  $$= 2.419.$$

- *Comment.* Note that this procedure leads to K_p. If we want an equilibrium constant expressed in concentrations we have to modify the expression. Use $[X] = n_X/V = p_X/RT$ in the expression for K_p and obtain
  $$K_p = \{(p_{Na}/p^\ominus)^2/(p_{Na_2}/p^\ominus)\}_e = (RT/p^\ominus)\{[Na]^2/[Na_2]\}_e.$$

Further reading

Elementary statistical thermodynamics. L. K. Nash; Addison-Wesley, Reading, Mass., 1968.

Statistical thermodynamics. B. J. McClelland; Wiley, New York, 1973.

Statistical mechanics. N. Davidson; McGraw-Hill, New York, 1962.

An introduction to statistical mechanics. T. L. Hill; Addison-Wesley, Reading, Mass., 1960.

Statistical mechanics, thermodynamics, and kinetics. O. K. Rice; Freeman, San Francisco, 1967.

Statistical thermodynamics. R. H. Fowler and E. A. Guggenheim; Cambridge University Press, 1965.

Statistical mechanics. A. Münster; Springer, Berlin, 1974.

Problems

21.1. Estimate the rotational partition function of HCl at (a) 100 K, (b) 298 K, (c) 500 K, on the basis of the high-temperature approximation.

21.2. The pure rotational, microwave spectrum of HCl has absorption lines at the following wavenumbers (in cm^{-1}): 21.19, 42.37, 63.56, 84.75, 105.93, 127.12, 148.31, 169.49, 190.68, 211.87, 233.06, 254.24, 275.43, 296.62, 317.80, 338.99, 360.18, 381.36, 402.55, 423.74, 444.92, 466.11, 487.30, 508.48. Calculate the rotational partition function at (a) 100 K, (b) 298 K by direct summation.

21.3. On the basis of the high-temperature approximation, calculate the rotational partition function of the water molecule at 298 K using the following rotational constants: $A = 27.878$ cm^{-1}, $B = 14.509$ cm^{-1}, $C = 9.287$ cm^{-1}. Above what temperature is the high temperature approximation valid?

21.4. The methane molecule is a spherical top with bond length of 109 pm. Calculate its rotational partition function at (a) 298 K, (b) 500 K, using the high-temperature approximation in each case ($\sigma = 12$).

21.5. Calculate the rotational partition function of methane by direct summation of the rotational energy levels.

21.6. The Sackur–Tetrode equation (p. 685) gives the theoretical entropy of a monatomic gas. Derive the corresponding expression for a gas confined to move in two dimensions. Hence find an expression for the molar entropy of condensation of a gas to form a mobile surface film. What would be the change of entropy if the surface film were not mobile?

21.7. Calculate the rotational entropy of benzene that is free to rotate in three dimensions at 362 K. Its moments of inertia are $I_A = 2.93 \times 10^{-38}$ $g\,cm^2$, $I_B = I_C = 1.46 \times 10^{-38}$ $g\,cm^2$, and its symmetry number is 12. What would be the change of rotational entropy if the molecule were adsorbed on to a surface and could rotate only about its six-fold axis?

21.8. In an experimental study of the thermodynamics of adsorption of organic molecules on graphite at 362 K (D. Dollimore, G. R. Heal, and D. R. Martin, *J. chem. Soc. Faraday Trans. I* 1784 (1973)) a change in entropy of -111 $J\,K^{-1}$ mol^{-1} was observed when there was only little surface coverage, but it dropped to 52 $J\,K^{-1}\,mol^{-1}$ for complete coverage. Use the results of the two preceding Problems to propose a model of the motion of the benzene molecules on the surface.

21.9. The *symmetry number* σ can be calculated simply by counting the number of indistinguishable orientations of the molecule that can be reached by rotational symmetry operations. What is the value of σ in the case of (a) N_2, (b) NO, (c) benzene, (d) methane, (e) chloroform?

21.10. Calculate the room temperature (25 °C) entropy of ClO_2 (a bent molecule with an unpaired electron): OClO angle 118.5°, ClO bond length 149 pm, C_{2v} symmetry, $^2B_1$ electronic state.

21.11. Calculate and plot the equilibrium constant for the reaction $CD_4 + HCl \rightleftharpoons CHD_3 + DCl$ in the gas phase from the following data on the vibrational energy levels of the species: $\tilde{v}(CHD_3)$: 2993(3), 2142(1), 1003(3), 1291(2), 1036(2) cm^{-1}; $\tilde{v}(CD_4)$: 2109(1), 1092(2), 2259(3), 996(3) cm^{-1}; $\tilde{v}(HCl)$: 2991 cm^{-1}; $\tilde{v}(DCl)$: 2145 cm^{-1}. (Numbers in brackets are degeneracies.) $B(HCl) = 10.59$ cm^{-1}, $B(DCl) = 5.445$ cm^{-1}. Take 300 K $\leqslant T \leqslant$ 1000 K.

21.12. The exchange of deuterium between acid and water is an important type of equilibrium, and we can deal with it on the basis of vibrational data on two types of molecule. Calculate the equilibrium constant at (a) 298 K, (b) 800 K for the gas-phase exchange reaction $H_2O + DCl \rightleftharpoons HDO + HCl$ on the basis of the following vibrational and rotational data: $\tilde{v}(H_2O)$: 3656.7, 1594.8, 3755.8 cm^{-1}; $\tilde{v}(HDO)$: 2726.7, 1402.2, 3707.5 cm^{-1}; rotational constants: H_2O 27.88, 14.51, 9.29 cm^{-1}; HDO 23.38, 9.102, 6.417 cm^{-1}; HCl 10.59 cm^{-1}; DCl 5.449 cm^{-1}.

21.13. The equilibrium constant for the $I_2 \rightleftharpoons 2I$ equilibrium at 1000 K was treated from an experimental point of view in Problem 9.21. The spectroscopic data for I_2 is as follows: $B = 0.0373$ cm^{-1}, $\tilde{v} = 214.36$ cm^{-1}, $D_e = 1.5422$ eV. The iodine atoms have $^2P_{3/2}$ ground states, implying 4-fold degeneracy. Calculate a statistical thermodynamic value of the equilibrium constant at 1000 K.

21.14. Iodine atoms have quite low-lying excited states at 7603 cm^{-1} (2-fold degenerate). At 2000 K these are significantly populated. What effect does their inclusion have on the value of the equilibrium constant calculated in the last Problem?

21.15. Now return to the simpler version of the calculation in Problem 21.13. Can a magnetic field have an appreciable effect on the equilibrium constant of the dissociation equilibrium? Investigate this possibility by calculating the effect of a magnetic field on the partition functions of the atoms. What magnetic field would be needed to affect the equilibrium constant by 1 per cent?

21.16. The harmonic oscillator plays a specially important role in statistical thermodynamics (as well as in other subjects) because closed forms of the partition function and thermodynamic properties may be obtained; furthermore, molecular vibrations are normally well approximated by harmonic motion, and so the vibrational contribution to various thermodynamic properties may be calculated very easily. As a first step, deduce expressions for the internal energy, enthalpy, entropy, Helmholtz function, and Gibbs function of a harmonic oscillator, and plot the results as a function of $x = \hbar\omega/kT$.

21.17. The Giauque function $\Phi_0(T)$ is $[G_m(T) - H_m(0)]/T$, and its applications were explored in Chapter 9. Find an expression for $\Phi_0(T)$ of a harmonic oscillator.

21.18. Now put the harmonic oscillator calculation to use. Calculate the vibrational contribution at 1000 K to the Giauque function of (a) ammonia, which has vibrational modes of frequencies 3336.7(1), 950.4(1), 3443.8(2), 1626.8(2) cm^{-1}, (b) methane, with vibrational modes at 2916.7(1), 1533.6(2), 3018.9(3), 1306.2(3) cm^{-1}. (Numbers in brackets are degeneracies.) The easiest procedure is to use the graphs constructed in the earlier Problems, and to find x for each mode.

21.19. The Giauque function is well suited for equilibrium calculations under conditions where data have not been tabulated. Calculate the total values of $\Phi_0(1000 \text{ K})$ for (a) H_2, (b) Cl_2, (c) NH_3, (d) N_2, (e) NO using rotational and vibrational data given in Table 17.2. Use $B(NO) = 1.7406$ cm^{-1}, $\tilde{v}(NO) = 1904$ cm^{-1}.

21.20. Find the equilibrium constant for the $N_2 + 3H_2 \rightleftharpoons 2NH_3$ equilibrium at 1000 K on the basis of the data in the last Problem.

21.21. Although expressions like $d \ln q/d\beta$ are useful for formal manipulations in statistical thermodynamics, and for arriving at neat expressions for thermodynamic quantities, they are sometimes more trouble than they are worth in practical applications. If you are presented with a table of energy levels it is often much more convenient to evaluate the following sums directly: $q = \sum_j \exp(-\beta\varepsilon_j)$, $\dot{q} = \sum_j (\varepsilon_j/kT) \exp(-\beta\varepsilon_j)$, $\ddot{q} = \sum_j (\varepsilon_j/kT)^2 \exp(-\beta\varepsilon_j)$. As a first step in seeing how these sums are employed, find expressions for U, S, and C_V in terms of q, $\dot{q}$, and $\ddot{q}$.

21.22. Practical applications can be simplified still further if we separate the internal modes of the molecule from its translation. Show that $U = U_{ex} + U_{int}$, and $S = S_{ex} + S_{int}$, and deduce expressions for U_{int}, S_{int}, and $C_{V,int}$ in terms of q, $\dot{q}$, and $\ddot{q}$ for the internal molecular modes.

21.23. The thermodynamic properties of monatomic gases may be calculated from a knowledge of their electronic energy levels obtained from spectroscopy. Calculate the electronic contribution to (a) the enthalpy $H_m(T) - H_m(0)$, (b) the Giauque function $\Phi_0(T)$, (c) the electronic contribution to C_V of magnesium vapour at 5000 K from the data given below. Use the direct summation procedures outlined in the last two Problems.

Term:	^1S	$^3P_0$	$^3P_1$	$^3P_2$	$^1P_1$	^3S
Degeneracy:	1	1	3	5	3	3
Energy/cm^{-1}:	0	21850	21870	21911	35051	41197

21.24. A rich source of data on atomic energy levels is *Atomic Energy Levels*, C. E. Moore, N.B.S. Circ. No. 476 (1949). Calculations of this kind are ideal for programming on to a computer or programmable calculator, especially when many electronic states are accessible at the temperature of interest. If you have access to such a calculator, calculate the value of $F_0(T)$ from room temperature to 5000 K using the data in the table overleaf which has been reproduced from Moore's collection. If you have no such calculator, evaluate the elebtronic $\Phi_0(T)$ and C_V at 3000 K.

21.25. Compute the electronic contribution to the heat capacity of monatomic sodium vapour from room temperature to 5000 K.

21.26. Sodium boils at 1163 K, and the vapour consists of both monomers and dimers. Calculate the equilibrium constant for the dimerization at this temperature and the proportion of dimers in the vapour at the boiling point on the basis of the Giauque function $\Phi_0(1163 \text{ K})$ for the atoms and molecules. (The molecule has a singlet electronic state, $B = 0.1547 \text{ cm}^{-1}$, and $\tilde{v} = 159 \text{ cm}^{-1}$.) What experiment could be done to confirm the prediction of this calculation?

21.27. Use equipartition theory to predict the likely heat capacities of the following molecules at room temperature: (a) I_2, (b) H_2, (c) CH_4, (d) benzene vapour, (e) water vapour, (f) carbon dioxide.

21.28. Plot $C_{V,m}$ as a function of $x = \hbar\omega/kT$ for a harmonic oscillator, and predict the heat capacities of ammonia and methane at (a) 298 K, (b) 500 K. (Vibrational data in Problem 21.18.)

21.29. The fundamental frequencies of the seven normal modes of acetylene (ethyne) are 612, 612, 729, 729, 1974, 3287, 3374 cm^{-1}. What is the heat capacity of the gas at (a) 298 K, (b) 500 K?

21.30. Obtain an expression for the heat capacity of a system in which there are only two levels separated by Δ. Draw the temperature dependence as a function of Δ/kT.

21.31. The NO molecule has a doubly degenerate ground state and a doubly degenerate electronically excited state 121.1 cm^{-1} above. Calculate the electronic contribution to the heat capacity of this molecule at (a) 50 K, (b) 298 K, (c) 500 K.

21.32. Can an applied magnetic field modify the heat capacity of a molecule? Investigate the question by deducing an expression for the heat capacity of NO_2

Na I

Config.	Desig.	J	Level	Interval
3s	3s ^{2}S	$\frac{1}{2}$	0.000	
3p	3p ^{2}P°	$\frac{1}{2}$	*16 956.183*	*17.1963*
		$1\frac{1}{2}$	*16 973.379*	
4s	4s ^{2}S	$\frac{1}{2}$	25 739.86	
3d	3d ^{2}D	$2\frac{1}{2}$	29 172.855	−0.0494
		$1\frac{1}{2}$	29 172.904	
4p	4p ^{2}P°	$\frac{1}{2}$	*30 266.88*	*5.63*
		$1\frac{1}{2}$	*30 272.51*	
5s	5s ^{2}S	$\frac{1}{2}$	33 200.696	
4d	4d ^{2}D	$2\frac{1}{2}$	34 548.754	−0.0346
		$1\frac{1}{2}$	34 548.789	
4f	4f ^{2}F°	$\left\{\begin{array}{c}2\frac{1}{2}\\3\frac{1}{2}\end{array}\right\}$	34 588.6	
5p	5p ^{2}P°	$\frac{1}{2}$	*35 040.27*	*2.52*
		$1\frac{1}{2}$	*35 042.79*	
6s	6s ^{2}S	$\frac{1}{2}$	36 372.647	
5d	5d ^{2}D	$2\frac{1}{2}$	37 036.781	−0.0230
		$1\frac{1}{2}$	37 036.805	
5f	5f ^{2}F°	$\left\{\begin{array}{c}2\frac{1}{2}\\3\frac{1}{2}\end{array}\right\}$	*37 057.6*	
5g	5g ^{2}G	$\left\{\begin{array}{c}3\frac{1}{2}\\4\frac{1}{2}\end{array}\right\}$	37 060.2	
6p	6p ^{2}P°	$\frac{1}{2}$	*37 296.51*	*1.25*
		$1\frac{1}{2}$	*37 297.76*	
7s	7s ^{2}S	$\frac{1}{2}$	38 012.074	
6d	6d ^{2}D	$2\frac{1}{2}$	38 387.287	−0.0124
		$1\frac{1}{2}$	38 387.300	

(which has a single unpaired electron) in an applied magnetic field. What change of heat capacity is brought about by a 5.0 T field at (a) 50 K, (b) 298 K?

21.33. The energy levels of a methyl group attached to a larger molecule are given by the expression for a particle on a ring so long as it is rotating freely. What is the high-temperature contribution to the heat capacity and the entropy of such a freely rotating group? (Its moment of inertia is 5.341×10^{-47} kg m^2.)

21.34. As a result of the Pauli principle, molecular hydrogen exists in two varieties distinguished by the different relative orientations of the two proton spins. If the proton spins are opposed to each other the variety is called *para*-hydrogen. A characteristic of this species is that the rotational quantum number can take only even values ($J = 0, 2, 4, \ldots$). In a sample of pure *para*-hydrogen at low temperatures the main contribution to the heat capacity comes from excitations from $J = 0$ to $J = 2$. Calculate the temperature dependence of the heat capacity of *para*-hydrogen on the basis that its rotational levels constitute, in effect, a two-level system (but note the degeneracy of the $J = 2$ state). Use $B = 60.864$ cm^{-1} and sketch the heat capacity curve. The experimental heat capacity does in fact show a hump at low temperatures.

21.35. The heat capacity of a gas determines the speed of sound. Since we can calculate heat capacities from molecular data it follows that we also have a route to the calculation of this speed. We need to know that $c_s = (\gamma RT/M_m)^{1/2}$, where $\gamma = C_p/C_V$ and M_m is the molar mass. Deduce an expression for c_s for an ideal gas of diatomic molecules (a) at high temperatures (translation, rotation, but no vibration), (b) for *para*-hydrogen at low temperatures.

21.36. Use the result of the last Problem to estimate the speed of sound in air at room temperature.

21.37. What is the residual entropy of a crystal in which the molecules can adopt (a) 3, (b) 5, (c) 6 orientations of equal energy at absolute zero?

21.38. The hexagonal molecule $C_6H_nF_{6-n}$ might have a residual entropy on account of the similarity of the hydrogen and fluorine atoms. What might be the residual value of the entropy for each value of n and for each isomer?

21.39. The thermochemical entropy of nitrogen gas at 298 K was discussed in the *Example* on p. 151 and the value 192.06 J K^{-1} mol^{-1} was obtained. On the basis that the rotational constant $B = 1.9987$ cm^{-1}, and the vibrational wavenumber is 2358 cm^{-1}, calculate the statistical thermodynamic value of the entropy at this temperature. What does the value suggest about the nature of the crystal at absolute zero?

22 Determination of molecular structure: diffraction methods

Learning objectives

After careful study of this chapter you should be able to

(1) Derive the *Bragg condition* for diffraction (eqn (22.1.1)).

(2) Specify crystal faces and the planes in a crystal lattice by their *Miller indices* (p. 729).

(3) Describe the *Laue method*, the *Bragg method*, and the *Debye–Scherrer method* of X-ray structural analysis of crystals (p. 732).

(4) *Index* reflections, and identify the nature of the unit cell from the *systematic absences* in the diffraction pattern (p. 734).

(5) Relate the structure of simple lattices to the *intensities* of X-ray reflections (p. 737).

(6) Define the *structure factor* (eqn (22.3.4)) and relate it to intensities and electron densities (eqn (22.3.7)).

(7) Describe the procedure for an X-ray *structure analysis* (p. 744).

(8) Outline the significance of and the methods of overcoming the *phase problem* (p. 741).

(9) Describe *hexagonal, cubic, and body-centred close-packed* arrangements of identical spheres (p. 746).

(10) State the *radius-ratio* rule for ionic crystals (p. 750).

(11) Draw the *rock salt, wurtzite,* and *caesium chloride* lattices (p. 749).

(12) Explain how the *absolute configurations* of molecules may be determined (p. 750).

(13) Indicate what information is available from a *neutron diffraction* experiment (p. 752).

(14) Describe the *electron diffraction* experiment (p. 753).

(15) Use the *Wierl equation* to deduce bond lengths and bond angles (p. 755).

Introduction

The diffraction of waves is the basis of several powerful methods for the determination of molecular structure. Sound waves and light waves are diffracted by objects with dimensions comparable to the wavelength of the radiation, and since X-rays have wavelengths comparable to the spacing of atoms in crystals (0.1 nm, or 100 pm) they are diffracted by crystal lattices. From the intensity pattern of the diffracted X-rays it is possible to draw up a detailed picture of the positions of atoms in molecules even as complex as proteins. It is even possible to determine the electron density distribution in individual bonds.

Electrons moving with a velocity of $20\,000\,\mathrm{km\,s^{-1}}$ (after acceleration by a potential difference of 4 kV) have a wavelength of 40 pm, and so they too may be used in diffraction studies. Neutrons generated in a nuclear reactor, and then slowed to thermal energies, have wavelengths of about the same magnitude, and are also extensively used.

22.1 General features of diffraction

A diffraction pattern can be regarded as arising either from the wave nature or from the corpuscular nature of the radiation. Both aspects give valuable insight into the nature of the diffraction experiment and its interpretation. The wave picture is treated here and the particle picture is described in the Appendix (p. 756). The wave interpretation of diffraction is based on the constructive and destructive interference that occurs when waves are superimposed. If the amplitudes are in-phase at some point, they augment each other and the intensity there is enhanced; where the amplitudes are out-of-phase they cancel, and the intensity is decreased. If the waves start from a common source their relative phase at a point depends on their path lengths. For instance, the diffraction pattern of the Young's slit experiment (Fig. 22.1) can easily be explained in terms of the path difference of the two rays, different points of the screen corresponding to places where the waves from the two slits are successively

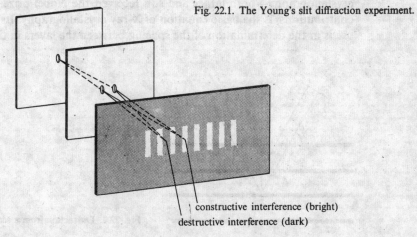

Fig. 22.1. The Young's slit diffraction experiment.

constructive interference (bright)
destructive interference (dark)

in phase (bright) and out of phase (dark).

Consider now a stack of reflecting layers arranged as shown in Fig. 22.2. The path-length difference for two rays reflected from two neighbouring layers is

$$AB + BC = 2d \sin \theta,$$

where d is the layer spacing and θ the angle defined in Fig. 22.2. For many choices of θ this path-length difference is not an integral number of wavelengths: in these cases the superposition of the two rays is out-of-phase, and the intensity is decreased. In other words, a viewer might observe no intensity when the source and detector are arranged at a general angle θ. At some angles, however, the path-length difference $AB + BC$ is an exact integral multiple of the wavelength, the amplitudes of all the waves deflected by the stack of layers are in phase, and the intensity is considerable. If the wavelength is λ, this constructive interference occurs when $AB + BC = n\lambda$, n an integer. Therefore the angle for constructive interference is given by the

(22.1.1)   *Bragg condition:* $n\lambda = 2d \sin \theta.$

Example (Objective 1). The separation of the layers of atoms in a crystal is 404 pm. At what angle will a reflection occur in a diffractometer using CuK_α X-rays (wavelength 154 pm).

- *Method.* Substitute directly into eqn (22.1.1). Use $n = 1$ for the first-order reflection.

- *Answer.* From eqn (22.1.1),

  $\theta = \sin^{-1}(\lambda/2d) = \sin^{-1}(154/808) = \sin^{-1}(0.191) = 10° 59'.$

- *Comment.* The second-order reflection will occur at $22° 27'$. With MoK_α X-rays ($\lambda = 70.8$ pm) the first-order reflection is at $5° 2'$: notice that the shorter the wavelength the smaller the diffraction angle.

The Bragg condition (named after one of the pioneers of X-ray crystallography—both father and son received the Nobel prize for their contributions) is the basic equation of X-ray crystallography. Its primary use is in the determination of the spacing between the layers in the lattice

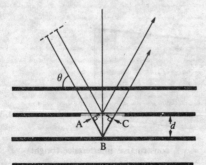

Fig. 22.2. Diffraction from a set of planes.

of a crystal, for by determining the angle θ at which maximum intensity is observed, d may readily be calculated if the incident wavelength is known.

22.2 Crystal lattices

The interpretation of diffraction patterns depends on knowing how atoms can be arranged in space. In Chapter 16 we saw the qualitative aspect of this problem when we investigated the structure of crystals on the basis of the symmetry of the way their components, the unit cells, were stacked together. In this section we turn to a quantitative discussion of crystal structure. We then return to the problems of measuring the lattice and locating the atoms by X-ray diffraction.

If the sites of atoms or ions are denoted by points, when they are stacked together in a crystal their positions define a *lattice*. The *unit cell* is a selected small regular figure which, by continuous replication, generates the entire lattice. Early evidence that crystals can be regarded as a lattice of atomic sites came from the *law of rational indices*. This states that *the intercepts of crystal planes on three suitably chosen axes set in the crystal can be expressed as integral multiples of three basic dimensions*. This points very clearly to the construction of a crystal from uniform building blocks, as illustrated in Fig. 16.30, and as discussed from the viewpoint of symmetry in Chapter 16.

Labelling the planes: the Miller indices. The planes of lattice points form the arrays that cause diffraction. Even in a rectangular lattice a large number of different planes can be selected from the array of lattice points, Fig. 22.3, and it is important to have a labelling scheme.

First consider a simple two-dimensional rectangular lattice formed from unit cells of side a,b, Fig. 22.3. In the figure four sets of planes have been selected, and it is clear that they can be distinguished by the distances along the axes where one representative member of each set of planes

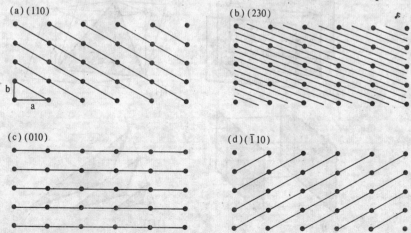

Fig. 22.3. Miller indices of some planes in a two-dimensional lattice.

intersects them. One labelling scheme is to denote each set by the distances along the two axes to the points of intersection and to choose as the representative member the plane that gives the least distances. The four sets shown could then be denoted respectively $(1a, 1b)$; $(3a, 2b)$; $(-a, b)$; and $(\infty a, 1b)$. If we agree always to quote distances along the unit cell axes in terms of the length of the cell in that direction, these planes can be denoted more simply as $(1, 1)$, $(3, 2)$, $(-1, 1)$, and $(\infty, 1)$. Furthermore, if the lattices as drawn are the top view of the three-dimensional rectangular array where the unit cell is of length c in the z-direction, all the three sets of planes intersect z at ∞c, and so the full labels of the sets of planes could be $(1, 1, \infty)$, $(3, 2, \infty)$, $(-1, 1, \infty)$, and $(\infty, 1, \infty)$. The appearance of ∞ is inconvenient, and a way of avoiding it would be to deal with the *reciprocals* of the indices. This turns out to have further advantages, as we shall see. The *Miller indices* are the reciprocals of the numbers in the brackets, with fractions cleared. For example, all the planes parallel to $(1a, 1b, \infty c)$, the $(1, 1, \infty)$ planes, become (110) in the Miller system. Similarly the $(3a, 2b, \infty c)$ planes in Fig. 22.3b turn from $(3, 2, \infty)$ to $(\frac{1}{3}, \frac{1}{2}, 0)$ when reciprocals are taken, to $(2, 3, 0)$ when fractions are cleared, and are therefore referred to as (230) planes in the Miller system. The Miller indices of the planes in Fig. 22.3c are (010). Negative indices (as for the $(-1, 1, \infty)$ planes shown in Fig. 22.3d) are written with a bar above the number, e.g., $(\bar{1}10)$.

One point to note is that a Miller index like (110) denotes a *set* of parallel planes; that is why we refer to the '(110) planes'. Sometimes it is necessary to refer to individual planes in a set; then we can refer to the '(110) plane' as the one that cuts the a, b, and c axes at distances a, b, and ∞ from some origin. The (220) plane would then be a plane lying parallel to the (110) plane, but cutting the axes at $\frac{1}{2}a$, $\frac{1}{2}b$, and ∞ from the same origin. A further use of the Miller indices (and the one for which they were originally introduced) is to denote a crystal *face*. Thus we might

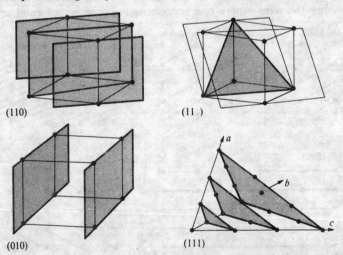

(110) (111)

(010) (111)

Fig. 22.4. Some planes and their indices.

refer to the (110) face as one parallel to a plane cutting the a, b, and c axes at distances a, b, and ∞.

Two things should be remembered about the Miller indices in order to see quickly which planes are under discussion. Refer to Fig. 22.4. First, the smaller the number h in the index (hkl) the more parallel are the planes to the a-axis. The same applies for k and the b-axis and l and the c-axis. When h is zero the intercepts of the planes with the a-axis are at infinity, and so the planes are then parallel to the a-axis. Similarly $k = 0$ indicates planes parallel to b, and $l = 0$ planes parallel to c. Secondly, the Miller index system is not confined to crystals with orthogonal (perpendicular) unit cell axes. Figure 22.4 also shows one indexing of a triclinic system: this is based on the same procedure as before, but the axes are no longer orthogonal.

One advantage of the Miller indices is that in a cubic crystal it is a simple matter to express the distances between neighbouring planes in terms of the indices h, k, l themselves. Referring to Fig. 22.5, it is easily confirmed that in two dimensions the separation is

$$d_{hk} = a/(h^2 + k^2)^{1/2},$$

and, by extension to three dimensions, that the separation of the (hkl) planes of a cubic crystal is

(22.2.1) $$d_{hkl} = a/(h^2 + k^2 + l^2)^{1/2}.$$

This is as far as we need go with labelling planes of atoms. We are now in a position to apply X-ray diffraction to the measurement of the separations d_{hkl}.

Example An orthorhombic unit cell has the following parameters: $a \doteq 50\,\text{pm}$ (5 Å), $b = 100\,\text{pm}$ (10 Å), $c = 150\,\text{pm}$ (15 Å). What is the spacing of the (123) planes?

- *Method.* The (hkl) planes are spaced by a distance d_{hkl} given by $1/d_{hkl}^2 = (h/a)^2 + (k/b)^2 + (l/c)^2$.

- *Answer.* $1/d_{123}^2 = (1/50\,\text{pm})^2 + (2/100\,\text{pm})^2 + (3/150\,\text{pm})^2 = 3(1/50\,\text{pm})^2$. Therefore

$$d_{123} = (50\,\text{pm})/\sqrt{3} = 29\,\text{pm}.$$

- *Comment.* The expression used is a generalization of eqn (22.2.1) and is applicable to any orthorhombic ($\alpha = \beta = \gamma = 90°$) cell.

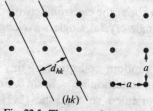

Fig. 22.5. The separation between lattice planes. (21) is illustrated.

22.3 X-ray crystallography

X-rays are produced by bombarding a metal with high-energy electrons. When the electrons plunge into the material two contributions to the overall X-radiation are generated. First there is a continuous background of radiation spanning a range of wavelengths. This continuum is called the *Bremsstrahlung* because it arises from the rapid deceleration of the electrons when they sink into the metal (*Bremsse* is German for brake, *Strahlung* for ray). Superimposed on this continuum are a few sharp peaks of high intensity. These peaks arise from interactions of the electrons with the electrons of the inner shells of the atoms of the metal. A collision knocks an electron out of an inner electron shell, and an electron in a higher energy shell drops into the vacancy, emitting the excess energy as a photon of high energy.

On the basis of fragments of evidence that the newly discovered X-rays might have wavelengths comparable to atomic spacings in crystals, von Laue suggested (in 1912) that they might be diffracted when passed through a crystal. This was confirmed almost immediately by Friedrich and Knipping, and so the technique of X-ray crystallography was created.

Laue's method consisted of passing a beam of X-rays of a broad band of wavelengths into a single crystal, and then recording the diffraction pattern on a photographic plate. The idea behind this approach is that a single crystal might not be suitably orientated to act as a diffraction grating for a single wavelength, but if the beam contains a wide range of wavelengths the diffraction condition will be satisfied by every orientation because a wavelength that satisfies the appropriate Bragg condition occurs in the beam.

Bragg's modification of the Laue procedure was to simplify the diffraction pattern by using a monochromatic beam. This was formed by passing a beam through a metal foil filter which eliminated the broad background frequencies but permitted the passage of one of the high-intensity monochromatic peaks.

An alternative to the Laue and Bragg methods (which both used single crystals) was introduced by Debye and Scherrer and by Hill. They used monochromatic radiation but a powdered sample. Since a powder is a collection of a large number of minute crystals packed together at random orientations, a proportion of them will always satisfy the diffraction condition even though the radiation is monochromatic. This technique is useful for a qualitative analysis of the material, and for an initial identification of the dimensions and symmetry of the unit cell, but it is unable to provide the detailed information about the electron density distribution which is available from the Bragg single-crystal, single-wavelength method.

The powder method. When the sample is a random collection of tiny crystals, a proportion of them satisfy the Bragg condition $n\lambda = 2d \sin \theta$. For example, some of the crystallites will be oriented so that the (111) planes, of spacing d_{111}, give rise to a diffracted intensity at the angle 2θ to the

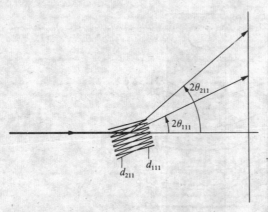

Fig. 22.6. Diffraction from two sets of lattice planes of a single crystal.

incident beam, Fig. 22.6. The (111) planes of other crystallites may be at an angle θ to the incoming beam, but at an arbitrary angle about the line of its approach. It follows that the diffracted beams lie on the surface of a cone with apex 4θ. Other crystallites will be oriented so that, for example, their (211) planes, with spacing d_{211}, satisfy the diffraction condition and give rise to a deflection through an angle $2\theta'$: this accounts for another cone of diffracted radiation, Fig. 22.6. In principle every plane (*hkl*) gives rise to a diffraction cone because some members of the randomly oriented sample are able to diffract the incoming monochromatic beam.

The original *Debye–Scherrer method* is illustrated in Fig. 22.7. The beam of monochromatic X-rays enters the powdered sample held in a capillary tube and rotated in order to ensure that the crystallites are distributed at random with respect to the incoming beam. The deflected rays describe cones, and are detected as arcs of circles where they cut the strip of photographic film wrapped round the camera. Typical powder photographs are shown in Fig. 22.8. The diffraction angle (2θ) for each cone

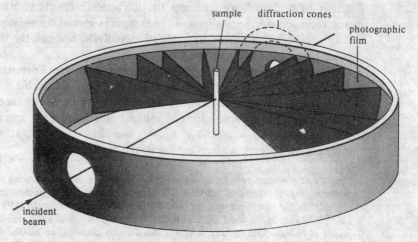

Fig. 22.7. The Debye–Scherrer X-ray diffraction method.

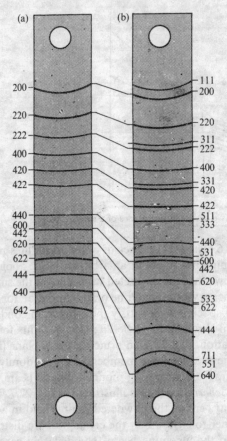

Fig. 22.8. X-ray powder photographs for (a) KCl and (b) NaCl.

can be determined from the position of the photographic image. If the cone can be ascribed to a particular (hkl) plane—this is called *indexing the reflection*—the value of d_{hkl} can be determined from the Bragg condition. In modern diffractometers, the diffraction intensities are monitored and recorded electronically, and the sample is spread on a flat plate. The major application now is to qualitative analysis, because the diffraction pattern is a kind of fingerprint.

The crux of the analysis is the indexing of the pattern. Fortunately some types of unit cell give characteristic and easily recognizable patterns of lines, and the type of cell and its dimensions can be determined without any difficulty. In a cubic lattice of unit cell dimension a the spacing of the (hkl) planes is $d_{hkl} = a/(h^2 + k^2 + l^2)^{1/2}$; therefore the angle at which the (hkl) planes diffract is given by the Bragg condition as*

---

* Note that $n = 1$ has been used, and so it may appear that we are confining attention to *first-order* ($n = 1$) reflections (reflection is a name often used for a diffraction spot). This is not actually the case, because a *second-order* ($n = 2$) reflection from the (110) planes, for example, occurs at the same angle as a first-order reflection from (220) planes which are planes of half the spacing: $d_{220} = \frac{1}{2}d_{110}$. Therefore by allowing the indexes (hkl) to run over all values we automatically accommodate the higher-order reflections.

(22.3.1) $\sin \theta_{hkl} = (\frac{1}{2}\lambda/a)(h^2 + k^2 + l^2)^{1/2}.$

In order to predict the diffraction pattern for the cubic lattice all we need do is insert the permitted values of h, k, and l into the last expression. This gives for $\sin^2 \theta$ the values (in multiples of $(\frac{1}{2}\lambda/a)^2$)

(hkl)	(100)	(110)	(111)	(200)	(210)	(211)	(220)	(300)	(221)	(310)...
$h^2 + k^2 + l^2$	1	2	3	4	5	6	8	9	9	10 ...

Notice that 7 is missing from the second row (as is 15, and others) because the sum of three squares of integers cannot equal 7 (or 15, etc.). Therefore the pattern has omissions which are characteristic of the primitive cubic structure.

Example (Objective 4). A powder diffraction photograph of KCl gave lines at the following distances from the centre spot when Mo X-rays (wavelength 70.8 pm) were used in a camera of radius 5.74 cm: 13.2, 18.4, 22.8, 26.2, 29.4, 32.2, 37.2, 39.6, 41.8, 43.8, 46.0, all in mm. Index the lines, identify the kind of unit cell, and determine its size.

- *Method.* From eqn (22.3.1) we have $\sin^2 \theta = (\lambda/2a)^2 (h^2 + k^2 + l^2)$. Therefore convert distances to θ and then to $\sin^2 \theta$, find the common factor $A = (\lambda/2a)^2$, then find $h^2 + k^2 + l^2$. Express this as (hkl). For the cell dimension, solve $A = (\lambda/2a)^2$ for a.

- *Answer.* The distance from the centre spot, D, is related to the diffraction angle (in radians) by $\theta = D/2R$. Convert to degrees by multiplying by $360/2\pi$. With $R = 57.4$ mm, we have $\theta/\text{degrees} = \frac{1}{2}(D/57.4 \text{ mm})(360/2\pi) = \frac{1}{2}D/\text{mm}$, and so the conversion from D to θ is very simple. Draw up the following table:

D/mm	13.2	18.4	22.8	26.2	29.4	32.2	37.2	39.6	41.8	43.8	46.0
θ/deg	6.6	9.2	11.4	13.1	14.7	16.1	18.6	19.8	20.9	21.9	23.0
$100 \sin^2 \theta$	1.32	2.56	3.91	5.14	6.44	7.69	10.17	11.47	12.73	13.91	15.27

The common divisor is 1.32/100: divide through to identify $h^2 + k^2 + l^2$

$h^2 + k^2 + l^2$	1	2	3	4	5	6	8	9	10	11	12

and attempt to express these in the form (hkl):

(hkl)	(100)	(110)	(111)	(200)	(210)	(211)	(220)	(300)	(310)	(311)	(222)

This has indexed the lines. Note an absence between 6 and 8: this indicates that the cell is primitive cubic. From $(\lambda/2a)^2 = 0.0132$ we find $a = 308$ pm.

- *Comment.* The same type of procedure can be used for any powder spectrum. Later we shall see that additional information comes from the intensities of the lines. Note that a cunning choice of R saves a lot of work.

This example confirms the features of the analysis just given and enables the cell size in KCl to be determined. When the method is applied to the closely analogous NaCl crystal the diffraction pattern is surprisingly different. The reason for this is that the K^+ and Cl^- ions, having the same number and configuration of electrons, act almost identically as scatterers of X-rays, and so although the lattice contains two sorts of ions the X-rays scatter as though it were a cubic lattice composed of identical ions. In the case of NaCl the scattering power of the Na^+ ions is less than that of the Cl^- ions, and the analysis presented so far has to be modified to allow for the effects of two types of ion in the same crystal.

This problem can be discussed qualitatively by regarding the NaCl structure as being composed of two interpenetrating cubic lattices, one of Na^+ ions and the other of Cl^- ions. Some reflections from the Na^+ lattice are out of phase with the reflections from the Cl^- lattice and interfere destructively. For other orientations the two sets of reflections are in phase, and intense lines appear. Cancellation is complete only if the ions have identical scattering power, but in the NaCl case the cancellation is incomplete, and an alternation of lines, some intense, some faint, appears. This accounts for the general appearance of the NaCl diffraction pattern shown in Fig. 22.8.

The nature of the diffraction pattern can be discussed quantitatively by considering the two-dimensional lattice shown in Fig. 22.9. This is composed of two interpenetrating lattices of atoms of type A and B. Their scattering powers are measured by the quantities f_A and f_B; these are called the *scattering factors* of the atoms. Since electrons are responsible for the scattering process the scattering factor is roughly proportional to atomic number; it also depends on the angle through which the scattering takes place, and is less for sidewards than for forward scattering.

Consider now a single unit cell and its overall contribution to the diffraction pattern. We concentrate on a cell in which there is an A atom at the origin, and a B atom at the position (xa, yb). When the incident beam is at a diffracting angle θ with respect to the (hk) plane of A atoms it is also at the same angle to the (hk) plane containing the B atoms, Fig. 22.9. The A and B planes are some distance d'_{hk} apart, and so reflections from the B planes contribute waves with phases shifted from those due to the A planes. Reference to Fig. 22.9 shows that the path-length difference between waves deflected by the A and B planes is $2d'_{hk} \sin\theta$. Since θ is a diffraction angle, it is given by $\sin\theta = \lambda/2d_{hk}$, and so the A,B path-length

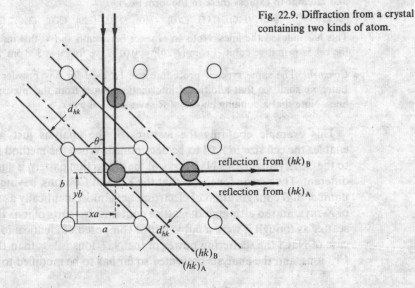

Fig. 22.9. Diffraction from a crystal containing two kinds of atom.

reflection from $(hk)_B$

reflection from $(hk)_A$

difference is $d'_{hk}\lambda/d_{hk}$. Simple trigonometric arguments show that $d'_{hk} = (hx+ky)d_{hk}$, and so the path difference is $(hx+ky)\lambda$. It follows that the A,B phase difference is $[(hx+ky)\lambda](2\pi/\lambda)$, or $2\pi(hx+ky)$. In three dimensions this becomes

(22.3.2)   A,B phase difference $= 2\pi(hx+ky+lz)$,

where the atom B is at the point (xa, yb, zc) in the unit cell and A is at the origin. (The last equation illustrates another result that can be expressed in a simple way by using the Miller indices.)

When the A, B phase difference is 180° (π radians) the amplitudes of the two waves cancel, and *if the atoms have equal scattering power* the intensity is completely annihilated. For example, if the unit cells are body-centred, each one contains an atom at $x = y = z = \frac{1}{2}$, and so

A,B phase difference $= \pi(h+k+l)$.

Therefore, all reflections corresponding to odd values of the sum $h+k+l$ vanish because the phases are 180° displaced. This means that *the pattern of scattering from a body-centred cubic lattice can be constructed from the primitive lattice simply by striking out all the reflections with $h+k+l$ odd.* Recognition of these *systematic absences* in a powder spectrum, Fig. 22.10, immediately indicates a body-centred lattice. A similar calculation can be done for the face-centred lattice, and the pattern of lines is also indicated in Fig. 22.10.

If the amplitude of the wave scattered from the A atoms is f_A at the detector, the amplitude of the wave scattered from B is $f_B \exp[2\pi i(hx+ky+lz)]$ because of the extra phase difference. Therefore the total amplitude at the detector is the sum

$$F = f_A + f_B \exp[2\pi i(hx+ky+lz)]$$

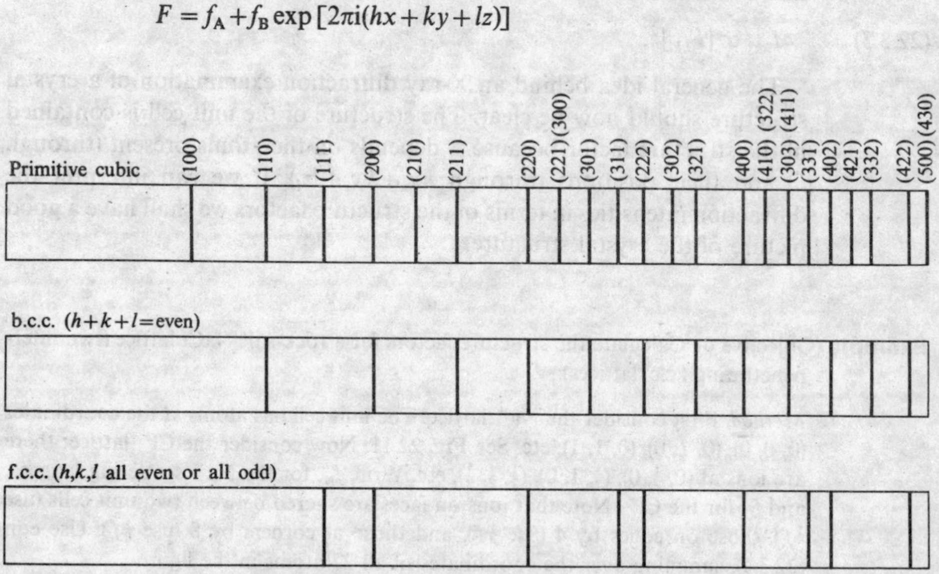

Fig. 22.10. Patterns and absences in powder photographs.

and the intensity there is

$$I_{hkl} \propto F^*F = |f_A + f_B \exp\{2\pi i T(hkl)\}|^2$$

because intensity of radiation is proportional to the square modulus of the amplitude of the wave. We have used the abbreviation $T(hkl) = hx + ky + lz$ for simplicity. This expression expands into

$$I_{hkl} \propto f_A^2 + f_B^2 + f_A f_B[\exp\{2\pi i T(hkl)\} + \exp\{-2\pi i T(hkl)\}]$$

(22.3.3) $$\propto f_A^2 + f_B^2 + 2f_A f_B \cos\{2\pi T(hkl)\}.$$

The cosine term either adds to or subtracts from $f_A^2 + f_B^2$ depending upon the value of $T(hkl)$, and so there is a variation in the intensity of the lines of different index (hkl) in the diffraction pattern. This is exactly what is observed in NaCl, and since the atomic scattering factors are known, it is a reasonably straightforward matter to extract the unit cell geometry from the intensities of the reflections via eqn (22.3.3).

The determination of crystal structures. If the unit cell contains several atoms with scattering factors f_i and coordinates $x_i a$, $y_i b$, and $z_i c$ within the unit cell, the overall amplitude of a wave from the (hkl) planes can be written as a simple extension of what has been done already:

(22.3.4)   total amplitude $= \sum_{i \atop \text{(unit cell)}} f_i \exp\{2\pi i(hx_i + ky_i + lz_i)\}.$

This sum is called the *structure factor*, and denoted F_{hkl}. The intensity of the reflection from the (hkl) planes of the crystal with such a composite unit cell is therefore

(22.3.5) $$I_{hkl} \propto |F_{hkl}|^2.$$

The general idea behind an X-ray diffraction examination of a crystal structure should now be clear. The structure of the unit cell is contained in the structure factor because it depends on the atoms present (through f_i) and their positions (through $hx_i + ky_i + lz_i$). If we can interpret the diffraction intensities in terms of the structure factors we shall have a good picture of the crystal structure.

Example (Objective 6). Calculate the structure factors for a rock-salt NaCl lattice (two interpenetrating f.c.c. lattices).

- *Method.* First consider the Na^+ lattice. The unit cell has atoms at the coordinates $(0, 0, 0)$, $(0, 1, 0)$, $(0, 1, 1)$, etc. See Fig. 22.11. Now consider the Cl^- lattice: there are ions at $(0, \frac{1}{2}, 0)$, $(\frac{1}{2}, 1, 0)$, $(\frac{1}{2}, \frac{1}{2}, \frac{1}{2})$, etc. Write f_{Na} for the Na^+ scattering lengths, and f_{Cl} for the Cl^-. Note that ions on faces are shared between two unit cells (use $\frac{1}{2}f$), those on edges by 4 (use $\frac{1}{4}f$), and those at corners by 8 (use $\frac{1}{8}f$). Use eqn (22.3.4), summing over the coordinates of all 27 atoms in the figure.

- *Answer.* In order to save space we shall write only a few terms of each line:

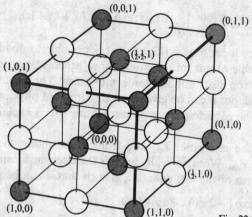

Fig. 22.11. A structure factor calculation.

$$F_{hkl} = f_{Na}[\tfrac{1}{8} + \tfrac{1}{8}\exp\{2\pi il\} + \ldots + \tfrac{1}{8}\exp\{2\pi i(h+k+l)\} + \ldots +$$
$$+ \tfrac{1}{2}\exp\{2\pi i(\tfrac{1}{2}h + \tfrac{1}{2}k + l)\}]$$
$$+ f_{Cl}[\exp\{2\pi i(\tfrac{1}{2}h + \tfrac{1}{2}k + \tfrac{1}{2}l)\} + \tfrac{1}{4}\exp\{2\pi i(\tfrac{1}{2}h)\} + \ldots + \tfrac{1}{4}\exp\{2\pi i(\tfrac{1}{2}h + l)\}].$$

By using $\exp(2\pi ih) = \exp(2\pi ik) = \exp(2\pi il) = 1$, because all h, k, l are integers and $\exp(2\pi i) = 1$, this simplifies considerably, and gives

$$F_{hkl} = f_{Na}[1 + \cos(h+k)\pi + \cos(h+l)\pi + \cos(k+l)\pi]$$
$$+ f_{Cl}[(-1)^{h+k+l} + \cos k\pi + \cos l\pi + \cos h\pi].$$

This may also be simplified because $\cos h\pi = (-1)^h$, and so

$$F_{hkl} = f_{Na}[1 + (-1)^{h+k} + (-1)^{h+l} + (-1)^{l+k}]$$
$$+ f_{Cl}[(-1)^{h+k+l} + (-1)^h + (-1)^k + (-1)^l].$$

This is a sufficient simplification for our purpose. Note that if

h, k, l are all *even*: $F_{hkl} = f_{Na}(1+1+1+1) + f_{Cl}(1+1+1+1) = 4(f_{Na} + f_{Cl})$;

if h, k, l are all *odd*: $F_{hkl} = 4(f_{Na} - f_{Cl})$;

if one odd, two even, or vice versa, $F_{hkl} = 0$.

- *Comment.* Note that the (hkl)-all odd are less intense than the (hkl)-all even lines. If $f_+ = f_-$ (as in KCl), the former have zero intensity, corresponding to the systematic absences of simple cubic unit cells.

This analysis can be taken further. We might not wish to divide up the unit cell into atoms, each with a contribution of strength f_i to the scattering pattern (in covalent structures it is not even possible to do this, except as an approximation). Instead we may think of a small volume $d\tau$ situated at the point $\mathbf{r} = (xa, yb, zc)$ in the unit cell, and in that volume an amount $\rho(\mathbf{r})\,d\tau$ of electronic charge. The amount of scattering from that volume element is proportional to the number of electrons it contains, and so instead of obtaining the structure factor F_{hkl} by summing over all the atoms present we can obtain it by integrating over all the volume elements. Equation (22.3.4) is then replaced by

(22.3.6)   total amplitude $= F_{hkl} \propto \int_{\substack{\text{unit}\\\text{cell}}} d\tau \rho(\mathbf{r}) \exp\{2\pi i(hx+ky+lz)\}.$

This expression reduces to eqn (22.3.4) if it is possible to divide the unit cell into identifiable atoms, but is much more general because it lets us explore the electron distribution inside molecules. The equation looks quite complicated, but it can be given a straightforward physical interpretation if we use the particle picture of diffraction. This is explained in the Appendix.

We measure the intensities I_{hkl} and from them obtain the observed structure factors F_{hkl}. What we really want is the electron density distribution $\rho(\mathbf{r})$. The form of the last equation is such that it can be inverted:

(22.3.7) $\rho(\mathbf{r}) = \sum_{hkl} F_{hkl} \exp\{-2\pi i(hx+ky+lz)\}.$

This is called a *Fourier synthesis* of the electron density. The sum may be evaluated in order to find the electron density at the point $\mathbf{r} = (xa, yb, zc)$, and therefore throughout the unit cell, if we know the structure factors for all the reflections of the crystal.

Example (Objective 5). Consider the $(h00)$ planes of a crystal extending indefinitely in the x-direction. In an X-ray structural analysis the structure factors for the planes were found to be as follows: 16(0), $-10(1)$, 2(2), $-1(3)$, 7(4), $-10(5)$, 8(6), $-3(7)$, 2(8), $-3(9)$, 6(10), $-5(11)$, 3(12), $-2(13)$, 2(14), $-3(15)$: the figures in brackets are the values of h, and $F_h = F_{-h}$. Construct a plot of the electron density along the x-axis of the cell.

- *Method.* Use eqn (22.3.7), noting that, since $F_h = F_{-h}$,

    $\rho(x) = F_0 + 2 \sum_{h>0} F_h \cos(2\pi hx).$

Evaluate the sum for $x = 0, 0.1, 0.2, \ldots 1.0$.

- *Answer.* The results of the evaluation of the 16-term sum at the 11 positions along the unit cell are plotted in Fig. 22.12 (full line).

- *Comment.* The positions of two or three atoms can be discerned very readily. The more terms there are in the sum the more accurate the density plot: terms corresponding to high values of h account for finer details of the variation of the electron density; broad features come from low values of h.

The problem at this stage is to obtain the intensities of all the diffraction spots, and to know which (hkl) planes are responsible for them. This is a considerable problem, and single-crystal techniques have to be used.

The basic single crystal technique uses the *oscillation camera*, in which the crystal (which might typically be of side 0.1 mm) is set on a mount called a *goniometer head* and oscillated through about 10°. The diffraction pattern is recorded on a cylindrical film surrounding the sample. The pattern can be simplified by placing a screen in front of the film so that only one *layer line* of spots is exposed, and gearing the film to the oscillating goniometer head so that it moves parallel to the axis of oscillation of the

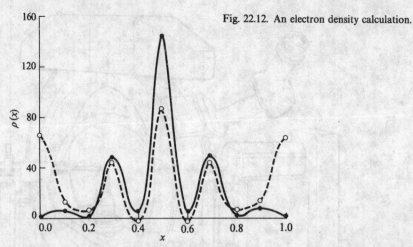

Fig. 22.12. An electron density calculation.

crystal: this is the *Weissenberg technique*. The advantage is that the indexing is greatly simplified, but the disadvantage is that the photographs are severely distorted. A refinement which overcomes this difficulty is to have a different coupling between the motions of the crystal and the screened film. In the *precession camera* the pattern is undistorted and can be indexed virtually at a glance.

Computing techniques are now available that lead not only to automatic indexing but also to the automatic determination of the shape, symmetry, and size of the unit cell. The most sophisticated technique in use at present uses the *four-circle diffractometer*, illustrated in Fig. 22.13. The crystal is set in an arbitrary orientation on a goniometer head. Since its unit cell properties will already have been determined (using a precession camera, for instance) the settings of the diffractometer's four angles that are needed to observe any particular *hkl* reflection can be computed. The computer controls the settings, and moves the diffractometer to each one in turn. There the diffraction intensity is monitored (using some kind of crystal detector or photomultiplier) and background intensities are assessed by making measurements at slightly different settings. The accumulated data for all the *hkl* reflections monitored are then converted into the structure factors F_{hkl}.

The phase problem. From the measured I_{hkl} we get the F_{hkl}, and then do the sum in eqn (22.3.7) to find $\rho(\mathbf{r})$. Unfortunately I_{hkl} is proportional to the *square modulus* $|F_{hkl}|^2$, and so we cannot say whether we should use $+|F_{hkl}|$ or $-|F_{hkl}|$ in the sum that gives $\rho(\mathbf{r})$. The situation is in fact worse than this suggests, because if we write F_{hkl} as the complex number $|F_{hkl}|e^{i\varphi}$, where $|F_{hkl}|$ is the real magnitude of F_{hkl} and φ is its phase, then the intensity lets us determine $|F_{hkl}|$ but reveals nothing about φ. This indeterminance of the complete form of F_{hkl} is the infamous *phase problem* of the X-ray method. Some way must be found to assign phases to the structure factors, otherwise the sum for $\rho(\mathbf{r})$ could not be evaluated and the X-ray method would be useless.

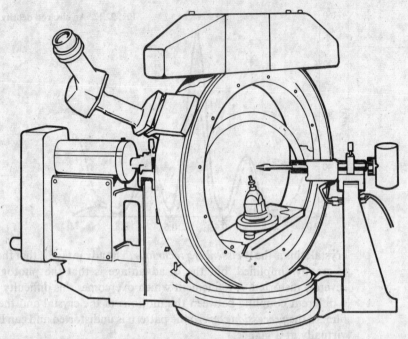

Fig. 22.13. The four-circle diffractometer (Nicolet XRD Corporation).

Example (Objectives 6, 7). Repeat the last *Example*, but suppose that all the structure factors after $h = 6$ have the same, positive, phase.

- *Method*. Repeat the evaluation of the 16-term sum, but with positive values of F_h after $h = 6$.

- *Answer*. The resulting electron density is plotted in Fig. 22.12 (p. 741) with a broken line.

- *Comment*. This emphasizes the importance of the phase problem: different phase choices give quite different electron densities.

The phase problem has been overcome to some extent by a variety of methods, the principal ones being the following:

(i) *Trial and error*. When the unit cell is simple we can be guided by a guess at the structure. If we guess the phases of the F_{hkl} and end up with an electron density that is obviously absurd (e.g., atoms overlapping) another guess is made until the structure is reasonable. This technique is now applied particularly to the study of refractory materials.

(ii) *Isomorphous replacement*. In some complex molecules, such as proteins, it is possible to implant a heavy atom without appreciably distorting the form of the crystal structure: this is *isomorphous replacement* (isomorphous means 'same form'). The advantage of the presence of a heavy atom is that its many electrons dominate the scattering, and the broad principles of the crystal structure may be worked out by considering them alone in the first instance. The finer details of the structure can then be discovered by examining how the other atoms modify the intensities of the heavy-atom reflections, and so build up a picture of relative phases. For example, if a spot is diminished in intensity by the presence of another atom in the unit cell, the phase of the extra contribution is negative relative to the first.

(iii) *Patterson synthesis*. Patterson investigated the possibility of using the intensities directly in a Fourier synthesis of the electron density. It is found that if I_{hkl} is inserted in the place of F_{hkl} in the expansion in eqn (22.3.7), then the quantity produced is a map of *separations* of scattering density (e.g., a map of separations of atoms in the unit cell). For example, if the unit cell has the structure shown in Fig. 22.14a the Patterson synthesis would give the map shown in Fig. 22.14b, where the distance of every spot from the origin gives the orientation and separation of all the pairs of atoms in the original structure. The disadvantage of this method is that as there are $\frac{1}{2}N(N-1)$ pairs of atoms in a structure of N atoms, the pattern rapidly gets very complex. Nevertheless, if one of the atoms happens to be heavy it dominates the Patterson map, and it may be possible to unravel the features of the pattern that depends on the heavy atom. Figures 22.14c,d show how the presence of a heavy atom affects the Patterson synthesis. This technique may be used in conjunction with the isomorphous replacement method, and is particularly useful for the investigation of organometallic compounds and highly symmetrical organic molecules.

(iv) *Direct methods*. Modern X-ray crystallography has largely discarded the techniques described above in favour of *direct methods* based on statistical procedures. This procedure is especially useful for studying complex organic molecules where the atom positions are virtually randomly located (from the radiation's point of view). It is possible to deduce relations between some structure factors and sums (and sums of squares) of others

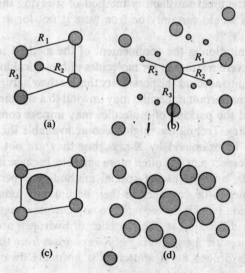

(a)

(b)

(c)

(d)

Fig. 22.14. (a) A unit cell and (b) its Patterson synthesis. When the central atom is replaced by a heavier element the cell in (c) has the Patterson synthesis shown in (d).

which have the effect of constraining the phases to have particular values, at least with high (99.7 per cent) probability. Since the relations can be built into computer programs, the determination of a crystal structure from a collection of observed intensities has become almost completely automated. Furthermore, since the data gathering is itself automatic, the entire procedure, at least in medium size natural products, from crystal mounting to electron density plotting, has become an automated technique.

The procedure for a structural analysis. The technique depends on the availability of single crystals of the material of interest, and this has led to a great deal of effort to obtain crystals of large biochemically significant molecules, such as enzymes and proteins. The indexing of the reflections and the determination of the unit cell parameters may be carried out using the precession camera technique described earlier. With the crystal randomly oriented in a goniometer head it is transferred to a diffractometer, e.g., a four-circle instrument of the kind already described. Then the intensities of all the diffraction reflections are interpreted as structure factors, their phases being assigned by the direct statistical technique. The electron density so calculated can then be drawn out as a contour diagram.

Various techniques of *refinement* may then be employed. For example, the atoms vibrate about their mean position in the structure, and this spreads out their electron density. It is possible to estimate the magnitude of this effect, and to eliminate it almost completely from the electron density map. In this way the contour maps of electron density are constructed, and from them equilibrium bond lengths, bond angles, and electron density distributions may be determined. One is shown in Fig. 22.15. On the basis of such maps models of the molecular structure may be built, and Fig. 0.3 on p. 6 shows an example.

The diagrams illustrated suggest, because they are so detailed, that X-ray analysis is the most satisfactory method of studying the structure of molecules. This would certainly be true were it not for a number of limitations.

The primary limitation is the confinement of the method to the solid state, which is a severe restriction for molecules of biological interest. It is important to take into account the possibility that, in their natural environment, biologically important molecules may unwind to a significant extent, and that in a solid the packing of molecules may impose constraints on their stereochemistry. Techniques might become available that enable *in vivo* molecules to be examined by X-rays, but they are not at present available. In this respect n.m.r. is often more suitable, because it can study molecules in fluids, and the use of special methods has permitted the structure of enzymes to be examined in their natural environment.

Another limitation (which makes n.m.r. complementary) is the lack of response of X-ray diffraction to the presence of hydrogen atoms in the structure. This is because the scattering of X-rays arises from the electrons present, and the hydrogen atom scattering is normally dwarfed by the

Fig. 22.15. The electron density of nickel phthalocyanine (J. M. Robertson, *Organic crystals and molecules*, Cornell University Press, 1953).

other atoms. Very careful refinement reveals the hydrogen atoms, but the technique is difficult, and n.m.r. can be of considerable assistance.

Finally, the cost of a complete X-ray analysis of a compound can be considerable. The economics of an X-ray analysis should take into account the search for a suitably crystalline sample as well as the extensive amount of computing on large-capacity machines.

22.4 Information from X-ray analysis

The bonds between the components of crystals may be of various kinds. Simplest of all are the *ionic bonds* between charged ions: the crystal is held together by the Coulombic interaction between opposite charges. The crystal structure is governed by the geometrical problem of packing the ions together in the energetically most favourable way. The opposite extreme of this type of bonding is *covalent bonding*, where chemical bonds of a definite spatial orientation link the atoms together. The stereochemical demands of valence override the simple geometrical problem of packing spheres together, and elaborate and extensive structures may be formed. A famous example of this is diamond, with the structure shown in Fig. 22.16. A crystal of diamond is an enormous single molecule with every carbon atom linked through sp^3-hybrid bonds to four neighbours in a tetrahedral disposition. Materials formed in this way are often tough and unreactive.

Many organic, and a number of inorganic, molecules are neither ionic nor have valencies free to form further covalent bonds. These species retain their individuality when they condense into crystals, and stick together by virtue of various kinds of *van der Waals* interactions of the same kind as give rise to gas imperfections (p. 782). These interactions may be dipole–dipole interactions, if the species are polar, or dispersion forces arising from instantaneous fluctuations in their electron distributions (see p. 784) if they are non-polar. The form of the crystal structure is then a solution of the problem of condensing unsymmetrical objects together to form an aggregate of minimum energy (actually minimum Gibbs function, p. 146), and the prediction of the favoured form is very difficult and almost impossible. The problem is made more complicated by the role of *hydrogen bonds*, where a proton acts as a link between two electronegative atoms. In some cases the H-bonds dominate the form of the crystal structure, as in ice, Fig. 22.17, but in others (e.g., phenol) they distort a structure which is largely determined by various van der Waals interactions.

Finally there are the *metallic bonds*, where a sea of electrons floods between arrays of cations and binds the whole together into a rigid structure. The crystal structure is then largely determined by the way spherical metal cations can pack together into an orderly array.

Packing of identical spheres: metal crystals. The simplest types of crystalline form are expected for metals because all the building blocks are identical and

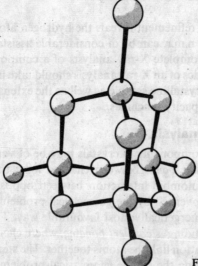

Fig. 22.16. The structure of the diamond lattice.

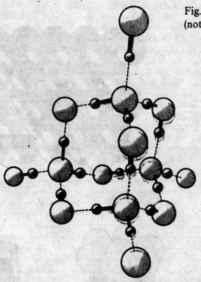

Fig. 22.17. The crystal structure of one form of ice (notice the underlying diamond structure).

spherical. It is found that most metals crystallize in one of three simple forms, two of which can be explained in terms of organizing spheres into the closest possible packing. See Fig. 22.18.

A close-packed layer of identical spheres can be formed as shown in Fig. 22.18a, and a similar layer can be laid on the depressions in this lattice, Fig. 22.18b. A third layer can be laid on top of this layer in two different ways. If it is placed so that it reproduces the first layer (Fig. 22.18c, so that the layer structure could be represented ABABAB... as the process continued) the crystal would be *hexagonal close-packed* (h.c.p.), Fig. 22.18d. The alternative arrangement places the third layer so that it does not reproduce the first (Fig. 22.18e, the layer structure is then ABCABC...): this is the *cubic close-packed* (c.c.p.) arrangement, Fig. 22.18f. The figure also indicates that this c.c.p. arrangement gives rise to face-centred unit cells, and so we may also denote it as f.c.c.

Both the h.c.p. and c.c.p. structures are closely packed, dense structures. A measure of this is the number of atoms surrounding any selected atom: the *coordination number* is 12 in each case. Metals showing these structures are indicated in Fig. 22.18.

Example Calculate the proportion of a hexagonally close-packed unit cell that is empty space.

- *Method*. Assume the ions to be perfect spheres of radius R. Refer to Fig. 22.19, and use some trigonometry.

- *Answer*. Area of the base of the unit cell (tinted in grey) is $3^{1/2}R \times 2R = (2\sqrt{3})R^2$. The height is obtained as follows. The next layer of atoms lies above the point marked B. This is at a distance $R/\cos 30° = 2R/\sqrt{3}$ from the centre of the neighbouring ion. The overlying ion has a centre a distance $2R$ from the ion just referred to, and so it is at a height $\sqrt{[(2R)^2-(2R/\sqrt{3})^2]} = 2(2/3)^{1/2}R$. The height of the

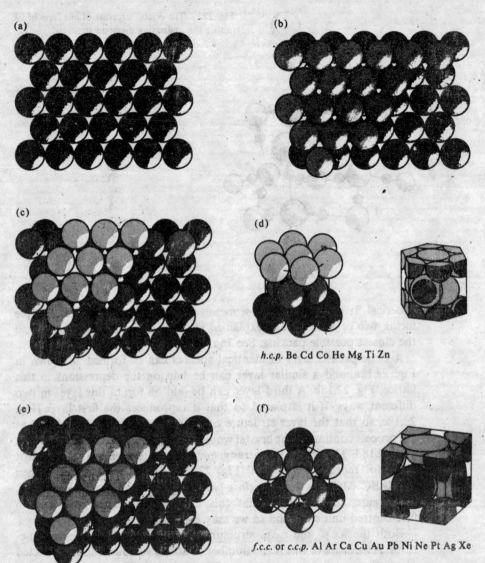

h.c.p. Be Cd Co He Mg Ti Zn

f.c.c. or *c.c.p.* Al Ar Ca Cu Au Pb Ni Ne Pt Ag Xe

Fig. 22.18. Close packing of identical spheres. (a) The A layer; (b) the AB layers: (c) ABA leads to hexagonal close packing; (d) detail of h.c.p.; (e) ABC leads to cubic close packing (or face-centred cubic); (f) detail of c.c.p.

unit cell itself is twice this, or $4(2/3)^{1/2}R$. The volume of the unit cell is (area) × (height) $= (2\sqrt{3})R^2 \times 4(2/3)^{1/2}R = (8\sqrt{2})R^3$. There are two complete ions per unit cell, and so the volume available to each is $(4\sqrt{2})R^3$. The volume of a spherical ion is $(4/3)\pi R^3$. Therefore the fraction of 'full' space is $(4/3)\pi R^3/(4\sqrt{2})R^3 = \pi/3\sqrt{2} = 0.74$; therefore 26 per cent is free space.

• *Comment*. The same type of argument may be applied to other cells. In an f.c.c. cell the free volume is 26 per cent, in a b.c.c. cell the free volume is 32 per cent, and in a simple cubic lattice 48 per cent. The *packing fractions* are 0.74, 0.68, and 0.52 respectively.

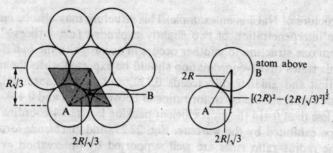

Fig. 22.19. Calculating the packing fraction.

In the third arrangement shown by a number of common metals (Ba, Cs, Cr, Fe, K, W) the atoms of the first layer are less closely packed than in the other structures, and the next layer lies in the dips on the first layer. The third layer reproduces the first in this arrangement, and so the structure is ABABAB.... Inspection of the unit cell shows that it corresponds to a *body-centred cubic* (b.c.c.) arrangement, and that the coordination number is 8.

Ionic crystals. In ionic crystals the problem of attaining the lowest energy is complicated by the fact that there are two kinds of spheres to pack together: not only do the spheres have different radii but they also have opposite charges. If, by some chance, the ions are the same size, it is still impossible to achieve dense 12-coordination. This is because such packing cannot be done in a way that surrounds every positive ion by the same number of negative ions, bearing in mind that every negative ion must also be surrounded by positive ions. The best that can be achieved is the 8-coordination of the caesium chloride structure, which is illustrated in Fig. 22.20a.

When the sizes of the ions become more disparate than in CsCl the packing reverts to a 6-coordination of the type shown in Fig. 22.20b: the

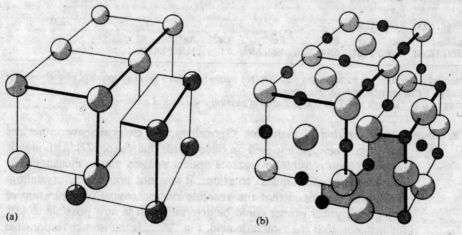

(a) (b)

Fig. 22.20. (a) The caesium chloride lattice and (b) the sodium chloride lattice.

structure of NaCl is an example. This structure may also be envisaged as the interpenetration of two slightly expanded f.c.c. lattices. The switch from one structure to another occurs in accord with the *radius–ratio rule*. This states that 8-coordination should be expected when the ratio of the cation and anion radii exceeds 0.732, and that 6-coordination (of the NaCl type) should occur for ratios of between 0.732 and 0.414. For ratios of less than 0.414 the most efficient packing leads to 4-coordination of the type exhibited by the *wurtzite*, Fig. 22.21, and *zincblende* forms of ZnS. The radius–ratio rules are well supported by observation even though they are founded on simple geometrical considerations. In fact, deviation of a structure from the predicted form indicates the shift from ionic towards covalent bonding.

The *ionic radii* to use in estimating the radius ratio, and wherever else it is important to know the size of the ions, can be established by collecting a large number of X-ray measurements of crystal structures. It is found that the distances between ions can be expressed as the sum of two contributions which can be transferred between different compounds. Thus Na^+ contributes to NaX separations an almost constant amount independent of the nature of X. This analysis can be extended to ions of valence higher than 1, and a list of values is given in Table 22.1. The same kind of analysis can be done on van der Waals molecular crystals, and a list of *van der Waals* radii of molecules and atoms built up. This is tabulated in Table 22.2 and illustrated pictorially in the Introduction (p. 7).

Table 22.1. Ionic radii, R/pm

Li^+	Be^{2+}			O^{2-}	F^-
59(4) 74(6)	27(4)			135(2) 138(4)	128(2) 131(4)
Na^+	Mg^{2+}		Al^{3+}	S^{2-}	Cl^-
102(6) 116(8)	58(4) 72(6)		39(4) 53(6)	184	181
K^+	Ca^{2+}				Br^-
138(6) 151(8)	100(6) 112(8)				196
Rb^+	Sr^{2+}				I^-
149(6) 160(8)	116(6) 125(8)				220
Cs^+	Ba^{2+}	Fe^{2+}	Fe^{3+}	Cu^{2+}	Ag^+
170(6) 182(8)	136(6) 142(8)	63(4) 78(6)	49(4) 65(6)	73	115(6) 130(8)

Numbers in brackets are coordination numbers. No number implies a value independent of coordination number. Fe values refer to high-spin complexes.
Source: G. W. C. Kaye and T. H. Laby, *Tables of physical and chemical constants*. Longmans.

The absolute configuration of molecules. Optically-active molecules are molecules that cannot be superimposed on their mirror images (p. 527). Although it had long been possible to separate optical isomers, and to measure their equal and opposite optical rotations, it was not until X-ray crystallography was developed that the absolute stereochemical configuration of any given optical isomer could be determined. It is now possible to say, for instance, that the L-tartaric acid, Fig. 22.22, is the isomer responsible for rotating light in a clockwise sense, denoted (+), and that D-tartaric

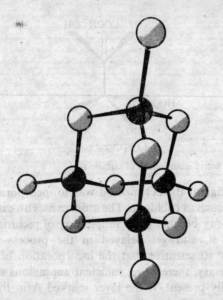

Fig. 22.21. The Wurtzite lattice (notice the underlying diamond structure).

acid rotates in an anticlockwise sense ($-$). The X-ray method is not trivial, because a crystal of one isomer gives an identical intensity pattern to the crystal of its mirror image isomer, its *enantiomer*. The information about the absolute configuration is contained in the phase of the diffracted radiation, and a special technique, due to Bijvoet, has to be used to extract it.

Table 22.2. Van der Waals radii, R/pm

		H	He
		120	120
N	O	F	Ne
150	140	135	234[a]
			320[b]
P	S	Cl	Ar
190	185	180	286[a]
			383[b]
As	Se	Br	
200	200	195	
Sb	Te	I	
220	220	215	

[a] Gas viscosity.   [b] Close packing.

Source: Principally *Handbook of chemistry and physics*, Chemical Rubber Co.

Consider first the diagrams shown in Fig. 22.23a. These represent an idealized crystal and its mirror image. Each plane of atoms gives rise to a scattered wave, and the superpositions are as shown. Note that the two superpositions have the same amplitude but differ in phase. The diffraction pattern therefore has the same intensity for each enantiomer, and cannot be used to distinguish them, as already mentioned.

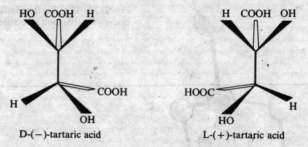

Fig. 22.22. The two optically active enantiomers of tartaric acid.

The essence of the original Bijvoet method was to incorporate in the molecule a heavy atom, such as rubidium. The atom causes an extra phase shift in the scattered X-ray because (as a simple way of picturing it) the X-rays tend to excite it, and get delayed in the process. Modern diffractometers are now so sensitive that the incorporation of a heavy atom is no longer necessary, there being sufficient anomalous scattering from the atoms normally present. If the layer marked A in the crystal contains these anomalous scatterers, the scattered waves are as shown in Fig. 22.23b. The essential point is that the anomaly in the scattering now produces superpositions that differ in amplitude, not just phase, and so the diffraction intensities are different in the two cases. Therefore the enantiomers can be distinguished.

22.5 Neutron diffraction

A neutron, generated in an atomic reactor, and slowed to thermal energies by repeated collisions with some *moderator* (such as graphite) until it is travelling at about $3.9 \, \text{km s}^{-1}$ has a momentum of $5.2 \times 10^{-24} \, \text{kg m s}^{-1}$,

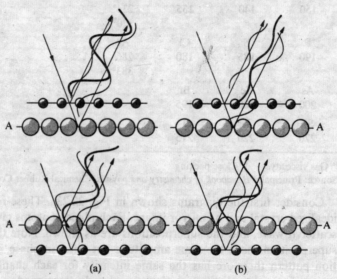

Fig. 22.23. (a) Normal scattering; (b) anomalous scattering.

and therefore a wavelength of 100 pm. This is comparable to the X-ray wavelengths used in X-ray diffraction, and so similar diffraction phenomena can be expected.

The scattering of X-rays is caused by the driven oscillations in the electrons of the atoms: as the incident radiation forces them into oscillation, so they radiate. The scattering of neutrons is a nuclear phenomenon: the neutrons pass through the electronic structure of the molecule, and interact with the nuclei through the forces that are responsible for binding nucleons together. The consequence is that neutrons are scattered much more evenly by all atoms (X-rays are scattered more by the electron-rich heavy atoms). Neutron diffraction therefore shows up the positions of protons in the molecular structure, whereas X-rays reveal them only weakly.

The difference in sensitivity can have a pronounced effect on the measurement of C–H bond lengths. Since X-rays respond to accumulations of electrons, the weak peaks from H-atoms in an X-ray diffraction experiment represent the location of the bulk of the electron density, and in a bond this may be shifted towards the C-atom. Such measurements on sucrose give C–H bond lengths of about 96 pm. Neutron scattering, however, responds to the position of the nuclei themselves, and C–H bond lengths of 109.5 pm have been reported for the same compound. The O–H bond lengths in sucrose show a similar variation: 79 pm by X-rays, but 97 pm from neutron studies.

The other attribute of neutrons that distinguishes them from X-ray photons is their possession of a spin magnetic moment. This magnetic moment can couple to the magnetic fields of the ions in the crystal lattice (if they have unpaired electrons) and modify the diffraction pattern. This is another reason why neutron diffraction experiments are done even though the sources are expensive and (because a monochromatic, single velocity ray has to be selected) the beam weak.

A simple example of this *magnetic scattering* is provided by the diffraction pattern from metallic chromium. The lattice is body-centred cubic, and so a series of systematic absences in the diffraction pattern is expected wherever $h+k+l$ is an odd number (as described for X-rays on p. 737). In fact the systematic absences are not observed in a neutron diffraction experiment, because the structure is such that atoms at the centres of the cubes have magnetic moments opposite to those at the vertices, and the structure is better regarded as two interpenetrating lattices, Fig. 22.24. Since the alternating layers are different as far as neutrons are concerned (but identical for X-rays) the basis of the systematic absences is lost, and spots appear in the diffraction pattern. Neutron diffraction is especially important for investigating these *magnetically ordered* lattices.

22.6 Electron diffraction

Electrons can be accelerated to precisely controlled energies by a known potential difference. When accelerated through 10 keV they acquire a wavelength of 12 pm, and this makes them suitable for diffraction studies

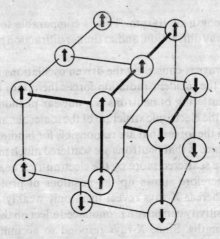

Fig. 22.24. Two interpenetrating magnetic lattices.

of molecules. Electrons are scattered strongly by interaction with the charges of the electrons and the nuclei of the molecule. As a consequence they are unsuitable for studying bulk samples. Nevertheless they are very suitable for precise measurements on molecules in the gas phase and for studies of surfaces and thin films.

A typical electron diffraction apparatus is illustrated in Fig. 22.25. Electrons are boiled off the hot filament on the left and accelerated through the potential difference. They then pass through the stream of gaseous sample, and on to a fluorescent screen.

The sample, being gaseous, presents all possible orientations of the scattering atom–atom separations to the beam, and the diffraction pattern is a superposition of the scattering due to the entire range of orientations. This appears in the diffraction picture as a series of concentric undulations on a decreasing background, Fig. 22.26a. The undulations can be ascribed to the sharply defined scattering from the nuclear positions, and the background ascribed to the much less well-defined contribution from the continuous distribution of electron density in the molecule. One way of eliminating the unwanted background is to insert a rotating, heart-shaped disc in front of the screen: this exposes the outer parts more than the

vacuum pump

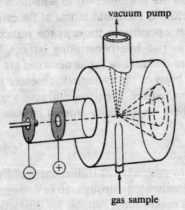

gas sample   Fig. 22.25. An electron diffraction apparatus.

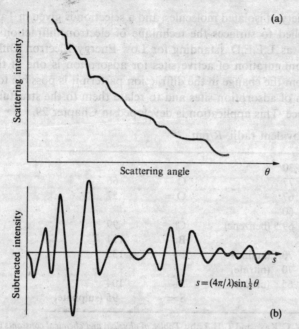

Fig. 22.26. (a) Electron diffraction from ferrocene vapour and (b) the appearance after eliminating unwanted background.

inner and helps to emphasize the undulations, Fig. 22.26b.

The scattering from a pair of atoms separated by a distance R_{ij} and orientated at some angle to the incident beam can be calculated. The overall diffraction pattern for the entire sample must then be calculated by allowing for contributions from all possible orientations of this pair of atoms. In other words we integrate over all orientations. This leads to the expression

$$I_{ij}(\theta) = 2f_e^2\{1 + \sin(sR_{ij})/sR_{ij}\}, \qquad s = (4\pi/\lambda)\sin\tfrac{1}{2}\theta,$$

where λ is the wavelength of the electron beam and θ is the scattering angle. The electronic scattering factor f_e is a measure of the intensity of the scattering power of the atoms. When the molecule consists of a number of atoms of scattering power f_i the total intensity has an angular variation given by the *Wierl equation*

$$(22.6.1) \qquad I(\theta) = \sum_{i,j} f_i f_j \sin(sR_{ij})/sR_{ij}.$$

The electron diffraction pattern can be ascribed to distances between all the possible pairs of atoms in the molecule and not just to those bonded together. The analysis of a many-atom molecule may therefore be very complicated. When there are just a few atoms the peaks can be analysed reasonably quickly, and the analysis proceeds by assuming a geometry and calculating the intensity pattern on the basis of the Wierl equation. The best fit is taken as the appropriate geometry.

Electron diffraction has provided precise bond distances and angles for a wide variety of isolated molecules, and a selection is given in Table 22.3. When applied to surfaces the technique of electron diffraction is often referred to as L.E.E.D. (standing for Low-Energy Electron Diffraction) and the configuration of active sites for adsorption is one of the main quests. From the change in the diffraction pattern it is possible to extract the pattern of adsorption sites and to relate them to the structure of the clean surface. This application is developed in Chapter 29.

Table 22.3. Typical covalent radii, R/pm

H—	30		
C—	77	O—	66
C=	67	O=	57
C≡	60		
C=	69.5 (benzene)	Cl—	99
		Br—	114
N—	70 (amino)	I—	133
N—	70 (nitrate)		
N=	65 (nitrate)	S—	104
		S=	95 (sulphate)

Source: G. W. C. Kaye and T. H. Laby, *Tables of physical and chemical constants* (11th edn.), Longmans.

Appendix: diffraction as a particle property

The underlying principle for dealing with diffraction as a particle property is the conservation of momentum in collisions. We begin by deriving the Bragg condition on this basis, and then pass on to a generalization which leads to an interpretation of eqn (22.3.7) on p. 740. This Appendix draws on a number of the ideas of quantum theory, introduced in Chapter 13. The aim is to show that the Fourier synthesis discussion of X-ray diffraction can be interpreted in a straightforward, physical way.

Consider the elastic (translational energy conserving) collision between a particle and a stationary wall as depicted in Fig. 22A.1. Conservation of linear momentum gives two conditions:

conservation of horizontal component: $p \cos \theta = p' \cos \theta'$

conservation of vertical component: $p \sin \theta = p_W - p' \sin \theta'$,

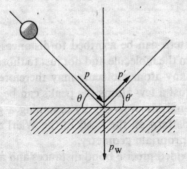

Fig. 22A.1. Photon picture of reflection.

p_W being the final momentum of the wall. Since the collision is elastic the magnitude of the particle's momentum is the same before and after the collision, and so $p' = p$. From the first condition we get $\theta = \theta'$, and from the second

$$2p \sin \theta = p_W.$$

According to the de Broglie relation (p. 396), the momentum of a particle (in this case the photon) may be expressed as a wavelength through $p = h/\lambda$, and so the last equation can be put into a form that already closely resembles the Bragg condition:

(22A.1) $$2 \sin \theta = p_W \lambda/h.$$

If the wall could take up all momenta this condition could be satisfied for all values of θ; if we can show that p_W is restricted to some integral multiple of h/d we shall have recovered the Bragg condition.

The key to understanding why p_W should be quantized can be obtained by examining the structure of the wall. Consider the vertical row of atoms in Fig. 22A.2. The variation of electron density down the column is also depicted. In the case of X-rays the collisions of interest are between the photon and the electrons of the solid's structure. Therefore the p_W in eqn (22A.1) must represent the momenta that the electrons may have up and down the column of atoms. The momentum of any electron in the structure is determined by its wavefunction's wavelength through the de Broglie relation $p_W = h/\lambda_e$, and so finding the permitted values of the wavelength λ_e of the wavefunctions will be enough to complete the derivation of the Bragg equation.

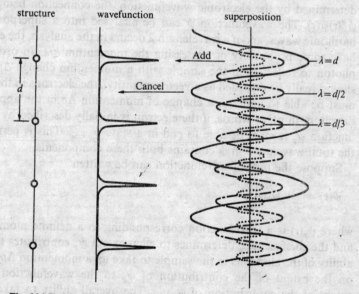

Fig. 22A.2. The structure of a line of atoms represented as a superposition of a series of waves.

In order to resolve this problem we have to realize that the wavefunction shown in Fig. 22A.2 can be regarded as a superposition of many wavefunctions with wavelengths d, $d/2$, $d/3$.... Therefore the electrons in the column of atoms can possess momenta h/d, $2h/d$, $3h/d$,..., and so p_W must be some integral multiple of h/d. Consequently, on reflection of the photon from the column, the momentum may *change* only by an integral multiple of h/d, or $p_W = nh/d$. It follows that

$$2 \sin \theta = [n(h/d)](\lambda/h) = n\lambda/d,$$

which is the Bragg condition.

The wavefunction for the electrons which has led to the Bragg condition is the superposition of waves:

(22A.2) $$\psi = \cos(2\pi z/d) + \cos(4\pi z/d) + \ldots = \sum_{l=1}^{\infty} \cos(2l\pi z/d).$$

Such a wavefunction, a superposition of an infinite number of harmonic waves, is a series of sharp spikes separated by a distance d. In an actual crystal the electron density is distributed with varying density throughout the whole of space, and the wavefunction for the electrons is less sharply peaked and localized. The breakdown into harmonic waves differs from the simple form given in the last equation, and so the uptakes of momenta that are allowed are also modified. This is the basic principle of the examination of crystals by diffraction: *the diffraction pattern represents the momenta that the electronic structure can accommodate, and the allowed momenta depend on the distribution of the electron density.*

This point can be developed as follows. The electrons are distributed with a density $\rho(\mathbf{r})$, where $\mathbf{r}$ is some point in the crystal. This density is determined by the electronic wavefunction, the connection being $\rho(\mathbf{r}) = \psi^*(\mathbf{r})\psi(\mathbf{r})$. The wavefunction ψ can be analysed into a superposition of harmonic waves, and if a wavelength λ occurs in the analysis, the electrons have some probability of possessing the momentum h/λ. In order for a photon to reflect from the sample with a momentum change Δp (which should really be regarded as a vector quantity), the electrons in the crystal must be able to undergo a change of momentum Δp in the appropriate direction. In other words, if the electron is initially described by a wavefunction ψ_p it must be able to end in a state $\psi_{p+\Delta p}$. This is permitted if the total wavefunction ψ contains both these components.

Suppose the overall wavefunction can be written

$$\psi(\mathbf{r}) = \sum_p c_p \psi_p(\mathbf{r})$$

where $\psi_p(\mathbf{r})$ is a wavefunction corresponding to a definite momentum p and the coefficient c_p determines to what extent ψ_p contributes to ψ. The ability of the electrons in the sample to take up a momentum Δp depends on the extent of the contribution of ψ_p to the wavefunction *and* the extent of the contribution of $\psi_{p+\Delta p}$. The overall ability to take up this momentum, and therefore the amplitude F of the scattering through the corresponding angle, can therefore be expressed as the sum

(22A.3)
$$F \propto \sum_p c_p c_{p+\Delta p}^*$$

(we have allowed for the possibility that the coefficients may be complex). The sum is taken because transitions $p \to p + \Delta p$ from all components ψ_p must be considered.

We now need to find the values of the coefficients c_p. The wavefunction for a state of definite momentum p in the x-direction is (p. 398)

$$\psi_p(x) \propto \exp(ipx/\hbar),$$

and so the equation defining c_p for this one-dimensional system is

$$\psi(x) = \sum_p c_p \exp(ipx/\hbar).$$

If this is multiplied by $\exp(-ip'x/\hbar)$ and integrated over x from one edge of the crystal (at $x = -\frac{1}{2}L$) to the other (at $x = \frac{1}{2}L$) we get

(22A.4)
$$\int_{-L/2}^{L/2} dx\,\psi(x)\,e^{-ip'x/\hbar} = \sum_p c_p \int_{-L/2}^{L/2} dx \exp[i(p-p')x/\hbar].$$

Both sides of the last equation can be simplified considerably. In the first place the wavefunction $\psi(x)$ repeats itself from cell to cell, and so instead of integrating it over the whole of the crystal it is sufficient to integrate it over the unit cell, and to get the total integral by multiplication by the number of unit cells. If the cell is of size a in the x-direction, the number of cells is L/a and the left-hand side of the last expression becomes

$$\int_{-L/2}^{L/2} dx\,\psi(x)\,e^{-ip'x/\hbar} = (L/a) \int_0^a dx\,\psi(x)\,e^{-ip'x/\hbar}.$$

The integral on the right in eqn (22A.4) is the superposition of a large number of waves, and they superimpose to give zero amplitude. The only case when this does not occur is when $p = p'$, for then the integral is

$$\int_{-L/2}^{L/2} dx = L.$$

Therefore the only value of p that contributes to the sum on the right-hand side of eqn (22A.4) is p', and so the sum reduces simply to $Lc_{p'}$. Now we have a simple expression for the coefficient $c_{p'}$ (or, changing p' to p, for c_p):

$$c_p = (1/a) \int_0^a dx\,\psi(x)\,e^{-ipx/\hbar}.$$

An integral of this form is called a *Fourier transform* of the periodic function $\psi(x)$.

The scattering amplitude can be expressed in terms of the wavefunction $\psi(x)$. Using the last equation in eqn (22A.3) leads to

$$F = (K/a^2)\sum_p \int_0^a dx\psi(x)\exp(-ipx/\hbar)\int_0^a dx'\psi^*(x')\exp\{i(p+\Delta p)x'/\hbar\}$$

$$= (K/a^2)\sum_p \int_0^a dx \int_0^a dx'\psi(x)\psi^*(x')\exp\{ip(x'-x)/\hbar\}\exp(i\Delta px'/\hbar),$$

where K is some coefficient of proportionality. The sum over p can now be done because it occurs only in the simple exponential term. This sum, $\sum_p \exp[ip(x'-x)/\hbar]$, is a superposition of waves of a wide variety of wavelengths, and so annihilation occurs, and it vanishes. If, however, $x = x'$ the sum does not vanish, because the oscillating exponential term is reduced to unity. Therefore, the only term that survives in the above expression is the one with $x = x'$, and so

$$F \propto \int_0^a dx\psi(x)\psi^*(x)\exp(i\Delta px/\hbar).$$

Now we recognize that $\psi(x)\psi^*(x)$ is the electron density at x, which is $\rho(x)$; therefore we arrive at the important result that the scattering is proportional to the Fourier transform of the electron density:

(22A.5) $$F \propto \int_0^a dx\rho(x)\exp(i\Delta px/\hbar).$$

. This equation already closely resembles eqn (22.3.6) on p. 740.

Next we insert the value for Δp. When a photon of wavelength λ collides with the column of electron density and is deflected through the angle 2θ, the change of momentum is $(2h/\lambda)\sin\theta = (4\pi\hbar/\lambda)\sin\theta$ in the x-direction. If θ is arranged so that an $(h00)$ reflection is being observed, since $d_{h00} = a/h$ in the present lattice, $\sin\theta = \lambda/2d_{h00} = h\lambda/2a$. Therefore the value of Δp for this arrangement is $\Delta p = 2\pi h\hbar/a$ (h is the index, $\hbar$ is Planck's constant divided by 2π), and the amplitude of radiation from the entire unit cell when the $(h00)$ reflection is being observed is

(22A.6) $$F_{h00} \propto \int_0^a dx\rho(x)\exp(i2\pi hx/a)$$

and the intensity of this radiation is proportional to $|F|^2$.

The development just given can be extended without difficulty to the three-dimensional case. Then the electron density in the unit cell of dimension a, b, c, depends on x, y, z, and the amplitude of an (hkl) reflection is

$$F_{hkl} \propto \int_0^a dx \int_0^b dy \int_0^c dz\, \rho(x,y,z)\exp\left\{2\pi i\left(\frac{hx}{a}+\frac{ky}{b}+\frac{lz}{c}\right)\right\}.$$

This is the central equation of X-ray crystallography and is the same as that obtained on p. 740 (except that x/a here is x there).

The interpretation of the crystal structure analysis is, therefore, that the scattering amplitude indicates the momenta that the lattice can take up, and that the Fourier integrals are the analysis of the electron density to find out these acceptable momenta.

Further reading

The third dimension in chemistry. A. F. Wells; Clarendon Press, Oxford, 1956 (reissued 1968).

The determination of molecular structure (2nd ed.) P. J. Wheatley; Clarendon Press, Oxford, 1968.

Crystal structure analysis. J. P. Glusker and K. N. Trueblood; Oxford University Press, New York, 1972.

Molecular structure: the physical approach. J. C. D. Brand and J. C. Speakman; Arnold, London, 1960.

Elementary crystallography. M. J. Buerger; Wiley, New York, 1956.

Introduction to crystal geometry. M. J. Buerger; McGraw-Hill, New York, 1971.

X-ray crystal structure analysis. W. N. Lipscomb and R. A. Jacobson; in *Techniques of chemistry* (A. Weissberger and B. W. Rossiter, eds.), Vol. IIID, 1, Wiley-Interscience, New York, 1972.

X-ray structure determination, a practical guide. G. H. Stout and L. H. Jensen; Macmillan, New York, 1968.

An introduction to X-ray crystallography. M. M. Woolfson; Cambridge University Press, 1970.

Neutron scattering in chemistry. G. E. Bacon; Butterworths, London, 1977.

Electron diffraction by gases. L. S. Bartell; in *Techniques of chemistry* (A. Weissberger and B. W. Rossiter, eds.), Vol. IIID, 125, Wiley-Interscience, New York, 1972.

Electron diffraction. T. B. Rymer; Chapman and Hall, London, 1970.

Tables of interatomic distances and configurations of molecules. L. E. Sutton (ed.); Chem. Soc. Special Publication, No. 11, 1958 (Supplement, Special Publication, No. 18, 1965).

Structural inorganic chemistry (4th ed.). A. F. Wells; Clarendon Press, Oxford, 1975.

Crystal structure (5 sections and supplements). R. W. G. Wycoff; Wiley-Interscience, New York, 1959.

Problems

22.1.   The first few Problems will give familiarity with the calculations that can be done on simple lattices. As a first step, draw an array of points as a two-dimensional rectangular lattice based on a unit cell of sides a, b. Mark planes with Miller indices (10), (01), (11), (12), (23), (41).

22.2.   Redraw the lattice in the last Problem so that the b axis makes an angle of 60° to the a axis. Mark the same planes as before.

22.3.   Calculate the distances between the (11) planes for the lattices in the preceding Problems.

22.4.   Crystal planes cut through the crystal axes at $(2a, 3b, c)$, (a, b, c), $(6a, 3b, 3c)$, $(2a, -3b, -3c)$. What are their Miller indices?

22.5.   Draw an orthorhombic unit cell and mark the (100), (010), (001), (011), (101), and (111) planes.

22.6.   Draw a triclinic unit cell and mark on it the same planes.

22.7.   What are the separations of the planes with indices (111), (211), and (100) in a crystal in which the cubic unit cell is of side 432 pm (100 pm = 1.00 Å).

22.8.   In the early days of X-ray crystallography there was an urgent need to know X-ray wavelengths. One technique was to measure the diffraction angle from a mechanically ruled grating, with the X-rays approaching it at a glancing angle. Another method was to estimate the separation of lattice planes from the measured density of a crystal. The density of NaCl is 2.17 g cm^{-3} and the (100) reflection using palladium K_α radiation occurred at $6° 0'$. What is the wavelength of the X-rays?

22.9. In their book of *X-rays and crystal structures* (which begins 'It is now two years since Dr. Laue conceived the idea...') the Braggs give a number of simple examples of X-ray analysis. For instance, they report that the first-order reflection from (100) planes of KCl occurs at $5° 23'$, but for NaCl it occurs at $6° 0'$ for the rays of the same wavelength. If the side of the NaCl unit cell is 564 pm, what is the size of the KCl cell? The densities of KCl and NaCl are 1.99 g cm^{-3} and 2.17 g cm^{-3} respectively: do these values support the X-ray analysis?

22.10. Potassium nitrate single crystals have orthorhombic unit cells of dimensions $a = 542 \text{ pm}$, $b = 917 \text{ pm}$, $c = 645 \text{ pm}$. Calculate the diffraction angle for first-order reflections from the (100), (010), and (111) planes using Cu K_α radiation (154.1 pm).

22.11. Copper(I) chloride (cuprous chloride) forms cubic crystals with four molecules per unit cell. The only reflections present in a powder photograph are those either with all even indices or with all odd. What is the nature of the unit cell? Note that this type of question about the form of the unit cell can be answered without knowing anything about the wavelength of the radiation: it is determined by symmetry, not size.

22.12. A powder diffraction photograph from tungsten shows lines which index as (110), (200), (211), (220), (310), (222), (321), (400),.... What type of unit cell is present?

22.13. Rock salt (NaCl) forms crystals with a simple cubic lattice of side 564 pm. What are the distances between the (100) planes, the (111) planes, and the (012) planes?

22.14. Show that the separation of the (hkl) planes in a crystal with an orthorhombic unit cell of sides a, b, and c is given by $1/d_{hkl}^2 = (h/a)^2 + (k/b)^2 + (l/c)^2$.

22.15. We often need to know the volumes of unit cells. Show that the volume of a triclinic unit cell of sides a, b, and c and angles α, β, and γ is $V = abc[1 - \cos^2\alpha - \cos^2\beta - \cos^2\gamma + 2\cos\alpha\cos\beta\cos\gamma]^{1/2}$. Note that by making the appropriate choice of angles and lengths this expression includes monoclinic and orthorhombic cells.

22.16. Naphthalene forms a monoclinic crystal with two molecules in each unit cell. The sides are in the ratio 1.377:1:1.436, and $\beta = 122° 49'$. The specific gravity is 1.152. What are the dimensions of the cell?

22.17. The type cf calculation used in the last Problem can be turned round, and X-ray analysis combined with density measurements to find a value of Avogadro's constant. The density of LiF is 2.601 g cm^{-3}. The (111) reflection occurs at $8° 44'$, when 70.8 pm X-rays from molybdenum are used. Find a value of Avogadro's constant on the basis that there are four LiF in a unit cell.

22.18. How many atoms are there in (a) a primitive cubic unit cell, (b) a b.c.c. unit cell, (c) an f.c.c. cell, (d) a unit cell for the diamond lattice?

22.19. Show that the maximum proportion of the available volume that may be filled by hard spheres is (a) 0.52 for a simple cubic lattice, (b) 0.68 for a b.c.c. lattice, (c) 0.74 for an f.c.c. lattice. These numbers are the *packing fractions* of the lattices. What method of packing is most economical for shipping (hard) oranges?

22.20. What is the packing fraction for long, straight cylindrical molecules (or pipes)?

22.21. The only known example of a metal that crystallizes into a primitive cubic lattice is polonium. The unit cell is of side 334.5 pm. Predict the form of the powder diffraction pattern using 154 pm X-rays. What do you predict for the density of polonium? Suppose it packed into an f.c.c. lattice, but with the same atomic radius: what would be its density?

22.22. The separation of the (100) planes of lithium metal is 350 pm. Its density is $0.53 \, \mathrm{g \, cm^{-3}}$. Is the lattice f.c.c. or b.c.c.?

22.23. Copper crystallizes in an f.c.c. lattice of side 361 pm. Predict the form of the powder diffraction pattern using 154 pm X-rays. What is the density of copper?

22.24. The coefficient of linear thermal expansion of copper is $1.67 \times 10^{-5} \, \mathrm{K^{-1}}$ at room temperature. What is the separation of the angles of the (111) reflections taken with 70.9300 pm X-rays at 100 K and 300 K? What other change in the appearance of the diffraction pattern may be expected when the sample is heated?

22.25. Calculate the coefficient of thermal expansion of diamond given that the (111) reflection shifts from $22° 2' 25''$ to $21° 57' 59''$ on heating the crystal from 100 K to 300 K and 154.0562 pm X-rays are used.

22.26. Chromium crystallizes in a b.c.c. lattice of side 288 pm. Predict the form of the powder diffraction pattern using 154 pm X-rays.

22.27. Calcium oxide is cubic and the unit cell of side 480 pm contains four molecules of CaO. What is the density of the crystal?

22.28. A powder diffraction pattern of KCl in a camera of radius 28.7 mm gave lines at the following distances from the centre spot of the film: 14.5, 20.6, 25.4, 29.6, 33.4, 37.1, 44.0, 47.5, 50.9, 54.4, 58.2, 62.1, 66.4, 78.1 mm. Index the lines. The radiation was $\mathrm{Cu} \, K_a$ (154 pm). What is the size of the unit cell?

22.29. In the same apparatus with the same wavelength radiation, LiF gave diffraction lines at the following distances from the central spot: 18.9, 22.1, 31.9, 38.3, 40.4, 48.9, 55.0, 58.0, 68.8, 83.3 mm. Deduce what you can about the nature and size of the unit cell.

22.30. The powder diffraction patterns of (a) tungsten, (b) copper are shown in the Figure. Both were obtained with 154 pm X-rays and the scale of the plate is indicated. Identify the nature of the cubic cell in each case, and find the lattice spacing. What are the crystal radii of the tungsten and copper atoms?

22.31. Mercury(II) chloride is orthorhombic and with $\mathrm{Cu} \, K_a$ radiation the (100), (010), and (001) reflections occur at $7° 25'$, $3° 28'$, and $10° 13'$ respectively. The

(a) W

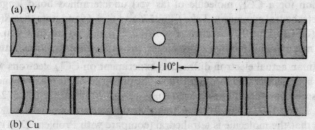

(b) Cu

Fig. 22.27. Powder diffraction patterns of (a) tungsten, (b) copper.

density of the crystal is 5.42 g cm^{-3}. What are the dimensions of the unit cell, and how many $HgCl_2$ units does it contain?

22.32.   The last Problem can be taken a little further. What do the following four observations imply about the symmetry of the unit cell?: (a) the (hkl) reflections show no general absences, (b) $(h0l)$ are missing for hl odd, (c) $(hk0)$ missing for k odd, (d) $(0kl)$ show no systematic absences.

22.33.   Genuine pearls consist of concentric layers of calcite crystals in which the trigonal axes are oriented along the radii. The nucleus of a cultured pearl is a piece of mother-of-pearl which has been worked to a sphere on a lathe, the oyster then deposits concentric layers of calcite on this central seed. Devise an X-ray method for distinguishing between real and cultured pearls.

22.34.   The coordinates, in units of a, of the atoms in a simple cubic lattice are $(0, 0, 0)$, $(0, 1, 0)$, $(0, 0, 1)$, $(0, 1, 1)$, $(1, 0, 0)$, $(1, 1, 0)$, $(1, 0, 1)$, and $(1, 1, 1)$. Calculate the structure factors F_{hkl} on the basis that all the atoms are the same. F is defined in eqn (22.3.4).

22.35.   The coordinates of the atoms in a b.c.c. unit cell are the same as in the last Problem for species A with scattering length f_A, and atom B, with scattering length f_B lies at $(\frac{1}{2}, \frac{1}{2}, \frac{1}{2})$. Find the scattering factors F_{hkl} and predict the form of the powder diffraction pattern in the cases when (a) $f_A = f, f_B = 0$, (b) $f_B = \frac{1}{2} f_A$, (c) $f_A = f_B = f$.

22.36.   Refer to Fig. 22.8. Why is the (311) reflection much less intense than the (222) reflection? You will need to think about structure factors.

22.37.   The *Patterson synthesis* (p. 743) gives spots of intensity density corresponding to the length and direction of the vectors joining scattering centres in the unit cell. Sketch the Patterson pattern that corresponds to (a) a planar, triangular BF_3 molecule, (b) the carbon atoms in benzene.

22.38.   The table (facing) shows the computed electron densities (in arbitrary units) in one section through a unit cell of LiCN. By visual inspection, draw contours that display the Li and CN groups. Choose steps of about 20–30 units for the contour lines. (Data from J. Bijvoet, N. H. Kolkmeyer, and C. H. MacGillary, *X-ray analysis of crystals*, Butterworths (1951)).

22.39.   What is the wavelength of electrons that have been accelerated through (a) 1 kV, (b) 10 kV, (c) 40 kV?

22.40.   What velocity should neutrons have if they are to have a wavelength of 50 pm? What is the wavelength of neutrons that have been *thermalized* by collision with a moderator (heavy water) at 300 K?

22.41.   As a first step in doing an actual electron diffraction analysis, take the Wierl equation, eqn (22.6.1) and calculate the expected electron diffraction intensity distribution for a CCl_4 molecule of (as yet) undetermined bond length but of known tetrahedral symmetry. Take $f_{Cl} = 17f$ and $f_C = 6f$ and note that $R(Cl, Cl) = (8/3)^{1/2} R(C, Cl)$. Plot I/f^2 against $x = sR(C, Cl)$, and notice that you obtain a series of maxima and minima representing the expected diffraction pattern.

22.42.   In an actual electron diffraction experiment on CCl_4 electrons were used that had been accelerated through 10 kV. The positions of the maxima and minima were observed at the following scattering angles. Maxima: $3° 10'$, $5° 22'$, $7° 54'$; Minima: $1° 46'$, $4° 6'$, $6° 40'$, $9° 10'$. Check that this pattern of maxima and minima confirms that the molecule is tetrahedral (compare with Problem 22.41) and then calculate the bond length of CCl_4.

```
 0   0   0   0   0   0   0   0   0   0   0   0   0   0   0   0   0   0   0   0   0   0   0   6  12  11   9   6   7  10
 1   0   0   0   0   0   0   0   0   0   0   0   0   0   0   0   0   0   0   0   0   7  20  39  42  32  16  10  12
16   5   0   0   0   0   0   0   0   0   0   0   0   0   0   0   0   0   0   6  23  60  99 107  86  40  16  17
47  30   7   0   0   0   0   0   0   0   0   0   0   0   0   0   0   0   0  10  20  40  81 124 145 102  32  42
50  34   5   0   0   0   0   0   0   0   0   0   0   0   0   0   0   0   9  30  40  50  62  84  85  67  40  34  46
26   6   0   0   0   0   0   0   0   0   0   0   0   0   0   4  34  61  87  78  60  40  39  26  17  17  30
 0   0   1   1   0   0   0   0   0   0   0   0   0   0   0   5  30  79 104  90  57  24  14   6   0   0   8
 0   0   7  16  18  12   0   0   0   0   0   0   0   0   2  21  42  62  58  22   8   0   0   0   0   0
 0   2  21  42  62  58  22   8   0   0   0   0   0   0   7  16  18  12   0   0   0   0   0   0   0
 0   5  30  79 104  90  57  24  14   6   0   0   8   0   0   1   1   0   0   0   0   0   0   0   0
 0   4  34  61  87  78  60  40  39  26  17  17  30  26   6   0   0   0   0   0   0   0   0   0   0
 0   0   9  30  40  50  62  84  85  67  40  34  46  50  34   5   0   0   0   0   0   0   0   0   0   0
 0   0   0   0  10  20  40  81 124 145 102  32  42  47  30   7   0   0   0   0   0   0   0   0   0   0
 0   0   0   0   6  23  60  99 107  86  40  16  17  16   5   0   0   0   0   0   0   0   0   0   0
 0   0   0   0   0   7  20  39  42  32  16  10  12   1   0   0   0   0   0   0   0   0   0   0   0
 0   0   0   0   0   0   6  12  11   9   6   7  10   0   0   0   0   6   5   0   0   0   0   0   0
 0   0   0   0   0   0   5   6   0   0   0   0   0  10   7   6   9  11  12   6   0   0   0   0   0
 0   0   0   0   0   0   0   0   0   0   0   0   1  12  10  16  32  42  39  20   7   0   0   0   0
 0   0   0   0   0   0   0   0   0   0   0   5  16  17  16  40  86 107  99  60  23   6   0   0   0
 0   0   0   0   0   0   0   0   0   0   7  30  47  42  32 102 145 124  81  40  20  10   0   0   0
 0   0   0   0   0   0   0   0   0   0   5  34  50  46  34  40  67  85  84  62  50  40  30   9   0   0
 0   0   0   0   0   0   0   0   0   0   6  26  30  17  17  26  39  40  60  78  87  61  34   4   0
 0   0   0   0   0   0   0   0   1   1   0   0   8   0   0   6  14  24  57  90 104  79  30   5   0
 0   0   0   0   0   0   0  12  18  16   7   0   0   0   0   0   0   8  22  58  62  42  21   2   0
 0   0   0   0   0   8  22  58  62  42  21   2   0   0   0   0   0   0   0  12  18  16   7   0   0
 8   0   0   6  14  24  57  90 104  79  30   5   0   0   0   0   0   0   0   0   0   0   1   1   0   0
30  17  17  26  39  40  60  78  87  61  34   4   0   0   0   0   0   0   0   0   0   0   0   6  26
46  34  40  67  85  84  62  50  40  30   9   0   0   0   0   0   0   0   0   0   0   0   5  34  50
42  32 102 145 124  81  40  20  10   0   0   0   0   0   0   0   0   0   0   0   0   7  30  47
17  16  40  86 107  99  60  23   6   0   0   0   0   0   0   0   0   0   0   0   0   0   5  16
12  10  16  32  42  39  20   7   0   0   0   0   0   0   0   0   0   0   0   0   0   0   0 - 0   1
10   7   6   9  11  12   6   0   0   0   0   0   0   0   0   0   0   0   0   0   0   0   0   0   0   0
```

23 The electric and magnetic properties of molecules

Learning objectives

After careful study of this chapter you should be able to

(1) Define the *dipole moment* of a molecule (p. 767) and the *polarization* of a dielectric (p. 769).

(2) Relate the *electric susceptibility* to the *relative permittivity* of a sample (eqn (23.1.5)).

(3) Define the molecular *polarizability* (eqn (23.1.10)) and *polarizability volume* (p. 772).

(4) Relate the permanent dipole moment to the relative permittivity by the *Debye equation* (eqn (23.1.11)) and define the *molar polarizability* (eqn (23.1.13)).

(5) Relate the *refractive index* of a sample to its polarizability, account for its frequency dependence (p. 776), and define *molar refractivity* (eqn (23.1.16)).

(6) Indicate the basis of *optical activity* (p. 779).

(7) Explain the source of, and assess the magnitude of, *dipole–dipole molecular interactions* (eqn (23.2.1)) and *dipole–induced-dipole interactions* (eqn (23.2.2)).

(8) Explain the source of *dispersion forces* and use the *London formula* to assess their strength (eqn (23.2.3)).

(9) Describe the *Lennard-Jones intermolecular potential* (eqn (23.2.5)).

(10) Define the *Madelung constant* (eqn (23.3.3)) explain its significance, and use it to calculate the internal energy of ionic lattices (p. 787).

(11) Describe the formation of a *molecular beam* (p. 789).

(12) Define *impact parameter* and *differential scattering cross-section* (p. 790).

(13) Explain the terms *glory* and *rainbow scattering*, and describe how molecular beam experiments yield information about intermolecular forces (p. 791).

(14) Describe the connection between *intermolecular forces* and *virial coefficients* of real gases (p. 792).

(15) Define the *radial distribution functions* of molecules in liquids and indicate their connection with thermodynamic properties (p. 795).

(16) Explain the temperature dependence of the *viscosity* of liquids (eqn (23.3.10)).

(17) Define the *magnetic susceptibility* and the *magnetizability* of a

material (eqns (23.4.1) and (23.4.3)) and explain how they are measured.
(18) Explain the source of the *paramagnetism* of molecules with unpaired spins (p. 791) and calculate their susceptibilities (eqn (23.4.9)).

Introduction

In this chapter we examine some of the electric and magnetic properties of molecules and interpret them in terms of their electronic structure. These properties include the dipole moment, polarizability, and magnetizability. The polarizability is related to refractive index, optical activity, and intermolecular forces, and the magnetizability is related to the magnetic susceptibility. We shall see how all such properties are related to molecular structure.

23.1 Electric properties

We shall discuss both the permanent electric dipole moment of a polar molecule, and the dipole moment induced in a molecule when it is in an electric field. The applied field distorts the electronic structure and changes the equilibrium positions of the nuclei and gives rise to a dipole moment of a size which depends on the strength of the field and the responsiveness of the molecule: this responsiveness is its *polarizability*.

The permanent dipole moment. When two charges q and $-q$ are separated by a distance R they constitute an electric *dipole* of magnitude qR. The dipole has a direction as well as a magnitude, and it is conventional to regard it as being directed from the negatively charged end towards the positive, Fig. 23.1. Dipole moments are generally quoted in *debye* (D):

$$1\,\text{D} = 3.336 \times 10^{-30}\ \text{C m}$$

(this unit is named after one of the pioneers of the study of polar molecules). When the charge q has the magnitude of the electronic charge and the separation is 100 pm (0.1 nm, 1 angstrom) the magnitude of the dipole is

$$p = (1.6022 \times 10^{-19}\ \text{C}) \times (1 \times 10^{-10}\ \text{m}) = 1.6022 \times 10^{-29}\ \text{C m}.$$

$$\approx 4.8\ \text{D}.$$

The magnitudes of dipole moments of small molecules are roughly of this order (but usually somewhat smaller).

Dipole moments are used to give information about molecular structure and also to test calculated wavefunctions. In practice, however, they are important for determining the suitability of a solvent for a solid, for the ability of a solvent to disrupt an ionic crystal depends on its ability to solvate the ions and to reduce the electrostatic interactions holding the crystal together. Polar molecules play a dual role in this process. One end of their dipole may be attracted electrostatically to the ion of opposite charge, and this lowers the potential energy of the ion in solution, Fig. 23.2. Their other role is to reduce the strength of the Coulombic interaction

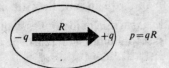

Fig. 23.1. The convention for the dipole moment.

between ions in the solution. This comes about as follows. When two ions are separated by a distance r their potential energy of interaction is proportional to $1/4\pi\varepsilon_0 r$ when the medium separating them is a vacuum, but when they are immersed in a solvent it is reduced to $1/4\pi\varepsilon r$, where ε is the permittivity. Usually one writes $\varepsilon = \varepsilon_0 \varepsilon_r$, where ε_r is the *relative permittivity* (or *dielectric constant*). The value of ε_r is determined in part by the polar nature of the solvent molecules, and it can have a very significant effect on the strength of the Coulombic interaction. For instance, water has $\varepsilon_r \approx 78$, which means that the Coulombic potential is reduced by nearly two orders of magnitude. The role of the permittivity of the solvent in governing the thermodynamic properties of ions in solution was discussed in Chapter 11.

Determination of dipole moments. Dipole moments can be measured in several ways. When the rotational spectrum of a molecule is examined in the gas phase it is found that the lines shift if the sample is exposed to a strong electric field. The magnitude of this *Stark effect* depends on the molecule's dipole moment, and so rotational spectra can be used to obtain very accurate data. When the molecule cannot be examined by rotational spectroscopy (when it is too complex, involatile, or unstable in the gas phase), the measurement is usually based on the determination of the relative permittivity ε_r of a bulk sample. The next paragraphs show how ε_r may be measured, and then how it may be related to the dipole moment.

We need to define the capacitance of a capacitor. Consider first the two plane, parallel plates depicted in Fig. 23.3. The charges on the plates are q and $-q$, and if their areas are A, this can be expressed as $\pm\sigma A$, where σ is the *surface charge density* (charge per unit area). When the charges have been established there is a potential difference $\Delta\phi$ between the plates. The *capacitance* of this arrangement is defined as

$$C = q/\Delta\phi.$$

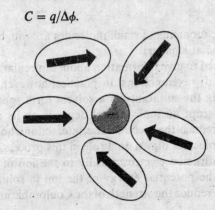

Fig. 23.2. Solvation of an ion by a polar solvent.

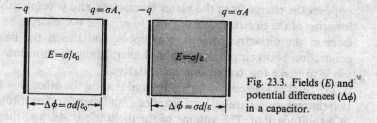

Fig. 23.3. Fields (E) and potential differences ($\Delta\phi$) in a capacitor.

Therefore in the present case, $C = \sigma A/\Delta\phi$. The potential difference can be written in terms of the charge density σ and the separation of the plates d. First we accept the standard result of elementary electrostatics that the electric field in the region between the plates is σ/ε_0 if the medium is a vacuum, but σ/ε if it is a dielectric. The presence of a field of strength E indicates the existance of a potential difference of magnitude Ed between the two plates (the field is the negative slope of the potential, and so $E = -\Delta\phi/d$ when the potential varies linearly). Therefore in the absence of a dielectric the capacitance is

$$C_0 = \sigma A/(\sigma d/\varepsilon_0) = \varepsilon_0 A/d,$$

but in the presence of the dielectric the capacitance is

$$C = \sigma A/(\sigma d/\varepsilon) = \varepsilon A/d.$$

Therefore a way of determining the relative permittivity is to measure the capacitance with and without the sample:

(23.1.1) $\qquad C/C_0 = \varepsilon/\varepsilon_0 = \varepsilon_r.$

Capacitance is easily measured by one of the standard types of experiments (e.g., using an a.c. bridge) and commercial instruments are available.

The next step is to relate ε_r to a molecular property. This is achieved by introducing the *polarization*, P, of the dielectric. The polarization is the *charge per unit area* of the sample. It is also the *average dipole moment per unit volume*. That these are equivalent may be seen by reference to Fig. 23.4, which shows how the individual dipoles of the medium align with the applied field and give rise to a charge on the surface of the dielectric but cancel in the bulk. The surface charge (of magnitude PA)

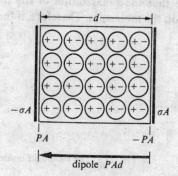

Fig. 23.4. Polarization of the molecules constituting a medium and the corresponding dipole.

opposes the charge from the plates themselves, and is responsible for the lowering of the electric field from σ/ε_0 to σ/ε. The two charges on opposite faces of the dielectric constitute a dipole, and this is the basis of the connection between polarization and average dipole moment. The next job is to make this connection quantitative.

The total charge on one face is PA and that on the other is $-PA$. These two charges are separated by a distance d, and therefore constitute a dipole of magnitude PAd. The volume of the dielectric is Ad, and therefore the average dipole moment per unit volume of dielectric is PAd/Ad, or P, as was to be shown. This equivalence indicates the two prongs of the calculation we have to make: one is to connect P with the measurable quantity ε_r, and the other is to relate it to the dipole moment of the individual molecules.

In order to connect P and ε_r we proceed as follows. In the absence of the dielectric the electric field between the plates is $E_0 = \sigma/\varepsilon_0$. In the presence of the dielectric, but for the same amount of charge on the plates, the field is changed to E. There are two ways of writing E. One expresses it in terms of the relative permittivity:

$$E = \sigma/\varepsilon = \sigma/\varepsilon_0\varepsilon_r.$$

The other assumes that the effect of the medium is simply to reduce σ to $\sigma - P$. Therefore:

$$E = (\sigma - P)/\varepsilon_0.$$

Eliminating σ from this pair of equations gives

(23.1.2) $P = \varepsilon_0(\varepsilon_r - 1)E,$

which relates the polarization to the field inside the dielectric.

We now concentrate on a single molecule somewhere inside the dielectric. This molecule experiences the field E together with a field that arises from the charges on the surface of the cavity that surrounds it, Fig. 23.5. Calculation shows that this extra contribution to the overall local field has a magnitude $\frac{1}{3}P/\varepsilon_0$ if the cavity is assumed to be spherical and the medium continuous. Therefore the total field at the molecule is

(23.1.3) $E^* = E + \frac{1}{3}P/\varepsilon_0 = \frac{1}{3}P(\varepsilon_r + 2)/\varepsilon_0(\varepsilon_r - 1).$

In the final step we suppose that the polarization of the dielectric is proportional to the field acting on the molecules. Therefore, if the constant

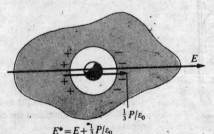

$\frac{1}{3}P/\varepsilon_0$

$E^* = E + \frac{1}{3}P/\varepsilon_0$

Fig. 23.5. The local field inside a dielectric.

of proportionality is written $\varepsilon_0\chi_e$ we write

(23.1.4) $P = \varepsilon_0\chi_e E^*$.

χ_e is called the *electric susceptibility*, and is dimensionless. If this expression for P is substituted into the last equation the E^* cancels and we find a relation connecting the susceptibility and the permittivity:

(23.1.5) $\chi_e = 3(\varepsilon_r - 1)/(\varepsilon_r + 2)$.

Example   (Objective 2). The capacitance of an empty sample cell was 5.0 pF; when it was filled with a sample of camphor its capacitance was 57.1 pF. What is the relative permittivity (dielectric constant) and electric susceptibility of camphor at room temperature?

● *Method.* Use eqn (23.1.1) to find ε_r and eqn (23.1.5) to find χ_e. The relative permittivity of air is virtually unity.

● *Answer.* From eqn (23.1.1), $\varepsilon_r = (57.1 \text{ pF})/(5.0 \text{ pF}) = 11.4$. From eqn (23.1.5), $\chi_e = 3(11.4 - 1)/(11.4 + 2) = 2.33$.

● *Comment.* We take this *Example* further below. Note that both ε_r and χ_e are dimensionless.

The susceptibility depends on the nature of the molecules because it represents how they acquire a net dipole moment in the presence of some electric field E^*. Even if the molecules have a permanent dipole moment p, in the absence of an applied field the *mean* dipole moment is zero in a fluid sample because their rotational motion averages the net moment to zero. When a field is present some orientations are energetically more favourable than others. In that case the average dipole moment of the sample differs from zero to an extent determined by the competition of the aligning influence of the field and the disruptive, stirring influence of the thermal motion in the sample. The net moment can be evaluated by using the Boltzmann distribution, p. 669, for a sample at a temperature T.

The energy of a molecule with dipole moment of magnitude p making an angle θ with an electric field of strength E^* is

$\mathscr{E} = -pE^*\cos\theta$.

When this energy is used in the Boltzmann distribution, the calculation of the average dipole moment gives

(23.1.6) $p_{\text{average}} = p\mathscr{L}(x)$

where x is pE^*/kT and $\mathscr{L}(x)$ is the *Langevin function*

$$\mathscr{L}(x) = \frac{e^x + e^{-x}}{e^x - e^{-x}} - \frac{1}{x}.$$

We shall be interested only in the value of the Langevin function at small values of x. This is because at reasonable temperatures and dipole moments, x is very much less than unity. For instance, if $p \approx 1\,\text{D}$ and $T \approx 300\,\text{K}$ the ratio pE^*/kT exceeds 0.01 only when the field strength exceeds $100\,\text{kV cm}^{-1}$: most conventional measurements are done at much smaller field strengths. When $x \ll 1$ the exponentials in $\mathscr{L}(x)$ can be expanded, and the largest term that survives in the expansion is

$$\mathscr{L}(x) = \tfrac{1}{3}x + \dots .$$

Therefore the average molecular dipole moment is

(23.1.7) $p_{\text{average}} \approx p^2 E^*/3kT.$

The last result can be used in the present problem. If the number density of molecules (the number per unit volume) is $\mathscr{N}$ the net dipole moment per unit volume in a field E^* is

(23.1.8) $P = \mathscr{N} p_{\text{average}} = \mathscr{N} p^2 E^*/3kT.$

This has the form of eqn (23.1.4) and so χ_e can be identified as $\mathscr{N} p^2/3\varepsilon_0 kT$. Consequently the relative permittivity of the solution is related to the permanent dipole moment of the molecules and the temperature by

(23.1.9) $\mathscr{N} p^2/3\varepsilon_0 kT = 3(\varepsilon_r - 1)/(\varepsilon_r + 2).$

Measurements of ε_r and the density (for $\mathscr{N}$) immediately give a value of the dipole moment. (But see below before using this expression.)

Polarizability. Non-polar molecules may acquire a dipole moment when they are exposed to an electric field. This is because their electron distributions and nuclear configurations become distorted and the centres of positive and negative charge, which were originally coincident, are separated. The magnitude of this induced dipole moment is proportional to the strength of the field (so long as the strength is not too great) and we write

(23.1.10) $p_{\text{induced}} = \alpha E.$

The coefficient of proportionality α is called the *polarizability* of the molecule. In general, it depends on the orientation of the molecule. If the field becomes very strong the induced moment also varies as E^2, and the coefficient of proportionality β in βE^2 is called the *hyperpolarizability*. We shall ignore this complication, which becomes important only at very high field strengths, such as are found in laser beams.

When the polarizability is defined as in eqn (23.1.10) it has the units $\text{J}^{-1}\,\text{C}^2\,\text{m}^2$ when E is expressed in V m^{-1} and p in C m. This can be checked by simple dimensional analysis (because $[p] = \text{C m}$ and $[E] = \text{V m}^{-1} = \text{J C}^{-1}\text{m}^{-1}$). That is awkward. Usually polarizabilities are expressed as a volume by dividing by $4\pi\varepsilon_0$ (which also has the units $\text{J}^{-1}\,\text{C}^2\,\text{m}^{-1}$), and from now on we shall often refer to *polarizability volumes*, $\alpha' = \alpha/4\pi\varepsilon_0$. Quite often α' is quoted in units of cm^3 or (angstrom)^3: see Table 23.1. Typical values are $0.20 \times 10^{-24}\,\text{cm}^3$ for He, rising to $10.5 \times 10^{-24}\,\text{cm}^3$ for CCl_4. The molecular volumes of these two species are about 40×10^{-24} cm³ and 230×10^{-24} cm³ respectively, and the parallelism of the polarizability volumes and the molecular volumes should be noted.

Table 23.1. Dipole moments (p), polarizabilities (α), and polarizability volumes (α')

	$p/10^{-30}$ C m	$p/$D	$\alpha'/10^{-24}$ cm³	$\alpha/10^{-40}$ J^{-1} C² m²
H_2	0	0	0.819	0.911
N_2	0	0	1.77	1.97
CO_2	0	0	2.63	2.93
CO	0.390	0.117	1.98	2.20
HF	6.37	1.91	0.51	5.67
HCl	3.60	1.08	2.63	2.93
HBr	2.67	0.80	3.61	4.01
HI	1.40	0.42	5.45	6.06
H_2O	6.17	1.85	1.48	1.65
NH_3	4.90	1.47	2.22	2.47
CCl_4	0	0	10.5	11.7
$CHCl_3$	3.37	1.01	8.50	9.46
CH_2Cl_2	5.24	1.57	6.80	7.57
CH_3Cl	6.24	1.87	4.53	5.04
CH_4	0	0	2.60	2.89
CH_3OH	5.70	1.71	3.23	3.59
CH_3CH_2OH	5.64	1.69		
C_6H_6	0	0	10.4	11.6
$C_6H_5CH_3$	1.20	0.36		
$o\text{-}C_6H_4(CH_3)_2$	2.07	0.62		
He	0	0	0.20	0.22
Ar	0	0	1.66	1.85

Source: *Handbook of chemistry and physics* (for p); principally *Theory of electric polarization*, C. J. F. Böttcher and P. Bordewijk, Elsevier; $\alpha' = \alpha/4\pi\varepsilon_0$.

Inside the sample the molecule experiences a field E^*, and the dipole induced is αE^*. This dipole contributes an amount $p_{induced} = \alpha E^*$, to the polarization of the medium. The total polarization is therefore

$$P = \mathcal{N}(\alpha + p^2/3kT)E^*.$$

Only the α term survives if the molecules have no permanent moment, but because molecules that are polar are also polarizable, both terms contribute for molecules with permanent moments. The susceptibility χ_e is now obtained by comparing the last equation with $P = \varepsilon_0\chi_e E^*$, and then eqn (23.1.5) gives

(23.1.11) $$(\mathcal{N}/\varepsilon_0)(\alpha + p^2/3kT) = 3(\varepsilon_r - 1)/(\varepsilon_r + 2),$$

which is called the *Debye equation*. The same expression, but lacking the contribution from the permanent dipole moment, is called the *Clausius-Mossotti equation*.

The Debye equation can be written in terms of the density ρ and the molar mass of the molecules, M_m, through $\mathcal{N} = \rho L/M_m$. Then

(23.1.12) $\quad \alpha + p^2/3kT = 3\varepsilon_0(M_m/L\rho)\{(\varepsilon_r - 1)/(\varepsilon_r + 2)\}.$

Quite often this expression is written in terms of the *molar polarizability* P_m:

(23.1.13a) $\quad P_m = (M_m/\rho)[(\varepsilon_r - 1)/(\varepsilon_r + 2)].$

Then

(23.1.13b) $\quad P_m = (L/3\varepsilon_0)\{\alpha + p^2/3kT\}.$

According to eqn (23.1.13) both the polarizability and dipole moment of the molecules can be measured by determining the temperature dependence of the molar polarizability. If P_m is plotted against $1/T$ the slope of the lines gives $p^2/3k$ and the intercept at $1/T = 0$ gives the polarizability α. The physical reason behind this result is that at very high temperatures the disordering effect of the thermal motions causes the permanent dipole to rotate so rapidly that its contribution to the polarization is averaged out, leaving the induced dipole alone. This induced dipole lies in the direction of the field that is inducing it, and so it remains in the same direction even though the molecule itself might be rotating; therefore it is not averaged to zero by the thermal motion, and survives to contribute to the permittivity even at the highest temperatures.

Example (Objective 4). A series of measurements on camphor in the same sample cell as in the last *Example*, but at a variety of temperatures, were made and are reported below. Use the data to find the dipole moment and the polarizability of the molecule.

$t/°C$	0	20	40	60	80	100	120	140	160	200
$\rho/\text{g cm}^{-3}$	0.99	0.99	0.99	0.99	0.99	0.99	0.97	0.96	0.95	0.91
C/pF	62.6	57.1	54.1	50.1	47.6	44.6	40.6	38.1	35.6	31.1

- *Method.* Find ε_r at each temperature; form $(\varepsilon_r - 1)/(\varepsilon_r + 2)$. In order to use eqn (23.1.12) multiply by $(3\varepsilon_0 M_m/\rho L)$ noting that ρ depends on the temperature. For camphor $M_m = 152.23 \text{ g mol}^{-1}$. Plot the r.h.s. of eqn (23.1.12) against $1/T$. The intercept gives α and the slope $p^2/3k$.

- *Answer.* At $0\ °C$, when the density is 0.99 g cm^{-3},

$$a \overset{\text{def}}{=} 3\varepsilon_0 M_m/\rho L$$
$$= 3 \times (8.8542 \times 10^{-12} \text{ J}^{-1} \text{ C}^2 \text{ m}^{-1}) \times (6.0221 \times 10^{23} \text{ mol}^{-1}) \times (0.99 \text{ g cm}^{-3})$$
$$= 6.78 \times 10^{-39} \text{ J}^{-1} \text{ C}^2 \text{ m}^2.$$

At higher temperatures its value can be found by multiplying by $\rho(0\,°C)/\rho(t)$. Draw up the following table:

$t/°C$	0	20	40	60	80	100	120	140	160	200
$(1/T)/10^{-3}\text{ K}^{-1}$	3.66	3.41	3.19	3.00	2.83	2.68	2.54	2.42	2.31	2.11
ε_r	12.5	11.4	10.8	10.0	9.5	8.9	8.1	7.6	7.1	6.2
$\{(\varepsilon_r - 1)/(\varepsilon_r + 2)\}$	0.793	0.776	0.766	0.750	0.739	0.725	0.703	0.688	0.670	0.634
$a\{(\varepsilon_r - 1)/(\varepsilon_r + 2)\}/$ $10^{-39} \text{ J}^{-1} \text{ C}^2 \text{ m}^2$	5.39	5.26	5.19	5.09	5.01	4.91	4.87	4.81	4.73	4.68

The plot is shown in Fig. 23.6. The intercept lies at 3.65, and so

$$\alpha = 3.65 \times 10^{-39} \text{ J}^{-1} \text{ C}^2 \text{ m}^2$$

$$\alpha' = \alpha/4\pi\varepsilon_0 = 3.65 \times 10^{-39} \text{ J}^{-1} \text{ C}^2 \text{ m}^2/4\pi \times (8.854 \times 10^{-12} \text{ J}^{-1} \text{ C}^2 \text{ m}^{-1})$$

$$= 3.28 \times 10^{-29} \text{ m}^3 = 3.28 \times 10^{-23} \text{ cm}^3.$$

The slope is 0.480, and so

$$p^2/3k = 0.480 \times (10^{-39} \text{ J}^{-1} \text{ C}^2 \text{ m}^2)/(10^{-3} \text{ K}^{-1}) = 0.480 \times 10^{-36} \text{ J}^{-1} \text{ C}^2 \text{ m}^2 \text{ K}.$$

Consequently

$$p = \sqrt{\{3 \times (1.380\,66 \times 10^{-23} \text{ J K}^{-1}) \times (0.480 \times 10^{-36} \text{ J}^{-1} \text{ C}^2 \text{ m}^2 \text{ K})\}}$$

$$= 4.46 \times 10^{-30} \text{ C m} \triangleq 1.34 \text{ D}.$$

• *Comment.* The odd thing about this result is that camphor does not melt until 175 °C, and so the data show that the spherical molecule is rotating even in the solid.

The polarizability of a molecule is governed by the strength with which the nuclear charges control the electrons and prevent their distortion by the applied field. If the molecule contains only a few electrons their distribution is tightly controlled by the nuclear charge and the polarizability is low. If the molecule contains large atoms with many electrons some distance from the nuclei, the degree of nuclear control is less, the electron distribution flabbier, and the polarizability greater.

The polarizability can be calculated from the wavefunction for the molecule, but it is possible to estimate its magnitude without resorting to detailed computation. Consider an atom of atomic number Z. It has no permanent dipole moment, but a dipole may be induced if an electric field is applied. Inducing a dipole moment involves shifting the electronic charge inside the atom; but that is another way of saying that the atom has to be excited to some extent. We can guess that the success of the field in inducing a dipole moment is proportional to the ratio of its energy of interaction with the electrons to the energy that has to be overcome in

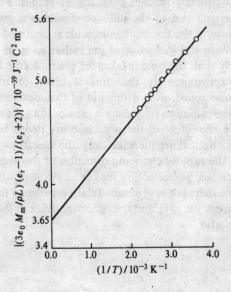

Fig. 23.6. Debye equation plot for camphor.

order to excite them. The instantaneous dipole moment of the atom is of the order of $-er$, where r is its radius, and so the energy of interaction with the applied field of strength E is of the order of erE. If the excitation energies are of the order of half the ionization energy I, the dipole moment induced is of the order of $-er(erE/\frac{1}{2}I)$ because the instantaneous dipole moment $-er$ is frozen into the atom to an extent determined by the ratio of the two energies. This suggests that the polarizability is of the order $2e^2r^2/I$. If there are Z electrons we can anticipate that

$$\alpha \approx 2Ze^2r^2/I, \text{ or } \alpha' \approx Ze^2r^2/2\pi\varepsilon_0 I.$$

This shows that the polarizability does indeed increase with increasing atomic number, atomic size, and ease of excitation of the atom, as the experimental results confirm.

Polarizability at high frequencies: refractive index. So far we have concentrated on molecules in static fields. When the applied field oscillates the Debye equation may lose its validity because when the frequency is very high the permanent dipole moments may be unable to reorientate themselves sufficiently rapidly to line up with the ever-changing direction of E. At such high frequencies the permanent dipole moment makes no contribution to the polarization of the medium, and so it drops out of the Debye equation. Since a molecule requires about 10^{-12} s to rotate in a fluid the dipolar contribution vanishes when the relative permittivity is measured above about 10^{11} Hz (in the microwave region). Inverting this argument suggests that it might be possible to measure the rate of molecular reorientation in liquids by observing the variation of the polarization and permittivity with frequency. This can be done, and the technique of *dielectric relaxation* is based on this idea.

The polarizability contribution survives at microwave frequencies, but it is modified as the frequency reaches even higher values. Part of the molecular distortion giving rise to the induced dipole is a geometrical distortion of the positions of the nuclei: the molecule is bent and stretched by the applied field. When the frequency of the radiation is so high that it changes more quickly than the time it takes for the molecule to change its shape—which is approximately the time it takes to undergo a vibration—this *distortion polarization* drops out of the contribution, Fig. 23.7. Therefore, when the radiation frequency is greater than the molecular vibration frequency (in the infrared) the polarizability drops to another value. At even higher, optical frequencies only the electrons are light enough to respond to the rapidly changing direction of the applied field, and then only the *electronic polarizability* makes any contribution.

At optical frequencies there is a very simple relation between the relative permittivity at the frequency v, $\varepsilon_r(v)$, and the *refractive index* of the medium at the same frequency $n_r(v)$:

(23.1.14) $\varepsilon_r(v) = n_r^2(v).$

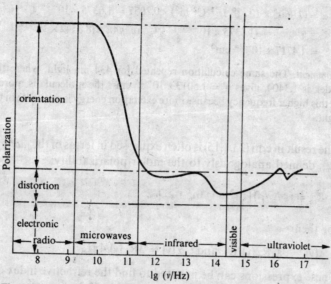

Fig. 23.7. The dependence of the polarization on the frequency.

The polarizability at optical frequencies can therefore be determined simply by measuring the refractive index of the sample, Table 23.2, and inserting the last result into the Clausius–Mossotti equation, eqn (23.1.12):

(23.1.15) $\alpha(v) = (3\varepsilon_0 M_m/L\rho)\{(n_r^2 - 1)/(n_r^2 + 2)\}.$

Table 23.2. Refractive index relative to air at 20 °C

	$\lambda = 434$ nm	589 nm	656 nm
Water	1.3404	1.3330	1.3312
Benzene	1.5236	1.5012	1.4965
Carbon tetrachloride	1.4729	1.4607	1.4579
Carbon disulphide	1.6748	1.6276	1.6182
KCl	1.5050	1.4904	1.4873
KI	1.7035	1.6664	1.6581

Source: *American Institute of Physics handbook*, McGraw-Hill.

Example (Objective 5). The refractive index of water at 20 °C is 1.3330 for light of wavelength 589 nm. What is the polarizability of the molecule at this frequency?

- *Method.* Use eqn (23.1.15) with $M_m = 18.015$ g mol^{-1} and $\rho = 0.9982$ g cm^{-3}.

- *Answer.* $(n_r^2 - 1)/(n_r^2 + 2) = (1.3330^2 - 1)/(1.3330^2 + 2) = 0.20$

$3\varepsilon_0 M_m/\rho L = 3 \times (8.8542 \times 10^{-12}$ J^{-1} C^2 m$^{-1}) \times (18.015$ g mol$^{-1})/(6.022 \times 10^{23}$ mol$^{-1}) \times (0.9982$ g cm$^{-3})$

$= 7.960\,48 \times 10^{-40}$ J^{-1} C^2 m^2.

Hence

$$\alpha = (7.9605 \times 10^{-40} \text{ J}^{-1} \text{ C}^2 \text{ m}^2) \times 0.2057 = 1.6375 \times 10^{-40} \text{ J}^{-1} \text{ C}^2 \text{ m}^2.$$

$$\alpha' = \alpha/4\pi\varepsilon_0 = (1.6375 \times 10^{-40} \text{ J}^{-1} \text{ C}^2 \text{ m}^2)/4\pi \times (8.8542 \times 10^{-12} \text{ J}^{-1} \text{ C}^2 \text{ m}^{-1})$$

$$= 1.4717 \times 10^{-24} \text{ cm}^3.$$

• *Comment.* The same calculation repeated for 434 nm light, where the refractive index is 1.3404, gives $\alpha' = 1.5013 \times 10^{-24} \text{ cm}^3$: the molecule is more polarizable at this higher frequency because more excitation energy is available in the incoming light.

The result in eqn (23.1.15) is often expressed in terms of the *molar refractivity* R_m, defined analogously to the molar polarizability:

(23.1.16) $$R_m = (M_m/\rho)\,[(n_r^2 - 1)/(n_r^2 + 2)],$$

for then

(23.1.17) $$\alpha(v) = 3\varepsilon_0 R_m(v)/L \quad \text{and} \quad \alpha'(v) = 3R_m(v)/4\pi L.$$

These expressions can be inverted to find the refractive index of a medium from tables of refractivities (such as Table 23.3):

(23.1.18) $$n_r = \left\{ \frac{V_m + 2R_m}{V_m - R_m} \right\}^{1/2}, \qquad V_m = M_m/\rho.$$

Table 23.3. Molar refractivities at 589 nm, $R_m/\text{cm}^3 \text{ mol}^{-1}$

C—H	1.65	C—C	1.20	C—O	1.41
O—H	1.85	C=C	2.79	C=O	3.34
		C≡C	4.79	C≡N	4.69
				He	0.5

Li$^+$	Be^{2+}		O^{2-}	F$^-$	Ne
0.07	0.20		7	2.65	1.00
Na$^+$	Mg^{2+}	Al^{3+}		Cl$^-$	Ar
0.46	0.24	0.17		9.30	4.14
K$^+$	Ca^{2+}			Br$^-$	Kr
2.12	1.19			12.12	6.26
Rb$^+$				I$^-$	Xe
3.57				18.07	10.16

Source: E. A. Moelwyn-Hughes, *Physical chemistry*, Pergamon.

Example (Objective 5). Estimate the refractive index of acetic acid for yellow sodium light.

• *Method.* Use the information in Table 23.3. The molecule CH_3COOH consists of $3(C—H) + (C—C) + (C=O) + (C—O) + (O—H)$. Then use eqn (23.1.18).

• *Answer.* $R_m/\text{cm}^3 \text{ mol}^{-1} = 3 \times 1.65 + 1.20 + 3.34 + 1.41 + 1.85 = 12.75.$

In order to use eqn (23.1.18) we require

$V_m = M_m/\rho = (60.05 \text{ g mol}^{-1})/(1.046 \text{ g cm}^{-3}) = 57.41 \text{ cm}^3 \text{ mol}^{-1}.$

Therefore, from eqn (23.1.18),

$n_r = [(57.41 + 25.50)/(57.41 - 12.75)]^{1/2} = 1.36.$

- *Comment.* The experimental value is 1.37, and so this technique for calculating n_r is quite good.

The refractive index is the ratio of the speed of light in a vacuum to the speed in the medium, $n_r = c/v$. The reason why the speed is affected by the polarizability of the medium can be explained by visualizing the propagation of the light as proceeding by the incident light distorting the molecules of the medium, and therefore inducing in them a dipole moment that oscillates with the incident frequency. This oscillating dipole generates radiation of the same frequency, but the process delays the phase of the propagating light wave. This delaying of the phase corresponds to the slowing of the speed of light in the medium. If the molecules are not polarizable the light does not interact, and no phase-delay is introduced; if the molecules are highly polarizable the interaction is strong and the phase lag considerable. Light of high frequency carries enough energy to distort the electronic distributions of the molecules in its path more effectively than light of lower frequency. Therefore, at optical frequencies the polarizability increases with the frequency of the incident light. This implies that the refractive index increases with frequency in the optical region. That is the basis of the well-known phenomenon of the dispersion of light by a prism: the refractive index is greater for blue light than for red, and therefore blue rays are bent more than red rays by a prism in their path. *Dispersion* is a name carried over from this phenomenon to mean the variation of the refractive index, or any other property, with the frequency.

Optical activity. The phenomenon of optical activity, in which a sample rotates the plane of polarization of a polarized light beam passing through it, can be related to a special property of the refractive index. The plane of polarization is rotated if the refractive index for *left-handed* circularly polarized light (n_L) is different from that for *right-handed* circularly polarized light (n_R). (We use the convention that in right-handed light the electric vector rotates clockwise from the viewpoint of the observer facing the oncoming beam, Fig. 23.8.)

That optical rotation occurs when $n_L \neq n_R$ may be seen by referring to Fig. 23.8. Before entering the medium the beam is plane-polarized at an angle $\theta = 0°$ to some axis. This plane-polarized light can be resolved into the superposition of two counter-rotating circularly-polarized components, Fig. 23.8. On entering the medium one component propagates faster than the other if the refractive indices are different, and if the sample is of length l the difference in passage time is $l/v_R - l/v_L$, where v_R and v_L are the velocities of the components. This may be expressed in terms of the

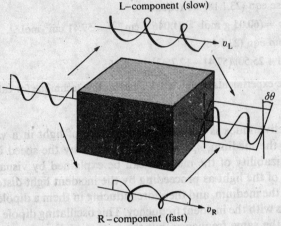

L–component (slow)

v_L

$\delta\theta$

v_R

R – component (fast)

Fig. 23.8. Linearly polarized light regarded as the superposition of two counter-rotating components which propagate with different velocities in an optically active medium.

refractive indices as $(n_R - n_L)l/c$, because $n = c/v$. The electric vectors are in phase at a different angle when they leave the sample, and so their superposition leads to a beam plane-polarized at a new angle $\delta\theta$. The angle of rotation is

(23.1.19) $\delta\theta = (n_L - n_R)2\pi l/\lambda,$

where λ is the wavelength of the light.

In order to explain why the refractive indices differ for the two polarizations of light we must explain why the polarizability of the molecule depends on the handedness of the light. The full theory of optical rotation is quite complicated, but one interpretation it permits is that if a molecule has a *helical structure* its polarizability depends on whether or not the electric field of the probing radiation follows the spiral of the molecule. If the molecule has this spiral structure, we can expect its interaction with the light to depend on the handedness of the light. Molecules having a helical structure are not superimposible on their mirror image, and this is the criterion for deciding whether a molecule is optically active (it was discussed in terms of symmetry on p. 527).

If optically inactive molecules solidify into a helical arrangement, optical activity can also be expected on the grounds that the polarizability of the solid depends upon the handedness of the light. This is confirmed by the structure of quartz, which has long columns of spiralling SiO_4 chains, Fig. 23.9, and is strongly optically active.

The variation of the optical activity with the frequency of the light is called *optical rotatory dispersion*. It can be interpreted in terms of the variation of the polarizability of the molecules in light of different frequencies, and is used to investigate the stereochemical configuration of large molecules. Associated with the differences of refractive index (the *optical birefringence*) is a difference of absorption intensities for left and

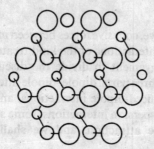

Fig. 23.9. The structure of quartz.

right circularly polarized light. This is called *circular dichroism*, and is used to obtain structural information.

Additive properties. To some extent it is possible to resolve a dipole moment into various contributions in a complex molecule, and to obtain the overall dipole moment by the addition (the vector addition) of these components. This can be illustrated by the case of the chlorobenzene molecules, Fig. 23.10. Chlorobenzene itself has a dipole moment of magnitude 1.57 D as a result of the presence of the chlorine atom, while *p*-dichlorobenzene has a zero moment because of the cancellation of the two equal but opposing moments. *o*-Dichlorobenzene has a dipole moment in a direction and with a magnitude that is approximately the resultant of two monochlorobenzene dipole moments arranged at 60°. *m*-Dichlorobenzene can be accounted for similarly, but with an angle of 120° between the contributions. This technique of vector addition can be applied with fair success to estimate the dipole moment of various molecules.

Polarizabilities are approximately additive properties of molecules, and so too are molar refractivities, eqn (23.1.16). If values of molar refractivities of ions are available, the refractive index of a crystal can be estimated by adding the respective refractivities. The same analysis can be made for groups in organic molecules, and the refractive index of molecules in homologous series, and the effect of adding various groups, can be estimated. Refractivities of ions and various groups are listed in Table 23.3.

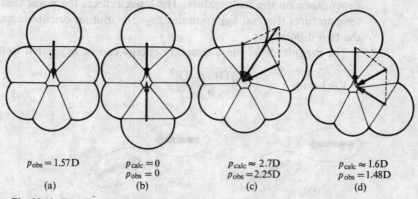

$p_{obs} = 1.57$ D

$p_{calc} = 0$
$p_{obs} = 0$

$p_{calc} \approx 2.7$ D
$p_{obs} = 2.25$ D

$p_{calc} \approx 1.6$ D
$p_{obs} = 1.48$ D

(a)   (b)   (c)   (d)

Fig. 23.10. The dipole moments of chlorobenzene molecules.

23.2 Intermolecular forces

In this section we consider the attractive, cohesive forces between molecules which can be traced to electrostatic interactions of the kind treated in the earlier part of the chapter. It is convenient to divide the discussion of these *van der Waals interactions* into three parts: dipole–dipole interactions between polar molecules, dipole–induced-dipole interactions, and finally induced-dipole–induced-dipole, or dispersion interactions. Ionic solids are held together by simple Coulombic attraction, and we shall discuss these too.

Dipole–dipole interactions. When two polar molecules are close to each other their dipoles interact, and so a force acts between them. In a liquid or gas one of the molecules rotates through all angles relative to the other, and so it might be expected that the overall force of interaction averages to zero because the attractive forces (when the dipoles are head to tail, Fig. 23.11a) are cancelled by the repulsive forces (when the dipoles are head to head, Fig. 23.11b). One important feature has been omitted from this argument. This is the tendency of one dipole to align the other into a favourable arrangement. In the fluid solution the attractive orientations, being energetically more favourable, slightly outweigh the repulsive orientations, and so the attractive forces slightly dominate the repulsive.

The extent to which the orientating effect of the dipoles overcomes the randomizing, thermal motion can be calculated on the basis of the Boltzmann distribution. (This calculation recalls the one used to obtain the net polarization of a fluid sample (p. 771).) When the calculation is carried through, and the approximation made that the energy of interaction is less than kT (a good approximation in most fluids) the average energy of interaction of two molecules with permanent dipole moments p_1 and p_2 is found to be

(23.2.1) $$V(R) = -2(p_1 p_2/4\pi\varepsilon_0)^2 (1/R^6)(1/3kT),$$

where R is their separation. The important point to notice is the dependence of this energy on the *inverse sixth power* of R, and its inverse dependence on the temperature. The latter reflects the point that at high temperatures thermal agitation destroys the mutual orienting effects of the two dipoles.

The magnitude of the interaction energy can be estimated from

$$V/\text{kJ mol}^{-1} \approx -\frac{(p_1/D)^2(p_2/D)^2}{1025(R/\text{nm})^6}$$

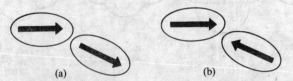

(a)   (b)

Fig. 23.11. The interaction between two polar molecules.

at room temperature ($T \approx 300$ K). The interaction energy for a collection of molecules with $p \approx 1$ D is therefore about -1.4 kJ mol^{-1} when the separation is 0.3 nm.

Dipole–induced-dipole interactions. The presence of a polar molecule in the vicinity of another molecule (which may itself be polar or non-polar) has the effect of polarizing the second molecule. The induced dipole can then interact with the dipole moment of the first molecule, and the two molecules are attracted together. The magnitude of the effect depends on both the size of the permanent dipole moment of the polar molecule and the polarizability of the second. Since the induced dipole moment follows the direction of the inducing dipole (Fig. 23.12) we do not need to take account of the effects of thermal motion (because it is unable to disorientate the induced moment from the direction of the inducing moment). The effect therefore survives even when the polar molecule is rotating rapidly in the vicinity of the polarizable molecule. The average interaction energy for the two molecules is

(23.2.2) $$V(R) = -p_1^2\alpha_2/(4\pi\varepsilon_0)^2 R^6 = -p_1^2\alpha_2'/4\pi\varepsilon_0 R^6,$$

where p_1 is the permanent dipole moment of molecule 1, α_2 is the polarizability of molecule 2, and α_2' its polarizability volume.

The magnitude of the interaction energy can be estimated from the following expression:

$$V/\text{kJ mol} \approx -\frac{(p_1/D)^2(\alpha_2'/10^{-24}\text{ cm}^3)}{1.66 \times 10^4\ (R/\text{nm})^6}.$$

For a molecule with $p \approx 1$ D (such as HCl) in the neighbourhood of a molecule of polarizability volume $\alpha' \approx 10 \times 10^{-24}$ cm^3 (such as benzene, Table 23.1) the interaction energy is about -0.8 kJ mol^{-1} when the separation is 0.3 nm.

Induced-dipole–induced-dipole interactions. Consider two non-polar molecules separated by a distance R from each other. Although they have no permanent moments their electron clouds are fluctuating, and we may consider them as having an instantaneous dipole moment which is constantly changing in magnitude and direction, Fig. 23.13. Suppose one molecule flickers into an electronic arrangement which gives it an instantaneous dipole p_1^*. This dipole polarizes the other molecule and induces in it an instantaneous dipole p_2^*. The two dipoles stick together, and so the molecules have an attractive interaction between them. Although the

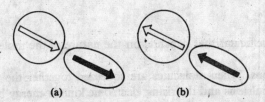

(a)   (b)

Fig. 23.12. The dipole-induced-dipole interaction: a polar molecule (dark arrow) can induce a dipole (light arrow.)

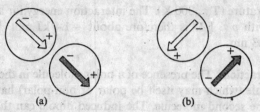

Fig. 23.13. The induced-dipole–induced-dipole interaction: an instantaneous dipole (shaded arrow) can induce a dipole (light arrow).

first molecule will go on to change the direction of its dipole, the second will follow it, and because of this correlation the attractive effect does not average to zero.

The magnitude of the energy depends on the polarizability of the first molecule, because the size of the instantaneous dipole p_1^* depends on the looseness of the nuclear potential's control over its outer electrons. It also depends on the polarizability of the second molecule, because the size of p_2^* depends on the extent to which it can be induced. The actual calculation of the magnitude of these induced-dipole–induced-dipole, or *dispersion forces* is quite involved, but a reasonable approximation to the interaction energy is given by the

(23.2.3)   *London formula* $V(R) \approx -\left(\dfrac{3I_1 I_2}{2(I_1 + I_2)}\right)\left(\dfrac{\alpha_1' \alpha_2'}{R^6}\right).$

In this expression the I_1 and I_2 are the ionization energies of the two molecules.

The magnitude of the dispersion energy for two methane molecules can be estimated from the last equation. Since $\alpha' \approx 2.6 \times 10^{-24}$ cm^3 and $I \approx 7$ eV (670 kJ mol^{-1}) we find

$V/\text{kJ mol}^{-1} \approx -3.4 \times 10^{-3}/(R/\text{nm})^6,$

and so $V \approx -4.7$ kJ mol^{-1} when the separation is about 0.3 nm (as in a liquid). A very rough check on this is provided by the enthalpy of vaporization of liquid methane, which is 8.2 kJ mol^{-1}. (The comparison is very rough because the enthalpy of vaporization is a many-body effect.)

The London formula also applies to interactions between polarizable polar molecules, and so the total interaction energy between molecules is given by the sum of eqns (23.2.1, 2, and 3). All three energies are negative, implying a lowering of energy as the molecules approach, and so they indicate the presence of attractive forces between the molecules. All three vary as R^{-6}, and this is the reason why attractive intermolecular energies are normally written in the form

(23.2.4) $V(R) = -C_6/R^6,$

C_6 being some coefficient that depends on the nature of the molecules.

Repulsive and total interactions. When molecules are squeezed together the nuclear and electronic repulsions and the rising electronic kinetic energy begin to dominate the attractive forces. Thus, although two helium atoms attract each other weakly at large distances, when pressed together they do not form a stable He_2 molecule. The reasons were explained in more detail in Chapter 15 (p. 476). The repulsive interactions rise very steeply with decreasing separation, and so the intermolecular energy is expected to follow a sharply rising curve like that shown in Fig. 23.14.

The behaviour at short distances is very complicated, and depends on the nature and electronic structure of the species. Nevertheless it has been found that quite a good approximation to the overall curve is given by the

(23.2.5)

$$\text{Lennard-Jones potential: } V(R) = C_n/R^n - C_6/R^6$$

where n is a large integer. When $n = 12$ we have the Lennard-Jones (12,6)-potential, and this is quite generally used because of its mathematical convenience. At small values of R, R^{12} is much smaller than R^6, and so the positive (repulsive) C_{12}/R^{12} term dominates the negative (attractive) $-C_6/R^6$ term. The opposite is true at large separations. Quite often the (12,6)-potential is written in the form

(23.2.6)

$$V(R) = 4\varepsilon\{(\sigma/R)^{12} - (\sigma/R)^6\}.$$

The parameter ε is the depth of the minimum of the curve, Fig. 23.14, which occurs at $R_e = 2^{1/6}\sigma$. Typical values are listed in Table 23.4.

Table 23.4. Lennard-Jones (12,6)-potential parameters

	$(\varepsilon/k)/K$	σ/pm
He	10.22	258
Ne	35.7	279
Ar	124	342
Xe	229	406
H_2	33.3	297
O_2	113	343
N_2	91.5	368
Cl_2	357	412
Br_2	520	427
CO_2	190	400
CH_4	137	382
CCl_4	327	588
C_2H_4	205	423
C_6H_6	440	527

Source: J. O. Hirschfelder, C. F. Curtiss, and R. B. Bird, *Molecular theory of gases and liquids*, Wiley.

The coefficients C_n and C_6 can be measured experimentally in a number of ways, for example by studying gas imperfections and the scattering of molecules in molecular beams: these methods and the applications of the results are described below.

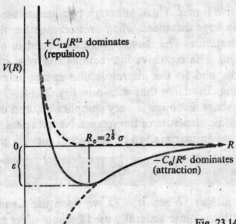

Fig. 23.14. The Lennard-Jones (12,6)-potential.

23.3 The role of intermolecular forces

Intermolecular forces hold substances together. Sometimes their average values are weak, as in gases, and their presence impairs the mobility of the molecules only a little. In crystals the interactions between atoms and ions are strong and make the structures rigid. Somewhere in between lie liquids: in them the kinetic energies of the molecules are comparable to their potential energies of interaction.

Ionic lattices. A knowledge of the separation of ions in crystals can be used to estimate the binding energy of ionic crystals. This was examined from a thermochemical viewpoint in Chapter 4, where it was shown that its magnitude could be determined by constructing a Born–Haber cycle. That dealt with the measurement of the lattice energy (the energy difference between the lattice MA and the gaseous ions M^+ and A^-): this section examines how it can be calculated theoretically.

The viability of this calculation depends on the simplicity of the forces that dominate the crystal structure in an ionic lattice. The attractive contribution to the cohesive energy can be supposed to arise principally from the Coulombic attraction between oppositely charged ions. This attraction, which tends to collapse the crystal down to a point, is opposed by the repulsive forces arising both from the interaction between like ions, and from the forces that arise when species are pushed together into the same region of space. The lattice spacing is determined by the balance of these effects.

The potential energy V of a pair of ions of charges $z_i e$ and $z_j e$ separated by a distance R_{ij} is

(23.3.1) $V_{ij} = z_i z_j e^2 / 4\pi\varepsilon_0 R_{ij}.$

The potential energy of a single cation is a result of its interaction with all the other anions and the cations in the lattice:

$$V^{(+)} = (z_+ e^2 / 4\pi\varepsilon_0) \sum_i z_i / R_{+i},$$

where R_{+i} is the distance of the selected cation of charge z_+e from the ion i of charge z_ie. If the shortest cation–anion distance in the crystal is R_0 we may take it as a unit of length and write $R_{+i} = \rho_{+i}R_0$; then

$$V^{(+)} = (z_+e^2/4\pi\varepsilon_0 R_0)\sum_i (z_i/\rho_{+i}).$$

Likewise a single anion may be selected, and its potential energy written as

$$V^{(-)} = (z_-e^2/4\pi\varepsilon_0 R_0)\sum_i (z_i/\rho_{-i}),$$

where $\rho_{-i}R_0$ is the distance of the ion i from the anion selected. The total energy of the crystal of unit amount of material arising from Coulombic interactions between ions is therefore

(23.3.2) $\qquad V = \tfrac{1}{2}L(V^{(+)} + V^{(-)}) = (Lz_+z_-e^2/4\pi\varepsilon_0 R_0)\mathscr{M}$

where

(23.3.3) $\qquad \mathscr{M} = \tfrac{1}{2}\sum_i \left(\frac{(z_i/z_-)}{\rho_{+i}} + \frac{(z_i/z_+)}{\rho_{-i}} \right).$

The factor $\tfrac{1}{2}$ in V arises because we must take care not to take into account every ion–ion interaction twice (which would happen if we just added V^+ and V^- and multiplied by the number of ions present in the crystal, L). The quantity $\mathscr{M}$ plays a special role in the theory of crystal lattices, and is called the *Madelung constant*. Its importance arises because it depends on the crystal type and not on the dimensions of the individual lattice. It has been evaluated for various lattices, as indicated in Table 23.5. We can therefore select the lattice type, and state its Coulombic potential energy simply in terms of the single parameter R_0 which governs the overall size of the lattice.

Table 23.5. Madelung constants

Lattice	$\mathscr{M}$
Rock salt	1.747 56
CsCl	1.772 67
Zinc blende	1.638 05
Wurtzite	1.641 32
Fluorite	2.519 39
Rutile	2.408
Cuprite	2.221 24
Corundum	4.1719

Source: *Handbook of chemistry and physics*, Chemical Rubber Co.

The calculation of the lattice energy of a given crystal lattice is now quite straightforward. The total energy is the sum of the attractive energy just calculated, and the contribution from the repulsive interactions between the ions which switch on when they are pressed together under the influence of the Coulombic collapse. This repulsive term is strong, but of short range, and its general form may be taken to be an exponential function of range R^*. Then the total energy of the lattice (for an MA species, with $z_+ = 1$, $z_- = -1$) is

$$U(R_0) = -(Le^2/4\pi\varepsilon_0)(\mathcal{M}/R_0) + K\exp(-R_0/R^*).$$

The minimum value of this expression occurs at a value of R_0 for which $dU/dR_0 = 0$, and this condition gives K as

$$K = \frac{Le^2\mathcal{M}R^*\exp(R_0/R^*)}{4\pi\varepsilon_0 R_0^2}.$$

Therefore the molar internal energy is

(23.3.4) $$U_m = -(Le^2\mathcal{M}/4\pi\varepsilon_0 R_0)(1 - R^*/R_0).$$

The only remaining variable in this expression is R^*, the range of the repulsive interaction. This can be determined from the resistance of the crystal to compression, that is, from the compressibility (p. 85), and so the lattice energy can be calculated by using one of the available values of the Madelung constant and the measured values of R_0 and R^*. Experimental values are recorded in Table 23.6. (Note that eqn (23.3.4) gives the internal energy: the enthalpy of the reaction

$$M^+(g) + A^-(g) \rightarrow MA(s)$$

can be obtained from it very simply by setting $pV = nRT$ for each gaseous component and using $H = U + pV$, $\Delta H_m^\ominus = U_m - 2RT$.)

Table 23.6. Lattice enthalpies at 25 °C, $\Delta H_m^\ominus/\text{kJ mol}^{-1}$

LiF	-1031						
NaF	-911.7	NaCl	-772.8	NaBr	-741.0		
KF	-810.0	KCl	-702.5	KBr	-678.2	KI	-637.6
RbF	-799.9	RbCl	—	RbBr	-658.6	RbI	-621.7

Source: J. O'M. Bockris and A. K. N. Reddy, *Modern electrochemistry*, Plenum.

Example (Objective 10). Calculate the internal energy of a sodium chloride crystal. The range of the repulsive interaction R^* is 32.1 pm, and the nearest neighbour separation 282 pm.

- *Method.* The Madelung constant for a rock-salt lattice is 1.747 558; substitute into eqn (23.3.4).

- *Answer.*
$$U_m = -\left\{\frac{(6.0221 \times 10^{23}\,\text{mol}^{-1}) \times (1.6022 \times 10^{-19}\,\text{C})^2 \times (1.747\,558)}{4\pi \times (8.8542 \times 10^{-12}\,\text{J}^{-1}\,\text{C}^2\,\text{m}^{-1}) \times (282 \times 10^{-12}\,\text{m})}\right\} \times \left(1 - \frac{32.1}{282}\right)$$

$$= -763\,\text{kJ mol}^{-1}.$$

- *Comment.* The way to find R^* is explored in Problem 23.27. The experimental value of U_m is 768 kJ mol^{-1} at 25 °C, and so the agreement is excellent.

Molecular interactions in beams. In recent years a notable advance in the experimental study of intermolecular forces has come from the development of *molecular beams*. A molecular beam is what its name implies: a narrow stream of molecules in an otherwise empty vessel. The beam of molecules is directed towards other molecules, and the scattering that occurs on impact is related to the intermolecular forces. The speed of the molecules in the beam may be controlled, and so their behaviour as they collide with the target molecules with different kinetic energies can be investigated.

A typical molecular beam apparatus is shown in Fig. 23.15. The velocity-selection device is often a set of slotted wheels which rotate in the path of the beam and allow only those molecules having the appropriate speed to pass through the slots. There are also other more sophisticated devices for *generating* molecules with the desired velocity: they have the advantage of giving a beam with greater intensity than can be obtained by blocking the passage of, and therefore discarding, molecules with the wrong velocity. Other types of selection are also possible: e.g., electric fields may be used to deflect polar molecules with various orientations; this results in a beam of aligned molecules.

The velocity-selected beam passes into the body of the apparatus. This is maintained under high vacuum so that the beam molecules do not collide and, by randomizing their motion, reach thermal equilibrium again. They pass on to the target gas, which may be inside a scattering chamber at a known pressure, or which may itself be in the form of a molecular beam crossing the path of the first beam. The latter *crossed beam* technique gives a lot of information because the states of both target and projectile molecules may be controlled.

Collisions scatter the beam molecules away from their initial direction, and the angle of scatter is the major piece of experimental information. In order to determine the intensity of scattered molecules at different angles, various kinds of detectors are used. These may consist of a chamber fitted with a sensitive pressure gauge, or an ionization detector, in which the incoming molecule is first ionized and then detected electrically. The state of excitation of the scattered molecules may be determined: there may be interest, for example, in the vibrational or rotational populations of the species after collision. In this case, vibrational or microwave spectroscopy on the scattered molecules can be used to determine the population of the energy levels.

Two of the principal parameters in discussions of molecular beam results

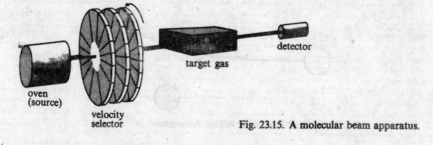

Fig. 23.15. A molecular beam apparatus.

oven
(source)

velocity
selector

target gas

detector

are the *impact parameter b*, and the *differential scattering cross-section dσ*. The impact parameter is the initial vertical separation of the paths of the colliding species, Fig. 23.16. The differential scattering cross-section measures the extent of scattering through different angles. Consider the infinitesimal solid angle $d\Omega$ at some angle Ω (Ω is capital Greek omega), Fig. 23.17. The ratio of the number of molecules scattered into this solid angle to the number in the incident beam is the value of dσ at that angle:

$$d\sigma(\Omega) = \frac{\text{number scattered into } d\Omega \text{ at } \Omega \text{ per unit time}}{\text{incident beam flux}}.$$

(The incident flux is the number of molecules per unit area per unit time: consequently σ has the dimensions of area.)

The differential scattering cross section depends on the impact parameter and the form of the intermolecular potential. This is most easily seen in terms of the collision of two hard spheres, Fig. 23.18. If the impact parameter is zero, the lighter, mobile ball is on a trajectory that leads to a head-on collision, and so all the scattering intensity is in the solid angle $d\Omega$ at the angle 180°. If the impact parameter is so great that the spheres do not come into contact, ($b > R_A + R_B$), then there is no scattering, and the scattering cross-section is zero for all angles. Glancing blows, with b greater than zero but less than $R_A + R_B$, give scattering intensity in cones around the forward direction, Fig. 23.18, and the differential cross-section peaks at directions lying on the cone.

This elementary discussion has to be modified in order to treat collisions of real molecules. In the first place real molecules are not hard spheres, and the scattering pattern depends on the radial dependence of their interaction, and on any anisotropy (angular dependence) that may be present if the molecules are non-spherical. Furthermore, the scattering depends on the relative speed of approach of the two species, for a very fast molecule might pass through a strong interaction zone without significant deflection, while a slow molecule of the same kind and with the same impact parameter might be captured by the intermolecular potential and the deflection might be considerable, Fig. 23.19. This implies that the dependence of the scattering cross-section on the relative speed and approach should give information about the strength and range of the intermolecular potential. This is why it is important to be able to control the relative speeds of the colliding beams.

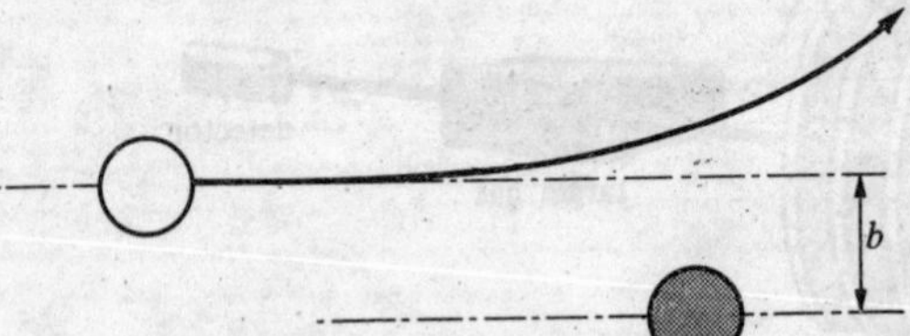

Fig. 23.16. Definition of the impact parameter.

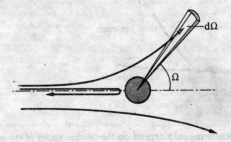

Fig. 23.17. Definition of the differential scattering cross-section.

The second modification recognizes that the outcome of collisions is controlled not by classical mechanics but by quantum mechanics. The major difference is the appearance of wave-like interference phenomena. Just as in a Young's slit type of interference experiment, p. 726, some idea of the quantum scattering pattern can be obtained by drawing all the classical trajectories that take the molecule from source to detector, and then allowing for interference between them.

Two quantum-mechanical effects are of particular importance. In the first place, a molecule with a certain impact parameter might approach the attractive part of the intermolecular potential in such a way that it is deflected towards the repulsive core, Fig. 23.20, which then repels it out through the attractive region so that it continues its flight in the forward direction along with molecules with such large impact parameters that they were undeflected. The two paths, the undeflected, large b, path and the forward-scattered, low b, path, interfere quantum mechanically, and the intensity in the forward direction is modified. This is called *glory scattering* because it is the same phenomenon that accounts for the optical glory effect where a bright halo can sometimes be seen surrounding an illuminated object (the coloured rings around the shadow of an aircraft

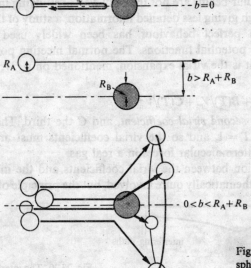

Fig. 23.18. The collision of two hard spheres with different impact parameters.

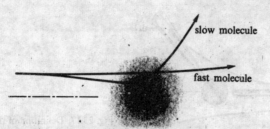

slow molecule

fast molecule

Fig. 23.19. The extent of scattering may depend on the relative speed of the molecules as well as the impact parameter.

cast on clouds by the sun, and often seen in flight, are an example of a glory). The other phenomenon is the observation of strongly enhanced scattering in some non-forward direction: this is called *rainbow scattering* because it arises for the same reason as an optical rainbow. Figure 23.21 illustrates the basic mechanism. As the impact parameter is reduced there comes a point where the scattering angle passes through a maximum: the interference then results in a strongly scattered beam. The *rainbow angle* is the angle for which $d\theta/db = 0$ and the scattering is strong.

A detailed study of the unravelling of all scattering data need not concern us, but the main point should be clear: the intensity distribution of the scattered molecules can be related to the intermolecular potential, and a detailed picture built up of its radial and angular dependences. One of the outcomes of this approach is the determination of the C_6 parameter in the van der Waals interaction (p. 784) and the testing of the Lennard-Jones expression (p. 785), for the potential. It is found that although the R^{-6} part of this potential is often found to be a good representation of the attractive potential, but that the repulsive part is often not at all well described as depending on R^{-12}.

Gas imperfections. Molecular beams are not the only way of studying intermolecular forces. Although giving less detailed information, a study of the deviations of gases from perfect behaviour has been widely used to examine intermolecular potential functions. The normal meeting point of theory and experiment is the virial expansion, mentioned on p. 37. This has the form

(23.3.5) $$pV_m/RT = 1 + B(T)/V_m + C(T)/V_m^2 + \dots$$

where B is the *second virial coefficient*, and C the third. The perfect gas satisfies $pV_m/RT = 1$, and so the virial coefficients must arise from the effects of the intermolecular forces in a real gas.

The connection between the virial coefficients and the intermolecular potential is mathematically quite involved, but the principle of the method,

interfering paths

Fig. 23.20. The interference of paths leading to glory scattering.

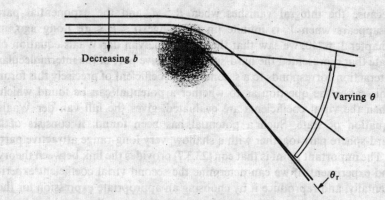

Fig. 23.21. The interference of paths leading to rainbow scattering. θ_r is the maximum scattering angle, the turning point of θ with respect to b.

and the expression for $B(T)$, is straightforward. In the chapter on statistical thermodynamics we saw (p. 704) that the pressure of a system is related to the canonical partition function by

$$(23.3.6) \qquad p = -(\partial A/\partial V)_T, \qquad A - A(0) = -kT \ln Q.$$

In the absence of intermolecular interactions the canonical partition function can be expressed in terms of the number of molecules present and the individual molecular partition functions through $Q = q^N/N!$. We saw on p. 704 that this leads to $p = nRT/V$. In the presence of intermolecular interactions the canonical partition function cannot be expressed so simply because the molecules are not independent and their energy depends on their separation. If the intermolecular potential energy is $V(R)$ an expression for the canonical partition function can be deduced, and eqns (23.3.5) and (23.3.6) used to derive an expression for B in terms of $V(R)$. The result is

$$(23.3.7) \qquad B(T) = 2\pi L \int_0^\infty \{1 - e^{-V(R)/kT}\} R^2 \, dR.$$

We see that B vanishes if V is everywhere zero, which it is in a perfect gas. B depends, through $V(R)$, on the separation of a *pair* of molecules, and so the second virial coefficient can be interpreted in terms of the effects due to collisions of pairs of molecules in the real gas. The third virial coefficient depends on the relative positions of three molecules, and so its role is to take into account the effects of transitory clusters of three molecules.

The integral in the last equation is difficult to evaluate except in a few cases. For example, if the true potential is approximated by a hard sphere potential in which $V(R)$ rises sharply from zero to infinity when two molecules comes into contact at $R = d$, we find

$$B(T) = 2\pi L \int_0^d R^2 \, dR = \tfrac{2}{3}\pi L d^3$$

because the integral vanishes when $R > d$, and the exponential part disappears when V is infinite (in the range $0 \leqslant R \leqslant d$). Long ago, in Chapter 1, p. 40, we saw that the part of the van der Waals equation of state that represented the hard-sphere repulsive part of the intermolecular interaction corresponds to a second virial coefficient of precisely this form. This raises the question as to whether a potential can be found which, when the virial coefficients are evaluated, gives the full van der Waals equation of state. Such a potential has been found: it consists of a hard-sphere part together with a shallow, very long-range attractive part.

The important point is that eqn (23.3.7) provides the link between theory and experiment. If we can determine the second virial coefficient experimentally, and reproduce it by choosing an appropriate expression for the intermolecular potential and evaluating the integral, then we have found the form of the intermolecular potential. Often a Lennard-Jones potential is chosen, and the parameters varied until agreement is reached. Some values found in this way were quoted in Table 23.4.

Example (Objective 14). Take the Lennard-Jones potential for argon in the form of eqn (23.2.6) with $\varepsilon/k = 120$ K and $\sigma = 340$ pm, and evaluate the second virial coefficient at 298 K.

- *Method.* Evaluate the integral in eqn (23.3.7). Do this numerically (the technical term is by *numerical quadrature*), using Simpson's rule. Simpson's rule is to evaluate the integrand at an even number of points (call these values y_m), with an interval h, then form the sum $(h/3)[y_0 + 4(y_1 + y_3 + \ldots + y_{m-1}) + 2(y_2 + y_4 + \ldots + y_{m-2}) + y_m]$. Since the integrand rises from zero up to a maximum, then crosses zero at 340 pm, and has a long tail out to several 1000 pm, it is sensible to do the integration in two ranges. One, from $R = 0$ to 340 pm, the second from 340 pm out to about 5000 pm. A small programmable calculator was used to evaluate the integral using 1 pm steps over the whole range, but you would be justified in using bigger steps in the second range.

- *Answer.* We write two fragments of the complete numerical table of values of the integrand:

R/pm	0	10	20	30	...	340	440	540
integrand/pm^2	0	100	400	900	...	0	-7.3×10^4	-2.88×10^4

Then from Simpson's rule, in the first range

$B/2\pi L = 1.129\,80 \times 10^7$ pm^3

and in the second range,

$B/2\pi L = -0.516\,29 \times 10^7$ pm^3;

and so the complete integral has the value $B/2\pi L = 6.135 \times 10^6$ pm^3; so that $B = 2.321 \times 10^{-5}$ m^3 mol^{-1}, or 23.2 cm^3 mol^{-1}.

- *Comment.* Calculations like this are ideal for a computer, or, as was used in this case, for a small programmable calculator. The two domains reflect the places where repulsions ($B > 0$) and then attractions ($B < 0$) are dominating.

Gas imperfections appear in other ways, and a particularly fruitful way of studying them is by measuring the transport properties of gases, especially their viscosities. The mobility of molecules in a gas, which determines the viscosity, depends on the way they stick together, and so deviations from perfect behaviour give information about the intermolecular potential. Results from this kind of study are also included in Table 23.4.

The structure, such as there is, of liquids. The starting point for a discussion of solids is the well-ordered form of perfect crystals. The starting point for a study of gases is the totally chaotic distribution of the particles of a perfect gas. The starting point for a study of liquids is somewhere between these two extremes, where there is some structure and some chaos. The theoretical description and the experimental examination of liquids remain in their infancy, although enough progress has been made for us to be able to outline some of their features.

The molecules of a liquid are held together by intermolecular attractions, but their kinetic energies are comparable to the depths of the potential wells. As a result, the whole structure, even though it is a definite phase (being separated by phase transitions from solids and gases), is very mobile. Just as the virial expansion gave the link between theory and experiment in gases, some concept is needed to do the same for liquids. This is provided by the *radial distribution functions*. The *pair distribution function* $g(R)$ is the most important and will be at the centre of our attention.

The pair distribution function gives the probability that another molecule will be found at some distance R from some selected molecule. More precisely, $g(R)dR$ is the probability that a molecule will be found in the range dR at a distance R from another. In a crystal this function is a periodic array of sharp spikes, representing the certainty (in the absence of defects and thermal motion) that an atom will be found at each of the lattice points. This regularity continues indefinitely, and so a crystal is said to have *long-range order*. When the crystal melts, the long-range order is lost; and wherever we look there is the same probability of finding a second molecule. Close to the first molecule, though, there may be a remnant of order because the nearest neighbours of the original molecule might still adopt approximately their original positions, and if they are displaced by newcomers, the new molecules will occupy their vacated positions. It may still be possible to detect a sphere of nearest neighbours at a distance R_1, and perhaps beyond them a sphere of next-nearest neighbours at a distance of about R_2 from the central molecule. The existence of this *short-range order* means that the pair-distribution function can be expected to have an oscillatory behaviour close to the central molecule with a peak at R_1, a smaller peak at R_2, and perhaps some more structure beyond that.

The pair distibution function (and, in more advanced descriptions, other distribution functions for triples, quadruples, etc.) is as close to the description of the structure of a liquid as we approach. The questions then are: (a) can $g(R)$ be measured experimentally?, (b) can it be calculated, and therefore used to test theories of liquid structure?, and (c) can $g(R)$

be used to discuss the properties of liquids?

The pair distribution function can be measured experimentally by X-ray diffraction. We saw in Chapter 22 that X-rays are diffracted by the distribution of electron density, and that the diffraction pattern leads to the determination of the structure of crystals. Liquids do have structure, even though it is only local, and so we can expect blurred diffraction patterns which should reveal the form of $g(R)$. This turns out to be the case. If there were only an amorphous, formless distribution of molecules, the picture would be a total blur, but the presence of definite rings of interference shows that there is some structure in the liquid, and that the pair distribution function has oscillations at short range. The diffraction pattern can be analysed in much the same way as that for a solid, and the intensity distribution used to find the pair distribution function itself: the one obtained for water at various temperatures is shown in Fig. 23.22, and the shells of local structure are unmistakable. Closer analysis shows that the central water molecule is surrounded, at least in the first shell, at the corners of a tetrahedron. This is just as in ice (p. 747), and the intermolecular forces, in this case largely the hydrogen bonds, are strong enough to govern the local structure right up to the boiling point.

The oscillatory behaviour arises in two ways. Even if the molecules were hard spheres without attractive interactions, an oscillatory $g(R)$ with a strong initial peak would be obtained. This shows that one of the factors influencing, and sometimes dominating, the structure of a liquid is simply the geometrical problem of stacking a lot of hard spheres together. The attractive well of the potential also plays a role, and its effect is to gather and trap molecules into its vicinity. One of the reasons for the difficulty in describing liquids theoretically is the importance of both the attractive and repulsive, hard-core, parts of the potential.

Once the radial distribution function has been determined it can be used to assess the thermodynamic properties of liquids. For example, the

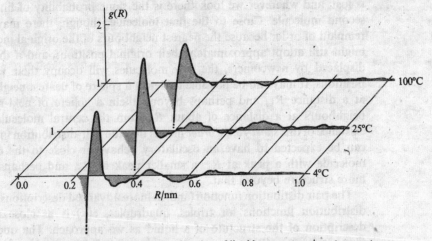

Fig. 23.22. The radial distribution function of liquid water at various temperatures (A. H. Norten, M. D. Danford, and H. A. Levy, *Discuss. Faraday Soc.*, **43**, 97 (1967)).

internal energy arising from the intermolecular potential is given by the integral

(23.3.8)
$$U_{\text{intermol.}} = \tfrac{1}{2}\mathcal{N}^2 V \int_0^\infty g(R)V(R)\,\mathrm{d}R$$

where $\mathcal{N}$ is the number of molecules per unit volume. This expression is the average of the intermolecular potential $V(R)$ weighted by the probability $g(R)\,\mathrm{d}R$ that a pair of molecules will be found with a separation R. Likewise, the pressure is given by the integral

(23.3.9)
$$p = \mathcal{N}kT - (\mathcal{N}^2/6) \int_0^\infty g(R)\,[R\,(\mathrm{d}V/\mathrm{d}R)]\,\mathrm{d}R$$

where the quantity in square brackets, $R\,(\mathrm{d}V/\mathrm{d}R)$, is called the *virial*. Since the first term on the right is the pressure exerted by a perfect gas under the same conditions, and since p of a liquid is very much less than that of a gas of the same density near the boiling point, the integral on the right must almost cancel the first term. This very fine balance between the two terms is yet another reason why calculations on liquids are so difficult.

Another more fundamental difficulty is the assumption that the interaction between three molecules can be expressed as the sum of the interaction between the three pairs. In fact it has been found that intermolecular energies do not sum in this simple way, and that the *three-body interactions*, the difference between the actual intermolecular energy of three molecules and the pairwise sum of energies, make significant contributions to the energy.

Another difficulty is the possibility that molecules have strongly anisotropic interactions. An extreme case of this is provided by a long, thin, molecule such as *p*-azoxyanisole,

When the solid melts some aspects of the long-range order are maintained above the melting point, and so, since the materials are then liquid (having incomplete long-range order) but are also in some respects crystals (having some elements of long-range order) they are called *liquid crystals*. One type of retained long-range order gives a *smectic* phase (from the Greek for soapy). In this phase the molecules align themselves in layers, Fig. 23.23a. Other materials, and some smectic liquid crystals at higher temperatures, do not possess a layered structure, but retain a parallel alignment: this is the *nematic phase* (from the Greek for thread), Fig. 23.23b. Liquid crystals have strongly anisotropic optical, electric, and magnetic properties, and one of their applications is in data displays in calculators and watches.

A discussion restricted to $g(R)$ ignores the dynamical structure of the

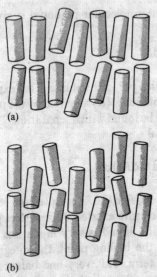

Fig. 23.23. The arrangement of molecules in (a) the smectic phase and (b) the nematic phase of a liquid crystal.

(a)

(b)

phase, whereas the mobility of liquids is one of their major characteristics. There are various experimental methods of determining the time-dependent properties of the local structure of liquids and one of the most effective is *inelastic neutron scattering*. This is a kind of neutron diffraction experiment, but interest centres on how much energy neutrons give up, or collect, from a liquid sample as they pass through. The analysis of the energy exchange process can be interpreted in terms of the detailed motion of molecules in the fluid.

Much more mundane are viscosity measurements, Table 23.7. Since, in order to move from one point to another, a molecule has to shed its neighbours, it has to acquire a minimum energy before it can move. The probability that it can acquire this energy is given by the Boltzmann distribution and is proportional to $\exp(-E_a/RT)$. The temperature dependence of the *mobility* of the fluid is therefore expected to follow this exponential Boltzmann factor. Since the viscosity η is inversely proportional to the mobility, this suggests that

Table 23.7. Viscosities of liquids at 25 °C, $\eta/(\text{kg m}^{-1} \text{ s}^{-1} \times 10^{-3})$

Water†	0.8909	Benzene	0.6010
n-pentane	0.224	Mercury	1.53
cyclohexane	0.895	Carbon tetrachloride	0.909

† The viscosity of water over its entire liquid range is represented with less than 1 per cent error by the expression

$$\lg(\eta_{20}/\eta_t) = \frac{1.370\,23[(t/°C)-20]+8.36\times10^{-4}[(t/°C)-20]^2}{109+(t/°C)}$$

Convert kg m⁻¹s⁻¹ to centipoise (cP) by multiplying by 10^3 (so that $\eta \approx 1$ cP for water).

Source: *American Institute of Physics handbook*, McGraw-Hill.

(23.3.10) $\eta \propto \exp(E_a/RT)$,

and so the viscosity is expected to decrease with increasing temperature: Fig. 23.24 shows this to be true. Once again the intermolecular forces govern the magnitude of E_a, but the problem of calculating it is very great because in order for one molecule to move several others may have to adjust their positions.

At this point we have arrived at time-dependent phenomena; but *change* is the topic of Part 3, and so the immensely difficult, and still largely unsolved, problem of liquid structures and properties will be put aside, and its simpler aspects taken up again there.

23.4 Magnetic properties

The magnetic properties of molecules are analogous to their electrical properties in two senses. First, some molecules possess permanent magnetic dipole moments. Second, the effect of a magnetic field is to induce a further contribution to the overall magnetic moment. The analogue of the polarization is the *magnetization M*, and the extent of magnetization brought about by the application of a magnetic field of strength H is called the *magnetic susceptibility*, κ (kappa):

(23.4.1) $M = \kappa H$.

The magnetic susceptibility is often denoted χ: this is the *mass magnetic susceptibility*. It is related to κ, the *volume magnetic susceptibility*, by $\chi = \kappa/(\rho/\text{kg m}^{-3}) = 1000\,\kappa/(\rho/\text{g cm}^{-3})$ where ρ is the density. This point is developed below. The *magnetic flux density* is denoted B, and is related to the applied field strength, and the magnetization it induces, by

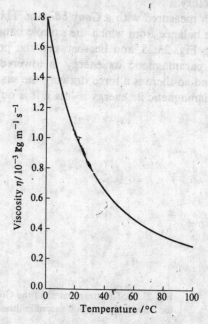

Fig. 23.24. Temperature dependence of the viscosity of water.

(23.4.2) $B = \mu_0(H + M) = \mu_0(1 + \kappa)H.$

μ_0 is a fundamental constant, the *vacuum permeability*; it has the value $4\pi \times 10^{-7}$ N A^{-2}, or $4\pi \times 10^{-7}$ J C^{-2} s^2 m^{-1}. B can be envisaged as a measure of the number of magnetic lines of force permeating the medium, and the energy of the dipole is determined by this density. The density of lines of force is increased if M adds to H (κ then being positive) but is decreased if M opposes H (κ then being negative). Materials in which κ (or χ) is positive are called *paramagnetic*, and those for which κ is negative are called *diamagnetic*.

The polarization is the electric dipole moment per unit volume. By analogy, M is the *magnetic* dipole moment per unit volume. The molecules might possess a permanent magnetic moment m. In that case the magnetization has a contribution proportional to $m^2/3kT$, as in the electrical case (eqn (23.1.8)). An applied field can also induce a magnetic moment to an extent that is governed by the molecular *magnetizability* ξ (a similar effect was encountered in the discussion of n.m.r.: see the discussion of chemical shifts on p. 646). The analogue of eqn (23.1.11) is therefore:

(23.4.3) $\kappa = \mathcal{N}\mu_0(\xi + m^2/3kT),$

and so the magnetic susceptibility can be related to the magnetizability, the permanent moment, the density, and the temperature. The point of introducing the mass magnetic susceptibility should now be clear: the presence of the density in $\chi = \kappa/\rho$ cancels the density ($\rho \propto \mathcal{N}$) in the expression for κ, and so simplifies the expressions by eliminating the density dependence of the susceptibility.

The susceptibility is often measured with a *Gouy balance*. This instrument consists of a sensitive balance from which the sample hangs in the form of a narrow cylinder, Fig. 23.25, and lies between the poles of a magnet. If the sample is paramagnetic its energy is lowered if it is inside the magnetic field, and so there is a force drawing the sample into the field. If the sample is diamagnetic its energy is less if it is outside the

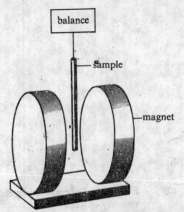

Fig. 23.25. The arrangement of the Gouy balance for measuring magnetic susceptibilities.

field, and so there is a force pushing it out of the field. The force is proportional to the susceptibility, and so determining the balance point allows κ to be determined. The apparatus is normally calibrated against a sample of known susceptibility. A typical paramagnetic susceptibility is of the order of 10^{-3} (notice that it is dimensionless in the unit system we are using), and a typical diamagnetic susceptibility is about $(-)10^{-5}$, Table 23.8.

Table 23.8. Volume magnetic susceptibilities, κ, and mass magnetic susceptibilities, $\chi^{\dagger}$, at 25 °C

	$\kappa/10^{-6}$	$\chi/10^{-8}$		$\kappa/10^{-6}$	$\chi/10^{-8}$
Water	−90	−0.90	$CuSO_4 . 5H_2O$	+ 176	+ 7.7
Benzene	− 7.2	−0.82	$MnSO_4 . 4H_2O$	+2639	+81.2
Cyclohexane	− 7.9	−1.02	$NiSO_4 . 7H_2O$	+ 416	+20.1
Carbon tetrachloride	− 8.9	−0.54	$FeSO_4(NH_4)_2SO_4 . 6H_2O$	+ 755	+40.6
NaCl	−13.9	−0.64	Al	+ 22	+ 0.82
Cu	− 9.6	−0.107	Pt	+ 262	+ 1.22
S	−12.9	−0.62	Na	+ 7.3	+ 0.75
Hg	−28.5	−0.21	K	+ 5.6	+ 0.65

† Mass and volume susceptibility are related by $\chi = \kappa/(\rho/\mathrm{kg\,m}^{-3})$. The values of χ quoted in the table are mass susceptibilities per kilogram. The values in CGS units per gram are obtained by forming $1000\chi/4\pi$, and in CGS units per gram mole by forming $1000\chi M_r/4\pi$. The *Example* on p. 802 elucidates the units.
Source: G. W. C. Kaye and T. H. Laby, *Tables of physical and chemical constants*, Longmans.

The permanent magnetic moment. The permanent magnetic moment can be extracted from susceptibility measurements by plotting κ against $1/T$. The *Curie Law* behaviour

(23.4.4)
$$\kappa = A + C/T$$

is often obeyed, and comparison with eqn (23.4.3) enables the slope C to be identified with $\mathcal{N}\mu_0 m^2/3k$.

The source of the permanent magnetic moment is the unpaired spin of the electrons on the molecules. We have seen already (p. 463) that an electron spin interacts with an applied field as though it had a magnetic moment proportional to its angular momentum:

(23.4.5)
$$m = 2\gamma_e s$$

where $\gamma_e = -e/2m_e$. Since in quantum mechanics the value of s^2 is $s(s+1)\hbar^2$,

(23.4.6)
$$m^2 = 2^2\gamma_e^2 s^2 \to 4\gamma_e^2\hbar^2 s(s+1).$$

If we introduce the Bohr magneton (p. 462) $\mu_B = e\hbar/2m_e$ ($\mu_B = 9.274 \times 10^{-24}$ J T^{-1}) we find

(23.4.7)
$$m^2 = 4s(s+1)\mu_B^2.$$

If there are several electron spins in the molecule they may combine to give a total electron spin angular momentum with a magnitude given by the quantum number S. In this case

(23.4.8) $m^2 = 4S(S+1)\mu_B^2.$

Consequently the contribution to the volume magnetic susceptibility is

(23.4.9) $\kappa = 4\mathcal{N}\mu_0\mu_B^2 S(S+1)/3kT.$

A practical form of this expression in terms of the mass magnetic susceptibility, in which $\mathcal{N}$ is cancelled by ρ (because $\rho = M_m\mathcal{N}/L$) is

(23.4.10) $\chi = 6.2856 \times 10^{-3}\{S(S+1)/M_r(T/K)\}.$

κ and χ are *positive* quantities, and so *the permanent spin moment contributes to the paramagnetic susceptibility* of the system. Note that this contribution to the susceptibility vanishes at high temperatures because the thermal motion randomizes the orientations of the individual spin magnetic moments, and the magnetization from this source is eliminated. In practice a paramagnetism also arises from the orbital angular momentum: we have discussed the *spin-only* contribution.

Example (Objective 18). Calculate the room temperature (25 °C) paramagnetic susceptibility of a metal complex with three unpaired electrons.

● *Method.* Use eqn (23.4.9); three unpaired electrons corresponds to $S = 3/2$. Take the density of the sample as 3.24 g cm^{-3} and $M_m = 200$ g mol^{-1}. Find both the volume susceptibility (κ) and the mass susceptibility (χ) *via* $\chi = \kappa/(\rho/\text{kg m}^{-3})$.

● *Answer.* The number density is $\mathcal{N} = L\rho/M_m$, hence

$\kappa = 4L\rho\mu_0\mu_B^2 S(S+1)/3kTM_m$

$\chi = \kappa/(\rho/\text{kg m}^{-3}) = \{4L\mu_0\mu_B^2 S(S+1)/3kTM_m\}$ kg m^{-3}.

Note that at 25 °C,

$4L\mu_0\mu_B^2/3kT = 4 \times (6.022\,05 \times 10^{23}\ \text{mol}^{-1}) \times (4\pi \times 10^{-7}\ \text{J s}^2\ \text{C}^{-2}\ \text{m}^{-1})$

$\times \dfrac{(9.274\,08 \times 10^{-24}\ \text{J T}^{-1})^2}{3 \times (1.380\,66 \times 10^{-23}\ \text{J K}^{-1}) \times (298.15\ \text{K})}$

$= 2.108\,21 \times 10^{-8}\ \text{m}^3\ \text{mol}^{-1}$

where we have used $1\text{T} = 1\text{J C}^{-1}\text{ s m}^{-2}$. It follows that

$\kappa = (2.108\,21 \times 10^{-8}\ \text{m}^3\ \text{mol}^{-1}) \times (\rho/M_m)S(S+1)$

$= (2.108\,21 \times 10^{-8}\ \text{m}^3\ \text{mol}^{-1}) \times (3.24 \times 10^3\ \text{kg m}^{-3}) \times$

$\times (15/4)/(200 \times 10^{-3}\ \text{kg mol}^{-1})$

$= 1280.7 \times 10^{-6}$

$\chi = (2.108\,21 \times 10^{-8}\ \text{m}^3\ \text{mol}^{-1}) \times (\text{kg m}^{-3}/M_m)S(S+1)$

$= (2.108\,21 \times 10^{-8}\ \text{kg mol}^{-1})S(S+1)/M_m$

$= (2.108\,21 \times 10^{-5}\ \text{g mol}^{-1})S(S+1)/M_m$

$= 2.108\,21 \times 10^{-5}\ S(S+1)/M_r$

since $M_r = M_m/\text{g mol}^{-1}$. Therefore, in the present case,

$$\chi = 2.108\,21 \times 10^{-5} \times (15/4)/200 = 29.53 \times 10^{-8}.$$

- *Comment.* Note how the density is not needed for the calculation of the mass susceptibility. Both χ and κ are dimensionless and (in this case) positive, indicating paramagnetism. Note their magnitudes ($\kappa \approx 10^{-3}$, $\chi \approx 10^{-7}$); diamagnetic susceptibilities are negative, and typically $-\kappa \approx 10^{-5}$.

At low temperatures a paramagnetic solid may sometimes show a phase transition to a state in which large domains of spins align with parallel orientations. This cooperative alignment gives rise to a very strong magnetization of the sample, and is the phenomenon of *ferromagnetism*, Fig. 23.26a. In other cases the cooperative effect might favour an alternation of spin orientations, so that they are locked into a low magnetization arrangement, Fig. 23.26b. This is known as *antiferromagnetism*, already mentioned in connection with neutron scattering, p. 753. The ferromagnetic transition occurs at the *Curie temperature*, and that for the antiferromagnetic transition occurs at the *Néel temperature*.

Induced magnetic moments. The application of a magnetic field to a molecule distorts its electron distribution, but whereas the electric field stretches the molecule and moves the positive and negative centres apart, the magnetic field twists the molecule. A twisting distortion of the molecule is equivalent to the stimulation of a circulation of current, and this current gives rise to a magnetic moment which usually opposes the applied field; in these cases the susceptibility is diamagnetic. In a few cases the induced moment is in the same direction as the applied field, and then the susceptibility is paramagnetic.

The great majority of molecules show a diamagnetic susceptibility. This is because the flow of paramagnetic current depends on the availability of low-lying excited electronic states of the molecule, and these are not always present. The diamagnetic flows of current occur within the ground state of the molecule, and so do not require the availability of excited states; they are therefore much more common. The induced paramagnetism when it occurs can be distinguished from the spin paramagnetism because it contributes via the ξ term in eqn (23.4.4), and is therefore temperature independent. The *temperature-independent paramagnetism* (t.i.p) is a consequence of the stimulation of orbital motion through the molecule, and should not be confused with the permanent (but temperature dependent) magnetic moment, which arises from the spins.

(a)

(b)

Fig. 23.26. Arrangement of spins in (a) a ferromagnetic material and (b) an antiferromagnetic material.

We can summarize the position by saying that all molecules have a diamagnetic component in their total susceptibility, but this is dominated by spin paramagnetism if the molecule has unpaired electrons. In a few cases the t.i.p is strong enough to make the molecules paramagnetic even though their electrons are paired.

Further reading

The determination of molecular structures (2nd ed.). P. J. Wheatley; Clarendon Press, Oxford, 1968.

Physical methods in advanced inorganic chemistry. H. A. O. Hill and P. Day (eds.); Interscience, London, 1968.

Electric dipole moments. J. W. Smith; Butterworths, London, 1955.

Dielectric behaviour and structure. C. P. Smyth; McGraw-Hill, New York, 1955.

Determination of dipole moments. C. P. Smyth; in *Techniques of chemistry* (A. Weissberger and B. W. Rossiter, eds.), Vol. IV, 397, Wiley-Interscience, New York, 1972.

Determination of dielectric constant and loss. W. E. Vaughan, C. P. Smyth, and J. C. Powles; in *Techniques of chemistry* (A. Weissberger and B. W. Rossiter, eds.), Vol. IV, 351, Wiley-Interscience, New York, 1972.

Tables of experimental dipole moments. A. L. McClellan; Freeman, San Francisco, 1963.

Molecular forces. B. Chu; Wiley-Interscience, New York, 1967.

Intermolecular forces: their origin and determination. G. C. Maitland, M. Rigby, E. B. Smith, and W. A. Wakeham; Clarendon Press, Oxford, 1981.

Experimental magnetochemistry. M. M. Schieber; Wiley, New York, 1967.

Physical techniques in chemistry. R. S. Drago; Saunders, Philadelphia, 1977.

Instrumentation and techniques for measuring magnetic susceptibility. L. N. Mulay; in *Techniques of chemistry* (A. Weissberger and B. W. Rossiter, eds.), Vol. IV, 431, Wiley-Interscience, New York, 1972.

Introduction to magnetochemistry. A. Earnshaw; Academic Press, New York, 1968.

Magnetism and the chemical bond. J. B. Goodenough; Interscience, New York, 1963.

Molecular beams in chemistry. M. A. D. Fluendy and K. P. Lawley, Chapman and Hall, London, 1974.

Molecular beams. J. Ross (ed.); *Adv. chem. Phys.* **10**, (1966).

The liquid state. J. A. Pryde; Hutchinson, London, 1966.

The structure of liquids. J. S. Rowlinson; in *Essays in chemistry* (J. N. Bradley, R. D. Gillard, and R. F. Hudson, eds.), 1, 1, Academic Press, London, 1970.

The molecular theory of gases and liquids. J. O. Hirschfelder, C. F. Curtiss, and R. B. Bird; Wiley, New York, 1954.

The virial coefficients of gases and mixtures: a critical compilation. J. H. Dymond and E. B. Smith; Clarendon Press, Oxford, 1980.

Optical rotation; experimental techniques and physical optics. W. Heller and H. G. Curmè; in *Techniques of chemistry* (A. Weissberger and B. W. Rossiter, eds.), Vol. IIIC, Wiley-Interscience, New York, 1972.

Problems

23.1. The dipole moment of toluene is 0.4 D; estimate the dipole moments of the three xylenes. Which answer can you be sure about?

23.2. The dipole moment of water is 1.85 D. Regard this as being the resultant of two bond dipoles at an angle of 104.5°, and predict the dipole moment of hydrogen peroxide as a function of the azimuthal angle between two OH groups (assume the OOH angle is 90°). The experimental dipole moment is 2.13 D: to what angle does this correspond?

23.3. What are the magnitudes of the fields that atoms and molecules are subjected to? Allow a water molecule ($p = 1.85$ D) to approach an ion. What is the favourable orientation of the molecule when the ion is an anion? Calculate the electric field experienced by the ion when the centre of the water dipole is at (a) 1.0 nm, (b) 0.3 nm, (c) 30 nm from its centre. Express your answer in $V\ m^{-1}$.

23.4. A water molecule is aligned by an external electric field of strength $1.0\ kV\ m^{-1}$ and an argon atom ($\alpha' = 1.66 \times 10^{-24}\ cm^3$) is brought up slowly from one side. At what separation is it favourable for the water molecule to flip over and point towards the approaching argon atom?

23.5. The relative permittivity of gaseous hydrogen halides is given by the expression $\varepsilon_r = 1 + \Delta/v$ in the range $0 \lesssim t/°C \lesssim 300$, the values of Δ being given below and v being a relative specific volume equal to unity at 273.15 K and 1 atm, so that $v = T/273.15$ K. What are their dipole moments and static polarizability volumes?

$t/°C$	0	100	200	300
$10^3\Delta(HCl)$	4.3	3.5	3.0	2.6
$10^3\Delta(HBr)$	3.1	2.6	2.3	2.1
$10^3\Delta(HI)$	2.3	2.2	2.1	2.1

23.6. The polarizability volume of a water molecule at optical frequencies is $1.5 \times 10^{-24}\ cm^3$. Estimate the refractive index of water. The experimental value is 1.33: what may be the explanation of the discrepancy?

23.7. The relative permittivity of water vapour at 1 atm pressure is given by $\varepsilon_r = 1.00785 - 1.6 \times 10^{-4}(t/°C - 140)$ in the range 140–150 °C. What is the dipole moment and polarizability volume of the molecule? Estimate the value of ε_r for water liquid on the basis of this information. The experimental value at 20 °C is 80; why is there a discrepancy?

23.8. The relative permittivity of chloroform was measured over a range of temperatures with the following results:

$t/°C$	−80	−70	−60	−40	−20	0	20
ε_r	3.1	3.1	7.0	6.5	6.0	5.5	5.0
$\rho/g\ cm^{-3}$	1.65	1.64	1.64	1.61	1.57	1.53	1.50

The melting point is −64 °C. Account for these results, and find the polarizability and dipole moment of the molecule.

23.9. The relative permittivity of methanol (m.p. −95 °C) is given below. What molecular information can be deduced from these values?

$t/°C$	−185	−170	−150	−140	−110	−80	−50	−20	0	20
ε_r	3.2	3.6	4.0	5.1	67	57	49	42	38	34

23.10. Show that in a gas (where the refractive index is close to unity) the refractive index depends on the pressure as $1 + (2\pi\alpha'/kT)p$. Inversely, show how to deduce the polarizability volume from a measurement of refractive index of a gas.

23.11. The refractive index of benzene is constant (at 1.51) from 0.4 GHz up to about 0.55 GHz (note that these are microwave frequencies), but then shows a series of oscillations between 1.47 and 1.54. Throughout the same frequency range toluene shows a greater refractive index (of about 1.55), but it also shows oscillations in the same place as benzene, and additional oscillations between 1.52 and 1.56 in the vicinity of 0.4 GHz. Account for these observations.

23.12. The dipole moment of chlorobenzene is 1.57 D and its polarizability is 1.23×10^{-23} cm^3. Estimate its relative permittivity at 25 °C, taking its density as 1.1732 g cm^{-3}.

23.13. Estimate the polarizabilities and polarizability volumes of the rare gases on the basis of the ionization energies and radii (Tables 4.9 and 22.2). The experimental values are given in Table 23.1.

23.14. Use eqn (23.1.15) to calculate and plot the refractive index of water between 0 °C and 100 °C using the following density data:

$t/°C$	0	20	40	60	80	100
$\rho/\text{g cm}^{-3}$	0.999 87	0.998 23	0.992 24	0.983 24	0.971 83	0.958 41

Take $\alpha' = 1.50 \times 10^{-24}$ cm^3.

23.15. What is the refractive index of steam at 100 °C, 1 atm pressure? Estimate the refractive index of crystals of CaCl$_2$, NaCl, and solid Ar from the data in Table 23.3. The densities are 2.15, 2.163, and 1.42 g cm^{-3} respectively. α' (H$_2$O) $\approx 1.5 \times 10^{-24}$ cm^3.

23.16. In his classic book on *Polar molecules*, Debye reports some early measurements on the molar polarizability of ammonia. From the selection below, find the dipole moment and polarizability volume of the molecule.

T/K	292.2	309.0	333.0	387.0	413.0	446.0
$P_m/\text{cm}^3\text{mol}^{-1}$	57.57	55.01	51.22	44.99	42.51	39.59

23.17. The refractive index of ammonia at 1 atm pressure and 273 K is 1.000 379 (this is for yellow sodium light). What is the molar polarizability of the gas at (a) this temperature, (b) 292.2 K? Combine this information with the static molar polarizability at this temperature and deduce a value for the molecular dipole moment from these two measurements alone.

23.18. Show that the mean interaction energy of 1 mol of atoms of diameter d interacting with a potential energy of the form $-C_6/R^6$ is given by $U = -2\pi L^2 C_6/3Vd^3$, where V is the volume to which they are confined, and all effects of clustering are ignored. Hence find a connection between van der Waals parameter a (p. 40) and C_6 from $n^2a/V^2 = (\partial U/\partial V)_T$. Investigate the consistency of this calculation in the case of argon showing that $C_6 \approx \frac{3}{4}I\alpha'^2$.

23.19. The expression for the second virial coefficient B in terms of the intermolecular potential $V(R)$ is given by eqn (23.3.7). Suppose that the atoms have a distance of closest approach d, and outside that range they are attracted together by a $-C_6/R^6$ potential. Suppose further, that when the atoms are not in contact, V/kT is so small that the exponential e^{-x} can be approximated by $1-x$. Find an expression for B in terms of C_6 and d.

23.20. Relate B to the polarizability volume of the gas atoms by connecting C_6 with α. Then estimate the value of B at 298 K for argon using data in Tables 4.9, 22.2, and 23.1.

23.21. Throughout the chapter we ignored totally the effect of gravitational forces between molecules. Is that justifiable? Estimate their relative importance for two argon atoms almost in contact. Take the gravitational constant as $G = 6.67 \times 10^{-11}$ N m^2 kg^{-2}.

23.22. The *cohesive energy density* is defined as $-U/V$, where U is the mean potential energy of attraction within the sample and V is its volume. Show that this quantity may be written as $-\frac{1}{2}\mathcal{N}^2\int V(R)\,d\tau$ where $\mathcal{N}$ is the number density of molecules and $V(R)$ their attractive potential energy, the integral ranging from d to infinity. Go on to show that the cohesive energy density of a uniform distribution of molecules that are held together by a van der Waals attraction of the form $-C_6/R^6$ is given by $(2\pi/3)(L^2/d^3M^2)\rho^2C_6$ where ρ is the density of the solid and M_m the molar mass of the molecules.

23.23. The cohesive energy density is approximately equal to the enthalpy of vaporization per unit volume. Estimate the molar enthalpy of vaporization of carbon tetrachloride on the basis that $\rho \approx 1.594$ g cm^{-3} and that the molecular polarizability volume is 10.5×10^{-24}.

23.24. Long ago, in Chapter 6, we saw that $(\partial U/\partial V)_T = n^2a/V^2$ for a van der Waals gas. If we identify U with the average cohesive energy we can use this relation to find a relation between a and C_6. Find this relation, and then express the critical constants in terms of d and C_6.

23.25. Estimate the critical constants of CCl_4 on the basis of its polarizability.

23.26. The Madelung constant is defined in eqn (23.3.3). One simple case where it can be calculated without much effort is in the case of a line of alternating singly charged positive and negative ions separated by R. What is its value for this array?

23.27. The internal energy of an ionic crystal, as given by eqn (23.3.4), requires some knowledge of the parameter R^*. This can be obtained from a knowledge of the isothermal compressibility of the crystal, and this Problem explores this relation. First, from the thermodynamic equation of state $(\partial U/\partial V)_T = T(\partial p/\partial T)_V - p$, evaluated at absolute zero, show that $1/\kappa$ is proportional to $\partial^2 U/\partial V^2$, and then deduce R^* in terms of V, κ, and R_0. (At some stage you may find it helpful to write $V = cR_0^3$, where c is a constant which can be eliminated at a later stage.)

23.28. Set up a Born–Haber cycle for determining the cohesive energy of KCl and evaluate it from the data in Chapter 4 and $\Delta H_{sub}(K) = 82.6$ kJ mol^{-1}.

23.29. Calculate the cohesive energy of KCl from eqn (23.3.4), using $\kappa = 1.1 \times 10^{-5}$ atm^{-1} for the isothermal compressibility and $\rho = 1.984$ g cm^{-3}.

23.30. The ionic radii of Na$^+$ and F$^-$ ions are 95 pm and 136 pm respectively. What is the molar internal lattice energy of NaF? Use $R^* \approx 29$ pm.

23.31. The ionic radii of Cs$^+$ and Cl$^-$ are respectively 182 pm and 181 pm. What is the molar internal lattice energy of CsCl? Use $R^* \approx 40$ pm.

23.32. Molecular beams are an important way of obtaining information about intermolecular forces, and this and the next Problem give a very simple introduction to the complicated task of unravelling the richness of information they provide and extracting the piece required. Consider the collision between a hard-sphere molecule of radius R_1 and mass m_1 and an infinitely heavy spherical atom of radius R_2 which is also an impenetrable sphere. (Imagine them as H and I, for example.) Plot the scattering angle θ as a function of the impact parameter b. Do the calculation on the basis of simple geometrical considerations.

23.33. The scattering characteristics of atoms depend on the energy of the collision. We can model the situation as follows. Let both atoms always behave as impenetrable hard spheres, but let the effective radius of the heavy atom vary with the relative speed of approach of the light atom. Suppose its effective radius depends on the speed v as $R_2 \exp(-v/v^*)$, where v^* is some constant. With this form tne effective radius of the heavy atom is R_2 at slow speeds of approach of the other atom, and less when the speed of approach is great. Take $R_1 = \frac{1}{2}R_2$ for simplicity, and an impact parameter $b = \frac{1}{2}R_2$ and plot the scattering angle as a function of (a) speed, (b) kinetic energy of approach.

23.34.   The magnetizability ξ and then the mass and volume magnetic susceptibilities (χ and κ) can be calculated from the wavefunctions of molecules. For instance, for a 1 electron atom $\xi \approx -(e^2/6m_e)\langle r^2 \rangle$. Calculate ξ and χ for hydrogen atoms.

23.35.   Calculate the paramagnetic and total susceptibilities (χ) for a hydrogen atom at 25 °C.

23.36.   Show that the spin-only mass magnetic susceptibility of a molecule of R.M.M. M_r and spin S at a temperature T is

$$\chi^P = 6.2856 \times 10^{-3}\{S(S+1)/M_r(T/K)\}.$$

Evaluate this expression for molecules with (a) two unpaired spins and (b) 5 unpaired spins at 4 K and 298.15 K.

23.37.   The strength of the ligand field affects the number of unpaired d-electrons in transition metal complexes. On the supposition that only the spin magnetic moments contribute to the susceptibility (this is the *spin-only* approximation) tabulate the expected mass paramagnetic susceptibilities of d^n octahedral complexes ($n = 1 \rightarrow 10$) for weak and strong ligand fields in terms of their R.M.M. and temperatures. (Refer to Problem 15.31.)

23.38.   As we saw in Chapters 20 and 21, the NO molecule is peculiar in so far as it has thermally accessible electronically excited states. It also has an unpaired electron, and so may be expected to be paramagnetic. In fact, its ground state is not paramagnetic. This is because the magnetic moment due to the orbital motion of the electron around the molecular axis cancels the spin magnetic moment. The first excited electronic state (at 121.1 cm^{-1}) is paramagnetic because the orbital moment augments, rather than cancels, the spin moment, and its magnetic moment is 2 Bohr magnetons. Since this upper state is thermally accessible, the paramagnetic susceptibility of the molecule will show a peculiar temperature dependence even at room temperature. Calculate the mass paramagnetic susceptibility of NO and plot it as a function of temperature. What is its value at 298 K?

24 The structures and properties of macromolecules

Learning objectives

After careful study of this chapter you should be able to

(1) List the techniques available for the determination of the shapes and sizes of macromolecules (p. 810).

(2) Use the measurements of *osmotic pressure* to determine the molar masses of macromolecules (p. 812).

(3) Explain why solutions of macromolecules are strongly non-ideal (p. 813) and estimate the size of the deviations (eqn (24.1.7)).

(4) Define the terms θ-*temperature*, *monodisperse*, *polydisperse*, *polyampholyte*, *polyelectrolyte*, and *dialysis* (p. 816).

(5) Distinguish between *number average* and *mass average* molar mass (p. 816).

(6) Account for the *Donnan effect* (p. 817) and deduce expressions for the effect of ionic charge on osmotic equilibria (eqns (24.1.12) and (24.1.14)).

(7) Describe the operation of an *ultracentrifuge* (p. 819) and relate the *rate of sedimentation* to the shapes and molar masses of macromolecules (p. 821).

(8) Describe how to obtain molar masses from observations on *sedimentation equilibria* (p. 824).

(9) Explain the basis and describe the applications of *electrophoresis* and *gel filtration* (p. 824).

(10) Define *intrinsic viscosity* (eqn (24.1.22)), describe how it is measured (p. 825), and explain its significance (p. 826).

(11) Discuss the application of *light scattering* measurements to the determination of the shapes and molar masses of macromolecules (p. 827).

(12) Describe how n.m.r. and e.s.r. are applied to the study of macromolecules (p. 833) and explain the use of *shift reagents* and *spin-labelling*.

(13) Explain the meaning of the terms *primary*, *secondary*, *tertiary*, and *quaternary structure* (p. 835) and *denaturation* (p. 836).

(14) Describe the structure of a *random coil* molecule (p. 837), relate its size to its composition (eqn (24.2.2), etc.), and explain the role of the *conformational entropy* (p. 838).

(15) Describe the nature of the *peptide link* and its role in the determination of the secondary structure of proteins (p. 839).

(16) Define the term *colloid* and distinguish between *sols*, *aerosols*, and *emulsions* and between *lyophilic* and *lyophobic* varieties (p. 842).

(17) Describe the *preparation and purification* of colloids (p. 843).

(18) Explain the terms *micelle* and *critical micelle concentration*, (p. 845) and state why micelles are important (p. 845).

(19) Account for the *stability of colloids* (p. 846) in terms of the *electric double layer* (p. 846), and state and account for the *Schultze–Hardy rule* (p. 846) and the *isoelectric point* (p. 847).

(20) Define *surface excess* (eqn (24.4.3)) and derive and use the *Gibbs surface tension equation* (eqn (24.4.5)).

Introduction

There are macromolecules everywhere, inside us and outside us. Some are natural: they include the proteins, DNA, cellulose, and so on. These macromolecules constitute, stabilize, energize, and in a sense control us, and they do the same for all the other forms of life. Others are man-made: they are the synthetic polymers formed by stringing together and cross-linking smaller units. Molecules and atoms sometimes swarm together under the influence of intermolecular forces, and the large conglomerates behave like macromolecules: these are the colloids. Varieties of them—the sols, emulsions, and foams—are intimately involved in the processes of the world around and within us. Life in all its forms, from its intrinsic nature to its technological interaction with its environment, is the chemistry of macromolecules.

Physical chemistry provides methods for discovering the shapes and sizes of macromolecules, for investigating the response of these character-istics to the environment, and for relating properties to constitution. Once constitution and structure are known, a deeper understanding of function may be accessible. Synthetic macromolecules are generally simpler than natural ones, although now that we are succeeding in synthesizing proteins this distinction is becoming old-fashioned. Even though synthetic polymers are relatively simple, it is still necessary to know the nature of the material at least to the extent of knowing the lengths of the polymer chains. We also ought to be able to relate physical properties, such as elasticity, to aspects of chemical constitution, such as chain length and the degree of cross-linking. Finally, the principal question about colloids is their stability: why do they form and why are they stable for long periods?

24.1 Size and shape

X-ray diffraction, Chapter 22, can reveal the position of almost every atom in even highly complex molecules. Why then, is it necessary to use any other technique? This can be answered at a variety of levels. In the first place the sample might be a mixture of polymers of different chain length and extent of cross-linking, in which case a detailed X-ray diffraction analysis would give some kind of blurred average. A related problem is that even if all the molecules are identical it might not be possible to obtain a single crystal, and so the X-ray procedure could hardly be started. Next, the technique is expensive, time consuming, and requires sophisticated computing facilities. Furthermore, although the work on haemoglobin, enzymes, and DNA has shown how immensely stimulating the data can be, the information is incomplete. For instance, what can be said about the shape of the molecule in its natural environment, the cell? What can be said about the way its shape changes in response to the environment? Shape and function are intimately related, and it is crucial to know how biological macromolecules, which often carry both acidic and basic groups, respond to the pH of the medium. Likewise it is of interest to investigate the process by which the conformation of a natural macromolecule changes from a regular to a less regular form. *Denaturation* to a less organized conformation is often accompanied by a loss of function, but sometimes it is an essential step in the fulfilment of function, as in the replication of DNA.

Another direct method for the observation of the size and shape of large molecules is the electron microscope. Resolution of about 500 pm (5 Å) may now be achieved, and so in principle the fine detail of the shape of a macromolecule can be discerned. Nevertheless there are severe limitations: the state of the sample is highly unnatural, for in order to obtain an image some kind of replica of the molecule has to be formed. The replica is often a cast taken by spraying the sample with metal atoms either directly or after coating the macromolecule with small protein or detergent molecules. Figure 24.1 was obtained in the first way. The *scanning electron microscope* differs from the transmission microscope in that it

Fig. 24.1. Scanning electron microscope photograph of (deformed) haemoglobin at a magnification of 2.75×10^6 diameters (A. V. Crewe, with permission).

obtains an image by scanning a metal cast with a spot-focused electron beam, and monitoring the intensity of the electrons ejected by and scattered from the beam. All these methods, however dramatic the pictures and useful the information they reveal, suffer from the limitation that the sample is modified.

Techniques are available for the determination of the sizes and shapes of macromolecules and colloidal particles in fluid solutions. Some give the molar masses or the relative molecular masses (R.M.M., or, more colloquially, 'molecular weights'). Others give an indication of the geometrical size and shape of the species, and indicate whether the molecule is rod-like or globular (spherical, egg shaped), and so on. The techniques can also be used to determine whether the chain is organized into a definite array, or whether it is simply a random coil. It is also becoming possible to use n.m.r. to find the disposition of atoms in a molecule in solution. We shall look at these methods in the next few subsections.

Osmosis and dialysis. The classical methods of determining the relative molecular masses of molecules are based on the colligative properties, Chapter 8. In the case of macromolecules, where the *number* of molecules in solution may be very small even though their total mass may be appreciable, only osmotic measurements are sufficiently sensitive.

The basic relation is the *van't Hoff equation* (p. 236):

$(24.1.1)°$ $\Pi V = n_p RT.$

Π is the osmotic pressure and n_p the amount of solute in a volume V of solution. Since the concentration of the solute, a macromolecule P, is $n_p/V = [P]$, a simpler form is

$(24.1.2)°$ $\Pi = RT[P].$

Furthermore, since the concentration $[P]$ is related to the mass concentration c_P through $[P] = c_P/M_m$, where M_m is the molar mass, another form of the equation is

$(24.1.3)°$ $\Pi/c_P = RT/M_m.$

The R.M.M., M_r, of the macromolecule is then related to the molar mass by $M_r = M_m/\text{g mol}^{-1}$ (that is, the R.M.M. is the numerical value of the molar mass when the latter is expressed in g mol^{-1}). It follows that a determination of the osmotic pressure of a solution containing a known mass concentration of the macromolecule gives the latter's R.M.M. A typical osmometer is illustrated in Fig. 24.2: the osmotic pressure is measured from the excess height h of the column of solution using $\Pi = \rho g h$, ρ being the solution density and g the acceleration due to gravity.

As always, there are complications. Three are of particular importance: the deviations from ideality, the presence of a range of R.M.M.s in the sample, and the presence (and possibility of its variability) of charge on the macromolecule.

Macromolecules give strongly non-ideal solutions. This is partly be-

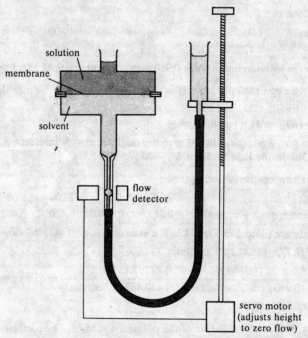

Fig. 24.2. Osmometer used in the determination of the molar masses of macromolecules.

cause, being so large, they displace a large quantity of solvent and do not dissolve by replacing individual solvent molecules with negligible disturbance. Furthermore, their great bulk means that there is a large excluded volume: one molecule is unable to swim freely through the solution since it is excluded from the regions occupied by the others. In thermodynamic terms this implies that the entropy change is important when a macromolecule goes into solution. There may also be a significant enthalpy of solution on account of the interaction of the solvent with a large number of the component monomer units in the polymer.

Deviations from ideality are normally taken into account by extending the van't Hoff equation in the same way as the perfect gas equation is extended to real gases; that is, by writing a virial expansion. The dependence of the osmotic pressure on the concentration of the macromolecule is written

(24.1.4a) $$\Pi/[\text{P}] = RT\{1 + B[\text{P}] + \ldots\}$$

(24.1.4b) $$\text{or} \quad \Pi/c_P = (RT/M_m)\{1 + (B/M_m)c_P + \ldots\}.$$

Then by plotting Π/c_P against c_P and extrapolating to zero concentration, M_m can be obtained from the intercept, and B, the *osmotic virial coefficient*, can be inferred from the slope.

Example (Objective 2). An osmometer was used to measure the osmotic pressure of a solution of polyvinylchloride (P.V.C.) in cyclohexanone at 25 °C. The data on the height of solution generated by the flow of solvent through the semipermeable membrane were as follows:

$c_P/\text{g dm}^{-3}$	1.00	2.00	4.00	7.00	9.00
h/cm	0.472	0.926	1.776	2.940	3.627

The average solution density was $0.980\,\text{g cm}^{-3}$. Find the R.M.M. of the polymer.

- *Method.* Use eqn (24.1.4b) with $\varPi = \rho g h$ and $g = 9.81\,\text{m s}^{-2}$. Since the equation may be written

$$h/c_P = (RT/\rho g M_m)\{1 + (B/M_m)c_P + \dots\}.$$

A plot of h/c_P against c_P should give a straight line with intercept $RT/\rho g M_m$ at $c_P = 0$. Obtain the R.M.M. from $M_r = M_m/\text{g mol}^{-1}$.

- *Answer.* Draw up the following table:

$c_P/\text{g dm}^{-3}$	1.00	2.00	4.00	7.00	9.00
$(h/c_P)/(\text{cm/g dm}^{-3})$	0.472	0.463	0.444	0.420	0.403

The points are plotted in Fig. 24.3. The intercept is at 0.482; it follows that

$$M_m = (RT/\rho g)/(0.482\,\text{g}^{-1}\,\text{cm dm}^3)$$

$$= \frac{(8.314\,\text{J K}^{-1}\,\text{mol}^{-1}) \times (298.15\,\text{K})}{(0.980\,\text{g cm}^{-3}) \times (9.81\,\text{m s}^{-2}) \times (0.482\,\text{g}^{-1}\,\text{cm dm}^3)} = 53.5\,\text{kg mol}^{-1}.$$

Hence $M_r = 53\,500$.

- *Comment.* This is a mean value of the polymer's R.M.M. The coefficient B can be obtained by equating the slope of the plot with $(RT/\rho g M_m)B/M_m$. This gives $B/M_m = -0.0195\,\text{g}^{-1}\,\text{dm}^3$.

The thermodynamic justification for the virial expansion, and the interpretation of B, can be traced by returning to the basic expressions for the osmotic pressure in terms of the chemical potential. Refer back to eqns (8.3.6) and (8.3.8) on p. 236. When the solution is non-ideal $RT \ln x_A$ in the latter has to be replaced by $RT \ln a_A$, where a_A is the activity of the solvent. If the calculation is then repeated as in Chapter 8 one arrives at

$$-RT \ln a_A = \int_p^{p+\varPi} V_m^* \, dp$$

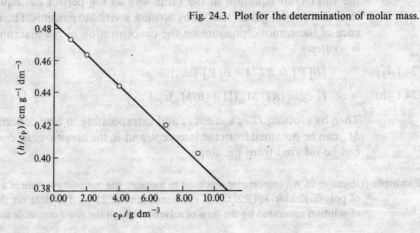

Fig. 24.3. Plot for the determination of molar mass.

in place of eqn (8.3.8). The integral is ΠV_m^* if the solvent is assumed to be incompressible, and so the osmotic pressure is given by

(24.1.5) $\qquad \Pi V_m^* = -RT \ln a_A.$

The activity is unity when the solvent is pure, and is equal to x_A when the solution is ideal. In the latter case we write $x_A = 1 - x_P$, where x_P is the mole fraction of macromolecule, and approximate $\ln (1 - x_P)$ by $-x_P$. In other words, we make the approximation $\ln a_A \approx -x_P$, and this leads to the van't Hoff equation. When the solution is non-ideal we suppose that the term $\ln a_A = -x_P$ is the start of a series:

(24.1.6) $\qquad \ln a_A = -\{x_P + B'x_P^2 + \ldots\};$

then substitution of this expansion in the equation for the osmotic pressure and modification of the units of B leads to eqn (24.1.4). The virial expansion for $\ln a_A$ is not a trivial result, although it might at first glance seem the obvious form. The *MacMillan–Meyer theory* of solutions, however, confirms that the polynomial expansion is appropriate to non-electrolytes, but because the activities of electrolyte solutions depend on the square root of concentrations (recall the Debye–Hückel theory), they are not captured by an expansion of the form given in eqn (24.1.6).

The coefficient B arises largely from the effect of excluded volume. This is reminiscent of the van der Waals gas, for there the virial coefficient B is equal to $b - a/RT$, which reduces to $B \approx b$ when the effect of excluded volume is dominant. If we imagine a solution of a macromolecule being built by the successive addition of macromolecules, each one being excluded from the space occupied by the ones that preceded it, the value of B turns out to be

(24.1.7) $\qquad B = \tfrac{1}{2} L v_P,$

where v_P is the excluded volume due to a single molecule ($v_P \approx 8 v_{\text{molecule}}$ for spheres).

It may happen that for a particular solvent and at a particular temperature the virial coefficient B is zero. This occurs when the excluded volume terms are compensated by any attractive terms tending to draw the macromolecules together. The temperature is called the θ-*temperature* (theta-temperature), and the solution is a θ-*solution*. This is the analogue of the Boyle temperature in gases, p. 38. The fortuitous cancellation of effects at the θ-temperature means that the solution then behaves virtually ideally, and its thermodynamic properties are easier to describe. As an example, the θ-temperature for polystyrene in cyclohexane is around 306 K, the exact value depending on its molar mass.

The second complication is the possibility that the sample is a mixture of macromolecules with different molar masses. A pure protein is a well-defined species with a single, definite molar mass (it is *monodisperse*) although there might be small variation (e.g., one amino acid in place of another) depending upon the source. On the other hand, a synthetic polymer is a mixture of various chain lengths, and the sample contains a

range of molar masses (it is *polydisperse*). Osmotic measurements then
yield an average molar mass. There are various ways of taking averages,
and the one given by osmotic pressure measurements is the *number average*
R.M.M., $\langle M_r \rangle_N$. This is defined as follows. Suppose there are N_i molecules
with R.M.M. M_{ri}, and N molecules in all. Then the number average
R.M.M. is the R.M.M. of each weighted by the numerical proportion
having that R.M.M.:

(24.1.8) $$\langle M_r \rangle_N = \sum_i (N_i/N) M_{ri} = (1/N) \sum_i N_i M_{ri}.$$

This is the same type of average taken when calculating the mean height
of a population, the mean speed of cars, and so on. The reason why
osmosis gives the number average R.M.M. and not something more
complicated is that it is a colligative property—a property that depends
on number not nature.

The third complication is the presence of charge on some types of
macromolecule. Some polymers are strings of acid groups (as in poly-
acrylic acid, $-(CH_2CHCOOH)_n-$), or strings of bases (such as nylon
$-[NH(CH_2)_6NHCO(CH_2)_4CO]_n-$), and proteins have both acid and
base groups. Polymers may therefore be *polyelectrolytes*, and, depending
upon their state of ionization, *polyanions*, *polycations*, and *polyampholytes*
(of mixed anionic and cationic nature).

One consequence of dealing with polyelectrolytes and naturally occurring
ionizable macromolecules is that it is necessary to know the extent of
ionization before osmotic data can be interpreted. For example, suppose
the sodium salt of a polyelectrolyte dissociates into v sodium ions and
the single polyanion P^{v-}, then the van't Hoff equation reads

(24.1.9)° $$\Pi/c_P = (v+1)RT/M_m,$$

where c_P is the mass concentration of the added Na_vP. If we guess that
$v = 1$ when in fact $v = 10$ the estimate of the molar mass will be seriously
wrong. The way out of this difficulty can be found by considering the
other consequence of dealing with charged macromolecules.

Suppose that the solution of the protein or macromolecule contains
additional salt (so that the solution of Na_vP also contains added NaCl,
for example), and that it is in contact through a membrane such as
cellophane (or a cell wall) with another salt solution, and that the membrane
is permeable to the solvent and the salt ions, but not to the polyelectrolyte
ion itself. This arrangement, which is referred to as *dialysis*, is one that
actually occurs in living systems, where osmosis is an important feature
of cell operation. What effect does the salt have on the osmotic pressure?
We shall see that answering this question leads to a resolution of the
laboratory problem of coping with the extent of ionization of the macromolecule. The reason why we anticipate an effect to arise from the presence
of the added salt is that the anions and cations cannot migrate through
the membrane in arbitrary amounts because electrical neutrality has to

be preserved on both sides: if an anion migrates in one direction, a cation must accompany it, and vice versa. The general term of the effect of added salt on dialysis equilibria is the *Donnan effect*.

In order to see the implications of the Donnan effect, consider what happens when the polyelectrolyte $(Na^+)_\nu P^{\nu-}$ is at a concentration $[P]$ on one side of a membrane and NaCl is added to both sides. On the left there are $P^{\nu-}$, Na^+, and Cl^- ions, and on the right there are Na^+ and Cl^-. The condition for equilibrium is that the chemical potential of NaCl should be the same on both sides of the membrane, and so a net flow of Na^+ and Cl^- ions occurs until the condition $\mu(NaCl, \text{left}) = \mu(NaCl, \text{right})$ is established. This requires

(24.1.10)
$$\mu^{\ominus}(NaCl) + RT\ln\{a(Na^+)a(Cl^-)\}_{\text{left}}$$
$$= \mu^{\ominus}(NaCl) + RT\ln\{a(Na^+)a(Cl^-)\}_{\text{right}}.$$

If activity coefficients are disregarded, and if it is assumed that the standard values $\mu^{\ominus}(NaCl)$ are the same on both sides of the barrier, then this is equivalent to requiring that

(24.1.11)
$$\{[Na^+][Cl^-]\}_{\text{left}} = \{[Na^+][Cl^-]\}_{\text{right}}.$$

The sodium ions are supplied by the polyelectrolyte as well as the added salt. Furthermore, if a sodium ion migrates through the membrane it has to be accompanied by a chloride ion in order to maintain electrical neutrality on both sides of the membrane (the macromolecule is too big to migrate). It follows that, at equilibrium,

$$[Na^+]_{\text{left}} = [Cl^-]_{\text{left}} + \nu[P]$$

$$[Na^+]_{\text{right}} = [Cl^-]_{\text{right}}.$$

These constraint equations can be combined with eqn (24.1.11) to obtain expressions for the differences in ion concentrations across the membrane:

(24.1.12a)°
$$[Na^+]_{\text{left}} - [Na^+]_{\text{right}} = \frac{\nu[P][Na^+]_{\text{left}}}{[Na^+]_{\text{left}} + [Na^+]_{\text{right}}} = \frac{\nu[P][Na^+]_{\text{left}}}{2[Cl] + \nu[P]}$$

(24.1.12b)°
$$[Cl^-]_{\text{left}} - [Cl^-]_{\text{right}} = \frac{-\nu[P][Cl^-]_{\text{left}}}{[Cl^-]_{\text{left}} + [Cl^-]_{\text{right}}} = \frac{-\nu[P][Cl^-]_{\text{left}}}{2[Cl]}.$$

In obtaining these expressions use has been made of the relations

$$[Cl] \overset{\text{def}}{=} \tfrac{1}{2}\{[Cl^-]_{\text{left}} + [Cl^-]_{\text{right}}\}$$
$$= \tfrac{1}{2}\{[Na^+]_{\text{left}} + [Na^+]_{\text{right}} - \nu[P]\},$$

the first being a definition of $[Cl]$, which can be measured by analysing the two solutions, and the second then coming from the first electroneutrality equation.

The final step is to note that the osmotic pressure depends on the difference of the numbers of particles on either side of the membrane at equilibrium, and so the van't Hoff equation $\Pi = RT[\text{solute}]$ becomes

$$\Pi = RT\{[P] + [Na^+]_{left} - [Na^+]_{right} + [Cl^-]_{left} - [Cl^-]_{right}\}$$

(24.1.13)° $$= RT[P]\{1 + v^2[P]/(4[Cl] + v[P])\}.$$

The second line comes from algebraic manipulations of the results outlined above.

There are several points that can now be made. First, when the amount of salt that has been added is so great that the condition $[Cl] \gg (v/4)[P]$ is satisfied, eqn (24.1.13) simplifies to

(24.1.14)° $$\Pi \approx RT[P]\{1 + (v^2/4[Cl])[P]\}$$

and the presence of the salt contributes to a kind of virial coefficient. When the salt is at such a high concentration that the ratio $v^2[P]/4[Cl]$ is much smaller than unity, the last equation reduces to $\Pi \approx RT[P]$, and the osmotic pressure is independent of the value of v. This is the first principal result we have been seeking: it means that if osmotic pressures are measured in the presence of high concentrations of salt (as is commonly the case for naturally occurring macromolecules) the molar mass may be obtained unambiguously. The underlying reason for this simplification is that when there is a great deal of salt present the fact that the polyelectrolyte provides additional cations (or anions if it is a polycation) is barely noticeable. In passing you should also notice that if no salt is added, so that $[Cl] = 0$, eqn (24.1.13) reduces to $\Pi = (1 + v)[P]RT$, the form encountered originally, eqn (24.1.9).

The second point arises from a consideration of eqn (24.1.12). There is often interest in the extent to which ions are bound by macromolecules, especially when a membrane (such as a cell wall) separates two regions. The equations show, however, that cations will dominate the anions in the compartment containing the polyanion (the concentration difference is positive for Na^+, negative for Cl^-, eqn (24.1.12)) simply as a result of the equilibrium and electroneutrality requirements, and so it would be wrong to conclude without further evidence that Na^+ tended to stick to the macromolecule.

Example (Objective 6). Two equal volumes of $0.200\,mol\,dm^{-3}\,NaCl$ aqueous solutions are separated by a membrane. A macromolecule with $M_r = 55000$, which cannot penetrate the membrane, is added as its sodium salt Na_6P to a concentration of $50\,g\,dm^{-3}$ to the left hand compartment. What are the equilibrium concentrations of Na^+ and Cl^- in each compartment?

- *Method.* Use eqn (24.1.12a) to find the concentration differences, and the following equation to find the sum of Na^+ on the left and right. Solve the equations for the individual concentrations. The chloride concentrations can then be found from the equations preceding eqn (24.1.12a). Use $[P] = c_P/M_m$.

- *Answer.* $[P] = (50\,g\,dm^{-3})/(55000\,g\,mol^{-1}) = 9.0909 \times 10^{-4}\,mol\,dm^{-3}$. Equation (24.1.12a) gives

$$[Na^+]_L - [Na^+]_R = \frac{6 \times (9.0909 \times 10^{-4}\,mol\,dm^{-3} \times [Na^+]_L}{2 \times (0.200\,mol\,dm^{-3}) + 6 \times (9.0909 \times 10^{-4}\,mol\,dm^{-3})}$$

$$= 0.01345\,[Na^+]_L$$

$$[Na^+]_L - [Na^+]_R = 2[NaCl] + 6[P] = (0.400 + 6 \times 9.0909 \times 10^{-4})\,mol\,dm^{-3}$$

$$= 0.40545\,mol\,dm^{-3}$$

Hence $[Na^+]_L = 0.2041\,mol\,dm$, $\qquad [Na^+]_R = 0.2014\,mol\,dm^{-3}$.

$$[Cl^-]_R = [Na^+]_R = 0.2014\,mol\,dm^{-3}$$

$$[Cl^-]_L = [Na^+]_L - 6[P] = 0.1986\,mol\,dm^{-3}.$$

- **Comment.** Note how the chloride accumulated slightly in the compartment not containing the macromolecule. The condition $[Cl] \gg (z/4)[P]$ is well satisfied, and so eqn (24.1.14) can be used to estimate the osmotic pressure. It gives $\Pi \approx 1.0409\,RT[P]$.

Sedimentation. In a gravitational field heavy particles settle towards the bottom of a column of solution. The rate at which this happens depends not only on the strength of the field, but also on both the masses and shapes of the particles. When the system is at equilibrium not all the particles are on the floor of the container because the effect of the gravitational field has to compete with the stirring effects of thermal motion. The particles are dispersed over a range of heights in accord with the Boltzmann expression for the distribution of populations. The spread of heights depends on molecular masses, and so the analysis of the equilibrium distribution is another way of determining molar mass. In the next few paragraphs we shall look at ways of assessing molar masses from sedimentation equilibria and both shapes and molar masses from sedimentation rates.

Sedimentation is normally a very slow process. The rate can be increased dramatically by replacing the gravitational field by a centrifugal field. This can be achieved in an *ultracentrifuge*, which is essentially a cylinder that can be rotated at high speed about its axis; the sample is held in a cell close to the periphery, Fig. 24.4. Modern centrifuges rotate so fast that accelerations about 10^5 that of gravity can be produced. Initially the sample is uniform, but in the course of the experiment the 'top' boundary of the solute moves outwards as sedimentation proceeds. The rate of

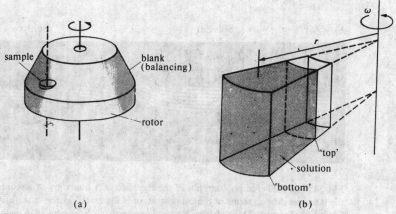

(a)　　　　　　　　　　　　　　(b)

Fig. 24.4. (a) An ultracentrifuge head, (b) details of the sample cavity.

movement of the boundary may be used to find the molar mass of the solute.

The principal problem is how to monitor the concentration of the sample at different radii when it is rotating at thousands of revolutions per minute. Spectroscopy is one possibility, but it is more common to detect the changing concentration profile by making use of its effect on the refractive index of the sample. The boundary between the sinking 'top' surface of the solute and the solvent it leaves behind as it sediments gives rise to a moderately abrupt change in the refractive index between the two parts of the sample. That boundary moves outwards. Since there is a refractive index gradient across the sample it behaves like a prism, and bends any light passing through it. The *Schlieren optical system* turns a refractive index gradient into a dark image, Fig. 24.5a; the alternative *interference system* monitors the refractive index gradient through its effect

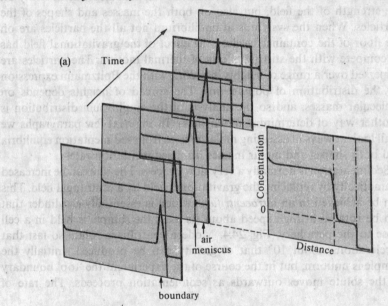

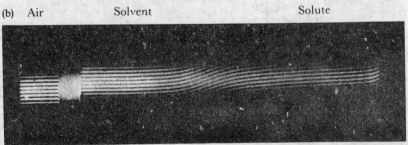

Fig. 24.5. (a) Schlieren photographs of the recession of a boundary at a sequence of times; the interpretation in terms of the concentration is illustrated in one case. (b) An interference photograph (D. Freifelder, *Physical biochemistry*, Freeman, 1976, with permission).

on the interference of two beams of light, one coming through the sample and the other through a blank, Fig. 24.5b. (Schlieren optics are also used to monitor air flow patterns in wind tunnels and shock tubes.)

The rate of sedimentation. Consider a solute particle of mass m in the solution. On account of the displacement of the liquid we have to take into account the buoyancy of the medium, and deal with the *effective mass* $m_{eff} = (1 - \rho v_s)m$, where ρ is the solution density and v_s is the specific volume (actually the partial specific volume), the volume per unit mass, of the solute. For simplicity we shall abbreviate the last expression to $m_{eff} = bm$, where b is the buoyancy correction factor. The solute particle experiences a centrifugal force $m_{eff}r\omega^2$ when it is at a distance r from the axis of the rotor and is travelling with an angular velocity ω. The acceleration outwards it induces is countered by a frictional force proportional to the speed, s, of the particle through the medium. This force is written fs, where f is the *friction coefficient*. The particle adopts a *drift speed* which is found by equating the two forces: $m_{eff}r\omega^2 = fs$ implies

(24.1.15) $s = m_{eff}r\omega^2/f = mbr\omega^2/f.$

The drift speed depends on the angular velocity and the radius of the rotor; we know both, and so it is convenient to concentrate on the ratio $s/r\omega^2$, which is called the *sedimentation constant*, S. Furthermore, since the mass of an individual molecule is related to the molar mass through $m = M_m/L$, we arrive at

(24.1.16) $S = s/r\omega^2 = M_mb/fL.$

Example (Objective 7). The sedimentation of bovine serum albumin was monitored at 25 °C. The initial radius of the solute surface was 5.500 cm, and during centrifugation at 56 850 r.p.m. it receded as follows:

t/s	0	500	1000	2000	3000	4000	5000
r/cm	5.50	5.55	5.60	5.70	5.80	5.91	6.01

Calculate the sedimentation constant.

- *Method.* Use eqn (24.1.16), identifying s with dr/dt. Integrate the equation $dr/dt = r\omega^2S$ finding $\ln\{r(t)/r(0)\} = \omega^2St$. A plot of $\ln\{r(t)/r(0)\}$ against t gives ω^2S from the slope. Remember that $\omega = 2\pi\nu$, where ν is in cycles/second.

- *Answer.* Draw up the following table:

t/s	0	500	1000	2000	3000	4000	5000
$\ln\{r(t)/r(0)\}$	0	0.0090	0.0180	0.0357	0.0531	0.0719	0.0887

The plot has a slope of 1.788×10^{-5}, hence $\omega^2S = 1.788 \times 10^{-5}\,s^{-1}$. Therefore

$$S = (1.788 \times 10^{-5}\,s^{-1})/[2\pi \times (56\,850/60)s^{-1}]^2 = 5.04 \times 10^{-13}\,s.$$

- *Comment.* We shall develop this result below. Note that the unit 10^{-13} s is sometimes called the *svedberg* and denoted S. Hence in this case the sedimentation constant is 5.04 S. Accurate results are obtained when concentrations are extrapolated to zero.

In order to make any progress, something has to be known about the friction constant f. For a spherical particle of radius a in a solvent of viscosity η it is given by *Stokes' relation,* $f = 6\pi a\eta$. Therefore, *for spherical molecules.*

(24.1.17) $S = bM_m/6\pi\eta aL,$

and S can be used to determine either M_m or the molecule's radius a. If the molecules are not spherical, the connection between f and their dimensions may still be made, and some of the expressions that arise in the case of rod-like molecules, prolate ellipsoids (like cigars), and oblate ellipsoids (like pancakes) are given in Table 24.1. You should note that care has to be taken to extrapolate the sedimentation data to zero concentration, because interference between macromolecules can cause severe complications.

Table 24.1. Frictional coefficients and molecular geometry

Sphere; radius a, $c = a$	f_0
Prolate ellipsoid; major axis $2a$, minor axis $2b$, $c = (ab^2)^{1/3}$	$\left\{ \dfrac{(1-b^2/a^2)^{1/2}}{(b/a)^{2/3} \ln\{[1+(1-b^2/a^2)^{1/2}]/(b/a)\}} \right\} f_0$
Oblate ellipsoid; major axis $2a$, minor axis $2b$, $c = (a^2b)^{1/3}$	$\left\{ \dfrac{(a^2/b^2-1)^{1/2}}{(a/b)^{2/3} \arctan[(a^2/b^2-1)^{1/2}]} \right\} f_0$
Long rod; length l, radius a, $c = (3a^2l/4)^{1/3}$	$\left\{ \dfrac{(l/2a)^{2/3}}{(3/2)^{1/3}\{2\ln(l/a)-0.11\}} \right\} f_0$

In each case $f_0 = 6\pi\eta c$ with the appropriate value of c.

For prolate ellipsoid:

a/b	2	3	4	5	6	7	8	9	10	50	100
f/f_0	1.04	1.11	1.18	1.25	1.31	1.38	1.43	1.49	1.54	2.95	4.07

For oblate ellipsoid:

a/b	2	3	4	5	6	7	8	9	10	50	100
f/f_0	1.04	1.10	1.17	1.22	1.28	1.33	1.37	1.42	1.46	2.38	2.97

Source: K. E. Van Holde, *Physical biochemistry*, Prentice Hall.

The shape of a molecule affects its rate of sedimentation, spherical (and compact molecules in general) sedimenting faster than rod-like or extended ones. For instance, DNA helices sediment much faster when they are denatured to a random coil. Consequently sedimentation velocity can be used to monitor denaturation.

In order to use the sedimentation rate to determine the molar mass it would appear that it is necessary to know the molecular radius a, or in general the friction coefficient f. Fortunately this problem can be avoided by drawing on a standard relation between f and the *diffusion coefficient* D. The diffusion coefficient, Table 24.2, is a measure of the rate at which molecules spread across a concentration gradient: it can be measured by observing the rate at which a boundary spreads, or the rate at which a more concentrated solution diffuses into another. There are also methods based on light scattering, as we shall see later in the chapter. The crucial equation for our present purposes is

(24.1.18) $\quad f = kT/D.$

(A special case of its derivation is given in Chapter 26.) The important point is that the relation is independent of the shape of the species. It follows from eqn (24.1.16) that

(24.1.19) $\quad M_m = fSL/b = SLkT/bD = SRT/bD.$

Therefore in order to obtain M_m we combine measurements of sedimentation and diffusion rates (for S and D respectively).

Table 24.2. Diffusion coefficients of macromolecules in aqueous solution at 20 °C

	M_r	$D/10^{-7} \, cm^2 \, s^{-1}$
Sucrose	342	45.86
Ribonuclease	13 683	11.9
Lysozyme	14 100	10.4
Serum albumin	65 000	5.94
Haemoglobin	68 000	6.9
Urease	480 000	3.46
Collagen	345 000	0.69
Myosin	493 000	1.16

Source: C. Tanford, *Physical chemistry of macromolecules*, Wiley.

Example (Objective 7). Use the result from the last *Example* in combination with the data below to find values for the R.M.M. of bovine serum albumin, and estimate its axial ratio on the basis that it is prolate ellipsoidal. Take $D = 6.97 \times 10^{-7} \, cm^2 \, s^{-1}$, $\rho = 1.0024 \, g \, cm^{-3}$, $v_s = 0.734 \, cm^3 \, g^{-1}$, $\eta = 0.890 \times 10^{-3} \, kg \, m^{-1} \, s^{-1}$, and the temperature as 25 °C.

- *Method.* Use eqn (24.1.19) for M_m, with $S = 5.04 \times 10^{-13} \, s$. In order to find the axial ratio, refer to Table 24.1. First find f from eqn (24.1.18), and f_0 from the assumption that the molecule is a sphere of radius c. Obtain c from v_s via $v_{mol} = (4/3)\pi c^3$. Then $f_0 = 6\pi\eta c$. Finally find a value of b/a from Table 24.1 which gives the observed value of f/f_0.

- *Answer.*

$$M_m = \frac{(5.04 \times 10^{-13} \, s) \times (8.314 \, J \, K^{-1} \, mol) \times (298.15 \, K)}{(1 - 1.0024 \times 0.734) \times (6.97 \times 10^{-11} \, m^2 \, s^{-1})}$$

$\quad = 67.9 \, kg \, mol^{-1} = 67\,900 \, g \, mol^{-1}.$

Hence $M_r = 67\,900.$

$f = kT/D = (1.381 \times 10^{-23} \, J \, K^{-1}) \times (298.15 \, K)/(6.97 \times 10^{-11} \, m^2 \, s^{-1})$

$\quad = 5.91 \times 10^{-11} \, kg \, s^{-1}.$

$v_{mol} = (0.734 \, cm^3 \, g^{-1}) \times (67\,900 \, g \, mol^{-1})/(6.022 \times 10^{23} \, mol^{-1})$

$\quad = 8.28 \times 10^{-26} \, m^3.$

$$c = [(3/4\pi)v_{mol}]^{1/3} = 2.70 \times 10^{-9}\,\text{m}$$
$$f_0 = 6\pi\eta c = 6\pi \times (0.890 \times 10^{-3}\,\text{kg m}^{-1}\text{s}^{-1}) \times (2.70 \times 10^{-9}\,\text{m})$$
$$= 4.54 \times 10^{-11}\,\text{kg s}^{-1}.$$

Therefore, $f/f_0 = (5.91 \times 10^{-11}\,\text{kg s}^{-1})/(4.54 \times 10^{-11}\,\text{kg s}^{-1}) = 1.30$. Reference to Table 24.1 shows that this indicates an axial ratio of slightly less than 6.

● *Comment.* The ellipsoid is like a cigar, six times longer than it is broad. More accurate calculations (based on extrapolations to zero concentration) give the axial ratio as about 4.4.

Sedimentation equilibria. The difficulty with using sedimentation rates to get a molar mass lies in the inaccuracies inherent in the determination of diffusion coefficients: the boundary is blurred by convection currents. The problem of needing to know D can itself be avoided by allowing the solution to settle into thermal equilibrium. Since the number of solute molecules with any given potential energy E is proportional to $\exp(-E/kT)$, the ratio of concentrations at two different heights (or radii in a centrifuge) can be used to find their masses. All we need to know is that the potential energy of a molecule of mass m_{eff} when it is being swung round at a radius r with an angular velocity ω is $\frac{1}{2}m_{eff}r^2\omega^2$. The ratio of concentrations at radii r_1 and r_2 is therefore

$$c_P(r_1)/c_P(r_2) = N(r_1)/N(r_2) = \exp\{-E(r_1)/kT\}/\exp\{-E(r_2)/kT\}$$

$$= \exp\{-\tfrac{1}{2}mb\omega^2(r_1^2 - r_2^2)/kT\},$$

(24.1.20)   or $$M_m = \frac{2RT\ln[c_P(r_2)/c_P(r_1)]}{(r_1^2 - r_2^2)b\omega^2}.$$

In order to use this technique, the centrifuge is run more slowly than in the sedimentation rate method because it is no use having all the solute pressed in a thin film against the bottom of the cell.

Electrophoresis. Many macromolecules carry an electric charge, and so they may be induced to move under the influence of an electric field. The phenomenon is termed *electrophoresis*. The solution may be supported on paper, but in *gel electrophoresis* the electrophoretic migration takes place through a cross-linked polyacrylamide gel. The mobility of the molecules depends, like their mobility in sedimentation experiments, on their masses and their shapes. One way of avoiding the problem of knowing neither the hydrodynamic shape of the species nor their total charge is to denature the material in a controlled manner. The detergent sodium dodecylsulphate has been found to be particularly useful in this respect. In the first place it denatures proteins into a rodlike shape by forming a complex with them. Therefore all proteins, whatever their initial form, are denatured into the same shape. Furthermore, most proteins have been found to bind a constant amount of the detergent per unit mass, and so the charge per protein molecule is well regulated. The determination of the molar mass of a protein may now be carried out by comparing its mobility in the rod-like complexed form with standard samples of known molar mass.

Gel filtration. Beads of porous polymeric material about 0.1 mm in diameter can capture molecules selectively, according to their size. Thus if a solution is filtered through a column, the small molecules require a long *elution time* while the larger ones, which are not captured, pass through rapidly. The molar mass of a macromolecule may therefore be determined by observing its elution time for a column calibrated using macromolecules of known molar mass. The range of the molar masses that can be determined can be altered by selecting columns made from polymers with different degrees of cross-linking. Elution times also depend on shape in a complicated way, and the technique is really suitable only for globular macromolecules.

Viscosity. The presence of macromolecules affects the viscosity of the medium, and so its measurement can be expected to give information about size and shape. The effect is large even at low concentrations, because the big molecules affect the surrounding fluid's flow over a long range.

The first step is to define the quantity required. At low concentrations of solute the viscosity of the solution η is expected to be related to the viscosity of the pure solvent by

$$\eta = \eta^* + Ac_P + \ldots$$

where A is some constant and η^* is the viscosity of the pure solvent. This constant is normally written $A = \eta^*[\eta]$, where $[\eta]$, the *intrinsic viscosity*, is the analogue of a virial coefficient. Then

(24.1.21) $\eta = \eta^* + \eta^*[\eta]c_P + \ldots = \eta^*\{1 + [\eta]c_P + \ldots\}.$

It follows that the intrinsic viscosity can be found experimentally by taking the following limit:

(24.1.22) $[\eta] = \lim\limits_{c_P \to 0} \{[(\eta/\eta^*) - 1]/c_P\}.$

$[\eta]$ has the dimensions of inverse concentration (e.g., $dm^3\,g^{-1}$).

Viscosity may be measured in a variety of ways. A common method is based on the *Ostwald viscometer*, which is essentially a capillary joining two reservoirs, Fig. 24.6a. The time taken for the solution to flow through the capillary is measured and compared with a standard sample. The method is particularly suitable for obtaining $[\eta]$ because the ratio of the viscosity of the solution and the pure solvent is proportional to the drainage times t_{drain} (if a correction for different densities ρ and ρ^* is made):

$$\eta/\eta^* = (t_{drain}/t^*_{drain})(\rho/\rho^*).$$

The value of this ratio can then be used directly in eqn (24.1.23). Care has to be taken to ensure that the temperature is not only constant but uniform. Viscometers in the form of rotating concentric cylinders are also employed, Fig. 24.6b. The viscosity is measured by monitoring the torque on the

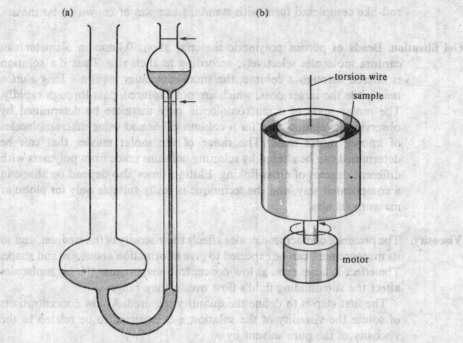

Fig. 24.6. Two types of viscometer: (a) Ostwald, (b) rotating drum.

inner cylinder while the outer one is rotated. Such viscometers have the advantage over the Ostwald type in so far as the shear gradient between the cylinders is much simpler in form than in the capillary, and so non-Newtonian behaviour (of the sort shortly to be described) is easier to study.

There are many complications in interpreting viscosity measurements and much (but not all) of the work is based on empirical observations. The measurement of viscosities of solutions made from calibrated solutes are used for assessing molar masses. For instance, it is found that linear polymers that have rolled up into a random, spherical coil in θ-solvents obey $[\eta] \propto M_r^{\frac{1}{2}}$, and in general $[\eta] = K M_r^a$ where K and a are constants, Table 24.3, that depend on the solvent and the type of macromolecule. Some theoretical justification can be given for the general form of this relation. For solid spheres it turns out that $a = \frac{1}{2}$, while for rigid rods $a = 2$. Therefore, by determining the value of a, information can be obtained on the shape of the species in solution. As an example, solutions of poly(γ-benzyl-L-glutamate) in its rigid, rod-like form has a limiting viscosity number about 4 times greater than when it is denatured and the rods collapse into random coils. Conversely, solutions of natural ribonuclease are less viscous than the denatured form: this suggests that the natural protein has a more compact structure than the denatured one.

Example (Objective 10). The viscosities of a series of solutions of polystyrene in toluene were measured at 25 °C with the following results:

$c_P/(\text{g dm}^{-3})$	0	2.0	4.0	6.0	8.0	10.0
$\eta/10^{-4}\,\text{kg m}^{-1}\text{s}^{-1}$	5.58	6.15	6.74	7.35	7.98	8.64

Find the intrinsic viscosity and estimate the R.M.M. on the basis that the viscosity of the system obeys $[\eta] = KM_r^a$ with $K = 3.80 \times 10^{-5}\,\text{dm}^3\,\text{g}^{-1}$ and $a = 0.63$.

- *Method.* The intrinsic viscosity is the limit of $[(\eta/\eta^*)-1]/c_P$ as c_P tends to zero. Therefore form this ratio and extrapolate to $c_P = 0$.

- *Answer.* Draw up the following table:

$c_P/(\text{g dm}^{-3})$	0	2.0	4.0	6.0	8.0	10.0
η/η^*	1	1.102	1.208	1.317	1.430	1.549
$[(\eta/\eta^*)-1]/(c_P/\text{g dm}^{-3})$	–	0.0511	0.0520	0.0528	0.0538	0.0549

The points are plotted in Fig. 24.7. The extrapolated intercept at $c_P = 0$ is 0.0504, and so $[\eta] = 0.0504\,\text{dm}^3\,\text{g}^{-1}$. The R.M.M. is

$$M_r = ([\eta]/K)^{1/a} = (0.0504\,\text{dm}^3\,\text{g}^{-1}/3.80 \times 10^{-5}\,\text{dm}^3\,\text{g}^{-1})^{1/0.63}$$
$$= 1326^{1.59} = 90450.$$

- *Comment.* This is an average R.M.M. Note that $\ln(\eta/\eta^*) = \ln[1+(\eta-\eta^*)/\eta^*] \approx (\eta-\eta^*)/\eta^* = (\eta/\eta^*)-1$, when $\eta \approx \eta^*$. Therefore the intrinsic viscosity can also be defined as the limit of $(1/c_P)\ln(\eta/\eta^*)$ as $c_P \to 0$. The intercept can be identified more precisely by plotting both functions.

Table 24.3. Intrinsic viscosity and R.M.M.

Macromolecule	Solvent	$t/°C$	$K/10^{-2}\,\text{cm}^3\,\text{g}^{-1}$	a
Polystyrene	Benzene	25	0.95	0.74
	Cyclohexane	34^θ	8.1	0.50
Polyisobutylene	Benzene	24^θ	8.3	0.50
	Cyclohexane	30	2.6	0.70
Amylose	$0.33\,\text{mol dm}^{-3}\,\text{KCl}$	25^θ	11.3	0.50
Various proteins†	Guanidine hydrochloride			
	$+\beta$-mercaptoethanol	—	0.716	0.66

θ: theta temperature.
†: Use $[\eta] = KN^a$, N the number of amino acid residues.
Source: K. E. van Holde, *Physical biochemistry*, Prentice-Hall.

One complication encountered is that in some cases it is found that the viscosity decreases when the rate of flow of the solution is increased. This is an example of *non-Newtonian behaviour*, and it indicates the presence of long rod-like molecules that are becoming orientated by the flow so that they slide past each other more freely. In some cases the stresses set up by the flow are so great that long molecules are ripped apart, with further consequences for the viscosity.

Light scattering. When light falls on an object it drives the latter's electrons into oscillation and they in turn radiate. If the medium is perfectly homogeneous (for instance, a perfect crystal) all the secondary waves interfere destructively except in the original direction of propagation. Therefore an observer will

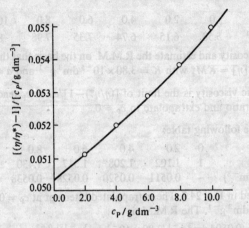

Fig. 24.7. Determination of the intrinsic viscosity.

see the beam only if he is looking exactly along the propagation direction. In contrast, if there are inhomogeneities in the medium, as in an imperfect crystal or in a solution containing foreign bodies (macromolecules in a solvent, smoke in air, and so on) scattered radiation in other directions survives and can be detected. An onlooker, even if he is not looking along the initial propagation direction, may see light. A familiar example is the light scattered by specks of dust in a sunbeam (and in advertisers' photographs of laser beams).

Consider first the features of scattering from particles much smaller than the wavelength (λ) of the incident light. This is called *Rayleigh scattering*. The intensity of the scattering depends on the wavelength as $1/\lambda^4$, so that short wavelengths are scattered more than long. This dependence is familiar, qualitatively at least, from the blue of the daytime sky, which arises from the predominant scattering of blue components in the sun's light by atoms and molecules in the atmosphere. The intensity also depends on the angle of observation, θ, and is proportional to $1 + \cos^2 \theta$ when the incident light is unpolarized. Therefore maximum intensity occurs in the forward ($\theta = 0$) and backward ($\theta = 180°$) directions, Fig. 24.8. In practice it turns out to be easier to make observations in a non-forward direction. The intensity also depends on the strength of the interaction of the light with the molecules: the interaction is large if the

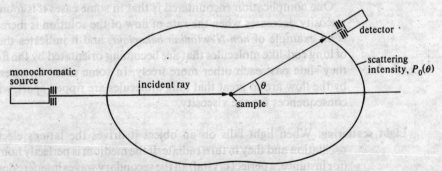

Fig. 24.8. Rayleigh scattering from point-like particles.

polarizability is large. This is a reason why light scattering is so useful in the study of macromolecules: they are large and much more polarizable than the surrounding medium, and so they dominate the scattering.

When all these aspects are combined into a quantitative theory, it turns out that the scattering intensity, $I(\theta)$, at the angle θ is

$(24.1.23)°$ $I(\theta) = AI_0 c_P M_r (1 + \cos^2 \theta),$

where I_0 is the intensity of the incident light, c_P the concentration of the solute, M_r its R.M.M., and A is a constant that depends in a known way on the refractive index of the solution, the wavelength, and the distance of the detector from the sample cell. This is an 'ideal' result in the sense that it ignores the complications that arise from the interaction of the macromolecules, and in an actual experiment it is important to extrapolate to zero concentration.

The obvious application of light scattering of the type described so far is to the determination of the molar mass of the solute. The intensity $I(\theta)$ is measured at a series of concentrations, and then M_r is obtained from the limiting value of $I(\theta)/c_P$. In the case of a polydisperse sample this procedure gives an average molar mass. It turns out that the average is not the number average encountered in osmotic measurements, but a different average, the *mass average R.M.M.*, $\langle M_r \rangle_M$. This is defined in terms of the masses of the individual molecules weighted according to their masses (as distinct from their numbers). Suppose that the total mass of the solute is M and that the mass of molecules with R.M.M. M_{ri} is M_i, then the mass average R.M.M. of the sample is

$(24.1.24)$ $\langle M_r \rangle_M = \sum_i (M_{ri}/M) M_{ri} = (1/M) \sum_i M_i M_{ri}.$

The reason why light scattering delivers the mass-weighted R.M.M. is that the intensity of scattering is greater for larger particles, and so the average is weighted more heavily in their favour.

Example (Objective 5). A sample of polymer consists of two components present in equal masses, one having $M_r = 30\,000$ and the other $M_r = 12\,000$. What are the values of the mass average and number average R.M.M.s?

- *Method.* Use eqn (24.1.8) for the number average and eqn (24.1.24) for the mass average. We have $M = M_1 + M_2$, $M_1 = M_2$. The proportions by mass are $M_1/M = \frac{1}{2}$, $M_2/M = \frac{1}{2}$; the proportions by number are

$$N_i/N = (M_i/M_{ri})/\{(M_1/M_{r1}) + (M_2/M_{r2})\}$$

or $N_1/N = M_{r2}/(M_{r1} + M_{r2})$

$$N_2/N = M_{r1}/(M_{r1} + M_{r2})$$

because $M_1 = M_2$.

- *Answer.* From eqn (24.1.8),

$$\langle M_r \rangle_N = \sum_i (N_i/N) M_{ri} = 2 M_{r2} M_{r1}/(M_{r1} + M_{r2})$$
$$= 2 \times 12\,000 \times 30\,000/(12\,000 + 30\,000) = 17\,143.$$

From eqn (24.1.24),

$$\langle M_r \rangle_M = \sum_i (M_i/M) M_{ri} = \tfrac{1}{2}(M_{r1} + M_{r2})$$

$$= \tfrac{1}{2}(12\,000 + 30\,000) = 21\,000.$$

• *Comment.* Note that the two averages give widely different results, the ratio being about 1.2. Since $N_i = M_i L/M_{ri}$ the mass average R.M.M. may also be written as $\langle M_r \rangle_M = (1/\sum_i N_i M_{ri}) \sum_i N_i M_{ri}^2$, and so is a (number) *mean square* R.M.M. Sedimentation experiments give yet another R.M.M., the *Z-average* R.M.M., $\langle M_r \rangle_Z = (1/\sum_i N_i M_{ri}) \sum_i N_i M_{ri}^3$, a *mean cubic* R.M.M.

The example indicates that while at first sight it might appear troublesome to have two types of average, the observation that $\langle M_r \rangle_N$ and $\langle M_r \rangle_M$ differ gives additional information about the range of molar masses in the sample. In the determinations of protein molar masses we expect the two averages to be the same because the sample is monodisperse (unless there has been degradation). In synthetic polymers there is normally a range of molar masses, and so the two averages are expected to be different, the range of the contributions being indicated by the magnitude of the difference. Typical synthetic materials have $\langle M_r \rangle_M/\langle M_r \rangle_M$ close to 3; the term 'monodisperse' is conventionally applied to synthetic polymers in which the ratio is smaller than about 1.1.

Since the solute is responsible for scattering light away from the forward direction of the beam it follows that the transmitted intensity is reduced. The transmitted intensity is given by a type of Lambert–Beer law (p. 605): if the initial intensity is I_0, the intensity that survives after the beam has travelled through a length l of solution is

(24.1.25)　　　$I_l = I_0 \exp(-\tau l),$

where τ is the *turbidity*. The turbidity increases with the scattering power of the solution, and its explicit form can be obtained from eqn (24.1.23) by evaluating the total amount of light scattered away from the initial direction. This implies that $\tau \propto c_P M_r$, and so the turbidity depends on the molar mass of the solute as well as its concentration. Typical values of τ are 10^{-5}cm^{-1} for pure transparent liquids (so that the sample would have to be 1 km long before the intensity had dropped to $1/e$ of its initial value by virtue of scattering alone: at such low turbidities absorption would normally dominate the reduction of intensity), 10^{-3}cm^{-1} for polymers at 1 per cent concentration, and 10cm^{-1} for milk (which, loosely interpreted, means that you can see about 1 mm into a glass of milk).

We now turn to a consideration of the information that can be obtained from light-scattering experiments when the wavelength is not greatly different from the size of the molecules responsible for the scattering. When the size of the particle is comparable to the wavelength, scattering occurs from different components, Fig. 24.9. This effect is also a common feature of the everyday world. It accounts, for instance, for the whiteness of clouds. Like the sky we see clouds by the light that scatters from them, but unlike

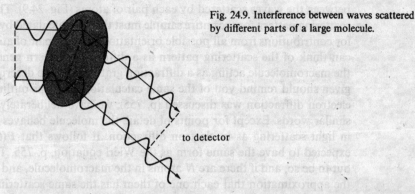

Fig. 24.9. Interference between waves scattered by different parts of a large molecule.

to detector

the sky they do not appear blue. The reason is that the water molecules group together into droplets of a size comparable to and greater than the wavelength of light, and scatter cooperatively. Although blue light scatters more strongly, more molecules can contribute cooperatively when the wavelength is longer (as in red light), and so white light scatters as white light from clouds. This paper is white for the same reason. Cigarette smoke is blue before it is inhaled, but brownish after it is exhaled because the particles aggregate in the lungs.

A major feature of larger particle size is that the intensity pattern is distorted from the $1 + \cos^2 \theta$ form characteristic of small particle, Rayleigh scattering, Fig. 24.10. An analysis of the angular dependence of the scattered intensity should then reveal information about the shape of the macromolecule in the solution. The analysis concentrates on the quantity $P(\theta)$, the ratio of the observed intensity at the angle θ to the intensity that would be scattered on the basis of the assumption that the entire structure of the molecule is concentrated into an infinitesimal region (so that it behaves as a Rayleigh scatterer):

(24.1.26) $$P(\theta) = I_{obs}(\theta)/I_{Rayleigh}(\theta)$$

If we think of the molecule as being formed from a collection of atoms at distances R_i from some convenient origin, then interference may occur

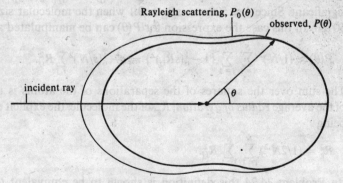

Rayleigh scattering, $P_0(\theta)$

observed, $P(\theta)$

incident ray

θ

Fig. 24.10. Distortion of the scattered light intensity distribution when the wavelength is comparable to the molecular diameter.

between the waves scattered by each pair of atoms (Fig. 24.9). The overall scattering pattern for the entire sample must then be calculated by allowing for contributions from all possible orientations of each pair of atoms. You can think of the scattering pattern as a diffraction pattern generated by the macromolecule acting as a diffraction grating, and the description just given should remind you of the same calculation that was outlined when electron diffraction was discussed (p. 755); we have deliberately adopted similar words. Except for points of detail, the molecule behaves the same in light scattering as in electron diffraction. It follows that $P(\theta)$ can be expected to have the same form as the Wierl equation, p. 755. This turns out to be so, and if there are N atoms in the macromolecule, and we make the approximation that each one of them has the same scattering power, then

(24.1.27) $$P(\theta) = (1/N)^2 \sum_{i=1}^{N} \sum_{j=1}^{N} \sin(sR_{ij})/sR_{ij}, \qquad s = (4\pi/\lambda)\sin\tfrac{1}{2}\theta.$$

(Compare eqn (22.6.1).) R_{ij} is the distance between atoms i and j, and λ is the wavelength of the light. The observed scattering is simply $I_{\text{Rayleigh}}(\theta)P(\theta)$, with I_{Rayleigh} given by eqn (24.1.23).

There are two simple developments of the last equation. In the first place, suppose that the dimensions of the molecule are much smaller than the wavelength of the light, then the product sR_{ij} will be much smaller than unity (because s is proportional to $1/\lambda$). The function $\sin(sR_{ij})$ expands as $sR_{ij} + \ldots$, and only the first term need be retained in the present approximation. Therefore

$$P(\theta) \approx (1/N)^2 \sum_{i=1}^{N} \sum_{j=1}^{N} (sR_{ij})/sR_{ij} = (1/N)^2 \sum_{i=1}^{N} \sum_{j=1}^{N} 1 = 1$$

In other words, Rayleigh scattering results when the molecule has dimensions much smaller than the wavelength of the light ($I_{\text{obs}}(\theta) = I_{\text{Rayleigh}}(\theta)$ when $P(\theta) = 1$).

Now consider the form of the scattering when the size of the molecule is still much less than the wavelength of the light but no longer vanishingly small. That is, sR_{ij} is still much smaller than unity but large enough for the second term in the expansion of $\sin(sR_{ij}) = sR_{ij} - \tfrac{1}{6}(sR_{ij})^3 + \ldots$ to be significant. Since $\lambda \approx 500\,\text{nm}$, $sR_{ij} \approx 0.1$ when the molecular size is about $50\,\text{nm}$. In this case the expression for $P(\theta)$ can be manipulated as follows:

$$P(\theta) \approx (1/N)^2 \sum_{i=1}^{N} \sum_{j=1}^{N} \{1 - \tfrac{1}{6}(sR_{ij})^2\} = 1 - \tfrac{1}{6}(s/N)^2 \sum_{i,j} R_{ij}^2.$$

The sum over the squares of the separations of the atoms is the square of the average *radius of gyration*, R_g, of the molecule, the explicit expression being

(24.1.28) $$R_g^2 = (1/2N^2) \sum_{i=1}^{N} \sum_{j=1}^{N} R_{ij}^2.$$

(In Problem 24.24 this definition is shown to be equivalent to another and more easily visualized one in the case of a chain of atoms or groups of identical mass: the radius of gyration is the *average root mean square distance* of the atoms from the centre of mass.) Then

(24.1.29) $P(\theta) \approx 1 - \frac{1}{3}s^2 R_g^2 = 1 - (16\pi^2/3\lambda^2)R_g^2 \sin^2 \frac{1}{2}\theta.$

The last equation shows than an analysis of the deviation of the scattering intensity from the Rayleigh form, $1 + \cos^2 \theta$, leads to the value of the radius of gyration of the molecule in the solution. That in turn can be interpreted in terms of the dimensions of the molecule if its shape is known. For instance, if the molecule is known to be spherical its radius of gyration is related to its radius R by $R_g = (3/5)^{1/2}R$, while if it is a long thin rod its length L is related to R_g by $R_g = L/2\sqrt{3}$. Once again, it has to be emphasized that the analysis must be performed on data obtained by extrapolating scattering intensities to zero concentration. The fact that both θ and concentration are variables leads to a special extrapolation procedure known as a *Zimm plot*. This is described in the references in the Bibliography. Some reported values are listed in Table 24.4.

Table 24.4. Radius of gyration of some macromolecules

	R.M.M.	R_g/nm
Serum albumin	66 000	2.98
Myosin	493 000	46.8
Polystyrene	3.2×10^6	49.4 (in poor solvent)
DNA	4×10^6	117.0
Tobacco mosaic virus	39×10^6	92.4

Source: C. Tanford, *Physical chemistry of macromolecules*, Wiley.

An example of the distortion of the intensity that size introduces is shown in Fig. 24.10. This has been calculated on the basis that the radius of gyration of the macromolecule is 30 nm and the wavelength is 500 nm. The form of the intensity distribution is obtained by calculating $P(\theta)(1 + \cos^2 \theta)$.

The development of the laser has led to further refinements in the application and interpretation of light scattering from solutions of macromolecules. A particularly significant shift of emphasis is towards the investigation of the time-dependence of the positions and orientation of macromolecules in solution. Their time behaviour can be studied by monitoring the frequency shifts that occur when initially monochromatic light is scattered by moving molecules, the general technique being called *dynamic light scattering*. In particular laser light scattering can be used in direct determinations of the diffusion characteristics of macromolecules, and it provides a direct and quick method for the measurement of diffusion coefficients.

Magnetic resonance. In recent years both n.m.r. and e.s.r. (Chapter 19) have been applied with success to the elucidation of the structure of biologically important macromolecules in solution. It is now possible to determine not only the general shape of the molecules but also the positions of their atoms and aspects of their motion. Since biological molecules operate in variable environments (for instance, under conditions where the pH of the surrounding medium may change) it becomes possible, through the use of magnetic resonance, to examine how their shapes respond.

The basic data from magnetic resonance spectra are line positions and line shapes. The positions of the spectral lines in an n.m.r. spectrum depend on the local magnetic fields at the protons or ^{13}C nuclei (and other magnetic nuclei) being examined. These fields are modified if there are extra sources of magnetic fields, such as may occur if a paramagnetic transition metal ion is deliberately incorporated into the molecule. The resonant nuclei experience a magnetic field additional to that applied externally and, as its strength depends on distance as $1/R^3$, the distances of the nuclei from the ion may be determined. One complication is that in a rotating molecule the shift gives the distance and not the direction, but the difficulty can be resolved by substituting the ion into different places and constructing a three-dimensional array of positions from separate experiments.

The *shift reagents* that are used may be of several kinds. The technique was developed using lanthanide ions, but it is not even necessary to use paramagnetic species, because the externally applied field can induce magnetic moments in sufficiently susceptible aromatic groups attached to the molecule, and these then act as sources of the magnetic field. One difficulty is the problem of knowing the site of the ion or group acting as the source of the shift field; but this can be overcome by having an initial idea of the structure (such as from solid-state X-ray studies). Another is the problem of knowing whether the substitution has distorted either the geometrical or the electronic structure.

Width is the other observable feature of a magnetic resonance line. Width is related to relaxation time (p. 570), and modern techniques measure the latter directly. As explained in Chapter 19 (p. 643), the relaxation time depends upon both the strength of the magnetic field causing the relaxation and on the rapidity with which the molecule is moving in the solution. The relaxing efficiency of a magnetic dipole attached to the molecule depends upon distance as $1/R^6$, and so by studying the relaxation times of resonant nuclei in the presence of a paramagnetic substituents their separation can be inferred. As in the case of shift probes, if this is done for a variety of substitution sites the network of separations can be interpreted as a three-dimensional map of nuclear positions. If two closely related lanthanides are taken, one of which is a shift probe (e.g., Eu(III)) and the other a broadening probe (e.g., Gd(III)), and which can reasonably be supposed to bind to the same sites, then since the former explores $1/R^3$ and the latter $1/R^6$, a detailed map of magnetic nuclei can be established.

The paramagnetic centre itself, rather than its influence on the surrounding nuclei, can be examined by e.s.r. The electron spin relaxation times, and hence the spectral linewidths, depend on the rotation rate of the molecule. Since this is not free rotation in solution it is normally referred to as molecular *tumbling*. The size of the macromolecule can therefore be inferred from the linewidth of its e.s.r. spectrum. If it is supposed that the molecule rotates like a sphere of radius a in a medium of viscosity η, then the time it requires to tumble through about 1 rad (57°) is given by the expression

$$\tau_{rot} = 4\pi a^3 \eta / 3kT.$$

For instance, a sphere of radius 2 nm in water at room temperature has $\tau_{rot} \approx 8$ ns. The e.s.r. technique is sensitive to processes on this time-scale.

It is also possible to determine the overall shape of a molecule bearing a paramagnetic centre—when the centre is a nitroxide, —R(NO)R′, these are called *spin-labelled molecules*—because the molecule may be able to rotate more rapidly about one axis than others. This anisotropic rotational behaviour can be detected from the shape of the e.s.r. spectrum of a spin-labelled species, and interpreted in terms of the geometrical shape of the species. The same technique is used to analyse the mobility of groups within the macromolecule, for some groups can wave around loosely while others are trapped in rigid conformations as a result of steric interactions. The use of spin labels therefore lets one investigate not only the size and shape of the entire molecule but also the flexibility of its structure.

24.2 Conformation and configuration

In this section we examine some of the factors that influence the shapes adopted by macromolecules. A convenient division of the discussion is into primary structure, secondary structure, and so on.

By *primary structure* is meant the sequence of chemical segments making up the chain. In the case of a synthetic polymer virtually all the segments are identical, and it is sufficient to name the monomer involved in the synthesis. Thus the primary structure of polyethylene (poly(ethene)) is the —CH_2CH_2— unit, and the chemical constitution of the chain is specified by denoting it as —$(CH_2CH_2)_n$—. Substituted polyethylenes have a primary structure of the form —$(CH_2CHR)_n$—; polystyrene has R = phenyl. The concept of primary structure ceases to be trivial in the case of biological macromolecules, for these are often chains of different molecules. Proteins are *polypeptides*, the name signifying chains formed from numbers of different amino acids (about twenty occur naturally) strung together by means of the *peptide link*, —CO—NH—. The determination of the primary structure is then a highly complex problem of chemical analysis. Fortunately, proteins are *linear* chains of amino acids, and so the *sequencing problem* (the determination of the order of the amino acids in the chain) is much less complicated than would have been the case if two or three dimensional nets had been involved. The linear nature is related to the fact that the biosynthesis of natural molecules is governed by DNA, which is itself a linearly sequenced molecule.

The *secondary structure* of macromolecules refers to the (often local) spatial, well characterized arrangement of the basic structural units. The secondary structure of an isolated molecule of polyethylene is a random coil, while that of a protein is a highly organized arrangement determined largely by hydrogen bonds and takes the form of helices or sheets in particular segments of the molecule. We examine both structures below. When the protein hydrogen bonds are destroyed (for instance by heat, as when you cook an egg) the structure collapses (*denatures*) into a random coil.

The difference between primary and secondary structures is closely related to the difference between the configuration and the conformation of a chain. The *configuration* refers to the aspects of structure that can be changed only by breaking bonds and re-forming new ones. Thus the chain —A—B—C— has a configuration different from —A—C—B—. The *conformation* of a chain refers to the spatial arrangement of its different parts, and so one conformation of a chain can be transformed into another simply by rotating one part of the chain round the bond joining it to another. The denaturation of a protein is a conformational, not a configurational, change. The secondary structure of a polymer like polyethylene is mobile because one conformation can change into another as a result of thermal motions in the chain itself or as a result of interactions with the surroundings, such as the solvent it might be dissolved in. In contrast, the conformation adopted by a protein is stabilized by hydrogen bonds, and the secondary structure is a vital aspect of its biological function.

The *tertiary structure* of a protein refers to the overall three-dimensional structure in the molecule. For instance, many proteins have a helix as their secondary structure, but in many this helix is bent in various places, and the result is a globular protein. The globular, folded helix arrangement is then the protein's tertiary structure.

The term *quaternary structure* refers to the way in which some molecules are formed by the combination of simpler chains into a composite structure. Haemoglobin is a famous example: it consists of four subunits of two types, they are referred to as the α and β chains, and are closely related to myoglobin.

Coils, helices, and sheets. As the first step in unravelling the various aspects of a macromolecule's structure, consider the most likely conformation of a chain of identical units which are incapable of forming hydrogen bonds or any other type of specific bond. The basic example is polyethylene, but the ideas and conclusions also apply to denatured proteins. We shall aim at assessing the size of the randomly coiled structure because this is a feature that can be determined experimentally and is what determines properties such as mobility.

The simplest model of a randomly coiled polymer chain is the *freely-jointed chain*, where any bond is free to make any angle with respect to the preceding one, Fig. 24.11a. This is obviously a great simplification because a bond is actually constrained to a cone of angles around the direction of the preceding one, Fig. 24.11b. Nevertheless, it turns out that the model can be adapted to the real case.

The main point to notice is that the model is the exact analogue of the

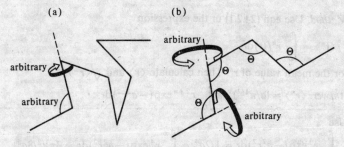

Fig. 24.11. (a) Freely jointed chain; (b) chain restricted to a fixed valence angle Θ but arbitrary azimuth.

classic problem of the *three-dimensional random walk*, each bond representing a step taken in a random direction. The mathematics of the random walk will be dealt with in Chapter 26 (p. 909): for our present purposes all we need is the central result that the probability of the ends of the chain (the net distance travelled by the random walker) lying in the range r to $r + dr$ is $f(r)dr$, where

(24.2.1) $\qquad f(r) = (a/\pi^{1/2})^3 4\pi r^2 \exp(-a^2 r^2), \qquad a^2 = 3/2Nl^2$

where N is the number of bonds (the number of paces) and l the bond length (the length of each pace).

It is clear from the last equation that in some coils the ends may be separated by great distances, while in others the separation will be small: the expression gives the proportion of each kind. An alternative interpretation is to regard each coil as writhing continually from one conformation to another, then $f(r)dr$ is the probability that at any moment it will be found with its ends lying between the separations r and $r + dr$.

We need some measure of the average separation of the ends of the random coil. The *root mean square separation*, R_{rms}, of the ends of the chain is one such measure: it may be calculated by averaging the separation R over its possible values weighted according to the probability expressions eqn (24.2.1):

(24.2.2) $\qquad R_{rms} = \left\{ \int_0^{\infty} R^2 f(R)\, dR \right\}^{1/2} = N^{1/2} l.$

Another measure of the bulk of the molecule is its *radius of gyration*, defined on p. 832. This too can be calculated on the basis of eqn (24.2.1) with the result that

(24.2.3) $\qquad R_g = N^{1/2} l/\sqrt{6}.$

The essential feature of these results is the dependence on N: as the number of monomer units is increased, the size of the random coil increases as $N^{1/2}$ (and its volume as $N^{3/2}$).

Example (Objective 14). Calculate the mean separation and the root mean square separation of the ends of a freely-jointed polymer chain of N bonds of length l.

- *Method.* Use eqn (24.2.1) in the expression

$$\langle r^n \rangle = \int_0^\infty r^n f(r) \, dr$$

for the mean value of r^n. Then calculate $\langle r \rangle$ and $\sqrt{\langle r^2 \rangle}$.

- *Answer.* $\langle r^n \rangle = (a/\pi^{1/2})^3 4\pi \int_0^\infty r^{2+n} \exp(-a^2 r^2) \, dr$.

Use

$$\int_0^\infty r^3 \exp(-a^2 r^2) \, dr = 1/2a^4; \qquad \int_0^\infty r^4 \exp(-a^2 r^2) \, dr = 3\pi^{1/2}/8a^5$$

to find

$$\langle r \rangle = 2/a\pi^{1/2} = (8/3\pi)^{1/2} N^{1/2} l$$
$$\langle r^2 \rangle = 3/2a^2 = Nl^2, \quad \text{or} \quad \sqrt{\langle r^2 \rangle} = N^{1/2} l.$$

- *Comment.* When the chain is not freely jointed these results have to be multiplied by a factor: see below.

Before making use of these conclusions we have to remove the obvious absurdity of allowing bonds to make all angles to each other. This turns out to be very simple in the case of long chains, for it is possible to take groups of several neighbouring bonds and to think about the direction of their resultant. Although the individual bonds are constrained to a single cone of angle Θ, the resultant of several lies in a random direction. By concentrating on groups rather than individuals it turns out that the final result is the same as in eqns (24.2.2) and (24.2.3), but there is an additional factor of $\{(1 - \cos\Theta)/(1 + \cos\Theta)\}^{1/2}$ multiplying each. In the case of tetrahedral bonds, where the angle $\Theta = 109.5°$, this factor is $2^{1/2}$. For such bonds it follows that

(24.2.4) $$R_{rms} = 2^{1/2} N^{1/2} l, \qquad R_g = N^{1/2} l/3^{1/2}.$$

An idea of the magnitudes involved can be obtained by taking a polyethylene chain with $M_r = 56\,000$, corresponding to $N = 4000$. Since $l = 154 \, pm$ for a C—C bond, we find $R_{rms} = 4.4 \, nm$ and $R_g = 1.78 \, nm$. Such a molecule is sketched in Fig. 24.12.

The model of a randomly coiled molecule is still an approximation even after the problem about the angle restriction has been resolved. This is because it does not take into account the impossibility of two atoms occupying the same place. Such self-avoidance tends to swell the coil, and so it is better to regard R_{rms} and R_g as lower bounds to the actual values. Furthermore, the model totally ignores the role of the solvent. A poor solvent will tend to cause the coil to tighten so that solvent–macromolecule contacts are minimized, while a good solvent does the opposite.

The random coil is the least structured conformation of a polymer chain, and so it corresponds to the state of maximum *conformational entropy*. Any stretching of the coil introduces order and so corresponds to reducing the entropy; conversely, the formation of a random coil from a more extended form is a natural, spontaneous process (so long as enthalpy contributions do not interfere with it). The elasticity of a perfect rubber

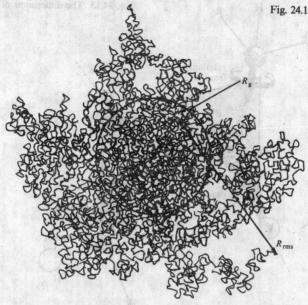

Fig. 24.12. A random coil.

(one in which the internal energy is independent of the extension) may be discussed in these terms, and the calculation is described in the Appendix on p. 851. The random coil model is also a helpful starting point for discussing the order of magnitude of the hydrodynamic properties of polymers and denatured macromolecules in solution.

The other extreme type of structure to consider is the natural conformation of macromolecules like proteins and DNA. Natural macromolecules need a precisely maintained conformation, for otherwise they would be unable to function. This is the major remaining problem in protein synthesis, for although the primary structure can be achieved, the product is inactive because the secondary structure still remains elusive.

The fundamental features of the secondary structures of proteins (we shall confine attention to these) are summarized in the set of rules formulated by Pauling and Corey. The essential feature is the stabilization of the structure by hydrogen bonds arising from the peptide link. The latter can act both as a donor of the H atom (the NH part of the link) and as an acceptor (the CO part). The rules are as follows:

(a) The atoms in the peptide link lie in the same plane, Fig. 24.13.

(b) The N, H, and O atoms of a hydrogen bond lie in a straight line, (with displacements of the H atom tolerated up to not more than $30°$ from the N–O vector).

(c) Every NH group and every CO group is engaged in bonding.

The implication of the rules is the existence of two possible conformations. One, where the hydrogen bonds form between peptide links of the same

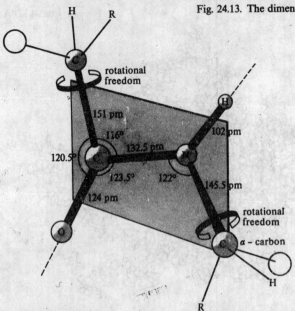

Fig. 24.13. The dimensions of the peptide link.

chain, is the *α-helix*. The other, where the hydrogen bonds link different polypeptide chains, is the *β-pleated sheet*; it is the secondary structure of the protein fibroin, the constituent of silk.

The α-helix is illustrated in Fig. 24.14. Each turn of the helix contains 3.6 amino acid residues, and so the period of the helix corresponds to 5 turns or 18 residues. The pitch of a single turn is 544 pm (5.44 Å). The N—H $\cdots$ O bonds lie parallel to the axis and link every fourth group (so that residue i is tied to residues $i-4$ and $i+4$). There is freedom for the helix to be arranged as either a right- or a left-handed screw, but the overwhelming majority of natural polypeptides are right-handed on account of the preponderance of the L-configuration of the naturally occurring amino acids and the thermodynamic stability of the right-handed helix they are able to form. The reason for the preponderance of L-amino acids in nature is uncertain.

Helical polypeptide chains are folded into a tertiary structure if there are other bonding influences between the residues of the chain that are strong enough to overcome the hydrogen bonds responsible for the secondary structure. The folding influences include disulphide links (—S—S—), ionic interactions (which depend on the pH of the environment), and stronger hydrogen bonds (such as O—H $\cdots$ O⁻). This is illustrated by the structure of myoglobin (which featured on p. 279 in the discussion of respiration). The full structure of the molecule has been determined by X-ray diffraction, and all 2600 atoms have been located. The folding as a result of the disulphide links between different residues of the fundamental α-helix can easily be discerned, Fig. 24.15; about 77 per cent of the structure is α-helix, the rest is involved in the folds.

Proteins with M_r exceeding about 50 000 are often found to be aggregates

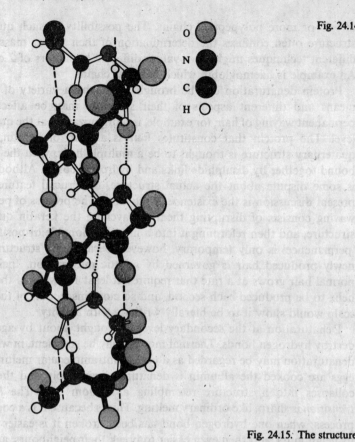

Fig. 24.14. The α-helix.

O
N
C
H

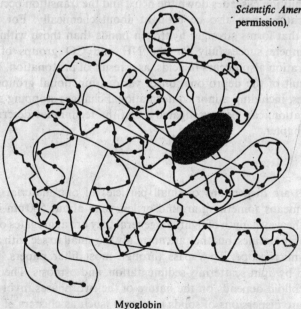

Fig. 24.15. The structure of myoglobin (based on M. F. Perutz, copyright *Scientific American*, 1964, with permission).

Myoglobin

of two or more polypeptide chains. The possibility of such quaternary structure often confuses the determination of their molar masses (since different techniques might give values differing by factors of 2 or more). An example is haemoglobin which has four chains.

Protein denaturation can be brought about by a variety of different means and different aspects of their structure may be affected. The permanent waving of hair, for example, is reorganization at the quaternary level. The protein that constitutes hair is a form of keratin, and its quaternary structure is thought to be a multiple helix, with the α-helices bound together by disulphide links and hydrogen bonds. Although there is some dispute about the actual structure, the crucial feature for the present discussion is the existence of the links. The process of permanent waving consists of disrupting them, unravelling the keratin quaternary structure, and then reforming it into a more fashionable disposition. The 'permanence' is only temporary, however, because the structure of the newly produced hair is governed by genetic information. Incidentally, normal hair grows at a rate that requires at least 10 twists of the keratin helix to be produced each second, and so close inspection of the human scalp would show it to be literally writhing with activity.

Denaturation at the secondary level is brought about by agents that destroy hydrogen bonds. Thermal motion may be sufficient, in which case denaturation may be regarded as a type of intramolecular melting. When eggs are cooked the albumin is denatured irreversibly and the protein collapses into a structure resembling a random coil. The *helix–coil transition* is sharp, like ordinary melting. This is because it is a cooperative process: when one hydrogen bond has been broken it is easier to break its neighbours, and then even easier to break their neighbours, and so on. The disruption cascades down the helix, and the transition occurs sharply. Denaturation may also be brought about chemically. For instance, a solvent that forms stronger hydrogen bonds than those within the helix will compete successfully for the NH and CO groups of the links. Denaturation also occurs in acids, as a result of protonation, or in bases, as a result of the deprotonation of various functional groups. All these processes, including minor conformational changes stopping short of full denaturation, can be investigated using the techniques described earlier in the chapter.

24.3 Colloids

Colloids are dispersions of small particles of one material in another. 'Small' means something around or less than about 500 nm in diameter (about the wavelength of light). In general they are aggregates of numerous atoms or molecules, but they are normally too small to see with an ordinary optical microscope. They pass through most filter papers, but can be detected by light scattering, sedimentation, and osmosis. The name given to the colloid depends on the nature of the two phases involved:

Sols are dispersions of solids in liquids (such as clusters of gold atoms in water), or of solids in solids (such as ruby glass, which is a gold in glass sol, and achieves its colour by scattering).

Aerosols are dispersions of liquids in gases (like fog and many sprays) and of solids in gases (such as smoke). The particles are often large enough to be seen under a microscope.

Emulsions are dispersions of liquids in liquids (such as milk). Sometimes *foams*, which are dispersals of gases in liquids (as in beer) or of gases in solids (such as pummice) are also included in the classification, but their inclusion is tidy rather than helpful. Aspects of foams were treated in Chapter 8.

A secondary classification of colloids is into *lyophilic* (solvent attracting) and *lyophobic* (solvent repelling) varieties. When the solvent is water the appropriate terms are *hydrophilic* and *hydrophobic* respectively. Lyophobic colloids include the metal sols; lyophilic colloids generally have some chemical identity with the solvent, such as hydroxyl groups, and so on, which are able to form hydrogen bonds. A *gel* is a semi-rigid mass of a lyophilic sol in which all the dispersion medium has been absorbed by the sol particles.

Preparation and purification. Preparation can be as simple as sneezing (which produces an aerosol). Laboratory and commercial methods make use of a variety of techniques. Material may be ground in the presence of the dispersion medium (for instance, colloidal dispersions of quartz may be produced in this way). Electrical methods are available. For instance, passing a heavy current through an electrolytic cell may lead to the crumbling of the electrode into colloidal particles, and arcing between metal electrodes beneath the surface of the support medium can do so too. Alkali metal sols in organic solvents may be prepared in this way. Chemical precipitation sometimes leads to the formation of a colloidal precipitate. Likewise, a precipitate may be dispersed by the addition of a third substance, a *peptizing agent*. Silver iodide may be dispersed into a colloidal suspension by peptization with either potassium iodide or silver nitrate. Clays may be peptized by alkalis, the OH^- ion being the active agent.

Emulsions are normally prepared by shaking the two components together, although some kind of emulsifying agent has to be used in order to stabilize the product. The emulsifier is either a soap (a long chain fatty acid), a detergent, or a lyophilic sol able to form a protective film around the dispersed phase. In the case of milk, which is an emulsion of fats in water, the emulsifying agent is casein, a protein containing phosphate groups. That casein is not wholly successful in stabilizing milk is apparent from the formation of cream on the surface: the dispersed fats coalesce into oily droplets which rise to the surface. This may be prevented by ensuring that the emulsion is dispersed very finely in the first place: violent agitation with ultrasonics brings this about, and the product is known as homogenized milk.

Aerosols are formed when a spray of liquid is torn apart under the influence of a jet of gas. The dispersal is aided if a charge is applied to the liquid, for then electrostatic repulsions blast the jet apart into tiny droplets. This procedure may also be used to produce emulsions, for the charged liquid phase may be squirted into another liquid, where it is shattered electrostatically.

Colloids are often purified by dialysis. The aim is to remove much (but not all, for reasons explained later) of the ionic material that may have accompanied their formation. As in the discussion of the Donnan effect, a membrane is selected which is permeable to solvent and ions but not to the bulky colloidal particles. Cellulose is often used. Dialysis is very slow, and so it is normally accelerated by making use of the electric charge carried by many colloids, the technique is then known as *electrodialysis*.

Structure, surface, and stability. The crucial feature of colloids is the very great surface area of the dispersed phase in comparison with the same amount of ordinary material. For instance, a 1 cm cube of material has a surface area of $6 \, cm^2$; but when it is dispersed as little cubes of side 10 nm the total surface area of the resulting 10^{18} smaller cubes is $6 \times 10^6 \, cm^2$. This dramatic increase in area means that surface effects are of dominating importance in colloid chemistry.

The first point to note is that colloids are thermodynamically unstable with respect to the bulk phase. This stems from the presence of the enlarged surface, for we saw in Chapter 8 that surface tension favours small surface areas ($dG = \gamma \, d\sigma$, which is negative for decreasing surface area, $d\sigma < 0$). The apparent stability must therefore be a consequence of the kinetics of collapse; colloids are kinetically but not thermodynamically stable.

At first sight even the kinetic argument seems to fail. This is because colloidal particles attract each other over great distances, and so there is a long-range force tending to collapse them down into a single blob. The reasoning behind this remark is as follows. The energy of interaction between two individual atoms, one in each colloid particle, varies with their separation as $1/R_{ij}^6$ (this is the van der Waals dispersion energy described on p. 783). But the sum of all these pairwise interactions gives the total interaction energy. When the sum is evaluated it turns out that the interaction energy falls off only as $1/R^2$, where R is the separation of the centres of the colloid particles, and this is of much greater range than the $1/R^6$ characteristic of individual atoms and small molecules.

There are various factors working against the long-range dispersion interaction. For instance, there may be a protective film at the surface of the colloid particle which stabilizes the interface and which cannot be penetrated when the two particles touch. For instance, the surface atoms of a platinum sol in water react chemically and are turned into $-Pt(OH)_3H_3$, and this layer encases the particle. A fat can be emulsified by a soap because the long hydrocarbon tails penetrate the oil droplet but the $-CO_2^-$ head groups (or other hydrophilic groups in detergents) surround the surface and form hydrogen bonds with the water, and if possible a shell of negative charge which repels a possible attack from another similarly charged particle.

Soap molecules can group together even in the absence of grease droplets, for their hydrophobic tails tend to congregate and their hydrophilic heads provide protection. The hundred or so molecules that accumulate together in this way constitute a *micelle*, Fig. 24.16. Micelles form only above a concentration which is typical of the system: this lower limit is called the *critical micelle concentration* (C.M.C.). They form, however, only if the temperature is above a minimum, the *Krafft temperature*, a characteristic of the system.

Non-ionic detergent molecules may cluster together in swarms of 1000 or more, but ionic species tend to be disrupted by the electrostatic repulsions between the head groups and are normally limited to groups of between 10 and 100 molecules. The micelle population is often polydisperse, and the shape of the individual micelles varies with concentration. For instance, while spherical micelles do occur, they are more commonly ellipsoidal with axial ratios not exceeding about 6:1 (i.e., they are somewhat flattened spheres) at concentrations close to the C.M.C., but they are often rod-like at higher concentrations. The interior of a micelle is like a droplet of oil, and magnetic resonance shows that the hydrocarbon chains are mobile, but slightly more restricted than in the bulk.

Micelles have an important role in both industry and biology on account of their solubilizing function: matter can be transported by water after it has dissolved in their hydrophobic interiors. For this reason micellar systems are used as detergents, drug carriers, and for organic synthesis, froth flotation, and petroleum recovery.

The thermodynamics of micelle formation shows that the enthalpy of

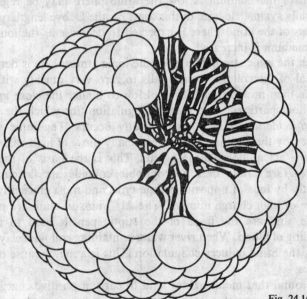

Fig. 24.16. A spherical micelle.

formation in aqueous systems is probably positive (that is, they are endothermic). The value of ΔH_m is in the region of 1 or 2 kJ per mole of detergent species. That they do in fact form above the C.M.C. indicates that their entropy of formation must then be positive, and measured values suggest that ΔS_m lies in the vicinity of $140 \, \mathrm{J \, K^{-1} \, mol^{-1}}$ at room temperature. That the entropy change is positive even though molecules are clustering together indicates a contribution to the entropy from the solvent, for it can move more freely once the detergent has herded into small clusters.

Apart from the physical stabilization of the colloid particles, a major source of kinetic stability is the existence of an electric charge on their surfaces. On account of this charge, ions of opposite charge tend to cluster nearby, and an ionic atmosphere is formed, as was described for normal ions in Chapter 11. Two regions of charge must be distinguished.

First, there is a fairly immobile layer of ions that stick tightly to the surface of the colloidal particle, and which may also include water molecules (if that is the support medium). The radius of the unit from the centre of the colloid particle up to the sphere that captures this rigid layer is called the *radius of shear*, and it is a major factor in determining the mobility of the particle. The electric potential at the radius of shear relative to its value in the distant, bulk medium is called the ζ-*potential* (zeta-potential) or the *electrokinetic potential*.

The charged central unit attracts an oppositely charged ionic atmosphere, and so there is also a diffuse cloud of opposite charge. The inner shell of charge and the outer atmosphere is referred to as the *electric double layer*. The structure of the atmosphere can be described in the same way as in the Debye–Hückel model of ionic solutions. When the ionic strength of the medium is low the atmosphere may be regarded as a spherically symmetric haze of thickness r_D, the Debye length (p. 319). The thickness of the atmosphere is expected to decrease as the ionic strength of the medium is increased.

When the ionic strength is large the ionic atmosphere is dense and the potential of the colloid particle falls to zero very quickly with distance. There is then no electrostatic repulsion to hinder the close approach of two colloid particles. As a result, coagulation (for which the alternative name *flocculation* is also used) readily occurs. The ionic strength is increased by the addition of ions, especially those of high valence, and so such ions act as flocculating agents. This is the basis of the empirical *Schultze–Hardy rule*, that hydrophobic colloids are flocculated more efficiently by ions of opposite charge type, and most efficiently when the ions are of high charge number. The Al^{3+} ions in alum are particularly effective, and are the basis of the stiptic pencils used to induce the congealing of blood. When river water containing colloidal clay flows into the sea, the brine induces coagulation. This is a major cause of silting in estuaries.

It is found that metal oxides tend to carry a positive charge. Sulphur and the noble metal sols tend to carry negative charges. Naturally occurring macromolecules also acquire a charge when they are dispersed in water (as in the cell, not only the laboratory). An important feature of proteins (and of other natural macromolecules) is that their overall charge depends on the pH of the medium. For instance, in acid environments protons tend to attach to basic groups, and the net charge on the macromolecule is positive; in basic media the opposite occurs and the molecule acquires a net negative charge. At the *isoelectric point* the pH is such that there is no net charge, and the molecule is neutral overall.

Example (Objective 19). The mobility of bovine serum albumin in aqueous solution was monitored at several values of pH. The data are listed below. What is the isoelectric point of the protein?

pH	4.20	4.56	5.20	5.65	6.30	7.00
mobility/$\mu m\,s^{-1}$	0.50	0.18	-0.25	-0.65	-0.90	-1.25

● *Method.* The protein has zero electrophoretic mobility when it is uncharged. Therefore the isoelectric point is the pH at which the protein does not move in an electric field. Plot mobility against pH and find, by interpolation, the pH of zero mobility.

● *Answer.* The data are plotted in Fig. 24.17. There is zero mobility when pH = 4.8, hence this is the isoelectric point.

● *Comment.* Sometimes the isoelectric point has to be obtained by extrapolation because the macromolecule might not be stable over a wide enough pH range. The isoelectric point takes note of the net charge arising from the protein and whatever other ions are present (as buffers). When there are no foreign ions present, except those arising from dissociation of the solvent, the solution is said to be *isoionic.*

The primary role of the electric double layer is to confer kinetic stability on colloid particles and macromolecules. Encountering particles break through the double layer and coalesce only if the collision is sufficiently energetic to disrupt the layers of ions and the solvating molecules, or if thermal motion has stirred away the surface accumulation of charge. This

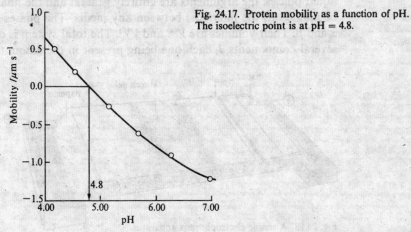

Fig. 24.17. Protein mobility as a function of pH. The isoelectric point is at pH = 4.8.

may happen at high temperatures, and so it is expected, and observed, that sols will precipitate when they are heated; furthermore, the protective role of the double layer explains why it is important not to remove all the ions when a colloid is being purified by dialysis. A further consequence is that since the net charge carried by proteins can be varied by changing the pH we also expect to modify the stability of their solutions. This is the reason why proteins coagulate most readily at their isoelectric point, when there is no overall charge.

The presence of charge on colloidal particles and natural macromolecules not only protects them but also permits us to exercise control over their motion. Applications that have been mentioned already are in electrodialysis and electrophoresis. Apart from its role in the determination of molar mass, electrophoresis has a number of analytical and technological applications. One analytical application is to the separation of different macromolecules, and a typical apparatus is illustrated in Fig. 24.18. A technical application is to the painting of objects by airborne charged paint droplets. Rubber molecules bear a charge when dispersed in a medium, and may be electrodeposited on anodes formed into the shape of the desired product. This electrophoretic rubber forming process is used to make articles where impermeability and sensitivity are demanded, such as, for instance, surgical gloves.

24.4 Surface tension and detergents

The general term for a species that is active in the interface between two phases is a *surfactant* or *surface-active agent*. Detergents are a good example, for they operate at the interface between hydrophilic and hydrophobic phases. In detergent action one expects the species to accumulate at the interface, and as a result to modify the surface tension. Thermodynamics lets us arrive at a precise relation between the accumulation of a species and its effect on surface tension.

Consider two phases in contact. Although we shall concentrate on two liquid phases, the arguments are entirely general and the final result may be applied to the interfaces between any media. The phases are labelled α and β. Their volumes are $V^{(\alpha)}$ and $V^{(\beta)}$. The total system is composed of several components J, each one being present in the amount n_J.

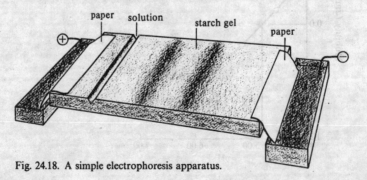

Fig. 24.18. A simple electrophoresis apparatus.

The total Gibbs function for the system is G. If the components were distributed uniformly through the phases right up to the interface, which is regarded as a well-defined surface of area σ, the total Gibbs function would be $G^{(\alpha)} + G^{(\beta)}$. But the components are not uniform, and the sum of the two Gibbs functions differs from G by an amount called the *surface Gibbs function*, $G^{(\sigma)}$:

(24.4.1)
$$G^{(\sigma)} = G - \{G^{(\alpha)} + G^{(\beta)}\}.$$

Similarly, if the bulk contains an amount $n_J^{(\alpha)}$ of J in phase α and an amount $n_J^{(\beta)}$ in β, both phases being regarded as homogeneous right up to the hypothetical dividing surface, the total amount of J differs from their sum by the amount

(24.4.2)
$$n_J^{(\sigma)} = n_J - \{n_J^{(\alpha)} + n_J^{(\beta)}\}.$$

This excess amount of substance is expressed as an amount per unit area of the surface by introducing the *surface excess* Γ_J through

(24.4.3)
$$\Gamma_J = n_J^{(\sigma)}/\sigma.$$

Note that both $n_J^{(\sigma)}$ and Γ_J may be either positive (accumulation of J at the interface relative to the bulk) or negative (relative deficiency).

Now we can embark upon the thermodynamic argument. A general change in G is brought about by changes in T, p, σ, and the n_J:

$$dG = -S\,dT + V\,dp + \gamma\,d\sigma + \sum_J \mu_J\,dn_J.$$

When this is applied to G, $G^{(\alpha)}$, and $G^{(\beta)}$ and eqn (24.4.1) is invoked, one arrives at

(24.4.4)
$$dG^{(\sigma)} = -S^{(\sigma)}\,dT + \sigma\,d\gamma + \sum_J \mu_J\,dn_J^{(\sigma)}$$

because at equilibrium the chemical potentials of the components are the same in every phase ($\mu_J^{(\alpha)} = \mu_J^{(\beta)} = \mu_J^{(\sigma)}$). Just as in the discussion of partial molar quantities in Chapter 8 (p. 217), this expression integrates at constant temperature to

$$G^{(\sigma)} = \sigma\gamma + \sum_J \mu_J n_J^{(\sigma)}.$$

We are seeking a connection between $d\gamma$, the change in surface tension, and the change in composition. Therefore we apply an argument that led in Chapter 8 to the Gibbs–Duhem equation (p. 219). This time, comparing

$$dG^{(\sigma)} = \sigma\,d\gamma + \gamma\,d\sigma + \sum_J n_J^{(\sigma)}\,d\mu_J + \sum_J \mu_J\,dn_J^{(\sigma)}$$

with eqn (24.4.4) with the temperature held constant ($dT = 0$) leads to the conclusion that

$$\sigma\,d\gamma + \sum_J n_J^{(\sigma)}\,d\mu_J = 0, \quad \text{at constant } T.$$

Division by σ then gives the

(24.4.5)   Gibbs surface tension equation: $d\gamma = -\sum_J \Gamma_J d\mu_J$.

The Gibbs equation can be put into a simpler form by the following argument. Suppose there is some surfactant, a detergent D, distributed in a two-phase system. We make the approximation that the oil and water phases are separated by a geometrically perfect surface. The detergent accumulates at the surface. Since the surface excesses of the water and the oil, Γ_{water} and Γ_{oil}, are zero, the Gibbs equation reduces to

$$d\gamma = -\Gamma_D d\mu_D,$$

where μ_D is the chemical potential of the detergent. For dilute solutions $d\mu_D = RT d\ln c_D$, where c_D is the concentration of the detergent. It follows that

(24.4.6)° $d\gamma = -(RT/c_D)\Gamma_D dc_D$,   or $(\partial\gamma/\partial c_D)_T = -RT\Gamma_D/c_D$.

The implication of this equation is as follows. If the detergent tends to accumulate at the interface its surface excess is positive, and so $(\partial\gamma/\partial c_D)_T$ is negative. That is, the surface tension *decreases* when a solute accumulates at a surface. Conversely, if the concentration dependence is known, the surface excess may be calculated from the last equation. The predictions of this type of equation have been tested by the simple (but technically elegant) procedure of slicing thin slices off the surfaces of solutions and analysing their compositions for the excess or deficiency of the surfactant. Note that the result summarized in the last equation is based on the assumption of ideal behaviour: there may be marked deviations at the concentrations of detergents employed in practice.

This section (and this Part) can be rounded off by linking the behaviour of the ideal surface phase as expressed by eqn (24.4.6) with the conception of the structure of ideal solutions introduced in Part 1 and with the work on the description of perfect gases that began Part 1. At small concentrations of detergent the surface tension can be expected to vary linearly with the concentration, and so

$$\gamma = \gamma^* - Kc_D,$$

where K is some constant. Then, from eqn (24.4.6),

$$\Gamma_D = Kc_D/RT = (\gamma^* - \gamma)/RT.$$

If now the difference $\gamma^* - \gamma$ is denoted π, the *surface pressure*, this equation becomes

(24.4.7)° $\pi\sigma = n_D^{(\sigma)}RT$,

which is the equation for a two-dimensional perfect gas. The excess solute at the interface of dilute, ideal solutions can therefore be regarded as behaving in the same way as the molecules of a perfect gas confined to a two-dimensional surface.

Appendix: the elasticity of rubber

Consider a one-dimensional freely-jointed polymeric chain. The three-dimensional case is more realistic, but the one-dimensional exhibits the principal feature without being too involved. The conformation of the chain can be expressed as the number of bonds pointing to the right, N_R, and the number to the left, N_L. The distance between the ends of the chain is $(N_R - N_L)l$, where l is the length of an individual bond. We write $n = N_R - N_L$. The total number of bonds is $N = N_R + N_L$. The number of ways of forming a chain with a given end to end distance nl is the number of ways of having N_R right-pointing and N_L left-pointing bonds. This is given by the binomial coefficient

$$W(n) = N!/N_R!N_L! = N!/\{\tfrac{1}{2}(N+n)\}!\{\tfrac{1}{2}(N-n)\}!,$$

The conformational entropy of the chain is then simply $S = k \ln W(n)$.

$$S(n)/k = \ln N! - \ln N_L! - \ln N_R!.$$

Since the factorials are all large, Stirling's approximation

$$\ln x! \approx \ln (2\pi)^{1/2} + (x + \tfrac{1}{2})\ln x - x$$

may be used (this is a more accurate form than $\ln x! \approx x \ln x - x$). This gives

$$S(n)/k = -\ln (2\pi)^{1/2} + (N+1)\ln 2 + (N + \tfrac{1}{2})\ln N \\ - \tfrac{1}{2}\ln\{(N+n)^{N+n+1}(N-n)^{N-n+1}\}.$$

The most probable conformation of the chain is the one with the ends close together ($n = 0$) as may be confirmed by differentiation. Therefore the maximum entropy is

$$S(0)/k = -\ln (2\pi)^{1/2} + (N+1)\ln 2 - \tfrac{1}{2}\ln N.$$

The change of entropy when the chain is stretched from this most probable conformation to a point such that the distance between the ends is nl is

$$\Delta S = S(n) - S(0)$$
$$= \tfrac{1}{2}k\{\ln N^{N+1}N^{N+1} - \ln (N+n)^{N+1+n}(N-n)^{N+1-n}\}$$
$$= -\tfrac{1}{2}kN \ln\{(1+v)^{1+v}(1-v)^{1-v}\}$$

where $v = n/N$. The entropy change for unit amount of chains is therefore

$$\Delta S_m = -\tfrac{1}{2}NR \ln\{(1+v)^{1+v}(1-v)^{1-v}\}.$$

(Remember that N is the number of links in the chain.) This is plotted in Fig. 24.19a. The entropy change is negative for all extensions, and so we conclude that contraction of the chain to its fully coiled state is the spontaneous process.

The calculation can be taken further. The work done on a piece of rubber when it is extended through a distance dx is $f dx$, where f is the restoring force. The First Law is therefore $dU = TdS - pdV + f dx$. It follows that

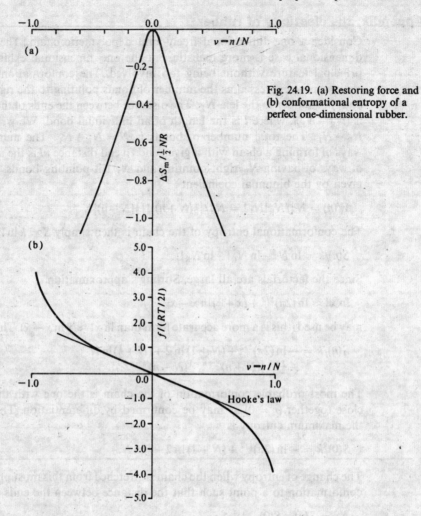

Fig. 24.19. (a) Restoring force and (b) conformational entropy of a perfect one-dimensional rubber.

$$(\partial U/\partial x)_{T,V} = T(\partial S/\partial x)_{T,V} + f.$$

In a perfect rubber, as in a perfect gas, the internal energy is independent of the dimensions at constant temperature, and so the l.h.s. of this expression is zero. The restoring force may therefore be identified as

$$f = -T(\partial S/\partial x)_{T,V}.$$

If the statistical expression for the entropy is introduced (and we evade problems arising from the constant volume constraint by assuming that the sample contracts laterally to an extent that maintains its volume constant when it is stretched), this becomes

$$f = -(T/l)(\partial S/\partial n)_{T,V} = -(T/lN)(\partial S/\partial v)_{T,V}$$

$$= (RT/2l)\ln\{(1+v)/(1-v)\},$$

which is plotted in Fig. 24.19b. At low extensions ($v \ll 1$), $f \approx RTn/Nl$, and so the sample exhibits Hooke's Law behaviour (restoring force $\propto$ displacement) but departs from it at higher extensions.

Further reading

Physical chemistry of surfaces. A. W. Adamson; Wiley, New York, 1976.

Colloid science. A. E. Alexander and P. Johnson; Clarendon Press, Oxford, 1949.

The dynamical character of adsorption. J. de Boer; Clarendon Press, Oxford, 1953.

The structure and action of proteins. R. E. Dickerson and I. Geiss; Benjamin, Menlo Park, 1969.

Physical biochemistry. K. E. van Holde; Prentice-Hall, Englewood Cliffs, 1971.

Physical biochemistry. D. Freifelder; W. H. Freeman, San Francisco, 1976.

Basic physical chemistry for the life sciences. V. R. Williams, W. L. Mattice, and H. B. Williams; W. H. Freeman, San Francisco, 1978.

Principles and problems in physical chemistry for biochemists (2nd edn). N. C. Price, and R. A. Dwek; Clarendon Press, Oxford, 1979.

Biochemistry. L. Stryer; W. H. Freeman, San Francisco, 1981.

Physical chemistry of macromolecules. C. Tanford; Wiley, New York, 1961.

Principles of polymer chemistry. P. Flory; Cornell University Press, 1953.

The handbook of biochemistry. H. Sober (ed.); Chemical Rubber Co., Cleveland, 1968.

Problems

24.1. The osmotic pressures of solutions of polystyrene in toluene were measured at 25 °C, with the following results:

$c_P/\text{mg cm}^{-3}$	3.2	4.8	5.7	6.9	7.8
h/cm	3.11	6.22	8.40	11.73	14.90

h is the height of the solution of density $0.867\,\text{g cm}^{-3}$ corresponding to the osmotic pressure. Obtain the molar mass and R.M.M. of the polymer by plotting h/c_P against c_P.

24.2. The osmotic pressure of a fraction of polyvinylchloride in a ketone solvent was measured at 25 °C. The density of the solvent (which is virtually equal to the density of the solution) was $0.798\,\text{g cm}^{-3}$. Calculate the molar mass and the virial coefficient B of the fraction from the following data:

$c_P/(\text{g}/100\,\text{cm}^3)$	0.200	0.400	0.600	0.800	1.000
h/cm	0.48	1.12	1.86	2.76	3.88

24.3. Calculate the number average R.M.M. and the mass average R.M.M. of a mixture of equal amounts of two polymers, one having $M_r = 62\,000$ and the other $M_r = 78\,000$.

24.4. A polymerization process produced a Gaussian distribution of polymers in the sense that the proportion of molecules having an R.M.M. in the range M_r to $M_r + dM_r$ was proportional to $\exp\{-(M_r - \bar{M}_r)^2/2\Gamma\}\,dM_r$. What are the number and mass average R.M.M.s when the distribution is narrow?

24.5. Calculate the excluded volume in terms of the molecular volume on the basis that the molecules are spheres of radius a. (The calculation is much more difficult in the case of rigid rods: see Tanford in *Further reading*.) Evaluate the osmotic virial coefficient in the case of bushy stunt virus, $a \approx 14.0\,\text{nm}$ and haemoglobin, $a \approx 3.2\,\text{nm}$.

24.6. Evaluate the percentage contribution to the osmotic pressures of $1.0\,\text{g}/$

$100 \, \text{cm}^3$ solutions of bushy stunt virus ($M_r \approx 1.07 \times 10^7$) and haemoglobin ($M_r \approx 66\,500$).

24.7.   The effective radius of a random coil a_{eff} is related to its radius of gyration R_g by $a_{\text{eff}} \approx \gamma R_g$ with $\gamma \approx 0.85$. Deduce an expression for the osmotic virial coefficient B in terms of the number of chain units for (a) a freely jointed chain, (b) a chain with tetrahedral interbond angles. Evaluate B for $l = 154 \, \text{pm}$ and $N = 4000$.

24.8.   Estimate the osmotic virial coefficient B for a randomly coiled polyethylene chain of R.M.M. M_r and evaluate it for $M_r = 56\,000$.

24.9.   Confirm that eqn (24.1.12) for the differences in ion concentrations on either side of a membrane follows from the preceding equations.

24.10.   Consider the effect of adding a salt $(M^+)_2 X^{2-}$ to a solution of a polyelectrolyte $(M^+)_\nu P^{\nu-}$. Find an expression for the differences of ion concentrations on either side of a membrane permeable to everything except the polyanion.

24.11.   Show that the ratio $[\text{Na}^+]_L/[\text{Na}^+]_R$ in the Donnan equilibrium is equal to $x + (1 + x^2)^{1/2}$, where $x = \nu[P]/2[\text{Na}^+]_R$ and sketch the ratio as a function of the polyelectrolyte concentration.

24.12.   A polyelectrolyte $\text{Na}_{20}P$ with $M_r = 100\,000$ at a concentration $1 \, \text{g}/100 \, \text{cm}^3$ was equilibrated in the presence of $0.001 \, \text{mol dm}^{-3}$ aqueous NaCl (i.e., $[\text{Na}^+]_R = 0.001 \, \text{mol dm}^{-3}$). What is the value of $[\text{Na}^+]_L$ at equilibrium?

24.13.   Discuss the role of the deviations from ideality ($\gamma_{\pm} \neq 1$) in determining the ratio of ion concentrations in the Donnan equilibrium.

24.14.   Investigation of the composition of the solutions used to study the osmotic pressure due to a polyelectrolyte with $\nu = 20$ showed that at equilibrium the concentrations corresponded to $[\text{Cl}] \approx 0.02 \, \text{mol dm}^{-3}$. Calculate the value of the osmotic virial coefficient for $\nu = 20$. Does it dominate the effect of excluded volume?

24.15.   Calculate the radial acceleration (as so many g) in a cell placed at $6.0 \, \text{cm}$ from the centre of rotation in an ultracentrifuge operating at $80\,000$ r.p.m.

24.16.   Calculate the speed of operation (in r.p.m.) of an ultracentrifuge needed in order to obtain a readily measurable concentration gradient in a sedimentation equilibrium experiment. Take that to be a concentration at the bottom of the cell about 5 times greater that at the top. Use $r_{\text{top}} = 5.0 \, \text{cm}$, $r_{\text{bottom}} = 7.0 \, \text{cm}$, $M_r \approx 10^5$, $\rho v_s \approx 0.75$, $T = 25 \, ^\circ\text{C}$.

24.17.   In an ultracentrifuge experiment at $20 \, ^\circ\text{C}$ on bovine serum albumin the following data were obtained: $\rho = 1.001 \, \text{g cm}^{-3}$, $v_s = 1.112 \, \text{cm}^3 \, \text{g}^{-1}$, $\omega/2\pi = 322 \, \text{Hz}$,

r/cm	10	11	12	13	14
$c/\text{g dm}^{-3}$	0.5354	0.4695	0.4067	0.3479	0.2940

What is the value of M_r?

24.18.   Sedimentation studies on haemoglobin in water gave a sedimentation constant $S = 4.5 \times 10^{-13} \, \text{s}$ at $20 \, ^\circ\text{C}$. The diffusion coefficient is $6.3 \times 10^{-7} \, \text{cm}^2 \, \text{s}^{-1}$ at the same temperature. Calculate the molar mass of haemoglobin using $v_s = 0.75 \, \text{cm}^3 \, \text{g}^{-1}$ for its partial specific volume and $\rho = 0.998 \, \text{g cm}^{-3}$ for the density of the solution.

24.19.   Estimate the effective radius of the haemoglobin molecule by combining the data in the last Problem with the information that the viscosity of the solution is $1.00 \times 10^{-3} \, \text{kg m}^{-1} \, \text{s}^{-1}$.

24.20. The diffusion coefficient for bovine serum albumin, a prolate ellipsoid, is $6.97 \times 10^{-7} \, \text{cm}^2 \, \text{s}^{-1}$ at $20 \, °\text{C}$, its partial specific volume is $0.734 \, \text{cm}^3 \, \text{g}^{-1}$, and its sedimentation constant is $5.01 \times 10^{-13} \, \text{s}$ in a solution of density $1.0023 \, \text{g} \, \text{cm}^{-3}$ and viscosity $1.00 \times 10^{-3} \, \text{kg} \, \text{m}^{-1} \, \text{s}^{-1}$. Estimate its dimensions.

24.21. The rate of sedimentation of a recently isolated protein was monitored at $20 \, °\text{C}$ and with a rotor speed of 50000 r.p.m. The boundary receded as follows:

t/s	0	300	600	900	1200	1500	1800
r/cm	6.127	6.153	6.179	6.206	6.232	6.258	6.284

Calculate the sedimentation constant S and the molar mass of the protein on the basis that its partial specific volume (which was measured in a pyknometer) is $0.728 \, \text{cm}^3 \, \text{g}^{-1}$ and its diffusion coefficient is $7.62 \times 10^{-7} \, \text{cm}^2 \, \text{s}^{-1}$ at $20 \, °\text{C}$, the density of the solution then being $0.9981 \, \text{g} \, \text{cm}^{-3}$.

24.22. Suggest a shape for the protein considered in the last Problem on the basis that the viscosity of the solution is $1.00 \times 10^{-3} \, \text{kg} \, \text{m}^{-1} \, \text{s}^{-1}$ at $20 \, °\text{C}$.

24.23. The viscosities of solutions of polyisobutylene in benzene were measured at $24 \, °\text{C}$ (the θ-temperature for the system) with the following results:

$c/(\text{g}/100 \, \text{cm}^3)$	0	0.2	0.4	0.6	0.8	1.0
$\eta/10^{-3} \, \text{kg} \, \text{m}^{-1} \, \text{s}^{-1}$	0.647	0.690	0.733	0.777	0.821	0.865

On the basis of the information in Table 24.3 deduce the R.M.M. of the polymer.

24.24. The radius of gyration is defined in eqn (24.1.28). Show that an equivalent definition is that R_g is the average root mean square distance of the atoms or groups (all assumed to be of the same mass); that is, that $R_g^2 = (1/N) \sum_j R_j^2$, where R_j is the distance of atom j from the centre of mass.

24.25. Use eqn (24.2.1) to deduce expressions for (a) the root mean square separation of the ends of the chain, (b) the mean separation of the ends, and (c) their most probable separation. (Integrals over Gaussians are given in Table 25.1 on p. 866.) Evaluate these three quantities for an $N = 4000$, $l = 154 \, \text{pm}$ fully flexible chain.

24.26. Construct a two-dimensional random walk either using tables of random numbers (e.g., *Handbook of mathematical functions*, Abramowitz and Stegun) or using a random number generating program for a calculator. Construct a walk of 50 and 100 steps. If there are many people working on the problem, investigate the mean and most probable separations in the plots by direct measurement. Do they vary as $N^{1/2}$?

24.27. Evaluate the radius of gyration of (a) a solid sphere of radius a, (b) a long straight rod of radius a and length l. (Use the expression deduced in Problem 24.24.) Show that in the case of a solid sphere of specific volume v_s, $R_g/\text{nm} \approx 0.056902\{(v_s/\text{cm}^3 \, \text{g}^{-1})M_r\}^{1/3}$. Evaluate R_g for species with $M_r = 100000$, $v_s = 0.750 \, \text{cm}^3 \, \text{g}^{-1}$, and, in the case of the rod, a radius $0.5 \, \text{nm}$.

24.28. On the basis of the information below and the expression for R_g of a solid sphere derived in the last Problem, classify the species below as globular or rod-like:

	M_r	$v_s/\text{cm}^3 \, \text{g}^{-1}$	Measured R_g/nm
serum albumin	66×10^3	0.752	2.98
bushy stunt virus	10.6×10^6	0.741	12.0
DNA	4×10^6	0.556	117.0

24.29. In formamide as solvent, poly(γ-benzyl-L-glutamate) is found by light scattering experiments to have a radius of gyration proportional to M_r; in contrast, polystyrene in butanone has R_g proportional to $M_r^{1/2}$. Present arguments to show that the first polymer is a rigid rod, while the second is a random coil.

24.30. Evaluate eqn (24.1.27) for $P(\theta)$ in the case of a long rigid rod of N identical scattering units with separation l, and assume that the units are so closely spaced that sums may be replaced by integrals.

24.31. Plot $P(\theta)$ as calculated in the last Problem as a function of θ on polar graph paper (like Fig. 24.10) in the case $L \approx \lambda$. The integral $\int_0^z (\sin x / x) dx$ is the 'sine-integral' $Si(z)$, and is tabulated in Abramowitz and Stegun, *Handbook of mathematical functions.*

24.32. Evaluate the rotational correlation time for serum albumin in water at $25\,°C$ on the basis that it is a sphere of radius $3.0\,nm$. What is the value for a CCl_4 molecule in carbon tetrachloride at $25\,°C$? (Viscosity data in Table 23.7; take $a(CCl_4) \approx 250\,pm$.)

24.33. We now turn attention to the thermodynamic description of stretching rubber. The observables are the tension t and length l (like p and V for gases). Since $dw = t\,dl$, the basic equation is $dU = T\,dS + t\,dl$. (pdV terms are supposed negligible throughout.) If $G = U - TS - tl$, find expressions for dG and dA and deduce the Maxwell relations $(\partial S/\partial l)_T = -(\partial t/\partial T)_l$ and $(\partial S/\partial t)_T = (\partial l/\partial T)_t$.

24.34. Continue the thermodynamic analysis by deducing the equation of state for rubber, $(\partial U/\partial l)_T = t - T(\partial t/\partial T)_l$.

24.35. On the assumption that the tension required to keep a sample at a constant length is proportional to the temperature (the analogue of $p \propto T$), show that the tension can be ascribed to the dependence of the entropy on the length of the sample. Account for this result in terms of the molecular nature of the sample.

24.36. The energy of interaction between two spheres of radius a composed of atoms at a number density $\mathscr{N}$ which individually interact as $-C_6/R^6$ is $V(R) = -\frac{1}{6}\pi^2 C_6 \mathscr{N}^2 f(s)$, with $f(s) = 2/(s^2 - 4) + 2/s^2 + \ln(1 - 4/s^2)$ and $s = R/a$; see A. W. Adamson, *Physical chemistry of surfaces*, p. 318. Plot $f(s)$ from $s = 4$ up to $s = 10$ in order to appreciate the range of the interaction energy.

24.37. The surface tensions of a series of aqueous solutions of a surfactant were measured at $20\,°C$, with the following results:

$[A]/mol\,dm^{-3}$	0	0.10	0.20	0.30	0.40	0.50
$\gamma/mN\,m^{-1}$	72.8	70.2	67.7	65.1	62.8	59.8

Calculate the surface excess concentration.

24.38. Evaluate the surface pressure π exerted by the surfactant in the last Problem, and investigate whether eqn (24.4.7) is satisfied.

24.39. The surface tensions of aqueous salt solutions are normally greater than that of water itself. Does the salt accumulate at the surface?

24.40. The surface tensions of solutions of salts in water at concentration c can be expressed in the form $\gamma = \gamma^* + (c/mol\,dm^{-3})\Delta\gamma$. The values of $\Delta\gamma$ at $20\,°C$ in the vicinity of $c = 1\,mol\,dm^{-3}$ are as follows:

$$\Delta\gamma/mN\,m^{-1} = 1.4(KCl), \quad 1.64(NaCl), \quad 2.7(Na_2CO_3).$$

Calculate the surface excess concentrations when the bulk concentrations are $1\,mol\,dm^{-3}$.

Part 3 · Change

25 Molecules in motion: the kinetic theory of gases

Learning objectives

After careful study of this chapter you should be able to

(1) Specify the *kinetic model* of a perfect gas (p. 860).

(2) Use the kinetic theory to calculate the *pressure* exerted by a perfect gas (eqn (25.1.2)).

(3) Define the *mean value* of discrete (eqn (25.1.5)) and continuous distributions (eqn (25.1.7)).

(4) Use probability arguments to derive the *Maxwell–Boltzmann distribution* of molecular velocities (eqn (25.1.11) and the *Maxwell distribution* of speeds (eqn (25.1.13)).

(5) Calculate the *mean speed* (eqn (25.1.15)), the *root mean square speed* (eqn (25.1.14)), and the *most probable speed* (eqn (25.1.16)) of gas molecules.

(6) Define *collision cross-section* (p. 871) and calculate the *collision frequency* (eqn (25.2.5)) and *mean free path* (eqn (25.2.7)) in a gas.

(7) Calculate the frequency of collisions of gas molecules with a surface (eqn (25.2.9)).

(8) Explain the term *transport property* (p. 874) and define *flux* (p. 875).

(9) State and use *Fick's First Law of Diffusion* (eqn (25.3.1)).

(10) Calculate the *rate of effusion* of a gas through a hole, state *Graham's Law*, and use *Knudsen's method* to determine vapour pressures (p. 877).

(11) Derive Fick's Law and calculate the *diffusion coefficient* for an ideal gas on the basis of kinetic theory (eqn (25.3.8)).

(12) Calculate the *coefficient of thermal conductivity* (eqn (25.3.10) and the *viscosity* (eqn (25.3.12)) of a gas from kinetic theory and explain their temperature and pressure dependences (p. 882).

(13) Describe how the viscosity of a gas is measured (p. 883).

Introduction

This chapter deals with the translational motion of collections of atoms and molecules in the gaseous state, and examines how the properties of a gas can be discussed on the basis of the continual translational motion of its components. We concentrate on the free translational motion of the molecules and ignore the details of the interactions between them. This means that we shall be investigating the *kinetic theory* of gases. The full problem, when molecules move under the influence of their mutual interactions, is called the *dynamic theory*; some aspects of this (single atom collisions) were met in Chapter 23; some more will be met in Chapter 28.

The rudiments of the kinetic theory of gases were outlined in the Introduction (pp. 21–2). There we saw that the view that a perfect gas is composed of a swarm of particles allows us to calculate a number of its properties. That brief outline is the starting point for a more detailed development in this chapter, and it should be looked at again before going further. In this chapter we shall sharpen some of the arguments used in the introduction, indicate some of the deficiencies of the calculations, and go on to find simple expressions for some of the more interesting properties of perfect gases. In particular we concentrate on the *transport properties* of gases: these are properties such as thermal conductivity, viscosity, and diffusion which have the common feature of involving the transport of some property from one region in the system to another.

The model on which the kinetic theory is based makes three assumptions.

(1) The gas consists of a swarm of particles of mass m in continual random motion.

(2) The particles have negligible size, in the sense that their diameters are much smaller than the average distance they travel between collisions.

(3) The particles do not interact except in so far as they undergo elastic collisions. An *elastic* collision means that the total translational kinetic energy of a colliding pair is the same before and after the collision: no energy is left in one of the colliding particles as rotational energy, or vibrational energy, etc.

Collisions ensure that the particles constantly change their speed and direction. The number of collisions made by a single particle, the *collision frequency z*, plays an important role in the discussion of transport properties and chemical reactions in the gas phase. We shall calculate its value on the basis of the kinetic theory. The average distance each particle travels between collisions is called the *mean free path λ* (lambda), and it plays a vital role in the discussion of transport phenomena because it indicates how far a molecule carries a certain property before colliding. The mean free path is a gauge of whether the kinetic theory is applicable to a real system. For example, if we know the molecules of a real gas have a diameter d, then the kinetic theory should give quite a good description of its properties if d/λ is small ($d/\lambda \ll 1$) but fail if it is not. You should also note that any mechanical quantity can be expressed as a combination of quantities with dimensions of mass, length, and time. These are the dimensions of m, λ, and $1/z$. Therefore, once λ and z have been calculated, any mechanical quantity can be obtained by taking a suitable combination of them.

25.1 The basic calculations

Putting the kinetic theory of gases on a quantitative basis involves two basic calculations. The first is to arrive at an expression for the pressure, and the second is to deduce details of the distribution of velocities of the component molecules.

The pressure exerted by a gas. The swarm of particles makes a large number of collisions with the walls of the containing vessel, and the impacts occur so often that the wall experiences a virtually constant force. Force per unit area is *pressure*, and so the kinetic theory can account qualitatively for the pressure of a gas. The kinetic theory accounts quantitatively for it as follows.

Consider the system illustrated in Fig. 25.1. When a molecule of mass m collides with the wall perpendicular to the x-axis its component of momentum along the x-axis changes from mv_x to $-mv_x$, the other components remaining unchanged. The total change of momentum on each collision is of magnitude $|2mv_x|$. The number of collisions in a time interval Δt is equal to the number of molecules able to reach the wall in that time. The distance a molecule of velocity v_x can travel in a time Δt is $|v_x|\Delta t$, and so all molecules lying within a distance $|v_x|\Delta t$ of the wall will strike it if they are travelling towards it. If the cross-section of the container has area A, all molecules lying in a volume $A|v_x|\Delta t$ will reach the wall (if they are travelling towards it). If $\mathcal{N}$ is the number of molecules per unit volume, $\mathcal{N}A|v_x|\Delta t$ is the number in the volume of interest. On the average half of these are moving to the right, and half to the left, and so the average number of collisions in the interval Δt is $\frac{1}{2}\mathcal{N}A|v_x|\Delta t$. The total momentum change imparted in that interval is

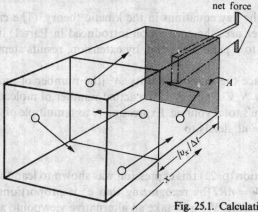

Fig. 25.1. Calculating the pressure exerted by a gas.

momentum change

$$= \text{(number of collisions)} \times \text{(momentum change on collision)}$$

$$= (\tfrac{1}{2}\mathcal{N}A|v_x|\Delta t)(2m|v_x|)$$

$$= m\mathcal{N}Av_x^2\Delta t,$$

and the *rate of change of momentum* is $m\mathcal{N}Av_x^2$. According to Newton's Law the rate of change of momentum is the *force*, and so the force exerted by the gas on the wall is $m\mathcal{N}Av_x^2$. The *pressure* is force per unit area; therefore, since the wall is of area A, the pressure is $m\mathcal{N}v_x^2$.

Not all molecules travel with the same velocity, and so the detected pressure is the average of the quantity just calculated. If the average value of v_x^2 is written $\langle v_x^2 \rangle$, we arrive at the expression

$$p = \mathcal{N}m\langle v_x^2 \rangle.$$

Since the motion of the molecules is random, the average of v_x^2 is the same as the average of the corresponding quantities in the y and z directions:

$$\langle v_x^2 \rangle = \langle v_y^2 \rangle = \langle v_z^2 \rangle.$$

Since the magnitude of the velocity, the speed, of an individual molecule is given by $v^2 = v_x^2 + v_y^2 + v_z^2$, it follows that we may write the average of v^2 as

$$\langle v^2 \rangle = \langle v_x^2 \rangle + \langle v_y^2 \rangle + \langle v_z^2 \rangle = 3\langle v_x^2 \rangle.$$

We shall write the average value of v^2 as c^2:

(25.1.1) $$c^2 = \langle v^2 \rangle$$

and call c^2 the *mean square speed* of the molecules. The square root of this quantity is called the *root mean square speed*, $c = \sqrt{\langle v^2 \rangle}$. The final expression for the pressure is obtained by collecting these remarks:

(25.1.2)° $$p = \tfrac{1}{3}\mathcal{N}mc^2.$$

This is one of the key equations in the kinetic theory. (The circle on the equation number uses the convention introduced in Part 1: it signifies a result confined to a perfect gas and, by extension, results stemming from kinetic theory.)

We can replace the number density $\mathcal{N}$ (the number of molecules per unit volume) by N/V, where N is the actual number of molecules present and V the system's total volume. Expressing N as a multiple of Avogadro's constant L, $N = nL$, leads to

(25.1.3)° $$pV = \tfrac{1}{3}nLmc^2.$$

In the Introduction (p. 21) this expression was shown to lead to the perfect gas equation $pV = nRT$ by recognizing that c^2 is proportional to T, the absolute temperature. We now take an alternative viewpoint, and develop the equation by examining the distribution of velocities according to the kinetic theory. We first deal with mean values of properties, and then specialize to the case of mean values of velocity components and speeds. It turns out that only one unknown parameter ζ is needed to describe the mean value of the molecular speeds (and any other property), and we shall derive the result that $pV \propto 1/\zeta$. Since we know from experiment that $pV = nRT$ we can obtain a value of this single unknown parameter ζ. With that established we can use the method to obtain the mean value of a wide range of properties, and develop the kinetic theory to take into account collisions and transport processes.

Mean values and distributions. Suppose we want the average value of a property X which may take any of the values $X_1, X_2, \ldots, X_Z$ (we call these the possible *outcomes* of the observation), and in a series of N measurements we find that X_1 occurs N_1 times, X_2 occurs N_2 times, and so on. The *mean value* $\langle X \rangle$ of the property X is the weighted sum of all the outcomes divided by the total number of observations:

$$\langle X \rangle = \frac{N_1 X_1 + N_2 X_2 + \ldots + N_Z X_Z}{N_1 + N_2 + \ldots + N_Z} = \frac{N_1 X_1 + N_2 X_2 + \ldots + N_Z X_Z}{N}$$

(25.1.4)
$$= \sum_{i=1}^{Z} (N_i/N) X_i.$$

The formula for the mean value can be expressed in a different way if we introduce the *probability* that an outcome X_i is obtained in an experiment. In the present example the probability that X_1 is the outcome is N_1/N, and similarly for all the other outcomes. We write these probabilities P_i (so that $P_1 = N_1/N$, etc., and in general $P_i = N_i/N$). Then the mean value can be written

(25.1.5)
$$\langle X \rangle = \sum_i P_i X_i$$

where the sum ranges over all possible outcomes. Note that from the definition of P_i

(25.1.6)
$$\sum_i P_i = (1/N) \sum_i N_i = 1.$$

Now suppose that the outcome of an experiment may take one of a *continuous* range of values. This is the case, for example, when we are interested in the mean height of a population, or in the mean velocity of a gas molecule. The definition of the mean value has to be modified.

We divide up the continuous range of possible outcomes into segments, Fig. 25.2. We count 1 every time a measurement has an outcome which falls anywhere in a given segment. For instance, consider the segment of length ΔX starting at X. If in a series of 300 experiments an outcome lying in this range is obtained in 6 experiments, we write $N(X) = 6$. If the total number of experiments is N, the possibility that the outcome of any single experiment lies in the range X to $X + \Delta X$ is $P(X) = N(X)/N$; which in this case is 1/50. The value of $N(X)$, and therefore the value of $P(X)$, is proportional to the length of the segment X to $X + \Delta X$ (if ΔX is small).

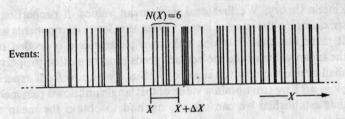

Fig. 25.2. A sequence of events.

We recognize this by writing $P(X) = f(X)\Delta X$.

So far we appear, by an approximation that groups together different outcomes, to have turned the continuous problem into the discrete one already described. This is indeed so, and we now proceed as we did in that case.

By analogy with the discussion of the truly discrete case we can calculate the *approximate* average value of X by taking its value for each segment, multiplying it by the probability $P(X)$ that the observations lie in that segment, and then summing over the segments:

$$\langle X \rangle \approx \sum_{\text{segments}} XP(X) = \sum_{\text{segments}} Xf(X)\Delta X.$$

This is only an approximation at this stage, because X might vary appreciably over the width of the segment. The relation can be made exact by allowing each segment to become infinitesimal, for then X will be constant over its range. We therefore replace the finite range ΔX by the infinitesimal range dX. The effect of this is to turn the sum into an integral:

(25.1.7)
$$\langle X \rangle = \int Xf(X)\,dX.$$

This is a very important expression: it lies at the heart of the development of this chapter.

The range of integration in the last equation is over all possible values of the observable X. For example, if the mean height h of a population is being determined, the range of heights is from 0 to ∞:

(25.1.8a)
$$\langle h \rangle = \int_0^\infty hf(h)\,dh,$$

because only positive heights are possible. If we were considering the mean velocity of a particle the appropriate integral would be

(25.1.8b)
$$\langle v_x \rangle = \int_{-\infty}^\infty v_x f(v_x)\,dv_x$$

because the x-component of velocity may have either sign, depending on the particle's direction.

The function $f(X)$ is called the *distribution* of the property X. From its definition $P(X) = f(X)\Delta X$ we see that it gives the probability that the property X lies somewhere in the range X to $X + \Delta X$. In the limit of an infinitesimal range dX, $f(X)$ gives the probability that an outcome lies in the infinitesimal region X to $X + dX$. For instance, $f(h)$ tells us the probability of a measured height falling in the range h to $h + dh$, and the distribution $f(v_x)$ tells us the probability of the x-component of velocity lying in the range v_x to $v_x + dv_x$. Thus, if $f(180 \text{ cm}) = 0.12 \text{ cm}^{-1}$ we would know that the probability of the height of a sample of the population falling in the range 180–181 cm is approximately 0.12, and that for the range 180–182 cm is approximately 0.24. (Note that the range is not truly infinitesimal, but approximately so.) By contrast, if $f(200 \text{ cm}) = 0.001 \text{ cm}^{-1}$ we would know that the probability of the height of a sample lying in the range 200–201 cm was only about 0.001.

The other property of probability distributions that concerns us relates to the question of considering several types of property simultaneously. For instance, it might be important to know the probability that a system has both the value X_i of some discrete property X *and* the value Y_j of some other discrete property Y. *If the properties are independent of each other*, the probability of the system having *both* the value X_i of the property X *and* the value Y_j of the property Y is

$$P(X_i, Y_j) = P(X_i)P(Y_j),$$

where $P(X_i)$ and $P(Y_j)$ are the individual probabilities. For example, if the probability of a person being a *man* is 0.495, and the probability of a *person* (man or woman) being left-handed is 0.110, then the probability of selecting a *left-handed man* by random choice from a crowd is $(0.110) \times (0.495) = 0.054$, or 1 in 18.5. If, however, left-handedness were a male characteristic this calculation would be false.

The same technique can be used for continuous properties. If the probability of X lying in the range dX at X is $f(X)dX$, and the probability that an *independent* property lies in the range Y to $Y + dY$ is $f(Y)dY$, then the probability of X lying in the range X to $X + dX$ *and* Y in Y to $Y + dY$ is the product of probabilities: $f(X)f(Y)dX\,dY$. For example, the probability that a molecule has a velocity component v_x in the range v_x to $v_x + dv_x$ *and* a velocity component v_y in the range v_y to $v_y + dv_y$ is $f(v_x)f(v_y)dv_x\,dv_y$, because these velocities are independent of each other (except in some contrived cases).

The distribution of molecular velocities. We are now in a position to set up the distribution function for the velocity components of the particles in a perfect gas according to the kinetic theory. The three components of velocity are independent of each other, and so the probability $F(v_x, v_y, v_z)dv_x\,dv_y\,dv_z$ that a molecule has a velocity with components in the range v_x to $v_x + dv_x$, v_y to $v_y + dv_y$, and $v_z + dv_z$ is the product of the individual probabilities of each component being in that range:

$$F(v_x, v_y, v_z)dv_x\,dv_y\,dv_z = f(v_x)f(v_y)f(v_z)dv_x\,dv_y\,dv_z.$$

We now assume that the probability of the molecules having a particular range of velocity components is independent of the orientation of the direction of flight, but that it does depend on the speed. So, for example, the probability of having velocity components $v_x = 1 \, \text{km s}^{-1}$, $v_y = 2 \, \text{km s}^{-1}$, $v_z = 3 \, \text{km s}^{-1}$ (and therefore a speed $\sqrt{14} \, \text{km s}^{-1}$) is the same as the probability of having the components $2 \, \text{km s}^{-1}, 3 \, \text{km s}^{-1}, 1 \, \text{km s}^{-1}$ respectively (same speed) or any other direction of motion such that the speed is $\sqrt{14} \, \text{km s}^{-1}$. This means that the distribution $F(v_x, v_y, v_z)$ can depend only on the speed v, where $v^2 = v_x^2 + v_y^2 + v_z^2$, and not on the individual components. Therefore F can be written as the function $F(v_x^2 + v_y^2 + v_z^2)$ and the last equation becomes

$$F(v_x^2 + v_y^2 + v_z^2) = f(v_x)f(v_y)f(v_z).$$

Only an exponential function satisfies this equation (because $e^{a+b+c} = e^a e^b e^c$) and so we write

$$f(v_x) = K \exp(\pm \zeta v_x^2),$$

where K and ζ (zeta) are constants. This is a satisfactory and unique form, and

$$f(v_x)f(v_y)f(v_z) = K^3 \exp(\pm \zeta [v_x^2 + v_y^2 + v_z^2]) = F(v_x^2 + v_y^2 + v_z^2),$$

as required.

The $\pm$ ambiguity in the exponent can be resolved on physical grounds. The probability of extremely high velocities must be very small; therefore the negative sign must be taken.

The value of the constants K and ζ can be established by two further arguments. First, the total probability that a velocity component lies in the range $-\infty \leqslant v_x \leqslant +\infty$ must be unity (it must have *some* velocity in that range). Therefore

$$\int_{-\infty}^{\infty} f(v_x) \, dv_x = 1.$$

Substituting the exponential form of $f(v_x)$ gives

$$\int_{-\infty}^{\infty} f(v_x) \, dv_x = K \int_{-\infty}^{\infty} \exp(-\zeta v_x^2) \, dv_x = K(\pi/\zeta)^{1/2}.$$

Therefore $K = (\zeta/\pi)^{1/2}$.

The value of ζ can be obtained by calculating the mean square speed of the molecules, and then using eqn (25.1.3) to equate it to the measurable quantity pV. We concentrate on calculating $\langle v_x^2 \rangle$, knowing that the other components have the same mean square value.

From the general expression for a mean value we can write, with $X = v_x^2$,

$$\langle v_x^2 \rangle = \int_{-\infty}^{\infty} v_x^2 f(v_x) \, dv_x = (\zeta/\pi)^{1/2} \int_{-\infty}^{\infty} v_x^2 \exp(-\zeta v_x^2) \, dv_x.$$

The integral on the right is standard, Table 25.1 (and may be derived very simply from the one encountered in the calculation of K above), and has the value $\frac{1}{2}(\pi/\zeta^3)^{1/2}$. It follows that

$$\langle v_x^2 \rangle = \tfrac{1}{2}(\zeta/\pi)^{1/2}(\pi/\zeta^3)^{1/2} = 1/2\zeta.$$

Therefore the mean square speed, $c^2 = \langle v^2 \rangle = \langle v_x^2 \rangle + \langle v_y^2 \rangle + \langle v_z^2 \rangle$, is

(25.1.9) $c^2 = 3/2\zeta.$

Incorporating this important result in eqn (25.1.3) gives

$$pV = \tfrac{1}{3}nLmc^2 = \tfrac{1}{2}nLm/\zeta.$$

But we also know that a perfect gas obeys the equation

$$pV = nRT = nLkT.$$

Equating the two equations allows ζ to be related to the temperature of the system:

(25.1.10)° $\tfrac{1}{2}nLm/\zeta = nLkT$ implies $\zeta = \tfrac{1}{2}m/kT.$

Therefore the complete form for the velocity distribution is

(25.1.11)° $f(v_x) = (m/2\pi kT)^{1/2} \exp(-\tfrac{1}{2}mv_x^2/kT).$

It can be seen that the right-hand side has the form of a Boltzmann distribution (p. 11) for the proportion of particles with a kinetic energy $\tfrac{1}{2}mv_x^2$ arising from their motion along the x-axis, and an alternative route to this equation is through the Boltzmann expression. For this reason the distribution $f(v_x)$ is called the *Maxwell–Boltzmann distribution* of molecular velocities, representing both Maxwell's contribution (he derived it originally) and Boltzmann's (he proved it rigorously).

Table 25.1. Integrals over $x^n \exp(-ax^2)$

$$I_n = \int_0^\infty x^n \exp(-ax^2)dx$$

$n =$	0	1	2	3	4	5	6
$I_n =$	$\tfrac{1}{2}(\pi/a)^{1/2}$	$1/2a$	$\tfrac{1}{4}(\pi/a^3)^{1/2}$	$1/2a^2$	$\tfrac{3}{8}(\pi/a^5)^{1/2}$	$1/a^3$	$\tfrac{15}{16}(\pi/a^7)^{1/2}$

The integral $(2/\sqrt{\pi})\int_0^z \exp(-x^2)dx$ is called the *error function* and denoted erf z. This function is listed in numerical tables, and a small selection is given below.

$z = 0$	0.20	0.40	0.60	0.80	0.90
erf $z = 0$	0.2227	0.4284	0.6039	0.7421	0.7969
$z = 1.0$	1.20	1.40	1.60	1.80	1.90
erf $z = 0.8427$	0.9103	0.9523	0.9763	0.9891	0.9928

Source: M. Abramowitz and I. A. Stegun, *Handbook of Mathematical functions*, Dover.

Example (Objective 5). A sample of caesium is heated to 500 °C in an oven. In one wall there is a small hole and the atoms emerge to form an atomic beam. What is their average velocity?

● *Method.* The beam is one-dimensional. Since only the atoms that are moving towards the right (positive x) are in the beam we know that $0 \leqslant v_x \leqslant \infty$. The mean velocity is $\langle v_x \rangle$, and so use eqn (25.1.8b), but integrate only from $v_x = 0$ up to

infinity. Use eqn. (25.1.11) for $f(v_x)$, but modified so that $\int_0^\infty f(v_x)\,dv_x = 1$: simply
multiply the right-hand side of eqn (25.1.11) by 2.

• *Answer.* From eqns (25.1.8b) and (25.1.11),

$$\langle v_x \rangle = \int_0^\infty v_x f(v_x)\,dv_x = 2(m/2\pi kT)^{1/2} \int_0^\infty v_x \exp(-mv_x^2/2kT)\,dv_x$$

$$= 2(m/2\pi kT)^{1/2}(kT/m) = \sqrt{(2kT/\pi m)}.$$

For caesium at 500 °C:

$$\langle v_x \rangle = \left\{ \frac{(2/\pi) \times (1.381 \times 10^{-23}\,\mathrm{J\,K^{-1}}) \times (773\,\mathrm{K})}{(132.9) \times (1.6605 \times 10^{-27}\,\mathrm{kg})} \right\}^{1/2} = 175.5\,\mathrm{m\,s^{-1}}.$$

• *Comment.* This is the *mean* velocity of the emergent atoms: the spread in velocities
may be very wide, and an atomic beam apparatus usually uses a velocity-selector
(p. 789).

Before discussing the content of the Maxwell–Boltzmann distribution
we complete the derivation by considering its three-dimensional form. The
probability that the velocity has components lying in the volume element
$dv_x\,dv_y\,dv_z$ at the point (v_x, v_y, v_z) is

$$dF(v_x, v_y, v_z) = f(v_x)f(v_y)f(v_z)\,dv_x\,dv_y\,dv_z$$

(25.1.12)° $$= (m/2\pi kT)^{3/2} \exp(-mv^2/2kT)\,dv_x\,dv_y\,dv_z.$$

Suppose we were only interested in the *speed* of the molecules, and were
not interested in their direction of flight. The probability that the molecule
has the speed v irrespective of direction is the sum of all the probabilities
given by the last equation over all orientations of the velocity (for the
specified v). The sum of all the volume elements $dv_x\,dv_y\,dv_z$ as the point
(v_x, v_y, v_z) is slid all over the surface of a sphere of constant radius v is
simply the volume of a spherical shell of radius v and thickness dv, Fig.
25.3. This volume is $4\pi v^2\,dv$. Therefore, the probability that the speed lies
in the range v to $v + dv$ irrespective of direction of motion, is

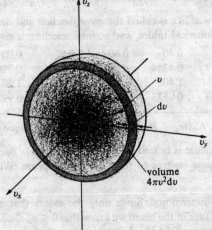

Fig. 25.3. The distribution of molecular velocities.

$(25.1.13)°$

$$dF(v) = 4\pi(m/2\pi kT)^{3/2} v^2 \exp(-mv^2/2kT)dv.$$

This important expression is the *Maxwell distribution of speeds*.

Example (Objective 5). What is the mean speed of the caesium atoms inside the oven of the last *Example*?

- *Method.* The mean speed is obtained from the expression $\bar{c} = \int_0^\infty vF(v)dv$, where $F(v)dv$ is the right-hand side of the last expression.

- *Answer.* From the last expression

$$\bar{c} = \int_0^\infty vF(v)dv = 4\pi(m/2\pi kT)^{3/2} \int_0^\infty v^3 \exp(-mv^2/2kT)dv$$

$$= 4\pi(m/2\pi kT)^{3/2} \int_0^\infty v^3 \exp(-mv^2/2kT)dv$$

$$= 4\pi(m/2\pi kT)^{3/2} \tfrac{1}{2}(2kT/m)^2 = (8kT/\pi m)^{1/2}.$$

For Cs at 500 °C,

$$\bar{c} = \left\{ \frac{8 \times (1.381 \times 10^{-23} \, \text{J K}^{-1}) \times (773 \, \text{K})}{\pi \times (132.9) \times (1.6605 \times 10^{-27} \, \text{kg})} \right\}^{1/2} = 351.0 \, \text{m s}^{-1}.$$

- *Comment.* Note that this average is significantly higher than the one-dimensional atomic beam in the last *Example*.

Some of the properties of the Maxwell distribution were examined in the Introduction (p. 14). The main features are recorded in Fig. 25.4 which shows how the spread of speeds broadens as higher temperatures are attained, and how the most probable speed, marked c^* in the figure, shifts to higher values as the temperature is raised. (It is also clear from Fig. 0.11 that, for the same temperature, light molecules move at higher speeds than heavier ones.)

We have to be careful to distinguish several ways of quoting the mean speed of the molecules. The *root mean square* (r.m.s.) *speed c* is the square root of the average value of v^2

$(25.1.14)°$ $c = (3kT/m)^{1/2}.$

The *mean speed $\bar{c}$* is the mean of the speeds calculated using the Maxwell distribution in eqn (25.1.13):

$(25.1.15)°$ $\bar{c} = \langle v \rangle = \int_0^\infty vF(v)dv = (8kT/\pi m)^{1/2}$

and the *most probable speed c^** is the speed at which the Maxwell distribution passes through a maximum:

$(25.1.16)°$ $c^* = (2kT/m)^{1/2}.$

They differ by small numerical factors ($c \approx 1.225c^*$, $\bar{c} \approx 1.128c^*$).

The Maxwell distribution has received extensive direct and indirect experimental verification, and so we may take eqn (25.1.13) to be firmly established. Molecular speeds can be measured directly with a velocity-

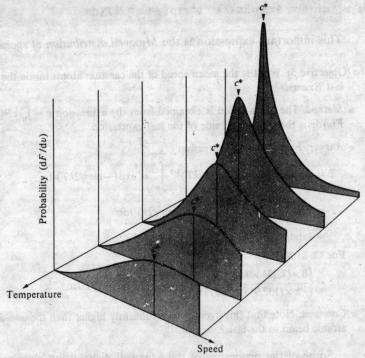

Fig. 25.4. The Maxwell distribution of speeds and its dependence on the temperature.

selector of the sort shown in Fig. 25.5. The spinning discs have slots which permit the passage of only those molecules moving through them at the appropriate speed, and the number of molecules can be determined by collecting them in a detector. The indirect method makes use of the Doppler effect on the wavelength of light emitted from a moving object and was discussed on p. 568. Some typical average speeds are given in Table 25.2.

Table 25.2. Average molecular speeds at 25 °C, $\bar{c}/\text{m s}^{-1}$

He	1256	H_2	1770	H_2O	592	C_6H_6	284
Ar	398	N_2	475	NH_3	609	Hg	177
		O_2	444	CO_2	379	Air	466
		Cl_2	298				

Source: Eqn (25.1.15) in the form $\bar{c}/\text{m s}^{-1} = 145.51 \, (T/K)^{1/2}/M_r^{1/2} = 2512.48/\sqrt{M_r}$ at 25 °C.

25.2 Collisions

We now calculate the frequency of intermolecular collisions. The principal features of the calculation were outlined in the Introduction. Here we summarize that calculation, and then improve it.

Intermolecular collisions. We count 'hit' whenever the centres of two molecules come within some distance d of each other, where d might be taken as their

detector

selector

source

Fig. 25.5. The determination of molecular speeds.

diameters, Fig. 25.6. The simplest approach to the problem is to freeze the positions of all the atoms except the one of interest, and then to observe what happens as the mobile one travels through the gas with an average speed $\bar{c}$ for a time Δt. In doing so it sweeps out a 'collision tube' of area $\sigma = \pi d^2$, length $\bar{c} \Delta t$, and therefore of volume $\sigma \bar{c} \Delta t$. σ is called the *collision cross-section*. The number of molecules with centres inside this volume is $\sigma \bar{c} \Delta t \mathcal{N}$, where $\mathcal{N}$ is the number of molecules per unit volume, and so the number of hits scored is $\sigma \bar{c} \Delta t \mathcal{N}$. Consequently the number of hits per unit time, the *collision frequency*, is $\sigma \bar{c} \mathcal{N}$.

The error in this calculation arises from the supposition that all the molecules are frozen in position. The $\bar{c}$ in the equation should really be the average *relative* speeds of the colliding molecules. When this is taken into account it turns out that $\bar{c}$ should be replaced by $\sqrt{2}\bar{c}$, and so the collision frequency is

$(25.2.1)°$ $\qquad z = \sqrt{2}\sigma \bar{c} N/V.$

This is the number of collisions a *single* molecule makes. If we require the total number of molecular collisions it is multiplied by $\frac{1}{2}N$ (the factor $\frac{1}{2}$ ensures that the collisions A...A' and A'...A are counted as only one collision). Therefore the number of collisions per unit volume per unit time is

$(25.2.2)°$ $\qquad Z_{AA} = \frac{1}{2}zN/V = \sigma \bar{c}(N/V)^2/\sqrt{2}.$

The value of $\bar{c}$ to use here has already been calculated in eqn (25.1.15), and so we conclude that

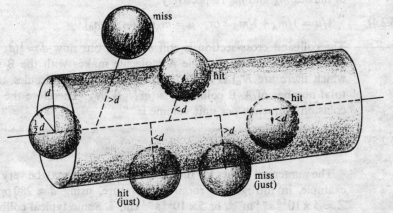

Fig. 25.6. The collision cross-section and the collision tube.

(25.2.3)° $Z_{AA} = \pi d^2 (4kT/\pi m)^{1/2}(N/V)^2.$

Example (Objective 6). We continue to explore the properties of the atomic beam apparatus
used in the last two *Examples*. How many collisions does a single Cs atom make
inside the oven in each second? How many collisions per second do all the atoms
inside the $50\,cm^3$ oven make?

- *Method.* Use eqn (25.2.1) for the collisions of a single atom and eqn (25.2.2) for
  the total number of collisions. Relate N/V to the pressure by $N/V = nL/V = p/kT$.
  The vapour pressure of Cs at $500\,°C$ is $80\,mmHg$. For the cross-section, use
  $d = 540\,pm$ (5.40 Å). Use $\bar{c}$ from the last *Example*.

- *Answer.* From eqn (25.2.1), $z = 2^{1/2}\sigma\bar{c}(p/kT)$;

  $\sigma = \pi d^2 = \pi \times (540 \times 10^{-12}\,m)^2 = 9.16 \times 10^{-19}\,m^2;$

  $p = (80/760) \times (1.013\,25 \times 10^5\,N\,m^{-2}) = 1.067 \times 10^4\,N\,m^{-2}.$

  Therefore

  $$z = \frac{2^{1/2} \times (9.16 \times 10^{-19}\,m^2) \times (351.0\,m\,s^{-1}) \times (1.067 \times 10^4\,N\,m^{-2})}{(1.3807 \times 10^{-23}\,J\,K^{-1}) \times (773\,K)}$$

  $= 4.54 \times 10^8\,s^{-1}.$

  From eqn (25.2.2), written in the form $Z_{CsCs} = \frac{1}{2}z(p/kT)$ we have

  $$Z_{CsCs} = \frac{\frac{1}{2} \times (4.54 \times 10^8\,s^{-1}) \times (1.067 \times 10^4\,N\,m^{-2})}{(1.3807 \times 10^{-23}\,J\,K^{-1}) \times (773\,K)}$$

  $= 2.27 \times 10^{32}\,s^{-1}\,m^{-3}.$

  Therefore, in a $50\,cm^3$ oven, the number of collisions per second is 1.14×10^{28}.

- *Comment.* Collisions will continue in the atomic beam, and they make it spread;
  after velocity-selection, collisions tend to return the velocity distribution back to
  the Boltzmann form. The effect can be retarded by working with a very weak beam.

If we were interested in the collision frequency of dissimilar molecules
the analysis would be modified only a little. In the first place the relative
speed is $(8kT/\pi\mu)^{1/2}$, where μ is the *reduced mass* of the colliding species
of masses m_A and m_B respectively:

(25.2.4) $1/\mu = 1/m_A + 1/m_B$ or $\mu = m_A m_B/(m_A + m_B).$

The collision cross-section is still $\sigma = \pi d^2$, but now $d = \frac{1}{2}(d_A + d_B)$. The
number of collisions a single A molecule makes with the B species (of
which there are N_B) is $\sigma\bar{c}N_B/V$. There are N_A A molecules, and so the
total number of A–B collisions is $(\sigma\bar{c}N_B/V)N_A$. Therefore the total A–B
collision frequency per unit volume is

(25.2.5)° $Z_{AB} = \pi d^2 (8kT/\pi\mu)^{1/2}(N_A N_B/V^2).$

The numerical values for the collision frequency may be very large. For
example, in nitrogen at room temperature, using $d \approx 280\,pm$ we find
$Z \approx 5 \times 10^{34}\,s^{-1}\,m^{-3}$, or $5 \times 10^{28}\,s^{-1}\,cm^{-3}$. Some typical collision cross-
sections are given in Table 25.3.

Table 25.3. Collision cross-sections, σ/nm^2

He	0.21	H_2	0.27	CO_2	0.52
Ne	0.24	N_2	0.43	SO_2	0.58
Ar	0.36	O_2	0.40	CH_4	0.46
		Cl_2	0.93	C_2H_4	0.64
				C_6H_6	0.88

Source: G. W. C. Kaye and T. H. Laby, *Tables of physical and chemical constants*, Longmans.

The calculation of the collision frequency allows us to determine an expression for the *mean free path* λ. If a molecule is travelling with a relative speed $\bar{c}$ and collides with a frequency z, it spends a time $1/z$ in free flight between collisions, and therefore travels a distance $\bar{c}/z$ between collisions. Therefore the mean free path is simply

$(25.2.6)°$
$$\lambda = \bar{c}/z = 1/\{\sqrt{2}\sigma(N/V)\}.$$

This can be expressed in terms of the pressure of the gas by writing $N = nL$, where n is the amount of gas, and then using $n/V = p/RT$ from the perfect gas equation:

$(25.2.7)°$
$$\lambda = (1/\sqrt{2}\sigma)(RT/Lp) = (1/\sqrt{2}\sigma)(kT/p)$$

and we see that the mean free path is inversely proportional to the pressure.

Example (Objective 6). Calculate the mean free path of Cs atoms in the oven of the last *Example* at 500 °C.

- *Method.* Use eqn (25.2.7) with $p = 80\,\text{mmHg}$, $\sigma = 9.2 \times 10^{-19}\,\text{m}^2$.
- *Answer.* From eqn (25.2.7),

$$\lambda = \frac{(1.381 \times 10^{-23}\,\text{J K}^{-1}) \times (773\,\text{K})}{2^{1/2} \times (9.2 \times 10^{-19}\,\text{m}^2) \times (1.067 \times 10^4\,\text{N m}^{-2})}$$

$$= 769 \times 10^{-9}\,\text{m, or } 769\,\text{nm}.$$

- *Comment.* The mean free path is about 1400 atomic diameters, and so we are in the range where the kinetic theory is likely to give reliable results.

Collisions with walls and surfaces. The estimation of the number of collisions with a given area of plane surface is straightforward, and underlies many of the calculations involved in the study of transport properties. In its simplest form the argument runs as follows.

Consider a wall of area A perpendicular to the x-axis. In the container the number density is $\mathcal{N}$ ($\mathcal{N} = N/V$). If a molecule has a velocity v_x lying between 0 and $+\infty$ it will strike the wall in a time Δt if it lies within a distance $v_x \Delta t$ of it (if it has a v_x lying between 0 and $-\infty$ it is moving in the wrong direction). Therefore all molecules in the volume $A v_x \Delta t$, with velocities in the right direction, will strike the wall in the interval Δt. The total number of collisions (on average) in this time is therefore

$$\text{No. of collisions} = \mathcal{N} A \Delta t \int_0^\infty v_x f(v_x)\, dv_x.$$

The integral can be evaluated easily using the explicit form of the velocity distributions:

$$\int_0^\infty v_x f(v_x)\, dv_x = (m/2\pi kT)^{1/2} \int_0^\infty v_x \exp(-mv_x^2/2kT)\, dv_x = (kT/2\pi m)^{1/2}.$$

Therefore the number of collisions per unit time per unit area is

(25.2.8)° $\qquad Z_W = (kT/2\pi m)^{1/2} N/V = \tfrac{1}{4}\bar{c} N/V.$

The number density N/V can be converted into an expression in terms of the pressure by writing $N/V = nL/V = p/kT$, and so

(25.2.9)° $\qquad Z_W = \tfrac{1}{4} p\bar{c}/kT = p/(2\pi mkT)^{1/2}.$

When the pressure is 1 atm $(1.01 \times 10^5\,\mathrm{N\,m^{-2}})$ and the temperature is 300 K, a container suffers about 3×10^{23} collisions per second per cm^2.

25.3 Transport properties

The flow of material from one region to another is an example of a *transport property*. For example, if a gas is confined to a container but open to a low-pressure region through a small hole, the gas will flow through the hole until the pressures are equal on both sides (this process is called *effusion*). When a gas is prepared with uniform density but non-uniform composition, the molecules *diffuse* through the system until the composition is uniform throughout. For example, the diffusion of a radioactive isotope into a non-radioactive gas is a transport process, and one type of molecule is transported past the other.

Transport processes occur in gases, liquids, and solids. A salt dissolves in a solvent, and the whole system becomes uniform by a process of diffusion. Two solids left in contact diffuse into each other at a very slow rate unless the temperature is high.

Transport processes are not confined simply to mass transfer. When a system is heated in one region the thermal motion is transported through it until its temperature is uniform. *Thermal conductivity* is the transport of energy in this way. *Electric conductivity* represents the transport of charge, either through the agency of the migration of electrons, as in metals, or by the migration of ions, as in ionic solutions or ionized gases. *Viscosity* is a transport property of a more subtle kind, and as we shall see, it depends on the transport of momentum.

In this section we shall examine some of the general aspects of transport, and then calculate the rates of some processes on the basis of the kinetic theory of gases. The subject is developed further in the next chapter, where we deal with transport processes in liquids, and especially with the motion of ions.

Flux. If matter, energy, charge, or momentum are being transported from one region
of a system to another we need a measure of the rate of flow. This is provided by the quantity called the *flux*, which is the amount of the property
passing through unit area in unit time. If mass is flowing we would speak
of a flux of so many $kg\,m^{-2}\,s^{-1}$; if energy were the property the flux
would be expressed in $J\,m^{-2}\,s^{-1}$, and so on. We shall denote flux by the
symbol J.

Experimental observations on the rates of various transport processes
show that the flux of a property is usually proportional to the gradient
of some other property of the system. For example, the rate of diffusion
(matter transport) is observed to be proportional to the concentration
gradient:

$$J_z(\text{matter}) \propto d\mathcal{N}/dz,$$

and the rate of thermal diffusion (energy transport) is observed to be proportional to the temperature gradient:

$$J_z(\text{energy}) \propto dT/dz.$$

The subscript on the flux indicates that the flow of the property along
the z-axis is proportional to the gradient along that axis.

We shall treat J as a vector, and *a positive sign indicates flow towards
increasing* z (to the right), *and a negative sign a flow to decreasing* z (to
the left). Mass flow occurs *down* a concentration gradient, and so if $d\mathcal{N}/dz$
is negative (concentration decreasing to the right, Fig. 25.7) J_z is observed
to be positive (flow to the right). Therefore the constant of proportionality
in the mass flux expression must be negative. If we write it as $-D$, where
D is the *diffusion coefficient*, we obtain

(25.3.1)   *Fick's First Law of Diffusion:* $J_z(\text{matter}) = -D(d\mathcal{N}/dz).$

Energy flows as heat down a temperature gradient, and so the same
reasoning leads to

(25.3.2) $J_z(\text{energy}) = -\kappa(dT/dz)$

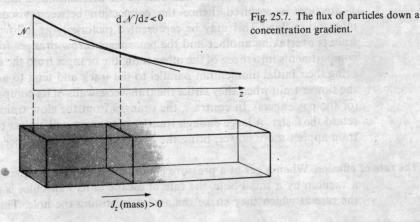

$d\mathcal{N}/dz < 0$

$J_z(\text{mass}) > 0$

Fig. 25.7. The flux of particles down a
concentration gradient.

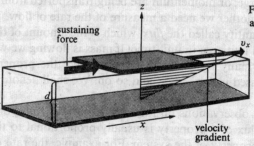

Fig. 25.8. Transport of momentum across a velocity gradient.

where κ (kappa) is the *coefficient of thermal conductivity*.

The connection between the flux of momentum and viscosity might appear odd at first, but it can be explained as follows. Consider a gas between two parallel plates, Fig. 25.8. In order to move the top plate relative to the bottom one a force must be applied in the x-direction, and the force must be maintained in order to sustain the motion. The velocity of the gas varies from zero (next to the bottom, stationary plate) to v_z (next to the upper plate), and in the region of *Newtonian flow* the velocity varies uniformly from one plate to the other. Observation shows that the force required to maintain the velocity of the top plate is proportional to the area of the plate (A) and the gradient of the velocity dv_x/dz ($= v_x/d$):

(25.3.3)   *required sustaining force* $\propto A(dv_x/dz)$,

(25.3.4) $F_x/A = \eta(dv_x/dz)$,

where η (eta) is the *coefficient of viscosity* (or simply the *viscosity*).

Now we concentrate on one of the layers of constant velocity into which we may imagine the gas to be divided, Fig. 25.9. The molecules in this layer move with some velocity v'_x, and so they each have a momentum mv'_x. There is a constant exchange of molecules between the layers, and molecules from above bring greater momenta (they come from a faster layer) and those from below bring smaller momenta. The net flow of momentum into unit area of the lamina of interest is J (momentum along x), and in order to sustain this change of momentum a force in the x-direction is required, hence the connection between viscosity and momentum flow. This may be represented pictorially as follows. If one train is overtaking another and the passengers throw oranges from their compartments into those of the other train, the oranges from the fast train bring their initial momentum parallel to the track and tend to accelerate the slower train when they strike the transverse walls of its compartments (or the passengers). In contrast, the oranges from the slow train tend to retard the faster. A large enough transfer of oranges will, unless the faster train applies more power, bring the two trains to the same speed.

The rate of effusion. When a gas at a pressure p and temperature T is separated from a vacuum by a small hole, the rate of escape of its molecules is equal to the rate at which they strike the area constituting the hole. This rate is

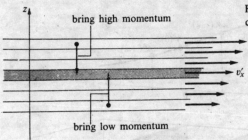

bring high momentum

z

v'_x

bring low momentum

Fig. 25.9. The transport involved in calculating the viscosity of a gas.

given by eqn (25.2.8). If the area of the hole is A_0 the number of molecules that emerge in unit time is $Z_W A_0$.

$(25.3.5)°$ $Z_W A_0 = pA_0/(2\pi mkT)^{1/2}.$

The rate of effusion is inversely proportional to the square-root of the relative molecular mass: this is *Graham's Law of Effusion*. This expression is also the basis of the *Knudsen method* for the determination of the relative molar mass, or, if that is known, the vapour pressure of a solid. Thus, if a solid has a vapour pressure p and is enclosed in a cavity equipped with a small hole, the rate of loss of mass from the container is proportional to p, and so p may be determined from eqn (25.3.5) if the mass loss is monitored.

Example (Objective 10). The atomic beam device used in the preceding *Examples* was used to measure the vapour pressure of Cs at 500 °C. The diameter of the hole in the wall was 0.5 mm, and in a period of 100 s the mass loss was 385 mg.

- *Method.* Use eqn (25.3.5) in the form $\delta m = Z_W A_0 \tau m$, where m is the mass of an atom, τ is the length of time for which effusion occurs, and δm is the loss of mass from the containers. Then $p = (2\pi kT/m)^{1/2}(\delta m/A_0 \tau)$.

- *Answer.* From the expression just quoted,

$$p = \left\{ \frac{2\pi \times (1.3801 \times 10^{-23} \times 773 \, \text{J})}{132.9 \times (1.6605 \times 10^{-27} \, \text{kg})} \right\}^{1/2} \left\{ \frac{(0.385 \times 10^{-3} \, \text{kg})}{\pi (0.25 \times 10^{-3} \, \text{m})^2 \times (100 \, \text{s})} \right\}$$

$$= 1.081 \times 10^4 \, \text{N m}^{-2}.$$

- *Comment.* Convert to mmHg by dividing by $1.013 \, 25 \times 10^5 \, \text{N m}^{-2}/\text{atm}$ and multiplying by 760 mmHg/atm: this gives 81.11 mmHg.

The calculation of the rate of effusion relies on the hole being much shorter than the mean free path of the gas molecules. If it is so long that they undergo many collisions inside the tube the transport is more like viscous flow, and the rate of effusion is governed by the viscosity. This aspect is taken up on p. 883.

The rate of diffusion. The problem now is to determine the value of the diffusion coefficient for a gas, and to provide a theoretical basis for Fick's Law, eqn (25.3.1).

We have to show that the flux of particles is proportional to the gradient

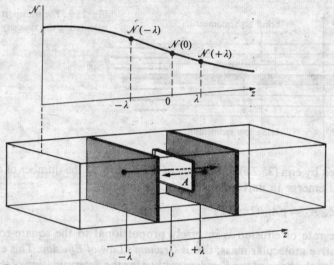

Fig. 25.10. Calculating the rate of diffusion of a gas.

of their concentration. In order to do this, consider the sample of gas shown in Fig. 25.10, and in which there is a variation of the concentration along the z-direction. We calculate the flux by finding the net flow of molecules per unit time through the imaginary window of area A set perpendicular to the z-axis. The key to this kind of calculation is the mean free path of the molecules. A molecule can be identified as originating from some point where the density of particles is $\mathcal{N}(z)$ only if the point z is no farther away from the window than some mean free path distance. If it is farther away it will probably have undergone a collision with another molecule.

On average, the molecules passing through the imaginary window have travelled about one mean free path, and so the average concentration in the region where they come from is about $\mathcal{N}(-\lambda) \approx \mathcal{N}(0) - \lambda(\mathrm{d}\mathcal{N}/\mathrm{d}z)_0$, where the subscript 0 indicates that the gradient is evaluated at the hole (where $z = 0$). Since the number of impacts on the window from the left in an interval Δt is on the average $\frac{1}{4}\mathcal{N}(-\lambda)\bar{c}A\,\Delta t$, the flux from left to right arising from the concentration on the left is

$$J(\mathrm{L} \to \mathrm{R}) \approx \tfrac{1}{4}\mathcal{N}(-\lambda)\bar{c}.$$

There is also a passage of molecules from right to left. On average the molecules that penetrate the window have originated from a distance λ to the right, and so the flux from right to left is

$$J(\mathrm{R} \to \mathrm{L}) \approx \tfrac{1}{4}\mathcal{N}(\lambda)\bar{c}.$$

The average concentration supplying the window from the right is $\mathcal{N}(\lambda) \approx \mathcal{N}(0) + \lambda(\mathrm{d}\mathcal{N}/\mathrm{d}z)_0$. The flow from the more concentrated region on the left dominates the back-wash, and so the net flux is positive (from left to right) and has the magnitude

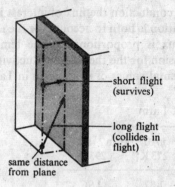

short flight (survives)

long flight (collides in flight)

same distance from plane

Fig. 25.11. Some molecules travel further in order to cross a plane.

$$J_z = \tfrac{1}{4}\{[\mathcal{N}(0) - \lambda(\mathrm{d}\mathcal{N}/\mathrm{d}z)_0] - [\mathcal{N}(0) + \lambda(\mathrm{d}\mathcal{N}/\mathrm{d}z)_0]\}\bar{c}$$
$$= -\tfrac{1}{2}\lambda\bar{c}(\mathrm{d}\mathcal{N}/\mathrm{d}z)_0$$

and is proportional to the gradient of the concentration, in accord with Fick's Law:

$$J_z = -D(\mathrm{d}\mathcal{N}/\mathrm{d}z)_0.$$

This calculation also provides an expression for the diffusion constant because D can be identified as $\tfrac{1}{2}\lambda\bar{c}$. It must be remembered, however, that the calculation is quite crude, and really amounts to an example of how to build an approximate model of the process of interest, and to assess a rough order of magnitude of the diffusion constant. One of the aspects of the model that has not been taken into account properly is illustrated in Fig. 25.11. This shows that although a molecule may have begun its journey very close to the window, it could have a long path before it penetrates. Since the path is long it stands a high probability of colliding before passing through the window, and so this molecule ought to be added to the graveyard of other molecules which have suffered a collision. As might be imagined, taking account of this effect introduces considerable labour into the calculation, but the net result turns out to be that the result we have derived is modified only to the extent of a factor of $\tfrac{2}{3}$, representing the lower flux because of the collisions suffered by the longer travelling particles.

The result of this modification is to permit the identification

(25.3.6)° $\qquad J_z = -\tfrac{1}{3}\lambda\bar{c}(\mathrm{d}\mathcal{N}/\mathrm{d}z)_0 \quad \text{with} \quad J_z = -D(\mathrm{d}\mathcal{N}/\mathrm{d}z)_0,$

which implies that

(25.3.7)° $\qquad D = \tfrac{1}{3}\lambda\bar{c}.$

Since the mean free path decreases with increasing pressure, and the mean speed increases with temperature, the diffusion coefficient has the same properties. From eqns (25.1.15) and (25.2.7) we find explicitly

(25.3.8)° $\qquad D = \tfrac{1}{3}(1/2^{1/2}\sigma)(kT/p)(8kT/\pi m)^{1/2} = \tfrac{2}{3}(1/\sigma p)(k^3T^3/m\pi)^{1/2},$

where σ is the collision cross-section πd^2.

Thermal conductivity. In the case of thermal conduction the flux of interest is that of energy, and the aim of the calculation is both to account for the empirical observation that the thermal flux is proportional to the temperature gradient, and to derive an expression for the thermal conductivity coefficient, eqn (25.3.2). Some experimental values of κ are given in Table 25.4.

Table 25.4. Transport properties of gases at 1 atm

	$\kappa/10^{-3}\,\mathrm{J\,cm^{-1}\,s^{-1}\,K^{-1}}$	$\eta/10^{-5}\,\mathrm{kg\,m^{-1}\,s^{-1}}$	
	273 K	273 K	293 K
He	1.4422	1.87	1.96
Ne	0.465	2.98	3.13
Ar	0.163	2.10	2.23
H_2	1.682	0.84	0.88
O_2	0.245	1.95	2.04
N_2	0.240	1.66	1.76
CO_2	0.145	1.36	1.47
Air	0.241	1.73	1.82

Source: G. W. C. Kaye and T. H. Laby, *Tables of physical and chemical constants*, Longmans.

Suppose every molecule carries an amount of energy ε (epsilon). When one passes through the imaginary window it transports that amount of energy. But the rate of transfer of molecules is the flux calculated in the last section, eqn (25.3.6). Therefore the flow of thermal energy is

$$J_z(\text{heat}) = \varepsilon J_z(\text{matter})$$
$$= -\tfrac{1}{3}\lambda \bar{c}\varepsilon(d\mathcal{N}/dz)_0$$
$$= -\tfrac{1}{3}\lambda \bar{c}(d\mathcal{E}/dz)_0,$$

where $\mathcal{E} = \varepsilon\mathcal{N}$ is the energy density (the amount of energy per unit volume). This shows that the energy flux is determined by the energy gradient. The energy gradient can be related to the temperature gradient by means of the heat capacity of the sample. If we let $C_{V,\mathrm{m}}$ denote the molar heat capacity at constant volume, the energy density gradient is

$$(d\mathcal{E}/dz)_0 = (n/V)C_{V,\mathrm{m}}(dT/dz)_0.$$

The flux is then

$$J_z(\text{energy}) \doteq -\tfrac{1}{3}\lambda \bar{c}(n/V)C_{V,\mathrm{m}}(dT/dz)_0$$

and so the coefficient of thermal conductivity is

$$(25.3.9)^\circ \qquad \kappa = \tfrac{1}{3}\lambda \bar{c}C_{V,\mathrm{m}}(n/V).$$

The feature to notice is that, because the mean free path is inversely proportional to n/V, κ is independent of the density. Explicitly

$$(25.3.10)^\circ \qquad \kappa = \tfrac{1}{3}\bar{c}C_{V,\mathrm{m}}/\sqrt{2}\sigma L.$$

This shows that κ is *independent of pressure*. The physical reason for this pressure independence is that the thermal conductivity is large when many molecules are available to transport the energy (hence the factor n in eqn (25.3.9)) but the presence of many molecules limits the mean free paths, and so the molecules are unable to carry the energy over great distances: these two effects balance. The thermal conductivity is found to be independent of pressure experimentally, except when the pressure is very low, when there is a linear dependence. The reason for this is that at very low pressures the mean free path is greater than the dimensions of the apparatus, and so the path length over which the energy is transported is determined by the size of the vessel and not by the other molecules present. The energy transported is proportional to the number of carriers, but the length of the journey is independent of their number, hence under these circumstances $\kappa \propto n/V \propto p$.

Example (Objective 12). Estimate the thermal conductivity of air at room temperature.

- *Method.* Use eqn (25.3.10). The molar heat capacity may be taken as $\frac{5}{2}R$, p. 713. Taking $M_r \approx 30$ for the mean R.M.M. of air is sufficiently accurate; so too is $\sigma \approx 0.42 \, \text{nm}^2$ (Table 25.3, q.v. O_2 and N_2). Estimate $\bar{c}$ from eqn (25.1.15).

- *Answer.* From eqn (25.1.15),

$$\bar{c} = \left\{ \frac{(8/\pi) \times (1.381 \times 10^{-23} \, \text{J K}^{-1}) \times (298.15 \, \text{K})}{30 \times (1.661 \times 10^{-27} \, \text{kg})} \right\}^{1/2} = 459 \, \text{m s}^{-1}.$$

Then from eqn (25.3.10),

$$\kappa = \frac{\frac{1}{3} \times (459 \, \text{m s}^{-1}) \times \frac{5}{2} \times (8.314 \, \text{J K}^{-1} \, \text{mol}^{-1})}{\sqrt{2} \times (0.42 \times 10^{-18} \, \text{m}^2) \times (6.022 \times 10^{23} \, \text{mol}^{-1})}$$

$$= 12.9 \times 10^{-3} \, \text{J K}^{-1} \, \text{m}^{-1} \, \text{s}^{-1}, \quad \text{or} \quad 0.129 \times 10^{-3} \, \text{J cm}^{-1} \, \text{s}^{-1} \, \text{K}^{-1}.$$

- *Comment.* If there were a temperature gradient of $10 \, \text{K cm}^{-1}$, then the flux of heat would be $(0.129 \times 10^{-3} \, \text{J K}^{-1} \, \text{cm}^{-1} \, \text{s}^{-1}) \times (10 \, \text{K cm}^{-1})$, or $1.29 \times 10^{-3} \, \text{J cm}^{-2} \, \text{s}^{-1}$.

The viscosity of gases. We have already seen that the coefficient of viscosity is determined by the rate at which momentum is transported. Molecules travelling downwards in Fig. 25.12 (from a fast layer to a slower) transport

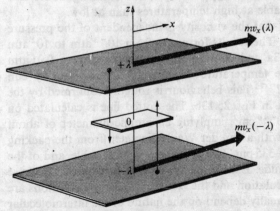

Fig. 25.12. Calculating the viscosity of a gas.

momentum $mv_x(\lambda)$ to their new layer at $z = 0$, and those travelling upwards transport $mv_x(-\lambda)$ to it. If the density of molecules is uniformly $\mathcal{N}$, the number of impacts on the imaginary window is $\frac{1}{4}\mathcal{N}\bar{c}$ per unit area per unit time. Those from above have an average x-component of momentum $mv_x(\lambda) = mv_x(0) + m\lambda(dv_x/dz)_0$ because on average they originate from the plane at a distance λ from the window. Those from below have a momentum $mv_x(-\lambda) = mv_x(0) - m\lambda(dv_x/dz)_0$. Therefore the net flux of x-momentum is

$$J_x(\text{momentum along } x) = \frac{1}{4}\mathcal{N}\bar{c}\{[mv_x(0) - m\lambda(dv_x/dz)_0]$$
$$- [mv_x(0) + m\lambda(dv_x/dz)_0]\}$$
$$= -\frac{1}{2}m\mathcal{N}\lambda\bar{c}(dv_x/dz)_0.$$

Comparison of this with the definition of the coefficient of viscosity, eqn (22.3.4) and the replacement of the factor $\frac{1}{2}$ by $\frac{1}{3}$ leads to

$(25.3.11)^\circ$ $\eta = \frac{1}{3}m\lambda\mathcal{N}\bar{c} = \frac{1}{3}m\lambda\bar{c}L(n/V).$

As in the case of thermal conductivity, the viscosity is independent of the pressure. This conclusion can be seen explicitly by introducing the expression for the mean free path:

$(25.3.12)^\circ$ $\eta = \frac{1}{3}m\bar{c}L(n/V)/2^{1/2}\sigma L(n/V) = \frac{1}{3}(\frac{1}{2})^{1/2}m\bar{c}/\sigma.$

The physical reason is the same as in the conductivity case: more molecules are available to transport the momentum, but fewer succeed because of their shortened mean free path. It should also be noticed that, according to the last equation, the viscosity *increases* with temperature because $\bar{c} \propto T^{1/2}$. This peculiar result arises because at higher temperatures the momentum is transported more rapidly through a given area, and so the force has to be increased to maintain the motion of the layers of gas (more oranges are thrown between the trains at higher temperatures). This viscosity behaviour is quite different from that shown by liquids, which flow more easily as temperature is increased. This difference stems from the fact that a liquid's viscosity is dominated by intermolecular forces: in order to flow, molecules need energy to escape from their neighbours, and this is more freely available at high temperatures than at low.

Experiments confirm that the viscosity is independent of the pressure over a very wide range. The results for argon from 10^{-3} atm to 10^2 atm are shown in Fig. 25.13a and confirm its invariance between 0.01 atm and about 50 atm. The temperature dependence is predicted by eqn (25.3.12) to depend on $T^{1/2}$. This behaviour is roughly confirmed by the results for argon shown in Fig. 25.13b. The dotted line is calculated on the basis of $\sigma = 22 \times 10^{-16}\ cm^2$, implying an atomic diameter of about 260 pm, in comparison with a van der Waals diameter (from the packing of atoms in the solid) of 335 pm. The agreement is reasonable, and of the correct order of magnitude. Exact equality cannot be expected because of the crudity of the calculation and the fact that molecular collisions are dynamic processes and really depend on the nature of the intermolecular

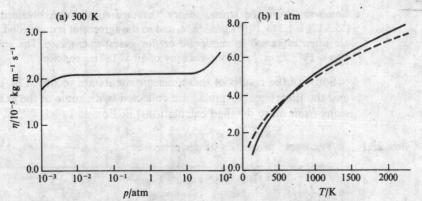

Fig. 25.13. The viscosity of argon as a function of (a) pressure and (b) temperature.

interactions.

Two principal methods for the determination of viscosities are in use. One depends on the observation of the rate of damping of the torsional oscillations of a disc surrounded by the gas. Analysis of the motion of the system shows that the amplitude of oscillation undergoes damped simple harmonic motion with a decay constant that depends on the viscosity of the gas and the details of construction of the apparatus.

The other method is based on *Poiseuille's formula* for the flow of fluid through a tube of radius r:

$$dV/dt = (p_1^2 - p_2^2)\pi r^4/16 l \eta p_0,$$

where V is the volume flowing, p_1 and p_2 are the pressures at each end of the tube of length l, and p_0 is the pressure at which the volume is measured. In order to determine the viscosity, the rate of flow under a known pressure difference is monitored and interpreted using this equation.

Example (Objective 13). In a Poiseuille's flow experiment to measure the viscosity of air at 298 K, the sample was allowed to flow through a 1 m tube of bore 1 mm. The high-pressure end was at 765 mmHg and the low-pressure end at 760 mmHg. The volume was measured at the latter pressure. In 100 s a volume of 90.2 cm³ passed through the tube. What is the viscosity of air at this temperature?

● *Method.* Substitute in the last equation. Convert pressures to $N\,m^{-2}$.

● *Answer.*

$$p_1^2 - p_2^2 = (765^2 - 760^2) \times (1.013\,25 \times 10^5\,N\,m^{-2})^2/(760)^2 = 1.355 \times 10^8\,N^2\,m^{-4}.$$

$$(\pi r^4/16 l p_0) = \frac{\pi \times (0.5 \times 10^{-3}\,m)^4}{16 \times (1\,m) \times (1.013\,25 \times 10^5\,N\,m^{-2})} = 1.211 \times 10^{-19}\,N^{-1}\,m^5.$$

$$dV/dt = (90.2\,cm^3)/(100\,s) = 9.02 \times 10^{-7}\,m^3\,s^{-1}.$$

Hence,

$$\eta = \frac{(1.211 \times 10^{-19}\,N^{-1}\,m^5) \times (1.355 \times 10^8\,N^2\,m^{-4})}{(9.02 \times 10^{-7}\,m^3\,s^{-1})} = 1.82 \times 10^{-5}\,kg\,m^{-1}\,s^{-1}.$$

• *Comment.* From the kinetic theory, the theoretical value, obtained from eqn (25.3.12), is $1.39 \times 10^{-5}\,\mathrm{kg\,m^{-1}\,s^{-1}}$, and so the agreement is quite good. Viscosities are often expressed in centipoise or (for gases) micropoise. The conversion is $1\,\mathrm{cp} = 10^{-3}\,\mathrm{kg\,m^{-1}\,s^{-1}}$: the viscosity of air is 182 micropoise.

Some of the results of these measurements are recorded in Table 25.4, and the theoretical formulae are collected (with some of the exact expressions from more detailed calculations) in Box 25.1.

Box 25.1

Transport properties of perfect gases

Property	Transported quantity	Simple kinetic theory	Refined theory	Units
Diffusion	Matter	$D = \frac{1}{3}\lambda\bar{c}$ $= kT\bar{c}/3p\sigma\sqrt{2}$	$\frac{3}{16}\lambda\bar{c}$	$\mathrm{m^2\,s^{-1}}$
Conductivity	Thermal energy	$\kappa = \frac{1}{3}\lambda\bar{c}C_{V,m}\mathcal{N}$ $= \bar{c}C_{V,m}/3\sigma L\sqrt{2}$	$(25\pi/64)\lambda\bar{c}C_{V,m}\mathcal{N}$	$\mathrm{J\,K^{-1}\,m^{-1}\,s^{-1}}$
Viscosity	Momentum	$\eta = \frac{1}{3}\lambda\bar{c}m\mathcal{N}$ $= m\bar{c}/3\sigma\sqrt{2}$	$0.499\lambda\bar{c}m\mathcal{N}$	$\mathrm{kg\,m^{-1}\,s^{-1}}$

$\mathcal{N}$ is the number density: $\mathcal{N} = N/V = nL/V = p/kT$.
λ is the mean free path: $\lambda = 1/\sigma\mathcal{N}\sqrt{2} = kT/\sigma p\sqrt{2}$.

Further reading

Kinetic theory of gases. W. Kauzmann; Benjamin, New York, 1966.
An introduction to kinetic theory. J. H. Hildebrand; Reinhold, New York, 1963.
Kinetic theory of gases. L. B. Loeb; Dover, New York, 1961.
Kinetic theory of gases. E. H. Kennard; McGraw-Hill, New York, 1938.
Kinetic theory of gases. R. D. Present; McGraw-Hill, New York, 1958.
Introduction to the kinetic theory of gases. J. H. Jeans; Cambridge University Press, 1940.
The molecular theory of gases and liquids. J. O. Hirschfelder, C. F. Curtiss, and R. B. Bird; Wiley, New York, 1954.
Transport phenomena. R. B. Bird, W. E. Stewart, and E. N. Lightfoot; Wiley, New York, 1960.

Problems

25.1. It is important to have some familiarity with the magnitudes of physical quantities involved in the discussion of gases, and the first few Problems give some practice. As a first step we become familiar with the magnitudes of *distances*. What is the *mean free path* in air ($\sigma \approx 0.43\,\mathrm{nm^2}$) at 25 °C and (a) 10 atm, (b) 1 atm, (c) 10^{-6} atm?

25.2. At what pressure does the mean free path of argon at 25 °C become comparable to the size of the $1\,\mathrm{dm^3}$ bulb that contains it? Take $\sigma \approx 0.36\,\mathrm{nm^2}$.

25.3. At what pressure does the mean free path of argon at 25 °C become comparable to the size of the atoms themselves?

25.4. At an altitude of 20 km the temperature is 217 K and the pressure 0.05 atm. What is the mean free path of nitrogen molecules? ($\sigma \approx 0.43\,\mathrm{nm^2}$.)

25.5. Now we turn to the time scale of events in gases. How many collisions does a single argon atom make in 1 s when the temperature is 25 °C and the pressure is (a) 10 atm, (b) 1 atm, (c) 10^{-6} atm?

25.6. What is the total number of molecular collisions in a 1 dm^3 sample of argon in the same conditions as in Problem 25.5?

25.7. The frequency of molecular collisions is an important quantity in atmospheric chemistry. How many collisions per second does an excited nitrogen molecule make at an altitude of 20 km? (See Problem 25.4.)

25.8. Calculate the number of collisions per cm^3 in air at 25 °C and 1 atm (a) between oxygen molecules, (b) between oxygen and nitrogen molecules. Take $R(O_2) \approx 178$ pm and $R(N_2) \approx 185$ pm.

25.9. How many collisions per cm^3 are made in the interior of the envelope of a Dewar flask where the remaining air is at a pressure of 1.2 mmHg?

25.10. What is the mean speed of (a) helium atoms, (b) methane molecules at (i) 77 K, (ii) 298 K, (iii) 1000 K?

25.11. What is the mean translational kinetic energy (in kJ mol^{-1}) of (a) hydrogen molecules, (b) iodine molecules in a gas at 300 K and 1 atm pressure?

25.12. In an experiment to measure the speed of molecules by a rotating slotted-disc experiment, the apparatus consisted of 5 coaxial 5 cm diameter discs separated by 1 cm, the slots in their rims being displaced by 2° between neighbours. The relative intensity I of the detected beam of krypton atoms for two different temperatures and at a series of rotation rates were as follows:

ν/Hz	20	40	80	100	120
I(40 K)	0.846	0.513	0.069	0.015	0.002
I(100 K)	0.592	0.485	0.217	0.119	0.057

Find the distribution of molecular velocities at these temperatures, and check that they conform to the theoretical prediction.

25.13. The Maxwell–Boltzmann distribution of velocities, eqn. (25.1.11), was derived from arguments about probability. But it was stated that it could also be derived from the Boltzmann distribution itself. Do so.

25.14. A specially constructed velocity-selector accepts a beam of molecules from an oven at a temperature T but blocks the passage of molecules with a speed greater than the mean. What is the mean speed of the emerging beam? (Note that this is essentially a one-dimensional problem.)

25.15. We encountered two ways of dealing with averages: one was for a discrete distribution, the other for a continuous distribution. Here is a chance to practise the manipulations involved. Cars are timed by police radar as they pass in both directions below a bridge. Their velocities (m.p.h., numbers of cars in brackets) to the east and west are as follows: 50 E (40), 55 E (62), 60 E (53), 65 E (12), 70 E (2); 50 W (38), 55 W (59), 60 W (50), 65 W (10), 70 W (2). What are (a) the mean velocity, (b) the mean speed, (c) the root mean square speed?

25.16. A population consists of people of the following heights (in feet and inches, numbers of individuals in brackets): 5′ 5″ (1), 5′ 6″ (2), 5′ 7″ (4), 5′ 8″ (7), 5′ 9″ (10), 5′ 10″ (15), 5′ 11″ (9), 6′ 0″ (4), 6′ 1″ (0), 6′ 2″ (1). What is (a) the mean height, (b) the root mean square height of the population.

25.17. What is the proportion of gas molecules having (a) more than, (b) less than the root mean square speed? What are the proportions having speeds greater and smaller than the mean speed ($\bar{c}$)?

25.18. What are the proportions of molecules in a gas that have a speed in a range at the speed nc^* relative to those in the same range at c^* itself? This calculation can be used to obtain an idea of the proportion of very energetic molecules (which is important when we turn to reactions). Evaluate the ratio for $n = 3$ and $n = 4$.

25.19. What is the *escape velocity* from the surface of a planet of radius R? What is the value for (a) the Earth, $R = 6.37 \times 10^6$ m, $g = 9.81$ m s^{-2}, (b) Mars, $R = 3.38 \times 10^6$ m, $m_{Mars}/m_{Earth} = 0.108$. At what temperatures do hydrogen, helium, and oxygen have *mean* speeds equal to their escape speeds? What proportion of the molecules have enough speed to escape when the temperature is (a) 240 K, (b) 1500 K? Calculations of this kind are very important in considering the composition of planetary atmospheres.

25.20. A manometer was connected to a bulb containing carbon dioxide under slight pressure. The gas was allowed to escape through a small pinhole, and the time for the manometer reading to drop from 75 cm to 50 cm was 52 s. When the experiment was repeated using nitrogen ($M_r = 28$) the same fall took place in 42 s. What is the R.M.M. of carbon dioxide?

25.21. A particular make of electric light bulb contains argon at a pressure of 50 mmHg and has a tungsten filament of radius 0.1 mm and length 5 cm. When operating, the gas close to the filament has a temperature of about 1000 °C. How many collisions are made with the filament in each second?

25.22. A space vehicle of internal volume 3 m^3 is struck by a meteor and a hole 0.1 mm radius is formed. If the oxygen pressure within the vehicle is initially 0.8 atm and its temperature 298 K, how long will it take to fall to 0.7 atm?

25.23. The *Knudsen cell* makes use of the expression for the frequency of collisions with an area in order to measure the vapour pressure of slightly volatile solids. A weighed amount of the sample is heated inside a container, in the wall of which there is a small hole. The mass loss over a period of time can be related to the vapour pressure at the temperature of the experiment. If Δm is the mass lost in time t through a circular hole of radius R, find an expression relating the vapour pressure p to Δm and t.

25.24. A Knudsen cell was used to determine the vapour pressure of germanium at 1000 °C. During an interval of 2 hr the mass loss through a hole of radius 0.50 mm amounted to 4.3×10^{-2} mg. What is the vapour pressure of germanium at 1000 °C?

25.25. In a study of the catalytic properties of a titanium surface it was necessary to maintain the surface free from contamination. Calculate the number of collisions per cm^2 of surface made by oxygen at a pressure of (a) 1 atm, (b) 10^{-6} atm, (c) 10^{-10} atm and a temperature of 300 K. Estimate the number of collisions made with a single surface atom in each second. This underlines the importance of working at very low pressures in order to study the properties of uncontaminated surfaces. Take the nearest neighbour distance as 291 pm.

25.26. The nucleus $^{244}_{97}$Bk (berkelium) decays by producing α-particles which capture electrons and form helium atoms. Its half-life is 4.4 hours. A 1 mg sample was placed in a 1 cm^3 container which was impermeable to α-radiation, but there was also a hole of radius 0.002 mm in the wall (that is not quite a paradox). What is the pressure of helium inside the container after (a) 1 hr, (b) 10 hr?

25.27. A spherical glass bulb of diameter 10 cm has a tube of radius 3 mm attached to it. While the bulb is maintained at 300 K the tube is at 77 K (it is immersed in liquid nitrogen). Initially the gas in the bulb is damp, the partial pressure of the water vapour being 1 mmHg. Estimate the time required for condensation in the tube to lower the partial pressure to 10^{-5} mmHg.

25.28. An atomic beam is designed to function with (a) cadmium, (b) mercury. The source is an oven maintained at 380 K, there being a small slit of dimensions 1 cm $\times 10^{-3}$ cm. The vapour pressure of cadmium is 1.0×10^{-3} mmHg and that of mercury is 1138 mmHg at this temperature. What is the atomic current (the number of atoms per unit time) in the beams?

25.29. The reaction $H_2 + I_2 \rightarrow 2HI$ depends on collisions between a variety of species in the reaction mixture, and we shall look at it in more detail in Chapter 26. Calculate the collision frequencies for the encounters (a) $H_2 + H_2$, (b) $I_2 + I_2$, (c) $H_2 + I_2$ for a gas with partial pressures of 0.5 atm for both species, the temperature being 400 K. $\sigma(H_2) \approx 0.27$ nm^2, $\sigma(I_2) \approx 1.2$ nm^2.

25.30. What is the viscosity of air at (a) 0 K, (b) 298 K, (c) 1000 K? Take $\sigma \approx 0.40$ nm^2. (The experimental value is 1.82×10^{-5} kg m^{-1} s^{-1} at 20 °C and 3.94×10^{-5} kg m^{-1} s^{-1} at 600 °C.)

25.31. Calculate the thermal conductivities of (a) argon, (b) helium at 300 K and 10^{-3} atm. The gases are confined in a cubic vessel of 10 cm side, one wall being at 310 K and the one opposite at 295 K. What is the rate of flow of heat from one wall to the other?

25.32. The viscosity of carbon dioxide was measured by comparing its rate of flow through a long narrow tube (recall Poiseuille's formula) with that of argon. For the same pressure differential, the same volume of carbon dioxide passed through the tube in 55 s as argon in 83 s. The viscosity of argon at 25 °C is 208 μpoise (2.08×10^{-5} kg m^{-1} s^{-1}); what is the viscosity of carbon dioxide? Estimate the molecular diameter of carbon dioxide.

25.33. Calculate the thermal conductivity of argon ($C_{V,m} = 12.5$ J K^{-1} mol^{-1}, $\sigma \approx 0.36$ nm^2) and air ($C_{V,m} = 21.0$ J K^{-1} mol^{-1}, $\sigma \approx 0.40$ nm^2) at room temperature.

25.34. What is the ratio of the thermal conductivities of gaseous hydrogen at 300 K to gaseous hydrogen at 10 K? Be circumspect.

25.35. In a double-glazed window the panes of glass are separated by 5 cm. What is the rate of transfer of heat by conduction from the warm room (25 °C) to the cold exterior (-10 °C) through a window of area 1 m^2? What power of heater is required to make good the loss of heat?

25.36. Calculate the diffusion constant D for argon at 25 °C and (a) 10^{-6} atm, (b) 1 atm, (c) 100 atm. If a pressure gradient of 0.1 atm cm^{-1} is established in a pipe, what is the flow of gas due to diffusion?

25.37. The rate of growth of droplets of lead from a vapour has been studied in the laboratories of an oil company (J. B. Homer and A. Prothero, *J. chem. Soc. Faraday Trans. I* **69**, 673 (1973)). Virtually all the lead condensed within 0.5 ms of the initiation of the run, and the concentration in the gas phase was no more than about 3×10^{15} atoms/cm^3. Find an expression for the rate of growth of the radius of the spherical particles. Take $T = 935$ K and assume that every atom sticks to the growing surface.

25.38. A cosmological cataclysm had the effect of destroying a huge amount of matter somewhere distant in the universe. The consequence was that the inertial mass of all remaining atoms was reduced by 25 per cent. Could the event be detected by examining the properties of a perfect gas? Would the equation $pV = nRT$ remain unchanged?

26 Molecules in motion: ion transport and molecular diffusion

Learning objectives

After careful study of this chapter you should be able to

(1) Define the *conductivity* (eqn (26.1.1)) and *molar conductivity* (eqn (26.1.2)) of solutions and explain how they are measured.

(2) Describe the *concentration dependence* of the molar conductivity of strong electrolytes (p. 891).

(3) State and use *Kohlrausch's law of the independent migration of ions* (eqn (26.1.4)).

(4) State and use *Ostwald's Dilution Law* (eqn (26.1.5)) for the molar conductivity of weak electrolytes.

(5) Define the *drift velocity* (eqn (26.1.9)) and the *mobility* (eqn (26.1.10)) of ions.

(6) Relate molar conductivity to ion mobility (eqn (26.1.14)).

(7) Define the *transport number* of an ion, and describe how it is measured (p. 899 and p. 914).

(8) List the factors that affect the mobility of ions, and the basis of the *Debye–Hückel–Onsager equation* (p. 900).

(9) Describe the *Wien effect* and the *Falkenhagen effect* (p. 902).

(10) Define a *thermodynamic force* (eqn (26.2.1)).

(11) Derive *Fick's Law of Diffusion* from the chemical potential (p. 903).

(12) Deduce and use the *Einstein relation* between mobility and diffusion coefficient (eqn (26.2.4)).

(13) Deduce and use the *Nernst–Einstein relation* between molar conductivity and ionic diffusion coefficients (eqn (26.2.5)).

(14) Deduce and use the *Stokes–Einstein relation* between viscosity and diffusion coefficients (eqn (26.2.6)).

(15) Derive the *diffusion equation* (eqn (26.2.7)) and use it to describe the diffusion of a solute into a solvent.

(16) Express the diffusive motion of a particle as a one-dimensional *random-walk* (p. 909).

(17) Deduce and use the *Einstein–Smoluchowski relation* relating step length and jump time to the diffusion coefficient (eqn (26.2.15)).

Introduction

In the last chapter we investigated a simple type of molecular motion, molecules moving freely as a gas. In this chapter we begin by examining what at first sight might appear to be an entirely different problem, the motion of charged ions in solution under the influence of an electric field produced by a pair of electrodes. When the average separation of the ions is large, as in very dilute solutions, their motion is determined (apart from the impacts of the solvent molecules) by the applied electric field, and their net motion is a uniform drift towards one or other of the electrodes. So long as we do not enquire too closely into the details of the motion, the overall drift can be discussed very simply. We shall also have to take into account the role of the solvent and of the other ions, but this can also be done quite simply if the analysis is not pressed too far.

There is, moreover, a very deep connection between the motion of ions and the motion of uncharged molecules in liquids and gases. This relation can be expressed thermodynamically, and the connection is examined in the second half of the chapter. The thermodynamic discussion presented there requires some of the material of Part 1, but the first sections of this chapter are independent of it.

26.1 Ion transport

The most direct evidence for the existence of ions in solution is the observation that the solution can conduct an electric current. This is accounted for by the motion of positively charged ions (*cations*) towards negatively charged electrodes, and of negatively charged ions (*anions*) to positive electrodes.

Migration of ions is a transport process of the kind considered in the last chapter. The two principal differences between ion transport and transport in gases are that the ions drift under the influence of an external electrical field, and that they are supported by a solvent. The presence of the field is a simplification because the average motion of the ions is easy to describe. The presence of the solvent, together with the ion–ion interactions, makes a detailed description quite difficult. We shall examine both in turn, but avoid the detailed calculations that underlie the theoretical model.

Ionic motion: the empirical facts. The simplest way of studying the motion of ions in solution is through their *conductivity*, their ability to conduct electricity. Determining conductivity involves measuring the electrical resistance of the solution, and the standard method consists of incorporating a cell of the kind shown in Fig. 26.1 into one arm of a bridge and searching for the balance point, as explained in standard texts on electricity. The major complication of the method results from the need to use an alternating current. A direct current would lead to an electrolysis reaction at the electrodes, and so the nature of the solution in their vicinity would change as the measurement proceeded; this is called *polarization* of the electrodes.

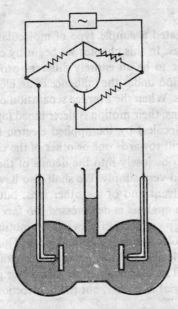

Fig. 26.1. A conductivity cell forming one arm of an a.c. bridge.

The use of a.c. eliminates the net reaction, for what polarization occurs as current flows in one half of the cycle is undone when the current flows in the opposite direction during the subsequent half. Furthermore there are capacitative effects at the solution–electrode interface, and the impedance they give rise to is minimized. A frequency of about 1 kHz is normally employed.

The resistance of a material increases with its length l but decreases with its cross-section $A: R \propto l/A$. The proportionality coefficient is called the *resistivity*, or *specific resistance*, and written ρ (rho). Then

$$R = \rho l/A.$$

The *conductivity* κ (kappa) is the inverse of the resistivity, and so

(26.1.1) $R = (1/\kappa)l/A$ or $\kappa = l/RA$.

Since the resistance is measured in ohm (Ω, capital omega) the units of κ are $\Omega^{-1}\,m^{-1}$ or $\Omega^{-1}\,cm^{-1}$. (Ω^{-1}, reciprocal ohm, is sometimes called mho, and sometimes the *siemens*: $1\ S = 1\ \Omega^{-1}$.)

Once the resistance of the sample has been measured the conductivity can be calculated from a knowledge of the cell dimensions. This, however, is an unreliable procedure because of a variety of complications, including the presence of stray currents in the cell. It is usual to calibrate the cell against a sample of known conductivity. If the resistance of a standard solution (often KCl in water) is R^* when measured in a particular cell and the resistance of the test sample is R in the same cell, writing the known conductivity of the standard as κ^* gives the conductivity of the sample as $\kappa = (R^*/R)\kappa^*$.

The conductivity depends on the number of charged carriers (ions) present, and so it is normal to express the conductivity as a molar quantity.

If the molar concentration is c, the *molar conductivity* is*

(26.1.2) $\Lambda_m = \kappa/c$.

The *equivalent conductivity* is also sometimes used. Then the conductivity is expressed in terms of the number of individual charges that are being carried. For example, in a solution of NaCl each ion carries a unit charge in each direction, and so the equivalent conductivity is the same as the molar conductivity. In $CuSO_4$, by contrast, each ion transports two units of charge, and the equivalent conductivity is half the molar conductivity. We shall always deal with molar conductivities.

Example (Objective 1). A cell contains 0.10 mol dm^{-3} aqueous KCl, which at that concentration has a molar conductivity of 129 Ω^{-1} cm^2 mol^{-1}. The measured resistance was 28.44 Ω. When the same cell was filled with 0.05 mol dm^{-3} NaOH aqueous solution the resistance was 31.6 Ω. Find the molar conductivity of aqueous NaOH at that concentration.

- *Method*. Calibrate the cell using $\kappa = C/R$, where R is the measured resistance, C the *cell constant*, and κ the conductivity. Determine C from the KCl data, then find κ for NaOH. Finally, convert to molar conductivities.

- *Answer*. For KCl the conductivity of the solution is

  $\kappa = (129\ \Omega^{-1}\ cm^2\ mol^{-1}) \times (0.10\ mol\ dm^{-3})$

  $= 12.9\ \Omega^{-1}\ (cm^2/dm^3) = 1.29 \times 10^{-2}\ \Omega^{-1}\ cm^{-1}$.

  $C = (1.29 \times 10^{-2}\ \Omega^{-1}\ cm^{-1}) \times (28.44\ \Omega) = 0.367\ cm^{-1}$.

  It follows that for NaOH:

  $C/R = (0.367\ cm^{-1})/(31.6\ \Omega) = 0.0116\ \Omega^{-1}\ cm^{-1}$.

  Therefore

  $\Lambda_m = (0.0116\ \Omega^{-1}\ cm^{-1})/(0.05\ mol\ dm^{-3}) = 232\ \Omega^{-1}\ cm^2\ mol^{-1}$.

- *Comment*. It is a good idea to use cells calibrated with standard solutions having similar conductances.

Measurements of molar conductivities reveal two classes of behaviour. Substances of one class have conductivities that depend only weakly on the concentration of the solute. From Fig. 26.2 we see that for solutions the molar conductivity rises slightly as the concentration falls. These solutes are called *strong electrolytes* (or, more strictly, *ionophores*). As the concentration of solute decreases, the molar conductivity rises to a limit which is called the *molar conductivity at infinite dilution*, Λ_m^0.

In an extensive series of measurements on strong electrolytes Kohlrausch showed that the molar conductivities of dilute solutions at concentration c obeyed

---

* The units of Λ_m are often quoted as Ω^{-1} cm^2 mol^{-1}; but some care is then needed with the interpretation of the concentration units. If κ is expressed in Ω^{-1} cm^{-1} and c in mol dm^{-3}, the ratio κ/c should be *multiplied by* 1000 to obtain Λ_m in Ω^{-1} cm^2 mol^{-1}. Explicitly:

$$\Lambda_m/\Omega^{-1}\ cm^2\ mol^{-1} = \frac{1000(\kappa/\Omega^{-1}\ cm^{-1})}{(c/mol\ dm^{-3})}.$$

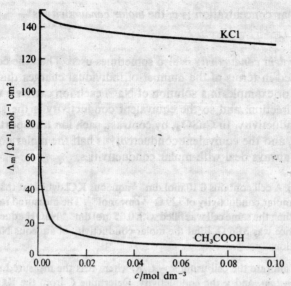

Fig. 26.2. The concentration dependence of molar conductivities.

(26.1.3) $$\Lambda_m(c) = \Lambda_m^0 - \mathscr{K} c^{1/2}.$$

$\mathscr{K}$ is a coefficient which depends more on the nature of the salt (e.g., whether it is of the form MA, M_2A, etc.) than its specific identity.

Kohlrausch was also able to confirm that the value of Λ_m^0 for any salt could be expressed as the sum of contributions from the individual ions. If the molar conductivity of a cation at infinite dilution is λ_+^0 and that for an anion is λ_-^0, then his law of the *independent migration of ions* is

(26.1.4) $$\Lambda_m^0 = v_+\lambda_+^0 + v_-\lambda_-^0$$

where v_+ and v_- are the numbers of cations and anions needed to form one molecule of the salt (e.g., $v_+ = v_- = 1$ for NaCl and $CuSO_4$, but $v_+ = 1$, $v_- = 2$ in the case of $MgCl_2$). This simple result, which can be understood on the grounds that ions behave independently when the solution is infinitely dilute, lets us predict the conductivity of any solution of ionophore using the data recorded in Table 26.1.

Example (Objective 3). Predict the molar conductivities of aqueous LiBr and $BaCl_2$ at infinite dilution at 25 °C.

● *Method.* Use the law of independent migration of ions, eqn (26.1.4), and the data in Table 26.1.

● *Answer.* For LiBr,
$\lambda_+^0 = 38.7 \ \Omega^{-1} \ cm^2 \ mol^{-1}$; $\lambda_-^0 = 78.1 \ \Omega^{-1} \ cm^2 \ mol^{-1}$
$\Lambda_m^0 = \lambda_+^0 + \lambda_-^0 = 116.8 \ \Omega^{-1} \ cm^2 \ mol^{-1}$.

For $BaCl_2$,

$$\lambda_+^0 = 127.2 \ \Omega^{-1} \ cm^2 \ mol^{-1}; \ \lambda_-^0 = 76.3 \ \Omega^{-1} \ cm^2 \ mol^{-1}$$
$$\Lambda_m^0 = \lambda_+^0 + 2\lambda_-^0 = 279.8 \ \Omega^{-1} \ cm^2 \ mol^{-1}.$$

- *Comment.* If we were asked for *equivalent conductances*, then the value for LiBr would be 116.8 Ω^{-1} cm^2 equiv^{-1}, while the value for BaCl$_2$ would be 139.9 Ω^{-1} cm^2 equiv^{-1} because twice the amount of charge is transferred in the latter case for every mole of material moved.

Table 26.1. Ionic conductivities at infinite dilution and 25 °C, $\lambda_\pm^0/\Omega^{-1}$ cm^2 mol^{-1}

Cations		Anions	
H$^+$	349.6	OH$^-$	199.1
Li$^+$	38.7	Cl$^-$	76.3
Na$^+$	50.1	Br$^-$	78.1
K$^+$	73.5	I$^-$	76.8
Ag$^+$	61.9	NO$_3^-$	71.4
NH$_4^+$	73.6	CH$_3$CO$_2^-$	40.9
Mg^{2+}	106.0	SO$_4^{2-}$	160.0
Ca^{2+}	119.0		
Cu^{2+}	107.2		
Zn^{2+}	105.6		
Ba^{2+}	127.2		

Source: G. W. C. Kaye and T. H. Laby, *Tables of physical and chemical constants*, Longmans.

Electrolytes of the second class, the *weak electrolytes* (or the *ionogens*), have molar conductivities that depend markedly on the concentration. The value of Λ_m remains low until very low concentrations are reached, when it sweeps up to values comparable to those of strong electrolytes. This is also illustrated by the data plotted in Fig. 26.2.

The reason for this behaviour can be traced to the existence of an equilibrium between the ionized and un-ionized forms of the ionogen,

$$MA \rightleftharpoons M^+ + A^-,$$

for this shifts to the right as the dilution is increased. The conductivity reflects the number of ions in the solution, and this depends on the equilibrium constant for the dissociation. If the *degree of ionization* is denoted α, the equilibrium constant (writing $c \equiv c/\text{mol dm}^{-3}$) is

$$K = c_{M^+}c_{A^-}/c_{MA} = (\alpha c)(\alpha c)/(1-\alpha)c = \{\alpha^2/(1-\alpha)\}c$$

where c is the concentration of added ionogen and c_{MA} is the concentration of the un-ionized form. The conductivity depends on the degree of ionization because the concentration of each ion is proportional to α. If the *measured* molar conductance is Λ_m', its relation to the molar conductance of the fully ionized solution, Λ_m, is

$$\Lambda_m' = \alpha \Lambda_m,$$

and so we arrive at *Ostwald's Dilution Law*

(26.1.5)
$$K = \left\{ \frac{(\Lambda'_m/\Lambda_m)^2}{1-(\Lambda'_m/\Lambda_m)} \right\} (c/\text{mol dm}^{-3})$$

connecting concentration and conductivity. This expression can be used to determine the dissociation equilibrium constant by measuring Λ'_m and estimating Λ_m from eqn (26.1.4). It can also be inverted to give

(26.1.6)
$$\Lambda'_m = \tfrac{1}{2}(K/c)\{(1+4c/K)^{1/2}-1\}\Lambda_m$$

for the explicit dependence of the molar conductivity on the concentration (with c understood as $c/\text{mol dm}^{-3}$).

Example (Objective 4). The resistance of an 0.01 mol dm^{-3} aqueous solution of acetic acid was measured in the same cell as in the *Example* on p. 891, and found to be $2220\ \Omega$. Find the value of the dissociation constant K_a, the value of pK_a, and the degree of dissociation of the acid at that concentration.

- *Method.* Find Λ'_m from the value of the cell-constant determined in the earlier *Example* ($C = 0.367$ cm^{-1}). Find Λ_m from the data in Table 26.1. Use eqn (26.1.5) to find K_a, then use $pK_a = -\lg K_a$. Finally, obtain the degree of dissociation from $\Lambda'_m = \alpha \Lambda_m$.

- *Answer.* From Table 26.1, $\Lambda^0_m(\text{CH}_3\text{COOH}) = \lambda^0_+(\text{H}^+)+\lambda^0_-(\text{CH}_3\text{COO}^-) = (349.6+40.9)\ \Omega^{-1}$ cm^2 mol^{-1} $= 390.5\ \Omega^{-1}$ cm^2 mol^{-1}. The observed conductivity is $\kappa = C/R = (0.367$ cm$^{-1})/(2220\ \Omega) = 1.653 \times 10^{-4}\ \Omega^{-1}$ cm^{-1}. Therefore the observed molar conductivity is

$$\Lambda'_m = (1.653 \times 10^{-4}\ \Omega^{-1}\ \text{cm}^{-1})/(0.01\ \text{mol dm}^{-3}) = 16.5\ \Omega^{-1}\ \text{cm}^2\ \text{mol}^{-1}.$$

The degree of dissociation is $\alpha = \Lambda'_m/\Lambda_m \approx \Lambda'_m/\Lambda^0_m = 0.0423$. Then from eqn (26.1.5),

$$K_a = \{(0.0423)^2/(1-0.0423)\} \times 0.01 = 1.86 \times 10^{-5}.$$

Therefore

$$pK_a = -\lg(1.86 \times 10^{-5}) = 4.73.$$

Since $\alpha = 0.0422$, 4.22 per cent of the acid is dissociated.

- *Comment.* Some corrections have been ignored (for example, we have identified Λ_m with Λ^0_m). The thermodynamic value of pK_a could be obtained by repeating the measurement at different concentrations, and extrapolating to zero concentration.

Convincing evidence for the view that ions exist in solution also came from the measurement of the colligative properties of dilute solutions (p. 228), in particular the osmotic pressure. This evidence was discussed in Chapter 8.

The experimental evidence summarised above can be used to obtain insight into the structure of ionic solutions and the way the ions move in them. We want to know the reasons why the mobilities of ions differ, why ions have different values of $\lambda^0_\pm$, and why the conductivities of strong electrolyte solutions depend on the concentration in the way indicated by Kohlrausch's Law, eqn (26.1.3).

The mobility of ions. An ion in solution is subject to a variety of forces. One force arises from the random bombardment of the ions by the solvent. Since the directions and the magnitudes of these collisions are both random, they do not cause any net motion of the ions in a uniform solution. If there are concentration gradients, the ions spread from the regions of high concentration into those of low. We shall disregard this complication for the present, and deal with uniform solutions, but the matter is taken up again on p. 903. The ions also respond to an electrical force. When two electrodes a distance l apart are at a potential difference $\Delta\phi$ there is a potential gradient of magnitude $\Delta\phi/l$ between them. Positive ions slide down the gradient, and negative ones slide up it, Fig. 26.3. This motion can be interpreted as the result of an electrical force operating on the ions, the magnitude of the force being proportional to the potential gradient.

We can make these ideas quantitative in the following way. First we note that if the charge on an ion is ze (where z is positive for cations and negative for anions) and it is in an electric field E, it experiences a force

(26.1.7) $$\mathscr{F} = zeE.$$

An electric field is the negative of the gradient of a potential:

(26.1.8) $$E = -d\phi/dx.$$

This fits the preceding discussion, because if the potential difference across the cell is $\Delta\phi$, its gradient is $\Delta\phi/l$, the field is $-\Delta\phi/l$, and so the force is

$$\mathscr{F} = -ze\Delta\phi/l.$$

If $\Delta\phi = \phi_R - \phi_L$ is negative (ϕ_R and ϕ_L being the potentials of the right and left hand electrodes), the force $\mathscr{F}$ is positive for positive ions (z positive) and so it pushes them in the direction of the more negative electrode, Fig. 26.3; the opposite applies to negative ions (z negative).

The force acting on an ion accelerates it, but as it rubs through the solvent a frictional force retards it. It is therefore accelerated only to some limiting velocity, which depends on the strength of the applied field and the viscosity of the solvent. This terminal velocity is called the *drift velocity* of the ion in solution, and is denoted v. If it is assumed that Stokes's formula for the viscous retarding force on a spherical object of radius a

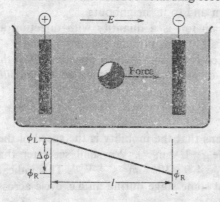

Fig. 26.3. A cation in a potential gradient during one cycle of an a.c. experiment.

in a medium of viscosity η applies even on a microscopic scale (and independent evidence suggests that it often gives the right order of magnitude), the frictional force depends on the velocity as $6\pi a\eta v$. The drift velocity is established when the accelerating force zeE is balanced by the retarding force; and so $6\pi a\eta v \approx zeE$ implies

(26.1.9) $v \approx zeE/6\pi a\eta$.

Since the drift velocity governs the rate at which current may be conducted it follows that we should expect the conductivity to decrease with increasing solution viscosity and increasing ion size.

Experimental results confirm the first of these predictions, but not the second. For example, the molar conductivities of the alkali metal ions *increase* on going from Li^+ to Cs^+ (actual values are given in Table 26.1) even though the ion sizes are known to increase markedly (a table of ionic radii obtained from X-ray diffraction studies is given on p. 750). This discrepancy can be explained by taking the solvation of the ions into account. The solvent molecules cluster around the ion and increase its effective size (which is called its *hydrodynamic radius*). Small ions are the source of stronger electric fields than large ions (this is a result of electrostatic theory, which shows that the electric field at the surface of a sphere of radius R is proportional to ze/R^2) and so the solvation is more extensive in the case of small ions. They have larger hydrodynamic radii than larger ions, lower drift velocities, and therefore lower conductivities.

The proton, however, although it is very small, has a very large molar conductivity of $350 \, \Omega^{-1} \, cm^2 \, mol^{-1}$. Even though we expect it to be strongly solvated it appears to be able to move rapidly through the solution. The hydrodynamic argument that led to the idea of a viscous drag cannot apply in its case, and another mechanism operates. This mechanism takes note both of the low mass of the proton and of the extensively hydrogen-bonded structure of water. Instead of a single, identifiable proton moving through the solution, it is believed that there is an *effective* motion of a proton, which involves the formation and destruction of bonds through a long chain of water molecules, Fig. 26.4. The conductivity is limited by the rate at which a water molecule (such as the one on the right in the diagram) can reorientate in order to accept and deliver protons from and to its neighbours.

The drift velocity of an ion is a quantity with direction as well as magnitude. We shall call its magnitude the *drift speed* $s_\pm$, so that $s_\pm = |v_\pm|$. The drift speed is proportional to the strength of the applied field, and the constant of proportionality is called the *mobility* of the ion, $u_\pm$:

(26.1.10) $s_\pm = u_\pm E$.

This definition also means that the mobility is the speed of the ion in a field of unit strength (e.g., $1 \, V \, m^{-1}$). We shall see later that typical mobilities lie in the range 4–$9 \times 10^{-8} \, m^2 \, s^{-1} \, V^{-1}$. This indicates that a potential drop of 1 V at some stage during an a.c. cycle across 1 cm of

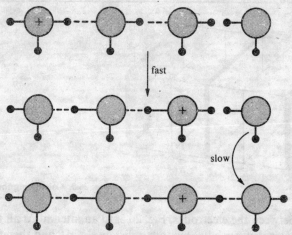

Fig. 26.4. The mechanism of conduction in water.

fast

slow

solution (so that $E = 10^2$ V m^{-1}) gives rise to a drift speed in the range $4\text{-}9 \times 10^{-6}$ m s^{-1}, or $4\text{-}9 \times 10^{-3}$ mm s^{-1}. This might seem slow; but not when expressed on a molecular scale, for it corresponds to the ion passing about 10 000 solvent molecules each second. Ionic mobilities are listed in Table 26.2 and their determination is described on p. 914.

Table 26.2. Ionic mobilities in aqueous solution at 25 °C, $u/10^{-4}$ cm^2 s^{-1} V^{-1}

Cations		Anions	
H$^+$	36.23	OH$^-$	20.64
Li$^+$	4.01	F$^-$	5.70
Na$^+$	5.19	Cl$^-$	7.91
K$^+$	7.62	Br$^-$	8.09
Rb$^+$	7.92	I$^-$	7.96
Ag$^+$	6.42	NO$_3^-$	7.40
NH$_4^+$	7.63	CO$_3^{2-}$	7.46
Ca^{2+}	6.17	SO$_4^{2-}$	8.29
Cu^{2+}	5.56	CH$_3$CO$_2^-$	4.24
La^{3+}	7.21		

Source: Principally Table 26.1 via $u = \lambda/zF$, eqn (26.1.15) below.

The usefulness of ionic mobilities is that they provide a link between quantities that can be calculated and quantities that can be measured. The mobilities can both be discussed in terms of the dynamics of the ions, and be related to the conductivity of the solution. We now go on to deal with these connections.

First we find the connection between mobility and conductivity. We take a solution of a salt $M_{\nu+}X_{\nu-}$ of concentration c (in amount per unit volume) so that it contains $\nu_+ cL$ cations and $\nu_- cL$ anions per unit volume. The cations have charge z_+e and the anions z_-e. Consider first the cations, and their motion through an imaginary window of area A set in the

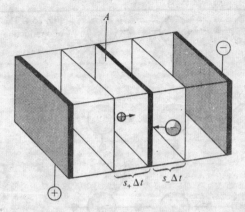

Fig. 26.5. Calculating the charge flux.

solution between the electrodes, Fig. 26.5. In an interval Δt all the cations (which have a drift speed s_+) within a distance $s_+\Delta t$, and therefore within the volume $s_+\Delta t A$ will pass through the window; and so the number that pass through is $(s_+\Delta t A)v_+cL$. (This is the same type of calculation as that done for gases in the last chapter; e.g., p. 861.) Each ion carries a charge z_+e, and so the flux of positive charge, the charge per unit area per unit time, is

$$J(\text{charge}) = (s_+\Delta t A)v_+cLz_+e/A\Delta t = v_+s_+z_+ceL.$$

Current is also transported by the anions: they move in the opposite direction but carry the opposite charge. Their effect is to augment the current. The total flux is therefore

(26.1.11) $\qquad J = v_+s_+z_+ceL + v_-s_-|z_-|ceL = (u_+v_+z_+ + u_-v_-|z_-|)cFE,$

where we have replaced the drift speed by the mobilities using eqn (26.1.10), and eL by F, the Faraday constant ($1 \, F = 9.65 \times 10^4 \, \text{C mol}^{-1}$).

The quantity J is the charge flux: the current through an area A is $I = JA$. If the field strength E arises from a potential difference $\Delta\phi$ across a solution of length l, its magnitude is $|\Delta\phi|/l$, and we find

$$I = (v_+u_+z_+ + v_-u_-|z_-|)cFA|\Delta\phi|/l.$$

But the current and the potential difference are also related to the conductivity through Ohm's law:

$$I = |\Delta\phi|/R = \kappa|\Delta\phi|A/l.$$

Comparison of these two expressions lets us identify the conductivity as

(26.1.12) $\qquad \kappa = (v_+u_+z_+ + v_-u_-|z_-|)cF.$

It follows that the molar conductivity is

(26.1.13) $\qquad \Lambda_m = (v_+u_+z_+ + v_-u_-|z_-|)F.$

and for a 1:1 salt with $z_+ = |z_-| = z$ this reduces to

(26.1.14) $\qquad \Lambda_m = z(u_+ + u_-)F.$

At this stage we can also see that the individual conductivities can be related to the individual mobilities:

(26.1.15) $$\lambda_\pm = u_\pm |z_\pm| F.$$

This is the connection we have been looking for between the mobilities and the conductivities of the ions. By measuring the conductivities we can find the mobilities through the last two equations. On p. 305 we shall see how to relate the mobilities to the size of the ions and the viscosity of the solution.

The fraction of the total current carried by each type of ion is directly related to their mobilities and if they are very mobile they carry a large proportion of the total current. The name *transport number* is given to the proportion of the current carried, and its definition (for a 1:1 salt) is

(26.1.16) $$t_+ = \frac{u_+}{u_+ + u_-}, \qquad t_- = \frac{u_-}{u_+ + u_-}.$$

Obviously $t_+ + t_- = 1$.

We can work backwards from these definitions and express $t_\pm$ in terms of the molar ion conductivities; using $\lambda_\pm = zu_\pm F$ (for a 1:1 salt) gives

$$t_\pm = \frac{\lambda_\pm}{\lambda_+ + \lambda_-} = \lambda_\pm / \Lambda_m,$$

or

(26.1.17) $$\lambda_\pm = \Lambda_m t_\pm.$$

Therefore, if we have an independent way of measuring transport numbers, the individual ion conductivities, and then the mobilities, may be established. Several techniques are available, and are summarized in the Appendix.

Conductivities and ionic interactions. The remaining problem is to account for the $c^{1/2}$ concentration dependence of the molar conductivities in dilute solution, eqn (26.1.3). In Chapter 11 on the structure of ionic solutions we met the $c^{1/2}$ dependence of the thermodynamic properties of solutions. We saw that ions fell into groups according to their valencies rather than their specific identities, and that these properties could be understood quantitatively in terms of the effect of the ionic atmosphere (p. 316) surrounding the ion of interest. Without drawing on the details of that chapter we can state the main qualitative features of the situation, and develop them for the present dynamical problem.

An ion tends to gather into its vicinity a cluster of ions of opposite charge. The accumulation is very faint, but it is significant because of the long range of the Coulombic interaction. The charge cloud is called the *ionic atmosphere*, and on the average it is spherically symmetrical. The presence of the charge cloud affects the electrostatic energy of the ion, and this affects its thermodynamic properties.

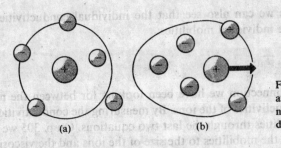

Fig. 26.6. (a) The ionic atmosphere when the ions have no net motion and (b) the distortion arising from motion.

When an external electric field is applied to a solution of ions, each with its tenuous atmosphere of opposite charge, the picture becomes dynamic, and we have to think of the cloud as a structure with motion. The first property to concentrate on is the time it takes to assemble and disassemble the atmosphere. The ions do not cluster together infinitely quickly, and so if an ion is moving through the solution the atmosphere is incompletely formed in front of its motion, and it has not fully decayed in its wake, Fig. 26.6. The overall effect is to displace the centre of charge of the atmosphere a short distance behind the moving ion. As the charge of the ion and atmosphere are opposite the result is to retard the motion of the moving ion. This is called the *relaxation effect*, because the formation and decay of the atmosphere is a kind of relaxation into an equilibrium distribution of the ions.

The other effect that has to be taken into account is the *electrophoretic effect*. The central ion experiences a viscous drag as it moves through the solution, and this determines its drift velocity, and hence its conductivity. When the ions are not infinitely far apart the effect is enhanced because ions of opposite charge, each with their cluster of solvent molecules, are rubbing past each other. This enhances the viscous drag, and lowers the drift velocity and therefore the conductivity.

The quantitative formulation of these effects is far from simple, but an approximation similar in sophistication to the Debye–Hückel theory of static solutions has been achieved. The result is the *Debye–Hückel–Onsager equation* for molar conductivities. This has the form

$$(26.1.18) \qquad \Lambda_m = \Lambda_m^0 - (A + B\Lambda_m^0)c^{1/2}$$

where A and $\overline{B}$ are constants depending on the electrical properties and temperature of the solution. For a symmetrical electrolyte,

$$A = \frac{zeF}{3\pi\eta}\left(\frac{2z^2e^2L}{\varepsilon kT}\right)^{1/2}, \qquad B = \frac{e^2z^2q}{24\pi\varepsilon kT}\left(\frac{2z^2e^2L}{\pi\varepsilon kT}\right)^{1/2}$$

where ε is the permittivity of the medium (p. 340) and $q \approx 0.59$ for a (1,1)-electrolyte. Their numerical values for several solvents (because nothing in the calculation limits its application only to water) are given in Table 26.3.

Example (Objective 8). The molar conductivity of aqueous KCl varies with concentration in the way reported below. Find the Kohlrausch coefficient $\mathcal{K}$ for this

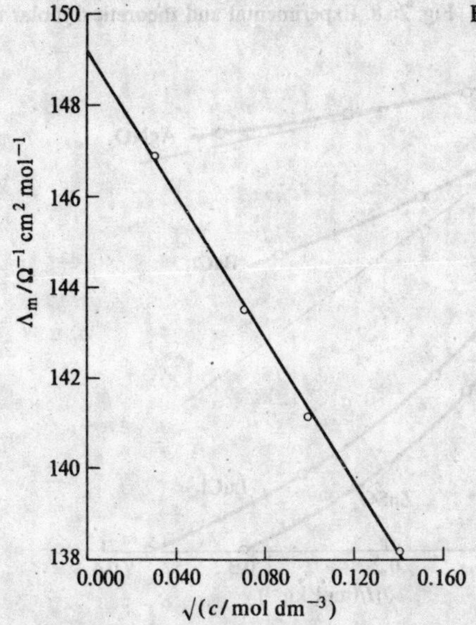

Fig. 26.7. The determination of $\mathscr{K}$ and Λ_m^0.

1,1-electrolyte, and confirm that its value is given by the Debye–Hückel–Onsager equation.

$c/\text{mol dm}^{-3}$	0.001	0.005	0.010	0.020
$\Lambda_m/\Omega^{-1}\text{ cm}^2\text{ mol}^{-1}$	146.9	143.5	141.2	138.2

- **Method.** Plot Λ_m against $\sqrt{(c/\text{mol dm}^{-3})}$; the slope is $-\mathscr{K}$. This slope is also given by eqn (26.1.18) as $-(A+B\Lambda_m^0)$. Take the value of A and B from Table 26.3 and that of Λ_m^0 from the intercept of the graph.

- **Answer.** The plot of $\Lambda_m/\Omega^{-1}\text{ cm}^2\text{ mol}^{-1}$ against $\sqrt{(c/\text{mol dm}^{-3})}$ is shown in Fig. 26.7. The intercept is at 149.3 and the slope is -79.14. Hence $\mathscr{K} = 79.14\ \Omega^{-1}\text{ cm}^2\text{ mol}^{-1}/(\text{mol dm}^{-3})^{1/2}$ (clumsy but convenient units) and $\Lambda_m^0 = 149.3\ \Omega^{-1}\text{ cm}^2\text{ mol}^{-1}$.

  From Table 26.3,

  $$A = 60.20\ \Omega^{-1}\text{ cm}^2\text{ mol}^{-1}/(\text{mol dm}^{-3})^{1/2}; \quad B = 0.229/(\text{mol dm}^{-3})^{1/2},$$

  and so $A+B\Lambda_m^0 = 60.20\ \text{cm}^2\text{ mol}^{-1}/(\text{mol dm}^{-3})^{1/2} + \{0.229/(\text{mol dm}^{-3})^{1/2}\} \times (149.5\ \Omega^{-1}\text{ cm}^2\text{ mol}^{-1}) = 94.4\ \Omega^{-1}\text{ cm}^2\text{ mol}^{-1}/(\text{mol dm}^{-3})^{1/2}$ Hence the theoretical value is

  $$\mathscr{K} \approx -94\ \Omega^{-1}\text{ cm}^2\text{ mol}^{-1}/(\text{mol dm}^{-3})^{1/2}.$$

Table 26.3. Debye–Hückel–Onsager coefficients for (1,1)-electrolytes at 25 °C

Solvent	a	b	Solvent	a	b
Water	60.20	0.229	Acetone	32.8	1.63
Methanol	156.1	0.923	Nitromethane	125.1	0.708
Ethanol	89.7	1.83	Nitrobenzene	44.2	0.776
Acetonitrile	22.9	0.716			

$a = A/(\Omega^{-1}\text{ cm}^2\text{ mol}^{-1}/(\text{mol dm}^{-3})^{1/2})$, $b = B/(\text{mol dm}^{-3})^{-1/2}$.
Source: J. O'M. Bockris and A. K. N. Reddy, *Modern electrochemistry*, Plenum.

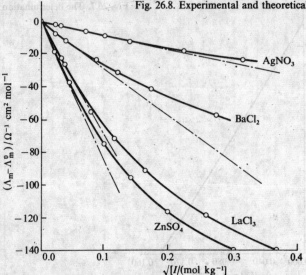

Fig. 26.8. Experimental and theoretical molar conductivities.

• *Comment*. The result is not at all bad considering the theoretical difficulties of dealing with a dynamical system of this complexity. Note that the plot of Λ_m against $\sqrt{c}$ is a useful way of making the extrapolation to find Λ_m^0.

Equation (26.1.18) has the same $c^{1/2}$ concentration-dependence as the empirical Kohlrausch expression, eqn (26.1.3). Furthermore, the slopes of the curves are predicted to depend on the valence type (z appears in the constants A and B). Some comparisons between theory and experiment are shown in Fig. 26.8, and this shows how well the theory accounts for the observations at low concentrations.

The success of the Debye–Hückel–Onsager equation suggests that the model of ion–ion interactions is substantially correct. A further test is obtained by investigating what happens when the effect of the ionic atmosphere is eliminated. This can be done in a variety of ways. In one the conductivities are measured at very high frequencies; then the central ion is moved backwards and forwards very rapidly, and the retarding effects of the ionic atmosphere ought to average to zero. This is the *Debye–Falkenhagen effect*, and the predicted increase in mobility at high frequencies has been observed. The other way of eliminating the effect of the atmosphere is to move the ions so rapidly that no atmosphere has time to build up. The *Wien effect* is the observation of higher mobilities at very high electric fields. (There are two Wien effects. The *first Wien effect* is the one just described; the *second Wien effect* is the enhancement of the degree of ionization of an ionogen, or weak electrolyte, by the applied field.)

This model of the interactions fails when the concentration becomes large, because ions tend to stick together in pairs, and even triples. This can be seen quite clearly from X-ray analysis of ionic solution, where peaks of scattering can be interpreted in terms of definite ion–ion distances.

26.2 Fundamental aspects of molecular transport

In this section we begin to draw together the threads of the discussion in this chapter and the last. We do so on the basis of thermodynamic and statistical principles and find that we can make a variety of important and useful connections between properties relating to the motion of molecules and ions in fluids.

Diffusion: the thermodynamic view. In Part 1 it was shown that the thermodynamic property that governs the direction of spontaneous change is the thermodynamic chemical potential (p. 182). When unit amount of solute is shifted from a region where its chemical potential is $\mu(1)$ to one where it is $\mu(2)$ the work required is $w = \mu(2) - \mu(1)$. Suppose the chemical potential depends on the position x in the system, then the work involved in transferring unit amount of material from x to $x + dx$ is

$$dw = \mu(x + dx) - \mu(x) = [\mu(x) + (d\mu/dx)dx] - \mu(x) = (d\mu/dx)dx.$$

(The derivatives ought strictly to be partial derivatives, and the transfer ought to be carried out under conditions of constant pressure and temperature—see p. 258 for details.) In classical mechanics the work required to shift an object through a distance dx against a force $\mathscr{F}$ is

$$dw = -\mathscr{F}\,dx.$$

By comparing the last two equations we see that *the gradient of the chemical potential acts like a force*. We shall therefore write

(26.2.1) $$\mathscr{F} = -(d\mu/dx).$$

There is no real force pushing molecules down the slope of chemical potential, for that is their natural drift as a consequence of the Second Law and the hunt for maximum entropy: nevertheless, thinking in terms of these phantom, effective, thermodynamic forces can be very useful, as we shall see.

In a solution where the concentration is c the chemical potential of an ideal solute is (p. 226).

(26.2.2)° $$\mu = \mu^{\ominus} + RT\ln(c/\text{mol dm}^{-3}).$$

If the concentration depends on position, the thermodynamic force acting is

(26.2.3)°
$$\mathscr{F} = -(d/dx)[\mu^{\ominus} + RT\ln(c/\text{mol dm}^{-3})]$$
$$= -(RT/c)(dc/dx)$$

because $\mu^{\ominus}$ is independent of position, and $d\ln f/dx = (1/f)df/dx$.

The form of the last equation lets us derive Fick's Law of Diffusion' (that flux is proportional to the concentration gradient, eqn (25.3.1)) from a thermodynamic viewpoint. We suppose that the flux is the response of the molecules to some force. If the force per unit amount is $\mathscr{F}$, the force per unit volume on the molecules present is $c\mathscr{F}$. If we assume that the flux of material, J(matter), is proportional to the impressed force, then J(matter) $\propto c\mathscr{F}$. But the effective force is given by eqn (26.2.3), and so

$$J(\text{matter}) \propto -RT(dc/dx) = -kT(d\mathscr{N}/dx)$$

and we have the flux as proportional to the concentration gradient, in accord with Fick's Law ($\mathscr{N}$ is the number density of molecules, $\mathscr{N} = cL$).

It is more convenient to develop a different line of argument, and to interpret the flux of particles as the product $v\mathscr{N} = vcL$, where v is their average velocity and c their concentration. Then Fick's Law,

$$v\mathscr{N} = J = -D(d\mathscr{N}/dx) = -DL(dc/dx)$$

reads

$$cv = -D(dc/dx)$$

or, using eqn (26.2.3),

$$v = -(D/c)(dc/dx) = (D/kT)\mathscr{F}.$$

Therefore, in response to a unit force, the molecules diffuse with a drift velocity of magnitude D/kT. We know, however, that the *mobility* of an ion is related to the electrical force on it. Since the mobility is defined through $s = uE$, and since the electrical force is ezE, it follows that

$$s = uE = (u/ze)(ezE) = (u/ze)\mathscr{F},$$

and therefore the drift speed under the influence of unit force is (u/ze). The nature of the force is irrelevant; therefore the two drift speeds $(D/kT)\mathscr{F}$ and $(u/ze)\mathscr{F}$ may be identified: then

$$u/ez = D/kT,$$

and so we arrive at the very important result, known as the

$(26.2.4)°$ *Einstein relation*: $D = ukT/ez$

connecting the diffusion constant and the mobility.

The last relation can be developed further in two directions. In the first place it can be used to relate the molar conductivity to the diffusion constants of the ions, D_+ and D_-. We write (for 1 : 1 salts)

$$\Lambda_m = zF(u_+ + u_-) = (z^2eF/kT)(D_+ + D_-)$$

and so arrive at the

$(26.2.5)°$ *Nernst–Einstein relation*: $\Lambda_m = (z^2F^2/RT)(D_+ + D_-)$.

One application of these expressions is to the determination of the ionic diffusion constants from conductivity measurements; the other is to the calculation of conductivities on the basis of models of ionic diffusion (see below).

The other direction of development of the Einstein relation is to relate the mobility to the viscosity. By combining the expressions

$$s = ezE/6\pi\eta a \quad \text{and} \quad s = uE,$$

the first being the expression for the drift speed defined in eqn (26.1.9), we are able to write

$$u = ez/6\pi\eta a.$$

Since the Einstein relation is $u = ezD/kT$, the two may be equated and combined into the

(26.2.6)° **Stokes–Einstein relation: $D = kT/6\pi\eta a$**

which connects the diffusion constant and the viscosity of liquids. An important feature of this result is that it is independent of the charge of the diffusing species, and therefore it also applies in the limit of vanishingly small charge, or neutral molecules. This means that we may use the Stokes–Einstein relation to estimate the diffusion constant from measurements of the viscosity. It must not be forgotten, however, that it is an approximation, being based on the assumption of the validity of the Stokes formula for the viscous drag. Some diffusion coefficients are listed in Table 26.4.

Table 26.4. Diffusion coefficients at 25 °C, $D/10^{-9}$ m^2 s^{-1}

I_2 in hexane	4.05	H_2^* in CCl$_4$	9.75
I_2 in benzene	2.13	N_2 in CCl$_4$	3.42
CCl$_4$ in n-heptane	3.17	O_2 in CCl$_4$	3.82
Glycine in water	1.055	Ar in CCl$_4$	3.63
Dextrose in water	0.673	CH$_4$ in CCl$_4$	2.89
Sucrose in water	0.521	Water in water	2.26
		Methanol in water	1.58
		Ethanol in water	1.24

Ions in water:

H$^+$	9.31	OH$^-$	5.30
Li$^+$	1.03	F$^-$	1.46
Na$^+$	1.33	Cl$^-$	2.03
K$^+$	1.96	Br$^-$	2.08
		I$^-$	2.05

Source: *American Institute of Physics Handbook*, McGraw-Hill, and (for the ions) eqn (26.2.4) and Table 26.2.

Example (Objective 12, 13, 14). Find the diffusion coefficient, the molar conductivity, and the effective hydrodynamic radius of the SO_4^{2-} ion in water at 25 °C.

● *Method.* In Table 26.2 the mobility of the ion is given as 8.29×10^{-4} cm^2 s^{-1} V^{-1}. Relate this to D through eqn (26.2.4). Then use eqn (26.2.5) to relate D (now written D_-) to λ_- (the anion contribution to Λ_m). Calculate the effective radius a_- from eqn (26.2.6). The viscosity of water at 25 °C is 1.00 cp $(1.00 \times 10^{-3}$ kg m^{-1} s$^{-1})$. Remember that J = C V and V A^{-1} = Ω.

- *Answer.* From eqn (26.2.4),

$$D_- = u_- kT/ez$$
$$= \frac{(8.29 \times 10^{-4}\,\text{cm}^2\,\text{s}^{-1}\,\text{V}^{-1}) \times (1.3807 \times 10^{-23}\,\text{J K}^{-1}) \times (298.15\,\text{K})}{2 \times (1.6022 \times 10^{-19}\,\text{C})}$$
$$= 1.065 \times 10^{-5}\,\text{cm}^2\,\text{s}^{-1}.$$

From eqn (26.2.5),

$$\lambda_- = \frac{2^2 \times (9.6485 \times 10^4\,\text{C mol}^{-1})^2 \times (1.065 \times 10^{-5}\,\text{cm}^2\,\text{s}^{-1})}{(8.3144\,\text{J K}^{-1}\,\text{mol}^{-1}) \times (298.15\,\text{K})}$$
$$= 160.0\,\Omega^{-1}\,\text{cm}^2\,\text{mol}^{-1}.$$

From eqn (26.2.6),

$$a_- = kT/6\pi\eta D_-$$
$$= \frac{(1.3807 \times 10^{-23}\,\text{J K}^{-1}) \times (298.15\,\text{K})}{6\pi \times (1.00 \times 10^{-3}\,\text{kg m}^{-1}\,\text{s}^{-1}) \times (1.065 \times 10^{-5}\,\text{cm}^2\,\text{s}^{-1})}$$
$$= 2.051 \times 10^{-10}\,\text{m} \quad \text{or} \quad 205\,\text{pm} \quad \text{or} \quad 2.05\,\text{Å}.$$

- *Comment.* The bond length in SO_4^{2-} is 144 pm, and so the radius calculated here (the radius of a sphere representing the molecule) is plausible and compatible with only a small degree of solvation.

Some experimental support for these ideas comes from conductivity measurements, because the empirical *Walden's rule* is that the product $\Lambda_m^0 \eta$ should be approximately constant for the same ions in different solvents. Since $\Lambda_m^0 \propto u$, and $u \propto 1/\eta$, we can see the theoretical basis of this rule. Its applicability is muddied by the role of solvation: different solvents solvate ions to different extents, and so the ions' effective hydrodynamic radii depend on their nature: both a and η vary with the solvent.

Diffusion as a time-dependent process. We turn now to the discussion of time-dependent diffusion processes, in which some distribution of concentration, or of temperature, etc., is established at some moment, and then allowed to disperse without replenishment. One example is a metal bar heated rapidly at one end and then allowed to reach equilibrium, and another is when a layer of solute is spread on the surface of a solvent and the concentration distribution in the solution changes as it dissolves.

In order to treat a time-dependent diffusion process we shall concentrate on the diffusion of matter, but the arguments are easily modified to apply to other properties. We fix our attention on a small slab of the system extending from x to $x + \Delta x$, and of cross-sectional area A, Fig. 26.9. Let the concentration at x be $\mathcal{N}(x, t)$ at the time t, then the increase in concentration inside the slab (of volume $A\Delta x$) by virtue of the flux from the left is

$$\partial \mathcal{N}(x, t)/\partial t = J(x, t)A/A\Delta x = J(x, t)/\Delta x$$

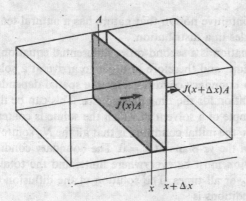

Fig. 26.9. The diffusion of material into a region.

because JA is the number of particles that enter through a window of area A in each unit time interval. There is also a flow out of the right-hand window; if the flux is $J(x+\Delta x)$ the concentration inside the slab changes due to this efflux with a rate

$$\partial \mathcal{N}(x,t)/\partial t = -J(x+\Delta x,t)A/A\Delta x = -J(x+\Delta x,t)/\Delta x,$$

the negative sign appearing because the concentration in the slab decreases when the flow is to the right (J positive). Therefore the total rate of change of concentration is

$$\partial \mathcal{N}(x,t)/\partial t = J(x,t)/\Delta x - J(x+\Delta x,t)/\Delta x.$$

The fluxes can now be related to the concentration gradients at each window. Using Fick's Law we can write

$$J(x,t) - J(x+\Delta x,t) = \{-D[\partial \mathcal{N}(x,t)/\partial x)]\} - \{-D[\partial \mathcal{N}(x+\Delta x,t)/\partial x)]\}$$

$$= -D\left(\frac{\partial \mathcal{N}(x,t)}{\partial x}\right) + D\frac{\partial}{\partial x}\left[\mathcal{N}(x,t) + \left(\frac{\partial \mathcal{N}(x,t)}{\partial x}\right)\Delta x\right]$$

$$= D(\partial^2 \mathcal{N}(x,t)/\partial x^2)\Delta x.$$

Substituting this back into the expression for the rate of change of the concentration in the slab leads to the

(26.2.7)   *diffusion equation:* $(\partial \mathcal{N}(x,t)/\partial t) = D(\partial^2 \mathcal{N}(x,t)/\partial x^2).$

This is also sometimes called *Fick's Second Law.*

First, a word about the general form of this equation. We see that the rate of change of the concentration is proportional to the curvature (the second-derivative) of the concentration dependence on the distance. If the concentration changes rapidly from point to point the rate at which the concentration changes with time is correspondingly rapid. If the curvature is zero, the concentration does not change with time. For example, if the concentration falls linearly with distance, the concentration at any point remains constant because the inflow of concentration is balanced by the outflow. The diffusion equation can be regarded as a mathematical formulation of the intuitive notion that nature has a natural tendency to eliminate the wrinkles in a distribution.

The diffusion equation is a second-order differential equation in space and first-order in time, and therefore in order to arrive at a solution we have to specify two boundary conditions for the spatial dependence and a single initial condition for the time dependence. This can be illustrated by the specific example of a solvent in which the solute is coated on one surface. At time zero the initial condition is that all the N_0 solute particles are concentrated on the yz-plane at $x = 0$. The boundary conditions are that the concentration must be everywhere finite, and the total number present must be N_0 at all times. The solution of the diffusion equation having these as conditions is

(26.2.8) $\mathcal{N}(x, t) = \{N_0/A(\pi Dt)^{1/2}\} \exp(-x^2/4Dt)$

as may be verified by direct substitution. The form of the result at different times is shown in Fig. 26.10, and it is clear that the concentration of particles spreads through the material as time advances. The use of the diffusion equation is that, with its aid, the concentration can be predicted at any point in the system at any time.

A number of important features of the diffusion process can be explained on the basis of the diffusion equation, and in particular with the help of the simple solution just quoted. For example, we can ask what is the *mean distance* through which the solute has spread after a time t. The number of molecules in the slab of thickness dx at x is $\mathcal{N}(x, t)A\,dx$, and so the probability that any of the N_0 molecules is there is $\mathcal{N}(x, t)A\,dx/N_0$. If the molecule *is* there it has travelled a distance x from the origin; therefore the mean distance travelled is

$$\langle x \rangle = \int_0^\infty x\mathcal{N}(x, t)A\,dx/N_0 = (1/\pi Dt)^{1/2} \int_0^\infty x\,e^{-x^2/4Dt}dx$$

(26.2.9) $= 2(Dt/\pi)^{1/2}.$

The average distance varies as the *square root* of the time lapse. This is an important general result which we shall return to later. If we use the Stokes–Einstein relation for the diffusion constant the mean distance covered in a solvent of viscosity η by particles of radius a is

(26.2.10) $\langle x \rangle = (2kT/3\pi^2\eta a)^{1/2}\sqrt{t}.$

The *root mean square distance* covered is $\bar{x} = \sqrt{\langle x^2 \rangle}$, and its value is

(26.2.11) $\bar{x} = \sqrt{\langle x^2 \rangle} = \sqrt{\int_0^\infty x^2 \mathcal{N}(x, t)A\,dx/N_0} = (2Dt)^{1/2}.$

This is a valuable measure of the spread of the particles when they are allowed to migrate in both directions (for then $\langle x \rangle = 0$). The value of $\bar{x}$ for molecules having $D = 5 \times 10^{-6}$ cm^2 s^{-1} is shown in Fig. 26.11: you can see how long it takes for diffusion to increase $\bar{x}$ to about 1 cm in an unstirred solution. The *proportion* of particles which remain within a distance $\bar{x}$ of

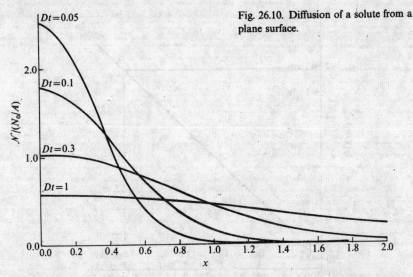

Fig. 26.10. Diffusion of a solute from a plane surface.

the origin is a useful number to have, because the mean might not convey enough information. Since the number in the slab at x is $\mathcal{N}(x,t)A\,dx$, the number in all the slabs up to the one at $\bar{x}$ is the sum (integral)

$$N(x \leqslant \bar{x}, t) = \int_0^{\bar{x}} \mathcal{N}(x,t)A\,dx = (0.68\ldots)(N_0/A),$$

where $\bar{x}$ has been replaced by $(2Dt)^{1/2}$, and the integral evaluated numerically.* It follows that the proportion of molecules inside the range $0 \leqslant x \leqslant \bar{x}$ is 0.68. Therefore, over two-thirds of the molecules are still clustered around the origin, and only 32 per cent have escaped beyond $\bar{x}$ (but do not forget that the value of $\bar{x}$ grows with time).

The diffusion equation can be solved for more complicated arrangements, for example, when ions are continuously generated at a plane electrode dipped in the solution, or when ions are deposited on an electrode and withdrawn from the solution. Calculations like these play an important part in the discussion of rates of reactions at electrodes (Chapter 30).

Diffusion: the statistical view. An intuitive picture of the mechanism of diffusion is one in which particles move in a series of small steps, and gradually migrate away from their original position. We shall build a model of diffusion on the basis that the particles can jump through a distance d, and do so in a time τ. This means that the distance covered by a molecule in a time t is $(t/\tau)d$. This does *not* mean that the particle will be found at

---

* The integral can be simplified by substituting $y = x/2(Dt)^{1/2}$, for then it becomes

$$N(x \leqslant \bar{x}, t) = \{N_0/A(\pi Dt)^{1/2}\} \int_0^{\bar{x}} dx\,e^{-x^2/4Dt} = (N_0/A)(2/\pi^{1/2}) \int_0^{1/2^{1/2}} e^{-y^2}dy.$$

It happens that the integral $(2/\pi^{1/2})\int_0^z e^{-y^2}\,dy$ is a standard mathematical form known as the *error function*, and written erf z. Standard tables of these are available, Table 25.1, and the value of $\mathrm{erf}(1/2)^{1/2}$, which is what we require, is 0.68.

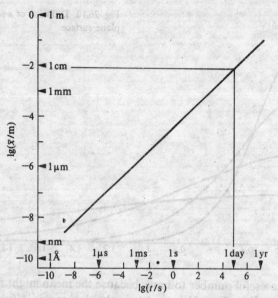

Fig. 26.11. Root mean square distance covered by particles with $D = 5 \times 10^{-6}$ cm^2 s^{-1}.

that distance from the origin. The direction of the steps is different on each occasion and so the net distance of diffusion must take this into account. We shall simplify the discussion by allowing the particle to move only along a straight line, the x-coordinate, but we must not forget that in a real system a particle is free to move in three dimensions. We shall also confine our attention to a model in which the particle can jump with equal probability through a distance d to the right or d to the left. This is called the *one-dimensional random walk*; we first met it in Chapter 24.

Our task is to find the probability that a molecule will be found at a distance x from the origin at a time t. During that time interval it will have taken t/τ steps: we shall write $n = t/\tau$. Many of these steps were steps to the right; many were steps to the left. If n_R is the number of steps to the right and n_L the number to the left, not only can we write the total number of steps as $n = n_R + n_L$, but we can also write the net distance travelled as $x = n_R d - n_L d$.

The probability of being at x after n steps of length d is the probability that of the n steps, n_R occurred to the right, n_L occurred to the left, and $n_R - n_L = x/d$.

What is the *total number of possibilities* for left or right steps? Since each step may occur in either of two directions (left or right) the total number of possibilities is 2^n.

How many ways are there of taking n_R *of the n steps to the right?* This is the same as the number of ways of choosing n_R objects from n possibilities, irrespective of the order: this is $n!/n_R!(n-n_R)!$. We can check this in the case of 4 steps, and ask what is the number of ways of taking 2 right steps. There are 2^4 possible step sequences:

LLLL LLLR LLRR LRRR RRRR
LLRL LRLR RLRR
LRLL LRRL RRLR
RLLL RLLR RRRL
RLRL
RRLL

and clearly there are 6 ways of taking 2 steps to the right and 2 to the left, which tallies with the expression $4!/2!2! = 6$. The probability that the particle is at the origin after 4 steps is therefore 6/16. The probability that it is at $x = 4d$ is 1/16 because, in order to be there, all four steps must be towards the right, and there is only one way of organizing that.

Returning now to the general case we see that the probability of being at x after n steps, each of length d, is

$$P(x) = n!/n_R!(n-n_R)!2^n,$$

with $n = n_R + n_L$ and $x/d = n_R - n_L$. Since

$$n_R = \tfrac{1}{2}(n + x/d), \qquad n - n_R = \tfrac{1}{2}(n - x/d),$$

it follows that

(26.2.12) $\qquad P(x) = n!/\{[\tfrac{1}{2}(n+s)]![\tfrac{1}{2}(n-s)]!2^n\},$

where $s = x/d$.

This expression does not seem to resemble the Gaussian distribution of probability, such as eqn (26.2.8), and so it looks as though the model of a random walk underlying a diffusion process is quite wrong. This, however, is not the case: the last equation becomes *identical* to the Gaussian distribution when we examine the limit in which the number of steps becomes very large.

The algebraic manipulation of this equation is based on the approximate formula for factorials of large numbers first used in Chapter 20 (p. 668). When N is a large number it is possible to use *Stirling's approximation*:

(26.2.13) $\qquad \ln N! \approx (N + \tfrac{1}{2})\ln N - N + \ln(2\pi)^{1/2}.$

This is a more accurate form of the approximation than the one used earlier. Even when N is quite small this expression is quite good. For example, instead of $10! = 3.629 \times 10^6$ it gives $10! \approx 3.60 \times 10^6$; when larger numbers are involved we can be very confident indeed about the results it gives. Taking logarithms of eqn (26.2.12), and then allowing n to be large, leads (after quite a lot of algebra) first from

$$\ln P = \ln n! - \ln([\tfrac{1}{2}(n+s)]!) - \ln([\tfrac{1}{2}(n-s)]!) - n\ln 2$$

to

$$\ln P \approx \ln(2/\pi n)^{1/2} - \tfrac{1}{2}(n+s+1)\ln(1+s/n) - \tfrac{1}{2}(n-s+1)\ln(1-s/n).$$

If we allow s/n to be a small number (so that x must not be a great distance from the origin) we can use the approximation $\ln(1+y) \approx y$, and obtain

$$\ln P \approx \ln(2/\pi n)^{1/2} - s^2/2n,$$

or

$$P \approx (2/\pi n)^{1/2} \exp(-s^2/2n),$$

which is already of a Gaussian form.

Now replace s by x/d and n by t/τ. We obtain

(26.2.14) $$P(x,t) = (2\tau/\pi t)^{1/2} \exp(-x^2\tau/2td^2),$$

and this has precisely the form of $\mathcal{N}(x,t)/N_0$ given in eqn (26.2.8) as a solution of the diffusion equation. (The differences of detail arise from allowing the particle to migrate in both directions away from $x = 0$, and letting it be found only at discrete points separated by d instead of being anywhere on a continuous line.) Therefore we can be confident that the diffusion *can* be interpreted as the result of a very large number of small steps in random directions. This also indicates the region of invalidity of the diffusion equation: we should not expect it to apply at times so short that the particles have had time to take only a few steps.

Finally we can make use of the identity of form of the two distributions to obtain yet another expression for D. Comparison of the two exponents leads to the identification

$$2d^2/\tau = 4D,$$

and therefore we come to the

(26.2.15)   *Einstein–Smoluchowski relation:* $D = \frac{1}{2}d^2/\tau.$

Example (Objective 17). Suppose that the SO_4^{2-} ion jumps through about its own diameter every time it makes a move in aqueous solution. How often does it change position?

- *Method.* The diffusion coefficient was found in the last *Example* to be 1.065×10^{-5} cm^2 s^{-1}, and the effective radius was found there to be 205 pm. Find τ from eqn (26.2.15).

- *Answer.* From eqn (26.2.15),

$$\tau = d^2/2D$$
$$= (2 \times 205 \times 10^{-12} \text{ m})^2/2 \times (1.065 \times 10^{-9} \text{ m}^2 \text{ s}^{-1}) = 7.89 \times 10^{-11} \text{ s}.$$

- *Comment.* The big, heavy SO_4^{2-} ion jumps through its diameter in about 8×10^{-11} s. If the ion were imagined as jumping through a distance equal to the diameter of a water molecule (≈ 150 pm) the jump time would be about 1.1×10^{-11} s.

The Einstein–Smoluchowski relation is a central connection between the microscopic properties of the size (d) and rate ($1/\tau$) of a molecular jump and the macroscopic properties of diffusion constant and viscosity (via the Stokes–Einstein relation eqn (26.2.6)). This also brings the discussion full circle and back to the properties of gases. For if d/τ is interpreted as a mean velocity of the molecules undergoing diffusion, and the jump length d is called a mean free path and written λ, the Einstein–Smoluchowski equation reduces to $D = \frac{1}{2}\lambda\bar{c}$, which is the same as that obtained for the diffusion constant from the kinetic theory of gases. This shows that the diffusion of a perfect gas can also be interpreted as a random walk through an average path length λ.

Summary of the general conclusions. The chapter began by examining various aspects of the motion of ions in solution. We saw that the conductivity could be expressed in terms of the *mobility* of the ions. We also saw that any species could be regarded as moving under the influence of an *effective force* $\mathscr{F}$ if its chemical potential varied from place to place, and we identified $\mathscr{F}$ with $-d\mu/dx$, eqn (26.2.1). The thermodynamic force led to the construction of *Fick's First Law of Diffusion*. We saw on quite general arguments, that if the particle was subjected to a unit force, it acquired a drift speed D/kT. This led to the *Einstein relation* between D and the mobility, eqn (26.2.4), and the *Nernst–Einstein relation*, eqn (26.2.5), between conductivity and D, see Box 26.1. Incorporation of the Stokes frictional force into the argument led to the *Stokes–Einstein relation*, eqn (26.2.6), between D and the viscosity, valid for molecules of any charge (including zero). We next set up equations for dealing with time-dependent diffusional processes, and derived the basic *diffusion equation*, eqn (26.2.7). The solutions of this equation could be reproduced, we found, if we modelled the diffusion process as a series of small steps of length d occurring with a frequency

---

Box 26.1   **Transport properties in solution**

Einstein–Smoluchowski relation between jump size d and jump time τ:

$$D = d^2/2\tau.$$

Stokes–Einstein relation between diffusion coefficient D and solution viscosity η:

$$D = kT/6\pi\eta a.$$

Einstein relation between diffusion coefficient and ion mobility $u_\pm$:

$$D_\pm = u_\pm kT/ez = u_\pm RT/zF.$$

Nernst–Einstein relation between diffusion coefficient and ion conductivity $\lambda_\pm$:

$$\lambda_\pm = (z^2 F^2/RT)D_\pm.$$

$1/\tau$; the solutions became the same when $\frac{1}{2}d^2/\tau$ was identified with the diffusion constant D: this is the *Einstein–Smoluchowski relation*, eqn (26.2.15). With this connection established we can interpret viscosity, ionic mobility, conductivity, and diffusion processes in general in terms of the microscopic, dynamical parameters d and τ.

Appendix: the measurement of transport numbers

The following are brief summaries of the three methods used to measure transport numbers of ions and, through them, individual ion conductivities and mobilities.

(1) *Moving boundary method*. Let MX be the salt of interest. Pour the solution of MX into the lower half of a narrow vertical tube. Select a salt NX where N is less mobile than M; prepare a solution of NX, and pour it on top of the MX solution so that there is a clear boundary. Pass a current I for a time t. The X^- move towards the anode (downwards) and the M^+ and N^+ move towards the cathode (upwards). The amount of cations transported for this amount (It/F) of electricity is $t_+(It/z_+F)$, their charge being z_+. If they are at a concentration c, the volume swept out is $t_+(It/z_+cF)$. But if the cross-section of the tube is S, and the distance moved by the boundary is x, the volume is also equal to xS. Therefore $t_+(It/z_+cF) = xS$, and monitoring the progress of the boundary for a series of times gives t_+.

Example (Objective 7). The transport numbers of H^+ and SO_4^{2-} were measured in a moving boundary experiment. The apparatus consisted of a tube of bore 6.40 mm containing aqueous sulphuric acid at a concentration of 0.015 mol dm^{-3}. A steady current of 1.23 mA was passed, and the boundary advanced as follows:

t/s	40	80	120	160	200
x/mm	0.860	1.722	2.586	3.450	4.309

Find t_+ and t_-.

- *Method*. Use the equations set out above.

- *Answer.* $t_+ = (ScF/I)(x/t)$

$$= \pi \times (3.20 \times 10^{-3} \text{ m})^2 \times (0.015 \text{ mol dm}^{-3}) \times (9.6485 \times 10^4 \text{ C mol}^{-1})$$
$$\times (1/1.23 \times 10^{-3} \text{ A}) \times (x/t)$$
$$= 3.785 \times 10^4 \text{ m}^{-1} \text{ s } (x/t) = 37.85 \, [(x/mm)/(t/s)].$$

Draw up the following table:

t/s	40	80	120	160	200
$10^4(x/mm)/(t/s)$	215	215.3	215.5	215.6	215.5

Average: $0.021\,54$ mm s^{-1}. Therefore $t_+(H^+) = 0.815$; and so $t_{\pm}(SO_4^{2-}) = 1 - 0.815 = 0.185$.

- *Comment*. These results can be used to recover the mobilities and the individual ionic conductivities, and are the basis of the data in the last *Example*.

(2) *The Hittorf method*. A cell is divided into three compartments and

an amount It of electricity is passed. An amount It/z_+F of cations are discharged at the cathode, but an amount $t_+(It/z_+F)$ of cations move into the cathode region. The net change in the amount of cations near the cathode is $-(It/z_+F)+t_+(It/z_+f) = -(1-t_+)(It/z_+F) = -t_-(It/z_+F)$. Therefore, measuring the change in composition in the cathode compartment gives t_-, the anion transport number. Likewise, the change in composition of anions at the anode is $-t_+(It/|z_-|F)$.

(3) *E.m.f. measurements.* The e.m.f. of a cell with transference having an electrode reversible with respect to anions is related to the e.m.f. of the cell with the same net reaction but without transference by $E_t = 2t_+E$ (the argument is similar to that used for the Hittorf method). Therefore set up two cells and compare their e.m.f.s.

Further reading

Conductimetry. T. Shedlovsky and L. Shedlovsky; in *Techniques of chemistry* (A. Weissberger and B. W. Rossiter, eds.), Vol. IIA, 163, Wiley-Interscience, New York, 1971.

Determination of transference numbers. M. Spiro; in *Techniques of chemistry* (A. Weissberger and B. W. Rossiter, eds.), Vol. IIA, 205, Wiley-Interscience, New York, 1971.

The principles of electrochemistry. D. A. MacInnes; Dover, New York, 1961.

The physical chemistry of electrolytic solutions. H. S. Harned and B. B. Owen; Reinhold, New York, 1958.

Ionic solution theory. H. L. Friedman; Wiley-Interscience, New York, 1962.

Electrolyte solutions. R. A. Robinson and R. H. Stokes; Academic Press, New York, 1959.

Electrolytic conductance. R. M. Fuoss and F. Accascina; Wiley-Interscience, New York, 1959.

Treatise on electrochemistry. G. Kortum; Elsevier, Amsterdam.

Experimental methods for studying diffusion in liquids, gases, and solids. P. J. Dunlop, B. J. Steel, and J. E. Lane; in *Techniques of chemistry* (A. Weissberger and B. W. Rossiter, eds.), Vol. IV, 205, Wiley-Interscience, New York, 1972.

Diffusion in solids, liquids, and gases. W. Jost; Academic Press, New York, 1960.

Problems

26.1. Conductivities are often measured by comparing the resistance of a cell filled with the sample to its resistance when filled with some standard solution, such as aqueous potassium chloride. The conductivity of water is $7.6 \times 10^{-4}\ \Omega^{-1}\ cm^{-1}$ at $25\ °C$ and the conductivity of 0.1 mol dm^{-3} aqueous KCl is $1.1639 \times 10^{-2}\ \Omega^{-1}\ cm^{-1}$. A cell had a resistance of 33.21 Ω when filled with 0.1 mol dm^{-3} KCl solution and 300 Ω when filled with 0.1 mol dm^{-3} acetic acid. What is the molar conductivity of acetic acid at that concentration and temperature?

26.2. In conductivity measurements it is common to write $\kappa = C/R$, where R is the measured resistance of the sample in the cell and C, the *cell constant*, is a characteristic of the particular cell in use. Since $\kappa = l/RA$ we see that $C = l/A$. Both l and A may be difficult to determine directly, and so it is normal to calibrate the cell with a sample of known conductivity. In several of the Problems that follow we shall refer to 'the cell'. We shall always mean the cell of this Problem, and so now we obtain its constant. A 0.020 mol dm^{-3} aqueous KCl solution has a molar conductivity of 138.3 Ω^{-1} cm^2 mol^{-1} at 25 °C, and in the cell its resistance was found to be 74.58 Ω. Find the cell constant C.

26.3. The resistances of a series of aqueous NaCl solutions, formed by successive dilution of a sample, were measured in the cell. The following resistances were found:

$c(NaCl)/$mol dm^{-3}	0.0005	0.001	0.005	0.010	0.020	0.050
R/Ω	3314	1669	342.1	174.1	89.08	37.14.

Check that the molar conductivity follows the Kohlrausch equation, eqn (26.1.3) and find the values of the molar conductivity at infinite dilution. Determine the coefficient $\mathcal{K}$. (The latter's theoretical value will be calculated in Problem 26.23.)

26.4. Use the value of $\mathcal{K}$ (which should depend only on the nature, not the identity of the ions) from the last Problem and the information that $\lambda^0_+(Na^+) = 50.1 \ \Omega^{-1}$ cm^2 mol^{-1} and $\lambda^0_-(I^-) = 76.8 \ \Omega^{-1}$ cm^2 mol^{-1}, to predict (a) the molar conductivity, (b) the conductivity, (c) the resistance it would show in the cell, of an 0.01 mol dm^{-3} aqueous solution of NaI at 25 °C.

26.5. The molar conductivities at infinite dilution of KCl, KNO$_3$, and AgNO$_3$ are 149.9 Ω^{-1} cm^2 mol^{-1}, 145.0 Ω^{-1} cm^2 mol^{-1}, and 133.4 Ω^{-1} cm^2 mol^{-1} respectively (all at 25 °C). What is the molar conductivity of AgCl at infinite dilution at this temperature.

26.6. The conductivity of a saturated solution of AgCl in water at 25 °C was found to be $1.887 \times 10^{-6} \ \Omega^{-1}$ cm^{-1}. Use the results of the last Problem to find the solubility and the solubility product at this temperature.

26.7. The molar conductivities at infinite dilution of aqueous sodium acetate, hydrochloric acid, and sodium chloride are 91.0 Ω^{-1} cm^2 mol^{-1}, 425.0 Ω^{-1} cm^2 mol^{-1}, and 128.1 Ω^{-1} cm^2 mol^{-1} respectively at 25 °C. What is the molar conductivity at infinite dilution of acetic acid?

26.8. When the resistance of an 0.020 mol dm^{-3} solution of acetic acid was measured in the cell its resistance was found to be 888 Ω. What is the degree of dissociation of the acid at this concentration?

26.9. What is the pH of the acid in the last Problem? Calculate this in two ways. First ignore activity coefficients. Then estimate them from the Debye–Hückel limiting law (p. 325).

26.10. Ostwald's Dilution Law was quoted in eqn (26.1.5). In order to check its validity and to put it to use, the resistances of aqueous acetic acid solutions were measured at 25 °C in the cell with the following results:

$c/$mol dm^{-3}	0.000 49	0.000 99	0.001 98	0.015 81	0.063 23	0.2529
R/Ω	6146	4210	2427	1004	497	253

The molar conductivity at infinite dilution was found in Problem 26.7. Draw the appropriate graph to show that eqn (26.1.5) is obeyed for concentrations of less than about 0.01 mol dm^{-3} and obtain a value for the acid dissociation constant K_a and for pK_a.

26.11. What is the molar conductivity, the conductivity, and the resistance (in the cell) of an 0.040 mol dm^{-3} solution of acetic acid at 25 °C.

26.12. At 25 °C the ionic conductivities of Li$^+$, Na$^+$, and K$^+$ are 38.7 Ω^{-1} cm^2 mol^{-1}, 50.1 Ω^{-1} cm^2 mol^{-1}, and 73.5 Ω^{-1} cm^2 mol^{-1} respectively. What are their mobilities?

26.13. What are the drift speeds of the ions in the last Problem when a potential difference of 10 V is applied across a 1 cm conductivity cell? How long would it take an ion to move from one electrode to the other? In conductivity measurements it is normal to use alternating current: what are the displacements of the ions in (a) cm, (b) solvent diameters, *c.* 300 pm, during a half cycle of 1 kHz applied potential?

26.14. The *transport number* of an ion in a mixture is defined as the fraction of the total current that it carries. Show that this definition reduces to eqn (26.1.16) in the case of a solution of a single 1:1 salt. Then show how the ratio of two transport numbers t', t'' for two cations in a mixture depends on their concentrations c', c'', and their mobilities u', u''.

26.15. The mobilities of H^+ and Cl^- at 25 °C in water are 3.23×10^{-3} cm² s⁻¹ V⁻¹ and 7.91×10^{-4} cm² s⁻¹ V⁻¹ respectively. What proportion of the current is carried by the protons in 10^{-3} mol dm⁻³ hydrochloric acid? What fraction do they carry when the NaCl is added to the acid so that it is 1.0 mol dm⁻³ in the salt? Note how concentration as well as mobility governs the transport of current.

26.16. In the *Hittorf method* for the determination of transport numbers (Appendix) a known amount of electricity is passed through a solution and the change in composition in the regions close to the electrodes is monitored. For unit amount of electrons passed through a solution of KCl the amount of KCl transferred from the anode compartment to the cathode compartment is t_+. In an experiment, 187.2 mg of silver were deposited in a silver coulometer in series with a cell that contained an aqueous 0.04 mol kg⁻¹ KCl solution. After the passage of this amount of electricity it was found that 118.42 g of the solution from the 250 cm³ cathode compartment contained 165.3 mg KCl while 114.11 g of solution from the 250 cm³ anode compartment contained 159.8 mg KCl. A 112.62 g sample from the central compartment contained 185 mg KCl before the experiment, and a 115.66 g sample contained 191 mg KCl after the experiment. What are the transport numbers of the K^+ and Cl^- ions?

26.17. Transport numbers may also be measured by the *moving boundary method* (Appendix). In a particular experiment on KCl the apparatus consisted of a tube of bore 4.146 mm, and it contained aqueous KCl at a concentration of 0.021 mol dm⁻³. A steady current of 1.82 mA was passed, and the boundary advanced as follows:

t/s	20	40	60	80	100
x/mm	0.64	1.28	1.92	2.54	3.18

Find the transport numbers of K^+ and Cl^-.

26.18. From the information in the preceding Problem, find the mobilities of K^+ and Cl^- ions in aqueous solutions, and then their ionic conductivities.

26.19. A third technique for measuring transport numbers (and therefore ion mobilities) is by an e.m.f. method. Show that the e.m.f. of the cell Ag,AgCl|HCl(c_1)|HCl(c_2)|AgCl,Ag with transference E_t is related to its e.m.f. without transference E by $E_t = 2t_+ E$. By the same argument show that for electrodes reversible with respect to the cation, the e.m.f. is given by $E_t = 2t_- E$.

26.20. In the case of the cell written in the last Problem, when $c_1 = 0.01$ mol dm⁻³ and $c_2 = 0.02$ mol dm⁻³ the e.m.f. was measured as -29.2 mV at 25 °C. What are the transport numbers and mobilities of the H^+ and Cl^- ions?

26.21. The proton possesses abnormal mobility in water, but does it behave normally in liquid ammonia? In order to investigate this question, a moving-boundary technique was used to determine the transport number of NH_4^+ in liquid ammonia (the analogue of H_3O^+ in liquid water) at $-40\,°C$ (J. Baldwin, J. Evans, and J. B. Gill, *J. chem. Soc. (A)*, 3389 (1971)). A steady current of 5 mA was passed for 2500 s, during which time the boundary formed between mercury(II) iodide and ammonium iodide ammoniacal solutions moved 286.9 mm in a 0.013 65 mol kg^{-1} solution and 92.03 mm in a 0.042 55 mol kg^{-1} solution. Calculate the transport number of NH_4^+ at these concentrations, and comment on the mobility of the proton in liquid ammonia.

26.22. The conductivity of the purest water prepared is $5.5 \times 10^{-8}\ \Omega^{-1}\ cm^{-1}$. What would be the resistance measured for a sample of this water in the conductivity cell of Problem 26.2? What is the dissociation constant of water? What are the values of pK_w and pH for pure water? (Take the mobilities of H^+ and OH^- from Table 26.2.)

26.23. In Problem 26.3 we found an experimental value for $\mathscr{K}$ in the Kohlrausch expression for the conductivity of NaCl. Calculate the values of A and B that appear in the Debye–Hückel–Onsager equation, eqn (26.1.18), and find a theoretical value for $\mathscr{K}$.

26.24. A dilute solution of potassium permanganate in water at $25\,°C$ was prepared. The solution was in a horizontal 10 cm tube, and at first there was a linear gradation of intensity of the purple solution from the left (where the concentration was 0.10 mol dm^{-3}) to the right (where the concentration was 0.05 mol dm^{-3}). What is the magnitude and sign of the thermodynamic force acting on the solute (a) close to the left face of the container, (b) in the middle, (c) close to the right face. Give the force per mole and force per molecule in every case.

26.25. The diffusion coefficient for sucrose in water at $25\,°C$ is $5.2 \times 10^{-6}\ cm^2\ s^{-1}$. Suppose that the permanganate in the last Problem were replaced by sucrose and we had some way (what, for instance?) of monitoring its concentration. What is the drift velocity of the sugar at the three points (left, middle, right)? What effect on the concentration distribution do these different drift velocities bring about? Sketch the concentration distribution at several later times. What is the drift velocity after an infinite time?

26.26. Estimate the effective radius of a sugar (sucrose) molecule in water at $25\,°C$ on the basis that its diffusion coefficient is $5.2 \times 10^{-6}\ cm^2\ s^{-1}$ and that the viscosity of water is 1.00 centipoise ($10^{-3}\ kg\ m^{-1}\ s^{-1}$).

26.27. The diffusion coefficient for molecular iodine in benzene is $2.13 \times 10^{-5}\ cm^2\ s^{-1}$. How long does a molecule take to jump through about one molecular diameter (approximately the fundamental jump length for translational motion)?

26.28. What is the root mean square distance travelled by (a) an iodine molecule in benzene, (b) a sucrose molecule in water at $25\,°C$ in 1 s?

26.29. About how long, on average, does it take for the molecules in the last Problem to drift to a point (a) 1 mm, (b) 1 cm from their starting points?

26.30. Estimate the diffusion coefficients and the effective hydrodynamic radii of the alkali metal cations in water from their mobilities at $25\,°C$ (Table 26.2).

26.31. Estimate the approximate number of water molecules that are dragged along by the cations in the last Problem. Crystal radii are given in Table 22.1.

26.32.   Nuclear magnetic resonance can be used to determine the mobility of molecules in liquids. A set of measurements on methane in carbon tetrachloride showed that its diffusion coefficient is 2.05×10^{-5} cm^2 s^{-1} at 0 °C and 2.89×10^{-5} cm^2 s^{-1} at 25 °C. Deduce what information you can about the mobility of methane in carbon tetrachloride.

26.33.   Confirm that eqn (26.2.8) is a solution of the diffusion equation with the correct initial value.

26.34.   A concentrated sucrose solution is poured into a 5 cm diameter cylinder. Take it as 10 g of sugar in 5 cm^3 of water. A further 1 dm^3 of water is then poured very carefully on top of the layer, without disturbing it. Ignore gravitational effects, and pay attention only to diffusional processes. Find the concentration at 5 cm above the lower layer after a lapse of (a) 10 s, (b) 1 year.

26.35.   The diffusion equation is valid when many elementary steps are taken in the time interval of interest; but the random walk calculation lets us discuss distributions for short times as well as for long. Use eqn (26.2.12) to calculate the probability of being 6 paces from the origin (that is, at $x = 6d$) after (a) 4, (b) 6, (c) 12 steps.

26.36.   Write a program for calculating $P(x)$ in a one-dimensional random walk, and evaluate the probability of being at $x = 6d$ for $n = 6, 10, 14, \ldots, 60$. Compare the numerical value with the analytical value in the limit of a large number of steps. At what value of n is the discrepancy no more than 0.1 per cent?

26.37.   In a series of observations on the displacement of rubber latex spheres of radius 2.12×10^{-5} cm the mean square displacements after selected time intervals were on average as follows:

t/s	30	60	90	120
$10^8 \langle x^2 \rangle / $cm^2	88.2	113.5	128	144

These results were originally used to find the value of Avogadro's constant (a remarkably eye-straining procedure for such a large quantity!) but there are now better ways of determining it, and so the data can be used to find another quantity. Find the effective viscosity of water at the temperature of this experiment (25 °C).

27   The rates of chemical reactions

Introduction

In this chapter we look into how chemical reactions occur. The principal aspect we examine is the *rate* of a reaction, and we shall see how it depends on the temperature and the concentrations of the species that are present.

There are two main reasons for studying the rates of reactions. The first is the practical importance of being able to predict how quickly a reaction mixture will move to its equilibrium state: the rate might depend on a number of factors under our control, such as the temperature, the pressure, and the presence of a catalyst, and, depending on our aims, we may be able to make the reaction proceed at an optimum rate. For instance, in an industrial process it might be economical for the reactions to proceed very rapidly; but not so rapidly as to produce an explosion. By contrast, in a biological process it may be appropriate for a reaction to proceed only slowly, and to be switched on and off at the demand of some activity.

The second reason for studying reaction rates (which, as we shall see, is closely bound up with the first) is that the study of rates can reveal the *mechanisms* of reactions. The term 'mechanism' has two connotations in this context. The first is the analysis of a chemical reaction into a sequence of elementary steps. For example, we might discover that the reaction of hydrogen and bromine proceeds by a sequence of steps involving the fission of Br_2 into two bromine atoms, the attack of one of these atoms on H_2, and so on. *The statement of all the elementary steps constitutes the statement of the mechanism of the reaction.* The other meaning of mechanism relates to the individual steps themselves, and concerns their detailed nature. In this sense 'mechanism' concerns what happens as a bromine atom approaches and attacks a rotating, vibrating, hydrogen molecule.

The first type of analysis of mechanism is the central feature of classical chemical kinetics, and we concentrate on it in this chapter. The second type of analysis, called *chemical dynamics,* had to await the technological advances that made available molecular beams for the study of individual molecular collisions, and is discussed in the next chapter. The dividing line between chemical kinetics and chemical dynamics is not clear cut: crude models of individual reaction steps were built on the basis of kinetic analyses, and we see something of this in the present chapter.

27.1. Empirical chemical kinetics

The basic data of chemical kinetics are the concentrations of the reactants and products as functions of time. The method selected for monitoring the concentrations depends on the nature of the species involved in the reaction, and on its rapidity.

Many reactions go to completion (that is, attain thermodynamic equilibrium) over a period of minutes or hours, and may be monitored by classical techniques. One of the following methods is often chosen.

(1) *Pressure changes.* A reaction in the gas phase might result in a change of pressure, and so its progress may be monitored by recording the pressure as a function of time. An instance of this is the decomposition of nitrogen(V) oxide, N_2O_5, according to

$$2N_2O_5(g) \rightarrow 4NO_2(g) + O_2(g).$$

For every mole of N_2O_5 destroyed, 5/2 moles of gaseous products are formed, and so the pressure of the system increases during the course of the reaction. This method is inappropriate for reactions that leave the overall pressure unchanged, and for reactions in solution.

(2) *Spectroscopy.* A technique that is available even when no pressure change occurs is the spectroscopic analysis of the mixture. For instance, the reaction

$$H_2(g) + Br_2(g) \rightarrow 2HBr(g)$$

can be followed by monitoring the intensity of absorption of visible light by the bromine.

(3) *Polarimetry.* When the optical activity of a mixture changes in the course of reaction, it can be monitored by measuring the angle of optical rotation. This is a historically important method because its application to the hydrolysis of sucrose was the first significant study of the rate of a reaction (by Wilhelmy in 1850).

(4) *Electrochemical methods.* When a reaction changes the number or nature of ions present in a solution, its course may be followed by monitoring the conductivity of the solution. One very important class of reactions consists of those occurring at electrodes, and we examine them in Chapter 30.

(5) *Miscellaneous methods.* Other methods of determining composition include mass spectrometry and chromatography. In order to employ these techniques, a small amount of the reaction mixture is bled from the reacting system at a series of times after the initiation of the reaction, and then analysed.

There are three main ways of applying these analytical techniques.

(1) *Real time analysis.* In this method the composition of the system is analysed while the reaction is in progress.

(2) *Quenching.* In this method the reaction is frozen after it has been allowed to proceed for a certain time, and then the composition is analysed by any suitable technique. The quenching can normally be achieved by

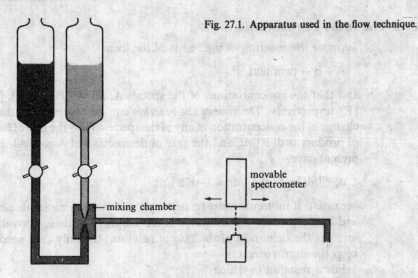

Fig. 27.1. Apparatus used in the flow technique.

movable
spectrometer

— mixing chamber

lowering the temperature suddenly, but this is suitable only for reactions that are slow enough for there to be little reaction during the time it takes to cool the mixture.

(3) *Flow method*. In this method, solutions of the reagents are mixed as they flow together into a chamber, Fig. 27.1. The reaction continues as the thoroughly mixed solutions flow through the outlet tube, and observation of the composition at different positions along the tube (for example, by spectroscopy) is equivalent to observing the reaction mixture at different times after mixing. Reactions that are complete within a few milliseconds can be observed with this technique, but its principal disadvantage is that large volumes of solutions are necessary. The method has been improved, and a modification, the *stopped-flow method* (p. 959), is in wide use.

The rates of chemical reactions have been found to depend very strongly on the temperature, and many follow the *Arrhenius rate law*, that the rate is proportional to $\exp(-E_a/RT)$, where E_a is the *activation energy* of the reaction. The implication of this observation for experimental studies is that the temperature of the reaction mixture must be held as constant as possible throughout the course of the reaction, otherwise the observed reaction rate will be a meaningless average of rates at different temperatures. This requirement puts severe demands on the design of the experiment. Gas phase reactions, for instance, are often carried out in a vessel held in contact with a substantial block of metal, and liquid phase reactions, including flow reactions, must be carried out in an efficient thermostat.

The general result of these experiments is that the rates of chemical reactions depend on the composition of the reaction mixture, and most depend exponentially on the temperature. The next few sections look at these observations in more detail.

27.2 The rates of reactions

Suppose the reaction of interest is of the form

   $A + B \rightarrow$ products, P

and that the concentrations of the species A, B, and P are [A], [B], and [P] respectively. The *rate of the reaction* can be expressed as the rate of change of the concentration of any of the species. Thus the rate of formation of product is $d[P]/dt$, and the rate of destruction of A is $d[A]/dt$. In the present case

   $d[P]/dt = -d[A]/dt = -d[B]/dt,$

because a B molecule must be destroyed for every A molecule destroyed, and in the process a P molecule is formed. Any of these derivatives can serve as the definition of the rate of reaction; the only care needed is to keep the signs correct.

For a reaction in which

   $A + 2B \rightarrow 3C + D$

the situation is less clear cut. In this case the concentration of B changes at twice the rate of change of the concentration of A, while the concentration of C increases at three times the rate of decrease of A. We may choose any of the time derivatives for the rate of the reaction, and they will be related by

   $-d[A]/dt = -\tfrac{1}{2}d[B]/dt = \tfrac{1}{3}d[C]/dt = d[D]/dt.$

The implication is that we cannot simply speak of the 'rate' of a reaction without saying exactly what we mean: we must state that we have chosen *rate* $= d[A]/dt$, or *rate* $= d[C]/dt$, and quote the chemical equation for the reaction. In this way there is never any ambiguity.

A consistent and succinct definition of the rate of a specified reaction can be given in terms of the *advancement* ξ (xi) of the reaction. This quantity was introduced in the thermodynamic discussion of equilibrium, p. 261, where we saw that if the reaction $A + 2B \rightarrow 3C + D$ proceeds by $d\xi$ the amount of A changes by $-d\xi$, the amount of B changes by $-2d\xi$, and so on. The *true rate* is defined as

(27.2.1)   *true rate* $= d\xi/dt,$

the rate of change of advancement of the reaction. This definition reduces to the elementary expressions for the rate of reaction on introducing the appropriate values of the stoichiometric coefficients.*

---

*The precise interpretation of true rate requires the reaction

   $A + 2B \rightarrow 3C + D$

to be written in the form

   $0 = -A - 2B + 3C + D,$

and in general,

$[I]_0/mol\,dm^{-3}$ 1.0×10^{-5} 2.0×10^{-5} 4.0×10^{-5} 6.0×10^{-5}

$(d[I_2]/dt)_0/mol\,dm^{-3}\,s^{-1}$ (a) 8.7×10^{-4} 3.48×10^{-3} 1.39×10^{-2} 3.13×10^{-2}

(b) 4.35×10^{-3} 1.74×10^{-2} 6.96×10^{-2} 1.57×10^{-1}

(c) 8.69×10^{-3} 3.47×10^{-2} 1.38×10^{-1} 3.13×10^{-1}

(a) corresponds to an argon concentration of $1.0 \times 10^{-3}\,mol\,dm^{-3}$, (b) to $5.0 \times 10^{-3}\,mol\,dm^{-3}$, and (c) to $10.0 \times 10^{-3}\,mol\,dm^{-3}$. Find the orders of reaction with respect to the iodine and argon atom concentrations, and the rate coefficient.

- *Method.* Plot $\lg(d[I_2]/dt)_0$ against $\lg[I]_0$ for a given $[Ar]$, and against $\lg[Ar]_0$ for a given $[I]_0$. Intercepts give $\lg k$, gradients the orders.

- *Answer.* The gradients are 2 (for the plot against $\lg[I]_0$) and 1 (for the plot against $\lg[Ar]_0$). Therefore the reaction is second order in I, first order in Ar, and third order overall. The intercept is at 9.94, and so $k_3 = 8.7 \times 10^9\,dm^6\,mol^{-2}\,s^{-1}$.

- *Comment.* A third-order reaction arises from the need for a three-body collision: the third body mops up some of the energy of the colliding I atoms and so prevents the I_2 molecule from dissociating as soon as it is formed.

Unfortunately the initial slope might not reveal the full rate law, for in a complex reaction the products themselves might be involved in intermediate steps. In the case of the HBr synthesis, for example, the true rate law involves the concentration of HBr, but initially no HBr is present, and a study of the initial rate as a function of the amount of hydrogen and bromine added would give incomplete information and be misleading about the overall reaction mechanism. In order to avoid this difficulty the rate law ought to be fitted to the data throughout the course of the reaction. Since the rate laws are differential equations giving the rate of change of concentrations at any stage of the reaction, integration will give an expression for the actual concentrations at any time. Different rate laws may give rise to different time dependences of the concentrations, and so the actual rate law can be found by fitting the various predictions to the observed concentrations.

The simplest rate law is for a first-order reaction:

(27.2.5) $-d[A]/dt = k_1[A].$

This rearranges into

$$-(1/[A])\,d[A] = k_1\,dt,$$

which can be integrated directly. Noting that at $t = 0$ the concentration of A is $[A]_0$, and at a later time t it is $[A]_t$, we have

$$\int_{[A]_0}^{[A]_t} (-1/[A])\,d[A] = \int_0^t k_1\,dt,$$

or

$$-(\ln[A]_t - \ln[A]_0) = k_1 t.$$

This expression can be reorganized into two useful forms:

(27.2.6a) $\ln([A]_t/[A]_0) = -k_1 t$

and

(27.2.6b) $[A]_t = [A]_0 \exp(-k_1 t).$

The second of this pair shows that the concentration of A falls exponentially with time with a rate determined by k_1. The first of the pair shows what to plot in order to confirm first-order behaviour and to determine the value of k_1: if $\ln([A]_t/[A]_0)$ is plotted against t the gradient of the straight line is $-k_1$. This is illustrated in the following *Example*, and some experimental values of k_1 are given in Table 27.1.

Table 27.1. Kinetic data for some first-order reactions

Reaction	Phase	$t/°C$	k_1/s^{-1}	$t_{1/2}$
$N_2O_5 \rightarrow NO_2 + NO_3$	Gas†	25	3.14×10^{-5}	6.1 h
		55	1.42×10^{-3}	8.2 m
$C_2H_6 \rightarrow 2CH_3$	Gas†	700	5.46×10^{-4}	21.2 m
Cyclopropane $\rightarrow$ propene	Gas†	500	6.71×10^{-4}	17.2 m
$N_2O_5 \rightarrow NO_2 + NO_3$	HNO_3 solution	55	9.27×10^{-5}	125 m
$N_2O_5 \rightarrow NO_2 + NO_3$	Br_2 solution	55	2.08×10^{-3}	333 s

† High-pressure limit
Sources: Principally K. J. Laidler, *Chemical kinetics*, McGraw-Hill; M. J. Pilling, *Reaction kinetics*, Oxford; J. Nicholas, *Chemical kinetics*, Harper and Row.

Example (Objective 7). The partial pressure of azomethane, $CH_3N_2CH_3$, was followed as a function of time at 600 K; the results are given below. Confirm that the decomposition $CH_3N_2CH_3 \rightarrow CH_3CH_3 + N_2$ is first-order in azomethane and find the value of the rate coefficient at this temperature.

t/s	0	1000	2000	3000	4000
$p(CH_3N_2CH_3)/$ mmHg	8.2×10^{-2}	5.72×10^{-2}	3.99×10^{-2}	2.78×10^{-2}	1.94×10^{-2}

- *Method.* Plot $\ln(p/p_0)$ against t. Get a straight line for a first-order reaction; its slope is $-k_1$.

- *Answer.* The data are plotted in Fig. 27.2. The plot gives a straight line with slope -3.6×10^{-4}, and so $k_1 = 3.6 \times 10^{-4}$ s. (The coefficient of determination of the least squares plot is 1.000.)

- *Comment.* First-order gas-phase reactions have some peculiar features: see below (p. 945).

On turning to *second-order reactions* it is necessary to distinguish between rate laws of the form

(27.2.7a) $-d[A]/dt = k_2[A]^2$

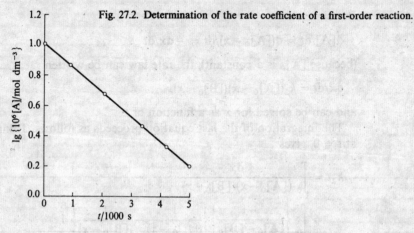

Fig. 27.2. Determination of the rate coefficient of a first-order reaction.

and

(27.2.7b) $-d[A]/dt = k_2[A][B].$

In the first case the integration is straightforward:

$$\int_{[A]_0}^{[A]_t} (-1/[A]^2)\, d[A] = \int_0^t k_2\, dt.$$

Hence

(27.2.8a) $(1/[A]_t) - (1/[A]_0) = k_2 t,$

which rearranges into

(27.2.8b) $[A]_t = \dfrac{[A]_0}{1 + k_2 t [A]_0}.$

The first of this pair shows what to plot in order to test for second-order behaviour: a straight line should be obtained when $1/[A]_t$ is plotted against t; if this is so, the slope is the second-order rate coefficient k_2. The second of the pair lets us predict the concentration of A at any time after the start of the reaction once we know the rate coefficient and the initial concentration of A.

The second type of second-order reaction, eqn (27.2.7b), can be integrated once the concentration of B is related to the concentration of A. This depends on the stoichiometry of the reaction, and for simplicity we consider one in which $A + B \rightarrow$ products. If the initial concentrations of A and B are $[A]_0$ and $[B]_0$, then when the concentration of A drops to $[A]_0 - x$ the concentration of B drops to $[B]_0 - x$, because every molecule of A that disappears entails the disappearance of one molecule of B. It follows that

$$-d[A]/dt = k_2[A][B] = k_2([A]_0 - x)([B]_0 - x).$$

But since

$$d[A]/dt = d([A]_0 - x)/dt = -dx/dt$$

(because $[A]_0$ is a constant), the rate law can be written

$$dx/dt = k_2([A]_0 - x)([B]_0 - x),$$

and can be solved for x as a function of t.

The integration of the last equation proceeds as follows. Taking $x = 0$ at $t = 0$, gives

$$k_2 t = \int_0^{x_t} \frac{dx}{([A]_0 - x)([B]_0 - x)}$$

$$= \int_0^{x_t} \left\{ \frac{-1}{[A]_0 - [B]_0} \right\} \left\{ \frac{1}{([A]_0 - x)} - \frac{1}{([B]_0 - x)} \right\} dx$$

$$= \left\{ \frac{-1}{[A]_0 - [B]_0} \right\} \left\{ \ln\left(\frac{[A]_0}{[A]_0 - x_t} \right) - \ln\left(\frac{[B]_0}{[B]_0 - x_t} \right) \right\}.$$

This expression can be simplified by combining the two logarithms and recalling that $[A]_t = [A]_0 - x_t$ and $[B]_t = [B]_0 - x_t$, for then

(27.2.9) $$k_2 t = \left\{ \frac{1}{[A]_0 - [B]_0} \right\} \ln\left\{ \frac{[A]_t[B]_0}{[A]_0[B]_t} \right\}.$$

In order to confirm second-order behaviour throughout the course of the reaction, the right-hand side of this expression, when plotted against t, should give a straight line. If that is so, the rate coefficient can be determined from its slope. Some experimental values of k_2 are reported in Table 27.2.

Table 27.2. Kinetic data for some second-order reactions

Reaction	Phase	$t/°C$	$k_2/dm^3\,mol^{-1}\,s^{-1}$
$2NOBr \rightarrow 2NO + Br_2$	Gas	10	0.80
$2NO_2 \rightarrow 2NO + O_2$	Gas	300	0.54
$H_2 + I_2 \rightarrow 2HI$	Gas	400	2.42×10^{-2}
$D_2 + HCl \rightarrow DH + DCl$	Gas	600	1.41×10^{-1}
$I + I \rightarrow I_2$	Gas	23	7×10^9
$I + I \rightarrow I_2$	Hexane	50	18×10^9
$CH_3Cl + CH_3O^-$	Methanol	20	2.29×10^{-6}
$CH_3Br + CH_3O^-$	Methanol	20	9.23×10^{-5}
$H^+ + OH^- \rightarrow H_2O$	Water	25	1.5×10^{11}

Sources: Principally K. J. Laidler, *Chemical kinetics*, McGraw-Hill; M. J. Pilling, *Reaction kinetics*, Oxford; J. Nicholas, *Chemical kinetics*, Harper and Row.

The technique just described can be extended to other rate laws, and some results are quoted in Box 27.1. The integrated rate expressions

Box 27.1   **Integrated rate laws**

Order	Reaction	Rate law $(x = [P])$	Integrated form	$t_{1/2}$
0	$A \rightarrow P$	$dx/dt = k_0$	$k_0 t = x$ for $x \leqslant [A]_0$	$[A]_0/2k_0$
1	$A \rightarrow P$	$dx/dt = k_1[A]$	$k_1 t = \ln\left\{\dfrac{[A]_0}{[A]_0 - x}\right\}$	$(\ln 2)/k_1$
2	$A \rightarrow P$	$dx/dt = k_2[A]^2$	$k_2 t = \dfrac{x}{[A]_0([A]_0 - x)}$	$1/k_2[A]_0$
	$A + B \rightarrow P$	$dx/dt = k_2[A][B]$	$k_2 t = \left\{\dfrac{1}{[B]_0 - [A]_0}\right\}$ $\times \ln\left\{\dfrac{[A]_0([B]_0 - x)}{([A]_0 - x)[B]_0}\right\}$	
	$A + 2B \rightarrow P$	$dx/dt = k_2[A][B]$	$k_2 t = \left\{\dfrac{1}{[B]_0 - 2[A]_0}\right\}$ $\times \ln\left\{\dfrac{[A]_0([B]_0 - 2x)}{([A]_0 - x)[B]_0}\right\}$	
	$A \rightarrow P$ with autocatalysis	$dx/dt = k_2[A][P]$	$k_2 t = \left\{\dfrac{1}{[A]_0 - [P]}\right\}$ $+ \ln\left\{\dfrac{[A]_0([P]_0 + x)}{([A]_0 - x)[P]_0}\right\}$	
3	$A + 2B \rightarrow P$	$dx/dt = k_3[A][B]^2$	$k_3 t = \left\{\dfrac{1}{2[A]_0 + [B]_0}\right\}$ $\times \left\{\dfrac{2x}{[B]_0([B]_0 - 2x)}\right\}$ $+ \left\{\dfrac{1}{2[A]_0 - [B]_0}\right\}^2$ $\times \ln\left\{\dfrac{[A]_0([B]_0 - 2x)}{([A]_0 - x)[B]_0}\right\}$	
$n \geqslant 2$	$A \rightarrow P$	$dx/dt = k_n[A]^n$	$k_n t = \dfrac{1}{n-1}\left\{\left(\dfrac{1}{[A]_0 - x}\right)^{n-1}\right.$ $\left. -\left(\dfrac{1}{[A]_0}\right)^{n-1}\right\}$	$\dfrac{2^{n-1} - 1}{(n-1)k_n[A]_0^{n-1}}$

rapidly become complicated, but they can often be simplified by Ostwald's *isolation method*. This depends on the approximation that when a reactant is present in large excess its concentration is hardly changed during the course of the reaction. If the rate law is

$$-d[A]/dt = k_2[A][B],$$

and B is in large excess, the concentration [B] is virtually constant, $[B] \approx [B]_0$, and may be absorbed into the rate coefficient to give a new coefficient $k'_1 = k_2[B]_0$. Then the rate law simplifies to

$$-d[A]/dt \approx k'_1[A].$$

This is a *pseudo-first-order rate law*. Its integrated form has already been derived, and the pseudo-first-order rate constant can be found from the expressions already quoted, eqn (27.2.6). In the same way a third-order rate law of the form

$$d[P]/dt = k_3[A]^2[B]$$

can be turned into a pseudo-first-order law by having the component A in large and virtually constant excess so that the rate follows the law

$$d[P]/dt = k'_1[B],$$

or into a pseudo-second-order reaction by ensuring that the component B is in excess. This technique of isolating in turn the contributions of the various components can be a great help in unravelling a complicated reaction mechanism.

Half-lives. A simple indication of the rate of a chemical reaction is the time it takes for the concentration of a reagent to fall to half its initial value: this is called the *half-life* of the reaction, and is denoted $t_{1/2}$. The half-life depends on the initial concentration in a characteristic way for reactions of different orders, and so its measurement gives a guide to the order.

In the case of a first-order reaction the time for the concentration of A to fall from $[A]_0$ to $\frac{1}{2}[A]_0$ can be calculated from eqn (27.2.6a):

$$-k_1 t_{1/2} = \ln(\frac{1}{2}[A]_0/[A]_0) = \ln\frac{1}{2} = -\ln 2,$$

or

(27.2.10) $t_{1/2} = (1/k_1)\ln 2.$

There are two significant points about this. One is that $t_{1/2}$ may readily be determined from a plot of the time dependence of $[A]_t$, and so this is a very rapid method of measuring the first-order rate coefficient. The other is that $t_{1/2}$ of a first-order reaction is independent of the concentration. This means that if the concentration at some arbitrary stage of the reaction is $[A]'$, then the concentration will have fallen to $\frac{1}{2}[A]'$ after a further interval of $(1/k_1)\ln 2$. Some half-lives are recorded in Table 27.1.

When we turn to reactions of higher order the concentration independence of the half-life is lost. For instance, eqn (27.2.8a) can be used to predict the half-life of a second-order reaction. Substituting $t = t_{1/2}$ and $[A]_t = \frac{1}{2}[A]_0$ leads to

(27.2.11) $t_{1/2} = 1/(k_2[A]_0).$

The value of $t_{1/2}$ clearly depends on the initial concentration: the higher the initial concentration the less time it takes to drop to half its value. This conclusion points to another way of checking for second-order behaviour: if $t_{1/2}$ is determined for a series of different initial concentrations, a plot of its value against $1/[A]_0$ should be a straight line. If it is, the second-order rate coefficient can be determined from its slope.

Example (Objective 9). The results of the alkaline hydrolysis of ethyl nitrobenzoate at various times are reported below. Determine the order of the reaction by the half-life method, and find the rate coefficient.

t/s	0	100	200	300	400	500	600	700	800
$100[A]/\text{mol dm}^{-3}$	5.00	3.55	2.75	2.25	1.85	1.60	1.48	1.40	1.38

- *Method.* Plot $[A]$ against t. Choose a sequence of times to regard as 'initial' times and note the corresponding 'initial' concentrations. Then find the time for the 'initial' concentration to fall to half its value. If $t_{1/2}$ is independent of $[A]_0$ the reaction is first order; if not, try plotting $t_{1/2}$ against $1/[A]_0$, eqn (27.2.11).

- *Answer.* The data are plotted in Fig. 27.3a. Initial times are chosen at a, b,...,f. We find

	a	b	c	d	e	f
$[A]_0/\text{mol dm}^{-3}$	0.050	0.045	0.040	0.035	0.030	0.0275
$t_{1/2}/s$	240	270	300	345	400	450

Clearly, $t_{1/2}$ is not independent of $[A]_0$. So plot $t_{1/2}$ against $1/[A]_0$ as in Fig. 27.3b. This gives a straight line, and so the reaction is second order. The slope is 12.5. Therefore $k_2 = (1/12.5)\,\text{mol}^{-1}\,\text{dm}^3\,\text{s}^{-1} = 8.0 \times 10^{-2}\,\text{mol}^{-1}\,\text{dm}^3\,\text{s}^{-1}$. (The coefficient of determination of the least-squares plot is 0.9973.)

- *Comment.* If the plot had not given a straight line, $t_{1/2}$ against $1/[A]_0^2$ could have been tried.

In the case of a general nth-order reaction, the value of the half-life depends on the initial concentration as $1/[A]_0^{n-1}$. The general expression is quoted in Box 27.1.

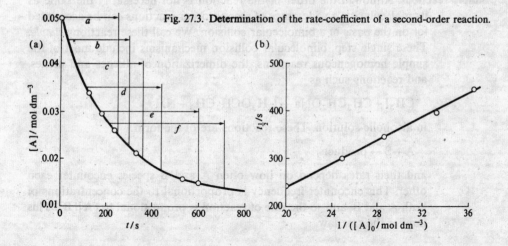

Fig. 27.3. Determination of the rate-coefficient of a second-order reaction.

A summary of the general approach. The kinetic investigation of a reaction sets out to establish the rate law and, following that, to measure the rate coefficient, often at a series of temperatures. The first step is to identify all the products of a reaction, and to investigate whether transient intermediates are formed in the course of the reaction. The isolation technique might then be used to examine the role of each component in turn, and to find the order of reaction with respect to each one. The initial slope procedure used in conjunction with the isolation technique gives an indication of the order, and a first check on this is to see if the half-life has the appropriate concentration dependence. If these two measurements agree, the concentration/time plot appropriate to the reaction order (e.g., the plots suggested by eqns (27.2.6a) or (27.2.8a)) can be drawn to check the indicated order, to confirm that it is maintained throughout the course of the reaction, and to find the value of the rate coefficient.

27.3 Accounting for the rate laws

We now move on to the second part of the analysis of kinetic data, and attempt to account for observed rate laws in terms of specific reaction mechanisms. Most reactions can be broken down into a sequence of steps that involve either a *unimolecular reaction* in which a single molecule shakes itself apart or into a new configuration, or a *bimolecular reaction* in which a pair of molecules collide and exchange energy, atoms, or groups of atoms. The *molecularity* of an individual reaction step is the number of molecules involved in it. We shall see that a reaction of some specified order can normally be accounted for in terms of a sequence of several unimolecular or bimolecular steps, and so, in general, the order of the reaction is quite distinct from the *molecularity* of the individual steps. *Order* is an empirical quantity, and obtained from the rate law; *molecularity* is characteristic of the underlying mechanism.

Simple reactions. Although the order of the reaction is not necessarily the same as its molecularity, a wide range of second-order reactions can be accounted for on the basis of a bimolecular collision. We call these reactions *simple*. These single step, bimolecular collision mechanisms include most of the simple homogeneous reactions, the dimerization of alkenes and dienes, and reactions such as

$$CH_3I + CH_3CH_2ONa \rightarrow CH_3OCH_2CH_3 + NaI$$

in alcoholic solution. These reactions are of the form

$$A + B \rightarrow products,$$

and their rates depend on how often A and B species encounter each other. This encounter frequency is proportional to the concentrations of both A and B, and so the rate of reaction is proportional to $[A][B]$. This implies the rate law

$$-d[A]/dt = k_2[A][B],$$

and therefore second-order kinetics.

The interpretation of a rate law is full of pitfalls, partly because a second-order rate expression can also result from a reaction scheme more complex than a simple bimolecular collision. We shall see examples of this in due course. For the present we emphasize that *if* the reaction is a simple bimolecular encounter process, then the kinetics will be second order, *but* if the kinetics are second order the reaction *might* be complex. The true mechanism can be determined only by detailed detective work on the system, and by investigating whether side products or intermediates appear during the course of the reaction. This was one of the ways, for example, in which the reaction

$$H_2 + I_2 \rightarrow 2HI$$

was shown to proceed by a complex mechanism after many years during which people had accepted on good, but insufficiently meticulous evidence, that it was a fine example of a simple bimolecular reaction in which atoms exchanged partners during a collision. Proof of a mechanism in chemical kinetics is less like mathematical proof and more like proof in a court of law.

The temperature dependence of the rates of simple reactions. The rates of most reactions increase as the temperature is raised. A good rule of thumb is that the rate doubles for every 10 K increase in temperature. There are exceptions, but most *simple* reactions fall somewhere in the range spanned by the hydrolysis of methyl ethanoate (methyl acetate), where the rate-constant at 35 °C is 1.82 times that at 25 °C, and the hydrolysis of sucrose, where the same increase in temperature changes the rate by a factor of 4.13. The temperature dependence of the rate coefficient has been found to fit the expression proposed by Arrhenius:

(27.3.1a) $$k_2 = A \exp(-E_a/RT).$$

The two parameters, the *pre-exponential factor* A (which is independent of temperature, or nearly so) and the *activation energy* E_a may be determined from a plot of $\ln k_2$ against $1/T$:

(27.3.1b) $$\ln k_2 = \ln A - E_a/RT,$$

the intercept is $\ln A$ and the slope is $-E_a/R$. This is called an *Arrhenius plot*, and a reaction giving a straight line is said to show *Arrhenius-type behaviour*. Once the activation energy of a reaction has been determined it is a simple matter to predict how the rate will respond to a change of temperature. Some experimental values are listed in Table 27.3.

Example (Objective 11). The rate of the second-order decomposition of acetaldehyde (ethanal) was measured over the temperature range 700–850 K, and the rate coefficients are reported below. Find the activation energy and the pre-exponential factor.

T/K	700	730	760	790	810	840	910	1000
$k_2/\text{mol}^{-1}\text{dm}^3\text{s}^{-1}$	0.011	0.035	0.105	0.343	0.789	2.17	20.0	145

- *Method.* Plot $\ln(k_2/\text{mol}^{-1}\text{dm}^3\text{s}^{-1})$ against $1/(T/\text{K})$, eqn (27.3.1), and analyse as described above.

- *Answer.* See Fig. 27.4. The intercept has the value 26.95, and so $A = 5.06 \times 10^{11}\,\text{dm}^3\text{mol}^{-1}\text{s}^{-1}$. The slope is -2.207×10^4, and so $E_a = (2.207 \times 10^4\,\text{K}) \times (8.314\,\text{J K}^{-1}\text{mol}^{-1}) = 184\,\text{kJ mol}^{-1}$. (The coefficient of determination for the least squares fit is 0.9986.)

Comment. This is a fairly typical reaction where a rise of temperature by about 10 K doubles the rate (in fact, on going from 700 K to 710 K the rate increases by a factor of 1.6). It may be more convenient to plot $\lg k_2$ in place of $\ln k_2$. The slope is then equal to $E_a/2.303R$. Note also how a $1/T$ plot leads to a cluster of points, and consequently a long, unreliable extrapolation, when only a narrow temperature range has been used in the experiment.

Table 27.3. Arrhenius parameters for rate coefficients

(1) First-order; k_1

Reaction	A/s^{-1}	$E_a/\text{kJ mol}^{-1}$
cyclopropane $\rightarrow$ propene	1.58×10^{15}	272
$CH_3NC \rightarrow CH_3CN$	3.98×10^{13}	160
cis CHD:CHD $\rightarrow$ trans CHD:CHD	3.16×10^{12}	256
cyclobutane $\rightarrow 2C_2H_4$	3.98×10^{15}	261
$C_2H_5I \rightarrow C_2H_4 + HI$	2.51×10^{13}	209
$C_2H_6 \rightarrow 2CH_3$	2.51×10^{17}	384
$N_2O_5 \rightarrow NO_2 + NO_3$	6.31×10^{14}	88
$N_2O \rightarrow N_2 + O$	7.94×10^{11}	250
$C_2H_5 \rightarrow C_2H_4 + H$	1.0×10^{13}	167

(2) Second-order; k_2

Reaction	$A/\text{dm}^3\text{mol}^{-1}\text{s}^{-1}$	$E_a/\text{kJ mol}^{-1}$
Gas phase:		
$O + N_2 \rightarrow NO + O$	1×10^{11}	315
$OH + H_2 \rightarrow H_2O + H$	8×10^{10}	42
$Cl + H_2 \rightarrow HCl + H$	8×10^{10}	23
$2CH_3 \rightarrow C_2H_6$	2×10^{10}	≈ 0
$NO + Cl_2 \rightarrow NOCl + Cl$	4×10^9	85
$SO + O_2 \rightarrow SO_2 + O$	3×10^8	27
$CH_3 + C_2H_6 \rightarrow CH_4 + C_2H_5$	2×10^8	44
$C_6H_5 + H_2 \rightarrow C_6H_6 + H$	1×10^8	≈ 25
Solution:		
$C_2H_5ONa + CH_3I$ in C_2H_5OH	2.42×10^{11}	81.6
$C_2H_5Br + OH^-$ in H_2O	4.30×10^{11}	89.5
Sucrose $+ H_2O$ in acid H_2O	1.50×10^{15}	107.9

Source: Principally J. Nicholas, *Chemical kinetics*, Harper and Row.

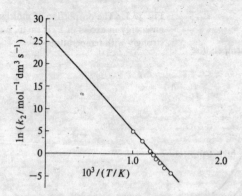

Fig. 27.4. Arrhenius plot for the decomposition of CH_3CHO.

The form of the Arrhenius expression can be obtained by a straightforward argument. If we suppose that the second-order rate coefficient is the result of a bimolecular reaction, there are two criteria to fulfil in order to have a reaction. First, the molecules must come together. In a gas we call this a *collision*, in a liquid we call it an *encounter*. The rate at which these collisions (or encounters) occur per unit volume is denoted Z. For the present we consider reaction in the gas phase, in which case Z can be identified as the *collision frequency*, p. 871. In a gas at a pressure of 1 atm the collision frequency is about $10^{28} \, s^{-1} \, cm^{-3}$ even at room temperature. If the occurrence of a collision were the only factor involved in determining whether a reaction occurred, all gas phase reactions at 1 atm pressure would be complete in about $10^{-9} \, s$, which is contrary to the facts. Furthermore, the collision frequency depends on the square root of the temperature (eqn (25.2.5), p. 872), and so as well as predicting an absurdly wrong rate, we also predict the wrong temperature dependence.

The missing factor comes from the realization that, in order to react, the molecules must collide with enough energy. A gentle collision does not lead to reaction; the collision has to be violent. If we suppose that the molecules must collide with at least an energy E_a for reaction to ensue, the collision frequency must be multiplied by the proportion of molecules colliding with at least the kinetic energy E_a along the line of approach. This proportion is given by the Boltzmann distribution (p. 11), and is $\exp(-E_a/RT)$ for a system at a temperature T. It follows that the predicted temperature dependence is

$$\text{rate} = Z \exp(-E_a/RT).$$

At normal temperatures the proportion of sufficiently energetic collisions is very small, Fig. 27.5, and so the model predicts a rate which is very much smaller than Z itself. Furthermore, the model predicts an exponential temperature dependence, because the proportion of sufficiently energetic collisions increases exponentially with temperature. The only snag, at this stage, is that the collision frequency itself is temperature dependent, whereas the experimental results seem to demand that the pre-exponential factor A is temperature independent. In fact the exponential temperature dependence is much stronger than the square-root dependence of Z, and

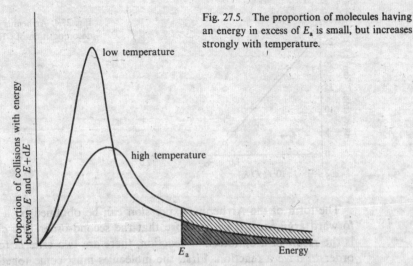

Fig. 27.5. The proportion of molecules having an energy in excess of E_a is small, but increases strongly with temperature.

it is a very difficult experimental problem to detect deviations from the exponential form. For example, for typical activation energies (of about 50–100 kJ mol^{-1}) the rate doubles for a 10 K rise in temperature but the collision frequency changes only by a factor of $(308\,K/298\,K)^{1/2} \approx 1.02$ for the same change at room temperature.

In order to be able to define an activation energy even when the rate coefficient has a non-Arrhenius form one uses the definition

(27.3.2) $E_a \overset{\text{def}}{=} RT^2 (\partial \ln k_2/\partial T)_V.$

If eqn (27.3.1) is substituted we find $(\partial \ln k_2/\partial T)_V = E_a/RT^2$, and so eqn (27.3.2) accords with the preceding discussion, but it is more general. For instance, if the rate of reaction is $Ze^{-E/RT}$, where Z is the collision frequency, then since $Z \propto T^{1/2}$ this definition leads to $E_a = \frac{1}{2}RT + E$. Experimental measurements are not normally accurate to within $\pm RT$, and so the difference between E and E_a is mainly of theoretical interest.

We shall adopt the view that collision theory reflects the principal features of the way that a bimolecular reaction takes place, and that the activation energy is a helpful parameter with some physical meaning. But clearly the theory of reaction rates requires a much stronger basis before it can be used to predict reaction rates reliably, and we return to the subject in the next chapter.

Reactions moving towards equilibrium. In this section we consider the first example of a reaction more complex than $A + B \rightarrow$ products, and see how the integrated rate laws are modified. We treat a reaction which proceeds towards some equilibrium, and in which both forward and backward reactions are first-order:

$$A \underset{k_{-1}}{\overset{k_1}{\rightleftharpoons}} B.$$

An example could be an isomerization reaction, such as cyclopropane $\rightleftharpoons$ propene.

The rate of change of the concentration of A has two contributions: A is depleted by the forward reaction at a rate $k_1[A]$, but is replenished by the backward reaction at a rate $k_{-1}[B]$. The total rate of change of the concentration of A is therefore

$$d[A]/dt = -k_1[A] + k_{-1}[B].$$

If the initial amount of A is $[A]_0$, and if initially there is no B present, then $[A] + [B] = [A]_0$ at all times. Therefore

$$d[A]/dt = -k_1[A] + k_{-1}([A]_0 - [A])$$
$$= -(k_1 + k_{-1})[A] + k_{-1}[A]_0.$$

The solution of this first-order differential equation is

(27.3.3) $[A]_t = [A]_0 \left\{ \dfrac{k_{-1} + k_1 \exp[-(k_1 + k_{-1})t]}{(k_1 + k_{-1})} \right\}.$

The properties of the solution accord with what we should expect. For example, if $k_{-1} = 0$ (no reverse reaction) the last equation reduces to the first-order solution already found, eqn (27.2.6).

What is the final state of the system? Letting t become infinite gives

$$[A]_\infty = k_{-1}[A]_0/(k_{-1} + k_1)$$
$$[B]_\infty = [A]_0 - [A]_\infty = k_1[A]_0/(k_{-1} + k_1).$$

After this very long time the system will have settled down into equilibrium, and the ratio of concentrations will be the *equilibrium constant K*. It follows that

(27.3.4) $K = ([B]/[A])_e = [B]_\infty/[A]_\infty = k_1/k_{-1}.$

In this way we arrive at a very important connection between the equilibrium constant and the rate coefficients of a simple reaction. The practical importance of the result is that if one of the rate coefficients is measured, the other can be obtained by combining it with a measurement of the equilibrium constant.

The same type of calculation may be made for other types of elementary equilibrium. For instance, in the case of the simple bimolecular, second-order reaction

$$A + B \underset{k_{-2}}{\overset{k_2}{\rightleftharpoons}} C + D,$$

(27.3.5) $K = \left\{ \dfrac{[C][D]}{[A][B]} \right\}_e = \dfrac{[C]_\infty[D]_\infty}{[A]_\infty[B]_\infty} = \dfrac{k_2}{k_{-2}}.$

Note that we emphasize *simple* reaction: the conclusion that $K = k_n/k_{-n}$ is valid for a simple, one-step reaction but is not necessarily valid for a general second-order reaction that is the consequence of several steps.

Consecutive reactions and the steady state. Some reactions proceed through the formation of an intermediate, as in the consecutive pair of first-order reactions

$$A \xrightarrow{k_1} B \xrightarrow{k_1} C.$$

An example, with half-lives of each stage, is the decay of a radioactive family, such as

$$^{239}_{92}U \xrightarrow[23.5 \text{ min}]{\beta^-} {}^{239}_{93}Np \xrightarrow[2.35 \text{ days}]{\beta^-} {}^{239}_{94}Pu.$$

The characteristics of this type of reaction can be obtained by setting up the rate equations for the formation and decay of all three species. A decays at a rate $k_1[A]$:

(27.3.6a) $d[A]/dt = -k_1[A]$.

The intermediate B increases on account of the decay of A, but it also decreases by decaying into the product C:

(27.3.6b) $d[B]/dt = k_1[A] - k'_1[B]$.

Finally, C is formed by first-order decay of B, and so

(27.3.6c) $d[C]/dt = k'_1[B]$.

We suppose that only A is present initially, and that its concentration is then $[A]_0$.

The first of the three rate equations corresponds to an exponential decay, and so the concentration of A follows

(27.3.7a) $[A]_t = [A]_0 \exp(-k_1 t)$.

If this result is inserted into the equation for B and the condition $[B]_0 = 0$ imposed, we arrive at

(27.3.7b) $[B]_t = [A]_0 \left\{ \dfrac{k_1}{k'_1 - k_1} \right\} (e^{-k_1 t} - e^{-k'_1 t})$.

The form of this function is plotted in Fig. 27.6. It shows that the intermediate's concentration rises from zero to a maximum, and then drops back to zero as A is depleted and C dominates in the mixture. The concentration of C can be obtained by inserting the solution for B into eqn (27.3.6c), and then integrating. Although this is quite straightforward there is a simpler route found by recognizing that at all times the concentrations are related by

$$[A]_t + [B]_t + [C]_t = [A]_0.$$

It follows that $[C]_t$ can be obtained directly from the solutions for $[A]$ and $[B]$:

(27.3.7c) $[C]_t = [A]_0 \left\{ 1 + \left(\dfrac{1}{k_1 - k'_1} \right) (k'_1 e^{-k_1 t} - k_1 e^{-k'_1 t}) \right\}$.

This expression is also plotted in Fig. 27.6, and we see how the concentration of the product grows until it reaches its final value $[A]_0$, when all A

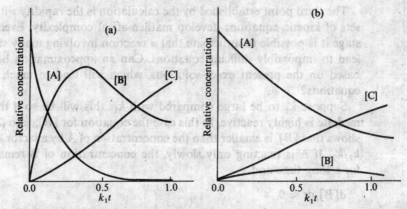

Fig. 27.6. The concentration of reactants, intermediates, and products for two consecutive first-order reactions (a) $k_1 = 10k_1'$, (b) $k_1 = 0.1k_1'$.

has changed into C.

This calculation brings into focus a number of points. The three equations (27.3.7) indicate how to analyse a reaction scheme consisting of two consecutive first-order reactions: some technique has to be found to measure the concentration of the intermediate B, as well as the concentration of the product, and the proposed mechanism has to be confirmed by checking that these equations are obeyed. If they are obeyed the values of the rate coefficients can be obtained.

The calculation also illustrates the meaning of 'rate-determining step'. Suppose the rate coefficient k_1' is very much greater than k_1, then whenever a B molecule is formed it decays quickly into C. The rate of formation of the product C is then determined almost wholly by the rate at which the B intermediates are formed. This can be confirmed by inspecting eqn (27.3.7c) for the case $k_1' \gg k_1$, for then $\exp(-k_1't)$ is much smaller than $\exp(-k_1t)$, and so may be neglected. This gives

$$(27.3.8) \qquad [C]_t \approx [A]_0 \left\{ 1 + \left(\frac{k_1'}{k_1 - k_1'} \right) e^{-k_1 t} \right\} \approx [A]_0 (1 - e^{-k_1 t}),$$

when k_1 in the denominator is neglected in comparison with k_1'. The formation of C is seen to depend only on the *smaller* rate coefficient, as anticipated. For this reason, the step with the slowest rate is called the *rate-determining step* of the reaction.

If the rate of formation of B is much faster than its decay into C, the rate determining step is the decay of B into products. The same argument as the one just described, but with $k_1' \ll k_1$ leads to

$$(27.3.9) \qquad [C]_t \approx [A]_0 (1 - e^{-k_1' t}).$$

As anticipated, the rate is governed by the coefficient for the slower, rate-determining, step of the overall reaction. This sequence, a fast step followed by a slow step, has been likened to building a six-lane highway up to a single-lane bridge: the traffic flow is governed by the rate of crossing the bridge.

The third point established by the calculation is the rapidity with which sets of kinetic equations develop mathematical complexity. Even at this stage it is possible to anticipate that a reaction involving many steps will lead to impossibly difficult equations. Can an approximation be found, based on the present exact solutions, which will lead to much simpler equations?

Suppose k_1' to be large compared with k_1: this will be so if the intermediate is highly reactive. In this case the equation for $[B]_t$, eqn (27.3.7b), shows that $[B]_t$ is smaller than the concentration of A by a factor of about k_1/k_2'. If A is reacting only slowly, the concentration of B remains low for a long time and

$$d[B]/dt \approx 0$$

for a reasonable length of time during the reaction, with the exception of the very beginning (when B's concentration has to build up from zero) and at the end (when it must drop to zero). The assumption that the major part of a reaction occurs with the reactive intermediates at virtually constant concentrations is called the *steady-state approximation*.

The steady-state approximation greatly simplifies the discussion of kinetic schemes. Since the intermediate is in a steady state, the rate equation for B, eqn (27.3.6b), reduces to

$$k_1[A] - k_1'[B] \approx 0,$$

so that

$$[B] \approx (k_1/k_1')[A].$$

Substituting this into the rate equation for the products, eqn (27.3.6c), gives

$$d[C]/dt = k_1'[B] \approx k_1[A].$$

This immediately indicates that C is formed by a first-order reaction from A, and so its solution can be written at once in the familiar exponential form.

A similar technique can be used to simplify the discussion of another type of consecutive reaction, one in which the intermediate attains an equilibrium with the reactants: this is called a *pre-equilibrium*. Consider a reaction scheme of the general form

$$A + B \underset{k_{-1}}{\overset{k_2}{\rightleftharpoons}} (AB) \xrightarrow{\ k_1\ } C,$$

where (AB) denotes some kind of intermediate. Suppose that the intermediate (AB) falls apart into C only very slowly in comparison with the rates at which it both forms from and decays back into A and B. Then as a first approximation k_1 may be neglected in the rate equation for the concentration of (AB), and so

$$d[(AB)]/dt \approx k_2[A][B] - k_{-1}[(AB)].$$

If the intermediate is in a steady state, the differential equation reduces to the algebraic equation

$$k_2[A][B] - k_{-1}[(AB)] = 0,$$

or

$$[(AB)] = (k_2/k_{-1})[A][B].$$

As we have seen already, the ratio of forward to backward rate coefficients of a simple reaction is the equilibrium constant K. Therefore

$$[(AB)] = K[A][B].$$

The rate of formation of product C is then given by

(27.3.10) $\quad d[C]/dt = k_1 K[A][B].$

Therefore the reaction has overall second-order kinetics.

Two examples of a consecutive reaction with pre-equilibrium illustrate how reaction mechanisms can be elucidated. The first reaction is the *third order* oxidation of nitrogen(II) oxide,

$$2NO + O_2 \xrightarrow{\; k_3 \;} 2NO_2.$$

One way of explaining third-order kinetics is to suppose that the reaction is termolecular. This, however, would require the simultaneous collision of three molecules, and such events are rare. Furthermore, the reaction rate *decreases* with increasing temperature. This is in contrast to the normal behaviour of reactions, and points to a reaction scheme of several steps.

In order to accommodate these remarks we investigate a mechanism consisting of the pre-equilibrium

$$NO + NO \rightleftharpoons N_2O_2, \quad \text{equilibrium constant } K,$$

together with a bimolecular reaction

$$N_2O_2 + O_2 \xrightarrow{\; k_2 \;} 2NO_2.$$

The existence of the pre-equilibrium implies that $[N_2O_2] = K[NO]^2$. The bimolecular reaction is second-order, and so the rate of formation of NO_2 is

$$d[NO_2]/dt = 2k_2[N_2O_2][O_2] = 2k_2 K[NO]^2[O_2].$$

This is a third-order rate law, as required by experiment, and the third-order rate coefficient can be identified as $k_3 = 2k_2 K$. The reason for the anomalous temperature dependence of the rate is also exposed by this analysis, for although k_2 probably behaves normally and increases with temperature, the equilibrium between NO and N_2O_2 shifts to the left (K decreases) as the temperature is raised, and the change is strong enough for $2k_2 K$ to decrease. (We know that K decreases with temperature because the dimerization reaction is exothermic; see the thermodynamic discussion on p. 266).

Another example of a pre-equilibrium reaction is the reaction mechanism proposed by Michaelis and Menten in 1913 for the mode of action of an enzyme. Let the enzyme be denoted E and the substance it acts on, the *substrate*, be denoted S. Then the overall reaction is

$$E + S \rightarrow P + E,$$

and in the conversion of S to a product P the enzyme undergoes no net change. Experiments show that the rate of formation of the product depends on the concentration of the enzyme, and so although the net reaction is simply $S \rightarrow P$, this must reflect an underlying mechanism with steps that involve the enzyme.

A simple mechanism is the following:

$$E + S \underset{k_{-1}}{\overset{k_2}{\rightleftharpoons}} \text{(ES)} \xrightarrow{k_1} P + E.$$

In this scheme (ES) denotes some bound, active combination of the enzyme and substrate, which may decay into products with the first-order rate coefficient k_1, or break up to give the original materials with the first-order rate coefficient k_{-1}. We require the rate of formation of product, which is given by

$$d[P]/dt = k_1[\text{(ES)}].$$

In order to deal with this equation it is necessary to know the concentration of bound substrate. Therefore we set up its rate equation:

$$d[\text{(ES)}]/dt = \underbrace{k_2[E][S]}_{\substack{\text{formation} \\ \text{from E,S}}} \underbrace{-k_{-1}[\text{(ES)}]}_{\substack{\text{decay to} \\ \text{E,S}}} \underbrace{-k_1[\text{(ES)}]}_{\substack{\text{decay to} \\ \text{product}}}.$$

The steady-state approximation then gives

$$k_2[E][S] - k_{-1}[\text{(ES)}] - k_1[\text{(ES)}] \approx 0$$

so that

$$[\text{(ES)}] \approx k_2[E][S]/(k_1 + k_{-1}).$$

The quantities [E] and [S] are the concentrations of the *free* enzyme and the *free* substrate. If $[E]_0$ is the total concentration of the enzyme we can write $[E] + [\text{(ES)}] = [E]_0$, a constant throughout the reaction. Since only a little enzyme was added, the total concentration of the substrate is approximately equal to the concentration of unbound substrate, $[S] + [\text{(ES)}] \approx [S]$. Therefore

$$[\text{(ES)}] \approx k_2\{[E]_0 - [\text{(ES)}]\}[S]/(k_1 + k_{-1})$$

or

$$[\text{(ES)}] \approx \frac{k_2[E]_0[S]}{k_1 + k_{-1} + k_2[S]}.$$

It follows that the rate of formation of products is

(27.3.11) $\qquad d[P]/dt \approx \dfrac{k_1 k_2 [E]_0 [S]}{k_1 + k_{-1} + k_2 [S]} = \dfrac{k_1 [E]_0 [S]}{K_M + [S]}$

where the *Michaelis constant* K_M is

$\qquad K_M = (k_1 + k_{-1})/k_2.$

According to eqn (27.3.11) the rate of enzymolysis depends linearly on the amount of enzyme added, and also on the amount of substrate present.

First-order reactions. The section summarizing the empirical observations on chemical kinetics showed that a number of gas phase reactions follow a first-order rate law. Whereas a second-order rate law can often be traced to a bimolecular collision process, how can we account for a first-order reaction in the gas phase? The difficulty is to account for first-order kinetics even though the molecule presumably acquires the energy necessary to change its form by collisions with other molecules. A reaction scheme must be found that leads to a rate of the form $k_1[A]$, but which involves bimolecular processes in order to accumulate enough energy in the molecule.

The first successful explanation of first-order reactions was provided by Lindemann in 1922. He supposed that a molecule A collided with another A molecule, and one became energetically excited at the expense of the other. Thus the initial bimolecular process in the scheme is

$\qquad A + A \xrightarrow{\ k_2\ } A^* + A.$

The excited molecule A^* does not lose its energy immediately, but might do so if it collides with another A molecule. Therefore the process

$\qquad A^* + A \xrightarrow{\ k_{-2}\ } A + A$

is also taking place. There is also the possibility that the excited molecule can shake itself apart, or into a new configuration, and release its energy by forming products according to the unimolecular decay

$\qquad A^* \xrightarrow{\ k_1\ } P.$

We shall suppose that P, which may be several species, cannot take part in the activating and deactivating collisions (e.g., it may be a solid). If it can take part, the discussion that follows has to be substantially modified. From the previous discussion we can anticipate that if the formation of A^* is fast, the rate-determining step in the overall scheme will be the unimolecular decay, and so first-order kinetics may emerge from this composite scheme.

The rate of formation of products is

$\qquad d[P]/dt = k_1 [A^*].$

It follows that an expression must be found for the concentration of the activated molecules. This is done by setting up their rate equation, and applying the steady-state approximation:

$$d[A^*]/dt = k_2[A]^2 - k_{-2}[A^*][A] - k_1[A^*] = 0.$$

This gives

(27.3.12) $$[A^*] = \frac{k_1[A]^2}{k_1 + k_{-2}[A]}.$$

The expression just obtained does not seem to give first-order kinetics. We must remember, though, that nothing has yet been said about the relative rates of the steps of the reaction. If the rate of deactivation of $A^* + A$ collisions is much greater than the rate of the unimolecular decay of A^*, we have

$$k_{-2}[A^*][A] \gg k_1[A^*] \quad \text{or} \quad k_{-2}[A] \gg k_1.$$

In this case the rate law becomes

(27.3.13) $$d[P]/dt \approx k_1 k_2[A]^2/k_{-2}[A] = (k_1 k_2/k_{-2})[A],$$

corresponding to first-order kinetics. This result conforms to the idea that led to the mechanism: if k_1 is small relative to $k_{-2}[A]$, the unimolecular decay is the rate-determining step, and we expect first-order kinetics.

The Lindemann mechanism has the virtue that it can be tested. If the pressure of A is reduced sufficiently, the condition $k_{-2}[A] \gg k_1$ fails, and is replaced by $k_{-2}[A] \ll k_1$. At these low pressures the rate law becomes

(27.3.14) $$d[P]/dt \approx k_2[A]^2,$$

and first-order kinetics are replaced by second-order kinetics. The physical reason for the change is that at low pressures the rate-determining step is the bimolecular formation of the excited molecules. This switch in the order of the reaction has been confirmed experimentally, and Fig. 27.7 shows the behaviour of the rate of isomerization of *trans*-$C_2H_2D_2$.

A more searching test of the Lindemann model is to investigate the quantitative characteristics of the change from first-order to second-order behaviour: does the order change with pressure in the way the model predicts? This can be answered as follows. First the rate law, eqn (27.3.13), is expressed as

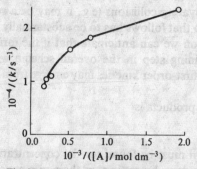

Fig. 27.7. The pressure-dependence of the first-order reaction rate for the isomerization of *trans*-$C_2H_2D_2$. (M. J. Pilling, *Reaction kinetics*, Clarendon Press, Oxford, 1975.)

(27.3.15) $d[P]/dt = k_{eff}^{\cdot}[A]$,

where the effective rate coefficient is

(27.3.16) $k_{eff} = \dfrac{k_1 k_2 [A]}{k_1 + k_{-2}[A]}$.

This can be rearranged into

(27.3.17) $1/k_{eff} = 1/k_2 [A] + k_{-2}/k_1 k_2$,

and so a plot of the observed rate coefficient $1/k_{eff}$ against $1/[A]$ should be a straight line. A typical result is shown in Fig. 27.8: it shows a pronounced curvature, corresponding to a larger value of k_{eff} at high pressures (low $1/[A]$) than would be expected by extrapolation of the low pressure results. Since the latter are dominated by the bimolecular process, this suggests that the unimolecular step has not been incorporated into the mechanism in a wholly realistic fashion.

One of the reasons for the discrepancy is that the model fails to recognize that a *specific* excitation of the molecule may be required before reaction occurs, whereas bimolecular collision gives a general, non-specific excitation. For example, in the first-order isomerization of cyclobutene

$$\square \rightarrow \wedge\!\!\wedge$$

the crucial step involves the stretching of one of the bonds, and occurs when that bond is highly excited vibrationally. In the collision leading to excitation, however, the energy of excitation is shared equally among all four bonds, and so the isomerization occurs only if the excitation energy has time to accumulate in the critical bond. This suggests that we ought to distinguish between the *energized molecule*, A*, where although there

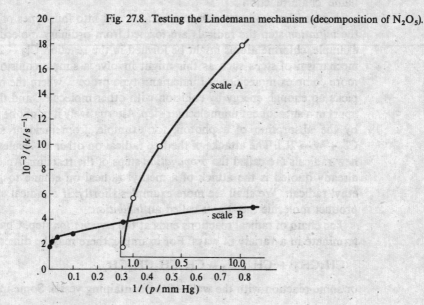

Fig. 27.8. Testing the Lindemann mechanism (decomposition of N_2O_5).

is enough energy for reaction, it is dispersed over a number of bonds, and the *activated state* of the molecule $A^{\ddagger}$, where the excitation is specific and the molecule is poised for reaction. The unimolecular part of the mechanism ought therefore to be modified to

$$A^* \xrightarrow{k_1} A^{\ddagger} \xrightarrow{k_1'} P$$

instead of the single step with a rate coefficient k_1. When this is incorporated into the mechanism, much of the discrepancy between the predicted and experimental pressure dependence of k_{eff} is eliminated.

27.4 Complex reactions

Reactions of technological and biological importance often involve a complex sequence of steps, or proceed at a useful rate only by virtue of the involvement of a catalyst. In this section we show how the ideas introduced earlier in this chapter can be developed to deal with such processes.

Chain reactions. Many reactions in the gas phase proceed through a series of steps involving *free radicals*. These molecular fragments can attack molecules and form a new radical, for example by extracting a hydrogen atom as in the reaction

$$\cdot CH_3 + CH_3CH_3 \rightarrow CH_4 + \cdot CH_2CH_3.$$

(A free radical is normally designated with a dot, indicating its unpaired electron.) The radicals so produced, in this case the ethyl radical $\cdot CH_2CH_3$, can go on to react either by producing yet another radical by attack on a molecule, or by meeting and combining with another free radical. This sequence of reactions, of radicals producing radicals, is the basis of the name 'chain reaction'.

In general a chain reaction can be analysed into four types of step. In the *initiation* step the radicals are formed from ordinary molecules. For example, chlorine atoms might be formed in the reaction $Cl_2 \rightarrow 2Cl$. The mechanism of steps such as this might involve a simple collision or, in more complex molecules, a Lindemann-type process, where the molecule picks up enough energy by collision with other molecules, and then falls apart in a subsequent unimolecular step. Alternatively it might be induced by the absorption of a photon, for example, from incident sunlight: $Cl_2 + h\nu \rightarrow 2Cl$. The attacks of the free radical on other molecules to give new radicals are called the *propagation* steps of the reaction. An example already quoted is the attack of a methyl radical on ethane to generate ethyl radicals. We shall see more examples shortly. If a radical attacks a product molecule the step is called an *inhibition*.

The chain of radical reactions ends at the *termination* step. Chains may terminate in a variety of ways. For example, there may be dimerization

$$CH_3CH_2\cdot + \cdot CH_2CH_3 \rightarrow CH_3CH_2CH_2CH_3,$$

or some reaction with the walls of the containing vessel. Some molecules react very rapidly with free radicals, and if introduced into the system are able to stop the reaction. The nitrogen(II) oxide molecule, NO, is one of these very efficient radical *scavengers*, and the quenching of a reaction when it is introduced is good evidence for a chain mechanism.

A chain reaction often, but not always, leads to a complicated rate law. As a first example we take the reaction between hydrogen and bromine. The overall reaction is

$$H_2 + Br_2 \rightarrow 2HBr$$

and the empirical rate law is

(27.4.1) $$d[HBr]/dt = \frac{k'[H_2][Br_2]^{1/2}}{1 + k''([HBr]/[Br_2])}.$$

This expression is complicated, which certainly suggests that a complex mechanism is involved. The mechanism that has been proposed is the following sequence of simple reactions:

Initiation: $Br_2 \xrightarrow{k_a} 2Br\cdot$

Propagation: $Br\cdot + H_2 \xrightarrow{k_b} HBr + H\cdot$

$H\cdot + Br_2 \xrightarrow{k_c} HBr + Br\cdot$

Inhibition: $H\cdot + HBr \xrightarrow{k_d} H_2 + Br\cdot$

Termination: $Br\cdot + Br\cdot \xrightarrow{k_e} Br_2.$

In order to obtain the rate law from this mechanism we have to find the rate of formation of HBr. Since it is generated in reactions (b) and (c), but removed in reaction (d), the net rate of formation is

$$d[HBr]/dt = k_b[Br][H_2] + k_c[H][Br_2] - k_d[H][HBr].$$

In order to develop this expression we must find the concentrations of bromine and hydrogen atoms. It is a simple matter to set up the two rate equations for the concentrations [H] and [Br], and to simplify them by imposing the steady state approximation:

$$d[H]/dt = k_b[Br][H_2] - k_c[H][Br_2] - k_d[H][HBr] \approx 0$$

$$d[Br]/dt = 2k_a[Br_2] - k_b[Br][H_2]$$

$$+ k_c[H][Br_2] + k_d[H][HBr] - 2k_e[Br]^2 \approx 0.$$

These two simultaneous equations may be solved for [H] and [Br] and the result substituted into the equation for d[HBr]/dt. This leads to

(27.4.2) $$d[HBr]/dt = \frac{2k_b(k_a/k_e)^{1/2}[H_2][Br_2]^{1/2}}{1 + (k_d[HBr]/k_c[Br_2])}.$$

This has the same form as the empirical rate law, and the two observed rate coefficients can be identified as

$$k' = 2k_b(k_a/k_e)^{1/2}, \qquad k'' = k_d/k_c.$$

Some chain reactions have simple kinetics, underlining yet again the care that has to be taken in analysing the data of chemical kinetics. For example, dehydrogenation of ethane to form ethene according to the net reaction

$$CH_3CH_3 \rightarrow CH_2{:}CH_2 + H_2$$

follows first-order kinetics

$$d[CH_2{:}CH_2]/dt = k_1[CH_3CH_3],$$

and the obvious interpretation is reaction by a Lindemann-type process. Further experiments, however, have shown that free radicals appear during the reaction, and so a more complex chain mechanism has to be found. The reaction sequence, known as the *Rice–Herzfeld mechanism*,

Initiation: $CH_3CH_3 \xrightarrow{k_a} 2CH_3{\cdot}$

Propagation: $CH_3{\cdot} + CH_3CH_3 \xrightarrow{k_b} CH_4 + CH_3CH_2{\cdot}$

$\qquad\qquad\quad CH_3CH_2{\cdot} \xrightarrow{k_c} CH_2{:}CH_2 + H{\cdot}$

$\qquad\qquad\quad H{\cdot} + CH_3CH_3 \xrightarrow{k_d} H_2 + CH_3CH_2{\cdot}$

Termination: $H{\cdot} + CH_3CH_2{\cdot} \xrightarrow{k_e} CH_3CH_3$

can be analysed in the way already outlined by setting up the appropriate rate equations and applying the steady state approximation to the concentrations of all the radical species. The conclusion is that

$$d[CH_2{:}CH_2]/dt = (k_ak_ck_d/k_e)^{1/2}[CH_3CH_3],$$

in accord with the observed first-order kinetics.

Explosions. Some reactions proceed explosively, either by design or by accident. There are two basic reasons for the occurrence of an explosion, and which one applies depends on the conditions as well as the type of reaction.

The basic reason for a *thermal explosion* is the exponential dependence of reaction rate on the temperature. If the energy of an exothermic reaction cannot escape, the temperature of the reaction system increases, and the reaction accelerates. This produces heat at an even greater rate; it cannot escape, and so the reaction goes even faster...catastrophically fast.

The other basic type of explosion depends on a chain reaction. If there are steps in the chain that *increase* the number of radicals in the system, the rate of reaction may cascade into an explosion. An example of a *branching chain reaction* is the hydrogen–oxygen reaction. Although the net reaction is very simple,

$$2H_2 + O_2 \rightarrow 2H_2O,$$

the mechanism is very complex, and has not yet been fully elucidated. It is known, however, that a free radical chain mechanism is involved, and the *chain carriers* (the radicals involved in the propagation steps) include $H\cdot$, $O\cdot$, $HO\cdot$, and $HO_2\cdot$. Some of the steps are

Initiation: $\quad H_2 + O_2 \rightarrow HO_2\cdot + H\cdot$

Propagation: $\quad H_2 + HO_2\cdot \rightarrow HO\cdot + H_2O$

$\qquad\qquad\quad H_2 + HO\cdot \rightarrow H\cdot + H_2O$

$\qquad\qquad\quad H\cdot + O_2 \rightarrow HO\cdot + O\cdot$ (branching)

$\qquad\qquad\quad O\cdot + H_2 \rightarrow HO\cdot + H\cdot$ (branching)

The last two reactions are branching reactions because they increase the number of carriers, and so the rate can increase very rapidly if they occur: this leads to the characteristic explosion when hydrogen and oxygen are sparked together.

The occurrence of an explosion depends on the temperature and pressure of the reacting system, and the regions of explosion for hydrogen and oxygen are depicted in Fig. 27.9. At low pressures the system is outside the explosion limits, and when the mixture is sparked reaction occurs without explosion. At these very low pressures the chain carriers produced in the branching reactions can reach the walls of the container, where they can combine and give up their excess energy (although the efficiency

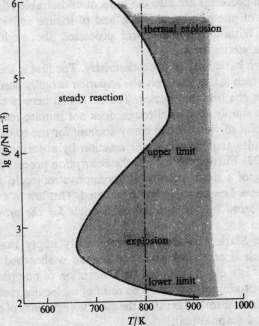

Fig. 27.9. Explosion regions for the $H_2 + O_2$ reaction.

of this depends on the composition of the walls). Raising the pressure takes the system through the *first explosion limit* (if the temperature exceeds about 400 °C). When now it is sparked, it explodes because the chain carriers react before reaching the walls and the branching reactions are explosively efficient. On increasing the pressure still further the system passes through the *second explosion limit* into a region where the reaction proceeds without explosion. The pressure is now large enough for the radicals produced in a branching reaction to combine in the gas, the molecules of the reaction mixture being able to soak up the excess energy. If the pressure is raised still further the system passes through the *third explosion limit*, and the reaction reverts to an explosion. This region appears to correspond to a thermal explosion.

Photochemical reactions. One of the ways that a chain reaction can be initiated is by the absorption of light: this was mentioned on p. 948 in connection with the reaction $Cl_2 \rightarrow 2Cl$. A wide variety of reactions, both chain and non-chain, can be initiated by the absorption of a photon, and the consequences are of inestimable importance to life on earth because photochemical processes are the primary mode of tapping the energy of the sun. Most of the currently available energy resources of the world can be ascribed to the sun's radiant energy which has been captured and accumulated by photochemical reactions. These reactions range from the heating of the atmosphere during daytime by absorption in the ultraviolet region of the spectrum through processes of the kind summarized in Fig. 27.10, to absorption in the red and blue by chlorophyll, and the subsequent deployment of the energy to bring about the synthesis of carbohydrates. Photochemical processes are also the basis of undesirable effects, such as the formation of photochemical smog, and of leisure activities, such as photography. Without photochemical processes, the world would be simply a warm, sterile, rock.

There are two basic laws of photochemistry. The first is the *Grotthuss-Draper Law*, which states that *only the radiation actually absorbed by the reacting system can initiate reaction*. The emphasis here is on the word *absorbed*: light simply passing through does not initiate reaction. From our modern point of view we can easily account for the basis of this law, because molecules acquire energy for reaction by absorbing photons of energy $h\nu$ (p. 603). Furthermore, since the absorption process corresponds to the trapping of a single photon by a receptive molecule, the basis of the *Einstein–Stark Law* can also be appreciated. This law asserts that *one quantum is absorbed by the molecule responsible for the primary photochemical process*.

The Einstein–Stark Law does not necessarily imply that only one product molecule is produced when one photon is absorbed: if the subsequent process is a chain reaction, the absorption of one photon might lead to several product molecules. The ratio of molecules reacting to the number of photons absorbed is called the *quantum efficiency* of the reaction and is denoted Φ (capital phi):

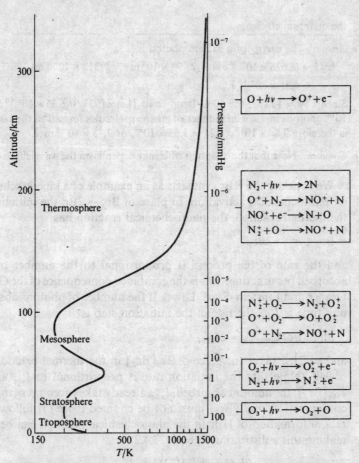

Fig. 27.10. Some of the photochemical processes in the atmosphere.

(27.4.3)
$$\Phi = \frac{\text{number of molecules that react}}{\text{number of photons absorbed}}.$$

In the photolysis of HI, for example, the processes are $HI + h\nu \to H + I$, $H + HI \to H_2 + I$, $2I \to I_2$, and as the absorption of one photon leads to the destruction of two HI molecules, the quantum efficiency is 2. In a chain reaction the quantum efficiency might be very large, and values of about 10^4 are quite common. In these cases the chain reaction acts as a chemical amplifier of the initial absorption step.

Example (Objective 22). The quantum efficiency for the formation of ethene from di-n-propyl-ketone (heptan-4-one) with 313 nm light is 0.21. How many molecules of ethene per second, and moles per second, are formed when the sample is irradiated with a 313 nm light operating at 50 W at that wavelength, and under conditions such that all the light is absorbed by the sample?

- *Method.* Calculate the number of photons emitted by the lamp in each second; all are absorbed, and so get the number of molecules decomposed by multiplying by the quantum efficiency.

● *Answer.* The energy of a 313 nm photon is

$$hc/\lambda = (6.626 \times 10^{-34}\,\text{J s}) \times (2.998 \times 10^8\,\text{m s}^{-1})/(313 \times 10^{-9}\,\text{m})$$

$$= 6.35 \times 10^{-19}\,\text{J}.$$

Since $1\,\text{W} = 1\,\text{J s}^{-1}$, the 50 W lamp emits $(1\,\text{s}) \times (50\,\text{W})/(6.35 \times 10^{-10}\,\text{J}) = 7.88 \times 10^{19}$ photons in 1 s. The number of ethene molecules formed in the same interval is therefore $7.88 \times 10^{19} \times 0.21 = 1.65 \times 10^{19}$, or $2.75 \times 10^{-5}\,\text{mol}$.

● *Comment.* Note that the quantum efficiency depends on the wavelength of the light.

We consider the HBr synthesis as an example of a kinetic scheme based on photochemical activation. In place of $Br_2 \rightarrow 2Br$, the initiation step in the thermal reaction, the photochemical reaction has

$$Br_2 + h\nu \rightarrow 2Br,$$

and the rate of the process is proportional to the number of photons absorbed per unit time (this is the combined consequence of the Grotthuss–Draper and Einstein–Stark Laws). If the number of photons absorbed per unit time is I_{abs}, the rate of the initiation step is

(27.4.4) $(d[Br]/dt)_{\text{initiation}} = 2I_{\text{abs}},$

and so $2I_{\text{abs}}$ takes the place of $2k_a[Br_2]$ in the thermal reaction scheme, p. 949. (Note that the initiation rate is proportional to I_{abs} and not to $I_{\text{abs}}[Br_2]$: the amount absorbed, I_{abs}, already takes account of the bromine concentration, and $[Br_2]$ must not be counted twice.) It follows that the rate of formation of HBr in the photochemical reaction can be found by making this substitution in eqn (27.4.2):

(27.4.5) $$d[HBr]/dt = \frac{2k_b(1/k_e)^{1/2}[H_2]I_{\text{abs}}^{1/2}}{1 + (k_d[HBr]/k_c[Br_2])},$$

and so the rate is predicted to increase as the square-root of the amount of light absorbed. This is confirmed experimentally.

Example (Objective 22). The number of photons absorbed can be determined by *chemical actinometry*. Careful experiments have shown that at 300 nm the quantum efficiency for the decomposition of uranyl oxalate is 0.570. In a particular experiment the incident light passing through an empty cell led to the decomposition of 6.201×10^{-3} mol of the oxalate in 2 hr. When the cell contained acetone and the irradiation continued for 10 hr, it was found that 1.40×10^{-3} mol of acetone were decomposed, and the light that passed through the cell and was not absorbed decomposed 2.631×10^{-2} mol of the oxalate. What is the quantum efficiency for the acetone decomposition?

● *Method.* Find the number of photons required to decompose the two amounts of oxalate: the former gives the incident rate, and so the amount absorbed can be obtained from the second.

● *Answer.* Number of photons needed to decompose 6.201×10^{-3} mol oxalate is

$(6.201 \times 10^{-3}\,\text{mol}) \times (6.022 \times 10^{23}\,\text{mol}^{-1})/(0.570) = 6.551 \times 10^{21}$. The photon flux is therefore $(6.551 \times 10^{21})/(2 \times 60 \times 60\,\text{s}) = 9.099 \times 10^{17}\,\text{s}^{-1}$. The number of photons incident in 10 hr is therefore 3.276×10^{22}. The number of photons not absorbed during the 10 hr irradiation is $(2.631 \times 10^{-2}\,\text{mol}) \times (6.022 \times 10^{23})/(0.570) = 2.78 \times 10^{22}$. Therefore the number of photons absorbed is $3.276 \times 10^{22} - 2.78 \times 10^{22} = 4.96 \times 10^{21}$. In the same interval $(1.40 \times 10^{-3}\,\text{mol}) \times (6.022 \times 10^{23}\,\text{mol}^{-1}) = 8.43 \times 10^{20}$ acetone molecules are decomposed. Therefore the quantum efficiency is $\Phi = (8.43 \times 10^{20})/(4.96 \times 10^{21}) = 0.17$.

- *Comment.* 1 mol of photons is called an *einstein*. When the quantum efficiency is 0.17, 1 einstein of 300 nm photons leads to the decomposition of 0.17 mol of acetone.

In a number of reactions the reactant molecule does not absorb because its electronic absorption spectrum lies outside the spectral range of the incident light: this may be the case, for example, when the incident light spans a very narrow range, as when it is generated in a laser or in a mercury or helium discharge lamp. Nevertheless, in such cases a photochemical reaction may still occur if a species is present that can both absorb and transfer its energy to the potentially reactive molecule. This process is called *photosensitization*.

An example of photosensitization by mercury atoms is their role in the synthesis of formaldehyde from carbon monoxide and hydrogen. When a mixture of carbon monoxide and hydrogen containing a trace of mercury vapour is irradiated with light from a mercury discharge, the mercury atoms are excited by absorption of the 254 nm light. The excited atoms collide with the other molecules present, and the transfer of energy between the mercury atom and a hydrogen molecule is efficient enough for the latter to acquire enough energy to dissociate. This initiates a radical reaction. The overall scheme is

Primary absorption: $\quad \text{Hg} + h\nu \rightarrow \text{Hg}^*$

Energy transfer: $\quad \text{Hg}^* + \text{H}_2 \rightarrow 2\text{H}\cdot + \text{Hg}$

Reaction: $\quad \text{H}\cdot + \text{CO} \rightarrow \text{HCO}\cdot$

$\quad\quad\quad\quad\quad\quad \text{HCO}\cdot + \text{H}_2 \rightarrow \text{HCHO} + \text{H}\cdot$

$\quad\quad\quad\quad\quad\quad 2\text{HCO}\cdot \rightarrow \text{HCHO} + \text{CO}.$

Some glyoxal, $\text{HCO}\cdot\text{HCO}$, is also formed by dimerization of the formyl radicals.

Photosensitization also plays an important role in solution kinetics, and molecules containing the carbonyl group, such as benzaldehyde ($\text{C}_6\text{H}_5\text{CHO}$) and benzophenone (diphenylmethanone, $\text{C}_6\text{H}_5\cdot\text{CO}\cdot\text{C}_6\text{H}_5$) are often used to trap incident light and transfer it to some potentially reactive species.

Catalysis. If the activation energy E_a is high, only a small proportion of molecular encounters are energetic enough to result in reaction, but if it is low a .

high proportion can react, and the rate coefficient is large. It follows that if the activation energy of the reaction can be lowered in some way, the reaction ought to proceed more rapidly. A *catalyst* is a substance which causes the reaction to go faster by lowering the activation energy of the rate determining steps. (The addition of a catalyst cannot change the position of equilibrium, p. 266: only the rate of attainment of equilibrium is increased.)

Some measure of the effectiveness of catalysts can be obtained by seeing what changes they bring about in the activation energies of various reactions. In the case of the decomposition of hydrogen peroxide in the absence of any catalyst the activation energy is $76 \, \text{kJ mol}^{-1}$, and decomposition is very slow at room temperature. When a little iodide is added, the same net reaction occurs, but the activation energy is only $57 \, \text{kJ mol}^{-1}$ and so at room temperature (when $RT \approx 2.5 \, \text{kJ mol}^{-1}$) the rate coefficient for the rate determining step is increased by a factor of

$$\frac{k(\text{catalyst})}{k(\text{no catalyst})} \approx \frac{\exp(-57 \, \text{kJ mol}^{-1}/2.5 \, \text{kJ mol}^{-1})}{\exp(-76 \, \text{kJ mol}^{-1}/2.5 \, \text{kJ mol}^{-1})} \approx e^{7.6} \approx 2000.$$

This accounts for the appearance of a stream of bubbles of oxygen when an iodide is added to hydrogen peroxide at room temperature. Even more dramatic is the change of activation energy when an enzyme is added to some biochemical systems. An enzyme is a biological molecule which has evolved to do a specific job with great efficiency. This is illustrated by the change in the activation energy for the hydrolysis of sucrose from $107 \, \text{kJ mol}^{-1}$ in the presence of H^+ to only $36 \, \text{kJ mol}^{-1}$ when a little of the enzyme saccharase is added: this corresponds to a change in rate of 22 orders of magnitude.

Two classes of catalyst may be distinguished. There are the *homogeneous catalysts*, which are in the same phase as the reaction mixture. This usually means that the reaction mixture is liquid, and the catalyst has been dissolved in it. In contrast there are the *heterogeneous catalysts* in which the catalyst and the reaction mixture are in different phases: this usually means that the reaction mixture is liquid or gaseous, and the catalyst solid. In this chapter we confine our attention to homogeneous catalysts, and postpone discussion of the others until we have discussed the structure and properties of surfaces (Chapter 29). The principle of their action, the lowering of the activation energy, is the same in each case, but it is brought about in different ways.

Some idea of the mode of action of homogeneous catalysts can be obtained by examining the kinetics of the bromine catalysed decomposition of hydrogen peroxide, the net reaction being

$$2H_2O_2 \rightarrow 2H_2O + O_2.$$

As usual, the statement of the net reaction gives no clue to the actual mechanism, and in this case, as in all catalysed reactions, the catalyst does not even occur in the net reaction although it must be involved in the mechanism. The reaction is believed to proceed through the following steps:

$$Br_2 + 2H_2O \rightarrow HOBr + H_3O^+ + Br^-$$

$$H_3O^+ + HOOH \rightarrow HOOH_2^+ + H_2O$$

$$HOOH_2^+ + Br^- \rightarrow HOBr + H_2O.$$

The net result of the last two reactions is

$$HOOH + H_3O^+ + Br^- \xrightarrow{k_a} HOBr + 2H_2O,$$

and this is the rate-determining step of the reaction. Once the HOBr has been formed it can attack the remaining peroxide by

$$HOOH + HOBr \rightarrow H_3O^+ + O_2 + Br^-.$$

This is a fast step, and so the rate is governed by the coefficient k_a, the rate law being

$$-d[H_2O_2]/dt = k_a[H_2O_2][H_3O^+][Br^-],$$

in agreement with the observed dependence of the rate on the bromide concentration and the hydrogen ion concentration (pH).

Important types of homogeneous catalysis are *acid catalysis* and *base catalysis*. A number of organic reactions proceed by one or other process, and sometimes both. Acid catalysis involves the transfer of a proton to the substrate:

$$BH + X \rightarrow B^- + HX^+, \quad HX^+ \text{ then reacts.}$$

This is the primary process in the solvolysis of esters, keto–enol tautomerism, and the inversion of sucrose. Base catalysis involves the transfer of a proton from substrate to catalyst:

$$XH + B \rightarrow X^- + BH^+, \quad X^- \text{ then reacts.}$$

This is the basis of the isomerization and halogenation of organic compounds, and of the Claisen and aldol reactions. The effectiveness of the catalysts depends on their ability to function as an acid or a base (in the general sense discussed in Chapter 12, p. 367). An important aspect of the study of acid and base catalysts is the relation of their activity to their strengths as measured by their pK_a or pK_b values (p. 368). This is a principal bridge between physical and organic chemistry, and is a major region of study in physical organic chemistry.

27.5 Fast reactions

All reactions are fast. At least, the individual steps of a reaction when molecules rearrange in a unimolecular step, or transfer atoms in a bimolecular encounter, occur on an atomic time scale and are complete in less than about 10^{-9} s. The slowness of the net reaction is due to the slowness with which molecules get activated, or come together; but even the net rate may become very fast when the activation energy can be provided very rapidly. This is the case, for example, when an explosion occurs: the system is then beginning to demonstrate just how fast molecular processes can be.

In recent years notable advances have been made in methods of studying fast reactions, which are reactions complete in less than about 1 s (and often much less), and the present thrust of modern chemical kinetics is to the study of processes occurring on ever shorter time scales. With special laser techniques it is now possible to observe processes occurring in a few picoseconds (1 ps $= 10^{-12}$ s).

Some of the techniques now available have been designed to monitor concentrations and to measure rate coefficients. These include flash photolysis, flow techniques, and relaxation methods. Others have been used to explore the dependence of the reaction rate on the state of vibrational excitation of molecules, or on the energy with which molecules collide, and the rate at which energy is changed from one form (e.g. vibrational) into another (e.g. rotational). The techniques then used include flash photolysis (again), shock tubes, and molecular beams. In this section we limit our attention to the first group, and see what kinetic information can be obtained; the other class of techniques is referred to in the next chapter.

Flash photolysis. The technique of flash photolysis depends on an initial photolytic step with a flash of light of very short duration. The contents of the reaction vessel are then monitored spectroscopically (e.g., by ultraviolet or visible absorption spectra, or magnetic resonance) as they evolve into stable products. The earlier techniques used photolytic flashes produced by discharging a bank of condensers through a gas. This is still done, but as the flash lasts for about 10^{-4} s, processes at very short times cannot be observed. Lasers are the basis of new flash techniques in the nanosecond $(10^{-9}$ s) region. The pulse from a laser has a duration of about 1 ns (and much less if special techniques, *mode locking*, are employed), and so the primary absorption step is complete by that time.

The remaining problem is to find a method of determining concentration and composition immediately after the flash. Spectroscopic techniques are usually employed. The normal method is to use a continuously operating white light source and to monitor the absorption electronically.

Nanosecond flash photolysis, like its slower forerunner, has been applied to numerous problems. As an example of the technique we consider its application to the photolysis of halogen molecules and their subsequent recombination. The incident flash generates $2X\cdot$ from the X_2 molecules. The only fate for the atoms is recombination, but if the atoms are to form X_2 the excess energy (the dissociation energy) must be removed, for otherwise the newly formed X_2 molecule would simply fly apart into atoms. One way of removing the excess energy is for the collision to involve three species:

$$X\cdot + X\cdot + M \rightarrow X_2 + M^*.$$

The third body, M, might be the wall of the container, or some added and possibly inert gas. One gas that is very efficient in the present case is nitrogen(II) oxide, NO. We have already discussed a third-order reaction in which NO took part in a pre-equilibrium, p. 943. The same kind of process occurs here, for NO can be expected to combine with a halogen atom (remember too that it is a good free-radical scavenger) to form NOX. Then the second halogen atom collides and displaces the NO and forms X_2. The NO acts as a sticky third body. When the spectrum of the photolysed halogen is observed immediately after the flash, and in the presence of NO, the spectrum of NOX can be detected.

Flow techniques. The basic idea behind a flow technique was described in Section 27.1. The disadvantage indicated there was that a conventional system depends on the availability of a large volume of solution. This is especially critical for fast reactions, because in order to spread the reaction over a length of tube the flow must be rapid. The modern development, the *stopped-flow technique*, avoids the disadvantage of the need for a large amount of material. The device is illustrated in Fig. 27.11. The two solutions are mixed very rapidly by injecting the material into a cavity. The cavity is fitted with a plunger that moves back as the fluids flood in. The flow ceases when the plunger reaches a stop, and the reaction continues in the thoroughly mixed solutions. Observations, usually spectroscopic, are made on the mixed sample as a function of time. The crucial aspect of the apparatus is that it brings a small amount of the reactants together in a thoroughly mixed form in a very brief time, then electronic techniques are used to resolve the time evolution of the reaction.

The ability of the method to monitor the behaviour of small samples has meant that it can be used to study biochemical processes, and a great deal of work has been done on the kinetics of enzyme action. We saw on p. 944 that the Michaelis–Menten mode of enzyme action leads to a rate law of the form

$$d[P]/dt = k_{eff}[E]_0,$$

with

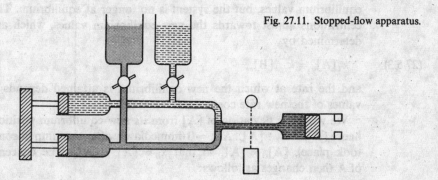

Fig. 27.11. Stopped-flow apparatus.

$$k_{\text{eff}} = k_1[S]/(K_M + [S]),$$

The expression for the effective rate coefficient can be rearranged into

(27.5.1) $$1/k_{\text{eff}} = 1/k_1 + (K_M/k_1)/[S].$$

It follows that a plot of $1/k_{\text{eff}}$ against $1/[S]$ will give the values of the rate coefficient k_1 and the Michaelis–Menten constant K_M, but not the values of the individual rate coefficients k_{-1} and k_2 that appear in K_M. The problem can be resolved with the stopped-flow technique because the rate of formation of the enzyme–substrate complex can be found by monitoring its concentration after mixing the enzyme and substrate. With the values of k_2 and k_1 established they can be combined with the known value of K_M to find $k_{-1} = k_2 K_M - k_1$.

Relaxation methods. The term 'relaxation' denotes the return of a system to equilibrium. In its application to chemical kinetics the term indicates that some externally applied influence has shifted the equilibrium position of a reaction, normally very quickly, and the reaction relaxes into the new equilibrium position.

We take as an example a simple equilibrium involving first-order reactions in both directions:

$$A \underset{k'_{-1}}{\overset{k'_1}{\rightleftharpoons}} B.$$

The primes have been added to the rate coefficients in order to denote their value under a particular set of conditions, for example, at some temperature. The rate of change of the concentration of A is

(27.5.2) $$d[A]/dt = -k'_1[A] + k'_{-1}[B].$$

When the system is at equilibrium, $d[A]/dt$ is zero. If the concentrations of A and B at equilibrium are denoted $[A]'_e$ and $[B]'_e$, we have

$$k'_1[A]'_e = k'_{-1}[B]'_e.$$

Now suppose that the condition of the system is suddenly changed (e.g. its temperature is raised) so that the rate coefficients change to k_1 and k_{-1}. The concentrations of A and B are still, momentarily, at their old equilibrium values, but the system is no longer at equilibrium. The concentrations adjust towards the new equilibrium values, which are now determined by

(27.5.3) $$k_1[A]_e = k_{-1}[B]_e,$$

and the rate at which the new equilibrium is attained depends on the values of the new rate coefficients.

We write the deviation of $[A]$ from its new equilibrium position as x, hence $[A]_t = x_t + [A]_e$. At $t = 0$ (immediately after the jump in conditions took place), $[A]_0 = [A]'_e$ so that $x_0 = [A]'_e - [A]_e$. The concentration of A then changes as follows:

$$d[A]/dt = -k_1(x_t + [A]_e) + k_{-1}(-x_t + [B]_e)$$
$$= -(k_1 + k_{-1})x_t$$

because of the cancellation of the two terms involving the equilibrium concentrations, eqn (27.5.3). Since the time derivative on the left is equal to dx/dt this is a simple first-order differential equation with the solution

(27.5.4) $x_t = x_0 \exp(-t/\tau)$,

where $1/\tau = k_1 + k_{-1}$. It follows that the concentration of A (and of B) relaxes into the new equilibrium at a rate determined by the sum of the two new rate coefficients. Since the equilibrium constant under the new conditions is $K = k_1/k_{-1}$, its measurement may be combined with a relaxation time measurement to find both k_1 and k_{-1}.

One of the most important relaxation techniques uses a *temperature jump*. The equilibrium is changed by causing a sudden change of temperature, and the concentrations are monitored as a function of time. One way of raising the temperature is to discharge an electric current through a sample which has been made conducting by the addition of ions. With a suitable choice of condensers, temperature jumps of between 5 and 10 K can be achieved in about 10^{-7} s.

Example A bank of condensers discharge 50 kV through a 10 cm³ aqueous solution in 20 μs. Calculate the temperature rise on the basis that the resistance of the solution is 40 Ω.

- *Method.* The power dissipated as heat is $\frac{1}{2}I^2R = \frac{1}{2}V^2/R$. In a time τ the energy dissipated as heat is $\frac{1}{2}V^2\tau/R$. If C is the heat capacity of the sample, the rise in temperature is $\frac{1}{2}V^2\tau/CR$. Take $C_m = 78\,\text{J K}^{-1}\,\text{mol}^{-1}$.

- *Answer.* From the expression just derived, noting that the amount of water in the sample is about (10/18) mol,

$$\Delta T = \frac{(50 \times 10^3\,\text{V})^2 \times (2 \times 10^{-5}\,\text{s})}{2 \times (40\,\Omega) \times (10/18) \times (78\,\text{J K}^{-1})} = 14\,\text{K}.$$

- *Comment.* The equipment can be designed to give the desired temperature jump by changing its size and modifying the resistance of the sample.

An important application of the temperature-jump technique has been to the determination of the rate of the reaction

$$H_3O^+ + OH^- \rightarrow 2H_2O.$$

A temperature jump changes the number of ions at equilibrium, and hence the conductivity of the solution. It was found that the relaxation time is $\tau \approx 40\,\mu\text{s}$ at room temperature, corresponding to $k_2 \approx 1.4 \times 10^{11}\,\text{dm}^3\,\text{mol}^{-1}\,\text{s}^{-1}$, making it one of the fastest solution phase reactions known (the reaction is even faster in ice, $k_2 \approx 8.6 \times 10^{12}\,\text{dm}^3\,\text{mol}^{-1}\,\text{s}^{-1}$).

Example (Objective 26). The $H_2O \rightleftharpoons H^+ + OH^-$ equilibrium relaxes in 37 μs at 25 °C. Find the rate coefficients for the forward and backward reactions.

- *Method.* Find an expression for τ in terms of k_1 (forward, first-order reaction) and k_2 (backward, second-order reaction). Do this in the same way as above, but make the approximation that the deviation from equilibrium is so small that x_r^2 can be neglected. Relate k_2 to k_1 through the dissociation constant $K_w = 1.0 \times 10^{-14}$. Solve for k_2, then for k_1.

- *Answer.* The forward rate is $k_1[H_2O]$, the backward rate is $k_2[H^+][OH^-]$. Following the same reasoning as above leads to

$$1/\tau = k_1 + k_2([H^+]_e + [OH^-]_e).$$

The equilibrium constant for the reaction is

$$K = \frac{([H^+]/\text{mol dm}^{-3})([OH^-]/\text{mol dm}^{-3})}{([H_2O]/\text{mol dm}^{-3})} = \frac{K_w}{([H_2O]/\text{mol dm}^{-3})}$$

$$= 1.0 \times 10^{-14}/55.5 = 1.8 \times 10^{-16}.$$

At equilibrium $k_1[H_2O]_e = k_2[H^+]_e[OH^-]_e$, and so

$$k_1 = k_2 K \,\text{mol dm}^{-3}.$$

Hence

$$1/\tau = k_2\{K \,\text{mol dm}^{-3} + [H^+]_e + [OH^-]_e\}$$
$$= k_2\{K + \sqrt{K_w} + \sqrt{K_w}\} \,\text{mol dm}^{-3} = 2 \times 10^{-7} k_2 \,\text{mol dm}^{-3}.$$

Therefore

$$k_2 = 1/(37 \times 10^{-6}\,\text{s}) \times (2 \times 10^{-7}\,\text{mol dm}^{-3}) = 1.4 \times 10^{11}\,\text{mol}^{-1}\,\text{dm}^3\,\text{s}^{-1}.$$

It follows that

$$k_1 = (1.4 \times 10^{11}\,\text{mol}^{-1}\,\text{dm}^3\,\text{s}^{-1}) \times (1.8 \times 10^{-16})\,\text{mol dm}^{-3} = 2.4 \times 10^{-5}\,\text{s}^{-1}.$$

- *Comment.* Notice how we have kept track of units: K and K_w are dimensionless, k_2 is expressed in $\text{mol}^{-1}\,\text{dm}^3\,\text{s}^{-1}$, and k_1 in s^{-1}.

The equilibrium position of a reaction is temperature dependent if the enthalpy of reaction differs from zero (p. 269); the temperature-jump method is therefore suitable, in principle, in all such cases. The equilibrium concentrations also depend on pressure if the volume of the system changes during the reaction. It follows that a *pressure-jump* technique ought also to be feasible. This is the case, but equilibria are much less responsive to pressure changes than to temperature changes, and so the technique is less widely used.

Relaxation techniques have been developed in a variety of ways, and one technique depends on observing the absorption of ultrasonic radiation by the system. Other relaxation techniques are those based on magnetic resonance: these were described in Chapter 19. Dielectric relaxation was described in Chapter 23.

Further reading

Chemical kinetics. K. J. Laidler; McGraw-Hill, New York, 1965.

Chemical kinetics. J. Nicholas; Harper and Row, London, 1976.

Rates and mechanisms of chemical reactions. W. C. Gardiner; Benjamin, New York, 1969.

Kinetics and mechanism. A. A. Frost and R. G. Pearson; Wiley, New York, 1961.

Homogeneous gas phase reactions. A. Maccoll; in *Techniques of chemistry* (E. S. Lewis, ed.), Vol. VIA, 47, Wiley–Interscience, New York, 1974.

Kinetics in solution. J. F. Bunnett; in *Techniques of chemistry* (E. S. Lewis, ed.), Vol. VIA, 129, Wiley–Interscience, New York, 1974.

Comprehensive chemical kinetics (Vols. 1–16). C. H. Bamford and C. F. Tipper (eds.); Elsevier, Amsterdam, 1969–76.

Photochemistry. R. P. Wayne; Butterworths, London, 1970.

Introduction to molecular photochemistry. C. H. J. Wells; Chapman and Hall, London, 1972.

Photochemistry. J. G. Calvert and J. N. Pitts; Wiley, New York, 1966.

Flash photolysis. G. Porter and M. A. West; in *Techniques of chemistry* (G. G. Hammes, ed.), Vol. VIB, 367, Wiley–Interscience, New York, 1974.

Fast reactions. J. N. Bradley; Clarendon Press, Oxford, 1974.

Rapid flow methods. B. B. Chance; in *Techniques of chemistry* (G. G. Hammes, ed.), Vol. VIB, 5, Wiley–Interscience, New York, 1974.

Temperature-jump methods. G. G. Hammes; in *Techniques of chemistry* (G. G. Hammes, ed.), Vol. VIB, 147, Wiley–Interscience, New York, 1974.

Pressure-jump methods. W. Knoche; in *Techniques of chemistry* (G. G. Hammes, ed.), Vol. VIB, 187, Wiley–Interscience, New York, 1974.

Photostationary methods. R. M. Noyes; in *Techniques of chemistry* (G. G. Hammes, ed.), Vol. VIB, 343, Wiley–Interscience, New York, 1974.

Kinetic data on gas phase unimolecular reactions. S. W. Benson and H. E. O'Neal; NSRDS-NBS-21, US Department of Commerce, Washington D.C., 1970.

Tables of bimolecular gas phase reactions. A. F. Trotman-Dickenson and G. S. Milne; NSRDS-NBS-9, US Department of Commerce, Washington D.C., 1967.

Problems

27.1. The rate constant for the first-order decomposition of N_2O_5 has the value $4.8 \times 10^{-4}\,s^{-1}$. What is the half-life of the reaction? What will be the pressure, initially 500 mmHg, after (a) 10 s, (b) 10 min after initiation of the reaction?

27.2. If the rate laws are expressed in (a) concentrations in $mol\,dm^{-3}$, (b) pressures in atmospheres, what are the units of the second-order and third-order rate co-efficients k_2 and k_3?

27.3. The half-life for radioactive decay of ^{14}C is 5730 yr (it emits β-rays with an energy of 0.16 MeV). An archaeological sample contained wood that had only 72 per cent of the ^{14}C found in living trees. What is its age?

27.4. One of the hazards of nuclear explosions is the generation of ^{90}Sr and its subsequent incorporation in place of calcium in bones. This isotope emits β-rays of energy 0.55 MeV, and has a half-life of 28.1 yr. Suppose 1 µg was absorbed by a newly born child. How much will remain after (a) 18 yr, (b) 70 yr?

27.5. The second-order rate coefficient for the reaction $AcOEt + NaOH(aq) \rightarrow AcONa + EtOH$ (where AcO is the acetate group) is $0.11\,mol^{-1}\,dm^3\,s^{-1}$. What will be the concentration after (a) 10 s, (b) 10 min when ethyl acetate is added to sodium hydroxide so that the initial concentrations are $[NaOH] = 0.05\,mol\,dm^{-3}$ and $[AcOEt] = 0.10\,mol\,dm^{-3}$?

27.6. On p. 929 an integrated expression for a second-order rate law was derived on the basis that the reaction was $A + B \rightarrow P$. Find the corresponding integrated form in the case when the stoichiometry is $2A + 3B \rightarrow P$.

27.7. Find the integrated form of a third-order rate law $-d[A]/dt = k_3[A]^2[B]$ in which the stoichiometry is $2A + B + \text{products}$ and the reactants are initially present in their stoichiometric proportions.

27.8. Repeat the last Problem, but with B present initially in twice the amount.

27.9. Find an expression for the half-life of the reaction in Problem 27.7 defined (a) as the time for A to halve its initial concentration, (b) defined as the time for B to halve its concentration, (c) defined as the time for the advancement of the reaction to reach $\xi = \frac{1}{2}$.

27.10. In an *autocatalysis reaction* the products contribute to the rate of the forward reaction, and so the rate is expected to increase, possibly very rapidly, as products are generated, and then to stop suddenly when the reagents are exhausted. This can be demonstrated by considering a simple reaction $A \rightarrow B$ having the rate law $-d[A]/dt = k_2[A][B]$. Solve this equation for the production of B taking initial concentrations of A and B as $[A]_0$ and $[B]_0$ respectively.

27.11. Show that $t_{1/2} \propto 1/[A]_0^{n-1}$ for a reaction that is nth-order in A.

27.12. Show that the ratio $t_{1/2}/t_{3/4}$, where $t_{1/2}$ is the half-life and $t_{3/4}$ is the three-quarter-life (the time for the concentration of A to drop to $\frac{3}{4}$ of its initial value, implying $t_{3/4} < t_{1/2}$) can be written as a function of n alone, and so it can be used as a rapid assessment of the order of a reaction.

27.13. The composition of a liquid phase reaction $2A \rightarrow B$ was followed as a function of time by a spectroscopic method with the following results:

t/min	0	10	20	30	40	∞
$[B]$/mol dm^{-3}	0	0.089	0.153	0.200	0.230	2.0

What is the order of the reaction? What is the value of the rate coefficient?

27.14. In an experiment to investigate the stability of substituted allyl radicals (A. B. Trenwith, *J. chem. Soc. Faraday Trans. I*, 1737 (1973)) the rate of formation of water in the reaction $CH_3CH(OH)CH:CH_2 \rightarrow H_2O + CH_2:CHCH:CH_2$ was monitored. At 810 K the results were as follows:

t/min	0.5	1.0	1.5	2.0	2.5	∞
V/cm^3	1.0	1.4	1.6	1.7	1.8	2.0

V is the volume of water produced. Find the order and rate coefficient. The C_4H_6 species did not appear to form as rapidly as the water: suggest a reason.

27.15. The experiment described in the last Problem was repeated at several temperatures and the following values of the rate coefficients were obtained:

T/K	773.5	786	797.5	810	810	824	834
k/units	1.63	2.95	4.19	8.13	8.19	14.9	22.2

What is the activation energy and pre-exponential factor A for the reaction? (Decide on the units and magnitude of k by referring to the previous Problem.)

27.16. Cyclopropane isomerizes into propene when heated to 500 °C in the gas phase. The amount converted for various initial pressures has been followed by gas chromatography by allowing the reaction to proceed for a time with various initial pressures:

p_0/mmHg	200	200	400	400	600	600
t/s	100	200	100	200	100	200
p/mmHg	186	173	373	347	559	520

where p_0 is the initial pressure and p is the final pressure of cyclopropane. What is the order and rate coefficient for the reaction under these conditions?

27.17. A second-order gas phase reaction is of the form $2A \rightarrow B$, both A and B being gases. Find an expression for the time-dependence of the total pressure of the reacting system. Let the initial pressure, when no B is present, be p_0; then plot p/p_0 as a function of $x = p_0 k_2 t$. What time is needed for the pressure to drop half way towards its final value? What is the advancement of the reaction at that time?

27.18. The composition of the gas phase reaction $2A \rightarrow B$ was monitored by measuring the total pressure as a function of time. The following are the results:

t/s	0	100	200	300	400
$p/mmHg$	400	322	288	268	256

What is the order of the reaction, and what is the value of the rate coefficient? At what time will the reaction be 99.99 per cent complete?

27.19. The radioactive decay of a family of nuclei was specified on p. 940. Calculate the abundance of the nuclides as a function of time, and plot the results as a graph.

27.20. In the isolated state of Malthusia, rigorous laws were imposed both on marriage and procreation. So harsh were the penalties that no husbands were unfaithful to their wives and every family generated one son and one daughter. These children, still under the severe regime, begat children at the same average rate. Set up and solve the rate laws for this society.

27.21. The addition of hydrogen halides to olefins has played a basic role in the investigation of organic reaction mechanisms. In one modern study (M. J. Haugh and D. R. Dalton, *J. Amer. chem. Soc.* **97**, 5674 (1975)) high pressures of hydrogen chloride (up to 25 atm) and propene (up to 5 atm) were examined over a range of temperatures and the amount of 2-chloropropane formed was determined by n.m.r. Show that if the reaction $A + B \rightarrow P$ proceeds for a short time Δt, the concentration of product follows $[P]/[A] = k_{m+n}[A]^{m-1}[B]^n \Delta t$ if the reaction is mth-order in A and nth-order in B. In a series of runs the ratio [chloropropane]/[propene] was independent of [propene] but the ratio [chloropropane]/[HCl] for constant amounts of propene depended on [HCl], and for $\Delta t \approx 100$ hr (which is short on the time scale of the reaction) the ratio rose from zero to 0.05, 0.03, 0.01 for $p(HCl) = 10$ atm, 7.5 atm, 5.0 atm. What are the orders of the reaction with respect to each reactant?

27.22. Show that the following mechanism can account for the rate law of the reaction in the last Problem:

$$2HCl \rightleftharpoons (HCl)_2 \qquad\qquad K_1$$
$$HCl + CH_3CH{:}CH_2 \rightleftharpoons complex \qquad K_2$$
$$(HCl)_2 + complex \rightarrow CH_3CHClCH_3 + 2HCl \qquad k_2, \text{ slow.}$$

What further tests could you apply to check this mechanism?

27.23. In the experiments described in the last two Problems, an inverse temperature dependence of the reaction rate was observed, the overall rate of reaction at 70 °C being roughly one-third that at 19 °C. Estimate the apparent activation energy and the activation energy of the rate determining step given that the enthalpies of the two equilibria are both of the order of -10.5 kJ mol^{-1}.

27.24. The second-order rate coefficients for the reaction of oxygen atoms with aromatic hydrocarbons have been measured (R. Atkinson and J. N. Pitts, *J. phys.*

Chem. **79**, 295 (1975)). In the reaction with benzene the rate coefficients are $1.44 \times 10^{+7}$ at 300.3 K, $3.03 \times 10^{+7}$ at 341.2 K, and $6.9 \times 10^{+7}$ at 392.2 K, all in $mol^{-1} dm^3 s^{-1}$. Find the pre-exponential factor and activation energy of the reaction.

27.25. In a study of the autoxidation of hydroxylamine (M. N. Hughes, H. G. Nicklin, and K. Shrimanker, *J. Chem. Soc. A* 3485 (1971)), the rate coefficient k_{obs} in the rate equation $-d[NH_2OH]/dt = k_{obs}[NH_2OH][O_2]$ was found to have the following temperature-dependence:

$t/°C$	0	10	15	25	34.5
$10^4 k_{obs}/mol^{-1} dm^3 s^{-1}$	0.237	0.680	1.02	2.64	5.90

What is the activation energy of the reaction? This analysis is taken further in Problem 27.28.

27.26. The equilibrium $A \rightleftharpoons B$ is first-order in both directions. Find an expression for the concentration of A as a function of time when the initial amounts of A and B are $[A]_0$ and $[B]_0$. What is the final composition of the system?

27.27. In Problem 27.16 the isomerization of cyclopropane over a limited pressure range was examined. If the Lindemann mechanism of first-order reactions is to be tested we also need data at low pressures. These have been obtained (H. O. Pritchard, R. G. Sowden, and A. F. Trotman-Dickenson, *Proc. Roy. Soc. A* **217**, 563 (1953)):

$p/mmHg$	84.1	11.0	2.89	0.569	0.120	0.067
$10^4 k_{1\,eff}/s^{-1}$	2.98	2.23	1.54	0.857	0.392	0.303

Test the Lindemann theory with these data (see eqn (27.3.17)).

27.28. The kinetics of autoxidation of hydroxylamine in the presence of EDTA (ethylenediamine tetra-acetic acid) have been studied in the hydroxide ion concentration range $0.5-3.2 \, mol \, dm^{-3}$ (reference in Problem 27.25). The reaction proceeds by the mechanism

$$NH_2OH + OH^- \rightarrow NH_2O^- + H_2O \quad \text{(fast)}$$

$$NH_2O^- + O_2 \rightarrow \text{products} \quad \text{(slow)},$$

with the rate laws

$$-d[(NH_2OH)]/dt = k_{obs}[(NH_2OH)][O_2]$$

$$-d[(NH_2OH)]/dt = k_2[NH_2O^-][O_2],$$

where $[(NH_2OH)]$ denotes the total hydroxylamine, $[NH_2OH] + [NH_2O^-]$, and $k_2 = k_{obs}/f$, where f is the fraction of hydroxylamine present as NH_2O^-. Show that a plot of $1/k_{obs}$ against $[H^+]$ should give a straight line, and that the acid dissociation constant of NH_2OH can be determined from the slope.

27.29. The data below relate to the reaction system described in the last Problem. Find pK_a for the hydroxylamine, and the amount of NH_2O^- present at each OH^- concentration.

$[OH^-]/mol\,dm^{-3}$	0.50	1.00	1.6	2.4
$10^4 k_{obs}/s^{-1}$	2.15	2.83	3.32	3.54

27.30. The *Rice-Herzfeld mechanism* for the dehydrogenation of ethane was set out on p. 950, and it was noted there that it led to first-order kinetics. Confirm this remark, and find the approximations that lead to the rate law quoted there. How may the conditions be changed so that the reaction shows different orders?

27.31. The following mechanism has been proposed for the thermal decomposition of acetaldehyde (ethanal):

$$CH_3CHO \rightarrow \cdot CH_3 + \cdot CHO, \qquad k_a$$

$$\cdot CH_3 + CH_3CHO \rightarrow CH_4 + \cdot CH_2CHO, \qquad k_b$$

$$\cdot CH_2CHO \rightarrow CO + \cdot CH_3, \qquad k_c$$

$$\cdot CH_3 + \cdot CH_3 \rightarrow CH_3CH_3, \qquad k_d.$$

Find an expression for the rate of formation of methane and the rate of disappearance of acetaldehyde.

27.32. The number of photons falling on a sample can be determined by a variety of methods, of which the classical one is *chemical actinometry*. The decomposition of oxalic acid, $(COOH)_2$, in the presence of uranyl sulphate, $(UO_2)SO_4$, proceeds according to the sequence

$$UO_2^{2+} + h\nu \rightarrow (UO_2^{2+})^*, \qquad (UO_2^{2+})^* + (COOH)_2 \rightarrow UO_2^{2+} + H_2O + CO_2 + CO$$

with a quantum efficiency of 0.53 at the wavelength used. The amount of oxalic acid remaining after exposure can be determined by titration (with $KMnO_4$) and the extent of decomposition used to find the number of incident photons. In a particular experiment, the actinometry solution consisted of 5.232 g oxalic acid in 25 cm^3 water (together with the uranyl salt). After exposure for 5 min the remaining solution was titrated with 0.212 mol dm^{-3} $KMnO_4$, and 17.0 cm^3 were required for complete oxidation of the remaining oxalic acid. What is the rate of incidence of photons at the wavelength of the experiment? Express the answer in photons/second and einstein/second, an *einstein* being 1 mol of photons.

27.33. The intensity of fluorescence or phosphorescence from an excited molecule M will depend on the efficiency of any competitive chemical quenching process. Consider the gas phase fluorescence from some excited molecule M* generated by $M + h\nu_i \rightarrow M^*$, the light intensity absorbed being I_a. This absorption is followed by $M^* + Q \rightarrow M + Q$, the second-order quenching reaction with rate coefficient k_q, in competition with the fluorescence intensity I_f arising from $M^* \rightarrow M + h\nu_f$, with rate coefficient k_f. Show that this scheme leads to the *Stern–Volmer relation*

$$1/I_f = (1/I_a)\{1 + (k_q/k_f)[Q]\}$$

between the intensity and the quencher concentration. Thus the ratio k_q/k_f can be determined from a plot of I_f against quencher concentration. Show, furthermore, that if the rate of disappearance of fluorescence is measured, k_q itself may be found.

27.34. When benzophenone is illuminated with ultraviolet light it is excited into a singlet state. This singlet changes rapidly into a triplet, which phosphoresces. Triethylamine, NEt_3, acts as a quencher for the triplet. In an experiment in methanol as solvent, the phosphorescence intensity varied with amine concentration as shown below. A flash photolysis experiment had also shown that the half-life of the fluorescence in the absence of quencher is 2.9×10^{-7} s. What is the value of k_q?

$[Q]/$mol dm^{-3}	0.001	0.005	0.010
$I_f/$arbitrary units	0.41	0.25	0.16

27.35. Find an expression for the rate of disappearance of a species A in a photochemical reaction where the mechanism is (a) initiation with light of intensity I, $A \rightarrow 2R\cdot$, (b) propagation, $A + R\cdot \rightarrow R\cdot + B$, rate coefficient k_p, (c) termination, $R\cdot + R\cdot \rightarrow R_2$, rate coefficient k_t. Hence show that rate measurements will give only a combination of k_p and k_t if a steady state is reached, but that both may be obtained if a steady state is not reached. The latter is the basis of the *rotating sector method* of studying photochemical reactions (see Further Reading).

27.36. The photochemical chlorination of chloroform in the gas has been found to follow the rate law $d[CCl_4]/dt = k_{1/2}[Cl_2]^{1/2}I_a^{1/2}$. Devise a mechanism that leads to this rate law when the chlorine pressure is quite high.

27.37. Conventional equilibrium considerations do not apply when a reaction is being driven by light absorption. Thus the steady-state concentration of products and reactions might differ significantly from equilibrium values: this is the *photostationary state*. For instance, suppose the reaction $A \rightarrow B$ is driven by light absorption, and that its rate is I_a, but that the reverse reaction $B \rightarrow A$ is bimolecular and second-order with a rate $k_2[B]^2$. What is the photostationary state concentration of B? Why does the photostationary state differ from the equilibrium state?

27.38. When anthracene is dissolved in benzene and exposed to ultraviolet light it dimerizes. In concentrated solutions the dimerization occurs with high quantum efficiency (and little fluorescence), but in dilute solutions fluorescence deactivates the excited molecules and the dimerization proceeds with low quantum efficiency. Calculate the quantum efficiency as a function of concentration, basing your derivation on the sequence $A + h\nu \rightarrow A^*$, $A^* + A \rightarrow A_2$, $A^* \rightarrow A + h\nu_f$, and account for the observations.

27.39. Studies of combustion reactions depend on knowing the concentrations of hydrogen atoms and hydroxyl radicals. Measurements on a flow system using e.s.r. for the detection of radicals gave information on the reactions

$$H + NO_2 \rightarrow OH + NO \quad k_2 = 2.9 \times 10^{10}\ dm^3\ mol^{-1}\ s^{-1}$$
$$OH + OH \rightarrow H_2O + O \quad k_2' = 1.55 \times 10^9\ dm^2\ mol^{-1}\ s^{-1}$$
$$O + OH \rightarrow O_2 + H \quad k_2'' = 1.1 \times 10^{10}\ dm^3\ mol^{-1}\ s^{-1}$$

(J. N. Bradley, W. Hack, K. Hoyermann, and H. G. Wagner, *J. chem. Soc. Faraday Trans. I*, 1889 (1973)). On the basis of initial hydrogen atom and NO_2 concentrations of $4.5 \times 10^{-10}\ mol\ cm^{-3}$ and $5.6 \times 10^{-10}\ mol\ cm^{-3}$ respectively, compute and plot curves showing the O, O_2, and OH concentrations as a function of time in the range 0–10 ns.

27.40. In a flow study of the reaction between oxygen atoms and chlorine (J. N. Bradley, D. A. Whytock, and T. A. Zaleski, *J. chem. Soc. Faraday Trans. I*, 1251 (1973)) at high chlorine pressures, plots of $\ln[O]_0/[O]$ against distances along the flow tube, where $[O]_0$ is the oxygen concentration at zero chlorine pressure, gave straight lines. Given the flow velocity as $6.66\ m\ s^{-1}$ and the data below, find the rate coefficient for the reaction $O + Cl_2 \rightarrow ClO + Cl$.

distance/cm	0	2	4	6	8	10	12	14	16	18
$\ln[O]_0/[O]$	0.27	0.31	0.34	0.38	0.45	0.46	0.50	0.55	0.56	0.60

$[O]_0 = 3.3 \times 10^{-8}\ mol\ dm^{-3}$, $[Cl_2] = 2.54 \times 10^{-7}\ mol\ dm^{-3}$, $p = 1.70\ mmHg$.

27.41. On p. 960 we examined the relaxation of a first-order reaction towards equilibrium. Show that the reaction $A \rightleftharpoons B + C$, first-order forwards, second-order backwards, also relaxes exponentially for small displacements from equilibrium. Find an expression for the relaxation time in terms of k_1 and k_2.

27.42. The initial rate of oxygen production by the action of an enzyme on a substrate was measured for a range of substrate concentrations; the data are below. What is the value of the Michaelis constant for the reaction?

$[S]/mol\ dm^{-3}$	0.050	0.017	0.010	0.005	0.002
rate/mm^3 min^{-1}	16.6	12.4	10.1	6.6	3.3

28   Molecular reaction dynamics

Learning objectives

After careful study of this chapter you should be able to

(1)   Specify the *collision theory* of bimolecular gas-phase reactions (p. 971).

(2)   Calculate the *second-order rate coefficient* from collision theory (eqn (28.1.1)).

(3)   Define, explain, and measure the *P-factor* and the *reactive cross-section* (p. 973).

(4)   Distinguish between *diffusion-controlled* and *activation-controlled* reactions in solution (p. 974).

(5)   Relate the second-order rate coefficient to the *diffusion coefficient* (eqn (28.1.5)) and the *viscosity* (eqn (28.1.6)).

(6)   Explain the terms *activated complex, reaction coordinate,* and *transition state* (p. 979).

(7)   Describe the *activated complex theory* of reaction rates (p. 980).

(8)   Derive the *Eyring equation* for reaction rates (eqn (28.2.7)).

(9)   Use the Eyring equation to calculate a rate coefficient (eqn (28.2.8)), and the magnitude of the *kinetic isotope effect* (eqn (28.2.9)).

(10)   Define and interrelate the *Gibbs function of activation, entropy of activation, enthalpy of activation,* and *internal energy of activation* (p. 986).

(11)   Explain the basis and derive the magnitude of the *kinetic salt effect* (eqn (28.2.16)).

(12)   Indicate how *molecular beams* are used to study reactive collisions (p. 990).

(13)   Sketch the *potential-energy surface* of a simple reaction (p. 992).

(14)   Describe the information available from measurements of the *angular distribution* of products (p. 994).

(15)   Distinguish between *attractive* and *repulsive surfaces* and explain how they determine the energy requirements of a reaction (p. 995).

Introduction

Now we are at the very heart of chemistry. Here we examine the detailed behaviour of molecules during the most crucial moments of reactions. Massive changes of form are occurring, energies of the size of dissociation energies are being redistributed among bonds, old bonds are being ripped apart and new bonds formed. The rate at which the molecules exchange atoms or groups of atoms, or a single molecule is switched into a different isomeric form, depends on the forces that operate at the climax of the reaction, and these in turn depend on the detailed disposition of all the charged particles of all the molecules involved in the step.

As may be imagined, the quantitative prediction of reaction rates, and the calculation of the bimolecular, second-order rate coefficient k_2 from first principles, is an extremely difficult undertaking. Nevertheless, like so many intricate problems, the broad features can be established very simply, and only when we enquire more deeply about the details do the complications emerge. In this chapter we look at three levels of approach to the interpretation and calculation of k_2. The simplest approach is in terms of *collision theory*, where attention is mainly confined to finding the rate of sufficiently energetic collisions. We shall look at some of the deficiencies of collision theory, and see that an improvement can be made by examining more closely the nature of the species formed in the collision: this is *activated complex theory*. Even that theory has serious imperfections, and one way of improving the description is to examine experimentally the individual molecular collisions involved in actual reactions. Modern technological progress has made this possible, and the use of molecular beams lets us follow in detail the course of reactive collisions, and to set up models of *molecular reaction dynamics*.

An understanding of the factors that govern the magnitude of k_2 is useful technologically as well as satisfying intellectually. One important technical application is in the development of chemical lasers: we saw on p. 668 that if molecules could be levered up into excited states, and then induced to emit their energy coherently, laser activity is obtained. Molecular beam studies have shown that the products of some reactions are produced in definite excited states: this means that the energy of these thermal reactions can be turned directly into coherent electromagnetic radiation. Setting up the appropriate conditions for the action of a chemical laser depends on a detailed knowledge of all the rate processes involved, and this can be obtained from the kind of observations and theory we shall describe. Another application concerns the possibility of enhancing the rates of chemical reactions by preparing the reactants in selected energy states. For example, some reactions go faster if the energy of the molecules is concentrated in their vibrational modes rather than in translation. If the translational kinetic energy can be switched into vibration, these reactions can be made to go faster. Knowledge of this kind depends on understanding how energy is released in a reaction, and how it is involved in reorganizing the bonds in the colliding molecules.

28.1 Molecular encounters

Chemical reaction depends on the encounter of two species. In the gas phase we imagine A banging into B at a rate equal to the collision frequency, and whether or not they react depends, among other things, on how much energy is involved in the collision. In solution we cannot so easily speak of 'collisions' because the relative migration of the species is diffusional and the solvent hinders their free flight, but the rate at which the potentially reactive species encounter each other in solution can be calculated and related to the diffusion constants of the dissolved species. In this section we concentrate on the frequencies of these two types of encounter, and see how they affect the rates of reactions.

Collision theory. The ideas underlying the collision theory of bimolecular reactions in gases were introduced in the last chapter, p. 937. The rate of a gas phase reaction in which A is destroyed is $-d[A]/dt$. Since $[A] = n_A/V = N_A/LV$, this rate can be written

$$-d[A]/dt = -(1/LV)dN_A/dt.$$

The collision rate of A with B per unit volume is Z_{AB}. Therefore the rate of change of the number of A molecules per unit volume, $(dN_A/dt)/V$, is Z_{AB} multiplied by the proportion of collisions that occur with a kinetic energy along the line of approach in excess of some threshold value E_a, the *activation energy* for the reaction:

$$(1/V)(dN_A/dt) = -Z_{AB}\exp(-E_a/RT).$$

Consequently,

$$-d[A]/dt = -(dN_A/dt)/LV = (Z_{AB}/L)\exp(-E_a/RT).$$

The collision frequency can be calculated on the basis of the kinetic theory of gases, Chapter 25:

$$Z_{AB} = \sigma(8kT/\pi\mu)^{\frac{1}{2}}(N_A/V)(N_B/V) = \sigma(8kT/\pi\mu)^{\frac{1}{2}}L^2[A][B],$$

where σ is the collision cross-section which, for hard sphere species of radii R_A and R_B, is $\pi(R_A+R_B)^2$, p. 870; μ is the reduced mass of the colliding species. It follows that

$$-d[A]/dt = \sigma L(8kT/\pi\mu)^{\frac{1}{2}}[A][B]\exp(-E_a/RT).$$

Since the second order rate coefficient is defined by the rate expression

$$-d[A]/dt = k_2[A][B],$$

the collision-theory expression for k_2 is

(28.1.1)
$$k_2 = \sigma L(8kT/\pi\mu)^{\frac{1}{2}}\exp(-E_a/RT).$$

The expression for the rate coefficient conforms qualitatively to the observed Arrhenius-type temperature dependence

(28.1.2) $k_2 = A \exp(-E_a/RT)$.

This certainly suggests that the collision theory takes into account the principal features of the reaction mechanism, that molecules have to collide (the pre-exponential term) but react only if the collisions are sufficiently energetic (the exponential term).

Qualitative agreement, though, is not enough, and we have to see whether the theory is quantitatively correct. The activation energy is a complicated property of the molecules, and so we might not expect to get good agreement between experimental values and calculations from first principles. The pre-exponential term, on the other hand, is well-defined, and apart from some doubt about what is meant by the collision cross-section, depends only on the masses of the species involved in the collision, and so it can easily be calculated.

The simplest procedure is to identify the cross-section with the cross-section for simple, non-reactive collisions. This can be estimated from tables of molecular radii, or determined experimentally from measurements of gas imperfections (Chapter 23). Table 28.1 compares some calculated values of A with experimental values obtained from Arrhenius plots of $\ln k_2$ against $1/T$. One of the pre-exponential factors shows good agreement between theory and experiment, but there are major discrepancies. Some pre-exponential factors are orders of magnitude smaller than the calculated values: this suggests that collision with sufficient energy is not the only criterion for reaction, and that some other factor, such as the relative orientation of the colliding species, has to be taken into account. The second deficiency is that one of the reactions in the Table has a pre-exponential factor greater than collision theory predicts. This seems to indicate that the reaction occurs more quickly than the molecules collide! In order to resolve this apparent paradox we can recall that the calculated value of A depends on the value chosen for the cross-section, and the values in the table are based on taking the value for a non-reactive collision cross-section. It may be the case that the distance of approach required for a reactive collision is quite different from the distance that leads simply to a deflection of the direction of motion.

Both effects, the orientational requirement and the distance requirement,

Table 28.1. Activation energies, pre-exponential factors for gas-phase reactions

Reaction	$A/\mathrm{dm^3\,mol^{-1}\,s^{-1}}$		$E_a/\mathrm{kJ\,mol^{-1}}$	$P = \sigma^*/\sigma$
	Experiment	Theory		
$2NOCl \rightarrow 2NO + Cl_2$	9.4×10^9	5.9×10^{10}	102.0	0.16
$2NO_2 \rightarrow 2NO + O_2$	2.0×10^9	4.0×10^{10}	111.0	5×10^{-2}
$2ClO \rightarrow Cl_2 + O_2$	6.3×10^7	2.5×10^{10}	0.0	2.5×10^{-3}
$K + Br_2 \rightarrow KBr + Br$	1.0×10^{12}	2.1×10^{11}	0.0	4.8
$H_2 + C_2H_4 \rightarrow C_2H_6$	1.24×10^6	7.3×10^{11}	180	1.7×10^{-6}

Source: Principally M. J. Pilling, *Reaction kinetics*, Oxford.

can be taken into account by replacing the collision cross-section σ in eqn (28.1.1) by the *reactive cross-section*, σ^*. A more conventional procedure takes the view that the cross-section for reaction can be expressed in terms of the collisional cross-section by introducing a factor P, such that $\sigma^* = P\sigma$; then the expression for the rate coefficient becomes

(28.1.3) $k_2 = P\{\sigma L(8kT/\pi\mu)^{\frac{1}{2}}\}\{\exp(-E_a/RT)\}.$

This displays very clearly the three contributions to k_2. The second factor is the *transport property* governing how rapidly the species move into each other's vicinity. The third factor, the exponential, is the *energy criterion*, governing the proportion of collisions that have enough energy to proceed to reaction. The first factor, P, incorporates the *local properties* of the reaction, the orientations required of the species, and the details of how close they have to come in order to react.

Example (Objective 3). Estimate the P-factor for the reaction $H_2 + C_2H_4 \rightarrow C_2H_6$ at $355\,^\circ C$ given that the experimental pre-exponential factor has the value $1.24 \times 10^6 \text{ dm}^3 \text{ mol}^{-1} \text{ s}^{-1}$.

- *Method.* Use eqn (28.1.3), identifying the theoretical pre-exponential factor as $\sigma LP(8kT/\pi\mu)^{\frac{1}{2}}$. Use $M_r(H_2) = 2.016$, $M_r(C_2H_4) = 28.05$, and estimate σ by taking the mean of the values given in Table 25.3, $\sigma(H_2) = 0.27 \text{ nm}^2$, $\sigma(C_2H_4) = 0.64 \text{ nm}^2$.

- *Answer.*

$$\mu = \left\{\frac{2.016 \times 28.05}{2.016 + 28.05}\right\} \times (1.6605 \times 10^{-27} \text{ kg}) = 3.12 \times 10^{-27} \text{ kg}$$

$$(8kT/\pi\mu)^{\frac{1}{2}} = \{8 \times (1.381 \times 10^{-23} \text{ J K}^{-1}) \times (628 \text{ K})/\pi \times (3.12 \times 10^{-27} \text{ kg})\}^{\frac{1}{2}}$$

$$= 2.661 \times 10^3 \text{ m s}^{-1}.$$

$$\sigma = \tfrac{1}{2}(0.27 \text{ nm}^2 + 0.64 \text{ nm}^2) = 0.46 \times 10^{-18} \text{ m}^2$$

$$A = \sigma PL(8kT/\pi\mu)^{\frac{1}{2}}$$

$$= (0.46 \times 10^{-18} \text{ m}^2) \times P \times (6.0221 \times 10^{23} \text{ mol}^{-1}) \times (2.661 \times 10^3 \text{ m s}^{-1})$$

$$= (7.37 \times 10^8 \text{ m}^3 \text{ mol}^{-1} \text{ s}^{-1})P = (7.37 \times 10^{11} \text{ dm}^3 \text{ mol}^{-1} \text{ s}^{-1})P.$$

Since $A_{\text{expt}} = 1.24 \times 10^6 \text{ dm}^3 \text{ mol}^{-1} \text{ s}^{-1}$, it follows that $P \approx 1.7 \times 10^{-6}$.

- *Comment.* This value accords with the one quoted in Table 28.1. As a rough guide, the more complex the molecules that are colliding, the smaller the value of P.

The theory would be all very well if we could calculate the steric factor P. In some cases we can. Take for example, the $K + Br_2$ reaction, for which $P = 4.8$, indicating an 'anomalously' large reaction cross-section, $\sigma^* \approx 4.8\sigma$. It has been proposed that the reaction proceeds by a *harpoon mechanism*. This brilliant name is based on a model of the reaction which pictures the K atom approaching the Br_2 molecule, and when the two are close enough, an electron (the harpoon) flips across to the Br_2. In place of two neutral species there are now two ions, and so there is a Coulombic attraction between them: this is the line on the harpoon. Under its influence the ions move together (the line is wound in), the reaction takes place, and $KBr + Br$ emerge. The harpoon extends the cross-section for the reactive encounter, and we greatly underestimate the reaction rate by taking for σ^* the mean of the values for simple $K + K$ and $Br_2 + Br_2$ collisions.

This qualitative argument accounts for $P > 1$, but we can go on to estimate its actual value. The calculation depends on estimating the distance between K atom and Br_2 molecule at which it is energetically favourable for the electron to leap from one to the other. The energy involved has three contributions. One is an investment of ionization energy in order to remove the electron from the K atom; this is $I(K)$. Energy is released as the electron enters the Br_2 to form Br_2^-, and the extent of lowering is the molecule's electron affinity $E_A(Br_2)$. The third contribution is due to the attractive Coulombic interaction between the two ions. When their separation is R the electrostatic energy is $-e^2/4\pi\varepsilon_0 R$. The total change of energy if an electron moves across when the two neutral species are separated by a distance R is therefore approximately

$$\Delta E = I(K) - E_A(Br_2) - e^2/4\pi\varepsilon_0 R.$$

$I(K)$ is bigger than $E_A(Br_2)$, and ΔE becomes negative only when R has decreased to less than some critical value R^* given by

$$e^2/4\pi\varepsilon_0 R^* = I(K) - E_A(Br_2).$$

When the species are at this distance the harpoon shoots across, and so the reactive cross-section can be identified as $\sigma^* = \pi R^{*2}$. This, in turn, means that the P-factor is

$$P = \sigma^*/\sigma = R^{*2}/d^2 \approx \{e^2/4\pi\varepsilon_0 d[I(K) - E_A(Br_2)]\}^2$$

where $d = R(K) + R(Br_2)$. Since $I(K) = 420$ kJ mol^{-1}, $E_A(Br_2) < 300$ kJ mol^{-1}, and $d \approx 310$ pm we arrive at $P < 12$, which is consistent with the experimental value of 4.8.

This example shows two things. First, that the P-factor is not a wholly useless quantity to introduce into the formalism, because in some cases it can be estimated. The other point is the converse: many reactions are much more complicated than $K + Br_2$, and we cannot hope to proceed in such a simple way to an estimation of P. What we want is a more powerful theory of P which lets us avoid simply having to guess its value.

Reactions in solution: diffusion control. The calculation of rates of reactions in solution can also be broken up into the problems of encounter rates and energy availability. Now the molecular motion is diffusional, in place of free flight, but the concepts of activation energy and steric requirements survive. Some measured activation energies are listed in Table 28.2. In this section we continue to concentrate on the encounter frequency.

Encounters in solution have a nature quite different from their counterparts in gases. Molecules have to jostle their way through the solvent, and so the encounter frequency is drastically less than in a gas. There is, however, another important factor. Since a molecule migrates only slowly into the region of a possible reaction partner, it also migrates only slowly away from it. In other words, the members of the encounter pair linger in each other's vicinity for much longer than in a gas, and so their chance of undergoing reaction is greatly enhanced. Furthermore, the activation energy of reaction is a much more complicated quantity in solution than in the gas, for the encounter pair is surrounded by solvent, and its energy is determined by all the interactions.

Table 28.2. Activation energies and pre-exponential factors for reactions in solution

Reaction	Solvent	$A/dm^3 mol^{-1} s^{-1}$	$E_a/kJ mol^{-1}$
$C_6H_5NH_2 + C_6H_5COCH_2Br$	benzene	9.1×10	34
tert-BuCl solvolysis	chloroform	1.4×10^4	45
	water	7.1×10^{16}	100
	methanol	3.2×10^{13}	107
	ethanol	3.0×10^{13}	112
	acetic acid	4.3×10^{13}	111
$CH_3I + C_2H_5O^-$	ethanol	2.42×10^{11}	81.6
$C_2H_5I + C_2H_5O^-$	ethanol	1.49×10^{11}	86.6
$CO_2 + OH^-$	water	0.15×10^{11}	38
$C_2H_5Br + OH^-$	ethanol	4.30×10^{11}	89.5
$CH_3I + S_2O_3^{2-}$	water	2.19×10^{12}	78.7

Source: A. A. Frost and R. G. Pearson, *Kinetics and mechanism*, Wiley, using
$$A = (kT/h)(RT/p^{\ominus}) \exp(\Delta S_m^{\ddagger}/R)$$
$$= (1.5291 \times 10^{14} dm^3 mol^{-1} s^{-1}) \times \exp(\Delta S_m^{\ddagger}/R) \text{ at } 25°C.$$

This complicated overall picture can be divided into simpler parts. First we set up a simple kinetic scheme for the overall process, taking into account the two steps of migration together followed by reaction. We suppose that the rate of forming the encounter pair is proportional to the concentrations of the components A and B:

$$A + B \xrightarrow{k_d} (AB), \qquad d[(AB)]/dt = k_d[A][B].$$

As we shall see, k_d is determined by the diffusion constants for the species. The encounter pair can break up without reaction or it can react to give products. Both processes might involve the solvent, but because it is present in great and constant excess it is not necessary to take its concentration into account explicitly (in the terms introduced in the last chapter, both the disruption and the reaction of the encounter pair are pseudo-first-order steps). The overall scheme is therefore

$$A + B \underset{k_{-d}}{\overset{k_d}{\rightleftharpoons}} (AB) \xrightarrow{k_1} \text{products, P.}$$

The steady-state concentration of (AB) can be found from the rate equation

$$d[(AB)]/dt = k_d[A][B] - k_{-d}[(AB)] - k_1[(AB)] = 0.$$

which gives

$$[(AB)] = \{k_d/(k_{-d} + k_1)\}[A][B],$$

and so the overall rate is

$$d[P]/dt = k_1[(AB)] = \left(\frac{k_1 k_d}{k_{-d} + k_1}\right)[A][B].$$

The overall rate law is second-order, and the second-order rate coefficient is

(28.1.4) $$k_2 = k_1 k_d/(k_{-d} + k_1).$$

Two limits can now be distinguished. If the rate of break-up of the encounter pair is much slower than the rate at which it forms products, then $k_{-d} \ll k_1$ and the expression for k_2 reduces to

$$k_2 \approx k_1 k_d/k_1 = k_d.$$

In this limit the rate of the reaction is determined by the rate at which the species diffuse together through the medium: this is called the *diffusion-controlled limit*, and the reaction is *diffusion controlled*. In due course we shall see that a signal that a particular reaction is diffusion controlled is that its rate coefficient is of the order of 10^9 dm^3 mol^{-1} s^{-1}, or greater. Because the combination of species with unpaired spins involves very little activation energy, radical and atom recombination reactions are often diffusion controlled.

The other limit arises when a substantial activation energy is involved in the reaction step, and then $k_1 \ll k_{-d}$. In this *activation-controlled* limit the overall rate constant is

$$k_2 = k_1(k_d/k_{-d}) = k_1 K,$$

where K is the equilibrium constant for the A, B, and (AB) species. In this case the rate depends on the accumulation of energy in the encounter pair as a result of its interaction with the solvent molecules.

We can arrive at the rates of diffusion-controlled reactions by calculating the rate at which the molecules diffuse together. The simplest approach is to consider a static A molecule immersed in a solvent also containing B molecules. Consider a sphere of radius r surrounding the static molecule. What is the total flow of B molecules through its surface, Fig. 28.1? Since the *flux J* of material is the amount of material passing through unit area in unit time (p. 927) the total *flow $\mathscr{J}$* through the spherical surface of area $4\pi r^2$ is $\mathscr{J} = 4\pi r^2 J$. From Fick's First Law of Diffusion (p. 927) the flux is proportional to the concentration gradient $L\,d[B]/dr$, and so

$$\mathscr{J} = 4\pi r^2 L(D_B\,d[B]/dr),$$

where D_B is the diffusion coefficient of type B molecules.

The overall concentration of B at any distance from A can be found by integrating the last equation. We need two pieces of information. (1) When $r \sim \infty$ the concentration of B is the same as in the bulk solution, $[B]$.

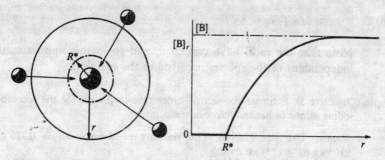

Fig. 28.1. Flux through a spherical surface, and the resulting concentration profile. Reaction occurs at R^*.

(2) The total flow through a shell is the same whatever its distance from A because no molecules are destroyed until A and B touch (Fig. 28.1): therefore $\mathscr{J}$ is a constant independent of r. It follows that

$$\int_{[B]_r}^{[B]} d[B] = \int_r^{\infty} (\mathscr{J}/4\pi r^2 D_B L) dr = (\mathscr{J}/4\pi D_B L) \int_r^{\infty} (1/r^2) dr.$$

Therefore

$$[B] - [B]_r = \mathscr{J}/4\pi r D_B L,$$

and the average concentration varies inversely with distance.

We now suppose that at some critical distance R^*, A and B 'touch', reaction takes place, and B is removed: that is, when $r = R^*$, $[B]_{R^*} = 0$. This can be inserted in the last equation to give an expression for the total rate of flow towards A:

$$\mathscr{J} = 4\pi R^* D_B L[B].$$

The rate of the diffusion-controlled reaction is the rate of flow of B species towards the A molecule. A is at a concentration $L[A]$, and so the total rate of reaction in the sample is $4\pi R^* D_B L^2 [A][B]$. It is unrealistic to suppose that the A molecules are static while the B molecules are mobile: but this is easily remedied by replacing the diffusion coefficient D_B by $D = D_A + D_B$. Then the diffusion-controlled rate coefficient can be identified as

(28.1.5) $k_2 = k_d = 4\pi R^* D L.$

The last expression can be taken further by incorporating the Stokes–Einstein relation between the diffusion coefficient and the viscosity of the medium (p. 913):

$$D_A = kT/6\pi\eta R_A, \qquad D_B = kT/6\pi\eta R_B,$$

where R_A and R_B are the effective hydrodynamic radii of A and B. Since this relation is quite crude, little extra error is introduced by limiting the discussion to the case where $R_A \approx R_B \approx \frac{1}{2}R^*$, which leads to

(28.1.6) $k_2 \approx 8LkT/3\eta = 8RT/3\eta$.

Note that the radii have cancelled, and so in this approximation k_2 is independent of the species involved in the reaction.

Example (Objective 5). Estimate the second-order rate coefficient for the recombination of iodine atoms in hexane at room temperature.

- *Method.* Use eqn (28.1.6). The viscosity of the solvent is 0.326 cP (3.26 × 10^{-4} kg m^{-1} s^{-1}) at 25 °C.

- *Answer.* From eqn (28.1.6).

$k_2 = (8/3) \times (8.3144$ J K^{-1} mol^{-1}) × (298.15 K)/(3.26 × 10^{-4} kg m^{-1} s^{-1})

$= 2.0 \times 10^7$ mol^{-1} m^3 s^{-1} or 2.0×10^{10} mol^{-1} dm^3 s^{-1}.

- *Comment.* The experimental value is 1.3×10^{10} mol^{-1} dm^3 s^{-1}; very good agreement considering the approximations involved.

Two further remarks may be made concerning diffusion-controlled reactions. The first concerns their temperature-dependence. Since the viscosity of a solvent obeys $\eta \approx A \exp(E_a/RT)$, p. 19, k_2 is Arrhenius-like:

$$k_2 \approx (8RT/3A)\exp(-E_a/RT).$$

Second, many diffusion-controlled reactions are between ions. If they are of the same sign of charge we should expect their encounter rate to be less than that calculated in this section, but more if they have opposite signs. The effect of charge can be taken into account, with the result that eqn (28.1.5) is replaced by

(28.1.7) $k_2 = 4\pi PR^*DL$

where the factor P is

$$P = \left(\frac{z_A z_B e^2}{4\pi \varepsilon_r \varepsilon_0 R^* kT}\right)\left(\frac{1}{\exp(z_A z_B e^2/4\pi \varepsilon_r \varepsilon_0 R^* kT) - 1}\right).$$

In this expression ε_r is the relative permittivity of the solvent and z_A and z_B the charges of the species. This predicts, as has been verified experimentally, that the rate should also depend on the relative permittivity (dielectric constant) of the medium. Note that we have here another example of the estimation of the P-factor.

28.2 Activated complex theory

We now turn from questions of molecular mobility to the discussion and calculation of the bimolecular rate constant from another point of view. This approach has the great advantage that it leads directly to the incorporation of the P-factor into the expression for the rate coefficient, and no artificial, *ad hoc* arguments need be made. That is not to say that the theory is complete, or even very reliable: it still remains an attempt to

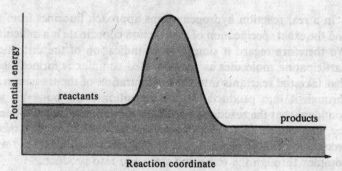

Fig. 28.2. The reaction profile. The abscissa is the reaction coordinate and the ordinate is the potential energy of the system.

identify the principal features that govern the magnitude of k_2 and takes little account of the time-dependent, dynamical features of the actual process.

The reaction coordinate and the transition state. The general features of how the energy of the species A and B changes as the reaction turns them into the products C + D are illustrated in Fig. 28.2. At first only A + B are present and their energy has some value. In the course of reaction, as A and B come into contact, distort, and begin to exchange or discard atoms, the energy rises to a maximum, then falls as the products separate, and attains a value characteristic of the products. The horizontal axis of the diagram represents the course of the reaction, and is called the *reaction coordinate*.

The climax of the reaction is at the peak separating reactants from products. Here a pair of reactants has been brought to the degree of closeness and distortion such that a small distortion in an appropriate direction will send the system in the direction of products. This crucial configuration is called the *transition state* of the reaction. If it passes through that crucial configuration it is inevitable that products will emerge from the encounter.

An example should make the idea of reaction coordinate and transition state as clear as is necessary for our purposes. We consider the approach of a hydrogen atom to a fluorine molecule, and for simplicity regard the atom as approaching along the F–F bond direction. At great distances the energy of the system is that of the isolated atom and molecule. When the atom comes to a point where its orbitals start to overlap the fluorine orbitals, the F–F bond begins to stretch, and to rise in energy, and an incipient bond forms between the atom and one of the fluorine atoms. The hydrogen atom moves closer (if it collides with enough energy), the F–F bond stretches even further, and the F–H bond gets stronger. There will be a point where the composite molecule, the *activated complex*, has a maximum energy, and is poised to pass through the transition state. An infinitesimal compression of the F–H bond takes it through the transition state: the other F–F bond then stretches further and breaks. The reaction coordinate represents the evolution of this set of events.

In a real reaction hydrogen atoms approach fluorines from all angles, and the exact specification of the reaction coordinate is a difficult problem. We therefore regard it simply as an indication of the distortions in the participating molecules as the activated complex is formed, the changes that take the reactants into the configuration of the transition state, and through it into products. At the transition state, however, we can be confident that the reaction coordinate corresponds to some bond stretching, and one of the central features of activated complex theory is the recognition that there is a single, special vibration that takes the activated complex through the transition state and into products.

The formation and decay of the activated complex. Activated complex theory (ACT) pictures a reaction between two species A and B as proceeding by the formation of an activated complex, $(AB)^{\ddagger}$, which falls apart into products at some rate $k^{\ddagger}$. The rate of disappearance of A, the rate of reaction, is the rate at which the activated complex produces products. For a reaction in the gas phase this is

(28.2.1) $$d[P]/dt = k^{\ddagger}[(AB)^{\ddagger}]$$

where $[(AB)^{\ddagger}]$ is the concentration of activated complex. There are two problems: to find $k^{\ddagger}$, and to find the concentration $[(AB)^{\ddagger}]$. The latter is likely to be proportional to the amount of reactants, and later we demonstrate explicitly that

(28.2.2) $$[(AB)^{\ddagger}] = K^{\ddagger}[A][B],$$

where $K^{\ddagger}$ is some constant of proportionality (with dimensions 1/concentration). It follows from eqn (28.2.1) that the rate of reaction is equal to $k^{\ddagger}K^{\ddagger}[A][B]$, and so the rate coefficient for this simple bimolecular reaction is

(28.2.3) $$k_2 = k^{\ddagger}K^{\ddagger}.$$

The coefficient $k^{\ddagger}$ can be related to a property of the activated complex by the following argument. The criterion for the complex passing into products is that it pass through a critical configuration, the transition state. If it does, products are formed. Once the activated complex has been formed, motion along the reaction coordinate corresponds to a distortion of some relevant bonds: passage of the activated complex through the transition state can therefore be identified with a particular mode of its vibration.

If the frequency of the crucial vibration of the complex is v, the frequency of passing through the transition state is also v. In fact it is possible that not every oscillation takes the molecule through the transition state: this is because the critical conformation of the molecule has a complicated dependence on the arrangement of all the atoms in the complex, and sometimes the overall conformation might not be appropriate for the formation of products. Another reason is the possibility that rotations also play a role in breaking up the activated complex, and so the transition state may also depend on the rotational state of the complex. In order to take these factors into account we assume that the rate of passage through the true transition state is *proportional* to the vibrational frequency v along the reaction coordinate, and write

(28.2.4) $k^{\ddagger} = \kappa v$

κ (kappa) is called the *transmission coefficient*; in many cases it is about unity.

The next step is the estimation of the concentration of the activated complex. The approach accepts that almost nothing is known about the nature of the complex, and in particular, that nothing is known about the population of its energy levels. Far from entailing defeat, ignorance of this kind can be used as a constructive principle. Since nothing is known about the distribution of energy among all the modes of the activated complex, we assume that all distributions of energy compatible with a given total energy are equally likely. This will be recognized as the hypothesis of equal *a priori* probabilities used in the construction of statistical thermodynamics in Chapter 20 (p. 680).

Figure 28.3 illustrates the principle. Superimposed on the energy levels of the reactants A and B are those of the activated complex. We want to know the number of molecules in the levels corresponding to $(AB)^{\ddagger}$ relative to the number in the levels corresponding to A and B, there being no bias in favour of particular types or identities of energy levels. This corresponds exactly to the problem we examined in the statistical thermodynamic approach to the calculation of the equilibrium constant, and Fig. 28.3 is essentially the same as Fig. 21.8 on p. 716. The answer to this problem can therefore be taken from the discussion based on Fig. 21.8, and the $K^{\ddagger}$ related to the partition functions of the three species involved. In the first place we have, since $[J] = p_J/RT$,

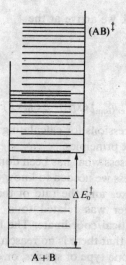

A + B   Fig. 28.3. Energy levels of reactants and the activated complex.

$$K^{\ddagger} = [(AB)^{\ddagger}]/[A][B] = RT\{p^{\ddagger}/p_A p_B\}$$

$$= (RT/p^{\ominus})\{(p^{\ddagger}/p^{\ominus})/(p_A/p^{\ominus})(p_B/p^{\ominus})\} = (RT/p^{\ominus})K_p^{\ddagger}$$

where $K_p^{\ddagger}$ is a (dimensionless) equilibrium constant expressed in terms of pressures. This is related to the molar partition functions by

(28.2.5) $\quad K_p^{\ddagger} = \{Lq_m^{\ddagger\ominus}/q_{A,m}^{\ominus}q_{B,m}^{\ominus}\}\exp(-\Delta E_0^{\ddagger}/kT),$

which is essentially eqn (21.3.25) on p. 719; $q_m^{\ddagger}$ is the molar partition function for the activated complex, $q_{A,m}$ and $q_{B,m}$ the molar partition functions for the reactant molecules, and $\Delta E_0^{\ddagger}$ is the energy separation of the complex and the reactants (Fig. 28.3).

In order to take the final step we focus attention on the partition function of the activated complex. We have already established that a vibration tips the complex through the transition state, and on to products. The partition function for this particular vibrational mode is $\{1 - \exp(-hv/kT)\}^{-1}$, where v is its frequency (the same frequency that determines $k^{\ddagger}$). This frequency will be much lower than for an ordinary molecular vibration because the motion corresponds to the falling apart of the activated complex. That being the case the exponential in the partition function can be expanded and only the leading term retained; this reduces it to kT/hv. We now can write

$$q_m^{\ddagger} = (kT/hv)\bar{q}_m^{\ddagger}$$

where $\bar{q}^{\ddagger}$ is the partition function for all the *other* modes of the activated complex.

All the parts of the calculation can now be combined into

$$k_2 = k^{\ddagger}K^{\ddagger} = k^{\ddagger}(RT/p^{\ominus})K_p^{\ddagger} = \kappa v(RT/p^{\ominus})(kT/hv)\bar{K}_p^{\ddagger}$$

where

(28.2.6) $\quad \bar{K}_p^{\ddagger} = \{L\bar{q}_m^{\ddagger\ominus}/q_{A,m}^{\ominus}q_{B,m}^{\ominus}\}\exp(-\Delta E_0^{\ddagger}/kT).$

Since the unknown frequencies v cancel, we arrive at the

(28.2.7a) *Eyring equation $k_2 = \kappa(kT/h)\bar{K}$,*

where

(28.2.7b) $\quad \bar{K} = (RT/p^{\ominus})\bar{K}_p^{\ddagger} = (RT/p^{\ominus})\{L\bar{q}_m^{\ddagger\ominus}/q_{A,m}^{\ominus}q_{B,m}^{\ominus}\}\exp(-\Delta E_0^{\ddagger}/kT).$

The last two equations are explicit expressions for calculating the rate of a simple bimolecular reaction from first principles.

It is well to pause at this point and assess what has been done. In order to find the rate of the bimolecular process we needed to know two things, the concentration of activated complexes and the rate of their passage through the transition state. The latter was related to a vibrational frequency along some relevant and critical coordinate. The former was found by making the single assumption that there is no reason for biasing the distribution of energy in favour of one type of species, or one type of energy level. In some treatments the same expression for the rate coefficient is obtained by making the assumption that the activated complex is in equilibrium with the reactants throughout the course of the reaction. On that basis we can go immediately to eqn (28.2.2), stating that $K^{\ddagger}$ is an equilibrium constant, use the conventional statistical thermodynamic expression, and so go at once to eqn (28.2.5). Although this route has the advantage of brevity and directness, the basic assumption that the rate process involves an equilibrium is often a stumbling block to its acceptance. Using the 'least prejudiced' approach, while ultimately equivalent to the equilibrium approach, indicates more fairly the nature of the approximation being made, and opens up more clearly the way to improving the model. For example, one might think that the model could be improved by biasing the population of the activated complex so that many of the vibrations retain the population distribution they had in the reactants.

How to use the Eyring equation. The possibility of using the Eyring equation to calculate the rate of a reaction depends on being able to calculate the partition functions for the species involved. The reactant partition functions can normally be calculated with confidence, either by using spectroscopic information about the energy levels or from the approximate expressions, Box 21.1, p. 703. The real difficulty lies in the determination of the partition function for the activated complex. This is not normally open to spectroscopic determination, and depends on assumptions made about its structure, including its size and shape.

As a first example of the calculation of a rate constant we take the simple case of two structureless particles A and B colliding to give an activated complex that can be treated as a diatomic molecule. This diatomic molecule then breaks up into products of some kind.

Since the reactants are structureless 'atoms' the only contributions to their partition functions are the translational terms:

$$q_{A,m}^{\ominus} = q_{A,m}^{t\ominus} = (2\pi m_A kT/h^2)^{3/2} V_m^{\ominus} = \tau_A V_m^{\ominus} = \tau_A (RT/p^{\ominus})$$

with a similar expression for q_B but with m_B in place of m_A. The activated complex is a diatomic molecule of mass $m_{AB} = m_A + m_B$ and moment of inertia I_{AB}. There is only one vibrational mode, but that corresponds to the reaction coordinate, and so it does not appear in $\bar{q}^{\ddagger}$. It follows that the molar partition function of the activated complex is

$$\bar{q}_m^{\ddagger\ominus} = \tau^{\ddagger}(2I_{AB}kT/\hbar^2)V_m^{\ominus}, \qquad \tau^{\ddagger} = (2\pi m_{AB}kT/h^2)^{3/2}.$$

Since the moment of inertia I_{AB} is related to the reduced mass μ_{AB} and the bond length of the diatomic by $I_{AB} = \mu_{AB}R_{AB}^2$ the expression for the rate coefficient is

$$k_2 = (RT/p^{\ominus})\left\{\frac{\kappa(kT/h)L\tau^{\ddagger}V_m^{\ominus}(2I_{AB}kT/\hbar^2)}{\tau_A\tau_B V_m^{\ominus 2}p^{\ominus}}\right\}\exp(-\Delta E_0^{\ddagger}/kT)$$

(28.2.8) $= \kappa(kT/h)(\tau^{\ddagger}/\tau_A\tau_B)(2I_{AB}kT/\hbar^2)\exp(-\Delta E_0^{\ddagger}/kT)$

             $= L(8kT/\pi\mu_{AB})^{\ddagger}\pi\kappa R_{AB}^2\exp(-\Delta E_0^{\ddagger}/kT),$

and by identifying $\kappa\pi R_{AB}^2$ as σ^*, the reactive cross-section, we arrive at precisely the same expression as obtained from simple collision theory, eqn (28.1.1).

Example (Objective 9). Obtain the order-of-magnitude of the *P*-factor for the reaction between two non-linear molecules.

- *Method.* Use the Eyring equation twice, first for two structureless molecules, and then for two molecules with internal structure (so that they can rotate and vibrate). The ratio of the two results is identified as *P*. For the order-of-magnitude estimate, assume all translational partition functions are the same, likewise for all rotational and all vibrational partition functions. An *N*-atomic molecule has 3 translational, 3 rotational, and $3N-6$ vibrational modes.

- *Answer.* For no internal modes of the colliding molecules,
$$q_{A,m}^{\ominus} = \tau V_m^{\ominus}, \qquad q_B^{\ominus} = \tau V_m^{\ominus}, \qquad \bar{q}_m^{\ddagger\ominus} = q_r^2 V_m^{\ominus}.$$
Therefore, from eqns (28.2.7) and (28.2.6), with $RT/p^{\ominus} = V_m^{\ominus}$,
$$k_2 \approx \kappa(kT/h)V_m^{\ominus}\{Lq_r^2\tau V_m^{\ominus}/\tau^2 V_m^{\ominus 2}\}\exp(-E_0^{\ddagger}/kT)$$
$$\approx \kappa(RT/h)(q_r^2/\tau)\exp(-E_0^{\ddagger}/kT).$$
When all the internal modes are active,
$$q_{A,m}^{\ominus} = \tau q_r^3 q_v^{3N_A-6}V_m^{\ominus}, \qquad q_{B,m}^{\ominus} = \tau q_r^3 q_v^{3N_B-6}V_m^{\ominus}, \qquad \bar{q}_m^{\ddagger\ominus} = \tau q_r^3 q_v^{3(N_A+N_B)-7}V_m^{\ominus}.$$
Note that one vibrational mode has been subtracted from the modes of the activated complex. It follows that
$$k_2^* \approx \kappa(RT/h)V_m^{\ominus}\{\tau q_r^3 q_v^{3(N_A+N_B)-7}V_m^{\ominus}/\tau^2 q_r^6 q_v^{3(N_A+N_B)-12}V_m^{\ominus 2}\}\exp(-E_0^{\ddagger}/kT)$$
$$\approx \kappa(RT/h)(q_v^5/\tau q_r^3)\exp(-E_0^{\ddagger}/kT).$$
Comparing the two results shows that
$$P = k_2^*/k_2 \approx q_v^5/q_r^5.$$
Since $q_v/q_r \approx 1/50$, $P \approx 1/50^5 = 3.2 \times 10^{-9}.$

- *Comment.* The calculation suggests that the reaction between two complex molecules will be much slower, both in the gas and in solution, than reactions between simple molecules, even though the activation energies may be similar.

Even the atom–atom example required an assumption about the structure of the activated complex, and the problem gets increasingly complicated as the number of atoms grows. There is, however, another simple but important illustration of its application. Consider the effect of *deuteration* on the rate of breaking a C–H bond. The reaction coordinate involves stretching the C–H bond, and the reaction profile is illustrated in Fig. 28.4. On deuteration the dominant change is a fall in the zero-point energy of the bond, on account of the isotope's greater mass. The whole reaction profile is not lowered, because the relevant vibration in the activated complex is very loose, and so there is very little zero-point energy in either the proton or the deuterium forms.

As a reasonable first assumption we assume that deuteration affects

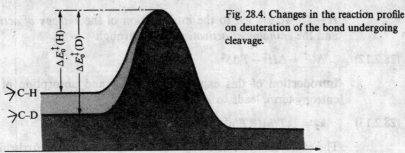

Fig. 28.4. Changes in the reaction profile on deuteration of the bond undergoing cleavage.

Reaction coordinate

only the reaction coordinate, and so the partition functions for all other internal modes remain the same. The translational partition functions change on deuteration, but the mass of the rest of the molecule is normally so great as to make the change insignificant. The value of $\Delta E_0^{\ddagger}$ changes because the lowest state of the reactants is lower when deuteration has taken place, but the lowest state of the activated complex remains the same: the change in $\Delta E_0^{\ddagger}$ is equal to the change in the zero-point energy of the C–H bond:

$$\Delta E_0^{\ddagger}(D) - \Delta E_0^{\ddagger}(H) \approx \tfrac{1}{2}\hbar\{\omega(C-H) - \omega(C-D)\}$$
$$\approx \tfrac{1}{2}\hbar k_f^{1/2}\{(1/\mu_{CH})^{1/2} - (1/\mu_{CD})^{1/2}\}.$$

k_f is the force-constant of the bond and μ is the reduced mass.

We can conclude that the rate coefficients for the undeuterated and deuterated species should be in the ratio

(28.2.9)
$$k_2(D)/k_2(H) \approx \exp\{\tfrac{1}{2}(\hbar k_f^{1/2}/kT)[(1/\mu_{CD})^{1/2} - (1/\mu_{CH})^{1/2}]\}.$$

This predicts that at room temperature cleavage should proceed 7 times as fast for a C–H bond as for a C–D bond, other conditions being equal. The reason as we have seen can be traced to the greater zero-point energy of the C–H bond, and therefore to its smaller activation energy.

Thermodynamic aspects. In the derivation of the expression for k_2 it was pointed out that the same result could be obtained by making the assumption that the activated complex was in thermodynamic equilibrium with the reactants. This analogy can be pursued by introducing the language of equilibrium thermodynamics.

In the first place a *Gibbs function of activation* can be defined through the expression

(28.2.10)
$$-RT\ln \bar{K}_p^{\ddagger} = \Delta G_m^{\ddagger}.$$

Then the rate coefficient $k_2 = \kappa(RT/p^{\ominus})(RT/h)\bar{K}_p^{\ddagger}$ becomes

(28.2.11)
$$k_2 = \kappa(kT/h)(RT/p^{\ominus})\exp(-\Delta G_m^{\ddagger}/RT).$$

Since a Gibbs function is related to the entropy and enthalpy by $G =$

$H - TS$ we are led to the introduction of the *entropy of activation*, $\Delta S^{\ddagger}$, and the *enthalpy of activation*, $\Delta H^{\ddagger}$ through

(28.2.12) $\Delta G^{\ddagger} = \Delta H^{\ddagger} - T\Delta S^{\ddagger}$.

Introduction of this expression into k_2, and absorption of κ into the entropy term, leads to

(28.2.13) $k_2 = (kT/h)(RT/p^{\ominus})\exp(\Delta S_m^{\ddagger}/R)\exp(-\Delta H_m^{\ddagger}/RT)$.

The expression just derived resembles very closely the Arrhenius rate law, and it is tempting to identify the A-factor as $(kT/h)(RT/p^{\ominus})\exp(\Delta S_m^{\ddagger}/R)$. This, however, neglects the fact that $\Delta H_m^{\ddagger}$ is not E_a. The connection between them can be established by using the general definition of the activation energy, eqn (27.3.2) on p. 938, which gives*

$$E_a = \Delta H_m^{\ddagger} + 2RT, \quad \text{or} \quad \Delta H_m^{\ddagger} = E_a - 2RT,$$

for a bimolecular gas phase reaction ($E_a = \Delta H_m^{\ddagger} + RT$ for a reaction in solution, when $\Delta H \approx \Delta U$). Consequently,

$$k_2 = (kT/h)(RT/p^{\ominus})\exp\{-\Delta H_m^{\ddagger}/RT + \Delta S_m^{\ddagger}/R\}$$
$$= (kTe^2/h)(RT/p^{\ominus})\exp(\Delta S_m^{\ddagger}/R)\exp(-E_a/RT),$$

and so

(28.2.14a) $A = (kTe^2/h)(RT/p^{\ominus})\exp(\Delta S_m^{\ddagger}/R)$.

(28.2.14b)   or $\Delta S_m^{\ddagger} = R\ln\{hAp^{\ominus}/L(ekT)^2\}$.

A practical form of this expression is given in the *Example* below. This expression makes contact with the collision theory approach because A was then determined by the occurrence of collisions in the gas, and collisions correspond to a decrease in entropy (they involve the coming together of two particles, and therefore a decrease in the randomness of the system). Furthermore, collisions with a well-defined relative orientation correspond to an even greater decrease in entropy than is brought about by collisions alone, and so the entropy of activation should then be more negative, and correspond to a value of P less than unity. The entropy of activation therefore corresponds to both the collision term Z and the steric factor P of simple collision theory.

* Write $(\partial\ln k_2/\partial T) = (1/k_2)(\partial k_2/\partial T)$, $RT/p^{\ominus} = V_m^{\ominus}$, and $k_2 = \kappa(kT/h)V_m^{\ominus}\bar{K}_p^{\ddagger}$, to obtain

$$E_a = RT^2(\partial\ln k_2/\partial T)_V = (RT^2/k_2)\left\{\frac{\partial}{\partial T}(\kappa(kT/h)V_m^{\ominus}\bar{K}_p^{\ddagger})\right\}_V$$

$$= (RT/\bar{K}_p^{\ddagger})\left\{\frac{\partial}{\partial T}(T\bar{K}_p^{\ddagger})\right\}_V = (RT/\bar{K}_p^{\ddagger})\left\{\bar{K}_p^{\ddagger} + T\left(\frac{\partial\bar{K}_p^{\ddagger}}{\partial T}\right)_V\right\} = RT + RT^2(\partial\ln\bar{K}_p^{\ddagger}/\partial T)_V.$$

Since $(\partial\ln K/\partial T)_V = \Delta U_m/RT^2$ it follows that $E_a = RT + \Delta U_m^{\ddagger}$, where $\Delta U_m^{\ddagger}$ is the *activation internal energy*. Since $\Delta H_m = \Delta U_m + \nu RT$ for perfect gases, and since for the formation of the activated complex $\nu = -1$ (because two species coalesce into one), we have $\Delta H_m^{\ddagger} = \Delta U_m^{\ddagger} - RT$. It follows that $E_a = \Delta H_m^{\ddagger} + 2RT$.

Example (Objective 10). Calculate the enthalpy, entropy, and Gibbs function of activation for the second-order hydrogenation of ethene at 355 °C.

- *Method*. From Table 28.1, the activation energy is 180 kJ mol^{-1} and $A = 1.24 \times 10^6$ mol^{-1} dm^3 s^{-1}. For a bimolecular rate-determining step, the activation energy E_a is related to the enthalpy of activation by $E_a = \Delta H_m^\ddagger + 2RT$. The entropy is obtained from eqn (28.2.14), and the Gibbs function of activation from eqn (28.2.12).

- *Answer*. $\Delta H_m^\ddagger = E_a - 2RT$

$$= 180 \text{ kJ mol}^{-1} - 2 \times (8.314 \text{ J K}^{-1} \text{ mol}^{-1}) \times (628 \text{ K})$$

$$= 170 \text{ kJ mol}^{-1}.$$

Inserting the values of the fundamental constants into the expression for $\Delta S_m^\ddagger$ gives an expression that is easy to use:

$$\Delta S_m^\ddagger = R \ln \left\{ \frac{(6.626\,18 \times 10^{-34} \text{ J s}) \times A \times (1.013\,25 \times 10^5 \text{ N m}^{-2})}{e^2 \times (6.022\,05 \times 10^{23} \text{ mol}^{-1}) \times (1.380\,66 \times 10^{-23} \text{ J K}^{-1})^2 \times T^2} \right\}$$

$$= R \ln \left\{ \frac{(7.9154 \times 10^{-11}) \times (A/\text{dm}^3 \text{ mol}^{-1} \text{ s}^{-1})}{(T/K)^2} \right\}.$$

It follows that in the present case, '

$$\Delta S_m^\ddagger = R \ln \{(7.9154 \times 10^{-11}) \times (1.24 \times 10^6)/(628 \text{ K})^2\} = -183.9 \text{ J K}^{-1} \text{ mol}^{-1}.$$

Then, from eqn (28.2.12),

$$\Delta G_m^\ddagger = (170 \text{ kJ mol}^{-1}) - (628 \text{ K}) \times (-183.9 \text{ J K}^{-1} \text{ mol}^{-1}) = 285 \text{ kJ mol}^{-1}.$$

- *Comment*. Note the large negative entropy of activation. The simple collision-theory magnitude of A (as calculated in the *Example* on p. 973), would correspond to a significantly smaller entropy of activation (-73 J K^{-1} mol^{-1}).

Reactions in solution. The application of activated complex theory to reactions in solution is very complicated because of the involvement of the solvent in the activated complex. We cannot hope to get very far by calculating partition functions for an activated complex surrounded by a cluster of solvent molecules. We shall accept this difficulty, but avoid coming to a dead stop by using the thermodynamic approach to k_2.

The *rate law* depends on the *concentrations* of the reactants A and B, but the *equilibrium constant* $K^\ddagger$ depends on the activities of the species. Therefore,

$$k_2[A][B] = k^\ddagger[(AB)^\ddagger]$$

should be combined with

$$K^\ddagger = a(AB)^\ddagger/a(A)a(B)$$

$$= \gamma_{(AB)^\ddagger}[(AB)^\ddagger]/\gamma_A[A]\gamma_B[B]$$

where the as are the activities and the γs the activity coefficients. This gives

(28.2.15) $\qquad k_2 = k^\ddagger K^\ddagger \{\gamma_A \gamma_B/\gamma_{(AB)^\ddagger}\} = k_2^0 \{\gamma_A \gamma_B/\gamma_{(AB)^\ddagger}\}.$

k_2^0 is the rate coefficient when all activity coefficients are unity.

The development can be taken a stage further by relating the activity coefficients to the ionic strength of the solution using the Debye–Hückel

Limiting Law derived in Chapter 11, p. 325:

$$\lg \gamma_i = -Az_i^2(I/\mathrm{mol\ kg}^{-1})^{1/2}$$

for then

$$\lg k_2 = \lg k_2^0 - A\{z_A^2 + z_B^2 - (z_A + z_B)^2\}\,(I/\mathrm{mol\ kg}^{-1})^{1/2}$$

(28.2.16) $$= \lg k_2^0 + 2Az_A z_B(I/\mathrm{mol\ kg}^{-1})^{1/2}.$$

Use has been made of the fact that if the charges of the reactants are z_A and z_B, the charge of the activated complex is $z_A + z_B$.

The last equation indicates that the rate coefficient for an ionic reaction in solution should depend on the ionic strength: this is the *kinetic salt effect*. If the ions are of like sign, increasing the ionic strength (by the addition, for example, of inert ions) increases the rate coefficient. This can be understood in terms of the formation of a single, highly charged activated complex from two ions of like charge: the process is favoured by a high ionic strength because of the favourable interaction of the new ion with its denser ionic atmosphere. Conversely, ions of opposite charge react more slowly in solutions of high ionic strength. This is because the

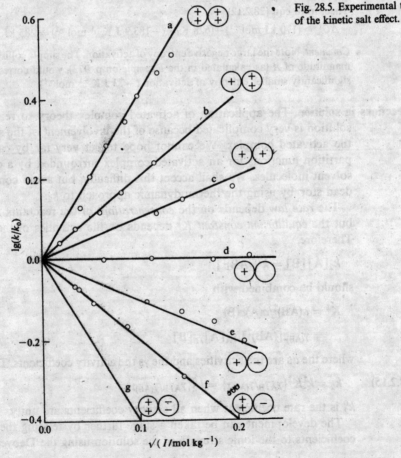

Fig. 28.5. Experimental test of the kinetic salt effect.

cancellation of charge on formation of the complex is unfavourable on account of the lessening of its interaction with the ionic atmosphere.

The kinetic salt effect can be checked experimentally by choosing ions of different charges, and plotting $\lg k_2$ against the square root of the ionic strength. The A-factor is known (for water at room temperature it has the value 0.509) and so the gradients of the plots can be compared with the theoretical value of $1.018 z_A z_B$: Fig. 28.5 shows how good is the agreement between theory and experiment. An alternative use of the theory is in the determination of the nature of the activated complex of a reaction: by finding the dependence of k_2 on the ionic strength we can deduce the charge of the species involved. This is illustrated in the following *Example*.

Example (Objective 11). The alkaline hydrolysis of the complex $[Co(NH_3)_5Br]^{2+}$ has a rate coefficient that depends on the ionic strength of the solution as tabulated below. What can be said about the nature of the activated complex in the rate-determining stage?

$I/\text{mol kg}^{-1}$	0.005	0.010	0.015	0.020	0.025	0.030
k_2/k_2^0	0.718	0.631	0.562	0.515	0.475	0.447

- *Method.* Plot $\lg(k_2/k_2^0)$ against $\sqrt{I}$. The gradient of the line will be $1.02\, z_A z_B$. The OH^- ion has $z_A = -1$, the complex ion has $z_B = +2$, and so if a line of slope -2.04 is obtained, we can conclude that the rate-determining step involves an activated complex formed from the two ions.

- *Answer.* Form the following table.

$I/\text{mol kg}^{-1}$	0.005	0.010	0.015	0.020	0.025	0.030
$\sqrt{(I/\text{mol kg}^{-1})}$	0.071	0.100	0.122	0.141	0.158	0.173
$\lg(k_2/k_2^0)$	-0.14	-0.20	-0.25	-0.29	-0.32	-0.35

The data are plotted in Fig. 28.6; the slope of the least-squares fitted line is -2.1, pointing to the involvement of the two ions in the activated complex. (The coefficient of determination for the fit is 0.9988.)

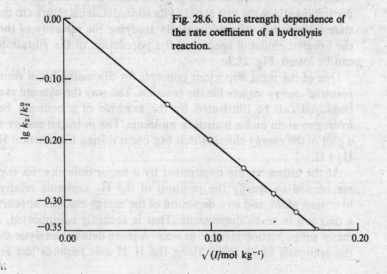

Fig. 28.6. Ionic strength dependence of the rate coefficient of a hydrolysis reaction.

● *Comment*. Another factor that influences the reaction rate is the relative permittivity of the medium; this is examined in Problem 28.28.

28.3 The dynamics of molecular collisions

We now come to the third level of examination of the factors that govern the rates of chemical reactions, and turn to the information about collisions between individual molecules made accessible by recent technological advances. *Molecular beam* techniques allow examination of collisions between molecules in preselected energy levels, and can be used to determine the energy states of the products of a reactive collision. Information of this kind is essential if a full picture of a reaction is to be built up, but we are still far from a complete theory: after all, most of the reactions we want to know about take place in solution, and our ability to study individual processes in liquids is still rudimentary.

Reactive encounters: the general situation. The experimental wing of attack depends on molecular beams, and especially crossed molecular beams (p. 789). The detector for the products of the collision of two beams can be moved to different angles, and so the technique can be used to determine the *angular distribution* of products. Furthermore, if the detector can distinguish between different energy states of the products, we are able to determine their states immediately after reaction. Since the molecules in the incoming beams can be prepared in different energy states (e.g. by selective vibrational excitation using lasers) and with different orientations (see p. 789), it is possible to study the dependence on final energy state, angular distribution, and probability of reaction for a range of initial states of the colliding molecules.

A helpful method of examining the final energy distribution in molecules is *infrared chemiluminescence*. If a product is formed in some non-equilibrium distribution of vibrational states (for example, if it is formed predominantly in the first excited vibrational state), it returns to the ground state by emitting infrared light. By studying the intensity of the lines in the infrared emission spectrum, the population of the vibrational states can be found, Fig. 28.7.

One of the most important concepts for discussing beam results is the *potential energy surface* for the reaction. The way the concept is normally employed can be illustrated by the example of a collision between a hydrogen atom and a hydrogen molecule. The potential energy surface is a plot of the energy changes/that can occur during the reaction $H + H_2 \rightarrow H_2 + H$.

At the outset we are confronted by a major difficulty: six coordinates are needed to specify the position of the H_2 molecule relative to the hydrogen atom, and so a depiction of the energy changes appears to need a diagram in seven dimensions. That is absurdly complicated, and so a major simplification is made at once. A more detailed analysis shows that the approach of an atom along the H–H axis requires less energy for

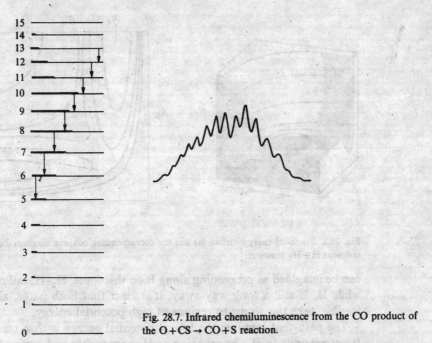

Fig. 28.7. Infrared chemiluminescence from the CO product of the O + CS → CO + S reaction.

reaction than any other approach, and so we confine our attention, for the moment, to a collinear collision. Now only two parameters are required to define the changes in atomic separations during the reaction: one is the H_a-H_b separation, R_{ab}, and the other the H_b-H_c separation, R_{bc}.

At the start of the collision R_{ab} is infinite and R_{bc} is the equilibrium bond distance of H_2; at the end of the collision (if reaction occurs) R_{bc} is infinite and R_{ab} has the H_2 equilibrium value. The energy of the three atoms depends on their relative separations, and can be found as in a normal molecular structure calculation. The plot of energy against R_{ab} and R_{bc} constitutes the potential-energy surface of the reaction, Fig. 28.8a. Since a three-dimensional drawing is difficult to depict successfully, a contour diagram is normally drawn, Fig. 28.8b.

If R_{ab} is maintained at a constant and large value, the changes of energy that occur with changing R_{bc} are those of an isolated H_2 molecule when its bond length changes. A section through the surface at $R_{ab} = \infty$ is the same as a molecular potential-energy curve of the type encountered in Chapter 15. At the edge of the diagram where R_{bc} is almost infinite (representing the case when the H_2 molecule is H_aH_b and atom H_c is far away), the profile of the surface has the form of the molecular potential energy curve of an isolated H_a-H_b molecule.

The surface can be used to trace the energy changes that take place as H_a approaches the H_b-H_c molecule. If the H_b-H_c bond length is held constant as H_a approaches, the potential energy would change as depicted by line A in Fig. 28.9. This shows that the system moves to a very high potential energy as H_a is pushed towards H_b-H_c, and then drops sharply as H_c breaks off and goes to a great distance. Alternatively, the reaction

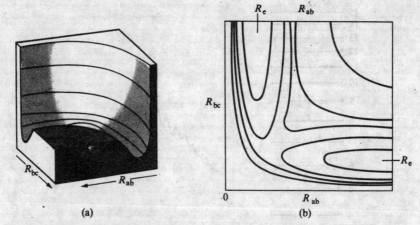

(a)   (b)

Fig. 28.8. Potential energy surface (a) and the corresponding contour diagram (b) of the collinear $H + H_2$ reaction.

can be imagined as proceeding along B on this route, $H_b–H_c$ falling apart while H_a is still a long way away. It is clear that both paths, although feasible, take the atoms to regions of high potential energy.

The path that involves the lowest potential energy is that marked C. It corresponds to R_{bc} lengthening as H_a approaches and begins to form a bond to H_b. The $H_b–H_c$ bond relaxes at the demand of the incoming atom, and although the energy rises, it climbs only as far as the *saddle point* marked $C^{\ddagger}$ in Fig. 28.9. The reaction path involving the least potential energy is the route C up the floor of the valley, through the saddle point, and down the floor of the other valley as H_c recedes and the new $H_a–H_b$ bond sinks to its equilibrium length.

The kind of questions investigated with molecular beams can be introduced by considering potential-energy surfaces of reactions. In the first place, in order to trace the path from reactants to products by making the journey along C, the incoming atom has to approach with sufficient kinetic energy to be able to climb up the potential-energy surface as far

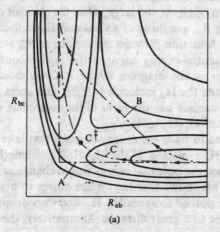

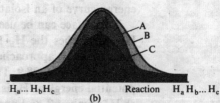

Fig. 28.9. Various trajectories (a) and the corresponding reaction profiles (b).

as the saddle point. By changing the speed of the incoming atom, and determining the energy at which reaction occurs, the accuracy of a calculated, or estimated, potential-energy surface can be tested. In the second place we can see that if the incoming atom brings a lot of kinetic energy the system may overshoot the saddle point, and follow a path that leads to a product molecule that vibrates in its well. By following the degree of vibrational excitation of the products the shape of the potential-energy surface can be built up, at least in the vicinity of the saddle point, and possibly for the whole space.

Another question that can be asked is whether it is better to smash the species together with a large amount of translational kinetic energy, or whether the reaction proceeds more efficiently if the energy is brought to the complex in the form of vibrations. For example, is trajectory C_2^* in Fig. 28.10, where the H_b–H_c molecule is initially in a vibrationally excited state, more efficient at leading to reaction than the trajectory C_1^* which follows the floor of the valley and corresponds to a gradual stretching of the bond?

How is the information given by molecular beam studies related to the value of the rate coefficient k_2? The connection is the realization that k_2 is a statistical quantity. In an ordinary gas-phase reaction the species

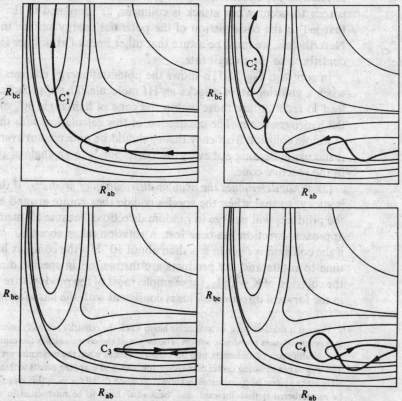

Fig. 28.10. Successful and unsuccessful encounters.

collide with a range of different energies and in a variety of different vibrational and rotational states. Each individual collision can be imagined as a trajectory on the potential-energy surface of the reaction. Some of these trajectories will be successful (C_1^* and C_2^* in Fig. 28.10) and some unsuccessful (C_3 and C_4) either because they lack sufficient energy or because the energy is distributed inappropriately. The rate of the reaction is an average over all these possible trajectories, and so, in order to calculate k_2, we have to calculate the form of a large number of trajectories, and then take the average. A major technique for doing this is the *Monte Carlo method*: the method is to sample a set of initial conditions by selecting them at random, but in such a way that overall the selection is characteristic of a thermal equilibrium system.

Some results of molecular beam studies. In this section we see how some of the questions raised in the last section can be answered, and how the study of collisions and the calculation of potential energy surfaces can give insight into the course of reactions.

(1) *Is collinear approach the least energetic route?* Figure 28.11a shows the results of a calculation of the change of energy as a hydrogen atom attacks a hydrogen molecule from different angles, and the bond is allowed to relax to the optimum length in each case. The least activation energy is seen to occur when attack is collinear, as presumed in the discussion that led to the construction of the potential energy surface in Fig. 28.8. Nevertheless, we must be aware that other lines of attack are feasible and contribute to the overall rate.

In contrast, Fig. 28.11b shows the potential-energy changes that occur when a chlorine atom attacks an HI molecule. The approaches that can lead to reaction are those within the cone of half-angle 30° surrounding the hydrogen atom. The connection of this calculation with the *P*-factor of simple collision frequency theory should be noticed: not every collision is successful because not every one corresponds to an angle of attack lying in the reactive cone.

(2) *What determines the angular distribution of products?* If the collision is sticky, so that when the species collide they rotate around each other, the products will emerge in random directions because all memory of the approach direction has been lost. A rotation takes about 10^{-12} s, and so if the collision is over in less than about 10^{-12} s the complex has not had time to rotate and the products are thrown off in specific directions. In the collision of K with I_2, for example, most of the products are thrown off in the forward direction*. This is consistent with the harpoon mechanism,

* There is a subtlety here. In molecular beam work the remarks normally refer to directions in a *centre-of-mass coordinate system*. The origin of the coordinates is the centre of mass of the colliding pair of molecules, and the collision occurs when the molecules are at the origin. The way of constructing centre of mass coordinates, and mapping events within them on to a laboratory system of coordinates, involves too much detail for us at this stage. All that need be remembered is that 'forward' and 'backward' have to be interpreted in an unconventional way.

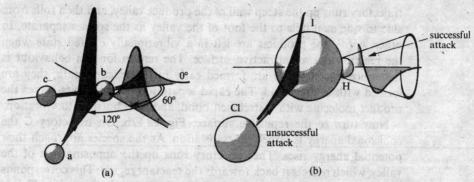

Fig. 28.11. Energy changes for various angles of attack: (a) $H + H_2$, (b) $Cl + HI$.

p. 973, because the collision takes place at long range. In contrast, the collision of K with methyl iodide, CH_3I, leads to reaction only if the species approach very closely. This resembles K bumping into a brick wall, and the KI product richocheting out in the backwards direction. The observation of this anisotropy in the angular distribution of products gives an indication of the distance and orientation of approach needed for reaction, as well as showing that the reaction must be complete in less than 10^{-12} s.

(3) *Is it best to have the collisional energy in translation or in vibration?* Some reactions are very sensitive to whether the energy has been predigested into a vibrational mode, or left in relative translation of the colliding species. For example, if two HI molecules are hurled together with more than twice the activation energy of the reaction, no reaction occurs because all the energy is in translation. In the case of the reaction $F + HCl \rightarrow Cl + HF$ it has been found that the reaction is about five times as efficient when the HCl is in its first vibrational excited state than when it is in its vibrational ground state.

The root of these requirements can be found by examining the potential-energy surfaces for reactions other than $H + H_2$. Two especially important cases are shown in Fig. 28.12. In the surface shown in Fig. 28.12a the highest point of the reaction trajectory occurs early on the path: this is called an *attractive* surface. In contrast Fig. 28.12b shows a surface where the saddle point occurs late: this is a *repulsive* surface.

Consider first the attractive surface. If the original molecule is vibrationally excited, a collision with an incoming species drives the system along the trajectory marked C: this tends to be bottled up in the region corresponding to reactants, and does not readily move over the saddle point. If, however, the same energy is present as simple translational energy, the system will tend to follow a trajectory close to C* when the collision occurs. This trajectory carries the system cleanly over the saddle point and into the region corresponding to products. We can conclude that reactions corresponding to an attractive potential surface proceed most efficiently if the energy is in a relative translational mode. Furthermore, the potential surface shows that once past the saddle point, the trajectory runs up the steep wall of the product valley, and then rolls from side to side as it falls to the foot of the valley as the species separate. In other words, the products are left in a vibrationally excited state when the reaction has an attractive surface. The reason for this behaviour is that since the products are formed early in the reaction path, they are formed with a long bond. The rapid separation of the species leaves the product molecule with a stretched bond, and so it bursts into vibration.

Now turn to the repulsive surface, Fig. 28.12b. On trajectory C the collisional energy is largely in translation. As the species approach their potential energy rises. The trajectory runs up the opposing face of the valley, which reflects it back towards the reactant region. This corresponds to an unsuccessful encounter, even though the total energy is sufficient for reaction. On the other trajectory, C*, some of the total energy is incorporated in the vibration of the molecule. The vibration causes the trajectory to weave from one side of the valley to the other as the system climbs to the saddle point on the approach of the other reactant. This motion is sufficient to tip the system round the corner and up the final slope to the saddle point. Once there it can run down into the product valley. In this case the product molecule is expected to be in an unexcited vibrational state. Another way of looking at the process is that the products do not separate until the new bond has been substantially formed and is virtually at its final equilibrium separation. The departing species is then shot out at a late stage, and carries away the excess energy. An example of a reaction with a repulsive potential surface is $H + Cl_2 \rightarrow HCl + Cl$, and a criterion for reaction is the necessity for the energy to be predigested as vibration.

The information in this section can be collected as follows. An *attractive surface* corresponds to a reaction that forms products in a vibrationally excited state and proceeds most efficiently if the energy of the collision is in the relative translational motion of the reactants. A *repulsive surface* indicates that the reaction proceeds most efficiently if the excitation energy

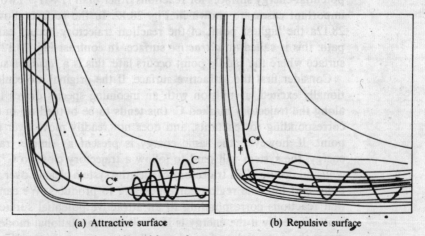

(a) Attractive surface   (b) Repulsive surface

Fig. 28.12. Attractive and repulsive surfaces.

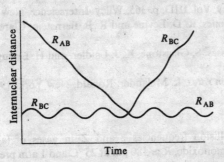

Fig. 28.13. Calculated trajectories for a reactive encounter between A and BC.

is brought up to the collision as a vibrational mode, and the products then appear in a vibrational ground state. A reaction that is attractive in one direction is repulsive in the reverse direction.

(4) *Can we calculate trajectories showing the reaction in progress?* A clear picture of the reaction can be obtained by using classical mechanics to calculate the trajectories of the atoms involved in a chemical reaction. Figure 28.13 shows the results of a classical calculation of the positions of three atoms in the reaction $A + BC \rightarrow AB + C$, the horizontal coordinate being time and the vertical one the separations involved. This shows clearly the vibration of the original species and the approach of the attacking atom. The reaction itself, the switch of atoms, takes place very rapidly. Then the new molecule shakes but settles down to steady, harmonic vibration as the expelled atom departs.

Although this type of calculation gives a good sense of what is going on during a reaction, its limitations should be understood. In the first place a real gas-phase reaction occurs with a wide variety of different speeds and angles of attack. In the second place the motion of the atoms, electrons, and nuclei, is governed by quantum mechanics. The concept of trajectory then fades and is replaced by the unfolding of a wavefunction that represents initially reactants and finally products. Nevertheless, recognizing these limitations should not be allowed to obscure the fact that recent advances in molecular dynamics have given us a first glimpse of the processes going on at the core of reactions.

Further reading

Reaction kinetics. M. J. Pilling; Clarendon Press, Oxford, 1974.

Gas phase reaction rate theory. H. S. Johnstone; Ronald, New York, 1966.

Activated complex theory: current status, extensions, and applications. R. A. Marcus; in *Techniques of chemistry* (E. S. Lewis, ed.) Vol. VIA, 13, Wiley–Interscience, New York, 1974.

The foundations of chemical kinetics. S. W. Benson; McGraw-Hill, New York, 1960.

Comprehensive chemical kinetics. (Vols. 1–16) C. H. Bamford and C. F. H. Tipper; Elsevier, Amsterdam, 1969–76.

Molecular beams in chemistry. M. A. D. Fluendy and K. P. Lawley; Chapman and Hall, London, 1974.

Molecular beams. J. Ross (ed.); *Adv. chem. Phys.* **10** (1966).

Molecular beams in chemistry. J. E. Jordan, E. A. Mason, and I. Amdur; in

Techniques of chemistry, Vol. IIID, p. 365, Wiley–Interscience, New York, 1972.

Molecular reaction dynamics. R. D. Levine and R. B. Bernstein; Clarendon Press, Oxford, 1974.

The theory of rate processes. S. Glasstone, K. J. Laidler, and H. Eyring; McGraw-Hill, New York, 1941.

Isotope effects on reaction rates. L. Melander; Ronald, New York, 1960.

Problems

28.1. Calculate the collision frequencies z and Z in gases of (a) ammonia, $R \approx 190$ pm, (b) carbon monoxide, $R \approx 180$ pm at 25 °C and 1 atm pressure. What is the percentage increase when the temperature is raised by 10 K at constant volume?

28.2. The Boltzmann distribution gives the number of molecules in the energy range dE at the energy E as proportional to $\exp(-E/kT)dE$. What is the constant of proportionality on the assumption of a uniform density of states? What is the proportion of molecules that have energies of *at least* E_a?

28.3. Collision theory depends on knowing the proportion of molecular collisions having at least the threshold energy E_a. What is this proportion when (a) $E_a = 10$ kJ mol^{-1}, (b) $E_a = 100$ kJ mol^{-1} at 200 K, 300 K, 500 K, 1000 K?

28.4. Calculate the percentage increase in the proportions in the last Problem when the temperature is raised by 10 K.

28.5. In the dimerization of methyl radicals at 25 °C the experimental pre-exponential factor is 2.4×10^{10} dm^3 mol^{-1} s^{-1}. What is (a) the reactive cross-section, (b) the P-factor for the reaction on the basis of a C–H bond length of 154 pm.

28.6. In Problem 27.24 on p. 965 the pre-exponential factor was determined. Estimate the P-factor and the reactive cross-section σ^* for the reaction of oxygen atoms with benzene at 340 K. Take $R(O) \approx R(Ne) \approx 78$ pm and $R(C_6H_6) \approx 265$ pm, Table 25.3.

28.7. Nitrogen dioxide reacts bimolecularly in the gas phase to give $2NO + O_2$. The temperature dependence of the second-order rate coefficient for the rate law $d[\text{Product}]/dt = k_2[NO_2]^2$ is given below. What is the P-factor and the reactive cross-section for the reaction?

T/K	600	700	800	1000
$k_2/\text{cm}^3\,\text{mol}^{-1}\,\text{s}^{-1}$	4.6×10^2	9.7×10^3	1.3×10^5	3.1×10^6

Take $\sigma \approx 0.60$ nm^2.

28.8. The diameter of the methyl radical is about 380 pm. What is the maximum rate coefficient in the expression $d[C_2H_6]/dt = k_2[CH_3]^2$ for second-order recombination of radicals at room temperature?

28.9. 10 per cent of a 1 dm^3 sample of ethane at 25 °C and 1 atm pressure is dissociated into methyl radicals. What is the minimum time for 90 per cent recombination?

28.10. Calculate the magnitude of the diffusion-controlled rate coefficient for a species in (a) water, (b) *n*-pentane, (c) *n*-decylbenzene. The viscosities are 1.00×10^{-3} kg m^{-1} s^{-1}, 0.22×10^{-3} kg m^{-1} s^{-1}, 3.36×10^{-3} kg m^{-1} s^{-1} (1.00 cP, 0.22 cP, 3.36 cP) respectively.

28.11. What is the magnitude of the P-factor in eqn (28.1.7) for ions in water at

25 °C? Evaluate the expression for $1:1$ ions of (a) like charge, (b) opposite charge, and take $R^* \approx 300$ pm.

28.12. One of the most important results derived in this chapter is the *Eyring equation*, eqn (28.2.7), and it is important to become familiar with the way it is applied. As a first step, evaluate the factor kT/h at (a) 0 °C, (b) 25 °C, (c) 1000 °C.

28.13. As a next step, estimate the orders of magnitude of the partition functions involved in the rate expression. State the order of magnitude of τ, q_{rot}, q_{vib}, q_{elec} for typical molecules. Check that in the collision of two structureless molecules the order of magnitude of the pre-exponential factor is of the same order as that predicted by collision theory.

28.14. Using the order-of-magnitude partition functions derived in the last Problem, estimate the P-factor for a reaction in which $A + B \rightarrow P$, and A and B are non-linear triatomic molecules.

28.15. The base-catalysed bromination of nitromethane-d_3 in water at room temperature proceeds 4.3 times more slowly than the bromination of the undeuterated material. Account for this difference. Use $k_f(C-H) \approx 450$ N m^{-1}.

28.16. Predict the order of magnitude of the isotope effect on the relative rates of displacement of (a) ^1H and ^3H (tritium), (b) ^{16}O and ^{18}O. Will raising the temperature enhance the difference? Take $k_f(CH) \approx 450$ N m^{-1}, $k_f(CO) \approx 1750$ N m^{-1}.

28.17. The major difficulty in applying ACT (and, it must be admitted, in devising straightforward problems to illustrate it) is to decide on the structure of the activated complex and to ascribe appropriate bond strengths and lengths to it. The following exercise gives some familiarity with the difficulties involved, yet leads to a numerical result for a reaction of some interest. Consider the attack of H on D_2, which is one step in the $H_2 + D_2$ reaction. Suppose that the H approaches D_2 from the side and forms a complex in the form of an isosceles triangle. Take the H–D distance as 30 per cent greater than in H_2 (74 pm) and the D–D distance as 20 per cent greater than in H_2. Let the critical coordinate be the antisymmetric stretching vibration in which one H–D bond stretches as the other shortens. Let all the vibrations be at about 1000 cm^{-1}. Estimate k_2 for this reaction at 400 K using the experimental activation energy of about 35 kJ mol^{-1}.

28.18. Now change the model of the activated complex in the last Problem so that it is linear. Use the same estimated molecular bond lengths and vibrational frequencies to calculate k_2 for this choice of model.

28.19. Clearly, there is much scope for modifying the parameters of the models of the activated complex in the last pair of Problems. If you have access to a computer, write and run a program that allows you to vary the structure of the complex and the parameters in a plausible way, and look for a model (or more than one model) that gives a value of k_2 close to the experimental value, 4×10^5 dm^3 mol^{-1} s^{-1}.

28.20. What is the entropy of activation for a collision between two essentially structureless particles at 25 °C?

28.21. Evaluate the entropies of activation for the $K + D_2$ reaction using the models described in Problems 27.17 and 27.18.

28.22. The Eyring equation can also be applied to physical processes. As an example, consider the rate of diffusion of an atom stuck to the surface of a solid. Suppose that in order to move from one site to another it has to reach the top of the barrier where it can vibrate classically in the vertical direction and in one horizontal direction, but vibration along the other horizontal direction takes it into the neighbouring site. Find an expression for the rate of diffusion, and evaluate it for tungsten atoms on a tungsten surface ($E_a = 60$ kJ mol^{-1}). Suppose that the vibration frequencies at the transition state are (a) the same as, (b) one half the value for the adsorbed atom. What is the value of the diffusion coefficient D at 500 K? (Take the site separation as 316 pm and $\nu \approx 10^{11}$ Hz.)

28.23. Suppose now that the adsorbed, migrating species is a spherical molecule, and that it can rotate classically as well as vibrate at the top of the barrier, but that at the adsorption site itself it can only vibrate. What effect does this have on the diffusion constant? Take the molecule to be methane. (Take note of one of the powerful aspects of ACT theory: if we accept the *unbiased* approach we predict one value of the rate; but we are also free to build a model with a bias.)

28.24. Electron spin resonance results show that a nitroxyl radical trapped in a solid was rotating at about 1.0×10^8 Hz at 115 K (see Problem 19.23). Use the Eyring equation to estimate the activation Gibbs function for the rotation.

28.25. The pre-exponential factor for the gas-phase decomposition of ozone at low pressures is 4.6×10^{12} dm^3 mol^{-1} s^{-1} and its activation energy is 10.0 kJ mol^{-1}. What is (a) the entropy of activation, (b) the rate coefficient at 25 °C, (c) the Gibbs function of activation?

28.26. What is the connection between $\Delta H_m^{\ddagger}$ and E_a? Show that for a bimolecular gas reaction $E_a = \Delta H_m^{\ddagger} + 2RT$.

28.27. The rates of thermolysis of a variety of *cis*- and *trans*-azoalkanes have been measured over a range of temperatures because of a controversy concerning the mechanism of the reaction. In ethanol an unstable *cis*-azoalkane decomposed at a rate that was followed by observing the nitrogen evolution, and this led to the rate coefficients listed below (P. S. Engel and D. J. Bishop, *J. Amer. chem. Soc.* **97**, 6754 (1975)). Find the enthalpy, entropy, energy, and Gibbs function of activation.

$t/°C$	-24.82	-20.73	-17.02	-13.00	-8.95
$10^4 \times k_1/s^{-1}$	1.22	2.31	4.39	8.50	14.3

28.28. If the activated complex is formed from ions of charges $z'e$ and $z''e$, and there is some characteristic distance $R^{\ddagger}$ between them in the activated complex, then the Gibbs function of activation will contain a term which is proportional to $z'z''/R^{\ddagger}\varepsilon_r$, where ε_r is the relative permittivity of the solvent. Deduce the expression $\ln k_2 = \ln k_2 - z'z''B/\varepsilon_r$, $B = e^2/4\pi\varepsilon_0 R^{\ddagger}kT$, for the dependence of the rate coefficient on ε_r. (Note that this Problem does not involve questions of ionic strength.)

28.29. The model just constructed can be tested on the basis of the following data. Bromophenol blue fades when OH$^-$ is added, the reaction rate being controlled by the step that can be symbolized as $B^{2-} + OH^- \rightarrow BOH^{3-}$. The reaction between an azodicarbonate ion (A^{2-}) and H$^+$ has a rate-determining step that may be symbolized as $A^{2-} + H^+ \rightarrow AH^-$. Both reactions were carried out in solvents of different relative permittivities, and the results are below. Do they support the model and calculation in the last Problem?

Bromophenol blue reaction:

ε_r	60	65	70	75	79
$\lg k_2$	-0.987	0.201	0.751	1.172	1.401

Azodicarbonate ion reaction:

ε_r	27	35	45	55	65	79
$\lg k_2$	12.95	12.22	11.58	11.14	10.73	10.34

28.30. In an experimental study of a bimolecular reaction in aqueous solution, the second-order rate coefficient was measured at 25 °C and at a variety of ionic strengths and the results are tabulated below. It is known that a singly charged ion is involved in the rate-determining step. What is the charge on the other ion involved?

$I/\text{mol kg}^{-1}$	0.0025	0.0037	0.0045	0.0065	0.0085
$k_2/\text{dm}^3 \text{mol}^{-1}\text{s}^{-1}$	1.05	1.12	1.16	1.18	1.26

28.31. Show that the intensities of a molecular beam before and after passing through a chamber of length I containing inert scattering atoms are related by $I(l) = I(0)\exp(-\mathcal{N}_s\sigma l)$ where σ is the collision cross-section and $\mathcal{N}_s$ the number density of scattering atoms.

28.32. In a molecular beam experiment to measure collision cross-sections it was found that the intensity of a CsCl beam was reduced to 60 per cent of its intensity on passage through CH_2F_2 at 10^{-5} mmHg, but that when the target was argon at the same pressure the intensity was reduced only by 10 per cent. What are the relative cross-sections of the two types of collision? Why is one much larger than the other?

28.33. The total cross-sections for reactions between alkali metal atoms and halogen molecules are given in the table below (R. D. Levine and R. B. Bernstein, *Molecular reaction dynamics*, Clarendon Press, Oxford, p. 72 (1974)). Assess the data on the basis of the harpoon mechanism (p. 973).

σ^*/nm^2	Cl_2	Br_2	I_2
Na	1.24	1.16	0.97
K	1.54	1.51	1.27
Rb	1.90	1.97	1.67
Cs	1.96	2.04	1.95

Electron affinities are approximately 1.3 eV (Cl_2), 1.2 eV (Br_2), and 1.7 eV (I_2), and ionization energies are 5.1 eV (Na), 4.3 eV (K), 4.2 eV (Rb), and 3.9 eV (Cs).

29 Processes at solid surfaces

Learning objectives

After careful study of this chapter you should be able to

(1) Describe how crystal surfaces grow (p. 1003).

(2) Explain the importance of a *screw dislocation* (p. 1004).

(3) Explain why *ultra-high vacuum techniques* must be used in order to study clean surfaces (p. 1006).

(4) Describe the *Auger effect* and *Auger spectroscopy* and explain how it is used to study surface composition (p. 1007).

(5) Explain how *photoelectron spectroscopy* is used in surface studies (p. 1007).

(6) Indicate the basis of *L.E.E.D.* techniques and discuss their application to surface studies (p. 1008).

(7) Describe *field emission microscopy* and *field ionization microscopy* (p. 1009).

(8) List the methods for determining the *rate of adsorption* and the *extent of surface coverage* (p. 1012).

(9) Describe the information available from *photoemission* and *electron loss spectroscopy* (p. 1014).

(10) Distinguish between *physisorption* and *chemisorption* (p. 1014).

(11) Define and measure the *desorption activation energy* (p. 1015).

(12) Describe the basis of *flash desorption spectroscopy* and explain how it may be used to study activation energies (p. 1018).

(13) Define *sticking probability* and *fractional coverage* (p. 1021).

(14) Derive and use the *Langmuir adsorption isotherm* (eqn (29.2.6)), the *Temkin isotherm* (eqn (29.2.15)), the *Freundlich isotherm* (eqn (29.2.16)), and the *Brunauer–Emmett–Teller isotherm* (eqn (29.2.12)).

(15) Describe the *Eley–Rideal* and *Langmuir–Hinshelwood mechanisms* of catalysed reactions (p. 1031).

(16) Express the *rate* of a catalysed reaction using the adsorption isotherm (p. 1031).

(17) Explain how *molecular beams* are employed in surface studies (p. 1032).

(18) Describe the processes involved in catalytic *hydrogenation* (p. 1034), *oxidation* (p. 1035), *cracking* (p. 1036), and *reforming* (p. 1036).

Introduction

Processes at surfaces govern most aspects of daily life, including life itself. Even if we limit our attention to solid surfaces, the importance and ubiquity of the processes are hardly reduced. Processes at solid surfaces govern the viability of industry, either constructively, as in the catalysis of chemical processes, or destructively, as in corrosion.

A fundamental dynamical aspect of a surface is its role as a region where a solid grows, evaporates, and dissolves. We therefore begin with a brief look at the growth of solids and the structure of simple surfaces. Then we turn to the way that surfaces are contaminated by the deposition of foreign material. This adsorption is the first step in heterogeneous catalysis, and we shall be examining this industrially important subject. A very special type of system relying on processes at surfaces is an electrode in contact with an electrolyte: that subject is treated in the next chapter.

29.1 The growth and structure of surfaces

In this section we see how surfaces are extended and crystals grow. This will let us picture the structures that are responsible for catalysis. We shall also see how surface properties may be monitored.

How crystals grow. A simple idea of the form of a perfect crystal surface is obtained by picturing it as a cobbled street. A gas atom that collides with the surface can be imagined as a ball bouncing erratically over the cobbles. The atom loses energy as it bounces, but it is likely to escape from the surface before it has lost enough kinetic energy to be trapped. The same applies to an ionic crystal in contact with a solution: there is very little energetic advantage for an ion in solution to shed a number of its solvating molecules and stick in an exposed position on the surface.

The picture changes when the surface has *defects*, for then there are ridges of incomplete layers of atoms or ions. A typical type of surface defect is a *step* between two otherwise flat *terraces*, Fig. 29.1. This step defect might itself be defective: it might have *kinks*.

When an atom settles on a terrace it bounces, and its bounce might lead it to a step or into a corner formed by a kink. Instead of interacting with a single terrace atom it now interacts with several, and the interaction

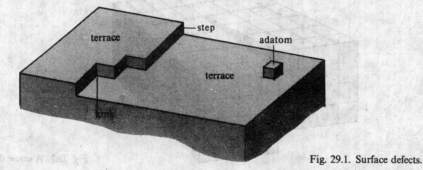

Fig. 29.1. Surface defects.

may be strong enough to stop and trap it. Likewise, in the case of deposition of ions from solution, the loss of the favourable solvation energy is compensated by a greater Coulombic interaction between the arriving ion and several ions in the vicinity of a surface defect.

This discussion shows the necessity of surface defects if deposition and growth are to occur. Yet this alone is not enough. As the process of settling into ledges and kinks on a surface continues, there comes a time when the entire lower terrace has been covered. At this stage the surface defects have been eliminated, and so growth will cease. For continuous growth a type of surface defect is needed that propagates itself as the crystal grows, and which is not eliminated by the process of growth. Such a defect can be found by examining the type of dislocations that exist in the bulk of a crystal.

Dislocations of crystal lattices can take various forms, and contribute to mechanical properties such as ductility and brittleness. They arise when atoms or ions are laid down in a manner that disrupts the regularity of the packing of the lattice. This might happen for a variety of reasons. One possibility is that the crystal is grown so rapidly that its atoms have insufficient time to settle into states of lowest potential energy before being trapped in position by the deposition of the next layer of atoms. Another reason may be that an impurity atom has distorted the lattice in its vicinity.

The lattice distortion called a *screw dislocation* can be depicted as in Fig. 29.2: imagine a cut in the crystal, with the atoms to the left of the cut being pushed up through the distance of a unit cell. The unit cells of the crystal now form a continuous spiral around the end of the cut, the *screw axis*. A path circulating about this axis spirals up to the top of the crystal. There the dislocation breaks through the surface and takes the form of a spiral ramp from the bottom of a step to the top, Fig. 29.2.

The surface defect formed by a screw dislocation is a step, possibly with some kinks, where growth can occur. The incoming atoms lie in ranks on the ramp, and successive ranks reform the step at an angle to its initial position. As deposition continues the step rotates around the screw axis, and is not eliminated. Growth can therefore continue indefinitely.

The whole length of the step might not grow at the same rate, and instead of covering the entire surface with new atoms, the rotating sequence

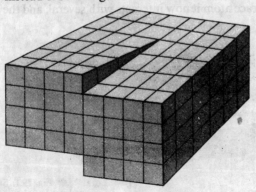

Fig. 29.2. A screw dislocation.

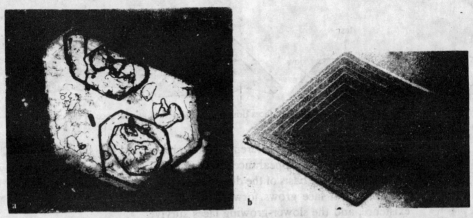

Fig. 29.3. Spiral growth; (a) cadmium iodide (H. M. Rosenberg, *The solid state*, 1975), (b) n-paraffin (B. R. Jennings and V. J. Morris, *Atoms in contact*, Clarendon Press, Oxford, 1974). (Courtesy of H. F. Kay.)

of deposition gradually forms a spiral. Several layers of deposition may occur, and so the edges of the spirals might be cliffs many atoms high. This type of growth of a crystal face into a well-defined spiral terrace is often observed, Fig. 29.3b.

Propagating spiral ledges can also lead to the formation of a series of flat terraces, giving a crystal face resembling a step pyramid, Fig. 29.3a. This comes about if growth occurs at neighbouring left- and right-handed screw dislocations, and Fig. 29.4 shows the building sequence. Deposition at the ledges of the neighbouring defects brings them into collision, but growth continues on the inside of the V-shaped cavity. Growth there propagates the two edges, and both make another complete rotation before colliding again. This process need not occur uniformly over the whole length of the edge, and so successive tables of atoms may decrease in size.

Terraces formed in these ways may fill up by further deposition at their edges, and more or less flat crystal planes can result. The rapidity of

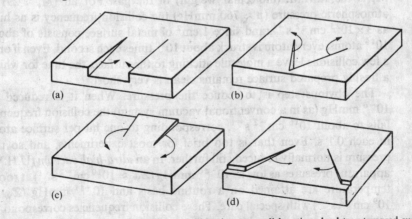

Fig. 29.4. Counter-rotating neighbouring screw dislocations lead to a terraced surface.

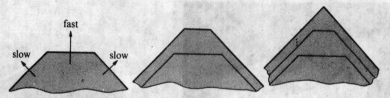

Fig. 29.5. The slower growing faces dominate the external appearance of the crystal.

growth often differs for different crystal planes, and the *slowest* growing faces dominate the appearance of the crystal. This conclusion can be understood on the basis of the diagrams in Fig. 29.5. We see that although the horizontal face grows forward most rapidly, it grows itself out of existence, and the slower-growing faces survive.

Experimental techniques: surface structure. We confine attention to the study of a solid surface in contact with a gas. It is necessary to know the nature of the clean surface before much can be said about its role in reactions, and so the first point is the preparation and characterization of clean surfaces. In this context clean means much more than scrubbing the sample and handling it with care. Under normal conditions a surface is constantly bombarded with gas molecules, and a freshly prepared surface is covered very quickly. Just how quickly can be estimated on the basis of the kinetic theory of gases. In Chapter 25 we derived the following expression for the number of collisions per unit time with unit area of surface when the gas pressure is p:

(29.1.1) $Z_{\mathrm{W}} = p/(2\pi mkT)^{\frac{1}{2}}$.

(This is eqn (25.2.9), on p. 874.) A practical form of the equation is

(29.1.2) $Z_{\mathrm{W}}/\mathrm{cm}^{-2}\,\mathrm{s}^{-1} \approx 3.51 \times 10^{22}\,(p/\mathrm{mmHg})/\{(T/\mathrm{K})M_{\mathrm{r}}\}^{\frac{1}{2}}$

$\approx 2.03 \times 10^{21}\,(p/\mathrm{mmHg})/\sqrt{M_{\mathrm{r}}}$ at 298 K.

M_{r} is the R.M.M. (molecular weight) of the gas. For air ($M_{\mathrm{r}} \approx 29$) at atmospheric pressure ($p \approx 760$ mmHg) the collision frequency is as high as 3×10^{25} cm^{-2} s^{-1}, and since 1 cm^2 of metal surface consists of about 10^{15} atoms, every atom is struck about 10^{10} times each second. Even if only a few collisions leave a molecule sticking to the surface, the time for which a freshly prepared surface remains clean is very short.

The obvious step is to reduce the pressure. When it is reduced to 10^{-6} mmHg (as in a conventional vacuum system) the collision frequency falls to about 10^{16} cm^{-2} s^{-1}, corresponding to one hit per surface atom in each 0.1 s. Even that is too brief for most experiments, and so the pressure is normally reduced still further. In an *ultra-high vacuum* (U.H.V.) apparatus pressures as low as 10^{-10} mmHg ($Z_{\mathrm{W}} \approx 10^{10}$ cm^{-2} s^{-1}) at room temperature are attained on a routine basis and 10^{-11} mmHg ($Z_{\mathrm{W}} \approx 10^{9}$ cm^{-2} s^{-1}) with special care. These collision frequencies correspond to each surface atom being hit once every 10^4–10^6 s, or about once a day.

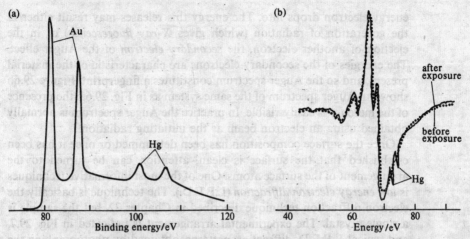

Fig. 29.6. (a) X-ray photoelectron and (b) Auger spectra of gold contaminated with mercury (M. W. Roberts and C. S. McKee).

The layout of a typical U.H.V. apparatus is such that the whole of the evacuated part can be heated to 200–300 °C for several hours in order to drive gas molecules from the walls. All the taps and seals are of metal in order to avoid contamination from greases. The sample is usually in the form of a thin foil, a filament, or a sharp point. Where there is interest in the role of specific crystal planes the sample is in the form of a single crystal with a freshly cleaved face. Initial cleaning of the surface is achieved either by heating electrically or by bombarding the surface with accelerated gaseous ions. The latter demands care because ion bombardment can shatter the surface structure, and leave it an amorphous jumble of atoms. High temperature annealing is then required to return the surface to an ordered state.

The *surface composition* (and, in particular, the detection of any remaining contamination after cleaning, or the detection of layers of material adsorbed later in the experiment) can be determined by a variety of ionization techniques. Their common feature is that the *escape depth* of the electrons is in the range of 0.1–1.0 nm, which ensures that only the surface characteristics are involved.

One technique is photoelectron spectroscopy (Chapter 18). X-p.e.s. seems to be better for the analysis of composition because it is able to fingerprint the materials present; U.V.-p.e.s is more suited to establishing the bonding characteristics and the details of valence shell electronic structures of materials on the surface. Figure 29.6. shows the X-ray photoelectron spectrum of a sample of gold foil on which has been deposited some mercury: the presence of the latter is clearly visible.

The most important technique, now widely used in the microelectronics industry, is *Auger spectroscopy*. The *Auger effect* is the emission of a second electron after high energy radiation has expelled another. Its mechanism is that the first electron leaves a hole in a low-lying orbital which a higher energy electron drops into. The energy this releases may result either in the generation of radiation (which gives *X-ray fluorescence*) or in the ejection of another electron, the *secondary electron* of the Auger effect. The energies of the secondary electrons are characteristic of the material present, and so the Auger spectrum constitutes a fingerprint. Figure 29.6b shows an Auger spectrum of the same system as in Fig. 29.6a: the presence of the mercury is easily visible. In practice the Auger spectrum is normally obtained using an electron beam as the initiating radiation.

Once the surface composition has been determined or once it has been established that the surface is clean, attention can be turned to the arrangement of the surface atoms. One of the most informative techniques is *low energy electron diffraction* (L.E.E.D.). The technique is basically the electron diffraction technique described in Chapter 22, but the sample is a single crystal. The experimental arrangement is indicated in Fig. 29.7, and typical L.E.E.D. diffraction patterns, obtained by photographing the fluorescent screen through the viewing port of the vacuum chamber, are shown in Fig. 29.8. The use of low energy electrons ensures that the principal features of the diffraction pattern relate to the surface characteristics of the sample rather than to its bulk.

The L.E.E.D. pattern represents the two-dimensional characteristics of the surface layers (although by studying how the diffraction intensities depend on the energy of the electrons it is also possible to infer some details about the vertical location of the atoms and to determine the thickness of the layers constituting the surface). The diffraction pattern is sharp if the surface is well-ordered for distances long compared with the wavelength of the incident electrons: in practice this means that sharp patterns are obtained for surfaces ordered for about 20 nm. Diffuse spots indicate either a poorly ordered surface structure or the presence of impurities. If the diffraction spots do not correspond to what can be predicted by extrapolating the bulk structure to the surface, then either a reconstruction of the surface has occurred (such as in LiF, where the Li^+ and F^- ions apparently lie on slightly different planes) or there is order in the arrangement of the adsorbed layer. One outcome of L.E.E.D.

Fig. 29.7. Apparatus for L.E.E.D.

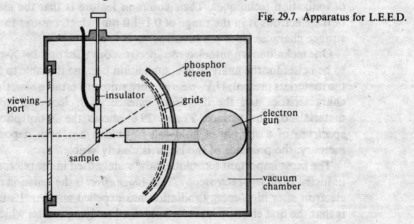

measurements is that they show how important it is not to assume that the surface of a crystal has exactly the same form as a slice through the bulk. As a general rule it is found that metal surfaces are simply truncations of the bulk lattice, although the distance between the top layer of atoms and the one below is contracted by around 5 per cent. Semiconductors generally have surfaces reconstructed to a depth of several layers.

The presence of terraces, steps, and kinks in the surface is shown in L.E.E.D. patterns, and their densities can be estimated. The importance of this will emerge later, when we see that the presence of a step or a kink may be crucial to the functioning of a catalyst. Three examples of the way that steps and kinks affect the L.E.E.D. pattern are shown in Fig. 29.9. (The samples may be prepared by cleaving a crystal at various angles to a plane of atoms: the cleavage has only terraces when the cut is parallel to the plane, and the density of steps increases as the angle of the cut increases.) The fact that in the stepped samples the L.E.E.D. patterns acquire additional structure (rather than merely blurring) indicates that the steps are arrayed regularly.

By far the most spectacular results are obtained from two closely related techniques. In *field emission microscopy* (f.e.m.) the sample is in the form of a filament etched to form a sharp tip. This is enclosed in an evacuated vessel equipped with a fluorescent screen. When a large potential difference is applied the electrons in the sample tend to be stripped out towards the screen. An electron emerging from the sample is accelerated from tip to screen, and gives a flash of light where it strikes. The ease with which electrons can escape from the metal depends on its surface structure (and, in particular, on the variation of the work function, p. 374, with the nature of the surface), and so the screen shows a variation of intensity corresponding to the variation of the surface. A typical result is shown in Fig. 29.10. The change in this pattern when material has been deposited can be used to detect the places where atoms are most likely to stick.

A development of f.e.m. is the *field ionization microscope* (f.i.m.). The apparatus is virtually the same, but the potential difference is reversed, with the fluorescent screen made negative relative to the tip. In the

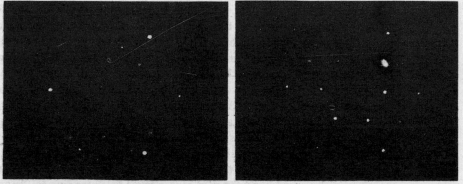

Fig. 29.8. The L.E.E.D. from (a) a clean Pt surface and (b) after exposure to propyne (G. A. Samorjai).

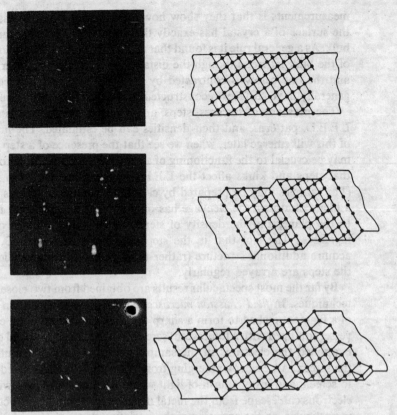

Fig. 29.9. L.E.E.D. patterns for a Pt surface with (a) low defect density, (b) regular steps spaced by about 6 atoms, (c) regular steps with kinks (G. A. Samorjai).

experiment a little gas, such as helium, is admitted. It strikes the tip and bounces over its cobbled, terraced, surface until it strikes a protruding atom, such as one of the rim of a ledge, Fig. 29.11. Protruding atoms are able to ionize the gas atom, and immediately the positive ion (e.g., He$^+$) is formed, the potential difference drags it off towards the screen, where its collision generates fluorescence.

The resolution of the f.i.m. technique is dependent on the transverse motion of the gas ions, and this can be reduced by cooling the tip to about 20 K, when the resolution is of the order of atomic dimensions. This means that the positions of individual atoms can be resolved, and Fig. 29.12 is an example: the small, bright spots are caused by individual atoms in the terraces of the tip. This remarkable picture shows the power of the f.i.m. technique, but we should not forget its limitations. In the first place ionization appears unequally at different atoms, and many atoms on the surface, even some at edges, are insufficiently exposed to give any ionization, and so do not lead to spots on the fluorescent screen. Furthermore, the sample has to be in the form of a tip, and be made of a material strong enough to withstand the very high electric fields tending to distort its structure. Despite these limitations the technique is remarkable for its

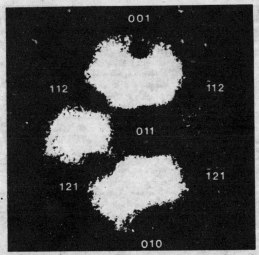

Fig. 29.10. A field emission photograph of a tungsten tip of radius 210 nm (M. Prutton, *Surface physics*, Clarendon Press, Oxford (1975)).

ability to portray the positions of individual atoms.

An elegant refinement of f.i.m. lets one determine the identity of individual atoms stuck to the surface of an otherwise clean tip. *Atom-probe f.i.m.* is the ultimate in surgery. The f.i.m. image of an adsorbed atom is brought into coincidence with a hole in the fluorescent screen. The imaging gas is eliminated, and a pulse of potential difference plucks off the atom (as an ion). It moves in the same direction as did the gas ions, and passes through the hole in the screen. Behind the screen there stands a mass spectrometer, and so the identity of the atom can be determined by determining its deflection in electrostatic and magnetic fields. Apart from being the ultimate analytical technique (as well as knowing exactly where the atom was in the sample, we also know what it is) events can be observed which on at atomic scale are really dramatic. For example, the analysing pulse lasts for about 2 ns, and during that time the evaporation of about ten atomic layers is sometimes observed. This corresponds to a rate of evaporation equivalent to the surface receding at about 1 m s^{-1}.

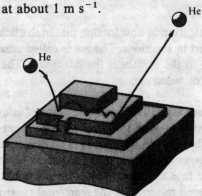

Fig. 29.11. Events leading to a f.i.m. image of a surface.

Fig. 29.12. A field ionization image of an iridium tip (E. W. Müller).

29.2 Adsorption at surfaces

We now turn to the investigation of how foreign materials stick to surfaces (*adsorb*) and then take part in reactions. The act is called *adsorption*. The material that adsorbs on to the surface is the *adsorbate*. The underlying material is the *adsorbent* or *substrate*.

The experimental analysis of the surface layer. The two basic items of information are the *extent of surface coverage* and the *rate of adsorption*. These are interdependent, because the rate can be inferred from observations of how the surface coverage changes with time, and the extent of coverage can be found if we know the rate by integration with respect to time. The former is normally expressed in terms of the *fractional coverage θ*,

$$\theta = \frac{\text{number of adsorption sites filled}}{\text{number of adsorption sites available}}.$$

Among the principal techniques are the following:

(1) *Flow methods.* The sample acts as a pump because adsorption removes molecules from the gas. One commonly used technique is therefore to monitor the rates of flow of the gas into and out of the system: the difference is the rate of uptake by the sample. Integration of this rate then gives the extent of coverage at any stage.

(2) *Thermal desorption.* In *flash desorption* the sample is suddenly heated (electrically) and the consequent rise of pressure in the system is interpreted in terms of the amount originally on the sample. The interpretation of the data is sometimes confused by the desorption of a compound (for instance WO_3 from tungsten covered with adsorbed oxygen).

(3) *Gravimetry.* The extent of adsorption can be followed directly by weighing the sample (on a microbalance) during the course of the experiment.

(4) *Radioactive tracer techniques.* The extent of adsorption can be measured by monitoring the radioactivity of the surface when it is exposed to an isotopically labelled gas.

While many important studies have been carried out simply by exposing a surface to a gas, modern work is increasingly making use of molecular beam techniques. One advantage is that the activity of specific crystal planes can be examined by directing the incoming beam on to an oriented surface with known structure and step and kink density (from L.E.E.D. investigations). Furthermore, if the sample undergoes a reaction of the surface, the products and their angular distribution may be analysed as they are ejected from the surface and pass into a mass spectrometer. Another advantage is that the time of flight of a molecule may be measured and interpreted in terms of the *residence time* of the molecule on the surface. In this way a very detailed picture can be constructed of the events involved in the interaction between molecules and surfaces and the energy characteristics of heterogeneous catalysis.

Apart from extent, rate, and crystallographic specificity, the other experimental information needed is the state of the adsorbed molecules and the nature of their bonds to the surface. One technique is photoelectron spectroscopy (especially U.V.-p.e.s., p. 622). In surface studies this is normally referred to as *photoemission spectroscopy*. Its particular strength lies in its ability to reveal which bonds are involved in the adsorption of the species. For instance, the principal difference between the photoelectron spectra of free benzene and benzene adsorbed on a palladium surface is in the energies of the π-orbitals: this is interpreted by regarding the molecule as lying parallel to the surface, and attached to it by its π-bonds. In contrast, pyridine is known to stand more or less perpendicular to the surface, and the attachment is by means of the σ-bond formed from the nitrogen lone pair.

Varieties of vibrational spectroscopy also reveal the nature of the adsorbed species, and in particular whether dissociation has occurred. Infrared and Raman spectroscopy have been improved by the development of laser and Fourier transform techniques, but still suffer from their low intensities on account of the low surface coverages normally encountered under laboratory conditions. Nevertheless, under conditions typical of industrial catalytic operations (high pressures and high coverages), infrared spectroscopy is one of the few viable techniques. Half way between photoemission spectroscopy and vibrational spectroscopy lies *electron loss spectroscopy*, in which the energy loss suffered by a beam of electrons when they are reflected from a surface is monitored. As in optical Raman spectroscopy, the spectrum of energy loss can be interpreted in terms of the vibrational spectrum of the adsorbed species. High resolution and sensitivity is attainable. As an example, Fig. 29.13 shows the electron loss spectrum of CO on a surface of a Pt crystal (the (111) plane) as the extent of coverage is increased. The major peak corresponds to CO attached perpendicular to the surface by a single Pt atom. As the coverage increases the smaller peak next to it grows in intensity. This peak corresponds to

CO at a *bridge site*, where it is attached to two Pt atoms, $\begin{matrix} M \\ M \end{matrix} \! \! > \! \! C \! \! = \! \! O$.

Physisorption and chemisorption. Molecules can stick to surfaces in two ways. In *physisorption*, which is a shortening of *physical adsorption*, there is a van der Waals interaction (for instance, dispersion or polar interactions)

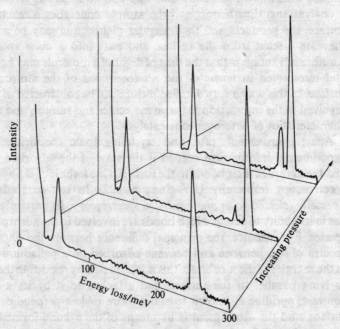

Fig. 29.13. Electron loss spectrum of CO on Pt(III) (H. Ibach).

between the surface and the adsorbed molecule (p. 782). This is a long-range but weak interaction and the amount of energy released when a molecule is physisorbed is of the order of the enthalpy of condensation. This energy can be absorbed as vibrations of the lattice and dissipated as heat, and a molecule bouncing across the cobbled surface will lose its kinetic energy and stick to the surface (this is the process of *accommodation*). The heat evolved in the physisorption can be determined by measuring the rise in temperature of the filament or foil, and enthalpy changes of about $20 \, kJ \, mol^{-1}$ are often observed (Table 29.1). These energies are insufficient to lead to bond breaking, and so in physisorption the molecule retains its identity, although it might be stretched or bent on account of the proximity of the surface.

Table 29.1. Enthalpies of physical adsorption, $-\Delta H_m/kJ \, mol^{-1}$

Maximum observed values:

H_2	84	CO	25	CH_4	21
O_2	21	CO_2	29	C_2H_2	38
N_2	21	H_2O	57	C_2H_4	34
Cl_2	36	NH_3	38		

Source: D. O. Hayward and B. M. W. Trapnell, *Chemisorption*, Butterworths.

A physisorbed molecule vibrates in its shallow potential well, and since the binding energy is low it may shake itself off the surface. This suggests that a molecule remains on the surface for only a short time before returning to the gas. The rate of departure can be expected to follow an Arrhenius-type law with a rate coefficient given by

$$k_{desorption} \approx A \exp(-E_a/RT),$$

and $1/k_{desorption}$ can be identified as the mean lifetime τ of a molecule on the surface. Then

(29.2.1) $$\tau \approx \tau_0 \exp(E_a/RT)$$

where $\tau_0 = 1/A$. Substituting $E_a \approx 25 \, kJ \, mol^{-1}$, the energy that must be acquired from the lattice in order to desorb, and guessing that the pre-exponential factor A is of the order of a vibrational frequency of the weak molecule–surface bond (of the order of $10^{12} \, s^{-1}$), lifetimes of about 10^{-8} s are predicted at room temperature. Lifetimes of the order of 1 s are obtained only when the temperature is lowered to about 100 K.

In *chemisorption*, which is a shortening of *chemical adsorption*, the molecules stick to the surface as the result of the formation of a chemical, and usually a covalent, bond and tend to find sites that maximize their coordination number with the substrate. The energy of attachment is very much greater than in physisorption, and typical values are in the region of $200 \, kJ \, mol^{-1}$ (measured, as before, by monitoring the rise in temperature of a sample of known heat capacity). A molecule undergoing chemisorption may be torn apart at the demand of the unsatisfied valencies of the surface atoms, and so it may lose its identity. The existence of molecular fragments on the surface as a result of the chemisorption of whole molecules is one of the reasons why surfaces can exhibit catalytic activity, and therefore the reason why the remainder of this section concentrates on chemisorption.

Chemisorbed species. Except in special cases, chemisorption must be exothermic if it is to be spontaneous. The argument behind this statement runs as follows. A spontaneous process requires a negative ΔG. Since there is a reduction in translational freedom when a species is adsorbed, ΔS is negative. Therefore, ΔH must be negative if $\Delta G = \Delta H - T\Delta S$ is to be negative, and a negative ΔH corresponds to an exothermic process. An exception may occur when the adsorbate dissociates and has high translational mobility on the surface. Thus hydrogen adsorbs endo-thermically ($\Delta H > 0$) on glass because there is a large increase of translational entropy accompanying the dissociation of the molecules, and this increase is not completely lost because the adsorbed atoms have high translational mobility. The net change of entropy for the process $\frac{1}{2}H_2(g) \rightarrow$ H(glass) is positive and $T\Delta S$ dominates the small positive ΔH.

A principal test to distinguish between chemisorption and physisorption was formerly the magnitude of the enthalpy of adsorption: ΔH_m for physisorption is rarely more negative than about -25 kJ mol^{-1}, while ΔH_m for chemisorption is usually more negative, and sometimes much more negative than -40 kJ mol^{-1}, Table 29.2. More direct methods are now available, because photoemission and vibrational spectroscopy can be used to determine the nature and state of the adsorbate.

Table 29.2. Enthalpies of chemical adsorption, $-\Delta H_m/\text{kJ mol}^{-1}$

Adsorbate	Ti	Ta	Nb	W	Cr	Mo	Mn	Fe	Co	Ni	Rh	Pt
H_2		188			188	167	71	134		117		
O_2						720					494	293
N_2		586						293				
CO	640							192	176			
CO_2	682	703	552	456	339	372	222	225	146	184		
NH_3				301				188		155		
C_2H_4		577		427	427			285		243	209	

Source: D. O. Hayward and B. M. W. Trapnell, *Chemisorption*, Butterworths.

The enthalpy of adsorption may **change** with the extent of surface coverage. A principal reason is the **interaction** between adsorbed molecules: if they repel (as for CO on Pd) the enthalpy of adsorption becomes less negative (less exothermic) as coverage increases. Furthermore, L.E.E.D. studies show that such species settle on the surface in a disordered way until the requirements of packing result in order. If the adsorbate molecules attract each other (as for O_2 on W) they tend to cluster into islands with growth occurring at the borders. These adsorbates also show order-disorder transitions when they are heated enough for thermal motion to overcome the attractive adsorbate–adsorbate interactions but not so much that they are desorbed.

The dependence of the energy of a molecule on its distance from the substrate surface is depicted in Fig. 29.14. As the molecule approaches the surface its energy drops as it becomes physisorbed (this physisorbed state is called the *precursor state* of chemisorption). Dissociation into fragments often takes place as the molecule moves into its chemisorbed state, and the energy of the molecule rises as the bonds stretch, and then decreases sharply as it falls into its chemisorbed state and the surface-adsorbate bonds reach their full strength. Even if the molecule does not fragment there is still likely to be a rise in energy as the surface atom adjusts its bonds with its neighbours in order to respond to the incoming molecule. In all cases, therefore, we can expect there to be an energy barrier separating the precursor physisorbed state from the chemisorbed state. This barrier, though, need not be very great, and it might not rise above the energy of the free molecule, Fig. 29.14a. In this case chemisorption from the gas is not an activated process. It can therefore be expected to proceed quickly, and many gas adsorptions on clean metal surfaces do in fact appear to be non-activated. In some cases the barrier between the precursor and final states may rise above the energy of the free molecule, in which case the process has an activation energy, Fig. 29.14b and it generally proceeds more slowly than non-activated chemisorption. An example is H_2 on Cu, which has an activation energy somewhere in the region of 20 to 40 kJ mol^{-1}.

One point that emerges from this discussion is that rates are not good criteria for distinguishing between physisorption and chemisorption. Chemisorption can be fast if the activation energy is small or zero, but it may be slow if the activation energy is large. Physisorption is usually fast,

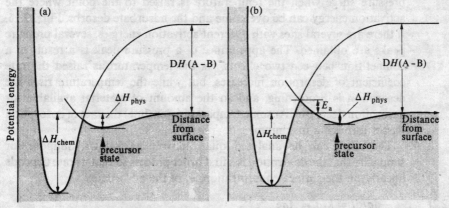

Fig. 29.14. Potential energy profiles for chemisorption. ΔH_{phys} is the enthalpy of physisorption, ΔH_{chem} that of chemisorption, and E_a the activation energy for chemisorption. (a) Zero adsorption activation energy, (b) non-zero adsorption activation energy.

but it can appear to be slow if adsorption is taking place on the surface of a porous medium.

The *desorption* of a chemisorbed species is always an activated process because the species has to be elevated from the foot of a well. Equation (29.2.1) can be used to estimate the residence time since now $E_a \approx$ 100 kJ mol^{-1} and guessing that $\tau_0 \approx 10^{-14}$ s (because the chemisorbed fragment has a stiffer bond to the surface than the physisorbed species) gives a residence time of as long as 3×10^3 s (about an hour) at room temperature, decreasing to 1 s only when the temperature has been raised to 370 K.

The activation energy for desorption can be measured in a variety of ways. The importance of knowing it lies not only in its role in discussions of the temperature dependence of catalytic processes, but also because it is a measure of the strength of the adsorbate–surface bond. In particular, if the chemisorption is non-activated, and if there is no dissociation, then the activation energy for desorption can be interpreted in terms of the adsorption enthalpy. The activation energy of desorption, however, is not as well defined as in homogeneous reactions because it may depend on the extent of surface coverage (for we have already seen that this is often true of the adsorption enthalpy), and therefore it may change during the course of the desorption. Furthermore, the transfer of the concepts of reaction order and the rate coefficient from bulk studies to surfaces is hazardous, and there are few examples of strictly first-order or second-order desorption kinetics.

If we disregard these complications it is clear that one way of measuring a desorption activation energy is to monitor the rate of pressure rise in the system when the surface is maintained at a series of temperatures, and then to attempt an Arrhenius plot of the rate coefficient. A more sophisticated technique is to determine the *flash desorption spectrum* of the sample. The basic observation is that in a pumped vessel there is a pressure surge when the temperature is raised to the point where the activation energy can be overcome and the adsorbate desorbs, Fig. 29.15. If there are several sites with different activation energies, several pressure peaks are obtained. The appearance of a pressure peak is a result of a competition between two effects. As the temperature is raised the rate coefficient of desorption increases, but while the temperature rises the desorption is proceeding, and so the amount of material available for desorption decreases. At some temperature (or time) the rate of desorption passes through a maximum.

The effect can be analysed quantitatively as follows. Suppose for simplicity that the desorption is a first order process so that its rate depends linearly on the surface concentration, σ, of the adsorbate:

(29.2.2) $-\mathrm{d}\sigma/\mathrm{d}t = k_1\sigma = A\sigma e^{-E_a/RT}.$

Furthermore, arrange for the temperature to rise with time according to

$T = T_i + \kappa t,$

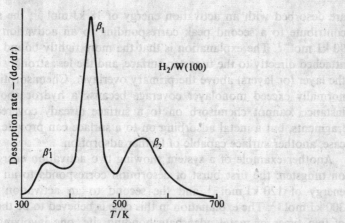

Fig. 29.15. Flash desorption spectrum of H_2 on W(100) (P. W. Tamm and L. D. Schmidt).

where κ depends on the rate at which the temperature is made to rise. Then the temperature, T^*, at which the rate goes through a maximum is the solution of

$$(d/dT)(-d\sigma/dt) = 0.$$

Since $dT = \kappa dt$, and using eqn (29.2.2), this condition can be developed as follows:

$$(d/dT)(-d\sigma/dt) = A(d\sigma/dT)e^{-E_a/RT} + (E_a A\sigma/RT^2)e^{-E_a/RT}$$

$$= (A/\kappa)k_1\sigma e^{-E_a/RT} + (E_a A\sigma/RT^2)e^{-E_a/RT}$$

$$= -(A^2/\kappa)\sigma e^{-2E_a/RT} + (E_a A\sigma/RT^2)e^{-E_a/RT}$$

$$= 0 \text{ at } T = T^*.$$

The temperature at which the pulse occurs is therefore the solution of

(29.2.3) $\quad (A/\kappa)e^{-E_a/RT^*} = E_a/RT^{*2}.$

This expression can be turned into a practical form by rearranging it into

(29.2.4) $\quad \ln(T^{*2}/\kappa) = \ln(E_a/RA) + E_a/RT^*.$

Therefore, in order to determine E_a, a series of different heating rates are taken and the temperature of desorption surge is determined. A plot of $\ln(T^{*2}/\kappa)$ against $1/T^*$ gives E_a from its slope and then A from its intercept at $1/T^* = 0$. A similar analysis can be applied to desorptions that are second-order in the amount of adsorbate.

In many cases only a single activation energy (and a single peak in the flash desorption spectrum) is observed. In some, though, the observation of several peaks indicates the presence of a variety of adsorption sites with different desorption activation energies. In some cases these correspond to adsorption on different crystal planes, although they may also indicate multilayer adsorption. For instance, cadmium on tungsten shows two activation energies. One site corresponds to loosely bound Cd atoms: they are desorbed with an activation energy of 18 kJ mol^{-1}; the second set contribute to a second peak corresponding to an activation energy of 90 kJ mol^{-1}. The explanation is that the more tightly bound atoms are attached directly to the tungsten surface, and the less strongly bound are the layer (or layers) above the primary overlayer. Chemisorption cannot normally exceed monolayer coverage because a hydrocarbon gas, for instance, cannot chemisorb on to a surface already covered with its fragments, but a metal adsorbing on to a surface can provide, as in this case, another surface capable of further adsorption.

Another example of a system showing two activation energies is CO on tungsten: the first burst of desorption corresponds to an activation energy of 120 kJ mol^{-1} and the second to an activation energy of 300 kJ mol^{-1}. The explanation in this case is believed to be the existence of two types of metal–adsorbate bonding site, one involving a simple M–CO bond and the other adsorption with dissociation (into individually adsorbed C and O atoms).

In some cases two desorption peaks may occur even though there is only one type of binding site. This complication arises when there are significant interactions between the adsorbate molecules so that at low surface coverages the enthalpy of adsorption is significantly different from its value at high surface coverages. The desorption spectrum then has a low temperature pulse corresponding to the removal of closely packed species, and a second pulse when the now-separated remaining adsorbate molecules are desorbed.

A further aspect of the strength of interaction between adsorbate and substrate is the former's mobility. The migration of species is often crucial to the function of catalysts, for a catalyst may be impotent if the reactant molecules adsorbed so strongly that they were unable to migrate over the surface. The activation energy for diffusion over a surface need not be the same as for desorption because the adsorbate can pass through valleys between potential energy peaks without completely leaving the surface. In general it turns out that the activation energy for migration is about 10–20 per cent of the energy of the surface–adsorbate bond, but it depends on the extent of surface coverage. Furthermore, the defect structure of the sample may play a dominant role because the adsorbed species might find it easier to skip across a terrace than to roll along the foot of a step, and it might get trapped in vacancies on an otherwise flat terrace. Diffusion may also be easier across one crystal face than another, and so surface mobility may depend on which lattice planes are exposed.

There are two very elegant methods for determining the diffusion characteristics of an adsorbate, and both make use of the f.i.m. technique. One involves depositing an adsorbate of a surface plane at low temperature, taking an f.i.m. image before and after the temperature is raised, and monitoring the migration of the boundary as the adsorbate floods across crystal surfaces at different rates. In a modification of the technique an individual atom is imaged, the temperature is raised, and then lowered after a definite interval. A new image is taken and the position of the

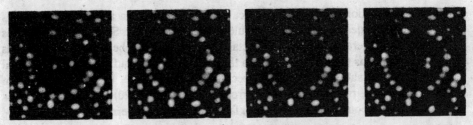

Fig. 29.16. FIM micrograph showing the migration of rhenium atoms on rhenium during 3 s intervals at 375 K (G. Ehrlich).

migrated atom is determined, Fig. 29.16. A sequence of pictures reveals that the atom makes a random walk across the surface, and the diffusion coefficient D can be inferred from the mean distance d travelled in the interval τ through the two-dimensional random walk expression $d \approx (D\tau)^{\frac{1}{2}}$. The value of D for different crystal planes at different temperatures can be determined directly in this way, and the activation energy for migration over each plane obtained from the Arrhenius type expression

(29.2.5) $$D = D_0 \exp(-E_a/RT).$$

Typical values for tungsten atoms on tungsten have E_a in the range 57–87 kJ mol^{-1} and $D_0 \approx 3.8 \times 10^{-7}$ cm^2 s^{-1}; for CO on tungsten E_a the activation energy is 144 kJ mol^{-1} at low surface coverage, but it drops to 88 kJ mol^{-1} when the coverage is high and the interactions between adsorbed molecules are significant.

The extent of adsorption: adsorption isotherms. The rate at which a surface is covered with adsorbate depends on a number of factors. One, as we have seen, is the activation energy; but as this is generally small or zero it rarely plays a dominant role. Another factor is the frequency of collisions: but this is very great under normal conditions, and becomes slow on the time-scale of normal experiments only when U.H.V. conditions are attained. The dominant factor can be introduced by the following argument. When a molecule crashes on to the surface it is trapped only if it is able to dissipate its energy into the thermal vibrations of the underlying lattice and do this sufficiently quickly. Otherwise it is simply reflected, or bounces over the surface until it reaches an edge where it returns to the gas phase. The proportion of collisions with the surface that lead to adsorption is called the *sticking probability, s*:

$$s = \frac{\text{rate of adsorption of molecules by the surface}}{\text{rate of collision of molecules with the surface}}.$$

Since the denominator of this expression can be calculated from kinetic theory once the pressure is known, and since the numerator can be measured by observing the rate of change of pressure, the sticking probability can be determined quite readily. Values of s vary widely. For instance, at room temperature carbon monoxide has s in the range 0.1–1.0 for several transition metal surfaces, but nitrogen has $s < 10^{-2}$ for adsorption on rhenium, indicating that more than a hundred collisions are needed before one molecule sticks successfully. Sticking probabilities are generally low on non-metal surfaces, but can be very low on metals too (thus O_2 on Ag has s less than 10^{-4}). Beam studies of specific crystal planes show pronounced specificity: for instance, for N_2 on tungsten sticking probabilities range from 0.74 (on the (320) faces) down to less than about 0.01 (on the (110) faces) at room temperature.

Example Calculate the time for 10 per cent of the sites on a (100) tungsten surface to be covered with nitrogen at 298 K when the pressure is 2.0×10^{-9} mmHg and the sticking probability is 0.55.

- *Method.* Begin by calculating the number of atoms per unit area of the (100) surface; the lattice constant of the b.c.c. unit cell is 316 pm. Then find Z_w from eqn (29.1.2).

- *Answer.* The distance between centres of atoms in the (100) face is 316 pm. Therefore each atom accounts for $(316)^2$ pm^2 of the surface. It follows that the number of sites per square metre is $1 \text{ m}^2/(316 \times 10^{-12} \text{ m})^2 = 1.00 \times 10^{19}$. The collision frequency is given by eqn (29.1.2):

$$Z_w/\text{cm}^{-2}\,\text{s}^{-1} \approx (2.03 \times 10^{21}) \times (2.0 \times 10^{-9})/\sqrt{28.02} = 7.67 \times 10^{11}.$$

so that $Z_w \approx 7.67 \times 10^{15} \text{ m}^{-2}\,\text{s}^{-1}$. Therefore, the time to cover 10 per cent of 1.00×10^{19} sites m^{-2} is

$$t = (1.00 \times 10^{18} \text{ m}^{-2})/(7.67 \times 10^{15} \text{ m}^{-2}\,\text{s}^{-1}) \times (0.55) = 237 \text{ s.}$$

- *Comment.* Covering the entire surface will take longer than 2370 s because the sticking probability falls with increasing coverage.

The sticking probability depends on how much of the surface is uncovered and as the surface sites get filled it drops to smaller values, Fig. 29.17. A simple model of surface effects would be to assume that the sticking probability is proportional to $(1-\theta)$, the fraction uncovered, and it is common to write $s = s_0(1-\theta)$, where s_0 is the sticking probability on a perfectly clean surface. The results in Fig. 29.17 do not fit this simple expression because they show that s remains close to s_0 until the coverage has risen to about 6×10^{13} molecules cm^{-2}, and then drops sharply to low values. The explanation of this behaviour is probably that the colliding molecule does not enter the chemisorbed state at once, but bounces over the surface until it comes into the vicinity of an empty site. The results in Fig. 29.17 also show the high selectivity of the sticking probability: adsorption is extremely slow on the (110) face.

The question that now arises is how the fractional coverage depends on the pressure of the gas above the surface. There is an equilibrium between molecules on the surface and molecules free in the gas phase, and the coverage depends on the pressure of the system. The dependence of θ on the pressure at a set temperature is called the *adsorption isotherm* of the system.

The simplest type of isotherm is based on the view that every adsorption site is equivalent, and the ability of the molecule to bind there is

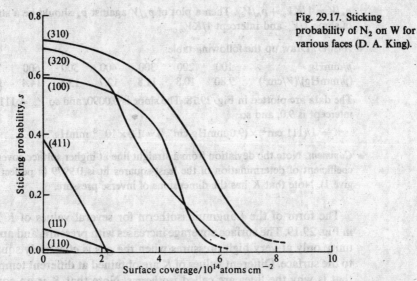

Fig. 29.17. Sticking probability of N_2 on W for various faces (D. A. King).

independent of whether or not the neighbouring sites are occupied. We suppose that the adsorbed molecules are in dynamic equilibrium with the free molecules, and write the rate coefficients for adsorption and desorption as k_a and k_d respectively:

$$A(g) + M(\text{surface}) \underset{k_d}{\overset{k_a}{\rightleftharpoons}} AM.$$

The rate of adsorption is proportional to the pressure of A, and the number of vacant sites on the surface, $N(1-\theta)$, where N is the total number of sites. Therefore

$$\text{rate of adsorption} = k_a p_A N(1-\theta).$$

The rate of desorption is proportional to the number of adsorbed species, $N\theta$:

$$\text{rate of desorption} = k_d N\theta.$$

At equilibrium the two rates are equal, and solving for θ gives the

(29.2.6)   *Langmuir isotherm:* $\theta = Kp_A/(1+Kp_A)$,

where $K = k_a/k_d$.

Example (Objective 14). The data below are for the adsorption of CO on charcoal at 273 K. Confirm that they fit the Langmuir isotherm, and find the constant K and the volume corresponding to complete surface coverage.

p/mmHg	100	200	300	400	500	600	700
V/cm^3	10.2	18.6	25.5	31.4	36.9	41.6	46.1

Mass of sample of charcoal: 3.022 g; in each case V has been corrected to 1 atm.

● *Method.* From eqn (29.2.6), $Kp_A\theta + \theta = Kp_A$. Write $\theta = V/V_\infty$, where V_∞ is the volume corresponding to complete coverage. The equation then becomes

$p_A/V = 1/KV_\infty + p_A/V_\infty$. Then a plot of p_A/V against p_A should be a straight line of slope $1/V_\infty$ and intercept $1/KV_\infty$.

- **Answer.** Draw up the following table:

p/mmHg	100	200	300	400	500	600	700
$(p/\text{mmHg})/(V/\text{cm}^3)$	9.80	10.8	11.8	12.7	13.6	14.4	15.2

The data are plotted in Fig. 29.18. The slope is 0.0090, and so $V_\infty = 111$ cm^3. The intercept is 9.0, and so

$$K = 1/(111 \text{ cm}^3) \times (9.0 \text{ mmHg cm}^{-3}) = 1.0 \times 10^{-3} \text{ mmHg}^{-1}.$$

- **Comment.** Note the deviation from a straight line at higher surface coverages. The coefficient of determination of the least-squares fit is 0.9979 (a perfect fit would give 1). Note that K has the dimensions of inverse pressure.

The form of the Langmuir isotherm for several values of K is shown in Fig. 29.19. The surface coverage increases with pressure, and approaches unity only at very high pressures when the gas is effectively squashed on to the surface. Different values of K are obtained at different temperatures: that is why the lines are called *isotherms*. Note that K is an equilibrium constant for the distribution of material between the surface and the gas phase. This indicates that its temperature dependence can be used to determine the *isosteric enthalpy of adsorption* (ΔH for a particular surface coverage) through the thermodynamic equation (p. 189)

(29.2.7) $(\partial \ln p/\partial T)_\theta = -\Delta H_m^{\text{des}}/RT^2 = \Delta H_m^{\text{abs}}/RT^2.$

Example The data below show the pressures of CO required for the volume of adsorption to be 10.0 cm^3 at each temperature (all volumes corrected to 1 atm and 273 K). The same sample was used as in the *Example* above. Find the enthalpy of adsorption at this surface coverage.

T/K	200	210	220	230	240	250
p/mmHg	30.0	37.1	45.2	54.0	63.5	73.9

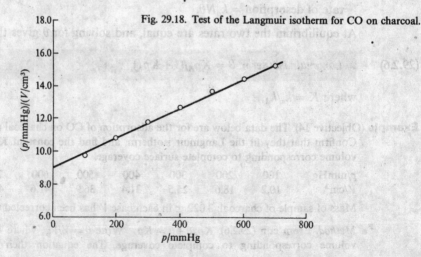

Fig. 29.18. Test of the Langmuir isotherm for CO on charcoal.

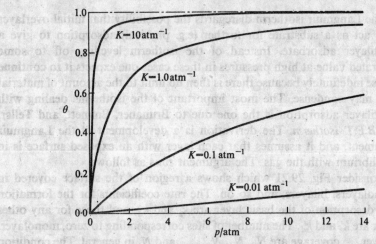

Fig. 29.19. The Langmuir isotherm, eqn (29.2.3), for several values of K.

● *Method.* From eqn (29.2.7) we have to plot $\ln p$ against $1/T$: the slope of the straight line is $\Delta H_m/R$.

● *Answer.* Draw up the following table:

T/K	200	210	220	230	240	250
$1000/(T/K)$	5.00	4.76	4.55	4.35	4.17	4.00
$\ln(p/\text{mmHg})$	3.40	3.61	3.81	3.99	4.15	4.30

These are plotted in Fig. 29.20. The slope of the least-squares fitted line is -904 and so $\Delta H_m = -904 \times (8.314 \text{ J K}^{-1} \text{ mol}^{-1}) = 7.5 \text{ kJ mol}^{-1}$.

● *Comment.* The value of K can be used to obtain a value of ΔG and then combined with the value of ΔH to obtain an entropy of adsorption, but there are some tricky features of interpretation.

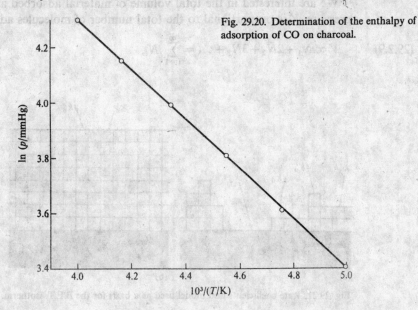

Fig. 29.20. Determination of the enthalpy of adsorption of CO on charcoal.

The Langmuir isotherm disregards the possibility that initial overlayer may act as a substrate for further (e.g. physical) adsorption to give a multilayer adsorbate. Instead of the isotherm levelling off to some saturated value at high pressures in these cases one expects it to continue to rise indefinitely because there is then no limit to the amount of material that may condense. The most important of the isotherms dealing with multilayer adsorption is the one due to Brunauer, Emmett, and Teller, the *B.E.T. isotherm*. The derivation is a development of the Langmuir argument, and it assumes that each layer with an exposed surface is in equilibrium with the gas. The argument runs as follows.

Consider Fig. 29.21 which shows a region of the surface covered in monolayers, bilayers, and so on. The rate coefficients for the formation and desorption of the basic layer are k_a and k_d, and those for any other layer are k_a' and k_d'. The number of sites corresponding to zero, monolayer, bilayer ... coverage are $N_0, N_1, N_2, \ldots$, and N_i in general. The condition for equilibrium of the initial layer is the equality of the rates of its adsorption and desorption: $k_a N_0 p = k_d N_1$. The condition for the next layer is $k_a' N_1 p = k_d' N_2$, and in general

$$k_a' N_{i-1} p = k_d' N_i, \qquad i = 2, 3, \ldots$$

The last expression may be expressed in terms of N_0 as follows:

$$N_i = (k_a'/k_d') p N_{i-1} = (k_a'/k_d')^2 p^2 N_{i-2} = \ldots = (k_a'/k_d')^{i-1} p^{i-1} N_1$$

$$= (k_a'/k_d')^{i-1}(k_a/k_d) p^i N_0.$$

Now write $k_a'/k_d' = x$ and $k_a/k_d = cx$; then with this simpler notation

(29.2.8) $$N_i = c(xp)^i N_0.$$

We are interested in the total volume of material adsorbed at a given pressure. V is proportional to the total number of molecules adsorbed:

(29.2.9) $$V \propto N_1 + 2N_2 + 3N_3 + \ldots = \sum_{i=1}^{\infty} i N_i,$$

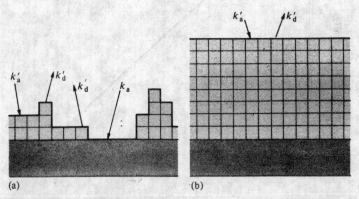

(a)   (b)

Fig. 29.21. Rate coefficients and model used as a basis for the B.E.T. isotherm.

because a monolayer site contributes one molecule, a bilayer contributes two, and so on, Fig. 29.21a. If there were complete monolayer coverage of the same sample of substrate the volume of adsorbate would be V_{mon}, where

(29.2.10) $$V_{mon} \propto N_0 + N_1 + N_2 + \ldots = \sum_{i=0}^{\infty} N_i$$

because each site in the actual sample then contributes only a single molecule to the total. It follows that

$$V/V_{mon} = \left\{ \sum_{i=1}^{\infty} iN_i \right\} \Big/ \left\{ \sum_{i=1}^{\infty} N_0 \right\}$$

$$= cN_0 \left\{ \sum_{i=1}^{\infty} i(xp)^i \right\} \Big/ \left\{ N_0 + cN_0 \sum_{i=1}^{\infty} (xp)^i \right\}.$$

Both sums can be evaluated using

$$\sum_{i=1}^{\infty} y^i = \{1/(1-y)\} - 1, \qquad \sum_{i=1}^{\infty} iy^i = y/(1-y)^2$$

with $y = xp$, and after a little rearrangement we arrive at

(29.2.11) $$V/V_{mon} = pxc/(1-xp)(1-xp+cxp).$$

The final step is to equate $1/x$ with the vapour pressure of the bulk liquid adsorbate, p^*. In order to see that this is so, consider the equilibrium between the gas and a sample in which the substrate surface is deeply and uniformly buried, Fig. 29.21b. The condition for equilibrium is now

$$k_a' Np = k_d' N, \quad \text{or} \quad k_a' p = k_d',$$

where N is the total number of sites. Such an equilibrium applies at the surface of the bulk liquid irrespective of whether or not a surface is deeply buried under it, and so p, the equilibrium pressure, can be identified with p^*, the bulk vapour pressure. It follows from the last equation that $x = k_a'/k_d' = 1/p^*$, as we set out to demonstrate. When this result is introduced into eqn (29.2.11) we obtain the

(29.2.12) $$\text{B.E.T. isotherm: } V/V_{mon} = c(p/p^*)/\{(1-p/p^*)[1-(1-c)(p/p^*)]\}.$$

This is often reorganized into

(29.2.13) $$z/(1-z)V = 1/cV_{mon} + (c-1)z/cV_{mon}, \qquad z = p/p^*.$$

cV_{mon} can therefore be obtained from the slope of a plot of the l.h.s. against z and $(c-1)/cV_{mon}$ can be found from the intercept; these results may be solved for c and V_{mon}. The form of the B.E.T. isotherm is drawn in Fig. 29.22: it reproduces the type of behaviour expected for multilayer adsorption. When the coefficient c is large the isotherm takes the simpler form

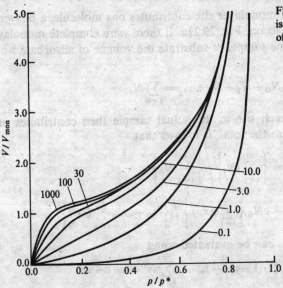

Fig. 29.22. The B.E.T. isotherms for various values of c.

(29.2.14) $V/V_{mon} \approx 1/(1 - p/p^*)$.

This is applicable to inert gases on polar surfaces, for then $c \approx 100$. The B.E.T. isotherm fits experimental observations moderately well over restricted pressure ranges, but it errs by underestimating the extent of adsorption at low pressures and overestimating it at high pressures.

Example (Objective 14). The data below relate to the adsorption of nitrogen on rutile at 75 K. Confirm that they fit a B.E.T. isotherm in the range of pressures involved, and find the constants V_{mon} and c.

p/mmHg	1.17	14.00	45.82	87.53	127.7	164.4	204.7
V/cm³	600.6	719.54	821.77	934.68	1045.75	1146.39	1254.14

$p^*(N_2) = 570$ mmHg at 75 K. The volumes have been corrected to 0 °C and 1 atm and refer to the volume of gas adsorbed per gram of substrate.

- *Method.* Use eqn (29.2.13), plotting $z/(1-z)V$ against z, finding $1/cV_{mon}$ from the intercept at $z = 0$ and $(c-1)/cV_{mon}$ from the slope. Use a least-squares procedure to analyse the data.

- *Answer.* Draw up the following table:

p/mmHg	1.17	14.00	45.82	87.53	127.7	164.4	204.7
z	0.0021	0.0246	0.0804	0.154	0.224	0.288	0.359
$\{z/(1-z)(V/\text{cm}^3)\}/10^{-4}$	0.0343	0.350	1.06	1.94	2.76	3.54	4.47

The points are plotted in Fig. 29.23. The intercept lies at 0.034, and so $1/cV_{mon} = 0.034 \times 10^{-4}$ cm⁻³. The slope of the line is 12.3, and so $(c-1)/cV_{mon} = 12.3 \times 10^{-4}$ cm⁻³. Solving these two equations gives $c-1 = 362$, or $c = 363$, and $V_{mon} = 810$ cm³ per gram of substrate.

- *Comment.* At 0 °C and 1 atm, 810 cm³ corresponds to 0.036 mol, or 2.18×10^{22} molecules. Since each molecule occupies an area of 0.16 nm², the surface area of the sample is 3488 m² per gram.

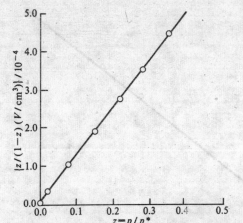

Fig. 29.23. B.E.T. isotherm plot.

Another assumption underlying the derivation of the Langmuir isotherm is the independence and equivalence of the adsorption sites. Intuition suggests that there is likely to be a variety of kinds of sites for adsorption, and it is likely that the energy of adsorption at any site depends on whether or not its neighbours are already occupied. Deviations from the Langmuir isotherm are widely observed, and the discrepancy can often be traced to the failure of these assumptions. For example, the enthalpy of adsorption often gets less negative as θ increases, and this points to the initial occupation of the energetically most favourable sites. Various attempts have been made to propose isotherms that take this variation into account. The

(29.2.15)   *Temkin isotherm:* $\theta = c_1 \ln(c_2 K p_A)$,

where c_1 and c_2 are constants, corresponds to supposing that the adsorption enthalpy changes linearly with the pressure, and the

(29.2.16)   *Freundlich isotherm:* $\theta = c_1 p_A^{1/c_2}$

corresponds to a logarithmic change. Different isotherms reproduce the experimental behaviour more or less well over restricted ranges of pressure, but they remain largely empirical. Empirical, however, does not mean useless; for if the parameters of a reasonably reliable isotherm are known, quite reliable results can be obtained for the extent of surface coverage under various conditions. This kind of information is vital for any discussion of heterogeneous catalysis.

Example (Objective 14). Examine whether the Freundlich isotherm is a better representation than the Langmuir isotherm for the data on CO adsorption, at 273 K.

- *Method.* If the Freundlich isotherm is applicable, a plot of $\ln V$ against $\ln p$ should give a straight line. Compare coefficients of determination of the least-squares fits.

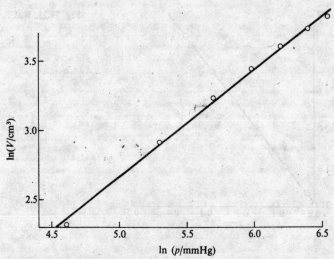

Fig. 29.24. Test of the Freundlich isotherm for CO on charcoal.

● *Answer*. Draw up the following table:

p/mmHg·	100	200	300	400	500	600	700
$\ln(p/\text{mmHg})$	4.61	5.30	5.70	5.99	6.21	6.40	6.55
$\ln(V/\text{cm}^3)$	2.32	2.92	3.24	3.45	3.61	3.73	3.83

The data are plotted in Fig. 29.24. The coefficient of determination of the least-squares fit to a straight line is 0.9968; this is further from unity than the coefficient for the Langmuir isotherm (0.9979) and so in this case the latter is the better representation over the pressure range considered.

● *Comment*. The Freundlich isotherm is often used in discussions of adsorption from fluid solutions. Then one writes $w = c_1 c^{1/c_2}$, where w is the mass of solute adsorbed per unit mass of adsorbent and c is the solute's concentration. This application is examined in the Problems.

29.3 Catalytic activity at surfaces

The reason for heterogeneous catalytic activity is the same as for homogeneous catalysis: the catalyst speeds the reaction by lowering the activation energy of the rate-determining step. Therefore, although it does not disturb the thermodynamically determined equilibrium composition of the reaction system, it increases the rate at which it is attained, Table 29.3.

Catalysis. The basis of much catalytic activity of surfaces is that chemisorption (and sometimes physisorption) organizes at least one of the reactant molecules into a form in which it can readily undergo reaction. Often this comes about because the chemisorption is accompanied by fragmentation. In that case the molecular fragments can be plucked off the surface by an incoming molecule or skip across the surface until they encounter some other fragment.

When a reaction proceeds by a gas phase species colliding with a

Table 29.3. Activation energies for catalysed reactions

Reaction	Catalyst	$E_a/\text{kJ mol}^{-1}$
$2HI \rightarrow H_2 + I_2$	None	184
	Au	105
	Pt	59
$2NH_3 \rightarrow N_2 + 3H_2$	None	350
	W	162
$2N_2O \rightarrow 2N_2 + O_2$	None	245
	Au	121
	Pt	134
$(C_2H_5)_2O$ pyrolysis	None	224
	I_2 (g)	144

Source: G. C. Bond, *Heterogeneous catalysis*, Oxford.

molecular fragment on the surface, the rate of formation of product is expected to be proportional to the pressure of the non-adsorbed gas, p_B, and to the extent of surface coverage by the adsorbed gas A, θ_A. This is the basis of the *Eley–Rideal mechanism*. The rate law is therefore expected to be

(29.3.1) $A + B \rightarrow \text{products}, \qquad -dp_A/dt = k_S p_A \theta_B,$

with k_S the rate coefficient. k_S might be much greater than for uncatalysed gas phase reactions because the catalyst has reduced the activation energy of the reaction (and the adsorption step itself is often non-activated). If the adsorption isotherm for the species B is known, then the rate law can be expressed in terms of its pressure in the gas, p_B. For instance, if B adsorbs according to the Langmuir isotherm in the range of pressure of interest, the overall rate law is

(29.3.2) $-dp_A/dt = k_S K p_A p_B/(1 + K p_A).$

When the pressure of A is high there is almost complete surface coverage, and the rate is approximately equal to $k_S p_B$ because the rate determining step is then the rate of collision of B with the adsorbed fragments. When the pressure of A has decreased, perhaps as a consequence of the reaction, $K p_A$ is small and the rate is equal to $k_S K p_A p_B$, and the extent of surface coverage becomes crucial.

Example (Objective 16). The decomposition of phosphine, PH_3, on tungsten is first order at low pressure and zeroth order at high pressures. Account for these observations.

- *Method.* The reaction is $PH_3 \rightarrow$ products. The rate is proportional to θ, the extent of surface coverage by PH_3. Substitute the Langmuir isotherm for θ and take the low and high pressure limits.

- *Answer.* The rate equation is

$-dp/dt = k_c \theta = k_c K p/(1 + K p),$

where p is the pressure of the phosphine.

When $Kp \ll 1$ (at low pressures), $-dp/dt \approx k_c Kp$. This is a first-order rate law, as required. The time dependence of the phosphine pressure is $p(0)\exp(-k_c Kt)$.

When $Kp \gg 1$ (at high pressure), $-dp/dt \approx k_c Kp/Kp = k_c$. Then the rate of reaction is independent of the phosphine pressure, and this corresponds to a zeroth-order reaction.

● *Comment.* Quite a few heterogeneous reactions are first order. In the Problems we shall see how to deal with reaction rates where the products adsorb and inhibit the action of the catalyst.

Many catalysed reactions, including olefin hydrogenation, involve collisions between fragments adsorbed on the surface, and so can be expected to be second-order in the extent of surface coverage. The rates of these *Langmuir–Hinshelwood* types of reaction follow expressions of the form

(29.3.3) $A + B \rightarrow products, \qquad -dp_A/dt = k_s \theta_A \theta_B.$

Insertion of the appropriate isotherms for the species then gives the rate in terms of their pressures. For instance, if both species follow a Langmuir isotherm in the pressure range of interest, then

$$\theta_A = K_A p_A/(1 + K_A p_A + K_B p_B), \qquad \theta_B = K_B p_B/(1 + K_A p_A + K_B p_B)$$

and the rate law is

(29.3.4) $-dp_A/dt = k_s K_A K_B p_A p_B/(1 + K_A p_A + K_B p_B)^2.$

(The integration of this rate law is taken up in Problem 29.22.) As the coefficients in the isotherms and the rate coefficient k_s are all temperature dependent, the overall temperature dependence of the rate may be strongly non-Arrhenius.

Molecular beam studies are able to give detailed information about mechanisms of reactions and their rates. In particular it has become possible to investigate how the catalytic activity of the surface depends on its structure as well as its chemical composition. For instance, the cleavage of C–H and H–H bonds appears to depend on the presence of steps and kinks, and a terrace often has only minimal catalytic activity. The reaction $H_2 + D_2 \rightarrow 2HD$ has been studied in some detail, and whereas terrace sites have no detectable activity, about one incident molecule in ten reacts when it strikes a stepped surface. While it may be the case that the step itself is the crucial feature, another explanation may be that the presence of a step merely exposes a more reactive crystal face (the face of the step). Likewise the dehydrogenation of cyclohexene to benzene on platinum appears to depend strongly on the density of steps in the exposed surface, but only weakly on the density of kinks in the steps. The hydrogenolysis of cyclohexane to n-hexane depends strongly on the kink density. Thus it appears that kinks are necessary if the aim is to cleave C–C bonds. These observations suggest a reason why even small amounts of impurities may poison catalysts: impurity species are likely to be attached at step and kink sites, and so the activity of the entire catalyst may be impaired if these crucial sites are blocked. A constructive aspect is that the extent of hydrogenolysis may be controlled relative to other types of reactions by seeking impurities which adsorb at kinks and act as specific poisons.

Examples of catalytic processes. Almost the whole of modern chemical industry depends on the development, selection, and application of catalysts, and all we can attempt to do in this section is to give an indication of the breadth of application rather than going into the details of technological applications, Table 29.4.

Table 29.4. Catalysts and their applications

Catalyst	Function	Example
Metals	Hydrogenation	Fe, Ni, Pt, Ag
	Dehydrogenation	
Semiconducting oxides and sulphides	Oxidation	NiO, ZnO, MgO
	Dehydrogenation	Bi_2O_3/MoO_3
	Desulphurization	
Insulating oxides	Dehydration	Al_2O_3, SiO_2, MgO
Acids	Polymerization	H_3PO_4, H_2SO_4
	Isomerization	SiO_2/Al_2O_3
	Cracking	
	Alkylation	

Source: G. C. Bond, *Heterogeneous catalysis*, Oxford.

The choice of a catalyst depends on the job to be done, the danger of it being poisoned by by-products or impurities in the reaction mixture, and economic considerations relating to its cost and lifetime.

The activity of a catalyst depends on the strength of chemisorption in a way that can be represented by a 'volcano' curve, Fig. 29.25. In order to have significant activity the catalyst should be extensively covered by the adsorbate, and this requires moderately strong adsorption. On the other hand, if the adsorption strength increases beyond what is required for significant amount of coverage, the strength of the adsorption bonds is so great that the catalytic activity declines either because the incoming species cannot react or because the adsorbed molecules are immobilized on the surface. This indicates that the activity increases with strength of adsorption (as measured, for example, by the enthalpy of adsorption), and then falls: the greatest activity is for catalysts lying in regions close to the mouth of the volcano.

Many metals are suitable for adsorbing gases, and the general sequence of adsorption strengths is O_2, C_2H_2, C_2H_4, CO, H_2, CO_2, N_2 with oxygen usually the most strongly adsorbing molecule, and nitrogen the least. Some of these molecules adsorb dissociatively (such as hydrogen). Transition metals such as iron, vanadium, and chromium show strong activity for all these gases, but metals like manganese and copper are unable to adsorb

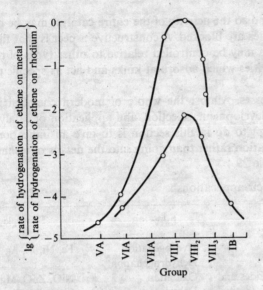

Fig. 29.25. A volcano curve of catalytic activity (G. C. Bond, *Heterogeneous catalysis*, Clarendon Press, Oxford, 1974). Lower curve: first row transition metals; upper curve: second and third row transition metals.

nitrogen and carbon dioxide, and adsorb hydrogen only weakly. Metals towards the left of the periodic table (such as magnesium and lithium), are able to adsorb only the most active gas, oxygen. These trends are summarized in Table 29.5.

Table 29.5. Chemisorption abilities

	O_2	C_2H_2	C_2H_4	CO	H_2	CO_2	N_2
Ti, Cr, Mo, Fe	+	+	+	+	+	+	+
Ni, Co	+	+	+	+	+	+	−
Pd, Pt	+	+	+	+	+	−	−
Mn, Cu	+	+	+	+	±	−	−
Al, Au	+	+	+	+	−	−	−
Li, Na, K	+	+	−	−	−	−	−
Mg, Ag, Zn, Pb	+	−	−	−	−	−	−

+ indicates strong chemisorption, ± weak chemisorption, − no chemisorption
Source: G. C. Bond, *Heterogeneous catalysis*, Oxford.

Hydrogenation. An example of catalytic action is the hydrogenation of olefins. In such cases the olefin adsorbs by forming two bonds from neighbouring carbon atoms:

$$CH_3CH_2CH{:}CH_2 \rightarrow CH_3CH_2$$

$$\begin{array}{c} \diagdown \\ CH{-}CH_2 \\ \diagup\diagup\diagup\diagup\diagup\diagup\diagup\diagup\diagup \end{array}$$

On the same metal surface there may be adsorbed hydrogen atoms, and in due course a collision occurs and one of the surface bonds will break:

$$CH_3CH_2 \quad CH_3CH_2CH_2 \quad CH_3CH_2 \quad CH_3$$
$$CH-CH_2+H \qquad CH_2 \text{ or } \qquad CH$$

Then another hydrogen atom moves to the neighbouring site, forms a bond with the attached molecule, and so releases the fully hydrogenated hydrocarbon, the thermodynamically favoured species.

The evidence for a two-step reaction is the observation of different isomeric olefins in the reaction mixture. This comes about because while the hydrocarbon chain is waving about over the surface of the metal, it might chemisorb again:

$$CH_3CH_2 \qquad CH_3CH_2 \quad CH_3 \quad H \quad CH_3 \qquad CH_3$$
$$CH-CH_2+H \qquad CH \qquad + \qquad CH-CH$$

This, when it desorbs, gives $CH_3CH:CHCH_3$. The new olefin would not be formed if the two hydrogen atoms reacted simultaneously.

A major industrial application of catalysed hydrogenation reactions is to the formation of edible fats from vegetable and animal oils. Raw oils obtained from sources such as the soya bean have the structure $CH(O_2CR_1)\cdot CH(O_2CR_2)\cdot CH(O_2CR_3)$, where R_1, R_2, and R_3 are long chain hydrocarbons with several double bonds. One disadvantage of the presence of many double bonds is that they make the oils susceptible to atmospheric oxidation, and so they turn rancid. The geometrical configuration of chains containing rigid double bonds is responsible for the liquid nature of the oil, and in many applications (such as on sandwiches) a solid fat is at least much better, and often necessary. Controlled, partial hydrogenation of an oil with a catalyst carefully selected so that hydrogenation is incomplete and so that the chains do not isomerize, is used on a wide scale to produce edible fats. The process, and the industry, is not made any easier by the seasonal variation of the number of double bonds in the oils.

Oxidation. Catalytic oxidation is also widely used in industry, and increasingly in pollution control. Although in some instances it is desirable to achieve complete oxidation, as in the elimination of nitrogen oxides from engine emissions, in others incomplete oxidation is the aim. For example, the complete oxidation of propene to carbon dioxide and water is wasteful, but its partial oxidation to acrolein ($CH_2:CHCHO$) is the start of important industrial processes. Likewise, the controlled oxidations of ethene to ethanol, ethanal (acetaldehyde), and, in the presence of acetic acid or chlorine, to vinyl acetate ($CH_2:CHCOOCH_3$) or vinyl chloride ($CH_2:CHCl$), are the initial stages of really important chemical industries.

Not all the reactions mentioned in the last paragraph proceed under the influence of simple metal catalysts, although platinum in exhaust emission control and palladium in the formation of vinyl acetate are two exceptions. A number depend on solid oxides of various kinds, and the physical chemistry of these complex surfaces is often very obscure. Some of the problems can be indicated by considering the oxidation of propene to acrolein. This is brought about by a bismuth molybdate surface which acts by adsorbing the propene with loss of a hydrogen atom to form the allyl radical CH_2CHCH_2. The oxygen in the surface can transfer to the allyl, leading to the formation of acrolein and its desorption from the surface. The hydrogen atom also escapes with a surface oxygen, with the formation of water. The surface is left with the charges of the oxide ions, and these are centres of attack for oxygen molecules in the surrounding atmosphere. These oxygens chemisorb on to the surface as oxide ions, and the catalyst is thereby reformed. This type of reaction obviously involves quite dramatic upheavals of the surface, and some materials break up under the stress of the conditions.

Cracking and reforming. Many of the small organic molecules used in the preparation of all kinds of chemical products, from pharmaceuticals to explosives and from perfumes to polymers, come from oil. Not only are chemicals constructed by catalytic oxidation, hydrogenation, chlorination, and polymerization, but the small building blocks are themselves cut from the long chain hydrocarbons squeezed out of the earth. The catalytically induced destruction of hydrocarbon chains to smaller, more volatile fragments, is called *cracking*, and has been brought about on silica-alumina catalysts. These operate by forming carbonium ions, which are unstable, fall apart, and isomerize to a more highly branched chain form. These branched chain molecules burn more smoothly and efficiently in internal combustion engines than their linear isomers, and are the basis of the higher 'octane' fuels.

Catalytic cracking has been largely superseded by *catalytic reforming* using a *dual-function catalyst*, such as a mixture of platinum and a treated alumina. The former is the *metal function*, and brings about dehydrogenation and hydrogenation. The latter is the *acidic function*, being able to form a carbonium ion. The sequence of events at this catalyst shows very clearly the complications that have to be unravelled if a reaction as important as this is to be fully understood, and the understanding used to develop an even better catalyst.

The first step is the attachment of the long chain hydrocarbon by chemisorption to the platinum: in the process first one, and then a second hydrogen atom is lost, and the olefin is formed. This olefin then migrates to an acidic site, where it accepts a proton, and attaches to the surface as a carbonium ion. This carbonium ion can undergo a variety of reactions. It can break in two, isomerize into a more highly branched structure (branched chain carbonium ions are more stable than straight chain ions), or undergo various types of ring closure. Then it loses a proton, escapes from the surface, and migrates as an olefin (possibly through the gas) to a metallic part of the catalyst where it is hydrogenated. We end up with a rich selection of smaller molecules which can be withdrawn, separated by fractionation, and then used as raw materials in other processes.

Further reading

Catalysis by metals. G. C. Bond; Academic Press, New York, 1962.

Chemisorption. D. O. Hayward and B. M. W. Trapnell; Butterworths, London, 1964.

Physical chemistry of surfaces. A. W. Adamson; Wiley-Interscience, New York, 1967.

Principles of surface chemistry. G. Somorjai; Prentice-Hall, Englewood Cliffs, N.J., 1972.

On physical adsorption. S. Ross and J. P. Oliver; Interscience, New York, 1964.

Chemistry of the metal–gas interface. M. W. Roberts and C. S. McKee; Clarendon Press, Oxford, 1978.

The solid state. H. M. Rosenberg; Oxford University Press, 1975.

Surface physics. M. Prutton; Oxford University Press, 1975.

Heterogeneous catalysis: principles and applications. G. C. Bond; Clarendon Press, Oxford, 1974.

Mechanism in heterogeneous catalysis. M. Boudart and R. L. Burwell; in *Techniques of chemistry* (E. S. Lewis, ed.), Vol. VI(A), 693, Wiley-Interscience, New York, 1974.

Introduction to the principles of heterogeneous catalysis. J. M. Thomas and W. J. Thomas; Academic Press, New York, 1967.

Problems

29.1.   Calculate the frequency of molecular collisions per cm^2 of surface in a vessel containing (a) hydrogen, (b) propane at $25\,°C$ when the pressure is (i) 1 mmHg, (ii) 10^{-7} mmHg.

29.2.   Knowing the frequency of collisions per unit area is not particularly informative unless we can translate it into collisions per atom on the surface. That can be done by drawing on some of the material of Chapter 22. Nickel is face-centred cubic with a unit cell of side 352 pm. What is the number of atoms per cm^2 exposed on a surface formed by (a) (100), (b) (110), (c) (111) planes?

29.3.   Tungsten crystallizes into a body-centred cubic form with a unit cell of side 316 pm. Calculate the number of atoms exposed per cm^2 on the same three surfaces as in the last Problem. In filaments of tungsten (110) and (100) planes predominate: estimate the average number of atoms exposed per unit area.

29.4.   Translate the frequencies in Problem 29.1 into frequencies per atom for the surfaces encountered in the last two Problems.

29.5.   For how long will a hydrogen atom remain on a surface at 298 K if its desorption activation energy is (a) 15 kJ mol^{-1}, (b) 150 kJ mol^{-1}? Take $\tau_0 = 10^{-13}$ s.

29.6.   For how long will the same atoms remain at 1000 K?

29.7.   What will be the corresponding times in the last two Problems if hydrogen is replaced by deuterium?

29.8.   The time for which an oxygen atom remains adsorbed to a tungsten surface is 0.36 s at 2548 K and 3.49 s at 2362 K. Find the activation energy for desorption. What is the pre-exponential factor for these tightly chemisorbed atoms?

29.9. The chemisorption of hydrogen on manganese is activated, but only weakly so. Careful measurements have shown that it proceeds 35 per cent faster at 1000 K than at 600 K. What is the activation energy for chemisorption?

29.10. The deposition of atoms and ions on a surface depends on their ability to stick, and therefore on the energy changes that occur. As an illustration, consider a two-dimensional square lattice of univalent positive and negative ions separated by 200 pm, and consider a cation approaching the upper terrace of this array from the top of the page. Calculate, by direct summation, its Coulombic interaction when it is in an empty lattice point directly above an anion. Now consider a high step in the same lattice, and let the approaching ion go into the corner formed by the step and the terrace. Calculate the Coulombic energy for this position, and decide on the likely settling point for a deposited cation.

29.11. The Langmuir isotherm is a moderately good representation of the adsorption behaviour of gases on a variety of surfaces, and it has the virtue of considerable simplicity. Quite a few of the Problems in this chapter will be based on it and they will give familiarity with the kind of calculations that can be done. As a first step, show that if a diatomic gas adsorbs as atoms the Langmuir isotherm becomes $\theta = \sqrt{(Kp)}/[1+\sqrt{(Kp)}]$.

29.12. How do we test the non-dissociative Langmuir isotherm? One way is to plot p/θ against p, but θ might not always be known. Show that a plot of p/V_a, where V_a is the volume of gas adsorbed (corrected here, and throughout the following, to a standard temperature and pressure) against p should give a straight line from which both V_a^0, the volume corresponding to complete coverage, and the constant K may be determined. Also show that for small surface coverages another test is to plot $\ln(\theta/p)$ against θ, when a line of gradient -1 should be obtained. What should the slope be when $\ln V_a/p$ is plotted against V_a at small coverages?

29.13. The data below are for the chemisorption of hydrogen on copper powder at 25 °C. Confirm that they fit the Langmuir isotherm. Then find the value of K for the adsorption equilibrium and the adsorption volume corresponding to complete coverage.

p/mmHg	0.19	0.97	1.90	4.05	7.50	11.95
V_a/cm^3	0.042	0.163	0.221	0.321	0.411	0.471

29.14. Suppose it is known that ozone adsorbs on a particular surface in accord with a Langmuir isotherm. How could you use the pressure dependence of the fractional coverage to distinguish between adsorption (a) without dissociation, (b) with dissociation into $O+O_2$, (c) with dissociation into $O+O+O$?

29.15. The data for the adsorption of ammonia on barium fluoride are reported below. Confirm that they fit a B.E.T. isotherm and find values of c and V_{mon}.

$T = 0 °C$, $p^* = 3222$ mmHg

p/mmHg	105	282	492	594	620	755	798
V/cm^3	11.1	13.5	14.9	16.0	15.5	17.3	16.5

$T = 18.6 °C$, $p^* = 6148$ mmHg

p/mmHg	39.5	62.7	108	219	466	555	601	765
V/cm^3	9.2	9.8	10.3	11.3	12.9	13.1	13.4	14.1

29.16. The *Freundlich isotherm* can be written $V_a = c_1 p^{1/c_2}$, where c_1 and c_2 are constants. The following data were obtained for the adsorption of methane on 10 g of carbon black at 0 °C. Which isotherm, the Langmuir or the Freundlich, fits the data better?

p/mmHg	100	200	300	400
V_a/cm^3	97.5	144	182	214

29.17. Carbon monoxide adsorbs on mica, and the data for 90 K are given below. Decide whether the Langmuir or the Freundlich isotherm is a better representation of the system. What is the value of K? Given that the total sample area is 6.2×10^3 cm^2, calculate the area occupied by each adsorbed molecule.

p/mmHg	100	200	300	400	500	600
V_a/cm^3	0.130	0.150	0.162	0.166	0.175	0.180

29.18. What volume of carbon monoxide would be adsorbed by the mica at 90 K when the pressure is 1 atm?

29.19. The designers of a new industrial plant wanted to use a catalyst code-named CR-1 in a step involving the fluorination of butadiene. As a first step in the investigation they determined the form of the adsorption isotherm. The volume of butadiene adsorbed per gram of CR-1 at 15 °C depended on the pressure as given below. Is the Langmuir isotherm suitable at this pressure?

p/mmHg	100	200	300	400	500	600
V_a/cm^3	17.9	33.0	47.0	60.8	75.3	91.3

29.20. Investigate whether the B.E.T. isotherm gives a better description of the adsorption of butadiene on CR-1. At 15 °C p^* (butadiene) = 1500 mmHg. Find V_{mon} and c.

29.21. Confirm the equation on p. 1032 for the extent of surface coverage in the presence of two adsorbing gases.

29.22. Integrate the rate law in eqn (29.3.4) in the case where $p_A(0) = p_B(0)$. Sketch the time-dependence of the pressure of the product in the case of $K_A \approx K_B \approx 1$ atm^{-1} and $p_A(0) \approx p_B(0) \approx 1$ atm.

29.23. Fluorine adsorbs on CR-2 in accord with the Langmuir isotherm, but butadiene adsorbs in accord with the Freundlich isotherm with $c_2 = 2$. Deduce an expression for the reaction rate on the basis that fluorination depends on encounters of F and butadiene on the catalyst surface. Assume that the adsorption of the two species occurs at different types of site.

29.24. Nitrogen gas adsorbed on charcoal to the extent of 0.921 cm^3 g^{-1} at a pressure of 4.8. atm and a temperature of 190 K, but at 250 K the same amount of adsorption was achieved only when the pressure was increased to 32 atm. What is the molar enthalpy of adsorption of nitrogen on charcoal?

29.25. In an experiment on the adsorption of oxygen on tungsten it was found that the same volume of oxygen was desorbed in 27 min at 1856 K, 2 min at 1978 K, and 0.3 min at 2070 K. What is the activation energy of desorption? How long would it take for the same amount to desorb at (a) 298 K, (b) 3000 K?

29.26. Ammonia was introduced into a bulb at a pressure 200 mmHg. At 856 °C it was found that a tungsten catalyst brought about a pressure change of 59 mmHg in 500 s, and 112 mmHg in 1000 s. What is the order of the catalysed decomposition? Account for the result.

29.27. In some catalytic reactions the species formed may adsorb more strongly than the reacting gas. This is the case, for instance, in the catalytic decomposition of ammonia on platinum at 1000 °C. As a first step in examining the kinetics of this type of process, show that the rate of ammonia decomposition should follow $-dp(NH_3)/dt = k_c p(NH_3)/p(H_2)$ in the limit of very strong adsorption of hydrogen.

Start by showing that when a gas J adsorbs very strongly, and its pressure is $p(J)$, that the fraction of uncovered sites is approximately $1/Kp(J)$.

29.28. Solve the rate equation for the catalytic decomposition of ammonia on platinum and show that a plot of $F(t) = (1/t)\ln(p/p_0)$ against $G(t) = (p-p_0)/t$ where p is the pressure of ammonia, should give a straight line from which k_c can be determined. Check the rate law on the basis of the data below, and find k_c for the reaction.

t/s	0	30	60	100	160	200	250
$p/mmHg$	100	88	84	80	77	74	72

29.29. Now suppose that the retarding gas is less weakly adsorbed and that the full form of the Langmuir isotherm has to be used. The reactant gas is still supposed not to adsorb. Deduce the rate law $-dp/dt = k_c p/(1+Kp')$, where p is the pressure of the reactant gas and p' that of the product. Integrate the rate law for the simple decomposition $A \rightarrow B+C$, where B adsorbs and inhibits but C does not.

29.30. An example of this type of reaction is the decomposition of dinitrogen oxide, N_2O, on platinum at $750\,^\circ C$. In this case the oxygen produced by the reaction adsorbs strongly and inhibits the catalyst. Confirm that the data below fit the integrated rate law and determine K and k_c for the reaction.

t/s	0	315	750	1400	2250	3450	5150
$p/mmHg$	95	85	75	65	55	45	35

29.31. Although the attractive van der Waals interaction between individual molecules varies as R^{-6}, the interaction of a molecule with a nearby solid (a collection of molecules) varies as R^{-3}, where R is its vertical distance above the surface. Confirm this assertion.

29.32. Calculate the interaction energy between an argon atom and the surface of sold argon on the basis of a Lennard–Jones (6,12)-potential. Estimate the equilibrium distance for an atom above the surface.

29.33. The adsorption of solutes on solids from liquids often follows a Freundlich isotherm. Check the applicability of this isotherm to the following data for the adsorption of acetic acid on charcoal at $25\,^\circ C$ and find the values of the parameters c_1 and c_2.

$[acid]/mol\ dm^{-3}$	0.05	0.10	0.50	1.0	1.5
w_a/g	0.04	0.06	0.12	0.16	0.19

w_a is the mass adsorbed per unit mass of charcoal. Investigate whether the adsorption of acetic acid on charcoal would be better represented by a Langmuir isotherm.

29.34. We now turn to the application of the Gibbs isotherm, p. 849, to the adsorption of gases. Show by a directly analogous argument to that used in Chapter 24 that the volume adsorbed per unit area of solid, V_a/σ, is related to the pressure of the gas by $V_a = -(\sigma/RT)(d\mu/d\ln p)$, where μ is the chemical potential of the adsorbed gas.

29.35. If the dependence of the chemical potential of the gas on the extent of surface coverage is known, the Gibbs isotherm can be integrated to give a relation between V_a and p, as in a normal adsorption isotherm. For instance, suppose that the change in the chemical potential of a gas when it adsorbs is of the form $d\mu = -c_2(RT/\sigma)dV_a$, where c_2 is a constant of proportionality: show that the Gibbs isotherm leads to the Freundlich isotherm in this case.

29.36. Finally we come full circle and return to the Langmuir isotherm. Find the form of $d\mu$ that, when inserted in the Gibbs isotherm, leads to the Langmuir isotherm.

30   Dynamic electrochemistry

Learning objectives

After careful study of this chapter you should be able to

(1)   Describe the structure of the *electric double layer* (p. 1042) and define *outer potential* and *inner potential* (p. 1043).

(2)   Relate the *rate* of an electrode process to the potential difference at the interface and the *transfer coefficient* (eqn 30.1.3)).

(3)   Calculate the *anodic* and *cathodic current densities* (p. 1046).

(4)   Define the *overpotential* of an electrode (eqn (30.1.11)) and the *exchange current density* (p. 1049).

(5)   Derive the *Butler–Volmer equation* for the current density at an electrode (eqn (30.1.7)).

(6)   Derive the low overpotential and high overpotential forms of the Butler–Volmer equation (p. 1050) and use the *Tafel plot* to obtain the exchange current density and the transfer coefficient (p. 1050).

(7)   Outline the experimental measurement of the overpotential (p. 1050).

(8)   Distinguish between *polarizable* and *non-polarizable* electrodes (p. 1052).

(9)   Describe *concentration polarization* and the *Nernst diffusion layer* (p. 1053), and derive an expression for the *limiting current density* (eqn (30.1.17)).

(10)   Explain the basis of *polarography* and the use of the *rotating disc electrode* (p. 1055).

(11)   Discuss the *deposition of metals* and the *evolution of gases* during electrolysis (p. 1058).

(12)   Explain how the *current* affects the potential of a cell (p. 1059), and relate the available *power* to the current drawn (eqn (30.3.1)).

(13)   Describe the operation of *fuel cells* (p. 1061).

(14)   List the factors involved in the *storage* of power (p. 1063).

(15)   Define the *corrosion potential* and the *corrosion current* (p. 1066).

(16)   Describe the processes involved in corrosion (p. 1065) and the techniques available for *passivation* and *protection* (p. 1067).

Introduction

In this chapter we turn from the properties of the gas–solid interface to the processes that occur at a metal surface in contact with an ionic solution. One of the most important examples of this type of process is the transfer of charge between one material and the other: the bulk of electrochemistry depends on processes of this kind, and the search for efficient forms of power generation and storage relies on a thorough knowledge of electrode reactions and the properties of the metal–solution interface.

The economic consequences of electrochemical reactions at surfaces are almost incalculable. Most of the modern methods of generating electricity are inefficient, and the development of fuel cells into commercially viable form could revolutionize our production and deployment of energy. Today we produce energy inefficiently to produce goods that decay by corrosion. Every step in this wasteful sequence could be improved by discovering more about the processes that go on at electrodes.

The problem at the root of the chapter is how rapidly ions can be discharged at electrodes. A measure of this rate is the *current density*, the current per unit area, at the electrode, and the main thrust of the discussion is towards discovering the factors that influence it. At equilibrium the flow of charge out of the electrode is equal to the flow towards it, and the potential difference at the interface of electrode and solution has its equilibrium value. When a cell is working there is a net flow of current, and the potential difference differs from its equilibrium value by an amount called the *overvoltage*. We shall set up an equation relating current flow to overvoltage, and one practical outcome will be a knowledge of the cell e.m.f. when it is actually producing current. Setting up a particular overvoltage (by connecting the electrode to an external source of e.m.f.) results in a current density, and so the same calculation will also allow us to discuss electrodeposition of metals.

30.1 Processes at electrodes

Chapter 11 explained how a potential difference arose at an interface, and showed how this difference could be discussed in terms of thermodynamics. In this section we return to a consideration of this potential difference, but deal with systems that are not at equilibrium, systems that are producing electric current, and systems where electrochemical processes such as electrodeposition and decay are in progress.

The double layer at the interface. An electric potential difference arises at the interface between an electrode and the surrounding solution (p. 330). When equilibrium properties are of interest there is no need to enquire about the structure of the charge separation. Now, however, we are interested in the passage of current between electrode and solution, and the description of this dynamic process depends on setting up a model of the microscopic structure of the interface.

The potential difference across the interface arises from a separation of charge. If electrons leave the electrode and reduce the cations in the solution, the electrode acquires a positive charge, and the solution loses its electroneutrality and becomes slightly negatively charged locally. The excess negative charge of the solution is confined to the region near the electrode, where there is an attractive interaction with the latter's net positive charge. The most primitive model of the vicinity of the electrode therefore consists of a sheet of positive charge at the surface of the electrode, and a sheet of negative charge at the neighbouring surface of the solution. This pair of charged sheets is called the *electric double layer*. (If the electrode has a negative charge, the polarity of the double layer is reversed.)

A more detailed model of the double layer can be constructed by speculating about the arrangement of the ions in the solution. The simplest view would be to suppose that the solvated ions range themselves along the surface of the electrode, and are held away from it only by the presence of their hydration spheres, Fig. 30.1. The position of the sheet of ionic charge can be identified as the plane running through the solvated ions, and is called the *outer Helmholtz plane*. This *Helmholtz model* of the double layer neglects the disrupting effect of thermal motion which tends to break up and disperse the rigid wall of charge. A similar effect is treated in the Debye–Hückel model of the ionic atmosphere of opposite charge surrounding an ion (p. 316); the difference is that the central ion is replaced by a planar, charged electrode. The charge distribution can be found in the same way as for the Debye–Hückel model but it is much easier because the surface is flat. In fact, this *Gouy–Chapman model* of the *diffuse double layer* preceded the Debye–Hückel model by thirteen years, and may have been the latter's inspiration.

We shall not go through the Gouy–Chapman derivation here, but illustrate the results. Figure 30.2 shows how the local concentrations of cations and anions differ from the bulk concentration: the clustering close to the electrode of ions of opposite charge, and the repulsion of ions of the same charge can easily be seen. This shows, incidentally, how misleading it might be to use activity coefficients characteristic of the bulk solution to discuss the thermodynamic properties of ions that lie very close to electrodes.

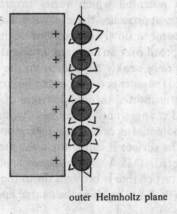

outer Helmholtz plane

Fig. 30.1. The outer Helmholtz plane at a positively charged electrode.

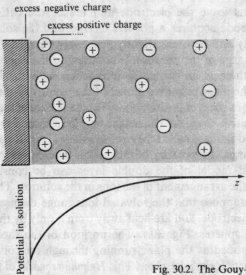

excess negative charge

excess positive charge

Fig. 30.2. The Gouy–Chapman model of the double layer.

Neither the Helmholtz nor the Gouy–Chapman model of the double layer is an exact representation of its structure. The former overemphasizes the rigidity of the ionic environment, and the latter overemphasizes its mobility. The two are combined in the *Stern model*. In this model, the ions closest to the electrode are constrained into a rigid Helmholtz plane, but outside that plane the ions are dispersed as in the Gouy–Chapman model.

The most important property of the double layer is the variation of potential in its vicinity. This potential governs the rate at which charges can be moved across from one side of the interface to the other, and so it controls the rate of the electrochemical processes at the electrode.

Consider the potential in the vicinity of an electrode carrying a positive charge. Imagine separating the metal from the solution but retaining the distribution of ions, solvent, and electrons. Now consider each half of the total system separately. At great distances from the metal a positive test charge experiences a Coulomb potential which varies inversely with distance, Fig. 30.3. As the charge approaches the electrode it reaches a region where the potential varies more slowly on account of the surface charge being not point-like but spread over an area, and at about 10^{-5} cm from the surface the potential is only weakly dependent on the distance. The potential in this region is called the *outer potential*, or the *Volta potential*, ψ. As the positive test charge is transported through the skin of electrons on the metal surface its potential changes until it reaches the inner, bulk metal environment. This extra potential is called the *surface potential*, χ (chi). The total potential inside the surface is called the *inner potential* or the *Galvani potential*, and is denoted $\phi(M)$. A similar sequence of changes in potential occurs as a positive test charge is brought up to the solution surface. The potential changes to its Volta value as the charge approaches

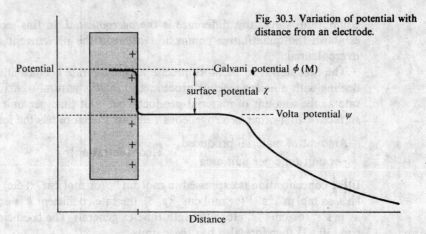

Fig. 30.3. Variation of potential with distance from an electrode.

the charged medium, and then to its Galvani value as it is taken through the orientated, dipolar solvent molecules and into the bulk. The inner potential of the solvent is denoted $\phi(S)$.

When the solution and electrode have been reassembled (but without any change in the charge distribution) the potential difference between points in the bulk metal and bulk solution is $\Delta\phi(M, S) = \phi(M) - \phi(S)$, the *Galvani potential difference*. This is the potential difference that lay at the centre of the discussion of electrode potentials in Chapters 11 and 12.

The rate of charge transfer. Processes at electrodes are dynamic because electrons are continually transferred in both directions. When the electrode is at equilibrium the transfers continue, but there is no net transfer. In contrast, when the cell is producing current there is an imbalance of transfers, and a net flow of electrons takes place either into the solution (accompanying reduction at the cathode) or into the electrode (accompanying oxidation at an anode).

In order for an ion (or a neutral molecule) to participate in charge transfer at an electrode, in some cases it has to discard its solvating molecules and migrate across the double layer to the electrode and in others it has to adjust its solvation sphere as it accepts or discards electrons. Similarly, a species already at the inner plane of the double layer has to be detached and migrate to the bulk. As a result both processes are activated and their rate coefficients are of the form

(30.1.1) $\qquad k = B \exp(-\Delta G_m^{\ddagger}/RT).$

$\Delta G_m^{\ddagger}$ is the molar Gibbs function of activation (p. 986) and B is some coefficient. The magnitude of $\Delta G_m^{\ddagger}$, and the flow of current in some direction, depends not only on the nature of the species but also on the potential difference between the initial state of the species and its transition state. These flows are in balance when the potential difference has its equilibrium value. The *net* flow of current to the electrode therefore depends on the extent to which the potential difference at the electrode differs from its equilibrium value: this difference is the *overpotential*. In this section we establish the quantitative connection between the net current and the overpotential.

The first step is to be more precise about the rate coefficient. We are dealing with a heterogeneous process, and so it is natural to express the rate as the amount of material produced per unit time per unit area of surface. A first-order heterogeneous rate law therefore has the form

$$\text{Amount of material produced per unit time per unit area} = k\{\text{concentration}\}.$$

If the concentration is expressed in mol dm^{-3} (or mol cm^{-3}, etc) and the l.h.s. as $\text{mol m}^{-2}\text{s}^{-1}$ (or $\text{mol cm}^{-2}\text{s}^{-1}$), the rate coefficient k is expressed as m s^{-1} (or cm s^{-1}, etc., or length/time in general). The coefficient B in eqn (30.1.1) therefore also has these units.

Now consider a reaction taking place at the electrode where a species is reduced by the transfer of a single electron in the rate determining step. (The last phrase is important: in the deposition of cadmium, for instance, only one electron is transferred in the rate determining step even though overall the deposition involves two.) Steps depending on the transfer of more than one electron can also be treated, but that introduces complications. Let the molar concentration of the oxidized species outside the double layer be [Ox] and the concentration of the reduced species be [Red]. Then the net current at the electrode is the difference of the currents arising from the reduction of Ox and the oxidation of Red. The rates are $k_{\text{red}}[\text{Ox}]$ and $k_{\text{ox}}[\text{Red}]$ respectively.

The reduction involves the transfer of one electron in each molecular step, and so the magnitude of the charge passed for unit amount of product generation is $eL = F$, Faraday's constant. The oxidation also involves the same transfer, but in an opposite direction. The *current densities* of the processes are the rates (expressed as amounts per unit time per unit area) multiplied by the charge associated with the material generated. Therefore there is a current density of magnitude $Fk_{\text{red}}[\text{Ox}]$ arising from the reduction and an opposite current density of magnitude $Fk_{\text{ox}}[\text{Red}]$ arising from the oxidation. The former is normally referred to as the *cathodic current* (recall that the cathode is defined as the seat of reduction), and the latter as the *anodic current*. For this reason k_{red} is normally denoted k_{c} and k_{ox} is denoted k_{a}. We shall henceforth adopt this notation.

It follows from the preceding discussion that the magnitude of the *net current density* at the electrode is the difference

$$j = Fk_{\text{c}}[\text{Ox}] - Fk_{\text{a}}[\text{Red}]$$

(30.1.2)
$$= B_{\text{c}}F[\text{Ox}]\exp(-\Delta G_{\text{m,c}}^{\ddagger}/RT) - B_{\text{a}}F[\text{Red}]\exp(-\Delta G_{\text{m,a}}^{\ddagger}/RT).$$

We have introduced eqn (30.1.1) to obtain the second line and have allowed for the possibility of the activation Gibbs functions and the pre-exponential factors being different for the cathodic (reduction) and anodic (oxidation) processes. That they are in general different is the central feature of the

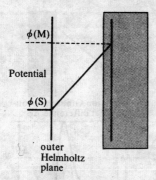

$\phi(M)$

Potential

$\phi(S)$

outer
Helmholtz
plane

Fig. 30.4. Electric potential in the vicinity of two charged plane surfaces.

remaining discussion.

At this stage we introduce the role of the potential difference $\Delta\phi = \phi(M) - \phi(S)$ across the double layer, Fig. 30.4. Consider the reduction step. In this step an electron is transferred from the electrode where the potential is $\phi(M)$ to the solution, where it is $\phi(S)$. There is an extra electrical contribution of magnitude $e\Delta\phi$ to the work. If the transition state corresponds to a state of the system in which the electron is transferred when the Ox species is essentially inside the double layer and close to the electrode, Fig. 30.5b, the molar Gibbs function is changed from $\Delta G_{m,c}^{\ddagger}$ when the potential difference is zero to $\Delta G_{m,c}^{\ddagger} + F\Delta\phi(M, S)$. If the electrode is at a higher (more positive) potential than the solution $\Delta\phi$ is positive and work has to be done in order to bring Ox to the transition state; and so the activation Gibbs function is increased. If, however, the transition state is a point on the reaction coordinate corresponding to the Ox species being very close to the outer plane of the double layer, Fig. 30.5c, the value of $\Delta\phi$ is irrelevant because the extra electrical work of attaining the transition state is the work involved in moving the Ox species from $\phi(S)$ to a closely similar value. In a real system the transition state is at some point intermediate between these two extremes, Fig. 30.5d, and so we write:

Activation Gibbs function for reduction $= \Delta G_{m,c}^{\ddagger} + \alpha F\Delta\phi$

where the parameter α is called the *transfer coefficient* or the *symmetry factor*. It lies in the range 0 to 1, and experimentally is often found to be about $\frac{1}{2}$.

Now consider the oxidation. In this case the Red species discards an electron to the electrode, and the extra work involved in attaining the transition state is about zero if the transition state lies close to the electrode (Fig. 30.5b), the full $-F\Delta\phi$ if it lies at the outer plane, and $\Delta G_{m,a}^{\ddagger} - (1-\alpha)F\Delta\phi$ in general. Thus if $\Delta\phi$ is positive, so that the potential of the electrode is greater than that of the solution, as in Fig. 30.5, the activation Gibbs function for oxidation is decreased and the rate of oxidation is greater:

Activation Gibbs function for oxidation $= \Delta G_{m,a}^{\ddagger} - (1-\alpha)F\Delta\phi$.

Now we can take the principal step in the argument: the two activation

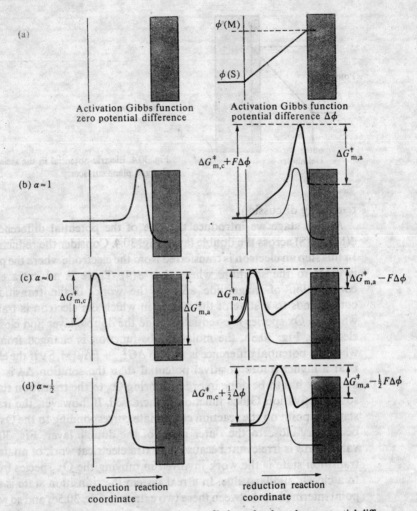

Fig. 30.5. Modification to the reaction profile brought about by a potential difference.

Gibbs functions just constructed can be inserted in place of the values in eqn (30.1.2), with the result that

$$(30.1.3) \qquad j = \{B_c F[\text{Ox}] \exp\left(-\Delta G_{m,c}/RT\right)\} \, e^{-\alpha F \Delta \phi / RT}$$
$$- \{B_a F[\text{Red}] \exp\left(-\Delta G_{m,a}/RT\right)\} \, e^{(1-\alpha)\Delta \phi / RT}$$

which is an explicit, if complicated, expression for the net current density in terms of the potential difference.

The appearance of the last equation can be simplified as follows. First we identify the individual cathodic and anodic current densities:

$$(30.1.4a) \qquad j_c = \{B_c F[\text{Ox}] \exp\left(-\Delta G_{m,c}/RT\right)\} \, e^{-\alpha F \Delta \phi / RT}$$

$$(30.1.4b) \qquad j_a = \{B_a F[\text{Red}] \exp\left(-\Delta G_{m,a}/RT\right)\} \, e^{(1-\alpha)F \Delta \phi / RT}$$

for then $j = j_c - j_a$. Now consider their values when $\Delta \phi$ has its equilibrium value $\Delta\phi_e$: this is the value we concentrated on in the equilibrium electro-chemistry chapters of Part 1. At equilibrium there is no net current, and so $j_{c,e} = j_{a,e}$, where $j_{c,e}$ and $j_{a,e}$ are the values obtained by substituting $\Delta\phi_e$ in the last two expressions. Now, when the potential difference differs from its equilibrium value by the *overpotential* η (eta),

(30.1.5)
$$\eta = \Delta\phi - \Delta\phi_e,$$

the currents are

(30.1.6a)
$$j_c = \{B_c F[Ox] \exp(-\Delta G_{m,c}/RT) e^{-\alpha F\Delta\phi_e/RT}\} e^{-\alpha\eta F/RT}$$
$$= j_{c,e} e^{-\alpha\eta F/RT}$$

(30.1.6b)
$$j_a = \{B_a F[Red] \exp(-\Delta G_{m,a}/RT) e^{(1-\alpha)\Delta\phi_e/RT}\} e^{(1-\alpha)\eta F/RT}$$
$$= j_{a,e} e^{(1-\alpha)\eta F/RT}.$$

And since we have already seen that $j_{a,e} = j_{c,e}$ we can drop the c,a subscript from j_e and write

(30.1.7)
$$j = j_c - j_a = j_e\{e^{-\alpha\eta F/RT} - e^{(1-\alpha)\eta F/RT}\}.$$

This is the *Butler–Volmer equation*. It lies at the heart of the remainder of the discussion, j_e is called the *exchange current density*. Remember that it is limited to electrodes where the rate determining step involves the transfer of only a single electron.

A small note may be added in passing. The expression for j_c (and j_a) may be used to calculate by how much the current density changes when the potential difference at an electrode is changed. Suppose $\Delta\phi$ is changed from 1 V to 2 V, then taking $\alpha = \frac{1}{2}$ gives

$$j_c \text{ (at 2 V)}/j_c \text{ (at 1 V)} = \exp\left\{\frac{-\frac{1}{2}\times(1\text{ V})\times(9.649\times10^4\text{ C mol}^{-1})}{(8.314\text{ J K}^{-1}\text{ mol}^{-1})\times(298\text{ K})}\right\}$$
$$= 2.9\times10^8.$$

This is a huge change of current density for a very small and easily applied change of conditions. Some insight into why it is so great can be obtained by expressing the change in the potential difference in terms of the change in the electric field acting on an ion in the outer plane of the double layer. A change in the potential difference of 1 V over a double layer 10^{-9} m thick (a few molecular diameters) corresponds to changing the electric field by 10^9 V m^{-1}. It is hardly surprising that huge changes in current result from such changes in the electric field. Innocuous as electrochemistry may appear from the outside, scrutiny at the molecular level reveals what great electrical stresses are involved as electrodes tear electrons from ions.

Overpotential. The magnitude of the net current density depends on two factors. The first is the equilibrium current density j_e for transport across the double layer in either direction when the electrode interface has its equilibrium potential difference. This quantity depends on the detailed structure of the double layer, and has a wide range of magnitudes for different electrodes and different conditions. For instance, the N_3^-, $N_2|Pt$ couple has $j_e \approx 10^{-76}\,A\,cm^{-2}$ whereas Hg^{2+}, $Hg|Pt$ has $j_e \approx 10^{10}\,A\,cm^{-2}$, a range of 86 orders of magnitude.

The content of the Butler–Volmer equation can be clarified by examining two of its limiting forms. When the overpotential is very small (less than about 0.01 V) the exponentials can be expanded using $e^x \approx 1 + x$. Then

(30.1.8) $j \approx j_e\{[1 - (\alpha\eta F/RT)] - [1 + (1 - \alpha)\eta F/RT]\} \approx -j_e\eta F/RT.$

(Since we are treating the Butler–Volmer equation as an equation for magnitudes we need not be troubled about overall signs. In a more detailed treatment the sign would signify the direction of current flow). In this case the current density is proportional to the overpotential, and so at low overpotentials the interface behaves like an Ohmic conductor (current $\propto$ potential difference).

Example (Objective 3). The exchange current density of a $Pt|H_2,H^+(aq)$ electrode is $0.79\,mA\,cm^{-2}$. What current flows through a standard electrode of total area $5\,cm^2$ when the potential difference across the electrode interface is $5\,mV$, the temperature $25\,°C$, and the proton activity unity?

- *Method.* The equilibrium potential is zero, and so the overpotential is $5\,mV$. Substitute in eqn (30.1.8) and then convert from current density to current. Use $RT/F = 25.68\,mV$.

- *Answer.* From eqn (30.1.8),

    $j = -(0.79\,mA\,cm^{-2}) \times (5\,mV/25.69\,mV) = -0.154\,mA\,cm^{-2}.$

    For an area of $5\,cm^2$ the magnitude of the total current is $(5\,cm^2) \times (0.154\,mA\,cm^{-2})$, or $0.77\,mA$.

- *Comment.* The current is in fact negative, indicating a flow away from the electrode. Note that $\alpha\eta F/RT = 0.097 \ll 1$, and so the linear approximation is valid.

When the overpotential is large (greater than about 0.1 V) the Butler–Volmer equation takes a different limiting form. When the large overpotential is positive, the first exponential in eqn (30.1.7) is much smaller than the second and may be neglected. Then ignoring overall signs, because we want only the magnitude of the current density,

$j \approx j_e \exp[(1 - \alpha)\eta F/RT],$

so that

(30.1.9) $\ln j \approx \ln j_e + (1 - \alpha)\eta F/RT.$

The plot of the logarithm of the current density against the overpotential, a *Tafel plot*, gives the exchange current density from its intercept and the symmetry parameter α from its gradient. When the overpotential is large but negative, the first exponential dominates the second:

$$j \approx j_e \exp(-\alpha \eta F / RT)$$

and the Tafel plot is based on

(30.1.10) $\ln j \approx \ln j_e - \alpha \eta F / RT.$

Example (Objective 6). The data below refer to the magnitude of the current through a $2\,\text{cm}^2$ platinum electrode in contact with a Fe^{2+}, Fe^{3+} solution at $25\,°C$. Find the exchange current density and the transfer coefficient for the electrode process.

η/mV	50	100°	150	200	250
I/mA	8.8	25.0	58.0	131	298

- *Method.* Use the Tafel equation, eqn (30.1.14) and plot $\ln j$ against η. The discharge reaction should be regarded as $(Fe^{2+})^+ + e^- \rightarrow Fe^{2+}$, and involves a one-electron transfer in its rate determining step.

- *Answer.* Draw up the following table:

η/mV	50	100	150	200	250
j/mA cm^{-2}	4.4	12.5	29.0	65.6	149
$\ln(j/\text{mA cm}^{-2})$	1.50	2.53	3.37	4.18	5.00

The high overpotential region gives a straight line of intercept 0.916 and slope 0.0163, Fig. 30.6. From the former it follows that $\ln(j_e/\text{mA cm}^{-2}) = 0.916$, so that $j_e = 2.5\,\text{mA cm}^{-2}$. From the latter, $(1-\alpha)F/RT = 0.0163\,\text{mV}$, so that $\alpha = 0.58$.

- *Comment.* Notice that the Tafel plot is non-linear for $\eta < 150\,\text{mV}$. This is the region where $\alpha \eta F / RT = 75/26 \approx 3$, which is not much greater than unity.

In order to draw a Tafel plot we need an experimental method of measuring the overpotential. A typical procedure makes use of three electrodes, and is illustrated in Fig. 30.7. The electrode of interest is called the *working electrode*, and the current flowing to or from it is controlled from an outside source. If its area is A and the current is I, the current density across its interface is I/A. The potential difference across the interface cannot be measured directly, but its potential relative to a third,

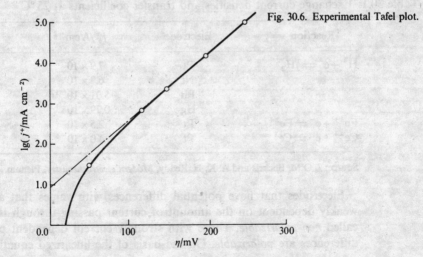

Fig. 30.6. Experimental Tafel plot.

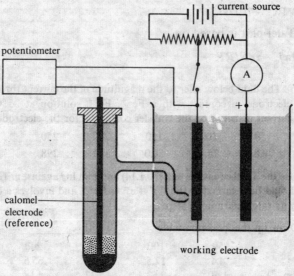

Fig. 30.7. Arrangement for measuring the overpotential.

reference electrode can be measured. The working and reference electrodes are brought into balance with a potentiometer, and no net current flows in that half of the circuit. Changing the amount of current flowing through the working circuit causes a change in potential of the working electrode. The potentiometer reading is used in the normal way to find its new potential relative to the reference.

A Tafel plot is illustrated in Fig. 30.6: there is quite good linearity in the high overpotential region, and from the gradient and extrapolated intercepts the symmetry factor and exchange current for the electrode can be found; some values are given in Table 30.1. Both parameters give valuable insight into the structure of the double layer.

Table 30.1. Exchange current densities and transfer coefficients at 25 °C

Reaction	Electrode	$j_e^+/A\,cm^{-2}$	α
$H^+ + e^- \rightleftharpoons \frac{1}{2}H_2$	Pt	7.9×10^{-4}	
	Ni	6.3×10^{-6}	0.58
	Pb	5.01×10^{-12}	
	Hg	0.79×10^{-12}	0.50
$Fe^{3+} + e^- \rightleftharpoons Fe^{2+}$	Pt	2.5×10^{-3}	0.58
$Ce^{4+} + e^- \rightleftharpoons Ce^{3+}$	Pt	4.0×10^{-5}	0.75

Source: J. O'M. Bockris and A. K. N. Reddy, *Modern electrochemistry*, Plenum.

Electrodes that have potential differences with values that are only weakly dependent on the amount of current passing through them are called *non-polarizable*. Those with strongly current dependent potential differences are *polarizable*. On the basis of the linearized equation, eqn

(30.1.8) written in the form

(30.1.11) $\qquad \eta \approx (RT/F)j/j_e$

it is clear that the criterion for low polarizability is high exchange current density (so that η, the deviation from $\Delta\phi_e$, may be small even though j is large). A non-polarizable electrode therefore has a double layer with a structure that permits a rapid adjustment of charge. The calomel and the H_2/Pt electrodes are both highly non-polarizable, which is one reason why they are so extensively used in equilibrium electrochemistry measurements. Equilibrium exchange current densities are also governed by the kinetics of the chemical processes involved. Many inorganic systems involve simple one-electron or successive one-electron steps; as a consequence, the electrode kinetics are sufficiently fast for thermodynamic data to be obtained readily. That is the underlying reason for the importance of cell e.m.f. measurements in inorganic redox chemistry. In contrast, organic redox systems (e.g., alcohol–aldehyde) often involve bond scission or formation. Such processes are slow. As a consequence, there is relatively little data on the thermodynamics of organic redox systems from electrochemical sources.

More aspects of deviations from equilibrium. One of the assumptions in the derivation of the Butler–Volmer equation is the uniformity of concentration in the vicinity of the electrode when the system is away from equilibrium. This assumption fails at high current densities because the diffusion towards the electrode from the bulk is too slow and may become rate-determining. The effect is called *concentration polarization*, a larger deviation from the equilibrium potential is then needed to bring about a given current, and the extra overpotential is called the *concentration overpotential*.

The concentration overpotential can be estimated as follows. We disregard the complication of having to think about the rate processes underlying the Butler–Volmer equation, and consider a case of pure concentration polarization. At equilibrium, when there is no net current density, the potential difference across the double layer is related to the activity of the ions in the solution by the Nernst equation (p. 355):

$$\Delta\phi_e = \Delta\phi^\ominus + (RT/zF)\ln a_{M^+}.$$

When the system is not at equilibrium the concentration just outside the double layer changes, the activity there changes to a'_{M^+}, and so the potential changes to

$$\Delta\phi = \Delta\phi^\ominus + (RT/zF)\ln a'_{M^+}.$$

The concentration overpotential, η^c, arising from this modification to the activities is

$$\eta^c = \Delta\phi - \Delta\phi_e = (RT/zF)\ln(a'_{M^+}/a_{M^+}).$$

We have already seen that there are severe difficulties in dealing with activities (there are concentration gradients, and solutions are often concentrated), and so we shall simply ignore the activity coefficients and work with the concentrations c. Then the concentration overpotential is given by

(30.1.12) $\eta^c = (RT/zF)\ln(c'/c)$.

The next problem is to make some estimate of the concentration just outside the double layer. The deviation of c' from the bulk depends on the size of the ionic diffusion coefficient: if the ions are mobile and D is large, the local concentration might be almost the same as the bulk even though a large current is passing.

The simplest approach to calculating c' starts with Fick's diffusion law, p. 875). This relates the flux of particles, the number crossing per unit area per unit time, to the concentration gradient:

(30.1.13) $J = -D(\partial \mathcal{N}/\partial x)$.

$\mathcal{N}$ is the concentration in number per unit volume, if c is expressed in amount per unit volume, $\mathcal{N} = cL$. We suppose, for simplicity, that the concentration of species has its bulk value up to a distance λ from the outer Helmholtz plane, and then falls linearly to c' at the plane itself. This *Nernst diffusion layer* is illustrated in Fig. 30.8. The concentration gradient is therefore

$$(\partial \mathcal{N}/\partial x) = (\mathcal{N}' - \mathcal{N})/\lambda = (c' - c)(L/\lambda),$$

and so the flux is

$$J = -(LD/\lambda)(c' - c).$$

The flux of species of charge ze corresponds to a current density of magnitude

(30.1.14) $j = zeJ$

and so combining the last two equations lets us express c' in terms of the current density at the double layer:

(30.1.15) $c' = c - (\lambda/zFD)j$.

Substituting this into the Nernst equation for the potential difference gives the concentration overpotential in the presence of a current density j as

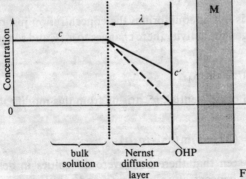

Fig. 30.8. The Nernst diffusion layer.

(30.1.16) $\eta^c = (RT/zF)\ln\{1-(j\lambda/zcFD)\}.$

The alternative form of this expression,

(30.1.16a) $j = (zcFD/\lambda)\{1-e^{zF\eta^c/RT}\},$

gives the current when there is an overpotential η^c (and when there are no other contributions to the deviation of $\Delta\phi$ from $\Delta\phi_e$ except concentration polarization).

The maximum rate of diffusion across the Nernst layer occurs when the gradient is the steepest, when $c' = 0$. This concentration occurs when an electron from an ion that diffuses across the Nernst layer is snapped over the activation barrier, through the double layer, and on to the electrode. No flow of ions can occur more rapidly than in this case, and so the magnitude of the limiting current density, j_L, is given by eqn (30.1.14) with $c' = 0$:

(30.1.17) $j_L = (zFD/\lambda)c.$

Example (Objective 9). Estimate the magnitude of the limiting current density at an electrode in which copper ions are at a concentration $0.01\,mol\,dm^{-3}$.

● *Method.* Use eqn (30.1.17). Take $\lambda = 0.5\,mm$ as a typical value. Estimate D from the ionic conductivity $\lambda_+ = 107\,\Omega^{-1}\,cm^2\,mol^{-1}$ and eqn (26.2.5).

● *Answer.* From eqn (26.2.5), $\lambda_+ = (z^2F^2/RT)D_+$, so that substituting this into eqn (30.1.21) leads to

$j_L = (zFD_+/\lambda)c_{M^+} = (RT\lambda_+/zF\lambda)c_{M^+}.$

Therefore

$j_L = (107 \times 10^{-4}\,\Omega^{-1}\,m^2\,mol^{-1}) \times (25.7\,mV) \times (0.01\,mol\,dm^{-3})/(2 \times 0.5\,mm)$

$= 2.75\,A\,m^{-2},$ or $0.275\,mA\,cm^{-2}.$

● *Comment.* This implies that the current towards a $1\,cm^2$ electrode cannot exceed $0.28\,mA$ in this solution. If the solution is stirred, or if the electrode surface is moving (as in polarography) the calculation of the limiting current is much more complicated.

The value of the limiting current density is determined by the charge, the diffusion coefficient, and the concentration of the ions, and monitoring it is the basis of the analytical method of *polarography* for determining the constitution of a solution. The technique consists of raising the potential between two electrodes and observing the current. The current climbs as the potential difference increases until the limiting current characteristic of the ion in the solution is reached. If several ions are present the current may rise to a sequence of limiting values and they are identified by determining their *half-wave potentials*, Fig. 30.9 and Table 30.2. The value of the limiting current is used to find the concentrations of the ions. One

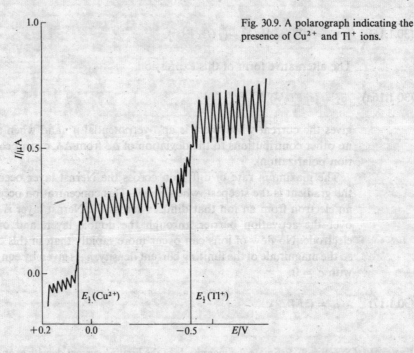

Fig. 30.9. A polarograph indicating the presence of Cu^{2+} and Tl^+ ions.

of the problems with this procedure is the contamination of the electrode surface. This can be avoided by using a *dropping mercury electrode*, Fig. 30.10, where the electrode is constantly regenerated as drops form at the end of the capillary, and then fall off. The growth of the drop, and therefore of the surface area of the electrode, accounts for the oscillations in the polarogram.

Table 30.2. Half-wave potentials for ions in $0.1\ mol\ kg^{-1}$ KCl at 25 °C, $E_{1/2}/V$ relative to the calomel electrode

Cd^{2+}	-0.599	Fe^{3+}	0.0
Co^{2+}	-1.4	Fe^{2+}	-1.3
Cr^{3+}	-0.91	Tl^+	-0.460
Cu^{2+}	$+0.04$	Zn^{2+}	-0.995

Source: G. W. C. Kaye and T. H. Laby, *Tables of physical and chemical constants*, Longmans.

The theoretical analysis of the dropping mercury electrode is far from simple because of its changing size and shape. Nevertheless, some moderately tolerable approximations may be made, and they lead to the *Ilkovic equation* for j_L. This indicates that $j_L \propto t^{1/6}$, and the peculiar time dependence reflects the competition between the growth of the drop, and the establishment of the limiting current. The problem of dealing with the complicated hydrodynamics of flow towards a growing electrode is eliminated by the *rotating-disc electrode*, Fig. 30.11. The electrode is a small, flat disc, set in a rotating axle. The rotation of its surface sets up a steady hydrodynamic flow which circulates the solution over its face. The flow

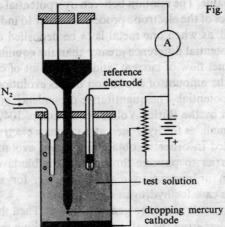

Fig. 30.10. The dropping mercury electrode.

reference
electrode

N_2

test solution

dropping mercury
cathode

pattern can be calculated quite readily, and the limiting current related
to the angular velocity of the disc.

Various modifications of the rotating-disc electrode can be used to study
the kinetics of electrode processes. One involves pulsing the potential
difference and then observing the growth of the current towards its limiting
value. Another development is the *ring-disc electrode* where the central
spinning disc is surrounded by an electrode in the form of a narrow ring.
As the disc rotates the flow lines bring the solution towards it and then
spread it over the ring. Analysis of the currents at the two electrodes, and
some knowledge of the time it takes for the products formed at the disc
to flow on to the ring and be detected, gives very detailed information
about the rates of electron transfer and other processes in solution.

30.2 Electrochemical processes

The discussion so far can be summarized as follows. The central concept
is the *overpotential*, the deviation of the potential difference at an electrode

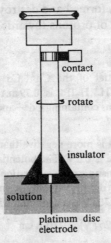

contact

rotate

insulator

solution

platinum disc
electrode

Fig. 30.11. The rotating-disc electrode.

from its equilibrium value. The relation between overpotential and current depends on the kinetics of the electrode process. In order to induce current to flow through a cell, as when one metal is to be deposited on another, we have to apply a potential difference greater than the equilibrium value in order to obtain a net flow of current, and the amount of current that flows, and therefore the amount of deposition or gas evolution obtained, depends on this overpotential. The quantitative dependence of current on overpotential is given by the Butler–Volmer equation, p. 1049. If the exchange current j_e is small, as for hydrogen at a mercury electrode, a large overpotential is needed in order to obtain significant evolution; if j_e is large because the barrier to passage through the double layer is small, appreciable evolution or deposition may occur even for small overpotentials. This is the case for hydrogen on platinum.

Current flows when a cell is used to generate power. Then the potential difference of a cell differs from the value when it is working reversibly by an amount called the *cell overpotential*. This overpotential arises from the shift of ions in the vicinity of both electrodes of the cell, the depletion of local concentrations of ions in the neighbourhood of the double layers, and the Ohmic drop in potential from one electrode to the other as a result of the flow of current within the cell.

How to discuss evolution and deposition at electrodes. An indication of the rate of gas evolution or metal deposition can be obtained from the Butler–Volmer equation and tables of exchange currents. The exchange current is strongly dependent on the nature of the electrode, and because this changes during electrodeposition of one metal on another, we shall look only at the simpler cases of metal on like metal and gas evolution.

A glance at Table 30.1 shows the range of exchange currents for a metal|hydrogen electrode. The most sluggish exchange currents occur for lead and mercury, and the equilibrium current of $10^{-12}\,\mathrm{A\,cm^{-2}}$ corresponds to the exchange of a monolayer of atoms (about $10^{15}\,\mathrm{cm^{-2}}$) in $10^8\,\mathrm{s}$ (about 5 years). In contrast, the exchange current on platinum is about $10^{-3}\,\mathrm{A\,cm^{-2}}$, and so a monolayer is formed and destroyed in 0.1 s. For a given overpotential the net currents of protons towards these electrodes differ by a factor of 10^9.

Example (Objective 11). The standard electrode potential of the Cd^{2+},Cd electrode is $-0.403\,\mathrm{V}$. The exchange current density of $H^+,H_2|Pt$ is $0.79\,\mathrm{mA\,cm^{-2}}$. Can cadmium be plated on to platinum from solutions containing Cd^{2+} and H^+ at unit activity?

- *Method.* The metal will deposit if the potential is more negative than $-0.403\,\mathrm{V}$; hydrogen evolves when the potential difference is 0.0 V. The potential difference drops to $-0.403\,\mathrm{V}$ only when the current density exceeds $j_e \exp(\alpha\eta F/RT)$ with $\eta = -0.403\,\mathrm{V}$. Take $\alpha = \frac{1}{2}$. Find the current density. If only a small current density is needed to bring the potential to $-0.403\,\mathrm{V}$ the cadmium will plate out.

- *Answer.* $j = (0.79\,\mathrm{mA\,cm^{-2}})\exp(0.403\,\mathrm{V}/2 \times 0.0257\,\mathrm{V}) = 2.01\,\mathrm{A\,cm^{-2}}$.

The potential difference drops to $-0.403\,$V only when the current density reaches $2\,A\,cm^{-2}$; for smaller current densities the hydrogen evolution dominates.

• *Comment.* For deposition on nickel the exchange current density is $3.2\,\mu A\,cm^{-2}$, and so a current density of only $8.1\,mA\,cm^{-2}$ is sufficient to drop the potential to the point where cadmium will plate out on to the nickel.

The exchange current also depends on which crystal face is exposed. For the deposition of copper on a copper electrode, the (100) face has an exchange current of $1 \times 10^{-3}\,A\,cm^{-2}$, and so for the same overpotential grows at 2.5 times the rate of the (111) face with its exchange current density of $4 \times 10^{-4}\,A\,cm^{-2}$. Put another way, a net current density of $10\,mA\,cm^{-2}$ can be achieved with an overpotential of $-125\,mV$ for the (100) face, but it requires $-185\,mV$ for the (111) face; the (110) face needs only $-85\,mV$ for the same net current density.

How the current affects the potential of a cell. We expect the e.m.f. of a cell to drop as current is withdrawn because it is then no longer operating reversibly and is capable of doing less than maximum work. This is found to be the case, and the connections between overpotential and current allows us to estimate the effect quantitatively.

Consider the cell $M|M^{+}(aq)||M'^{+}(aq)|M'$. We shall ignore all the complications that arise from any liquid junctions present. The e.m.f. of this cell, when no current is being drawn, is determined by the difference of the electrode potentials $E(M^{+},M)$ and $E(M'^{+},M')$. When current is drawn the potential difference between the electrodes can be written in terms of the potential differences at the interfaces, together with any potential difference across the solution itself:

$$\Delta\phi(M', M) = \phi(M') - \phi(M)$$

$$= \phi(M') - \phi(S') + \phi(S') - \phi(S) + \phi(S) - \phi(M)$$

$$= \Delta\phi(M', S') + \Delta\phi(S', S) - \Delta\phi(M, S)$$

where S is the solution in the vicinity of M, and S′ the solution in the vicinity of M′. From now on we shall write $\Delta\phi(M', M) = E$, the working e.m.f. of the cell.

The working e.m.f. can be expressed in terms of the equilibrium e.m.f., E_e, by introducing the equilibrium potential differences and the overpotentials:

$$\Delta\phi(M, S) = \Delta\phi_e(M, S) + \eta, \qquad \Delta\phi(M', S') = \Delta\phi_e(M', S') + \eta'.$$

Then

(30.2.1) $$E = E_e + \eta' - \eta + \Delta\phi(S', S),$$

and E can be found by using the Butler–Volmer equation to evaluate the overpotentials corresponding to the current I being drawn. The potential difference across the solution $\Delta\phi(S', S)$ has a magnitude IR_s where R_s is the solution's resistance. Since it is a contribution to the cell's irreversibility (it is a thermal dissipation term) the sign of IR_s is such as always tend to reduce the cell e.m.f., making E less positive if it is positive, but less negative if it is negative.

On the basis that the areas (A) of the electrodes are the same, that only one electron is transferred in the rate-determining steps at the two electrodes, that the transfer coefficients are both $\frac{1}{2}$, and that the high-overpotential form of the Butler–Volmer equation may be used, the connection between E and I is

$$(30.2.2) \qquad E \approx E_e - (4RT/F)\ln\{I/A\sqrt{(j_e j'_e)}\} - IR_s.$$

The concentration overpotential also reduces the cell e.m.f. If we adopt the simple Nernst diffusion layer model for the effect, we can use eqn (30.1.16) for η^c at each electrode. The change in the e.m.f. due to this effect can then be written at once:

$$(30.2.3) \qquad E = E_e + (RT/zF)\ln\{(1 - I/Aj'_L)(1 - I/Aj_L)\}$$

where j_L and j'_L are the limiting current densities at the two electrodes, eqn (30.1.17). This result can be added to eqn (30.2.2) to obtain a full (but still very approximate) expression for the e.m.f. of a cell when a current I is being drawn:

$$(30.2.4) \qquad E = E_e - IR_s - (2RT/zF)\ln f(I),$$

where

$$f(I) = \frac{[I/A\sqrt{(j_e j'_e)}]^{2z}}{[(1 - I/Aj_L)(1 - I/Aj'_L)]^{1/2}}.$$

This expression depends on a lot of cell parameters, but an example is sketched in Fig. 30.12: notice the very steep fall in working e.m.f. when the current is high and close to the limiting current density of one of the electrodes.

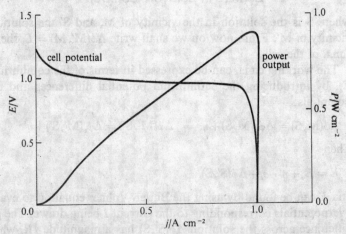

Fig. 30.12. Dependence of working cell e.m.f. and its power on the current drawn.

30.3 Power generation and storage

Potential difference, or e.m.f., is not the only criterion of usefulness of an electrochemical source: we should also be interested in *power*. The power supplied by a cell is IE, where I is the current and E is the e.m.f. at that current. If I is expressed in amperes and E in volts, the power P is given in watts (W). The expression $P = IE$ summarizes the need to study cells away from equilibrium: at equilibrium they produce no current, and although the e.m.f. may be greatest, I is zero, and so no power is produced.

An expression for the power produced by a cell can be obtained very simply by multiplying eqn (30.2.4) by I:

(30.3.1) $$P = IE_e - I^2 R_S - (2RTI/zF)\ln f(I).$$

The first term in this equation is the power that would be obtained if the cell produced its equilibrium e.m.f. even when delivering current; the second term is the power dissipated uselessly as heat in the electrolyte as a result of its resistance; and the third term is the correction taking account of the lowering of the electrode potential differences as a result of drawing current. The form of the dependence of the power on the current is illustrated in Fig. 30.12 for the e.m.f./current curves shown there. Notice how, in this case at least, maximum power is obtained just before the concentration polarization quenches the e.m.f. Information of this kind is essential if the optimum conditions for employing electrochemical devices are to be found, and their performance improved.

Power generation in fuel cells. Recent advances in electrochemistry and technology have led to the introduction of a compact source of power in which fuels are used to produce electricity without the intervention of thermal devices such as boilers, turbines, and generators. These *fuel cells* can be much more efficient than conventional thermal sources; they have already been employed in spacecraft, and may become the basis of pollution-free power for transport.

A fuel cell operates just like a conventional electrochemical cell with the exception that the materials to be oxidized and reduced at the electrodes are stored outside the cell rather than forming an integral part of its construction. A fundamental, but important, example of a fuel cell is the hydrogen|oxygen cell, illustrated in Fig. 30.13. If the electrolyte is acid, the reduction reaction at the cathode is $O_2 + 4H^+ + 4e^- \rightarrow 2H_2O$, and at the anode the reaction is $2H_2 \rightarrow 4H^+ + 4e^-$. The equilibrium e.m.f. of this cell is $(1.229\,V) - (0\,V) = 1.229\,V$, but as we are interested in power the current/e.m.f. characteristics of the cell are important.

Example (Objective 13). Calculate the e.m.f. of a propane|oxygen fuel cell operating reversibly.

- *Method.* The reaction is $C_3H_8 + 5O_2 \rightarrow 3CO_2 + 4H_2O$, and the standard Gibbs function of the reaction is $-2108\,kJ\,mol^{-1}$. As written, the reaction involves the transfer of 20 electrons. This number is arrived at by noting that the reaction can be expressed as the combination of $5O_2 + 20H^+ + 20e^- \rightarrow 10H_2O$ with $C_3H_8 + 6H_2O \rightarrow 20e^- + 3CO_2 + 20H^+$. Use eqn (12.3.1) with $v = 20$.

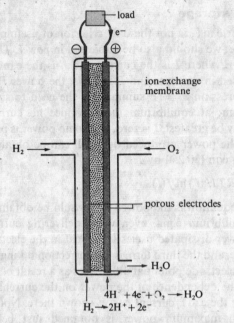

Fig. 30.13. A hydrogen|oxygen fuel cell.

$$4H^+ + 4e^- + O_2 \rightarrow H_2O$$
$$H_2 \rightarrow 2H^+ + 2e^-$$

- *Answer.* From eqn (12.3.1)

$$E^{\ominus} = -(-2.108 \times 10^6 \, J \, mol^{-1})/20 \times (9.648 \times 10^4 \, C \, mol^{-1}) = 1.092 \, V.$$

- *Comment.* The enthalpy of combustion at 25 °C is 2220 kJ mol^{-1}, and so in a cell working with perfect efficiency some heat is given out in the reaction (an amount equal to 2220 kJ mol^{-1} − 2108 kJ mol^{-1} = 112 kJ mol^{-1}).

One advantage of the hydrogen|oxygen system is the large exchange current of the hydrogen reaction. Unfortunately the oxygen reaction has an exchange current of only about 10^{-10} A cm^{-2}, and so this limits the current available from the cell. One way round this difficulty is to use a catalytic surface (to increase the exchange current density) and a very large surface area for the electrode in order to obtain a large current even though the current density is small. In the cell illustrated in Fig. 30.13 the electrodes are made of titanium coated with platinum, and the electrolyte is a cation-exchange resin.

Other types of fuel cell have been developed, making use of various fuels and technical devices for increasing the magnitude of the exchange current density, Table 30.3. The latter can be achieved, for example, by selecting an electrode surface with useful catalytic properties, or by heating the entire cell to improve the diffusional characteristics and the rate of crossing the double layers. Although oxygen is normally the oxidizing fuel (because of its easy availability) the fuel for the anode reaction may be any of a wide variety of materials. A particularly important group of anode fuels are the hydrocarbons. Hydrocarbon|air fuel cells normally use platinum electrodes, concentrated phosphoric acid electrolyte, and operate at temperatures in the region of 100 °C. Models now available can produce about 0.1 W per cm^2 of electrode surface.

Table 30.3. Thermodynamic characteristics of some possible fuel cells

Reaction	$\Delta G_m^{\ominus}/\text{kJ mol}^{-1}$	$\Delta H_m^{\ominus}/\text{kJ mol}^{-1}$	E/V	ε†
$H_2 + \frac{1}{2}O_2 \rightarrow H_2O$	-237.2	-258.9	1.229	0.83
$CH_4 + 2O_2 \rightarrow CO_2 + 2H_2O$	-818.0	-890.4	1.060	0.92
$CH_3OH + \frac{3}{2}O_2 \rightarrow CO_2' + 2H_2O$	-706.9	-764.0	1.222	0.93
$C + O_2 \rightarrow CO_2$	-137.3	-110.5	0.712	1.24

†$\varepsilon = \Delta G_m^{\ominus}/\Delta H_m^{\ominus}$, a measure of the relative efficiency of electrical and thermal energy resources.
Source: J. O'M. Bockris and A. K. N. Reddy, *Modern electrochemistry*, Plenum.

Power storage. Electricity generated in one device may be stored in another, and then transported to the site where it is to be used. The battery in a car is one example (although here the generator is also carried round by the vehicle), and rechargeable batteries in tape recorders, shavers, etc., are others. These two sets of examples indicate the range of performance required: the weight of a car battery is not of great importance, but it must be able to deliver massive currents for short periods: in contrast, the weight of batteries in tape recorders, shavers, and electrically propelled vehicles, is a significant part of the total, but they are required to provide moderate or small, steady currents over long periods.

Electricity storage cells operate like conventional cells while they are producing electricity, but their fuel is formed within them by passing a current from an outside source. This is in contrast to a simple cell where the fuels are built in during manufacture, and to a fuel cell where they are supplied from outside during the cell's operation. We shall look at only two of the many devices now available, and concentrate on the lead storage battery and the nickel–cadmium cell.

The lead storage battery is an old device, but one quite well suited to the job of starting cars. During charging, the cathode reaction is the reduction of Pb^{2+} and its deposition as lead on the lead electrode:

$$PbSO_4 + 2e^- \rightarrow 2Pb + SO_4^{2-}.$$

This occurs instead of the reduction of the acidic solution to hydrogen gas because of the latter's low exchange current density on lead.

The anode reaction driven during charging is the oxidation of Pb^{2+} to Pb^{4+}: this is hydrolysed to PbO_2 and deposited as the oxide on the electrode:

$$PbSO_4 + 2H_2O \rightarrow PbO_2 + 4H^+ + SO_4^{2-} + 2e^-.$$

On discharge of the cell these two reactions run in reverse. The exchange currents for the reactions are high, which is why the lead battery can produce large currents on demand.

The nickel–cadmium cell is used in many units requiring a small, steady current. One electrode is coated with $Cd(OH)_2$, the other with $Ni(OH)_2$, the electrolyte being KOH. When electrons are pumped into the cadmium electrode reduction occurs:

$$Cd(OH)_2 + 2e^- \rightarrow Cd + 2OH^-.$$

At the same time electrons are withdrawn from the nickel electrode, and the reaction

$$2Ni(OH)_2 + 2OH^- \rightarrow 2NiOOH + 2H_2O + 2e^-$$

is induced. Charging is complete when all the available $Cd(OH)_2$ and $Ni(OH)_2$ have been converted. When current is required, the reactions are run in reverse, electrons are supplied at the cadmium electrode, and accepted by the nickel.

30.4 Corrosion

It may appear unduly pessimistic to end this book with a section on the decay of materials by corrosion, the ultimate triumph of the Second Law. Nevertheless, the importance of corrosion is familiar to everyone, and an understanding of the underlying principles does at least give some indication of how it may be controlled, even though we cannot hope for ultimate victory. Since the annual cost of corrosion is about 1 per cent of gross national product (a figure with many powers of ten in any currency), the organization of our resources can have a considerable economical impact.

The thermodynamics of corrosion. The *thermodynamic* warning of the likelihood of corrosion can be stated quite simply on the basis of equilibrium electrochemistry, Chapter 12. The reaction

$$M + M'^+ \rightarrow M^+ + M'$$

has a tendency to proceed spontaneously to the right if

$$E_e(M'^+, M') > E_e(M^+, M),$$

where the Es are the (equilibrium) electrode potentials of the two half reactions $M^+ + e^- \rightleftharpoons M$ and $M'^+ + e^- \rightleftharpoons M'$. One of the most important half-reactions is

$$\tfrac{1}{2}Fe^{2+} + e^- \rightleftharpoons \tfrac{1}{2}Fe \qquad E_e^{\ominus}(Fe^{2+}, Fe) = -0.4402 \text{ V}$$

because, if it can be driven to the left by coupling to another half-reaction, it is responsible for the decay of steel artifacts. We shall often identify M with Fe in the subsequent discussion.

The reaction often responsible for driving the $M \rightarrow M^+$ oxidation is the reduction of protons or oxygen:

In acid solution:

(a) $H^+ + e^- \rightleftharpoons \tfrac{1}{2}H_2 \qquad E_e^{\ominus}(H^+, H_2) = 0$

(b) $H^+ + \tfrac{1}{4}O_2 + e^- \rightleftharpoons \tfrac{1}{2}H_2O \qquad E_e^{\ominus}(H^+, O_2) = 1.23 \text{ V}$

In alkaline solution:

(c) $\frac{1}{2}H_2O + \frac{1}{4}O_2 + e^- \rightleftharpoons OH^-$ $E_e^\ominus(O_2, OH^-) = 0.401\,V$.

Since all three electrode potentials are greater than $E^\ominus(Fe^{2+}, Fe)$, all three can drive the corrosion of iron. In contrast, the electrode potentials for the so-called noble metals, such as gold, are strongly positive ($E_e^\ominus(Au^{3+}, Au) = +1.50\,V$) and so they are not corroded in moist air in the absence of complexing ions such as Cl^-.

Note that we have quoted the *standard* electrode potentials: electrode potentials may be quite different when ion concentrations deviate from unit activity, and so corrosion may become thermodynamically feasible. In particular, the electrode potentials of the two acid reactions (a) and (b) vary with pH as

$$E_e(a) = E_e^\ominus(a) - (RT/F)\ln a_{H^+} = (-0.059\,V)\,pH$$

$$E_e(b) = E_e^\ominus(b) - (RT/F)\ln a_{H^+} = (1.23\,V) - (0.059\,V)\,pH,$$

and so are capable of driving a range of oxidation reactions, depending on the pH of the medium.

Example (Objective 16). Will a mild steel vessel corrode when in contact with a solution of pH = 3.0?

- *Method.* The electrode potential of the acid corrosion reaction is $-(0.059\,V)\,pH$. Adopt the criterion that the vessel has corroded if the Fe^{2+} concentration exceeds $10^{-6}\,mol\,dm^{-3}$. Find $E(Fe^{2+}, Fe)$ for this concentration, and decide whether $E(H^+, H_2) > E(Fe^{2+}, Fe)$ at the stated pH.

- *Answer.* $E(H^+, H_2) = (-0.059\,V) \times 3.0 = -0.177\,V$.

  $$E(Fe^{2+}, Fe) = (-0.440\,V) + \frac{1}{2}(0.059\,V)\lg 10^{-6} = -0.617\,V \approx -0.6\,V.$$

  Since the hydrogen couple exceeds that of the iron, the vessel will corrode beyond the concentration $10^{-6}\,mol\,dm^{-3}$.

- *Comment.* The corrosion by this reaction ceases to be thermodynamically favoured only when the pH rises beyond 10.1.

The kinetics of corrosion. A thermodynamic discussion of corrosion is interesting only in so far as it shows whether a particular conjunction of metals has a tendency to corrode, or whether the exposure of a metal to a particular environment is likely to lead to its decay. What is of great interest is the *kinetics* of corrosion. Given that a system is unstable, we need to know how rapidly it will corrode, and knowing that, whether we can decrease the rate. In the remainder of this section we consider some of the factors that influence the rate of a corrosion reaction which has a thermodynamic tendency to occur.

A model of the corrosion system is illustrated in Fig. 30.14a: it can be taken to be a drop of slightly acidic or alkaline water and containing some dissolved oxygen, in contact with a metal (M). The oxygen rich area accepts electrons and undergoes reduction; its area is A. Elsewhere the metal is oxidized over an area A'. The droplet acts as a short-circuited

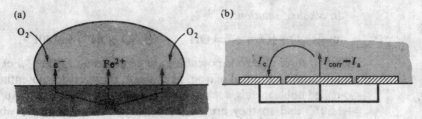

Fig. 30.14. (a) Corrosion under a droplet of water and (b) the model systems.

electrochemical cell, Fig. 30.14b, with the cathodic region acting as one electrode, and the anodic region the other.

The rate of corrosion can be measured by the current of M^+ (e.g. Fe^{2+}) ions leaving the metal surface. This is the *corrosion current*, I_{corr}, and can be identified with the anodic current, I_a. Since any current emerging from the anodic region must find its way to the cathodic region, the corrosion current must also be equal in magnitude to the cathodic current, I_c. In terms of the current densities at the metal oxidation site (j) and the oxygen reduction site (j') we have

(30.4.1) $$I_{corr} = jA = j'A' = \sqrt{(jj'AA')}.$$

The current densities can be expressed in terms of the Butler–Volmer equation at the relevant overpotential so long as we disregard concentration polarization effects. For simplicity we assume that the overpotentials are large enough to use the exponential form of eqn (30.1.7). Notational simplification is obtained by setting $\alpha = \frac{1}{2}$ and supposing that the rate determining step is a one electron transfer. All these assumptions can be removed, but that leads to more complicated expressions without showing anything new. Since the cell is short-circuited, there is zero potential difference between the cathodic and anodic regions. Moreover, if the droplet is small there is negligible potential difference between one part of the electrolyte surface and the other. Therefore the potential difference between the two regions and the solutions are the same. This common potential difference is called the *corrosion potential*, $\Delta\phi_{corr}$. The overpotentials for the two regions are therefore $\eta = \Delta\phi_{corr} - \Delta\phi_e$ and $\eta' = \Delta\phi_{corr} - \Delta\phi'_e$, and so the current densities are

(30.4.2a) $$j = j_e \exp(\eta F/2RT) = j_e \exp(\Delta\phi_{corr}/2RT)\exp[-\Delta\phi_e F/2RT],$$

(30.4.2b) $$j' = j'_e \exp(-\eta'F/2RT) = j'_e \exp(-\Delta\phi_{corr}/2RT)\exp[\Delta\phi'_e F/2RT].$$

This is very nearly the end of the calculation, because these two equations can be combined, using $jA = j'A'$, to find the corrosion potential:

(30.4.3) $$\exp(\Delta\phi^c F/RT) \approx \left(\frac{A'_e j'_e}{Aj_e}\right) \exp\{(F/2RT)[\Delta\phi_e + \Delta\phi'_e]\}.$$

This can be substituted into $I_{corr} = \sqrt{(jj'AA')}$ to give the final expression for the corrosion current:

(30.4.4) $$I_{corr} = (AA')^{1/2}(j_e j'_e)^{1/2} \exp\{(F/2RT)[E'_e - E_e]\}.$$

Several conclusions can be drawn from this approximate expression. In the first place, the rate of corrosion depends on the areas of the surfaces of the metal and the impurity: if no impurity is exposed, $A' = 0$, and, as we should expect, there is no corrosion. This points to a trivial, yet often effective, method of slowing the rate of corrosion: cover the surface with a coating, such as paint, so that there are no impurity areas open to the moist environment. (Paint also increases the effective solution resistance between the cathodic and anodic patches on the surface.) Other things being equal (i.e., for corrosion reactions with similar exchange currents) the rate is large when E_e is significantly larger than E'_e; that is, rapid corrosion may occur if the metal and the oxidizing couple have widely differing electrode potentials.

The role of the exchange currents in determining the corrosion rate can be seen by considering the case of iron in contact with an acidic solution. Thermodynamically the iron can be oxidized by reactions (a) and (b), p. 1064, and the latter is thermodynamically dominant in the sense that it couples with the iron reaction to give the greatest difference of electrode potentials. Nevertheless, its exchange current at an iron interface is only about $10^{-14}\,\mathrm{A\,cm^{-2}}$, whereas the exchange current for hydrogen evolution on iron is $10^{-6}\,\mathrm{A\,cm^{-2}}$. The latter therefore occurs more rapidly, and iron corrodes by virtue of the hydrogen evolution reaction in acid solution. Furthermore, the exchange currents depend on the pH of the solution (so do the electrode potentials) and the actual pH dependence of acidic corrosion is quite involved.

The inhibition of corrosion. Several techniques for inhibiting or preventing corrosion are available. Coating the surface with some impermeable layer, such as paint, may prevent the access of damp air. Unfortunately the protection fails disastrously if defects in the paint surface allow access of moist air. The oxygen has access to the exposed metal surface and can supply electrons there for consumption at the edges of the pinhole where the oxygen supply is less rich. Corrosion then takes place beneath the paintwork, and the extent of damage may be much greater than casual inspection of a blemish or a scratch might suggest. Almost every car owner suffers from this aspect of electrochemistry: so do people with submerged iron pipes or posts.

Another form of surface coating is provided by galvanizing: this consists of coating the iron object with zinc. Since the electrode potential of the reaction $\frac{1}{2}Zn^{2+} + e^- \rightleftharpoons \frac{1}{2}Zn$ is $-0.76\,\mathrm{V}$, and is more negative than the iron reduction, the zinc corrosion is thermodynamically favoured and the iron survives. In contrast, tin plating of iron leads to very rapid corrosion of the iron if the tin layer is scratched: the tin reduction potential is $-0.14\,\mathrm{V}$, and so it drives the iron oxidation. Some oxides are stable kinetically, in the sense that they adhere to the metal surface and form an impermeable layer over a fairly wide pH range: this is the reason why aluminium is stable in air even though its reduction potential is strongly negative $(-1.662\,\mathrm{V})$. This type of *passivation* can be viewed as cutting down the

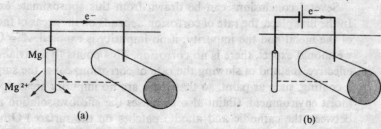

Fig. 30.15. (a) Cathodic and (b) impressed current cathodic corrosion protection.

exchange currents by sealing the surface.

Another method of protection is to change the potential of the object by pumping in electrons. It is then able to donate these excess electrons to the oxygen reduction reaction without needing to make the transition from M to M^+. One way of providing an excess pool of electrons is *cathodic protection*. This is achieved by connecting to the object a metal with a more negative electrode potential, such as magnesium ($-2.363\,V$). The magnesium acts as a *sacrificial anode*, supplying its own electrons to the iron object and decaying in the process, Fig. 30.15a. A block of magnesium, replaced occasionally, is often much cheaper than the ship, building, or pipeline for which it is being sacrificed. Another method of supplying electrons is *impressed current cathodic protection*, Fig. 30.15b. In this case a reaction in an external source provides the electrons and eliminates the need for the iron to transfer its own. In both cases, however, the Second Law is leading on, inexorably, to final corruption. All we can do is slow its operation, mitigate its effects, and not squander its power.

Further reading

Electrode processes. G. J. Hills; in *Essays in chemistry* (J. N. Bradley, R. D. Gillard, and R. F. Hudson, eds.), **2**, 19, Academic Press, London, 1971.

Experimental approach to electrochemistry. N. J. Selley; Edward Arnold, London, 1977.

Modern electrochemistry. J. O'M. Bockris and A. K. N. Reddy; Plenum, New York, 1970.

Electrode kinetics. W. J. Albery; Clarendon Press, Oxford, 1974.

Treatise on electrochemistry. G. Kortum; Elsevier, Amsterdam, 1965.

Reactions of molecules at electrodes. N. S. Hush; Wiley-Interscience, New York, 1971.

Potentiometry; Oxidation–reduction potentials. S. Wawzonek; in *Techniques of chemistry* (A. Weissberger and B. M. Rossiter, eds.), Vol. IIA, 1, Wiley-Interscience, New York, 1971.

Fuel cells: their electrochemistry. J. O'M. Bockris and S. N. Srinivasan; McGraw-Hill, New York, 1969.

Fuel cells. A. McDougall; Macmillan, London, 1976.

Problems

30.1. The key to the entire chapter is the Butler–Volmer equation, and so in the first few Problems we gain some familiarity with the way it can be used and manipulated. First, take $\alpha = \frac{1}{2}$ and plot the current density, j/j_e, as a function of the overpotential η at 25 °C.

30.2. A typical exchange current density, that for H^+ discharge at platinum, is 0.79 mA cm^{-2} at 25 °C. What is the current density at an electrode when its overpotential is (a) 10 mV, (b) 100 mV, (c) -5.0 V?

30.3. The exchange current density for a Pt|Fe^{3+}, Fe^{2+} electrode is 2.5 mA cm^{-2}. The standard electrode potential is $+0.771$ V. Calculate the current flowing through an electrode of surface area 1 cm^2 as a function of the potential of the electrode. Take unit activity for both ions.

30.4. Suppose that the electrode potential is set at 1.00 V. Calculate the current that flows for the ratio of activities $a(Fe^{2+})/a(Fe^{3+})$ in the range 0.1 to 10.0 and a temperature of 25 °C.

30.5. What overpotential is needed to sustain a 20 mA current at the Fe^{3+}, Fe^{2+}|Pt electrode in which both ions are at an activity $a = 0.1$?

30.6. How many electrons or protons are transported through the double layer in each second when the Pt|H_2, H^+, Pt|Fe^{3+}, Fe^{2+}, and Pb|H_2, H^+ electrodes are at equilibrium at 25 °C? (The exchange current density for the last is 5.0×10^{-12} A cm^{-2}; take the area as 1 cm^2 in each case.)

30.7. In order to appreciate the magnitude of these quantities on an atomic scale, estimate the number of times each second a single atom on the surface takes part in an electron transfer event.

30.8. What is the effective resistance at 25 °C of an electrode interface when the overpotential is small? Evaluate it for 1 cm^2 (a) Pt, H_2|H^+, (b) Hg, H_2|H^+ electrodes. (Data in Table 30.1.)

30.9. If $\alpha = \frac{1}{2}$, an electrode interface is unable to rectify alternating current because the current density curve is symmetrical about $\alpha = 0$ (Problem 30.1). When $\alpha \neq \frac{1}{2}$ the magnitude of the current density depends on the sign of the overpotential, and so some degree of rectification may be obtained: this is called *faradaic rectification*. Suppose that the overpotential varies as $\eta(t) = \eta_0 \cos \omega t$. Find an expression for the mean flow of current (averaged over a cycle) for general α, and confirm that the mean current is zero when $\alpha = \frac{1}{2}$. In each case work in the limit of small η_0, but you will have to work to second order in $\eta_0 F/RT$.

30.10. Calculate the mean direct current density at 25 °C for a 1 cm^2 hydrogen-platinum electrode with $\alpha = 0.38$ when the overpotential varies between ± 10 mV at 50 Hz.

30.11. Now suppose that the overpotential is in the high overpotential region at all times even though it is oscillating. What waveform will the current across the interface show if η varies linearly and periodically (as a sawtooth waveform) between η_- and η_+ around η_0? Take $\alpha = \frac{1}{2}$.

30.12. The Tafel plot, p. 1050, gives a means of determining both the exchange current density and the transfer coefficient α for electrodes, and so it lies at the root of the information we need when discussing electrode processes. In an experiment on the Pt|H_2, H^+ electrode in dilute H_2SO_4 the following current densities were observed at 25 °C. Find α and j_e for the electrode.

η/mV	50	100	150	200	250
$j^+/mA\,cm^{-2}$	2.66	8.91	29.9	100	335

30.13. How would the current density at this electrode depend on the over-potential of the same set of magnitudes but of opposite sign?

30.14. State what happens when a platinum electrode in an aqueous solution containing both cupric and zinc ions at unit activity is made the cathode of an electrolysis cell.

30.15. What are the minimum (thermodynamically determined) potentials at which (a) zinc, (b) copper can be deposited from aqueous molar solutions? What are the corresponding potentials when the concentrations are $0.01\,mol\,dm^{-3}$?

30.16. What are the conditions that allow a metal to be deposited from aqueous acidic solution before hydrogen evolution occurs significantly? Why may silver be deposited from aqueous silver nitrate? Why may cadmium be deposited from aqueous cadmium sulphate? (The overpotential for hydrogen evolution on cadmium is about $1\,V$ at current densities of $1\,mA\,cm^{-2}$.)

30.17. The exchange current density for H^+ discharge at zinc is about $5\times 10^{-11}\,A\,cm^{-2}$. Can zinc be deposited from a unit activity aqueous solution of a zinc salt?

30.18. The standard potential of the $Zn^{2+}|Zn$ electrode is $-0.763\,V$ at $25\,°C$. The exchange current density for H^+ discharge at platinum is $0.79\,mA\,cm^{-2}$. Can zinc be plated on to platinum at that temperature? (Take unit activities.)

30.19. Can magnesium be deposited on a zinc electrode from a unit activity acid solution at $25\,°C$?

30.20. Find an expression for the current density at an electrode where the rate process is diffusion controlled and η^c is known. Sketch the form of j/j_L as a function of η^c. What changes occur if anion currents are involved?

30.21. The limiting current density for the reaction $I_3^- + 2e^- \rightarrow 3I^-$ at a platinum electrode is $28.9\,\mu A\,cm^{-2}$ when the concentration of KI_3 is $6.6\times 10^{-4}\,mol\,dm^{-3}$ and the temperature $25\,°C$. The diffusion coefficient of I_3^- is $1.14\times 10^{-5}\,cm^2\,s^{-1}$. What is the thickness of the diffusion layer?

30.22. The limiting current density is given in terms of the ionic diffusion constant D_+ in eqn (30.1.12), but D can be related to the ionic conductivity λ_+ (p. 904). Derive an expression for the limiting current in terms of λ_+, the concentration c_+, and the thickness of the diffusion layer λ.

30.23. The ionic conductivity of Fe^{2+} is $40\,\Omega^{-1}\,cm^2\,mol^{-1}$. The limiting current at a platinum electrode of area $40\,cm^2$ dipping in to a solution of iron(II) chloride at $25\,°C$ was measured at various concentrations, the results are given below. What is the thickness of the diffusion layer at each concentration? (The answer you should get is a quite typical thickness.) Can you think of an independent non-electrochemical way of measuring λ?

$[FeCl_2]/mol\,dm^{-3}$	0.250	0.125	0.063	0.031
I/mA	215	107	49	23

30.24. The standard electrode potentials of lead and tin are $-126\,mV$ and $-136\,mV$ respectively at $25\,°C$, and there is little overvoltage for their deposition. What should be their relative activities in order to ensure simultaneous deposition from a mixture?

30.25. The standard electrode potentials of silver, copper, and zinc are 799 mV, 337 mV, and −763 mV respectively. When they are present in a mixture with cyanide ions they may be deposited simultaneously. Account for this observation.

30.26. We now turn to power generation. As a first step, find the maximum (reversible) potential difference of a nickel–cadmium cell, and the maximum possible power output when 100 mA is drawn, the temperature being 25 °C.

30.27. Calculate the thermodynamic limit to the e.m.f. of fuel cells operating on (a) hydrogen and oxygen, (b) methane and air. Use the Gibbs function information in Table 5.3, and take the species to be in their standard states at 25 °C.

30.28. Estimating the power output and e.m.f. of a cell under operating conditions is very difficult, but eqn (30.2.4) summarizes, in an approximate way, some of the parameters involved. As a first step in manipulating this expression, identify all the quantities that depend on the ionic concentrations. Express E in terms of the concentration and conductivities of the ions present in the cell.

30.29. Estimate the parameters in the last Problem for $Zn|ZnSO_4(aq)||CuSO_4(aq)|$ Cu. Take electrodes of area $5\,cm^2$ separated by $5\,cm$. Ignore both potential differences and resistance of the liquid junction. Take the concentration as $1\,mol\,dm^{-3}$, the temperature being 25 °C, and neglect activity coefficients. Plot E as a function of the current drawn.

30.30. On the same graph, plot the power output of the cell. What current corresponds to maximum power?

30.31. Consider a cell in which the current is activation controlled. Show that the current for maximum power can be estimated by plotting $\lg(I/I_e)$ and $c_1 - c_2 I$ against I (where $I_e^2 = A^2 j_e j'_e$ and c_1 and c_2 are constants), and looking for the point of intersection of the curves. Carry through this analysis for the cell in Problem 30.29 ignoring all concentration overpotentials.

30.32. If copper and mild steel are in contact (as in a badly designed domestic water system) which will tend to corrode?

30.33. Find the pH dependence of the electrode potential of the alkaline corrosion reaction, reaction (c) on p. 1065. The $H^+ + e^- \rightleftharpoons \frac{1}{2}H_2$ reaction has a more negative potential at any pH: why, then, does it ever play a role in corrosion?

30.34. Which of the following metals has a thermodynamic tendency to corrode in moist air at pH 7: Fe, Cu, Pb, Al, Ag, Cr, Co? Take as a criterion of corrosion a metal ion concentration of at least $10^{-6}\,mol\,dm^{-3}$.

30.35. Find the pH at which each of the metals in the last Problem will corrode at 25 °C.

30.36. Estimate the magnitude of the corrosion current for a patch of zinc of area $0.25\,cm^2$ in contact with a similar area of iron in an aqueous environment at 25 °C. Take the exchange current densities as $10^{-6}\,A\,cm^{-2}$ and the local ion concentrations as $10^{-6}\,mol\,dm^{-3}$.

Answers to problems

Chapter 1 (1.1) 10 atm. (1.2) (a) 15.1 m^3; (b) 151 m^3. (1.3) 0.5 m^3. (1.4) 1.47 kPa, 1.5 per cent.
(1.5) 30 K. (1.6) 29.7 lb in^{-2}. (1.7) 0.031 atm. (1.8) $p = \rho(RT/M_m)$, $M_r = 46.1$.
(1.9) 4.6×10^3 mol or 9.2 kg hydrogen, payload 129 kg. With He, payload 120 kg. At
30000 ft, payloads are 124 kg and 115 kg, respectively. (1.10) (a) 2.72×10^{-26} kg; (b)
$M_r = 16.4$, according to this experiment. (1.11) $M_r = 102$, CH_2FCF_3.
(1.12) 0.184 mmHg, 68.6 mmHg, 0.184 mmHg. (1.13) $\chi(H_2) = \frac{2}{3}$, $\chi(N_2) = \frac{1}{3}$, $p(H_2) =$
2 atm, $p(N_2) = 1$ atm, $p = 3$ atm. (1.14) $p(N_2) = \frac{1}{3}$ atm, $p(NH_3) = \frac{4}{3}$ atm, $p = \frac{5}{3}$ atm.
(1.15) No, 24.5 atm. (1.16) 21.6 atm. (1.17) (a) 1.00 atm, 821 atm; (b) 0.994 atm,
1468 atm. (1.18) (a) 8.73 atm, 3665 K; (b) 4.54 atm, 2619 K; (c) 0.177 atm, 47.0 K.
(1.19) 54.5 atm, 67.8 cm^3 mol^{-1}, 120 K. (1.20) $a = 1.333$ atm dm^6 mol^{-2},
$b = 0.0329$ dm^3 mol^{-1}, $v_{mol} \approx 0.055$ nm^3, $r_{mol} \approx 0.24$ nm. (1.21) He 226 pm, Ne 202 pm,
Ar 246 pm, Xe 287 pm. (1.22) $a = 5.66$ dm^6 atm mol^{-2}, $b = 59.5$ cm^3 mol^{-1}.
(1.23) $B(T) = b - a/RT$, $C(T) = b^2$. (1.24) $B(T) = b - a/RT$, $C(T) = b^2 - (ab/RT) + (a^2/2R^2T^2)$.
(1.25) $p_c = 39.0$ atm, $V_{c,m} = 104$ cm^3 mol^{-1}, $T_c = 131$ K (for van der Waals gas);
$p_c = 25.8$ atm, $V_{c,m} = 87.8$ cm^3 mol^{-1}, $T_c = 102$ K (for Dieterici gas). (1.26) $p_c = B^3/27C^2$,
$V_{c,m} = 3C/B$, $T_c = B^2/3RC$, $Z_c = \frac{1}{3}$. (1.27) $B' = B/RT$, $C' = (C - B^2)/R^2T^2$.
(1.28) 1001 K, $T_B = (a/Rb)$. (1.29) (a) $T_{B,r} \approx (27/8)$; (b) $T_{B,r} \approx 4$. (1.30) -0.257 dm^3 mol^{-1}.
(1.32) (a) 0.99998 p_0; (b) 0.95 atm. (1.33) Partial pressures of N_2, O_2, Ar, CO_2 in atm
at sea level: 0.782, 0.208, 0.009, 0.0003; corresponding values for (a) 0.747, 0.198, 0.008,
0.0003; (b) 0.289, 0.067, 0.002, 6×10^{-5}; (c) 1.10×10^{-5}, 5.41×10^{-7}, 9.3×10^{-10},
6.2×10^{-12}. (1.34) $k = 1.50 \times 10^{-23}$ J K^{-1}, $L = 5.5 \times 10^{23}$ mol^{-1}.

Chapter 2 (2.1) (a) 9.8 J; (b) 1.6 J. (2.2) 2.0 kJ. (2.3) 10 J, 10 J. (2.4) 200 kJ, 200 W, 68 °C.
(2.5) (a) $w = 2Fa/\pi$; (b) $w = 0$. (2.6) 101 J. (2.7) 4.9 J, 4.9 J. (2.8) The capacity of the
surroundings to do work is reduced by 4.9 J. (2.9) (a) -273 J; (b) -941 J.
(2.10) -8.09 kJ. (2.11) -8.09 kJ. (2.12) -1.53 kJ. (2.13) $w = -nRT\{\ln (V_f/V_i) -$
$nB[(1/V_f) - (1/V_i)] + \ldots\}$. (2.14) (a) -1.53 kJ; (b) -51 kJ; (c) -1.57 kJ.
(2.15) $w = -nRT\ln [(V_f - nb)/(V_i - nb)] + n^2a[(V_f - V_i)/V_fV_i]$.
(2.17) $w_r = -(8/9)nT_r \ln [(3V_{r,f} - 1)/(3V_{r,i} - 1)] + n[(V_{r,f} - V_{r,i})/(V_{r,f}V_{r,i})]$.
(2.18) (a) $(\partial p/\partial T)_V = p/T$; $(\partial p/\partial V)_T = -p/V$; (b) $(\partial p/\partial T)_V = [1 + (na/RTV)](p/T)$, $(\partial p/\partial V)_T =$
$[(na/RTV) - V/(V - nb)](p/V)$. (2.19) (a) $\delta p/p \approx 1$ per cent; (b) $\delta p/p \approx 2$ per cent;
(c) $\delta p/p \approx 3$ per cent. (2.20) (a) $(\frac{3}{2})nR$; (b) $3nR$. (2.21) 18.02 cal deg^{-1} mol^{-1},
75.4 J K^{-1} mol^{-1}. (2.22) 125 J, $w = 0$; (b) 208 J, $w = -83$ J. (2.23) 663 kJ; 663 s.
(2.24) 314 kJ, 5 min 14 s. (2.25) (a) -172 kJ; (b) 2.25×10^3 kJ; (c) 2.08×10^3 kJ;
(d) 2.25×10^3 kJ. (2.26) $w = -147$ kJ, $q = 2.44 \times 10^3$ kJ, $\Delta U = 2.29 \times 10^3$ kJ,
$\Delta H = 2.44 \times 10^3$ kJ. (2.27) 2.25×10^3 kJ, 2.44×10^3 kJ, 3.75 kW, 4.07 kW, 23 km, 25 km.
(2.28) 314 kJ. (2.29) $q = 22.2$ kJ, $w_{rev} = 1.7$ kJ, $\Delta H_{vap} = 22.2$ kJ, $\Delta U_{vap} = 20.5$ kJ,
$\Delta H_{vap,m} = 40.0$ kJ mol^{-1}. (2.30) $\Delta H_{vap,m} = 98.9$ kJ mol^{-1}, $\Delta U_{vap,m} = 96.2$ kJ mol^{-1}.

Chapter 3 (3.1) Mass, enthalpy, susceptibility, heat capacity. (3.2) 0, 0. (3.3) (a) -7.0 mm^3;
(b) -2.1 cm^3. (3.4) $\delta V_T = -2.6$ cm^3, combined effect is -2.6 cm^3 at 100 ft, and
-4.7 cm^3 at 5000 fath. (3.6) $d \ln p/dt = (1/\tau_V) - (1/\tau_T)$; p is independent of time when
$\tau_T = \tau_V$. (3.7) $(\partial p/\partial T)_V = nR/(V - nb)$, $(\partial p/\partial V)_T = -nRT/(V - nb)^2 + 2n^2a/V^3$.
(3.9) $\alpha = RV^2(V - nb)/[RTV^3 - 2a(V - nb)^2]$, $\kappa = V^2(V - nb)^2/[nRTV^3 - 2n^2a(V - nb)^2]$.
(3.10) $\mu_{JT}C_p = -\{[nbRTV^2 - 2an(V - nb)^2]/[RTV^3 - 2an(V - nb)^2]\}V$, $\mu_{JT}(Xe) = 1.46$ K atm^{-1}.
(3.11) $T_1 = 2(a/bR)(1 - nb/V)^2$, $T_{1,r} = (27/4)(1 - 1/3V_r)^2$, (a) 223 K, (b) 2053 K.
(3.12) -4.2 atm. (3.13) 1.3 K atm^{-1}. (3.14) $(\partial H/\partial p)_T = -T(\partial V/\partial T)_p + V$.
(3.15) 0.29 K atm^{-1}. (3.16) (a) 0.91 mm^3; (b) 0.015 mm^3. (3.17) $\lambda = 1.11$ $C_{p,m} - C_{V,m} \approx$
9.2 J K^{-1} mol^{-1}. (3.18) (a) 0.731 J K^{-1} mol^{-1}, 0.29 kJ; (b) 39.8 J K^{-1} mol^{-1}, 12.7 kJ.
(3.19) Cu 2.05×10^9 J m^{-3}, benzene 4.07×10^8 J m^{-3}. (3.20) 0.75 kJ for each, the

amounts differ by only $\alpha p V_m \Delta T \approx 5$ mJ. **(3.21)** $w = -20.2$ J; $q = 0$, $\Delta U = -20.2$ J, $\Delta H = -26.0$ J, $\Delta T = -0.35$ K. **(3.22) (a)** 226 K; **(b)** 234 K. **(3.23)** 31.6 J K^{-1} mol^{-1}. **(3.24)** 1.31, 41.3 J K^{-1} mol^{-1}. **(3.25)** $\Delta U_m = -1.57$ kJ, $\Delta H_m = -1.98$ kJ. **(3.27)** $\frac{5}{3}$, $\frac{4}{3}$. **(3.28)** $1 + \frac{2}{3}\lambda$, $1 + \frac{1}{3}\lambda$. **(3.29) (a)** 1.671; **(b)** 1.336. **(3.30) (a)** 1.02 km s^{-1}; **(b)** 346 m s^{-1}. **(3.31)** 1.24, 34.6 J K^{-1} mol^{-1}. **(.32)** 357 Hz (F above middle C).

Chapter 4 **(4.1) (a)** exothermic; **(b, c)** endothermic. **(4.2) (a)** -57.2 kJ mol^{-1}; **(b)** -176.0 kJ mol^{-1}; **(c)** -32.9 kJ mol^{-1}; **(d)** -55.8 kJ mol^{-1}. **(4.3)** 641.2 J K^{-1}. **(4.4) (a)** -2.802×10^3 kJ mol^{-1}; **(b)** -2802 kJ mol^{-1}; **(c)** -1.274×10^3 kJ mol^{-1}. **(4.5)** $\Delta H_m = 17.7$ kJ mol^{-1}, $\Delta H_f^{\ominus} = 72.5$ kJ mol^{-1}. **(4.6)** -2130 kJ mol^{-1}, -2130 kJ mol^{-1}, -1267 kJ mol^{-1}. **(4.7)** 78.6 kJ mol^{-1}. **(4.8)** -383 kJ mol^{-1}. **(4.9) (a)** -6.20 kJ mol^{-1}; **(b)** -7.73 kJ mol^{-1}; **(c)** 7.78 kJ mol^{-1}. **(4.10)** -97.8 kJ mol^{-1}. **(4.11) (a)** 2877 kJ mol^{-1}, 3536 kJ mol^{-1}, 5471 kJ mol^{-1}; **(b)** 49.50 kJ g^{-1}, 49.01 kJ g^{-1}, 47.91 kJ g^{-1}. **(4.12)** 49.40 kJ g^{-1}, 48.91 kJ g^{-1}, 48.17 kJ g^{-1}. **(4.13) (a)** 0.47 g; **(b)** 0.47 kg. **(4.14)** 11.3 kJ mol^{-1}. **(4.15)** 1.90 kJ mol^{-1}. **(4.16)** $\Delta H(T_2) = \Delta H(T_1) + \Delta a(T_2 - T_1) + (1/2)\Delta b(T_2^2 - T_1^2) - \Delta c[(1/T_2) - (1/T_1)]$. **(4.17)** -59.7 kJ mol^{-1}. **(4.18) (a)** -286.6 kJ mol^{-1}; **(b)** -283.5 kJ mol^{-1}. **(4.19)** -283.2 kJ mol^{-1}, -283.4 kJ mol^{-1}. **(4.20)** $\Delta U(T_2) = \Delta U(T) + \int_{T_n}^{T} \Delta C_V(T)\mathrm{d}T$. **(4.21) (a)** -1059 kJ mol^{-1}; **(b)** -2338 kJ mol^{-1}. **(4.22)** $\Delta H_{c,m} = -268$ kJ mol^{-1}, ΔH (to lactic acid) $= -5376$ kJ mol^{-1}. **(4.23)** 10 m. **(4.24)** 2.4×10^3 kJ, 155 g; 82 °F. **(4.25)** $\Delta H_c^{\ominus}$(liq) $= -2205$ kJ mol^{-1}, ΔU_c(liq) $= -2200$ kJ mol^{-1}. **(4.26)** -2203 kJ mol^{-1}, -2198 kJ mol^{-1}. **(4.27)** 150 kJ mol^{-1}. **(4.28)** $\Delta H_{sol,m}$(KF)/kJ mol^{-1} $= -35.4 + 6.2\ m$, $\Delta H_{sol,m}$(KFAcOH)/kJ mol^{-1} $= 3.1 + 0.9\ m$. **(4.29)** 1173 kJ mol^{-1}. **(4.30)** 22.1 kJ mol^{-1}. **(4.31)** -1890.3 kJ mol^{-1}. **(4.32)** -45.5 kJ mol^{-1}.

Chapter 5 **(5.1) (a)** 6.25×10^{-2}; **(b)** 9.77×10^{-4}; **(c)** $1.55 \times 10^{-2} \times 10^{-10^{23}} \approx 0$. **(5.2) (a)** 91.5 J K^{-1}; **(b)** 67.0 J K^{-1}. **(5.3)** 45.4 J K^{-1}. **(5.4) (a)** 0; **(b)** 51.2 J K^{-1}. **(5.5)** 152.68 J K^{-1} mol^{-1}. **(5.6) (a)** 200.7 J K^{-1} mol^{-1}; **(b)** 232.0 J K^{-1} mol^{-1}. **(5.7)** 2.83 J K^{-1}. **(5.8)** 17.0 J K^{-1}. **(5.9)** 36.0 J K^{-1}. **(5.10) (a)** $\Delta S_{sys} = 2.88$ J K^{-1}, $\Delta S_{surr} = -2.88$ J K^{-1}, $\Delta S_{univ} = 0$; **(b)** $\Delta S_{sys} = 2.88$ J K^{-1}, $\Delta S_{surr} = 0$, $\Delta S_{univ} = 2.88$ J K^{-1}; **(c)** $\Delta S_{sys} = 0$, $\Delta S_{surr} = 0$, $\Delta S_{univ} = 0$. **(5.11) (a)** 21.3 J K^{-1} mol^{-1}; **(b)** -109.7 J K^{-1} mol^{-1}. **(5.12)** -0.11 kJ mol^{-1}, 0.75 kJ mol^{-1}. **(5.13) (a)** 109 J K^{-1} mol^{-1}; **(b)** 87.3 J K^{-1} mol^{-1} (changes in systems); $\Delta S_{surr} = -\Delta S_{sys}$; $\Delta S_{univ} = 0$. **(5.14)** Water 11.8 J K^{-1} mol^{-1}, benzene 84.4 J K^{-1} mol^{-1}. **(5.15) (a)** 0.11 kJ mol^{-1}; **(b)** 0.11 kJ mol^{-1}. **(5.16)** 798 kJ mol^{-1}, 802.8 kJ mol^{-1}. **(5.17) (a)** 63.88 J K^{-1} mol^{-1}; **(b)** 66.08 J K^{-1} mol^{-1}. **(5.18)** 152.0 J K^{-1} mol^{-1}. **(5.19)** $w(1) = -RT_h \ln (V'/V_i)$, $w(2) = C_{V,m}(T_c - T_h)$, $w(3) = -RT_c \ln (V'''/V'')$, $w(4) = C_{V,m}(T_h - T_c)$, $q(1) = -w(1)$, $q(2) = q(4) = 0$, $q(3) = -w(3)$, $\Delta S(1) = R \ln (V'/V_i)$, $\Delta S(2) = \Delta S(4) = 0$, $\Delta S(3) = R \ln (V'''/V'')$, $\Delta U(1) = \Delta U(3) = 0$, $\Delta U(2) = w(2)$, $\Delta U(4) = w(4)$. **(5.23)** 0.11, 0.38. **(5.24)** 0.10 kg, 0.03 kg. **(5.25)** 6.9 km. **(5.27)** 22.6 J K^{-1}. **(5.28)** 0.95 J K^{-1} mol^{-1}. **(5.29)** 12.1 J K^{-1} mol^{-1}, -12.1 J K^{-1} mol^{-1}, 0. **(5.30) (a)** -153 J K^{-1} mol^{-1}; **(b)** $+33.3$ J K^{-1} mol^{-1}; **(c)** 511.9 J K^{-1} mol^{-1}. **(5.31) (a)** -178.7 kJ mol^{-1}; **(b)** -228.6 kJ mol^{-1}; **(c)** -5798 kJ mol^{-1}. **(5.32) (a)** -163.3 J K^{-1} mol^{-1}, -285.8 kJ mol^{-1}, -237.2 kJ mol^{-1}; **(b)** -360.8 J K^{-1} mol^{-1}, -205.2 kJ mol^{-1}, -97.7 kJ mol^{-1}; **(c)** -192.9 J K^{-1} mol^{-1}, -318.2 kJ mol^{-1}, -260.7 kJ mol^{-1}. **(5.33)** 96.864 J K^{-1} mol^{-1}, 76.9 J K^{-1} mol^{-1}. **(5.36)** 243 J K^{-1} mol^{-1}, 34.4 kJ mol^{-1}, -128 J K^{-1} mol^{-1}.

Chapter 6 **(6.4)** $(\partial H/\partial p)_T = -T(\partial V/\partial T)_p + V$. **(6.5) (a)** 0; **(b)** -8.2 J atm^{-1}, -8.2 J. **(6.6) (a)** 306 J m^{-3}; **(b)** 30.6 kJ m^{-3}. **(6.8)** $(\partial C_V/\partial V)_T = (RT/V^2)(\partial^2 BT/\partial T^2)_V$, 0.06 J K^{-1} mol^{-1}. **(6.10)** $\mu_J C_V = p - T\alpha/\kappa$. **(6.11)** $(\partial U/\partial V)_T = a/V_m^2$; **(a)** 0.23 kJ m^{-3}; **(b)** 23 kJ m^{-3}. **(6.12)** $(\partial U/\partial V)_T = anp/RTV$, $(\partial U_r/\partial V_r)_T = p_r/T_r V_r$. **(6.14)** $q = nRT \ln [(V_r - nb)/(V_i - nb)]$. **(6.15)** $T\mathrm{d}S = C_p \mathrm{d}T - T(\partial V/\partial T)_p \mathrm{d}p$, $q = -\alpha TV(p_f - p_i)$, -0.5 kJ. **(6.16) (a)** -17.0 kJ; **(b)** 8.87 kJ; **(c)** -8.1 kJ. **(6.17)** 1.00 kJ. **(6.18)** $G(p_f) \approx G(p_i) + V\Delta p - (1/2)\kappa V(\Delta p)^2$.

(6.19) (a) 72.13 J mol^{-1} (error $\approx$ 0.01 per cent); (b) 7.213 kJ mol^{-1}; (error $\approx$ 1 per cent).
(6.20) (a) 182.7 J mol^{-1}; (b) 13.79 kJ mol^{-1}. **(6.21)** $\Delta G(T_f) \approx \tau \Delta G(T_i) +$
$[\Delta H(T_i) - T_i \Delta C_p] - T_f \Delta C_p \ln \tau$, $\tau = T_f/T_i$. **(6.22)** -231.9 kJ mol^{-1}, -52.8 kJ mol^{-1}.
(6.23) (a) 10.7 kJ mol^{-1}; (b) 114 kJ mol^{-1}. **(6.24)** 5 kJ mol^{-1}. **(6.25)** 11.4 kJ.
(6.26) $G(\bar{p}) = G(0) + p^* V_0 \{1 - \exp[-\bar{p}/p^*]\}$. **(6.27)** 73.1 atm. **(6.28)** $f = p\gamma$,
$\ln \gamma = Bp/RT + \frac{1}{2}(C - B^2)p^2/R^2T^2$. **(6.29)** (a) 0.9991 atm; (b) 0.99999 atm.
(6.30) 1.00095 atm. **(6.31)** (a) 0.9894 atm; (b) 34.4 atm.
(6.32) $f = p \exp[\sqrt{(1 + 4pB/R)} - 1]/(1/2)[1 + \sqrt{(1 + 4pB/R)}]$.

Chapter 7 **(7.1)** 2.8666 kJ mol^{-1}. **(7.2)** (a) 2.8668 kJ mol^{-1}; (b) 2.8668 kJ mol^{-1}.
(7.3) (a) -22.0 J K^{-1} mol^{-1}; (b) -109 J K^{-1} mol^{-1}. **(7.4)** (a) -1.65×10^{-4} kJ mol^{-1}
atm^{-1}; (b) 3.04 kJ mol^{-1} atm^{-1}. **(7.5)** 110 J mol^{-1}. **(7.6)** 0.55 kJ mol^{-1}, 0.61 kJ mol^{-1}.
(7.7) $-7.5\,°C$. **(7.8)** $8.7\,°C$. **(7.9)** 0.36 K. **(7.10)** $d(\Delta H/T) = \Delta C_p d \ln T$.
(7.11) $\delta T = 0.07$ K. **(7.13)** 10.2 mmHg. **(7.14)** 63 kJ mol^{-1}, 490 K. **(7.15)** $22\,°C$.
(7.16) 24 g s^{-1}. **(7.17)** (a) 1.7 kg; (b) 30.8 kg; (c) 1.4 g. **(7.18)** (a) 17 mmHg; (b) 1.2 kg.
(7.19) (a) 357 K; (b) 34.9 kJ mol^{-1}. **(7.20)** 55 kJ mol^{-1}, $227.5\,°C$.
(7.21) $1/T_h = 1/T_b + (M_m g/T\Delta H_m)h$, 363 K. **(7.22)** $\Delta H_m/$kJ mol$^{-1} = a/1000$, 0.039 mmHg,
63.1 kJ mol^{-1}. **(7.23)** 3 mmHg. **(7.28)** (a) 13.7, (b) 8.8. **(7.29)** 6.10 kJ, 61 s.
(7.30) $w = C_p[(T_f - T_i) - T_h \ln(T_f/T_i)]$. **(7.31)** 6.85 kJ, 69 s, 6.85 kJ. **(7.32)** 3.7×10^{-7} J.
(7.33) 6.73×10^{-6} J. **(7.34)** 8×10^{-6} J, 8 s. **(7.35)** (a) 0; (b) 9.6 kJ. **(7.36)** 17 W.
(7.37) (a) 0.05; (b) 5×10^{-5}; (c) 5×10^{-7}. **(7.38)** 9.5 J, 9.5 J. **(7.39)** (a) 1.0002; (b) 1.020.
(7.40) 4.8 g, 4.9 g. **(7.41)** (a) 1.49 cm, 1.23 cm; (b) 14.9 cm, 12.3 cm.
(7.42) 5.8 cm, 440 N m^{-2}. **(7.43)** 15 cm. **(7.44)** $h = (2\gamma/\rho gr)\cos\theta$. **(7.45)** $(\partial V/\partial\sigma)_{p,T} = \frac{1}{2}r$.
(7.46) (a) 3.8 m s^{-1}, (b) 2.2 m s^{-1}.

Chapter 8 **(8.1)** 17.48 cm^3 mol^{-1} (NaCl), 18.07 cm^3 mol^{-1} (H_2O). **(8.2)** (a) -1.39 cm^3 mol^{-1};
(b) 18.04 cm^3 mol^{-1}. **(8.5)** 886.4 cm^3, 887.7 cm^3. **(8.6)** 45.7 g of each, 57.6 cm^3 ethanol
and 45.7 cm^3 water, 0.97 cm^3. **(8.8)** Show that $(\partial V_A/\partial x_B)/(\partial V_B/\partial x_B) = -x_B/(1 - x_B)$.
(8.9) Show that $\partial \ln p_B/\partial \ln x_B = 1$ if $p_A = x_A p_A^*$. **(8.10)** 80.36 cm^3 mol^{-1}.
(8.11) -351 J, 1.18 J K^{-1}. **(8.12)** -702 J, 2.36 J K^{-1}, -133 J. **(8.13)** 4.69 J K^{-1} mol^{-1}.
(8.14) -18.5 kJ, 62.0 J K^{-1}, 0. **(8.15)** (a) 1:1; (b) 1.00:0.860.
(8.16) (a) 3.4×10^{-3} mol dm^{-3}; (b) 3.4×10^{-2} mol dm^{-3}. **(8.17)** N_2 5.1×10^{-4} mol kg^{-1},
O_2 2.7×10^{-4} mol kg^{-1}. **(8.18)** 0.3 mol kg^{-1}. **(8.19)** 3.94×10^5 mmHg.
(8.20) 0.920:0.080, 0.968:0.032. **(8.21)** $K_f = 32$ K/(mol kg^{-1}), $K_b = 5.22$ K/(mol kg^{-1}).
(8.22) $-1.3\,°C$. **(8.23)** 82. **(8.24)** ± 0.05 K. **(8.26)** (104 K)x_B, (103 K)x_B.
(8.27) $-0.167\,°C$. **(8.28)** $CaCl_2$. **(8.30)** 5 or 6. **(8.31)** $x_B = 0.92$. **(8.32)** 24 g kg^{-1}.
(8.33) 87 000. **(8.34)** 14 000. **(8.35)** (a) $y_T = 0.36$; (b) $y_T = 0.82$. **(8.37)** $119\,°C$, $x_T = 0.410$,
liq:vap $= 0.73$. **(8.38)** 4. **(8.39)** 9.

Chapter 9 **(9.1)** a, c, e. **(9.2)** b, d. **(9.3)** (a) 53 kJ mol^{-1}; (b) -53 kJ mol^{-1}. **(9.4)** 9.5/ln K,
$100(K^{0.1} - 1)$. **(9.5)** (a) 775; (b) 6.01×10^5; (c) 1.29×10^{-3}. **(9.6)** -14.4 kJ mol^{-1}.
(9.7) 1.68×10^{-5}. **(9.8)** (a) 9.00; (b) -12.8 kJ mol^{-1}; (c) 161.6 kJ mol^{-1};
(d) 249 J K^{-1} mol^{-1}. **(9.9)** 5.64, 0.764, $+129$ kJ (mol dimes)$^{-1}$. **(9.10)** 1.8×10^{-10},
1.3×10^{-5} mol kg^{-1}. **(9.11)** 57 kJ mol^{-1}. **(9.12)** H_2 14.7 kJ mol^{-1}, CO 18.8 kJ mol^{-1}.
(9.13) (a) 299 K; (b) 301 K. **(9.14)** $x(H_2) = 0.01$, $x(I_2) = 0.11$, $x(HI) = 0.78$.
(9.17) $K = [16(2 - \alpha)^2\alpha^2/27(1 - \alpha)^4][1/(p/atm)^2]$. **(9.18)** $\alpha_e = 1 - \{1/[1 + (3/4)\sqrt{(3K_p/atm)}]\}$.
$x(N_2) = 0.0033$, $x(H_2) = 0.0100$, $x(NH_3) = 0.9867$. **(9.20)** -66.1 kJ mol^{-1},
-142 J K^{-1} mol^{-1}. **(9.21)** 157.6 kJ mol^{-1}. **(9.22)** 8.3, -5.2 kJ mol^{-1},
-169 J K^{-1} mol^{-1}. **(9.23)** K_p (mean) $= 104$, $K_p' = 760 K_p$. **(9.24)** -137 kJ mol^{-1}.
(9.25) -98.2 kJ mol^{-1}, -301 J K^{-1} mol^{-1}, 1.04×10^{-6}. **(9.26)** $(\partial \ln K_p/\partial p)_T \approx$
$[2a(HI)p(HI) - a(I_2)p(I_2) - a(H_2)p(H_2)]p/RT$, K_p (550 atm) $\approx 1.15 K_p$(500 atm).
(9.27) 212.7 kJ mol^{-1}, 218.7 kJ mol^{-1}. **(9.28)** 3.3×10^6 kJ, -9.4×10^5 kJ.
(9.29) $\Delta G(T_2) = \Delta G(T_1) + [(T_1 - T_2)/T_1][\Delta H(T_1) - \Delta G(T_1)] + (T_2 - T_1)(\Delta a - T_2\Delta b) +$
$(1/2)\Delta b(T_2^2 - T_1^2) - T_2 \Delta a \ln(T_2/T_1) - \Delta c[(1/T_2) - (1/T_1)] + \frac{1}{2}\Delta c T_2[(1/T_2^2) - (1/T_1^2)]$.
(9.30) -225.4 kJ mol^{-1}. **(9.31)** $S(T) = [H(T) - H(0)]/T - \Phi_0(T)$.
(9.32) (a) 125.9 kJ mol^{-1}; (b) -20.2 kJ mol^{-1}, 25.2 kJ mol^{-1}. **(9.33)** (a) 2.7×10^{-7};
(b) 132, 0.22. **(9.34)** $\Phi_{298}(T) = \Phi_{298}(298) + \int_{298K}^{T} C_p(T')[(T' - T)/TT'] dT'$.

Chapter 10 **(10.2) (a)** 2; **(b)** 2; **(c)** 3. **(10.4)** $C = 2$, $P = 2$. **(10.5)** $P = 2$, $C = 1$; $P = 2$, $C = 2$.
(10.6) $C = 2$, $P = 3$, $F = 1$. **(10.7)** $C = 2$, $P = 2$, $F = 2$. **(10.9) (a)** 2150 °C;
(b) liq + sol, $x(MgO) = 0.35$, $y(MgO) = 0.18$; liq : sol = 0.42; **(c)** 2650 °C.
(10.11) (a) liq : sol = 5; **(b)** liq : sol = 0. **(10.14)** $T_{uc} = 122$ °C, $T_{lc} = 8$ °C.
(10.15) One phase up to $w = 0.18$, two phases up to $w = 0.84$, then one phase.
(10.18) 215 °C. **(10.19) (a)** 80 % Ag; **(b)** decomposes; **(c)** 80% Ag; **(d)** 20% Sn.
(10.21) $MgCu_2$ deposited at ≈ 770 °C, solidified at 560 °C to give $MgCu_2 + Mg_2Cu$.
(10.25) 2 phases; add 81 g or remove 46 g. **(10.28) (a)** 2 phases; **(b)** 3 phases,
(c) 1 phase; **(d)** invariant point. **(10.29) (a)** 19.5 mol kg^{-1}; **(b)** 23.8 mol kg^{-1}.

Chapter 11 **(11.1)** $I = [(1 + \sum_j n_j M_{j,m})/2\rho_{soln}]\sum_j c_j z_j^2$. **(11.3)** 0.9 mol kg^{-1}. **(11.4)** 2.52 mol kg^{-1}.
(11.5) 0.25 mol kg^{-1}. **(11.7)** $a(KCl) = \gamma_\pm^2 m^2$, $a(MgCl_2) = 4\gamma_\pm^3 m^3$, $a(FeCl_3) = 27\gamma_\pm^4 m^4$,
$a(CuSO_4) = \gamma_\pm^2 m^2$, $a(Al_2(SO_4)_3) = 108\gamma_\pm^5 m^5$. **(11.8) (a)** -1.92 kJ; **(b)** -52.9 kJ.
(11.9) 2.9×10^{-23} N at all distances. **(11.10) (a)** 2.1×10^{-23} N; **(b)** 1.1×10^{-64} N;
(c) ≈ 0. **(11.11) (a)** 5.54 nm; **(b)** 5.30 nm. **(11.12)** 3.18 nm. **(11.13)** -3.95.
(11.14) 0.964, 0.949, 0.920, 0.889, 0.847. **(11.19)** 0.773. **(11.21)** $pK_a' \approx pK_a - 2A[K_a c^0(HA)]^{1/4} = 4.719$. **(11.22)** 1.80×10^{-10}, 9.04×10^{-7}. **(11.23)** 9.1×10^{-7} mol kg^{-1}.
(11.26) 4.99×10^{-11} mol kg^{-1}, 6.31×10^{-11} mol kg^{-1}. **(11.27)** $\lg m_{MX} = (1/2)\lg K_s + A\sqrt{I}$.
(11.28) (a) 2.3×10^{-10} mol dm^{-3}; **(b)** 2.3×10^{-9} mol kg^{-1}; **(c)** 1.51×10^{-5} mol kg^{-1}.
(11.28) $m \approx 1.36 \times 10^{-5}$ mol kg^{-1}. **(11.29)** $K_{sp} = 1.58 \times 10^{-10}$ mol kg^{-1}, $\gamma_\pm = 0.800$.
(11.30) 386 kJ mol^{-1}. **(11.31)** -0.83 V. **(11.32)** $\Delta\phi = \Delta\phi^{\ominus} + (RT/2F)\ln(f_{O_2}^{1/2}/a_{O^{2-}})$.
(11.33) $\Delta\phi = \Delta\phi^{\ominus} - (RT/F)\ln a_{OH^-}$, -37.6 mV. **(11.34)** $\Delta\phi = \Delta\phi^{\ominus} - (RT/6F) \times \ln(a_{Cr^{3+}}^2/a_{Cr_2O_7^{2-}} \cdot a_{H^+}^{14})$.

Chapter 12 **(12.1) (a)** R: $AgCl + e^- \rightleftarrows Ag + Cl^-$, L: $H^+ + e^- \rightleftarrows \frac{1}{2}H_2$, $AgCl + \frac{1}{2}H_2 \rightleftarrows Ag + HCl$;
(b) R: $\frac{1}{2}Sn^{4+} + e^- \rightleftarrows \frac{1}{2}Sn^{2+}$, L: $Fe^{3+} + e^- \rightleftarrows Fe^{2+}$, $\frac{1}{2}Sn^{4+} + Fe^{2+} \rightleftarrows \frac{1}{2}Sn^{2+} + Fe^{3+}$;
(c) R: $\frac{1}{2}MnO_2 + 2H^+ + e^- \rightleftarrows \frac{1}{2}Mn^{2+} + H_2O$, L: $\frac{1}{2}Cu^{2+} + e^- \rightleftarrows \frac{1}{2}Cu$, $\frac{1}{2}MnO_2 + 2H^+ + \frac{1}{2}Cu \rightleftarrows \frac{1}{2}Mn^{2+} + H_2O + \frac{1}{2}Cu^{2+}$; **(d)** R: $AgBr + e^- \rightleftarrows Ag + Br^-$, L: $AgCl + e^- \rightleftarrows Ag + Cl^-$,
$AgBr + Ag + Cl^- \rightleftarrows Ag + AgCl + Br^-$. **(12.2) (a)** $Zn|ZnSO_4(aq)||CuSO_4(aq)|Cu$;
(b) $H_2(g)|HCl(aq)|AgCl, Ag$; **(c)** $O_2(g)|HCl(aq)|H_2(g)$; **(d)** $Na|NaI(EtNH_2)|Na$,
$Hg|NaOH(aq)|H_2(g)$; **(e)** $H_2(g)|HI(aq)|I_2$. **(12.3)** (for cells of 12.2); **(a)** 1.100 V;
(b) 0.2223 V; **(c)** 1.229 V; **(d)** 1.884 V; **(e)** 0.535 V. **(12.4) (a)** -728 kJ mol^{-1};
(b) -202.2 kJ mol^{-1}; **(c)** -292 kJ mol^{-1}; **(d)** 123 kJ mol^{-1}. **(12.5) (a)** 1.2×10^{16};
(b) 1.29×10^{83}; **(c)** 7.3×10^{-7}. **(12.6) (a)** 1.100 V; **(b)** -212 kJ mol^{-1}; **(c)** 1.9×10^{37}.
(12.7) $3Sn^{4+} + 2Al \rightleftarrows 3Sn^{2+} + 2Al^{3+}$; **(a)** 1.865 V, 1.845 V; **(b)** -1068 kJ mol^{-1}; **(c)** 10^{184}.
(12.8) 1.884 V. **(12.9)** 0.978 V. **(12.10)** 17.6 mV, -12.0 mV. **(12.11)** 0.482 V.
(12.12) 1.000 atm, 1.000, 34.6 atm, 0.691, 1530 atm, 15.3. **(12.13)** $E = E^{\ominus} + (2RT/F)\ln(1 + \kappa t)$, $(\partial E/\partial t) = (2RT/F)/(1 + \kappa t)$. **(12.14)** 1.229 V, 237 kJ mol^{-1}.
(12.15) 2877 kJ mol^{-1}, 2738 kJ mol^{-1}, 2746 kJ mol^{-1}. **(12.17) (a)** 0.2223 V;
(b) 0.2223 V. **(12.18) (a)** 0.0796; **(b)** 0.7957; **(c)** 1.10.
(12.19) $\Delta G_m^{\ominus} = -21.458$ kJ mol^{-1}, $\Delta H_m^{\ominus} = -40.032$ kJ mol^{-1},
$\Delta S_m^{\ominus} = -62.296$ J K^{-1} mol^{-1}, $\Delta G_m^{\ominus} = -131.18$ kJ mol^{-1}, $\Delta H_m^{\ominus} = -167.06$ kJ mol^{-1},
$S_m^{\ominus} = 33.9$ J K^{-1} mol^{-1}. **(12.20)** 0.926. **(12.21)** 0. **(12.22) (a)** 9.19×10^{-9} mol kg^{-1};
(b) 8.44×10^{-17}. **(12.23)** $\gamma^* = 0.533$. **(12.24)** 0.26838 V, 0.99895. **(12.25)** $pK_w(20 °C) = 14.23$, $pK_w(25 °C) = 14.01$, $pK_w(30 °C) = 13.79$, $\Delta H_m^{\ominus} = 74.9$ kJ mol^{-1},
$\Delta S_m^{\ominus} = -17.1$ J K^{-1} mol^{-1}. **(12.26)** 0.2916 V. **(12.27)** $A_{calc} = 0.419$, $A_{exp} = 0.317$.
(12.28) $c_{H^+}^2 = K_w K_a/(K_a + c + c_{H^+} - K_w/c_{H^+})$. **(12.29)** pH $\approx \frac{1}{2}(pK_w - pK_b - \lg c)$.
(12.30) (a) 5.1; **(b)** 8.88; **(c)** 9.38; **(d)** 2.88. **(12.31)** 8.43, 9.43. **(12.32)** $c_{H^+} + c_s = (c_a + c_s)/[1 + (c_{H^+}/K_a)] + K_w/c_{H^+}$, pH $\approx pK_a + \lg(c_s/c_a)$. **(12.33)** 9.14. **(12.38)** -1.1 V.

Chapter 13 **(13.1) (a)** 3.31×10^{-19} J, 199 kJ mol^{-1}; **(b)** 3.61×10^{-19} J, 218 kJ mol^{-1};
(c) 4.97×10^{-19} J, 299 kJ mol^{-1}; **(d)** 9.93×10^{-19} J, 598 kJ mol^{-1}; **(e)** 1.32×10^{-15} J,
79.8×10^4 kJ mol^{-1}; **(f)** 1.99×10^{-23} J, 0.012 kJ mol^{-1}. **(13.2) (a)** 1.10×10^{-27} kg m s^{-1},
0.66 m s^{-1}; **(b)** 1.20×10^{-27} kg m s^{-1}, 0.72 m s^{-1}; **(c)** 1.66×10^{-27} kg m s^{-1},
0.99 m s^{-1}; **(d)** 3.31×10^{-27} kg m s^{-1}, 1.98 m s^{-1}; **(e)** 4.42×10^{-24} kg m s^{-1}, 2640 m s^{-1};
(f) 6.63×10^{-32} kg m s^{-1}, 3.96×10^{-5} m s^{-1}. **(13.3)** 21.1 m s^{-1}.

(13.4) (a) 2.77×10^{18} s^{-1}; (b) 2.77×10^{20} s^{-1}.   (13.5) (a) 1.60×10^{-33} J m^{-3};
(b) 2.52×10^{-4} J m^{-3}.   (13.6) $hc/5k$.   (13.7) 6.51×10^{-34} J s.   (13.8) 6000 K.
(13.9) (a) 19.7 J K^{-1} mol^{-1}; (b) 22.4 J K^{-1} mol^{-1}; (c) 24.4 J K^{-1} mol^{-1}, classical
value 24.9 J K^{-1} mol^{-1} at all temperatures.   (13.10) 341 K.   (13.11) (a) No ejection;
(b) 3.19×10^{-19} J (1.99 eV), 837 km s^{-1}.   (13.12) 6.93 keV.   (13.13) (a) 2.426 pm;
(b) 1.321×10^{-15} m.   (13.15) (a) 399 kJ mol^{-1}; (b) 39.9 kJ mol^{-1};
(c) 3.99×10^{-13} kJ mol^{-1}.   (13.16) (a) 6.626×10^{-29} m; (b) 6.626×10^{-36} m; (c) 73 pm;
(d) 123 pm; 38.8 pm; 3.88 pm.   (13.17) 1.06×10^{-28} m s^{-1}, 1.06×10^{-27} m.
(13.18) 5.3×10^{-25} kg m s^{-1}, 5.8×10^5 m s^{-1}.   (13.19) (a) 0.020; (b) 0.007; (c) 6.6×10^{-6};
(d) 0.5; (e) 0.61.   (13.20) 9.0×10^{-6}, 1.2×10^{-6}.   (13.21) (a) $\sqrt{(2/L)}$; (b) $1/c\sqrt{(2L)}$;
(c) $1/\sqrt{(\pi a_0^3)}$; (d) $1/\sqrt{(32\pi a_0^5)}$.   (13.22) (a) 1.80×10^{-19}, 109 kJ mol^{-1}, 1.13 eV,
9090 cm^{-1}; (b) 6.62×10^{-19}, 399 kJ mol^{-1}, 4.14 eV, 33 300 cm^{-1}.
(13.23) 7.48×10^{-13} kJ mol^{-1}, 2.24×10^9, 1.12×10^{-19} kJ mol^{-1}.   (13.24) $E_{lmn} =$
$(h^2/8m)[(l/L_X)^2 + (m/L_Y)^2 + (n/L_Z)^2]$, $E_{lmn} = (h^2/8mL^2)(l^2 + m^2 + n^2)$.   (13.28) $g = \frac{1}{2}\sqrt{(mk/h^2)}$,
$\frac{1}{2}\hbar\omega$, $\hbar\sqrt{(k/m)}$.   (13.29) (a) 3.3×10^{-34} J; (b) 3.3×10^{-33} J; (c) 2.2×10^{-29} J;
(d) 4.7×10^{-21} J, 2.8 kJ mol^{-1}; (e) 3.14×10^{-20} J, 1580 cm^{-1}.   (13.30) HI (314.2 N m^{-1}),
HBr (411.8 N m^{-1}), HCl (515.9 N m^{-1}), NO (1595 N m^{-1}), CO (1902 N m^{-1}).
(13.32) 78.23×10^{-3} kJ mol^{-1}, 6.541 cm^{-1}, 1.055×10^{-34} J s.   (13.34) 0, 13.18, 39.53,
79.05 cm^{-1}.   (13.39) $\theta = \arccos\{m/\sqrt{[l(l+1)]}\}$; 54° 44′.   (13.40) (a) Poles; (b) equator;
(c) equator.   (13.41) (a) $x = \frac{1}{2}L$; (b) π.   (13.42) (a) $L/2\sqrt{3}$; (b) $\pi/\sqrt{3}$.   (13.43) $\langle x \rangle = 0$
for $v = 0, 1, 2$, $[\delta x]_{v=0} = \sqrt{(\hbar/2m\omega)}$, $[\delta x]_{v=1} = \sqrt{(3\hbar/2m\omega)}$, $[\delta x]_{v=2} = \sqrt{(5\hbar/2m\omega)}$.
(13.44) (a) $\hbar k$; (b) 0; (c) 0; (d) 0.   (13.45) (a) ik; (c) 0.   (13.46) (a) $-k^2$; (b) $-k^2$;
(c) 0; (d) 0.   (13.47) 0, 0, 0, $\hbar i$, 0.   (13.48) No, yes, yes.
(13.49) $l_{x,op} = (\hbar/i)\{y(\partial/\partial z) - z(\partial/\partial y)\}$, etc., no.

Chapter 14 (14.1) $n_2 \to n_1 = 6$.   (14.2) 397.13 nm, 3.40 eV.   (14.3) 987 663 cm^{-1}.
(14.4) 137 175 cm^{-1}, 185 187 cm^{-1}.   (14.5) 3.3594×10^{-27} kg.   (14.6) 7616 cm^{-1},
10 282 cm^{-1}, 11 516 cm^{-1}, 6.80 eV.   (14.7) $E = -R_H/n^2$.   (14.8) Trajectory not defined;
angular momentum should be zero.   (14.9) $r^* = a_0/Z$; (a) 26 pm; (b) 5.9 pm.
(14.10) 0.42 pm.   (14.11) (a) 0; (b) 0; (c) $\hbar\sqrt{6}$; (d) $\hbar\sqrt{2}$; (e) $\hbar\sqrt{2}$.   (14.12) Closer, 698 pm.
(14.13) $E_0 = -Z^2 R_H$, -1102 eV.   (14.14) $z = \pm 2a_0/Z = \pm 106$ pm.
(14.15) (a) Nondegenerate; (b) 9; (c) 25.   (14.16) b, c, e.   (14.17) (a) 2; (b) 6; (c) 10; (d) 18.
(14.18) H 1s, He $(1s)^2$, Li $(1s)^2 2s$, Be $(1s)^2(2s)^2$, B $(1s)^2(2s)^2 2p_x$, C $(1s)^2(2s)^2 2p_x 2p_y$,
N $(1s)^2(2s)^2 2p_x 2p_y 2p_z$, O $(1s)^2(2s)^2(2p_x)^2 2p_y 2p_z$, F $(1s)^2(2s)^2(2p_x)^2(2p_y)^2 2p_z$,
Ne $(1s)^2(2s)^2(2p_x)^2(2p_y)^2(2p_z)^2 = KL$, Na KL$3s$, Mg KL$(3s)^2$, Al KL$(3s)^2 3p_x$, ...,
Cl KL$(3s)^2(3p_x)^2(3p_y)^2 3p_z$, Ar KL$(3s)^2(3p_x)^2(3p_y)^2(3p_z)^2 = KLM$.   (14.19) 987 663 cm^{-1},
122.5 eV.   (14.20) 43 505 cm^{-1}, 5.39 eV.   (14.21) 14.0 eV, 4.1 eV.   (14.22) $I_D = 1.000272\, I_H$.
(14.23) $j = \frac{7}{2}, \frac{5}{2}$.   (14.24) $\frac{41}{2}, \frac{39}{2}$.   (14.25) (a) $J = 8, 7, 6, 5, 4, 3, 2$; (b) Same,
$[J(J+1)]^{1/2}\hbar$.   (14.27) (a) 1, 0; 3, 0 states; (b) $\frac{3}{2}, \frac{1}{2}, \frac{1}{2}$; 4, 2, 2 states; (c) 2, 1, 1, 1, 0, 0;
5, 3, 3, 0, 0 states.   (14.28) (^{1}S) 0, 1 state, (^{2}P) $\frac{3}{2}, \frac{1}{2}$; 4, 2 states, (^{3}P) 2, 1, 0; 5, 3, 1 states,
(^{3}D) 3, 2, 1; 7, 5, 3 states, (^{2}D) $\frac{5}{2}, \frac{3}{2}$; 6, 4 states, (^{1}D) 2; 5 states, (^{4}D) $\frac{7}{2}, \frac{5}{2}, \frac{3}{2}, \frac{1}{2}$;
8, 6, 4, 2 states.   (14.29) ^{1}S none, ^{2}P$_{3/2}$ ↔ ^{2}D$_{5/2}$, ^{2}D$_{3/2}$, ^{2}P$_{1/2}$ ↔ ^{2}D$_{3/2}$, ^{3}P$_2$ ↔ ^{3}D$_3$, ^{3}D$_2$,
^{3}D$_1$, ^{3}P$_1$ ↔ ^{3}D$_2$, ^{3}D$_1$, ^{3}P$_0$ ↔ ^{3}D$_1$, ^{1}D none, ^{4}D none.   (14.30) (a) ^{2}S$_{1/2}$; (b) ^{2}P$_{3/2}$; ^{2}P$_{1/2}$;
(c) ^{2}D$_{5/2}$, ^{2}D$_{3/2}$; (d) ^{2}P$_{3/2}$, ^{2}P$_{1/2}$.   (14.31) 2.14T.   (14.32) (a) 1; (b) $\frac{4}{3}$; (c) $\frac{1}{3}$.
(14.33) Use $g(^3S) = 2$, $g(^3P_2) = \frac{3}{2}$, $g(^3P_1) = \frac{3}{2}$, $g(^3P_0) = 1$.

Chapter 15 (15.7) (a) 8.61×10^{-7}; (b) 8.61×10^{-7}; (c) 3.7×10^{-7}; (d) 4.86×10^{-7}.
(15.8) (a) 1.96×10^{-6}; (b) 1.96×10^{-6}; (c) 0; (d) 5.47×10^{-7}.   (15.9) (a) 1.91 eV, (b) 133 pm.
(15.11) $N = 1/\sqrt{[2(1+S)]}$.   (15.14) $R/a_0 = 2.1$.   (15.15) H$_2^-$ $(1s\sigma_g)^2(1s\sigma_u^*)$,
N$_2(1s\sigma_g)^2(1s\sigma_u^*)^2(2s\sigma_g)^2(2s\sigma_u^*)^2(2p\pi_u)^4(2p\sigma_g)^2$,
O$_2(1s\sigma_g)^2(1s\sigma_u^*)^2(2s\sigma_g)^2(2s\sigma_u^*)^2(2p\pi_u)^4(2p\sigma_g)^2(2p_x\pi_g^*)(2p_y\pi_g^*)$,
CO$(1s\sigma)^2(1s\sigma^*)^2(2s\sigma)^2(2s\sigma^*)^2(2p\pi)^4(2p\sigma)^2$, NO$(1s\sigma)^2(1s\sigma^*)^2(2s\sigma)^2(2s\sigma^*)^2(2p\pi)^4(2p\sigma)^2(2p\pi^*)$,
CN$(1s\sigma)^2(1s\sigma^*)^2(2s\sigma)^2(2s\sigma^*)^2(2p\pi)^4(2p\sigma)$.   (15.16) (a) C$_2$, CN; (b) NO, O$_2$, F$_2$.
(15.17) σ, π, δ, (a) $(d\sigma)^2$; (b) $(d\sigma)^2(d\pi)^3$; (c) $(d\sigma)^2(d\pi)^4(d\delta)^2$.   (15.18) Yes.
(15.19) (a) π_g^*; (b) inapplicable; (c) δ_g; (d) δ_u^*; (e) a$_{1u}$, e$_{2g}$, e$_{1u}$, b$_{1g}$.
(15.20) 472 kJ mol^{-1} (4.89 eV).   (15.21) Use $e^2/4\pi\varepsilon_0 R^* = I - E_A$.   (15.22) 1.
(15.23) Use $\cos\theta = \frac{1}{2}(3\cos^2\Phi - 1)$.   (15.24) CO$_2$, NO$_2^+$, H$_2$O^{2+}.   (15.25) NH$_3^{2+}$, CH$_3$,

NO_3^-, CO_3^{2-}. **(15.29)** $5(h^2/8m_eL^2)$. **(15.30)** 4.32×10^{-19} J (2.70 eV); absorbs blue; appears red. **(15.31) (a)** $t^1(1)$, $t^2(2)$, $t^3(3)$, $t^4(2)$, $t^5(1)$, $t^6(0)$, $t^6e(1)$, $t^6e^2(2)$, $t^6e^3(1)$, $t^6e^4(0)$; **(b)** $t^1(1)$, $t^2(2)$, $t^3(3)$, $t^3e(4)$, $t^3e^2(5)$, $t^4e^2(4)$, $t^5e^2(3)$, $t^6e^2(2)$, $t^6e^3(1)$, $t^6e^4(0)$. **(15.35) (a)** $\psi \approx H(1)F(2) + H(2)F(1)$; **(b)** $\psi \approx F(1)F(2)$; **(c)** $\psi = 0.89 \psi_{cov} + 0.45 \psi_{ion}$.

Chapter 16 **(16.1)** R_3, C_{2v}, D_{3h}, $D_{\infty h}$, $C_{\infty v}$, D_3, D_{6h}, C_{4v}, C_s. **(16.2)** C_{2v}, C_{3v}, C_{3v}, D_{2h}, C_{2v}, C_{2h}, D_{2h}, D_{2h}, C_{2v}. **(16.3)** D_{3d}, D_{3d}, D_{2h}, $D_{\infty h}$, D_3, D_{4d}. **(16.4)** *trans* CHCl : CHCl. **(16.5)** $C_2\sigma_h = i$. **(16.6)** NO_2, CH_3Cl, CCl_3H, *cis* CHCl : CHCl, C_6H_5Cl. **(16.7)** $Co(en)_3^{3+}$. **(16.11)** No, d_{xy} on S. **(16.12)** $x \to B_1$, $y \to B_2$, $z \to A_1$. **(16.13)** A_2. **(16.14) (a)** $E_1(x, y)$, $A_1(z)$; **(b)** $B_1(x)$, $B_2(y)$, $A_1(z)$. **(16.15)** $A_1 + T_2$, yes, d_{xy}, d_{xz}, d_{yz} transform as T_2. **(16.16) (a)** All d; **(b)** all except $d_{xy}(A_2)$. **(16.19) (a)** $2A_1$, A_2, $2B_1$, $2B_2$; **(b)** A_1, $3E$; **(c)** T_2, T_1, A_1; **(d)** T_{1u}, T_{2u}, A_{2u}. **(16.20) (a)** T_2, T_1, A_1; **(b)** T_{1u}, T_{2u}, A_{2u}. **(16.21) (a)** i, ii; **(b)** ii. **(16.23) (a)** $t(A_1 + B_1 + B_2)$, $r(A_2 + B_1 + B_2)$; **(b)** $t(A_1 + E)$, $r(A_2 + E)$; **(c)** $t(A_1 + E_1)$, $r(A_2 + E_1)$; **(d)** $t(T_{1u})$, $r(T_{1g})$. **(16.24) (a)** Only p and f; **(b)** yes. **(16.25) (a)** $A_1 \to A_2$, B_1, B_2; **(b)** $T_{2g} \to A_{2g}$, E_g, T_{1g}, T_{2g}; **(c)** A_{2g}. **(16.27) (a)** Yes; **(b)** no; **(c)** yes. **(16.28)** E, i, $3C_2$, $4C_3$, 3σ, $O_h(m3m)$. **(16.29)** E, $3C_2$, $4C_3$, 6σ, cubic, $T_d(\bar{4}3m)$. **(16.30)** Triclinic, $C_i(\bar{1})$. **(16.31)** Monoclinic. **(16.32)** Cubic, $O_h(m3m)$. **(16.34)** $\phi = 54° \, 44'$.

Chapter 17 **(17.1)** HCl, CH_3Cl, CH_2Cl_2, H_2O, H_2O_2, NH_3, NH_4Cl(gas). **(17.2)** HCl, CO_2, H_2O, CH_3CH_3, CH_4, CH_3Cl, N_3^-. **(17.3)** H_2, HCl, CH_3Cl, CH_2Cl_2, CH_3CH_3, H_2O. **(17.4)** All. **(17.5)** 0.999999925×660 nm; -6.36×10^7 m s^{-1} (or -7.02×10^7 m s^{-1} considering relativity). **(17.6)** 2.397×10^4 km s^{-1}, 8.351×10^5 K. **(17.7) (a)** 2.1×10^{-6}, 0.67 MHz, 0.006 cm^{-1}; **(b)** 9.7×10^{-7}, 3.3×10^{-3} MHz, 3.7×10^{-4} cm^{-1}. **(17.8) (a)** 5.3×10^{-11} s; **(b)** 5.3×10^{-12} s; **(c)** 1.6×10^{-9} s. **(17.9) (a)** 53 cm^{-1}; **(b)** 0.53 cm^{-1}. **(17.10)** $\tau = 1/z \approx 2.29 \times 10^{-10}$ s, 696 MHz, 0.73 mmHg. **(17.11) (a)** 0.052; **(b)** 0.067; **(c)** 0.200. **(17.12)** 30. **(17.13)** 6. **(17.14) (a)** 1.6266×10^{-27} kg, 2.6422×10^{-47} kg m^2; **(b)** 3.1622×10^{-27} kg, 5.1367×10^{-47} kg m^2; **(c)** 1.6291×10^{-27} kg, 2.6462×10^{-47} kg m^2. **(17.15)** 596 GHz, 19.9 cm^{-1}, 0.252 mm, 0.083 mm, ... (diminishing separation on λ scale). **(17.16)** $B(^{12}C^{16}O) = 1.9318$ cm^{-1}, $B(^{13}C^{16}O) = 1.8466$ cm^{-1}. **(17.17)** 2.728×10^{-47} kg m^2, 129.5 pm. **(17.18)** 42.86, 53.56, 64.16, 74.77, ..., cm^{-1}. **(17.20) (a)** 4.6016×10^{-48} kg m^2; **(b)** 7.4946×10^{-45} kg m^2. **(17.21)** 160.5 pm. **(17.22)** 0.10057 cm^{-1}, 218 pm. **(17.23)** 1.37998×10^{-45} kg m^2. **(17.24)** $R(CO) = 116.28$ pm, $R(CS) = 155.97$ pm. **(17.25)** See J. Kraitchman, *Am. J. Phys.* **21**, 17 (1953). **(17.26)** 215 pm. **(17.27)** 0.0459 cm^{-1}, 0.0184 cm^{-1}, 0.0103 cm^{-1}. **(17.28)** 967.1 N m^{-1}, 515.6 N m^{-1}, 411.8 N m^{-1}, 314.2 N m^{-1}. **(17.29)** 3002.3 cm^{-1}, 2143.7 cm^{-1}, 1885.9 cm^{-1}, 1640.1 cm^{-1}. **(17.32)** 4.13 eV. **(17.33)** 93.80 N m^{-1}, 142.81 cm^{-1}, 3.4 eV. **(17.34) (a)** 5.16 eV; **(b)** 5.21 eV. **(17.36)** 480.7 N m^{-1}, 128 pm, 130 pm. **(17.37)** 198.9 ppm. **(17.39) (a)** Yes; **(b)** no.

Chapter 18 **(18.1)** 449 cm^{-1} mol^{-1} dm^3. **(18.2)** 159 cm^{-1} mol^{-1} dm^3. **(18.3)** 0.090, 0.449, 0.896, 4.52; 0.32. **(18.4) (a)** 69%, 1.3×10^{-14}%; **(b)** 90.2%, 0.0032%. **(18.5) (a)** 87.6 cm; **(b)** 291 cm. **(18.6) (a)** 3.5×10^{18} dm^3 mol^{-1} cm^{-1} s^{-1}; **(b)** 1.7×10^{17} dm^3 mol^{-1} cm^{-1} s^{-1}. **(18.7) (a)** 0.219; **(b)** 0.011. **(18.9)** $\mathscr{A} = 3.32 \times 10^{15}$ dm^3 mol^{-1} cm^{-1} s^{-1}, $f = 2.08 \times 10^{-4}$, forbidden. **(18.10)** $f(280 \text{ nm}) \approx 2.5 \times 10^{-3}$, $f(430 \text{ nm}) \approx 3 \times 10^{-3}$. **(18.11)** 4.74×10^{-2}, forbidden, A_1, B_1, B_2. **(18.12)** $f_{\vec{n} \; n+1} = (64/3\pi^2)[n^2(n+1)^2/(2n+1)^3]$, $f_{\vec{n} \; n+2} = 0$. **(18.13)** 7350 cm^{-1}, infrared absorption, 1.4×10^5 mol^{-1} dm^3 cm^{-1}, 2.1×10^{-5} cm. **(18.14)** Lengthen, blue. **(18.16)** $R/a_0 = 2.10380$. **(18.17) (a)** Allowed; **(b)** forbidden. **(18.21)** pH $= pK_{In} + \lg\{\alpha/(1-\alpha)\}$. **(18.23)** 7.1 (7.9 at pH 9). **(18.24)** 0.848 eV. **(18.25)** 5.09 eV. **(18.26) (a)** $3+1$, $5+3$; **(b)** $4+4$, $2+2$. **(18.28)** 16.52 eV ionization of $2p\sigma$ electron, 15.65 eV ionization of $2p\pi$ electron, 9.21 eV ionization of $2p\pi^*$ electron.

Chapter 19 **(19.1)** 1.34 T (13.4 kG). **(19.2)** 1.66×10^{-11} mol dm^{-3}, $N_\alpha = 1.2492 \times 10^{14}$, $N_\beta = 1.2509 \times 10^{14}$. **(19.3) (a)** 0.0503; **(b)** 0.0007. **(19.4)** 2.00224. **(19.5) (a)** 331.9 mT; **(b)** 1.201 T. **(19.6) (a)** 13.4 G; **(b)** 48.5 G. **(19.7)** 1.9920, 2.0022. **(19.8)** 50.7 mT. **(19.9)** 2.3 mT. **(19.10)** 2.0025, 64.5 MHz. **(19.11)** 330.2 mT, 332.2 mT, 332.8 mT, 334.8 mT. **(19.12)** $\tau_J \lesssim 9.4 + 10^{-9}$ s. **(19.14)** 0.69 mT, 0.21 mT. **(19.15)** Quartet (1 : 3 : 3 : 1) of triplets (1 : 2 : 1), triplets replaced by 1 : 2 : 3 : 2 : 1 quintets. **(19.17)** 10%.

(19.18) 38%. **(19.19)** 0.48, 0.52, 3.8, yes, 131°. **(19.20) (a)** 6.697×10^{-4} T;
(b) 2.486×10^{-3} T. **(19.23)** 1.6×10^{-9} s. **(19.24) (a)** 1.409 T, 9.181 T, 5.605 T, 19.504 T,
1.498 T, 3.481 T; **(b)** 7.05 T, 45.90 T, 28.03 T, 97.52 T, 7.49 T, 17.41 T.
(19.25) (a) 7.663×10^{-5}, 3.831×10^{-4}, 1.788×10^{-3}; **(b)** 1.022×10^{-6}, 5.108×10^{-6},
2.384×10^{-5}. **(19.26)** 211 T, 2.14 mT. **(19.27)** 2.06 T, -4.84×10^{-6}, $\beta(m_s = -\frac{1}{2})$
is lower. **(19.28) (a)** -11.4 μT; **(b)** -53.2 μT. **(19.29) (a)** 456 Hz; **(b)** 2.28 kHz.
(19.31) 158 pm. **(19.32)** 77.6 μT. **(19.33)** When lifetime of each isomer $\lesssim 2$ ms.
(19.34) 56.2 kJ mol^{-1}. **(19.35) (a)** 81.3 m s^{-1}, 6.56×10^{10} Hz;
(b) 7.69×10^{-20} m s^{-1}, 6.2×10^{-11} Hz. **(19.37)** Ionicity of AuI < AuBr < AuCl.

Chapter 20 **(20.1)** 1. **(20.2)** (4, 0, 0, 0, 0, 1), 5; (3, 1, 0, 0, 1, 0), 20; (3, 0, 1, 1, 0, 0), 20; (2, 2, 0, 1, 0, 0), 30;
(2, 1, 2, 0, 0, 0), 30; (1, 3, 1, 0, 0, 0), 20; (0, 5, 0, 0, 0, 0), 1; most probable are
(2, 2, 0, 1, 0, 0) and (2, 1, 2, 0, 0, 0) jointly. **(20.4)** (4, 2, 2, 1, 0, 0, 0, 0, 0, 0) with
weight = 3780. **(20.5)** ≈ 160 K. **(20.6)** 103.8 K, $q = 1 + a$. **(20.9)** a, b, d, f; g involves
knowing about the Pauli principle. **(20.10) (a)** 4.76×10^{25}; **(b)** 2.45×10^{26}; **(c)** 4.76×10^{28};
(d) 1. **(20.11)** Unless $h^2/mX^2kT \ll 1$, 3.5×10^{-15} K, 7.53. **(20.12) (a)** 5.00; **(b)** 6.25.
(20.13) (a) 1.00, 0.80; **(b)** 6.5×10^{-11}, 0.12. **(20.14)** 64% lower, 36% upper.
(20.15) 0.518 kJ mol^{-1}. **(20.16) (a)** 1.0488; **(b)** 1.5553. **(20.17) (a)** 0.9535, 0.0450,
0.0021; **(b)** 0.6430, 0.2298, 0.0826. **(20.18) (a)** 126.3 J mol^{-1}; **(b)** 1.395 kJ mol^{-1}.
(20.19) 0.904, 0.999. **(20.20)** $q = 1 + e^{-x} + e^{-2x}$, $\langle \varepsilon \rangle = -g_n\mu_N B[(1 + e^{-2x})/(1 + e^{-x} + e^{-2x})]$,
where $x = g_n\mu_N B\beta$. **(20.21) (a)** -298 K; **(b)** -10 K; **(c)** -0 K.
(20.24) (a) 13.38 J K^{-1} mol^{-1}; **(b)** 18.07 J K^{-1} mol^{-1}. **(20.25) (a)** 11.2 J K^{-1} mol^{-1};
(b) 11.4 J K^{-1} mol^{-1}. **(20.26) (a)** 1.661 J K^{-1} mol^{-1}; **(b)** 8.352 J K^{-1} mol^{-1}.
(20.27) $S_m/R = xe^{-x}/(1 + e^{-x}) + \ln(1 + e^{-x})$. **(20.28)** 155 J K^{-1} mol^{-1}.
(20.29) (a) 7.66 J K^{-1} mol^{-1}; **(b)** 42.8 J K^{-1} mol^{-1}.

Chapter 21 **(21.1) (a)** 6.56; **(b)** 19.6; **(c)** 32.8. **(21.2) (a)** 6.9051; **(b)** 19.8893. **(21.3)** 43.10; $T \gg 40$ K.
(21.4) (a) 36.3; **(b)** 78.9. **(21.5)** 36.6055 (at 298 K); 79.1716 (at 500 K). **(21.6)** $\Delta S_m = $
$R \ln[(\sigma/V)(h^2\beta/2\pi me)^{1/2}]$. **(21.7)** 89 J K^{-1} mol^{-1}; -62 J K^{-1} mol^{-1}. **(21.9) (a)** 2;
(b) 1; **(c)** 12; **(d)** 12; **(e)** 3. **(21.10)** 256 J K^{-1} mol^{-1}. **(21.12) (a)** 3.886; **(b)** 2.414.
(21.13) 3.15×10^{-3}. **(21.14)** Multiply by 1.004. **(21.15)** ≈ 100 T. **(21.16)** $U - U(0) = $
$H - H(0) = Nh\omega e^{-x}/(1 - e^{-x})$, $S = Nk[xe^{-x}/(1 - e^{-x}) - \ln(1 - e^{-x})]$, $A - A(0) = G - G(0) = $
$NkT \ln(1 - e^{-x})$. **(21.17)** $R \ln(1 - e^{-x})$. **(21.18) (a)** -4.31 J K^{-1} mol^{-1};
(b) -6.52 J K^{-1} mol^{-1}. **(21.19) (a)** -136 J K^{-1} mol^{-1}; **(b)** -232 J K^{-1} mol^{-1};
(c) -203 J K^{-1} mol^{-1}; **(d)** -198 J K^{-1} mol^{-1}; **(e)** -217 J K^{-1} mol^{-1}.
(21.20) 4.13×10^{-7}. **(21.21)** $U = nRT(q'/q)$, $S = nR[(q'/q) + \ln(eq/N)]$, $C_V = nR[(q''/q) - $
$(q'/q)^2]$. **(21.22)** $U_{int} = nRT(q'_{int}/q_{int})$, $S_{int} = nR[(q'/q)_{int} + \ln q_{int}]$, $C_{V,int} = nR[(q''/q)_{int} - $
$(q'/q)_{int}^2]$. **(21.23) (a)** 4.322 kJ mol^{-1}; **(b)** -0.138 J K^{-1} mol^{-1}; **(c)** 5.405 J K^{-1} mol^{-1}.
(21.24) -5.771 J K^{-1} mol^{-1}, $0.060 R = 0.499$ J K^{-1} mol^{-1}. **(21.26)** 8.671,
$x(Na_2) = 0.095$, $x(Na) = 0.905$. **(21.27) (a)** 21 J K^{-1} mol^{-1}; **(b)** 21 J K^{-1} mol^{-1};
(c) 25 J K^{-1} mol^{-1}; **(d)** 25 J K^{-1} mol^{-1}; **(e)** 25 J K^{-1} mol^{-1}; **(f)** 21 J K^{-1} mol^{-1}.
(21.29) (a) 35.72 J K^{-1} mol^{-1}; **(b)** 46.44 J K^{-1} mol^{-1}. **(21.30)** $C_{V,m}/R = x^2 e^{-x}/(1 + e^{-x})^2$,
where $x = \Delta/kT$. **(21.31) (a)** 2.94 J K^{-1} mol^{-1}; **(b)** 0.654 J K^{-1} mol^{-1};
(c) 0.244 J K^{-1} mol^{-1}. **(21.32) (a)** 3.72×10^{-2} J K^{-1} mol^{-1}; **(b)** 1.06×10^{-3} J K^{-1} mol^{-1}.
(21.33) 4.16 J K^{-1} mol^{-1}, 15.1 J K^{-1} mol^{-1}. **(21.35) (a)** $c_s = (1.40 \, RT/M_m)^{1/2}$;
(b) $c_s = [5(1 + a)RT/(3 + 2a)M_m]^{1/2}$, $a = 2x^2 e^{-x}/(1 + 5e^{-x})^2$, $x = 6hcB/kT$.
(21.36) 346 m s^{-1}. **(21.37) (a)** 9.13 J K^{-1} mol^{-1}; **(b)** 13.4 J K^{-1} mol^{-1};
(c) 14.9 J K^{-1} mol^{-1}. **(21.39)** 191.4 J K^{-1} mol^{-1}; zero residual entropy.

Chapter 22 **(22.3)** $d = ab/\sqrt{(a^2 + b^2)}$, $d = \sqrt{3}ab/2\sqrt{(a^2 + b^2 - ab)}$. **(22.4)** (326), (111), (122), (3$\bar{2}\bar{2}$).
(22.7) 249 pm, 176 pm, 432 pm. **(22.8)** 58.9 pm. **(22.9)** 628 pm, yes. **(22.10)** 8° 10',
4° 49', 11° 46'. **(22.11)** Face-centred cubic. **(22.12)** Body-centred cubic. **(22.13)** 564 pm,
326 pm, 252 pm. **(22.16)** 834 pm, 606 pm, 870 pm. **(22.17)** 6.05×10^{23} mol^{-1}.
(22.18) (a) 1; **(b)** 2; **(c)** 4; **(d)** 8. **(22.19)** f.c.c. **(22.20)** 0.9069. **(22.21)** $\theta = 13°$ 17',
18° 59', 23° 28', ...; 9.32 g cm^{-3}, 13.2 g cm^{-3}. **(22.22)** b.c.c. **(22.23)** $\theta = 21°$ 41', 25° 15',
37° 06', ...; 8.97 g cm^{-3}. **(22.24)** $-3'$ 45", blurred lines. **(22.25)** $\beta_{vol} = 4.8 \times 10^{-5}$ K^{-1},
$\beta_{linear} = 1.6 \times 10^{-5}$ K^{-1}. **(22.26)** $\theta = 22°$ 00', 32° 00', 40° 00', ... **(22.27)** 3.37 g cm^{-3}.

(22.28) 312 pm. **(22.29)** f.c.c., $a = 412$ pm. **(22.30)** (a) b.c.c., $a = 316$ pm; (b) f.c.c., $a = 361$ pm. **(22.31)** 59 l pm, 1270 pm, 434 pm, 4. **(22.32)** (a) Primitive; (b) (010) is a glide plane with translation $= \frac{1}{2}$ diagonal length; (c) (001) is glide plane with transl. $= \frac{1}{2}$ axial length along b; (d) (100) is ordinary symmetry plane. **(22.33)** See J. M. Bijvoet, *X-ray analysis of crystals*, Butterworths, London, p. 56 (1951). **(22.35)** (a) $F_{hkl} = f$; (b) $\frac{1}{2} f_A$ if $h + k + l$ odd, $\frac{3}{2} f_A$ if $h + k + l$ even; (c) 0 if $h + k + l$ odd. **(22.36)** $F_{hkl} = 4(f_M - f_{Cl})$ h, k, l all odd. **(22.39)** (a) 38.8 pm; (b) 12.26 pm; (c) 6.132 pm. **(22.40)** 7912 m s^{-1}, 145 pm. **(22.42)** 177 pm.

Chapter 23 **(23.1)** 0 (*p*-xylene), 0.7 D (*o*-xylene), 0.4 D (*m*-xylene), *p*-xylene value from symmetry. **(23.2)** $p = (3.02 \text{ D}) \cos \frac{1}{2}\phi$, $\phi = 91°$. **(23.3)** (a) 1.12×10^8 V m^{-1}; (b) 4.11×10^9 V m^{-1}; (c) 4.11×10^3 V m^{-1}. **(23.4)** 22.7 nm. **(23.5)** HCl: 1.03 D, 3.52×10^{-24} cm^3, HBr: 0.80 D, 3.69×10^{-24} cm^3, HI: 0.36 D, 5.59×10^{-24} cm^3. **(23.6)** 1.34. **(23.8)** 1.19×10^{-23} cm^3, 0.9 D. **(23.9)** $\alpha' = 1.38 \times 10^{-23}$ cm^3, $p = 0.32$ D. **(23.12)** 13. **(23.13)** $\alpha'(\text{He}) = 1.7 \times 10^{-24}$ cm^3, $\alpha'(\text{Ne}) = 7.3 \times 10^{-24}$ cm^3. **(23.15)** 1.00019, 1.69, 1.65, 1.23. **(23.16)** 1.59 D, 1.98×10^{-24} cm^3. **(23.17)** (a) 5.661 cm^3 mol^{-1}; (b) same; 1.578 D. **(23.19)** $B(T) \approx (2/3)\pi L d^3 [1 - (C_6 / kT d^6)]$. **(23.20)** $B \approx 227$ cm^3 mol^{-1}. **(23.23)** 1.8 kJ mol^{-1}. **(23.24)** $V_{m,c} = (\pi/2)Ld^3$, $p_c = (8/9\pi)(C_6/d^9)$, $T_c = (32/27)C_6/kd^6$. **(23.25)** $V_{m,c} = 285$ cm^3 mol^{-1}, $p_c = 6.88$ atm, $T_c = 63$ K. **(23.26)** $2 \ln 2 = 1.38629\ldots$ **(23.27)** $R^* = (\frac{2}{3})R_0 + (18\pi/5)\varepsilon_0 R_0^2 V_m / \kappa L e^2 \mathcal{M}$. **(23.28)** -708.4 kJ mol^{-1}. **(23.29)** -150 kJ mol^{-1}. **(23.30)** -919 kJ mol^{-1}. **(23.31)** -605 kJ mol^{-1}. **(23.34)** $\xi = 3.9456 \times 10^{-29}$ C^2 m^2 kg^{-1}, $\chi = -2.9622 \times 10^{-8}$. **(23.35)** $\chi = 1.5686 \times 10^{-5}$, $\chi(\text{total}) = 1.5657 \times 10^{-5}$. **(23.36)** (a) $3.118 \times 10^{-3}/M_r$, $4.183 \times 10^{-5}/M_r$, (b) $1.364 \times 10^{-2}/M_r$, $1.830 \times 10^{-4}/M_r$.

Chapter 24 **(24.1)** $M_m = 88\,000$ g mol^{-1}, $M_r = 88\,000$. **(24.2)** 147 000 g mol^{-1}, 1.12×10^7 cm^3 mol^{-1}. **(24.3)** 70 000, 70 914. **(24.4)** $\bar{M}_r + \sqrt{(2\Gamma/\pi)}$, $\{\bar{M}^2 + 2(2\Gamma/\pi)^{1/2}\bar{M} + \Gamma\}/2(\bar{M} + \Gamma/\sqrt{\pi})$. **(24.5)** $V_p = 8v_{mol}$, 28 m^3 mol^{-1}, 0.33 m^3 mol^{-1}. **(24.6)** 2.8%, 5.0%. **(24.7)** $B \approx (4.22 \times 10^{23} \text{ mol}^{-1})l^3 N^{3/2}$; (a) 0.39 m^3 mol^{-1}; (b) 1.10 m^3 mol^{-1}. **(24.8)** 1.10 m^3 mol^{-1}. **(24.12)** 0.0024 mol dm^{-3}. **(24.14)** 5×10^{-2} m^3 mol^{-1}. **(24.15)** 4.3×10^5. **(24.16)** 3500 r.p.m. **(24.17)** 66 400. **(24.18)** 69 100. **(24.19)** 3.4 nm. **(24.20)** $a \approx 6.2$ nm, $b \approx 1.8$ nm. **(24.21)** $S = 5.193 \times 10^{-13}$ s, 60 750 g mol^{-1}. **(24.22)** oblate or prolate with $a/b \approx 2.8$. **(24.23)** 158 000. **(24.25)** (a) $N^{1/2}l$; (b) $(8N/3\pi)^{1/2}l$; (c) $(2N/3)^{1/2}l$; 9.7 nm, 9.0 nm, 8.0 nm. **(24.27)** (a) $(3a^2/5)^{1/2}$; (b) $(l^2/12)^{1/2}$, 0.92 nm, 46 nm. **(24.28)** sphere, sphere, rod. **(24.30)** $P(\theta) \approx (2/sL)\int_0^{sL} (\sin u/u) \, du - \{\sin(\frac{1}{2}sL)/(\frac{1}{2}sL)\}^2$, $s = (4\pi/\lambda) \sin \frac{1}{2}\theta$. **(24.32)** 2.45×10^{-8} s, 1.42×10^{-11} s. **(24.35)** $t = -T(\partial S/\partial l)_T$. **(24.39)** $\Gamma < 0$, not accumulate. **(24.40)** -5.74×10^{-3}, -6.73×10^{-3}, 1.11×10^{-2} mol cm^{-2}.

Chapter 25 **(25.1)** (a) 6.7 nm; (b) 67 nm; (c) 67 nm. **(25.2)** 8.0×10^{-7} atm. **(25.3)** 133 atm. **(25.4)** 0.97×10^{-6} m. **(25.5)** (a) 5.0×10^{10} s^{-1}; (b) 5.0×10^9 s^{-1}; (c) 5.0×10^3 s^{-1}. **(25.6)** (a) 6.2×10^{33} s^{-1} m^{-3}; (b) 6.2×10^{31} s^{-1} m^{-3}; (c) 6.2×10^{19} s^{-1} m^{-3}. **(25.7)** 4.2×10^8 s^{-1}. **(25.8)** (a) 3.33×10^{27} s^{-1}; (b) 2.67×10^{28} s^{-1}. **(25.9)** 1.4×10^{23} s^{-1}. **(25.10)** (a) 638 m s^{-1}, 1256 m s^{-1}, 2301 m s^{-1}; (b) 319 m s^{-1}, 628 m s^{-1}, 1150 m s^{-1}. **(25.11)** (a) 3.74 kJ mol^{-1}; (b) same. **(25.14)** $\langle v_x \rangle = 0.47(2kT/\pi m)^{1/2}$. **(25.15)** (a) 1.75 m.p.h.; (b) 56.2 m.p.h.; (c) 56.4 m.p.h. **(25.16)** (a) 5' 9½"; (b) 5' 9½". **(25.17)** (a) 0.39, 0.53; (b) 0.61, 0.47. **(25.18)** $n^2 \exp (1 - n^2)$, 3.02×10^{-3}, 4.9×10^{-6}. **(25.19)** $\sqrt{(2gR)}$, 11.2 km s^{-1}, 5.0 km s^{-1}, 11 900 K, 23 600 K, 188 900 K on Earth, 2430 K, 4810 K, 38 500 K on Mars, 2.3×10^{-27}, ≈ 0, ≈ 0 on Earth at 240 K, 1.3×10^{-5}, 7.4×10^{-11}, ≈ 0 on Mars at 240 K, 2.6×10^{-4}, 9.1×10^{-9}, ≈ 0 on Earth at 1500 K, 0.26, 0.046, 7.9×10^{-14} on Mars at 1500 K. **(25.20)** 43. **(25.21)** 2.45×10^{21} s^{-1}. **(25.22)** 1.09×10^5 s (30 hours). **(25.23)** $p = (\Delta m/St)(2\pi kT/m)^{1/2}$. **(25.24)** 7.3×10^{-3} N m$^{-2} \triangleq 7.2 \times 10^{-8}$ atm $\triangleq 5.5 \times 10^{-5}$ mmHg. **(25.25)** (a) 2.72×10^{23} s^{-1} cm^{-2}, 2.3×10^8 s^{-1}; (b) 2.72×10^{17} s^{-1} cm^{-2}, 230 s^{-1}; (c) 2.72×10^{13} s^{-1} cm^{-2}, 0.02 s^{-1}. **(25.26)** (a) 0.61 mmHg; (b) 0.18 mmHg. **(25.27)** 1.44 s. **(25.28)** (a) 1.7×10^{14} s^{-1}; (b) 1.4×10^{20} s^{-1}. **(25.29)** (a) 3.29×10^{34} m^{-3} s^{-1}; (b) 1.30×10^{34} m^{-3} s^{-1}; (c) 8.0×10^{34} m^{-3} s^{-1}.

(25.30) (a) No gas; **(b)** 2.69×10^{-5} kg m^{-1} s^{-1}; **(c)** 4.64×10^{-5} kg m^{-1} s^{-1}.
(25.31) (a) 5.41×10^{-3} J K^{-1} m^{-1} s^{-1}, 8.1×10^{-3} J s^{-1}; **(b)** 2.93×10^{-2} J K^{-1} m^{-1} s^{-1},
4.4×10^{-2} J s^{-1}. **(25.32)** 1.38×10^{-5} kg m^{-1} s^{-1}, 190 pm.
(25.33) 1.62×10^{-2} J m^{-1} s^{-1} K^{-1}, 2.88×10^{-2} J m^{-1} s^{-1} K^{-1}. **(25.34)** 9.1.
(25.35) 20 J s^{-1}, 20 W. **(25.36) (a)** 10.6 m^2 s^{-1}, 0.427 mol cm^{-2} s^{-1};
(b) 1.06×10^{-5} m^2 s^{-1}, 4.32×10^{-7} mol cm^{-2} s^{-1}; **(c)** 1.06×10^{-2} m^2 s^{-1},
4.32×10^{-9} mol cm^{-2} s^{-1}. **(25.37)** $dr/dt = (s\mathcal{N}/L\rho)(M_m RT/2\pi)^{1/2} \lesssim 6.9 \times 10^{-4}$ cm s^{-1},
or $\lesssim 4$ nm in 0.5 ms.

Chapter 26 (26.1) $5.3 \, \Omega^{-1}$ cm^2 mol^{-1}. **(26.2)** 0.2063 cm^{-1}. **(26.3)** 76.5 $(\Omega^{-1}$ cm^2 mol$^{-1})$/
(mol dm$^{-3})^{1/2}$. **(26.4) (a)** 119.2 Ω^{-1} cm^2 mol^{-1}; **(b)** $1.192 \times 10^{-3} \, \Omega^{-1}$ cm^{-1};
(c) 173.1 Ω. **(26.5)** 138.3 Ω^{-1} cm^2 mol^{-1}. **(26.6)** 1.36×10^{-5} mol dm^{-3}, 1.86×10^{-10}.
(26.7) 387.9 Ω^{-1} cm^2 mol^{-1}. **(26.8)** 0.03. **(26.9)** 3.22, 3.23. **(26.10)** 1.91×10^{-5}, 4.72.
(26.11) 8.38 Ω^{-1} cm^2 mol^{-1}, $3.35 \times 10^{-4} \, \Omega^{-1}$ cm^{-1}, 615 Ω. **(26.12)** 4.01×10^{-4}
cm^2 s^{-1} V^{-1}, 5.19×10^{-4} cm^2 s^{-1} V^{-1}, 7.62×10^{-4} cm^2 s^{-1} V^{-1}.
(26.13) Li$^+$: 4.01×10^{-3} cm s^{-1}, ≈ 4 min, 1.3×10^{-6} cm, 43 diameters, Na$^+$:
5.19×10^{-3} cm s^{-1}, ≈ 3 min, 1.7×10^{-6} cm, 55 diameters, K$^+$: 7.62×10^{-3} cm s^{-1},
≈ 2 min, 2.4×10^{-6} cm, 81 diameters. **(26.14)** $t'_+/t''_+ = c'_+ u'_+/c''_+ u''_+$. **(26.15)** 0.82, 0.003.
(26.16) 0.48, 0.50. **(26.17)** 0.48, 0.52. **(26.18)** 7.46×10^{-4} cm^2 s^{-1} V^{-1},
8.08×10^{-4} cm^2 s^{-1} V^{-1}, 72.0 Ω^{-1} cm^2 mol^{-1}, 78.0 Ω^{-1} cm^2 mol^{-1}. **(26.20)** 0.82, 0.18,
3.61×10^{-3} cm^2 s^{-1} V^{-1}, 7.92×10^{-4} cm^2 s^{-1} V^{-1}. **(26.21)** 0.407. **(26.22)** 3.75 MΩ,
1.01×10^{-14}, 14.0, 7.0. **(26.23)** 60.4 Ω^{-1} cm^2 mol^{-1}/(mol dm$^{-3})^{1/2}$, 0.13/(mol dm$^{-3})^{1/2}$,
76.8 Ω^{-1} cm^2 mol^{-1}/(mol dm$^{-3})^{1/2}$. **(26.24) (a)** 1.24×10^4 N mol^{-1},
2.06×10^{-20} N/molecule; **(b)** 1.65×10^4 N mol^{-1}, 2.75×10^{-20} N/molecule;
(c) 2.48×10^4 N mol^{-1}, 4.12×10^{-29} N/molecule. **(26.25) (a)** 2.60 nm s^{-1};
(b) 3.47 nm s^{-1}; **(c)** 5.19 nm s^{-1}. **(26.26)** 420 pm. **(26.27)** 2.1×10^{-11} s.
(26.28) (a) 6.5×10^{-5} m; **(b)** 3.2×10^{-5} m. **(26.29) (a)** 235 s, 962 s; **(b)** 2.35×10^4,
9.62×10^4 s. **(26.30)** Li$^+$: 1.03×10^{-5} cm^2 s^{-1}, 212 pm, Na$^+$: 1.33×10^{-5} cm^2 s^{-1},
164 pm, K$^+$: 1.96×10^{-5} cm^2 s^{-1}, 112 pm, Rb$^+$: 2.04×10^{-5} cm^2 s^{-1}, 107 pm.
(26.32) $a \approx 83$ pm, $E_a \approx 9.3$ kJ mol^{-1}. **(26.34) (a)** 0; **(b)** 0.06 mol dm^{-3}.
(26.35) (a) 0; **(b)** 0.016; **(c)** 0.054. **(26.37)** 1.2×10^{-3} kg m^{-1} s^{-1}.

Chapter 27 (27.1) 1.4×10^3 s; **(a)** 504 mmHg; **(b)** 688 mmHg. **(27.2) (a)** dm^3 mol^{-1} s^{-1},
dm^6 mol^{-2} s^{-1}; **(b)** atm^{-1} s^{-1}, atm^{-2} s^{-1}. **(27.3)** 2716 yr. **(27.4) (a)** 0.64 μg;
(b) 0.18 μg. **(27.5) (a)** 0.09 mol dm^{-3}; **(b)** 0.05 mol dm^{-3} (AcOEt). **(27.6, 27.7, 27.8,**
27.10) See Box 27.1. **(27.12)** $t_{1/2}/t_{3/4} = (2^{n-1} - 1)/[(4/3)^{n-1} - 1]$. **(27.13)** First,
5.57×10^{-4} s^{-1}. **(27.14)** First, 1.4×10^{-2} s^{-1}, probably is reactive. **(27.15)** 241 kJ mol^{-1},
1.1×10^{13} s^{-1}. **(27.16)** First, 7.2×10^{-4} s^{-1}. **(27.17)** $p = (1/2)p_0[(2 + k_2 t p_0)/(1 + k_2 t p_0)]$,
$t = 1/p_0 k_2$, $\xi = 1/2$. **(27.18)** Second, 1.61×10^{-5} mmHg^{-1} s^{-1}, 3.1×10^6 s.
(27.21) Third in HCl, first in propene. **(27.23)** -18 kJ mol^{-1}, 3 kJ mol^{-1}.
(27.24) 1.14×10^{10} mol^{-1} dm^3 s^{-1}, 16.7 kJ mol^{-1}. **(27.25)** 66 kJ mol^{-1}.
(27.28) $1/k_{obs} = 1/k_2 + [H^+]/k_2 K_a$. **(27.29)** p$K_a = 13.7$. **(27.31)** d[CH$_4$]/dt $=$
$k_b(k_a/2k_d)^{1/2}[CH_3CHO]^{3/2}$, d[CH$_3$CHO]/dt $= -k_a[CH_3CHO] - k_b(k_a/2k_d)^{1/2}$
$[CH_3CHO]^{3/2}$. **(27.32)** 1.9×10^{20} photons/s, 3.1×10^{-4} einstein s^{-1}.
(27.34) 5.1×10^8 mol dm^{-3} s^{-1}. **(27.35)** d[A]/dt $= -I_a - k_p(I_a/k_t)^{1/2}[A]$.
(27.36) Cl$_2$ + $h\nu \rightarrow$ 2Cl, Cl + CHCl$_3 \rightarrow$ CCl$_3$ + HCl, CCl$_3$ + Cl$_2 \rightarrow$ CCl$_4$ + Cl, 2CCl$_3$ + Cl$_2 \rightarrow$
2CCl$_4$. **(27.37)** $(I_a/k_2)^{1/2} \propto [A]^{1/2}$. **(27.38)** $\Phi = 2/\{(k_t/k_2) + [A]\}$.
(27.40) 4.96×10^7 mol^{-1} dm^3 s^{-1}. **(27.41)** $1/\tau = k_1 + k_2[B]_e + k_2[C]_e$.
(27.42) 0.01 mol dm^{-3}.

Chapter 28 (28.1) (a) 9.6×10^9 s^{-1}, 1.2×10^{29} s^{-1} cm^{-3}, 1.8%; **(b)** 6.7×10^9 s^{-1}, 8.3×10^{28} s^{-2} cm^{-3},
1.8%. **(28.2)** $N\beta$, exp $(-\beta E_a)$. **(28.3) (a)** 2.44×10^{-3}, 1.81×10^{-2}, 9.02×10^{-2},
3.00×10^{-1}; **(b)** 7.62×10^{-27}, 3.87×10^{-18}, 3.57×10^{-11}, 5.98×10^{-6}. **(28.4) (a)** 30.1%,
13.4%, 4.8%, 1.2%; **(b)** 301%, 134%, 48%, 12%. **(28.5) (a)** 4.3×10^{-20} m^2; **(b)** 0.14.
(28.6) 0.07, 0.026 nm^2. **(28.7)** 0.01, 4.0×10^{-21} m^2. **(28.8)** 2.5×10^{11} dm^3 mol^{-1} s^{-1}.
(28.9) 8.8 ns. **(28.10) (a)** 6.61×10^9 dm^3 mol^{-1} s^{-1}; **(b)** 3.00×10^{10} dm^3 mol^{-1} s^{-1};
(c) 1.97×10^9 dm^3 mol^{-1} s^{-1}. **(28.11) (a)** 0.24; **(b)** 2.62. **(28.12) (a)** 5.6915×10^{12} s^{-1};

(b) 6.2124×10^{12} s^{-1}; (c) 2.6528×10^{13} s^{-1}. (28.13) $\tau \approx 3.5 \times 10^{29}$ dm^{-3}, $q_{rot} \approx 10^2$ to 10^3, $q_{vib} \approx 1$, $q_{elec} \approx 1$. (28.16) (a) $k_2(T)/k_2(H) = \frac{1}{15}$; (b) $k_2(CO(18))/k_2(CO(16)) = 1.2$, no. (28.17) 1.4×10^6 dm^3 mol^{-1} s^{-1}. (28.18) 1.2×10^6 dm^3 mol^{-1} s^{-1}. (28.20) -76 J K^{-1} mol^{-1}. (28.21) -88 J K^{-1} mol^{-1}, -89 J K^{-1} mol^{-1}. (28.22) $k_1 \approx (\nu^3/\nu^{\ddagger 2}) \exp(-\beta \Delta E)$, (a) 2.7×10^{-11} cm^2 s^{-1}; (b) 1.1×10^{-10} cm^2 s^{-1}. (28.24) 9.6 kJ mol^{-1}. (28.25) (a) -46 J K^{-1} mol^{-1}; (b) 18.6 kJ mol^{-1}; (c) 8.2×10^{10} dm^3 mol^{-1} s^{-1}. (28.27) 88.2 kJ mol^{-1}, 17.8 J K^{-1} mol^{-1}, 90.7 kJ mol^{-1}, 82.9 kJ mol^{-1}. (28.30) 2 (of same sign as first ion). (28.32) $\sigma(CH_2F_2)/\sigma(Ar) = 4.9$.

Chapter 29 (29.1) (a) 1.44×10^{21} cm^{-2} s^{-1}, 1.44×10^{14} cm^{-2} s^{-1}; (b) 3.06×10^{20} cm^{-2} s^{-1}, 3.06×10^{13} cm^{-2} s^{-1}. (29.2) (a) 1.61×10^{15} cm^{-2}; (b) 1.14×10^{15} cm^{-2}; (c) 1.86×10^{15} cm^{-2}. (29.3) 1.00×10^{15} cm^{-2}, 1.41×10^{15} cm^{-2}, 1.16×10^{15} cm^{-2}, 1.20×10^{15} cm^{-2}. (29.5) (a) 4.2×10^{-11} s; (b) 1.9×10^{13} s (600 000 yrs). (29.6) (a) 6.1×10^{-13} s; (b) 7.3×10^{-6} s. (29.7) At 298 K, $\tau_D = 1.80 \tau_H$, at 1000 K, $\tau_D = 1.52 \tau_H$. (29.8) 611 kJ mol^{-1}, 1.07×10^{-13} s. (29.9) 3.7 kJ mol^{-1}. (29.12) $p/V_a = p/V_a^0 + 1/KV_a^0$, $\ln(\theta/p) \approx \ln K - \theta$ if $\theta \ll 1$, $-1/V_a^0$. (29.13) 0.52 mmHg^{-1}, 0.48 cm^3. (29.17) Langmuir, 0.017 mmHg^{-1}, 0.13 nm^2. (29.18) 0.186 cm^3. (29.19) No. (29.23) $k_2 p_F^{1/2} p_B^{1/2}/[1 + (Kp_F)^{1/2}]$. (29.24) -12.7 kJ mol^{-1}. (29.25) 700 kJ mol^{-1}, (a) 2.1×10^{104} min; (b) 50 μs. (29.26) Zeroth. (29.28) $G(t) = k_c + F(t)$, 0.02 mmHg s^{-1}. (29.29) $G(t) = (k_c/K) + \{(1 + Kp_0)/Kp_0\}F(t)$. (29.30) 0.033 mmHg^{-1}, 4.3×10^{-4} s^{-1}. (29.32) 295 pm. (29.33) 0.16, 2.4. (29.36) $d\mu = (RTV_a^0/\sigma)d[\ln(1-\theta)] = -\{(RT/\sigma)/(1-\theta)\} dV_a$.

Chapter 30 (30.2) (a) 0.31 mA cm^{-2}; (b) 5.41 mA cm^{-2}; (c) 1.39×10^{42} mA cm^{-2}. (30.5) 107 mV. (30.6) 4.93×10^{15} cm^{-2} s^{-1}, 1.56×10^{16} cm^{-2} s^{-1}, 3.12×10^7 cm^{-2} s^{-1}. (30.7) 3.8 s^{-1}, 12 s^{-1}, 2.4×10^{-8} s^{-1}. (30.8) RT/SjF, (a) 32.5 Ω; (b) 3.3×10^{10} Ω. (30.9) $\langle j \rangle = (\frac{1}{4})(2\alpha - 1)(F/RT)^2 j_e \eta_0^2$. (30.10) 7.2 μA. (30.12) 0.38, 0.78 mA cm^{-2}. (30.14) When E falls below 0.34 V, Cu deposited until limiting current density attained; then drop in potential difference until zinc begins depositing when E is below -0.76 V. (30.15) (a) -0.76 V, -0.82 V; (b) 0.34 V, 0.28 V. (30.17) Draw only 0.14 mA cm^{-2} (giving negligible hydrogen evolution) before potential drops to -0.76 V, so zinc may be deposited. (30.18) Will draw 2.2×10^3 A cm^{-2}, so there is massive hydrogen evolution before zinc can deposit. (30.19) Must draw 5.3×10^9 A cm^{-2}, giving massive hydrogen evolution, so no magnesium will be deposited. (30.20) $j = j_L\{1 - \exp(F\eta_c/RT)\}$. (30.21) 0.25 mm. (30.22) $j_L = (RT/zF)(\lambda_+/\lambda)c_{M^+}$. (30.23) 0.23 mm at all concentrations. (30.24) $a(Sn^{2+})/a(Pb^{2+}) \approx 2.2$. (30.26) -1.299 V, 0.13 W. (30.27) (a) 1.23 V; (b) 1.060 V. (30.32) Fe. (30.36) 13 mA.

Acknowledgements

The permissions of the respective copyright holders to reproduce tables of data and extracts from them is gratefully acknowledged as follows: Academic Press (4.10 [© 1969]), Addison Wesley (16.1, 16.2 [© 1962]), Butterworths Ltd. (29.1, 29.2 [© 1964]), Professor J. G. Calvert (18.1, 18.2 [© 1966]), Chapman and Hall Ltd. (17.3 [© 1975]), CRC Press Inc. (1.1, 4.4, 11.1, 12.1, 12.3, 22.2, 23.1, 23.5 [© 1979]), Cornell University Press (4.5, [© 1960]), Dover Publications Inc. (13.1, 25.1 [© 1965]), Elsevier Scientific Publishing Company (23.1 [© 1978]), Professor A. A. Frost (28.2), Longman (1.2, 2.2, 3.1, 4.1, 4.4, 4.8, 5.2, 5.3, 7.1, 8.2, 9.2, 12.2, 12.3, 22.1, 22.3, 23.8, 25.3, 25.4, 26.1, 26.2, 30.2 [© 1973]), McGraw-Hill Book Company (1.2, 1.3, 3.1, 3.2, 4.1, 4.2, 4.6, 4.7, 4.9, 5.1, 17.2, 23.2, 23.7, 26.4 [© 1972]; 4.3, 4.6, 6.1, 9.1 [© 1961]; 27.1, 27.2 [© 1965]; 13.2, 14.1, [© 1935]; 3.2 [© 1968]; 17.1 [© 1955]), Dr. J. Nicholas (27.1, 27.2, 27.3 [© 1976]), Oxford University Press (27.1, 27.2, 27.3 [© 1975]; 29.3, 29.4, 29.5 [© 1974] Pergamon Press Ltd. (23.3 [© 1961]), Prentice Hall Inc. (24.1, 24.3 [© 1971]), John Wiley and Sons Inc. (8.1 [© 1975]; 15.2 [© 1944]; 23.4 [© 1954]; 24.2, 24.4 [© 1961]). The sources of the material are quoted at the foot of each table.

The permission of the following individuals and journals for reproduction of illustrations is gratefully acknowledged: *Acta crystallogr.* (0.2), Professor B. J. Alder (0.14), Dr R. F. Barrow (13.7, 18.16), Professor G. C. Bond (29.25), Professor A. V. Crewe (24.1), Professor G. Ehrlich (29.16), Professor D. Freifelder (24.5b), Dr H. F. Kay (29.3), Professor D. A. King (29.17), Professor D. A. Long (17.19), Dr K. A. McLauchlan (19.2), Dr G. Morris (19.15), Professor E. W. Müller (0.1, 29.12), *Nature* (0.3), Nicolet XRD Corpn (22.13), Dr A. H. Norton (23.22), Open University Press (18.19 [© 1976]), Oxford University Press (19.2, 29.3, 29.10), Dr M. Prutton (29.10), Professor M. W. Roberts (29.6), Professor G. A. Samorjai (29.8, 29.9), *Scientific American* (10.1, 24.5b).

Table index

Table index

Subject index

PERIODIC TABLE

IA	IIA	IIIA	IVA	VA	VIA	VIIA		VIII		IB	IIB	IIIB	IVB	VB	VIB	VIIB	O
1 H 1.008																	2 He 4.003
3 Li 6.941	4 Be 9.012											5 B 10.81	6 C 12.01	7 N 14.01	8 O 16.00	9 F 19.00	10 Ne 20.18
11 Na 22.99	12 Mg 24.31											13 Al 26.98	14 Si 28.09	15 P 30.97	16 S 32.06	17 Cl 35.45	18 Ar 39.95
19 K 39.10	20 Ca 40.08	21 Sc 44.96	22 Ti 47.90	23 V 50.94	24 Cr 52.00	25 Mn 54.94	26 Fe 55.85	27 Co 58.93	28 Ni 58.71	29 Cu 63.55	30 Zn 65.37	31 Ga 69.72	32 Ge 72.60	33 As 74.92	34 Se 78.96	35 Br 79.90	36 Kr 83.80
37 Rb 85.47	38 Sr 87.62	39 Y 88.91	40 Zr 91.22	41 Nb 92.91	42 Mo 95.94	43 Tc (99)	44 Ru 101.1	45 Rh 102.9	46 Pd 106.4	47 Ag 107.9	48 Cd 112.4	49 In 114.8	50 Sn 118.7	51 Sb 121.8	52 Te 127.6	53 I 126.9	54 Xe 131.3
55 Cs 132.9	56 Ba 137.3	57 La 138.9	72 Hf 178.5	73 Ta 180.9	74 W 183.9	75 Re 186.2	76 Os 190.2	77 Ir 192.2	78 Pt 195.1	79 Au 197.0	80 Hg 200.6	81 Tl 204.4	82 Pb 207.2	83 Bi 209.0	84 Po (210)	85 At (210)	86 Rn (222)
87 Fr (223)	88 Ra (226)	89 Ac (227)															

Lanthanides

58 Ce 140.1	59 Pr 140.9	60 Nd 144.2	61 Pm (147)	62 Sm 150.4	63 Eu 152.0	64 Gd 157.3	65 Tb 158.9	66 Dy 162.5	67 Ho 164.9	68 Er 167.3	69 Tm 168.9	70 Yb 173.0	71 Lu 175.0

Actinides

90 Th 232.0	91 Pa (231)	92 U 238.0	93 Np (237)	94 Pu (242)	95 Am (243)	96 Cm (248)	97 Bk (247)	98 Cf (251)	99 Es (254)	100 Fm (253)	101 Md (256)	102 No (254)	103 Lw (257)

MASSES (m/u)* AND NATURAL ABUNDANCES (per cent)† OF SELECTED NUCLIDES

H

^{1}H(99.985)1.0078
^{2}H(0.015)2.0141

He

^{3}He(0.00013)3.0160
^{4}He(100)4.0026

Li

^{6}Li(7.42)6.0151
^{7}Li(92.58)7.0160

B

^{10}B(19.78)10.0129
^{11}B(80.22)11.0931

C

^{12}C(98.89)12.0000
^{13}C(1.11)13.0034

N

^{14}N(99.63)14.0031
^{15}N(0.37)15.0001

O

^{16}O(99.76)15.9949
^{17}O(0.037)16.9991
^{18}O(0.204)17.9992

F

^{19}F(100)18.9984

Ne

^{20}Ne(90.92)19.9924
^{21}Ne(0.257)20.9938
^{22}Ne(8.82)21.9914

Na

^{23}Na(100)22.9898

P

^{31}P(100)30.9738

S

^{32}S(95.0)31.9721
^{33}S(0.76)32.9715
^{34}S(4.22)33.9679
^{36}S(0.014)35.9671

Cl

^{35}Cl(75.53)34.9688
^{37}Cl(24.4)36.9651

Ar

^{36}Ar(0.337)35.9675
^{38}Ar(0.063)37.9627
^{40}Ar(99.60)39.9623

Br

^{79}Br(50.54)78.9183
^{81}Br(49.46)80.9163

Kr

^{78}Kr(0.35)77.9204
^{80}Kr(2.27)79.9164
^{82}Kr(11.56)81.9135
^{83}Kr(11.55)82.9141
^{84}Kr(56.90)83.9115
^{86}Kr(17.37)85.9106

I

^{127}I(100)126.9045

* $1u = 1.66056 \times 10^{-27}$ kg; $m = M_r u$

† Figures in parentheses.